COMPUTER SCIENCE/INTERNET

CONSUMER

CONSTRUCTION/TRADE INDUSTRY

EDUCATION

ENGLISH/GRAMMAR STRUCTURE

GAMES

GENERAL INTEREST

(continued on next page)

GENERAL SCIENCE

GOVERNMENT

HEALTH/MEDICAL

HISTORICAL

LAW ENFORCEMENT

LEISURE

MATH-RELATED

(*continued on inside back cover*)

Fundamentals of Mathematics

Fundamentals of Mathematics

TENTH EDITION

William M. Setek, Jr.
Monroe Community College

Michael A. Gallo
Florida Institute of Technology

Upper Saddle River, New Jersey 07458

Library of Congress Cataloging-in-Publication Data

Setek, William M.
Fundamentals of mathematics / William M. Setek, Jr.,
Michael A. Gallo. – 10th ed.
p. cm.
Includes index
ISBN 0-13-113941-X
1. Mathematics. I. Gallo, Michael A, II. Title.
CIP data available

Executive Acquisitions Editor: *Petra Recter*
Editor-in-Chief: *Sally Yagan*
Vice President/Director of Production and Manufacturing: *David W. Riccardi*
Production Editor: *Debbie Ryan*
Senior Managing Editor: *Linda Mihatov Behrens*
Assistant Managing Editor: *Bayani Mendoza de Leon*
Executive Managing Editor: *Kathleen Schiaparelli*
Assistant Manufacturing Manager/Buyer: *Michael Bell*
Manufacturing Manager: *Trudy Pisciotti*
Marketing Manager: *Krista Bettino*
Marketing Assistant: *Rachel Beckman*
Art Director: *Heather Scott*
Interior Designer/Cover Designer: *Susan Anderson*
Creative Director: *Carole Anson*
Director of Creative Services: *Paul Belfanti*
Art Editor: *Thomas Benfatti*
Editorial Assistant: *Joanne Wendelken*
Cover Image: © *StockFood/Schieren*
Art Studio: Laserwords

Pearson Prentice Hall
Pearson Education, Inc.
Upper Saddle River, NJ 07458

Printed in the United States of America

10 9 8 7 6 5 4 3

ISBN 0-13-113941-X

Pearson Education, Ltd., *London*
Pearson Education Australia PTY. Limited, *Sydney*
Pearson Education Singapore, Pte., Ltd
Pearson Education North Asia Ltd, *Hong Kong*
Pearson Education Canada, Ltd., *Toronto*
Pearson Education de Mexico, S.A. de C.V.
Pearson Education – Japan, *Tokyo*
Pearson Education Malaysia, Pte. Ltd

For two very special people:

Jillian Marie Setek & Rachel Jennie Setek

W. M. S.

To Janie . . . my wife and cutie π

M.A.G.

Contents

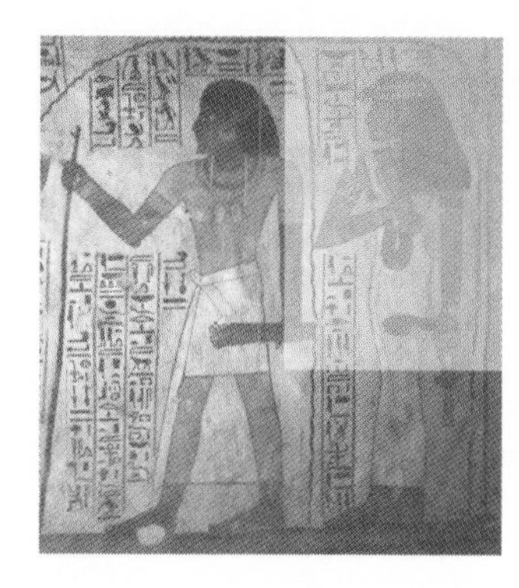

7 Systems of Numeration 355

8 Sets of Numbers and Their Structure 417

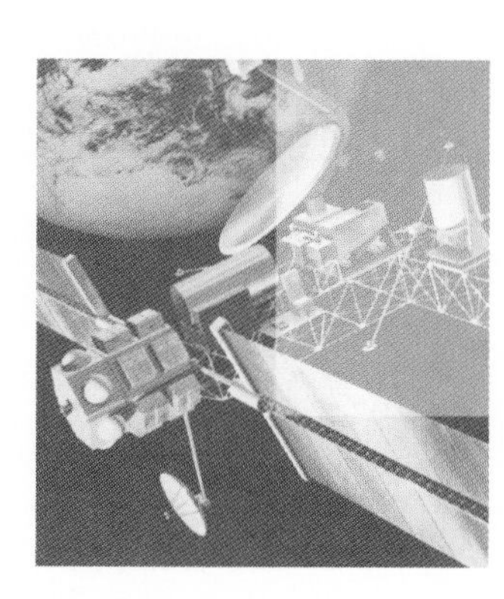

9 An Introduction to Algebra 489

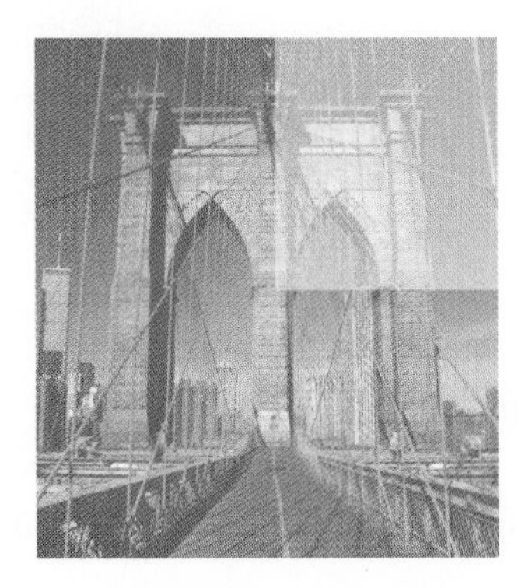

10 Selected Topics in Algebra 549

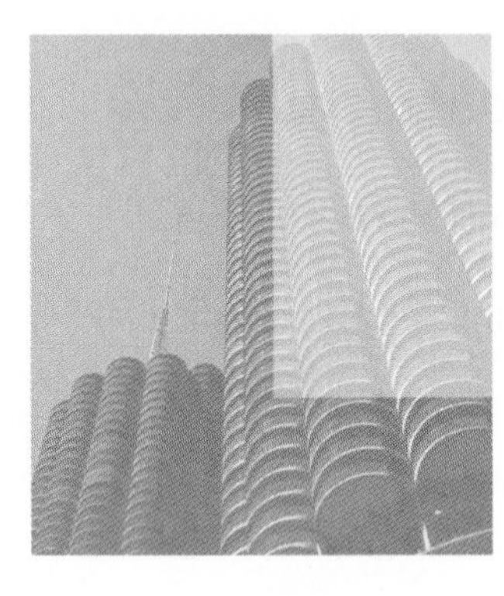

11 An Introduction to Geometry 593

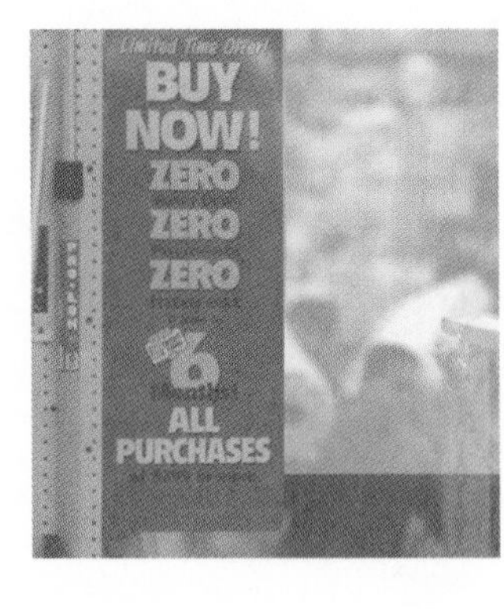

12 Consumer Mathematics 659

FOR
15
km

To the Instructor

The tenth edition of *Fundamentals of Mathematics* reflects our enduring commitment to providing liberal arts mathematics students and teachers the best possible textbook for maximizing student success. Using a patient, friendly, and straightforward approach, this tenth edition builds on its nearly 30-year history by presenting topics and concepts in a simple and uncomplicated manner. Its focus continues to be "user friendly," and the exposition is presented from an intuitive perspective. The success and longevity of this text can be summed up in two words: **It Works!**

Changes to the Tenth Edition

In writing this tenth edition, several substantive changes were made that helped fine-tune the presentation of certain material. A summary of the more salient ones follows:

- *Over 20% of the Examples and Exercises, many involving sourced data* are either new or updated. These examples and exercises link real-world data to the mathematical concepts being presented.
- A "Now Work" exercise follows each example. "Now Work" suggests an end of section exercise that is similar in style and concept to the preceding example. This gives students the opportunity to check and confirm their understanding. Answers to the "Now Work" exercises are found in the Answers Section in the back of the text.
- *Math Connections* have replaced the *Math Notes* from the ninth edition. This change in name better reflects the extension of mathematics to other related areas. Additionally, over a dozen new *Math Connections* have been added in this edition.
- *Chapter Overviews,* which highlight the key topics of a chapter, have been written. Corresponding to these new chapter *overviews*, *introductions* have been revised.
- A new optional section on *mathematical modeling* has been written in Section 9.6 that extends the application of linear equations in two variables.
- Chapter 10 (Selected Topics in Algebra) has been reorganized, and the material on quadratic equations has been rewritten so that it flows more logically. The entire chapter has been reorganized so it proceeds from introductory to more difficult topics.
- *Chapter Review* materials were revised. Each *Chapter Summary* is now organized by sections and the summary information was changed from narrative to bulleted list form to facilitate recall. The *Vocabulary Check* was retitled as *Key Terms and Concepts*, and page numbers are now included for each item to reference where in the text the item can be found. Finally, the review exercises are organized by chapter section.
- All chapters were carefully reviewed and, where appropriate, the prose was modified to improve flow and understanding. Most notable among these are Chapters 8, 9, 10, and 11.
- An *index of applications* that provides readers with a single reference list of real-world applications was prepared.

Design Changes and Features

Several new design changes have also been incorporated into this edition. Key among them are:

- *Updated artwork*
- *Example labels* that identify specific topics or concepts being illustrated.
- *Exercise labels* that specify real-world applications.

Retained Features

Although several content and design changes have been incorporated into this tenth edition, we were also careful to retain the hallmarks of *Fundamentals of Mathematics* that have made it an effective and engaging resource for liberal arts mathematics students since 1976. These include:

- **Intuitive Approach.** Throughout the book, the material is written specifically for nonmath majors. The exposition is clear, patient, and friendly. We present the mathematics in a non-intimidating manner, without clutter, and limit the introduction of new material to one or two concepts at a time.
- **Development.** The material is carefully and thoroughly developed. The prose contains complete explanations to facilitate student understanding and meaningful learning, and "signals" are provided throughout the presentations to help students identify important terms and concepts. Many students have commented that the textbook reads as if it were a transcript of a classroom lesson.
- **Examples.** The text contains over 600 completely worked-out examples with systematic step-by-step solutions; there are no gaps or "magic" solutions.
- **Exercises.** The exercises are plentiful (don't let the numbering system fool you). Moreover, they are graded in difficulty (from basic to moderately difficult) and parallel in nature so that an instructor may independently assign either odd- or even-numbered problems to cover the material in a section. More than 6,500 problems appear in this edition, far more than any competitor offers.
- **Illustrations.** There are more than 800 illustrations and photos that help students visualize mathematics and use that visualization in solving mathematical problems.
- **Biographies, Historical Notes, Notes of Interest, and Just for Funs.** Throughout the text, specially designed sidebars or problems supplement the exposition. *Biographies* provide vignettes about people who have made important contributions to mathematics. *Historical Notes* embellish a discussion by providing a historical context for a particular math concept or topic. *Notes of Interest* contain adjunct information about a topic that students might find interesting, and *Just for Funs* are problems that range from serious extensions of mathematical ideas to light-hearted puzzles, recreational mathematics, or brain teasers.
- **Glossary.** A 250-word glossary of key mathematical terms or phrases is provided at the end of the book.

In the back of the text, answers for all odd-numbered exercises, as well as answers for all review exercises, chapter quizzes and *Just for Funs*, are provided.

Fully worked out solutions for the odd-numbered problems are found in the *Student Study Guide and Solutions Manual*, and the even-numbered answers are provided in the *Instructor's Solutions Manual*. For a closer look at the textbook's many features, turn to the text walkthrough on pages xx–xxiii.

Instructor Supplements

All instructor resources can be downloaded from a website (URL and password can be obtained from your PH Sales Representative), or ordered individually.

Instructor's Solutions Manual. Contains complete step-by-step worked out solutions to all exercises in the textbook.

Test Item File. With over 1000 questions, this text bank allows easy creation of full tests and quizzes for your students.

TestGen EQ (computerized test generator—cross platform). This computerized test bank allows you to view and edit test bank questions, transfer them to exams, and print in a variety of formats. The program offers many options for organizing and displaying test banks and tests, and a built-in random number and test generator allows instructors to create multiple versions of exams.

Companion Website – **www.prenhall.com/setek** This offers an interactive study guide, matched to each chapter of the text, technology help, quizzes, objectives and other valuable internet resources.

Student Supplements

Student Study Guide and Solutions Manual. This combined Study Guide and Solutions Manual contains step-by-step worked out solutions to all the odd-numbered exercises in the textbook and a comprehensive study guide.

Tutor Center offers text-specific tutoring for students, by trained mathematicians. Students can access tutors by toll free phone, fax and e-mail. For more information, see www.prenhall.com/tutorcenter

Companion Website – www.prenhall.com/setek This offers an interactive study guide, matched to each chapter of the text, technology help, quizzes, objectives and other valuable internet resources.

Acknowledgments

This book evolved from an idea into final form through the efforts of many individuals. We are grateful and indebted to those users of the previous editions who provided us with many valuable suggestions and constructive comments—in particular, our students and colleagues at Monroe Community College and Florida Institute of Technology. Our thanks to those who took the time to express their comments in writing.

ALEXANDER ADDONA
Monroe Community College

EVAN B. ALDERFER
Ocean County College

JEAN M. ALLIMAN
Hesston College

AVIYADASA ALUTHGE
Marshall University

GERALD BEER
California State University—Los Angeles

JOHN BEKLE
Harris Stone State College

GARY JESSEE
Mountain Empire Community College

DOUG JOHNSTON
Wallace College-Dothan

ANNE F. LANDRY
Duchess Community College

GARY LIANO
Northland Pioneer College

MYRA LIEBHART
Illinois Valley Community College

DIANE LYNCH
Niagara Communithy College

Tate Bennett
Madisonville Community College

Prem Bhatia
Coppin State College

Paul Brunetto
Medaille College

Gail Bulster
Niagara County Community College

Jane Cappello
Dutchess Community College

Gayle Childers
J. Sargaent Reynolds Community College

Jeanne Christensen
Vennard College

Lawrence Clar
Monroe Community College

Robert Cole
Davidson County Community College

Dale Craft
South Florida Community College

Sharon Daulk
Faulkner University

Tom Dellaquila
Monroe Community College

Amy J. Dianiska
Ursuline College

Quiang S. Dotzel
University of Missouri-St. Louis

Patrick Downs
Monroe Community College

Tom Drew
Fullerton Community College

Chuck Ellison
Bessemer State Technical College

Donald Fama
Cayuga Community College

Mary Ellen Ferraro
St. Thomas Aquinas College

Daniel Fitzgerald
Holyoke Community College

Daniel N. Fitzgerald
Holyoke Community College

Marvel Froemming
Moorhead State University

Gene Grace
Hopkinsville Community College

Robert A. Gullo
Monroe Community College

Darlene Harding
Medaille College

Maxwell M. Hart
University of Mary Hardin-Baylor

Steven Hatfield
Marshall University

William Wassey
Dickinson State University

Bob Maynard
Rockingham Community College

Wanzie A. McAuley
Columbia State Community College

Tania McDuffie
Spartanburg Technical College

Tim McKenna
University of Michigan-Dearborn

Carol M. Meehan
Dutchess Community College

Sister Marie Mueller, O.P.
Caldwell College

Agashi Nwogbaga
Wesley College

Ron Orban
Fayetteville Technical Community College

Leon R. Pepper
Frank Phillips College

Anne M. Peterson
Iowa Lakes Community College

Verna Phillips
Southern West Virginia Community College

Mazine S. Porter
Paul D. Camp Community College

Susan Poss
Spartanburg Technical College

Marvin L. Proctor
Pratt Community College

Russell Rainbolt
Ovachita Baptist University

Robert C. Riswanger
Columbia College

Barbara Ruslander
Maria College

John J. Salera
Delaware Valley Cottage

Leonore Senfeld
Miami-Dade College

Mark Sereeransky
Camden County College

Felice Shore
Baltimore City Community College

Rose M. Jenkins
Midlands Technical College

Walter Sizer
Moorhead State University

John Smashey
Southwest Baptist University

Pat Stringer, O.P.
Caldwell College

Carole Summerville
Richard Bland College

SISTER ANN DOMINIC TASSONE
BETH HENKLE
Avila College

LIBBY HIGGINS
Greenville Technical College

RUTH E. HOFFMAN
Toccoa Falls College

CALVIN HOLT
Paul D. Camp Community College

MARY JANE HUTCHINS
Ovachita Baptist University

DUANE JACOBS
Frank Phillipes College

ALLISON SUTTON
Camden County College

ALEX THANNIKKARY
Halifax Community College

MARIAN VANVLEET
Saint Mary College

STAN WIECHERT
Washburn University

KAREN ANN YOHO
Fairmont State College
Washburn University

A sincere thank you to the editors and designers at Prentice Hall, particularly Sally Yagan, Petra Recter, and Debbie Ryan, for their enthusiastic support throughout the project.

For their direct contributions to this and previous editions of the text, we also thank the following persons: Dr. Hester Lewellen of Baldwin Wallace College, Professor Jane Edgar of Brevard Community College, Dr. W. Michael Goho, Professor Angel Andreu, and Professor Florence Whittaker, all of Monroe Community College, and Kim Query of Lindenwood Community College.

To Pam Dretto, thank you for excellent work.

A special tribute to Addie Setek, who helped in so many ways. This edition, and the others, would not have been possible without her support, understanding and patience. Addie, we are forever grateful to you.

W.M.S.

M.A.G.

To the Student

This book is designed to help you learn some mathematics, regardless of your mathematical background. It is written so that you can understand, appreciate, and even enjoy areas of mathematics to which you may or may not have been exposed. But, in order for this to occur, you must use this book. Someone once said:

I hear and I forget.

I see and I remember.

I do and I understand.

Mathematics is not a spectator subject; it is a participation sport—you must actively use the text. Here are some hints we give our students at the beginning of the course:

1. Read the text with pencil in hand—*before the lecture*. Knowing what to expect and what is in the book, you can take fewer notes and spend more time listening and understanding the lecture.
2. Work the illustrative examples. There are more examples in this text than in any other of this nature. Their purpose is to help you understand the material and learn by doing. Make use of the wide margins—they are designed for "scratch" work.
3. After a concept has been introduced and an example given, you will see *Now Work Problem xx.* Go to the exercise at the end of the section, work the problem cited, and check your answer in the back of the book. If you get it right, you can be confident in continuing on in the section. If you do not get it right, go back over the explanations and examples to see what you might have missed, then rework the problem. Ask for help if you miss it again.
4. If you are confused about something, visit your instructor during office hours immediately, before you fall behind. Bring your attempted solutions to problems with you to show your instructor where you are having trouble.
5. To study for the exam, begin by reviewing each chapter and your notes, then concentrate on the end-of-chapter problems.
6. Use the Student Study Guide and Solutions Manual. It will provide help.
7. Make use of the companion Website. It is designed to complement and expand upon the text. Visit www.prenhall.com/setek

We encourage you to look through the following text walkthrough and become familiar with the different learning aids that will appear and how they can be used to understand the material.

We hope you will find reading and using this book a worthwhile and enjoyable endeavor. Good luck! We welcome any and all comments. Feel free to write and let us know your thoughts and reactions to this text.

WILLIAM M. SETEK, JR.
Monroe Community College
Rochester, New York 14623
wsetek@monroecc.edu

MICHAEL A. GALLO, PH.D.
Florida Institute of Technology
Melbourne, FL 32901
gallo@fit.edu

HOW TO USE THIS BOOK:

Designed for your success...

Over 400 Interesting and Diverse Applications...

- Represent a wide range of disciplines.
- Feature unique and sourced data that illustrate how mathematics is applied in many settings.
- Using real world data to emphasize that mathematics is a tool used to understand the world around us.

Real-World Data

Internet Spam *Use Figure 1.2 to answer questions 76–78.*

Year	Spam messages (in billions)
1999	1.0
2000	2.3
2001	4.0
2002	5.6
2003	7.3[1]
2004	8.8[1]

1 - estimate

Source: IDC   By Quin Tian, USA TODAY

Figure 1.2
Worldwide spam messages sent daily.

Page 15

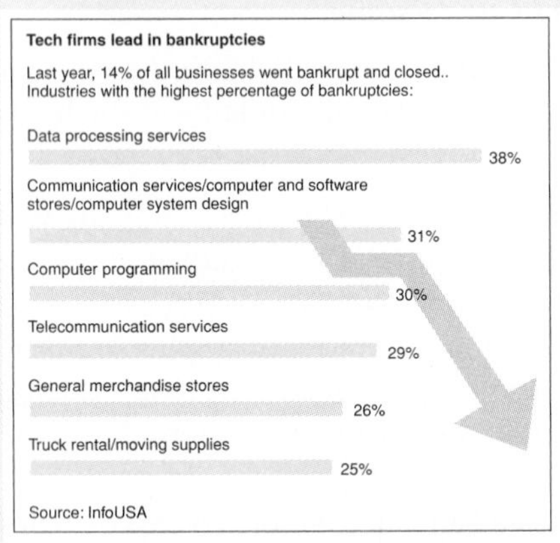

Page 78

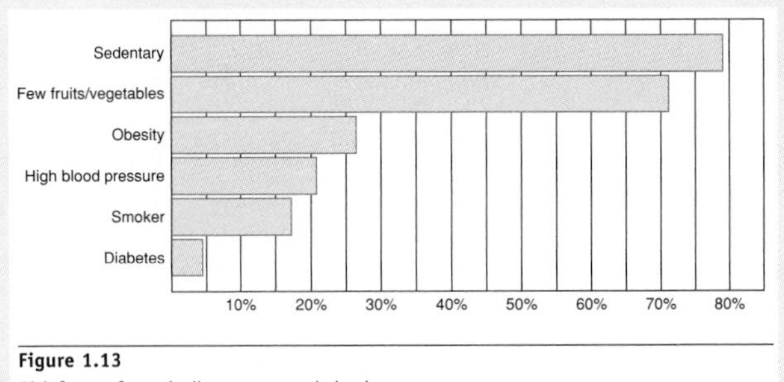

Figure 1.13
Risk factors for early (i.e., premature) death.

Page 28

Unique Chapter Openers

- Each chapter opens with a photo that is representative of the chapter content, a list of learning objectives, a chapter outline, and frequently used symbols. All of these prepare you for the chapter ahead.

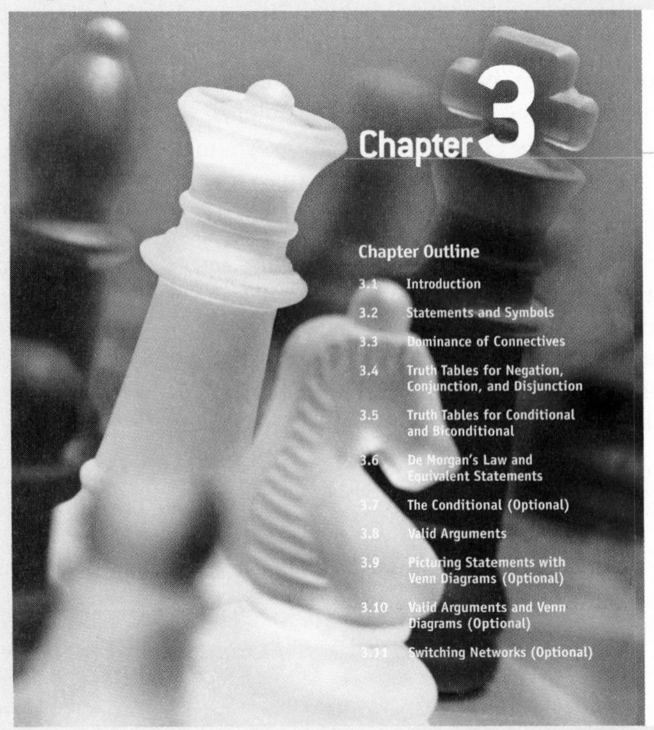

Logic

After Studying This Chapter, You Will Be Able to Do the Following:

1. Distinguish between simple and compound statements, and determine if a compound statement is a negation, conjunction, disjunction, conditional, or biconditional.
2. Write English sentences in symbolic form and express symbolized statements as English sentences.
3. Construct truth tables.
4. Determine if two statements are logically equivalent, consistent, or inconsistent.
5. Identify the forms of a conditional statement.
6. Determine if an argument is valid or invalid.
7. Represent statements using Venn diagrams.
*8. Express a switching network in symbolic form and construct a switching network given a corresponding symbolic statement.

Note: *indicates optional material.

Notation frequently used in this chapter

$P \wedge Q$	P and Q
$P \vee Q$	P or Q
$P \veebar Q$	P or Q but not both
$P \rightarrow Q$	if P, then Q
$P \leftrightarrow Q$	P if and only if Q
$\sim P$	not P; it is false that P; it is not the case that P
iff	if and only if
$\equiv$	is the same as

Pages 112-113

Over 600 Clear Examples and "Now Work" Problems

- A step-by-step approach to reaching the solution with precise explanations built into each step.
- Abundant in number because it is easier to learn by example.
- "Now Work" Problems follow examples and suggest end-of-section problems that are similar to the corresponding examples. This is a good way to check your understanding as you work through the chapter. The answers to all "Now Work"problems can be found either in the back of the book or in the *Student Solutions Manual.*

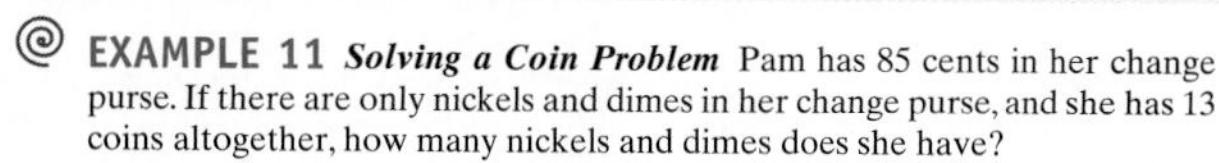

EXAMPLE 11 ***Solving a Coin Problem*** Pam has 85 cents in her change purse. If there are only nickels and dimes in her change purse, and she has 13 coins altogether, how many nickels and dimes does she have?

Solution Let x = number of dimes; then $13 - x$ = the number of nickels. Now if Pam has x dimes in her purse, then the value of these dimes in cents is $10x$. For instance, if she had three dimes, then the value is $10 \cdot 3$, or 30 cents. Similarly, the value of the nickels in Pam's purse is $5(13 - x)$. The total value of the coins in her purse is 85 cents, and hence we have the equation

$$10x + 5(13 - x) = 85$$

Solving the equation,

$10x + 65 - 5x = 85$	*distributive property*
$5x + 65 = 85$	*combining like terms*
$5x + 65 - 65 = 85 - 65$	*subtracting 65 from both sides*
$5x = 20$	*combining like terms*
$\frac{5x}{5} = \frac{20}{5}$	*dividing both sides by 5*
$x = 4$	*number of dimes*
$13 - x = 13 - 4 = 9$	*number of nickels*

Check:

$x = 4$, and $13 - x = 9$, so it checks that Pam possesses 13 coins. But we must also check the value of the coins; that is, $4 \cdot 10 = 40$ (the value of the dimes), $9 \cdot 5 = 45$ (the value of the nickels), and $40 + 45 = 85$ cents (the total value of the coins). Our solution is verified.

NW ***Now Work Problem 19.***

Pages 519-520

Thorough and Extensive Exercise Sets

- End-of-Section Exercises are extensive and are organized within five categories: Practice Exercises, Writing Mathematics, Challenge Exercises, Group/Research Activities and Real-World Data.
- Exercises are graded in difficulty, from basic to moderately difficult.

Writing Mathematics

9. Describe what is meant by a universal set and provide at least two examples.

10. Explain the difference between the inclusive and exclusive use of the word *or*. Describe how it is typically used in mathematics.

11. Describe what is meant by the intersection of two sets. Provide at least two examples. What is a practical analogy to the concept of set intersection?

12. Describe what is meant by the union of two sets. Give at least two examples. What is a practical analogy to the concept of set union?

Challenge Exercises

In addition to intersection and union, another set operation is difference. Specifically, the difference between sets A and B, denoted $A - B$ (read "a minus b"), represents the set of all elements that belong to set A but not to set B. Use the concept of set difference for Exercises 13 through 18.

13. Express $A - B$ in set-builder notation.

14. Represent the set $A - B$ using intersection and complement.

15. Given a universal set U and a set B where $B \subseteq U$, define the complement of B using set difference.

16. Let $U = \{1, 2, 3, \ldots, 10\}$, $A = \{1, 2, 3, 4, 5\}$, and $B = \{1, 3, 5, 7, 9\}$. Find $A - B$.

17. Do you think that $A - B = B - A$? Why? Give an example to defend your answer.

18. What does $A - \emptyset$ equal?

19. What does $\emptyset - A$ equal?

20. Still another set operations is the *symmetric difference* of two sets, denoted $A \Delta B$, and defined as $A \Delta B = (A - B) \cup (B - A)$. Repeat Exercises 16 through 18 using set symmetric difference.

Group/Research Activity

21. The concepts of set difference and set symmetric difference are basic set operations used for sorted sequences in computer science. Research these two concepts and prepare a report that describes a practical application of their use. As part of your report, summarize the set difference and set symmetric difference algorithms and demonstrate the result of these algorithms for two sets A and B, where $A = \{1, 2, 3, 4, 5\}$ and $B = \{3, 4, 5, 6, 7\}$.

22. Research the concept of a *truth set* and prepare a report on its meaning and use in mathematics.

Real-World Data

Fat Content of Foods *Figure 2.3 shows the fat content of various foods. Use this figure to represent in roster form the elements of the sets in Exercises 23–30.*

23. The set that contains foods with more than 3 grams of saturated fat or has 0 grams of polyunsaturated fat.

Page 86

Math Connections

Math Connections boxes appear in abundance throughout the text They:

- Enhance your learning by helping make connections among a variety of topics.
- Facilitate your understanding of how mathematics can be applied to other disciplines.

Math Connections

Genetics

One application of probability is in the field of genetics. For example, every expectant mother thinks about what her child will look like when it is born. Will it have blue eyes and dark hair, or will it have brown eyes and light hair? These are questions of probability (e.g., What is the probability that my child will have blue eyes?). To answer these type of questions, we first need to introduce some basic biological concepts.

The genetic constitution of an individual is called **genotype**. An individual's genotype for a specific trait such as eye color is determined by the genes of the person's parents. From the moment of conception (i.e., when the sperm and egg unite), the genotype of the baby is determined. If the sperm carries a brown-eyed gene (B) and the egg possesses a blue-eyed gene (b), then the baby's genotype consists of one brown-eyed gene and one blue-eyed gene (Bb). In the case of eye color, we say that brown is the **dominant characteristic** and blue is the **recessive characteristic**, which implies that when both genes are present, the baby's eye color will be the dominant one, namely, brown. The physical appearance of an individual trait is known as the **phenotype**. For our simple eye color example, three possible genotypes and two possible phenotypes are

Note that there are four possibilities for the baby's genotype, two of which are bb. Thus, for parents who possess these genotypes, the probability that their baby will have blue eyes is $\frac{2}{4} = \frac{1}{2}$. The Bb entries in the Punnett square mean that the baby is a brown-eyed hybrid (heterozygous).

What do you think the probability is of a baby being born with blue eyes if both parents have brown eyes? To examine this probability, we need to know if the parents genotype is pure dominant (i.e., BB) or hybrid (i.e., Bb). From the preceding table, note that there are three different arrangements in which this can occur:

- Both parents are pure dominant (BB and BB).
- Both parents are hybrid (Bb and Bb).
- One parent is pure dominant and the other parent is hybrid (BB and Bb).

Thus, a Punnett square will have to be constructed for each arrangement. Doing so, leads to the following probabilities:

- If both parents are pure dominant, then P(blue-eyed baby) = 0.

Page 229

Optional Enrichment Essays

Biographies provide vignettes about people who have made important contributions to mathematics.

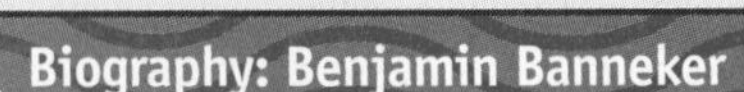

Biography: Benjamin Banneker

Benjamin Banneker (1731–1806), an African-American born of a free mother and a slave father, was eventually able to buy his own freedom. He had very little formal education, so he was for the most part self-taught in mathematics, astronomy, geology, and physics. From 1792 through 1799 he published *Benjamin Banneker's Almanac and Ephemeris*. One feature of the almanac was his reporting and calculation of the seventeen-year cycle for the locust plagues. His almanac was the first scientific work published by an African-American. Through his scientific work, Mr. Banneker gained a respected reputation that won him a position on the survey commission of six men to assess the land that was to become the District of Columbia. When the head of the commission, Major L'Enfant, resigned and returned to France with all the plans, Mr. Banneker's excellent memory was able to reconstruct the plans in their entirety.

Page 558

Notes of Interest contain information about a topic related to mathematics.

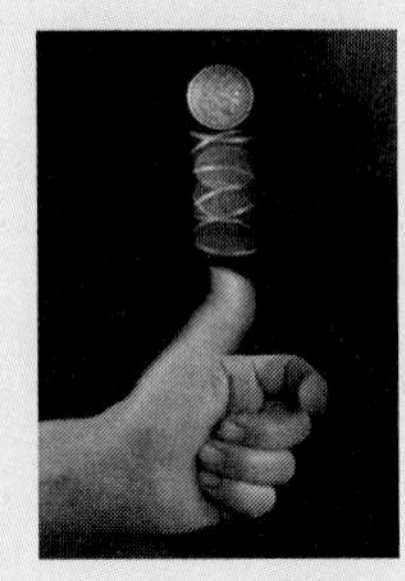

Note of Interest

No one expects a coin to fall heads once in every two tosses; however, in a large number of tosses the results tend to even out. For a coin to fall heads 50 consecutive times it would take 1 million people tossing coins 10 times a minute for 40 hours per week. This is so unlikely to happen that it would occur only once every nine centuries.

Page 200

Historical Notes enhance a discussion by providing historical context for a particular concept or topic.

Historical Note

The laws of chance as we know them today began with interests in gambling. A French nobleman, the Chevalier de Méré, asked a mathematician friend of his, Blaise Pascal (1623–1662), "how to split the pot in a dice game that has to be discontinued."

Pascal pondered the problem for some time and then relayed it to another mathematician friend, Pierre de Fermat (1601–1665). From the correspondence and research of Pascal and Fermat regarding various gambling situations, the theory of probability has evolved.

Page 199

Just for Fun present problems that range from serious extensions of mathematical ideas to light-hearted puzzles, recreational mathematics, or brainteasers. Answers to these are found in the back of the book.

Just for fun

How many cubes are shown in the figure?

Page 223

Chapter Review

- Summary reinforces the key concepts presented in the chapter.
- Key Terms and Concepts lists important terms with requisite page numbers so that they can be easily located within the chapter.
- Review Exercises provide the opportunity to practice the concepts presented within the chapter.
- Chapter Quiz enables you to test your understanding of the chapter concepts.

CHAPTER REVIEW MATERIALS

Summary / Chapter 3

3.2 Statements and Symbols

- In logic, *statements*, also called *propositions*, are either true or false, but not both. Statements are either simple or complex.
- *Simple statements* do not contain any connectives.
- *Complex statements* contain the connectives *not, and, or, if then* . . . , and *if and only if*. These statements respectively are called *negation, conjunction, disjunction, conditional*, and *biconditional*.

. . .

Key Terms and Concepts

biconditional, *118*
conclusion, *165*
conditional, *117*
conjunction, *117*
consistent, *174*
contradiction, *144*
contrapositive, *158*
converse, *158*
De Morgan, *154*
disjunction, *117*
iff, *118*
implication, *121*
inclusive "or", *121*
inconsistent, *174*
invalid, *166*
inverse, *158*
logically equivalent, *144*
negation, *117*
paradox, *115*
premise, *171*
statement, *115*
syllogism, *171*
tautology, *144*
valid, *165*

Review Exercises

3.2 Statements and Symbols

1. Identify each sentence as a simple statement, compound statement, or neither. If the statement is compound, then classify it as a negation, conjunction, disjunction, conditional, or biconditional.
   - **a.** Either Hugh sold his car, or he traded it in.
   - **b.** It is raining?
   - **c.** Yesterday was Friday.
   - **d.** Scott stayed home, but Joe went to the show.
   - **e.** Carlos did not bring his notes to class.
   - **f.** If Mary went swimming, then she did not study.
2. By means of the appropriate connectives and parentheses, symbolize each statement, using the given symbols for the simple statements:

3.4 Truth Tables for Negation, Conjunction, and Disjunction/3.5 Truth Tables for Conditional and Biconditional

5. Construct a truth table for each statement.
   - **a.** $\sim P \rightarrow Q$
   - **b.** $P \vee \sim Q$
   - **c.** $P \vee Q \leftrightarrow P$
   - **d.** $P \vee Q \rightarrow R$
   - **e.** $P \wedge \sim Q \rightarrow Q \vee R$
   - **f.** $\sim(P \wedge \sim Q) \rightarrow \sim(\sim P \vee Q)$

Use Figure 3.67 to determine the truth value of the given statements in Exercises 6–9.

Chapter Quiz

Indicate whether each statement is true or false.

1. The sentence "How old are you?" is considered to be a statement in logic.
2. "Jim is late or he is absent" is an example of a disjunction.
3. "Dana will graduate, if she passes the exam," is an example of a conditional.
4. Given P = *Today is Monday* and Q = *I am tired*, then $\sim P \wedge Q$ can be written as "Today is not Monday, but I am tired."
5. The answer column of the truth table for $P \rightarrow \sim Q$ is *T, F, T, F*.
6. $P \wedge (Q \vee \sim P)$ is a tautology.
7. "If you cut class, then you missed the exam" is logically equivalent to "Either you did not cut class or you missed the exam."
8. "It is false that you did not study and that you passed the exam" is logically equivalent to "You did study or you did not pass the exam."
9. The following argument is valid:

   If I work, then I will be paid.
   I did not work.
   Therefore, I was not paid.
10. A contradiction is a statement that is always false.
11. Statements that cannot be true at the same time are inconsistent.
12. The following argument is valid:

    All ants are busy creatures.
    All busy creatures are creative.
    Therefore, some ants are creative.

14. "If you work hard, then you will succeed" is the contrapositive of "If you don't succeed, then you didn't work hard."
15. "If you attend class, then you will pass" is the inverse of "If you did not pass, then you did not attend class."
16. "It snows only if it is cold" is logically equivalent to "If it snows, then it is cold."
17. The biconditional is the most dominant connective and the conjunction is the least dominant connective.
18. The disjunction is a more dominant connective than the conjunction.
19. Given $\sim(P \wedge Q \rightarrow \sim P)$, the answer column in the truth table for this statement would be found under the arrow.
20. Given P = *Today is Monday* and Q = *I am tired*, then $\sim P \rightarrow Q$ can be written as "Today is not Monday if I am tired."
21. The answer column of the truth table for $P \vee Q \rightarrow \sim R$ is *F, T, F, T, F, T, T, T*.
22. $P \wedge \sim Q \rightarrow Q \vee R$ is a contradiction.
23. The statement $\sim P \wedge Q \rightarrow \sim R$ is an example of a negation.

*24. Given the two premises

All students are industrious.
No politicians are industrious.

A conclusion that will make the argument valid is

Some students are politicians.

*25. Given the symbolic statement

$(P \wedge Q) \vee (\sim P \wedge \sim Q)$

the corresponding network would be described as a series

Pages 190, 192, 194

Fundamentals of Mathematics

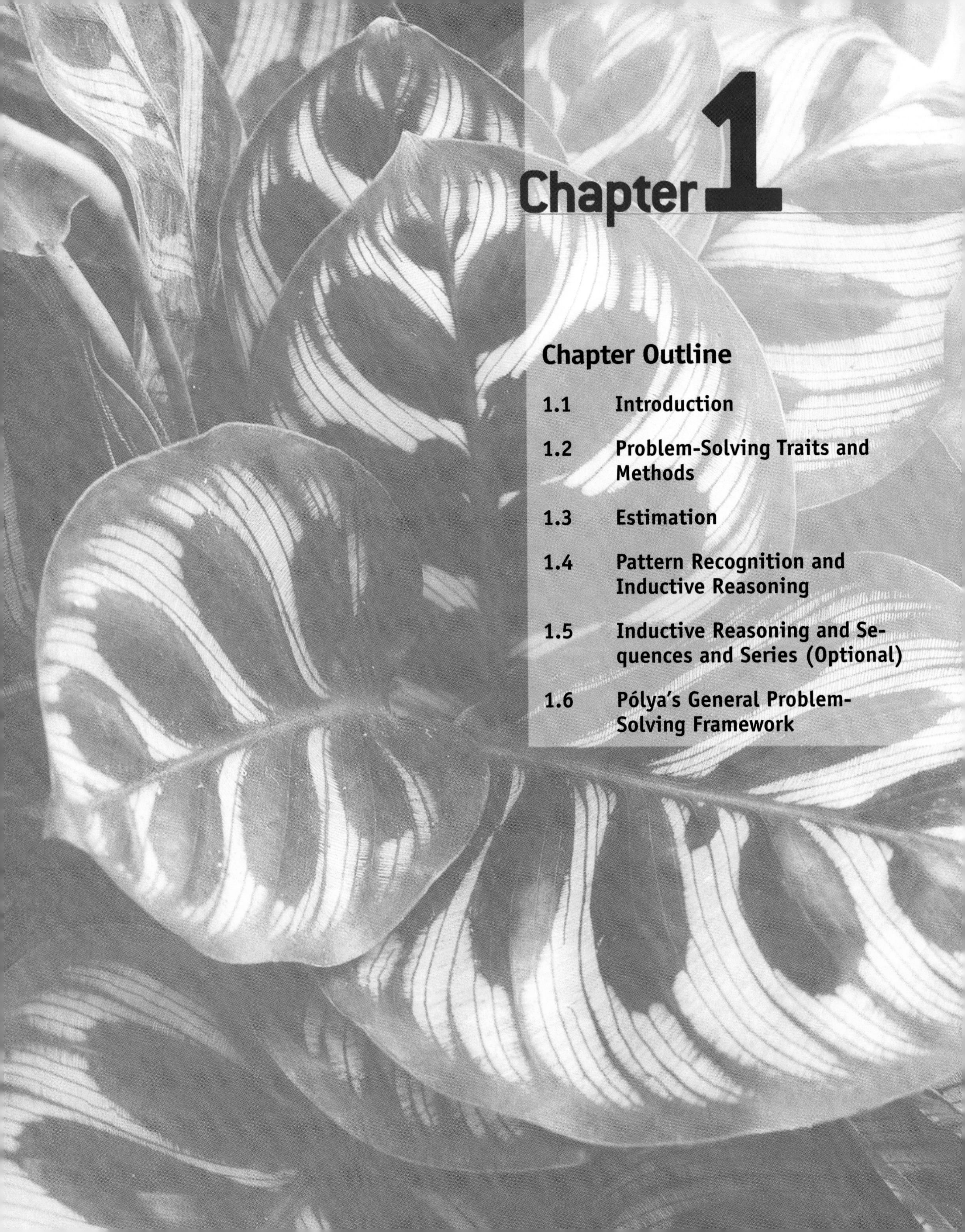

Chapter 1

Chapter Outline

Fundamentals of Problem Solving

After Studying This Chapter, You Will Be Able to Do the Following:

1. Identify and use problem-solving strategies such as reasoning, pattern recognition, and questioning.
2. Use estimation to approximate the solution to problems.
3. Understand the difference between inductive and deductive reasoning.
4. Use pattern recognition and inductive reasoning for prediction.
5. Apply Pólya's problem-solving model.

Note: *indicates optional material.

WWW Visit the PH Companion Web site at www.prenhall.com/setek

OVERVIEW

Problem solving is a part of nearly everything we do. Solving problems, however, is not always an easy task and rarely does it involve a single step. This chapter discusses the concept of problem solving and provides various strategies that can help make us better problem solvers. Key among these strategies is pattern recognition. To quote English social historian, philosopher, and essayist Sir Isaiah Berlin (1909–1997), "To understand is to perceive patterns." Patterns are all among us—it is up to us to discover and apply them to help solve problems.

Patterns often occur in nature. An example appears in the photo in the chapter opener page. In Section 1.4 we discuss the concept of pattern recognition and in Section 1.5, we introduce a pattern known as the Fibonacci sequence, which is found throughout nature.

1.1 INTRODUCTION

Every day we face opportunities that challenge our ability to solve problems. For example, suppose your car does not start when you leave for work. What do you do? In your mathematics class, if you do not understand a homework assignment, what do you do? If you do not get a dial tone when you pick up your telephone handset to call someone, what do you do? Each of these situations illustrates everyday problems that require solutions.

This chapter explains some basic concepts related to problem solving. In Section 1.2, we consider various human characteristics that foster problem solving. We also provide an overview of several different problem-solving methods. These include reasoning, finding patterns, understanding and questioning, thinking critically, and estimating. In the remaining sections of the chapter, we examine many of these methods in greater detail. For example, Section 1.3 discusses estimation, and Sections 1.4 and 1.5 extend the concepts of logical reasoning and pattern recognition to help us arrive at conclusions. Finally, Section 1.6 introduces a framework, developed by George Pólya in his book *How to Solve It*, which can be used as a general guide for solving problems. As part of our discussion in Section 1.6, we incorporate many of the methods presented in Section 1.2 to demonstrate how mixing and matching various strategies can facilitate the problem-solving process.

1.2 PROBLEM-SOLVING TRAITS AND METHODS

Problem solving is a "hands-on, minds-on" activity. It requires developing a concrete model that can be physically manipulated (hands-on), and it requires thinking actively about a problem and its possible solution (minds-on). Problem solving is not a one-time event that uses a single strategy. Instead, good problem-solving skills incorporate several different human traits as well as a variety of different techniques. Human characteristics that are critical to problem solving include patience, observation, and experience. A brief discussion of these traits follows.

Patience

Before acting, good problem solvers always take time to think through a problem. Patience is a virtue here because it gives us time to (1) think about the problem, (2) fully understand the problem and what it involves (including constraints and special conditions), (3) develop a "game plan" for problem resolution, and (4) assemble various strategies that can be used to solve the problem.

EXAMPLE 1 ***Patience*** At the end of a discussion on divergent thinking, Mrs. Pannell asks her students to consider why manhole covers are round. One of her students, Clay, goes home and immediately searches the World Wide Web for the answer but cannot find it. He asks his friends, parents, and siblings. No one knows the answer. After three hours of working on the problem he still does not have the answer. What can he do?

Solution Perhaps the best strategy Clay could use is to "sleep on it." In other words, he should set the problem aside and simply stop thinking about it. This concept of "sleeping on it" is called *incubation.* Incubation provides us with a period of time that (1) enables our "subconscious mind" to work on the problem; (2) allows us to stop making incorrect assumptions about the problem, which might be preventing us from solving it; and (3) permits new information to emerge that might provide the basis for a solution. Most of us have probably experienced at least one occasion where the solution to a problem or insight into its solution occurred when we were not thinking about the problem. This is the power of patience. In Clay's case, for example, his "Aha!" moment might come when he brings the trash pail (which is cylindrical) to the curb the next morning and realizes that the cover cannot fall into the pail.

Observation

A second human trait that is key to problem solving is observation. Human observation usually occurs at one of three levels. The first level, which we call **functional observation**, involves viewing an object as nothing more than an object. The second level, **recognition observation**, entails seeing an object in just enough detail so that we are able to identify a feature or attribute of the object or name it. The highest level of observation is **active observation**, which demands concentration and focus so we can provide detailed information about an object.

EXAMPLE 2 ***Recognizing Objects*** In each of the following descriptions, identify the type of observation illustrated.

a. A police officer asks the victim of a purse snatching to describe her assailant. Her response is, "He was a white male."

b. An art teacher at a local community college asks his class if there are any trees planted in the median of the expressway his students drive on en route to the college. No one can say for certain although they are sure something is planted in the median.

c. At the end of a house-warming party, Kay asks Lester if he noticed what Bernice was wearing. Lester says, "Sure. She was wearing clothes." Kay then says, "Men! She was wearing a new dress. It was an Armani and it was a very flattering blue. She also had on a new pair of shoes that matched her outfit."

Solution

a. The woman is operating at the recognition level of observation because she was able to identify the person as a white male but could not provide any further details.

b. Students in the class are operating at the functional level of observation. They know something is in the median but they are unable to identify it.

c. Lester's response illustrates functional observation; Kay's comment illustrates active observation.

NW ***Now Work Problem 3.***

Although good observation skills can help us in our problem-solving endeavors, we must be careful not to rely on it solely in a decision-making process. For example, in deciding if weather conditions are acceptable to go for a bicycle ride, we might go outside and observe the current conditions. Through our observations we might conclude that it looks like it might rain, or that it feels too cold, or that it is too windy and decide not to ride. Without getting additional input, however (e.g., watching the local forecast on The Weather Channel), our decision not to ride might be incorrect. Nevertheless, effective problem solvers are active observers. By thoroughly and carefully examining a problem, they can identify common themes or recognize distinct patterns that can provide insight into solving a problem or making a decision. In subsequent sections of this chapter, we will help sharpen your observation skills by presenting examples that facilitate problem solving through pattern recognition.

Experience

A third human trait effective problem solvers possess is experience. The more problems we solve, the better we become at problem solving. This is because the knowledge we gain from solving a problem in one setting can be used to help us solve a different problem in another setting.

EXAMPLE 3 ***Using Experience*** Sue Ann discovered on several occasions that whenever a lamp was not working it was because the bulb was "bad," the lamp's cord was not plugged into the wall socket, the switch was not turned on, or because the circuit breaker that serviced the electrical outlet "tripped." In what other settings can Sue Ann apply this knowledge?

Solution Sue Ann's experience can be applied to situations involving other electrical devices. For example, if her car radio isn't working, it could be because it hasn't been switched on or it might be because the fuse that protects the electrical circuit to which the radio is connected burned out. If her computer printer isn't working it might be because it either isn't plugged in or switched on. There is no guarantee that these are correct solutions, but **knowledge transfer** does provide a *probable solution.* This ability to transfer knowledge from a past experience to a later one is a cornerstone to problem solving.

In addition to human traits, good problem solvers also possess a "bag of tricks." That is, through their problem-solving experiences they develop a repertoire of different problem-solving methods. Some of the more common ones include reasoning, finding patterns, understanding and questioning, thinking critically, and estimating. A brief discussion of each method follows. Then, in later sections, we examine these techniques in more detail.

Reasoning

The obvious goal of problem solving is to solve the problem. In our attempts to find a solution we often make decisions or conclusions based on some *input* we receive. Those decisions are usually the result of some reasoning process. There are two general methods of reasoning we use to arrive at a conclusion: inductive and deductive.

Inductive reasoning is a process that involves making a decision, prediction, or conclusion based on an observed pattern. The conclusion we make from our observations is called a **conjecture**. For example, an examination of five lamps that do not work might reveal that all have defective light bulbs, which might lead to the conclusion: All nonworking lamps have bad bulbs.

When arriving at a conjecture, we must be careful not to over generalize. That is, we should not claim or infer our conjecture as fact or truth. This is because with inductive reasoning it is not feasible to examine every conceivable case; it takes only one *counterexample* to show that a conjecture is not always true. For example, in the preceding lamp illustration, it is possible that after examining a sixth lamp we find it is not working because it isn't plugged into the electrical socket. Thus, our previous conjecture that all nonworking lamps have bad bulbs no longer holds.

EXAMPLE 4 ***Inductive Reasoning—The Effectiveness of a Flu Shot***
Assume that over the past five flu seasons, 2 million people were given a flu shot. If none of these 2 million people got the flu, can we conclude that getting a flu shot will prevent us from getting the flu? If one person of the 2 million people who received the vaccine got the flu, can we conclude that getting a flu shot will prevent us from getting the flu?

Solution Most people will answer yes to the first question. In the second question, does the one "failure" disprove our conclusion that getting a flu shot prevents us from getting the flu? Probably not. This is because we do not expect the vaccine to work in every case, but it will work in most cases. In other words, anyone who gets a flu shot will *probably* not get the flu. This logical reasoning is an example of inductive reasoning. We are arriving at a conclusion based on specific observations.

Inductive reasoning is an excellent strategy to help identify *probable solutions*. Most of us use inductive reasoning on a fairly regular basis to make decisions. Inductive reasoning is also used in science to help explain scientific phenomena. Nevertheless, we still must be cautious about making any bold generalizations resulting from an inductive reasoning process because an underlying presumption of inductive reasoning is that the present or future will resemble the past. Consequently, any conclusions are at best probable and not necessarily true. We examine inductive reasoning in more detail in Sections 1.4 and 1.5.

Math Connections

The Scientific Method

Two of the problem-solving methods discussed in this section—namely, observation and inductive reasoning—serve as the foundation of the *scientific method*, which is a process used by scientists to construct an accurate representation of the world. In its most simple form, the scientific method involves the following steps:

1. Observe a particular phenomenon or some aspect of our world or universe.
2. Develop a tentative description (i.e., a *hypothesis*) to explain what was observed.
3. Use the hypothesis to make conjectures or predictions about the existence of other phenomena, or to predict the outcome of new observations.
4. Test the hypothesis experimentally and modify the initial hypothesis according to the results of the experiments. (For example, if we find a counterexample, then the hypothesis no longer holds and hence must be modified.)
5. Continue repeating steps 3 and 4 until there are no discrepancies between what was hypothesized and the results of experiments or subsequent observations.

Once step 5 is completed, consistency has been obtained between the hypothesis and experimental results and the hypothesis is now considered a *theory* or a *law*. Thus, a theory or law represents a hypothesis (or a group of hypotheses) that has been confirmed through repeated experimental tests. This theory then serves as the framework that provides us with an explanation of our observations. It also enables us to make predictions within this framework. Consider, for example, the law of gravity, which explains why objects fall to the ground. The scientific method can be applied to everyday problem solving. For example, if our telephone doesn't work, we could formulate hypotheses that explain the problem (e.g., the inside wiring might be defective, the phone company could have a problem with its equipment, or the phone itself might be defective) and then test these hypotheses. The results of our tests will determine whether we modify the initial hypothesis.

A second form of logical reasoning is **deductive reasoning**, which is a process that involves making decisions based on previously accepted and proven facts. Deductive reasoning begins with statements that are considered true and shows that other statements logically follow from them. By correctly applying accepted statements of truth to a specific example, we can produce a conclusion that must be true.

EXAMPLE 5 ***Deductive Reasoning*** If we agree that today is Thursday, can we conclude that tomorrow is Friday?

Solution The answer is yes. Because we agree that the first statement is true (i.e., that today is indeed Thursday), then the second statement is a logical consequence of the first statement. In a deductive reasoning process, statements that are assumed to be true are called *hypotheses* or *premises*, and the statement that logically follows the premises is called the *conclusion*.

The concept of deduction is extremely important in mathematics. One form of deductive reasoning we will study in Chapter 3 is the concept of a *syllogism*, which consists of two premises and a conclusion.

EXAMPLE 6 ***Deductive Reasoning—Logical Arguments*** If we assume that the following two statements are true, what conclusion can we draw from them?

All fish can swim.
A shark is a fish.

Solution If we agree that the first two statements are true, then since a shark is a fish and all fish can swim, a logical conclusion is that *a shark* can *swim*. Collectively, these three statements comprise an *argument*. The first two statements are called the premises and are assumed to be true; the statement that logically follows from the premises is called the conclusion. One of our goals in Chapter 3 will be to determine if a given argument is valid or invalid. This is done by examining an argument's conclusion to see if it follows logically from its premises. If it does, then the argument is valid; otherwise it is invalid.

When using logical reasoning methods as a problem-solving strategy, it is important that we be able to distinguish between inductive and deductive reasoning. The key to making this distinction correctly is to determine if the conclusion is a necessary condition or simply a probable one.

In inductive reasoning, the conclusion is never more than probable. In deductive reasoning, the conclusion is inescapable if the hypothesis is accepted as true.

Finding Patterns

Searching for and analyzing patterns are an integral part of problem solving. Finding patterns that are inherent within a problem can help us discover the solution to the problem. For example, in trying to solve a string of burglaries, an investigating detective might focus on days of the week in which the burglaries occurred, time of day, or location. If a pattern exists among the individual burglaries, the detective might then be able to predict when or where the next one will occur, or identify possible suspects who might fit a certain profile. The key to finding patterns is observation.

EXAMPLE 7 ***Finding Patterns*** Consider the following four words:

indigo *onions* *forms* *tomatoes*

a. What is common among these words?

b. Which of the following words do you think could be added to the first set of words so that the pattern or common theme is maintained?

feisty *wither* *plums* *bypass*

Solution

a. After carefully studying the first set of words, notice that each word begins with a preposition, namely, *in, on, for,* and *to*.

b. Once we identify the pattern, we know that there are two words in the second set, *wither* and *bypass*, which can be included with the first set of words.

Pattern recognition is not always easy. It requires active observation as well as good reasoning skills. It is, however, a very efficient method of problem solving. In Sections 1.4 and 1.5, we examine pattern recognition in more detail as part of our discussion on inductive reasoning.

NW ***Now Work Problem 21.***

Understanding and Questioning

Problem solving is an active process that involves understanding *all* aspects of a given problem. To make a decision about something we must first understand exactly what is being asked, what the constraints are (if any), and what certain components mean for a given situation (i.e., context). One way to acquire this understanding is to ask questions. If we do not understand a given problem or something about it, then a questioning strategy is needed. Various questioning strategies are available, and there are advantages and disadvantages to each strategy. One questioning strategy is to follow the lead of a litigator who tries to arrive at a conclusion by asking a series of "yes–no" questions.

EXAMPLE 8 ***Investigating An Auto Accident*** Develop a series of yes–no questions that might help a police officer arrive at a conclusion about the cause of an automobile accident.

Solution Many different questions can be asked. Following is one possible set of questions.

- Were you going fast?
- Were you listening to the radio, a tape, or CD?
- Were you talking on your cell phone?
- Were you distracted by something at the time of the accident?
- Did you drink any alcoholic beverages before you started driving?
- Was anyone else in the car with you?
- Were you wearing your corrective lenses?

Another questioning strategy is to ask open-ended questions instead of yes–no questions. Open-ended questions tend to provide us with "rich" information, which can provide greater insight or depth to a particular problem.

EXAMPLE 9 ***Investigating a Computer Problem*** Kelly discovers that her computer is not working. Develop three open-ended questions that will solicit from Kelly responses that could lead to solving the problem.

	0	@	P	`	p
!	1	A	Q	a	q
"	2	B	R	b	r
#	3	C	S	c	s
$	4	D	T	d	t
%	5	E	U	e	u
&	6	F	V	f	v
'	7	G	W	g	w
(	8	H	X	h	x
)	9	I	Y	i	y
*	:	J	Z	j	z
+	;	K	[	k	{
,	<	L	\	l	|
-	=	M	]	m	}
.	>	N	^	n	~
/	?	O	_	o	

Collectively known as *alphanumeric characters*, these symbols represent the foundation of written communication in the English language.

Solution Many different questions can be asked. Following is one possible set of questions.

- When was the computer last working?
- What changes occurred between the last time it worked and now?
- What possible solutions have you attempted to date and what were their outcomes?

NW ***Now Work Problem 27.***

Another strategy that helps us understand a problem is to study the *context* of the problem. In our language, for example, we rely on various symbols for written communication. These include the 26 letters of the alphabet (both uppercase and lowercase), the digits 0 through 9, and various other symbols, such as commas, dots, question marks, and parentheses. Each symbol is unique in form and can be used to represent different ideas or concepts. For example, the letter A can be used to represent the first letter of the English alphabet, or to signify the first item in a sequence or group, or to denote excellence (e.g., receiving an A on an exam), or to represent a particular sound (e.g., the musical note A). What enables us to distinguish among the various meanings is its context. Once its context is known, we can then interpret the symbol's meaning. Note that the interpretation (or meaning) of a symbol is also unique to a particular context. If it weren't, then there would be ambiguity. When solving problems, it is necessary to first identify the context in which symbols are being used before we assign meaning to them.

EXAMPLE 10 ***Interpreting Lyrics*** In the song "Wooden Ships" by David Crosby and Stephen Stills, the lyrics begin with the following sentence:

If you smile at me I will understand, 'cause that is something everyone does in the same language.

Using the concept of context, interpret the meaning of this sentence.

Solution Key to understanding the meaning of this statement is knowing what is meant by the word *smile*. As with many words, the word *smile* can be interpreted in more than one way. For example, a smile can be symbolic of good will, kindness, friendliness, or compassion. Or a smile can imply mischievous amusement (e.g., an impish grin), or a smirk or sneer, which would imply sarcasm (e.g., "He sneered at his mother as she disciplined him."). To distinguish among these different meanings, we need to understand the *context* in which the word is used. The context in which the word *smile* is used in the example leads us to interpret a smile as a symbol of benevolence. As a result, this statement means that regardless of who you are or where you come from, a smile represents a sign of friendship.

EXAMPLE 11 ***Identifying Word Context*** Consider the word that is spelled i-n-v-a-l-i-d. Identify two contexts in which this word has different meanings depending on how it is pronounced.

Solution

a. If the word is pronounced as IN-va-lid, then it would imply ill health.

b. If the word is pronounced in-VAL-id, then it would mean something that is not valid.

TABLE 1.1 Common Mathematical Symbols

Symbol	English Translation
$=$	Is equal to
$\neq$	Is not equal to
$<$	Is less than
$>$	Is greater than
$\leq$	Is less than or equal to
$\geq$	Is greater than or equal to
$\sim$	Not
$\approx$	Is approximately equal to

Mathematics, like our English language, also uses symbols as a form of written communication. (See Table 1.1 for a list of some common mathematical symbols and their meanings.) Mathematical symbols are used as a special kind of shorthand notation to represent expressions or model problems. The interpretation of mathematical statements also involves recognizing the context in which the symbols are being used. Thus, it is important that we first identify the context in which these symbols are used before we assign meaning to them.

EXAMPLE 12 ***Assigning Context*** Is the mathematical statement $9 + 5 = 2$ a "true" statement?

Solution At first glance, our initial reaction to this statement might be, "That's wrong. Nine plus five is equal to fourteen, not two." However, without knowing the context in which these symbols are being used we cannot determine whether or not this statement is true or false or interpret its meaning correctly. For example, if the digits represent "hours" of a 12-hour clock system, then the statement implies that 5 hours past 9 is 2, which is true. Alternatively, if the digits represent the months of the year numbered consecutively beginning with one from January to December, then the statement implies that five months past September is February, which is also true. If, on the other hand, the digits represented distance in meters, then the statement is obviously false because adding 5 meters to 9 meters does not yield 2 meters. Once again, note the importance of first understanding the context in which symbols are being used in a given situation *before* making any interpretations or conclusions. The ability to understand symbols and the context in which they are used in a given statement is an important component of problem solving.

EXAMPLE 13 ***Assigning Context*** Given the statement A-B-D-C-C-F = *C*, provide a context that would make this a true statement.

Solution The statement would be true if the letters represented grades earned for courses taken by a student. Notice that if we assigned the numbers 4, 3, 2, 1, 0, respectively, to the letters A, B, C, D, and F, then a student who received these grades would have a grade point average (GPA) of 2.0, which is a C. Also, in the given statement the symbol – does not mean subtraction; it is used merely to separate the letter grades. It is important first to understand the context in which symbols are being used before interpreting the statement containing them.

NW ***Now Work Problems 29 and 43.***

Thinking Critically

Critical thinking involves investigating a problem from different perspectives. This involves two general approaches: top-down processing and bottom-up processing. **Top-down processing** is a systematic method of reducing a large or complicated problem into smaller but related independent ones, which can then be solved more readily. Depending on the nature of the problem, some of the smaller components may need to be reduced further into still smaller units. This process is sometimes called a *modular approach* to problem solving or the "divide and conquer" method because we deal with a large task by completing it one small step at a time. These small steps or modules are solved independently of each other and then joined together to form the solution to the problem. Collectively, the entire set of components forms a *hierarchy,* which is organized in a treelike fashion that comprises a top level (the original problem), followed by second-level components, followed by third-level components, and so forth.

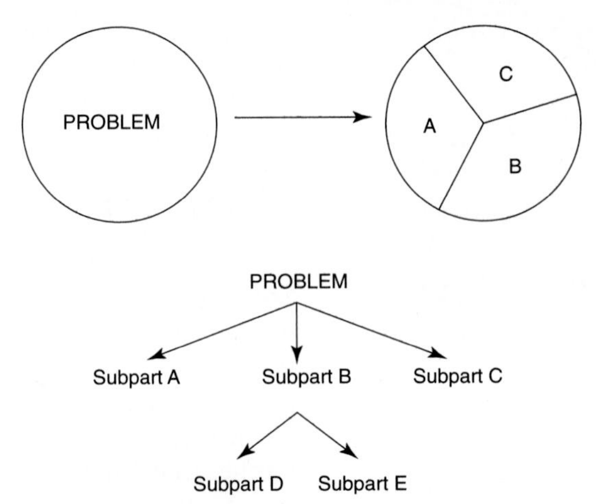

Top down processing involves dividing a problem into easier-to-solve smaller problems. This partitioning generates a hierarchy of problem components (i.e., subparts) that are first solved separately and then reassembled to solve the overall problem.

EXAMPLE 14 ***Organizing a Deck of Playing Cards*** Use top-down processing to order numerically a standard shuffled deck of 52 playing cards.

Solution We begin by first separating the cards by suits (i.e., clubs, hearts, diamonds, and spades). Working with each suit independently, we next arrange the suits numerically (i.e., ace through 10 followed by jack, queen, and king). Finally, with all suits numerically ordered, we combine them to form a completely numerically arranged deck of cards.

In Example 14, notice how we reduced a complicated problem into four independent simpler ones that were less difficult to solve. This approach is similar to the way a textbook author prepares a book. The textbook is first divided into chapters, chapters are divided into sections, and sections are further partitioned into subsections if necessary. The concept is to begin with the ultimate goal or task and then divide it into smaller, more manageable components.

Bottom-up processing, on the other hand, relies on intuition or creativity. For example, instead of numerically arranging a standard shuffled deck of 52 cards as in Example 14, we might have arranged the cards

- by suit without regard to numerical order
- by grouping like cards (e.g., all aces, all twos, etc.)
- by grouping picture cards versus nonpicture cards

Math Connections

Working Backward

The method of working backward is an extremely powerful problem-solving approach that has served as the foundation on which various products are designed in the computer industry. Using a process known as *reverse engineering*, manufacturers are able to take a product whose design is protected by patent or copyright laws and develop their own product that performs in exactly the same manner as the initial product without violating any patents or copyrights. Many professional Web page designers also use the concept of reverse engineering in their profession. A designer will first sketch out the "look and feel" of a client's Web page based on client input, and then, once the initial page is designed, the corresponding computer code is written.

Bottom-up processing also involves other strategies, including *working backward* and *symbolic manipulation*. Working backward is the process of starting at the end and working toward the beginning. Symbolic manipulation involves rearranging information so that we can view it from a different perspective.

EXAMPLE 15 ***Determining a Test Grade*** Shane, a student in Mrs. Edgar's College Algebra class, has a test average of 74 based on the six unit exams Mrs. Edgar has given to date. Mrs. Edgar is planning on giving only one more test and Shane needs an 80 average to get a B for the course. Is it possible for Shane to get an 80 average, and if so, what must his final test score be?

Solution Although many students with an algebra background might solve this problem algebraically, we can also use a working-backward approach:

- If Shane has a 74 average after 6 tests, then he has a total of $74 \times 6 = 444$ points.
- If Shane's average after 7 tests is 80, then he will have a total of $80 \times 7 = 560$ points.
- Thus, Shane must get $560 - 444 = 116$ points on the final test.

Assuming that the final test is worth 100 points and that Mrs. Edgar is not giving a bonus problem worth at least 16 points, then Shane cannot get an 80 average.

EXAMPLE 16 ***Constructing Vanity Plates*** Using no more than eight characters, rearrange the symbols of English and mathematics to construct automobile "vanity plates" for the given expressions:

a. Two for tea and tea for two
b. I see you are the one
c. Why are you lazy?
d. For once I am greater than you

Note: Unlike automobile license plates, which are limited to letters and numbers, we will include mathematical symbols, punctuation marks, and other symbols found on a keyboard in our construction of vanity plates.

Math Connections

Creating Strong Computer Passwords

The concept of "vanity plates" is often used to create strong passwords on computer systems. Unlike weak passwords—which are easy to guess or "crack" and include a person's name, phone number, dictionary word, or birthdate—strong passwords are difficult to crack because they include a mix of uppercase and lowercase letters, digits, punctuation symbols, and special characters such as =, *, ^, and @. Given that most computer systems accept only the first eight characters of a password, vanity plate expressions can effectively serve as strong passwords. Furthermore, since vanity plate expressions represent cute or meaningful phrases, they are also easy to remember and hence, if they are used as passwords, we do not have to write them down. A password committed to memory is much more secure than one that is written down.

In addition to vanity plates, two other suggestions for creating strong passwords include the following:

- Intermix the first letters of an easy to remember (short) phrase with digits, punctuation symbols, or special characters. For example, *It was 20 years ago today* can be represented by intermixing each word's initial letters, IWTYAT, with the special characters $ and ^ to produce the password: Iw$ty^aT. Note that we also chose to use a mix of upper- and lowercase letters because most computer systems are case sensitive.
- Combine two relatively short words with a special character, digit, or punctuation symbol. For example, *buzz* and *off* combined with the tilde character produces the password: BuzZ ~ OfF.

Given today's highly computerized and networked world, security is paramount and we should all do our part to prevent unauthorized access to our systems. Remember: *A single user with a weak password can compromise the security of an entire computer system or network.*

Solution

a. Two for tea and tea for two: *24T&T42*
b. I see you are the one: *Icurthe1*
c. Why are you lazy?: *Yrulaz?*
d. For once I am greater than you: *41cIm > u* 

EXAMPLE 17 ***Translating Character Strings to English*** Write an English expression for the given character strings:

a. Dont4get
b. URgr8
c. 1c&4all
d. cu@62day

Solution

a. Dont4get: *Don't forget*
b. URgr8: *You are great*
c. 1c&4all: *Once and for all*
d. cu@62day: *See you at 6 today* 

NW ***Now Work Problems 49 and 55.***

Estimation

The last problem-solving method we discuss is estimation, which involves forming an approximate response or solution to a given situation. For example, a general contractor might provide an estimate of the *cost* to remodel a house, or we might estimate the *distance* between two cities, or we might estimate the *time* it will take to complete a project. In each case, the cost, distance, and time estimates are only *approximate* values and not exact answers.

When we approximate numbers, we frequently use a process called *rounding*. To round a number means to give a ball park figure. In other words, we find another number that is about equal to the given number but not as precise. In its most general form, rounding is achieved by the rule, "Five or more, round up; less than five, round down." To implement this rule, we have to first identify which part of the number is being rounded. For example, given the number 1248,

- If we are rounding to the *nearest tens* (48), then we round up to 12**50**.
- If we are rounding to the *nearest hundreds* (248), then we round down to 1**200**.
- If we are rounding to the *nearest thousands* (1248), then we round down to **1000**.

EXAMPLE 18 ***Using Rounding to Estimate an Answer*** In preparing for its twelfth annual Intracoastal Waterway Century (a 100-mile bicycle ride), the Spacecoast Freewheelers Association is planning on supporting four rest stops. Last year, each rest stop, respectively, used 196, 322, 555, and 977 gallons of water. If Irv is in charge of ordering water for this year's event, approximately how many gallons of water should Irv order?

Solution If Irv follows the general rule of rounding, then 196 is rounded to 200, 322 is rounded to 300, 555 is rounded to 600, and 977 is rounded to 1000. Thus, Irv estimates that he should order $200 + 300 + 600 + 1000 = 2100$ gallons of water.

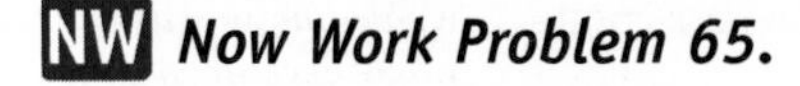

Estimation is a very handy problem-solving method. It is also an extremely important one given our increasing dependence on technology. This is because technology is not infallible. For example, hardware problems can result in errors, correct data that are entered incorrectly can result in errors, and incorrect data entered correctly can also lead to errors. Estimation skills enable us to verify that computations are, at the very least, reasonable answers to a given problem. We will discuss estimation in more detail in Section 1.3.

After reading this section, you might be asking yourself, "Which problem-solving method should I use" or "Which method is the best?" The answer to both questions is, "It depends." Effective problem solvers do not always rely on a single strategy. Instead, they achieve greater success by mixing and matching various strategies to serve their needs. In Section 1.6, we will examine this "mix and match" approach in detail. For the time being, however, you are encouraged to follow the lead of a professional golfer who is about to make a critical putt. Instead of simply stepping up to the ball and putting, the pro will study the situation from different perspectives: by walking around the green, by crouching down at different locations to examine the green at ground level, by talking to the caddy for additional insight. In short, the pro gains as much information as possible before attempting the putt. Before you attempt to solve a problem, you, too, need to acquire as much information as possible about the problem.

Exercises for Section 1.2

1. Identify a problem that you were able to solve after you set it aside for a while or "slept on it."
2. Give an illustration of how one of your past experiences helped you to solve a problem.

In Exercises 3–5, identify the type of observation strategy (functional, recognition, active) that is needed and why.

3. NW **Secret Service** A U.S. secret service agent who is assigned to guarding the president of the United States
4. **Catching a Flight** A businessman who is running through an airport to get to the departure gate on time (Assume that the businessman is familiar with the airport.)
5. **Going to the Movies** A movie theatre patron who leaves her seat prior to the start of the movie to get popcorn but will return to her seat after the movie has started

Using Reasoning Skills

In Exercises 6 through 15, state whether the reasoning process being used is inductive or deductive. Also determine whether the conclusion given is justified. Explain your answer.

6. Tim earns an A on his first two mathematics exams. He concludes that he will get an A on the remaining three exams of the semester.
7. NW **Pet Dogs** Melinda adopted a female rottweiler and discovered that the dog is extremely smart. She decides that all rottweilers are smart and encourages her mother to get one.
8. **Eating Dinner** Ralph and Elaine have dinner at a new restaurant and are dissatisfied with the food. They conclude the restaurant is "bad" and decide never to go there again.
9. **Car Problems** Jane discovered that the reason her car would not start was because of a new safety feature that linked the clutch to the ignition: In order to start the car the clutch pedal has to be depressed. Jane concludes that whenever her car won't start it is because of the clutch.
10. A visitor at a local college walks into the physics lab and observes that all of the students present are males. He concludes that only males take physics.
11. Every student taking Mr. Ketes's mathematics course at a local community college works full time during the day. Gerlando is taking Mr. Ketes's college night mathematics class. Therefore, Gerlando works full time during the day.
12. **Asphalt Paving** When it is raining outside, employees of Bennett Asphalt Paving Company do not work. Employees of Bennett's are not working today. Therefore, it must be raining.
13. **Riding a Bicycle** Mike rides his bicycle with Jane every Saturday and Sunday. Mike rides his bicycle today. Therefore, today must be Saturday or Sunday.
14. Item a is larger than item b. Item b is larger than item c. Therefore, a is larger than c.
15. NW All Florida residents are also residents of the United States. Jesse is not a resident of Florida. Therefore, Jesse is not a resident of the United States.

Finding Patterns

16. If we write the consecutive odd numbers as a triangular list as shown and then add each horizontal row, an interesting pattern emerges. Identify the pattern.

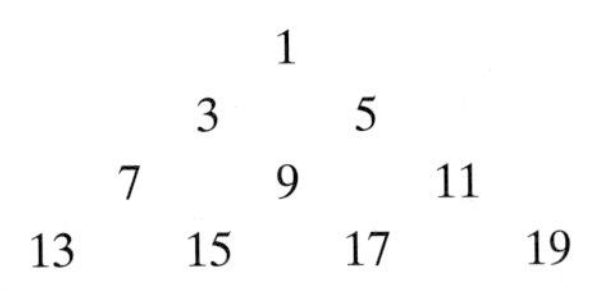

17. **Pascal's Triangle** Another interesting triangular list of numbers is *Pascal's triangle.* Identify the pattern of this list and write the next two rows.

1
1 1
1 2 1
1 3 3 1
1 4 6 4 1

In Exercises 18–20, find the next three terms in the given list of numbers.

18. 2, 4, 6, 8, 10, 12
19. 3, 6, 12, 36, 39, 78
20. 1, 1, 2, 6, 6, 12

In Exercises 21 and 22, find the next three figures.

NW **21.**
22.

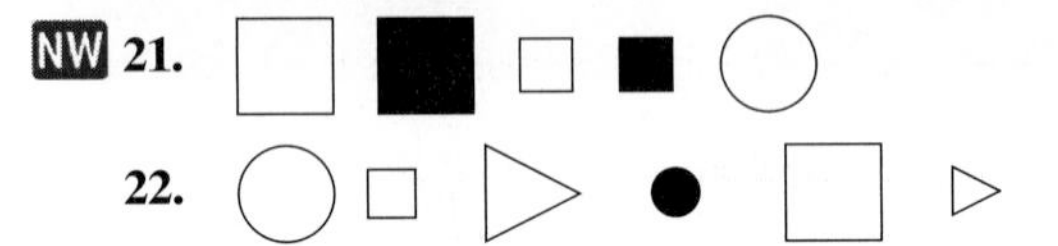

Asking Questions

In Exercises 23–27, prepare a set of questions (either yes–no or open ended) that can be asked to help solve the given problem.

23. Your car does not start.

24. **Watching Television** When you power on your television set you get static on your screen with sound, but no picture.

25. **Classroom Lecture** A teacher is trying to decide whether to use a chalk and chalkboard, transparencies and an overhead projection unit, or Microsoft Corporation's *PowerPoint* and a computer for her lecture.

26. **Auto Purchase** You are trying to decide whether to purchase a car or a sport utility vehicle.

NW **27.** **Internet Connection** You make a connection to the Internet and your computer freezes up.

Understanding Context

Given the symbols in Exercises 28 through 37, (a) state at least two contexts in which each symbol can be used; (b) name and define each symbol as it would be used for each context; and (c) give an example of its use.

28. # NW **29.** – **30.** + **31.** ' **32.** x **33.** !
34. ( ) **35.** e **36.** , **37.** :

Each word given in Exercises 38 through 47 can be pronounced in more than one way. Write a sentence using the word for each pronunciation, and state the context in which the word is being used.

38. Project **39.** Wind
40. Lead **41.** Bow
42. Number NW **43.** Tear
44. Wound **45.** Live
46. Present **47.** Refuse

48. The following sign is posted in a restaurant window: "Breakfast $199." What does this sign mean to you? Explain.

NW **49.** The following sign is posted outside a theater window: "Tickets purchased before June 10th are $10.00; after June 10th, $12.00." What is wrong with this sign?

50. **Vanity Plates** Using the symbols of English and mathematics, construct 10 vanity plates that contain no more than eight characters. Write an English statement equivalent to each of the 10 vanity plates.

51. When writing, how do you distinguish between a lowercase L (l) and the numeral one (1)?

52. When writing, how do you distinguish between the letter Oh (O) and the numeral zero (0)?

53. Read the following telephone number: 773-8012. How did you read this number:

"Seven-seven-three-eight-zero-one-two"?

or

"Seven-seven-three-eight-Oh-one-two"?

Do you think it's important to distinguish between these two characters (0 vs. O) when either writing or reading passages that contain them? Explain your answer. Provide an illustration where you do not think it is necessary to make such a distinction.

54. Provide a context that will make the statement 12 + 3 = 3 true.

55. NW Provide a context that will make the statement 2200 = 10 true.

56. Provide a context that will give the expression :-) meaning.

57. Provide a context that will give the expression XXX meaning.

Working Backward

In Exercises 58–60, work backward to solve the given problem.

58. **Airlines-Airports** Jane and Michael arrived at Orlando International Airport to take a flight to Scotland. When they arrived, Jane looked at her watch and noted that they had one hour until their flight was scheduled to depart. The moment she looked at her watch, an announcement was made that their flight would be delayed two hours. While waiting, a second announcement was made that there was going to be an additional one-and-one-half hour delay. The flight finally departed at 8 P.M. What time did Jane and Michael arrive at the airport?

59. Fill in the blanks with the correct numbers so that the given solution is maintained.

$$\_\_\_ + 42 = \_\_\_ \times 2 = \_\_\_ - 39 = \_\_\_ \div 25 = 3$$

60. **Playing Darts** Consider the modified dart board shown in Figure 1.1. If the objective is to get a total of 95, what is the fewest number of darts that need to be tossed?

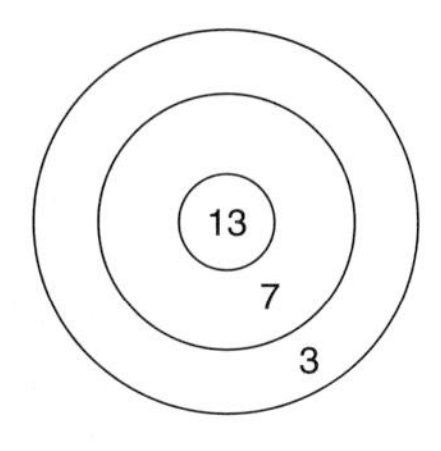

Figure 1.1

Estimation

In Exercises 61–65, use estimation techniques to solve the given problem.

61. **Patents** At the end of 1999, the U.S. Patent and Trademark Office issued patent number 6,000,000. If the first patent was issued in 1836, and patent numbers 1 million, 2 million, 3 million, 4 million, and 5 million were issued in 1911, 1936, 1961, 1976, and 1991, respectively, approximate the year when patent number 7 million will be issued.

62. Approximate the total number of seconds you have been alive.

63. **Heart Beats** Estimate the number of times your heart has beaten since your birth.

64. **Downloading a File** Estimate how long it takes to download a 5-Megabyte graphics file from the Internet if you are using a 56,000 bit per second modem. (*Note:* 1 byte = 8 bits and 1 Megabyte = 1,000,000 bytes.)

NW **65.** **Disney World** The Albergs are planning their annual trip to Disney World in Orlando, Florida, with their three children. Last year one airline ticket cost $283 and a single 5-day pass cost $64. Approximately how much should the Albergs budget for tickets and passes this year?

Writing Mathematics

66. Write a report that describes a problem-solving activity that you were involved in and what you did to solve it. Include in your report the strategy you used, your thought processes, and any errors you might have made.

67. Explain the difference between inductive and deductive reasoning.

68. Describe what you perceive to be the single most important problem-solving strategy you have learned from this section and why.

Challenge Exercises

In Exercises 69–72, identify the pattern that exists between each row of numbers and then use this pattern to write a rule that will generate the next two numbers. For example, if the x list contains the numbers 1, 2, 3, 4, 5, 6 and the y list contains the numbers 2, 4, 6, 8, then the rule is to multiply each number in the x list by 2 to get the corresponding number in the y list. This rule is written as y = 2x.

69.

x:	0	1	2	3	4	5
y:	1	3	5	7	?	?

70.

x:	0	1	2	3	4	5
y:	3	5	7	9	?	?

71.

x:	1	3	5	7	9	11
y:	0	6	12	18	?	?

72.

x:	2	4	6	8	10	12
y:	3	4	5	6	?	?

73. Make up a list of six numbers and suggest different patterns that can be used to generate the next three terms.

Group/Research Activities

74. **Using Analogies** A problem-solving strategy that was not discussed in this section is an analogy, which is based on previous experience. An analogy involves identifying a similarity between two seemingly unlike entities. For example, Gutenberg's invention of the printing press was influenced by the analogy he made to the wine press and the punches used for making coins. Research the concept of analogies and write a report that discusses how it can be used as a problem-solving strategy.

75. **Scientific Method** Write a report on the scientific method (see page 5) and discuss the role of inductive reasoning. As part of your report, discuss why Galileo, in his study of falling bodies, modified his initial hypothesis.

Real-World Data

Internet Spam *Use Figure 1.2 to answer questions 76–78.*

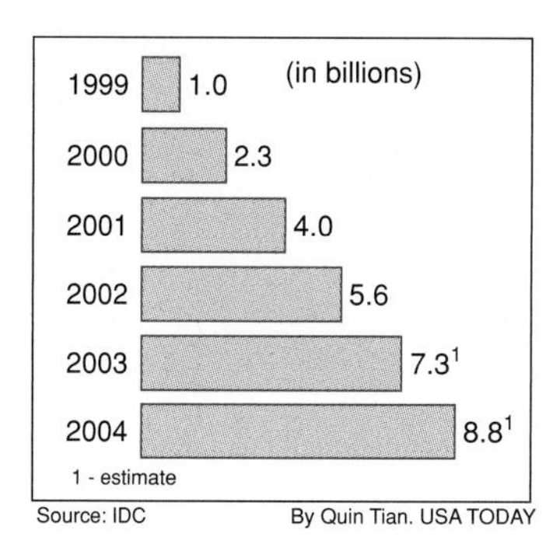

Figure 1.2
Worldwide spam messages sent daily.

76. Determine the *general* pattern of daily worldwide spam messages.

77. Assuming that the 2003 and 2004 estimates are correct and that the pattern were to continue, estimate the number of daily worldwide spam messages in 2005.

78. Do you think this pattern will continue? If so, why? If not, why not? As part of your response, state the reasoning strategy you are using to base your decision.

Student Loan Rates *Use Figure 1.3 to answer questions 79–81.*

79. Determine the *general* pattern of the federal student loan interest rates.

80. Assuming that the 2003–2004 projected rate is correct and that the pattern were to continue, estimate the federal student loan interest rates for 2004–2005.

81. Do you think this pattern will continue? If so, why? If not, why not? As part of your response, state the reasoning strategy you are using to base your decision.

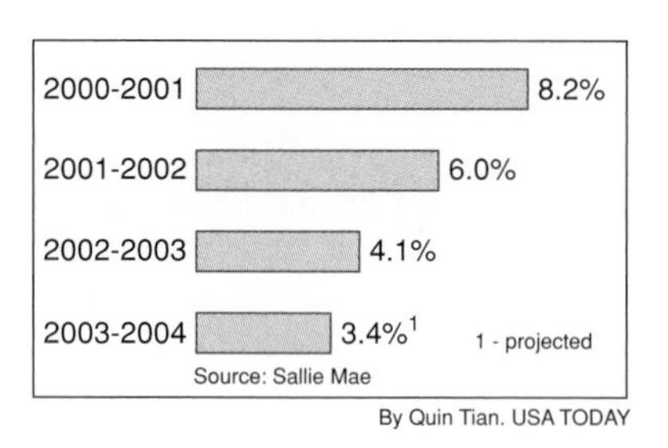

Figure 1.3
Federal student loan interest rates since 2000.

Just for fun

A car, traveling at a constant speed of 50 mph in a 25-mph school zone, approaches a parked police car at point *A*. The instant the speeder passes the police car, the police officer begins pursuing the speeder and eventually catches and stops him at point *B*. If the speeder maintains a constant speed and the police car accelerates at a constant rate, will the average speed of the police car be less than, greater than, or equal to the average speed of the speeder between points *A* and *B*?

1.3 ESTIMATION

As noted in Section 1.2, estimation is a process that involves forming an approximate response or solution to a given situation. Estimation is an important problem-solving technique because it enables us to mentally determine an approximate solution to a problem; it also can help us determine if a given solution is reasonable. Several techniques are available for estimating an answer. In Section 1.2, we briefly examined one of them: rounding. In this section we expand on this technique and introduce others.

Estimation by Rounding

In Section 1.2, we provided a general rule for rounding: "Five or more, round up; less than five, round down." The key to rounding is to examine the digit to the immediate right of the digit being rounded. With this in mind, we now introduce a more formal rule:

1. If the digit to the immediate right of the digit being rounded is 5 or greater, then we increase the digit being rounded by 1 and replace all digits to its right with 0s.
2. If the digit to the immediate right of the digit being rounded is less than 5, then we leave the digit being rounded unchanged and replace all digits to its right with 0s.

EXAMPLE 1 ***Using Rounding at the Grocery Store*** Jane makes a quick trip to the grocery store and purchases the following items: a half-gallon of freshly squeezed orange juice at $2.45, a half-gallon of skim milk at $1.19, four containers of fat-free yogurt at 69¢ each, two energy bars at $1.49 each, and one loaf of whole wheat bread at $2.69. Jane is in a hurry and wants to check out at the express lane because there is no waiting line. However, a sign at this lane reads "No more than 10 items and no more than $10."

a. Use rounding to estimate if Jane's purchases total more than $10.
b. If the total bill is $23.56, should Jane challenge it?

Solution

a. Before we answer this question, we have to decide if we will round to the nearest cent, dime, or dollar. In this situation, the quickest way for Jane to estimate her purchases is to estimate to the nearest dollar. (Why?) Doing so, yields a total of $12. Thus, her estimate is greater than $10.

Item	Original Price	Rounded to Nearest Dollar
Orange Juice	2.45	2.00
Milk	1.19	1.00
Yogurt	0.69	1.00
Yogurt	0.69	1.00
Yogurt	0.69	1.00
Yogurt	0.69	1.00
Energy Bars	1.49	1.00
Energy Bars	1.49	1.00
Bread	2.69	3.00

b. Based on Jane's estimate, $23.56 is unreasonable and she should challenge it. (The actual total is $12.07.)

Estimation by Compatible Numbers

In addition to rounding, we can also use *compatible numbers* to help us estimate an answer. This method is useful in problems involving products or quotients. What we do is substitute for the numbers making up a problem with special numbers that will make the calculation simpler. As an example, consider estimating the product of 137×9 by rounding and compatible numbers:

Rounding: $137 \times 9 \approx 140 \times 10 = 1400$

Compatible Numbers: $137 \times 9 \approx 137 \times 10 = 1370$

Note that with the compatible numbers technique, we simply substituted 9 with 10 because multiplying by 10 is relatively simple. This made it unnecessary to modify 137.

EXAMPLE 2 ***Using Compatible Numbers to Estimate Weight*** A sign mounted on a panel in an elevator reads, "Elevator Capacity: 3300 pounds (17 persons)."

a. Use compatible numbers to provide an estimate of the largest acceptable average weight of the 17 people.
b. Do you think the elevator can support 22 first-grade children? Why?

Solution

a. One estimate can be obtained by recognizing that 33 and 17 are close to the compatible numbers 32 and 16, which are multiples of each other (i.e., $16 \times 2 = 32$). Thus, we substitute 3300 with 3200 and 17 with 16. We can now divide mentally: 3200 divided by 16 is 200. Therefore, the approximate largest acceptable average weight of the 17 people is 200 pounds.

b. Most people will answer yes to this question. This is because elevator capacities are usually based on adults. Thus, a logical reasoning method will follow this train of thought: If 17 adults with an estimated average weight of 200 pounds each can fit in the elevator, then it is conceivable that 22 first-graders who might have an average weight of 50 pounds will fit also.

EXAMPLE 3 *Using Compatible Numbers to Estimate Gasoline Mileage*

Julia purchased a new Honda Accord and traveled 356 miles before refueling.

a. If she needed 15.6 gallons of gasoline to fill the car's tank, use compatible numbers to estimate her gasoline mileage.

b. If the cost of gasoline was \$1.43 per gallon, use compatible numbers to estimate the total cost of the gasoline.

Solution

a. Note that this is a division problem. We must divide the number of miles traveled by the number of gallons of gasoline needed. Also recognize that 15 and 300 are multiples of each other. Thus, if we substitute 356 with 300 and 15.6 with 15, then we get an estimate of $\frac{300}{15} = 20$ miles per gallon.

b. Note that 15.6 and 1.43 are both relatively close to 15. This is a key observation because $15 \times 15 = 225$. Thus, if we substitute 15.6 with 15 and \$1.43 with \$1.50, then we get an estimate of \$22.50.

In Example 3, part (b), the reason we chose to round 15.6 to 15.0 and 1.43 to 1.50 was so that we could take advantage of a special multiplication "trick" that enables us to multiply 15×15 readily easily. This trick is demonstrated in the examples below:

Note the pattern: Given any 2 two-digit numbers where the rightmost digits (i.e., the units place) add to 10 and the leftmost digits (i.e., the tens place) are the same, the product will always be composed of two parts. The left part is the product of the tens' digits plus one of these digits; the right part is the product of the units' digits. Thus, when working with compatible numbers, it is helpful to apply this "trick" if it is appropriate to round the numbers so they have the described relationship.

$17 \times 13 = 221$: $1 \times 1 + 1 = 2$; $7 \times 3 = 21$; 2 | 21

(a)

$26 \times 24 = 624$: $2 \times 2 + 2 = 6$; $6 \times 4 = 24$; 6 | 24

(b)

$39 \times 31 = 2709$: $3 \times 3 + 3 = 12$; $9 \times 1 = 09$; 12 | 09

(c)

$85 \times 85 = 7225$: $8 \times 8 + 8 = 72$; $5 \times 5 = 25$; 72 | 25

(d)

Front-End Estimation

Another estimation technique is a process known as *front-end estimation*, which involves tallying the leading digits of the numbers in an addition problem. This method is very simple and straightforward: Add the leading digits to estimate an answer. The only thing we need to be sensitive to is to make sure we combine "like" digits. For example, we can combine hundreds to hundreds, but we cannot combine hundreds to tens.

EXAMPLE 4 ***Using Front-End Estimation at the Grocery Store*** Use front-end estimation to approximate the cost of the groceries given in Example 1.

Solution Front-end estimation involves adding the leading digits of the original prices. This method yields an estimate of $2 + 1 + 0 + 0 + 0 + 0 + 1 + 1 + 2 = \$7$.

Item	Original Price	Front-End Estimation
Orange Juice	2.45	**2**.45 → 2
Milk	1.19	**1**.19 → 1
Yogurt	0.69	**0**.69 → 0
Yogurt	0.69	**0**.69 → 0
Yogurt	0.69	**0**.69 → 0
Yogurt	0.69	**0**.69 → 0
Energy Bars	1.49	**1**.49 → 1
Energy Bars	1.49	**1**.49 → 1
Bread	2.69	**2**.69 → 2

EXAMPLE 5 ***Estimating Populations*** Using the data from the following chart, estimate the Western Hemisphere's total population by rounding, using compatible numbers, and front-end estimation.

Region	Population
North America	306,364,000
Central America	136,439,000
Caribbean	36,026,000
South America	344,787,000

Solution

Region	Population	Rounding (Nearest Million)	Compatible Numbers	Front-End Estimation (Leading Digit)	Front-End Estimation (Leading 2 Digits)
North America	306,364,000	306,000,000	300,000,000	**3**06,364,000	**30**6,364,000
Central America	136,439,000	136,000,000	100,000,000	**1**36,439,000	**13**6,439,000
Caribbean	36,026,000	36,000,000	36,000,000	36,026,000	**3**6,026,000
South America	344,787,000	345,000,000	300,000,000	**3**44,787,000	**34**4,787,000
Total:	823,616,000	823,000,000	736,000,000	**7**00,000,000	**8**00,000,000

Now Work Problems 9 and 15.

Estimation Accuracy

A key aspect of any estimation technique is *accuracy*. That is, how accurate is an estimation? Given the three different estimation techniques discussed so far, which do you think provides the most accurate approximation? Furthermore, which estimation method should you use? If we were to compare the results shown in Example 5, we might be inclined to say rounding yields the most accurate estimation and hence we should always use it.

Before committing to any particular answer, though, let's examine this situation a little more closely. (Remember: This is a key feature of effective problem solving.) Notice what happens if we round the data in Example 5 to the nearest 100 million and then to the nearest 10 million. Our estimates are 700,000,000 and 830,000,000, respectively.

Region	Population	Rounding to Nearest 100 Million	Rounding to Nearest 10 Million	Rounding to Nearest Million
North America	306,364,000	300,000,000	310,000,000	306,000,000
Central America	136,439,000	100,000,000	140,000,000	136,000,000
Caribbean	36,026,000	0	40,000,000	36,000,000
South America	344,787,000	300,000,000	340,000,000	345,000,000
Total:	823,616,000	700,000,000	830,000,000	823,000,000

Comparing these results to those given in Example 5, rounding to the nearest 100 million is less accurate than compatible numbers and front-end estimation using the leading two digits, but the same as front-end estimation using only the leading digits. Rounding to the nearest 10 million, however, is more accurate than any of the other methods except rounding to the nearest million. This example reveals the following fact: *When rounding, the smaller (i.e., finer) the unit, the more accurate the estimate.* Thus, the answer to the question of which method yields the most accurate estimate is, "It depends." Similarly, the answer to the question of which estimation technique we should use is also, "It depends." A general rule of thumb is as follows:

- If we want a relatively accurate approximation, then we should use rounding and we should round to as small a unit as possible.
- If we simply want a ballpark figure and are not concerned with accuracy, then any of the estimation techniques are sufficient.

@ **EXAMPLE 6** ***Home Remodeling*** Larry Wright Construction is estimating the price of a home remodeling project. As part of this estimate Larry needs to consider costs involving labor, subcontractors (e.g., plumber, electrician), material and supplies, permits, painting, and profit.

a. If the homeowner requests a fixed-cost contract, which estimation technique do you think Larry should use to prepare his bid?

b. If the homeowner agrees to a variable-cost contract, which estimation technique do you think Larry should use to prepare his bid?

Solution

a. A fixed-cost contract implies that there is no deviation in the price of the job. Thus, Larry should prepare his bid by rounding and he should round to the smallest unit possible. Accuracy is important.

b. A variable-cost contract implies that deviation in the costs are permitted without penalty. Since Larry has some latitude, he could use any of the estimation techniques. Accuracy is less important.

Overestimates and Underestimates

Our discussion on the accuracy of an estimate brings forth another aspect of estimation, namely, underestimation and overestimation. How do we know if an estimate is less than or greater than the exact answer? In what situations is it better to underestimate, and in what situations is it better to overestimate?

Estimating by compatible numbers and front-end estimation generally tends to yield estimations that are less than the exact answer. Hence, these techniques *underestimate*. Rounding, on the other hand, is a bit more problematic because within a problem some numbers might get rounded up and others might get rounded down. From a strictly mathematical perspective, though, rounding will usually produce an *overestimate* because the general rule says to round up at the halfway point. For example, to round 15, we round up to 20 and hence have an overestimation. However, if we want to be *guaranteed* of an overestimate when rounding, then we should disregard the rule and round everything up. Similarly, if we want to be guaranteed of an underestimate when rounding, we should round everything down.

EXAMPLE 7 ***Estimating College Costs*** David is preparing for his second semester at Rollins College. During his first semester, he registered for 17 hours and paid $5435 for tuition, $420 for books, and $1125 for gasoline expenses driving between home and campus.

a. Should David overestimate or underestimate the cost of his second semester? Why?

b. David plans on taking 12 hours instead of 17 hours during his second semester and estimates that he only needs to save $5000. Is this a reasonable amount? If not, how much should he save?

Solution

a. David should overestimate his second-semester costs. If not, then he might be short of money.

b. We need to first determine the hourly tuition cost. Rounding hours and tuition (to the nearest thousand dollars) "up," tuition is approximately $\frac{\$6000}{20\text{ hours}} = \$300$ per hour. If David is going to register for 12 hours, then he will need approximately $3600 for tuition. We now round up his book and gasoline costs. Rounding to the nearest hundred dollars, David will need approximately $500 for books and $1,200 for gasoline expenses. In total, he will need approximately $\$3600 + \$500 + \$1200 = \$5300$. As a result, David's savings of $5000 is a reasonable amount.

To be really safe, though, David might want to round up to the nearest thousand dollars. Although he is taking less hours, his books for the new term might be more expensive or his gasoline expenses might increase. In this case, then, he will need approximately $3600 for tuition, $1000 for books, and $2000 for gasoline expenses for a total of $6600. Based on this estimate, his savings of $5000 is not sufficient.

EXAMPLE 8 ***Overestimate versus Underestimate*** Give one example of where it is better to underestimate than overestimate.

Solution In most situations it is better to overestimate. There are, however, some cases where an underestimate is better. For example, consider the situation where a police officer is monitoring vehicle speeds along a highway with a radar unit. We "know" that the officer will permit some latitude above the posted speed limit. What we do not know, though, is how much: 2 miles an hour more? 5 miles an hour more? 10 miles an hour more? Here is an illustration of where it is prudent to underestimate the amount of latitude the officer is allowing rather than overestimate it.

EXAMPLE 9 ***Overestimate versus Underestimate*** The graph in Figure 1.4 shows the recording times of the songs on The Rolling Stones' CD *Let It Bleed*. Patti wants to make a tape recording of this CD for her friend Jane who does not have a CD player and intends to use a 60-minute cassette tape (30 minutes per side). She also wants to record all the songs on a single side of the tape.

a. Should Patti overestimate or underestimate the recording time? Why?

b. Is a 60-minute tape sufficient? If not, what size tape should she use?

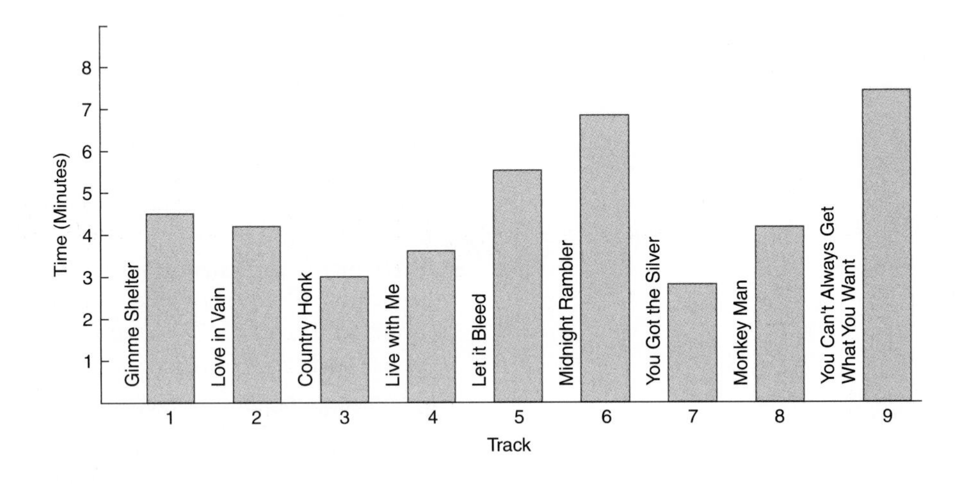

Figure 1.4

Recording times of songs from The Rolling Stones' CD Let It Bleed.

Solution

a. Patti should overestimate the recording time. If she underestimates, then she might not be able to record all of the songs on one side of the tape.

b. Reading from the graph, the approximate recording times for each track are given in the following table. To overestimate these times, we round up to the next minute. Based on the estimated total, it appears that a 60-minute tape is not sufficient. Patti should use a 90-minute tape (45 minutes per side) or use the 60-minute tape and tell Jane that the songs are recorded on both sides of the tape.

Track	Approximate Time	Recording Time Rounded Up to Next Minute
1	4 minutes 30 seconds	5 minutes
2	4 minutes 20 seconds	5 minutes
3	3 minutes	3 minutes
4	3 minutes 30 seconds	4 minutes
5	5 minutes 30 seconds	6 minutes
6	6 minutes 50 seconds	7 minutes
7	2 minutes 50 seconds	3 minutes
8	4 minutes 10 seconds	5 minutes
9	7 minutes 30 seconds	8 minutes
	Estimated Total	46 minutes

NW ***Now Work Problem 35.***

Estimation when Exact Answers Cannot Be Obtained

At the beginning of this section we stated that estimation serves two useful purposes: It enables us to determine mentally (and quickly) an approximate solution to a problem, and it enables us to verify if an exact answer to a problem is reasonable. Sometimes, though, there are problem-solving situations that do not lend themselves to exact answers and approximations are all that we can achieve. For example, it is sometimes impractical to count all of the people who participate in a protest march or who attend a free outdoor concert. In these instances, estimation skills are a necessity.

EXAMPLE 10 ***Attending a Concert*** A free outdoor concert was held Sunday afternoon in Highland Park. No seating was provided. Those who attended stood and sat on the ground, or brought lawn chairs to sit on. The audience area was completely full for the duration of the concert and an aerial snapshot was taken of the crowd at one point during the concert. Using this photo, estimate the number of people who attended the concert.

Solution To estimate the size of the audience, we apply the concept of top-down processing, which we discussed in the previous section. Specifically, we partition the problem into smaller, independent regions of equal areas (see Figure 1.5), count the number of people within one region (or a region that appears to be the most representative of all regions), and then multiply this number by the number of equal regions. Thus, if we partition the audience area into 60 equal

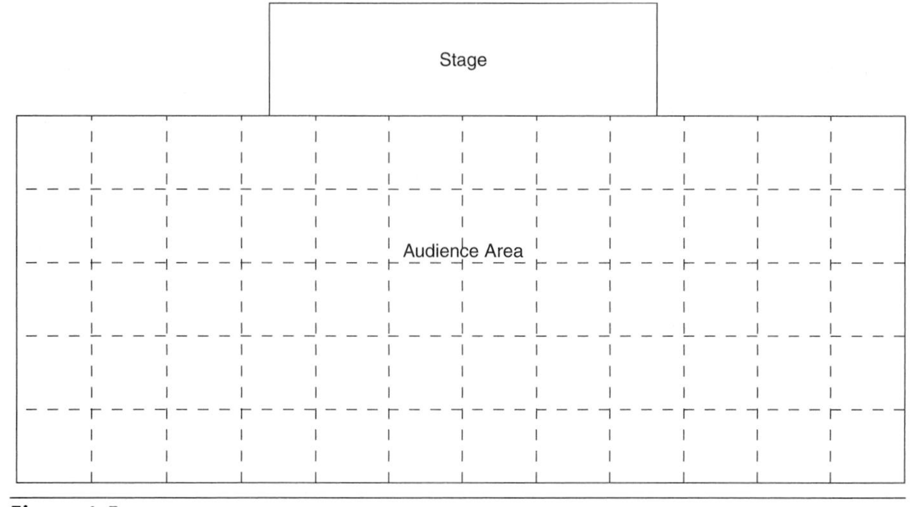

Figure 1.5

square regions and there is an average of 50 people in one region, then a reasonable estimate is that 3,000 people attended the concert. Once again, note the concept of accuracy. The more regions we have (i.e., the finer the unit), then the more accurate the estimate.

Estimations Based on Percentages

Frequently, data are presented in circle graphs (sometimes called pie charts) using percentages. That is, the circle graph is partitioned into sectors with each sector representing a percentage of the overall circle. To find the data of a particular sector, we must be given the overall total of whatever the graph represents. Once this is known, we then multiply the given total by a sector's percentage, which is expressed as a decimal. This involves moving the decimal point two places to the left and dropping the percent sign. Example 11 illustrates these concepts.

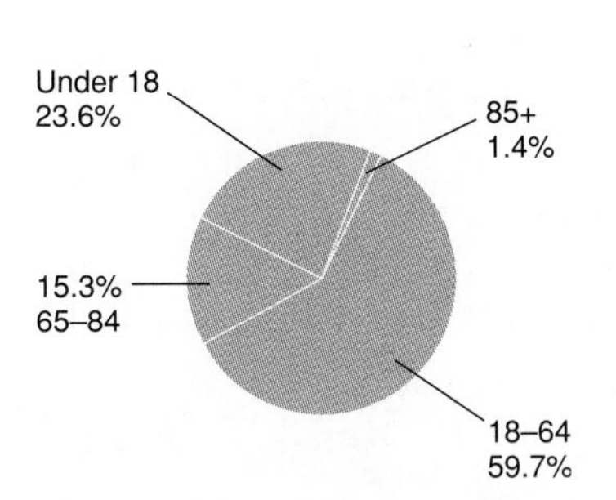

Figure 1.6
Population of Brevard County, FL, by age

EXAMPLE 11 ***Estimating Population*** The pie chart in Figure 1.6 shows the population of Brevard County, Florida, based on age groups. If the total population of the county is 460,977, approximately how many residents are under 18 years of age?

Solution To estimate the number of residents under the age of 18, we first approximate the total population by rounding 460,977 down to 460,000 and then rounding 23.6% up to 25%. Next, we multiply 460,000 by 0.25. Thus, there are approximately $460{,}000 \times 0.25 = 115{,}000$ Brevard County residents under the age of 18. Note that another estimate can be acquired using compatible numbers. This requires recognizing that 23.6% is approximately 1/4. Now, by rounding 460,977 to 440,000, we have the simple problem of taking 1/4 of 440,000, which is 110,000.

Exercises for Section 1.3

In Exercises 1–10, estimate the answers to the given problems by (a) rounding to the nearest whole number, (b) using compatible numbers, and (c) using front-end estimation.

1. $7.59 + 3.8 + 24.37$
2. $19.25 + 7.06 + 67.28$
3. 2196×23
4. 1937×44
5. $3.97 \times 42.8 \times 0.663$
6. $17.8 \times 5.26 \times 0.085$
7. $\dfrac{4659.53}{16.38}$
8. $\dfrac{7902.81}{59.83}$
9. NW $(199.95 \times 3.98) + \dfrac{0.004}{23848}$
10. $(2958 + 899.67) \times \dfrac{999.99}{0.99999}$

In Exercises 11–20, read each problem carefully and estimate the answer.

11. **School Travel** The distance between Brad's house and school is 8.25 miles. Estimate the total number of miles Brad drives between home and school in one month. (Assume Brad does not drive between home and school on weekends.)
12. **Gasoline Costs** Using the information from Exercise 11, let's assume that Brad's car gets 18 miles per gallon and that the capacity of his car's gasoline tank is 11 gallons. Approximately how much money does Brad spend per month on gasoline if he only uses his car to travel between home and school and the cost of gasoline is \$1.54 per gallon?
13. **Light Bulb Life Expectancy** If the life span of a light bulb is 2500 hours, approximately how many weeks can you keep this light on 24 hours per day?
14. Every morning during "orange season," Robert uses 3.5 Florida oranges to make himself a 12-oz glass of freshly squeezed orange juice. (He then eats the remaining half.) Approximately how many oranges will Robert need if he wants to make 1 liter of orange juice? (*Note:* 1 fl oz = 30 milliliters and 1 liter = 1000 milliliters.)
15. NW **Airline Travel** In a flight from Los Angeles to New York, Evan and Lisa each had two carry-on bags with them that averaged 22 pounds per bag. Evan noticed that most people boarding the plane also had two carry-on bags. If the flight had 167 passengers and each passenger brought onboard two carry-on bags that were comparable in weight to those of Evan's and Lisa's bags, what is the approximate weight of all the carry-on bags?
16. **Bridge Weight** If the weight limit of Mathers Bridge is 10 tons, approximately how many cars can be on the bridge at

the same time if each car has an average weight of 2300 pounds? (*Note:* 1 ton = 2000 lb.)

17. Bicycle Ride In the annual Cross Florida bicycle ride, bicyclists ride from Cocoa Beach to Pine Island—a distance of 170 miles—in one day. Located throughout the ride are 7 rest stops where bicyclists can take a break and get something to eat or drink. If the ride begins at 6 A.M. and Mike and Bill average 16.6 miles per hour, approximately what time will Mike and Bill finish the ride if they average 12 minutes at each rest stop and encounter no problems while on the road?

18. Presidential Election In the 2000 U.S. presidential election, several Florida counties recounted their ballots by hand. If 431,650 ballots were recounted in one county, approximately how many days did it take if 50 two member teams worked in two 7-hour shifts daily and each team reviewed an average of 7 ballots per minute?

19. Internet Service An Internet service provider contractually guarantees a 99.99% "uptime" of its network to all of its customers. (a) If this guarantee is measured on a daily basis, approximately how many minutes per day can the network be contractually unavailable (i.e., "down")? (b) If this guarantee is measured on a monthly basis, approximately how many minutes per month can the network be down? (c) If this guarantee is measured on a yearly basis, approximately how many minutes per year can the network be down?

20. Hotel Rates The nightly room rate at the Forte Posthouse in Edinburgh, Scotland is 127.44 pounds sterling. If the monetary exchange rate between the U.K. pound sterling and the U.S. dollar is 1.56 (i.e., 1 pound sterling = $1.56), approximately how much in U.S. dollars will a three-night stay cost at the Forte Posthouse?

21. Age Groups In an online survey, participants were asked to report their age. The data collected are shown in the bar graph in Figure 1.7. Use these data to answer the following questions:

- **a.** Approximately how many people are younger than 21?
- **b.** Approximately how many people are older than 50?
- **c.** Approximately how many people reported their age?
- **d.** What age group is missing from this set of data?
- **e.** If you were an advertiser, to which age group would you try to focus your advertisements? Why?

22. Internet Service The bar graph in Figure 1.8 on p. 26 shows the expected number of people (in millions) who will subscribe to either cable modem service or DSL (digital subscriber line) service.

- **a.** Which service will be more popular during the 5-year period of 2000–2004?
- **b.** Approximately how many more subscribers will choose cable modem service over DSL service during the given 5-year period?
- **c.** Does the gap between the number of subscribers between the two services appear to be increasing, decreasing, or remaining the same?
- **d.** Based on the pattern shown in the chart, in what year would you project that DSL service will overtake cable modem service?

23. Music Copyrights The circle graph in Figure 1.9 on p. 26 represents the results of a telephone survey in which participants were asked the question, "Should Internet companies be held responsible for the illegal activities (piracy of copyrighted material) their software might facilitate?" If 1022 people were surveyed, approximately how many people (a) had an opinion? (b) felt strongly against holding such companies responsible?

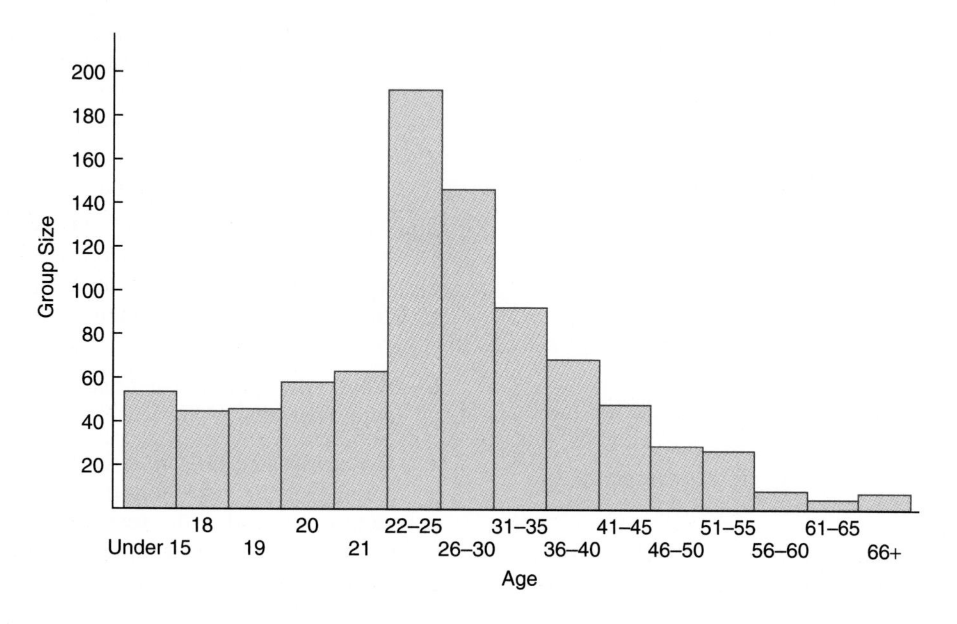

Figure 1.7

Number of Internet users by age groups.

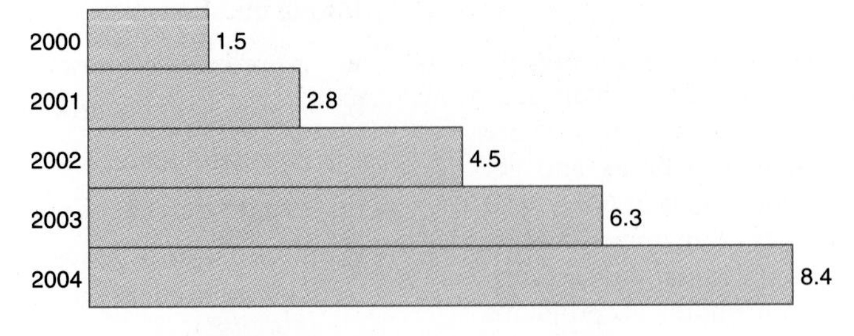

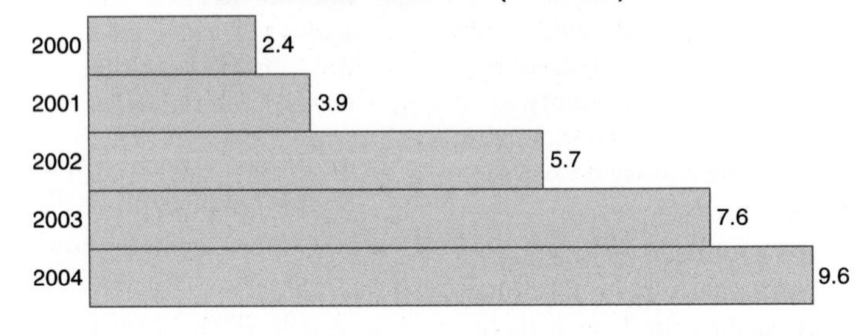

Figure 1.8
Source: The Yankee Group

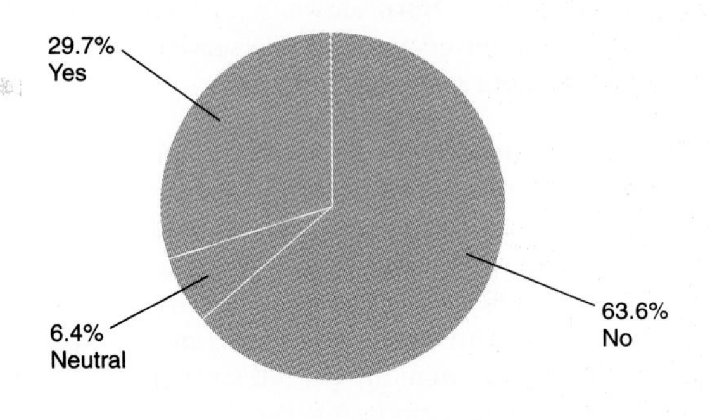

Figure 1.9
Should Internet companies be held responsible for the illegal activities (piracy of copyrighted material) their software might facilitate?

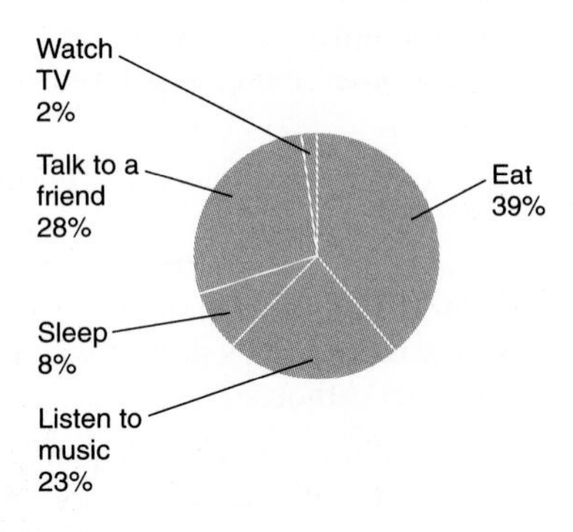

Figure 1.10
What people do when they feel upset or worried.

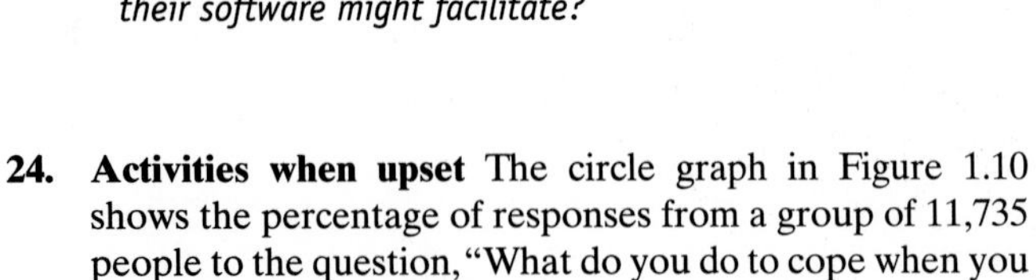

24. Activities when upset The circle graph in Figure 1.10 shows the percentage of responses from a group of 11,735 people to the question, "What do you do to cope when you feel upset and worried?"

- **a.** Approximately how many of the people polled eat when they feel upset?
- **b.** What is the approximate percentage of the top three activities combined?
- **c.** Estimate the number of people who responded in each category.

25. Stock Quotes The line graph in Figure 1.11 tracks the stock prices between an Internet company (**www.addie.com**) and a traditional "bricks and mortar" company (JB Industries) during a one-year period.

- **a.** Approximately at what point during the year was the difference in stock prices the greatest? Estimate this difference.
- **b.** What is the approximate price per share of each company at the end of the year?
- **c.** If you invested $10,000 in each company at the beginning of the year, determine if your total investment at the end of the year increased, decreased, or remained the same and approximate what the difference is (if any).

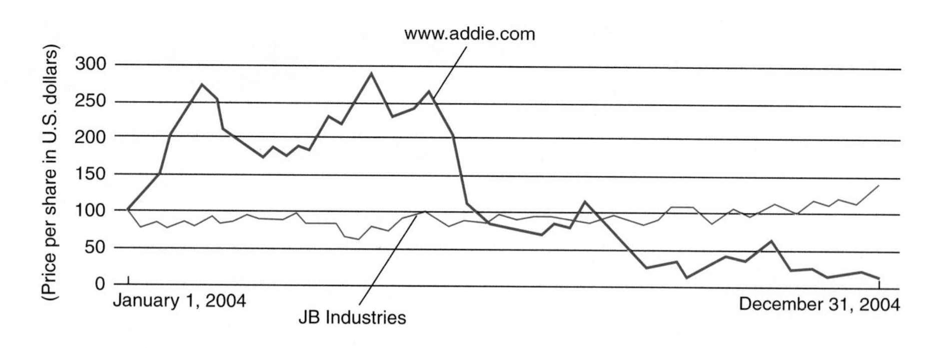

Figure 1.11

26. **Vehicle Costs** A random sample of 10 individuals were asked to report how much they paid for their current vehicle and the most money they spent buying a new vehicle. The results are shown in the line graph in Figure 1.12.

    a. Approximately how much money did Beth spend for her current vehicle?

    b. Estimate the most amount of money Sean spent on a vehicle.

    c. Is the difference between the price of their current vehicle and the most money they spent on a vehicle greater, less than, or about the same between Samantha and Ally?

27. **Best Male Comedian** The first chart on p. 28 reflects the results of a survey in which respondents were asked to vote for their favorite male comedian.

    a. Approximately how many people voted?

    b. Were the combined votes of Jerry Seinfeld and Robin Williams approximately more, less, or about the same when compared to the votes of all the other comedians? Estimate this difference (if any).

28. **Annual Salaries** The second chart on p. 28 shows the median annual salaries for various information technology (IT) professions for two geographical regions of the United States (West Coast and Midwest).

    a. Approximately how much more money does a West Coast network administrator earn than a West Coast telecommunications specialist?

    b. Approximately how much more money does a West Coast network administrator earn than a Midwest network administrator?

    c. continued on page 28

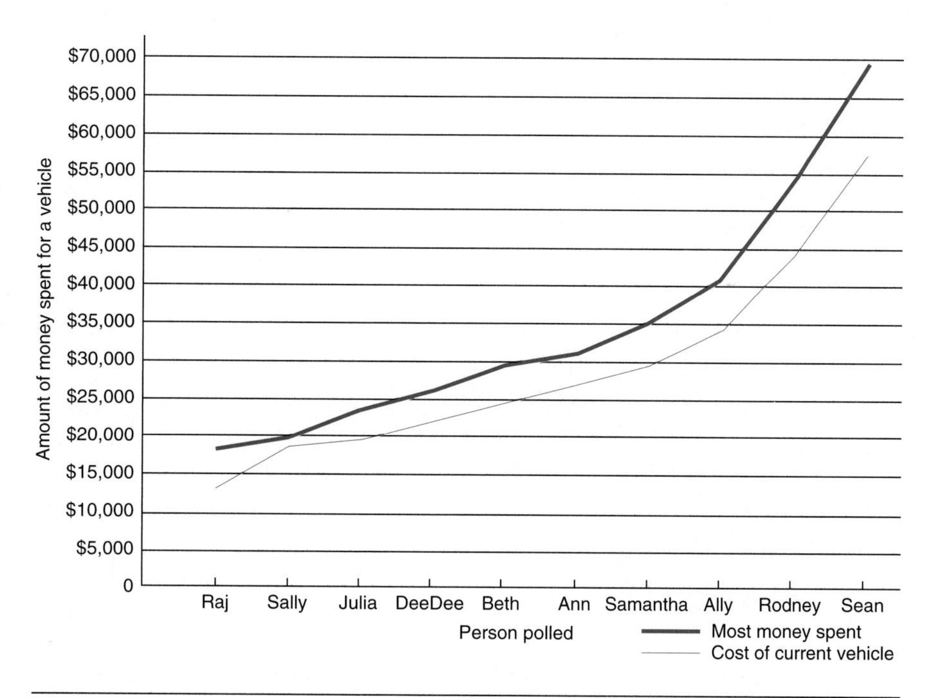

Figure 1.12

Comedian	Number of Votes in Favor
Tim Allen	411
George Carlin	503
Bill Cosby	437
Billy Crystal	190
Rodney Dangerfield	111
Arsenio Hall	261
Andy Kaufman	75
Jay Leno	175
David Letterman	283
Richard Lewis	27
Bill Mahr	51
Steve Martin	240
Dennis Miller	318
Eddie Murphy	328
Conan O'Brien	143
Richard Pryor	86
Chris Rock	308
Jerry Seinfeld	807
Robin Williams	1076

IT Profession	Median Salary (West Coast)	Median Salary (Midwest)
Senior Software Engineer	$82,900	$78,700
Software Engineer	71,700	64,900
Sr. Database Analyst/Admin.	80,400	80,900
Post Y2K Analyst	84,200	78,700
Object-Oriented/GUI Developer	79,200	73,600
WWW/Internet Developer	82,000	70,700
Network Administrator LAN/WAN	74,300	69,800
Sr. Systems Analyst Programmer	75,400	67,800
Systems Analyst Programmer	62,000	61,300
Sr. Systems Admin./Unix	80,000	67,900
Sr. Client Server Prog./Analyst	73,800	71,700
Client-Server Prog./Analyst	67,800	65,000
Sr. Mid/MF Programmer Analyst	72,800	67,100
Mid/MF Programmer Analyst	60,700	58,100
Telecommunications Specialist	64,100	59,000
PC Applications Specialist	53,900	48,700
Quality Assurance Analyst	69,800	66,900
Security Specialist	83,500	80,900

GUI = graphical user interface; WWW = World Wide Web; LAN = local area network; WAN = wide area network; MF = mainframe. Thus, Mid/MF Programmer Analyst refers to a Middle Manager/Mainframe Programmer Analyst.

28. **c.** Which two West Coast IT professions have the biggest difference in salary?

d. Overall, approximately how much more money do West Coast IT professionals earn annually than Midwest IT professionals regardless of the profession?

29. **Health Risk Factors** The bar graph in Figure 1.13 shows the risk factors of premature death (i.e., before their life expectancy) and the corresponding percentages for individuals living in Fairfield County, Connecticut, which has an approximate population of 833,315.

a. Which risk factor will be the result of the most early deaths?

b. Estimate the number of people in Fairfield County who will die early from not eating a sufficient amount of fruits and vegetables.

c. Estimate the total number of people in Fairfield County who will die early from either smoking or diabetes.

30. **Education Courses** The bar graph in Figure 1.14 shows the approximate distribution of weekly classes during the first year of secondary education in Kuwait. Assume there are approximately 423,875 such classes throughout the country.

a. Which classes have a distribution rate of more than 25%?

b. Approximately how many more Arabic language classes are there compared to the number of English classes?

c. Estimate the total number of math and science classes.

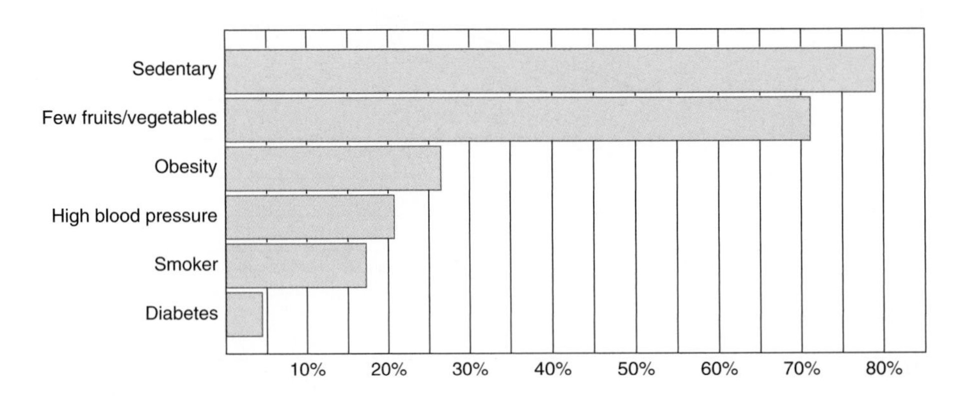

Figure 1.13
Risk factors for early (i.e., premature) death.

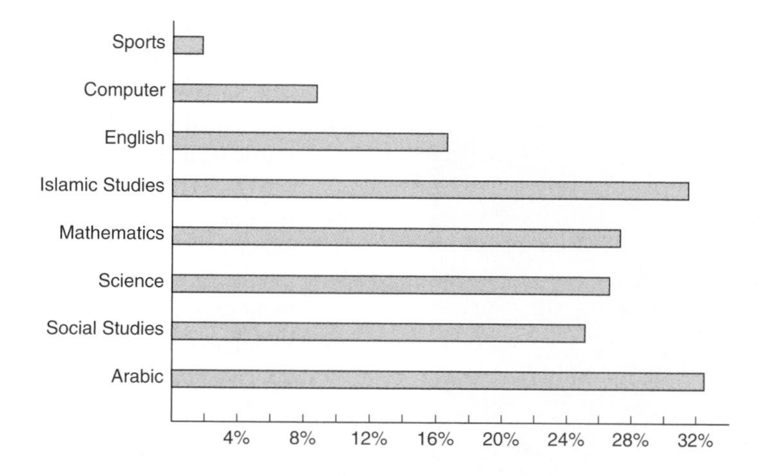

Figure 1.14

Distribution of weekly classes in the first year of secondary education in Kuwait.

31. Driving Distances The map in Figure 1.15 shows the south Florida area. See the legend in Figure 1.16.

a. Estimate the distance (in miles) between Boca Raton and Hollywood.

b. Estimate the distance (in miles) between the cities of Coral Springs and Pembroke Pines.

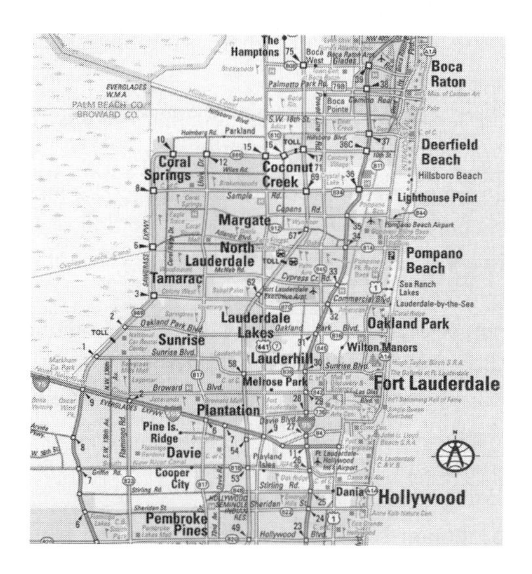

Figure 1.15

Source: Map © by RMC, 04-S-24, www.randmcnally.com

32. Driving Distances The map in Figure 1.16 shows the San Francisco area.

a. Estimate the distance (in miles) between Pacifica and San Mateo.

b. Approximately how far (in miles) is Palo Alto from San Francisco?

33. Crowd Estimates Estimate the number of people shown in the photo at the right.

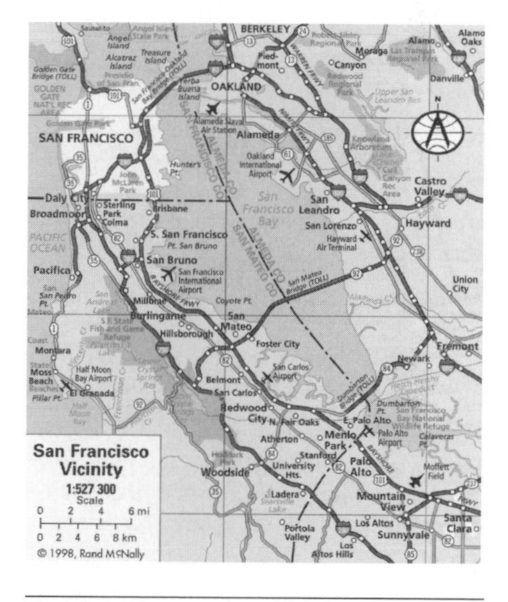

Figure 1.16

Source: Map © by RMC, 04-S-24, www.randmcnally.com

34. **Automobile Estimate** Estimate the number of cars on the freeway shown in the photo.

NW 35. **Travel Costs** In planning for a one-week vacation to Italy, Claudia and Scott want to estimate their travel costs based on a similar trip they took a year earlier. Should they underestimate or overestimate their costs? Why?

36. **Space Shuttle Mission** In planning for a shuttle mission, Helen noted that the seven astronauts consumed an average of 10 food packs a day. Should Helen overestimate or underestimate the number of food packs the next shuttle mission requires? Why?

37. **Television Game Show** In a popular daytime television show, *The Price Is Right*, contestants estimate the price of a product and the person who is closest to the actual retail price without going over it wins. Should contestants overestimate or underestimate? Why?

38. **Diet and Calories** In estimating the number of calories various foods contain, should you overestimate or underestimate your daily calorie intake if you are on a diet? Why?

39. Is it better to overestimate or underestimate the amount of money you owe in taxes? Why?

40. Is it better to overestimate or underestimate the amount of gasoline that remains in your automobile's fuel tank. Why?

41. **Room Dimensions** The dimensions of most ceiling tiles are either 2 ft by 2 ft or 2 ft by 4 ft. Find a room in your school that contains ceiling tiles and then estimate the dimensions of the room using these tiles. Now find the actual room dimensions using a tape measure and check the accuracy of your estimate.

42. **Floor Dimensions** Find a room that has a tiled floor and measure the dimensions of one tile. Use this measurement to estimate the dimension of the room. Now find the actual room dimensions using a tape measure and check the accuracy of your measurement.

43. Estimate how long you can hold your breath and then use a stopwatch to time yourself to check the accuracy of your estimate.

44. Estimate the number of seconds you have to wait at a red traffic light and then use a watch to time the light so you can check your accuracy.

45. Estimate the number of paperclips that will fill a small paper cup. To check the accuracy of your measurement, fill the cup with paperclips and then count the actual number of paperclips used.

46. Estimate the number of books that can be placed on a bookshelf that is 10 feet long. Now fill the shelf with books and count the number of books used to check your accuracy.

Writing Mathematics

47. **Movie Ticket Sales** Every Monday morning, various newspapers report the estimated revenue current movies grossed over the previous weekend. Explain how you think these estimates are determined.

48. Most new car reports generally provide an estimate of a car's mileage. Explain how you think these estimates are determined.

49. Explain why there is never a "correct" answer when it comes to estimation.

50. Describe a situation where, using the same set of criteria, one estimate can be "bad" whereas another estimate can be "good."

In Exercises 51–58, explain if an estimate is either sufficient or insufficient for the given situation.

51. A waitress uses estimation to determine the amount of tax that should be charged to your dinner bill.

52. A waitress uses estimation to determine your total dinner bill.

53. A customer uses estimation to determine if a dinner bill is correct.

54. A customer uses estimation to determine how much of a tip to leave.

55. A sports reporter estimates the number of people who attended a football game.

56. An accountant estimates the gross revenue from ticket sales to a football game.

57. Your employer estimates your weekly salary.

58. A bank estimates the amount of interest or dividends it pays its depositors.

59. In this section, we described the general rounding rule as "five or more, round up; less than five, round down." An alternative to this rule is, "if the amount is even, round up; if the amount is odd, round down." Which rule do you think provides a more accurate estimate? Why? Give an example to support your answer.

Challenge Exercises

An important concept in higher mathematics involves finding the area under a curve. Although the mathematics needed to do this is beyond the scope of this book, we can still estimate the area under a curve using "approximating rectangles." This technique involves (a) subdividing the region into rectangles of equal bases, (b) finding the areas of these rectangles, and then (c) calculating the sum of these areas. One such approach uses inscribed rectangles, which implies that each rectangle's height is equal to the minimum height for a particular subregion. This is illustrated in Figure 1.17.

Area Under A Curve

Use the concept of approximating rectangles to answer Exercises 60–63.

60. In the example just given, is it necessary to use rectangles of equal bases? Why? If rectangles of unequal bases were used, would the estimate be better, worse, or about the same? Explain.

61. Is the approximate area of the region shown in Figure 1.17 an overestimate or underestimate? Why?

62. If your answer to Exercise 61 is overestimate, then explain what we can do to get an underestimate. If your answer to question 61 is underestimate, then explain what we can do to get an overestimate.

63. When using approximating rectangles, what is producing the error in our estimate and how do you think we can reduce this error?

Group/Research Activities

64. Do a Web search on the Internet for the term *estimation* and prepare a report on the way in which estimation is used in at least three different fields.

65. Interview a building contractor and prepare a report on the estimation techniques he or she uses to estimate the cost of a project.

66. Working with your classmates, select 25 people and estimate the number of hairs each person has on his or her scalp. Now use these data to provide a general estimate of the average number of hairs on the human scalp. How accurate do you think your estimate is? Why? Do you think your result is representative of the human population? Why?

67. Discuss with your classmates situations in which you used estimation to help solve a problem.

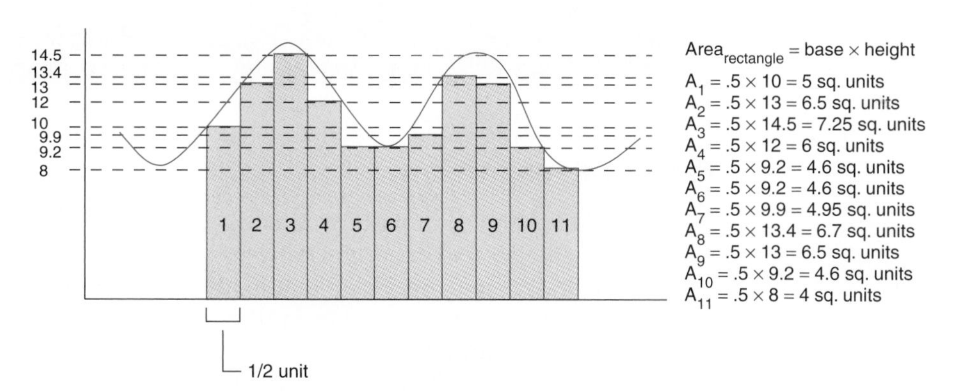

The area of the shaded region is approximately

$$A_1 + A_2 + A_3 + A_4 + A_5 + A_6 + A_7 + A_8 + A_9 + A_{10} + A_{11}$$
$$= 5 + 6.5 + 7.25 + 6 + 4.6 + 4.6 + 4.95 + 6.7 + 6.5 + 4.6 + 4$$
$$= 61 \text{ sq. units}$$

Figure 1.17

Just for fun

A scale contains three consecutive weights that register 54 kilograms. What weight is the lightest of the three?

1.4 PATTERN RECOGNITION AND INDUCTIVE REASONING

In Section 1.2, we discussed briefly the concepts of finding patterns and reasoning. In this section, we examine these two concepts in more detail.

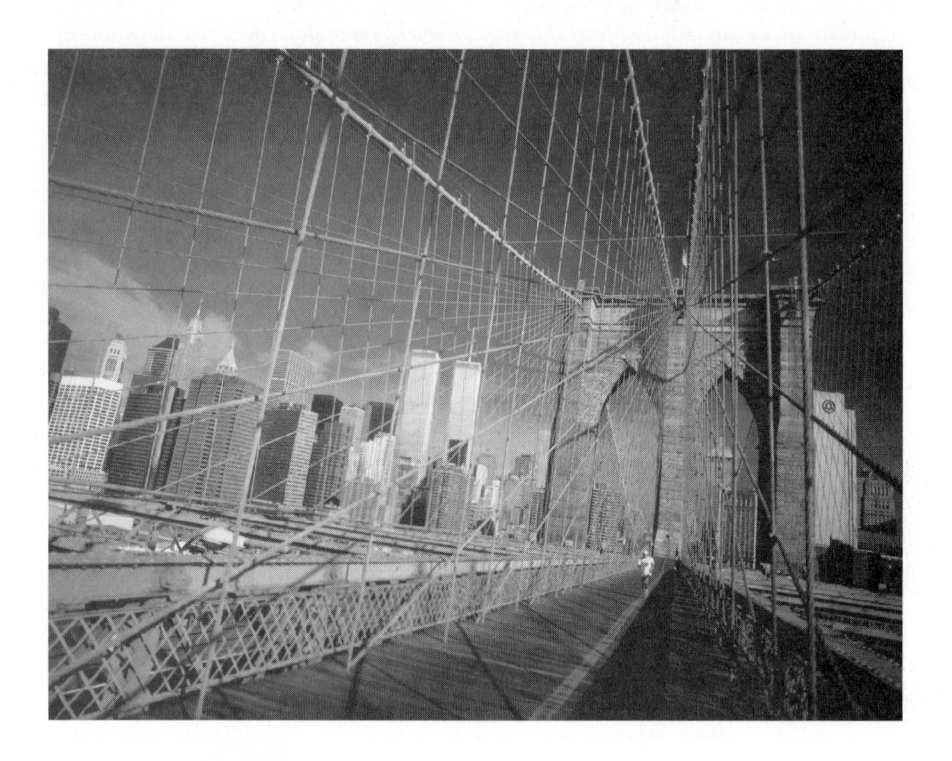

Pattern Recognition

In our world, examples of patterns are everywhere, including in music, in art, in science, in medicine, and in nature. Patterns can also be found throughout mathematics. In fact, someone once said that mathematics is the science of pattern and order. Whether in the real world or in the study of mathematics, our challenge is to discover these patterns and order and use them to further our knowledge. To illustrate this, consider the multiplication table for 9:

$$\begin{array}{lll} 9 \times 1 = 9 & 9 \times 2 = 18 & 9 \times 3 = 27 \\ 9 \times 4 = 36 & 9 \times 5 = 45 & 9 \times 6 = 54 \\ 9 \times 7 = 63 & 9 \times 8 = 72 & 9 \times 9 = 81 \end{array}$$

There are actually two patterns that exist in these products. The obvious one is the sum of the digits that make up the product is 9 (e.g., $9 \times 4 = 36$ and $3 + 6 = 9$). A not so obvious pattern, however, is that the leading digit of each product is one less than the number being multiplied by 9. For example,

$$9 \times 4 = 36 \qquad\qquad 9 \times 8 = 72$$

3 is one less than 4   7 is one less than 8

By combining this pattern with the previous one, a rule for multiplying a single-digit number by 9 can now be established: The product of a single-digit number (n) and 9 is a two-digit number where the individual digits of the product add to

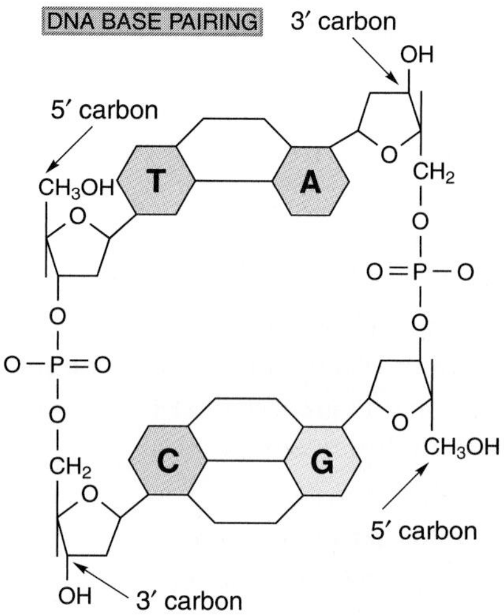

DNA BASE PAIRING

9 and the leading digit of the product is one less than the number being multiplied $(n - 1)$.

Sometimes the absence of a pattern is the pattern. For example, in studying the fingerprints of millions of people, it was concluded that no pattern exists and that no two people have exactly the same fingerprints. A similar "pattern" was discovered with DNA, which led to a similar conclusion—namely, no two people have exactly the same DNA.

As illustrated by the previous examples, pattern recognition is often used to generate rules, make decisions, make conclusions, or make predictions. Furthermore, as we learned in Section 1.2, active observation can help us identify patterns or common themes that might exist for individual cases. If a pattern exists and is recognized, we can then predict the next case. This concept is demonstrated in the following examples.

EXAMPLE 1 ***Recognizing Patterns for Prediction*** Given the following list of characters, use pattern recognition to predict the next two characters:

1, Z, 3, W, 9, T, 27, Q, 81

Solution Observe that the first character and every other character after it is a number. Further observe that the second character and every other character thereafter is a capital letter. Focusing only on the numbers, it appears that the numbers are increasing in value and every succeeding number is three times more than the previous number. Similarly, focusing on the letters, it appears they are in reverse order from their standard alphabetic position, and each succeeding letter is "three removed" from the previous one. Based on our observations and the patterns we identified, the next two characters are probably N and 243.

EXAMPLE 2 ***Recognizing Patterns: Bubble Sort*** Given the following *columns* of numbers, predict the next two columns.

$$\begin{array}{cccc} 2 & 2 & 2 & 2 \\ 6 & 4 & 4 & 4 \\ 4 & 6 & 6 & 1 \\ 8 & 8 & 1 & 6 \\ 1 & 1 & 8 & 8 \end{array}$$

Solution To solve this problem it may help if we can identify the changes being made from one column to the next: Moving from the first column to the second, 6 and 4 are switched; moving from the second column to the third, 8 and 1 are switched; and moving from the third column to the fourth, 6 and 1 are switched. Unfortunately, this observation alone is not helpful. We need more information.

Specifically, we need to identify the reason behind interchanging two numbers. In column 1, the only logical reason for switching 6 and 4 is because 6 is greater than 4. Notice that this reasoning holds true for the remaining columns as well. If this is indeed the pattern, then the fifth column should show 4 and 1 switched:

$$\begin{array}{ccccc} 2 & 2 & 2 & 2 & \mathbf{2} \\ 6 & 4 & 4 & 4 & \mathbf{1} \\ 4 & 6 & 6 & 1 & \mathbf{4} \\ 8 & 8 & 1 & 6 & \mathbf{6} \\ 1 & 1 & 8 & 8 & \mathbf{8} \end{array}$$

The sixth column should show 2 and 1 switched:

2	2	2	2	2	**1**
6	4	4	4	1	**2**
4	6	6	1	4	**4**
8	8	1	6	6	**6**
1	1	8	8	8	**8**

(*Note:* This example is an illustration of an algorithm known as the Bubble Sort, which is studied in computer science. Working down the list of numbers in a column, successive pairs of numbers are compared. If the first is greater than the second, then the numbers are interchanged. The result is a numerically ordered list.)

EXAMPLE 3 ***Recognizing Patterns for Prediction*** For the four shapes shown in Figure 1.18, identify the pattern that exists and use that pattern to predict the next figure in the sequence.

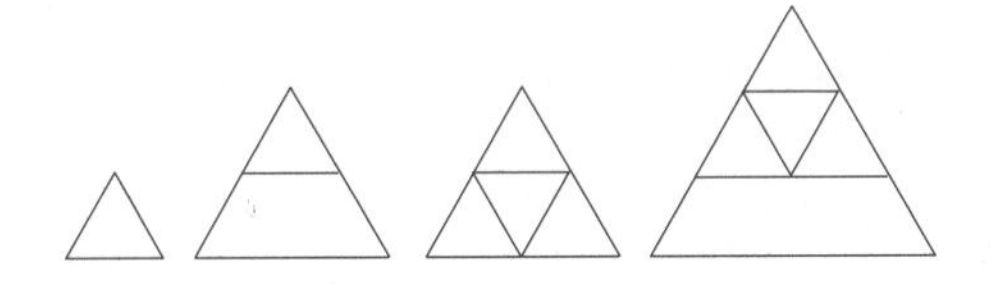

Figure 1.18

Solution The pattern that appears to be emerging is as follows: Begin with a triangle; extend the sides of the triangle to create a larger, outer triangle; place an inverted triangle within the new outer triangle. If we were to continue this pattern the picture would be as shown in Figure 1.19.

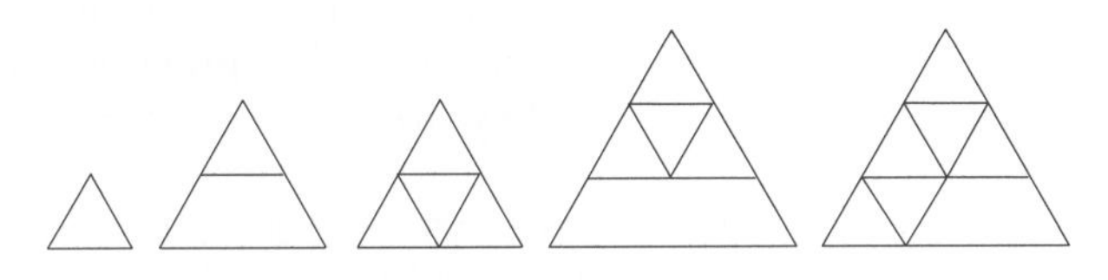

Figure 1.19

EXAMPLE 4 ***Recognizing Patterns: Consumer Price Index*** The Consumer Price Index (CPI) is a measure of the average change over time in the prices paid by urban consumers for a market basket of consumer goods and services. The chart in Figure 1.20 shows the CPI for the six-month period between

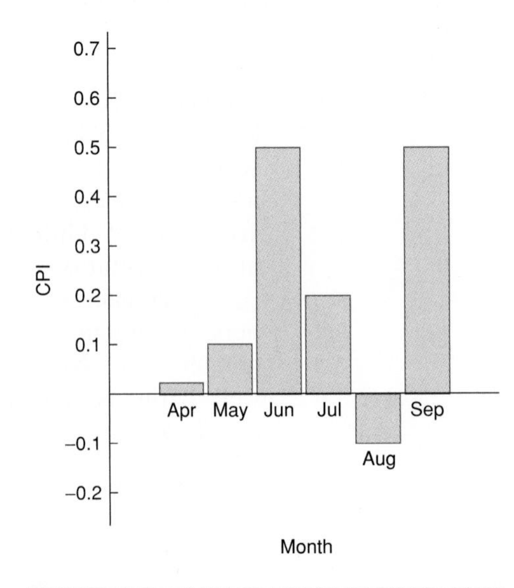

Figure 1.20
Consumer Price Index: April–September.

April and September of a given year. Predict whether the CPI will increase or decrease in October. What do you think it will be?

Solution The pattern that appears to be emerging is that there were three consecutive months where the CPI increased (April, May, and June), followed by two consecutive months where the CPI decreased (July and August), followed by an increase in September. Based on this pattern, it appears that the CPI will probably increase in October and that it will be more than 0.5. Note that we could also argue that the CPI's highest value was 0.5 in June and then decreased to 0.2 the following month. Given this pattern, the October CPI might decrease to 0.2 in October since its September value was 0.5.

NW ***Now Work Problem 1.***

Inductive Reasoning

Recognizing patterns or common themes is the foundation of inductive reasoning. In Section 1.2, we discussed briefly the concept of inductive reasoning, and explained that it involves first observing individual events and then making a general statement based on those events. We now examine this subject in more detail. A formal definition follows:

> ***Inductive reasoning is the process of arriving at a general conclusion based on the results of specific cases.***

As noted in the preceding definition, the process of induction involves a transition from specific cases to general principles or conclusions. To illustrate this, consider the following scenario: Suppose you have a pet cat who likes to eat turkey. Further suppose that a friend of yours also has a pet cat who likes to eat turkey. Based on these two specific observations you might conclude that *all* cats like to eat turkey. This conclusion was arrived at inductively, namely, a generalization (*All cats like to eat turkey*) was made based on two specific observations. As noted in Section 1.2, we must be careful when using induction to arrive at conclusions because we cannot be certain that *every instance* of a certain phenomenon holds true. All that is needed to prove the cat–turkey generalization false is one cat who does not like to eat turkey.

When applying inductive reasoning to problem-solving situations, we typically rely on active observation to identify patterns or themes that might exist among individual cases. We also must remember that any results obtained from an inductive reasoning process are at best probable.

EXAMPLE 5 ***Inductive Reasoning and Number Sequences*** Use inductive reasoning to determine the next probable number in the following list of numbers:

$$1, \frac{1}{2}, \frac{1}{3}, \frac{1}{4}, \frac{1}{5}$$

Solution Using only the method of observation, note that the succeeding denominators of each fraction are being increased by one but that their respective numerators remain constant. It appears that the next probable number of the list is $\frac{1}{6}$.

In searching for a pattern that might exist in a list of numbers, a strategy that can be employed is to subtract each succeeding number by the number that immediately precedes it. For example, in the list of numbers 2, 4, 6, 8, each such subtraction yields a *common difference* of 2.

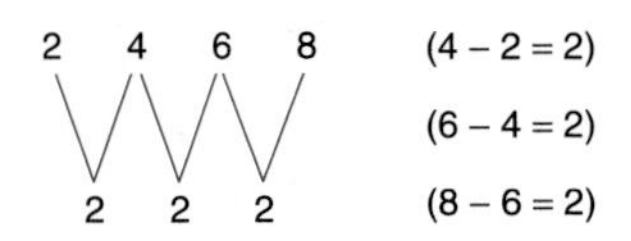

A list of numbers that has a common difference is called an **arithmetic progression**. In the preceding example we might conclude that the next number in this list is 10. Once again you are reminded that without any additional information about this problem our conclusion is neither *right* nor *wrong*, only *probable*.

EXAMPLE 6 ***Arithmetic Progression*** Use inductive reasoning to determine the next probable number in the list of numbers 1, 3, 5, 7.

Solution The common difference among the numbers is 2. Thus the next probable number is 9.

EXAMPLE 7 ***Arithmetic Progression*** Use inductive reasoning to determine the next probable number in the list of numbers $\frac{2}{3}, 1, \frac{4}{3}, \frac{5}{3}, 2$.

Solution The common difference among the numbers is $\frac{1}{3}$. Thus the next probable number is $\frac{7}{3}$.

NW ***Now Work Problem 11.***

The common differences in Examples 6 and 7 were discovered by doing a *surface-level analysis*. For some lists of numbers, it might be necessary to perform a *multilevel analysis* before a common difference is discovered. This is illustrated in Examples 8 and 9.

EXAMPLE 8 ***Arithmetic Progression and Multilevel Analysis*** Determine the next probable number in the list of numbers 20, 28, 40, 56 using inductive reasoning.

Solution A surface-level examination of this list of numbers does not immediately yield a common difference. However, by using top-down processing we can continue our analysis at subsequent levels. We first perform a level 1 analysis, as we did in the previous two examples, by subtracting each term by its predecessor. The common differences are shown and labeled level 1.

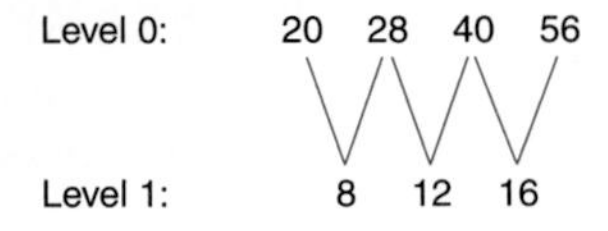

Note that at Level 1 we still do not have a common difference. As a result we proceed to a second level by subtracting successive number pairs in level 1.

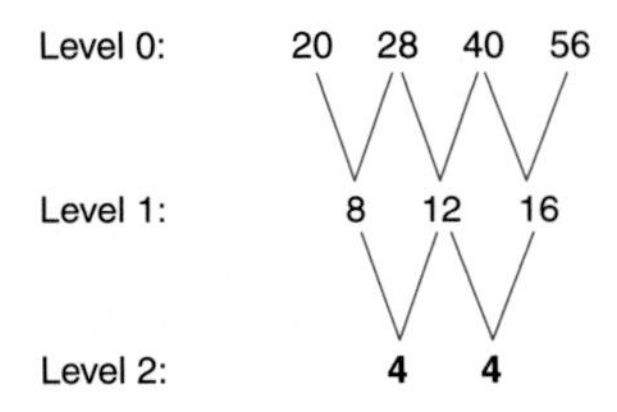

A second level analysis yields a common difference of 4. The next probable number in the list can now be determined by working backwards. Specifically, we first add 4 to 16 at level 1, which yields 20. We then add 20 to 56 at level 0, which yields 76. Thus the next probable number is 76. A complete analysis is shown.

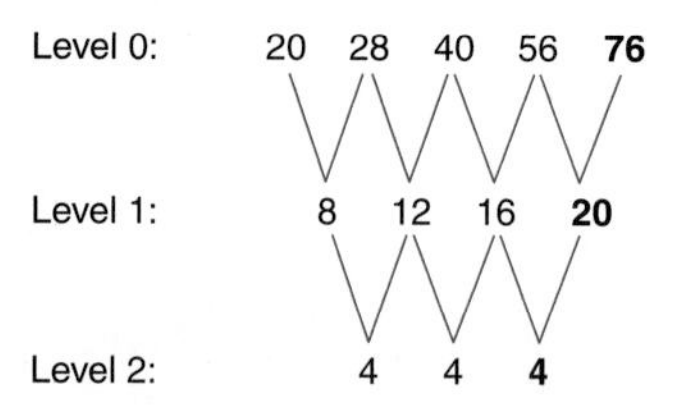

EXAMPLE 9 ***Arithmetic Progression and Multilevel Analysis*** Determine the next probable number in the list of numbers 8, 26, 56, 100, 160, 238 using inductive reasoning.

Solution This problem requires a three-level analysis. The completed analysis is shown. The next probable number in the given list is 336.

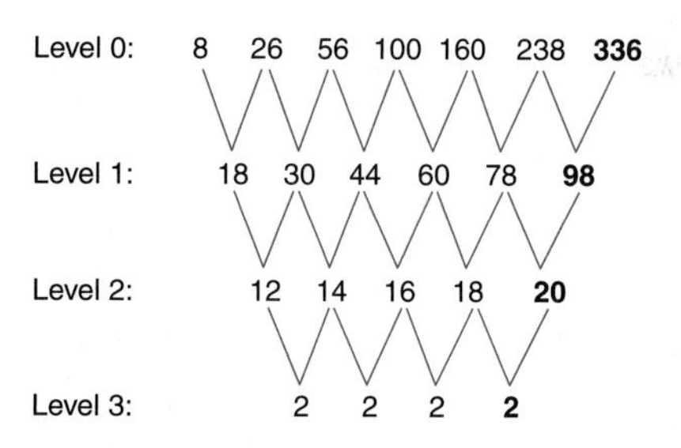

NW ***Now Work Problem 17.***

It is important to note that a multilevel analysis will not always yield a common difference. As stated earlier, problem solving involves developing a "bag of tricks," or strategies that can be employed in our pursuit toward solving a problem. The method of common differences is only one strategy within our "bag of tricks" we can try using when searching for a common theme or pattern in a list of numbers.

The lists of numbers given in Examples 6 through 9 are known as *arithmetic progressions* because successive terms were found by adding common differences.

Some lists of numbers have *common factors* rather than common differences. That is, successive terms are found by multiplying by a fixed number. Just as common differences were found by subtracting a succeeding term by its predecessor, common factors are found by dividing a succeeding term by its predecessor. A list of numbers that have a common factor is called a **geometric progression**, as illustrated in Example 10.

EXAMPLE 10 ***Geometric Progression*** Determine the next probable number in the list of numbers 2, 6, 18, 54 using inductive reasoning.

Solution A multilevel analysis of this list (as was done in Examples 8 and 9) does not yield a common difference. However, if we divide each successive number by the number that immediately precedes it, then a common factor of 3 is obtained:

$$6 \div 2 = 3; \quad 18 \div 6 = 3; \quad \text{and} \quad 54 \div 18 = 3.$$

Thus the next probable number in this list is $54 \times 3 = 162$.

NW ***Now Work Problem 33.***

Pattern recognition is a powerful tool in problem solving. By identifying common themes or patterns existing among sets of data, we are often able to make sense of what might appear initially to be nonsensical information. Pattern recognition helps us to acquire insight into a problem. It facilitates the decision-making process by making us more knowledgeable and informed about a specific situation before we make a decision or arrive at a conclusion. We end this section with examples from three other branches of mathematics: graph theory, geometry, and algebra.

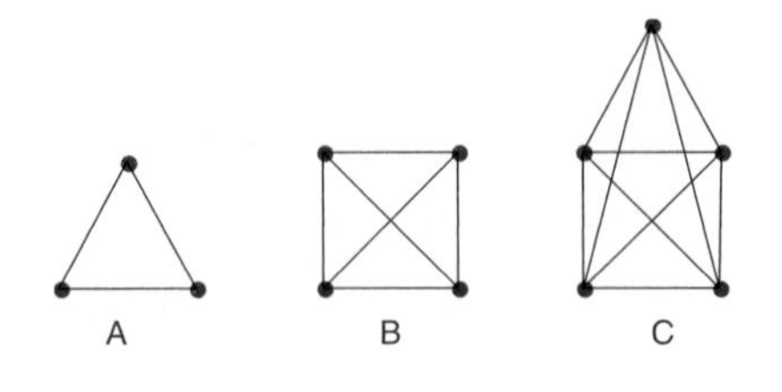

Figure 1.21

EXAMPLE 11 ***Graph Theory*** For parts A, B, and C as shown in Figure 1.21, identify the probable relationship between the number of sides each figure has and the number of "lines" that meet at each vertex (denoted by a •) using inductive reasoning. (*Note:* The number of "lines" that meet at a vertex is called the *degree of the vertex*, and the "lines" are called *edges*. Also, each figure shown is *fully connected* because all the vertices of each figure are connected to one another.)

Solution

a. A is a triangle and consists of three vertices and three sides. Each vertex in part A has degree 2 since two edges meet at each vertex.

b. Part B is a four-sided figure (called a *quadrilateral*) and consists of four vertices. The degree of each vertex is 3.

c. Part C is a five-sided figure (called a *pentagon*) and has five vertices. The degree of each vertex is 4.

Organizing our analysis into tabular form, we can readily identify the probable relationship that exists between the number of sides of the figures and the degree of the vertices:

Number of Sides	Degree of Vertex
3	2
4	3
5	4

The degree of a vertex for a fully connected "graph" is one less than the total number of sides.

EXAMPLE 12 ***Geometric Relationships*** After studying figures A, B, and C in Figure 1.22, identify what the probable value of A is for figure D using inductive reasoning.

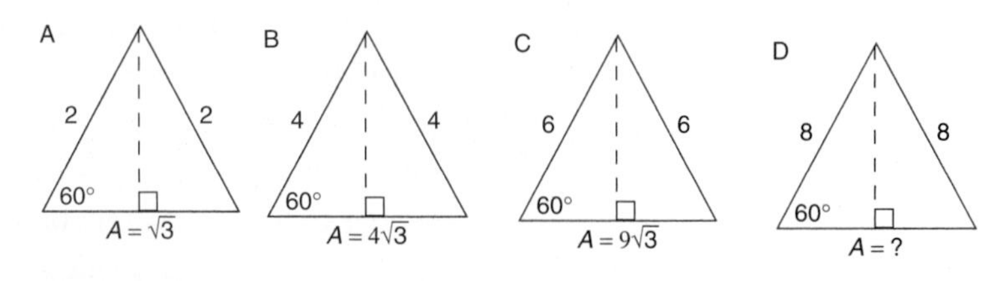

Figure 1.22

Solution The common theme existing among figures A, B, and C is that the coefficient of $\sqrt{3}$ is equal to one-half of the given side squared. For example, in Figure A the given side is 2. One-half of 2 is 1, and $1^2 = 1$. In Figure B the given side is 4. One-half of 4 is 2, and $2^2 = 4$. In Figure C the given side is 6. One-half of 6 is 3, and $3^2 = 9$. Following this pattern for Figure D, $\frac{8}{2} = 4$, and $4^2 = 16$. As a result of this analysis, $A = 16\sqrt{3}$.

EXAMPLE 13 ***Algebraic Pattern*** Study the pattern that exists between the two equations shown. Using that pattern, generate an expression for $(a + b + c + d)^2$.

$$(a + b)^2 = a^2 + b^2 + 2ab$$
$$(a + b + c)^2 = a^2 + b^2 + c^2 + 2ab + 2ac + 2bc$$

Solution Focusing on the right side of the equals sign, three common themes exist between the two equations: Each term within parentheses is squared; each pair of terms within parentheses is multiplied by 2; and all terms are separated by + signs. As a result of this analysis we have the following equation:

$$(a + b + c + d)^2 = a^2 + b^2 + c^2 + d^2 + 2ab + 2ac + 2ad + 2bc + 2bd + 2cd$$

NW ***Now Work Problem 57.***

EXAMPLE 14 ***Time Spent in Hospital*** Study the pattern of hospital stays given in Figure 1.23 and estimate the average length of time a patient stays in the hospital in 2010. Do you think this is an accurate projection? Why or why not?

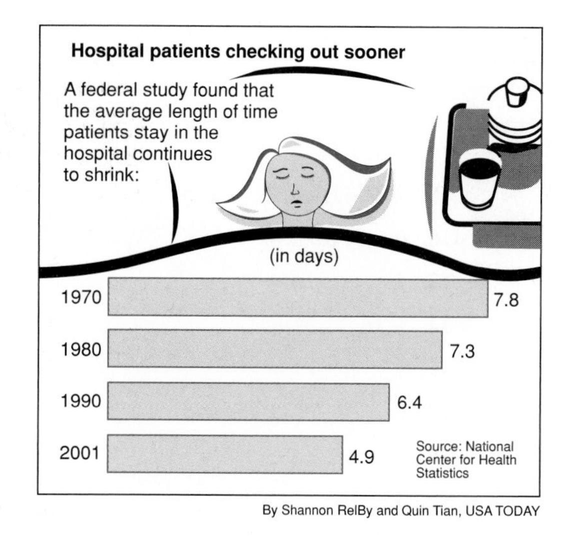

Figure 1.23

Average length of time (in days) patients stay in the hospital.

Solution The average length of time patients stay in the hospital has decreased steadily since 1970. From 1970 to 1980, there was a decrease of 0.5 day; from 1980 to 1990, there was a decrease of 0.9 day; and from 1990 to 2001, there was a decrease of 1.5 days: The pattern is

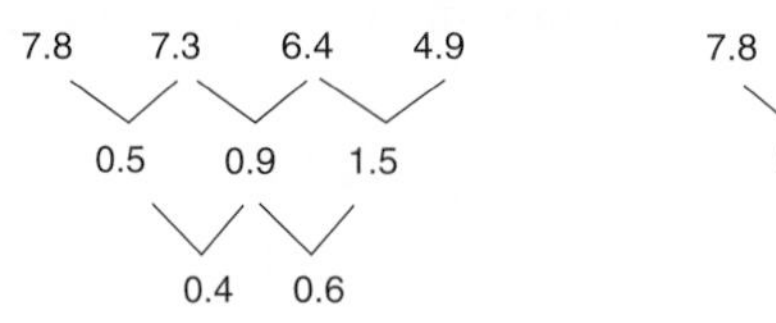

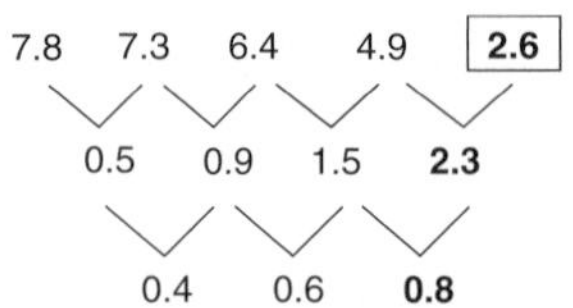

If this pattern were to continue, there will be an average decline of 2.3 days between 2001 and 2010. Thus, the projected average hospital stay will be $4.9 - 2.3 = 2.6$ days. Given the advances in medical technology, healthier lifestyles, and increased health insurance rates, it is reasonable to think that this trend will continue.

Exercises for Section 1.4

In Exercises 1 and 2, use pattern recognition to determine the next two iterations of the given pattern.

NW **1.** ○ △ ○○ △ ○○○ △ ___?___ ___?___

2. | ∠ △ □ ⊠ ___?___ ___?___

In Exercises 3 through 40, use pattern recognition and inductive reasoning to write the next probable number. Also state whether the given list of numbers forms an arithmetic or geometric progression (if appropriate).

3. 2, 5, 8, 11

4. 1, 7, 13, 19

5. 2, 4, 3, 9, 4

6. 1, 4, 9, 16

7. $1, \frac{3}{2}, 2, \frac{5}{2}$

8. $2, \frac{13}{2}, 11, \frac{31}{2}$

9. 1, 1, 2, 3 5

10. 0, 5, 5, 10, 15

NW **11.** 5, 10, 16, 23, 31

12. 4, 5, 8, 13, 20

13. 2, 29, 59, 92, 128

14. 3, 16, 34, 57, 85

15. 4, 7, 12, 20, 32, 49

16. 1, 2, 4, 9, 19, 36

NW **17.** 1, 10, 20, 37, 67, 116

18. 0, 5, 10, 20, 35, 65

19. $19, 10, 2, -5, -11$

20. $7, 2, -1, -2, -1$

21. 1, 3, 9, 27

22. 1, 6, 36

23. $1, \frac{1}{3}, \frac{1}{9}, \frac{1}{27}$

24. $1, \frac{1}{2}, \frac{1}{4}, \frac{1}{8}$

25. $\frac{1}{9}, \frac{1}{3}, 1$

26. $\frac{1}{4}, \frac{1}{2}, 1$

27. $64, -16, 4$

28. $27, -9, 3$

29. $1, -\frac{1}{2}, \frac{1}{4}$

30. $1, -\frac{1}{3}, \frac{1}{9}$

31. $(x - y), x, (x + y)$

32. $x, (x + z), (x + 2z), (x + 3z)$

NW **33.** $x, 2x, 4x, 8x$

34. $x, \frac{x}{2}, \frac{x}{4}, \frac{x}{8}$

35. x, x^2, x^3

36. $\frac{1}{x}, \frac{1}{x^2}, \frac{1}{x^3}$

37. $(x + y), (x + y)^2, (x + y)^3$

38. $(x + y), (x^2 + y^2), (x^3 + y^3)$

39. $x, 2x^2, 4x^3, 8x^4$

40. $(x + 1), x(x + 1), x^2(x + 1), x^3(x + 1)$

If the reciprocals of a list of numbers have a common difference, then the original numbers are said to form a harmonic progression. *In Exercises 41 through 48, use inductive reasoning to show that the lists of numbers form a harmonic progression.*

41. $1, \frac{1}{2}, \frac{1}{3}, \frac{1}{4}$

42. $\frac{1}{2}, \frac{1}{5}, \frac{1}{8}, \frac{1}{11}$

43. $\frac{1}{2}, \frac{1}{4}, \frac{1}{6}, \frac{1}{8}$

44. $\frac{1}{3}, \frac{1}{6}, \frac{1}{9}, \frac{1}{12}$

45. $\frac{2}{3}, \frac{1}{2}, \frac{2}{5}, \frac{1}{3}$

46. $\frac{3}{4}, \frac{3}{5}, \frac{1}{2}, \frac{3}{7}$

47. $3, \frac{6}{5}, \frac{3}{4}, \frac{6}{11}$

48. $1, \frac{4}{7}, \frac{2}{5}, \frac{4}{13}$

49. What number can never be part of a harmonic progression? Explain your answer.

50. Use the pattern established in Example 13 to write an expression for $(a + b + c + d + e)^2$.

In Exercises 51 and 52, determine the pattern and then generate the next two iterations.

51. $37 \times 3 = 111$
$37 \times 6 = 222$
$37 \times 9 = 333$

52. $37037 \times 3 = 111111$
$37037 \times 6 = 222222$
$37037 \times 9 = 333333$

Use the patterns from Exercises 51 and 52 to complete the multiplication problems in Exercises 53 through 55. Confirm your results.

53. $37037037 \times 3 = ?$

54. $37037037037 \times 6 = ?$

55. $37037037037037 \times 9 = ?$

In Exercises 56 through 59, identify the pattern and write the next two iterations. Confirm your results where possible.

56.
$$1 + 2 = 3$$
$$4 + 5 + 6 = 7 + 8$$
$$9 + 10 + 11 + 12 = 13 + 14 + 15$$

NW **57.**
$$1 \times 8 + 1 = 9$$
$$12 \times 8 + 2 = 98$$
$$123 \times 8 + 3 = 987$$

58.
$$0 \times 9 + 8 = 8$$
$$9 \times 9 + 7 = 88$$
$$98 \times 9 + 6 = 888$$
$$987 \times 9 + 5 = 8888$$

59.
$$\left(1 + \frac{1}{2}\right) \times 3 = \frac{9}{2}$$
$$\left(1 + \frac{1}{3}\right) \times 4 = \frac{16}{3}$$
$$\left(1 + \frac{1}{4}\right) \times 5 = \frac{25}{4}$$

60. High School Dropouts The line graph in Figure 1.24 shows the percentage of 15- to 24-year-olds who dropped out of grades 10–12. Use inductive reasoning to predict the percent of people who will drop out of grades 10–12 for 2004.

61. High School Dropouts The line graph in Figure 1.25 shows the percentage of 16- to 24-year-olds who were high school dropouts. Use inductive reasoning to predict the percent of high school dropouts for 2004.

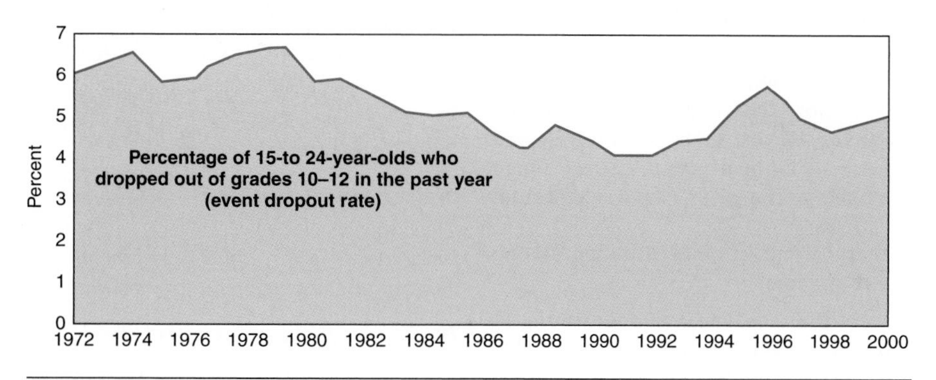

Figure 1.24

Source: U.S. Department of Commerce, U.S. Census Bureau, Current Population Survey, October (1972–2000). Available: http://nces.ed.gov/pubs2002/droppub_2001/Figs.asp

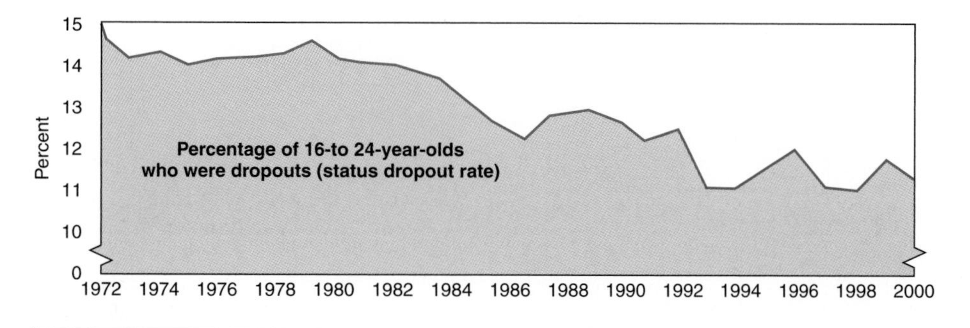

Figure 1.25

Source: U.S. Department of Commerce, U.S. Census Bureau, Current Population Survey, October (1972–2000). Available: http://nces.ed.gov/pubs2002/droppub_2001/Figs.asp

62. Cigarette Smoking The table on p. 42 shows the percentage of high school seniors and eighth-graders in the United States who answered yes when asked if they had smoked a cigarette within the past month.

a. What is the general trend in cigarette smoking among seniors and eighth-graders?

b. Use pattern recognition and inductive reasoning to predict the percent of cigarette-smoking seniors for 2002–2004.

c. Use pattern recognition and inductive reasoning to predict the percent of cigarette-smoking eighth-graders for 2002–2004.

	All Seniors	All 8th-Graders
1991	28.3	14.3
1992	27.8	15.5
1993	29.9	16.7
1994	31.2	18.6
1995	33.5	19.1
1996	34.0	21.0
1997	36.5	19.4
1998	35.1	19.1
1999	34.6	17.5
2000	31.4	14.6
2001	29.5	12.2

Source: Determinants and Measures of Health: Table 65. Use of selected substances (cigarettes) by high school seniors and eighth-graders: United States, selected years 1980–2001. Available: http://www.cdc.gov/nchs/data/hus/tables/2002/02hus065.pdf

63. **Binge Drinking** The following table shows the percentage of high school seniors and eighth-graders in the United States who answered yes when asked if they had consumed five or more alcoholic drinks in a row ("binge drinking") at least once in the prior two-week period.

a. What is the general trend in binge drinking among seniors and eighth-graders?

b. Use pattern recognition and inductive reasoning to predict the percent of binge-drinking seniors for 2002–2004.

c. Use pattern recognition and inductive reasoning to predict the percent of binge-drinking eighth-graders for 2002–2004.

	All Seniors	All 8th-Graders
1991	29.8	12.9
1992	27.9	13.4
1993	27.5	13.5
1994	28.2	14.5
1995	29.8	14.5
1996	30.2	15.6
1997	31.3	14.5
1998	31.5	13.7
1999	30.8	15.2
2000	30.0	14.1
2001	29.7	13.2

Source: Determinants and Measures of Health: Table 65. Use of selected substances (binge drinking) by high school seniors and eighth-graders: United States, selected years 1980–2001. Available: http://www.cdc.gov/nchs/data/hus/tables/2002/02hus065.pdf

Writing Mathematics

64. Explain in your own words the relationship between pattern recognition and inductive reasoning.

65. Review current newspapers or magazine articles that contain data presented in chart or graph form and write a summary of how these data are presented.

66. Compose a pattern recognition problem and explain how to solve it.

67. Compose an inductive reasoning problem and explain how to solve it.

68. Explain the concept of a counterexample relative to inductive reasoning. Of what importance is a counterexample?

Challenge Exercises

69. Generate a pattern that appears to hold true "forever" based on inductive reasoning but eventually fails when the pattern is extended beyond a certain number of iterations.

A **magic square** *consists of an equal number of rows and columns and is an arrangement of numbers where the sum of the vertical columns, horizontal rows, and diagonals are equal. In the following three-by-three magic square, each row, column, and diagonal has a sum of 15.*

8	1	6
3	5	7
4	9	2

In Exercises 70 through 73, find the missing numbers so that the resulting arrangement forms a magic square.

70.

4	9	2
3	5	7
8		

71.

	21	26
23		27
24		22

72.

15	8	13
	12	14
11	16	

73.

16	2	3	
5	11		8
9	7	6	12
	14	15	

74. Study the following magic square closely. What unique feature can you discover about it?

96	11	89	68
88	69	91	16
61	86	18	99
19	98	66	81

Group/Research Activities

75. Rent the film *Donald in Mathmagic Land* and view it with your classmates. Discuss the various ways in which patterns

are presented in the movie and their applications in the world. Also discuss what you found to be the most interesting aspect of the film.

76. Review the book *Pattern* (*Math Counts*) (Children's Press, New York) by Henry Pluckrose and write a report on the various ways in which patterns are found throughout our world.

77. Leonhard Euler discovered an interesting relationship among the number of vertices (V), number of faces (F), and number of edges (E) of ordinary polyhedra. Research this relationship and prepare a report that describes it.

Real-World Data

Counterfeit Money Arrests *Use the following table to answer questions 78–80.*

Foreign arrests

Locations of foreign arrests made in seizures of counterfeit money:

	2000	2001	2002	Total
Paris	66	150	94	310
Bogota, Colombia	87	96	122	305
Bangkok, Thailand	56	25	50	131
Pretoria, South Africa	49	21	7	77
Milan, Italy	35	7	21	63
Rome	16	35	12	63
Berlin	5	24	18	47
Moscow	11	15	20	46
San Juan, Puerto Rico	17	11	12	40
Hong Kong	7	1	17	25
Manila	25	0	0	25
London	8	4	4	16
Guam	0	0	11	11
Montreal	4	5	0	9
Frankfurt, Germany	0	0	6	6

Source: U.S. Secret Service

78. Use pattern recognition and inductive reasoning to predict the number of arrests that will be made in Bogota, Columbia in 2003–2005.

79. Use pattern recognition and inductive reasoning to predict the number of arrests that will be made in Pretoria, South Africa in 2003–2005.

80. Organize the cities listed in the table by the following regions: Europe, South America, and Asia. (*Note*: Some cities will not be placed in any of these categories.) Review these data and determine what the overall trend appears to be for each region. Provide a plausible explanation for these trends.

Computers Used for Counterfeit *Use Figure 1.26 to answer questions 81–84.*

81. What is the overall trend of all U.S. counterfeit activities?

82. What is the overall trend of using computers to make counterfeit money?

83. Use pattern recognition and inductive reasoning to predict the amount of computer-generated counterfeit money that will be made in 2003–2005.

84. Between 1998 and 2002, has the *percentage* of computer-generated money to all U.S. counterfeit activities increased, decreased, or remained the same?

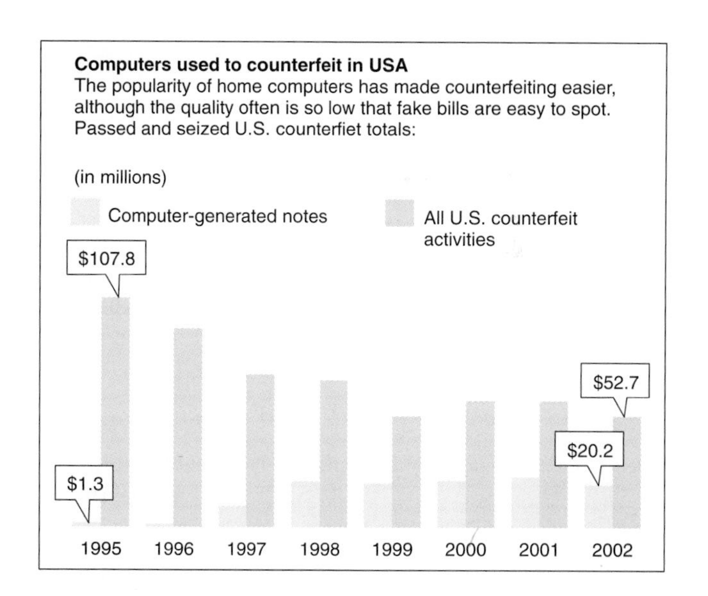

Figure 1.26

Computers used to counterfeit in the United States. Passed and seized U.S. counterfeit totals (in millions).
Source: U.S. Secret Service

Just for fun

Can you think of at least seven three-letter words that begin with the letters BO?

1.5 INDUCTIVE REASONING AND SEQUENCES AND SERIES (OPTIONAL)

In Section 1.4, we presented several examples that explored the concept of inductive reasoning and learned that central to the inductive process is pattern recognition. We also learned that any result we obtain through an inductive process is at best probable because we cannot examine every individual case to confirm that an observed pattern holds true all the time. In this section, we extend our discussion of inductive reasoning by examining it from a purely mathematical perspective. Specifically, our discussion focuses on sequences and series, which are useful and important concepts in advanced courses in mathematics, physics, and engineering.

Sequences

Without getting too formal, a **sequence** is simply a list of ordered objects. For example, the list of numbers

$$1, 1, 2, 3, 5, 8$$

represents a sequence of numbers that contains six elements. The first element is 1, the second element is l, the third element is 2, the fourth element is 3, the fifth element is 5, and the sixth element is 8. The individual elements of a sequence are called *terms*. A sequence also can be either *finite* or *infinite*. A **finite sequence** contains a specific number of terms. The sequence given above is a finite sequence; it contains a total of six terms. An **infinite sequence** has an unlimited number of terms (i.e., it is impossible to count the total number of terms because they are *unbounded*). We also use an *ellipsis* (. . .) in a sequence to denote that an observed pattern of the terms is to be continued. An example of an infinite sequence is the list

$$2, 4, 6, \ldots$$

This sequence consists of three terms and has a common difference of two between any two adjacent terms. The ellipsis indicates that this pattern is to continue indefinitely. Thus, based on the observed pattern, it appears that each succeeding term is found by adding 2 to the immediately preceding term. Assuming that this pattern is correct, then the next three terms of the sequence are 8, 10, and 12. Once again, note how we are using inductive reasoning to make this decision.

EXAMPLE 1 ***Finite and Infinite Sequences*** Given the sequences in **a** through **e**, determine if the sequence is finite or infinite, state the number of terms the sequence contains, and identify the established pattern that exists between terms.

a. $1, 2, 3$ **b.** $1, 2, 3, \ldots$ **c.** $2, 4, 6, \ldots$
d. $1, 3, 5, \ldots$ **e.** $2, 4, 6, \ldots, 50$

Solution

a. The sequence is finite; it contains three terms; and there is a common difference of one between any two adjacent terms.

b. The sequence is infinite; it contains an infinite number of terms; and there is a common difference of one between any two adjacent terms. This sequence is not the same as the sequence given in part a. Furthermore, the terms of this sequence correspond to the set of *counting numbers*, which is mathematically represented as $\{1, 2, 3, \ldots\}$.

c. The sequence is infinite; it contains an infinite number of terms; and there is a common difference of two between any two adjacent terms. The terms of this sequence correspond to the set of *even counting numbers* $\{2, 4, 6, \ldots\}$.

d. The sequence is infinite; it contains an infinite number of terms; and there is a common difference of two between any two adjacent terms. The terms of this sequence correspond to the set of *odd counting numbers* $\{1, 3, 5, \ldots\}$.

e. The sequence is finite; it contains 25 terms; and there is a common difference of two between any two adjacent terms. This is not an infinite sequence because we can count all of the terms. This sequence represents the first 25 even counting numbers.

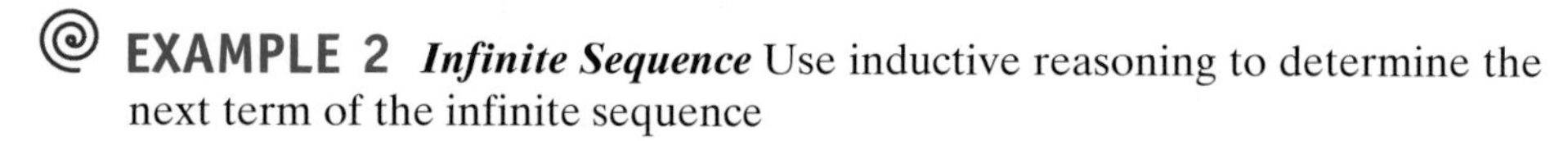

EXAMPLE 2 ***Infinite Sequence*** Use inductive reasoning to determine the next term of the infinite sequence

$$\frac{1}{2}, \frac{1}{3}, \frac{1}{4}, \ldots$$

Solution The pattern that is emerging from these three terms is as follows: The numerator remains constant (1), and the denominator is being increased by one for each subsequent term. Based on this pattern the next term should be $\frac{1}{5}$.

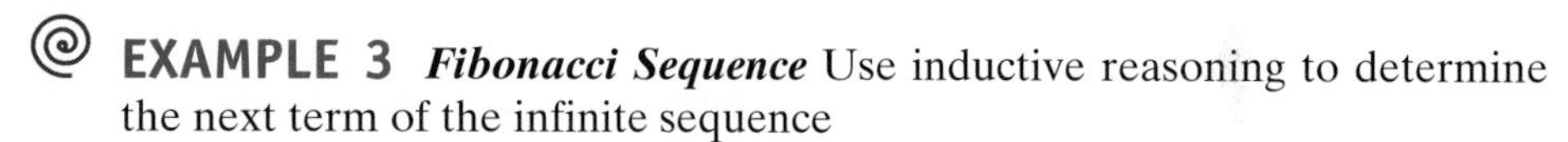

EXAMPLE 3 ***Fibonacci Sequence*** Use inductive reasoning to determine the next term of the infinite sequence

$$1, 1, 2, 3, 5, 8, \ldots$$

Fibonacci numbers can be seen in the arrangement of seeds on flower heads.

Solution The pattern that is emerging from these six terms is as follows: Beginning with the second term, each subsequent term is the sum of the two immediate previous terms. For example, the third term (2) is the sum of the first two terms $(1 + 1 = 2)$. The fourth term (3) is the sum of its immediate two previous terms $(1 + 2 = 3)$. Continuing in this fashion, $5 = 2 + 3$, and $8 = 3 + 5$. Thus the next term is $(5 + 8 = 13)$. This sequence is known as the **Fibonacci sequence**, named after Leonardo Fibonacci, a thirteenth-century Italian mathematician. See the biography on the next page.

In Example 2, we *conjectured* that the fourth term of the given sequence was $\frac{1}{5}$. Similarly, in Example 3, we conjectured that the seventh term of the given sequence was 13. In both instances our conclusions were based on inductive reasoning—we examined the given terms to determine a pattern and then used this pattern to generate the next term. When working with infinite sequences, a technique known as *mathematical induction* is used to prove that an observed pattern is valid for all of the terms of the sequence. For example, it can be mathematically proven that the sequence $2, 4, 6, \ldots$ is equal to $\{2n\}$, where n is equal to the set of counting numbers $\{1, 2, 3, \ldots\}$. The expression $\{2n\}$ is called the **general term** of the sequence $2, 4, 6, \ldots$. By substituting specific values for n in the general term we can generate the terms of the sequence. For example, if n is equal to 1, 2, and 3, then $2n$ is equal to 2, 4, and 6, respectively, which are the first three terms of the sequence.

$$\text{When } n = 1: \quad 2(1) = 2$$
$$\text{When } n = 2: \quad 2(2) = 4$$
$$\text{When } n = 3: \quad 2(3) = 6$$

Using the general term, this sequence can be written as

$$2, 4, 6, \ldots, 2n, \ldots$$

When the general term is written in this context, it is referred to as the ***nth term***.

Although the derivation of a sequence's general term is usually done using mathematics that are not covered in this textbook, we can still, nevertheless, derive *probable* general terms using inductive reasoning. The reason we use the word *probable* is because we will be basing our results on inductive reasoning rather than deductive methods, which include formal mathematical theorems and proofs. Since it will be impossible to examine every individual case, any general term we derive can only be regarded as probable. For example, consider the expression

$$2n + n(n - 1)(n - 2)(n - 3)$$

(where n is a counting number). This expression also can be used to represent the general term of the sequence 2, 4, 6, ... because it too generates the first three terms of this sequence. However, if the first four terms of the sequence were given as 2, 4, 6, 8, ..., then this second rule could not be considered because its fourth term is 32 and the observed pattern (common difference of 2) would not apply. This illustrates two important points about inductive reasoning. First, the more cases we are able to test and find true for a general rule or principle derived from inductive reasoning, then the more likely our rule will be correct. (It is particularly important to try to identify and test "special" cases if any exist.) Second, a general rule derived from inductive reasoning can readily be proven false by a single case, called a *counterexample*. Consequently, unless proven otherwise, any general rule, principle, or conjecture arrived at or made through inductive reasoning can be considered only probable at best.

EXAMPLE 4 ***Finding the nth term of a Sequence*** Given the sequences in **a, b,** and **c**, develop a probable general term for each sequence and write each sequence using this general term.

a. $1, 3, 5, 7, 9, \ldots$ **b.** $2, 4, 8, 16, 32, \ldots$ **c.** $1, \frac{1}{2}, \frac{1}{3}, \frac{1}{4}, \frac{1}{5}, \ldots$

Biography: Leonardo Fibonacci

The Granger Collection, New York

Leonardo de Pisa, better known as Fibonacci (Fibonacci coming from Filius Bonacci, son of Bonacci) was born 1175 in Pisa and died after 1240. Fibonacci was the only mathematical genius during the Middle Ages in Europe. In his most influential work, *Liber Abaci* ("a book about the abacus"), first published in 1202 with a second edition in 1228, he introduced Europe to the Hindu-Arabic numeral system. He did extensive travels in Egypt, Syria, Greece, and Sicily, absorbing all the knowledge he could from the various people about mathematics. In his book he proposed a problem about a rabbit population and its solution. It is stated as,

"Someone placed a pair of rabbits in a certain place, enclosed on all sides by a wall, to find out how many pairs of rabbits will be born there in the course of one year, it being assumed that every month a pair of rabbits produces another pair, and that rabbits begin to bear young two months after their own birth."[†]

The solution produces the following sequence: 1, 2, 3, 5, 8, 13, 21, 34, 55, 89, ... During the nineteenth century the French mathematician Edouard Lucas named the above sequence the Fibonacci sequence. Scientists and mathematicians study the many applications of the Fibonacci sequence (now written: 0, 1, 1, 2, 3, 5, 8, 13, ...) such as the number spirals of a sunflower, the scale patterns of pine cones, the ancestry of male bees, and the bud arrangement on a stem. There currently is a journal devoted to the Fibonacci sequence and related mathematics, called *The Fibonacci Quarterly*.

Fibonacci also wrote *Practica Geometriae* (1220; "Practice of Geometry") and *Liber quadratorum* (1225; "Book of Square Numbers"). The latter is considered his masterpiece even though his *Liber Abaci* influenced mathematicians for centuries to come.

[†]Vorob'ev, N. N., *Fibonacci Numbers*, Blaisdell Publishing Co. New York, © 1961 p. 2.

Solution

a. The given terms of the sequence are the first five odd counting numbers. A probable general term is $\{2n - 1\}$, where n is a counting number. Assuming this general term to be correct, we would write this sequence as $1, 3, 5, 7, 9, \ldots, 2n - 1, \ldots$

b. The given terms are powers of 2:

$$2^1 = 2; \quad 2^2 = 4; \quad 2^3 = 8; \quad 2^4 = 16; \quad 2^5 = 32$$

A probable general term is $\{2^n\}$, where n is a counting number. Assuming this to be true, the sequence would be written as $2, 4, 8, 16, 32, \ldots, 2^n, \ldots$.

c. The common pattern that exists among the given terms of this sequence is the denominator of each term is being increased by one. Thus a probable general term is $\{\frac{1}{n}\}$, where n is a counting number. Assuming this to be true, the sequence can be written as $1, \frac{1}{2}, \frac{1}{3}, \frac{1}{4}, \frac{1}{5}, \ldots, \frac{1}{n}, \ldots$.

NW ***Now Work Problem 13.***

EXAMPLE 5 ***Fibonacci Sequence and Recursion*** Develop a probable general term for the Fibonacci sequence

$$1, 1, 2, 3, 5, 8, \ldots$$

Solution A probable general term for this sequence can be developed as follows:

$$\begin{aligned} &\text{Let } T_n = \text{the } n\text{th term of the sequence} \\ &\text{Let } T_1 = 1 \\ &\text{Let } T_2 = 1 \end{aligned}$$

Thus when $n = 1$, the first term of the sequence is $T_1 = 1$, and when $n = 2$, the second term of the sequence is $T_2 = 1$. The general term can now be expressed as

$$\{T_{n-1} + T_{n-2}\}$$

where n is a counting number greater than 2. This can be confirmed as follows:

When $n = 3$,

$$\begin{aligned} T_{3-1} + T_{3-2} &= T_2 + T_1 \\ &= 1 + 1 \\ &= 2 \end{aligned}$$

Thus the third term of the sequence (T_3) is equal to 2.

When $n = 4$,

$$\begin{aligned} T_{4-1} + T_{4-2} &= T_3 + T_2 \\ &= 2 + 1 \\ &= 3 \end{aligned}$$

Thus the fourth term of the sequence (T_4) is equal to 3.

A similar procedure would be used to confirm the fifth and sixth terms. Note that in order to define a general term for this sequence we had to first define the first two terms specifically and then develop a general term that referred to earlier versions of itself. This concept of first defining specific terms (called *initial conditions*) and then giving a rule for successively defining additional terms by using those terms previously defined is known as **recursion** (and the general term is called the *recurrence relation*). Recursion and recursive algorithms have important applications in various branches of mathematics and computer science.

Math Connections

The Fibonacci Sequence and the Golden Ratio

The terms of the Fibonacci sequence can be visually demonstrated by developing a growing pattern of squares. This is shown in Figure 1.27. We begin with a small square with a side of 1 unit (a). We then append to this square another square, which also must have a side of 1 unit (b). In (c) we attach a square to the figure shown in (b). Note that this square must have a side of 2 units. Following this pattern, we continue attaching new, larger squares onto each previous design. Thus, in (d), the new square must have a side of 3 units; in (e), the new square must have a side of 5 units; in (f), the new square must have a side of 8 units; and in (g), the new square must have a side of 13 units. Once again, note that the sequence of the lengths of the sides of each new square forms the Fibonacci sequence 1, 1, 2, 3, 5, 8, 13, ...

Another interesting observation about the Fibonacci sequence is that the ratio of successive terms approaches the number 1.618. That is, when we divide each successive term by its preceding term, the quotient gets closer and closer to 1.618. This is illustrated using the first 11 terms of the Fibonacci sequence.

This number, 1.618, is called the *golden ratio*. Based on this observation, note that the ratio of the sides of the rectangles formed by attaching a new square to a previous design in Figure 1.27 approximates the golden ratio. For example, the sides of the rectangle in (d) are of the ratio 5 to 3. Any rectangle whose sides form the golden ratio is called a *golden rectangle*. Ancient Greek architecture, as well as contemporary art and architecture, frequently incorporates golden rectangles. In fact, rectangular figures whose sides were in the ratio of 5 to 3 were regarded by the Greeks as "divinely inspired" and the most pleasing to the eye.

The golden ratio has even been used in plastic surgery. It appears that good looks has a lot to do with how symmetrical you are. This symmetry is grounded in the golden ratio. For example, the "perfect" ratio between the width of a good looking person's nose and the width of the person's mouth is 1 : 1.618. To exploit

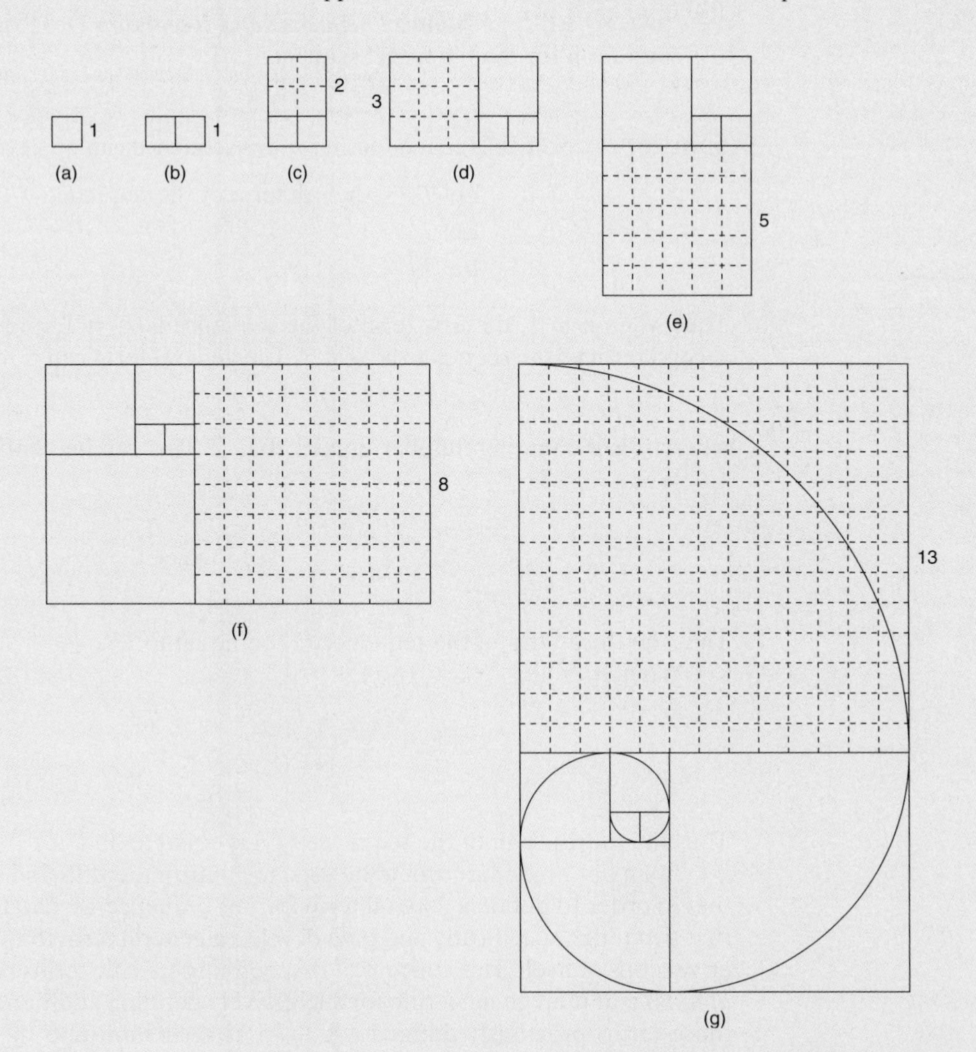

Figure 1.27

(continued)

(continued)

this link between the golden ratio and good looks, Stephen Marquardt, a retired California plastic surgeon, designed a mask that applies the golden ratio to the human face. The mask can be used to gauge how attractive a person is—the closer a person's face is aligned with the mask, the more attractive the person is perceived to be. For a closer look at Dr. Marquardt's mask, go to **http://www.beautyanalysis.com/mba_youandthemask_page.htm**.

Adjacent terms of the Fibonacci sequence occur in many different living entities. For example, a daisy contains 21 spirals in a clockwise direction and 34 counterclockwise. In pine cones, 5 spirals go one way and 8 the other. Pineapples, too, have opposing spirals of 8 and 13. Finally, the spiral shown in (g) is the same spiral that is found in the shell of the chambered nautilus.

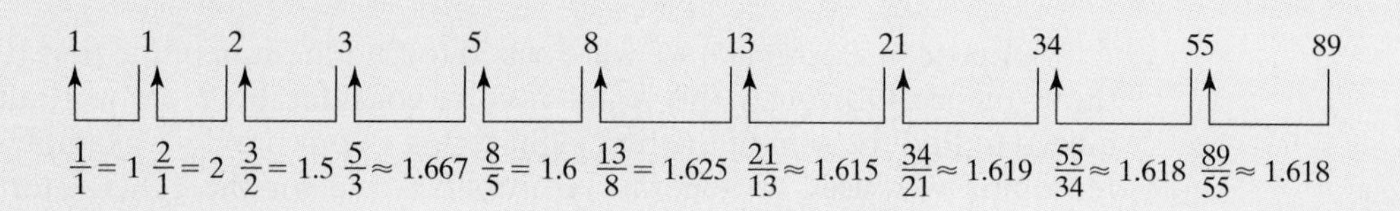

Series

Another concept that is closely related to infinite sequences is **infinite series**. An infinite series (commonly referred to as simply *series*) represents the sum of the terms of an infinite sequence. Although we may not realize it, most of us have already been introduced to the concept of series. For example, when we write the fraction $\frac{1}{3}$ in decimal form

$$\frac{1}{3} = 0.33333\ldots$$

we are really writing an infinite series:

$$\frac{1}{3} = 0.3 + 0.03 + 0.003 + 0.0003 + 0.00003\ldots$$

Note that when written in this manner, we have a **sequence of partial sums**, which is the sum of a finite number of terms. For example, given the series for $\frac{1}{3}$,

- The sum of the first term, denoted S_1, is 0.3.
- The sum of the first two terms, denoted S_2, is $0.3 + 0.03 = 0.33$.
- The sum of the first three terms, denoted S_3, is $0.3 + 0.03 + 0.003 = 0.333$.

S_1 is called the *first partial sum*; S_2 is called the *second partial sum*; and S_3 is called the *third partial sum*. Furthermore, if the nth term of an infinite sequence is given, then the sum of the first n terms is called the *nth partial sum*.

EXAMPLE 6 ***Infinite Series*** Given the infinite sequence 2, 4, 6, 8, 10, ... $2n$, ...,

a. Write the corresponding series.
b. Find the third partial sum of the series.
c. Find the eighth partial sum of the series.
d. Write the nth partial sum of the series.

Solution

a. The corresponding series is $2 + 4 + 6 + 8 + 10 + \cdots + 2n + \cdots$.
b. The sum of the first three terms is $2 + 4 + 6 = 12$. Thus, $S_3 = 12$.

c. Since we are only given the first five terms of the sequence, we first need to generate the sixth, seventh, and eighth terms. Using the general term $2n$, when $n = 6, 2n = 12$; when $n = 7, 2n = 14$; and when $n = 8, 2n = 16$. Therefore, the sum of the first eight terms is

$$2 + 4 + 6 + 8 + 10 + 12 + 14 + 16 = 72$$

Hence, $S_8 = 72$.

d. The nth partial sum of the series is written as follows: $S_n = 2 + 4 + 6 + 8 + 10 + \cdots + 2n$.

Suppose in Example 6 we were asked to find the sum of the first 100 terms of the sequence. Obviously, this would involve considerably more work than was required to find the sum of the first eight terms. In our previous examples inductive reasoning was used to generate a probable rule for the general term of a sequence. In a similar fashion, we can use inductive reasoning to help generate a probable formula for adding the terms of a series. This is demonstrated in the following two examples.

EXAMPLE 7 ***Infinite Series and the General Formula*** We wish to develop a general formula for adding the first n terms of the set of counting numbers $\{1, 2, 3, 4, 5, \dots\}$. To do this we will use bottom-up processing, pattern recognition, and inductive reasoning. Study these three equations and then answer the questions that follow.

$$1 + 2 = \frac{2 \times 3}{2}$$

$$1 + 2 + 3 = \frac{3 \times 4}{2}$$

$$1 + 2 + 3 + 4 = \frac{4 \times 5}{2}$$

a. What is $1 + 2 + 3 + 4 + 5 = ?$

b. Write a general rule based on these cases.

c. Use the rule from **b** to find the sum of the first 25 counting numbers; that is,

$$1 + 2 + 3 + 4 + 5 + \cdots + 23 + 24 + 25 = ?$$

Solution

a. Focusing on the right-hand side of the given equations, two observations are noted. First, in each case the denominator of the fraction remains constant at 2. Second, the numerator consists of two factors: The first factor is the last term of the expression on the left side of the equals sign; the second factor is 1 more than the first factor. This analysis leads to the following equation:

$$1 + 2 + 3 + 4 + 5 = \frac{5 \times 6}{2}$$

b. A probable general rule is

$$1 + 2 + 3 + 4 + 5 + \cdots + n = \frac{n(n + 1)}{2}$$

where n is a counting number

c. To find the sum of the first 25 counting numbers we set $n = 25$ and $(n + 1) = 26$:

$$\frac{n(n+1)}{2} = \frac{25(26)}{2} = \frac{650}{2} = 325$$

In Example 7, notice how we used bottom-up processing by expressing the sum of each addition, using a different perspective. For example, rather than expressing the sum of $1 + 2$ as 3, we chose to express it as $\frac{2 \times 3}{2}$. Similarly, $1 + 2 + 3$ was expressed as $\frac{3 \times 4}{2}$ rather than 6, and $1 + 2 + 3 + 4$ was expressed as $\frac{4 \times 5}{2}$ rather than 10. This perspective enabled us to identify a common pattern that existed among the individual cases that would not have been recognized otherwise. Once a common pattern was identified, we then used inductive reasoning to *generalize* what we observed from the individual cases. (*Note:* The generalized rule from Example 7 is a formula for adding the first n counting numbers.)

EXAMPLE 8 ***Infinite Series and the General Formula*** After studying the following equations, use pattern recognition and inductive reasoning to answer the questions.

$$1 + 3 = 2^2$$
$$1 + 3 + 5 = 3^2$$
$$1 + 3 + 5 + 7 = 4^2$$

a. What is $1 + 3 + 5 + 7 + 9 = ?$

b. Write a general rule based on these equations.

c. Use the rule from part **b** to find the following sum:

$$1 + 3 + 5 + 7 + \cdots + 21 + 23 + 25$$

d. Interpret the meaning of this rule.

Solution

a. A relationship that exists among the three equations is that the *number of terms* given on the left-hand side of the equation is the number being squared on the right-hand side of the equation. (The first equation has two terms and 2 is being squared; the second equation has three terms and 3 is being squared, and so forth.) Also, the terms consist of the odd counting numbers. Based on this analysis, we have

$$1 + 3 + 5 + 7 + 9 = 5^2$$

b. A probable general rule is

$$1 + 3 + 5 + 7 + 9 + \cdots + (2n - 1) = n^2$$

where n is a counting number.

c. The given expression contains 13 terms. Thus the sum is $13^2 = 169$.

d. This rule is a formula for finding the sum of the first n consecutive odd counting numbers.

NW ***Now Work Problem 23.***

Math Connections

Inductive Reasoning, Visual Proofs, and Math Induction

In mathematics, as well as in the sciences, we frequently make generalizations based on specific observations or collected data. Unfortunately, such generalizations are made by inductive reasoning and hence are always open to revision. Sometimes, though, we can visually demonstrate these generalizations. As an example, consider the set of odd counting numbers. In Example 4a, we generalized that the *nth* term of the sequence $1 + 3 + 5 + 7 + \cdots$ was $(2n - 1)$, and then in Example 8, we generalized that the sum of the first n consecutive odd counting numbers was n^2, that is, $1 + 3 + 5 + 7 + \cdots + (2n - 1) = n^2$. This arithmetic relationship, although derived from an inductive reasoning process, can be "proven" visually using square blocks, as shown in Figure 1.28.

Unfortunately, not all mathematical proofs involving infinite sequences or series lend themselves to visual "proofs." Instead, we rely on algebraic methods. One such method used in mathematics to prove infinite relationships is a process known as *mathematical induction*, which uses exactly two steps to prove an infinite number of steps. Although the specific manipulations of math induction require a background in algebra, we can nevertheless demonstrate this process intuitively to "prove" that $1 + 3 + 5 + 7 + \cdots + (2n - 1) = n^2$. To understand this intuitive reasoning process, we use the concept of partial sums as discussed in this section. That is, S_1 = the sum of the first term, S_2 = the sum of the first two terms, S_3 = the sum of the first three terms, and S_n = the sum of the first n terms. Thus

$$S_1 = 1 = 1^2 = 1$$

$$S_2 = 1 + 3 = 2^2 = 4$$

$$S_3 = 1 + 3 + 5 = 3^2 = 9$$

$$S_n = 1 + 3 + 5 + \cdots + (2n - 1) = n^2$$

We now intuitively reason through the process of mathematical induction:

1. We first show that S_1 is true. That is, $1 = 1^2$, which is indeed true.
2. We next assume that S_k is true for an arbitrary number k and then show that S_{k+1} is also true.

The first step of math induction is the initialization stage. We show that the relationship holds for the first term. The second step is the induction stage. By assuming that the relationship holds for the first k terms, if we can then show that the relationship also holds when we include the very next term, that is, the

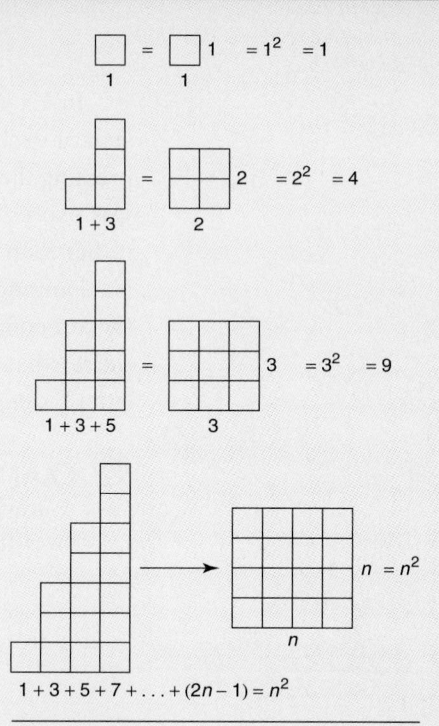

Figure 1.28

$k + 1$ term, then we have a deductive reasoning process that essentially says the following:

> *If S_1 is true, then this implies that S_2 is true. If S_2 is true, then this implies that S_3 is true, which implies that S_4 is true, which implies that S_5 is true, and so on.*

To illustrate this second step, let $k = 3$. That is, let's assume that the sum of the first three terms, S_3, is 3^2. That is, $1 + 3 + 5 = 3^2$. If we can now show that the sum of the first four terms, that is, S_{3+1}, is equal to 4^2, then we know that the relationship holds for the first four terms. As another example, if $k = 50$, then we assume that the sum of the first 50 terms, S_{50}, is 50^2. If we can now show that the sum of the first 51 terms is 51^2, then we know that the relationship holds for the first 51 terms.

Note how mathematical induction relies on inductive and deductive reasoning. We use inductive reasoning to first observe a pattern and make a generalization from our observation. We then use deductive reasoning by assuming that the observed pattern is true for the first k terms and then showing that the result we get for $k + 1$ terms follows logically from our assumption. Through mathematical induction, we effectively collapse an infinite number of steps into only the two steps given previously.

Exercises for Section 1.5

Given the sequences in Exercises 1 through 20, use pattern recognition and inductive reasoning to develop a probable general term. Write the sequences using this ***nth*** *term.*

1. $1, \frac{1}{\sqrt{2}}, \frac{1}{\sqrt{3}}, \frac{1}{2}, \ldots$

2. $1, \frac{1}{4}, \frac{1}{9}, \frac{1}{16}, \ldots$

3. $0, \frac{1}{2}, \frac{2}{3}, \frac{3}{4}, \ldots$

4. $\frac{1}{2}, \frac{1}{4}, \frac{1}{6}, \frac{1}{8}, \ldots$

5. $2, \frac{3}{2}, \frac{4}{3}, \frac{5}{4}, \frac{6}{5}, \ldots$

6. $\frac{1}{2}, \frac{4}{3}, \frac{9}{4}, \frac{16}{5}, \ldots$

7. $\frac{1}{2}, \frac{1}{3}, \frac{1}{4}, \frac{1}{5}, \ldots$

8. $0, \frac{1}{3}, \frac{2}{4}, \frac{3}{5}, \frac{4}{6}, \ldots$

9. $2, 1, \frac{2}{3}, \frac{2}{4}, \frac{2}{5}, \ldots$

10. $1, \frac{4}{3}, \frac{6}{4}, \frac{8}{5}, \ldots$

11. $\frac{2}{3}, \frac{3}{4}, \frac{4}{5}, \ldots$

12. $1, \frac{2}{3}, \frac{3}{5}, \ldots$

NW **13.** $\frac{1}{2}, \frac{2}{3}, \frac{3}{4}, \ldots$

14. $1, \frac{3}{4}, \frac{3}{5}, \frac{3}{6}, \ldots$

15. $2, \frac{4}{3}, \frac{6}{5}, \frac{8}{7}, \ldots$

16. $2, \frac{1}{2}, \frac{2}{9}, \frac{2}{16}, \frac{2}{25}, \ldots$

17. $\frac{1}{\pi}, \frac{1}{\pi^2}, \frac{1}{\pi^3}, \frac{1}{\pi^4}, \ldots$

18. $1, e, e^2, e^3, \ldots$

19. $1, -1, 1, -1, 1, -1, \ldots$

20. $-1, 1, -1, 1, -1, 1, \ldots$

21. Consider the following two sequences:

(i) $1, \frac{1}{2}, \frac{1}{3}, \frac{1}{4}, \ldots, \frac{1}{n}, \ldots$ **(ii)** $2, 1, \frac{2}{3}, \frac{2}{4}, \frac{2}{5}, \ldots, \frac{2}{n}, \ldots$

a. What do you notice about the numerators and denominators of the terms of the sequences?

b. As n increases, are the terms of the sequences getting smaller or larger?

c. Using your observations from **a** and **b**, and inductive reasoning, what can you conjecture about sequences that have this pattern?

22. We want to develop a general formula for adding the first n terms of the following sequence:

$$2, 4, 6, \ldots, 2n, \ldots$$

a. Study the equations and then write a general rule based on these equations using inductive reasoning.

$$2 + 4 = 2(3)$$
$$2 + 4 + 6 = 3(4)$$
$$2 + 4 + 6 + 8 = 4(5)$$

b. Use the rule from part **a** to find the sum of the first 99 terms.

NW **23.** We want to develop a general formula for adding the first n terms of the following sequence:

$$1^3, 2^3, 3^3, \ldots, n^3, \ldots$$

a. Study the following equations and then write a general rule based on these equations, using inductive reasoning. (*Hint:* Refer to Example 7.)

$$1^3 = 1 = (1)^2$$
$$1^3 + 2^3 = 9 = (1 + 2)^2$$
$$1^3 + 2^3 + 3^3 = 36 = (1 + 2 + 3)^2$$
$$1^3 + 2^3 + 3^3 + 4^3 = 100 = (1 + 2 + 3 + 4)^2$$

b. Use the rule from part **a** to find the sum of the first 10 terms.

Given the equations in Exercises 24 through 26, use inductive reasoning to develop a general rule.

24.
$$37 \times 3 = 111$$
$$37 \times 6 = 222$$
$$37 \times 9 = 333$$

25.
$$37037 \times 3 = 111111$$
$$37037 \times 6 = 222222$$
$$37037 \times 9 = 333333$$

26.
$$\left[1 + \frac{1}{2}\right] \times 3 = \frac{9}{2}$$
$$\left[1 + \frac{1}{3}\right] \times 4 = \frac{16}{3}$$
$$\left[1 + \frac{1}{4}\right] \times 5 = \frac{25}{4}$$

27. Use inductive reasoning to develop a general rule for the equations shown for parts **a** and **b**. What can you conclude from parts **a** and **b**?

a.
$$(1 + 1) \times 2 = \frac{4}{1}$$
$$\left(1 + \frac{1}{2}\right) \times 3 = \frac{9}{2}$$
$$\left(1 + \frac{1}{3}\right) \times 4 = \frac{16}{3}$$
$$\left(1 + \frac{1}{4}\right) \times 5 = \frac{25}{4}$$

b.
$$(1 + 1) + 2 = \frac{4}{1}$$
$$\left(1 + \frac{1}{2}\right) + 3 = \frac{9}{2}$$
$$\left(1 + \frac{1}{3}\right) + 4 = \frac{16}{3}$$
$$\left(1 + \frac{1}{4}\right) + 5 = \frac{25}{4}$$

28. Use inductive reasoning to generalize the following relationship. [*Note:* The expression $\binom{n}{j}$ is a special symbol used to denote the number of unordered arrangements of j things taken from n things. It is read as "n on j".]

$$3 + \binom{3}{2} = 3 + 3 = 6$$
$$4 + \binom{4}{2} = 4 + 6 = 10$$

$$5 + \binom{5}{2} = 5 + 10 = 15$$

$$6 + \binom{6}{2} = 6 + 15 = 21$$

29. **Arithmetic Progression** In Section 1.4 we learned that a list of numbers that contains a common difference is referred to as an *arithmetic progression*. By adding this common difference to one term of the sequence we can obtain the next term. If we let a = the first term and d = the common difference, then an arithmetic progression can be written as

$$a, (a + d), (a + 2d), (a + 3d), \ldots$$

Use pattern recognition and inductive reasoning to develop the nth term of this sequence.

30. **Geometric Progression** In Section 1.4 we learned that a list of numbers that contains a common factor is referred to as a *geometric progression*. If we multiply this common factor by one term of the sequence, we can obtain the next term. If we let a = the first term and r = the common factor, then a geometric progression can be written as

$$a, ar, ar^2, ar^3, \ldots$$

Use pattern recognition and inductive reasoning to develop the nth term of this sequence.

31. **Recursion** Develop an initial condition and a recurrence relation for the following sequence (see Example 5).

$$1, 2, 4, 8, 16, \ldots, 2^n, \ldots$$

32. **Recursion** Develop an initial condition and a recurrence relation for the following sequence (see Example 5).

$$1, 3, 9, 27, \ldots, 3^n, \ldots$$

Writing Mathematics

33. Explain the difference between an infinite sequence and infinite series.
34. Explain the concept of a partial sum.
35. Explain the concept of recursion. As part of your explanation, give an example of where recursion is used.

Challenge Exercises

36. Refer to the Math Connections on inductive reasoning, visual proofs, and math induction. Construct a similar visual "proof" for $1 + 2 + 3 + 4 + 5 + \cdots + n = \frac{n(n + 1)}{2}$.
37. Find the Fibonacci sequence that is embedded in Pascal's triangle, which was discussed in Exercise 17 in Section 1.2.
38. **Fibonacci Sequence** Divide each of the first 30 numbers of the Fibonacci sequence by 4 and record the remainders. Now determine the emerging pattern of these remainders.
39. **Fibonacci Sequence** Note the following about the Fibonacci sequence:

$$1 + 1 + 2 = 4$$
$$1 + 1 + 2 + 3 = 7$$
$$1 + 1 + 2 + 3 + 5 = 12$$

   a. Determine the pattern that appears to be emerging.
   b. Use this pattern and inductive reasoning to write the sum of the first 20 terms of the Fibonacci sequence.

Group/Research Activities

40. **Lucas Numbers** Another sequence similar to the Fibonacci sequence is called Lucas numbers, named after nineteenth century French mathematician Edouard Lucas (1842–1891). Research this sequence and report on what it is and its application.
41. In higher-level mathematics, a sequence is said to be *monotonic* if successive terms are nondecreasing, nonincreasing, decreasing, or increasing. Research the concept of monotonic sequences and prepare a report (with examples) of monotonically increasing and monotonically decreasing sequences.
42. Research what it means to say that a series *converges* or *diverges*.

Just for fun

A boat is tied up at a dock, and its ladder is hung over the side. Four of its 12 rungs are above water. The rungs are 8 inches apart. If the tide goes out at a constant rate of 7 inches per hour, how many rungs will be above water after 4 hours?

1.6 PÓLYA'S GENERAL PROBLEM-SOLVING FRAMEWORK

Thus far we have studied several techniques of problem solving. Some of the strategies included asking questions to better understand a problem, identifying common patterns or themes, using inductive reasoning, recalling past experiences or solutions, looking at a problem from different perspectives, and breaking down a problem into smaller parts (top-down processing). In this section we consolidate that information into a coherent, logical framework that can be used as a general guide for solving problems.

In his book *How to Solve It*, George Pólya lists four main phases, or steps, to problem solving.

Pólya's Four Steps for Problem Solving

1. **Understand the problem:** To solve a problem we must first understand what is being asked. Techniques or strategies that can facilitate such understanding include reading the problem carefully (sometimes more than once), listing specific facts that are present or details that you know to be true, identifying key phrases, identifying unknowns or missing information, selecting important and pertinent information, and identifying specific conditions or constraints that apply to a particular problem.
2. **Devise a plan:** Once a problem is fully understood, the next step is to decide on an approach or method that can be used to solve the problem. This can include searching for a pattern or common theme, using top-down or bottom-up processing, using prior knowledge to frame an appropriate technique, using inductive reasoning, drawing a sketch or making a table or chart, using trial and error, working backward, and restating the problem if necessary.
3. **Carry out the plan:** During this phase of Pólya's four-step process the technique selected in step 2 is implemented.
4. **Look back:** The final step is to review the solution to the problem. This involves checking the results for correctness and reasonableness, confirming that all conditions of the problem have been addressed, and adding the problem and its method of solution to your "bag of tricks" that can be used for subsequent problems.

Biography: George Pólya (1887–1985)

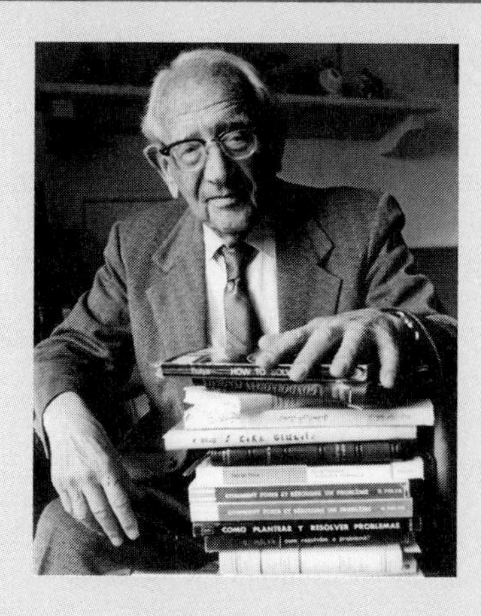

Known in teaching circles as the author of *How to Solve It* (which has been translated into fifteen languages), Pólya came to mathematics in a roundabout way. He first entered college to study law, but he could not stand it. Then he studied languages and literature and received his teaching certificate for Latin and Hungarian. However, he never made use of that certificate. Later he studied philosophy, physics, and mathematics. He was more interested in philosophy and physics, but eventually chose mathematics because, as he put it, "I thought I am not good enough for physics and I am too good for philosophy. Mathematics is in between." His many books are about the art of solving (mathematical) problems. When asked how he became interested in this topic, he mentioned Descartes's *Regulae* and Ernst Mach's book on the history of mechanics. In the latter book, the theme is that in order to really understand a theory you need to know how it was discovered. These influences and others are what led Pólya to develop a systematic approach to problem solving. As he told his students, "Guess and Test" and when trying to solve a difficult problem, try solving one that is simpler and similar first, get insight, and move up to the difficult problem.

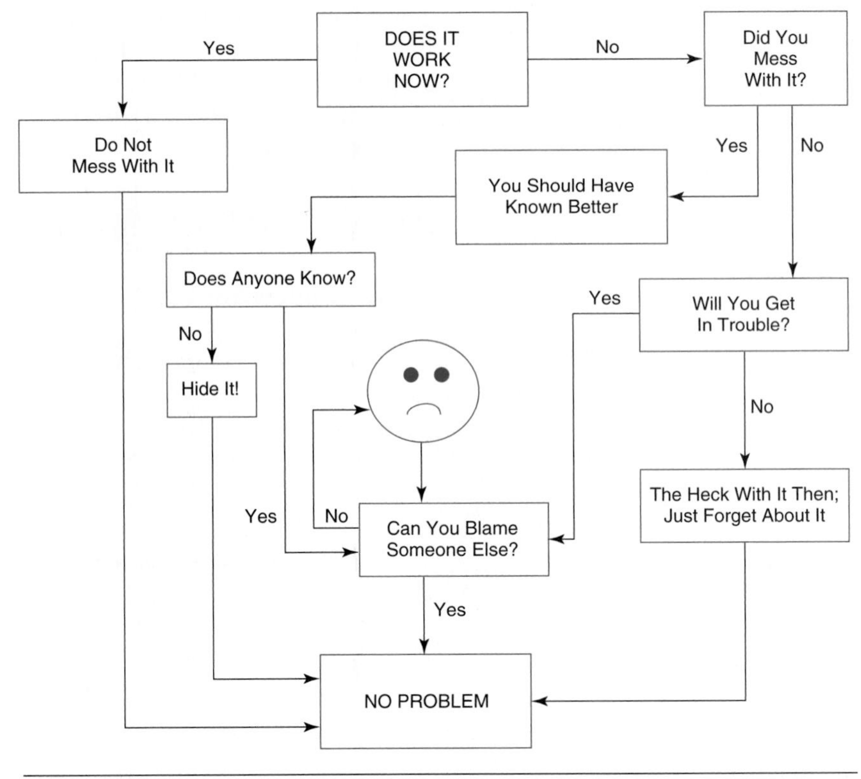

Figure 1.29

A humorous (and sometimes practical) approach to problem solving.

The "look back" step can also involve solving the problem using an alternative strategy to confirm the initial result.

An important characteristic of Pólya's problem-solving framework is its generality; the framework can be applied to many different types of problems in many different situations. Do not be misled by the framework's simplicity. Problem solving is a nontrivial activity. It takes time, perseverance, and practice to gain experience and confidence. In the examples that follow we use Pólya's four-step process to solve specific problems.

Math Connections

Variations of Pólya

In addition to Pólya's general problem-solving framework, several other models have been devised that can help guide the problem solving process. One such model is called the IDEAL Problem Solver[1]:

Identify a problem.
Define the problem.
Explore alternative approaches.
Act on a plan.
Look at the effects.

Identifying a problem is often an ignored component in most problem-solving models. Frequently, many people are quick to accept or maintain the status quo without thinking about how to improve something. If life presents us with an inconvenience or an unpleasant situation, we should strive to improve it. Problem identification is the source of many inventions, including caller ID, the fax machine, and the umbrella. Once a problem is identified, we next **define the problem**. Correctly defining the problem is important because the approaches that we consider to solving it will depend on how the problem is defined. When considering possible solutions to a problem, we should not restrict ourselves to only one. Instead, we should **explore alternative approaches**, such as using top-down processing to break the problem into parts, working backward, or using past experiences (see Section 1.2). We next **act on a plan**. That is, we implement our approach. Finally, we

(continued)

(continued)

look at the effects of our action. This involves checking the solution to see if it works and to see if it satisfies all the necessary conditions and stipulations of the original problem.

To illustrate this process, consider the following situation. Guests at a high rise hotel began complaining that the elevators were too slow shortly after the hotel completed its renovations of the upper floors. Using the IDEAL problem-solver model,

- The problem is identified: Elevators are too slow.
- The hotel manager defines the problem: "The problem is with the elevators."
- Various approaches are explored, including trying to speed up the elevators and installing more elevators. None of the approaches are feasible.
- A bell boy redefines the problem: "The problem is not with the elevators; it is how to take the hotel guests' minds off of time while they are waiting for the elevators."
- A solution is to install mirrors on each floor in front of the elevators.
- This solution is deemed feasible and implemented.
- After the mirrors were installed, guests' complaints stopped. Thus, the problem was solved.

In this illustration, note how important it is to define the problem correctly. The way we define a problem strongly influences the approaches we will consider. As a result, we need to be cognizant of some of the barriers to problem definition and solutions, including jumping to immediate action, not using all available information, misusing given information, and being locked into unsuccessful attempts.

[1]*Source:* Bransford, J. D., and Stein, B. S. (1984). *The IDEAL Problem Solver.* New York: W. H. Freeman and Company.:

EXAMPLE 1 ***Handshakes*** Professor Horton's educational statistics class consists of 10 students. If every student is to shake hands with each of the other students, how many handshakes will occur among the students?

Solution ***Understand the problem.*** In this example we are asked to find the number of handshakes that will occur among 10 students. We are told that each student will shake hands with each of his or her peers. We do not know and do not care if Professor Horton will be shaking hands with any of his students; we are interested only in the number of handshakes among his 10 students. It is also important to note that a handshake between two students is counted as one handshake regardless of who initiates it. Thus, if student A initiates a handshake with student B, and then student B initiates a handshake with student A, the second handshake is not counted because it is between the same two people who have already shaken hands.

Devise a plan. Rather than trying immediately to solve this problem for 10 students, let us first examine the problem by counting the number of handshakes that occur with one, two, three, and four persons to see if a pattern emerges. In short, we reduce the problem to smaller related ones.

Carry out the plan. Using the letters of the alphabet to represent the students and the symbol $\leftrightarrow$ to represent a handshake between two students, we have the following:

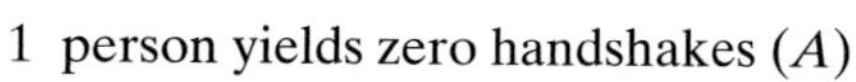
1 person yields zero handshakes (A)
2 people yield 1 handshake ($A \leftrightarrow B$)
3 people yield 3 handshakes
($A \leftrightarrow B, A \leftrightarrow C, B \leftrightarrow C$)
4 *people yield 6 handshakes* ($A \leftrightarrow B, A \leftrightarrow C, A \leftrightarrow D, B \leftrightarrow C,$)
$B \leftrightarrow D, C \leftrightarrow D$)

We use the following chart to organize our work and search for a pattern:

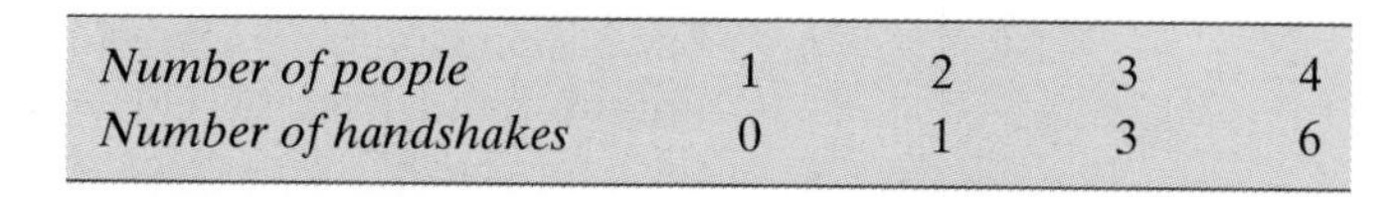

Number of people	1	2	3	4
Number of handshakes	0	1	3	6

Using the strategy of common differences from Section 1.4, notice that the number sequence associated with the number of handshakes is a level 2 common difference problem:

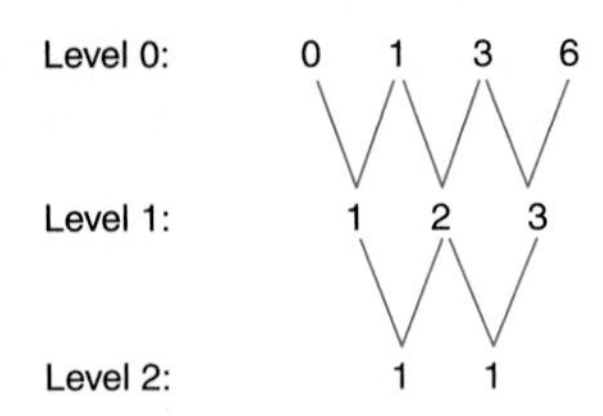

If we were to continue this pattern for 10 students, we would have a total of 45 handshakes.

Number of people	1	2	3	4	5	6	7	8	9	10
Number of handshakes	0	1	3	6	10	15	21	28	36	45

(*Note:* There are at least two additional patterns that can be extracted from this table. Can you identify them?)

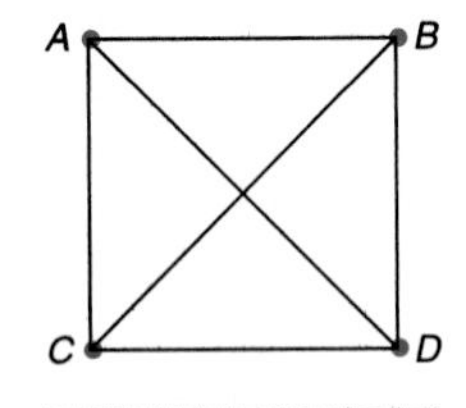

Figure 1.30

Look back. From the pattern that emerged from the preceding example, our solution of 45 handshakes appears reasonable. There are a few things that we can do to confirm this result. First, we can extend our plan to five, six, and seven students and see if we get the same result. Second, we can repeat steps 2 and 3; that is, we can devise and carry out a different plan and see if we obtain the same result. For example, another way that this problem can be solved is to *model* the problem by using a graph. If we use vertices to represent students and line segments (that connect the vertices) to represent handshakes, then Figure 1.30 is a model of the number of handshakes among four students.

From our model we "see" that a total of 12 handshakes occurs (three each for A, B, C, and D). However, observe that we have counted a handshake between two people twice. To remove this redundancy we divide the total number of handshakes depicted in this model by 2. Then the total number of handshakes among four students is $12 \div 2 = 6$, which corresponds to our earlier result. Since, using two different strategies, our results are the same for four people we now are more confident that our answer to the original problem is correct.

By combining this model with a little prior knowledge, it is not too difficult to develop a general rule that can be used to determine the number of handshakes that would occur among any number of people. First, notice that the model is a *fully connected* graph (see Example 11 of Section 1.4). Note further that each vertex is of degree three (there are three line segments joining at each vertex), and the degree of a vertex is equal to one less than the total number of vertices. Finally, if we multiply the degree of a vertex by the total number of vertices we obtain the total number of handshakes with redundancy. By dividing this product by two, we determine the total number of unique handshakes. Thus, the

$$\begin{pmatrix}\text{Total number of} \\ \text{unique handshakes}\end{pmatrix} = \frac{(\text{Degree of vertex}) \times (\text{Number of vertices})}{2}$$

As a result, a fully connected graph with 10 vertices has degree 9; the total number of handshakes with redundancy then is $10 \times 9 = 90$; and the total number of unique handshakes is $90 \div 2 = 45$. This answer matches the answer from our first approach.

The strategy used to solve the problem in Example 1 included reducing it to smaller (but related) parts, making a chart, and searching for a pattern. In the next example we use the strategy of logical reasoning.

NW ***Now Work Problem 1.***

EXAMPLE 2 ***Basketball Tournament*** A "single-elimination" basketball tournament, consisting of 10 teams, is being held at the local Grant Community Center. How many games must be played to determine the winner of the tournament?

Solution ***Understand the problem.*** In this example we are asked to find the number of basketball games that must be played to determine a winner. Important information includes: A basketball game involves two teams; the term *single elimination* means that a team is eliminated if it loses a game; and 10 teams will be playing.

Devise and carry out the plan. Using logical reasoning, if there is to be one winning team out of 10 teams, then there must be nine losing teams. Given the fact that if a team loses a game it cannot be a winner (by virtue of the "single elimination" rule), then only nine games need to be played.

Look back. This answer of nine games is reasonable. To check our work we can draw a tournament chart (Figure 1.31) that shows the winner of each game played. Notice that a total of nine games are played before the tournament winner is declared.

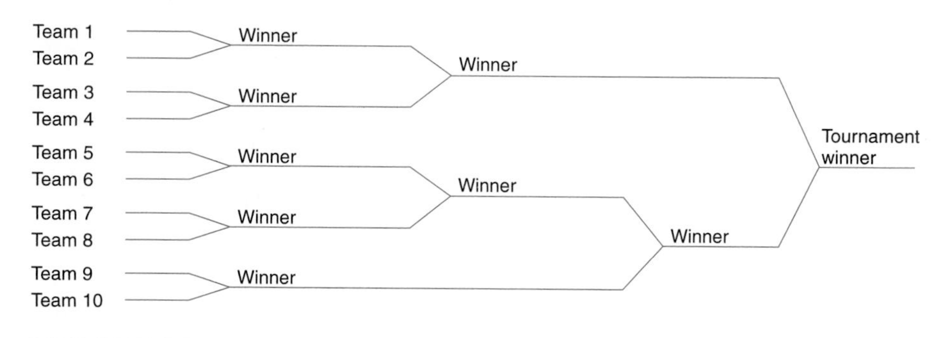

Figure 1.31

NW ***Now Work Problem 13.***

Our next example illustrates the method of trial and error. (In the remaining examples of this section we do not itemize each step of Pólya's framework as we did in Examples 1 and 2.)

EXAMPLE 3 ***Partitioning a Region*** Sally, an animal researcher, wishes to design a square cage to house seven mice. She wants each mouse to be e closed in its own separate area. Dr. Nenno, Sally's former mathematics professor, tells her that the interior region of a square can be partitioned into exactly seven separate areas by three intersecting lines. How can this be accomplished?

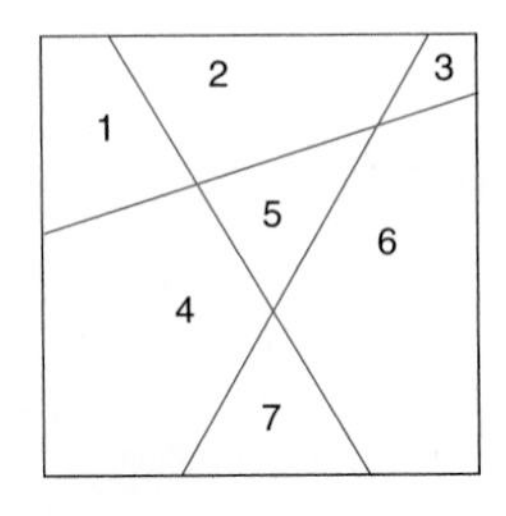

Figure 1.32

Solution The solution to this problem can be arrived at through trial and error. Our job is to draw three intersecting line segments within a square so that seven separate regions are formed. One solution is shown in Figure 1.32.

For future reference the general formula

$$\frac{n(n+1)}{2} + 1$$

where n is the number of intersecting lines, yields the maximum number of separate regions produced by the intersecting lines.

In the next example we use a *matrix* and the method of elimination to solve a problem.

EXAMPLE 4 ***Using Matrices to Model a Problem*** Three women, Janie, Elaine, and Leah, live in separate but adjacent apartments. Janie lives in the middle apartment. The "occupations" of the three women are grandmother, math teacher, and socialite. We know the following about these women:

a. The socialite watches Janie's apartment when Janie goes on vacation.

b. The math teacher has to pound on Leah's wall when Leah has the television on too loud.

Determine Janie's occupation.

Solution The solution can be facilitated by using a three-by-three matrix. The names of the women are placed at the top and their "occupations" are placed along the left side. Since Janie lives in the middle apartment, we put Janie's name in the middle column:

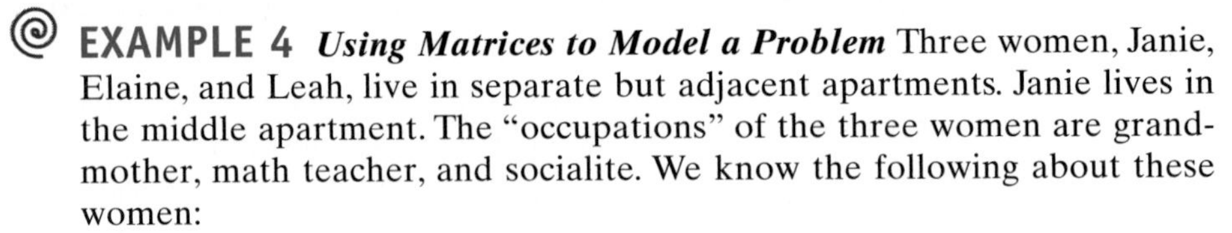

	Leah	Janie	Elaine
Socialite			
Math Teacher			
Grandmother			

The next step is to place an X in each "cell" to indicate that a particular "occupation" cannot be associated with any of these women. We are guided in our efforts by the information contained within the problem: We know from **a** that Janie cannot be the socialite. Similarly, from **b** we know that Leah cannot be the math teacher. Also from **b**, Elaine cannot be the math teacher since she cannot pound on Leah's wall. As a result of this analysis, we place an X in each of the appropriate cells.

	Leah	Janie	Elaine
Socialite		X	
Math Teacher	X		X
Grandmother			

From the chart we can now see that, through a process of elimination, Janie must be the math teacher since it is not possible for either Leah or Elaine to be the math teacher.

Can we conclude anything else from this matrix? Explain your answer.

In the preceding examples we demonstrated Pólya's general framework for problem solving. The various strategies used can now be added to our "bag of tricks," which we can draw from to help us solve future problems. These strategies can also serve as a catalyst for generating new ideas or plans to improve and refine our problem-solving abilities. The problems that follow will provide ample opportunities to develop your problem-solving skills. Read each problem carefully and make sure that you understand it completely before trying to solve it. You are also encouraged to employ more than one strategy, as we did in some of the examples.

Exercises for Section 1.6

NW **1. Music** A phonograph record has a diameter of 12 inches. The record has two nonrecorded tracks. The first is a 1-inch gap that exists on the outer margin of the record; the other is a 4-inch gap at the center. If the average number of grooves in the record is 120 per inch, how far must the turntable needle travel when the record is played?

2. Water Rationing In a remote village on the island of Sicily, many houses have running water only from 8 A.M. to 10 A.M. During those hours Sarafina fills several containers of different sizes with water, which her family will use for the day. One afternoon Sarafina needed exactly 4 liters of water. If she had an 8-liter container filled with water and two empty containers of 5 and 3 liters, respectively, how can she measure exactly 4 liters of water by using these containers?

3. Prescription Drugs Your doctor gives you a prescription instructing you to take one pill every 15 minutes. How many pills will you take during the first half-hour?

4. Pet Cat Julia, the cat, wants to climb a tree that is 73 feet high. Every day she climbs 11 feet, but then during the night she comes back down 7 feet. Assuming that Julia does not fall, how long will it take her to reach the top of the tree?

5. Internet Connection Marshal purchased a DSL modem and a special cable for a total of \$210. If the modem costs \$200 more than the cable, how much did Marshal pay for each?

6. Graph Theory In Figure 1.33 on p. 61 the vertices represent people and the line segments connecting the vertices represent a parent–child relationship. For example, the segment from A to E means that *A is the biological parent of E*. Given this information, determine if the graph is valid. Explain your reasoning.

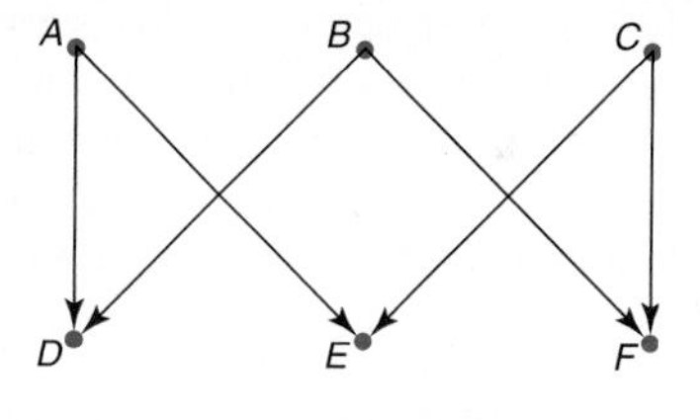

Figure 1.33

7. Keeping Time As an experiment, Howard programmed three of his many clocks to keep time as follows: At noon on July 1, 1998 all three clocks displayed the correct time. Immediately thereafter, though, one clock will continue to show the correct time, but the other two clocks gained and lost 1 minute respectively every 24 hours. Determine the date (month, day, and year) that all three clocks will show the same time at noon.

8. **Athletic Endeavors** Cousins Sue Ann, Claudia, and Jane are athletes. One of them is playing golf on Sanibel Island, a second is in the Blue Ridge Mountains, and the third is at Aspen. One cousin is bicycling, another is skiing, and the third is playing golf. Sue Ann is not on Sanibel Island, Jane is not at Aspen, and the one who is playing golf is not in the Blue Ridge Mountains. If the golfer is not Jane, who is playing golf and where?

9. **Classroom Desk Arrangement** Some students feel that Professor Chen is old-fashioned and regimented. For example, the desks in his class are in straight rows with the same number of desks in each row. He allows no deviations from this pattern. If all of the students in his Survey of Mathematics class are present, then every desk will be occupied. John, one of the students, is assigned the desk that is located in the second row from the front and the fourth row from the back. The desk is also the third one, counting from the left side of the row, and the fifth one, if you count from the right side of the row. How many students are in Professor Chen's Survey of Mathematics class?

10. **Logical Reasoning** Assume that you have only one match. Further assume that you are in a room containing only three items: a kerosene lamp, an oil burner, and a wood-burning stove. Which would you light first?

11. **Distance–Time Relationship** It is important that Julio arrive at exactly 5 P.M. for a meeting. Julio determined that if, riding his bicycle, he averaged 15 miles per hour, he would arrive an hour too soon. On the other hand, if he averaged 10 miles per hour, he would arrive an hour too late. Based on his calculations, what *distance* must Julio ride his bicycle to arrive on time at the meeting?

12. **Distance–Time Relationship** A freight train 1 mile long is traveling at a constant rate of 60 miles per hour. How long will it take for the train to pass completely through a 1-mile long tunnel?

NW 13. **NCAA Basketball Tournament** The annual NCAA Basketball Tournament is a "single-elimination" tournament comprising 64 teams. How many games must be played to determine the winner?

14. Using the first 11 counting numbers, place a number in each of the 11 circles in Figure 1.34 so that the numbers in any three circles that are part of the same line segment have a sum of 18.

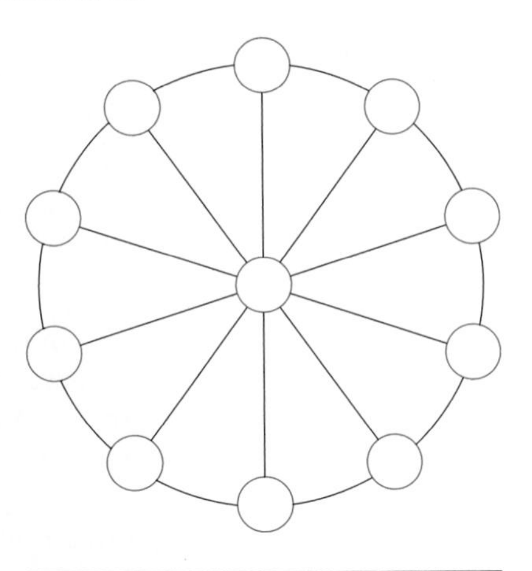

Figure 1.34

15. Dr. Bill says that he has created a liquid that can dissolve anything. Do you believe him? Explain your reasoning.

16. Given the set of standard arithmetic symbols and four 7s, write an expression that equals 100. For example, four 5s can be arranged as

$$(5 + 5) \times (5 + 5),$$

which equals 100.

17. **Land Parcels** A building contractor wants to subdivide a square parcel of land that contains 11 trees. He wants each lot to have one tree. What is the minimum number of fences he must erect? How would he do it? (See Example 3.)

18. Why are street manhole covers circular in shape?

19. **Holiday Gift Sharing** Every year the eight members of the Computing Services department at a local college exchange holiday gifts. What is the total number of gifts exchanged if all members of the department participate?

20. **Crossing a River** Two adults and two children need to cross a river. Their boat has a load maximum of 150 pounds. Two of the individuals each weighs 150 pounds; the other two each weighs 75 pounds. Using the boat, how can this group get across the river? How many boat crossings are required?

21. There are 20 pigeons sitting on a telephone wire in a remote field. A young boy fires a shotgun blast at the pigeons and shoots a fourth of them. How many pigeons remain?

22. **Used Auto Purchase** Kelly purchased a used auto for \$1500 and then sold it for \$1600. She then bought the same car back again for \$1700 but later sold it for \$1800. What was Kelly's overall profit from these transactions?

23. **The Fugitive** Let's assume that you are Richard Kimble, *The Fugitive*, and that you are walking along a rural road. Let's further assume that Lt. Girard is in a police car and that he is driving toward you. Under what circumstances would you run *toward* Lt. Girard's car instead of away from it when you see him?

Writing Mathematics

24. Summarize the four steps of Pólya's problem-solving framework.
25. The IDEAL Problem Solver model was presented as a Math Note as an alternative to Pólya's model. Explain the similarities and differences between these two models.
26. Another alternative to Pólya's general framework is a problem-solving model from the business world developed by the worldwide management consulting firm, Kepner-Tregoe (see **http://www.kepner-tregoe.com/**). This model involves the following five steps: (1) recognize a situation, (2) take interim action, (3) find the cause, (4) choose an action, and (5) implement the plan. Compare this model to Pólya's and write a report that describes the differences and similarities between the models.

Challenge Exercises

27. **Presidential Election** In a hotly contested presidential election, the declared winner of Florida won the presidency. Throughout the weeks that followed this election, many people were hoping that one of the candidates would step forward and say that in the interest of the nation, he was conceding to his opponent. Although this did not occur, let's assume that X was more of a statesman than Y and said to the Y campaign, "I am willing to allow the outcome of this election to be determined by chance. You may place two pieces of paper in a hat. One piece shall have 'winner' written on it and the other shall have 'loser' written on it. I will not look at either piece of paper. In a public ceremony, I will then select one piece of paper from the hat and accept the outcome of the drawing." After much deliberation, Y agreed to the challenge. During the morning of the drawing, the head of Y's campaign asked Y why he was willing to do this. "Don't worry about it," said Y. "I am going to write 'loser' on both pieces of paper." Not to be outwitted, X also had a strategy and the outcome of the drawing declared him the winner. What did X do?
28. Go to the Web site **http://www.greylabyrinth.com/puzzles.htm**, and see if you can solve one of the given problems. (*Note:* Also check out the plethora of puzzles and problems that are available in the archive component of this site.)

Group/Research Activities

29. Two frequent barriers to effective problem solving are making incorrect or unnecessary assumptions about a problem and imposing unnecessary constraints on a given situation that make the problem impossible to solve. An excellent example that demonstrates this is the story of Copernicus, a sixteenth-century scientist who studied the movement of the planets in our solar system and is recognized as the founder of modern astronomy. Research this story and prepare a report that explains what Copernicus did to help explain the movement of the planets.
30. One collaborative problem-solving activity is known as *brainstorming*. Research this concept and prepare a report that specifies the rules of brainstorming. Then identify a problem of interest and use brainstorming to help identify possible solutions.
31. Read Pólya's book *How To Solve It*, and prepare a report that summarizes its salient aspects.

Just for fun

How many planets are in our solar system? Ranging from nearest to farthest from the sun, what are the names of these planets?

CHAPTER REVIEW MATERIALS

Summary/Chapter 1

1.2 Problem-Solving Traits and Methods

- Human traits or characteristics that lend themselves to problem solving include patience, observation, experience, and reasoning.
- Specific methods or strategies used to solve problems include pattern recognition, understanding and questioning, critical thinking, and estimation.

1.3 Estimation

- Estimation provides a means for mentally determining an approximate solution to a problem.
- Estimation can be used to determine if the final solution to a problem is reasonable.
- Methods of estimation include rounding, compatible numbers, and front-end estimation.
- When estimating the solution to a problem, we must be sensitive to the accuracy of the estimate. Depending on the technique, the result may lead to an overestimate or an underestimate.
- Estimation is appropriate when it is impossible to obtain an exact answer (e.g., the size of a crowd at a peace rally).

1.4 Pattern Recognition and Inductive Reasoning

- Pattern recognition involves identifying a specific order or arrangement of objects.
- The presence or absence of a pattern can help predict the solution to a problem.
- Inductive reasoning is a mental process that enables us to make predictions by developing generalizations based on a limited

number of specific observations or experiences. In contrast to inductive reasoning is deductive reasoning, which is based on making predictions from general observations or experiences.

- Inductive reasoning is commonly the result of pattern recognition.
- A prediction resulting from an inductive reasoning process is considered probable.

1.5 Inductive Reasoning and Sequences and Series

- A sequence is a list of ordered objects. Sequences can be finite (they have a specific number of terms) or infinite (they have an unlimited number of terms).
- Inductive reasoning and pattern recognition can be used to help determine the general term of a sequence. This general term can then be used to generate the terms of a sequence.
- The concept of recursion involves defining the general term of a sequence by first defining the sequence's first two terms and then developing a rule for specifying subsequent terms by using previously defined terms. The first two terms represent the initial conditions and the general term is called the recurrence relation. A well-known recursive sequence is the Fibonacci sequence.
- A series represents the sum of the terms of an infinite sequence.
- Inductive reasoning, pattern recognition, and bottom-up processing can be used to determine the general formula of a series. This general formula can then be used to find the sum of the series.

1.6 Pólya's General Problem-Solving Framework

- George Pólya was a mathematician who developed a general framework for problem solving.
- Pólya identified four main phases, or steps, to solving problems: understand the problem, devise a plan, carry out the plan, and look back.

Key Terms and Concepts

Review Exercises

1.2 Problem-Solving Traits and Methods

In Exercises 1 through 3, state whether the reasoning process used is inductive or deductive.

1. **Making Generalizations** After eating Lemon Chicken for the first time, Ralph comments to his wife Elaine that he likes it. Elaine decides to make and serve Lemon Chicken to Ralph every week.
2. **Flying** When it is cloudy outside Marshal does not go flying. Marshal is not flying today; therefore it must be cloudy outside.
3. **Homeless** The results of a survey of 15 homeless men in South Florida indicate that the majority of those interviewed had beards. Using the results of this survey a newspaper reporter, writing a story on the homeless in America, states that the majority of homeless men have beards.
4. **Identifying Context** Give examples of at least two contexts in which the symbol * can be used.
5. **Interpreting Statements** The following statement appears on a sign announcing a new lottery game: "You must be 18 to play." Interpret the meaning of this statement.

1.3 Estimation

In Exercises 6 through 10, use estimation to answer the given problems.

6. **Estimating A Bill and Tax** Lynn and Sue went to breakfast at Sonny's one morning and ordered the items shown in the following table.
   a. Estimate their bill without tax.
   b. If the tax is 8%, approximately how much is their total bill with tax?
   c. If Lynn and Sue want to leave between a 15% and 20% tip, approximately how much should they leave?

Qty	Item	Unit Price ($)
2	Coffee	0.65
2	Orange Juice	1.50
2	Muffin	0.95
1	Liver Mush	2.35
1	Ham & Eggs	2.25
1	Oatmeal	1.95

7. **Marijuana** The line graph in Figure 1.35 shows the percentage of high school seniors who answered yes when asked if they had experimented with marijuana at least once in a prior two-week period. The results were part of an annual survey conducted during the 1990s. Based on these data, approximate the percentage of high school seniors who will answer yes to this question in 2004.
8. **Job Stress** The circle graph in Figure 1.36 shows the percentage of workers who were asked to report their job stress level. If 2095 people were polled, approximately how many workers reported being very stressed?
9. **Job Satisfaction** The circle graph in Figure 1.37 shows the percentage of workers who were asked to report their level of job satisfaction via an online Web survey. If 1783 people participated, approximately how many reported being *less than satisfied* with their current job?
10. **Corporate Revenue** The bar graph in Figure 1.38 shows earned revenue in billions of dollars of a major United

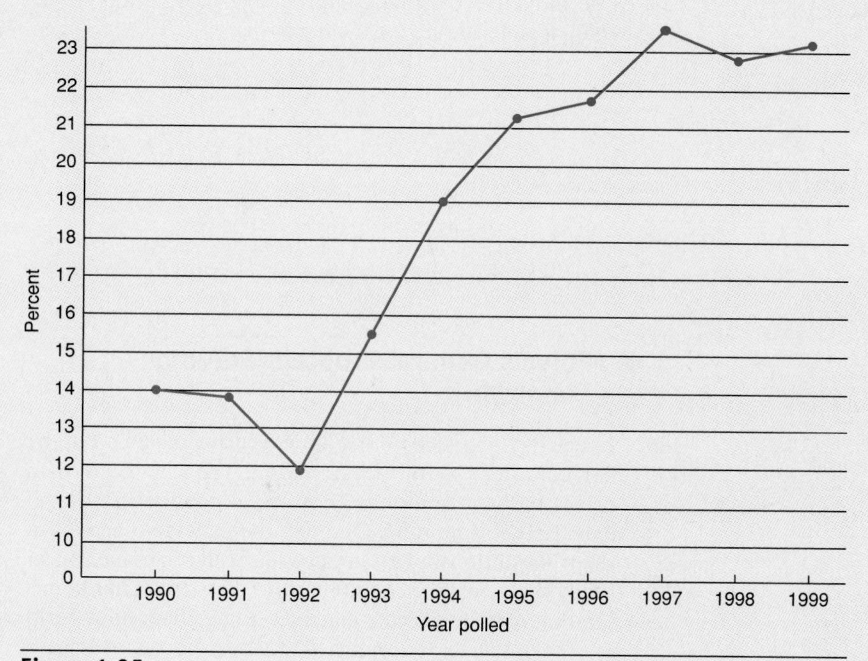

Figure 1.35

Percentages of high school seniors who experimented with marijuana.

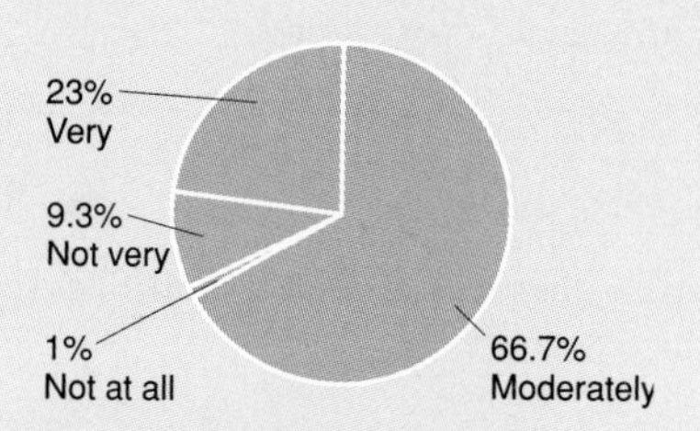

Figure 1.36

Job stress level

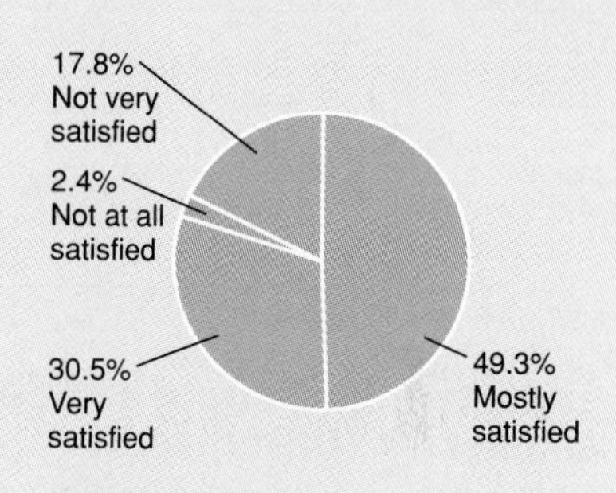

Figure 1.37

Job satisfaction level

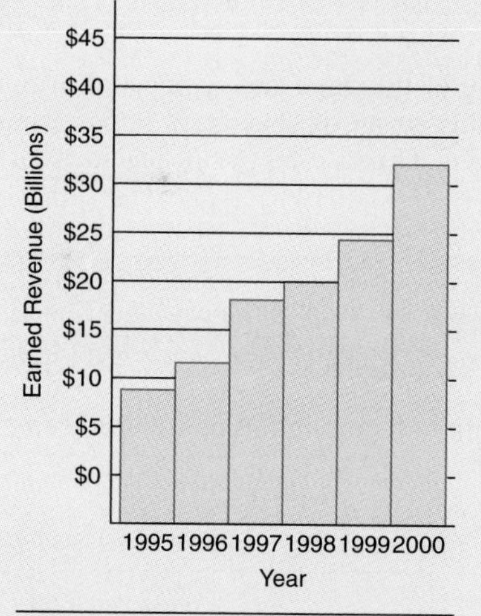

Figure 1.38

Earned revenue for Jillian Enterprises

10. continued

States corporation during the period 1995–2000. Based on these data, estimate the amount of revenue this corporation will earn in 2005.

1.4 Pattern Recognition and Inductive Reasoning

In Exercises 11 through 13, identify the pattern and use inductive reasoning to determine the next probable iteration.

11. $10_2 = 2$

$10_3 = 3$

$10_4 = 4$

12. $10^2 = 100$

$10^3 = 1000$

$10^4 = 10{,}000$

13. $\frac{S}{N} = 10$ implies 10 decibels

$\frac{S}{N} = 100$ implies 20 decibels

$\frac{S}{N} = 1000$ implies 30 decibels

14. Figure 1.39 shows a relationship between the number of points on a circle and the number of regions created when the points are connected by line segments.

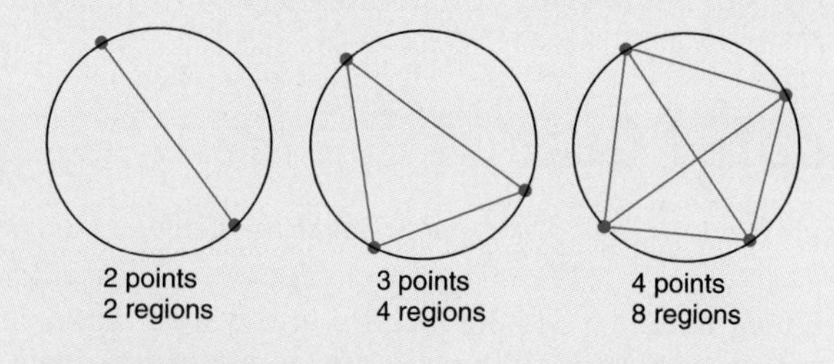

Figure 1.39

a. Determine the next iteration and draw a sketch of the figure.

b. Based on this pattern, how many regions will six fully connected points produce? How would you justify your answer? Is your answer based on inductive or deductive reasoning?

15. What is the next character in the following pattern?

Z, O, T, T, F, F

(*Note:* The second character is the uppercase letter "Oh.")

In Exercises 16 through 20, determine the pattern that exists and use inductive reasoning to write the next probable number. State whether the given list of numbers forms an arithmetic or geometric progression (or neither).

16. 2, 7, 12, 17 **17.** $3, 1, \frac{1}{3}, \frac{1}{9}$ **18.** 4, 10, 18, 28, 40

19. 144, 121, 100, 81, 64 **20.** 1, 1, 2, 3, 5

21. Study Figure 1.40 below and find the value of c in the third triangle.

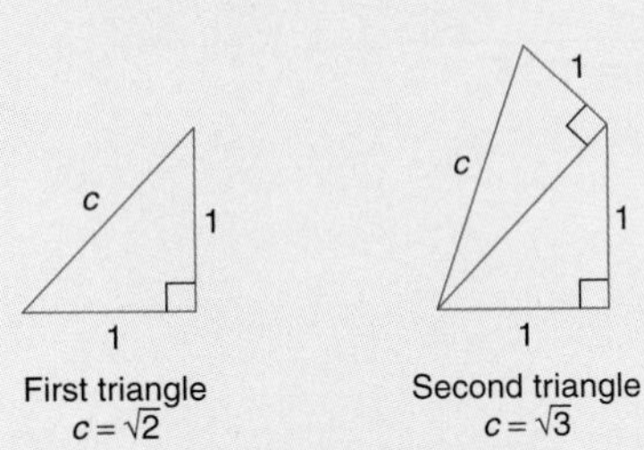

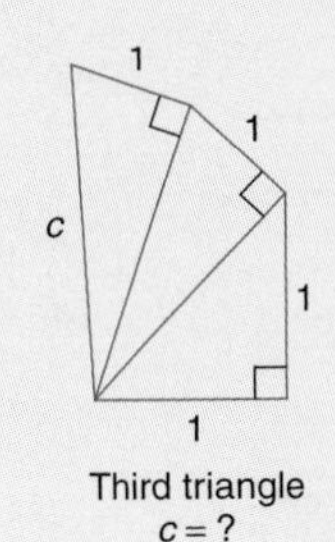

Figure 1.40

1.5 Inductive Reasoning and Sequences and Series (Optional)

In Exercises 22 and 23, use pattern recognition and inductive reasoning to develop a probable general term.

22. $\frac{1}{x}, \frac{1}{\sqrt[2]{x}}, \frac{1}{\sqrt[3]{x}}, \frac{1}{\sqrt[4]{x}}, \ldots$

23. $\left(1 + \frac{1}{2}\right) \times 3, \left(1 + \frac{1}{3}\right) \times 4, \left(1 + \frac{1}{4}\right) \times 5, \ldots$

24. Use inductive reasoning and pattern recognition to develop a general rule for the following:

$$\left(1 + \frac{1}{2}\right) \times 3 = \frac{9}{2}$$

$$\left(1 + \frac{1}{3}\right) \times 4 = \frac{16}{3}$$

$$\left(1 + \frac{1}{4}\right) \times 5 = \frac{25}{4}$$

1.6 Pólya's General Problem-Solving Framework

25. Every day at exactly 5 P.M. an executive is picked up by her chauffeur at the train station. The chauffeur then drives the woman home. They arrive home every day at exactly the same time. One day the executive arrives at the train station at 4 P.M. and begins walking home. Eventually she meets the chauffeur, who then drives her home. On that day the executive arrives home 20 minutes earlier. Using logical reasoning, what time did the executive meet the chauffeur on her walk home?

26. Given the arrangement

$$1 \quad 2 \quad 3 \quad 4 \quad 5 \quad 6 \quad 7 \quad 8 \quad 9 = 100$$

use any of the signs and symbols of arithmetic to transform this string of characters into a true mathematical statement. Do not alter the arrangement of the characters.

Just for fun

If one person out of every 16 can catch a certain type of flu, then how many people in a population of 100,000 can catch that flu?

Chapter Quiz

In Exercises 1 through 3, indicate whether the statement is true or false.

1. **Doing Laundry** Sorting laundry into whites, darks, sheets, towels, and work clothes is an example of top-down processing.
2. Inductive reasoning involves developing a general statement or conclusion based on small, specific cases.
3. A geometric progression consists of a sequence of numbers that have a common factor.
4. Find the next two most probable terms from the following list: 2, 1, 3, 4, 7.
5. Use problem-solving skills to find the following sum.

$$\frac{1}{2} + \frac{1}{2^2} + \frac{1}{2^3} + \frac{1}{2^4} + \cdots + \frac{1}{2^n} + \cdots$$

(*Hint:* Try to identify a pattern by focusing on "partial sums.")

6. **Geometric Relationship** Using Figure 1.41, identify the relationship that exists between the number of interior line segments and the total number of triangles generated.

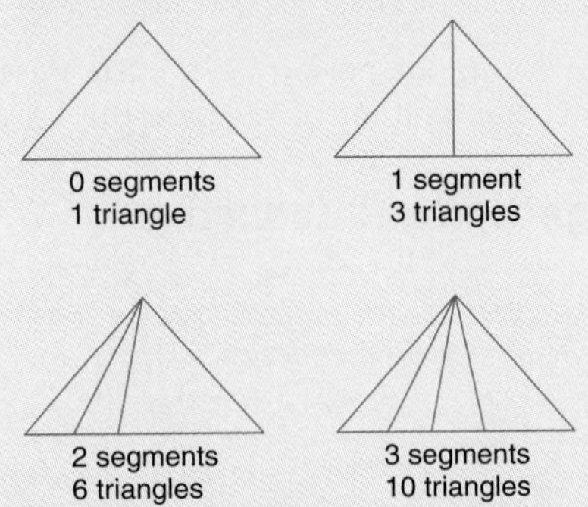

Figure 1.41

7. **Pólya** State Pólya's four-step problem-solving process.
8. The cost of two items is $200. If the first item is $100 more than the second, how much does each item cost?

9. Use logical reasoning to determine the amount of dirt in a hole that is 10 feet long, 5 feet wide, and 3 feet deep.

10. **Favorite Fruits** Laura, Susan, Claudio, and Pedro were each asked to name their favorite fruit. The fruits were bananas, peaches, plums, and apples. No two individuals had the same favorite fruit. One of the boys likes apples while the other likes peaches. Also, Laura claimed she does not like plums, whereas Pedro said his favorite fruit is spelled with the same first letter as his first name. Using this information identify each person's favorite fruit.

11. **Bar Mitzvah** Ralph is planning to attend Andrew's and Marc's bar mitzvah, which begins at 7 P.M. If it takes Ralph 25 minutes to get to the temple; 40 minutes to shower, shave, and dress; and 15 minutes to pick up his date, what time should Ralph begin getting ready? What problem-solving strategy did you use to answer this question?

12. **Holiday Retail Sales** The circle graph in Figure 1.42 shows various time periods of the Christmas holiday season, which is between Thanksgiving and the end of the year, and the average percentage of holiday retail sales per time period. If Rachel's Christmas Boutique generated $285,000 during last year's Christmas holiday season, approximately how much money do you think Rachel's will generate this year during the last two weeks of the holiday season?

13. **Company Stock Values** The line graph in Figure 1.43 shows the end-of-year stock closings for two companies: RBN Fiber and The Wireless Group.

    a. Approximately how much more was The Wireless Group's stock worth per share than RBN Fiber's stock at the close of 2003?

    b. In what year did the two stocks' price per share differ the greatest and what was the approximate difference?

    c. Based on these data, estimate what you think these two companies' end-of-year stock closing will be in 2005.

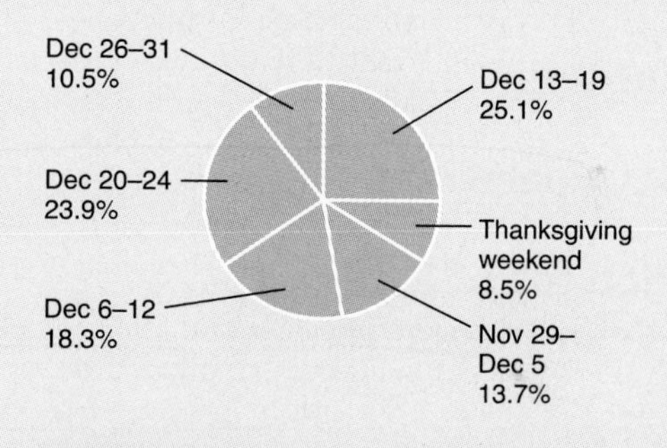

Figure 1.42

Average percentage of retail sales during the Christmas holiday season.

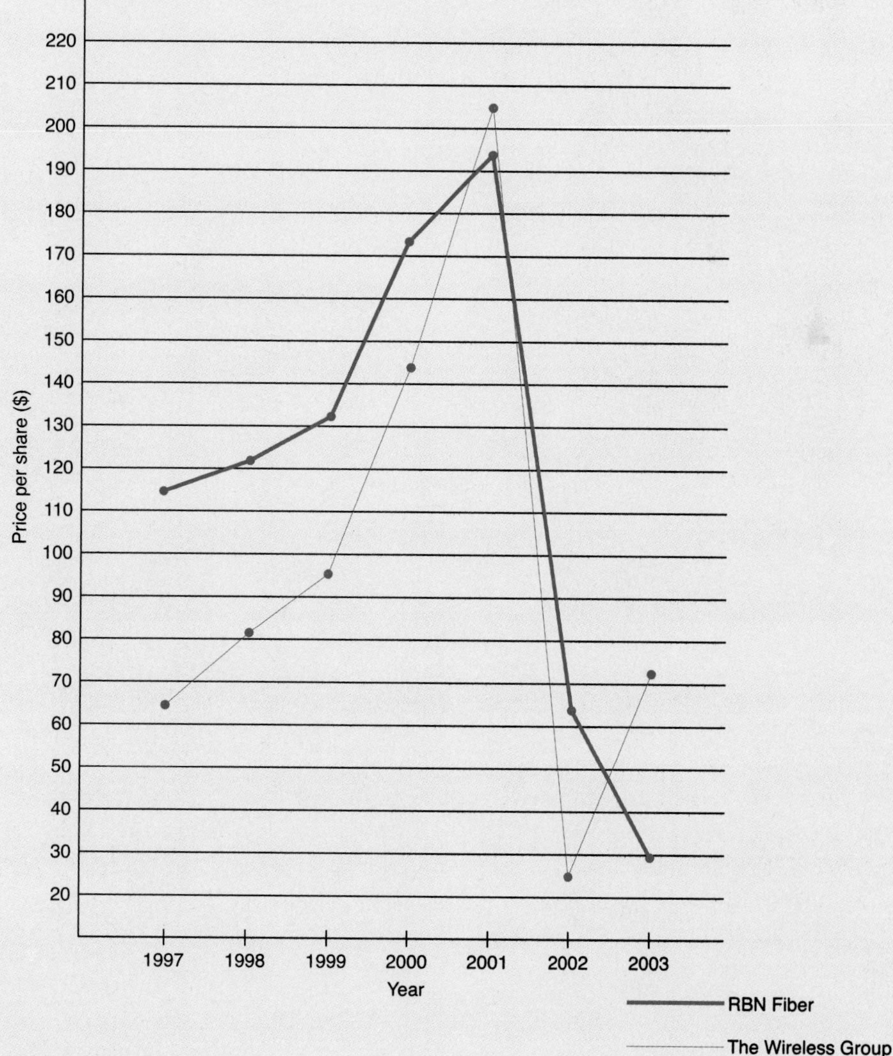

Figure 1.43

End-of-year stock closing prices.

Chapter 2

Chapter Outline

Sets

After Studying This Chapter, You Will Be Able to Do the Following:

1. Describe the meaning of the word **set**, and write a given set in two ways: **roster form** and **set-builder notation.**
2. Identify **well-defined sets, finite sets, infinite sets, equal sets, equivalent sets**, and **disjoint sets.**
3. Find the **subsets** and **proper subsets** of a given set, and determine the **intersection** and **union** of two or more sets.
4. Identify a **universal set** and find the **complement** of any set contained in a given universal set.
5. Draw **Venn diagrams** to show the relationship between sets, and use Venn diagrams to solve **survey problems.**
6. Show a **one-to-one correspondence** between any two equivalent sets, and find the **cardinality** of sets.
7. Find the **Cartesian product** ($A \times B$) of two sets A and B.

Note: *indicates optional material.

Notations Frequently Used in This Chapter

{ }	braces, used to enclose members of a set
$\in$	"is an element of"
$\notin$	"is not an element of"
. . .	proceed in the indicated pattern
$\varnothing$	the empty set, also denoted by { }
$\subset$	"is a proper subset of"
$\subseteq$	"is a subset of"
$\not\subseteq$	"is not a subset of"
U	the universal set
A'	the complement of A
$\cap$	intersection
$\cup$	union
$n(A)$	the cardinal number of set A
(a, b)	the ordered pair a and b
=	"is equal to"
$A \times B$	the Cartesian product of sets A and B
$a \mid b$	a such that b
$n!$	n factorial $n! = n \times (n - 1) \times \cdots \times 3 \times 2 \times 1$

Visit the PH Companion Web site at www.prenhall.com/setek

OVERVIEW

The concept and properties of a **set** are often uses in mathematics. Students study some of set theory at all levels of mathematics, from grade school through college. It has been said that the use of sets is the one unifying idea that unites all branches of mathematics.

It is not uncommon for us to use the concept of sets in everyday experiences. For example, you or someone you know may have purchased a *set of tires*, a *set of golf clubs*, or a *set of dishes*. Note that each of these sets contains a certain number of *elements*. For example, a set of tires would typically contain four elements. In Section 2.2 we will learn how to list and describe sets more formally. We can also identify *subsets* based on certain attributes. For example, if we were to consider the set of students at your college, we could identify subsets of students based on various characteristics. One such subset might be all female students who work full time. Another subset might be all students that are enrolled as part-time students. This concept of subsets is discussed in Section 2.3.

2.1 INTRODUCTION

A ***set*** is an intuitive concept. This is similar to the study of geometry where, you might recall, we do not define what we mean by a point or line. We may think of a set as a collection of objects. These objects are called **elements** or **members** of the set. In addition to the everyday experiences we may have with sets (see chapter overview), the concept of sets is also used throughout mathematics. For example, you may have discussed some of the following:

The set of counting numbers

The set of integers

A set of points

A set of solutions for an equation

A set of ordered pairs

Although a set is an intuitive concept, a set has specific properties and may be manipulated by performing various operations. In this chapter, we discuss the concept of sets, and examine some basic properties, rules, and operations that pertain to sets. We begin in Section 2.2, where we introduce notation used to represent sets as well as some of the vocabulary associated with sets. We then show how we can describe sets using this notation. In Section 2.3, we learn about subsets, which contain some of the elements of a particular set. We also discuss the concept of a universal set and the complement of a set. In Section 2.4, we learn how to manipulate sets using the set operations of intersection and union. We extend this discussion of set operations in Section 2.5, where we learn how to represent sets pictorially using Venn diagrams. In Section 2.6, we apply our knowledge of set operations and Venn diagrams to help solve survey problems. Finally, in Section 2.7, we discuss Cartesian products, which is another set operation.

2.2 NOTATION AND DESCRIPTION

We do not have to describe a set by a name as we did in the introduction; we can also describe it by listing or naming its elements. Braces, { }, are used to enclose the members of a set when we list them. Note that the only correct fence or enclosure is braces, not parentheses, ( ), nor brackets, [ ].

If we are talking about the set of vowels in the English alphabet, we may denote this set of vowels as $\{a, e, i, o, u\}$. The listing of the elements in a different order does not change the set: We could also write the set as $\{i, o, u, e, a\}$, and it would still be the set of vowels in the English alphabet. When we list the elements in a set, as we have just done, it is called the **roster form** of the set.

Sets are usually denoted by capital letters such as A, B, or C. Therefore, we could write

$$A = \{a, e, i, o, u\}$$

to indicate that set A contains the elements a, e, i, o, u.

We shall use $\in$ to indicate that elements are members of a set. The symbol $\in$ is read "is a member of" or "is an element of," and the notation $\notin$ is read "is *not* an element of." Using our previous example $A = \{a, e, i, o, u\}$, we may say that $a \in A, e \in A, u \in A, 2 \notin A$, and $z \notin A$.

If a set contains many elements, we often use three dots, ..., called an **ellipsis**, to indicate that there are elements in the set that have not been written down. The following are some examples of sets where we list some elements and then use an ellipsis to indicate that the pattern is to be continued indefinitely:

$$N = \{1, 2, 3, 4, \dots\}, \quad A = \{5, 10, 15, \dots\}, \quad W = \{0, 1, 2, 3, \dots\}$$

The set N is called the set of **counting numbers** (or **natural numbers**), whereas the set W is called the set of **whole numbers** (note the addition of the element zero). Using the sets N, W, and A, we can say that $2 \in N, 2 \in W, 2 \notin A$, and $0 \notin N, 0 \in W, 0 \notin A$.

We can also use an ellipsis in another manner when listing elements in a set, to indicate that some elements are missing in the listing. Consider the set K consisting of those counting numbers from 1 through 100. We can make use of the ellipsis to list the set K as

$$K = \{1, 2, 3, \dots, 98, 99, 100\}$$

This notation tells us the first element in the set, some of the succeeding elements, and the last element in the set.

When we use three dots in the roster form of listing elements, we must list some of the elements so that the pattern can be determined. Remember that the ellipsis means that the listing of the elements will continue in the indicated pattern.

If we use the three dots to indicate that the pattern continues indefinitely, as in $N = \{1, 2, 3, 4, \dots\}$, then we have what is known as an **infinite set** because the set has an unlimited number of elements. The pattern is unending, and the list of elements goes on and on. If we have a set such as $A = \{1, 2, 3, \dots, 10\}$—where the ellipsis shows an indicated pattern, and we know the last element of the pattern and how many elements are in the set—then we have a **finite set**. An infinite set has an unlimited number of elements; a finite set has a last element, *and* we can count the number of elements in the set.

Example 1 ***Type of Set*** Is set $A = \{1, 2, 3, 4, \dots\}$ an infinite set or a finite set?

Solution Set $A = \{1, 2, 3, 4, \dots\}$ is an infinite set. It has an unlimited number of elements.

Is the set of contented lions a well-defined set?

Example 2 ***Type of Set*** Is set $B = \{a, b, c, d, \ldots, x, y, z\}$ an infinite set or a finite set?

Solution Set $B = \{a, b, c, d, \ldots, x, y, z\}$ is a finite set. From the pattern indicated, we know that set B is the set of letters in the alphabet, and that there are 26 elements in the set.

We should be able to determine whether any given element is a member of any given set; that is, any set that we consider should be **well defined**. There should not be any ambiguity as to whether an element belongs to a set. The following are some sets that are not well defined. Why?

The set of interesting courses you can take
The set of nice people in your class
The set of good instructors in your school

None of these sets is well defined because there is no common agreement as to what is meant by "interesting courses," "nice people," or "good instructors."

Biography: Georg Cantor (1845–1918)

Georg Ferdinand Ludwig Philipp Cantor was born in St. Petersburg, Russia, on March 3, 1845. During his youth, Cantor was initially home-schooled by a private tutor before attending primary school in St. Petersburg. At the age of 11, his family moved to Germany, where he remained for the rest of his life. Cantor studied at the Realschule in Darmstadt and graduated in 1860; he was recognized for his exceptional skills in mathematics, particularly trigonometry. Although he initially began his university studies at the Höheren Gewerbeschule in Darmstadt as an engineering student to please his father, Cantor eventually, with his father's permission, entered the Polytechnic of Zurich in 1862 to study mathematics. After his father died in 1863, Cantor returned to Germany and attended the University of Berlin, where he ultimately completed his dissertation on number theory in 1867. Two years later, Cantor received his habilitation, which is an extra postdoctoral qualification required to lecture at a German university, and began teaching at the university in Halle.

In 1873, Cantor proved that the set of rational numbers is countable, and by the end of that year, he also proved that the set of real numbers was not countable. His findings were ultimately published in a paper in 1874, which was the first documented description of the one-to-one correspondence concept. Later, in 1878, Cantor published another paper that precisely explains this concept. Cantor continued his research and made several other important contributions to the field of mathematics, including geometry, and even surprised himself at some of his own discoveries. It is reported, for example, that upon proving that a one-to-one correspondence exists between the points on the interval [0, 1] and the points in p-dimensional space, Cantor wrote, "I see it, but I don't believe it!"

Cantor is considered by many to be the father of set theory because he first developed this branch of mathematics. His ideas on set theory, which were initially published in a series of six papers between 1879 and 1884, were often met with public ridicule, particularly his idea of a set with an infinite number of elements. This was considered revolutionary by mathematicians of his time.

Throughout his adult life, Cantor suffered from periods of depression and died in a mental institution in Germany at the age of 73. At one time, many believed that his mental breakdowns were caused by attacks other mathematicians made on his work. However, given our recent (and improved) understanding of mental illness, it is now widely believed that Cantor's mathematical anxieties and his often difficult relationships with his colleagues were most likely exacerbated by his depression, but they were not its source. (For additional information about Cantor, see **http://www-groups.dcs.st-and.ac.uk/~history/References/Cantor.html.**)

Example 3 ***Well-Defined Set*** Is the set of big people a well-defined set?

Solution No, it is an ill-defined set. By what characteristic are big people defined—height, power, money? The word *big* is ambiguous.

Example 4 ***Well-Defined Set*** Given $A = \{2, 4, 6, 8, \dots\}$, is set A a well-defined set?

Solution Yes, set A is a well-defined set. Set A is the set of even counting numbers beginning with 2 and proceeding on. From the given description we can ascertain that $100 \in A$. Note that set A is a well-defined infinite set.

NW ***Now Work Problems 3a, d, e.***

It is sometimes cumbersome to write a word description of a set, and it also is sometimes awkward to describe a set by listing all of its elements in roster form. There is another method that we can use to describe a set, called **set-builder notation**.

Consider set $A = \{2, 4, 6, 8, \dots\}$. We have described set A as the set of even counting numbers. We could say that A is the set of all even counting numbers. We can also say A = the set of all x's (or any other letter) such that x is an even counting number. We can refine this to

$$A = \{x\text{'s such that } x \text{ is an even counting number}\}$$

Math Connections

Roster versus Set-Builder Notation

When defining a set, we must ensure that the set contains an objective test for membership. For example, given the set of all the smart people in the world, it is difficult to determine the elements of this set because the word *smart* is ill defined. If we redefine this set to the set of all people in the world with an IQ of at least 120, then we now have an objective method for determining its elements. One method might be to administer an IQ test such as the Stanford-Binet and anyone who scores at least 120 becomes a member of the set. This redefined set effectively transforms the word *smart* from a subjective state to an objective one, which in turn makes the new set well defined.

When describing a set using the roster method, note that the objective test for membership is *explicit*, that is, the actual members are listed. Thus, to check membership, we simply see if an element is listed. Although this might be acceptable for finite sets, this is not always practical for infinite sets. In contrast to this, set-builder notation specifies membership *implicitly*; that is, instead of listing specific members, it provides us with a *membership rule* that we can apply to test for membership. Thus, to determine if an element is a member of a set, we apply the test (or rule) to the element. If it passes the test, then the element is a member of the set; otherwise it is not.

Of the two set description methods, the roster method introduces the potential for ambiguity or uncertainty, particularly when specifying infinite sets. For example, given the set

$$A = \{0, 2, 4, 6, \dots\}$$

we would be inclined to believe that this set specifies the set of even numbers and that even though it is not shown, 8 is the next element of this set. However, making this conclusion is pure speculation on our part because we do not know set A's membership rule. On the other hand, the set defined as

$$B = \{x \mid x \in 2n + n(n-1)(n-2)(n-3), \text{ where } n \text{ is a whole number}\}$$

provides us with a membership test. Furthermore, we can show that 0, 2, 4, and 6 are elements of set B by evaluating its membership rule, namely, $2n + n(n-1)(n-2)(n-3)$, for $n = 0, 1, 2,$ and 3. Unlike set A, though, we also know without any doubt that the next element of set B is *not* 8 because when n assumes the value of 4, the given rule evaluates to 32.

Given this illustration, we should observe that set-builder notation has a distinct advantage over the roster method when describing infinite sets. Although it implicitly specifies its elements, set-builder notation explicitly specifies the rule to follow, which eliminates any ambiguity or uncertainty relative to whether an element is or is not a member of a set. In a certain sense, this discussion lends itself to the central theme of mathematics: *Mathematics is an exact and precise discipline.* Generalizing further, the set-builder notation concept suggests that whenever we encounter problems involving mathematics, it is preferable first to establish a well-defined set of criteria and then test possible solutions by seeing if they satisfy the criteria. This idea has tremendous ramifications not only in mathematics but also in the broad spectrum of everyday life.

In set notation we use a vertical line (|) to stand for "such that." Hence, we now have $A = \{x | x \text{ is an even counting number}\}$. This is read as "$A$ is the set of all x such that x is an even counting number." The set-builder notation is commonly used in mathematics when discussing sets.

Example 5 ***Set Notation*** What does $A = \{x | x \text{ is a Great Lake}\}$ mean?

Solution This set is described in set-builder notation. Set A is the set of all x such that x is a Great Lake. The vertical line after the x stands for "such that." The x after the first brace tells us that we are considering all x's—that is, all of the Great Lakes.

Example 6 ***Set Notation*** List the elements of $\{x | x$ is a vowel in the word *Westhampton*$\}$.

Solution We have the set of all x such that x is a vowel in the word *Westhampton*. The vowels are a, e, i, o, u, and those that appear in *Westhampton* are e, a, and o, that is $\{e, a, o\}$.

NW ***Now Work Problems 5a, b and 9a, b.***

A set does not have to contain elements that are related. It may be that the only thing that the elements in a set have in common is that they are in the same set. For example, consider the sets $A = \{\Delta, \square, a, 2\}$ and $B = \{\text{red, blue, 1, 1,000, XII}\}$.

Sets may contain a definite number of elements or an unlimited number of elements. It is also true that a set may contain *no* elements. If a set does not contain any elements, it still contains a definite number of elements—namely, zero elements. Consider the set of lobsters that live in Lake Ontario. This set has no elements: There are no lobsters living in Lake Ontario. If we were to list the elements for this set, we would have to put nothing between the braces. We would have $\{\ \}$. This is an **empty set**, also called the **null set**. The empty set is usually denoted by $\{\ \}$, but another common symbol is $\varnothing$.

When you denote the empty set you may use either symbol, but do not use both together.

$\{\ \}$ or $\varnothing$ correct

$\{\varnothing\}$ *wrong!*

The notation $\{\varnothing\}$ is incorrect because the set $\{\varnothing\}$ is not empty: The symbol $\varnothing$ is inside the braces. It is false that there is nothing inside the braces; hence the set cannot be empty.

Math Connections

The Empty Set

When designating the empty set, we can use the symbol $\varnothing$, or empty braces, $\{\ \}$. Frequently, though, many students want to write the empty set as $\{\varnothing\}$, which is incorrect. Note that this representation specifies the set that contains the symbol for the empty set and has one element, namely, the symbol for the empty set. To help clarify this concept, think of a set as an envelope. If the set is empty, then the envelope is empty. On the other hand, if the set is not empty—that is, it contains at least one element—then there are items in the envelope. One such item can be another envelope. Using this analogy, the symbol $\{\varnothing\}$ specifies an empty envelope contained within another envelope. Thus, the set has one element and hence is not empty.

Example 7 *Empty Set* List the elements in the following set:

The set of months containing 33 days

Solution There are no months that have 33 days; therefore, there are no elements in the set. Hence, the set of months containing 33 days is the empty set, or null set, denoted by $\{\ \}$ or $\varnothing$.

Two sets that contain exactly the same elements are said to be **equal sets**. If we are given $A = \{a, e, i, o, u\}$ and $B = \{i, o, u, a, e\}$, then we can say that $A = B$. These two sets contain exactly the same elements and therefore they are equal.

Two sets that contain exactly the same number of elements are **equivalent sets**. If we are given $A = \{a, e, i, o, u\}$ and $B = \{1, 2, 3, 4, 5\}$, we say that A is equivalent to B. Both sets contain five elements and hence they are equivalent—but they are not equal.

Example 8 *Equal Sets* Are the following sets equal?

a. $A = \{d, a, b\}$; $B = \{b, a, d\}$

b. $C = \{1, 2, 3, 4, \dots\}$; $D = \{5, 10, 15, \dots\}$

Solution

a. Yes, sets A and B are equal sets because they contain exactly the same elements. The order of the listing of the elements does not change the set.

b. No, sets C and D are not equal sets. They do not contain exactly the same elements. For example, from the set descriptions we know that $6 \in C$, but $6 \notin D$, because set D contains only those counting numbers that are multiples of 5.

Example 9 *Equivalent Sets* Are the following sets equivalent?

a. $A = \{$T. Woods, M. Jordan, D. Jeter$\}$; $B = \{$golf, basketball, baseball$\}$

b. $C = \{1, 2, 3, \dots\}$; $D = \{l, o, v, e\}$

Solution

a. Sets A and B are equivalent because they each contain three elements.

b. Set C contains an unlimited number of elements while D contains four elements. Hence, sets C and D are not equivalent.

NW ***Now Work Problems 3b, c, f.***

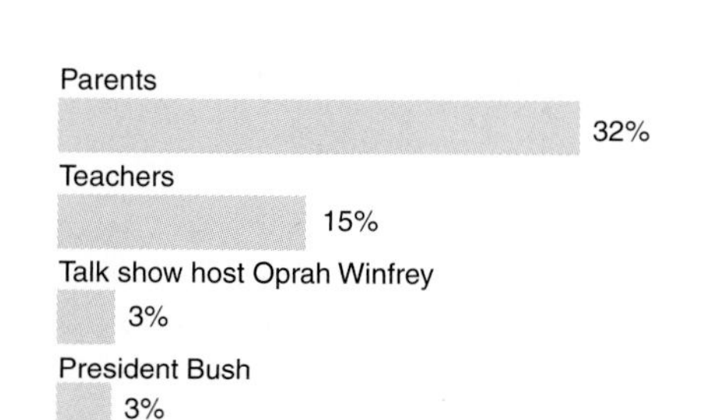

Figure 2.1

Teens' top role models.
Source: Junior Achievement/Harris Interactive Poll of 624 teens, ages 13 to 18.

Example 10 *Representing Sets from Graphical Data* The bar graph in Figure 2.1 shows teenagers' (ages 13–18) top role models. Use this figure to list the elements of each of the following sets in roster form:

a. $A =$ The set of role models who were selected by more than 10% of teens.

b. $B =$ The set of role models who were selected by less than 5% of teens.

c. $C =$ The set of role models who were selected by at least 50% of teens.

d. $D = \{x \mid x$ is a role model who is an entertainer$\}$.

e. $E = \{x \mid x$ is a role model who is a founder of a company$\}$.

Solution

a. From Figure 2.1, only parents and teachers were selected by more than 10% of teens. Therefore,

$$A = \{\text{Parents, Teachers}\}$$

b. The role models with less than 5% of the votes were

$$B = \{\text{Oprah Winfrey, President Bush, Bill Gates, Condoleezza Rice, Jennifer Lopez, Michael Dell}\}$$

c. This is the empty set. Thus, $C = \{\ \}$.

d. The only role model listed as an entertainer is Jennifer Lopez. Therefore, $D = \{\text{Jennifer Lopez}\}$. Note that some people might want to include Oprah Winfrey in this set. However, if we were to do so, then set D would not be well defined based on the descriptions given in Figure 2.1.

e. $E = \{\text{Bill Gates, Michael Dell}\}$ because these are the only role models whose names are associated with being a founder of a company. As in part d, note once again that some people might want to include Oprah Winfrey and Jennifer Lopez because they too have their own businesses. We do not include them in this set, however, because they are not listed in Figure 2.1 as a founder of a company. Once again, this underscores the need for well-defined sets.

Exercises for Section 2.2

1. Are the following statements true or false?

a. $4 \in \{1, 2, 3, 4\}$
b. $8 \in \{2, 4, 6, \dots\}$
c. $\{e, a, t, s\} = \{s, e, a, t\}$
d. $\{1, 2, 3\}$ is equivalent to $\{4, 5, 6\}$.
e. $12 \in \{x | x \text{ is a counting number}\}$
f. $0 \in \{\ \}$

2. Are the following statements true or false?

a. $4 \notin \{1, 2, 3, 4, 5\}$
b. $D \in \{a, b, c, d, \dots, z\}$
c. $\{\ \} = \emptyset$
d. $\{1, 2, 3\}$ is both equivalent to and equal to $\{3, 1, 2\}$.
e. March $\in \{x | x \text{ is a month of the year}\}$
f. $\{1, 2\} = (1, 2)$

3. Are the following statements true or false?

NW **a.** $\{a, b, c, \dots, x, y, z\}$ is a finite set.
NW **b.** Equal sets are equivalent sets.
NW **c.** $\{0\}$ is an empty set.
NW **d.** The set of students enrolled in this course is a finite set.
NW **e.** The set of big cars manufactured in the United States is a well-defined set.
NW **f.** Equivalent sets are equal sets.

4. Are the following statements true or false?

a. $\{n, a, m, e\} = \{m, a, n, e\}$
b. $\{a, b, c\}$ is equivalent to $\{x, y, z\}$.
c. $\{a, b, c, \dots, z\}$ is a well-defined set.
d. Equal sets are well-defined sets.
e. $6 \in \{x | x \text{ is a whole number}\}$
f. $95 \in \{5, 10, 15, \dots\}$

5. List the elements of each set in roster form.

NW **a.** The set of Great Lakes
NW **b.** The set of states whose names begin with the word *New*
c. The set of states whose names begin with the letter *A*
d. The set of states whose names begin with the letter *B*
e. The set of states whose names end with the letter *o*

6. List the elements of each set in roster form.

a. The set of seasons of the year
b. The set of days with names containing the letter *s*
c. The set of days with names containing the letter *x*

d. The set of months containing 31 days

e. The set of months containing 32 days

7. List the elements of each set in roster form.

a. $\{x|x$ is a letter in the English alphabet$\}$

b. $\{x|x$ is a Great Lake$\}$

c. $\{x|x$ is the capital of Canada$\}$

d. $\{x|x + 3 = 9\}$

e. $\{x|x + 1 = x\}$

8. List the elements of each set in roster form.

a. $\{x|x$ is a counting number less than $5\}$

b. $\{x|x$ is a counting number greater than $6\}$

c. $\{x|x$ is a whole number less than $5\}$

d. $\{x|x$ is a month whose name contains an $r\}$

e. $\{x|x$ is a month whose name does not contain an $r\}$

9. Write each set in set-builder notation.

NW **a.** {Monday, Tuesday, Wednesday, Thursday, Friday, Saturday, Sunday}

NW **b.** $\{a, e, i, o, u\}$

c. $\{1, 3, 5, 7, \dots\}$

d. $\{2, 3, 5, 7, 11, 13, 17, \dots\}$

10. Write the following in set-builder notation.

a. {CBS, ABC, NBC, FOX}

b. $\{1, 2, 3, \dots\}$

c. $\{2, 4, 6, \dots\}$

d. $\{a, b, c, \dots, z\}$

11. Is the empty set a well-defined set? Explain your answer.

12. If $A = \{1, 2, 3, 4, \dots\}$ and $B = \{5, 10, 15, 20, \dots\}$, is A equivalent to B?

Writing Mathematics

13. What is a set?

14. What is the empty set?

15. Describe the difference between equal sets and equivalent sets.

16. Describe the difference between finite sets and infinite sets.

Challenge Exercises

17. Write the empty set using set-builder notation.

18. Define an empty set using set-builder notation. For example, $\{x|x \in W$ and $x < 0\}$ specifies an empty set because there is no such element that is both a whole number and is less than zero.

Group/Research Activity

19. In 1742, Christian Goldbach had observed that every even integer, except 2, seemed representable as the sum of two primes. That is, $4 = 2 + 2, 6 = 3 + 3, 8 = 5 + 3, \dots, 18 = 11 + 7, \dots, 100 = 97 + 3$, etc. This became known as **Goldbach's conjecture**. Thus far, it has not been proven. Each group member should research Goldbach and describe his life and work. This research should result in a presentation made by the group to the entire class.

20. For any set to be useful and functional, it must be well defined. We must be able to determine whether an element is a member of a particular set. Discuss in a group each of the following and then present ways of determining whether or not a set is well defined.

a. Three players on the women's basketball team at Brevard Community College were described as tall.

b. Jillian described Professor Gallo as her best mathematics instructor.

c. Joe's Pizza Palace advertises that it sells the biggest "large" pizza in town.

Real-World Data

Baseball *Table 2.1 shows the name of active baseball players, their team, the number of home runs (HR) and stolen bases (SB) by these players, and the players' age as of 10/01/03. Use this table to represent in roster form the elements of the sets in Exercises 21–30.*

TABLE 2.1

Player, Team	HR	SB	Age
Barry Bonds, Giants	658	500	39
Rickey Henderson, Dodgers	297	1406	44
Kenny Lofton, Cubs	115	538	36
Roberto Alomar, White Sox	206	474	35
Eric Young, Giants	73	436	36
Marquis Grissom, Giants	203	425	36
Sammy Sosa, Cubs	539	233	34
Rafael Palmero, Rangers	528	93	39
Fred McGriff, Dodgers	491	72	39
Ken Griffey, Reds	481	177	33
Juan Gonzalez, Rangers	429	26	34

21. The set of baseball players who have more than 1000 stolen bases.

22. The set of baseball players who are in the 500–500 club (i.e., at least 500 home runs and at least 500 stolen bases).

23. The set of baseball players who are younger than 35.

24. The set of baseball players who play for the Yankees.

25. The set of baseball players who have hit less than 100 home runs.

26. $\{x|x$ is a baseball player whose first and last initials are the same$\}$.

27. $\{x|x$ is a baseball player who is in his 40s$\}$.

28. $\{x|x$ is a baseball player who does not play for a major league team$\}$.

29. $\{x|x$ is a baseball team that has more than one player listed$\}$.

30. $\{x|x$ is a baseball player who is in the 200–200 club (i.e., at least 200 home runs and at least 200 stolen bases)$\}$.

Bankruptcies *Figure 2.2 lists the name of industries with the highest percentage of bankruptcies. Use this figure to represent in roster form the elements of the sets in Exercises 31–34.*

31. The set industries that are *not* technology related.

32. The set of industries that have at least a 30% bankruptcy rate.

33. $\{x|x$ is service-oriented business$\}$.

34. $\{x|x$ is a grocery-related business$\}$.

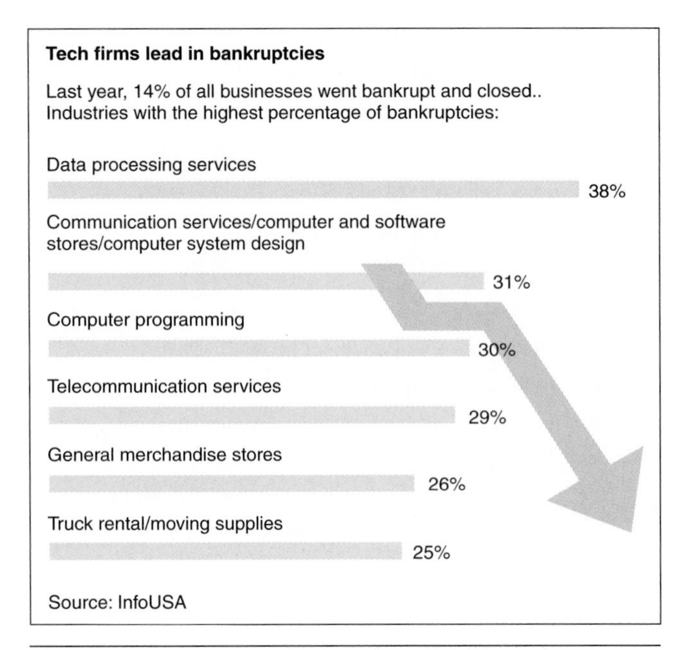

Figure 2.2

Just for fun

We have considered the set $\{a, e, i, o, u\}$ in our discussion. Can you find a word that contains all of these vowels in the order they are listed? (The vowels may be separated by other letters.)

2.3 SUBSETS

In many situations two or more sets contain some, but not all, of the same elements. For example, consider the set of positive even whole numbers, $A = \{2, 4, 6, 8, \dots\}$, and the set of positive whole numbers, $B = \{1, 2, 3, 4, \dots\}$. We can see that $4 \in A$ and $4 \in B$; similarly, we note that $10 \in A$ and $10 \in B$. In fact, every element that is in set A is also contained in set B. Therefore, we can say that set A is **contained in** set B, or, symbolically,

$$A \subseteq B$$

When a set A is contained in another set B, we say that A is a *subset* of B.

> ***Given any two sets A and B, if every element in A is also an element in B, then A is a subset of B.***

Since every set is a subset of itself, we have to be careful with our notation. A subset of a given set that is *not* the set itself is called a *proper subset.* If set A is a proper subset of set B, then two conditions must be satisfied: First, A must be a subset of set B; second, set B must contain at least one element that is not found in set A. If A is a proper subset of B, then we say that A is **properly contained in *B***, and we write

$$A \subset B$$

If A is a subset of B and there is at least one element in B that is not contained in A, then A is a proper subset of B.

In other words, A is a proper subset of B if and only if A is a subset of B and A does not equal B.

If A is a subset of B, but not necessarily a proper subset of B, then we denote this by

$$A \subseteq B$$

How many subsets does the set of rowers in this boat have?

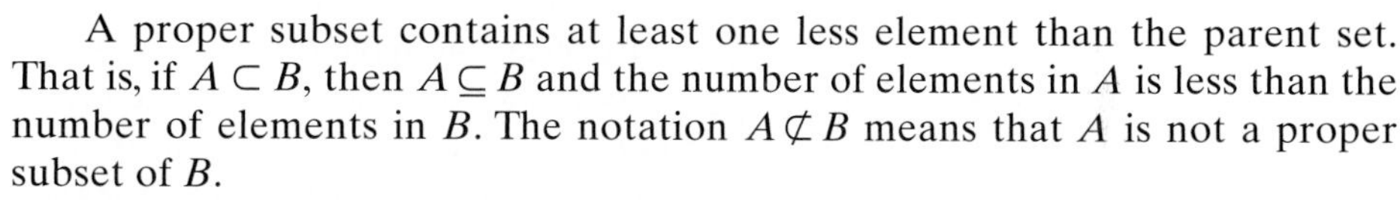

A proper subset contains at least one less element than the parent set. That is, if $A \subset B$, then $A \subseteq B$ and the number of elements in A is less than the number of elements in B. The notation $A \not\subset B$ means that A is not a proper subset of B.

Consider the sets $A = \{1, 2, 3\}$ and $B = \{3, 2, 1\}$. We can state that $A = B$ and $A \subseteq B$ because each element in A is also an element in B. But we cannot say $A \subset B$. That is, $A \not\subset B$ because there must be at least one element in set B that is not contained in set A.

Consider the empty set $\{\ \}$. The empty set has no elements. This means that it is impossible to find an element in the empty set that is not in set A. Since the empty set has no elements, there are none that can fail to be elements of A. Hence the empty set is a subset of A. By the same reasoning the empty set is a subset of set B. In fact, the empty set is a subset of every set.

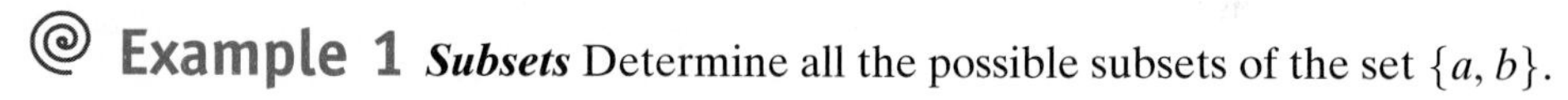

Example 1 ***Subsets*** Determine all the possible subsets of the set $\{a, b\}$.

Solution From our discussion on subsets, we know that every set is a subset of itself; thus $\{a, b\}$ is a subset. We also know that the empty set is a subset of all sets, so we have $\{\ \}$. Are there any others? Yes—namely, $\{a\}$ and $\{b\}$. It appears that the complete list of subsets of $\{a, b\}$ is

$$\{a, b\}, \{\ \}, \{a\}, \{b\}$$

Example 2 ***Subsets*** Determine all the possible subsets of $\{a, b, c\}$.

Solution We know that we can list the set itself, $\{a, b, c\}$, and the empty set, $\{\ \}$. But what about the others? If we are to proceed in a manner that has some order, then we should probably first consider the subsets that would be obtained by taking the elements one at a time, then two at a time, and so on. Note that taking zero elements at a time gives us the empty set.

Since the set $\{a, b, c\}$ contains three elements, we would have the following:

Zero at a Time	One at a Time	Two at a Time	Three at a Time
$\{\ \}$	$\{a\}$	$\{a, b\}$	$\{a, b, c\}$
	$\{b\}$	$\{a, c\}$	
	$\{c\}$	$\{b, c\}$	

There are eight subsets for the given set. Remember that, since $\{a, b\}$ is the same set as $\{b, a\}$, we do not list both of these, because we would not wish to list the same subset twice.

NW ***Now Work Problem 5c.***

The sets in the two examples we just considered contained two and three elements, and it was not too difficult to determine all of the subsets for each set. In order to determine the number of subsets for any given set with n elements, we may use the following rule:

If a set contains n elements, then it has 2^n subsets.

The set $\{a, b\}$ contains two elements, so it has 2^2, or $2 \times 2 = 4$, possible subsets. The set $\{a, b, c\}$ contains three elements, so it has 2^3, or $2 \times 2 \times 2 = 8$, possible subsets. Note that this rule also works for the empty set $\{ \}$. The empty set contains zero elements, so we have 2^0, which is equal to 1. The subsets of the empty set are the set itself, and the empty set, which is again the set itself; hence, we list it only once, and have only one subset.

When we discuss sets, we often refer to some general set that contains all the other sets under consideration. If we are discussing a set of numbers, this set could be generated from many other sets—the set of counting numbers, the set of whole numbers, the set of integers, and so on. To avoid confusion, it is necessary to know what general set the elements are taken from. This general set is called the **universal set**, and it contains all of the elements being considered in the given discussion or problem. The universal set can change from problem to problem, depending on the nature of the set being discussed. We usually denote the universal set in any problem by the capital letter U.

For example, the universal set $U = \{0, 1, 2, 3, 4, 5, 6, 7, 8, 9\}$ contains the digits 0 through 9. In a discussion using this universal set, we would consider only those sets whose elements are members of U. For example, $A = \{0, 2, 4\}$ might be discussed, but $C = \{2, b, c\}$ would not be, because not all the elements of C are elements of U.

The **complement** of a set A is the set of all the elements in the given universal set, U, that are not in the set A. The notation for the complement of A is A'. Some texts use the notation $\overline{A}$ for the complement of A, but we shall use the "prime" notation. In order to find the complement of a set, we must be given a universal set U.

Example 3 ***Complement of a Set*** Given $U = \{a, e, i, o, u\}$ and $A = \{i, o, u\}$, find A'.

Solution The set A', the complement of A, is the set of elements that are in U, but not in A. These elements are a and e. Hence, we have $A' = \{a, e\}$.

Example 4 ***Complement of a Set*** Given $U = \{a, e, i, o, u\}$, $B = \{a, i\}$, and $C = \emptyset$, find

a. B' **b.** C'

Solution

a. The complement of B, B', is the set of elements that are in U, but not in B. These elements are e, o, and u. Therefore, $B' = \{e, o, u\}$.

b. Set C is the empty set and therefore contains no elements, and its complement is the set of elements in U that are not in C. In this case, the complement of C is all the elements in the universal set. Therefore, $C' = \{a, e, i, o, u\} = U$.

NW ***Now Work Problem 7.***

Exercises for Section 2.3

1. Tell whether each statement is true or false.

a. $\{4, 5\} \subset \{1, 2, 3, 4, 5\}$ **b.** $0 \subset \emptyset$
c. $\{1, 3, 5\} \subset \{5, 3, 1\}$ **d.** $\{i, o\} \subseteq \{o, i\}$
e. $\{0\} \subset \emptyset$ **f.** $\{\emptyset\} = \{\ \}$

2. Tell whether each statement is true or false.

a. $\{x, y, z\} \subset \{x, y, z\}$
b. $\{5, 10\} \subset \{1, 2, 3, \dots\}$
c. $\{1, 3, 5, \dots\} \subset \{1, 2, 3, \dots\}$
d. $\{5, 10\} \subset \{2, 4, 6, \dots\}$
e. $\{\emptyset\} \subset \{\ \}$
f. $\{2, 4, 6\} \subseteq \{6, 2, 4\}$

3. Tell whether each statement is true or false.

a. $5 \subset \{1, 3, 5, \dots\}$ **b.** $\{b, a, t\} \subset \{t, a, b\}$
c. $\{2\} \in \{2, 4, 6\}$ **d.** $\{d\} \subset \{a, b, c, \dots, z\}$
e. $\{2\} \subseteq \{2, 4, 6\}$ **f.** $\{A\} \subset \{a, b, c\}$

4. List all possible subsets of each set.

a. $\{5, 10\}$ **b.** $\{b, a, s, e\}$
c. $\{x, y, z\}$ **d.** $\{a, e, i, o, u\}$

5. List all possible subsets of each set.

NW **a.** $\{15, 20\}$ **b.** $\{m, a, t, h\}$
NW **c.** $\{1, 3, 5\}$ **d.** $\emptyset$

6. If $U = \{0, 1, 2, 3, 4, 5, 6, 7, 8, 9\}$, find

a. $\{0, 1, 2\}'$ **b.** $\{0, 2, 4, 6, 8\}'$
c. $\{0, 5\}'$ **d.** $\{\ \}'$
e. $\{1, 3, 5, 7, 9\}'$ **f.** U'

NW **7.** If $U = \{m, e, t, r, i, c\}$, find

a. $\{m, e, t\}'$ **b.** $\{r, i, c\}'$
c. $\{m, e, i, c\}'$ **d.** $\{e, r, i, c\}'$
e. $\{\ \}'$ **f.** $\{m, e, t, r, i, c\}'$

8. If any two sets A and B are equal, is it true that A is a proper subset of B? Explain your answer.

9. Mr. Reed has a nickel, a dime, a quarter, and a half-dollar in his pocket. How many different sums of money may he select as a tip for the paper carrier? What is the largest tip he can give? What is the smallest amount of money he can choose to give?

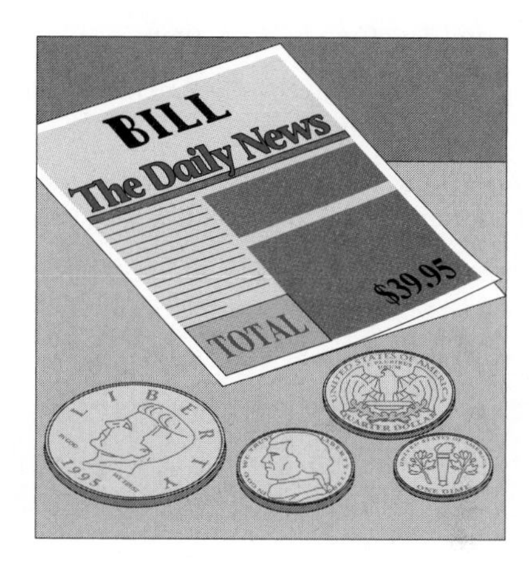

10. If you have three bills—a one-dollar bill, a five-dollar bill, and a ten-dollar bill—and you are allowed to select any number of these bills, how many different sums of money can you select? What is the greatest amount of money you can select? What is the least amount of money you can select?

11. A set contains six elements. How many subsets does it contain? How many proper subsets does it contain?

12. A set has 255 proper subsets. How many elements are in the set?

In Exercises 13–18, find an appropriate universal set to which the given elements belong.

13. AltaVista, Ask Jeeves, Excite, Goto, Hotbot, LookSmart, Lycos, Google

14. AOL, Yahoo, MSN, Netscape

15. Columbia, Endeavor, Challenger, Discovery

16. Oak, Maple, Hickory, Cherry, Walnut

17. Red, Orange, Yellow, Green, Blue, Indigo, Violet

18. UPS, FedEx, Avery, Airborne Express

Writing Mathematics

19. What is a subset?

20. What is the difference between a proper subset and a subset?

21. Describe the formula for finding the number of subsets for a given set. Give at least two examples.

22. Explain the difference between the symbols $\in$ and $\subseteq$. Give examples of how each symbol is used.

23. If a set A has n elements, then write a mathematical expression that specifies the number of proper subsets of A.

Challenge Exercises

In Exercises 24 and 25, state the conclusion you can make from the given statement.

24. If $A \subseteq B$ and $B \subseteq A$, then _____.

25. If $A \subseteq B$ and $B \subseteq C$ and $C \subseteq A$, then _____.

26. Which of the following two statements is true: $\emptyset \subset A$ or $\emptyset \subseteq A$? Defend your answer.

27. Determine if the following statement is true or false and defend your answer: If $A \subseteq B$ and $B \subseteq C$, then $A \subseteq C$.

The **power set** of a set A, denoted $P(A)$, is the set whose elements are all possible subsets of A. For example, if $A = \{a, b, c\}$, then $P(A) = \{\emptyset, \{a\}, \{b\}, \{c\}, \{a, b\}, \{a, c\}, \{b, c\}, A\}$. Thus, the members of $P(A)$ are sets themselves, specifically the subsets of A. Use the concept of a power set to answer Exercises 28–32.

28. When working with sets whose members are sets, what distinction should you make between using the symbols $\in$ and $\subseteq$? Give an example.

29. Write the definition of $P(A)$ using set-builder notation.

30. If set A has n elements, then how many elements does $P(A)$ contain?

31. Is it possible for a given set A such that $P(A) = \emptyset$?

32. Let's denote the power set of S as 2 raised to the S power. What is $2^{\{a, b\}}$ equal to?

Group/Research Activities

33. We defined a proper subset. Research the concept of an **improper subset** and report back to your class.

34. Research **Russell's paradox**, which involves calling any collection of objects a set. Discuss this paradox with your class and then answer the following questions: Given a set A, if $A \notin A$, then A is called an *ordinary set*. For example, if A is the set of all automobiles, then $A \notin A$ because A is not an automobile. Let $X = \{x | x \text{ is an ordinary set}\}$. Is $X \in X$? Is $X \notin X$? What can we say about the collection of ordinary sets?

Just for fun

Can you find two words in the English language that rhyme with the word orange?

2.4 SET OPERATIONS

In the preceding sections, we discussed sets to some extent and considered various properties of sets and subsets. Now we are ready to examine set operations that will enable us to combine sets.

In arithmetic, we have operations such as addition and subtraction that enable us to combine numbers. In this section, we shall consider the intersection and union of sets, and we shall also do some more work with the complement of a set.

Intersection: *If we have two sets A and B, then the intersection of A and B is a set of elements that are members of both A and B. The notation for A intersection B is* $A \cap B$.

In other words, the *intersection* of sets A and B results in another set, denoted by $A \cap B$. This resulting set is composed of elements that are common to both A and B.

Example 1 ***Intersection of Sets*** Given $A = \{1, 2, 3, 4, 5\}$ and $B = \{2, 4, 6, 8\}$, find $A \cap B$.

Solution The elements that are in A and also in B are 2 and 4. Hence, $A \cap B = \{2, 4\}$.

Example 2 ***Intersection of Sets*** Given $A = \{1, 2, 3, \ldots, 10\}$ and $B = \{2, 4, 6, \ldots\}$, find $A \cap B$.

Solution We see that set A contains the whole numbers 1 through 10, whereas set B contains the even whole numbers. The elements common to both sets are 2, 4, 6, 8, and 10; therefore, $A \cap B = \{2, 4, 6, 8, 10\}$.

Example 3 ***Intersection of Sets*** Given $A = \{2, 4, 6, 8\}$ and $B = \{a, e, i, o, u\}$, find $A \cap B$.

Solution Examining sets A and B, we see that there are no elements common to both. Therefore, the intersection of these two sets is the empty set. Hence, $A \cap B = \{\ \}$, or $A \cap B = \emptyset$.

Two sets whose intersection is the empty set (as was the case in example 3) are said to be **disjoint**. Disjoint sets have no elements in common.

Example 4 ***Multiple Set Operations*** Given $U = \{a, e, i, o, u\}$, $A = \{a, e, o\}$, and $B = \{e, i, o\}$, find $(A \cap B)'$.

Solution We are asked to find $(A \cap B)'$. As in arithmetic and algebra, when we have parentheses, we first do the work inside the parentheses, and then perform the operation(s) outside the parentheses. So we must find $A \cap B$, and then find the complement of our answer. $A \cap B = \{e, o\}$. The complement of $\{e, o\}$ is the set of all elements that are in U, but not in $A \cap B$. Therefore, the complement of $\{e, o\}$ is $\{a, i, u\}$, and

$$(A \cap B)' = \{e, o\}' = \{a, i, u\}$$

Example 5 ***Multiple Set Operations*** Given sets $A = \{1, 2, 4, 5, 6, 7\}$, $B = \{1, 3, 5, 7, 9\}$, and $C = \{2, 4, 6, 7, 8\}$, find $(A \cap B) \cap C$.

Solution First we find $(A \cap B)$, then find the intersection of that with C. Since $(A \cap B) = \{1, 5, 7\}$,

$$(A \cap B) \cap C = \{1, 5, 7\} \cap C = \{1, 5, 7\} \cap \{2, 4, 6, 7, 8\} = \{7\}$$

The *union* of sets A and B is the set of all the elements that are members of either set A or set B, or both. When we list the elements in the union of two sets, we list all of the elements in set A and all of the elements in set B, but if an element is in both sets, we list it only once. Therefore, the union of sets A and B is the set of elements that are elements of at least one of the two sets. The notation for A union B is $A \cup B$.

Union: *If we have two sets A and B, then the union of A and B, $A \cup B$, is the set of elements that are members of A, or members of B, or members of both A and B.*

Example 6 ***Union of Sets*** Given $A = \{1, 2, 3, 4, 5\}$ and $B = \{2, 4, 6, 8\}$, find $A \cup B$.

Solution The elements of A are 1, 2, 3, 4, 5. The elements of B are 2, 4, 6, 8. The union of A and B is the set of all these elements, because all of these elements belong either to set A or to set B, or to both. Therefore, $A \cup B = \{1, 2, 3, 4, 5, 6, 8\}$.

Note that 2 and 4 are members of both sets, but we list each only once in our solution. Also, it is not necessary to list the elements in order, but it does provide a means of checking that all elements are included.

Example 7 ***Union of Sets*** Given $A = \{1, 3, 5, \dots\}$ and $B = \{2, 4, 6, \dots\}$, find $A \cup B$.

Solution We have the set of odd counting numbers and the set of even counting numbers. Therefore, the union of these two sets is the set of counting numbers: $A \cup B = \{1, 2, 3, 4, 5, 6, \dots\}$.

NW ***Now Work Problem 1.***

Example 8 ***Multiple Set Operations*** Given $U = \{a, e, i, o, u\}$, $A = \{a, e, o\}$, and $B = \{e, i, o\}$, find $A' \cup B'$.

Solution We want to find the union of two sets, A' and B'. First we must find the complement of each set, and then the union of the two complements. $A' = \{i, u\}$ and $B' = \{a, u\}$. Therefore, $A' \cup B' = \{i, u, a\}$ or $\{a, i, u\}$.

Example 9 ***Multiple Set Operations*** Given $U = \{a, e, i, o, u\}$, $A = \{a, e, o\}$, and $B = \{e, i, o\}$, find $(A \cup B)'$.

Solution This problem is a little different from the previous example. This time we want $(A \cup B)'$, which is the complement of $A \cup B$. Recall that we must first perform the operation inside the parentheses, and then perform the operation outside the parentheses. First we find $A \cup B = \{a, e, i, o\}$; then we find the complement of $\{a, e, i, o\}$, which is $\{u\}$. Hence, $(A \cup B)' = \{a, e, i, o\}' = \{u\}$.

NW ***Now Work Problem 3.***

Note that Examples 8 and 9 show that $A' \cup B' \neq (A \cup B)'$. It is important to remember this, and to exercise care in reading problems and computing solutions. It is also true that $A' \cap B' \neq (A \cap B)'$. As we shall see later, $(A \cap B)' = A' \cup B'$ and $(A \cup B)' = A' \cap B'$.

There are many different combinations of operations in set theory. However, if you do each operation as it is indicated, you will be able to do any problem involving set operations.

It should be noted that the words *and*, *or*, and *not* correspond to operations in set theory. The intersection of two sets A and B is the set of elements in set A *and* set B. The union of two sets A and B is the set of elements in set A *or* set B *or* both. The complement of set A is the set of elements in the universal set U, but *not* in set A. These operations are summarized using set-builder notation:

$$A \cap B = \{x \mid x \in A \text{ and } x \in B\}$$
$$A \cup B = \{x \mid x \in A \text{ or } x \in B\}$$
$$A' = \{x \mid x \in U \text{ and } x \notin A\}$$

Math Connections

Set Laws

Set operations obey various mathematical laws. A summary of the most salient ones follows.

Empty Set Laws

$$A \cap \emptyset = \emptyset$$
$$A \cup \emptyset = A$$

Idempotency Laws

$$A \cap A = A$$
$$A \cup A = A$$

Commutative Laws

$$A \cap B = B \cap A$$
$$A \cup B = B \cup A$$

Associative Laws

$$A \cap (B \cap C) = (A \cap B) \cap C$$
$$A \cup (B \cup C) = (A \cup B) \cup C$$

Distributive Laws

$$A \cap (B \cup C) = (A \cap B) \cup (A \cap C)$$
$$A \cup (B \cap C) = (A \cup B) \cap (A \cup C)$$

Absorption Laws

$$A \cap (A \cup B) = A$$
$$A \cup (A \cap B) = A$$

DeMorgan's Laws

$$A - (B \cap C) = (A - B) \cup (A - C)$$
$$A - (B \cup C) = (A - B) \cap (A - C)$$

Exercises for Section 2.4

NW **1.** For each pair of sets, find $A \cap B$ and $A \cup B$.

- **a.** $A = \{2, 4, 6, 8\}$, $B = \{1, 3, 4, 6, 7\}$
- **b.** $A = \{f, u, n\}$, $B = \{t, e, a\}$
- **c.** $A = \{a, e, i, o, u\}$, $B = \{a, e, i\}$
- **d.** $A = \{m, i, s, t, e, r\}$, $B = \{p, o, s, t\}$
- **e.** $A = \{5, 10, 15, \ldots\}$, $B = \{10, 20, 30, \ldots\}$
- **f.** $A = \{1, 3, 5, 7, \ldots\}$, $B = \{2, 4, 6, 8, \ldots\}$

2. Find $A \cap B$ and $A \cup B$ for each pair of sets.

- **a.** $A = \{\text{duck, dog, skunk}\}$, $B = \{\text{duck, fish, bird}\}$
- **b.** $A = \{1, 2, 3, \ldots\}$, $B = \{2, 4, 6, 8, \ldots\}$
- **c.** $A = \{¢, \&, \#\}$, $B = \{¢, *, \&\}$
- **d.** $A = \{a, b, c\}$, $B = \{x, y, z\}$
- **e.** $A = \{1, 3, 5\}$, $B = \emptyset$
- **f.** $A = \{m, a, t, h\}$, $B = \{e, a, s, y\}$

NW **3.** Given the sets $U = \{2, 3, 4, 5, 6, 7, 8\}$, $A = \{3, 5, 7, 8\}$, and $B = \{2, 4, 5, 6, 7\}$, find

- **a.** A' **b.** B'
- **c.** $A' \cap B'$ **d.** $A' \cup B'$
- **e.** $(A \cap B)'$ **f.** $(A \cup B)'$

4. Given the sets $U = \{a, b, c, d, e\}$, $A = \{a, b, e\}$, and $B = \{c, d, e\}$, find

- **a.** A' **b.** B'
- **c.** $A' \cap B'$ **d.** $A' \cup B'$
- **e.** $(A \cap B)'$ **f.** $(A \cup B)'$

5. If $U = \{0, 1, 2, 3, \ldots, 9\}$, $A = \{1, 2, 3, 4\}$, $B = \{3, 4, 5, 6, 7\}$, and $C = \{3, 5, 6, 7, 8\}$, find

- **a.** $A' \cap B'$ **b.** $A' \cup B'$
- **c.** $(A \cap B)'$ **d.** $(A \cup B)'$
- **e.** $(A \cap B) \cap C$ **f.** $A \cup (B \cap C)$
- **g.** $(A \cap B)' \cup C$ **h.** $A' \cap (B' \cap C')$

6. If $U = \{0, 1, 2, 3, 4, 5, 6, 7, 8, 9\}$, $A = \{1, 3, 4, 5, 7\}$, $B = \{2, 3, 4, 5, 6\}$, and $C = \{0, 2, 4, 6, 8, 9\}$, find

- **a.** $(A \cap B)'$ **b.** $A' \cap B'$
- **c.** $(A \cup B)'$ **d.** $A' \cup B'$
- **e.** $(A \cap B) \cup C$ **f.** $(A \cup B) \cap C$
- **g.** $A' \cap (B \cup C)$ **h.** $(A \cup B)' \cap C'$

7. If $U = \{n, i, c, k, e, l\}$, $A = \{n, i, c\}$, $B = \{i, e, l\}$, and $C = \{n, c, l\}$, is each of the following statements true or false?

- **a.** $B \cup C = U$ **b.** $(A \cup B) \subset U$
- **c.** $A' \cap B' = \emptyset$ **d.** $(B \cap C) \subset A$
- **e.** $(A \cap B)' = A' \cap B'$ **f.** $(A \cup B)' = A' \cup B'$

8. If $U = \{a, e, i, o, u\}$, $A = \{i, o, u\}$, $B = \{e, i, o\}$, and $C = \{a, i, o\}$, is each of the following statements true or false?

- **a.** $A \cap B = C$ **b.** $A \cup B = U$
- **c.** $B \cup C = U$ **d.** $A' \cap B' = \emptyset$
- **e.** $(A \cap B) \subset C$ **f.** $(B \cap C) \subseteq A$

Writing Mathematics

9. Describe what is meant by a universal set and provide at least two examples.
10. Explain the difference between the inclusive and exclusive use of the word *or*. Describe how it is typically used in mathematics.
11. Describe what is meant by the intersection of two sets. Provide at least two examples. What is a practical analogy to the concept of set intersection?
12. Describe what is meant by the union of two sets. Give at least two examples. What is a practical analogy to the concept of set union?

Challenge Exercises

In addition to intersection and union, another set operation is difference. Specifically, the difference between sets A and B, denoted A − B (read "a minus b"), represents the set of all elements that belong to set A but not to set B. Use the concept of set difference for Exercises 13 through 19.

13. Express $A - B$ in set-builder notation.
14. Represent the set $A - B$ using intersection and complement.
15. Given a universal set U and a set B where $B \subseteq U$, define the complement of B using set difference.
16. Let $U = \{1, 2, 3, \ldots, 10\}$, $A = \{1, 2, 3, 4, 5\}$, and $B = \{1, 3, 5, 7, 9\}$. Find $A - B$.
17. Do you think that $A - B = B - A$? Why? Give an example to defend your answer.
18. What does $A - \varnothing$ equal?
19. What does $\varnothing - A$ equal?
20. Still another set operations is the *symmetric difference* of two sets, denoted $A \Delta B$, and defined as $A \Delta B = (A - B) \cup (B - A)$. Repeat Exercises 16 through 18 using set symmetric difference.

Group/Research Activity

21. The concepts of set difference and set symmetric difference are basic set operations used for sorted sequences in computer science. Research these two concepts and prepare a report that describes a practical application of their use. As part of your report, summarize the set difference and set symmetric difference algorithms and demonstrate the result of these algorithms for two sets A and B, where $A = \{1, 2, 3, 4, 5\}$ and $B = \{3, 4, 5, 6, 7\}$.
22. Research the concept of a *truth set* and prepare a report on its meaning and use in mathematics.

Real-World Data

Fat Content of Foods *Figure 2.3 shows the fat content of various foods. Use this figure to represent in roster form the elements of the sets in Exercises 23–30.*

23. The set that contains foods with more than 3 grams of saturated fat or has 0 grams of polyunsaturated fat.

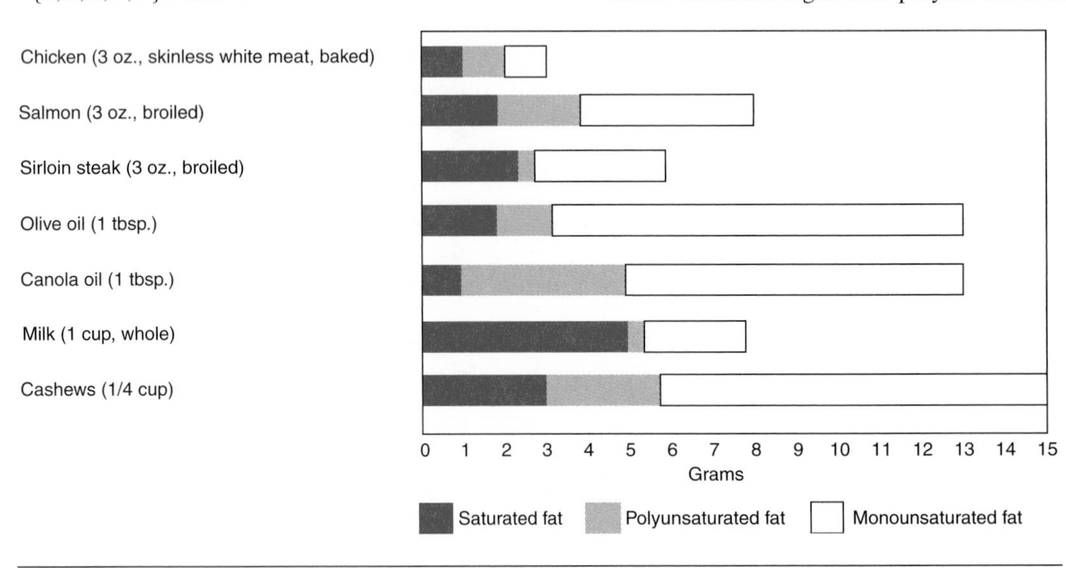

Figure 2.3
Fat content of selected foods.

24. The set that contains foods that have less than 2 grams saturated fat and a total fat content of less than 5 grams.
25. The set that contains foods with at least 5 grams of monounsaturated fat and no more than 1 gram of saturated fat.
26. $\{x \mid x$ is a food with 3 grams of saturated fat$\} \cup \{x \mid x$ is a food with 1 gram of polyunsaturated fat$\}$.
27. $\{x \mid x$ is a food with < 1 gram polyunsaturated fat$\} \cap \{x \mid x$ is a food with a total fat content of < 6 grams$\}$.
28. $\{x \mid x$ is a food with a total fat content of < 10 grams$\} \cup \{x \mid x$ is a food with 1 gram of saturated fat$\}$.
29. $\{x \mid x$ is a food with the highest saturated fat$\} \cup \{x \mid x$ is a food with the lowest polyunsaturated fat$\} \cup \{x \mid x$ is a food food with the lowest monounsaturated fat$\}$.
30. $\{x \mid x$ is a food with the lowest saturated fat$\} \cap \{x \mid x$ is a food highest polyunsaturated fat$\} \cup \{x \mid x$ is a food with the highest monounsaturated fat$\}$.

Just for fun

A snake is stuck at the bottom of a 30-foot well. It can climb up 3 feet every hour, but at the end of each hour it stops to rest and slips back 2 feet. At this rate, how long will it take for the snake to get out of the well?

2.5 PICTURES OF SETS (VENN DIAGRAMS)

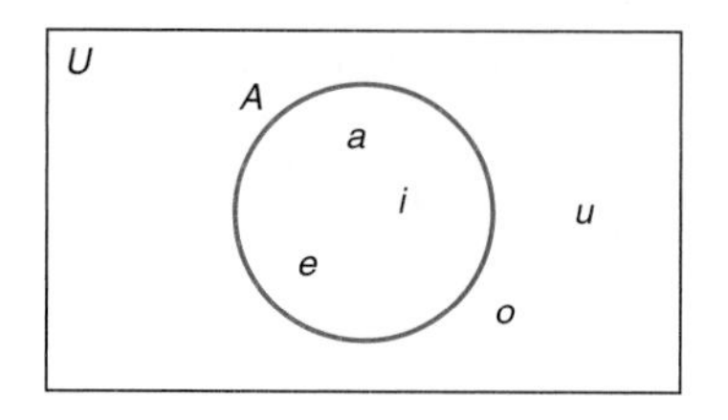

Figure 2.4

Note that the elements of the universal set that are in A are inside the circle, while those not in A are inside the rectangle, but outside the circle.

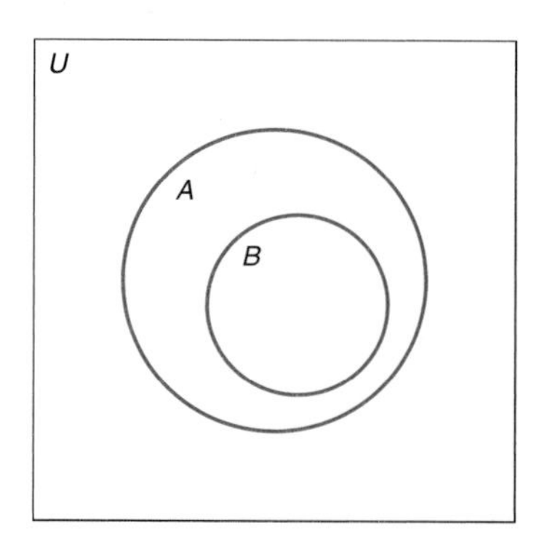

Figure 2.5

It is sometimes useful to represent sets and relationships between sets by means of a picture or diagram. This is almost always done by using a rectangle to represent the universal set and a circle or circles inside the rectangle to represent the set or sets being considered in the discussion. It is understood that the elements in the set are inside the circle that represents the set.

Example 1 ***Universal Set*** Let $U = \{a, e, i, o, u\}$ and $A = \{a, e, i\}$. Make a picture to show the relationship between A and U.

Solution The diagram that represents the given information is Figure 2.4.

Leonhard Euler (1707–1783), a Swiss mathematician, first used circular regions to represent sets. Known today as **Euler circles**, Euler arranged these regions so that each diagram portrayed the exact relationship among the subsets of a universal set. For example, Figure 2.5 shows the relationship $B \subset A$ (B is a proper subset of A). Note that this figure also shows the relationship $A \cap B = B$.

Example 2 ***Euler Diagram*** Use Euler circles to show the relationships given in **a** through **d**.

a. A and B are disjoint sets.

b. A is a proper subset of B.

c. A is equal to B.

d. A and B share common elements.

Solution

a. If A and B are disjoint then their intersection is the empty set. That is, $A \cap B = \emptyset$. See Figure 2.6a. (on p. 88)

b. If A is a proper subset of B, then all the elements of A are contained in B, and there is at least one element in B that is not contained in A. See Figure 2.6b.

c. If A and B are equal sets, then all the elements of A are contained in B, and all the elements of B are contained in A. See Figure 2.6c.

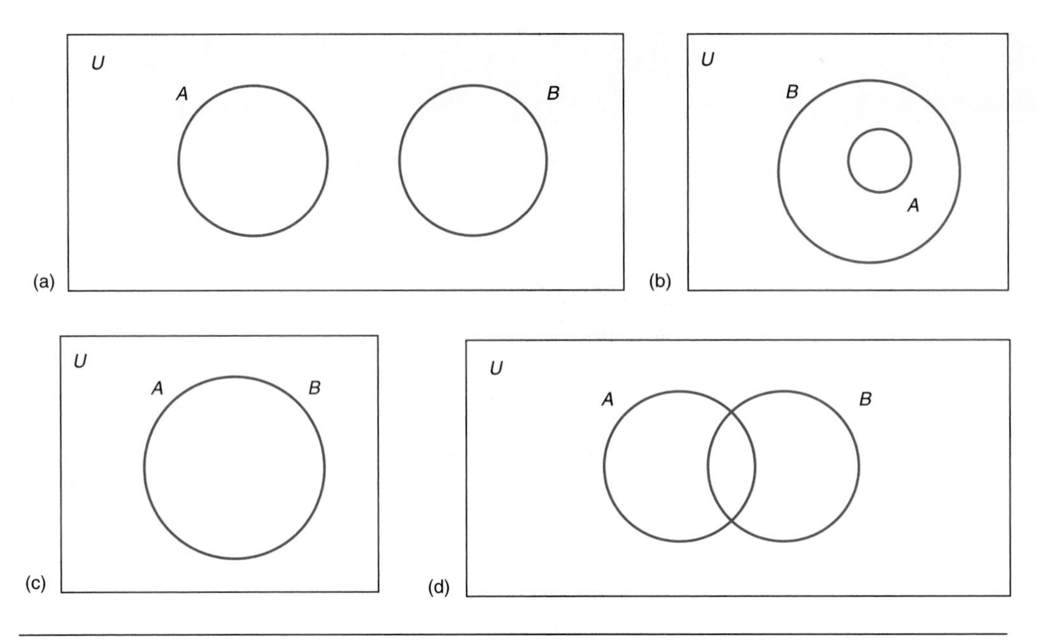

Figure 2.6

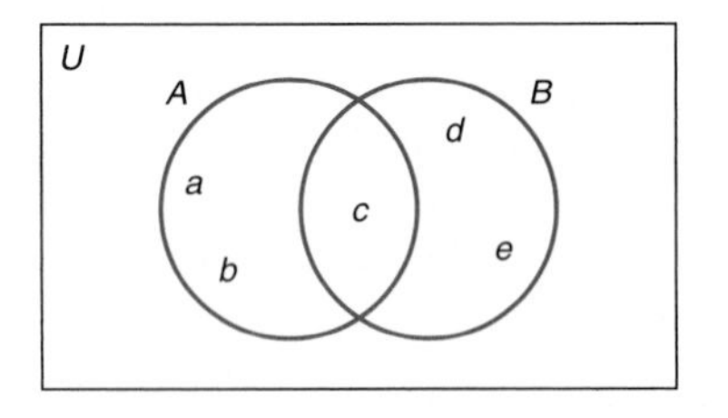

Figure 2.7

Sets A and B have the element c in common.

d. If A and B share common elements, then there are elements in A not contained in B, and elements in B not contained in A. See Figure 2.6d.

As an illustration of Example 2d, let $a = \{a, b, c\}$ and $B = \{c, d, e\}$. Here B and A have an element, c, in common. This is shown in Figure 2.7. Note that the two sets overlap, which allows us to illustrate that c is an element common to both set A and set B.

Euler circles were later refined by John Venn (1834–1923), who made great contributions to modern mathematics. His diagrams, known today as **Venn diagrams**, are used to picture relationships in set theory and logic.

Note of Interest

What does a set sound like?

The composer, architect, and mathematician Iannis Xenakis composes music that is based on mathematical models. In 1960–1961 he composed a piece of music titled *Herma*, which is based on using sets. This was done in the following way: Let *R* be the universal set of all the sounds of a piano, the 88 keys. Now define a subset *A* of *R* with a particular characteristic property. Two other subsets of *R, B* and *C*, are chosen in a similar fashion. Now by using the set operations of unions, intersections, and complements on the sets *A, B*, and *C*, and creating a stochastic (probabilistic) correspondence between the pitches and a set *T* of "the moments of occurrence," he was able to create a composition for piano. The composition was first performed in 1962 by the gifted Japanese pianist Yuji Takahashi in Tokyo.

Iannis Xenakis (b. 1922 Romania, d. 2001) was the originator of *musique stochastique*, music based on mathematical probability models and composed with the aid of computers. In 1966 he founded the School of Mathematical and Automatic Music. During the 1950s, Xenakis was associated with the famous French architect Le Corbusier. Prior to all of this, he fought in the Greek resistance movement during World War II, whereupon he lost sight in one eye. After the war he attended the Athens Polytechnic, graduating in 1947. He was exiled to France due to his political views; it was there that he later studied musical composition at the Paris Conservatory. In 1962 he published his book *Formalized Music, Thought and Mathematics in Music*, where he described many of his compositions and the mathematics behind them. For additional information, see **http://www.iannis-xenakis.org**

Biography: Leonhard Euler (1707–1783)

Leonhard Euler was born in Basel, Switzerland April 15, 1707, but was raised in nearby Riehen when his family moved there one year after his birth. Euler's early education was devoid of mathematics. His interest in the subject, however, was presumably piqued by his father, a Protestant minister, and Euler began a self-study of mathematics coupled with private math lessons.

Euler began attending the University of Basel in 1720, to prepare for the ministry. In 1723, he received his Master's degree in philosophy and began studying theology later that year. His interest and enthusiasm for theology, however, could not compare to his excitement about mathematics. With the persuasive assistance of his private math tutor, Johann Bernoulli, who was also a personal friend of his father, Euler received his father's permission to study mathematics and completed his studies in 1726.

Euler accepted a position at St. Petersburg (Russia) Academy of Science in 1727, where he was appointed to the division of mathematics and physics; he became professor of physics in 1730. Among his colleagues were Jakob Hermann, Daniel Bernoulli, and Christian Golbach. Euler later became the senior chair of mathematics when Daniel Bernoulli left this post in 1733 to return to Switzerland. During his tenure at the Academy, Euler gained recognition as a distinguished mathematics scholar through his many publications, including his book *Mechanica*.

In June 1741, at the personal invitation of Frederick the Great, Euler left St. Petersburg for Berlin and assumed a position at the Society of Sciences. In 1744, the Society was replaced by the new Berlin Academy of Science and Euler became director of mathematics. During his 25 years in Berlin, Euler established himself as a prolific author by writing nearly 400 articles as well as books on the calculus of variations, the calculation of planetary orbits, artillery and ballistics, shipbuilding and navigation, the motion of the moon, and lectures on differential calculus.

Euler returned to St. Petersburg in 1766, and soon after became totally blind. Nevertheless, with the assistance of his sons and other members of the St. Petersburg Academy of Science, Euler produced nearly half of his total works after his return to St. Petersburg. Euler's repository of unpublished work was so great that even after his death in 1783, the St. Petersburg Academy continued to publish his writings for nearly another 50 years. Among his many contributions, Euler is credited for the notation $f(x)$ for a function, e for the base of natural logarithms, i for the square root of -1, π for pi, and Σ for summation. (For additional information about Euler, see **http://www-groups.dcs.st-and.ac.uk/~history/Mathematicians/Euler.html.**)

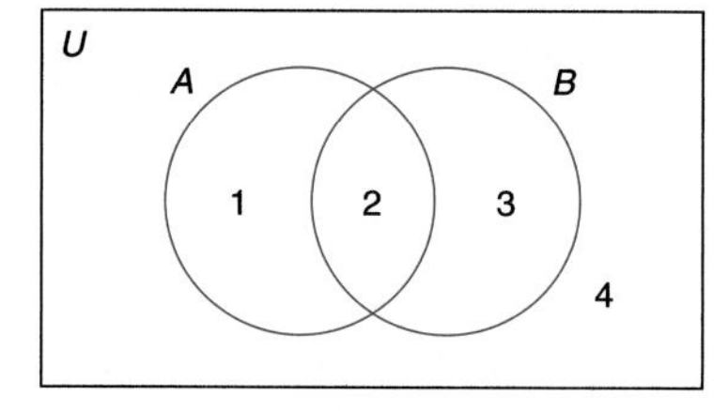

Figure 2.8

A Venn diagram that we will use quite often is shown in Figure 2.8. This type of illustration allows us to show all relations that might exist between the two sets A and B. The overlapping technique is used because it is the most efficient; it allows us to consider all of the possibilities that might exist. Although we have shown sets A and B overlapping, it does not mean that the two sets intersect.

We have assigned numbers to each region in the diagram in Figure 2.8. This enables us to discuss easily the diagram and its various parts. Region 2, for example, represents the intersection of A and B, $A \cap B$. In using the overlapping diagram, we are not concerned with whether sets A and B are disjoint; we just want to identify the region that represents $A \cap B$. The two most common methods of doing this are shading the region and using stripes (a type of shading) to distinguish the region. See Figures 2.9 and 2.10.

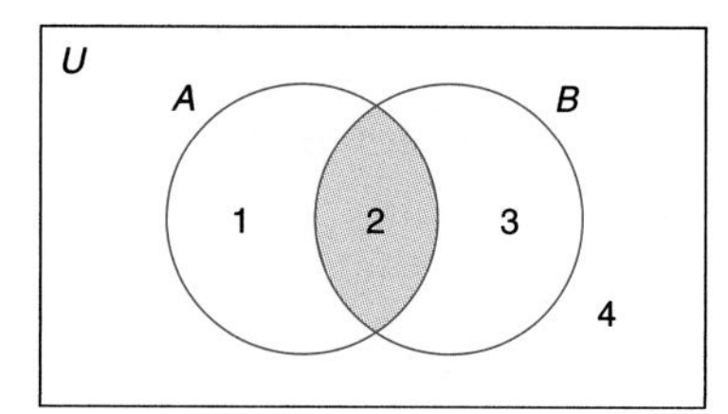

Figure 2.9

Region 2 represents $A \cap B$.

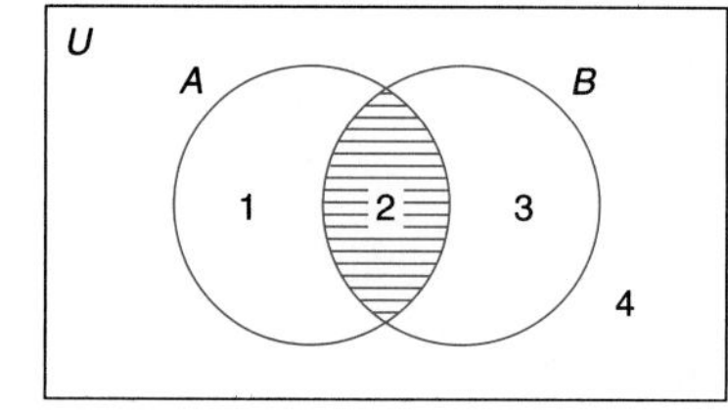

Figure 2.10

Region 2 represents $A \cap B$.

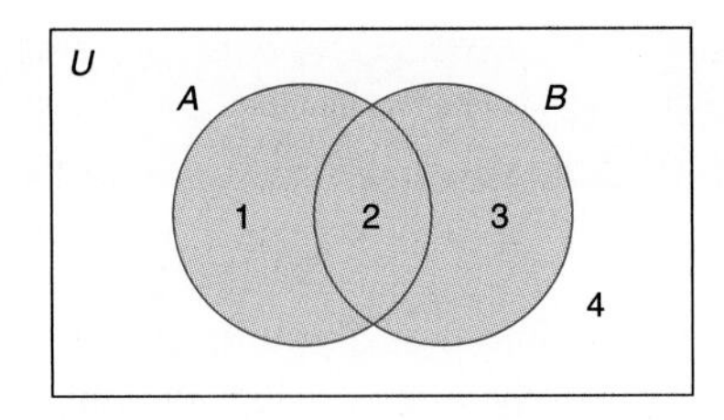

Figure 2.11
Regions 1, 2, and 3 represent $A \cup B$.

We know that region 2 represents $A \cap B$, but what region or regions represent $A \cup B$? Since $A \cup B$ is the set of elements that are elements of A or B or both, we must have all the elements in both sets. Therefore the Venn diagram showing $A \cup B$ would have regions 1, 2, and 3 shaded, as in Figure 2.11.

Example 3 *Venn Diagram* Use a Venn diagram to show $(A \cup B)'$.

Solution We must determine what region or regions represent $(A \cup B)'$. In the preceding discussion, we determined that regions 1, 2, and 3 represent $A \cup B$; we now want to determine $(A \cup B)'$, the complement of $A \cup B$. This region must be in the universal set, but not in $A \cup B$; hence it must be region 4. See Figure 2.12.

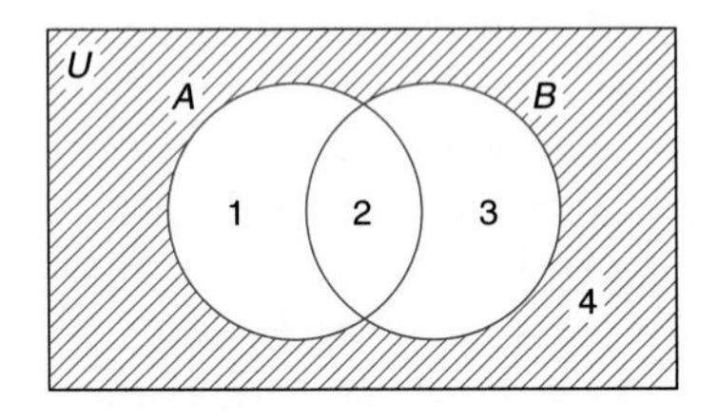

Figure 2.12
Region 4 represents $(A \cup B)'$.

Note that in Figure 2.12 we used stripes to identify a particular region in our Venn diagram. We are not concerned about what elements are in the region, or even whether there are any elements in it; we just want to distinguish it from the other regions in some way. If we have to identify more than one region in a diagram, we should use a different type of shading for each part of the problem. Each set of stripes should run in a different direction (such as horizontal or vertical), as shown in Example 4.

Example 4 *Venn Diagram* Use a Venn diagram to show $A \cap B$.

Solution In Figure 2.13, regions 1 and 2 are in A; we shade these with vertical stripes. Set B is made up of regions 2 and 3; we shade set B with horizontal stripes. Since we are looking for the intersection of A and B, we want the region with stripes that are common to both A and B. Note that this occurs in region 2, which is shaded with both horizontal and vertical stripes. Therefore, region 2 represents $A \cap B$.

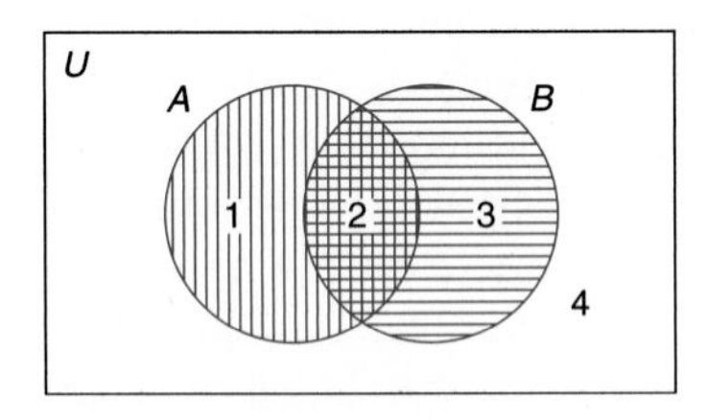

Figure 2.13
Region 2 represents $A \cap B$.

NW ***Now Work Problems 1a, c.***

In Example 4, the regions of A and B were shaded separately, and the intersection of the two sets was the shading common to both sets. For the union of A and B, $A \cup B$, we would have done the problem in the same manner, but we would have taken all of the shaded areas for our answer.

Example 5 *Venn Diagram* Use a Venn diagram to show that $(A \cap B)' = A' \cup B'$.

Solution In order to illustrate that the statement is true, we must use two Venn diagrams. We will let one diagram represent the left side of the equation and the second diagram represent the right side, and then show that the final results for each diagram have the same regions shaded.

The set $(A \cap B)'$ is the complement of $A \cap B$. The region that would satisfy the stated problem must be in U, but not in $A \cap B$. The region satisfying this consists of regions 1, 3, and 4 in Figure 2.14.

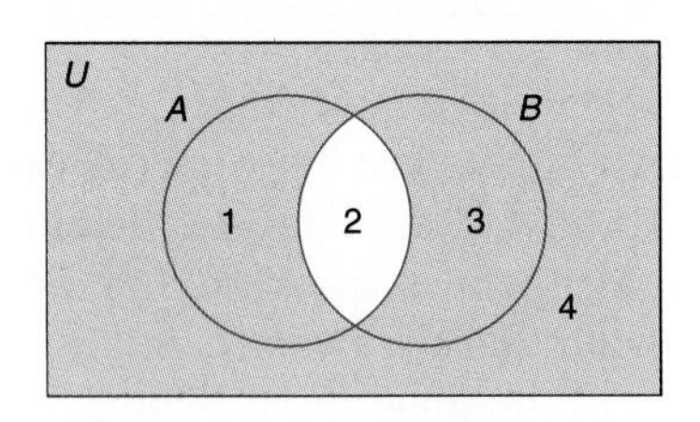

Figure 2.14
Regions 1, 3, and 4 represent $(A \cap B)'$.

The set A' is represented by regions 3 and 4 in Figure 2.15; we shade this with a vertical set of lines. The set B' is represented by regions 1 and 4, and we shade this with a horizontal set of lines.

The set $A' \cup B'$ is determined by shading the region for A' vertically, shading the region for B' horizontally, and then taking all of the shaded areas. See Figure 2.15.

Now note that the shaded region for $(A \cap B)'$ in Figure 2.14 is the same as the shaded region for $A' \cup B'$ in Figure 2.15. Hence, we have shown that $(A \cap B)' = A' \cup B'$.

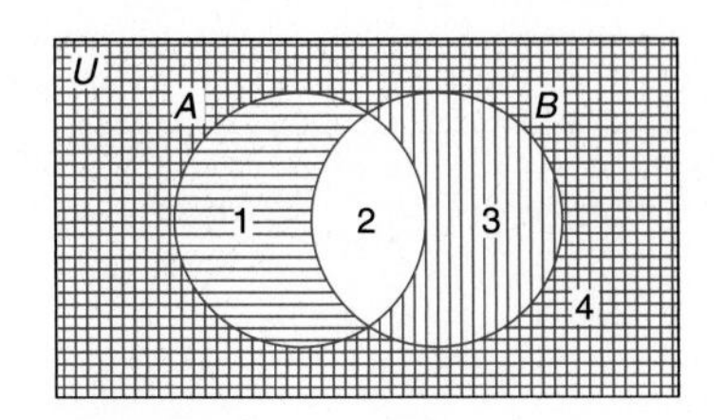

Figure 2.15
Regions 1, 3, and 4 represent $A' \cup B'$.

NW ***Now Work Problems 5a, c.***

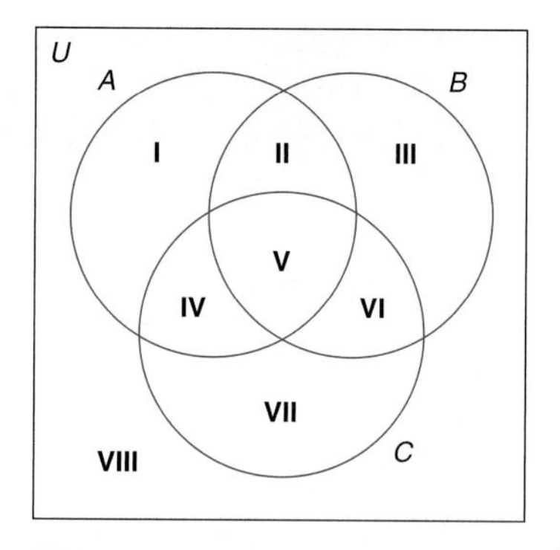

Figure 2.16

Venn diagrams may also be used when working with three sets. Figure 2.16 is a typical Venn diagram involving three sets, A, B, and C. We show the three sets overlapping as this allows us to see what relationships might exist between A and B, B and C, A and C, and so on. Because there are many possible combinations for such a problem, we have to use care in shading the proper region. For example, consider the following sets: $A' \cap (B \cup C)$ and $(A \cap B) \cup C'$. Before we attempt to illustrate such sets, let us agree that we shall number the separate regions as shown in Figure 2.16. Note that we used Roman numerals in numbering the regions. We could have used Arabic numerals $(1, 2, 3, \ldots)$, but, as you will see later, the use of Roman numerals enables us to avoid some confusion in discussing Venn diagrams. We shall use Roman numerals from now on.

Examine the Venn diagram shown in Figure 2.16. What regions represent $A \cup B$? We see that the regions comprising $A \cup B$ are I, II, III, IV, V, and VI. If we are asked to shade $A \cup B$, then these are the regions we should shade.

Regions II and V represent $A \cap B$; regions IV and V represent $A \cap C$; and regions V and VI represent $B \cap C$. Now what region represents $(A \cap B) \cap C$? That is, what region is common to all three sets? Examining Figure 2.16, we see that it is region V.

What does region VIII represent? It is the region inside U, but outside of all three sets. Hence, it must be the complement of the union of all three sets. Therefore, region VIII is the complement of $A \cup B \cup C$; we write this as $(A \cup B \cup C)'$.

Now that we are somewhat familiar with the different regions, let us consider some examples using shading.

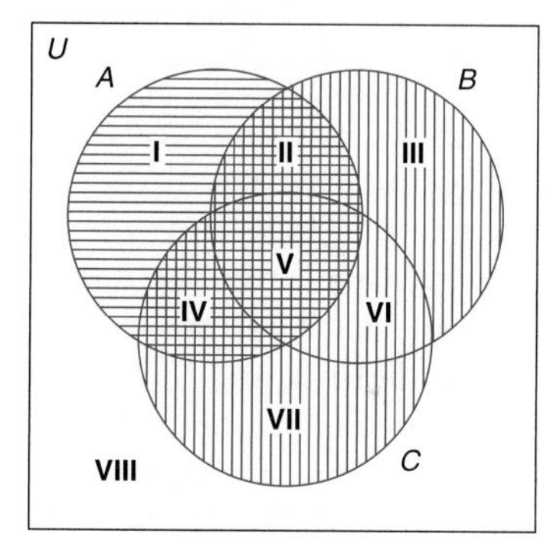

Figure 2.17

Regions II, IV, and V represent $A \cap (B \cup C)$.

Example 6 *Venn Diagram (Three Sets)* Use a Venn diagram to show $A \cap (B \cup C)$.

Solution We shall first shade $(B \cup C)$ with vertical lines. In Figure 2.17, $B \cup C$ is regions II, III, IV, V, VI, and VII. We then shade A with horizontal lines. This problem asks for $A \cap (B \cup C)$, the intersection of A with the union of B and C. Therefore, our answer appears in the regions with double lines—namely, regions II, IV, and V.

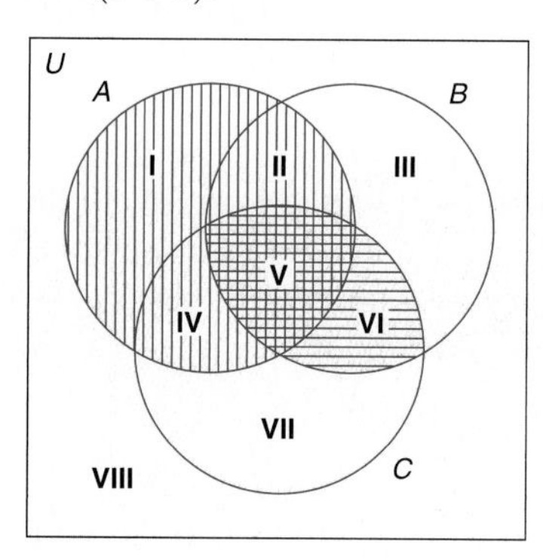

Figure 2.18

Regions I, II, IV, V, and VI represent $A \cup (B \cap C)$.

Example 7 *Venn Diagram (Three Sets)* Use a Venn diagram to show $A \cup (B \cap C)$.

Solution We shall shade $(B \cap C)$ with horizontal lines. In Figure 2.18, the regions shaded for $B \cap C$ are V and VI. Set A is shaded with vertical lines. Since we want the union of A with $(B \cap C)$, our answer is regions I, II, IV, V, and VI.

NW ***Now Work Problem 3a.***

Example 8 *Venn Diagram (Three Sets)* Show that $A \cup (B \cap C) = (A \cup B) \cap (A \cup C)$.

Solution We shall show that $A \cup (B \cap C) = (A \cup B) \cap (A \cup C)$ by comparing the corresponding Venn diagrams. Recall that if they both contain the same regions for a final answer, then the sets that the regions represent are equal. We have already determined the regions for $A \cup (B \cap C)$ (see Figure 2.18); they are I, II, IV, V, and VI.

Now we must represent $(A \cup B) \cap (A \cup C)$ by means of a Venn diagram. We first shade $(A \cup B)$ with horizontal lines and then shade $(A \cup C)$ with vertical lines, as shown in Figure 2.19. We want the intersection of these two, so our answer is regions I, II, IV, V, and VI. These are precisely the same regions that we had for $A \cup (B \cap C)$. Therefore, we have shown that $A \cup (B \cap C) = (A \cup B) \cap (A \cup C)$.

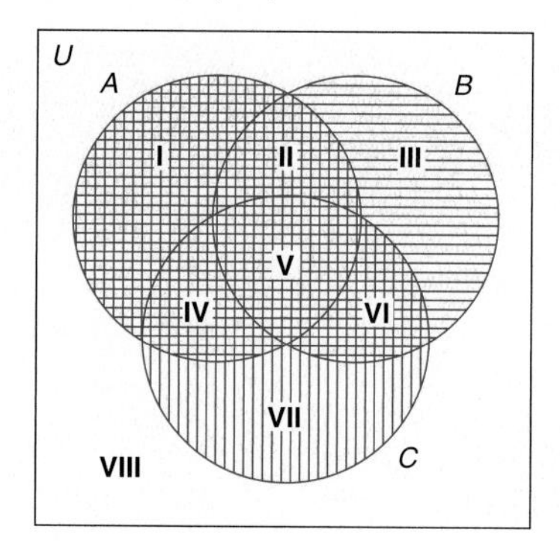

Figure 2.19

Regions I, II, IV, V, and VI represent $(A \cup B) \cap (A \cup C)$.

Math Connections

Countably Infinite Sets

Consider the sets

$$A = \{1, 2, 3, 4, \dots\} \text{ and}$$
$$B = \{2, 4, 6, 8, \dots\}$$

We can describe both A and B as infinite sets; each set has an unlimited number of elements. What is the cardinality of each set? At first glance, it may appear that A contains a greater number of elements than B, but does it? The cardinal number of the set of counting numbers is called "aleph-null" or "aleph-zero," and is designated by $\aleph_0$. Aleph is the first letter of the Hebrew alphabet. It can be shown that the set $B = \{2, 4, 6, \dots\}$ can be placed in a one-to-one correspondence with the set of counting numbers (set A), which implies that they contain the same number of elements. Thus, the cardinality of B is also $\aleph_0$. A and B are called countably infinite sets. The cardinal number of all countably infinite sets is $\aleph_0$. Thus, an infinite set that can be placed in a one-to-one correspondence with the set of natural numbers, N, is countably infinite; otherwise, it is uncountable.

The reason we used Roman numerals to designate the regions of a set is that many times we may want to use Arabic numerals to tell how many members are in a set. This is called the **cardinality** of a set. A **cardinal number** tells us "how many," as opposed to an **ordinal number**, which tells us "what position." (Examples of ordinal numbers are first, second, third, fourth, fifth, and so on.)

A cardinal number gives the number of elements in a set. The empty set contains no elements, so its cardinal number is zero. We may say that the *cardinality* of the empty set is zero. The set $A = \{a, e, i, o, u\}$ has five elements; hence its cardinality is 5.

We shall use the notation $n(A)$ to stand for the cardinality of A. Therefore, we may say $n(A) = 5$. But suppose set A is in a Venn diagram—how could we tell how many members it has? One solution is to list the elements in the circle. Another solution is to write the Arabic numeral 5 to indicate that set A contains five elements. See Figures 2.20 and 2.21.

Although you may think that neither technique is better than the other, suppose set A contained all the letters of the alphabet. Would you list all 26 letters separately in the circle? Carrying this idea a bit further, suppose the set contains all the students in your school. If we are concerned only with cardinality of this set, then it is much more efficient to write the Arabic numeral for that number than to list all of the elements.

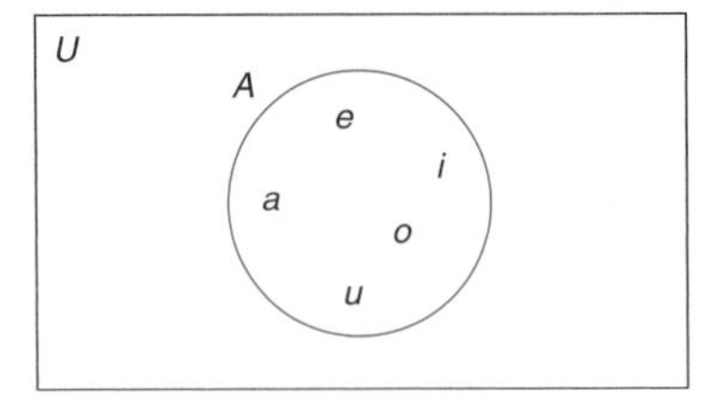

Figure 2.20

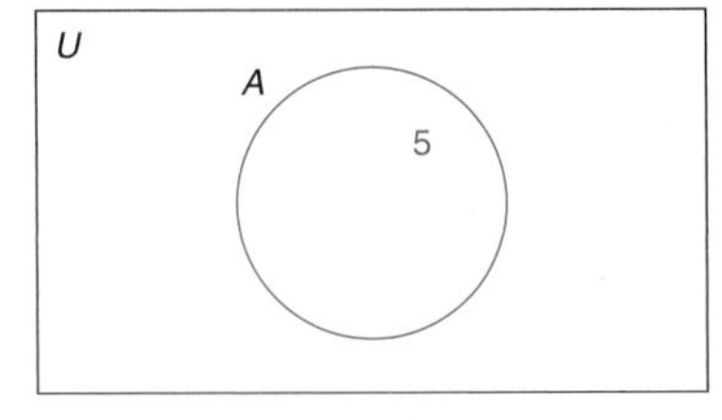

Figure 2.21

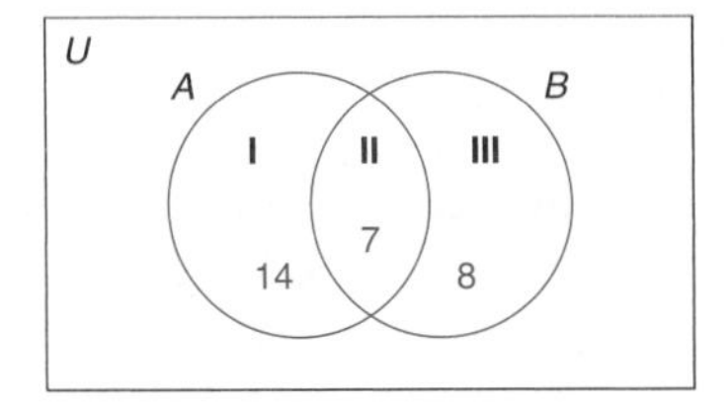

Figure 2.22

Example 9 ***Cardinality*** Use Figure 2.22 to find the cardinality of set A—that is, $n(A)$.

Solution In Figure 2.22, set A is composed of regions I and II. The number of elements in region I is 14, and the number of elements in region II is 7. The total number of elements in set A is 21. Therefore,

$$n(A) = 21$$

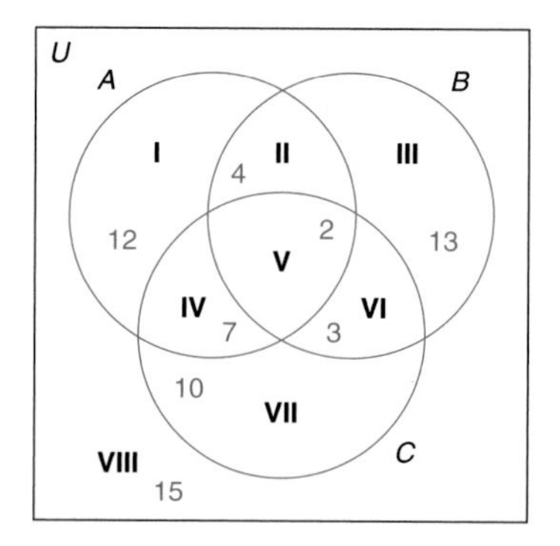

Figure 2.23

Example 10 ***Cardinality*** Use Figure 2.23 to find $n(A \cap B)$, $n(B \cup C)$, and $n(U)$.

Solution In Figure 2.23 we see that $A \cap B$ is composed of regions II and V. The total number of elements in these two regions is 6. Therefore,

$$n(A \cap B) = 6$$

For $B \cup C$, we use regions II, III, IV, V, VI, and VII. The total number of elements in all of these regions is 39. Recall that in doing set union problems, we list an element only once even if it appears in two sets. Similarly, we need to count the elements only once in a region even if the region is in both sets. Hence,

$$n(B \cup C) = 39$$

To find the cardinality of the universal set, we must find the total number of elements in all eight regions. Adding these numbers, we find that the cardinality of U is 66, or

$$n(U) = 66$$

NW ***Now Work Problem 7.***

Recall that two sets that contain the same number of elements are said to be equivalent sets. If two sets are equivalent, then they have the same cardinality, and the elements of one set can be paired with the elements of the other set. Consider the following two sets, A and B:

$$A = \{a, b, c\}$$

$$B = \{bat, cat, rat\}$$

The two sets have the same cardinality, so we can match the elements of the two sets with each other. We could pair *a* with *bat*, *b* with *cat*, and *c* with *rat*. We can show this by

$$\begin{array}{ccc} \{a, & b, & c\} \\ \updownarrow & \updownarrow & \updownarrow \\ \{bat, & cat, & rat\} \end{array}$$

This is not the only way that we could have paired the elements of the two sets; there are others. For example, we could pair *a* with *rat*, *b* with *bat*, and *c* with *cat*. A pairing of the elements of two sets is called a **one-to-one correspondence** between the sets. When two sets are in a one-to-one correspondence, it means that each element of the first set is paired with just one element of the second set, and each element of the second set is paired with just one element of the first set. Note that two sets that are in one-to-one correspondence must be equivalent sets and have the same cardinality.

Example 11 ***One-to-One Correspondence*** In how many ways can a one-to-one correspondence be established between $A = \{a, b, c\}$ and $B = \{bat, cat, rat\}$?

Solution We have already discussed two different one-to-one correspondences for these two sets, but we shall start from the beginning. The letter *a* can be paired with any of the three words *bat, cat,* or *rat* in set *B*. Now, if *a* is paired, then *b* can be paired with either of the two remaining words. With *a* and *b* paired, *c* has to be paired with the remaining word. Therefore, the total number of one-to-one correspondences is

$$3 \times 2 \times 1 = 6$$

Note: Some texts may use the notation $|A|$ to denote the cardinality of set A.

Here are the six different one-to-one correspondences:

$$\begin{array}{ccc} \{a, & b, & c\} \\ \updownarrow & \updownarrow & \updownarrow \\ \{bat, & cat, & rat\} \end{array} \quad \begin{array}{ccc} \{a, & b, & c\} \\ \updownarrow & \updownarrow & \updownarrow \\ \{cat, & bat, & rat\} \end{array} \quad \begin{array}{ccc} \{a, & b, & c\} \\ \updownarrow & \updownarrow & \updownarrow \\ \{rat, & bat, & cat\} \end{array}$$

$$\begin{array}{ccc} \{a, & b, & c\} \\ \updownarrow & \updownarrow & \updownarrow \\ \{bat, & rat, & cat\} \end{array} \quad \begin{array}{ccc} \{a, & b, & c\} \\ \updownarrow & \updownarrow & \updownarrow \\ \{cat, & rat, & bat\} \end{array} \quad \begin{array}{ccc} \{a, & b, & c\} \\ \updownarrow & \updownarrow & \updownarrow \\ \{rat, & cat, & bat\} \end{array}$$

Alternate Solution The two given sets are equivalent; they have the same number of elements. Therefore, a one-to-one correspondence can be established between A and B. The number of ways a one-to-one correspondence can be established between equivalent sets is $n!$, read ***n-factorial*** where $n! = n \times (n-1) \times (n-2) \times (n-3) \times \cdots \times 3 \times 2 \times 1$ and n is the number of elements in a given set. In this case $n = 3$ and we have

$$3! = 3 \times 2 \times 1 = 6$$

which is the same answer we obtained previously.

NW ***Now Work Problem 9.***

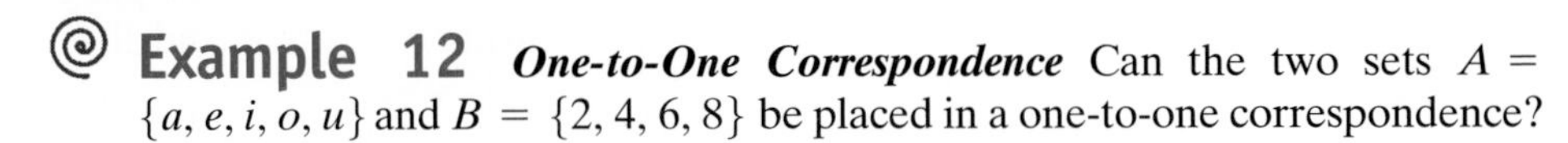

Example 12 ***One-to-One Correspondence*** Can the two sets $A = \{a, e, i, o, u\}$ and $B = \{2, 4, 6, 8\}$ be placed in a one-to-one correspondence?

Solution No, the two sets do not have the same cardinality, because the cardinality of set A is 5, while the cardinality of set B is 4. If we tried pairing the elements of the two sets, we would have an element left over, and it would have to be paired with an element that had already been matched.

$$\begin{array}{ccccc} \{a, & e, & i, & o, & u\} \\ \updownarrow & \updownarrow & \updownarrow & \updownarrow \swarrow & \\ \{2, & 4, & 6, & 8\} & \end{array} \quad \textit{not one-to-one}$$

Exercises for Section 2.5

1. Illustrate each of the following with a Venn diagram. Number the regions in your diagram as shown in Examples 3–5 and list the regions that make up your answer.
   - NW **a.** $A \cap B$ **b.** $A' \cap B'$
   - NW **c.** $A \cup B$ **d.** $A' \cup B'$
   - **e.** $(A' \cap B')'$ **f.** $(A' \cup B')'$

2. Illustrate each of the following with a Venn diagram. Number the regions in your diagram as shown in Examples 3–5 and list the regions that make up your answer.
   - **a.** $A \cap B'$ **b.** $A' \cup B$
   - **c.** $A' \cap B$ **d.** $(A \cup B)'$
   - **e.** $(A \cap B)'$ **f.** $A \cup B'$

3. Use a Venn diagram to illustrate each of the following. Number the regions in your diagram as shown in Examples 6–10 and list the regions that make up your answer.
   - NW **a.** $A \cap (B \cap C)$ **b.** $A \cup (B \cup C)$
   - **c.** $(A \cap B) \cup C$ **d.** $(A \cap B) \cup (A \cap C)$
   - **e.** $(A \cup B) \cap (A \cup C)$ **f.** $(A' \cap C') \cup B$

4. Use a Venn diagram to illustrate each of the following. Number the regions in your diagram as shown in Examples 6–10 and list the regions that make up your answer.
   - **a.** $(A \cap B) \cap C$ **b.** $(A \cup B) \cup C$
   - **c.** $(A \cup B) \cap C$ **d.** $(A \cap B) \cap (B \cap C)$
   - **e.** $(B \cup C) \cup (A \cap C)$ **f.** $(C' \cup A) \cap B$

5. Use Venn diagrams to show that each of the following statements is true.
   - NW **a.** $A \cap B = (A' \cup B')'$
   - **b.** $A \cup B = (A' \cap B')'$
   - NW **c.** $(A \cup B)' = A' \cap B'$

d. $(A \cap B)' = A' \cup B'$
e. $A \cap (B \cup C) = (A \cap B) \cup (A \cap C)$
f. $A \cup (B \cap C) = (A \cup B) \cap (A \cup C)$

6. Use Figure 2.24 to find the following cardinalities.

a. $n(A)$ **b.** $n(B)$
c. $n(A \cap B)$ **d.** $n(A \cup B)$
e. $n(A')$ **f.** $n(B')$
g. $n(A' \cap B')$ **h.** $n(A' \cup B')$

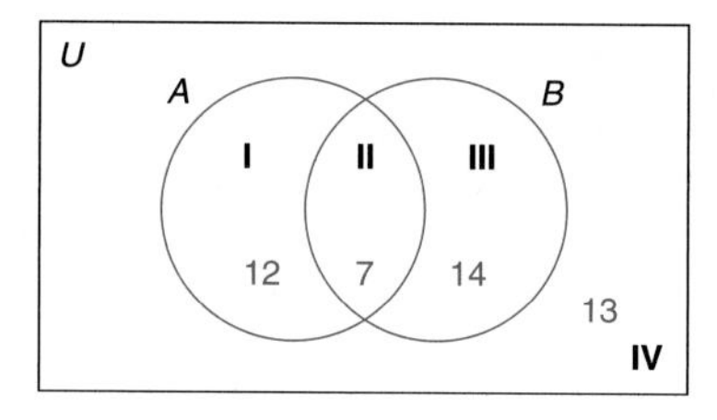

Figure 2.24

NW **7.** Use Figure 2.25 to find the following cardinalities.

a. $n(A)$ **b.** $n(B)$
c. $n(C)$ **d.** $n(U)$
e. $n(B \cup C)$ **f.** $n(B \cap C)$
g. $n(A \cap B \cap C)$ **h.** $n(A \cup B \cup C)$

8. Show a one-to-one correspondence between $A = \{10, 20\}$ and $B = \{15, 30\}$. In how many ways can you do this?

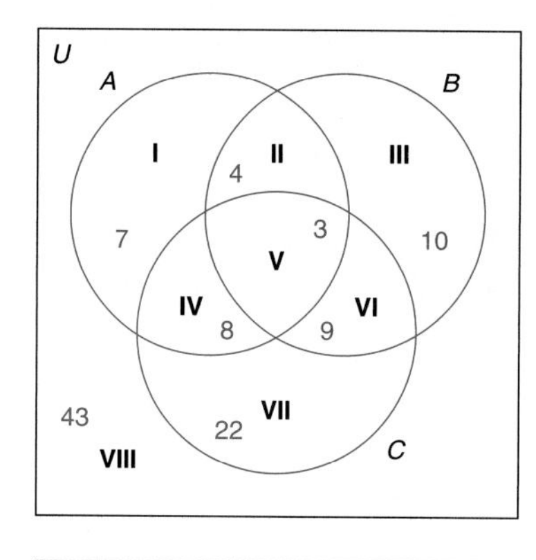

Figure 2.25

NW **9.** In how many ways can one show a one-to-one correspondence between $A = \{i, o, u\}$ and $B = \{a, b, c\}$? Show one of these correspondences.

10. Show a one-to-one correspondence between $A = \{Jill, Adam, Chris, Neil\}$ and $B = \{3, 6, 9, 12\}$. In how many ways can you do this?

11. In how many ways can a one-to-one correspondence be established between $A = \{m, a, t, h\}$ and $B = \{f, u, n\}$? Explain your answer.

12. For each of the diagrams in Figures 2.26a–f, use set notation to describe the situation shown. For example, Figure 2.26 shows $(A \cap B)$.

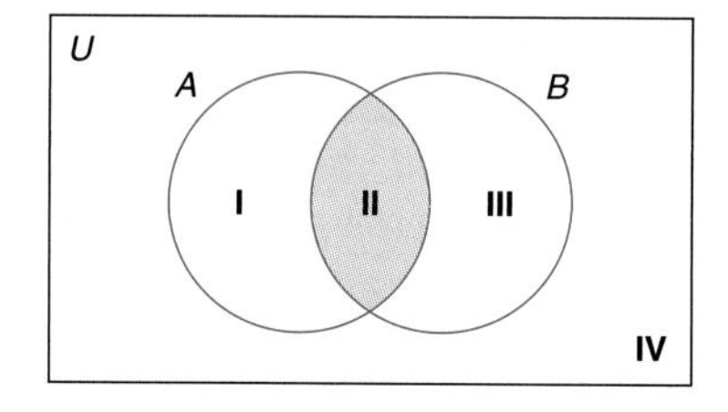

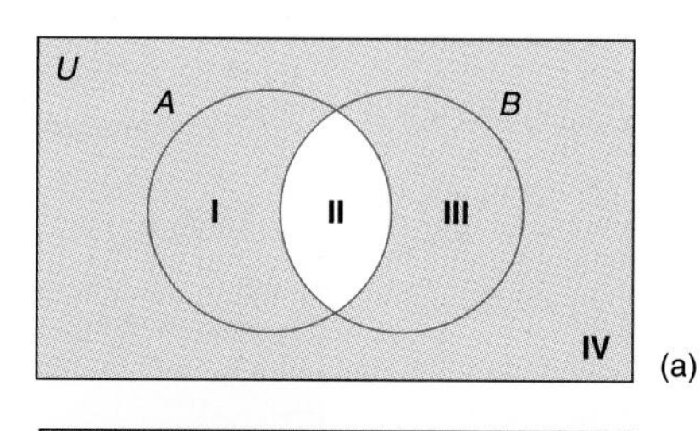

(a)

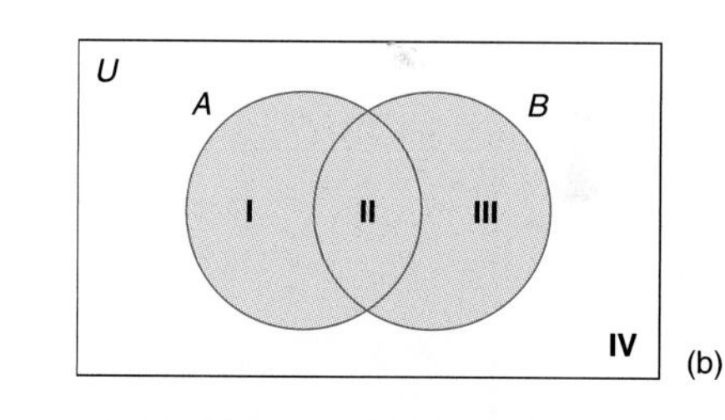

(b)

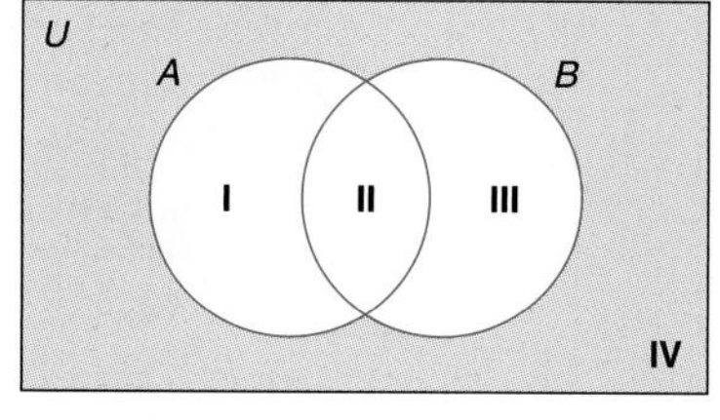

(c)

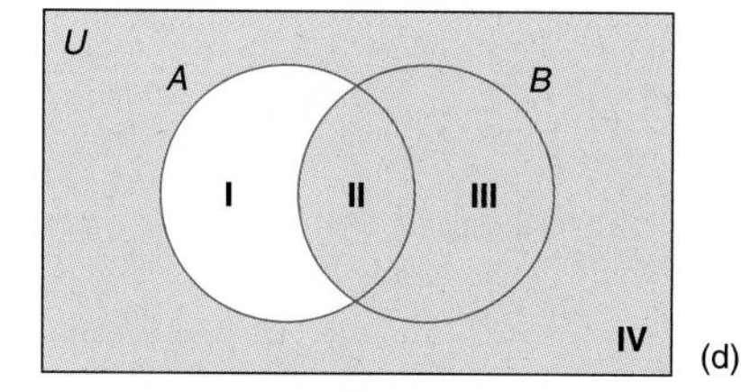

(d)

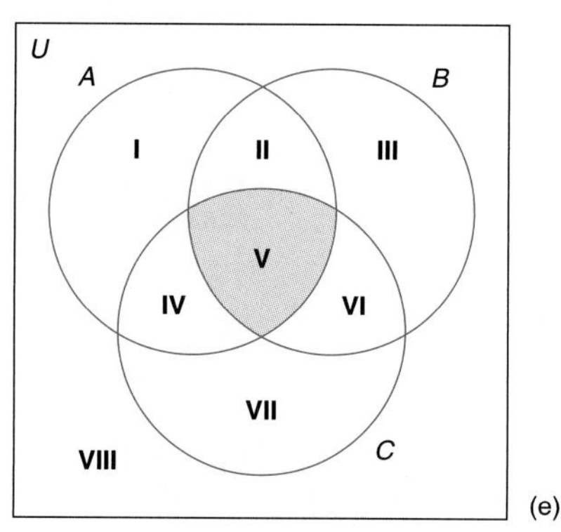

(e)

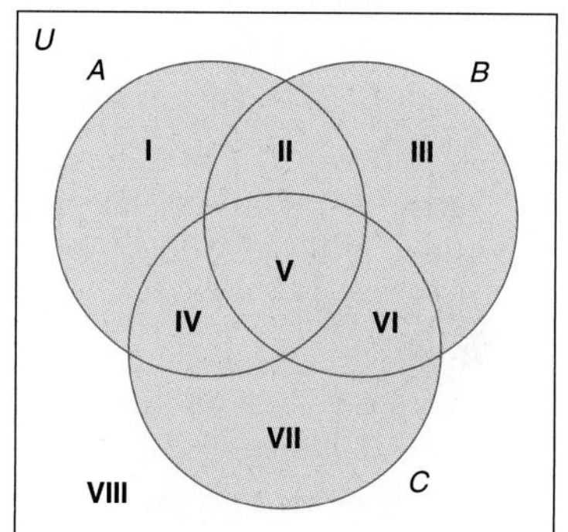

(f)

Figure 2.26

***13.** Use Figure 2.27 to determine if the expressions given in **a** through **h** are true or false.

a. $B \subseteq A$ **b.** $x \notin A$
c. $y \in U$ **d.** $(B \cap C) \subset A$
e. $z \in C$ **f.** $x \in (A \cap B \cap C)$
g. $y \in (B \cup C)'$ **h.** $w \in A'$

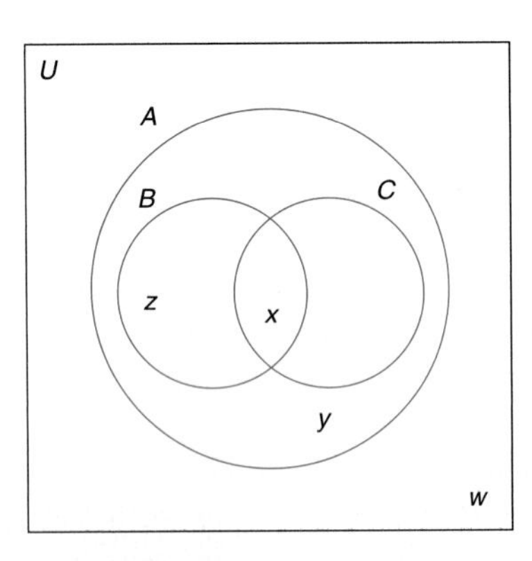

Figure 2.27

***14.** Use Figure 2.28 to determine if the expressions given in **a** through **h** are true or false.

a. $A \cap B = \emptyset$ **b.** $x \notin B'$
c. $B \cap C = C$ **d.** $y \in (B \cap C)$
e. $w \in (A \cup B)$ **f.** $A \cap C = \emptyset$
g. $(A \cap C) \subset B$ **h.** $w \in (A \cup C)'$

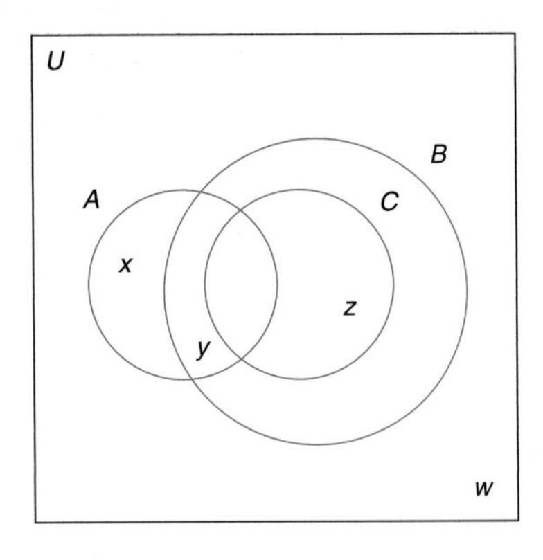

Figure 2.28

***15.** Use Figure 2.27 to answer the questions in **a** through **e** as true or false.

a. There is at least one element in C that is also contained in B (i.e., *Some C are in B*).
b. All elements in C are also contained in A (i.e., *All C are in A*).
c. There is at least one element in B that is not in A (i.e., *Some B are not in A*).
d. All elements in B are also in A (i.e., *All B are in A*).
e. There is at least one element in B that is also contained in A (i.e., *Some B are in A*).

***16.** Use Figure 2.28 to answer the questions in **a** through **e** as true or false.

a. All elements in C are also contained in B (i.e., *All C are in B*).
b. There is at least one element in C that is not in A (i.e., *Some C are not in A*).
c. There is at least one element in B that is not in A (i.e., *Some B are not in A*).
d. Some elements in A are also in B (i.e., *Some A are in B*).
e. There is at least one element in A that is not in B (i.e., *Some A are not in B*).

17. The following chart contains the results of a nationwide survey on factors that contribute to job dissatisfaction based on employees' age groups. Using the letters indicated, find the number of people in each of the given sets in **a** through **j**.

Age (Years)	Salary and Compensation (*S*)	Relationship with Management (*M*)	Hours Worked (*H*)
Under 30 (*T*)	197	210	188
30–50 (*F*)	328	375	419
Over 50 (*O*)	216	507	393

a. F **b.** M
c. $T \cap M$ **d.** $O \cup H$
e. $S \cap M$ **f.** $F \cap T$
g. $(F \cap S)'$ **h.** $(T \cup M)'$
i. $(T \cap S) \cup H$ **j.** $(O \cup M) \cap F$

18. The following chart contains the results of a nationwide survey on common types of employee compensation provided by employee status. Using the letters indicated, find the number of people in each of the given sets in **a** through **j**.

Employee Status	401K (*F*)	Bonus or Incentive (*B*)	Stock Options (*O*)
Full-time salaried (*S*)	306	781	569
Full-time hourly (*H*)	438	222	119
Part-time (*P*)	83	167	46

a. S **b.** O
c. $H \cap B$ **d.** $P \cup F$
e. $H \cap P$ **f.** $F \cap O$
g. $(F \cap P)'$ **h.** $(H \cup O)'$
i. $(S \cap B) \cup O$ **j.** $(H \cup F) \cap B$

Writing Mathematics

19. What is a Venn diagram, and how are Venn diagrams used?

20. Describe the Venn diagram for two disjoint sets. How does the diagram show that the sets have no elements in common?

21. What does the notation $n(A)$ mean in this section and how is it used?

22. What does the notation $n!$ mean in this section and how is it used?

Challenge Exercises

23. A Venn diagram with one set has two regions; a diagram with two sets has four regions; and one with three sets has eight regions. How many regions would a Venn diagram with four sets have? Try to construct such a diagram. *Hint:* The sets should not be circle shaped.

24. Use Venn diagrams to illustrate $A - (B \cap C)$. (See Section 2.4 Challenge Exercises.)

25. If $A \subset B$, use Venn diagrams to visually prove that $A \cup B = B$.

Use the concept of countably infinite (see the Math Connections in this section) to answer Exercises 26 and 27.

26. Given the set of integers $Z = \{\ldots, -3, -2, -1, 0, 1, 2, 3, \ldots\}$, is Z countably infinite? Why?

27. Given the set of real numbers, R (see Chapter 8), is R countably infinite? Why?

Group/Research Activity

28. As noted in the discussion, Venn diagrams are named after John Venn. Research the origin of Venn diagrams and prepare a report that describes the context in which they were first used.

29. Research the relationship between Euler circles and Venn diagrams.

30. Go to the Web site **http://forum.swarthmore.edu/dr.math/-problems/mcswain11.7.97.html** and read about the concept of **transfinite cardinal numbers.** Prepare a report to the class about what you learned.

Just for fun

There are many different ways to describe a group or set of animals. Some of these words have fallen into disuse but are still found in the dictionary. See how many of these you can match with the appropriate species.

band	clowder	drift	mob	sleuth
bed	colony	flock	murder	span
bevy	congregation	gaggle	pod	troop
brood	crash	herd	school	volery
cast	down	litter	skulk	yoke

2.6 AN APPLICATION OF SETS AND VENN DIAGRAMS

Our knowledge of sets and Venn diagrams is particularly useful in solving problems involving overlapping sets of data on individuals. Consider Example 1.

Example 1 ***Pizza*** In a certain group of 100 customers at Phil's Pizza Palace, 60 customers ordered cheese and pepperoni on their pizza. Altogether 80 customers ordered a pizza with cheese on it, and 72 customers ordered pizza with pepperoni on it.

a. How many customers ordered cheese on their pizza, but no pepperoni?

b. How many customers ordered pepperoni on their pizza, but no cheese?

c. How many customers in the group of 100 customers ordered neither cheese nor pepperoni on their pizza?

Solution Because we have two pizza toppings to consider, cheese and pepperoni, we can draw a Venn diagram illustrating these two sets. See Figure 2.29a. We label the sets C (cheese) and P (pepperoni) and also number the regions.

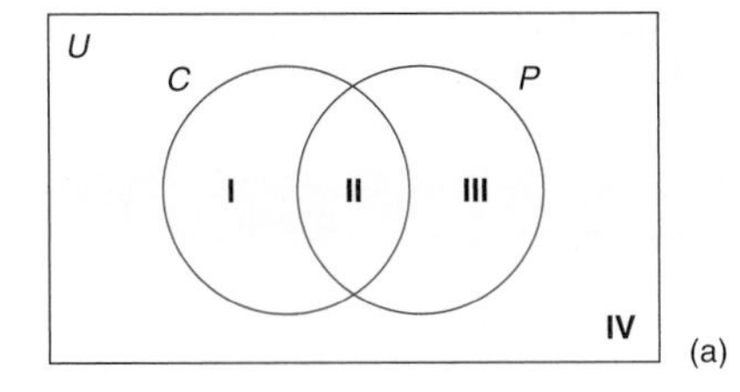

(a)

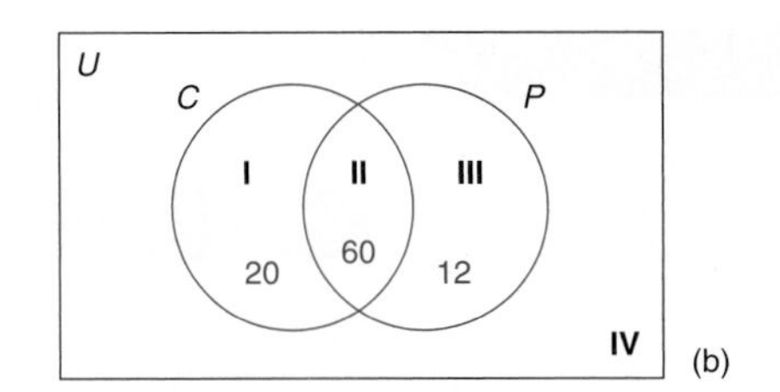

(b)

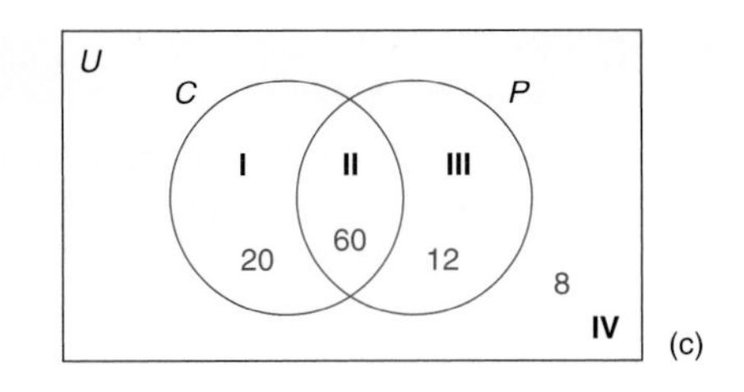

(c)

Figure 2.29

There were 60 customers who ordered pizza with cheese *and* pepperoni. The *and* is our clue that these customers belong in the intersection of C and P—namely, region II. Altogether 80 customers ordered a pizza with cheese on it, but we cannot place 80 in region I. If we put 80 in region I, since we already have 60 in region II, that would give us 140 customers in set C—and there are only 100 customers total. The 60 customers who ordered cheese and pepperoni on their pizza are also counted as part of the 80 customers who ordered cheese on their pizza. If there are 80 customers in set C, and we already have 60 customers in region II (part of C), then we have 20 in region I.

Seventy-two customers are in set P, and we know 60 of them are in region II; hence we write 12 in region III. We now have the diagram in Figure 2.29b.

Adding up the total number of people in regions I, II, and III, we see that there are only 92 customers in our diagram. We had a total of 100 customers in our universal set. Where are the other 8? As they are not contained in set P or set C, but are in the universal set, they must be in region IV. Therefore, we place 8 customers in region IV. The completed diagram is shown in Figure 2.29c.

Now that we have our completed diagram, we can answer the original questions.

a. How many of these customers ordered cheese on their pizza, but no pepperoni? These are the customers in region I and therefore the answer is 20.

b. How many of these customers ordered pepperoni on their pizza, but no cheese? These customers are in region III and the answer is 12.

c. How many customers in their group of 100 customers ordered neither cheese nor pepperoni on their pizza? These customers are in region IV and we see that the answer is 8.

NW ***Now Work Problem 1.***

Note that in determining the numbers that went in the different regions of the diagram in Example 1, we started with the most specific piece of information. We

Note of Interest

Albert Einstein (1879–1955) was born in Ulm, Germany. While growing up, he taught himself calculus and other areas of mathematics. After attending college in Switzerland, he worked in the Swiss Patent Office. During his stay at the Patent Office he wrote papers on physics and mathematics and, in 1905, his first published article on the theory of relativity introduced the famous formula $E = mc^2$. This led to the discovery of nuclear energy. The formula predicted how much energy would be released by a nuclear reaction similar to that of an atomic bomb explosion. It should be noted that Einstein's discoveries were based on mathematical reasoning rather than on experiments. Einstein's theories also illustrate the close relationship between science and mathematics.

first placed the 60 customers who were in the intersection of the two sets, and then proceeded with the completion of the diagram. Once a diagram is complete, we can examine it for information and answer the questions that are asked. But we must use care in entering the data in our Venn diagram. If the data are entered correctly, then we can answer the questions by reading directly from the Venn diagram.

Example 2 ***Course Registration*** In a certain group of 75 students, 16 students are taking psychology, geology, and English; 24 students are taking psychology and geology; 30 students are taking psychology and English; and 22 students are taking geology and English. However, 7 students are taking only psychology, 10 students are taking only geology, and 5 students are taking only English.

a. How many of these students are taking psychology?

b. How many of these students are taking psychology and English, but not geology?

c. How many students in this group are not taking any of the three subjects?

Solution This problem involves overlapping sets of data, so we may represent the data by means of a Venn diagram. There are three subjects, so we will have three circles; see Figure 2.30a. We label the sets P (psychology), G (geology), and E (English) and number the regions.

It is best to start with the most specific piece of information and work from it. The most specific information we have is that 16 students are taking all three subjects (the intersection of all three sets). These students would have to appear in region V. If 24 students are taking psychology and geology (the intersection of P and G), and we already know that 16 are in region V, then we need to place 8 of them in region II. Why not put 24 there? If we did that, then we would have 24 in region II and 16 in region V, which would give us a total of 40 students taking psychology and geology. That is not the case!

Since 30 students are in English and psychology, we place 14 in region IV. Similarly, we place 6 in region VI because 22 students are taking geology and English. Since 7 students are taking only psychology, they belong in region I; similarly, we place 10 students in region III and 5 students in region VII. We now have the diagram in Figure 2.30b.

Counting the total number of students in the circles, we get 66. But there are 75 students in our group and therefore we place 9 students in region VIII to complete our diagram. See Figure 2.30c.

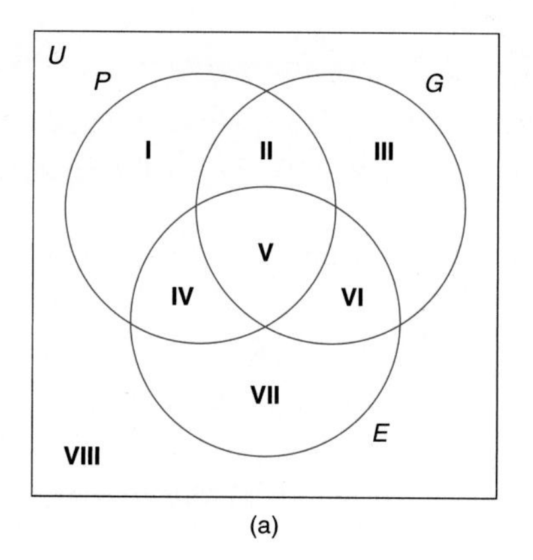

(a)

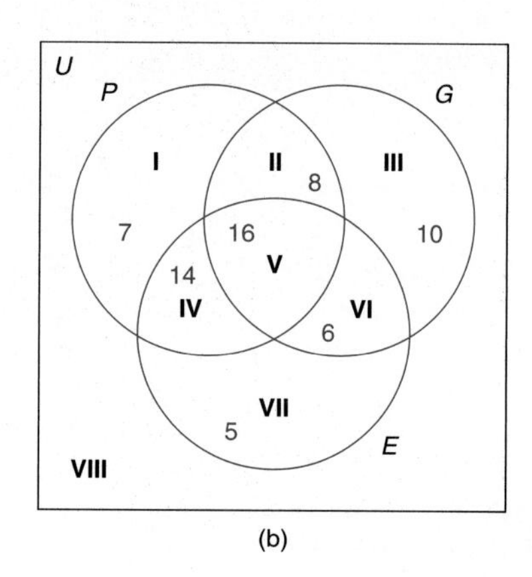

(b)

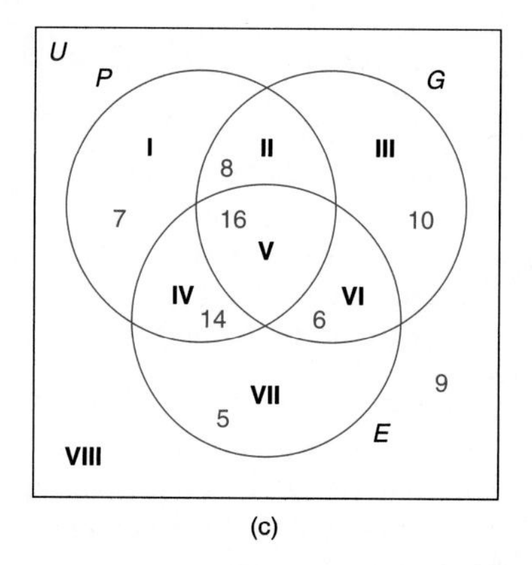

(c)

Figure 2.30

Now we can readily answer the questions by reading directly from the Venn diagram.

a. How many of these students are taking psychology? These students are in set P, so we add the numbers of students in regions I, II, IV, and V. The answer is 45.

b. How many of these students are taking psychology and English, but not geology? These are the students in region IV, and the answer is 14.

c. How many students in this group are not taking any of the three subjects? These students are in region VIII, and the answer is 9.

NW ***Now Work Problem 5.***

Example 3 ***Brands of Automobiles*** At a meeting of 50 car dealers, the following information was obtained: 12 dealers sold Buicks, 15 dealers sold Toyotas, 16 dealers sold Pontiacs, 4 dealers sold both Buicks and Toyotas, 6 dealers sold both Toyotas and Pontiacs, 5 dealers sold both Buicks and Pontiacs, and 1 dealer sold all three brands.

a. How many dealers sold Buicks and neither of the other two brands?

b. How many of the dealers at the meeting did not sell any of these cars?

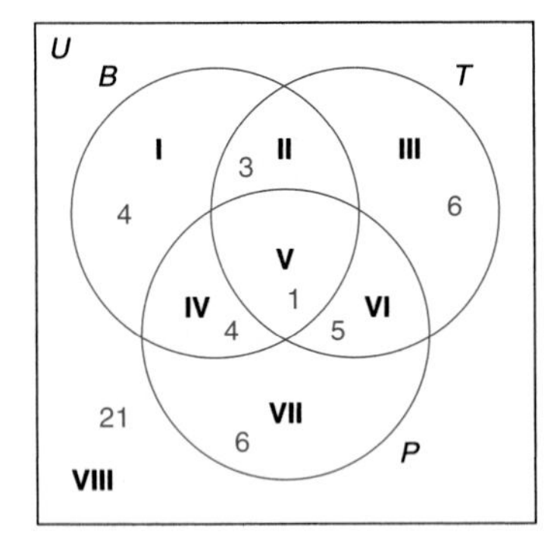

Figure 2.32
B = the set of Buicks,
T = the set of Toyotas, and
P = the set of Pontiacs.

Solution We may obtain answers to these questions by completing a Venn diagram and reading our answers directly from the figure; see Figure 2.32. It is important to remember to start with the most specific piece of information, that which belongs in the intersection of all three sets (in this case, the one dealer who sold all three brands of cars). After using this information, we proceed to the data that belong in the intersection of two sets, and continue working backward.

In our diagram, we labeled the circles Buick, Toyota, and Pontiac in that order, but we could have changed the order if we had so chosen. The resulting information would still be the same, and we would still be able to answer the questions.

Reading the information from the Venn diagram, we can answer the questions.

a. How many dealers sold Buicks and neither of the other brands? These are the dealers in region I, and the answer is 4.

b. How many of the dealers at the meeting did not sell any of these cars? These are the dealers in region VIII; they are not members of any of the three sets, and the answer is 21.

NW ***Now Work Problem 7.***

Math Connections

Finding the Intersection

In all of the examples discussed in this section, our strategy was to begin with the most specific piece of information, which represented the intersection of the given sets, and then work backward to complete the Venn diagram. Suppose, though, that the information representing the intersection was not given and we were asked to find it.

To illustrate this, let's revisit Example 1, in which 100 customers were at Phil's Pizza Palace. Instead of being told that 60 people ordered cheese and pepperoni pizzas, let's assume that all we know is that 20 customers ordered a pizza with only cheese on it, 12 people ordered a pizza with only pepperoni on it, and 8 people ordered neither cheese nor pepperoni on their pizza. To find out how many people ordered a pizza with cheese and pepperoni, which is the intersection, we do the following:

1. Create a Venn diagram to represent each set and then insert the appropriate information into regions I, III, and IV. We also place an x in region II because this information is unknown. This is illustrated in Figure 2.33.
2. Establish a mathematical model that will enable us to find x. Since the regions are *mutually exclusive*, it is clear from our Venn diagram that the sum of all four regions must equal 100. That is,

$$20 + x + 12 + 8 = 100$$

(continued)

(continued)

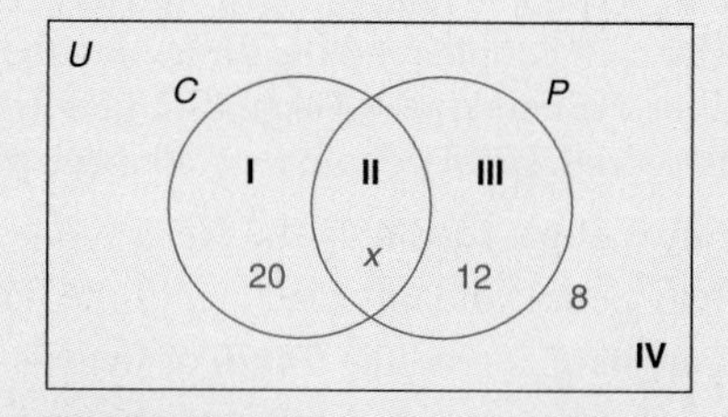

Figure 2.33

3. We now find x by solving the algebraic equation from step 2 (see Chapter 9 for information about solving algebraic equations):

$$20 + x + 12 + 8 = 100$$
$$20 + 12 + 8 + x = 100$$
$$40 + x = 100$$
$$x = 100 - 40$$
$$x = 60$$

Thus, 60 people ordered pizza with cheese and pepperoni.

Observe that the equation in step 2 could also be set up using logical reasoning. We begin our reasoning process as follows:

The cardinality of set C added to the cardinality of set P added to the cardinality of the complement of the two sets must equal 100.

That is,

$$n(C) + n(P) + n(C \cup P)' = 100$$
$$= (\text{Region I} + \text{Region II}) + (\text{Region II} + \text{Region III}) + \text{Region IV} = 100$$

This reasoning is partially flawed, though, because it leads to adding Region II twice. To remove this redundancy we simply subtract one occurrence of Region II, which leads to the equation

$$(\text{Region I} + \text{Region II}) + (\text{Region II} + \text{Region III}) + \text{Region IV} - \text{Region II} = 100$$
$$(20 + x) + (x + 12) + 8 - x = 100$$

This now simplifies to the equation

$$20 + x + 12 + 8 = 100$$

Exercises for Section 2.6

NW 1. **Magazines** In a certain group of 75 students at Monroe Community College, the following information was obtained: 44 students subscribe to *Sports Illustrated,* 26 students subscribe to *People,* and 14 subscribe to both magazines.

 a. How many students in this group do not subscribe to *People?*

 b. How many students do not subscribe to *Sports Illustrated?*

 c. How many students subscribe to neither magazine?

 d. How many students subscribe to *People* or *Sports Illustrated?*

2. **Pets** At the Lollipop Pet Shop, 150 customers were asked what type of pet they had. Fifty-one people in the group had a dog, 48 people had a cat, and 29 people had both a dog and a cat.

 a. How many of these people had only a dog?

 b. How many of these people did not have a cat?

 c. How many of these people had neither a dog nor a cat?

 d. How many of these people had a dog or a cat?

3. **Course Registration** In a survey of 125 students standing in a college registration line, the following data were collected: 70 students registered for a mathematics course, 60 students registered for a science course, and 40 students registered for a mathematics course and a science course.

 a. How many of the students registered only for a science course?

 b. How many of the students did not register for a mathematics course?

 c. How many of the students did not register for a science course?

 d. How many students registered for a mathematics course or a science course?

4. **Newspapers** In a certain group of 350 passengers at JFK Airport in New York City, the following information was obtained: 200 passengers purchased *The Wall Street Journal,* 140 passengers purchased *USA Today,* and 45 passengers purchased both newspapers.

 a. How many passengers in this group did not purchase *USA Today?*

 b. How many passengers purchased neither newspaper?

 c. How many passengers purchased only *The Wall Street Journal?*

 d. How many passengers purchased *USA Today* or *The Wall Street Journal?*

NW 5. **Course Selections** Seventy-five students participated in a survey at a local college and the following data were collected: There were 27 students taking accounting, 26 taking psychology, and 41 taking statistics. Twelve students were taking accounting and psychology, 13 students were taking accounting and statistics, and 17 were taking psychology and statistics. Five students were taking all three courses.

a. How many students were taking only psychology?

b. How many students in the group were not taking any of the three subjects?

c. How many students were taking accounting and statistics, but not psychology?

d. How many students were taking just one of these subjects?

6. Investments Ms. Commission, an investment advisor, analyzed the investments of her clients. She noted that 13 of her clients had invested in stocks, bonds, and mutual funds. Of the 53 clients who invested in mutual funds, 27 also had invested in bonds, while 28 had also invested in stocks. Also, of the 55 clients who had invested in bonds, 29 had invested in stocks. Ms. Commission also noted that she had 54 clients who had invested in stocks, and finally, she had 9 clients who had invested only in other areas.

a. How many clients does Ms. Commission have altogether?

b. How many clients have invested in stocks or bonds?

c. How many clients have not invested in stocks?

d. How many of the stock investors have not invested in mutual funds?

NW **7. Electronics** In a survey of 87 residents in a certain dormitory, the following data were collected: 45 residents had computers in their rooms, 34 residents had CD players in their rooms, and 38 residents had televisions in their rooms. Twenty-one residents had both CD players and televisions. 19 residents had televisions and computers, and 22 had computers and CD players in their rooms. Ten residents had computers, televisions, and CD players in their rooms.

a. How many residents had none of these items in their rooms?

b. How many residents did not have televisions in their rooms?

c. How many residents had only computers in their rooms?

d. How many residents did not have computers in their rooms?

8. Football In analyzing the scoring for college football teams in a particular conference, the following facts were gathered: 70 players had scored touchdowns, 44 players had scored points after touchdowns (PATs), and 32 players had scored field goals, while 19 players had scored both touchdowns and PATs, 16 players had scored touchdowns and field goals, and 21 players had scored both PATs and field goals. Six players had scored in all three ways.

a. How many players scored only touchdowns?

b. How many players scored field goals and PATs but not touchdowns?

c. How many players scored touchdowns or field goals?

d. How many players did not score a PAT?

9. Newspapers At a subway stop in New York City, 125 people were asked what newspaper they read. Forty-six people read the *Times,* 43 people read the *Post,* and 65 people read the *News*. Nineteen people read the *Times* and the *Post,* 18 people read the *Post* and the *News,* and 11 people read the *Times* and the *News*. Eight people read all three papers.

a. How many people read only the *News?*

b. How many people did not read any of the papers?

c. How many people read the *Times* or the *News?*

d. How many people read the *Post* and the *News*, but not the *Times?*

10. Banking At a local bank, 100 customers were asked what bank services they used. Fifty-two people in the group had savings accounts, 52 people had checking accounts, and 57 people had bank debit cards, while 23 people had savings and checking accounts, 25 people had checking accounts and bank debit cards, and 24 people had savings accounts and bank debit cards. Eleven people used all three types of services—checking accounts, savings accounts, and debit cards.

a. How many of these people had only checking accounts?

b. How many of these people did not have savings accounts?

c. How many of these people did not have bank debit cards?

d. How many of these people had savings accounts or checking accounts?

11. Games In the college union, 130 students were surveyed as to what board games they played. The following data were collected: 57 played backgammon, 89 played checkers, and 76 played chess, while 35 played checkers and backgammon, 50 played chess and checkers, and 30 played backgammon and chess. Twenty students played all three games.

a. How many of these students played only chess?

b. How many of these students did not play chess?

c. How many students in this group played at least one of these games?

d. How many students in this group played backgammon or checkers?

12. Communication At the Grand Central Terminal, a landmark railroad station in New York City, a pollster conducted a survey of 1000 commuters and the following data were collected: 680 people had a cellular phone, 600 people had a pager, and 560 people had a laptop computer, while 360 people had a cell phone and a pager, 380 people had a pager and a laptop computer, and 370 people had a cell phone and a laptop computer. Two hundred forty people had all three pieces of equipment.

a. How many of the commuters did not have a pager?

b. How many of the commuters had a cell phone or a pager?

c. How many of the commuters had only a laptop computer?

d. How many of the commuters had a cell phone and a pager, but not a laptop computer?

13. Clams Many fish markets also sell clams. Usually these stores sell three different sizes of clams: cherrystones, littlenecks, and chowders. During a certain week, the following data were collected regarding the sale of clams: 40 customers purchased cherrystone clams, 47 customers purchased littleneck clams, and 32 customers purchased chowder clams, while 18 customers purchased cherrystones and littlenecks, 14 customers purchased littlenecks and chowders, and 9 customers purchased cherrystones and chowders. Four customers purchased all three types of clams.

a. How many customers purchased clams during the week?

b. How many customers purchased only littlenecks?

c. How many customers purchased cherrystones or littlenecks?

d. How many customers purchased only one type of clam?

14. Advertising An advertising agency conducted a survey of 50 retail outlets and the following data were collected: 24 merchants advertised on radio; 20 merchants advertised on television, and 27 merchants advertised in the newspapers. Eleven merchants advertised on radio and television, 10 merchants advertised on television and in the newspapers, and 9 merchants advertised on radio and in the newspapers. Five merchants advertised through all three media.

a. How many merchants advertised only on television?

b. How many merchants did not use any of these media?

c. How many merchants advertised on radio or television?

d. How many merchants advertised on radio and television, but not in the newspapers?

15. Automobiles A used-car dealer must complete an inventory of the cars on his lot. He notes that he has 22 compact, two-door, standard-transmission cars. Of the 50 standard-transmission cars on the lot, 28 are classified as compact, while 30 are two-door. Also, of the 47 two-door cars on the lot, 31 are classified as compact. The dealer also notes that he has 44 compact cars on his lot and 15 large, four-door, automatic-transmission cars.

a. How many cars are there on the lot altogether?

b. How many compact cars have standard transmission, but are not the two-door type?

c. How many of the two-door cars with standard transmission are not compact?

d. How many of the standard-transmission cars are not compact?

16. Automobiles An automobile dealership completed an inventory of new cars on its lot. The inventory showed that there are 30 full-size, four-door, automatic-transmission cars on the premises. Of the 60 automatic-transmission cars, 50 are classified as full-size, while 35 are four-door. Also, of the 70 four-door cars on the premises, 55 are classified as full-size. There are 100 full-size new cars on its lot. The inventory also revealed that the dealership has 25 compact, two-door, standard-transmission cars on the premises.

a. How many of the automatic-transmission cars are not full-size?

b. How many of the four-door cars with automatic transmission are not full-size?

c. How many full-size cars have automatic transmission, but are not the four-door type?

d. How many new cars are on the inventory list?

17. Farming In a recent survey of 300 farmers in the Northeast regarding crops that they grew, the following information was gathered: 150 farmers grew strawberries, 150 farmers grew cauliflower, and 170 farmers grew potatoes, while 80 farmers grew strawberries and cauliflower, 65 farmers grew strawberries and potatoes, and 105 farmers grew potatoes and cauliflower. Fifty farmers grew all three crops.

a. How many farmers grew potatoes or cauliflower?

b. How many farmers grew *at least one* of these crops?

c. How many farmers grew *at most two* of these crops?

d. How many farmers did not grow strawberries?

18. Television Networks In a recent survey of 300 people regarding television programming, the following information was gathered: 160 people watched ABC, 150 people watched CBS, and 150 people watched NBC, while 90 people watched both ABC and CBS, 70 people watched CBS and NBC, and 100 people watched ABC and NBC. Forty people watched all three networks.

a. How many people watched ABC or NBC?

b. How many people watched only one of the networks?

c. How many people did not watch any of the networks?

d. How many people did not watch NBC?

19. Survey A statistician reported to his employer that he had gathered the following information: In a survey of 40 households in a certain tract, 36 households had a digital video disk (DVD) player, 36 had two cars, and 21 owned a camper, while 22 households had both a DVD player and two cars, 19 had a DVD player and a camper, and 17 had a camper and two cars. Six households had a DVD player, two cars, and a camper. The statistician was promptly fired by his employer. Why?

20. **Survey** An independent survey agency was hired by the Metropolitan Transit Authority (MTA) to find out how people commuted to their jobs. The agency interviewed 1000 commuters and submitted the following report: 631 people came to work by car, 554 people came to work by bus, and 759 came to work by subway. Also, 373 people came to work by a combination of car and bus, 301 people came to work by bus and subway, and 268 people came to work by car and subway, while 231 people used all three means of transportation to get to work. The MTA refused to accept the report, stating that it was inaccurate. Why?

Writing Mathematics

21. Explain why, when solving a survey problem with a Venn diagram using three circles, we usually complete region V first.
22. After constructing a Venn diagram with three sets and completing region V, we typically complete regions II, IV, and VI next. Explain why we should usually do this.

Challenge Exercises

23. **Bridge Partners** A group of people, Meisha, Jillian, Rachel, Sue, Andre, Ted, Joe, and Al, are going to play bridge. The players are divided into three different sets according to their likes and dislikes. The sets are

$$A = \{\text{Ted, Sue, Andre, Meisha}\}$$
$$B = \{\text{Al, Sue, Jillian, Ted}\}$$
$$C = \{\text{Joe, Andre, Ted, Al}\}$$

By means of a Venn diagram, determine the pairings of bridge partners that will result from the given information.

24. **Tennis Partners** A group of tennis enthusiasts, Amy, Jason, Jon, Keysha, Kim, Eric, Tania, and Jose, are going to play doubles. The instructor divided the group into three sets according to ability and compatibility. The sets are

$$A = \{\text{Amy, Jason, Keysha, Kim}\}$$
$$B = \{\text{Jason, Jon, Eric, Kim}\}$$
$$C = \{\text{Keysha, Eric, Tania, Kim}\}$$

Use a Venn diagram to determine the doubles partners that result from the given information.

25. If $A+B=(A \cap B') \cup (A' \cap B)$ and $A \oplus B=(A \cap B) \cup (A' \cap B')$, then draw the Venn diagram for $(R+S) \oplus T$.

Group/Research Activity

26. This problem is usually easier to solve if a group of students discuss and solve it together. Carefully read the problem and decide what the regions represent before trying to solve it.

In a co-ed volley ball class the following information was collected: 20 players were over 6 feet tall, 16 were blonde, 19 were male, 11 were blonde and over 6 feet tall, 7 were over 6 feet tall and male, 8 were blonde and male, and 4 were over 6 feet tall, male, and blonde.

a. How many students are not over 6 feet tall?
b. How many students are over 6 feet tall and male, but not blonde?
c. How many students are female?
d. How many female students are blonde and not over 6 feet tall?
e. How many students are blonde or over 6 feet tall?
f. How many female students in the class are not blonde?

Math Connections

Blood Typing and Donation

An interesting application of Venn diagrams involves blood typing and donation. Approximately 55% of the blood's total volume consists of *plasma*, which (among others) contains dissolved blood proteins. A large group of proteins called *immunoglobulins* defend the body against invasion by disease organisms or other foreign substances. These invaders are called *antigens*, which, when introduced into the body, trigger the production of antibodies. Antibodies usually react only with the specific antigen that triggered them, and once triggered, retain their ability to react to that antigen. For example, the first time you acquire a certain virus, it takes several days for your body to produce enough antibodies to react to the virus, and hence, you are sick. Any subsequent invasion by this virus, however, does not cause you to become sick because your body quickly produces the corresponding antibodies. Thus, you become immune to this particular virus.

Our body normally has to be exposed to an antigen before it develops an antibody to it. However, our blood plasma also contains antibodies to whatever antigens are not present in our red blood cells. The absence of these antigens from our red blood cells serves as the basis for blood typing. Two such antigens are symbolized A and B. Thus, blood that contains the A antigen is referred to as type A; blood that has the B antigen is referred to as type B; blood that contains both A and B antigens is referred to as type AB; and blood that contains neither A and B antigens is referred to as type O. In addition to A and B, a third antigen is symbolized Rh. Blood that contains the Rh antigen is referred to as positive (+), and if it does not have this antigen then it is referred to as negative (−). These three

(continued)

(continued)

antigens can be represented using a three-circle Venn diagram, as shown in Figure 2.31.

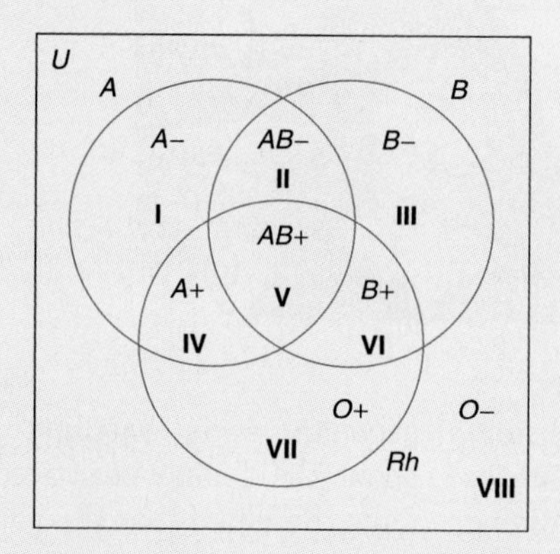

Figure 2.31

Note that irrespective of the Rh antigen, blood type A (regions I and IV) contains anti-B antigens; blood type B (regions III and VI) contains anti-A antigens; blood type O (regions VII and VIII) contains both anti-A and anti-B antigens; and blood type AB (regions II and V) contains neither anti-A nor anti-B antigens. This information is extremely important in blood transfusions. If you have type A blood and receive a transfusion from someone with type B blood, the anti-B antibodies in your type A plasma will attack the "foreign substance" and cause the red blood cells in the donated blood to clump together and disintegrate. The clumping could cause blockage of your small blood vessels, and the disintegration of the red blood cells could cause fatal kidney failure. Consequently, when a person donates blood, his or her antigens should be a subset of the receiver's antigens. Using the Venn diagram in Figure 2.31, we can construct the following donor table:

Receiver's Blood Type	Donor's Blood Type
AB+	AB+, AB−, A+, A−, B+, B−, O+, O −
AB−	AB−, A−, B−, O−
A+	A+, A−, O+, O−
A−	A−, O−
B+	B+, B−, O+, O−
B−	B2, O−
O+	O+, O−
O−	O−

Note that regardless of what blood type a person has, he or she can receive type O−, which has none of the three antigens. This is why a person with O− blood type is called a universal donor. As an exercise, have your blood typed and determine to whom you may donate blood.

Just for fun

Here is a problem about a set of people. See if you can solve it. An accountant, an attorney, an architect, and an author all belong to the same club. Their names, although not necessarily in order, are Jack, Joe, Sue, and Sharon. The following is known about these people: Jack and the attorney are not on speaking terms with Sue. Joe and the author are good friends. Sue and the architect live in the same apartment complex. The accountant is good friends with Sharon and the author.

Given this much information, determine the profession of each person.

2.7 CARTESIAN PRODUCTS

As the last set operation in this chapter, we shall consider the *Cartesian product* of two sets (named after René Descartes, a seventeenth-century French mathematician and philosopher). The Cartesian product is unique among set operations in that it produces new elements that are not members of the universal set. These new elements are called *ordered pairs.*

An **ordered pair** is a pair of objects where one element is considered first and the other element is considered second. Which element is first and which element is second is important. If we have a pair of socks, it does not matter which sock we put on first. A pair of socks is *not* an ordered pair. But if we

Biography: René Descartes

René Descartes (1596–1650) was a French mathematician–philosopher who revolutionized mathematical concepts and initiated modern mathematics with his creation of coordinate (plane analytic) geometry. Descartes also developed views of philosophy that were revolutionary in their break with the past and caused him to be known as the father of modern philosophy.

As a young boy he was sent to a Jesuit school. It was there that he developed (because of delicate health) his lifelong habit of lying in bed as late as he pleased. As an adult he slept 10 hours each night and never let anyone disturb him before noon.

Descartes lived in various countries in Europe before settling in Holland, where he pursued his mathematical studies and philosophical contemplations. In 1637 he published a treatise outlining his views on philosophy. It was in the appendix of this book that he introduced analytic geometry and the Cartesian plane, for which he is famous today.

According to legend, Descartes was led to the contemplation of analytic geometry while watching a fly crawl about the ceiling near a corner of his room. He noted that the path of the fly could be described, if he knew a relation connecting the fly's distance from the two adjacent walls. The technical terms *coordinate, abscissa,* and *ordinate* used in analytic geometry today were contributed by Leibniz in 1692.

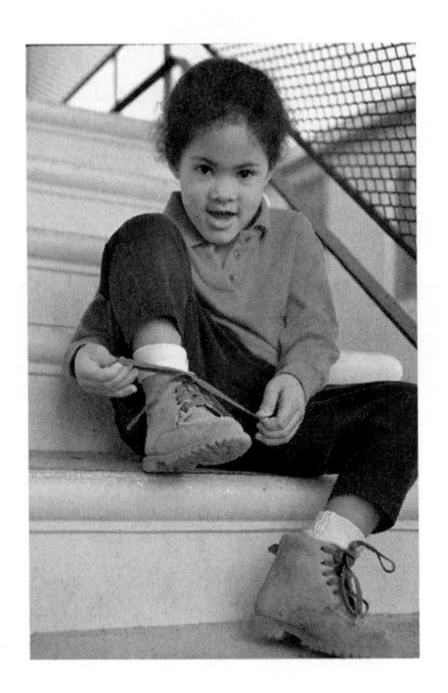

have a sock and a shoe, it does matter which we put on first: Try putting a sock on over a shoe!

In mathematics, if we wish to discuss an ordered pair of elements consisting of a and b, we use the notation (a, b). This notation tells us that we want to consider a first and b second. Suppose a stands for "walk a blocks north" and b stands for "walk b blocks west." Now consider the ordered pair (3, 2). This would mean "walk 3 blocks north and then walk 2 blocks west." This is certainly different from (2, 3), which would mean "walk 2 blocks north and then walk 3 blocks west." We would not arrive at the same place with these two ordered pairs.

Given two sets A and B, the Cartesian product of A and B, denoted by A × B (read "A cross B"), is the set of all possible ordered pairs such that the first element of the ordered pair is an element of A and the second element of the ordered pair is an element of B.

Example 1 ***Cartesian Product*** Given $A = \{\text{Joe, Scott}\}$ and $B = \{p, q\}$, find $A \times B$, the Cartesian product of A and B.

Solution The Cartesian product $A \times B$ is the set of all possible ordered pairs such that the first element is an element of A and the second element is an element of B. We must pair the elements in A, Joe and Scott, with the elements in B, p and q. Joe may be paired with p or q; hence, the possible ordered pairs with Joe are (Joe, p) and (Joe, q). Now we do the same thing for Scott; the possible ordered pairs are (Scott, p) and (Scott, q). Therefore, we have

$$A \times B = \{(\text{Joe}, p), (\text{Joe}, q), (\text{Scott}, p), (\text{Scott}, q)\}$$

Note that $A \times B$ gives us a set where the elements are ordered pairs. These are not members of the universal set, since the universal set consists of single elements such as p, q, Joe, and Scott.

Example 2 ***Cartesian Product*** Given $A = \{4, 8\}$ and $B = \{a, b, c\}$, find $A \times B$.

Solution The first element in our ordered pairs must come from A. Therefore, we pair 4 with a, then 4 with b, etc. Then we do the same for 8. Hence, we have

$$A \times B = \{(4, a), (4, b), (4, c), (8, a), (8, b), (8, c)\}$$

Example 3 ***Cartesian Product*** Given $A = \{4, 8\}$ and $B = \{a, b, c\}$, find $B \times A$.

Solution This is similar to the previous example, but here we want to find $B \times A$. Note that $B \times A$ is not the same as $A \times B$. In $B \times A$, we want the set of all possible ordered pairs such that the first element is an element of B and the second element is an element of A. Here we take the elements in B—a, b, and c—and pair them with the elements in A—4 and 8. Therefore,

$$B \times A = \{(a, 4), (a, 8), (b, 4), (b, 8), (c, 4), (c, 8)\}$$

NW ***Now Work Problem 3.***

By comparing the results of Examples 2 and 3, we see that $A \times B \neq B \times A$. The two sets contain different ordered pairs, as $(4, a)$ is not the same as $(a, 4)$.

Note that in Example 1 the number of elements $A \times B$ was 4 [that is, $n(A \times B) = 4$] and in Example 2, $n(A \times B) = 6$. If we want to determine the number of elements in a Cartesian product $A \times B$ (the cardinality of the set), we take the number of elements in A and multiply it by the number of elements in B. If set A has m elements in it and set B has n elements in it, then the number of elements in $A \times B$ is $m \times n$. In other words,

$$n(A \times B) = n(A) \times n(B)$$

This provides us with a handy check in computing $A \times B$ because we can use the cardinality to check whether we have all of the possible ordered pairs.

Example 4 ***Cartesian Product*** Find $A \times A$ if $A = \{1, 2, 3\}$.

Solution To find $A \times A$, we pair each element in A with every element in A. Therefore, we have

$$A \times A = \{(1, 1), (1, 2), (1, 3), (2, 1), (2, 2), (2, 3), (3, 1), (3, 2), (3, 3)\}$$

Since there are three elements in A, we should have $3 \times 3 = 9$ elements in $A \times A$. Checking our answer, we see that we do have nine different ordered pairs in $A \times A$.

NW ***Now Work Problem 5.***

The Cartesian product of two sets, $A \times B$, gives us a set of ordered pairs. This set of ordered pairs may be pictured by means of an **array** or **lattice**. Consider the following example.

Example 5 ***Cartesian Product*** Find $A \times B$ if $A = \{a, b, c\}$ and $B = \{d, e\}$.

Solution There are three elements in A and two elements in B; hence there are $3 \times 2 = 6$ ordered pairs in the Cartesian product $A \times B$. Pairing each element of

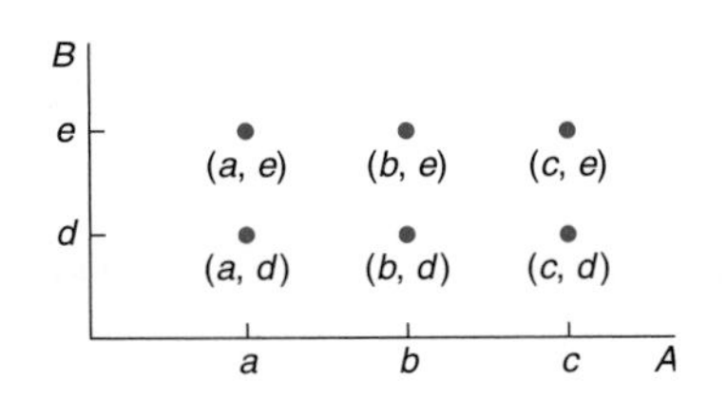

Figure 2.34

A with every element in B, we have

$$A \times B = \{(a, d), (a, e), (b, d), (b, e), (c, d), (c, e)\}$$

Figure 2.34 shows the array of ordered pairs for this example. In the lattice, a dot represents an ordered pair.

Note that the vertical axis represents set B with the elements d and e. The horizontal axis represents set A with elements a, b, and c. It is traditional to use the horizontal axis to represent the first element in an ordered pair.

One other way of picturing the formation of a Cartesian product is a *tree diagram*. A **tree diagram** consists of a number of branches that illustrate the possible pairings in $A \times B$, as in Figure 2.35.

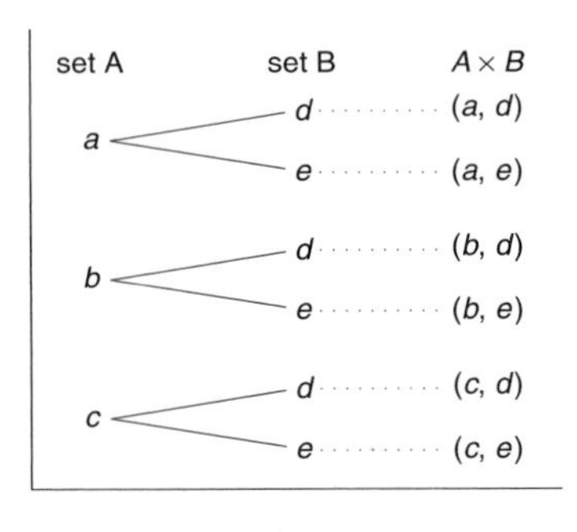

Figure 2.35

Exercises for Section 2.7

1. If $A = \{5, 10\}$ and $B = \{x, y, z\}$, find $A \times B$, $B \times A$, and $n(A \times B)$.

2. If $A = \{\$, \#, ¢\}$ and $B = \{d, e, f\}$, find $A \times B$, $B \times A$, and $n(A \times B)$.

NW 3. If $C = \{4, 6, 8\}$ and $D = \{3, 5, 7\}$, find $C \times D$, $D \times C$, and $n(C \times D)$.

4. If $V = \{a, e, i, o, u\}$ and $Z = \{x, y, z\}$, find $V \times Z$, $Z \times V$, and $n(V \times Z)$.

NW 5. Let $T = \{t, f\}$. Find $T \times T$ and $n(T \times T)$.

6. Let $A = \{d, e, f\}$ and $B = \{2, 4, 6\}$. Make a lattice showing $A \times B$.

7. Let $A = \{4, 5, 6\}$ and $B = \{x, y\}$. Make a lattice showing $A \times B$.

8. Let $A = \{a, b, c, d\}$ and $B = \{x, y, z\}$. Make a tree diagram showing $B \times A$.

9. If $A = \{x, y\}$ and $B = \{a, e, i, o, u\}$, make a tree diagram showing $A \times B$.

10. If $U = \{1, 2, 3, 4\}$, find $U \times U$ and $n(U \times U)$.

11. Find each of the following, given $A = \{a, b\}$, $B = \{b, c, d\}$, and $C = \{c, d, e\}$.

    a. $A \times B$ **b.** $A \times C$
    c. $n(B \times C)$ **d.** $(A \cap B) \times C$
    e. $C \times (A \cup B)$ **f.** $A \times (B \cap C)$

12. If $A = \{1, 2, 3\}$, $B = \{3, 4, 5, 6\}$, and $C = \{4, 5, 6, 7\}$, find each of the following:

    a. $A \times B$ **b.** $n(B \times C)$
    c. $(A \cap C) \times B$ **d.** $(B \cup C) \times A$
    e. $(B \cap C) \times A$ **f.** $(A \cap B) \times (B \cap C)$

13. Given that $A = \{1, 2\}$ and $B = \{3, 4\}$, does $A \times B = B \times A$? Why or why not?

Writing Mathematics

14. Using set-builder notation, we define a Cartesian product more formally as

$$A \times B = \{(a, b) | a \in A \text{ and } b \in B\}$$

Describe, in your own words, the concept of a Cartesian product.

15. Describe two different ways to obtain the Cartesian product, $A \times B$.

Challenge Exercises

16. Define the Cartesian product of three sets A, B, and C using set-builder notation.

17. Given sets A, B, and C, is $A \times B \times C$ different from $A \times (B \times C)$? Explain.

18. Find $A \times \emptyset$.

19. Given the two sets A and B,

a. How many distinct one-to-one correspondences are there if $n(A) = n(B) = 1$?

b. How many distinct one-to-one correspondences are there if $n(A) = n(B) = 2$?

c. How many distinct one-to-one correspondences are there if $n(A) = n(B) = 3$?

d. How many distinct one-to-one correspondences are there if $n(A) = n(B) = 4$?

e. How many distinct one-to-one correspondences are there if $n(A) = n(B) = 5$?

f. Based on your responses to parts **a** through **e**, write a mathematical statement that describes the relationship between the number of distinct one-to-one correspondences and two equivalent sets that have n elements.

Group Activity/Research

20. Among his many contributions to the field of mathematics, René Descartes is also well known for his philosophical statement, "I think, therefore I am." Conduct a Web search to research Descartes's contributions to the field of mathematics and philosophy. One place to begin your research is at **http://www.epistemelinks.com/Main/MainPers.asp**.

Just for fun

Given

$142857 \times 2 = 285714$, $142857 \times 3 = 428571$,

$142857 \times 4 = 571428$, $142857 \times 5 = 714285$,

$142857 \times 6 = 857142$, then $142857 \times 7 = ?$

CHAPTER REVIEW MATERIALS

Summary/Chapter 2

2.1 Sets

- A *set* may be thought of as a collection of objects. These objects are called *elements* or *members*, of the set.

2.2 Notation and Description

- *Equal sets* contain exactly the same elements.
- *Equivalent sets* have the same number of elements, that is, the same *cardinality*.
- The *empty set* or *null set* contains no elements; it is denoted by { } or Ø.

2.3 Subsets

- Given any two sets, A and B, if every element in A is also an element in B, then A is a *subset* of B, denoted by $A \subseteq B$.
- If A is a subset of B and there is at least one element in B that is not contained in A, then A is a *proper subset* of B, denoted by $A \subset B$.
- If a set contains n elements, then it has 2^n *subsets.*
- The *universal set*, denoted by U, contains all of the elements being considered in the given discussion or problem.
- The *complement* of set A, denoted by A', is the set of all the elements in the given universal set, U, that are not in the set A.

2.4 Set Operations

- If we have two sets A and B, then the *intersection* of A and B, denoted by $A \cap B$, is a set of elements that are members of both A and B.
- If we have two sets A and B, then the *union* of A and B, denoted by $A \cup B$, is the set of elements that are members of A, or members of B, or members of both A and B.

2.5 Pictures of Sets (Venn Diagrams)

- *Venn diagrams* are useful representations of sets and the relationships between sets. Typically, a rectangle is used to represent the universal set, and circles inside the rectangle represent the sets being considered.
- A pairing of the elements of two equivalent sets is called a *one-to-one correspondence* between the sets.

2.6 An Application of Sets and Venn Diagrams

- Venn diagrams are particularly useful in solving problems involving overlapping sets of data (survey problems).
- *Venn diagrams* may also be used to answer questions when we are given certain data and wish to determine more information from the data (survey problems).

2.7 Cartesian Products

- Given two sets, A and B, the *Cartesian product* of A and B, denoted by $A \times B$, is the set of all possible *ordered pairs* such that the first element of the ordered pair is an element of A and the second element of the ordered pair is an element of B.

Key Terms and Concepts

Review Exercises

2.2 Notation and Description

1. Describe in your own words what a *set* is.
2. Write each set in two ways.
   a. The set of states whose names begin with the letter *T*.
   b. The set of Great Lakes.
   c. All even whole numbers.
   d. All positive whole numbers.
   e. All counting numbers less than 8.
3. State whether each sentence is true or false.
   a. The set of pretty flowers is a well-defined set.
   b. $\{1, 2, 3, \ldots\}$ is a well-defined set.
   c. $\{2, 3, 5, 7, 11, 13, \ldots\}$ is not a well-defined set.
   d. The set of books in a bookstore is a finite set.
   e. $\{a, b, c, \ldots, z\}$ is an infinite set.
   f. $\emptyset$ is a finite set.
4. State whether each sentence is true or false.
   a. $\{t, e, a, m\} = \{m, e, a, t\}$
   b. $\emptyset = \{\ \}$
   c. $\{l, o, v, e\}$ is equivalent to $\{h, a, t, e\}$.
   d. $\{1, 2, 3, \ldots\}$ and $\{5, 10, 15, 20, \ldots\}$ are equivalent.
   e. If $A = \{1, 3, 5, \ldots\}$ and $B = \{2, 4, 6, \ldots\}$, then sets A and B are disjoint sets.
   f. $\{m, a, t, h, i, s, f, u, n\} \subset \{a, b, c, \ldots, z\}$

2.3 Subsets

5. State whether each of the following is true or false.
   a. $\{1, 7\} \subset \{1, 2, 3, \ldots\}$
   b. $\{a, b\} \subseteq \{a, b, c\}$
   c. $\{1, 3, 5, \ldots\} \subset \{1, 2, 3, \ldots\}$
   d. $\emptyset \subseteq \{a, b\}$
   e. Disjoint sets have at least one element in common.
   f. The set $\{m, a, t, h\}$ has 32 possible subsets.
6. List all the possible subsets of $\{d, e, f\}$.

2.4 Set Operations

7. Let $U = \{0, 1, 2, 3, 4, 5\}$, $A = \{0, 1, 3, 5\}$, $B = \{1, 3, 4, 5\}$, and $C = \{0, 2, 4, 5\}$. Find each of the following.
   a. $A \cap B$
   b. $B \cup C$
   c. $A' \cap B'$
   d. $B' \cup C'$
   e. $(A \cap B) \cup C$
   f. $A \cup (B \cap C)$
   g. $(A' \cap B')' \cup C$
   h. $(A \cap B)' \cup C'$

2.5 Pictures of Sets (Venn Diagrams)

8. Use a Venn diagram to illustrate each of the following sets. List the regions that make up your answer.
   a. $A \cup B$
   b. $(A \cap B)'$
   c. $A' \cap B'$
   d. $A \cap (B \cup C)$
   e. $A \cup (B \cap C)$
   f. $(A \cap C) \cup B'$
9. Show a one-to-one correspondence between $A = \{1, 2, 3\}$ and $B = \{5, 10, 15\}$. In how many ways can you do this?
10. Use Figure 2.36 to find each cardinality.

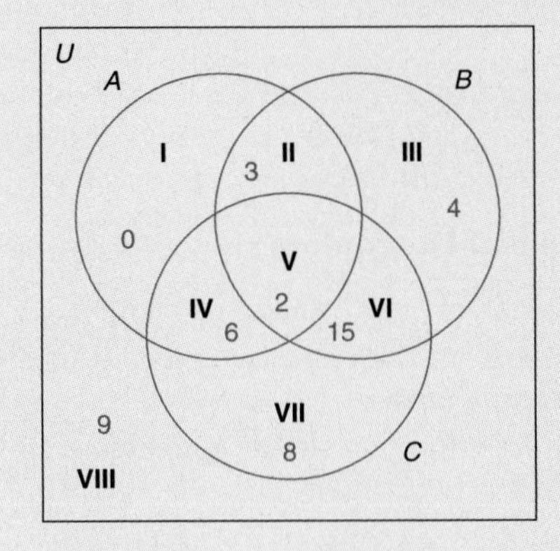

Figure 2.36

   a. $n(A)$
   b. $n(B)$
   c. $n(C')$
   d. $n(A \cap B)$
   e. $n(A \cup C)$
   f. $n(B \cup C)$
   g. $n(U)$
   h. $n(A' \cap B)$

2.6 An Application of Sets and Venn Diagrams

11. **Class Schedules** In a survey of 77 students at a local college, the following information was gathered: 45 students had a class at eight o'clock in the morning, 43 students had a class at nine o'clock in the morning, 47 students had a class at 10 o'clock in the morning. Twenty-six students had an 8:00 A.M. and a 9:00 A.M. class, 24 students had an 8:00 A.M. and a 10:00 A.M. class, and 25 students had a 9:00 A.M. and a 10:00 A.M. class. Eleven students had classes at 8:00, 9:00, and 10:00 A.M.

a. How many of the students did not have a class at any of the three times?

b. How many students had a class at 9:00 A.M. and 10:00 A.M., but not at 8:00 A.M.?

c. How many students did not have a class at 8:00 A.M.?

d. How many students had a class at 8:00 A.M. or 10:00 A.M.?

2.7 Cartesian Products

12. Let $A = \{m, a, t, h\}$ and $B = \{e, a, s, y\}$.

a. Show a one-to-one correspondence between A and B.

b. In how many ways can you do this?

c. List all possible subsets of A.

d. Find $A \times B$.

e. Find $n(A \times B)$.

Various Sections

13. When are the following statements true?

a. $A \cap B = A$ **b.** $A \cup B = A \cap B$

c. $(A \cap B)' = A' \cup B'$ **d.** $B \subseteq \emptyset$

e. $(A \cup B)' = A' \cap B'$

14. Explain the difference between a subset and a proper subset.

15. In this chapter we encountered the notations 2^n, $n!$, and $n(A)$. Explain how each is used in the study of sets.

Chapter Quiz

State whether each of the following is true or false.

1. $\{m, o, n, e, y\} = \{r, i, c, h\}$

2. $\{d, o, g\}$ is equivalent to $\{c, a, t\}$

3. $\{2, 4, 6\} \subseteq \{4, 6, 2\}$

4. $\{0\} = \emptyset$

5. $35 \in \{2, 4, 6, \dots\}$

6. $44 \notin \{x | x \text{ is a natural number}\}$

7. $44 \in \{x | x \text{ is a whole number}\}$

8. $A \times B = B \times A$

9. If $A = \{a, e, i, o\}$, then A has at most eight subsets.

10. If A and B are disjoint sets, then A and B have only one element in common.

11. If A and B are any two disjoint sets, then it is impossible to show a one-to-one correspondence between them.

Use Figure 2.37 to answer Questions 12–19.

12. $n(A) =$

13. $n(B \cup C) =$

14. $n(A' \cap B') =$

15. $n(B \cap C)' =$

16. $n(A \cup C) =$

17. $n(C') =$

18. $n(A \cap B)' =$

19. $n(A' \cup C') =$

Use the following information to answer Questions 20–25:

$$U = \{u, l, i, k, e, m, a, t, h\},$$
$$A = \{l, i, k, m, t\}$$
$$B = \{u, i, e, a, h\},$$
$$C = \{l, i, m, a\}$$

Find

20. $B \cap C$

21. $A \cup B \cup C$

22. $A' \cap B'$

23. $(A \cap B)'$

24. $C' \cup (A \cap B)$

25. $A' \cup B'$

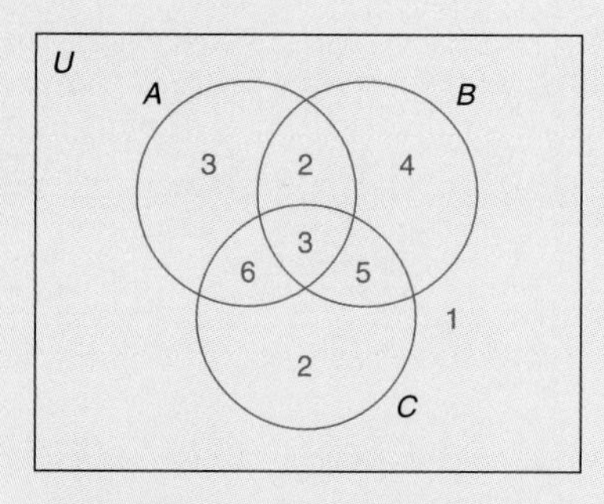

Figure 2.37

Chapter 3

Chapter Outline

Logic

After Studying This Chapter, You Will Be Able to Do the Following:

1. Distinguish between simple and compound statements, and determine if a compound statement is a negation, conjunction, disjunction, conditional, or biconditional.
2. Write English sentences in symbolic form and express symbolized statements as English sentences.
3. Construct truth tables.
4. Determine if two statements are logically equivalent, consistent, or inconsistent.
5. Identify the forms of a conditional statement.
6. Determine if an argument is valid or invalid.
7. Represent statements using Venn diagrams.

*8. Express a switching network in symbolic form and construct a switching network given a corresponding symbolic statement.

Note: *indicates optional material.

Notations Frequently Used in This Chapter

$P \wedge Q$	P and Q
$P \vee Q$	P or Q
$P \underline{\vee} Q$	P or Q but not both
$P \rightarrow Q$	if P, then Q
$P \leftrightarrow Q$	P if and only if Q
$\sim P$	not P; it is false that P; it is not the case that P
iff	if and only if
$\equiv$	is the same as

Visit the PH Companion Web site at www.prenhall.com/setek

OVERVIEW

Logic, which was first studied by Aristotle (384 B.C.–322 B.C.), is the science of thinking and reasoning correctly. Aristotle's collection of works, known as the "Organon," or instrument, essentially created the discipline of logic. Since the late nineteenth century, logic has taken on an increasingly well-defined, technical character. Key contributors include Leonhard Euler (1707–1783), George Boole (1815–1864), John Venn (1834–1923), and Bertrand Russell (1872–1970). Today, logic is a branch of mathematics, known as symbolic logic, and is primarily concerned with the structure of reasoning. Symbolic logic formally defines a symbol-based language that enables us to translate natural language sentences into symbolic form. This then allows us to investigate complex relationships among the elements of the sentences, including concepts such as validity, consistency, and contradiction. It is this kind of logic that we will discuss in this chapter.

3.1 INTRODUCTION

As indicated in the chapter overview, logic is the science of thinking and reasoning correctly. Many times false assumptions are made about things or people because the meaning of certain statements or actions are misconstrued, sometimes deliberately. For example, consider the following statements and think about how you would interpret them:

It's good to see you here on time!

I am glad to see you sober today!

Four dentists out of five said they used "Smiley" toothpaste.

One thousand heroin addicts admitted that they used marijuana.

Note that each sentence has hidden implications and as such may be misinterpreted. Thus, it is possible that the reader or listener understands something other than what the writer or speaker has written or said. We should always phrase statements so they express our meaning exactly. In short, we should strive to follow the guidelines of an old proverb that states, "Say what you mean and mean what you say." An understanding of logic and its uses will help us in this endeavor and increase our skills in analytical thinking.

This chapter explains the fundamental concepts of logic. In Section 3.2, we first discuss the basic building blocks of logic, namely, simple statements, and then

Biography: George Boole

It was not until the early twentieth century that logic came to be considered part of the study of mathematics.

George Boole (1815–1864), an English mathematician, was primarily responsible for this acceptance. His book *Laws of Thought* developed logic as an abstract mathematical system. It consisted of undefined terms, binary operations, and rules for using these operations on the terms. Today, we know the terms as propositions; the operations are the conjunction, disjunction, and negation.

Boole was able to use logic as a special kind of algebra. The system follows many of the rules of ordinary arithmetic. Today Boolean algebra is used extensively in the design of computers.

The advantage of Boole's symbolic notation is that many errors in reasoning are greatly reduced. The ambiguities of language are avoided by the use of symbols because once a problem is translated into symbolic notation, the solution becomes mostly mechanical.

show how these statements can be combined using connecting words to form compound statements. We also introduce notation that is used to symbolize compound statements. In Section 3.3, we expand on this discussion by examining compound statements that contain multiple connecting words. In Sections 3.4 and 3.5, we show how to determine the truth value of statements by analyzing them using truth tables, and then in Section 3.6, we learn how to form and test for logically equivalent statements. Section 3.7 examines "if ... then" statements (called conditional statements) in greater detail than previous sections. The next three sections focus on the concept of a logical argument. We introduce this concept in Section 3.8 and discuss how to determine if an argument is valid using truth tables. We then extend this concept to a special argument called a syllogism. We first show how to represent a syllogism graphically (Section 3.9), and then we use this graphical representation to determine the validity of a syllogism (Section 3.10). Finally, in Section 3.11, we apply the basic building blocks of logic to switching circuits.

3.2 STATEMENTS AND SYMBOLS

Simple and Compound Statements

In preparing for any task, we must first equip ourselves with the proper tools and the "know-how" required to do the job. The first thing that we must be able to do is identify and symbolize sentences. In logic we concern ourselves only with those sentences that are either true or false, but not both. A **statement** is a declarative sentence that is either true or false (but not both true and false).* We shall not concern ourselves with sentences that cannot be assigned a true or false value. (Sentences of this nature are usually questions or commands.)

Note that it is not possible to assign a true or false value to the following:

Did you do the assignment?
Hand in your paper.
Is it raining?
Close the door when you leave.
Stop the car!

The following statements are either true or false:

February has 30 days.
$4 + 2 = 3 \times 2$
Bill Clinton was President of the United States.
Ottawa is the capital of Canada.
Tomorrow is Saturday.

There are other types of sentences that cannot be assigned a true or false value. The sentence "I am lying to you" is one example. Suppose it is true that I am lying to you; then—if I am lying—the sentence is false. On the other hand, assume that the sentence is false. If that is the case, then I am not lying, so the sentence is true. This is known as a **paradox**.

An example of a visual paradox is shown in Figure 3.1. It cannot exist in three dimensions.

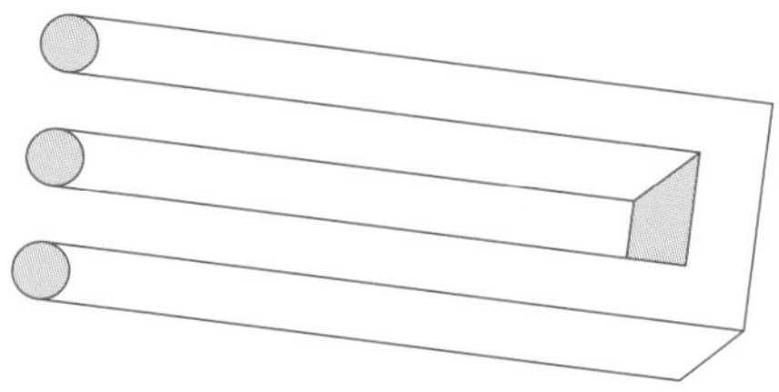

Figure 3.1
A visual paradox.

Another example of a paradox is "All rules have exceptions." This rule negates itself. It says that the rule itself must have exceptions and therefore cannot

Many texts distinguish between a proposition and a statement. A **proposition is defined as a statement that is either true or false, but not both. We shall use both terms interchangeably, as this is the only type of statement we shall consider.*

be true. Many people like the paradox about the little boy who is concerned about God. He has been told that God can do anything. The boy then asks, "If that is the case, then can God make a stone so big that he can't move it?"

Remember that in logic we concern ourselves with statements—that is, sentences that are true or false, but not both. The basic type of statement in logic is called a *simple statement*. A **simple statement** is a complete sentence that conveys

More Paradoxes

As indicated in the text, a paradox is a statement or proposition that is self-contradictory. Paradoxes can be found throughout literature and in various contexts. Following are some additional examples of paradoxes:

1. A fifth century (B.C.) philosopher, Zeno of Elea (490–425 B.C.), purportedly wrote a book that contained 40 paradoxes concerning the continuum. We say purportedly because there are no surviving works of Zeno. Much of what we know about Zeno's paradoxes comes from Plato, who, as quoted in *A History of Greek Mathematics* by T. L. Heath (Oxford, 1931), wrote that Zeno's book of 40 paradoxes was "a youthful effort, and it was stolen by someone, so that the author had no opportunity of considering whether to publish it or not. Its object was to defend the system of Parmenides by attacking the common conceptions of things." One of Zeno's paradoxes is called The Dichotomy, which effectively challenges the concept of motion:

   There is no motion because that which is moved must arrive at the middle of its course before it arrives at the end and so on ad infinitum.

   Zeno concluded that motion is impossible because it can never begin. He reasoned that before we can traverse a distance that is one unit in length, we must first get to the middle. However, before we can get to the middle, we must first get one-fourth of way, but before we get one-fourth of the way, we must get one-eighth of the way, and so on. Note that this reasoning is the reverse of the sum of the infinite series $1/2 + 1/4 + 1/8 + \cdots$. Although some people resolve this notion by showing that an infinite series can have a finite sum (see Section 1.5), the real point that Zeno is making is that if we assume these premises to be true, then we can logically reason that we can never get started because we are trying to build up this infinite sum from the opposite (i.e., "wrong" end).

2. Another paradox is the *paradox of the barber*. Consider the following:

   In a certain village there is a man who is a barber; the barber shaves all and only those men in the village who do not shave themselves.

   The question is, does the barber shave himself? Note that any man in the village is shaved by the barber if and only if he does not shave himself. Can you see the problem? The barber shaves himself if and only if he does not. Hence, we have another example of a paradox.

3. Economists sometimes talk about the *paradox of thrift*, which was introduced by John Maynard Keynes, and is familiar to most students of economics:

   The more consumers save, the less they spend. The less they spend, the worse the economy gets and the more they save.

   Assuming this statement is true, saving money is bad, yet we were always taught that saving money was good. Once again, can you see the paradox? On the flip side of this is the paradox of profits. What do you think this says?

4. Here's an example of a *word paradox*. Consider the word *cleave*, which means both to "join together" and "to break apart." Can you think of another word paradox?

5. Our last example is a *geometrical paradox*, which was one of Lewis Carroll's favorite puzzles (see Weaver, W., 1938, "Lewis Carroll and a Geometrical Paradox," *American Mathematical Monthly, 45*, 234–236). Take a standard chessboard, which is an 8×8 square, and cut it into four pieces as shown below. Now reassemble the pieces into a rectangle. Note that the original 64 squares have been rearranged into a 5×13 rectangle, which has 65 squares. Can you explain how this happened?

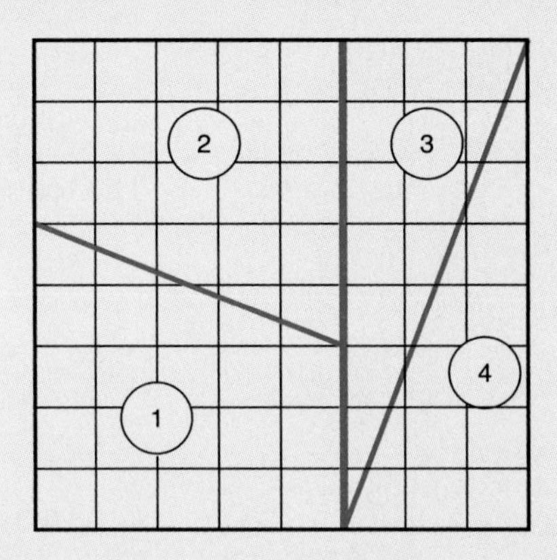

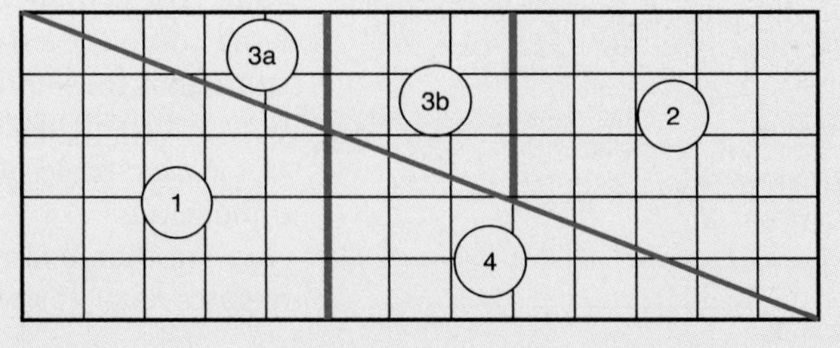

Paradoxes can be fun . You are encouraged to explore additional paradoxes.

one thought with no connecting words. The following are examples of simple statements.

> Five is a counting number.
> The Tampa Bay Buccaneers have won the Super Bowl.
> Sally was late for class.
> Today is Monday.
> George W. Bush won the 2000 presidential election.

Now, if we take simple statements and put them together using connecting words, we form sentences that are known as **compound** or **complex** statements. The basic connectives are *and, or, if ... then ..., if and only if,* and the negation *not.*

The word *not* does not connect two simple statements, but it is still thought of as a connective. It negates a simple statement. Some logicians do not like to call a negated simple statement a compound statement, but if it is no longer simple, then it must be compound. Therefore we shall think of a compound statement as a sentence that is formed by connecting one or more simple statements with a connective.

A simple statement such as "Today is Monday" is no longer simple if we say "Today is not Monday" or "It is false that today is Monday." The original simple statement has been negated, so we call the newly formed compound statement a **negation**.

When we connect two simple statements using the word *and,* we have a compound statement that is called a **conjunction**. The sentence "Today is Monday *and* tomorrow is Wednesday" is a conjunction. Remember that we are not concerned about the meaning of the sentence, only what type of statement it is. Consider

> ***All looms are booms and all booms are zooms.***

This statement is a conjunction, even though we cannot make too much sense out of it.

Sometimes the word *but* will be used in place of *and* in a sentence:

> ***Bonnie was early and Clyde was late.***

could be written as

> ***Bonnie was early, but Clyde was late.***

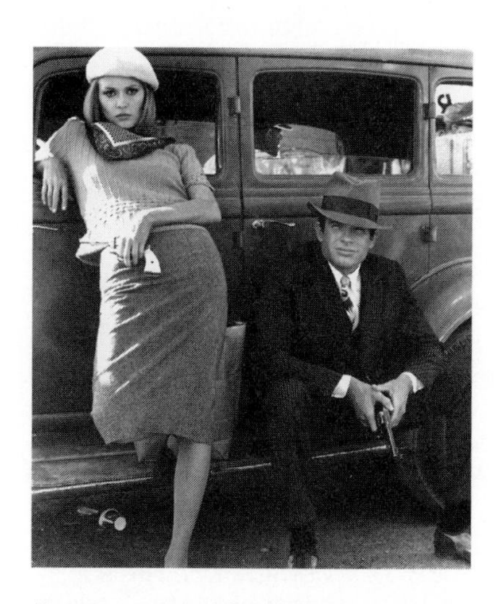

Faye Dunaway and Warren Beatty in the Bonnie and Clyde movie.

The connective *or* is used in forming a compound statement called a **disjunction**. The following are some examples of disjunctions:

> Either he took my coat or someone stole it.
> I will pass history or I will be sad.
> Today will be sunny or the weather forecast is wrong.

The connective *if ... then ...* is used in compound statements referred to as **conditionals**. An example of a conditional is

> ***If you do your homework, then you will pass the exam.***

The statement between the *if* and *then* ("you do your homework") is called the **antecedent** of the conditional. The part of the sentence that follows *then* ("you will pass the exam") is called the **consequent**. As with other connectives, there are variations in writing conditional statements. Two of the more common variations are illustrated by the following examples:

> If someone was late, it was Benny.
> We will win the game if Jackson doesn't play for them.

In the first sentence, *then* was omitted, but it is understood to be there. In the second sentence, we switched the two parts around and also omitted *then*. Nonetheless, both of these statements are conditional.

You may remember the following statement from the study of geometry:

If two sides of a triangle are equal, then two angles of the triangle are equal, and if two angles of a triangle are equal, then two sides of the triangle are equal.

It is the conjunction of two conditional statements where the antecedent and consequent of the first statement have been switched in the second. This type of sentence is usually stated as

Two sides of a triangle are equal if and only if two angles of the triangle are equal.

This type of statement is called a **biconditional.** It has the advantage of shortening the original statement. An abbreviation for *if and only if*, is ***iff***; we shall sometimes use this abbreviation.

Remember that a biconditional statement is the conjunction of two conditional statements where the antecedent and consequent of the first statement have been switched in the second.

You should be able to identify a simple statement or a compound statement. If the statement is compound, then it must be one of the following: negation, conjunction, disjunction, conditional, or biconditional.

A summary of the different types of statements is given in Table 3.1.

Beyonce Knowles and Destiny's Child

TABLE 3.1

Statement	*Defining Characteristic(s)*	*Example*
Simple	• Can be assigned a truth value (i.e., it is either true or false, but not both) • Has no connecting words.	• Beyonce Knowles is a member of Destiny's Child.
Negation	• A simple statement that has been negated. • Uses the word *not* or an equivalent expression such as *it is false that.*	• Beyonce Knowles is *not* a member of Destiny's Child. • *It is false that* Beyonce Knowles is a member of Destiny's Child.
Conjunction	• Combines two (or more) simple statements using the word *and.*	• Beyonce Knowles is a member of Destiny's Child *and* she is a solo artist.
Disjunction	• Combines two (or more) simple statements using the word *or.*	• Beyonce Knowles is a member of Destiny's Child *or* she is a solo artist.
Conditional	• Combines two simple statements into an *if then* form. • The statement following the word *if* is the *antecedent*; the statement following the word *then* is the *consequent.*	• *If* Beyonce Knowles is a member of Destiny's Child, *then* she is a solo artist. • *If* <antecedent> *then* <consequent>.
Biconditional	• Combines two conditional statements into a conjunction where the antecedent and consequent of the first statement have been switched in the second statement. • Uses the abbreviation *iff,* which means *if and only if.*	• *If* Beyonce Knowles is a member of Destiny's Child, *then* she is a solo artist *and if* she is a solo artist, *then* Beyonce Knowles is a member of Destiny's Child. • Beyonce Knowles is a member of Destiny's Child *if and only if* she is a solo artist.

Symbolizing Statements

In mathematics and the English language, many statements are lengthy and cumbersome. We would all tire quickly if we had to copy this page word for word. In logic, this problem is overcome by using symbols to represent simple statements.

It is traditional in algebra to use x, y, and z as symbols for variables and a, b, and c as symbols for constants. In logic, we normally use the letters P, Q, R, and S, and sometimes A and B, to represent statements. Other letters may also be used if needed.

Consider the statement "Today is Friday." We shall let the letter P represent this simple statement. We shall also let Q stand for the statement "I have a test." Hence, we have

$$P = \textit{Today is Friday}$$

$$Q = \textit{I have a test}$$

Symbolizing Conjunctions

If we combine the two statements to form a conjunction, "Today is Friday and I have a test," we would symbolize this as

$$P \text{ and } Q$$

It certainly seems odd to have the simple statements symbolized, but not the connective. However, we will find that there are symbols for each of the connectives.

Ampersand, &, is the typewriter symbol for *and*. In logic, it is more common to use $\wedge$ to represent *and*. Some of you may be familiar with this symbol as a *caret*. Using this symbol, the statement "Today is Friday and I have a test" can be completely symbolized as

$$P \wedge Q \quad \text{(conjunction)}$$

Symbolizing Negations

If we have a statement *not* P, and P stands for "Today is Friday," it is awkward to say

Not today is Friday.

We would be more comfortable if we said

It is not the case that today is Friday.

We would probably be even more comfortable if we said

Today is not Friday.

When we use the word *not* in a sentence, we are negating the original statement. The logical symbol most commonly used to show negation is a *tilde*, ~, a diacritical mark used in some languages. If we let P stand for "Today is Friday," then the statement "Today is not Friday" would be symbolized as

$$\sim P \quad \text{(negation)}$$

Remember that $\sim P$ may also be interpreted as

Not P
It is false that P
It is not the case that P

EXAMPLE 1 ***Symbolizing Negations and Conjunctions*** Let P = *Today is Monday* and Q = *I am tired*. Write each of the following statements in symbolic form:

a. Today is not Monday.
b. Today is Monday and I am tired.
c. Today is Monday and I am not tired.
d. Today is not Monday and I am tired.
e. Today is not Monday and I am not tired.

Solution

a. This statement is the negation of P, and hence it would be symbolized as $\sim P$.
b. The statement is a conjunction of P and Q and therefore would be symbolized as $P \wedge Q$.
c. This is also a conjunction, but here we have Q negated. The proper symbolization is $P \wedge \sim Q$.
d. This is similar to statement **c**, but now P is negated, and we have $\sim P \wedge Q$.
e. This time each part of the statement is negated. We would symbolize this as $\sim P \wedge \sim Q$.

We shall see later that statements such as $\sim P \wedge Q$ and $\sim(P \wedge Q)$ are not the same and must be interpreted differently.

Symbolizing Disjunctions

A disjunction is a compound statement consisting of two statements connected by the word *or*. The symbol for this connective is $\vee$. If we let

$$P = \textit{Today is Monday}$$
$$Q = \textit{Tomorrow is Wednesday}$$

then the statement

Today is Monday, or tomorrow is Wednesday.

is symbolized as

$$P \vee Q \quad \text{(disjunction)}$$

Consider the statement

Either two is not even, or three is not odd.

In this case we would let

$$P = \textit{Two is even}$$
$$Q = \textit{Three is odd}$$

and the compound statement would be symbolized as

$$\sim P \vee \sim Q$$

EXAMPLE 2 ***Symbolizing Negations and Conjunctions*** Let P = *Today is Monday* and Q = *Yesterday was Sunday*. Write each of the following statements in symbolic form:

a. Either today is Monday, or yesterday was Sunday.
b. Yesterday was not Sunday, or today is Monday.
c. Either today is not Monday, or yesterday was not Sunday.

Solution

a. This statement is a disjunction since we have the connective *or*; the word *either* also tells us that we have a disjunction. The correct symbolization is $P \vee Q$.

b. This is also a disjunction, but here we have the statements interchanged and Q is negated. Therefore, the statement should be symbolized as $\sim Q \vee P$.

c. Here each part of the compound statement is negated; we would symbolize this as $\sim P \vee \sim Q$.

It should be noted that the word *or* can be used in two different ways in a sentence. For example, consider the following statements:

The weather forecast calls for rain or snow.
I will get an A or B for this course.

The first statement illustrates the **inclusive** use of *or*, since it might rain, it might snow, or it might do both. The second statement illustrates the **exclusive** use of *or*, since it is not possible for both things to occur. That is, the grade for the course is an A or a B, but not both. The symbol commonly used for the exclusive *or* is $\underline{\vee}$. Hence, *P or Q but not both* is symbolized as $P \underline{\vee} Q$. Unless otherwise noted, we shall assume that *or* is used in the inclusive sense.

NW ***Now Work Problems 3a, b, and c.***

Symbolizing Conditionals

A conditional is a statement that implies something. The symbol used in mathematics for implication is $\rightarrow$. The statement $P \rightarrow Q$ is usually interpreted as

If P, then Q.

or, equivalently,

P implies Q.

Consider the following statement:

If the Browns win the championship, then I'll eat my hat.

If we let

P = *The Browns win the championship*
Q = *I'll eat my hat*

the compound statement is symbolized as

$$P \rightarrow Q \quad \text{(conditional)}$$

Let us examine a conditional statement where the antecedent and consequent are negated. If we let

P = *x is negative*
Q = *x is less than zero*

then $\sim P \rightarrow \sim Q$ is interpreted as

If x is not negative, then x is not less than zero.

Symbolizing Biconditionals

A biconditional is the conjunction of two conditional statements where the antecedent and consequent of the first statement have been switched in the second. The symbol for the connective in a biconditional is $\leftrightarrow$.

Consider the statement

Skating is permitted if and only if the ice is 6 inches thick.

Let

$$P = \textit{Skating is permitted}$$

$$Q = \textit{The ice is 6 inches thick}$$

The compound statement is symbolized as

$$P \leftrightarrow Q \quad \text{(biconditional)}$$

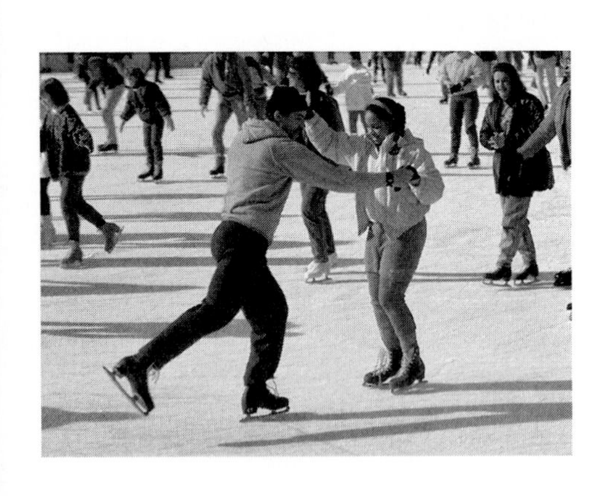

We know from the preceding discussion that this statement is the same as the conjunction of two conditionals, and therefore we are aware that if skating is permitted, then the ice is 6 inches thick, and if the ice is 6 inches thick, then skating is permitted; that is,

$$P \leftrightarrow Q \equiv (P \rightarrow Q) \wedge (Q \rightarrow P)$$

(The symbol $\equiv$ means *is the same as.*)

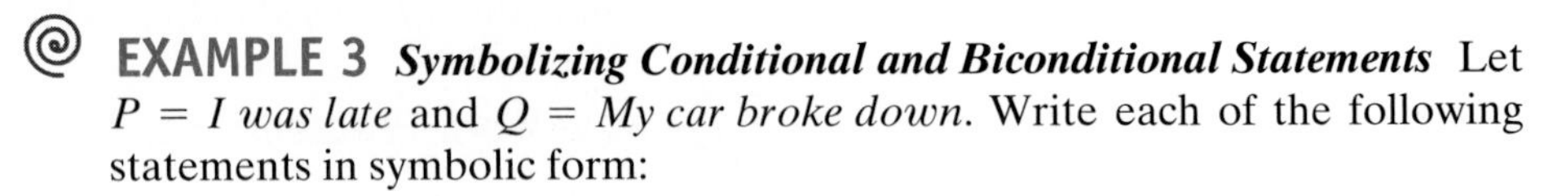

EXAMPLE 3 ***Symbolizing Conditional and Biconditional Statements*** Let $P = \textit{I was late}$ and $Q = \textit{My car broke down}$. Write each of the following statements in symbolic form:

a. If I was late, then my car broke down.
b. If my car broke down, then I was late.
c. I was late if and only if my car broke down.

Solution

a. This statement is a conditional; the key is the connective *if ... then ....* We would symbolize this as $P \rightarrow Q$.
b. This is similar to statement **a**, but here we have "my car broke down" as the antecedent, and therefore we symbolize the statement as $Q \rightarrow P$.
c. Statement **c** is a biconditional, as indicated by the phrase "if and only if," and the correct symbolization is $P \leftrightarrow Q$.

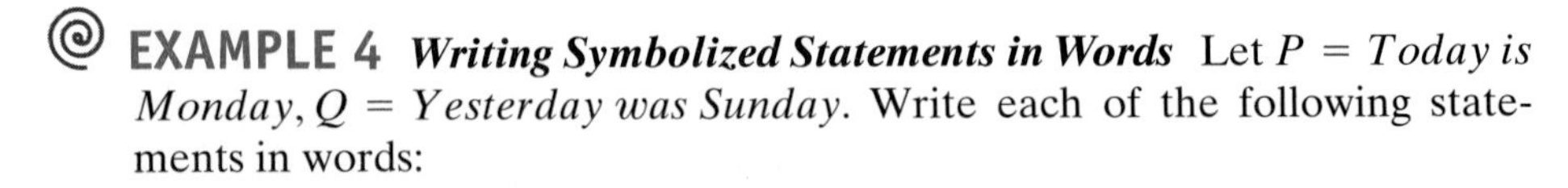

EXAMPLE 4 ***Writing Symbolized Statements in Words*** Let $P = \textit{Today is Monday}$, $Q = \textit{Yesterday was Sunday}$. Write each of the following statements in words:

a. $P \wedge Q$
b. $\sim P \vee Q$
c. $P \rightarrow \sim Q$
d. $\sim P \leftrightarrow \sim Q$
e. $\sim P \wedge \sim Q$
f. $P \rightarrow (P \wedge \sim Q)$

Solution

a. This is a conjunction and we may write this directly from the symbols: "Today is Monday and yesterday was Sunday."

b. This is a disjunction with the first part negated: "Today is not Monday, or yesterday was Sunday."

c. This is a conditional statement with the consequent negated: "If today is Monday, then yesterday was not Sunday."

d. This statement is a biconditional with both parts negated: "Today is not Monday if and only if yesterday was not Sunday."

e. This conjunction could be written as "Today is not Monday and yesterday was not Sunday." Another correct interpretation is "Neither is today Monday, nor was yesterday Sunday." A *neither–nor* statement is a conjunction where both parts are negated. That is, a sentence of the form "neither P nor Q" is symbolized as $\sim P \wedge \sim Q$.

f. This statement is a conditional whose consequent is a conjunction: "If today is Monday, then today is Monday and yesterday was not Sunday."

NW ***Now Work Problems 3d, e, and f.***

At this point, you should be familiar with the following types of statements and their connective symbols:

Type of Statement	Connective	Symbol	Name of Symbol
Negation	not	$\sim$	Tilde
Conjunction	and	$\wedge$	Wedge (caret)
Disjunction	or	$\vee$	Vee
Conditional	if ... then ...	$\rightarrow$	Arrow
Biconditional	if and only if iff	$\leftrightarrow$	Double arrow

Exercises for Section 3.2

1. Identify each of the following sentences as a simple statement, compound statement, or neither. Classify each compound statement as a negation, conjunction, disjunction, conditional, or biconditional.
   a. Two is a counting number.
   b. It is not the case that Scott has a new car.
   c. You may attend the conference iff you have paid the registration fee.
   d. Today is not a holiday.
   e. If Addie went swimming, then Julia went sailing.
   f. Where is your homework?
   g. Both Ruth and Florence are members of the band.
   h. Either Bill is here, or he did not come to school.

2. Identify each sentence as a simple statement, compound statement, or neither. Classify each compound statement as a negation, conjunction, disjunction, conditional, or biconditional.
   a. If $3 + 4 = 8$, then $9 - 2 = 6$.
   b. When in doubt, punt!
   c. A student may take Math 176 iff he has successfully completed Math 175.
   d. Neither Hugh nor Bill is here.
   e. Addie is not late, and Julia is not early.
   f. Mary went swimming or cycling.
   g. Yesterday was Sunday.
   h. Norma was not in class.

3. Let P = *Polly is good* and Q = *Johnny is good*, and let us agree that *bad* = *not good*. Write each of the following statements in symbolic form:
   NW **a.** Both Polly and Johnny are good.
   NW **b.** Either Johnny or Polly is good.
   NW **c.** Polly is not good, or Johnny is bad.
   NW **d.** If Polly is good, then Johnny is bad.
   NW **e.** It is false that Polly is bad.
   NW **f.** Johnny is good iff Polly is good.

4. If "Polly is good" and "Johnny is good" are true statements, which of the statements in Exercise 3 do you think are true?

5. Let P = *Algebra is difficult* and Q = *Logic is easy*. Write each of the following statements in symbolic form:
   a. Algebra is difficult, or logic is easy.
   b. Logic is not easy, and algebra is difficult.
   c. It is false that logic is not easy.
   d. Logic is easy iff algebra is difficult.

e. If algebra is difficult, then logic is easy.
f. Neither is algebra difficult nor is logic easy.

6. If "Algebra is difficult" and "Logic is easy" are true statements, which of the statements in Exercise 5 do you think are true?
7. Write the following statements in symbolic form using the letters in the parentheses:
a. Either I sink this putt, or I lose the match. (P, M)
b. Five is greater than zero, and 5 is positive. (G, P)
c. If you do not attend class, then you will be dropped from the course. (A, D)
d. Either the bus is late, or my watch is not working correctly. (B, W)
e. Two equals 1 iff 3 is greater than 4. (T, F)
f. Smith will raise taxes if he is elected. (R, E)
8. Write the following statements in symbolic form using the letters in parentheses:
a. It is not the case that Brian did not take the test. (B)
b. I do not have a ticket, but I am going to the show. (T, S)
c. Neither five nor six will win the horse race. (F, S)
d. If you like pasta, then Amiel will prepare it. (P, A)
e. I can go to the show iff I finish my homework. (S, H)
f. I will be on time, if I find a parking space. (T, S)
9. Let P = *I like algebra* and Q = *I like geometry*. Write each of the following statements in words:

a. $P \wedge Q$ b. $P \rightarrow \sim Q$
c. $P \vee Q$ d. $P \vee \sim Q$
e. $\sim P \wedge \sim Q$ f. $P \leftrightarrow Q$

10. Let P = *Tom is tired* and Q = *Today is Monday*. Write each of the following statements in words:

a. $P \vee \sim P$ b. $P \leftrightarrow \sim Q$
c. $\sim P \vee Q$ d. $\sim P \wedge \sim Q$
e. $\sim Q \rightarrow P$ f. $P \rightarrow \sim Q$

Writing Mathematics

11. What is a statement? Explain why each of the following are not statements: questions, commands, opinions. Give an example of each.
12. Describe the difference between a compound statement and a simple statement. Give an example of each.
13. What is a conjuction statement? Give an example.
14. What is a disjunction statement? Give an example.
15. What is a conditional statement? Give an example.
16. What is a biconditional statement? Give an example.

Challenge Exercises

17. **Spelling Rules** A common spelling rule that we are familiar with is "i before e, except after c," which leads us to another rule "All Rules Have Exceptions." Explain why this is a paradox.
18. Construct a sentence that is a paradox.
19. Write a compound statement (in words) containing at least two connectives. Write it in symbolic form.

Group/Research Activity

20. Find a newspaper or magazine article, or a legal document of some kind (warranties, tax form instructions, mortgage, mortgage papers, etc.) that contains the following connectives: *and, or, not, if ... then ..., if and only if.* Highlight each connective. Each member of the group should have a complete set of examples. The group should then present at least two of these examples explaining how the connectives are used and what the sentences mean.
21. One of the most well-known paradoxes is the *Achilles and the Tortoise* paradox of Lewis Carroll. Go to the Web site **http://www.mathacademy.com/ platonic_realms/encyclop/articles/carroll.html** to read about this paradox. Prepare a report and present your findings to your class. As part of your report, compare Carroll's *Achilles and the Tortoise* paradox to Zeno's *Paradox of the Tortoise and Achilles.*

Real-World Data

Famous Quotes

In Exercises 22–27 are quotes from various celebrities, politicians, or famous people. Classify each statement as a negation, conjunction, disjunction, conditional, or biconditional, and then write each statement in symbolic form using the letters in the parentheses.

22. "I am always ready to learn, but I do not always like being taught." (R, T) *Sir Winston Churchill*
23. "Education is not the filling of a pail, but the lighting of a fire." (P, F) *William Butler Yeats*
24. "If you want to make enemies, try to change something." (E, C) *Woodrow Wilson*
25. "I will not condemn you for what you did yesterday, if you do it right today." (C, R) *Sheldon S. Maye*
26. "If you aren't fired with enthusiasm, you will be fired with enthusiasm." (F, F) *Vince Lombardi*
27. "The secret of life is honesty and fair dealing. (H, D) If you can fake that, you've got it made." (F, M) *Groucho Marx*

Vince Lombardi

Just for fun

What month has 28 days?

Math Connections

Boolean Searches and the Internet

Most major Internet sites today, such as Yahoo!, AOL (America OnLine), Netscape, and MSN (Microsoft Network), enable users to search the Internet for specific information. A user's request for information, though, does not involve searching the Internet directly, but instead invokes a search engine that examines the contents of a database of Internet sites that have been compiled by a particular organization. Thus, "searching the Internet" really involves searching these databases. Several search engines are available and some of the most popular ones include Alta Vista **(http://www.altavista.com)**, Excite **(http://www.excite.com)**, Google **(http://www.google.com)** and HotBot **(http://www.hotbot.com).** There are also *metasearch engines*, which submit a user's search request to multiple search engines at the same time. Two popular ones are Dogpile **(http://www.dogpile.com)** and MetaCrawler **(http://www. metacrawler.com)**.

Most Internet database searching relies on the principles of Boolean logic and consists of the three logical operators, OR, AND, and NOT, all of which have exactly the same meaning as discussed in this section. When using an Internet search engine, though, the type of Boolean expressions permitted varies from one engine to another. Most search engines enable users to do a full Boolean search that requires Boolean logical operators. As an illustration, let's assume we are interested in finding information about learning disorders in children such as hyperactivity or attention deficit disorder:

- To find information about either one of these disorders our search would involve the OR operator: *hyperactivity OR attention deficit disorder*.
- To find information that addresses both disorders, our search would involve the AND operator: *hyperactivity AND attention deficit disorder*.
- To find information about one disorder but not the other, our search would involve the NOT operator: *hyperactivity NOT attention deficit disorder* (or *attention deficit disorder NOT hyperactivity*).

Search engines also support **implied Boolean logic**, in which symbols are used to represent Boolean logical operators. For example,

- A plus symbol (+) implies AND and is used when you want to include a keyword. Thus, $hotel + california$ will search for pages that contain both of the words "hotel" and "california."
- A minus symbol (−) implies NOT and is used when you want to exclude a keyword. Thus, $hotel - california$ will search for pages that contain the word "hotel" but do not contain the word "california."
- The absence of either a plus or minus symbol defaults to OR in most engines. Thus, *hotel california* will search for pages that contain either the word "hotel" or the word "california." (*Note:* Some sites use AND as its default for blank spaces. To find out which operator an engine is using as its default for a blank space, see the help files at the site itself.)

Most search engines also provide a **search template** that allows users to select the Boolean operator from a menu. Often the logical operator is expressed with substitute language instead of the operator itself. For example, as part of this menu, a user can select "all of the words," "exact phrase," or "Boolean." Search templates are usually available by selecting "advanced search," "super search," or "power search" from a site's search page. As an example see MSN's advanced search (**http://search.msn.com/advanced.asp**). You are also encouraged to experiment with a metasearch engine such as Dogpile.

3.3 DOMINANCE OF CONNECTIVES

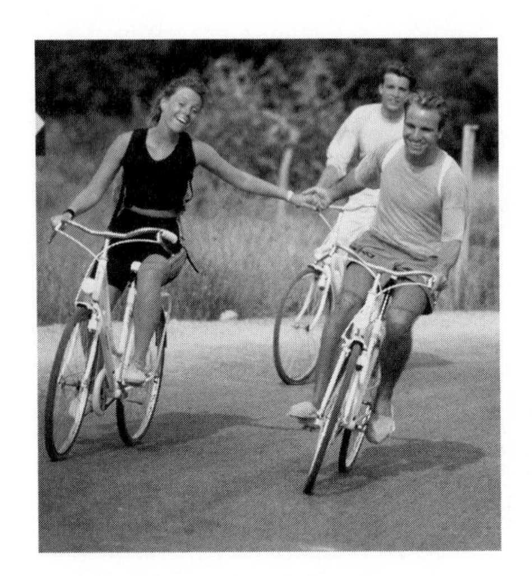

Up to this point in our discussion, except for some of the propositions in the exercises, we have encountered few compound sentences with multiple connectives. We have examined a few statements that contained a negation in addition to another connective, but that is all.

Suppose that we have a statement such as

A: I will go swimming or I will go cycling, and I will go to the movies.

This statement is a conjunction because the comma separates the sentence into two parts with the connective *and*. But if the comma were omitted, we would not know whether the statement was a disjunction or a conjunction. Some of us would interpret it as

B: I will go swimming, or I will go cycling and I will go to the movies.

whereas others would interpret it as it is stated in A.

We need punctuation marks in writing statements in order to make sense out of them. The following example points out this need. Try to make sense out of this statement.

Tom is here or Jim left and Bob came late.

Unless this statement is correctly punctuated, we do not know whether it is a disjunction or a conjunction.

Mathematics uses parentheses or other forms of grouping symbols to help us correctly evaluate expressions (see the Math Connections on calculator logic). Logic also uses parentheses to help us correctly identify the type of statement being considered.

Some statements do not need parentheses, whereas others do; some even need more than one set.

Math Connections

Calculator Logic

Many calculators have one of two types of logic: *serial logic or sums of products logic*. Calculators with serial logic perform all operations in exactly the sequence they are entered. Calculators with sums of products logic perform all multiplications and divisions first, and then additions and subtractions last. As an example, consider the problem $1 \times 2 + 3 \times 4$.

- Serial logic performs all operations as they are entered resulting in a final answer of 20.

$$\underbrace{1 \times 2} + 3 \times 4$$
$$\underbrace{2 + 3} \times 4$$
$$\underbrace{5 \times 4}$$
$$20$$

- Sums of products logic, working from left to right, performs all multiplications and divisions first and then all additions and subtractions last, resulting in a final answer of 14.

$$\underbrace{1 \times 2} + \underbrace{3 \times 4}$$
$$\underbrace{2 + 12}$$
$$14$$

Which answer is correct? The correct answer is 14. This is because in mathematics we follow a prescribed **order of operations**. Specifically, this order is as follows:

Working from left to right,

1. Perform all operations within parentheses or other grouping symbols.
2. Evaluate all exponential expressions.
3. Evaluate all multiplications and divisions.
4. Evaluate all additions and subtractions.

In logic, compound statements that involve multiple connectives also follow a prescribed order that helps us correctly identify the type of statement given. Without an agreed upon dominant order of the connectives, it is possible to misinterpret a given statement similar to the way we arrived at two different answers in the preceding example.

If we let

$$P = I\ will\ go\ swimming$$
$$Q = I\ will\ go\ cycling$$
$$R = I\ will\ go\ to\ the\ movies$$

then statement A would be symbolized as $(P \vee Q) \wedge R$, and statement B would be symbolized as $P \vee (Q \wedge R)$.

It is superfluous, but not wrong, to use parentheses in statements such as the following:

$$\sim P \qquad P \wedge Q \qquad P \vee Q \qquad P \rightarrow Q \qquad P \leftrightarrow Q$$

When we interpret a statement, we shall interpret it exactly as written. That will enable us to determine the symbolic form of the statement.

EXAMPLE 1 ***Identifying Statements with Multiple Connectives*** Identify each of the following statements as a negation, conjunction, disjunction, conditional, or biconditional.

a. Either the Jets won, or the Bills won and the Browns lost.

b. If John goes to college and Jack goes to art school, then their father will have to take a loan.

c. José is here or Larry is here, and it is hot.

d. It is false that if I took French II, then I had taken French I before.

Solution

a. This statement is a disjunction, as the comma indicates.

b. Statement **b** is a conditional statement because of the word *if*.

c. Statement **c** is a conjunction, as the comma indicates.

d. The phrase *it is false that* identifies statement **d** as a negation, even though it contains a conditional. *It is false that* negates everything that follows.

When we want to translate symbolic statements into words, we need parentheses. We will also adopt the convention of the dominance of connectives. In logic, some connectives are considered more dominant than others. The following is a list of the connectives in their dominant order, as found in most logic books; the most dominant connective is listed first:

1. Biconditional $\leftrightarrow$
2. Conditional $\rightarrow$
3. Conjunction $\wedge$, disjunction $\vee$
4. Negation $\sim$

Note that conjunction and disjunction are listed together. They are of equal value. If a compound statement contains both of these connectives and no others, we must use parentheses to designate it as a conjunction or a disjunction. Also, a symbol outside parentheses outranks any symbol inside the parentheses. For example, consider the statement symbolized as $\sim(P \rightarrow Q)$. At "face value" the arrow is the more dominant connective. However, because the arrow is contained within parentheses, it loses its dominance over the tilde. As a result, this statement is a negation and not a conditional. One further note: The connectives *and, or, if-then,* and *iff* are **binary operators** because they connect two statements. The connective *not*, however, is a **unary operator** because it operates on a single statement.

EXAMPLE 2 ***Identifying Statements by Their Dominant Connective*** Identify each symbolic statement.

a. $P \wedge Q \leftrightarrow R$ **b.** $(P \vee Q) \wedge R$
c. $\sim P \vee Q \rightarrow R \wedge S$ **d.** $\sim(P \wedge Q)$
e. $\sim(P \rightarrow Q \vee R)$ **f.** $P \vee (Q \rightarrow R)$

Solution

a. The statement $P \wedge Q \leftrightarrow R$ is a biconditional because the double arrow is the dominant connective and there are no parentheses.

b. The parentheses in $(P \vee Q) \wedge R$ separate the statement at the *and* connective; hence, it is a conjunction.

c. In $\sim P \vee Q \rightarrow R \wedge S$ there are no parentheses and none are needed, as the conditional arrow is stronger than any of the other connectives in the statement. The statement is a conditional. Note that $\sim P \vee Q$ is the antecedent and $R \wedge S$ is the consequent.

d. At first glance it might appear that $\sim(P \wedge Q)$ is a conjunction, but it is not because the negation sign is outside the parentheses. If the statement were $\sim P \wedge Q$, it would be a conjunction, but with the parentheses it is a negation.

e. The negation sign in $\sim(P \rightarrow Q \vee R)$ takes precedence because it is outside the parentheses and the arrow is inside. Therefore, it is a negation.

f. The parentheses divide the statement into two major parts and the connective outside the parentheses is dominant; hence, it is a disjunction.

NW ***Now Work Problems 1a, b, and c.***

EXAMPLE 3 ***Writing Symbolized Statements with Multiple Connectives into Words*** Let P = *I can go*, Q = *You can go*, and R = *Lew can go*. Write each of the following symbolic statements in words:

a. $P \wedge (Q \vee R)$ **b.** $P \vee Q \rightarrow R$
c. $\sim(P \wedge Q)$ **d.** $\sim P \vee \sim Q$

Solution

a. Because of the parentheses, this statement is a conjunction, so we write "I can go, and you can go or Lew can go."

b. The arrow is the dominant connective, so statement **b** is a conditional: "If I can go or you can go, then Lew can go."

c. Statement **c** is a negation; hence, we have "It is false that I can go and you can go."

d. This statement is a disjunction with each part negated. Therefore, we have "I cannot go or you cannot go."

NW ***Now Work Problems 7a, b, and c.***

Parentheses are used in logic to tell us what type of statement we are considering. If there are no parentheses, then we follow the convention of the dominance of connectives. The biconditional ($\leftrightarrow$) is the strongest connective, followed by the conditional ($\rightarrow$), and then conjunction ($\wedge$) and disjunction ($\vee$). The conjunction and disjunction are of equal value. The negation ($\sim$) is the weakest connective.

Exercises for Section 3.3

1. Add parentheses in each statement to form the type of compound statement indicated. If none are needed, indicate that fact.
   - NW **a.** Conditional: $P \wedge \sim Q \rightarrow R$
   - NW **b.** Conjunction: $P \wedge Q \leftrightarrow R$
   - NW **c.** Negation: $\sim P \vee Q \rightarrow R$
   - **d.** Disjunction: $\sim P \vee Q \wedge R$
   - **e.** Biconditional: $P \leftrightarrow Q \vee R$
   - **f.** Conjunction: $\sim P \wedge Q \vee R$
2. Add parentheses in each statement to form the type of compound statement indicated. If none are needed, indicate that fact.
   - **a.** Negation: $\sim P \rightarrow \sim Q$
   - **B.** Biconditional: $\sim P \leftrightarrow Q \vee R$
   - **c.** Disjunction: $P \vee Q \rightarrow R \wedge S$
   - **d.** Conditional: $P \vee Q \rightarrow R \wedge S$
   - **e.** Conjunction: $P \vee Q \rightarrow R \wedge S$
   - **f.** Negation: $\sim P \wedge Q \rightarrow \sim R$
3. Let P = *Algebra is difficult*, Q = *Logic is easy*, and R = *Latin is interesting*. Use appropriate connectives and parentheses to symbolize each statement.
   - **a.** If logic is easy and algebra is difficult, then Latin is interesting.
   - **B.** Latin is interesting and algebra is difficult, or logic is easy.
   - **c.** It is false that logic is easy and algebra is difficult.
   - **d.** Either logic is easy and Latin is interesting, or algebra is difficult.
   - **e.** Algebra is difficult iff Latin is interesting and logic is easy.
   - **f.** Neither is algebra difficult, nor is Latin interesting.
4. Let P = *Sue likes golf*, Q = *Ryan likes football*, and R = *David likes basketball*. Use appropriate connectives and parentheses to symbolize each statement.
   - **a.** If Sue likes golf and Ryan likes football, then David likes basketball.
   - **B.** Either Sue likes golf and Ryan likes football, or David likes basketball.
   - **c.** It is false that Sue likes golf and Ryan likes football.
   - **d.** Sue likes golf iff Ryan likes football and David likes basketball.
   - **e.** Sue likes golf or David likes basketball, and Ryan likes football.
   - **f.** If David likes basketball, then it is false that Sue likes golf and Ryan likes football.
5. Using the suggested notation, symbolize each statement completely.
   - **a.** If you study, then you will pass the exam and you will pass the course. (*S, E, C*)
   - **B.** Neither Jill nor Cindy is tall. (*J, C*)
   - **c.** It is not the case that Adam will not graduate. (*A*)
   - **d.** If Anthony cuts class, then he will miss the exam and receive a zero. (*C, E, Z*)
   - **e.** Either Joan lost her assignment, or it is in the computer and she is waiting for a printout. (*J, C, P*)
6. Using the suggested notation, symbolize each statement completely.
   - **a.** Today is not Friday iff tomorrow is Sunday. (*F, S*)
   - **B.** Paul likes strawberry ice cream and Juanita likes strawberry ice cream, or Connie made a mistake. (*P, J, C*)
   - **c.** We will be happy if Barbara is elected chairperson. (*H, B*)
   - **d.** It is false that I was late and missed the test. (*L, T*)
   - **e.** Today is Friday, and if today is Friday then I get paid. (*F, P*)
7. Let P = *Algebra is difficult*, Q = *Logic is easy*, and R = *Latin is interesting*. Write each symbolic statement in words.
   - NW **a.** $P \wedge (Q \vee R)$
   - NW **b.** $P \wedge Q \rightarrow R$
   - NW **c.** $P \vee (Q \wedge R)$
   - **d.** $P \wedge (Q \rightarrow R)$
   - **e.** $\sim(P \wedge Q)$
   - **f.** $\sim P \leftrightarrow Q \wedge \sim R$
8. Let P = *Harry studies*, Q = *Harry will pass*, and R = *Harry will succeed*. Write each symbolic statement in words.
   - **a.** $P \rightarrow Q \vee R$
   - **b.** $\sim(P \wedge Q)$
   - **c.** $(P \rightarrow Q) \vee R$
   - **d.** $Q \wedge R \leftrightarrow P$
   - **e.** $\sim P \wedge \sim Q$
   - **f.** $P \wedge (Q \vee \sim R)$

Writing Mathematics

9. Why do we need punctuation marks when writing compound statements?
10. In mathematics and logic, what are used as punctuation marks? Provide at least two examples for each category.
11. In logic some connectives are considered more dominant than others. List the connectives and their names in their dominant order; the most dominant connective should be listed first.
12. Discuss the difference between $\sim P \vee Q$ and $\sim(P \vee Q)$.
13. Do the two given symbolic statements $P \wedge Q \rightarrow R$ and $P \wedge (Q \rightarrow R)$ represent the same thing? Why or why not?

Challenge Exercises

14. Given the following set of words, see if you can punctuate it to make sense of it:

    time flies impossible they pass at such irregular intervals

    Be certain to explain your answer.

Group/Research Activities

15. Using the suggested notation and proper connectives, symbolize the paragraph completely. (on p. 130.)

Either the store is open or it is an official holiday. If it is an official holiday, then we will save our money. If the store is open, then we will not enjoy the holiday. Hence, we will not enjoy the holiday or we will save our money. (S, O, M, E)

16. Using the suggested notation and proper connectives, symbolize the paragraph completely.

If you pass algebra, then you may take trigonometry or statistics. If you take trigonometry, then you did not pass algebra. If you take history, then you cannot take statistics. Therefore, it is false that you passed algebra and that you took history. (A, T, S, H)

17. Working in two- or three-group teams, examine at least 25 different calculators to determine the type of logic they employ. You can test a calculator's logic by using the simple arithmetic expression given in the Math Connections of this section ($1 \times 2 + 3 \times 4$). Organize your data by the manufacturer and model number of the calculators and include the percentage of calculators that have serial logic and the percentage that have sums of products logic. Present your findings to your class.

18. Research the origin of the order of operations for evaluating arithmetic expressions to find out why multiplication and division precede addition and subtraction.

19. The logic we consider in this chapter is based on the two truth values of "truth" and "falsity." That is, either a statement is true or it is false. In other systems of logic, it is possible to have more than two truth values. For example, in a four-valued logic system, a statement might be considered as true, plausible, implausible, or false. Emil L. Post (1897–1954) proposed a logic system that consisted of n truth values. Prepare a report on Post and include a description of his logic system.

Real-World Data

20. Dinner Menus The dinner menu at the Olive Garden Restaurant includes the following statement at the end of its entrées listing: *Enjoy our freshly baked garlic breadsticks and your choice of homemade soup or garden-fresh salad with any entrée.*

a. How would you interpret this statement?

b. Rewrite this sentence using a comma to emphasize the statement's meaning.

c. Symbolize the statement.

Famous Quotes

In Exercises 21 and 22 are quotes from various celebrities, politicians, or famous people. Determine the type of statement given and then write each statement in symbolic form using the letters in parentheses.

21. "If a man can write a better book (B), preach a better sermon (P), or make a better mousetrap (T) than his neighbor, though he build his house in the woods, the world will make a beaten path to his door" (M). *Ralph Waldo Emerson*

22. "Either you think (T)—or else others have to think for you (O) and take power from you (F), pervert and discipline your natural tastes (P, D), civilize and sterilize you" (C, S). *F. Scott Fitzgerald*

Teens and School

The statements in Exercises 23–25 refer to Figure 3.2. Let D = high school dropouts, S = attending school, and W = working. Determine the type of statement given and then write each statement in symbolic form.

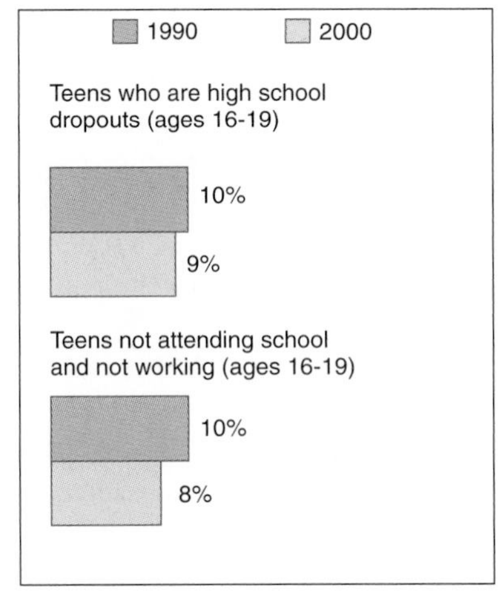

Figure 3.2

23. In 1990, 10% of teens were high school dropouts, and 10% of teens were not attending school and not working.

24. In 2000, 9% of teens were high school dropouts, or 8% of teens were not attending school and not working.

25. In 1990, 10% of teens were not attending school and not working, or in 2000, only 8% of teens were not attending school and not working.

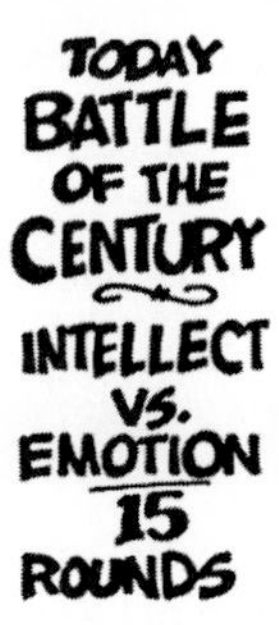

Just for fun

Even the least-interested sports fan knows that a football field is 100 yards long (120 if we count the end zones), but do you know how wide it is?

3.4 TRUTH TABLES FOR NEGATION, CONJUNCTION, AND DISJUNCTION

Because a statement in logic is one that may be true or false, we must be able to determine the **truth value** of a given statement; that is, we need to know under what conditions a statement is true or false. Some readers might think that this can be done rather easily by interpreting the statement within its given context. For example, if someone says "Today is Saturday" on Tuesday, then we know the statement is false. Unfortunately, determining the truth value of a statement by interpretation is not always possible or infallible because the meaning of certain statements can be misconstrued. Fortunately, we have another way to assess a statement's truth value. It is by constructing **truth tables**. This method is strictly mechanical but it leaves nothing to chance. Truth tables are constructed by arranging all possible truth values of a particular statement in tabular form. In this section we examine the truth tables for negations, conjunctions, and disjunctions. In the next section, we consider conditional and biconditional statements.

Negation ($\sim$)

TABLE 3.2 Negation

P	$\sim P$
T	F
F	T

If we are given a simple statement, P, we know that P must be either true or false, but not both. If P is a true statement, then we say that the truth value of P is *true*; if P is false, then its truth value is *false*. Now what happens if we negate P? If P is true, then $\sim P$ must be false, and if P is false, then $\sim P$ must be true. This analysis is summarized in Table 3.2, which is the truth table for a negation. Thus, a negation has two possible truth values.

Conjunction ($\wedge$)

TABLE 3.3a

P	Q	$P \wedge Q$
T	T	
T	F	
F	T	
F	F	

The conjunction $P \wedge Q$ contains two simple statements. P has two possible truth values—namely, true or false—and Q has two possible values. How many possible values does a compound statement such as $P \wedge Q$ have? We know that P and Q could both be true, or they could both be false. That gives us two possibilities. There is also the case where one could be true and the other false; P could be true and Q false, or Q could be true and P false. Therefore, we have four possibilities. This is summarized in Table 3.3a.

We will not go through this type of reasoning every time we want to construct a truth table. If there are n simple statements in a compound statement, then there are 2^n possible true–false combinations. The statement $P \wedge Q$ has two simple statements and, therefore, we have $2^2 = 2 \times 2 = 4$ possible true–false combinations. The statement $P \wedge Q \rightarrow R$ has $2^3 = 2 \times 2 \times 2 = 8$ possible true–false combinations.

Now let us examine a statement that will enable us to complete the truth table for a conjunction $P \wedge Q$. Consider the following statement.

Today is Tuesday and I have a math class.

Let

$$P = Today\ is\ Tuesday$$

$$Q = I\ have\ a\ math\ class$$

The statement is a conjunction and is symbolized as $P \wedge Q$. When is this compound statement true?

As noted in our previous discussion and in Table 3.3a, a conjunction has four possible *T–F* combinations:

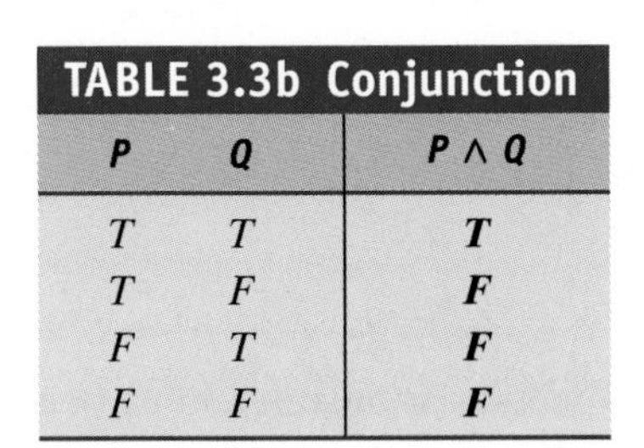

TABLE 3.3b Conjunction

P	Q	$P \wedge Q$
T	*T*	***T***
T	*F*	***F***
F	*T*	***F***
F	*F*	***F***

- ***Both statements are true (T–T)****:* If the person who makes this statement does so on Tuesday and he or she has a math class, then the statement is true. Thus, a *T–T* combination for a conjunction yields a true statement.
- ***The first statement is true and the second statement is false (T–F)****:* If the person who makes this statement does so on Tuesday but does not have a math class, then he or she is making a false statement. Thus, a *T–F* combination for a conjunction yields a false statement.
- ***The first statement is false and the second statement is true (F–T)****:* If the person makes this statement on any day other than Tuesday, then he or she is not telling the truth. Thus, a *F–T* combination for a conjunction yields a false statement.
- ***The first statement is false and the second statement is false (F–F)****:* This is similar to the *F–T* reasoning. Thus, a *F–F* combination for a conjunction yields a false statement.

The completed truth table appears in Table 3.3b. This table shows that a conjunction is true only when both parts are true; otherwise it is false.

Disjunction (∨)

Let's now consider the disjunction. When is a disjunction true? Let's say that Ron makes the following statement to his wife Randee:

I will buy you flowers or I will take you to dinner.

Let

$$P = \text{I will buy you flowers}$$

$$Q = \text{I will take you to dinner}$$

As was the case for the conjunction, a disjunction has four possible truth combinations:

- ***Both statements are true (T–T):*** If Ron does both activities (i.e., he buys Randee flowers and takes her to dinner), he made a true statement. This is because we are using the inclusive *or* in logic. Thus, a *T–T* combination for a disjunction yields a true statement.
- ***The first statement is true and the second statement is false (T–F):*** If Ron only bought Randee flowers, he still made a true statement. Thus, a *T–F* combination for a disjunction yields a true statement.
- ***The first statement is false and the second statement is true (F–T):*** Once again, by doing only one of the activities, Ron made a true statement. Thus, a *F–T* combination for a disjunction yields a true statement.

TABLE 3.4 Disjunction

P	Q	$P \vee Q$
T	T	T
T	F	T
F	T	T
F	F	F

- ***The first statement is false and the second statement is false (F–F):*** Now Ron is in trouble. He failed to do either activity and hence made a false statement. Thus, a *F–F* combination for a disjunction yields a false statement.

The completed truth table appears in Table 3.4. This table shows that a disjunction is true in all cases except when both parts are false.

Recall from Section 3.2 that the exclusive *or*—that is, one event *or* the other, but not both—is symbolized by $P \underline{\vee} Q$. This means we want P or Q, but not both at the same time. The truth table for the exclusive *or* is shown in Table 3.5. The statement $P \underline{\vee} Q$ is true only when exactly one part is true; otherwise, it is false.

TABLE 3.5 Exclusive or

P	Q	$P \underline{\vee} Q$
T	T	F
T	F	T
F	T	T
F	F	F

EXAMPLE 1 ***Truth Table with Negation and Disjunction*** Construct a truth table for $\sim(P \vee Q)$.

Solution Note that the parentheses tell us that the statement is a negation. The whole statement is to be negated; therefore, the negation is the final column completed in the table. The first thing we must do is list the truth values for the variables (we have numbered the columns so that you can follow, in order, the step-by-step process):

P	Q	$\sim$	$(P$	$\vee$	$Q)$
T	T		T		T
T	F		T		F
F	T		F		T
F	F		F		F
1	2		1		2

Since $\vee$ is the connective inside the parentheses, we complete the column for this connective, column 3:

P	Q	$\sim$	$(P$	$\vee$	$Q)$
T	T		T	T	T
T	F		T	T	F
F	T		F	T	T
F	F		F	F	F
1	2		1	3	2

Now, to complete the truth table, we must negate the statement inside the parentheses (the disjunction). To negate the statement, we negate column 3. This gives us column 4 and completes the table:

P	Q	$\sim$	$(P$	$\vee$	$Q)$
T	T	F	T	T	T
T	F	F	T	T	F
F	T	F	F	T	T
F	F	T	F	F	F
1	2	4	1	3	2

Note that we fill in the columns headed by variables before filling in those columns headed by connectives. We fill in the column of the least dominant connective before filling in the columns of the more dominant connectives.

NW ***Now Work Problem 3.***

EXAMPLE 2 ***Truth Table with Negation and Conjunction*** Construct a truth table for $\sim(P \wedge \sim Q)$.

Solution This problem is similar to Example 1, but it is a little more involved. The statement is a negation, so that is the last column completed in our truth table. Listing the truth values for the variables, we have the following table:

P	Q	$\sim$	$(P$	$\wedge$	$\sim$	$Q)$
T	T		T			T
T	F		T			F
F	T		F			T
F	F		F			F
1	2		1			2

The statement inside the parentheses is a conjunction, but before we can complete the column for the connective we must negate Q, because the conjunction is of P with *not* Q:

P	Q	$\sim$	$(P$	$\wedge$	$\sim$	$Q)$
T	T		T		F	T
T	F		T		T	F
F	T		F		F	T
F	F		F		T	F
1	2		1		3	2

Now, to figure out the value of the conjunction, we compare columns 1 and 3. This gives us column 4, which is the column that we negate to get column 5 and complete the table:

P	Q	$\sim$	$(P$	$\wedge$	$\sim$	$Q)$
T	T	**T**	T	F	F	T
T	F	**F**	T	T	T	F
F	T	**T**	F	F	F	T
F	F	**T**	F	F	T	F
1	2	5	1	4	3	2

The completed truth table (column 5) tells us that the statement $\sim(P \wedge \sim Q)$ is true in all cases except when P is true and Q is false.

Remember, a negation ($\sim$) changes the value of the statement to its opposite. That is, if you negate T, you get F; if you negate F, you get T. A conjunction ($\wedge$) is true only when both parts are true, and a disjunction ($\vee$) is true unless both parts are false.

EXAMPLE 3 ***Childrens' Free-Time Activities*** Use the bar graph in Figure 3.3 to determine the truth value of the following statements:

a. The percentage of children who watch TV is less than 30% and the percentage of children who play is more than 25%.

b. The percentage of children who study is 5% or the percentage of children who read is 5%.

c. It is false that more than 10% of the children play sports and less than 10% do household work.

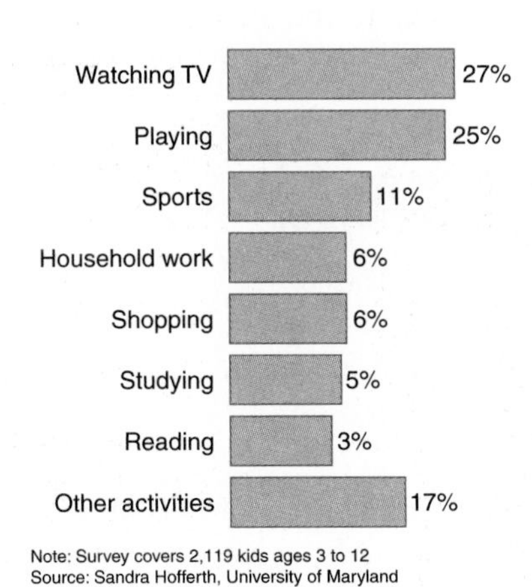

Figure 3.3

Solution

a. The given statement is a conjunction. Recall that a conjunction is true only when both parts are true. Let's examine the truth value of each statement separately:

- *The percentage of children who watch TV is less than 30%.*
  According to the bar graph, 27% of the children watch TV. Thus, this statement is true ($27\% < 30\%$).
- *The percentage of children who play is more than 25%.*
  The bar graph shows that 25% of the children play. This statement is false because 25% is not more than 25%.
  As a result, we have a *T–F* combination, which is false for a conjunction. Thus, the given statement is false.

b. The given statement is a disjunction, which is true in all cases except when both parts are false. We examine the truth value of each statement separately:

- *The percentage of children who study is 5%.*
  The bar graph shows that 5% of the children listed studying as an activity. Thus, this statement is true.
- *The percentage of children who read is 5%.*
  The bar graph shows that 3% of the children listed reading as an activity. Since $3\% \neq 5\%$, the statement is false. As a result, we have a *T–F* combination, which is true for a disjunction. Thus, the given statement is true.

c. This statement is a negation. It is negating a conjunction. This becomes more evident when we symbolize it. If we let S = *More than 10% of the children play sports* and H = *less than 10% do household work*, we get $\sim(S \wedge H)$. Examining the conjunction, once again recall that a conjunction is true only when both parts are true: S is *true* because $11\% > 10\%$, and H is *true* because $6\% < 10\%$. Thus, the conjunction is true. We must now negate this statement. The negation of a true statement is false. Thus, the given statement is false.

Alternative Method for Constructing Truth Tables

An alternative method that can be used to construct truth tables is to use *top-down processing* (see Chapter 1), in which a given statement is first reduced into smaller and separate components. (Recall that a compound statement is composed of simple statements.) Once the statement has been partitioned into separate parts we then use a building process to help "build" the truth table. A tool that is used to facilitate this process is a *tree diagram.*

The first two examples of this section are repeated here and solved using this alternative method.

EXAMPLE 1A ***Truth Table with Negation and Disjunction*** *(Alternative Method)* Construct a truth table for $\sim(P \vee Q)$.

Solution To partition a statement we always work from the most dominant connective to the weakest. This statement is a negation; it is negating the statement within parentheses, which is a disjunction. Furthermore, the disjunction connects the two simple statements P and Q. A tree diagram that pictures how this statement can be partitioned is shown in Figure 3.4.

Once we have a tree diagram we next begin the building process by working from the bottom level (Level 2) to the top level (Level 0). In doing so, note that

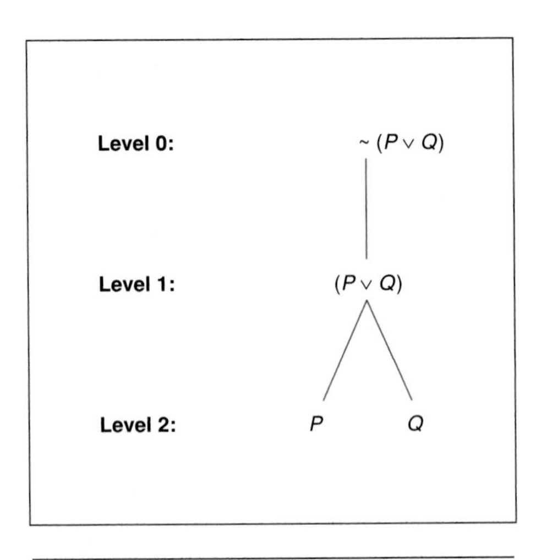

Figure 3.4

we start with the simple statements P and Q, which each have preestablished truth assignments. We then proceed to Level 1, where the statements are connected using a disjunction. Finally, at Level 0 we negate the result we get from Level 1. The individual steps of the truth table construction are now presented for this example.

Step 1 This step corresponds to Level 2 of the tree diagram. Assign truth values to the simple statements P and Q, and label these columns 1 and 2, respectively.

P	Q
T	T
T	F
F	T
F	F
1	2

Step 2 This step corresponds to Level 1 of the tree diagram. Moving up the tree from Level 2, we must combine P and Q into a disjunction. We complete a column for $P \vee Q$ by combining the truth assignments given in columns 1 and 2 under the rules of disjunction. Label this column 3.

P	Q	$P \vee Q$
T	T	T
T	F	T
F	T	T
F	F	F
1	2	3

Step 3 This step corresponds to Level 0 of the tree diagram. Moving up the tree from Level 1, we negate the statement $(P \vee Q)$. This is done by negating the truth assignments of column 3. Label this column 4. This last column represents the truth assignment for the given statement.

P	Q	$P \vee Q$	$\sim(P \vee Q)$
T	T	T	**F**
T	F	T	**F**
F	T	T	**F**
F	F	F	**T**
1	2	3	4

Note how the construction of this truth table followed directly from the tree diagram. Using the tree diagram, we were able to determine the order in which the columns of the truth table are to be completed. By working our way up from the bottom of the tree, we first completed those columns headed by the weaker connectives (those at lower levels) and finished with those columns headed by the strongest connectives (those at higher levels).

NW *Now Work Problem 3.*

EXAMPLE 2A ***Truth Table with Negation and Conjunction*** *(Alternative Method)* Construct a truth table for $\sim(P \wedge \sim Q)$.

Solution This statement is a negation; it negates the statement within the parentheses, which is a conjunction. The conjunction connects P and $\sim Q$. Finally, $\sim Q$ negates the simple statement Q. A tree diagram that pictures the partitioning of this statement is shown in Figure 3.5. (*Note:* P was carried down to Level 3 for completeness' sake.)

As in Example 1A, a truth table is built using the tree diagram as an aid.

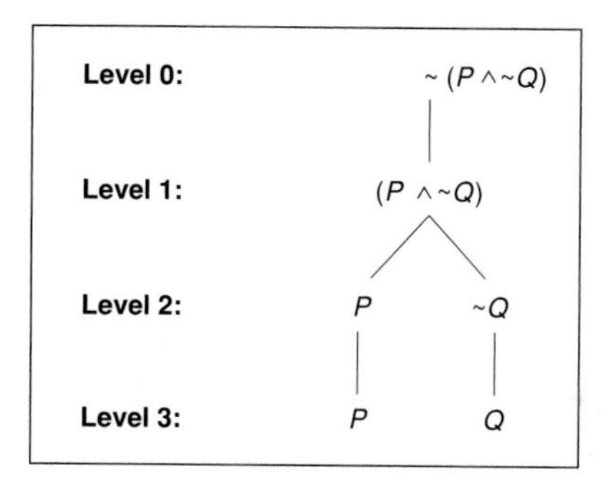

Figure 3.5

Step 1 This step corresponds to Level 3 of the tree diagram. Assign truth values to the simple statements P and Q, and label these columns 1 and 2, respectively.

P	Q
T	T
T	F
F	T
F	F
1	2

Step 2 This step corresponds to Level 2 of the tree diagram. Moving up the tree from Level 3, we must negate Q. This is done by negating the truth assignments given in column 2. Label this column 3. (Note that we do not have to do anything with P since that was taken care of in Level 3.)

P	Q	~Q
T	T	F
T	F	T
F	T	F
F	F	T
1	2	3

Step 3 This step corresponds to Level 1 of the tree diagram. Moving up the tree from Level 2, we must combine P and $\sim Q$ into a conjunction. This is done by combining the truth assignments of columns 1 and 3 under the rules of conjunction. Label this column 4.

P	Q	$\sim Q$	$(P \wedge \sim Q)$
T	T	F	F
T	F	T	T
F	T	F	F
F	F	T	F
1	2	3	4

Step 4 This step corresponds to Level 0 of the tree diagram. Moving up the tree from Level 1, we must negate $(P \wedge \sim Q)$. This is done by negating the truth values from column 4. Label this column 5. This last column represents the final answer.

P	Q	$\sim Q$	$(P \wedge \sim Q)$	$\sim(P \wedge \sim Q)$
T	T	F	F	$\boldsymbol{T}$
T	F	T	T	$\boldsymbol{F}$
F	T	F	F	$\boldsymbol{T}$
F	F	T	F	$\boldsymbol{T}$
1	2	3	4	5

Math Connections

Logic Gates

In 1938, as a student at Massachusetts Institute of Technology (MIT), Claude Shannon published the paper "A Symbolic Analysis of Relay and Switching Circuits," which was based on his master's thesis. This paper is noteworthy because it was the first published work to describe a relationship between George Boole's logic theories of the mid-1800s and current-day relay circuits. Shannon theorized that the on–off states of electronic relay circuits could be represented, respectively, as a series of 0s and 1s, which could lead to information being electronically processed via on–off switches. Shannon's theories effectively provided the mathematical foundation for designing digital electronic circuits, which are the basis for modern-day information processing. (Incidentally, Shannon also provided the first published account of the term *bit* to denote the abbreviation of *binary digit* as a representation of the individual information units a computer processes in his paper "Mathematical Theory of Communication," which was published in the *Bell System Technical Journal* in 1948.) In all, there are seven basic logic gates, which form the basic building blocks of digital circuits. These gates make logical decisions based on the functions NOT, AND, OR, NAND (NOT AND), XOR (exclusive OR), NOR (NOT OR), and XNOR (exclusive NOR).

At any given moment, a logic gate is in one of two binary conditions, namely, 0 or 1, which represent the state of a voltage variable called a *logic level*. In most logic gates, a "low" level is approximately 0 volts and a "high" level is approximately positive 5 volts. A 0 level is equivalent to a false or off state; a 1 level is equivalent to a true or on state. Logic gates accept binary-coded input signals, which are then combined to yield a specific output signal (i.e., a logical 0 or 1). The properties of logic gates can be represented using truth tables, which list the output that will be triggered by each of the possible combinations of input signals.

A summary of each gate and its corresponding truth table is shown in Figure 3.6. Note the parallel between these truth tables and their truth values for a given set of inputs to the truth tables discussed in this chapter. For example, an AND gate will yield a logical 1 output only when it receives a logical 1 signal through both its inputs. Similarly, an OR gate yields a logical 1 in every case except when both inputs are 0. Finally, the NOT gate is simply a logical inverter that inverts 0 to 1 or 1 to 0. Using combinations of logic gates, nearly any complex calculation can be performed. In theory, there is no limit to the number of gates that can be placed together in a single device. In practice, however, physical space constraints limit this number.

Recently, computer scientists succeeded in simulating the functions of logic gates using DNA. Thus, instead of responding to an electronic signal, DNA-based gates respond to nucleotide sequences. In DNA-based computing, commonly called *molecular computing*, switches, transistors, and logic

(continued)

(continued)

gates—the basic building blocks of computer chips—are constructed of individual molecules. Given a molecule that is designed to stop or start the flow of electricity, and connecting it to wires that are only a molecule or so wide, the creation of processors and memory chips millions of times smaller than the fingernail-sized silicon chips currently in production now becomes possible. This type of technology will not only lead to massively powerful processors the size of a grain of sand, but it will also revolutionize computing and networking as we currently know it.

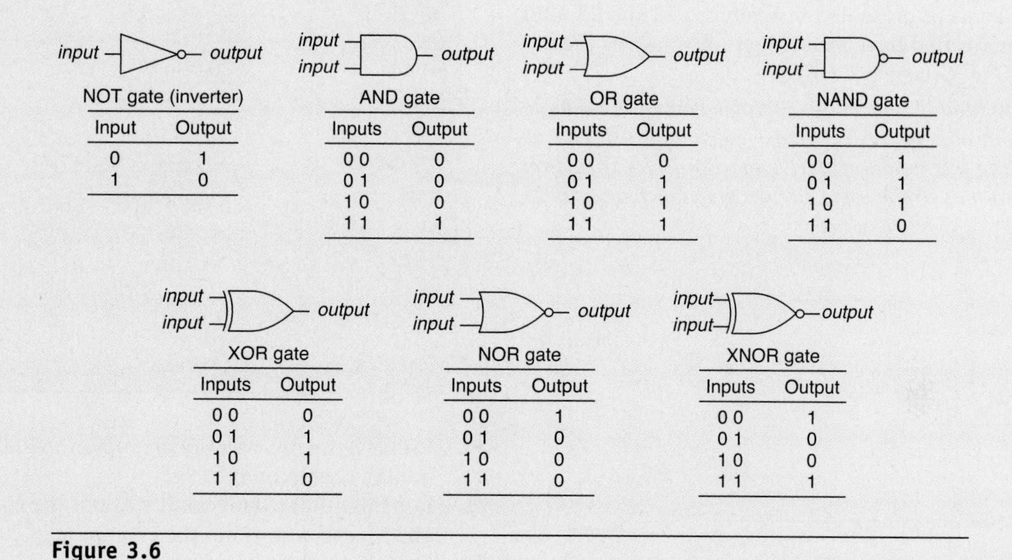

Figure 3.6

Exercises for Section 3.4

In Exercises 1–16, construct a truth table for each statement.

1. $\sim P \wedge \sim Q$

2. $\sim P \wedge Q$

 **3.** $\sim(P \wedge Q)$

4. $P \vee \sim P$

5. $P \wedge \sim P$

6. $\sim P \vee \sim Q$

7. $P \vee \sim Q$

8. $\sim P \vee Q$

9. $\sim(P \vee \sim Q)$

10. $\sim(\sim P \wedge \sim Q)$

11. $\sim P \vee (P \wedge \sim Q)$

12. $Q \vee (\sim P \wedge Q)$

13. $\sim P \underline{\vee} Q$

14. $\sim(P \underline{\vee} \sim Q)$

15. $\sim(\sim P \underline{\vee} \sim Q)$

16. $P \underline{\vee} \sim Q$

Writing Mathematics

17. If there are n simple statements in a compound statement, then there are 2^n possible true–false cases. How many cases must be listed in a truth table that contains two simple statements? List all the cases.

18. When is a conjunction true?

19. When is a conjunction false?

20. When is a disjunction true?

21. When is a disjunction false?

Challenge Exercises

Throughout this section, we constructed truth tables to find the truth values of given compound statements. In Exercises 22 and 23, your challenge is to do the reverse. That is, find the compound statement that has the given truth values as noted by its truth table. You can confirm if your derived statements are correct by constructing their corresponding truth tables and comparing the result to the given truth tables.

22.

P	*Q*	?
T	*T*	***F***
T	*F*	***T***
F	*T*	***F***
F	*F*	***T***

23.

P	*Q*	?
T	*T*	**F**
T	*F*	**F**
F	*T*	**F**
F	*F*	**F**

24. Using the concept presented in Exercises 22 and 23, find *two* statements that have the same truth values as the exclusive or, $P \veebar Q$ (i.e., $F\,T\,T\,F$).

Refer to the Math Connections on logic gates. Let's denote the NOT gate as x', the AND gate as xy, and the OR gate as $x + y$. Using this notation, derive the Boolean expression that corresponds to the logic circuits given in Exercises 25 and 26.

25.

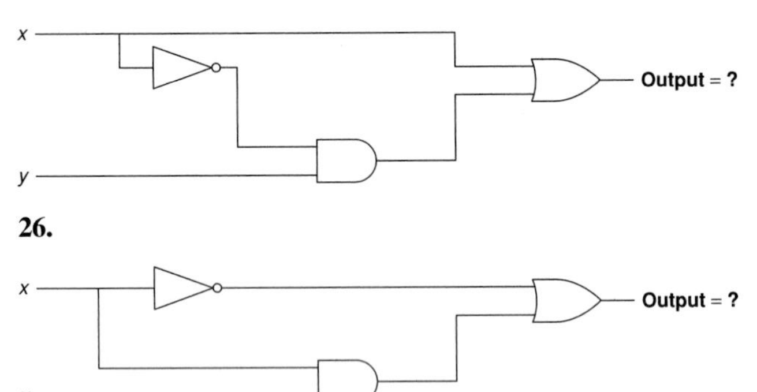

26.

27. Using the notation given in Exercises 25 and 26, write two Boolean expressions for the XOR gate and then design a gate implementation for each expression.
(*Hint:* See Exercise 24.)

Group/Research Activities

28. A compound statement that consists of simple statements using the *and* connective is called a *conjunctive* statement. Similarly, a compound statement that consists of simple statements using the *or* connective is called a *disjunctive* statement. Observe that the correct solutions to Exercises 22 through 24 consist of a disjunctive statement made up of conjunctions. A statement that has this form is referred to as being written in **disjunctive normal form (DNF)**. Note also that one of the solutions to Exercise 24 is a conjunctive statement made up of disjunctions. A statement of this form is referred to as being written in **conjunctive normal form (CNF)**. In logic, every compound statement can be expressed in disjunctive normal form or conjunctive normal form. The concepts of DNF and CNF have important implications and applications in computer science. Research how DNF and CNF are used.

29. An alternative method to truth tables is the **inverted trapezoid** method originally conceived by W. T. Parry and published in "A New Symbolism for the Propositional Calculus," *Journal of Symbolic Logic* (Vol. 19, No. 3, 1954, 161–168). Read this article and prepare a report on how this method is used to determine the truth values of a statement.

Real-World Data

Use Table 3.6 to determine the truth value of the statements given in Exercises 30–34.

TABLE 3.6 NCAA Division I Men's Basketball Champion

Year	*Champion*
2003	Syracuse
2002	Maryland
2001	Duke
2000	Michigan State
1999	Connecticut
1998	Kentucky
1997	Arizona
1996	Kentucky
1995	UCLA
1994	Arkansas

30. Either Syracuse was the champion in 2003 or Duke was not the champion in 2001.

31. It is false that Connecticut was not the champion in 1999 and Maryland was not the champion in 2002.

32. Duke was the champion in 2001 and Arizona was the champion in 1999.

33. UCLA was not the champion in 1995 or Kentucky was the champion in 1998.

34. Michigan State was not the champion in 2000 and Syracuse was not the champion in 2003.

Shopping Patterns

Use Figure 3.7 to determine the truth value of the statements given in Exercises 35–38.

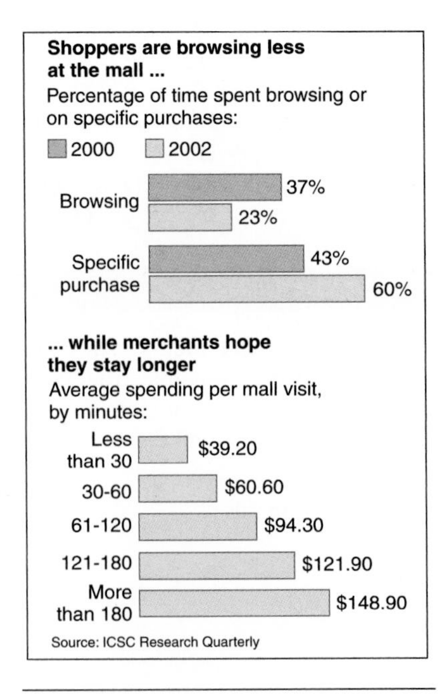

Figure 3.7

35. The percentage of time browsing in 2000 decreased by more than 10% in 2002 but the percentage of time making a specific purchase increased by more than 20% in the same period.
36. Either the percentage of time browsing decreased from 2000 to 2002 or those who spent at least 45 minutes per mall visit spent an average of more than $50.
37. The percentage of time making a specific purchase did not decrease between 2000 and 2002 and those who visited the mall for more than three hours failed to spend at least an average of $150.
38. It is not the case that those who visited the mall for an hour spent an average of less than $100 but those who visited the mall between one and two hours spent an average of at least $100.

Just for fun

Can you number the rest of the squares in the figure so that each row, column, and diagonal totals 15? No number may be used more than once.

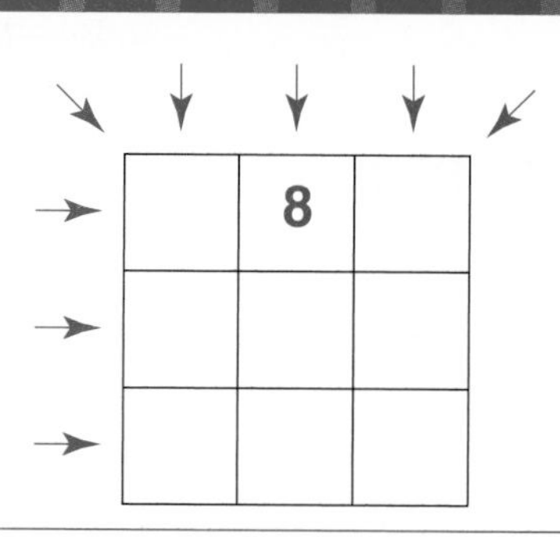

3.5 TRUTH TABLES FOR CONDITIONAL AND BICONDITIONAL

Conditional ($\rightarrow$)

We are often confronted with the *if ... then ...* type of statement. When is the statement $P \rightarrow Q$ true, and when is it false? We shall examine a completed truth table shortly. But let us first try to justify the entries by means of a discussion. Try to follow the discussion closely so that the resulting table appears natural.

Suppose you are told the following by a counselor:

If you pass biology, then you will graduate.

Now, the statement says that if you pass biology, then you will graduate, regardless of what you may have done previously, or will do later. Let

$$P = \textit{You pass biology}$$

$$Q = \textit{You will graduate}$$

Since we have two variables in the compound statement $P \rightarrow Q$, there are four possible true–false combinations.

Let's examine each case separately.

- ***Both statements are true (T–T):*** In this instance, you do pass biology and you do graduate. The counselor's statement must be a true one; it is certainly not a lie. (If the counselor had lied, then the statement would be false.) Thus, a *T–T* combination for a conditional yields a true statement.

- ***The first statement is true and the second statement is false (T–F):*** Suppose you do pass biology, but you do not graduate. This is the case where P is true and Q is false. Has the counselor told the truth? No, the original statement is not true. Therefore, the statement $P \rightarrow Q$ is false when the antecedent is true and the consequent is false.

- ***The first statement is false and the second statement is true (F–T):*** Now, let's consider the case where P is false, and Q is true. In this instance, we have the situation where you did not pass biology, but you still graduated. Examine the original statement: "If you pass biology, then you will graduate." Did the counselor lie to you? No, the counselor said what would happen if you did pass biology, not what would happen if you did not pass. There may be other ways of meeting the graduation requirements, such as taking another course. In this case, the counselor's original statement does not apply. A lie would be a false statement, but since the counselor did not lie to you, the original statement must be true! Therefore, the statement $P \rightarrow Q$ is true when P is false and Q is true.
- ***The first statement is false and the second statement is false (F–F):*** The last case to consider is when the antecedent and consequent are both false—that is, P is false and Q is false. You did not pass biology and you did not graduate. The original statement said, "If you pass biology, then you will graduate." In this instance, you did not pass biology and you did not graduate. Did the counselor lie to you? No, the counselor said what would happen if you *passed* biology. As we stated earlier, a lie would be a false statement; if the counselor did not lie, then the original statement must be true.

TABLE 3.7 Conditional

P	Q	P → Q
T	T	**T**
T	F	**F**
F	T	**T**
F	F	**T**

We are now aware that a conditional is true in all cases except one: A conditional is false when the antecedent is true and the consequent is false; otherwise it is true. The truth table for the conditional statement is shown in Table 3.7.

Consider the following statement:

If the moon is made of green cheese, then this is a mathematics book.

You may balk at the sentence as being silly or absurd. But the point here is that we want to use logic to determine whether it is true or false—without being influenced by our own emotions or prejudices. Is the statement true or false? It is true! Why? The statement is true in this particular case because the antecedent is false and the consequent is true. We see from the truth table for a conditional that F–T yields a T. Remember, the only time a conditional statement is false is when the antecedent is true and the consequent is false.

Another conditional statement to consider is

If $2 = 5$, then $6 = 9$.

Is this statement true or false? It is true. We can determine this from our truth table: Since the antecedent is false, the statement has to be true.

EXAMPLE 1 ***Truth Table with Conditional and Negation*** Construct a truth table for $P \rightarrow \sim Q$.

Solution The statement is a conditional; the arrow is the most dominant connective. Therefore, the last column completed in the truth table will be under the arrow. We first list the truth values for the variables:

P	Q	P	→	~	Q
T	T	T			T
T	F	T			F
F	T	F			T
F	F	F			F
1	2	1			2

Next, we fill in the column for the least dominant connective, the negation:

P	Q	P	$\rightarrow$	$\sim$	Q
T	T	T		F	T
T	F	T		T	F
F	T	F		F	T
F	F	F		T	F
1	2	1		3	2

Column 3 is derived from column 2. In order to figure out the truth values for the arrow, we must compare columns 1 and 3 (the truth values for the antecedent and consequent). Thus we obtain column 4, which completes the table:

P	Q	P	$\rightarrow$	$\sim$	Q
T	T	T	**F**	F	T
T	F	T	**T**	T	F
F	T	F	**T**	F	T
F	F	F	**T**	T	F
1	2	1	4	3	2

NW Now Work Problem 5.

EXAMPLE 2 ***Truth Table with Conjunction, Conditional and Negation***

Construct a truth table for $P \wedge Q \rightarrow \sim P$.

Solution This statement is a conditional statement since there are no parentheses and the arrow is the dominant connective. The antecedent is $P \wedge Q$ and the consequent is $\sim P$. The completed truth table is as follows:

P	Q	P	$\wedge$	Q	$\rightarrow$	$\sim$	P
T	T	T	T	T	**F**	F	T
T	F	T	F	F	**T**	F	T
F	T	F	F	T	**T**	T	F
F	F	F	F	F	**T**	T	F
1	2	1	3	2	5	4	1

Column 3 is the truth value for the antecedent and column 4 is the truth value for the consequent. We compare columns 3 and 4 (using the rules for the conditional) to obtain column 5. This tells us that the statement $P \wedge Q \rightarrow \sim P$ is true in all cases except when P and Q are both true.

NW Now Work Problem 9.

Biconditional ($\leftrightarrow$)

Recall from Section 3.2 that a biconditional is the conjunction of two conditional statements where the antecedent and consequent of the first conditional have been switched in the second. Therefore, a biconditional statement $P \leftrightarrow Q$ is the same as $(P \rightarrow Q) \wedge (Q \rightarrow P)$. Using this information, let us construct the truth table for the biconditional $P \leftrightarrow Q$ by constructing a table for the statement $(P \rightarrow Q) \wedge (Q \rightarrow P)$. If we figure out the truth table for the conjunction, we will know the truth values for $P \leftrightarrow Q$, since they are equivalent. The first steps in our truth table are shown in Table 3.8a.

So far we have worked out both sides of the conjunction. Note that each side is a conditional, but the parentheses tell us that $\wedge$ is the dominant connective.

TABLE 3.8a

P	Q	$(P$	$\rightarrow$	$Q)$	$\wedge$	$(Q$	$\rightarrow$	$P)$
T	T	T	T	T		T	T	T
T	F	T	F	F		F	T	T
F	T	F	T	T		T	F	F
F	F	F	T	F		F	T	F
1	2	1	3	2		2	4	1

TABLE 3.8b

P	Q	$(P$	$\rightarrow$	$Q)$	$\wedge$	$(Q$	$\rightarrow$	$P)$
T	T	T	T	T	**T**	T	T	T
T	F	T	F	F	**F**	F	T	T
F	T	F	T	T	**F**	T	F	F
F	F	F	T	F	**T**	F	T	F
1	2	1	3	2	5	2	4	1

Our final step is to compare columns 3 and 4 using the rules for a conjunction. This leads to column 5, which completes the truth table (see Table 3.8b).

Table 3.8b shows that the conjunction of $(P \rightarrow Q)$ and $(Q \rightarrow P)$ is true when P and Q are both true, and when P and Q are both false. Because this statement is equivalent to $P \leftrightarrow Q$ we also know the truth value of $P \leftrightarrow Q$. We may say that a biconditional is true when P and Q are both true or when P and Q are both false. That is, a biconditional is true when both parts have the same truth value; otherwise it is false. The truth table for the biconditional appears in Table 3.9.

TABLE 3.9 Biconditional

P	Q	$P \leftrightarrow Q$
T	T	**T**
T	F	**F**
F	T	**F**
F	F	**T**

EXAMPLE 3 ***Truth Table with Biconditional*** Construct a truth table for $(\sim P \vee Q) \leftrightarrow (P \rightarrow Q)$.

Solution

P	Q	$(\sim$	P	$\vee$	$Q)$	$\leftrightarrow$	$(P$	$\rightarrow$	$Q)$
T	T	F	T	T	T	**T**	T	T	T
T	F	F	T	F	F	**T**	T	F	F
F	T	T	F	T	T	**T**	F	T	T
F	F	T	F	T	F	**T**	F	T	F
1	2	3	1	4	2	6	1	5	2

As with the other truth tables, we list the possibilities for P and Q first (columns 1 and 2). Our next step is column 3. Then we compare columns 3 and 2 to get column 4 (the disjunction). On the right side, we compare 1 and 2 to get column 5 (the conditional). Our final step is to compare columns 4 and 5 to obtain column 6. Here we use the rule for the biconditional.

NW ***Now Work Problem 11.***

Note that column 6 in Example 3 contains only T's. This tells us that the compound statement $(\sim P \vee Q) \leftrightarrow (P \rightarrow Q)$ is true for all cases, regardless of the truth values of the variables P and Q. A compound statement that is true for any combination of truth values of the variables in the statement is called a **tautology**.

A statement that yields all F's—that is, a statement that is always false—is called a **contradiction**. A common example of a contradiction is $P \wedge \sim P$.

A biconditional that is a tautology tells us something else, too. In Example 3, the truth values for columns 4 and 5 are exactly the same. Columns 4 and 5 represent the truth values for the left and right members of the biconditional. Two statements are **logically equivalent** if they have identical truth tables. Therefore, a biconditional that is a tautology tells us that the left and right members are logically equivalent. In some logic texts, the biconditional is referred to as the **equivalence**, and therefore in Example 3 we have a tautological equivalence.

EXAMPLE 4 ***True-False Combinations for Three Simple Statements*** How many cases have to be considered in order to construct a truth table for the statement $(P \wedge Q) \rightarrow R$?

Solution Note that this compound statement contains three simple statements. How many true–false combinations are there? Recall that if there are n simple statements in a compound statement, then there are 2^n possible true–false combinations. So we have $2^3 = 2 \times 2 \times 2 = 8$ possible cases to consider.

 **EXAMPLE 5** ***Truth Table with Three Simple Statements*** Construct a truth table for $(P \wedge Q) \rightarrow R$.

Solution There are eight true–false possibilities to consider in our truth table. We should therefore have eight rows in our table so we can consider all possibilities for the three variables, P, Q, and R. We have to consider "all true," "two true and one false," "one true and two false," and "all false." Here is a listing of all such possibilities for P, Q, and R.

P	Q	R
T	T	T
T	T	F
T	F	T
T	F	F
F	T	T
F	T	F
F	F	T
F	F	F

After noting that a pattern occurs here, we can easily remember how to construct the possibilities. The first column has four T's followed by four F's. The second column then has two T's, two F's, two T's, two F's. The third column is T, F, T, F, and so on.

Next we complete the truth table:

P	Q	R	$(P$	$\wedge$	$Q)$	$\rightarrow$	R
T	T	T	T	T	T	**T**	T
T	T	F	T	T	T	**F**	F
T	F	T	T	F	F	**T**	T
T	F	F	T	F	F	**T**	F
F	T	T	F	F	T	**T**	T
F	T	F	F	F	T	**T**	F
F	F	T	F	F	F	**T**	T
F	F	F	F	F	F	**T**	F
1	2	3	1	4	2	5	3

Now Work Problem 19.

Remember, a conditional ($\rightarrow$) is true in all cases except when the antecedent is true and the consequent is false. A biconditional ($\leftrightarrow$) is true when both parts have the same truth value; otherwise it is false.

Finding the Truth Value of a Compound Statement without Using Truth Tables

A truth table gives us the truth value of a compound statement for each possible combination of the truth or falsity of the simple statements with the compound statement. If we want to find the truth value of the compound statement for any specific case when we know the truth values of the simple statements, we do not have to develop the entire table. For example, to develop the truth value for the statement

Ottowa is the capital of Canada and Buffalo is the capital of New York

Historical Note

The idea of a tautology in the current propositional sense was first developed in the early part of the twentieth century by the founder of American Pragmatism, a scientist and foremost a logician, Charles Sanders Peirce. The philosopher Ludwig Wittgenstein, who was one of the founding fathers of Linguistic Analysis, introduced the word tautology, giving it the current definition used in this book. From his book, *Tractatus Logical-Philosophicus* (1921): "In one of these cases the proposition is true for all the truth-possibilities of the elementary propositions. We say that the truth-conditions are tautological."

Charles Sanders Peirce (1839–1914) is now considered the most original and versatile intellect that this country has produced. In 1867 he was elected a fellow of the American Academy of Arts and Sciences and a member of the National Academy of Science in 1877. While a member of the latter, 1878 to 1911, he presented 34 papers covering areas such as logic, mathematics, physics, geodesy, spectroscopy, and experimental psychology. In 1880 he was elected a member of the London Mathematical Society.

Ludwig Wittgenstein (1889–1951), an Austrian-born English philosopher, is considered the most influential philosopher of the twentieth century. He introduced two influential and original philosophical systems of thought; his logical theories and his philosophy of language. The former was introduced in the

(continued on next page)

Tractatus. This book is a series of remarks that are numbered. Although only 75 pages long, it covers a wide range of topics: logic, language, ethics, philosophy, the self, will, death, good, evil, and the mystical. The last line of his book is good advice: "What we cannot speak about we must pass over in silence."

we let P = Ottowa is the capital of Canada and Q = Buffalo is the capital of New York. Now we can write the compound statement as $P \wedge Q$. We know that P is a true statement and Q is a false statement. Hence, we can substitute T for P and F for Q and evaluate the statement:

$$\begin{array}{c} P \wedge Q \\ T \wedge F \\ F \end{array}$$

Therefore, the compound statement is false.

EXAMPLE 6 ***Using Truth Values of Simple Statements*** Determine the truth value for each simple statement. Using these truth values, determine the truth values of the compound statement.

a. $4 + 5 = 9$ or $7 + 5 = 11$
b. $5 \times 4 = 20$ and $3 \times 3 = 6$
c. If the moon is made of green cheese, then water freezes at 0° Celsius.

Solution

a. Let P: $4 + 5 = 9$ and Q: $7 + 5 = 11$. We know that P is a true statement and Q is a false statement. Substitute T for P and F for Q and evaluate the statement.

$$\begin{array}{c} P \vee Q \\ T \vee F \\ T \end{array}$$

Therefore, "$4 + 5 = 9$ or $7 + 5 = 11$" is a true statement.

b. Let P: $5 \times 4 = 20$ and Q: $3 \times 3 = 6$. We know that P is a true statement and Q is a false statement. Substitute T for P and F for Q and evaluate the statement.

$$\begin{array}{c} P \wedge Q \\ T \wedge F \\ F \end{array}$$

Therefore, "$5 \times 4 = 20$ and $3 \times 3 = 6$" is a false statement.

c. Let P: The moon is made of green cheese. Q: Water freezes at 0° Celsius. We know that P is a false statement and Q is a true statement. Substitute F for P and T for Q and evaluate the statement.

$$\begin{array}{c} P \rightarrow Q \\ F \rightarrow T \\ T \end{array}$$

Therefore, "If the moon is made of green cheese, then water freezes at 0° Celsius" is a true statement.

NW ***Now Work Problem 33.***

EXAMPLE 7 ***2000 U.S. Presidential Election*** Use the bar graph in Figure 3.8 to determine the truth values of the given statements.

a. In the 2000 presidential election, more people younger than 30 voted for Bush and the majority of people 60 or older voted for Gore.

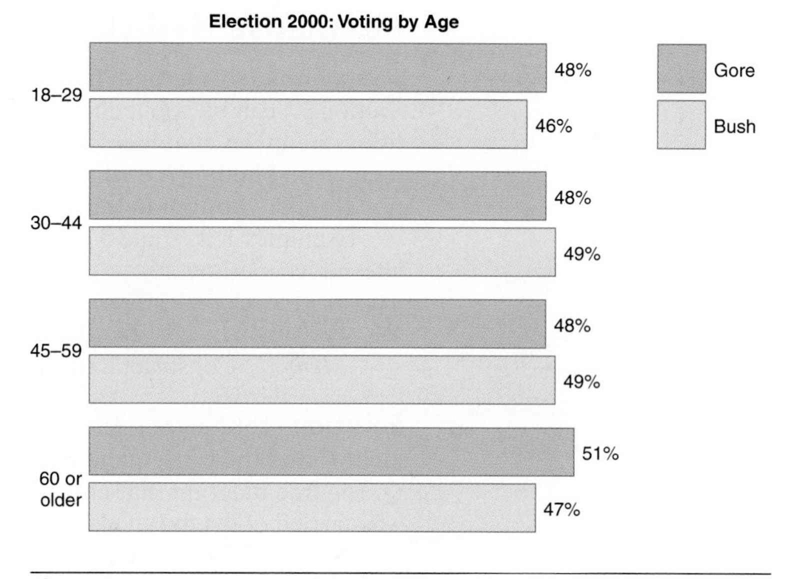

Figure 3.8
Source: Voter News Service

b. It is not the case that more 18–29 year-olds and fewer people between 45 and 59 voted for Bush.

c. If more people between 45 and 59 years old voted for Bush, then more people 60 or older did not vote for Bush.

Solution

a. If we let P = *more people younger than 30 voted for Bush* and Q = *the majority of people 60 or older voted for Gore*, the given statement can be symbolized as $P \wedge Q$. Reading the graph in Figure 3.8, observe that P is false because 48% of the 18–29 age group voted for Gore but only 46% voted for Bush. On the other hand, Q is true. Thus, we have

$$\begin{aligned} &P \wedge Q \\ &= F \wedge T \\ &= F \end{aligned}$$

Therefore, the given statement is false.

b. Let P = *more 18–29 year-olds voted for Bush* and Q = *fewer people between 45 and 59 voted for Bush*. The given statement is now symbolized as $\sim(P \vee Q)$. From the graph in Figure 3.8, observe that P is false and Q is false. Thus, we have

$$\begin{aligned} &\sim(P \wedge Q) \\ &= \sim(F \wedge F) \\ &= \sim(F) \\ &= T \end{aligned}$$

Therefore, the given statement is true.

c. Let P = *more people between 45 and 59 voted for Bush* and Q = *more people 60 or older voted for Bush*. The given statement is symbolized $P \rightarrow \sim Q$ and we have the following truth assignments:

$$\begin{aligned} &P \rightarrow \sim Q \\ &= T \rightarrow T \\ &= T \end{aligned}$$

Therefore, the given statement is true.

Alternative Method for Constructing Truth Tables

The alternative method for constructing truth tables that was presented in Section 3.4 can be applied to this section as well. As was done before, we first partition the given statement into separate components using a tree diagram. We then begin building a truth table for the statement by tracing the branches of the tree from the bottom to the top.

Examples 1, 2, 3, and 5 of this section are repeated here and solved using this alternative method.

EXAMPLE 1A ***Truth Table with Conditional and Negation*** *(Alternative Method)* Construct a truth table for $P \rightarrow \sim Q$.

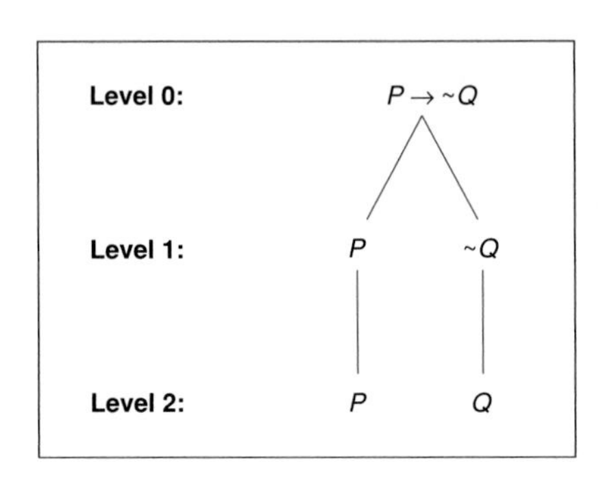

Figure 3.9

Solution This statement is a conditional. The antecedent is the simple statement P, and the consequent is $\sim Q$, which is the negation of the simple statement Q. The tree diagram that partitions this statement is given in Figure 3.9. (*Note:* P was carried down to Level 2 for completeness.)

We now begin building the truth table by working from the bottom level (Level 2) to the top level (Level 0) using the tree diagram as our guide.

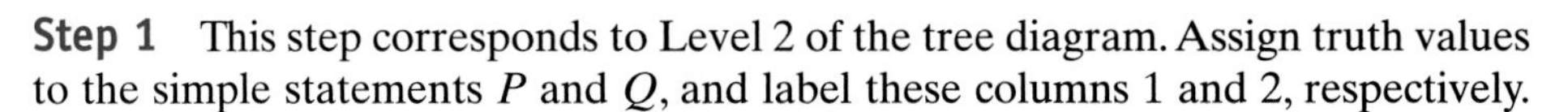

Step 1 This step corresponds to Level 2 of the tree diagram. Assign truth values to the simple statements P and Q, and label these columns 1 and 2, respectively.

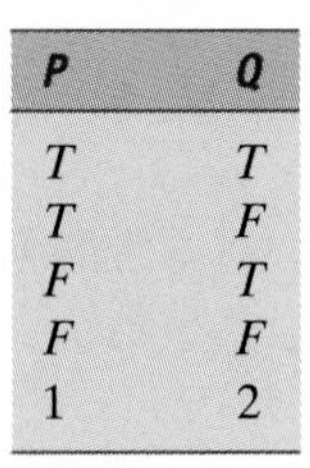

P	Q
T	T
T	F
F	T
F	F
1	2

Step 2 This step corresponds to Level 1 of the tree diagram. Moving up the tree from Level 2, we negate Q by negating the truth assignments given in column 2. Label this column 3. (Note that we do not have to do anything with P since that was taken care of in Level 2.)

P	Q	~Q
T	T	F
T	F	T
F	T	F
F	F	T
1	2	3

Step 3 This step corresponds to Level 0 of the tree diagram. Moving up the tree from Level 1, we combine P and $\sim Q$ into a conditional statement. This is done by combining the truth assignments of columns 1 and 3 under the rules of a conditional. Label this column 4. This last column represents the final answer, which is the truth assignments of the given statement.

P	Q	~Q	P → ~Q
T	T	F	**F**
T	F	T	**T**
F	T	F	**T**
F	F	T	**T**
1	2	3	4

NW ***Now Work Problem 5.***

EXAMPLE 2A *Truth Table with Conjunction, Conditional and Negation (Alternative Method)* Construct a truth table for $P \wedge Q \rightarrow \sim P$.

Solution This statement is a conditional. Its antecedent is the compound statement, $P \wedge Q$, which is a conjunction that consists of the simple statements P and Q. The consequent of the given conditional statement is the negation $\sim P$, which negates P. The tree diagram that partitions this statement is given in Figure 3.10. (*Note:* P and Q were carried down to Level 2 for completeness.)

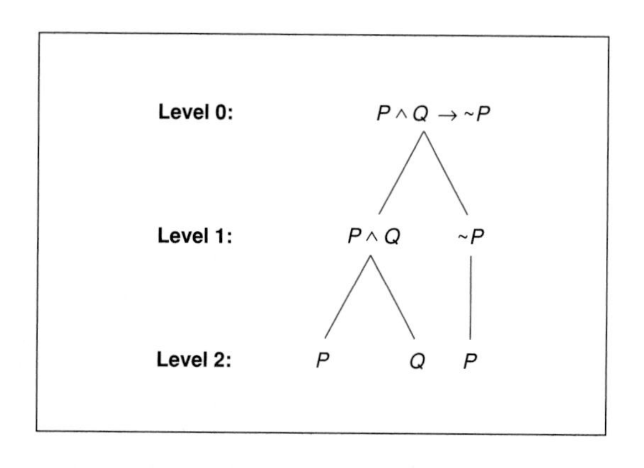

Figure 3.10

We now construct a truth table using the tree diagram as our guide. The completed truth table is

P	Q	$P \wedge Q$	$\sim P$	$P \wedge Q \rightarrow \sim P$
T	T	T	F	**F**
T	F	F	F	**T**
F	T	F	T	**T**
F	F	F	T	**T**
1	2	3	4	5

From the table, note the following: (1) Columns 1 and 2 correspond to Level 2 and are the preestablished truth assignments for the simple statements P and Q; (2) Column 3 corresponds to Level 1 and represents the truth assignments of columns 1 and 2 combined under the rules of conjunction; (3) Column 4 also corresponds to Level 1 and represents the negated truth assignments from column 1; and (4) Column 5 corresponds to Level 0 and represents the truth assignments of columns 3 and 4 combined under the rules of a conditional. Column 5 is the final column and represents the truth values for the given conditional statement.

NW ***Now Work Problem 9.***

EXAMPLE 3A *Truth Table with Biconditional (Alternative Method)* Construct a truth table for $(\sim P \vee Q) \leftrightarrow (P \rightarrow Q)$.

Solution This statement is a biconditional consisting of the following two parts: (1) the disjunction, $\sim P \vee Q$, and (2) the conditional, $P \rightarrow Q$. The disjunction is composed of $\sim P$ and the simple statement Q; $\sim P$ is the negation of the simple statement P. The conditional is composed of the simple statements P and Q, where P is the antecedent and Q is the consequent.

The tree diagram that partitions this statement is given in Figure 3.11.

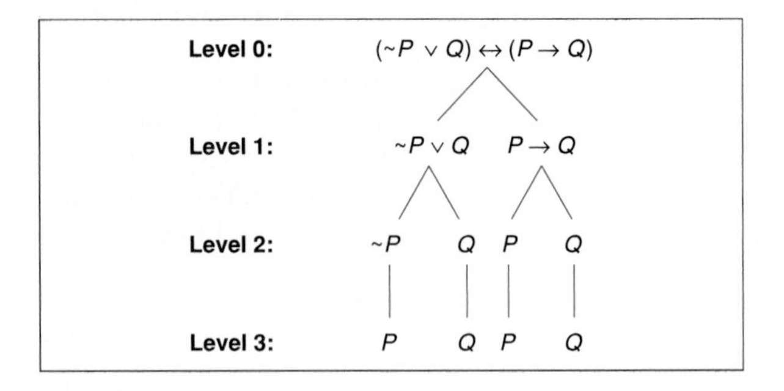

Figure 3.11

The completed truth table for this statement is as follows:

P	Q	$\sim P$	$\sim P \vee Q$	$P \to Q$	$(\sim P \vee Q) \leftrightarrow (P \to Q)$
T	T	F	T	T	**T**
T	F	F	F	F	**T**
F	T	T	T	T	**T**
F	F	T	T	T	**T**
1	2	3	4	5	6

Once again note the building process we use to construct this truth table. We begin with the simple statements at Level 3 (columns 1 and 2). We then move to Level 2 and negate P (column 3). Next we move to Level 1, where we complete the disjunction (column 4) by using columns 2 and 3, and complete the conditional (column 5) by using columns 1 and 2. Finally, we move to Level 0, where we complete the biconditional (column 6) by using columns 4 and 5.

NW ***Now Work Problem 11.***

EXAMPLE 4A ***Truth Table with Three Simple Statements*** *(Alternative Method)* Construct a truth table for $(P \wedge Q) \to R$.

Solution This statement is a conditional. Its antecedent is the conjunction $P \wedge Q$, which consists of the simple statements P and Q; its consequent is the simple statement R.

The tree diagram that partitions this statement is given in Figure 3.12

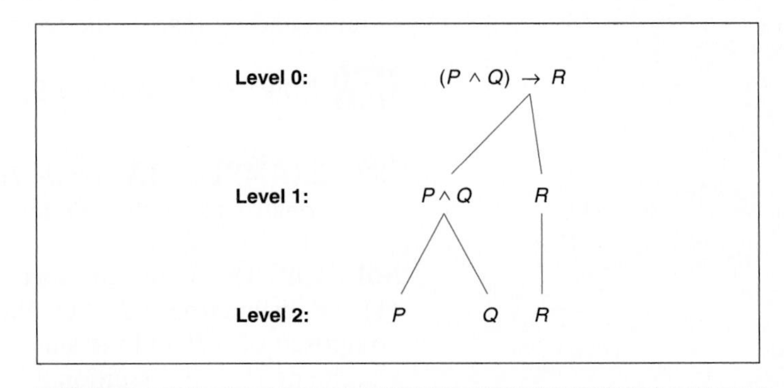

Figure 3.12

The completed truth table for this statement is as follows:

P	Q	R	$P \wedge Q$	$(P \wedge Q) \rightarrow R$
T	T	T	T	$\boldsymbol{T}$
T	T	F	T	$\boldsymbol{F}$
T	F	T	F	$\boldsymbol{T}$
T	F	F	F	$\boldsymbol{T}$
F	T	T	F	$\boldsymbol{T}$
F	T	F	F	$\boldsymbol{T}$
F	F	T	F	$\boldsymbol{T}$
F	F	F	F	$\boldsymbol{T}$
1	2	3	4	5

NW **Now Work Exercise 19.**

Fuzzy Logic

In Boolean logic, a statement has only one of two truth values—it is either true or false. Another way of saying this is that truth values of statements are *discrete,* that is, there is no "sort of true" or "sort of false." In some instances, however, this strict interpretation is not necessarily accurate. For example, in most situations a light bulb is either on or off. What happens, though, if we put a dimmer in the circuit? Although we can still say the light is on when we dim the light, it really isn't "fully" on. In other words, the dimmer creates a "degree of 'onness." ' That is, instead of asking if the light is on, we now ask "to what degree is the light on?" This concept of *degrees of truth* was initially introduced in the mid-1960s by Dr. Lotfi Zadeh of the University of California at Berkeley as a means to model the uncertainty of natural language. To address the concept of "partial truth," Zadeh described **fuzzy logic**, which is an extension of conventional (i.e., Boolean) logic. In fuzzy logic, the truth value of a statement is assigned 0 if the statement is false, 1 if the statement is true, and a number between 0 and 1 if the statement is partially true.

As an example, consider the set of old people from a classical set theory perspective (Chapter 2). As stated, this set is ill defined because there is no clear definition of what constitutes *old*. If we now define *old* as more than 30 years of age, then we have a well-defined set and can classify people as either old or not old (i.e., young). However, this rigid definition implies that on the day of a person's thirtieth birthday, he or she is considered young, but the very next day this person is considered old. In fuzzy logic, we can express the degree of oldness by letting age = 0 for all ages less than or equal to 30, age = 1 for all ages greater than or equal to 40, and age = a value between 0 and 1 for all ages between 30 and 40, exclusive. (*Note:* The age 40 was chosen arbitrarily.) This definition is expressed mathematically as

$$\text{old}(x) = \left\{ \begin{array}{ll} 0, & \text{if age}(x) \leq 30 \\ \dfrac{\text{age}(x) - 30}{10}, & \text{if } 30 < \text{age}(x) < 40 \\ 1, & \text{if age}(x) \geq 40 \end{array} \right\}$$

This is illustrated in the graph in Figure 3.13.

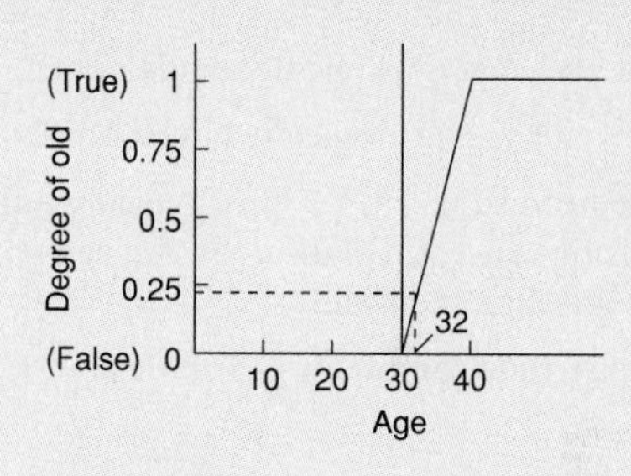

Figure 3.13

Given this definition, the statement *A 15-year-old person is old* is false (0) and the statement *A 40-year-old person is old* is true (1). However, the statement *A 32-year-old person is old* is "sort of true." It's degree of oldness is $(32 - 30)/10 = 0.2$. This is calculated by dividing the difference between a person's age and 30 by 10, which is the difference between the two end points (30 and 40).

Fuzzy logic has tremendous application in real-world settings, which are replete with "gray area" (i.e., vague) concepts such as fast, bright, tall, good, heavy, or cold. For example, Nissan Motors uses fuzzy logic to maintain efficient and stable control of its automobile engines. Nissan and Subaru employ fuzzy logic in their automobile cruise-control systems. Sony uses fuzzy logic as a means for recognizing handwritten symbols used in hand-held computers. Fuzzy logic is also used for backlight control in Sanyo camcorders and to compensate against vibrations in Matsushita camcorders. In short, fuzzy logic enables computers, which rely on discrete signals, to control devices that require analog interpretation or manipulation as part of their control.

Exercises for Section 3.5

In Exercises 1–20, construct a truth table for each symbolic statement.

1. $P \rightarrow Q$
2. $\sim(P \rightarrow Q)$
3. $\sim P \rightarrow \sim Q$
4. $P \rightarrow \sim Q$
5. NW $\sim P \rightarrow Q$
6. $\sim P \leftrightarrow Q$
7. $\sim P \leftrightarrow \sim Q$
8. $P \wedge Q \rightarrow P$
9. NW $P \vee Q \rightarrow \sim Q$
10. $P \rightarrow Q \vee \sim P$
11. NW $(P \rightarrow Q) \vee P \rightarrow Q$
12. $P \wedge (\sim P \vee Q) \rightarrow Q$
13. $P \wedge Q \leftrightarrow P \vee Q$
14. $\sim P \wedge \sim Q \leftrightarrow \sim(P \vee Q)$
15. $(P \vee Q) \wedge R$
16. $P \vee (Q \wedge R)$
17. $(P \wedge Q) \vee (P \wedge R)$
18. $P \wedge (Q \vee \sim R)$
19. NW $P \leftrightarrow Q \vee R$
20. $P \vee \sim(Q \wedge R) \rightarrow P \vee Q$
21. Determine whether $\sim(P \vee Q)$ is logically equivalent to $\sim P \wedge \sim Q$.
22. Determine whether $\sim(P \wedge Q)$ is logically equivalent to $\sim P \vee \sim Q$.
23. Are $P \wedge \sim Q$ and $\sim(\sim P \vee Q)$ logically equivalent?
24. Are $\sim P \vee Q$ and $\sim(P \wedge \sim Q)$ logically equivalent?
25. In Example 3 we found that $\sim P \vee Q$ is logically equivalent to $P \rightarrow Q$. Using this equivalence, rewrite each of the following statements. For example,

If today is Monday, then tomorrow is Tuesday.

can be rewritten as

Today is not Monday or tomorrow is Tuesday.

   a. If the tide is out, then we can go clamming.
   b. If Louise drove her new car, then Florence will accompany her.
   c. If today is Friday, then tomorrow is not Sunday.
   d. Either Chris does not toss the ball, or Gary punts.
   e. Carlos wasn't feeling well, or he will not go to New York City.

In Exercises 26–31, if P is true and Q is false and R is true, determine the truth value of the statement.

26. $P \wedge \sim Q$
27. $Q \rightarrow \sim R$
28. $P \wedge Q \rightarrow R$
29. $\sim P \vee Q \leftrightarrow \sim R$
30. $\sim(P \wedge \sim Q) \rightarrow R$
31. $\sim Q \vee R \leftrightarrow P \wedge \sim R$

In Exercises 32–40, determine the truth values for each simple statement. Use these truth values to determine the truth values of the compound statement.

32. $7 + 11 = 16$ or $3 \times 4 = 12$
33. NW $8 - 5 = 3$ and $4 - 1 = 3$
34. If $3 \div 6 = 0.5$ then $5 \times 2 = 9$.
35. $8 \times 2 = 10$ iff $3 + 1 = 2$
36. Tiger Woods plays golf or the cat barks.
37. If the Giants won the Super Bowl, then Bill Clinton was President of the United States.
38. Albany is the capital of New York, or Seattle is not in Washington and Baltimore is in Maryland.
39. If Elton John plays the piano, then Ringo plays the drums and Elvis played the violin.
40. Either Jackie Robinson was a Dodger and Babe Ruth played for the Red Sox, or Derek Jeter played for the Mets.

In Exercises 41–46, use the bar graph in Figure 3.14 to determine the truth values of the given statements.

41. Winter visitors averaged \$50.86 daily on food and entertainment, and they averaged less than \$30 per day on shopping and gifts.
42. Winter visitors averaged more than \$15 per day on other expenses or less than \$20 per day on gasoline/rental car expenses.

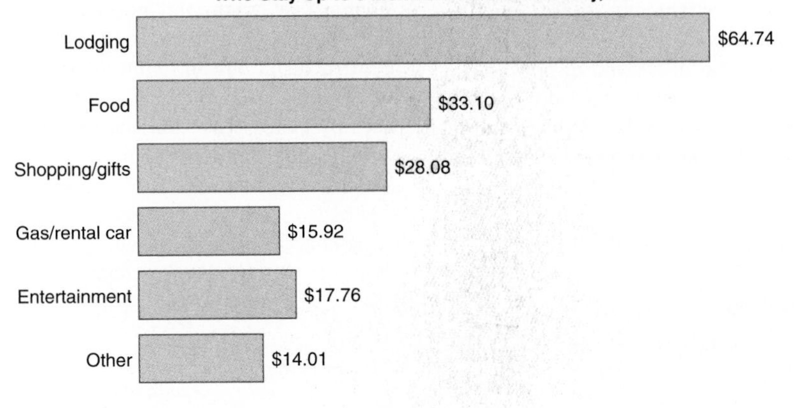

Figure 3.14
Source: 1998 Florida Space Coast Economic Impact Study

43. It is not the case that winter visitors averaged less than \$50 per day for lodging and they did not spend more than \$10 per day on gasoline/rental car expenses.

44. It is false that food or entertainment expenses were not more than \$50 per day or that shopping/gift expenses averaged more than \$25 per day.

45. If daily shopping/gift expenses averaged more than \$30 per day, then other expenses were less than \$20 per day.

46. If entertainment expenses were less than \$20 per day, then food expenses were not more than \$30 per day.

On Line Purchases *In Exercises 47–50, use the bar graph in Figure 3.15 to determine the truth values of the given statements.*

47. The two largest percentages of people intend to purchase music CDs and books, or more than 20% of the people polled intend to purchase toys/games as Christmas gifts via the Internet.

48. The largest percentage of people do not intend to purchase apparel and footwear, or the largest percentage of people intend to purchase wine/gift baskets or at least 50% of the people intend to purchase books.

49. If less than 25% of the people polled intend to purchase music CDs, then it is false that more than 50% of the people intend to purchase books or the percentage of people who intend to purchase toys/games is greater than the percentage of people who intend to make wine/gift basket purchases.

50. The largest percentage of people do not intend to buy books if and only if the percentage of people who intend to purchase music CDs is greater than the percentage of people who intend to purchase toys/games, and the percentage of people who intend to purchase wine/gift baskets is not less than 15%.

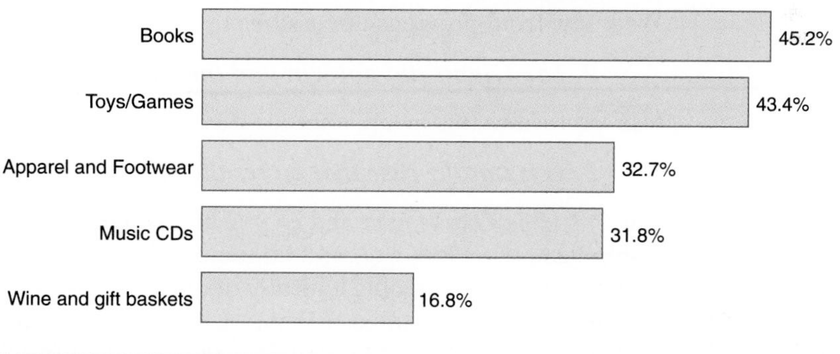

Figure 3.15
Source: Jupiter Media Metrix (www.JMM.com)

Writing Mathematics

51. When is a conditional true?
52. When is a conditional false?
53. When is a biconditional true?
54. When is a biconditional false?
55. What is a tautology? Give at least two examples.

Challenge Exercises

56. Construct a truth table for $P \wedge Q \rightarrow R \vee S$.

57. Construct a truth table for $\sim P \wedge Q \rightarrow R \wedge \sim S$.

Group/Research Activities

58. Using logic, solve the following problem. Each member of the group should be able to explain and defend the solution.

Pam, Dana, and Addie live next to each other in an apartment complex. One is an artist, one is a lawyer and one is a teacher. Dana lives in the middle. When Addie is away, the artist feeds her cat. The lawyer bangs on Pam's wall when the music is too loud. What is the occupation of each woman?

59. The problem given in Exercise 58 is referred to as a logic recreation problem. Similar logic recreations or logic puzzles can be found in the book *Games for the Super-Intelligent* by James Fixx (BBS Publishing, NY) in the chapter entitled "Those Wonderful Laws of Logic (and How They Can Fool You Every Time)." Review this chapter and try to solve some of the puzzles presented. Select one puzzle and present it to your class as an exercise.

Just for fun

If 3 cats kill 3 rats in 3 minutes, then how many cats will it take to kill 30 rats in 30 minutes?

3.6 DE MORGAN'S LAW AND EQUIVALENT STATEMENTS

In the preceding section, we were introduced to logically equivalent statements. Two statements that have exactly the same truth values were said to be *logically equivalent*. In Example 3 of Section 3.5, we showed that $P \rightarrow Q$ is logically equivalent to $\sim P \vee Q$. We can write this as

$$P \rightarrow Q \equiv \sim P \vee Q$$

We determined this by constructing the truth table for the statement $(\sim P \vee Q) \leftrightarrow (P \rightarrow Q)$, and obtaining all T's for the biconditional (a tautology). By means of this technique, we can always determine if one statement is logically equivalent to another.

Sometimes when we are given a statement we can create another statement that is logically equivalent to it. We can change a disjunction, a conjunction, or the negation of one of these to an equivalent statement by means of a rule known as **De Morgan's law**. To illustrate how this works, consider the statement

1. Neither the Giants nor the Browns won.

We know from previous discussions that this is the same as

2. The Giants did not win and the Browns did not win.

Can we restate this in still another way? Consider

3. It is not the case that either the Giants or Browns won.

Let P = *The Giants won* and Q = *The Browns won*; then statement 2 would be symbolized as $\sim P \wedge \sim Q$, and statement 3 would be symbolized as $\sim(P \vee Q)$. Are these two statements logically equivalent? We can determine if they are by means of a truth table (Table 3.10). Let's examine the truth table for $(\sim P \wedge \sim Q) \leftrightarrow \sim(P \vee Q)$. If it is a tautology, then we will know that the two statements are equivalent.

Biography: Augustus De Morgan

Augustus De Morgan (1806–1871) was born of British parents in India, but spent most of his life in and around London, England.

As a youngster De Morgan had to endure intense ridicule from his classmates as he had only one eye. Consequently, he developed an indifferent attitude toward social activities and a dislike for the outdoors, athletics, and animals. De Morgan, instead, turned to books. At an early age he was a bookworm, he spoke five languages fluently, played the flute masterfully, and was well on his way to becoming a mathematician.

De Morgan became known and respected as an energetic worker, a prolific writer, and an excellent instructor. His total dedication to his work won him an appointment as the first professor of mathematics at University College, London.

His attitudes on religious freedom were reflected on a number of occasions throughout his life. While in college, De Morgan was denied a candidacy for a fellowship and was also refused the right to continue on to the M.A. degree because he would not take and sign a theological test. Later in life he resigned his professorship in protest against the preference given to members of the Church of England in the selection of books for the college and the religious philosophy of the college itself. He was a man of integrity and unafraid of making personal sacrifices for a cause or a principle in which he believed.

TABLE 3.10

P	Q	$(\sim$	P	$\wedge$	$\sim$	$Q)$	$\leftrightarrow$	$\sim$	$(P$	$\vee$	$Q)$
T	T	F		F	F		**T**	F	T	T	T
T	F	F		F	T		**T**	F	T	T	F
F	T	T		F	F		**T**	F	F	T	T
F	F	T		T	T		**T**	T	F	F	F
1	2	3		5	4		8	7	1	6	2

Column 8 of Table 3.10 shows us that the given statement is a tautology, thereby verifying the fact that statements 2 and 3 are logically equivalent.

De Morgan's law enables us to create an equivalent statement when we are given a certain type of statement. To begin with, we must have some form of a disjunction or conjunction, or the negation of one of these. In order to create an equivalent statement using De Morgan's law, we must perform the following three steps:

1. Negate the whole statement.
2. Negate each statement that makes up the disjunction or conjunction.
3. Change the conjunction to a disjunction or the disjunction to a conjunction.

EXAMPLE 1 ***De Morgan's law and Conjunction*** Use De Morgan's law to create a statement equivalent to $P \wedge Q$.

Solution We first negate the whole statement, which gives us $\sim(P \wedge Q)$. Next we negate each part, and that yields $\sim(\sim P \wedge \sim Q)$. We then make the third and final change, changing $\wedge$ to $\vee$, which gives us $\sim(\sim P \vee \sim Q)$. Hence, $P \wedge Q \equiv \sim(\sim P \vee \sim Q)$.

NW ***Now Work Problem 1.***

EXAMPLE 2 ***De Morgan's law with Multiple Negations*** Use De Morgan's law to create a statement equivalent to $\sim(\sim P \vee Q)$.

Solution We are given $\sim(\sim P \vee Q)$. First we negate the whole statement, which therefore makes it $\sim\sim(\sim P \vee Q)$ or $(\sim P \vee Q)$. Now we negate each part, that is, $\sim\sim P \vee \sim Q$, which leaves us with $P \vee \sim Q$. (Note that $\sim\sim P \equiv P$; $\sim\sim P$ is called a **double negation.**) We apply the third step and change $\vee$ to $\wedge$, which yields $P \wedge \sim Q$. Finally, we have $\sim(\sim P \vee Q) \equiv P \wedge \sim Q$.

NW ***Now Work Problem 3.***

EXAMPLE 3 ***De Morgan's law with Negation and Conjunction*** Use De Morgan's law to write a statement equivalent to the statement "It is false that Allan abhors anatomy and that Lucy loves Latin."

Solution Let A = *Allan abhors anatomy* and L = *Lucy loves Latin*; then the given sentence may be symbolized as $\sim(A \wedge L)$. Now we create an equivalent statement using De Morgan's law: $\sim(A \wedge L) \equiv \sim A \vee \sim L$. Translating the new statement, we have "Either Allan does not abhor anatomy or Lucy does not love Latin."

NW ***Now Work Problem 11.***

EXAMPLE 4 ***De Morgan's law with Negation and Conjunction*** Use De Morgan's law to write a statement equivalent to the statement "Sally did not stay and Quincy quit."

Solution Let S = *Sally stayed* and Q = *Quincy quit.* The statement may be symbolized as $\sim S \wedge Q$. Using De Morgan's law, we have $\sim S \wedge Q \equiv \sim(S \vee \sim Q)$. Translating the equivalent statement, we have "It is not the case that Sally stayed or that Quincy did not quit."

EXAMPLE 5 ***De Morgan's law with Negation and Conditional*** Use De Morgan's law to create a statement equivalent to $\sim(P \rightarrow Q)$.

Solution At first glance, it may seem that we cannot do this problem, because we do not have a disjunction or conjunction. However, recall that in Example 3 of Section 3.5 we discovered that $P \rightarrow Q \equiv \sim P \vee Q$. Therefore, we can take the given statement and rewrite it as $\sim(\sim P \vee Q)$. Now we are ready to use De Morgan's law: $\sim(\sim P \vee Q) \equiv P \wedge \sim Q$. Therefore, $\sim(P \rightarrow Q) \equiv P \wedge \sim Q$.

Note that Example 5 verifies that a statement such as

It is false that if today is Friday, then I get paid.

is equivalent to

Today is Friday and I do not get paid.

It is interesting to note that the logical operation *and* corresponds to the set operation *intersection*. It is also the case that *or* corresponds to set *union,* and that *not* corresponds to set *complementation*. Two of the most common examples of De Morgan's law for equivalent statements are

$$\sim(P \wedge Q) \equiv \sim P \vee \sim Q$$
$$\sim(P \vee Q) \equiv \sim P \wedge \sim Q$$

The corresponding expressions in set theory are

$$(A \cap B)' = A' \cup B'$$
$$(A \cup B)' = A' \cap B'$$

These two statements can be shown to be true by means of Venn diagrams (see Example 5 and Exercise 5c in Section 2.5). Therefore, you can see that set operations are similar to some of those in logic. The basic difference is in the notation.

Remember that in logic we can use De Morgan's law only on some form of a disjunction, conjunction, or negation of one of these. Table 3.11 gives a list of logically equivalent statements. Their equivalence can be verified by means of truth tables.

Table 3.11 Equivalent Statements

De Morgan's law	$\sim(P \wedge Q) \equiv \sim P \vee \sim Q$
	$\sim(P \vee Q) \equiv \sim P \wedge \sim Q$
Implication	$P \rightarrow Q \equiv \sim P \vee Q$
Contraposition	$P \rightarrow Q \equiv \sim Q \rightarrow \sim P$
Biconditional	$P \leftrightarrow Q \equiv (P \rightarrow Q) \wedge (Q \rightarrow P)$
Association	$(P \wedge Q) \wedge R \equiv P \wedge (Q \wedge R)$
	$(P \vee Q) \vee R \equiv P \vee (Q \vee R)$
Distribution	$P \wedge (Q \vee R) \equiv (P \wedge Q) \vee (P \wedge R)$
	$P \vee (Q \wedge R) \equiv (P \vee Q) \wedge (P \vee R)$
Idempotent	$P \wedge P \equiv P$
	$P \vee P \equiv P$

Exercises for Section 3.6

In Exercises 1–10, use De Morgan's law to create a statement equivalent to each given statement.

NW 1. $P \vee Q$
2. $\sim(P \vee \sim Q)$
NW 3. $\sim(\sim P \wedge Q)$
4. $\sim P \vee Q$
5. $P \wedge \sim Q$
6. $\sim P \vee \sim Q$
7. $\sim(\sim P \vee \sim Q)$
8. $\sim(\sim P \wedge \sim Q)$
9. $\sim(P \rightarrow \sim Q)$
10. $\sim[P \vee (Q \wedge R)]$

In Exercises 11–22, use De Morgan's law to rewrite each statement.

NW 11. It is false that Larry is busy and cannot serve on the committee.
12. Neither Bill nor Mary is tall.
13. I did not pass the test, or I studied too much.
14. Either Julia studied or she failed the exam.
15. It is false that Pam stayed late or Angie left early.
16. Neither today nor tomorrow is a holiday.
17. Either the bus is late, or my watch is not working correctly.
18. You cannot go to the concert, but I have a ticket.
19. It is not the case that x is greater than zero and that x is negative.
20. Either two is a counting number or it is an integer.
*21. If the wind doesn't come up, then we can't sail.

*22. If you do not attend class, then you will be dropped from the course.

Use De Morgan's law on the set expressions in Exercises 23–30 to create equivalent expressions.

23. $(A \cap B)'$
24. $(A \cup B)'$
25. $A' \cap B'$
26. $A' \cup B$
27. $A \cap B$
28. $(A' \cap B)'$
29. $(A \cup B')'$
30. $(A \cap B')'$

Writing Mathematics

31. What are equivalent statements?
32. How can you determine whether two statements are equivalent?
33. Describe how to construct a logically equivalent statement using De Morgan's law.
34. Describe how to construct a statement logically equivalent to a conditional statement.

Challenge Exercises

35. Construct a statement that is logically equivalent to $\sim[\sim(\sim P \wedge Q)]$.
36. Construct a statement logically equivalent to $P \rightarrow \sim Q$ that is a negation.
37. Construct a statement that is logically equivalent to $\sim[\sim P \wedge (Q \vee \sim R)]$.

The implication rule states that $P \rightarrow Q \equiv \sim P \vee Q$. Using this equivalence, rewrite the sentences (in words) given in Exercises 38–43.

38. If today is Wednesday, then tomorrow is Friday.
39. If Horace did not play, then we lost.
40. Derrick did not pass the test or he is unhappy about something else.
41. Either two equals three or four does not equal six.
42. Tiger will win, if he sinks this putt.
43. We will go skating, if the pond freezes over.

Group/Research Activity

44. Prepare a report on De Morgan. As part of your report, include the contributions De Morgan made to the concepts of mathematical induction and complex numbers, as well as his fascination with odd numerical facts (e.g., he noted that he had the distinction of being x years old in the year x^2).

Real-World Data

Use Figure 3.16 to do the following for each given statement in Exercises 45–50:

a. Symbolize the statement using the letters in parentheses.
b. Determine if it is true or false.
c. Use De Morgan's law to write an equivalent statement in symbolized and word form.

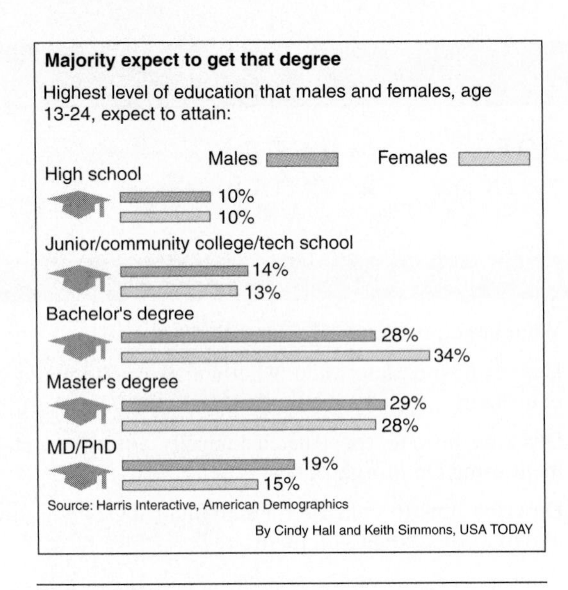

Figure 3.16

45. Thirty-four percent of women expect to earn a bachelor's degree (B) and 29% of women expect to earn a master's degree (M).

46. More than 50% of men and women expect to get a bachelor's degree (B), or less than 10% of women expect to get a doctorate (D).

47. If 10% of men expect to get a high school diploma (M), then 25% of women expect to get a master's degree (W).

48. It is not the case that if less than 20% of men expect to get a master's degree (M), then 10% of women expect to get a high school diploma (W).

49. If more than 50% of women expect to get a master's or doctorate degree (W), then 50% of men also expect to get a master's or doctorate degree (M).

50. If more than 50% of men expect their highest degree to be a bachelor's degree (M), then it is false that the percentage of women who expect to get a master's or doctorate is greater than those of men (W).

Just for fun

Which straight-line segment is longer?

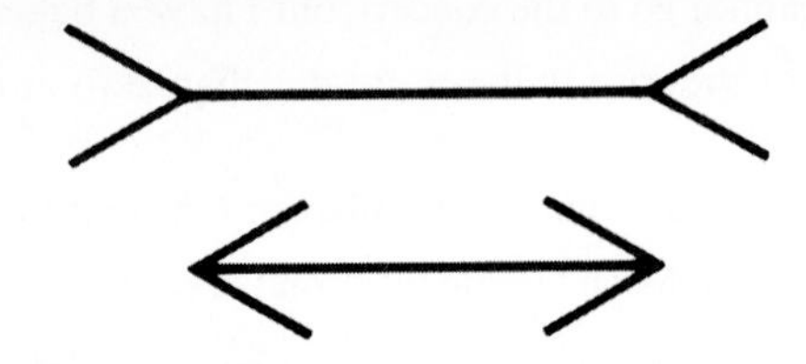

3.7 THE CONDITIONAL (OPTIONAL)

Forms of the Conditional: Converse, Inverse, and Contrapositive

All math students have, at one time or another, encountered rules or theorems stated as conditionals. One of the first theorems proved in high school geometry is this conditional: "If two sides of a triangle are equal, then the angles opposite these sides are equal."

In this section, we shall examine the conditional statement in detail. We shall concern ourselves with statements that are equivalent to *If P, then Q*, and the various forms that these statements may take.

Suppose we are given two statements, P and Q, and we are asked to construct a conditional statement using P and Q, or $\sim P$ and $\sim Q$. What are the possible variations? After some thought, you would probably come up with the following list (their names are included):

1. $P \rightarrow Q$ Statement
2. $Q \rightarrow P$ *Converse* of statement 1
3. $\sim P \rightarrow \sim Q$ *Inverse* of statement 1
4. $\sim Q \rightarrow \sim P$ *Contrapositive* of statement 1
5. $\sim(P \rightarrow Q)$ *Negation* of statement 1

Note, however, that statement 5, $\sim(P \rightarrow Q)$ is not a conditional statement but the *negation* of a conditional statement. Therefore, we will concern ourselves with only the first four statements. By comparing their truth tables, we can discover which statements are logically equivalent.

From Table 3.12 we see that a conditional statement is logically equivalent to its contrapositive. Hence, when we have a statement such as "If I study, then I shall pass math," we know that it is equivalent to "If I did not pass math, then I did not study."

TABLE 3.12 Different Forms of a Conditional

Statement			*Converse*			*Inverse*			*Contrapositive*		
P	$\rightarrow$	Q	Q	$\rightarrow$	P	$\sim P$	$\rightarrow$	$\sim Q$	$\sim Q$	$\rightarrow$	$\sim P$
T	**T**	T	T	**T**	T	F	**T**	F	F	**T**	F
T	**F**	F	F	**T**	T	F	**T**	T	T	**F**	F
F	**T**	T	T	**F**	F	T	**F**	F	F	**T**	T
F	**T**	F	F	**T**	F	T	**T**	T	T	**T**	T

The conditional statement is dangerous in the hands of the uninformed. A common error made by some people in reasoning about mathematics or everyday occurrences is to assume that the converse or inverse of a given statement has the same truth value as the given statement. This is particularly the case when the given conditional statement seems to be obviously true. Checking the truth tables in Table 3.12, we see that the inverse and converse of a statement are not always true when the given statement is true.

In some of the advertising that we see, hear, or read, the advertiser wants us to believe that the inverse or converse of a statement is true. Consider the following:

For a cleaner wash, insist on Sudsy Soap.

This translates to "If you want a cleaner wash, then use Sudsy Soap." The advertiser wants you to use Sudsy Soap, but he or she also wants you to believe that if you do not use Sudsy Soap, then your wash will not be clean; hence, you had better buy some Sudsy Soap.

Consider another conditional statement,

If x is greater than zero, then x is not negative.

This statement is true. Its converse is

If x is not negative, then x is greater than zero.

This statement is not true because x could be zero (zero is neither positive nor negative), and therefore x would not be greater than zero. Thus we see that an inverse or converse does not follow just because the given statement is true.

Recall that *then* is omitted in a conditional statement when it is thought to be obvious. Consider the following:

If anybody was late, it had to be Dave.

EXAMPLE 1 ***Writing Different Forms of a Conditional*** Write the converse, inverse, and contrapositive of the statement

If you use Sudsy Soap, then your clothes are clean.

Solution Let P = *You use Sudsy Soap* and Q = *Your clothes are clean.* We can symbolize the given statement as $P \rightarrow Q$.

The converse of $P \rightarrow Q$ is $Q \rightarrow P$, or "If your clothes are clean, then you use Sudsy Soap." Note that even if the given statement is true, the converse is not necessarily true.

The inverse of $P \rightarrow Q$ is $\sim P \rightarrow \sim Q$, which may be stated as "If you do not use Sudsy Soap, then your clothes are not clean." This statement is also not necessarily true.

The contrapositive of $P \rightarrow Q$ is $\sim Q \rightarrow \sim P$. "If your clothes are not clean, then you do not use Sudsy Soap." If the given statement is true, then the contrapositive is also true because it is logically equivalent to the given statement.

EXAMPLE 2 ***Writing Different Forms of a Conditional*** Write the converse, inverse, and contrapositive of the statement

If today is Sunday, then yesterday was not Friday.

Solution Let P = *Today is Sunday* and Q = *Yesterday was Friday.* The given statement is symbolized as $P \rightarrow \sim Q$.

The converse is $\sim Q \rightarrow P$, "If yesterday was not Friday, then today is Sunday."

The inverse of $P \rightarrow \sim Q$ is $\sim P \rightarrow \sim\sim Q$. Since we have a double negation, $\sim\sim Q$, we may therefore interpret it as Q; that is, $\sim\sim Q \equiv Q$. Hence the inverse $\sim P \rightarrow \sim\sim Q$ is the same as $\sim P \rightarrow Q$: "If today is not Sunday, then yesterday was Friday."

The contrapositive of $P \rightarrow \sim Q$ is $\sim\sim Q \rightarrow \sim P$, which is thus the same as $Q \rightarrow \sim P$: "If yesterday was Friday, then today is not Sunday."

NW ***Now Work Problems 1a and b.***

Additional Forms of the Conditional: Only If, Implies, Whenever, Sufficient, and Necessary

The conditional causes a great deal of confusion and trouble for many people because they do not understand it. We shall use an example to show the various ways that a conditional may be stated. You should already be aware that $P \rightarrow Q$ can be translated as

If P, then Q.
If P, Q.
Q if P.

In the following discussion we shall see that $P \rightarrow Q$ may also be translated in other ways. Consider the statement

If you are a citizen of Buffalo, then you are a citizen of New York State.

P = *You are a citizen of Buffalo*

Q = *You are a citizen of New York State*

Then the statement would be symbolized as $P \rightarrow Q$.

Now, suppose someone says "You are a citizen of Buffalo only if you are a citizen of New York State." What can we conclude if you live in New York State? We know that you live in the state of New York, but you may live in Buffalo, Rochester, Syracuse, New York City, or a thousand other places. But what can we conclude if you live in Buffalo? We know that you are a citizen of New York State because you are a citizen of Buffalo *only if you are a citizen of New York State. If* you live in Buffalo, *then* you must live in New York State.

Another statement that may help you understand P *only if* Q is

It snows only if it is cold.

If the thermometer tells us it's cold, do we know anything? We know it's cold, but we do not know if it will snow. But suppose we look out the window and see that it is snowing. Then we can deduce that it is cold. Why? Because we know that it snows only if it is cold. That is, *if* it snows, *then* it is cold. We would conclude that P *only if* Q may be symbolized as $P \rightarrow Q$.

Consider a variation of the citizenship statement, "Being a citizen of Buffalo is sufficient for being a citizen of New York State." This statement means that in order to be a citizen of New York State, it is enough to be a citizen of Buffalo, since every citizen of the city of Buffalo is also a citizen of the state of New York. Do you have to do anything else? No, because *if* you are a citizen of Buffalo, *then* you are a citizen of New York State.

Suppose an instructor tells you that doing your homework every day is sufficient for passing the course. What do you know? Do you have to do your homework every day? The instructor did not say that you had to do your homework every day; but he did say that if you did, then you would pass the course. You could pass the course some other way, perhaps by passing all of the exams. But you know that if you do your homework every day, then you will pass the course. Suppose you did not pass the course? We may conclude that you did not do your homework every day. That is, if you did not pass the course, then you did not do your homework every day. But this is the contrapositive of "If you do your homework every day, then you will pass the course." We may conclude from these discussions that P *is sufficient for* Q may be symbolized as $P \rightarrow Q$.

Another variation of "If you are a citizen of Buffalo, then you are a citizen of New York State" is the following sentence: "Being a citizen of New York State is necessary for being a citizen of Buffalo." This means that being a citizen of New York State is a necessary condition for being a citizen of Buffalo. Do you have to be a citizen of Buffalo in order to be a citizen of New York? No, but you have to be a citizen of New York State in order to be a citizen of Buffalo. Since you have to be a citizen of New York State, we might conclude that

If you are not a citizen of New York State, then you are not a citizen of Buffalo.

But that is the contrapositive of

If you are a citizen of Buffalo, then you are a citizen of New York State.

Therefore, we may conclude that Q *is necessary for* P may be symbolized as $P \rightarrow Q$.

Another example that may aid you in understanding Q *is necessary for* P is the following:

Oxygen is necessary for fire.

Given this statement, we know that we cannot have fire unless we have oxygen. That is, if we do not have oxygen, then we cannot have a fire. Again, we can apply the contrapositive to this, and we have the statement

If we have fire, then we have oxygen.

In summary, we have the fact that Q *is necessary for* P is symbolized as $P \rightarrow Q$.

Consider the statement

If you studied Algebra II, then you studied Algebra I.

Three other sentences that have the same meaning are

You studied Algebra II only if you studied Algebra I.

Studying Algebra II is sufficient for studying Algebra I.
Studying Algebra I is necessary for studying Algebra II.

When we have the conditional statement $P \rightarrow Q$, it is usually interpreted as *If P, then Q*, but three other common equivalent wordings are

P only if Q.
P is sufficient for Q.
Q is necessary for P.

The following is a list of equivalent wordings of the conditional statement $P \rightarrow Q$. Bear in mind that the list is a sampling and not intended to be a list of all possible variations, but all statements listed are interpretations of the original conditional statement.

1. *If P, then Q.*
2. *If P, Q.*
3. *Q if P.*
4. *P implies Q.*
5. *Q is implied by P.*
6. *Q whenever P.*
7. *P only if Q.*
8. *P is sufficient for Q.*
9. *Q is necessary for P.*

(All of the above: $P \rightarrow Q$)

The following examples are conditional statements using some of the variations in the previous list:

1. If the Mets won, then the Dodgers lost.
2. If the Mets won, the Dodgers lost.
7. The Mets won only if the Dodgers lost.
8. The Mets' winning is sufficient for the Dodgers' losing.
9. The Dodgers' losing is necessary for the Mets' winning.

EXAMPLE 3 ***Symbolizing Different Forms of a Conditional*** Let P = *Today is Saturday* and Q = *Tomorrow is Sunday*. Write each of the following statements in symbolic form:

a. If today is Saturday, then tomorrow is Sunday.
b. If tomorrow is not Sunday, then today is not Saturday.
c. Tomorrow is Sunday if today is Saturday.
d. Today's being Saturday is sufficient for tomorrow's being Sunday.
e. Tomorrow's being Sunday is necessary for today's being Saturday.
f. Today is not Saturday only if tomorrow is not Sunday.
g. It is false that today's being Saturday is sufficient for tomorrow's being Sunday.
h. Tomorrow's not being Sunday is sufficient for today's not being Saturday.

Solution

a. $P \rightarrow Q$ b. $\sim Q \rightarrow \sim P$ c. $P \rightarrow Q$
d. $P \rightarrow Q$ e. $P \rightarrow Q$ f. $\sim P \rightarrow \sim Q$
g. $\sim(P \rightarrow Q)$ h. $\sim Q \rightarrow \sim P$

NW ***Now Work Problems 3a, b, and c.***

Forms of the Biconditional: Necessary and Sufficient

Because a conditional statement may be written in various ways, the same is true for a biconditional statement. Consider this statement:

A number is positive if and only if that number is greater than zero.

From the discussion on truth tables in Section 3.5, we know that this statement may be rewritten as a conjunction:

If a number is positive, then the number is greater than zero, and if a number is greater than zero, then the number is positive.

Let P = *A number is positive* and Q = *A number is greater than zero*. Then the preceding statement would be symbolized as

$$(P \rightarrow Q) \wedge (Q \rightarrow P)$$

For the first part of the conjunction, we can say *P is sufficient for Q,* and for the second part, *P is necessary for Q*. Or we could switch them around: *P is necessary for Q* and *P is sufficient for Q,* or simply *P is necessary and sufficient for Q:*

$$P \leftrightarrow Q \begin{cases} P \textit{ if and only if } Q \\ P \textit{ is necessary and sufficient for } Q \end{cases}$$

EXAMPLE 4 ***Symbolizing Different Forms of a Biconditional*** Using the suggested notation, write the following in symbolic form:

a. You may play on this course if and only if you are a member of Oak Hill. (*C, O*)

b. Citizens become irate, whenever, and only whenever, their taxes are raised. (*C, T*)

c. Being a student here is a necessary and sufficient condition for obtaining a yearbook. (*S, Y*)

d. Quickness and coordination are necessary and sufficient for being a pole vaulter. (*Q, C, P*)

Solution

a. $C \leftrightarrow O$ **b.** $C \leftrightarrow T$ **c.** $S \leftrightarrow Y$ **d.** $(Q \wedge C) \leftrightarrow P$ 

Exercises for Section 3.7

1. Use the suggested notation to write the converse, inverse, and contrapositive for each of the following statements in symbolic form:

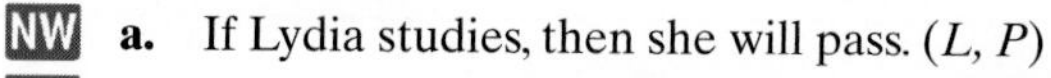

NW **a.** If Lydia studies, then she will pass. (*L, P*)

NW **b.** We will pick apples if they are ripe. (*P, R*)

c. If the Bills do not win, then the fans will be sad. (*B, F*)

d. If French is difficult, then mathematics is not easy. (*F, M*)

e. If today isn't your birthday, then there will not be a party. (*B, P*)

2. **Hurricane** "If the storm is a hurricane, then its winds are traveling at least 74 miles per hour." Given that this is a true statement, which of the following must also be true? (Use your knowledge of the conditional.)

a. If the storm is not a hurricane, then its winds are not traveling at least 74 miles per hour.

b. If the storm's winds are traveling at least 74 miles per hour, then the storm is a hurricane.
c. It is false that if the storm's winds are not traveling at least 74 miles per hour, then the storm is not a hurricane.
d. If the storm's winds are not traveling at least 74 miles per hour, then the storm is not a hurricane.
e. The storm is not a hurricane, or the storm's winds are traveling at least 74 miles per hour.
f. The storm's winds are not traveling at least 74 miles per hour, or the storm is not a hurricane.

3. Let P = *It is a logic course* and Q = *It is interesting*. Write each of the following statements in symbolic form:
NW a. It is a logic course only if it is interesting.
NW b. If it isn't interesting, then it isn't a logic course.
NW c. The fact that it is a logic course is sufficient for it to be interesting.
d. Only if it is interesting is it a logic course.
e. Being a logic course is necessary and sufficient for it to be interesting.

4. Use the suggested notation to write the following in symbolic form:
a. A student can be successful in mathematics only if he does his homework and pays attention in class. (S, H, P)
b. A sufficient condition for a student to flunk is that he not attend class. (S, A)
c. A necessary condition for me to go swimming is that the water must be warm or the air temperature must be 80° F. (S, W, T)
d. If a person is contributing to society, he is contributing to society only if he is helping his fellow man. (C, H)
e. Only exams are boring. (E, B)

5. Let P = *He works hard* and Q = *He is happy*. Write each of the following statements in symbolic form:
a. He works hard only when he is happy.
b. If he isn't working hard, then he isn't happy.
c. His working hard is necessary for his being happy.
d. He works hard iff he is happy.
*e. He is not happy unless he works hard.

6. Sneaky Sam made the following promise to Gullible Gary: "I will help you only if you give me five dollars." Gary gave Sam five dollars and Sam walked away. Did Sam break his promise? Why or why not?

7. An instructor told a student "I will pass you only if you come to class every day." The student came to class every day and still failed the course. Does the student have a legitimate gripe? Why or why not?

8. Let P = *It is a logic course* and Q = *It is interesting*. Write each of the following in symbolic form.
a. It is a logic course only if it is interesting.
b. Only if it is interesting, is it a logic course.
c. Being interesting is a necessary condition for it to be a logic course.
d. The fact that it is a logic course is a sufficient condition for it to be interesting.

Writing Mathematics

9. Describe how to obtain the contrapositive statement from a given conditional statement.
10. Describe how to obtain the implication statement from a given conditional statement.

Challenge Exercise

11. Do you think the following statement is a tautology? Why or why not? Confirm your answer by constructing a truth table. Explain verbally what this relationship implies.

$$[(P \wedge Q) \vee (\sim P \wedge Q)] \vee [(P \wedge \sim Q) \vee (\sim P \wedge \sim Q)]$$

Group/Research Activity

12. Each group member should find a minimum of five examples of deceptive advertising. For example, consider the statement "If you use Sudsy Soap, then you will be clean." The advertiser wants you to use Sudsy Soap, but also implied is that if you do not use Sudsy Soap, then you will not be clean. Also, consider endorsements, such as those by super stars that say you had better eat a certain cereal implying that you might be just like this famous person if you do so. The group should present five of the best examples and present them to the class.

Real-World Data

TV Viewership *Use Figure 3.17 to do the following for each given statement in Exercises 13–20:*
a. Symbolize the statement using the letters in parentheses.
b. Determine if it is true or false.
c. Write the converse, inverse, contrapositive, and negation in symbolic form (except in Exercise 20).

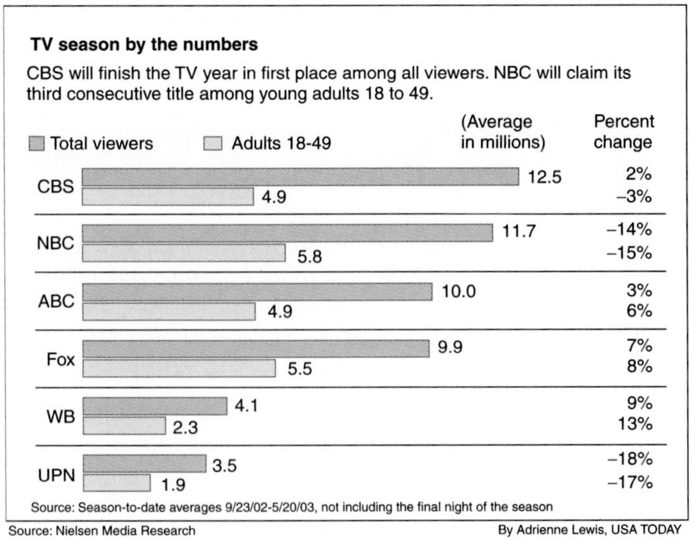

TV season by the numbers

CBS will finish the TV year in first place among all viewers. NBC will claim its third consecutive title among young adults 18 to 49.

Network	Total viewers (Average in millions)	Adults 18-49 (Average in millions)	Percent change: Total viewers	Percent change: Adults 18-49
CBS	12.5	4.9	2%	–3%
NBC	11.7	5.8	–14%	–15%
ABC	10.0	4.9	3%	6%
Fox	9.9	5.5	7%	8%
WB	4.1	2.3	9%	13%
UPN	3.5	1.9	–18%	–17%

Source: Season-to-date averages 9/23/02-5/20/03, not including the final night of the season

Source: Nielsen Media Research — By Adrienne Lewis, USA TODAY

Figure 3.17

13. If a TV program was on CBS (C), then an average of 4.9 million adults watched it (A).

14. If a TV program was on Fox (F), then an average of 9.9 total viewers watched it (T).
15. The average number of adults watching was less than 1 million (A) if the program was on the WB network (W).
16. The average total viewership was more than 10 million people (T) if a TV program was on ABC (A).
17. A TV program was on NBC (N) only if its average total viewership was at least 5 million people (T).
18. Having an average total viewership of at least 10 million people (T) is necessary for a TV program to be on one of the three major networks—CBS, NBC, and ABC (M).
19. A TV program on one of the two fledgling networks, WB and UPN, (F) implies that the average number of adults who watched it was less than 2 million (A).
20. Having an average viewership of at least 5 million adults (A) is a necessary and sufficient condition for broadcasting a TV program on Fox (F).

Just for fun

Let us agree that all special agents always tell the truth and all spies always lie. Once three people appeared in a court of law before a judge. The first accused person spoke, but the judge was unable to hear what the accused had said. The judge asked the second accused person what the first had said. Number 2 replied, "She said she is a spy." The judge then asked the third accused person whether number 2 was a special agent or a spy. Number 3 replied, "Number 2 is a special agent." Determine whether each accused person was a special agent or a spy.

3.8 VALID ARGUMENTS

We shall now review another application of the truth table, for by means of a truth table we can determine whether an argument is valid.

An **argument,** or **proof,** consists basically of two parts: the given statements, which are called premises, and the conclusion. You may recall proofs from geometry. A proof or argument is said to be **valid** if the conclusion follows logically from the premises. What does this mean? This is really another way of saying that an argument is valid if the premises imply the conclusion. We can take the premises of an argument and connect them using conjunction, and then use this compound statement as the antecedent of a conditional statement of which the conclusion is the consequent. If this conditional, or implication, is a tautology, then the argument is valid. A tautological implication tells us that the premises do indeed imply the conclusion.

Consider the following argument:

> If the given figure is a square, then it is a rectangle.
> (major premise)
> The given figure is a square.
> (minor premise)
> Therefore, it is a rectangle.
> (conclusion)

Let

$$P = \textit{The given figure is a square}$$
$$Q = \textit{It is a rectangle}$$

Then the argument would be symbolized as

$$\begin{array}{l} P \rightarrow Q \\ \underline{P \quad\quad} \\ Q \end{array}$$

If we connect the premises using conjunction, and then imply the conclusion, the corresponding conditional statement is as follows:

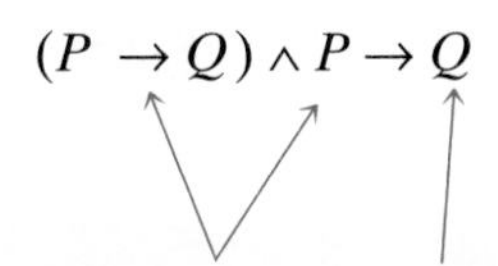

Is this argument valid? Does the conclusion follow logically from the premises? That is, is $(P \rightarrow Q) \wedge P \rightarrow Q$ true for all possible values of P and Q? Table 3.13 shows the truth table for $(P \rightarrow Q) \wedge P \rightarrow Q$.

Note: If a symbolized statement has more than one letter in it, then we enclose it in parentheses when we construct the corresponding conditional statement. We do this to avoid confusion as to which is the dominant connective.

Note that we skipped a couple of steps in Table 3.13. Our major concern is column 5, where we see that the implication is true in all cases, therefore verifying the fact that the premises imply the conclusion in all cases. Hence, the argument is valid.

TABLE 3.13

P	Q	(P → Q)	∧	P	→	Q
T	T	T	T	T	**T**	T
T	F	F	F	T	**T**	F
F	T	T	F	F	**T**	T
F	F	T	F	F	**T**	F
1	2	3	4	1	5	2

Maybe you had already guessed that. However, we are trying to convey what a valid argument is, and how to determine validity. It should be noted from Table 3.13 that a valid argument can have either a true or a false conclusion. The truth or falsity of the conclusion does not determine the validity of the argument. Also, the validity of an argument does not guarantee the truth of its conclusion. But if the premises are true, a valid argument has to have a true conclusion. If you have an argument in which all the premises are true, and the conclusion is false, then the argument is **invalid**.

EXAMPLE 1 ***Valid or Invalid?*** Determine whether the following argument is valid or invalid:

If I study, then I will pass math.
I didn't pass math.
Therefore, I didn't study.

Solution Let P = *I study* and Q = *I will pass math.* The argument in symbolic form would appear as

$$\begin{array}{l} P \rightarrow Q \\ \underline{\sim Q \quad\quad} \\ \sim P \end{array}$$

If we connect the premises using conjunction and imply the conclusion, the corresponding conditional statement is

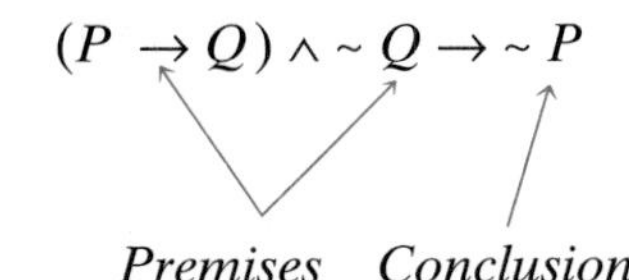

(Note that the first premise is in parentheses because it has more than one letter in it.) The truth table for this conditional statement is as follows:

P	Q	$(P \to Q)$	$\wedge$	$\sim Q$	$\to$	$\sim P$
T	T	T	F	F	$\mathbf{T}$	F
T	F	F	F	T	$\mathbf{T}$	F
F	T	T	F	F	$\mathbf{T}$	T
F	F	T	T	T	$\mathbf{T}$	T
1	2	3	6	4	7	5

Column 7 tells us that we have a tautology, and therefore the argument is valid. The premises do imply the conclusion.

The argument used in Example 1 illustrates one of the basic rules of inference in logic: the **law of contraposition**, or **modus tollens**. If the statement $P \to Q$ is true, and $\sim Q$ is known to be true, then $\sim P$ must be true.

NW ***Now Work Problem 1.***

EXAMPLE 2 ***Valid or Invalid?*** Determine whether the following argument is valid or invalid:

If you are healthy, then you are wealthy.
You are wealthy.
Therefore, you are healthy.

Solution Let P = *You are healthy* and Q = *You are wealthy*. The argument in symbolic form would be

$$\begin{array}{l} P \to Q \\ \underline{Q \quad\quad} \\ P \end{array}$$

If we connect the premises using conjunction and imply the conclusion, the corresponding conditional statement is

$$(P \to Q) \wedge P \to Q$$

Premises *Conclusion*

The corresponding truth table is as follows:

P	Q	$(P \to Q)$	$\wedge$	Q	$\to$	P
T	T	T	T	T	$\mathbf{T}$	T
T	F	F	F	F	$\mathbf{T}$	T
F	T	T	T	T	$\mathbf{F}$	F
F	F	T	F	F	$\mathbf{T}$	F
1	2	3	4	2	5	1

The resulting conditional statement is not a tautology (see column 5). The premises do not imply the conclusion in all cases, and therefore the argument is invalid. Note that in the third horizontal line of the truth table we have a case where the premises are true, but the conclusion is false. A valid argument does not allow us to go from true premises to a false conclusion.

An invalid argument is one in which the conclusion does not logically follow from the premises. That is, the form of the argument does not permit only true conclusions to be logically derived from true premises. The form of the invalid argument in Example 2 is

$$\begin{array}{l} P \rightarrow Q \\ \underline{Q \qquad} \\ P \end{array}$$

This form of incorrect reasoning is commonly called the **fallacy of affirming the consequent**. If the consequent Q of a conditional statement $P \rightarrow Q$ is true, it does *not* follow that the antecedent P must be true. The *statement* $P \rightarrow Q$ can still be true, even if the consequent Q is false.

NW ***Now Work Problem 3.***

Can an argument be valid if one or more of the premises is false and the conclusion is false? Consider the argument in Example 3.

EXAMPLE 3 ***Valid or Invalid?*** Determine whether the following argument is valid or invalid:

If the moon is made of green cheese, then two equals one.
The moon is made of green cheese.
Therefore, two equals one.

Solution Let P = *The moon is made of green cheese* and Q = *Two equals one*. The argument in symbolic form would appear as

$$\begin{array}{l} P \rightarrow Q \\ \underline{P \qquad} \\ Q \end{array}$$

If we connect the premises using conjunction and imply the conclusion, the corresponding conditional statement is

$$(P \rightarrow Q) \wedge P \rightarrow Q$$

Premises *Conclusion*

The corresponding truth table is as follows:

P	Q	(P→Q)	∧	P	→	Q
T	T	T	T	T	**T**	T
T	F	F	F	T	**T**	F
F	T	T	F	F	**T**	T
F	F	T	F	F	**T**	F
1	2	3	4	1	5	2

Column 5 consists of all T's, so the conditional resulting from the argument is a tautology. Hence, the argument is valid. The symbolic argument used in this example illustrates another of the basic rules in logic. It is called the **law of detachment**, or **modus ponens**.

We have just examined an example of a valid argument, parts of which are false (refer back to the truth table in Example 3). There is a difference between the validity of an argument and its truth: An argument is valid because of its form—that is, the manner in which the conclusion is derived from the premises—not because of the meaning of the statements in it.

EXAMPLE 4 ***Valid or Invalid?*** Determine whether the following argument is valid or invalid:

Either the bank is closed, or it is not after three o'clock.
It is not after three o'clock.
Therefore, the bank is not closed.

Solution Let P = *The bank is closed* and Q = *It is after three o'clock*. The argument in symbolic form is

$$\begin{array}{l} P \vee \sim Q \\ \underline{\sim Q \quad\quad} \\ \sim P \end{array}$$

If we connect the premises using conjunction and imply the conclusion, the corresponding conditional statement is

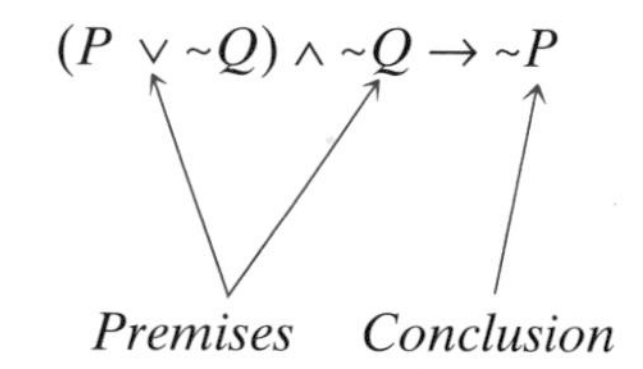

(Note that the first premise is in parentheses because it has more than one letter in it.) The corresponding truth table is as follows:

P	Q	$(P \vee \sim Q)$	$\wedge$	$\sim Q$	$\rightarrow$	$\sim P$
T	T	T	F	F	**T**	F
T	F	T	T	T	**F**	F
F	T	F	F	F	**T**	T
F	F	T	T	T	**T**	T
1	2	5	6	3	7	4

The resulting conditional statement is not a tautology (see column 7). The premises do not imply the conclusion in all cases. Therefore, the argument is invalid.

NW ***Now Work Problem 11.***

In order to test whether an argument is valid or invalid, we need consider only the one conditional statement whose antecedent is the conjunction of all the premises of the argument and whose consequent is the conclusion of the argument. If this conditional statement is a tautology, then the argument is valid. If it is not a tautology, then the argument is invalid.

Exercises for Section 3.8

Symbolize each of the following arguments (using the suggested notation) and, by means of a truth table, determine whether the argument is valid or invalid. State your answer.

NW **1.** If I pass the test, then I will quit coming to class. (P, Q)
I quit coming to class.
Therefore, I passed the test.

2. If Judy was on time, then her watch was correct. (J, W)
Judy was on time.
Therefore, Judy's watch was correct.

NW **3.** If I pass math, then I will graduate. (P, G)
I graduated.
Therefore, I passed math.

4. If Andre gets help, then Andre will pass. (H, P)
Andre did not get help.
Therefore, Andre did not pass.
5. I will graduate iff I pass math. (G, P)
I graduated.
Therefore, I passed math.
6. It will be sunny or cloudy today. (S, C)
It isn't sunny.
Therefore, it will be cloudy.
7. Addie and Bill will be at the party. (A, B)
Bill was at the party.
Therefore, Addie was at the party.
8. If one equals two, then four equals eight. (T, E)
Four does not equal eight.
Therefore, one does not equal two.
9. You may attend the concert iff you purchased a ticket. (A, T)
You purchased a ticket.
Hence, you may attend the concert.
10. If it snows, Benny will go skiing. (S, B)
It did not snow.
Therefore, Benny did not go skiing.

NW 11. Pat or Sandy will bring the doughnuts. (P, S)
Pat did not bring the doughnuts.
Therefore, Sandy did not bring the doughnuts.
12. You are healthy, if you eat apples. (H, A)
You are healthy.
Therefore, you eat apples.
13. If it rains, Bobby takes his umbrella to school. (R, B)
It did not rain.
Therefore, Bobby did not take his umbrella to school.
14. If you finish the exam early, you may leave early. (E, L)
You did not finish the exam early.
Therefore, you may not leave early.

*15. Paul did not study, or he is bluffing. (P, B)
If he is bluffing, then he will cut class. (B, C)
Therefore, Paul did not study, or he will cut class.

*16. If I get a job, then I will not be able to study. (J, S)
It is false that I will fail history and that I will not get a job. (F, J)
Therefore, if I study, then I will not fail history.

*17. If a is greater than b, then b is greater than c. (A, B)
If b is greater than c, then c is greater than d. (B, C)
Therefore, if a is greater than b, then c is greater than d.

*18. If the heavenly body is Mars, then it is near Venus. (M, V)
If it is not near Venus, then it is not close to Saturn. (V, S)
Therefore, either it is not close to Saturn, or the heavenly body is Mars.

Writing Mathematics

19. Describe what is meant by a valid argument.
20. Is it possible for a valid argument to have a conclusion that is false? Explain your answer.
21. Is it possible for an invalid argument to have a conclusion that is true? Explain your answer.
22. Is it possible for a valid argument to have all false premises and a false conclusion? Explain your answer.

Challenge Exercises

Symbolize each of the following arguments. Using the suggested notation, and by means of a truth table, determine whether the argument is valid or invalid. State your answer.

23. Either the store is open or it is an official holiday. If it is an official holiday, then we will save our money. If the store is open, then we will not enjoy the holiday. Hence, we will not enjoy the holiday or we will save our money. (S, O, M, E)
24. If the lease is not broken, then we may keep our office. They will vote to reach a decision or not evict us. It is false that we may keep our office and that they will vote to reach a decision. Therefore, if they evict us, then our lease is broken. (L, O, V, E)

Consider the following two statements:

If I pass logic, then I will graduate
If I graduate, then I will get a job.

If these statements are given as premises, then we may logically conclude the following:

If I pass logic, then I will get a job,

This argument is symbolized as follows, where P = I pass logic, Q = I will graduate, and R = I will get a job:

$$\begin{array}{c} P \rightarrow Q \\ \underline{Q \rightarrow R} \\ P \rightarrow R \end{array}$$

This result is a rule of inference known as the **law of syllogism** *and can be verified by showing that the argument is valid. In Exercises 25 through 27, use the law of syllogism to write the conclusion that can be logically derived from the given set of premises.*

25. If it rains today, then we can't go swimming. If we can't go swimming, then we will go bowling.
26. If Ralph sinks this putt, then he will advance to the final round. If Ralph advances to the final round, then he will be playing for the title.
27. If I do not get home soon, then I will be in trouble. If I get home soon, then I will be able to watch my favorite television show.
28. Given the following argument, find a valid conclusion.

All work was to be done in the blue book.
None of the work was written in black ink, except problems that required detailed solutions.
All of the work written in the blue book was graded.
If the work was graded, then it did not require a detailed solution.

Also, use the law of syllogism to reduce the conclusion to a single conditional statement.

Group/Research Activities

29. Find examples of valid (or invalid) arguments in advertising, magazine, or newspaper articles. Explain why the arguments are valid (or invalid).

30. Lewis Carroll posed many problems in his writings. Select one of his works, such as *Alice in Wonderland,* and prepare a report on how he used logic in the book. Give at least four specific examples.

Just for fun

You are given six line segments of equal length. You are to construct four triangles by using these six line segments. Can you do it?

3.9 PICTURING STATEMENTS WITH VENN DIAGRAMS (OPTIONAL)

Venn diagrams can also be used to determine whether certain kinds of arguments are valid or invalid. The arguments that we shall consider here are called *syllogisms*. A **syllogism** is an argument that contains three statements: the **major premise,** the **minor premise,** and the **conclusion.**

Recall that a valid argument is one in which the conclusion follows logically from the premises. The conclusion is derived from the premises according to the laws of logic. The following is an example of a syllogism that is a valid argument. The conclusion follows from the premises.

All mathematics students are ambitious. (*major premise*)
No ambitious people are lazy. (*minor premise*)
Therefore, no mathematics students are lazy. (*conclusion*)

Note that each of the statements in the example contains a quantifier such as *all* or *no*.

Biography: John Venn

John Venn (1834–1923) was an Englishman who devoted his life to the study and teaching of both mathematics and religion. He was a mathematical scholar and ordained a deacon in his church. Two of his better known works are *The Logic of Chance* and *Symbolic Logic.* At the age of 69, Venn was appointed president of Gonville and Caius College, near London.

John Venn developed his diagrams to give sensible illustrations of the relation of terms and properties to each other. In his own words, his purpose was "to save time, avoid unpleasant drudgery and to be sure to avoid mistakes and oversights." Venn devised his own diagrams because he found Euler diagrams inadequate and ineffective, even for the propositions of formal logic.

George Boole significantly influenced the development of Venn's diagrams, for they are founded on Boole's system of logic. However, Boole did not use such diagrams, nor did he indicate or suggest their introduction.

The syllogisms that we shall consider here contain quantified statements. There are four such types of statements:

1. The **universal affirmative** statement states that "All *A*'s are *B*'s." For example, "All students are scholars."
2. The **universal negative** statement states that "No *A*'s are *B*'s." For example, "No students are scholars."
3. The **particular affirmative** statement states that "Some *A*'s are *B*'s." For example, "Some students are scholars."
4. The **particular negative** statement states that "Some *A*'s are not *B*'s." For example, "Some students are not scholars."

Do you think that the following syllogism is a valid argument? Does the conclusion follow from the premises?

All golfers are swingers. (*major premise*)
No hackers are swingers. (*minor premise*)
Therefore, no hackers are golfers. (*conclusion*)

The preceding example syllogism is a valid argument. You will see why shortly, but first we must be able to symbolize (picture) a statement using Venn diagrams.

Picturing the Universal Affirmative

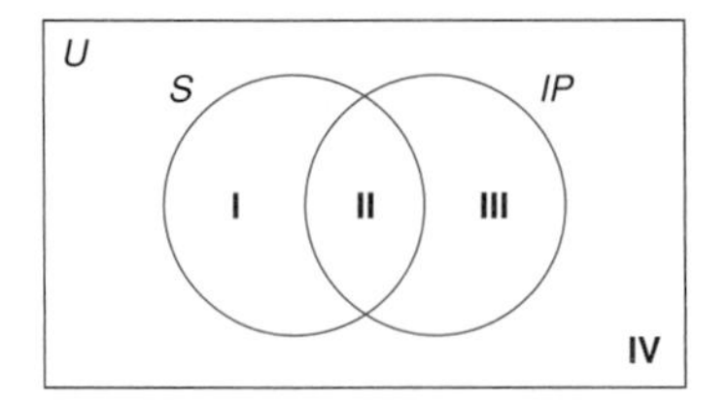

Figure 3.18

S = the set of students, and IP = the set of industrious people.

Consider the statement "All students are industrious." We have a set of students and a set of industrious people. Hence, we may draw two circles and label them (see Figure 3.18). It does not matter which circle is on the left, but it is convenient to take them in the order given.

Note that we overlap the circles so that all possibilities may be considered. What do we know from the statement "All students are industrious"? We know that if a person is a student, then he or she is industrious; that is, the set of students is a subset of the set of industrious people. It does not follow that if a person is industrious, then he or she is a student.

How are we going to picture our statement? The statement tells us that all the students are in the "industrious" circle; that is, all the elements in the set containing students are in the set containing industrious people. So they are in region II, and therefore region I is empty. Probably your first inclination is to shade region II—that is, the set $S \cap IP$—as shown in Figure 3.19, because all the elements in set *S* are also in set *IP*.

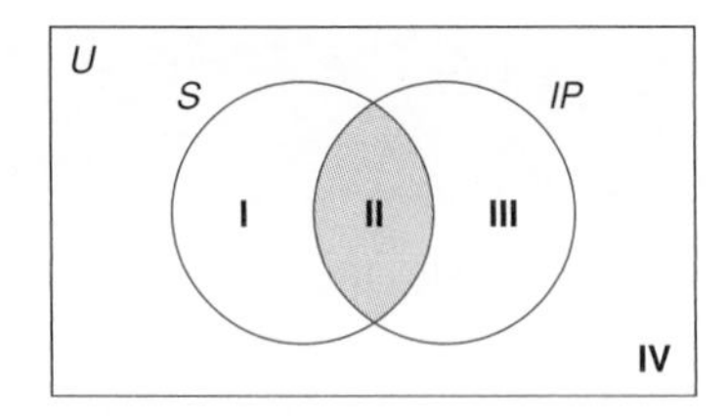

Figure 3.19

However, suppose one or more of the sets we are picturing do not contain any elements. For example, consider the statement "All purple people-eaters are one-eyed." This is a universal affirmative statement; it is of the form "All *A*'s are *B*'s." To picture this statement, we must consider the set of purple people-eaters, which is an empty set. The shading technique used in Figure 3.19 would not be appropriate for the statement "All purple people-eaters are one-eyed," because shading region II would indicate that the set of purple people-eaters has members, which is not true. Therefore, we must use another way of picturing statements in Venn diagrams, since we must have a method that works for all cases.

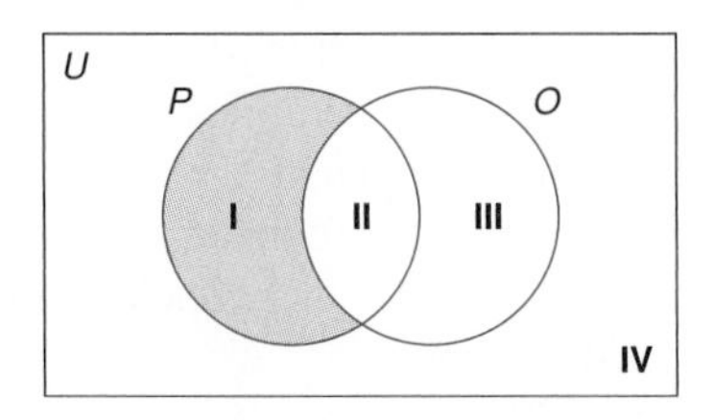

Figure 3.20

P = the set of purple people-eaters, and O = the set of one-eyed objects.

A technique that does work for all cases is to distinguish the sets (regions) that *cannot* have elements in them, according to the given statement. For example, in constructing a picture of the statement "All purple people-eaters are one-eyed," we note that if purple people-eaters exist, then they would also all be in the set of one-eyed objects (region II in Figure 3.20). But whether they exist or not, region I is known to be empty. According to the given statement, there are no purple people-eaters that are not one-eyed. Therefore, we can eliminate region

I—that is, we can cross it out. In Figure 3.20, we have indicated that region I is empty by shading it.

Now, back to the original statement under consideration, "All students are industrious." From this statement we know that there is nothing in the set of students that is not also in the set of industrious people. That is, there are no students in set *S* who are not also members of set *IP*. Hence, region I is empty and we can eliminate it, or cross it out. The statement "All students are industrious" may be pictured as in Figure 3.21.

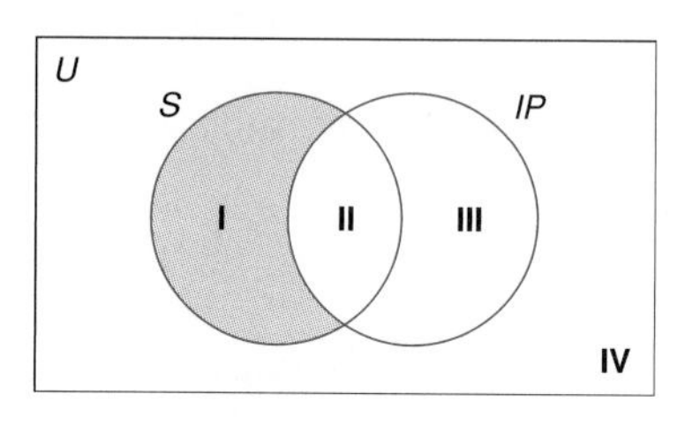

Figure 3.21

S = the set of students, and
IP = the set of industrious people.

Picturing the Universal Negative

We have discussed the diagram of the universal affirmative statement. Now let's examine the next one, the universal negative statement. Consider the statement "No students are industrious." Again we have two sets, *students* and *industrious people*, but what region do we eliminate? There are no elements in the set of students that are also in the set of industrious people. Therefore, region II in Figure 3.22 should be crossed out because it contains no elements. Hence, the statement "No students are industrious" would be pictured as in Figure 3.22.

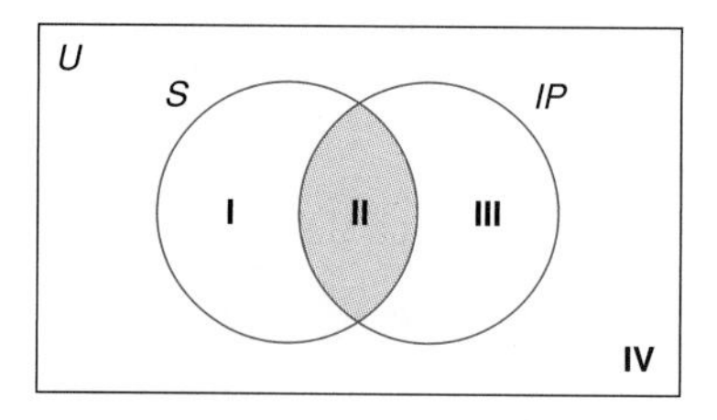

Figure 3.22

S = the set of students, and
IP = the set of industrious people.

If two statements have the same diagram, the two statements are logically equivalent. The statements have the same meaning because they have the same picture in the diagram and the same region eliminated. The statement "No *A*'s are *B*'s" is logically equivalent to "No *B*'s are *A*'s," but how do "All *A*'s are *B*'s" and "All *B*'s are *A*'s" compare? Are they logically equivalent? By checking diagrams of these two statements, you will see that they are not! (See Figures 3.23a and 3.23b.)

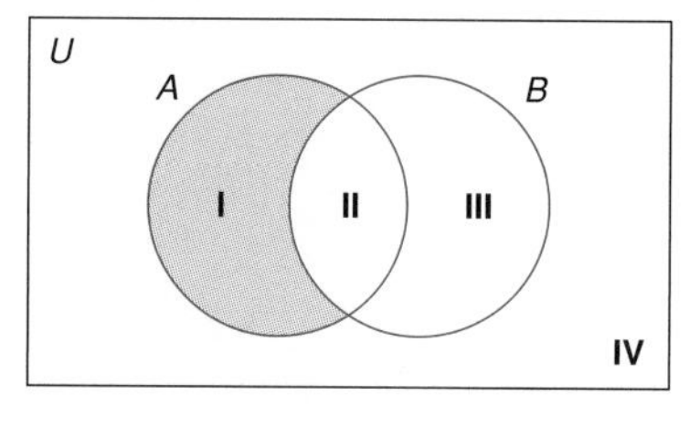

Figure 3.23a

All A's are B's.

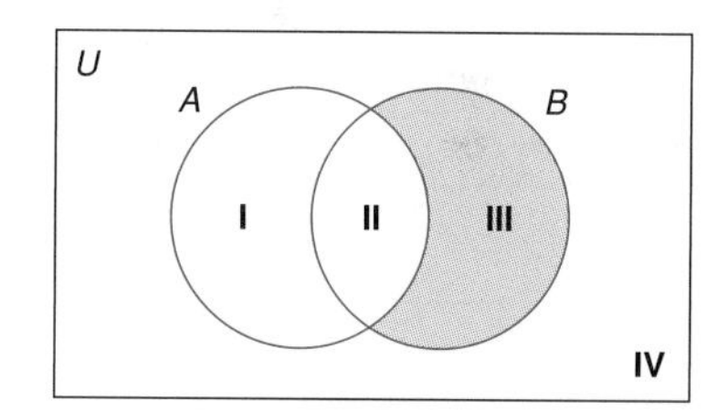

Figure 3.23b

All B's are A's.

Picturing the Particular Affirmative

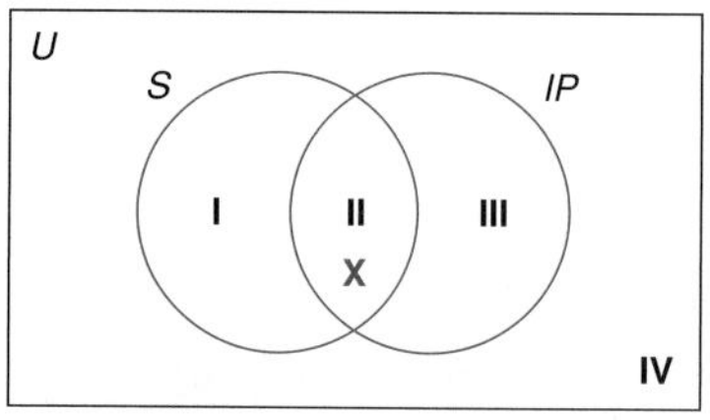

So far we have pictured the universal affirmative and universal negative statements. Consider the statement "Some students are industrious." We have the set of students and the set of industrious people, and from the given statement we know that there are some students that are industrious. We are discussing "some" students. In logic, the word *some* is interpreted to mean "There is at least one," but there could be more. The word *some* does not specify a certain number, so we just maintain that there is at least one. Since there are some students who are industrious, we know that there is at least one student who is industrious. This one student is common to both sets. We diagram this by showing that the intersection of the two sets is not empty. We place something in the intersection (region II) to show that it is not empty—namely, the symbol *x*. Therefore, the statement "Some students are industrious" would be pictured as in Figure 3.24. Note that when we cross out a region in a Venn diagram, then there are no elements in that region; but when we place the symbol *x* in a region, then there is at least one element in that region.

U
S
IP
I
II
X
III
IV

Figure 3.24

S = the set of students, and
IP = the set of industrious people.

Picturing the Particular Negative

Consider the particular negative statement "Some students are not industrious." This statement tells us that there is at least one element in the set of students that is not in the set of industrious people. Therefore the intersection of the set of students and the set of nonindustrious people is not empty. Hence, we would place the symbol x in region I because it contains at least one element that is a student and is not industrious. The proper diagram for "Some students are not industrious" is shown in Figure 3.25.

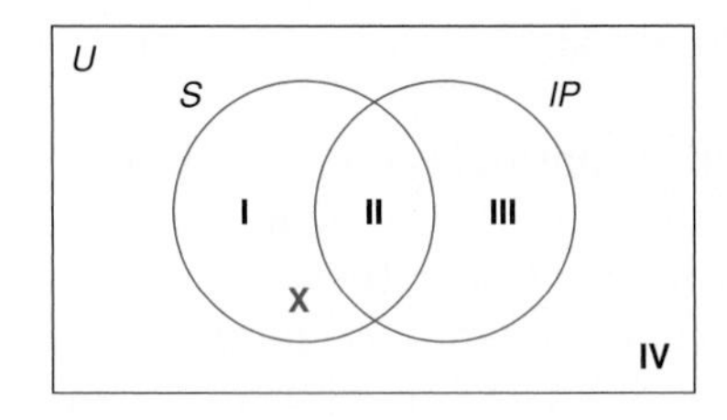

Figure 3.25

S = the set of students, and IP = the set of industrious people.

Using examples, we have diagrammed the four types of statements that may appear in a syllogism: universal affirmative, universal negative, particular affirmative, and particular negative (see Figures 3.26a, 3.26b, 3.26c, and 3.26d).

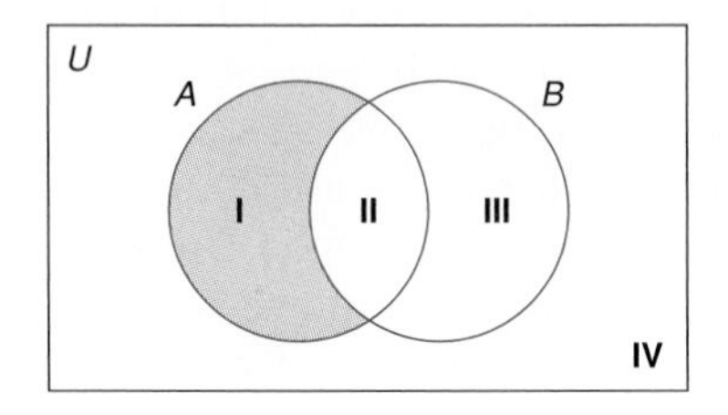

Figure 3.26a

Universal affirmative: All A's are B's.

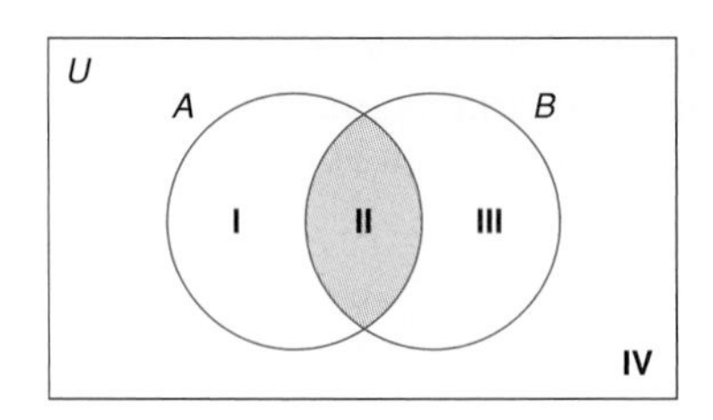

Figure 3.26b

Universal negative: No A's are B's.

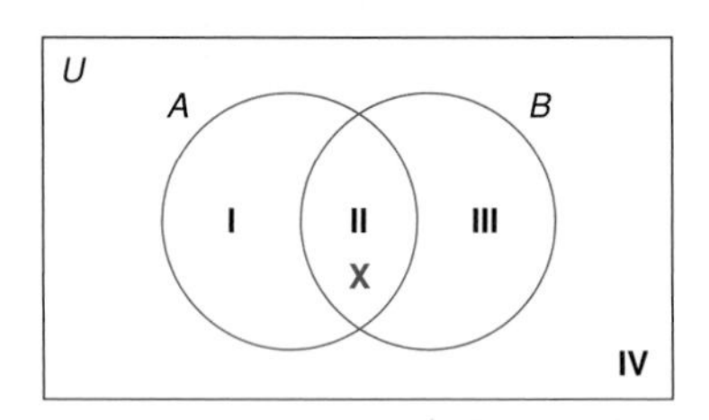

Figure 3.26c

Particular affirmative: Some A's are B's.

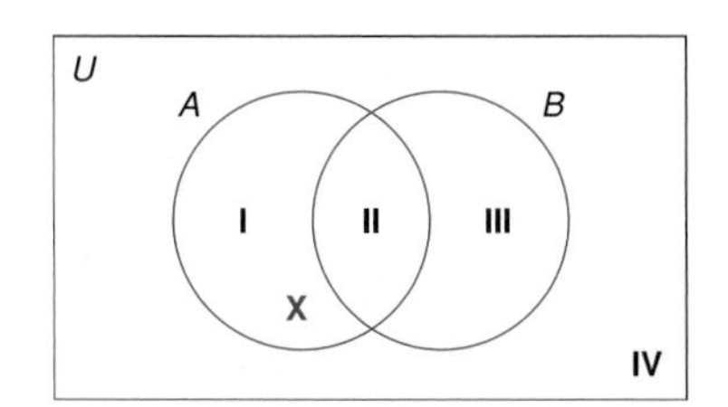

Figure 3.26d

Particular negative: Some A's are not B's.

EXAMPLE 1 ***Diagramming Quantified Statements*** Identify and diagram each of the following:

a. *All bugs are pesky.* **b.** *Some bugs are pesky.*

c. *No bugs are pesky.* **d.** *Some pesky things are not bugs.*

Solution Let B = the set of bugs, and P = the set of pesky things.

a. Universal affirmative; see Figure 3.27a.
b. Particular affirmative; see Figure 3.27b.
c. Universal negative; see Figure 3.27c.
d. Particular negative; see Figure 3.27d.

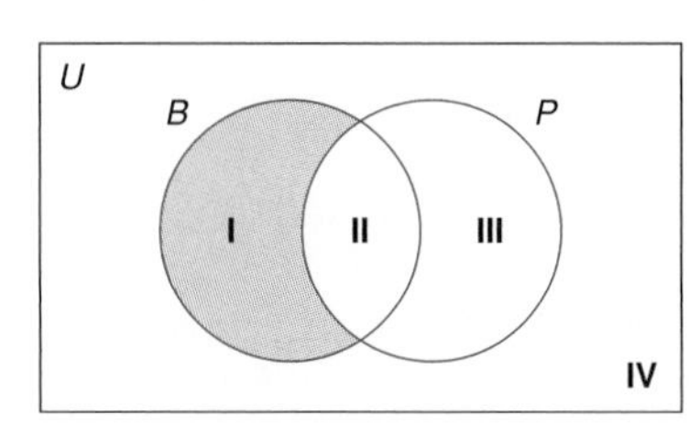

Figure 3.27a

All bugs are pesky.

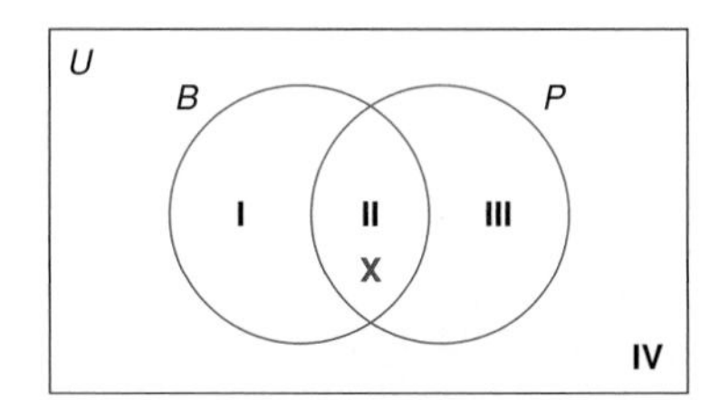

Figure 3.27b

Some bugs are pesky.

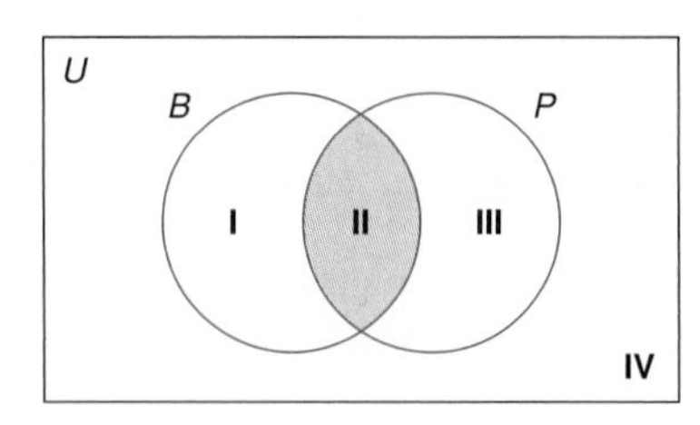

Figure 3.27c

No bugs are pesky.

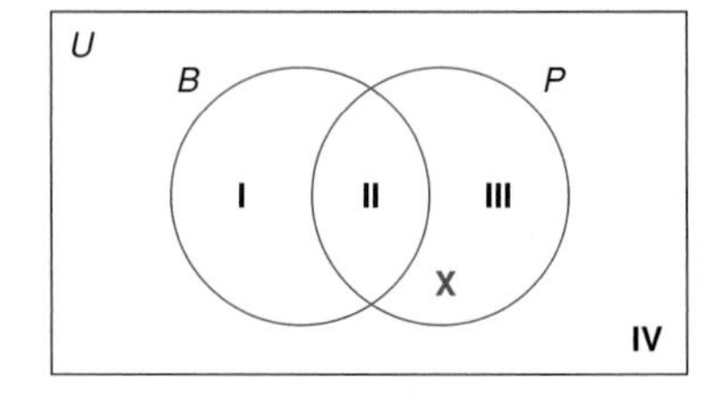

Figure 3.27d

Some pesky things are not bugs.

Statements that cannot be true together are said to be **inconsistent**. Given two statements, "Today is Monday" and "Today is not Monday," we see that these two statements contradict each other and cannot be true at the same time. Statements that *can* be true together are **consistent** statements; they do not contradict each other. Many times it is difficult to determine whether two statements are consistent. Let us see how we may use Venn diagrams to do this.

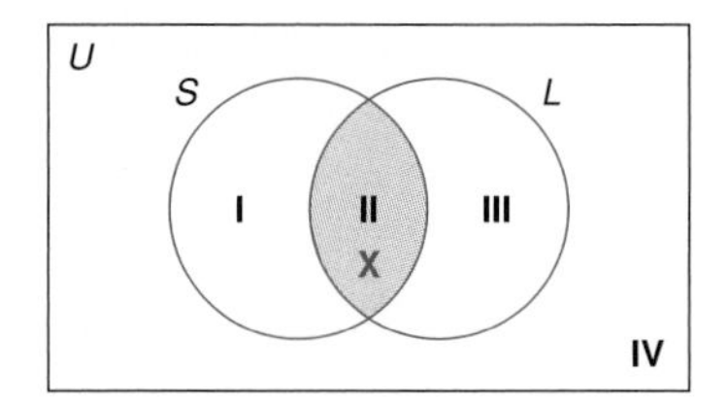

Figure 3.28

S = the set of students, and
L = the set of lazy people.

EXAMPLE 2 ***Consistent or Inconsistent?*** Determine whether the following pair of statements is consistent or inconsistent:

1. *Some students are lazy.*
2. *No students are lazy.*

Solution We construct a Venn diagram and diagram both statements in the same picture (see Figure 3.28). The shading in region II represents statement 2 (*No students are lazy*), and the x in region II represents statement 1 (*Some students are lazy*). We have both crossed out a region and placed an x in the region. When we cross out a region to picture a statement, it means that there are no elements in that particular region. If we place an x in a region, then there is at least one element in that region. This is a contradiction; the two statements cannot be pictured in the same diagram. They are inconsistent. Two statements that are not consistent cannot be true at the same time.

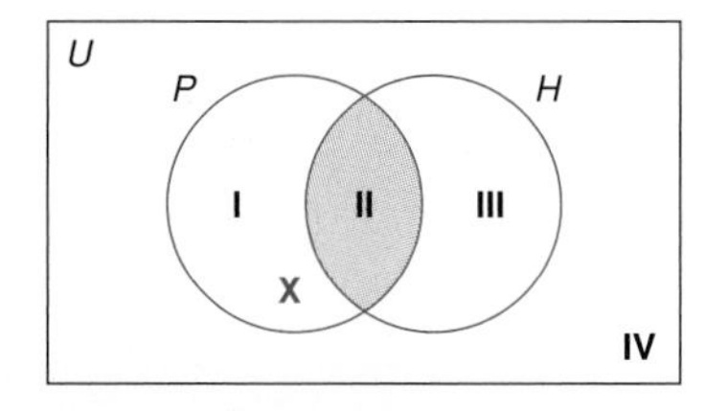

Figure 3.29

P = the set of politicians, and
H = the set of honest people.

EXAMPLE 3 ***Consistent or Inconsistent?*** Determine whether the following pair of statements is consistent or inconsistent:

1. *No politicians are honest.*
2. *Some politicians are not honest.*

Solution We construct a Venn diagram and diagram statements 1 and 2 in the same picture (see Figure 3.29). The shading in region II represents statement 1 (*No politicians are honest*), and the x in region I represents statement 2 (*Some politicians are not honest*). Because we are able to picture both statements in the same diagram without any contradiction, we may conclude that the statements are consistent.

Exercises for Section 3.9

Identify each statement as universal affirmative, universal negative, particular affirmative, or particular negative, and then diagram it by means of a Venn diagram.

1. No heroes are imps.
2. All neighbors are kind.
3. Some girls are not imps.
4. No doctors are cheerful.
5. Some taxpayers are happy.
6. All books are entertaining.
7. Some letters are not interesting.
8. All instructors are patient.
9. Some math courses are challenging.
10. All candidates are energetic.
11. Some politicians are not honest.
12. All Republicans are politicians.
13. No math teachers are compassionate.
14. All math teachers are compassionate.
15. Some cars are not gas eaters.
16. Some gamblers are losers.
17. Some losers are not gamblers.
18. All math teachers are dull.
19. No math teachers are boring.
20. Some students are procrastinators.

Use a Venn diagram to determine whether each pair of statements is consistent or inconsistent.

21. Some kind people are clever.
    No clever people are kind.
22. Some math courses are interesting.
    No math courses are interesting.
23. No math teachers are compassionate.
    All math teachers are compassionate.
24. All dogs are barkers.
    Some barkers are not dogs.
25. Some gamblers are losers.
    Some losers are not gamblers.
26. Some tests are not easy.
    All tests are easy.
27. No math courses are boring.
    Some math courses are boring.

28. All students are procrastinators.
No procrastinators are students.

29. All politicians are sly.
No sly people are politicians.

30. Some real numbers are rational.
Some rational numbers are real.

31. All logic students are gullible.
No logic students are gullible.

32. No Democrats are politicians.
Some politicians are not Democrats.

33. Some tests are long.
Some tests are not long.

34. Some girls are not imps.
No imps are girls.

35. Some houses are homes.
Some homes are houses.

Writing Mathematics

36. What are inconsistent statements? Give at least two examples.

37. What are consistent statements? Give at least two examples.

38. Explain how to use a Venn diagram to determine whether two statements are consistent or inconsistent.

Challenge Exercises

In mathematics, the symbol $\forall$ *is called the* **universal quantifier,** *and when it is written with the variable x,* $\forall x$*, it is read as "For all possible values of x." Expressions that imply the use of the universal quantifier include "any," "anything," "anyone," "everything," "everyone," "everybody," "for every x," and "for all x." Using this notation, statements such as "Everything is fine" can be symbolized by rewriting the statement first as "Every x is fine" and then further refining the statement to "For all x, x is fine. If we let F = the term "fine," then our original statement can now be symbolized as* $(\forall x)(Fx)$.

Express each of the statements in Exercises 39 through 44 using this new notation.

39. Everything is good.

40. Everybody loves Raymond.

41. Everything is not good.

42. Everybody does not love Raymond.

43. No one likes snakes.

44. No one likes math.

Consider the statement All flowers are beautiful. *Note that this statement does not imply that "For all x, x is a flower and it is beautiful" because the statement states that all flowers are beautiful, not all things. Thus, the correct interpretation is "For all x, if x is a flower then x is beautiful." Using the universal quantifier notation, this statement is symbolized as* $(\forall x)(Fx \rightarrow Bx)$, *where* Fx = *x is a flower and* Bx = *x is beautiful. Use this concept to symbolize the statements in Exercises 45 through 50.*

45. All integers are rational numbers.

46. Every square is a rectangle.

47. All triangles are not isosceles.

48. All counting numbers are not odd.

49. No kites are circles.

50. No politicians are honest.

Group/Research Activities

51. The four quantified statements were initially developed by Aristotle (384–322 B.C.). Research the history of these statements and prepare a report why Aristotle developed them. Include in your report what Aristotle referred to as the A-form, the E-form, the I-form, and the O-form and why these terms were used.

52. In addition to the universal quantifier, we have the **existential quantifier** denoted as $\exists$. Research the meaning of $\exists$ and give an example of how it is used in logic to symbolize statements. As part of your research, try to provide the English translation of the statement $\forall\exists\exists\forall$.

Just for fun

If it takes 5 seconds for a clock to strike 5, how long does it take to strike 10?

3.10 VALID ARGUMENTS AND VENN DIAGRAMS (OPTIONAL)

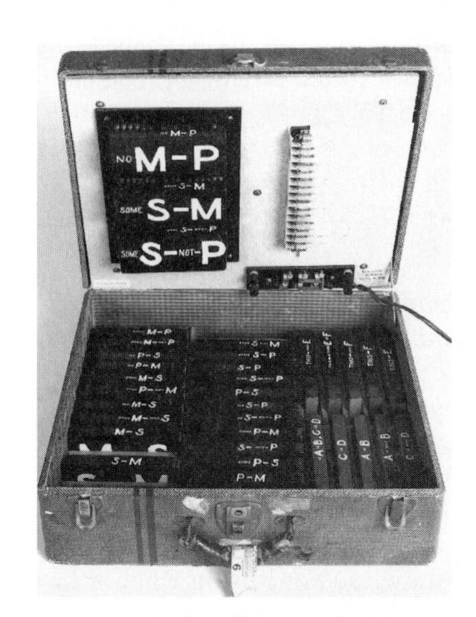

This machine, on display at the Smithsonian Institution in Washington, D.C., can be used to evaluate the validity of syllogisms.

An argument consists of premises and a conclusion. An argument is said to be valid if the conclusion follows from the premises. Arguments may contain more than two premises, but here we are going to consider syllogisms only. Recall that a syllogism contains a major premise, a minor premise, and a conclusion. An argument is valid if the conclusion follows from the premises.

You should be aware of the difference between truth and validity. An argument may be valid even though your own knowledge tells you that the conclusion is false. But if the conclusion follows from the premises, then the argument is valid. On the other hand, a conclusion may be true and the argument invalid. You may know that a certain conclusion is true, but if it does not follow from the premises, the argument is not valid.

Venn diagrams are useful to determine the validity of syllogisms. An argument is valid if the conclusion follows from the premises. How do we determine if a conclusion follows from the premises? We diagram the premises in a Venn diagram, and if the conclusion is shown in the diagram of the premises without any ambiguity, then the argument is valid. If it is possible to diagram the premises without at the same time showing the conclusion, then the argument is invalid. Let's examine some examples to see how this technique is used.

EXAMPLE 1 ***Valid or Invalid?*** *Using Venn Diagrams* Determine whether the following argument is valid:

All students are industrious.
No dropouts are industrious.
Therefore, no dropouts are students.

Solution First we construct a Venn diagram with three overlapping circles—one for the set of students, one for the set of industrious people, and one for the set of dropouts (see Figure 3.30). Next we diagram each of the premises in the argument. "All students are industrious" is pictured by crossing out regions I and IV. "No dropouts are industrious" is pictured by crossing out regions V and VI. You will note that the conclusion, "No dropouts are students," is already pictured in the diagram as a result of diagramming the premises. This tells us that the argument is valid. (If the conclusion is shown in the diagram of the premises without any ambiguity, the argument is valid.) See Figure 3.31 for the completed diagram.

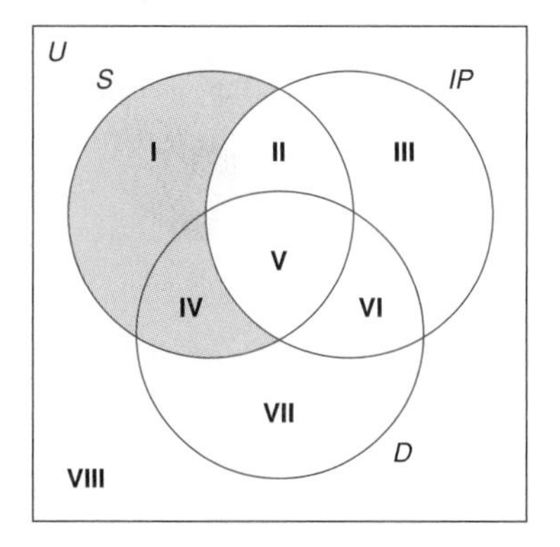

Figure 3.30

S = the set of students, IP = the set of industrious people, and D = the set of dropouts.

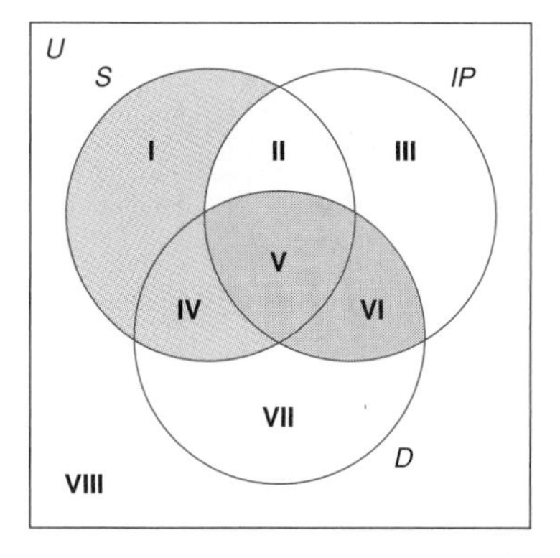

Figure 3.31

Biography: Charles Lutwidge Dodgson

Charles Lutwidge Dodgson (1832–1898), or Lewis Carroll as we know him, was born in England. As a child he was fascinated with math and solving puzzles. His mathematical career began with teaching at Oxford. In later years, most of his time was spent writing, solving puzzles, and lecturing at Christ Church. *Alice in Wonderland* and *Through the Looking Glass* are two of his most famous works.

Carroll posed many problems in his work. His poetry, puns, and novels have been read by many mathematicians and philosophers. They have studied the logic in his work extensively, and many of them have been inspired by Carroll.

Following is a puzzle typical of those found in his writings. See if you can solve it without looking at the answer, which is given below the rhyme.

Dreaming of apples on a wall,
and dreaming often dear,
I dreamed that, if I counted all,
—how many would appear?

The answer is 10 (dreaming of-ten).

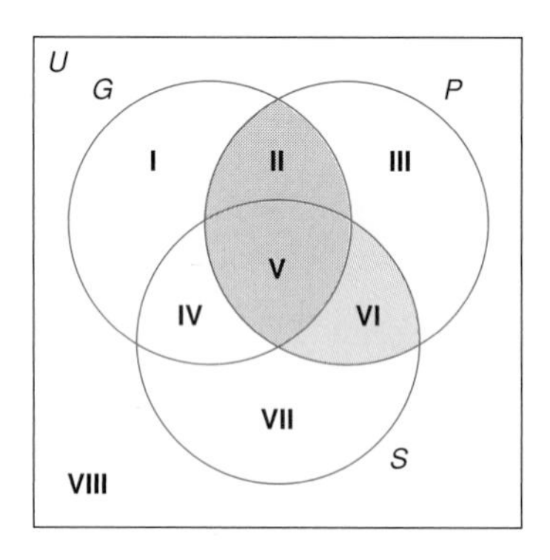

Figure 3.32

G = the set of generals, P = the set of pacifists, and S = the set of soldiers.

The premises of a valid argument *must* show the conclusion when they are diagrammed. We do not diagram the conclusion; its picture must be a result of diagramming the premises. If we are able to diagram the premises without showing the conclusion, then the argument is invalid.

NW ***Now Work Problem 5.***

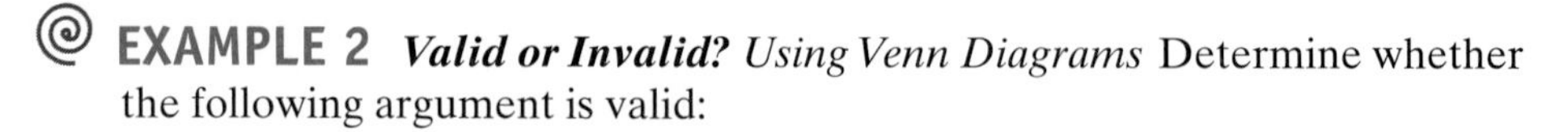

EXAMPLE 2 ***Valid or Invalid?*** *Using Venn Diagrams* Determine whether the following argument is valid:

No generals are pacifists.
No pacifists are soldiers.
Hence, no generals are soldiers.

Solution We diagram the premises in Figure 3.32 to see if the conclusion follows from the premises. We have three circles: generals, pacifists, and soldiers. "No generals are pacifists" is pictured by eliminating regions II and V. "No pacifists are soldiers" is pictured by eliminating regions V and VI. We have already crossed out region V, so we need eliminate only region VI. The conclusion, "No generals are soldiers," is not pictured in the diagram. If it were, then regions IV and V would be crossed out—and region IV is not. The conclusion does not follow from the premises; therefore the argument is invalid.

NW ***Now Work Problem 11.***

EXAMPLE 3 ***Valid or Invalid?*** *Using Venn Diagrams* Determine whether the following argument is valid:

All generals are soldiers.
Some generals are fighters.
Therefore, some soldiers are fighters.

U G S I II III V X IV VI VII F VIII

Figure 3.33

G = the set of generals, S = the set of soldiers, and F = the set of fighters.

Solution We diagram the premises to see if we also obtain a picture of the conclusion (see Figure 3.33). The statement "All generals are soldiers" is diagrammed by eliminating regions I and IV. The second premise, "Some generals are fighters," tells us that there are some elements in set G that are also in set F.

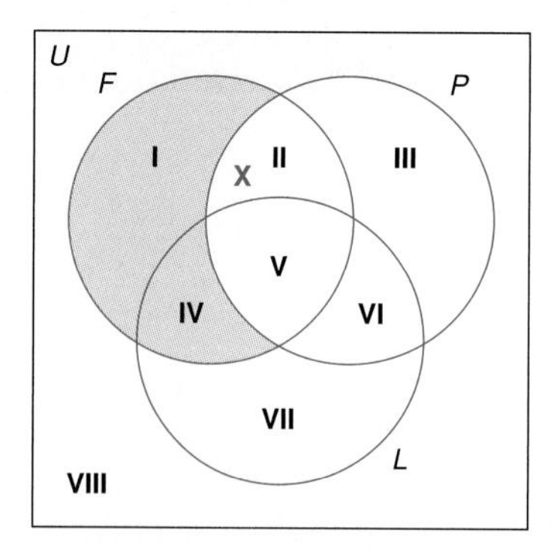

Figure 3.34

F = the set of fishermen, P = the set of patient people, and L = the set of liars.

We have a choice of placing an x in region IV or V. We must place the x in region V, since we have already eliminated region IV in diagramming the first premise. Now we see that the conclusion, "Some soldiers are fighters," is already diagrammed in Figure 3.33. The conclusion does follow from the given premises, so the argument is valid.

EXAMPLE 4 ***Valid or Invalid?*** *Using Venn Diagrams* Determine whether the following argument is valid:

All fishermen are patient.
Some fishermen are not liars.
Hence, some liars are not fishermen.

Solution The major premise, "All fishermen are patient," is diagrammed in Figure 3.34 by eliminating regions I and IV. "Some fishermen are not liars" tells us that there are some elements in set F that are not in set L. We must place the x in region II, since region I has already been eliminated. The conclusion, "Some liars are not fishermen," is not pictured in the diagram. This conclusion does not follow from the premises. Hence, the argument is invalid.

Note that by placing an x in region II we could have shown the statement "Some fishermen are not liars," but that is not the desired conclusion.

If it is possible to diagram the premises without showing the conclusion at the same time, then the argument is invalid. This may occur when we diagram one particular statement and have an option of placing an x in more than one region (only after the first premise has already been diagrammed).

To summarize, *an argument is valid if the conclusion is shown without any ambiguity as soon as the premises are diagrammed.* We do not diagram the conclusion; its picture must be a result of diagramming the premises. If it is possible to diagram the premises without at the same time showing the conclusion, then the argument is invalid.

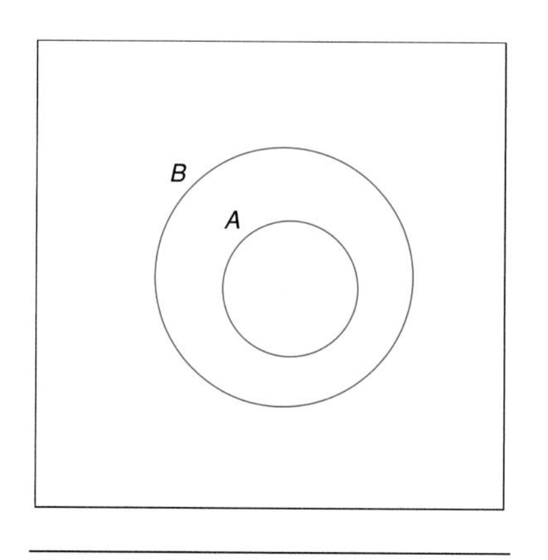

Figure 3.35a

Universal affirmative: All A's are B's.

NW ***Now Work Problem 15.***

Alternative Method for Determining the Validity of an Argument

In addition to Venn diagrams, Euler circles (see Chapter 2, Section 2.5) can be used as an alternative method for determining whether or not an argument is valid. Figures 3.35a, 3.35b, 3.35c, and 3.35d show the Euler circle representations of the universal affirmative, universal negative, particular affirmative, and particular negative, respectively.

The four examples of this section are repeated here using Euler circles.

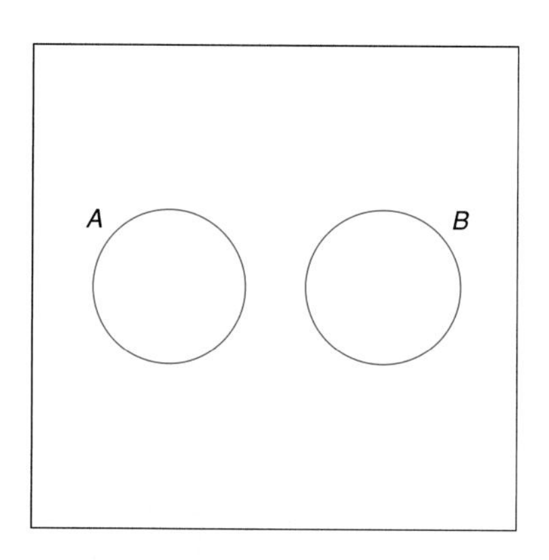

Figure 3.35b

Universal negative: No A's are B's.

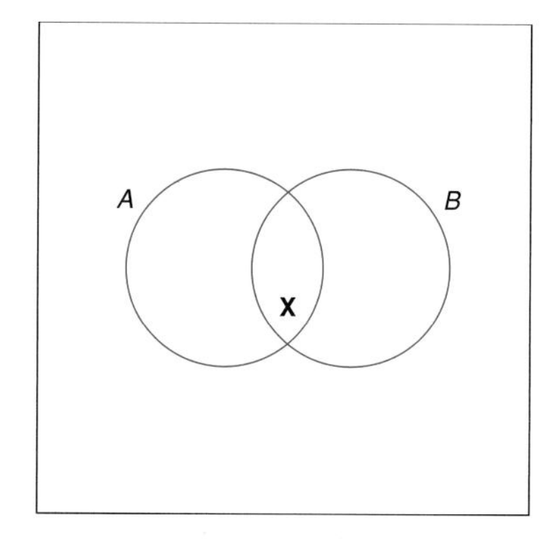

Figure 3.35c

Particular affirmative: Some A's are B's.

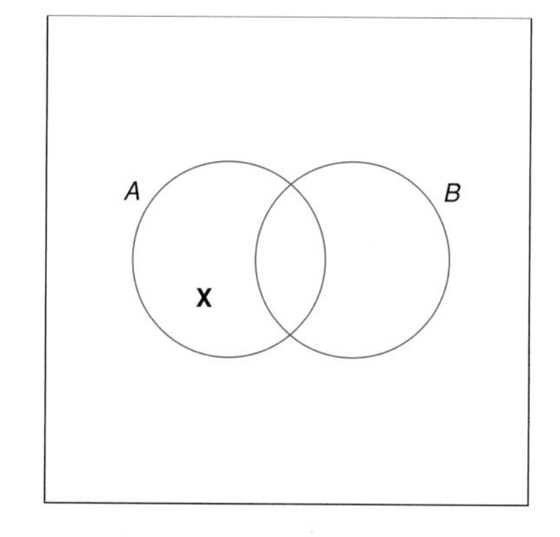

Figure 3.35d

Particular negative: Some A's are not B's.

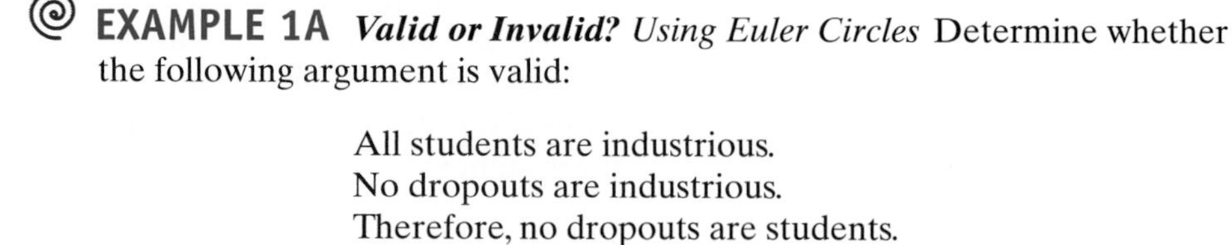

EXAMPLE 1A ***Valid or Invalid?*** *Using Euler Circles* Determine whether the following argument is valid:

All students are industrious.
No dropouts are industrious.
Therefore, no dropouts are students.

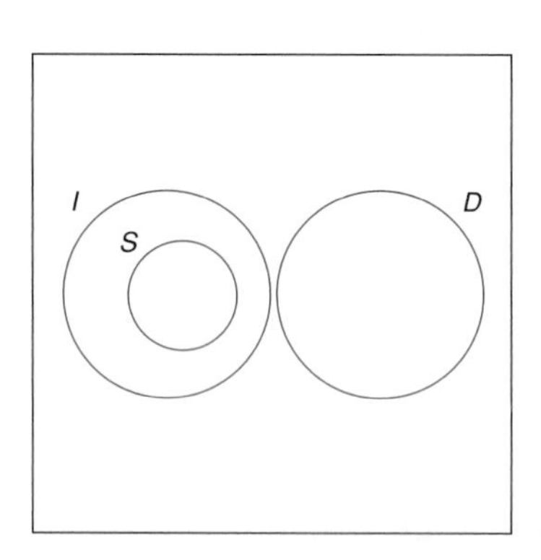

Figure 3.36

I = the set of industrious people, S = the set of students, and D = the set of dropouts.

Solution Let S = the set of students, I = the set of industrious people, and D = the set of dropouts. We now represent the premises using Euler circles. The first premise is a universal affirmative. It means that the set of students is a proper subset of the set of industrious people. The second premise is a universal negative. It means that the intersection of the sets of dropouts and industrious people is empty. These two premises are pictured together (i.e., within the same universal set) in Figure 3.36. Note that the conclusion is a universal negative and implies that the intersection of the sets of dropouts and students is empty. This statement is already pictured in Figure 3.36. Because the conclusion is shown in the diagram of the premises without any ambiguity, this argument is valid.

NW ***Now Work Problem 5.***

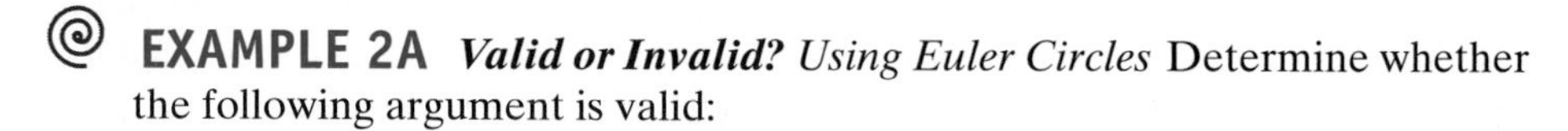

EXAMPLE 2A ***Valid or Invalid?*** *Using Euler Circles* Determine whether the following argument is valid:

No generals are pacifists.
No pacifists are soldiers.
Hence, no generals are soldiers.

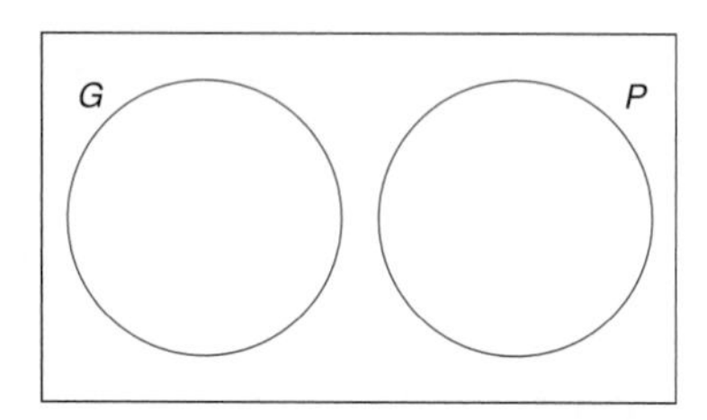

Figure 3.37a

Euler representation of the first premise where P = the set of generals and P = the set of pacifists.

Solution Let G = the set of generals, P = the set of pacifists, and S = the set of soldiers. We now represent the premises using Euler circles. The first premise is a universal negative. This means that the intersection of the sets of generals and pacifists is empty. The second premise also is a universal negative. Thus the sets of pacifists and soldiers also are disjoint. We now picture these two premises together and see if the conclusion results. In doing so, though, we have several choices in how the two premises can be represented. To see this we begin with picturing the first premise. This is shown in Figure 3.37a. We next picture the second premise within the same universal set. Here is where ambiguity arises. Figures 3.37b, 3.37c, and 3.37d are all proper Euler circle representations of the two premises. In all three figures (3.37b, 3.37c, and 3.37d), the intersection of sets P and S is empty, which is what the second premise implies. Of these three, only one figure, Figure 3.37b, also shows the conclusion, *No generals are soldiers*. The conclusion is not pictured in the other two figures. Since it is possible to diagram the premises

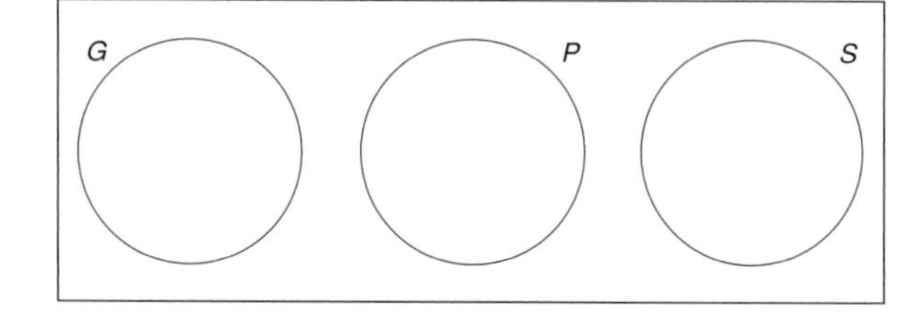

Figure 3.37b

One Euler representation of the first and second premises. G = the set of generals, P = the set of pacifists, and S = the set of soldiers. This also shows the conclusion.

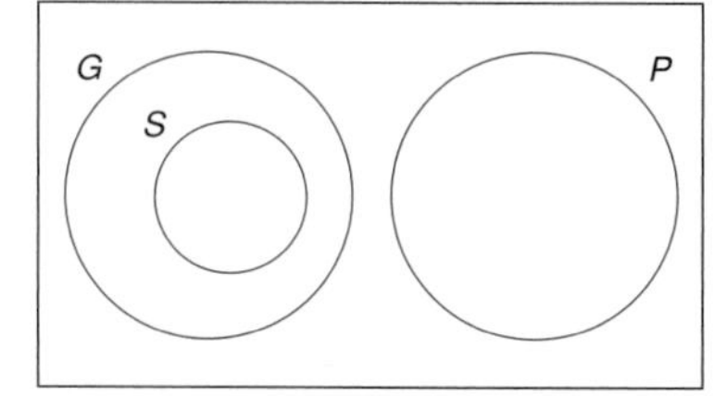

Figure 3.37c

A second Euler representation of the two premises. This does not show the conclusion.

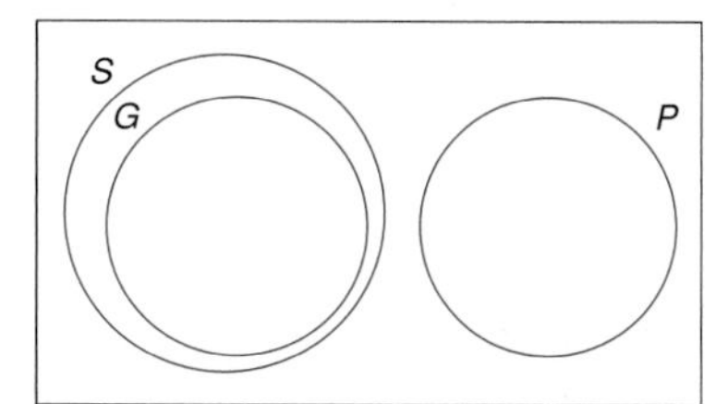

Figure 3.37d

A third Euler representation of the two premises. This does not show the conclusion.

without showing the conclusion at the same time (Figures 3.37c and 3.37d), the argument is invalid.

NW ***Now Work Problem 11.***

EXAMPLE 3A ***Valid or Invalid?*** *Using Euler Circles* Determine whether the following argument is valid:

All generals are soldiers.
Some generals are fighters.
Therefore, some soldiers are fighters.

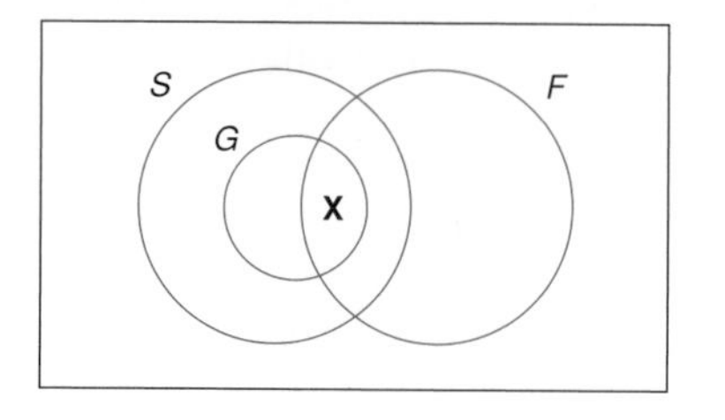

Figure 3.38

G = the set of generals, S = the set of soldiers, and F = the set of fighters.

Solution Let G = the set of generals, S = the set of soldiers, and F = the set of fighters. We now represent the premises using Euler circles. The first premise is a universal affirmative and implies that the set of generals is a proper subset of the set of soldiers. The second premise is a particular affirmative. There is at least one general who also is a fighter. These two premises are pictured together in Figure 3.38. The conclusion is a particular affirmative and implies that there is at least one soldier who also is a fighter. This statement is already pictured in Figure 3.38. Since the conclusion is shown in the diagram of the premises, this argument is valid.

To give some additional insight into why we diagrammed the premises as we did in Example 3A, it might be beneficial to examine the premises one at a time. Let's begin with the first premise, which is a universal affirmative. Its Euler circle representation is shown in Figure 3.39a. We next diagram the second premise, which is a particular affirmative. This is shown in Figure 3.39b. Since we must diagram the premises together, we keep set S in Figure 3.39b, but we de-emphasize it so we can focus only on the second premise. We now examine the conclusion, which is a particular affirmative, to see if it follows logically from the premises. More concretely, is the conclusion already pictured in the diagram of the premises? The answer is yes. This can be seen more clearly if we just focus on the Euler circle representation of the conclusion. This is shown in Figure 3.39c, where we de-emphasized set G. Note that the diagram shows that there is at least one soldier who is a fighter. That is, there is at least one element contained within the intersection of sets S and F. Since the conclusion is shown in the diagram of the premises without any ambiguity, the argument is valid.

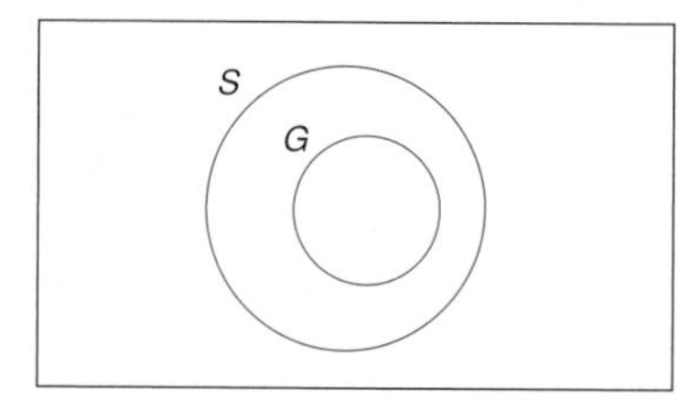

Figure 3.39a

Euler circle representation of the first premise.

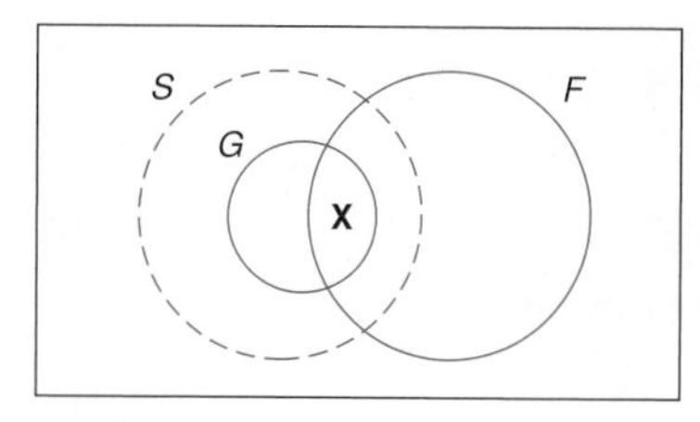

Figure 3.39b

Euler circle representation of the second premise.

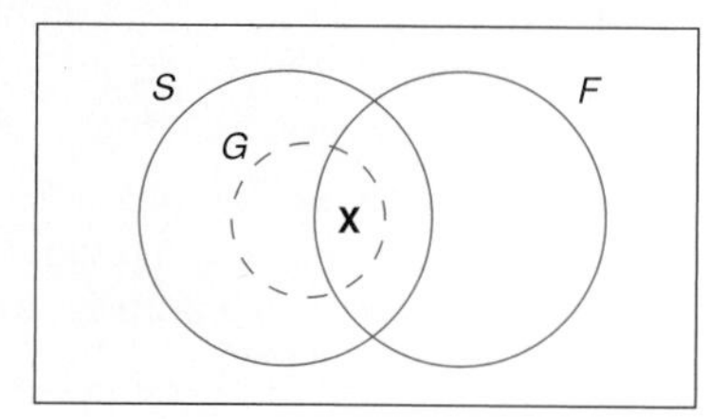

Figure 3.39c

Euler circle representation of the conclusion.

EXAMPLE 4A ***Valid or Invalid?*** *Using Euler Circles* Determine whether the following argument is valid:

All fishermen are patient.
Some fishermen are not liars.
Hence, some liars are not fishermen.

Solution Let F = the set of fishermen, P = the set of patient people, and L = the set of liars. We now represent the premises using Euler circles. The first premise is a universal affirmative and implies that the set of fishermen is a proper subset of the set of patient people. This is pictured in Figure 3.40a. The second premise is a particular negative. There is at least one fisherman who is not a liar. This is pictured in Figure 3.40b. We de-emphasize set P so we can concentrate on sets F and L. We now examine Figure 3.40b to see if the conclusion, which is a particular negative, is pictured. It is not. In order to show that there is at least one liar who is not a fishermen, there must be an x placed in the region outside of set F, but still contained within set L. There are two places where this can happen. We have labeled these regions as y_1 and y_2 in Figure 3.40c. Since there is no x in either of these regions, the conclusion is not pictured in the diagram of the premises. Thus the conclusion does not follow logically from the premises and hence is invalid.

NW ***Now Work Problem 15.***

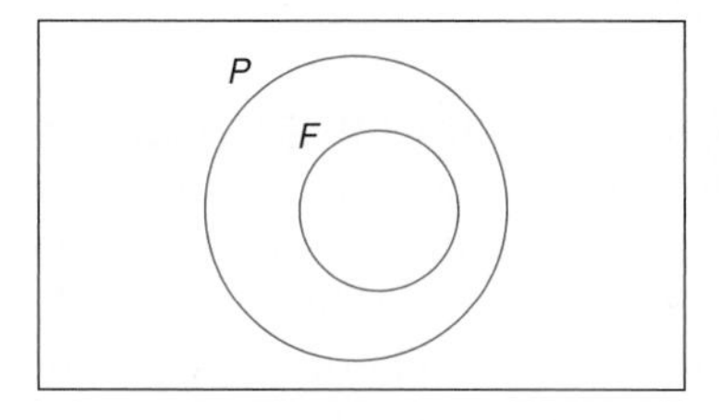

Figure 3.40a

Euler circle representation of the first premise, All fishermen are patient.

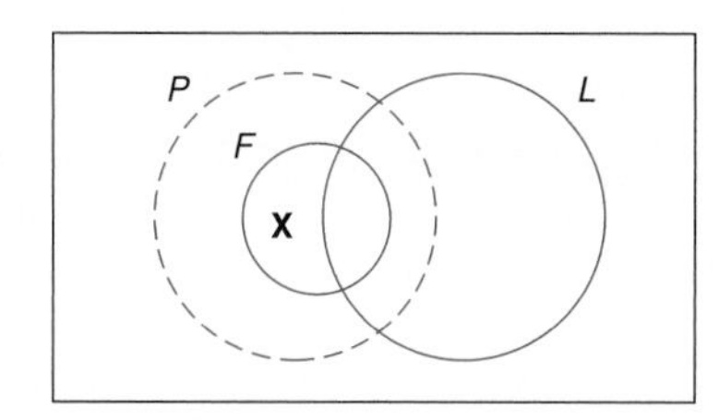

Figure 3.40b

Euler circle representation of the second premise, There is at least one fisherman who is not a liar.

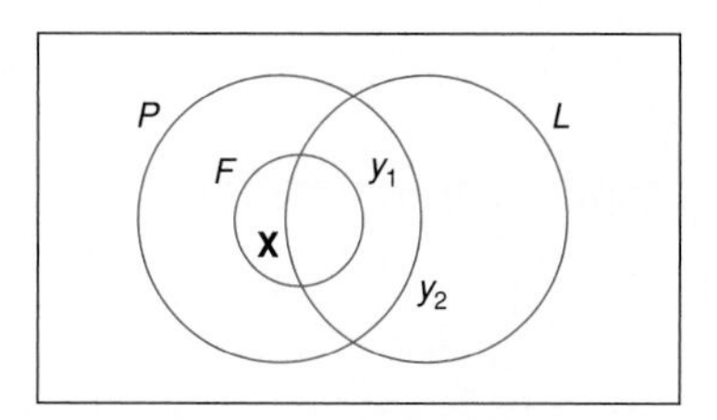

Figure 3.40c

An x needs to be placed in either of the regions labeled y_1 or y_2 in order for the conclusion to be accurately represented from the premises.

Exercises for Section 3.10

Use Venn diagrams to determine whether the following arguments are valid or invalid. State your answer.

1. No logic students are gullible.
   Some gullible people are superstitious.
   Hence, no logic students are superstitious.
2. All mechanics are intelligent.
   Some authors are intelligent.
   Therefore, some authors are mechanics.
3. All documents are fragile.
   Some documents are irreplaceable.
   Therefore, some fragile items are irreplaceable.
4. All tabloids are scandal mongers.
   Some scandal mongers are biased.
   So, some tabloids are biased.
5. **NW** All cars have four wheels.
   No bikes have four wheels.
   Therefore, no cars are bikes.
6. Some mowers are noisy.
   Some noisy items are choppers.
   So, some mowers are choppers.
7. All lawyers are aggressive.
   Some aggressive people are not pleasant.
   So, some lawyers are not pleasant.
8. All mathematics students are clever.
   Some history students are clever.
   Therefore, some mathematics students are history students.
9. All logic students are adroit.
   Some logic students are ambitious.
   So, some ambitious people are adroit.
10. All logic students are adroit.
    Some logic students are ambitious.
    Hence, some adroit people are ambitious.
11. **NW** All logic students are adroit.
    All adroit people are ambitious.
    Therefore, some ambitious people are logic students.
12. All gamblers are losers.
    Some quarterbacks are not losers.
    So, some quarterbacks are not gamblers.
13. All gamblers are losers.
    No losers are lucky.
    Hence, no gamblers are lucky.
14. No winners are losers.
    Some gamblers are losers.
    Therefore, some gamblers are not winners.
15. **NW** All winners are gamblers.
    Some gamblers are losers.
    Hence, some gamblers are winners.

16. No skyjackers are sane.
Some writers are sane.
Hence, some skyjackers are not writers.

17. Some gamblers are not kibitzers.
Some kibitzers are not winners.
Therefore, some gamblers are not winners.

18. All politicians are leaders.
Some politicians are lawyers.
Hence, some lawyers are leaders.

19. Some teachers are dull.
All dull people are boring.
Therefore, some teachers are boring.

20. Some instructors are understanding.
All understanding people are tolerant.
So, some instructors are tolerant.

21. All warriors are brave.
No cowards are brave.
Hence, no cowards are warriors.

22. No circles are squares.
No squares are triangles.
Therefore, no circles are triangles.

23. All squares are rectangles.
Some rectangles are parallelograms.
So, some squares are parallelograms.

24. All handball players are agile.
No weightlifters are agile.
Hence, no handball players are weightlifters.

25. All actors are vain.
Some vain people are not greedy.
So, some actors are not greedy.

26. All vain people are greedy.
Some actors are greedy.
Therefore, some actors are not vain.

Writing Mathematics

27. Explain the difference between truth and validity.
28. Explain how to use a Venn diagram to determine whether a syllogistic argument is valid or invalid.

Challenge Exercises

29. Supply the missing conclusion that will make this argument valid.

All flowers are pretty. Some plants are flowers. Hence, ...

30. Supply the missing conclusion that will make this argument valid.

Some television programs are stupid. All stupid things are useless. Therefore, ...

Group/Research Activity

31. John Venn and Leonhard Euler both made significant contributions to the development of logic. Do research and write a report on the life of one of these men. Be sure to include their contributions to logic and other areas that might be of interest. References may include history of mathematics books and the Internet.

Just for fun

Mike's car traveled 11 miles at a speed of 55 miles per hour. How many minutes did it take to travel 11 miles?

3.11 SWITCHING NETWORKS (OPTIONAL)

A practical application of logic is the design of switching networks. Switching networks may also be thought of as electrical circuits. Electrical circuits, such as those in light switches and computers, are familiar objects of everyday experience. You do not need any previous knowledge of electricity to understand the material presented in this section.

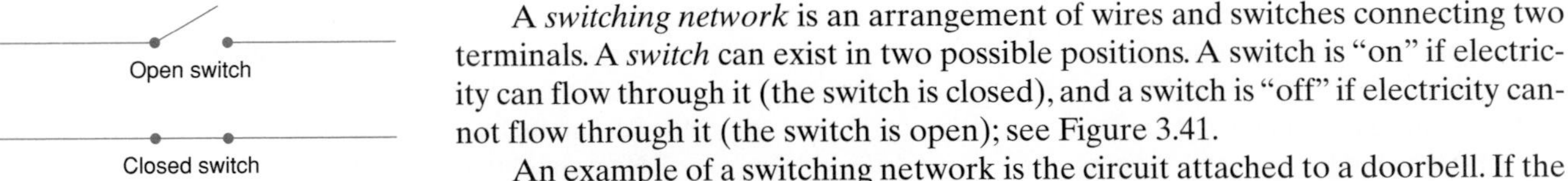

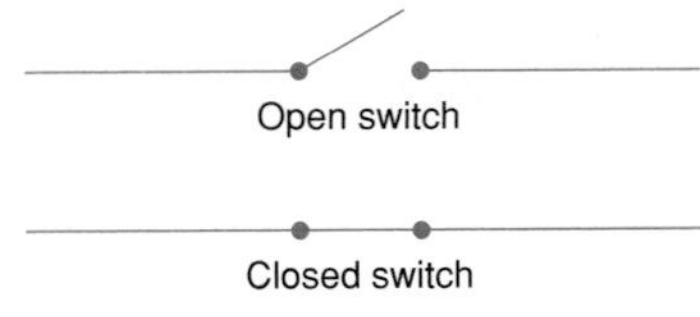

Figure 3.41

A *switching network* is an arrangement of wires and switches connecting two terminals. A *switch* can exist in two possible positions. A switch is "on" if electricity can flow through it (the switch is closed), and a switch is "off" if electricity cannot flow through it (the switch is open); see Figure 3.41.

An example of a switching network is the circuit attached to a doorbell. If the button is depressed, the switch becomes closed, electricity flows through it, and the doorbell rings. Normally the button is not depressed, the switch is open, and electricity cannot flow through it; hence the doorbell does not ring. A simple circuit for a doorbell appears in Figure 3.42.

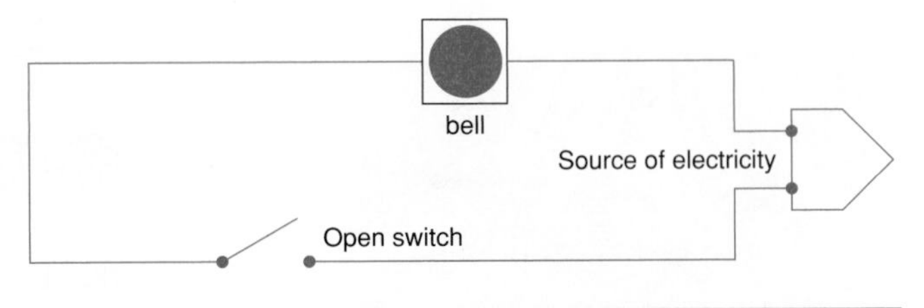

Figure 3.42

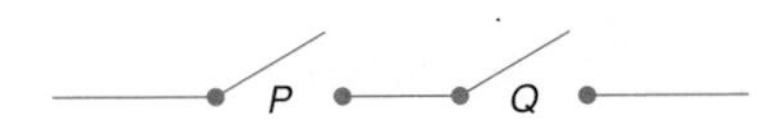

Figure 3.43
Series connection.

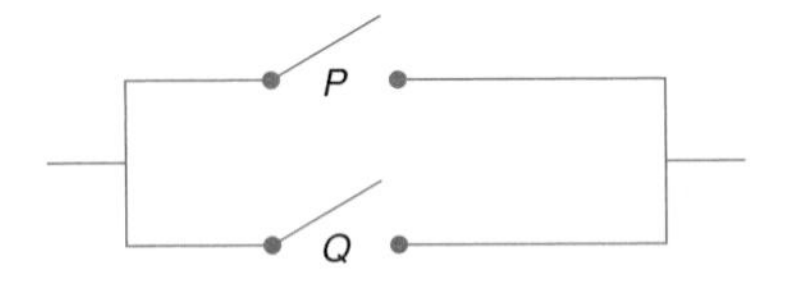

Figure 3.44
Parallel connection.

Figure 3.45

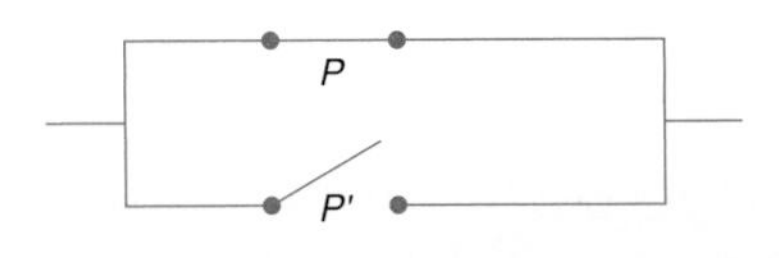

Figure 3.46

It should be noted that most switching networks encountered in everyday experience consist of more than one switch. Two switches can be arranged in a switching network in two different ways. The two arrangements are *series* connection and *parallel* connection. Figure 3.43 shows two switches, P and Q, connected in series. In a series connection, electricity will flow through the network only when both switches P and Q are closed. Electricity will not flow through the network if one or both of the switches are open. Does this remind you of anything in logic? Note that a series connection resembles the conjunction $P \wedge Q$. A conjunction is true only when both parts are true; otherwise, it is false. The series connection in Figure 3.43 will work (allow electricity to flow through) only when both switches are closed; otherwise, current will not pass through the network.

Figure 3.44 shows two switches in a parallel connection. In a parallel connection, electricity will flow through the network if P is closed, or if Q is closed, or if both P and Q are closed. The only time that current will not pass through this network is when both switches are open. A parallel connection corresponds to the disjunction $P \vee Q$. A disjunction is true when P is true, or Q is true, or when both P and Q are true. The only time that a disjunction is false is when both parts are false. The only time that current will fail to pass through a parallel connection is when both switches are open.

In some networks there are switches whose positions (open–closed) are determined by another switch. That is, when one switch is open, the other switch is in the opposite position (closed). Similarly, when one switch is closed, the other switch is open. Two switches that always have opposite positions are said to be *complementary switches*. If we have two complementary switches, we can name one of them P and its complement P'. Figures 3.45 and 3.46 illustrate complementary switches in series and parallel connections, respectively.

Note that in Figure 3.45 we have complementary switches in a series connection. If P is open, then P' is closed; and if P is closed, then P' is open. One of the two switches will always be open. Therefore, it is impossible for electricity to flow through this network. In Figure 3.46 we have complementary switches in a parallel connection. Electricity will always flow through this network. One of the switches will always be closed. Again, notice the similarity of complementary switches with a statement and its negation in logic. If P is true, then $\sim P$ is false, and if P is false, then $\sim P$ is true. A statement and its negation always have opposite truth values.

So far we have discussed series, parallel, and complementary switches. Let's examine these switches and their corresponding truth tables. But first let us agree

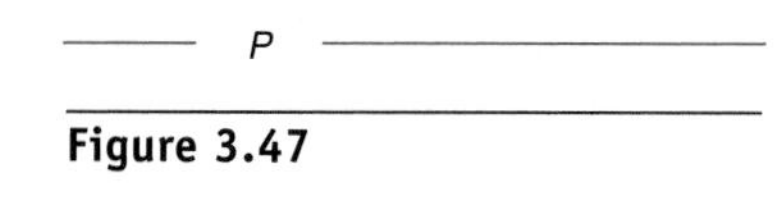

Figure 3.47

that the notation in Figure 3.47 will represent a switch whose position is not known. That is, we do not know if switch P is open or closed.

Table 3.14 shows the logic truth table corresponding to the switching network in Figure 3.48. Current will flow only when P and Q are closed.

Figure 3.48
Series connection.

TABLE 3.14

P	$\wedge$	Q	
T	$\boldsymbol{T}$	T	P closed, Q closed
T	$\boldsymbol{F}$	F	P closed, Q open
F	$\boldsymbol{F}$	T	P open, Q closed
F	$\boldsymbol{F}$	F	P open, Q open
1	3	2	

Figure 3.49
Parallel connection.

Table 3.15 shows the truth table corresponding to the switching network in Figure 3.49. Current will flow when P is closed, or Q is closed, or both P and Q are closed.

TABLE 3.15

P	$\vee$	Q	
T	$\boldsymbol{T}$	T	P closed, Q closed
T	$\boldsymbol{T}$	F	P closed, Q open
F	$\boldsymbol{T}$	T	P open, Q closed
F	$\boldsymbol{F}$	F	P open, Q open
1	3	2	

Figure 3.50
Series connection of complementary switches.

Figures 3.50 and 3.51 show series and parallel connection of complementary switches. In Figure 3.50, current cannot flow through; when P is closed, P' is open, and vice versa (see Table 3.16). In Figure 3.51, current will always flow; when P is closed, P' is open, and vice versa (see Table 3.17). Note that the logic symbol for P' is $\sim P$.

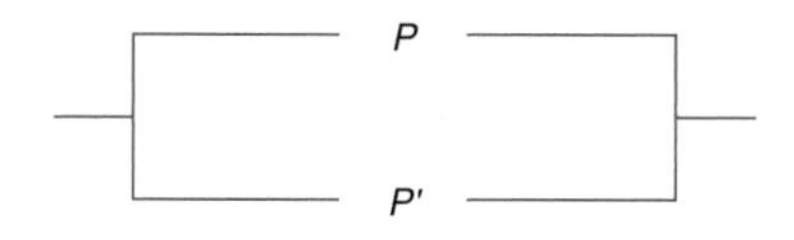

Figure 3.51
Parallel connection of complementary switches.

TABLE 3.16

P	$\wedge$	$\sim$	P
T	$\boldsymbol{F}$	F	T
F	$\boldsymbol{F}$	T	F
1	3	2	1

TABLE 3.17

P	$\vee$	$\sim$	P
T	$\boldsymbol{T}$	F	T
F	$\boldsymbol{T}$	T	F
1	3	2	1

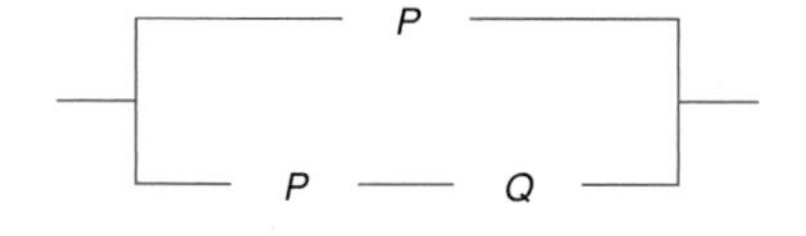

Figure 3.52

We have already discussed complementary switches, P and P'. These two switches will always have opposite positions to each other. When P is open, P' is closed, and when P is closed, P' is open. There also exist switches that are set up to operate together; when one is open the other is open, and when one is closed, the other is closed; they are always in the same position. When switches are set up to operate together in a network, we use the same letter to designate each such switch. Figure 3.52 shows a network that has two such switches. We have labeled each of these with the letter P.

You will note in Figure 3.52 that we have a network that is more involved than the previous examples we have considered. Earlier in this chapter we took simple statements and worked out truth tables for them; then we combined the simple statements into compound statements. We are about to do the same thing with networks. Figure 3.52 is a parallel connection, but within the parallel network is a series connection. The P, Q connection in the bottom wire is a series connection. From our previous discussion, we know that this can be represented as $P \wedge Q$. Since the top wire contains switch P and the whole network is parallel (current will flow through either the top wire or bottom wire), the entire network may be represented by $P \vee (P \wedge Q)$.

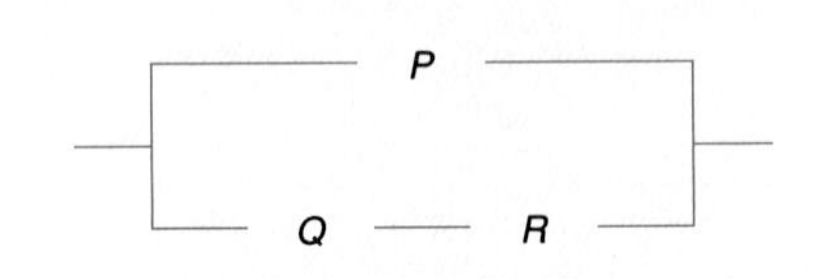

Figure 3.53

EXAMPLE 1 ***Writing a Symbolic Statement for a Network*** Write a symbolic statement for the network shown in Figure 3.53.

Solution The network shown is parallel. The bottom wire contains switches Q and R, and they are connected by a series connection, so we can write $Q \wedge R$. Switch P is in parallel with the rest of the network; therefore, the symbolic statement is $P \vee (Q \wedge R)$.

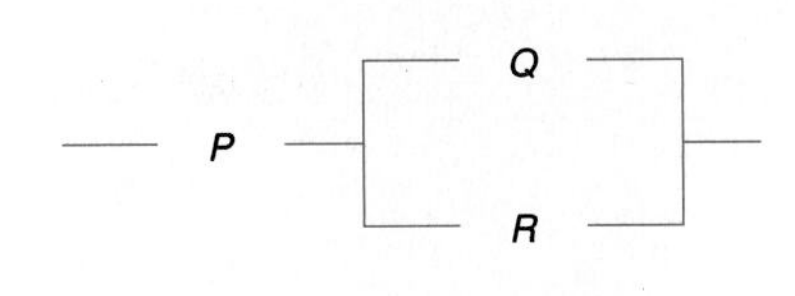

Figure 3.54

EXAMPLE 2 ***Writing a Symbolic Statement for a Network*** Write a symbolic statement for the network shown in Figure 3.54.

Solution Switches Q and R are connected in parallel so we write $Q \vee R$. Note that switch P is in a series connection with the rest of the network. Therefore, the symbolic statement is $P \wedge (Q \vee R)$.

NW ***Now Work Problem 3.***

EXAMPLE 3 ***Writing a Symbolic Statement for a Network*** Write a symbolic statement for the network shown in Figure 3.55.

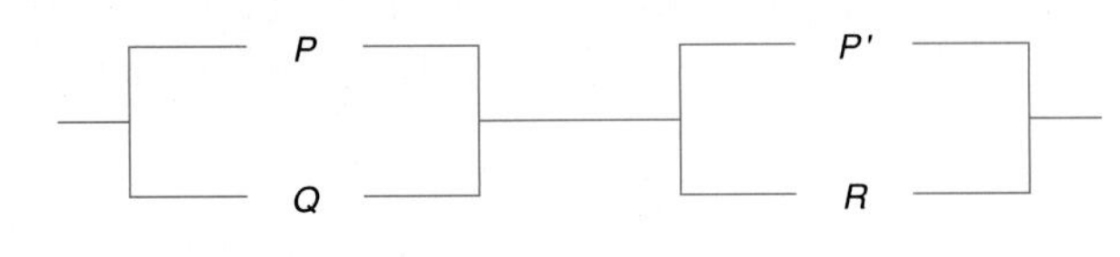

Figure 3.55

Solution We have two sets of parallel switches; P and Q are in parallel, as are P' and R. But note that the sets of parallel switches are in series connection with each other. Therefore, we can symbolize this network as $(P \vee Q) \wedge (\sim P \vee R)$. Note that we used the logic symbol $\sim P$ for the switch P'.

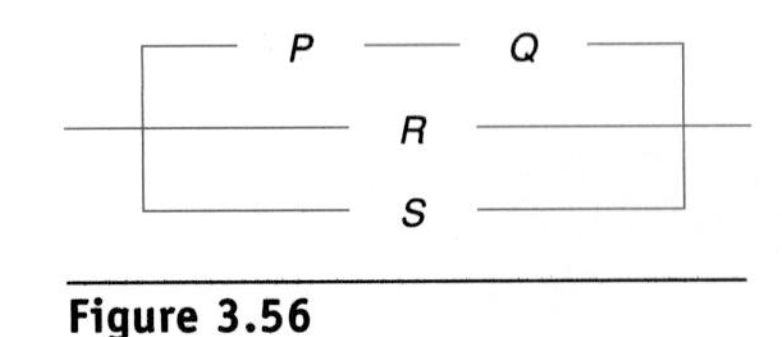

Figure 3.56

EXAMPLE 4 ***Writing a Symbolic Statement for a Network*** Write a symbolic statement for the network shown in Figure 3.56.

Solution Note that the given circuit is a parallel one. There are three wires that are parallel to each other. The top wire contains two switches P and Q that are set in a series connection, so we have $P \wedge Q$ for the top wire. Since each wire is in parallel with the others, we can symbolize this network as $(P \wedge Q) \vee R \vee S$.

NW ***Now Work Problem 7.***

So far we have been considering given networks and then writing symbolic statements for them. We can also construct (draw) networks that correspond to given symbolic statements. When we interpret such statements, we must remember that the connective $\vee$ means a parallel connection, the connective $\wedge$ means a series connection, and the negation symbol $\sim P$ means we have a switch that is complementary to switch P. In a circuit we denote this complementary switch by P'.

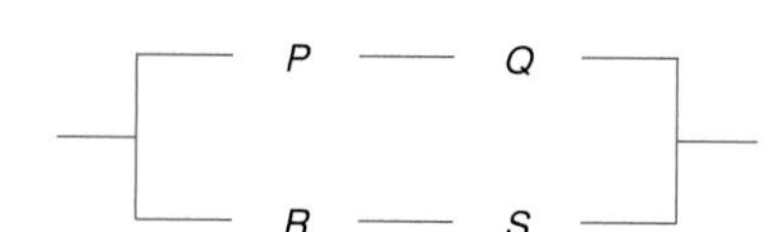

Figure 3.57

EXAMPLE 5 ***Drawing a Network from a Statement*** Draw a network representing the statement $(P \wedge Q) \vee (R \wedge S)$.

Solution The statement $(P \wedge Q) \vee (R \wedge S)$ is a disjunction. The connective $\vee$ is the dominant connective. Therefore, the circuit will contain two parallel wires. Each wire contains two switches set in a series connection, $P \wedge Q$ and $R \wedge S$. Hence, the network is shown in Figure 3.57.

EXAMPLE 6 ***Drawing a Network from a Statement*** Draw a network to represent the following statement:

$$(P \vee Q) \wedge [P \vee (R \wedge \sim Q)]$$

Solution The parentheses and brackets indicate that the statement

$$(P \vee Q) \wedge [P \vee (R \wedge \sim Q)]$$

is a conjunction. Therefore, the entire network will be in series. The first part, $(P \vee Q)$, is a parallel connection, as is $[P \vee (R \wedge \sim Q)]$. Inside of $[P \vee (R \wedge \sim Q)]$ we have a series connection, $R \wedge \sim Q$. The resulting network is shown in Figure 3.58. Note that $\sim Q$ appears in the network as Q'.

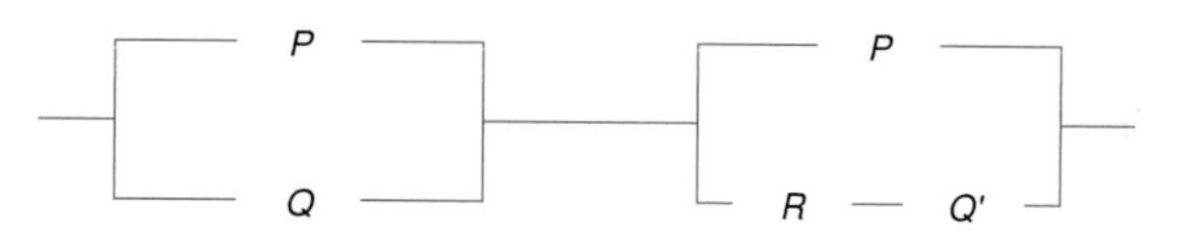

Figure 3.58

NW ***Now Work Problem 13.***

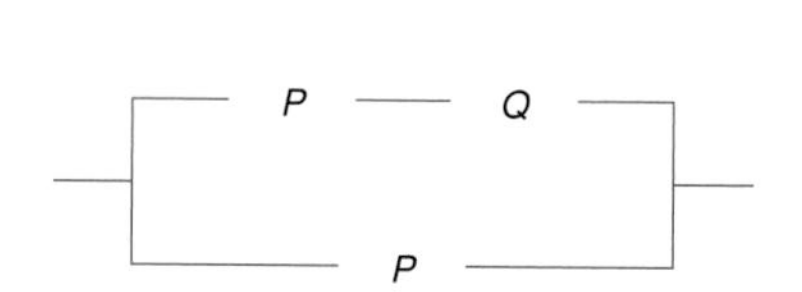

Figure 3.59

We have used logic to develop switching networks, and we may use logic to simplify a network. When we simplify a network, we reduce the number of switches in it. In practical terms, a person or company can save money if the number of switches required in a network is reduced. Two networks are considered to be equivalent when both networks function in the same manner. That is, if current flows through the first network, then it flows through the second, and if current does not flow through the first, then it does not flow through the second. Consider the network shown in Figure 3.59. The symbolic statement for this network is $(P \wedge Q) \vee P$. Constructing a truth table for the statement $(P \wedge Q) \vee P$, we have Table 3.18.

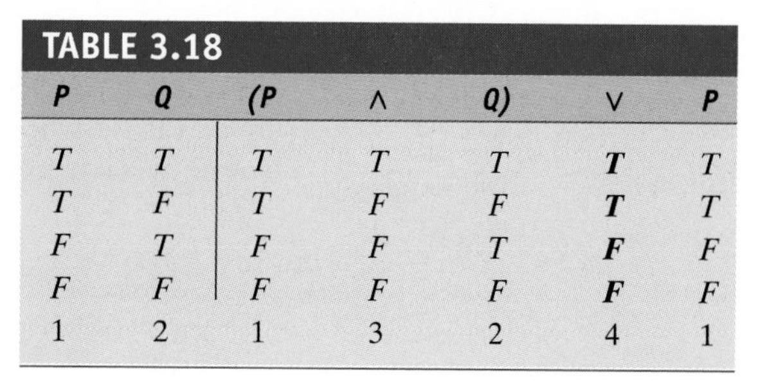

TABLE 3.18

P	Q	(P	∧	Q)	∨	P
T	T	T	T	T	**T**	T
T	F	T	F	F	**T**	T
F	T	F	F	T	**F**	F
F	F	F	F	F	**F**	F
1	2	1	3	2	4	1

The answer, or final result, is column 4. A switch is closed when it has a T value and open when it has an F value. Current will flow through this network whenever switch P is closed. In Table 3.18 rows 1 and 2 indicate that this is the case. Similarly, current will not flow through the network when P is open. Therefore, a simpler network need contain only switch P. Thus, the network $(P \wedge Q) \vee P$ in Figure 3.59 is equivalent to the network in Figure 3.60—namely, P.

P

Figure 3.60

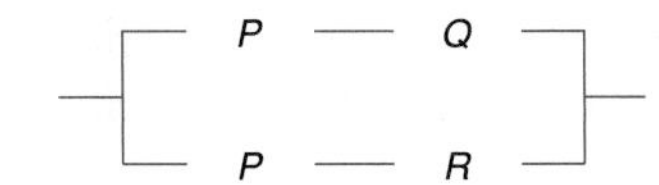

Figure 3.61

EXAMPLE 7 ***Simplifying a Network*** Find and draw a simplified network equivalent to the one shown in Figure 3.61.

Solution We must first write a symbolic statement for the network shown in Figure 3.61. It is a parallel network with series connections in the top and bottom wire. Therefore, the symbolic statement is $(P \wedge Q) \vee (P \wedge R)$.

P	Q	R	(P	∧	Q)	∨	(P	∧	R)
T	T	T	T	T	T	**T**	T	T	T
T	T	F	T	T	T	**T**	T	F	F
T	F	T	T	F	F	**T**	T	T	T
T	F	F	T	F	F	**F**	T	F	F
F	T	T	F	F	T	**F**	F	F	T
F	T	F	F	F	T	**F**	F	F	F
F	F	T	F	F	F	**F**	F	F	T
F	F	F	F	F	F	**F**	F	F	F
1	2	3	1	4	2	6	1	5	3

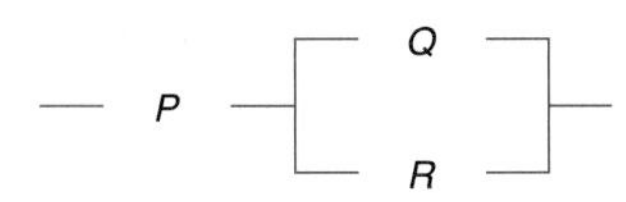

Figure 3.62

Next, we construct a truth table for this statement. Column 6 is our answer, and we see that the statement is true in the first three rows in our table. Current will flow through the network when switch P is closed and Q or R is closed. Hence, a symbolic statement for an equivalent network is $P \wedge (Q \vee R)$. This network is shown in Figure 3.62. We have reduced the number of switches by one. This could save a manufacturing company a lot of money.

NW ***Now Work Problem 21b.***

Alternative Solution Recall that in Section 3.6 we discussed logically equivalent statements. Two logical equivalences that we mentioned were the distributive properties.

$$P \wedge (Q \vee R) \equiv (P \wedge Q) \vee (P \wedge R)$$
$$P \vee (Q \wedge R) \equiv (P \vee Q) \wedge (P \vee R)$$

Note that the symbolic statement for the network illustrated in Figure 3.61 is $(P \wedge Q) \vee (P \wedge R)$. This is part of the first distributive property above. Therefore, $(P \wedge Q) \vee (P \wedge R)$ may be simplified to the equivalent statement $P \wedge (Q \vee R)$ by means of the distributive property. This is the desired result for the simplified, equivalent network shown in Figure 3.62.

EXAMPLE 8 ***Simplifying a Network Using the Distributive Property*** Given the network shown in Figure 3.63, find and draw a simplified, equivalent network using the distributive property.

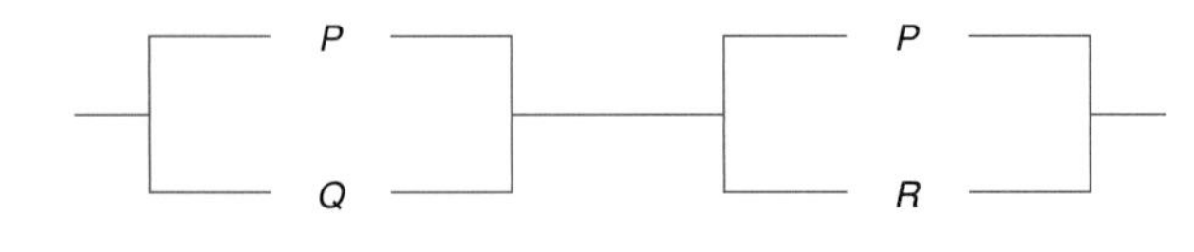

Figure 3.63

Solution We must first write a symbolic statement for the network of Figure 3.63. It is a series network with parallel switches in the series. The symbolic statement is $(P \vee Q) \wedge (P \vee R)$. By means of the distributive property of Section 3.6

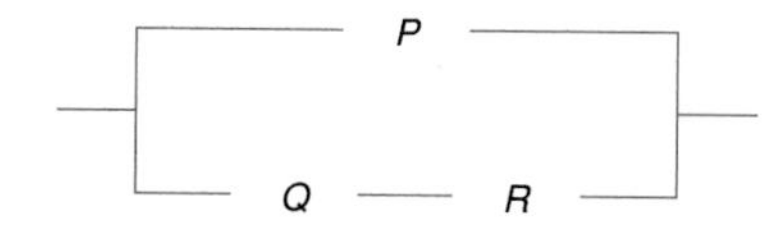

Figure 3.64

we have $(P \vee Q) \wedge (P \vee R) \equiv P \vee (Q \wedge R)$. The network corresponding to the simplified, equivalent statement is shown in Figure 3.64. ◎

The next problem is a traditional one in switching networks. It is usually called the "hall light problem." We have a hall light with a switch at each end of the hall. We want to control the light from either switch. That is, we want to be able to turn the light on (or off) by means of either switch. We want current to pass through the network when we operate either of the switches; we also want the current to cease when we operate either of the switches. That is, we want the value of the network to change whenever we change the position of one of the switches. Let's call the switches P and Q. Listing the possibilities for P and Q, we have Table 3.19.

TABLE 3.19

	P	*Q*
1.	*T*	*T*
2.	*T*	*F*
3.	*F*	*T*
4.	*F*	*F*

In the case represented by line 1 in Table 3.19, P and Q are both closed, and current could definitely flow. It would cease if we changed the position of P or Q, as in the cases corresponding to lines 2 and 3. Now, the fourth case (line 4 in Table 3.19) could be the result of changing the position of either P or Q in lines 2 and 3, and in this case we would want the current to flow through the network.

In other words, the hall light will be on when P and Q are both T, or when P and Q are both F. We need a parallel network that will function in this manner. (A series network will not work when a switch is open—that is, F.) If P and Q are both true, then $(P \wedge Q)$ is true, and if P and Q are both false, then $(\sim P \wedge \sim Q)$ is true. Thus, the symbolic statement that behaves in the desired manner is $(P \wedge Q) \vee (\sim P \wedge \sim Q)$, as shown in Table 3.20. The corresponding network is shown in Figure 3.65.

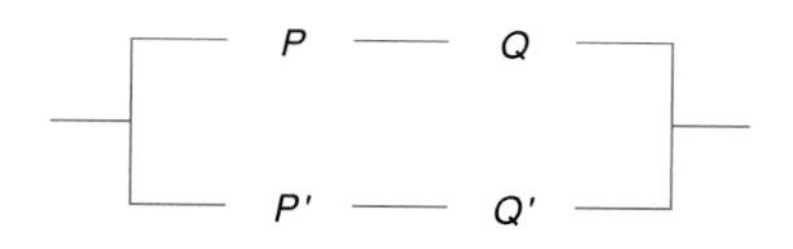

Figure 3.65

TABLE 3.20

P	*Q*	*(P*	∧	*Q)*	∨	*(~P*	∧	*~Q)*
T	*T*	*T*	*T*	*T*	***T***	*F*	*F*	*F*
T	*F*	*T*	*F*	*F*	***F***	*F*	*F*	*T*
F	*T*	*F*	*F*	*T*	***F***	*T*	*F*	*F*
F	*F*	*F*	*F*	*F*	***T***	*T*	*T*	*T*
1	2	1	3	2	7	4	6	5

Exercises for Section 3.11

For Exercises 1–10, write a symbolic statement for the networks shown.

1.

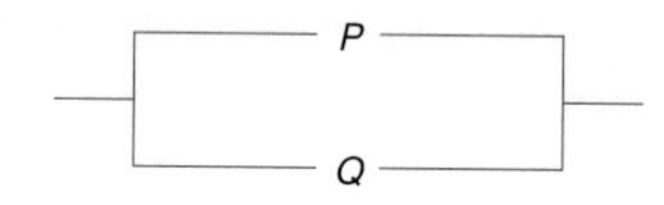

2.

3.

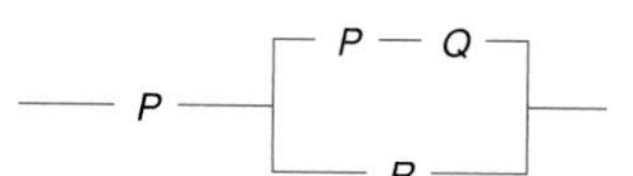

4.

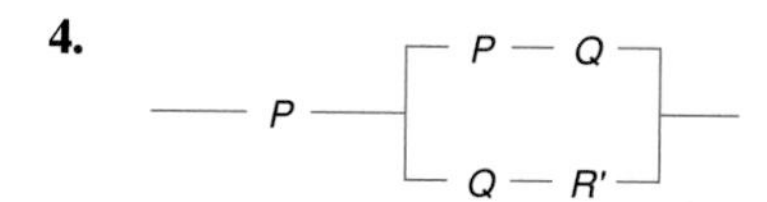

5.

6.

NW **7.**

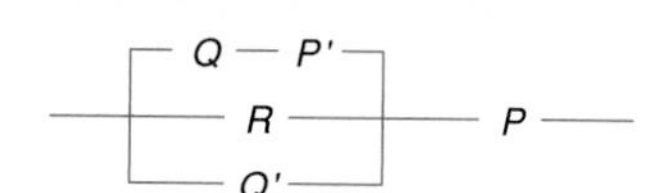

8.

9.

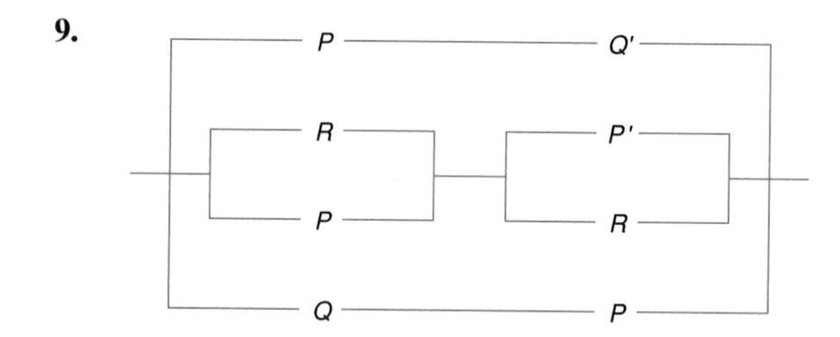

10.

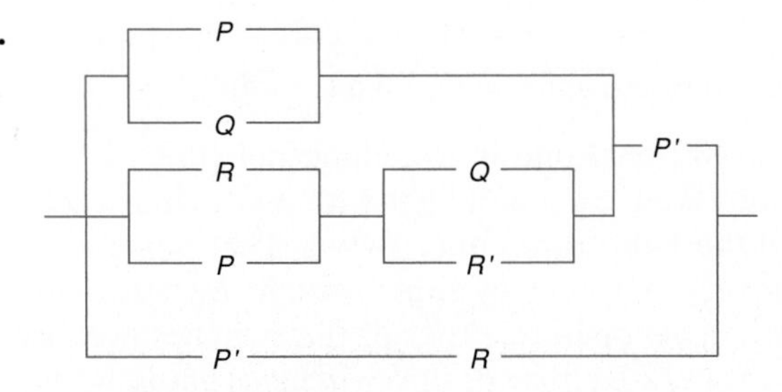

For Exercises 11–20, construct the switching networks that correspond to the given statements.

11. $(P \wedge Q) \vee R$

12. $[(P \wedge Q) \vee R] \wedge P$

NW **13.** $(P \vee Q) \wedge (P \vee R)$

14. $(P \wedge Q) \vee (P \wedge R)$

15. $(P \vee Q) \wedge (\sim P \vee R)$

16. $(P \wedge Q) \wedge [(P \wedge R) \vee (\sim Q \wedge P)]$

17. $P \vee [Q \wedge (P \vee \sim Q)]$

18. $(P \wedge \sim Q) \vee [(P \wedge Q) \vee (\sim P \wedge Q)]$

***19.** $P \rightarrow Q$

***20.** $P \rightarrow (Q \wedge R)$

21. Construct switching networks that are equivalent to and also simpler than the given networks.

a.

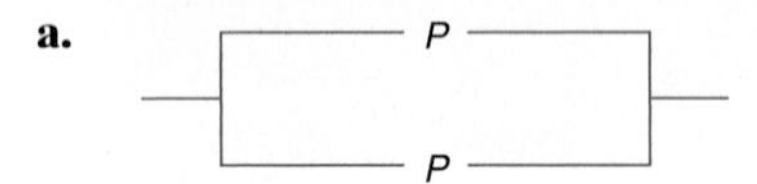

NW **B.**

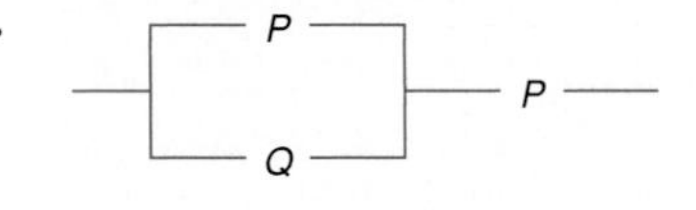

c.

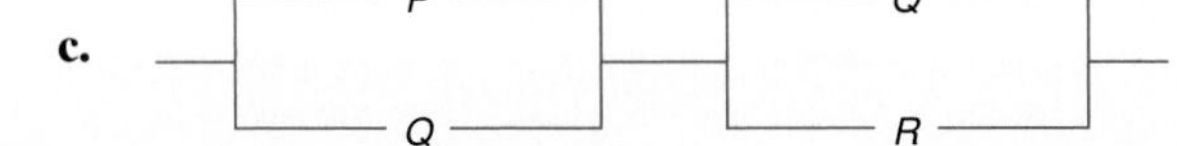

d.

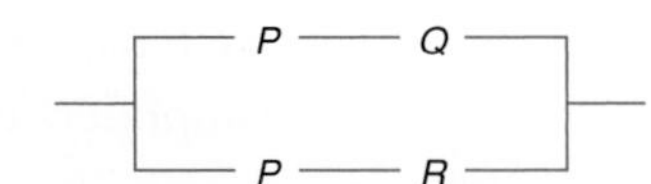

Writing Mathematics

22. Describe the difference between a parallel switching network and a series switching network.

23. What are complementary switches in a switching network and how are they used?

Challenge Exercises

24. A committee of three (Bob, Joe, and Fred) must vote yes or no on an important proposal. They want to do this secretly. Design a switching network for indicating a majority vote. Write the symbolic statement and draw the network.

25. An engineer designed the network of switches shown in Figure 3.66. Find a simpler but equivalent network.

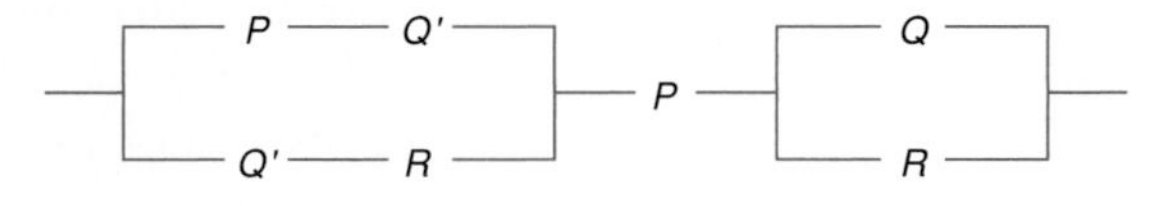

Figure 3.66

26. Draw a network that represents a tautology.

Group/Research Activity

27. The term *switching network* has a different meaning in the context of computer networks. Find out what a switching network is in the world of computer networks and how the concepts discussed in this section might relate to switched networks.

Just for fun

Which is bigger?

3 7

CHAPTER REVIEW MATERIALS

Summary / Chapter 3

3.2 Statements and Symbols

- In logic, *statements*, also called *propositions*, are either true or false, but not both. Statements are either simple or complex.
- *Simple statements* do not contain any connectives.
- *Complex statements* contain the connectives *not, and, or, if ... then ...*, and *if and only if.* These statements respectively are called *negation, conjunction, disjunction, conditional*, and *biconditional.*

- Letters such as P, Q, R, and S, and sometimes A, B, and C, are traditionally used to symbolize simple statements.
- The symbols used for the connectives are: $\sim$ (*not*), $\wedge$ (*and*), $\vee$ (*or*), $\rightarrow$ (*if ... then ...*), and $\leftrightarrow$ (*if and only if*).

3.3 Dominance of Connectives

- Parentheses are used to tell us the type of compound statement we are considering. In the absence of parentheses, we follow the convention of the dominance of the connectives.
- The biconditional ($\leftrightarrow$) is the most dominant connective, followed by the conditional ($\rightarrow$), and then the equally dominant conjunction ($\wedge$) and disjunction ($\vee$). The negation ($\sim$) is the least dominant connective.

3.4 Truth Tables for Negation, Conjunction, and Disjunction

- *Truth tables* show us all possible truth values for a given compound statement.
- The truth assignment for a negation is always the opposite of the original statement. Thus, if a given statement, P is true, then $\sim P$ is false; and if P is false, then $\sim P$ is true,
- A conjunction is true only when both of the simple statements that make up the conjunction are true. Thus, a *T–T* combination under a conjunction yields a true statement, but *T–F*, *F–T*, and *F–F* combinations yield a false statement.
- A disjunction is true in all cases except when both of the simple statements that make up the disjunction is false. Thus, *T–T*, *T–F*, and *F–T* combinations under a disjunction yield a true statement, but a *F–F* combination yields a false statement.
- A statement that is always true is called a *tautology* and a statement that is always false is called a *contradiction*.

3.5 Truth Tables for Conditional and Biconditional

- A *conditional* is true all cases except when the first statement is true and the second statement is false. Thus, a *T–F* combination under a conditional yields a false statement, but *T–T*, *F–T*, and *F–F* combinations under a conditional yield a true statement.
- A *biconditional* is true when both simple statements that make up the biconditional have the same truth value. Thus, *T–T* and *F–F* combinations under a biconditional yield a true statement, and *T–F* and *F–T* combinations yield a false statement.

3.6 De Morgan's Law and Equivalent Statements

- *De Morgan's law* is used to create logically equivalent statements from conjunctions and disjunctions. The procedure is to (1) negate the entire given statement, (2) negate each part of the given statement, and (3) change the *and* to *or* or change *or* to *and*.
- De Morgan's law can also be used to create logically equivalent statements from a conditional. To do so, though, we must first convert $P \rightarrow Q$ to its logically equivalent form, $\sim P \vee Q$.

3.7 The Conditional (Optional)

- Variations of a given conditional statement, $P \rightarrow Q$ include *converse* ($Q \rightarrow P$), *inverse* ($\sim P \rightarrow \sim Q$), and *contrapositive* ($\sim Q \rightarrow \sim P$).
- A conditional statement is logically equivalent to its contrapositive.
- A conditional statement may be worded in many different ways, all of which have the same meaning. The following are all equivalent to *If P, then Q: Q, if P; P only if Q; P is sufficient for Q; Q is necessary for P; P implies Q; and Q whenever P.*
- The biconditional, *P if and only if Q*, may also be expressed equivalently as *P is necessary and sufficient for Q*.

3.8 Valid Arguments

- In logic, an *argument* contains *premises*, which are assumed to be true, and a *conclusion*. An argument is considered *valid* if the conclusion follows logically from the premises; that is, the argument is valid if the premises imply the conclusion.
- To determine if an argument is valid, we (1) conjunct the premises (i.e., we connect them using a conjunction), (2) imply the conclusion (i.e., connect the newly formed conjunction statement to the premise using a conditional), and (3) complete the truth table for the resulting conditional statement. If the result is a tautology, then the argument is valid; otherwise it is *invalid*.

3.9 Picturing Statements with Venn Diagrams (Optional)

- A logic argument that contains three statements—a *major premise*, a *minor premise* and the *conclusion*—is called a *syllogism*. The syllogisms discussed were composed of quantified statements, which can be pictured using Venn diagrams.
- There are four types of quantified statements: *universal affirmative* (*All As are Bs*), *universal negative* (*No As are Bs*), *particular affirmative* (*Some As are Bs*), and *particular negative* (*Some As are not Bs*).
- Venn diagrams can also be used to determine if two statements are *consistent*. Two statements are consistent if can be true at the same time; otherwise they are *inconsistent*.

3.10 Valid Arguments and Venn Diagrams (Optional)

- Venn diagrams can be used to determine if a syllogism is valid.
- To use Venn diagrams to determine the validity of a syllogism, we first picture the premises in a Venn diagram. We next examine the Venn diagram to see if it contains the conclusion. If the conclusion is shown without any ambiguity as soon as the premises are drawn, then the argument is valid.
- If it is possible to diagram the premises without showing the conclusion at the same time, then the argument is invalid.
- An alternative method for determining the validity of a syllogism is to use *Euler circles*.

3.11 Switching Networks (Optional)

- A *switching network*, which is an application of logic principles, is an arrangement of switches and wires connecting two terminals.
- A switch is "on" (i.e., closed) if current can flow through it, and "off" (i.e., opened) if current cannot flow through it.
- Switches can be arranged in a *series connection*, a *parallel connection*, or a combination of series and parallel connections. Switches in a series connection correspond to a conjunction (*P and Q*), and switches in a *parallel connection* correspond to a disjunction (*P or Q*).
- *Complementary switches* are pairs of switches that are always in opposite positions. Such switches correspond to a statement and its negation (*P and not P*).

Key Terms and Concepts

Review Exercises

3.2 Statements and Symbols

1. Identify each sentence as a simple statement, compound statement, or neither. If the statement is compound, then classify it as a negation, conjunction, disjunction, conditional, or biconditional.

a. Either Hugh sold his car, or he traded it in.
B. It is raining?
c. Yesterday was Friday.
d. Scott stayed home, but Joe went to the show.
e. Carlos did not bring his notes to class.
f. If Mary went swimming, then she did not study.

2. By means of the appropriate connectives and parentheses, symbolize each statement, using the given symbols for the simple statements:

$$P = \textit{Jill is reticent},$$
$$Q = \textit{Robin is cautious},$$
$$R = \textit{Bob is successful}.$$

a. Jill is reticent, but Robin is cautious.
B. If Robin is not cautious, then Bob is not successful or Jill is reticent.
c. Neither is Jill reticent nor is Robin cautious.
d. Robin is cautious and Jill is reticent, if Bob is successful.
e. Either it is false that both Bob is successful and Robin is cautious, or Jill is reticent.
f. Jill is reticent if and only if Robin is cautious or Bob is successful.

3.3 Dominance of Connectives

3. Let P = *Bill is golfing*, Q = *Addie is sailing*, and R = *Flo is jogging*. Write each statement in words.

a. $P \wedge (Q \vee R)$
B. $(P \wedge Q) \vee R$
c. $P \vee Q \rightarrow R$
d. $P \vee (Q \rightarrow R)$
e. $R \wedge Q \rightarrow \sim P$
f. $\sim P \leftrightarrow Q \wedge \sim R$

4. Explain why it is necessary to have a ranking or dominance of connectives in symbolic logic.

3.4 Truth Tables for Negation, Conjunction, and Disjunction/3.5 Truth Tables for Conditional and Biconditional

5. Construct a truth table for each statement.

a. $\sim P \rightarrow Q$
B. $P \vee \sim Q$
c. $P \vee Q \leftrightarrow P$
d. $P \vee Q \rightarrow R$
e. $P \wedge \sim Q \rightarrow Q \vee R$
f. $\sim(P \wedge \sim Q) \rightarrow \sim(\sim P \vee Q)$

Use Figure 3.67 to determine the truth value of the given statements in Exercises 6–9.

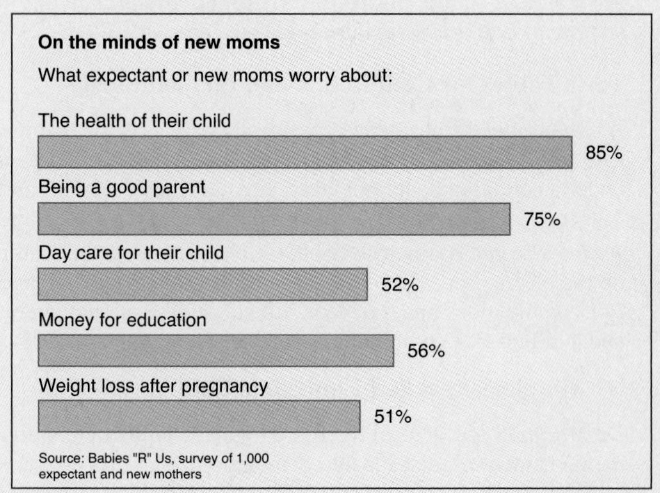

Figure 3.67

6. If 85% of expectant or new moms worry about the health of their child, then more than 50% worry about day care for their child or money for education.

7. It is false that more than 50% of expectant or new moms worry about weight loss after pregnancy and 100% worry about being a good parent.

8. If 56% of expectant or new moms do not worry about money for education, then more than three-fourths of these women worry about weight loss after pregnancy and less than 50% do not worry about day care for their child.

9. More than 75% of expectant or new moms worry about being a good parent if and only if more than 50% worry about weight loss after pregnancy.

3.6 De Morgan's Law and Equivalent Statements

10. Use De Morgan's law to rewrite each statement.

a. It is false that today is Monday and that tomorrow is Sunday.

B. Neal is not first, or David is second.

c. Hugh is painting or cutting the grass.

d. It is not the case that Norma went to the store or that Laurie went swimming.

e. If mathematics is difficult, then logic is easy.

11. Determine whether $\sim P \vee Q$ is logically equivalent to $\sim(P \wedge \sim Q)$.

3.7 The Conditional

***12.** Using the suggested notation, write in symbolic form the converse, inverse, and contrapositive of each conditional statement.

a. If I pass math, then I will graduate. (P, G)

B. If I pass the test, then I will not come to class. (P, C)

c. If you do not attend class, then you will not pass the course. (A, C)

***13.** A young man promised his girl friend, "I will marry you only if I get a job." He eventually got a job and then married someone else. Is the young man a heel? Why or why not? Explain your answer by using logic.

14. Given that "If I study logic then I will not be gullible" is a true statement.

a. Write another statement (in words) that you know is *true* from the above.

B. Write another statement (in words) that you know is *false* from the above.

***15.** Let P = *It is a gas eater* and Q = *It is a car*. Write each of the following statements in symbolic form:

a. If it is a gas eater, then it is a car.

B. Being a car is necessary for it to be a gas eater.

c. Not being a car is sufficient for its not being a gas eater.

d. It isn't a car only if it isn't a gas eater.

e. It isn't a gas eater unless it is a car.

Use Figure 3.68 to do the following for each statement in Exercises 16–19:

a. Symbolize the statement using the letters in parentheses.

B. Determine if the statement is true or false.

c. Write the converse, inverse, contrapositive, and negation of the statement in symbolic form.

***16.** If the SUV was a Jeep Wrangler (J), then it received a "P" rating for the front crash test (P).

***17.** An SUV received a marginal rating (M) for the front crash test only if it was a Ford Escape/Mazda Tribute (F).

How small SUVs rated

Results of the Insurance Institute for Highway Safety side and front crash tests of the smallest sport-utility vehicles. Ratings: G (good), A (acceptable), M (marginal), P (poor). Side air bag status: S (standard), O (optional), N (none).

	Side crash	Front crash	Side air bags
Subaru Forester	G	G	S
Hyundai Santa Fe	A	G	S
Honda CR-V	M	G	N
Ford Escape[1]/ Mazda Tribute	G	M	O
Jeep Wrangler	M	A	N
Honda Element	P	G	N
Saturn Vue[2]	P	G	O
Mitsubishi Outlander	P	G	N
Land Rover Freelander	P	A	N
Suzuki Grand Vitara	P	A	N
Toyota RAV4	P	A	N

1 – Escape tested with optional side air bag
2 – Vue tested without optional air bag
Source: Insurance Institute for Highway Safety

Figure 3.68

***18.** Receiving a "G" rating for the side crash test (G) implies that the SUV was not a Subaru Forester (S).

***19.** The Honda Element does not have side air bags (H) if the Toyota RAV4 does not have side air bags (T).

3.8 Valid Arguments

20. Use a truth table to determine whether the following arguments are valid or invalid. State your answer.

a. If Paula is not polite, then Sally is quiet. (P, S)
It is false that Sally is quiet.
Therefore, Paula is not polite.

B. If Bill is golfing, then Addie is sailing. (B, A)
If Addie is sailing, then Flo is jogging. (A, F)
Therefore, if Bill is golfing, then Flo is jogging.

***21.** Use Venn diagrams to determine whether each pair of statements is consistent or inconsistent.

a. No liars are virtuous.
Some liars are virtuous.

B. No liars are virtuous.
All liars are virtuous.

c. Some cars are gas guzzlers.
Some gas guzzlers are not cars.

d. All mathematics students are industrious.
Some mathematics students are not industrious.

22. A valid argument can have a true conclusion or a false conclusion. Explain how this is possible.

3.9 Picturing Statements with Venn Diagrams/ 3.10 Valid Arguments and Venn Diagrams

***23.** Use Venn diagrams to determine whether each argument is valid or invalid. State your answer.

a. All logic students are mathematics students.
No mathematics students are gullible.
Hence, no logic students are gullible.

b. All logic students are mathematics students.
Some logic students are not gullible.
Therefore, some gullible people are not mathematics students.

c. No golfers are sane.
All students are sane.
Therefore, no students are golfers.

d. Some television programs are stupid.
All stupid things are useless.
So, some television programs are useless.

e. All commercials are deceiving.
Some ads are deceiving.
Hence, some ads are commercials.

3.11 Switching Networks

***24.** Write a symbolic statement for each network.

a.

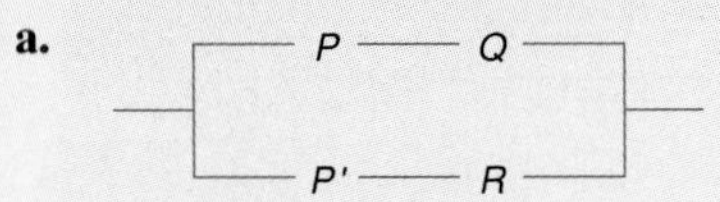

b.

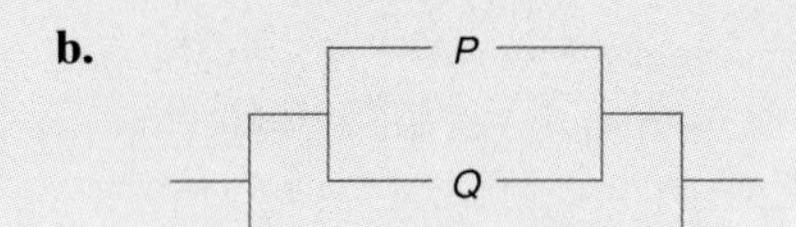

c.

R — Q
P
Q' — R'

d.

P
Q — R
P'

***25.** Construct the switching networks that correspond to the given statements.

a. $P \wedge (Q \vee R)$ **b.** $P \vee (Q \wedge R)$

c. $(P \wedge Q) \vee (P \wedge R)$ **d.** $(P \wedge Q) \vee [(P \wedge Q) \vee R]$

Chapter Quiz

Indicate whether each statement is true or false.

1. The sentence "How old are you?" is considered to be a statement in logic.

2. "Jim is late or he is absent" is an example of a disjunction.

3. "Dana will graduate, if she passes the exam," is an example of a conditional.

4. Given P = *Today is Monday* and Q = *I am tired*, then $\sim P \wedge Q$ can be written as "Today is not Monday, but I am tired."

5. The answer column of the truth table for $P \rightarrow \sim Q$ is *T, F, T, F.*

6. $P \wedge (Q \vee \sim P)$ is a tautology.

7. "If you cut class, then you missed the exam" is logically equivalent to "Either you did not cut class or you missed the exam."

8. "It is false that you did not study and that you passed the exam" is logically equivalent to "You did study or you did not pass the exam."

9. The following argument is valid:

If I work, then I will be paid.
I did not work.
Therefore, I was not paid.

10. A contradiction is a statement that is always false.

11. Statements that cannot be true at the same time are inconsistent.

12. The following argument is valid:

All ants are busy creatures.
All busy creatures are creative.
Therefore, some ants are creative.

13. "If I do not pass this course, then I cannot graduate" is the converse of "If I do pass this course, then I can graduate."

14. "If you work hard, then you will succeed" is the contrapositive of "If you don't succeed, then you didn't work hard."

15. "If you attend class, then you will pass" is the inverse of "If you did not pass, then you did not attend class."

16. "It snows only if it is cold" is logically equivalent to "If it snows, then it is cold."

17. The biconditional is the most dominant connective and the conjunction is the least dominant connective.

18. The disjunction is a more dominant connective than the conjunction.

19. Given $\sim(P \wedge Q \rightarrow \sim P)$, the answer column in the truth table for this statement would be found under the arrow.

20. Given P = *Today is Monday* and Q = *I am tired*, then $\sim P \rightarrow Q$ can be written as "Today is not Monday if I am tired."

21. The answer column of the truth table for $P \vee Q \rightarrow \sim R$ is *F, T, F, T, F, T, T, T.*

22. $P \wedge \sim Q \rightarrow Q \vee R$ is a contradiction.

23. The statement $\sim P \wedge Q \rightarrow \sim R$ is an example of a negation.

***24.** Given the two premises

All students are industrious.
No politicians are industrious.

A conclusion that will make the argument valid is

Some students are politicians.

***25.** Given the symbolic statement

$(P \wedge Q) \vee (\sim P \wedge \sim Q)$

the corresponding network would be described as a series connection.

Just for fun

Bill, the bewildered builder, discovered that a large picture window had been broken in his big new beautiful building. Bill knew that three workers were on the premises when the window was broken—Bob, Bart, and Barry. The workers' professions were painter, mason, and carpenter. But Bill did not know which man did which job, although he did know that one had committed the foul deed. He also knew the painter always told the truth, the mason never told the truth, and the carpenter always told one true statement and one false statement.

Barry said, **"Bart didn't do it."**
"Bob did it."
Bob said, **"I didn't do it."**
"Bart did it."
Bart said, **"I didn't do it."**
"Barry did it."

Using the true–false idea, help Bill discover the culprit's name and profession.

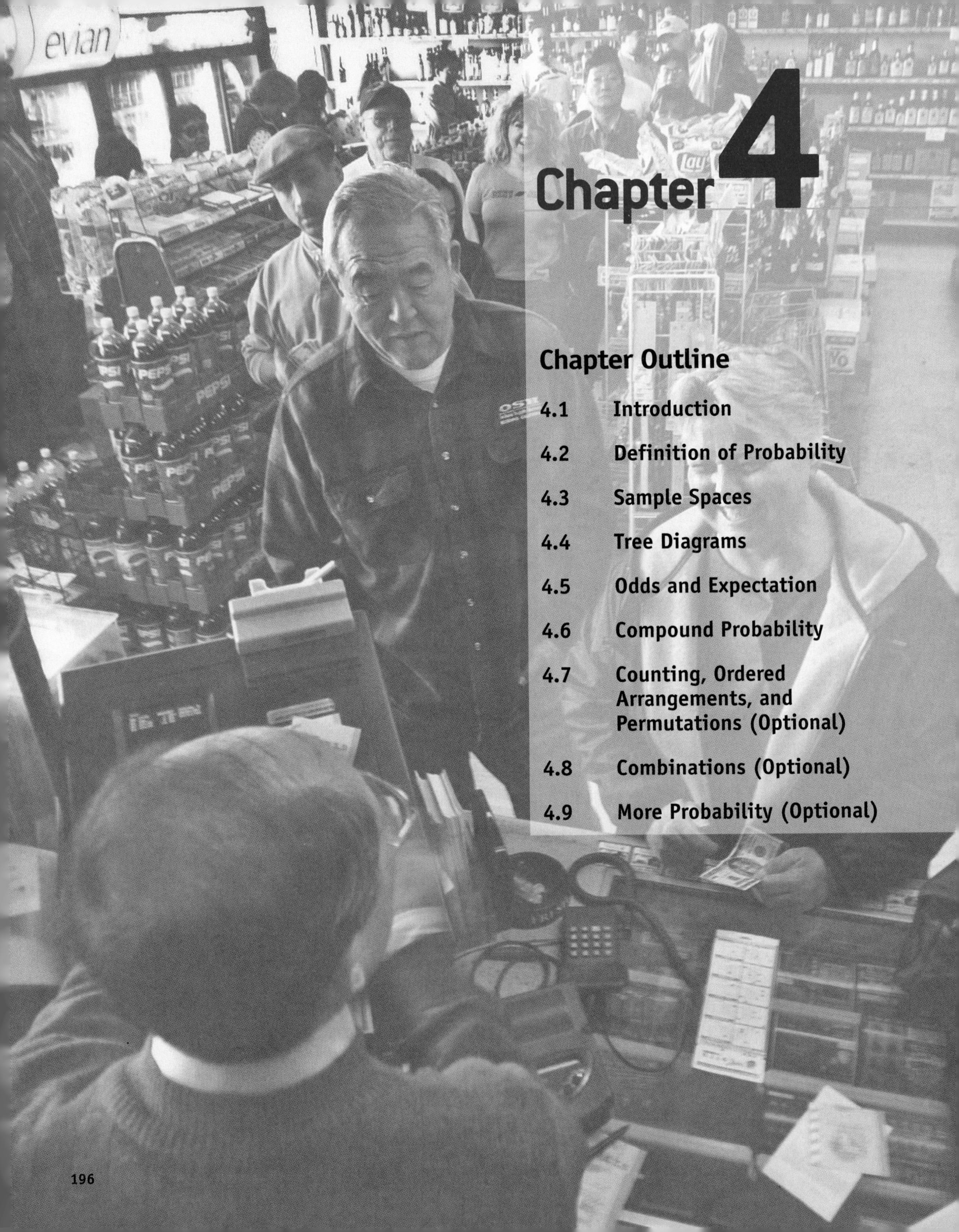

Chapter 4

Chapter Outline

Probability

After Studying This Chapter, You Will Be Able to Do the Following:

1. Compute the **probability** that an event *A* will occur, and given the probability of *A*, compute the probability of *not A*.
2. Use the **fundamental counting principle** to determine how many ways two or more events can occur together.
3. Construct a **sample space**, either by listing or by using a **tree diagram**, to show the possible outcomes of an experiment.
4. Compute the **odds in favor** of and **odds against** an event.
5. Compute the **mathematical expectation** for an event.
6. Determine whether two events are **dependent** or **independent**, and determine whether two events are **mutually exclusive**.
7. Compute the probability of the event *A and B* and the probability of the event *A or B*, given the probability of *A* and the probability of *B*.

*8. Determine the number of **permutations** that can occur for *n* things taken *r* at a time, determine the number of **combinations** that can occur for *n* things taken *r* at a time, and then use permutations and combinations to compute the probabilities of events involving unordered arrangements and a large number of possible outcomes.

Note: *indicates optional material.

Notations Frequently Used in This Chapter

$P(A)$	the probability that event A will occur
$A{:}B$, A *to* B, $\frac{A}{B}$	ratio of A and B
$n!$	n factorial $n! = n \times (n-1) \times \cdots \times 3 \times 2 \times 1$
${}_nP_r$	the number of permutations of n things taken r at a time
${}_nC_r$	the number of combinations of n things taken r at a time

Visit the PH Companion Web site at www.prenhall.com/setek

OVERVIEW

The remark "I'll bet you!" is a phrase that many of us have used at one time or another—most people have bet on something. It might have been on the results of an election or what the weather will be like tomorrow. Maybe you have bet with someone on the outcome of a World Series or Super Bowl game. Millions of people make some sort of a wager on major horse races such as the Kentucky Derby. You may even have bet with someone on how you would do on an exam.

Most people are introduced to the topic of probability through betting or games of chance. Even games that we played as small children—or play now—involve the use of a pair of dice. Even if you have never made a friendly bet or played any dice games (Remember Monopoly?), you have heard the weather forecaster mention that there is an "80% chance" of showers for the weekend. Probability is with us every day.

4.1 INTRODUCTION

Probability has been studied by mathematicians for a long time. The concept of probability was first formally studied in the sixteenth century. At that time, it was the outgrowth of a study on gambling and games of chance.

Today, probability is still used to help people understand games of chance such as blackjack, craps, and lotteries. But there are certainly other uses of probability. Insurance companies are concerned with the probable life expectancy of their policy holders. Surveyors of public opinion use probability to determine the results of their polls. The Harris and Gallup polls arrive at their results by means of probability, as do the various television polls. Biology (genetics), astronomy, and manufacturing are some other areas that make extensive use of probability theory.

The topics covered in this chapter will provide you with a basis for understanding probability and some everyday applications.

4.2 DEFINITION OF PROBABILITY

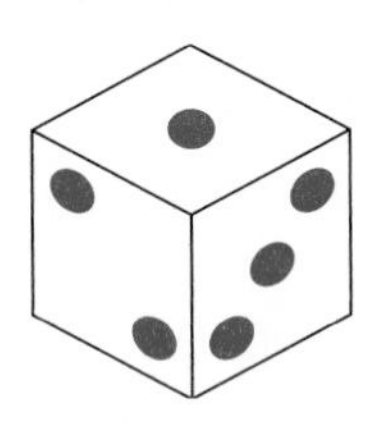

Figure 4.1

Suppose we are given one die from a pair of dice (Figure 4.1). Upon examination, we see that it is a six-sided solid cube in which one side has one dot, another side has two dots, and so on, until the last side has six dots.

When we toss a die, we are performing an **experiment** to see which set of dots, or number, will turn face up. The number of dots that are on the top surface when we toss the die is called the **outcome** of the experiment. Each number has an equal chance of occurring, so we say that each outcome—1, 2, 3, 4, 5, or 6—is **equally likely** to occur.

The set of all possible outcomes of an experiment is called a **sample space**. For example, when the experiment is tossing a die, the sample space is $\{1, 2, 3, 4, 5, 6\}$. An **event** is any subset of the sample space: $\{3\}$ and $\{1, 6\}$ are both events.

If we toss a die, what is the probability of obtaining a 3? If we are interested in the probability of obtaining a 3, then we are actually interested in the probability that the event $\{3\}$ will occur. Therefore, before we proceed any further, let us define what we mean by the probability that an event A will occur.

If an experiment has a total T of equally likely possible outcomes, and if exactly S of them are considered successful (or favorable)—that is, they are members of event A—then **the probability that event A will occur**, denoted by $P(A)$, is

$$P(A) = \frac{\text{Number of successful outcomes}}{\text{Total number of all possible outcomes}} = \frac{S}{T}$$

Historical Note

The laws of chance as we know them today began with interests in gambling. A French nobleman, the Chevalier de Méré, asked a mathematician friend of his, Blaise Pascal (1623–1662), "how to split the pot in a dice game that has to be discontinued."

Pascal pondered the problem for some time and then relayed it to another mathematician friend, Pierre de Fermat (1601–1665). From the correspondence and research of Pascal and Fermat regarding various gambling situations, the theory of probability has evolved.

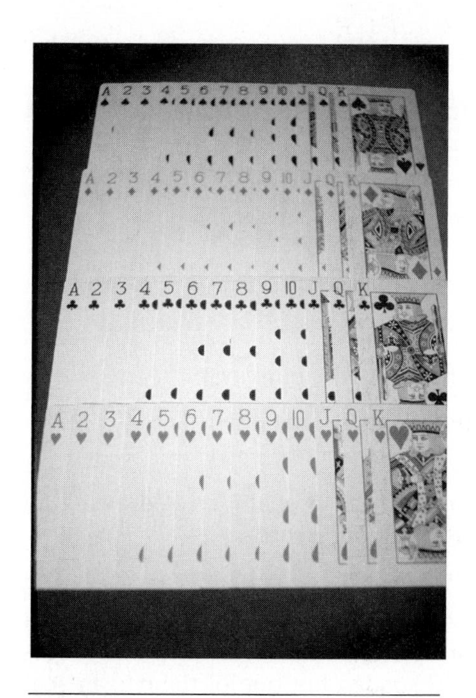

Figure 4.2
A standard deck of 52 cards.

A helpful way to remember this probability formula, $P(A) = S/T$, is to remember that to find the probability in favor of an event A, we find the number of *successful* outcomes and divide it by the *total* number of possible outcomes.

Hence, the probability of obtaining a 3 in a toss of one die is $\frac{1}{6}$. We see that there is only one possible successful outcome, a 3. There is a total of six possible outcomes—namely, 1, 2, 3, 4, 5, and 6. To compute the probability of obtaining a 3 for this experiment, we have

$$P(3) = \frac{1}{6} = \frac{\text{Number of successful outcomes}}{\text{Total number of possible outcomes}}$$

EXAMPLE 1 ***Flipping a Coin*** Find the probability of getting a head when you flip a quarter.

Solution There are two different possible outcomes (head, tail), which are equally likely, and only one of these outcomes (head) is successful. Hence, the probability in favor of obtaining a head is $\frac{1}{2}$:

$$P(\text{head}) = \frac{1}{2} = \frac{\text{Number of successful outcomes}}{\text{Total number of possible outcomes}}$$

EXAMPLE 2 ***Tossing a Die*** Find the probability of obtaining a number greater than 4 on a single toss of a die.

Solution There are six different possible outcomes and two of these outcomes, $\{5, 6\}$, are successful. Hence, the probability in favor of obtaining a number greater than 4 is $\frac{2}{6}$:

$$P(\text{number greater than } 4) = \frac{2}{6} = \frac{1}{3}$$

EXAMPLE 3 ***Picking a Card*** Find the probability of drawing a king (one pick) from a shuffled standard deck of 52 cards. (A standard deck of cards is the most common type of deck used in most card games. If you are not familiar with such a deck, see Figure 4.2, which shows all 52 cards.)

Solution There are 52 different possible outcomes. Four of these outcomes are successful: king of spades, king of hearts, king of clubs, king of diamonds. Therefore, the probability in favor of obtaining a king is $\frac{4}{52}$:

$$P(K) = \frac{4}{52} = \frac{1}{13}$$

EXAMPLE 4 ***Picking a Card*** Find the probability of drawing a jack or a queen from a shuffled standard deck of 52 cards.

Solution There are 52 different possible outcomes and eight of these are considered successful outcomes, because if a jack *or* a queen is picked, we are successful. Hence, the probability in favor of obtaining a jack or a queen is $\frac{8}{52}$:

$$P(J \text{ or } Q) = \frac{8}{52} = \frac{2}{13}$$

NW ***Now Work Problems 1a, b, and c.***

EXAMPLE 5 ***Zero Probability*** Find the probability of obtaining a 7 on a single toss of one die.

Solution There are six different possible outcomes and none of these outcomes would produce a 7. That is, zero of these outcomes would be successful. The probability in favor of obtaining a 7 on a single toss of one die is $\frac{0}{6}$, or 0:

$$P(7) = \frac{0}{6} = 0$$

When an event cannot possibly succeed, we say it is an **impossible event**. The probability of an impossible event is zero:

$$P(\text{impossible event}) = \frac{0}{T} = 0 \quad (T \neq 0)$$

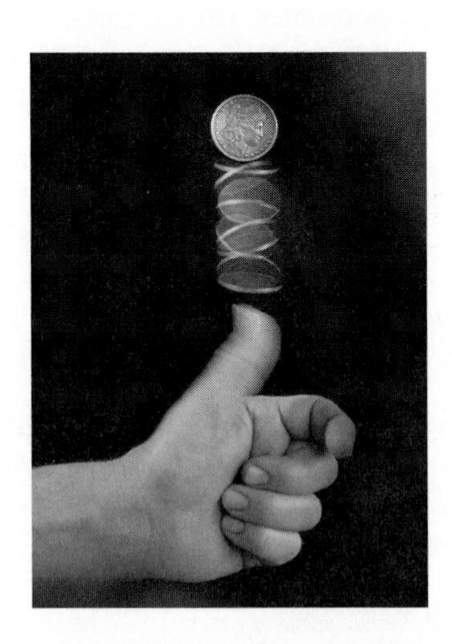

EXAMPLE 6 ***Perfect Probability*** Find the probability of getting heads or tails when you flip a quarter.

Solution You can't lose! There are two different outcomes (head, tail) and both of these outcomes are successful. Hence, the probability in favor of obtaining heads or tails is $\frac{2}{2}$, or 1:

$$P(\text{head or tail}) = \frac{2}{2} = \frac{1}{1} = 1$$

When an event is sure to occur, then success is inevitable, and we say that the probability is 1. This is sometimes called certainty:

$$P(\text{certain event}) = \frac{T}{T} = 1 \quad (T \neq 0)$$

For any event, $P(A) \geq 0$ and $P(A) \leq 1$, so we may write

$$0 \leq P(A) \leq 1$$

If an event can never occur, then its probability is 0.

If an event is certain to occur, then its probability is 1.

Note of Interest

No one expects a coin to fall heads once in every two tosses; however, in a large number of tosses the results tend to even out. For a coin to fall heads 50 consecutive times it would take 1 million people tossing coins 10 times a minute for 40 hours per week. This is so unlikely to happen that it would occur only once every nine centuries.

EXAMPLE 7 ***Redundant Outcomes*** Find the probability of drawing an ace or a spade from a shuffled standard deck of 52 cards.

Solution Your initial answer might be $\frac{17}{52}$, but this is not correct! There are 52 different possible outcomes, and there are 4 aces and 13 spades; hence, adding 4 and 13, we would have $\frac{17}{52}$. Why isn't this correct? If we examine the possible successful outcomes, we see there are indeed 13 spades, but one of them is an ace! We include the ace with our 13 spades, leaving 3 other aces to include in our number of successful outcomes. If you count the aces first, then you have included the ace of spades, and there are 12 spades left to include in the number of successful outcomes. Altogether, there are 16 successful outcomes, and the probability of drawing an ace or a spade is $\frac{16}{52}$, or $\frac{4}{13}$:

$$P(\text{ace or spade}) = \frac{16}{52} = \frac{4}{13}$$

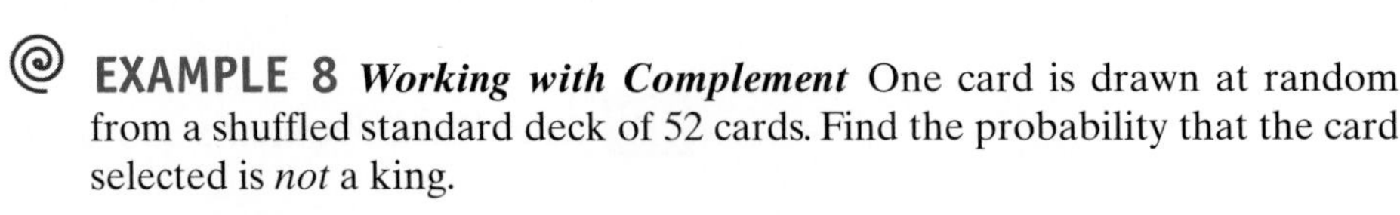

EXAMPLE 8 ***Working with Complement*** One card is drawn at random from a shuffled standard deck of 52 cards. Find the probability that the card selected is *not* a king.

Solution There are 52 different possible outcomes. There are 4 kings in a deck, so the other 48 cards are not kings, and these are the successful outcomes. Hence, the probability that the card selected is not a king is $\frac{48}{52}$:

$$P(not\ K) = \frac{48}{52} = \frac{12}{13}$$

The sum of the probability that an event will occur and the probability that it will not occur is 1. Therefore, we can solve Example 8 in another way.

Be sure to remember that if

$$P(A) = \frac{S}{T} \quad \text{then} \quad P(not\ A) = 1 - \frac{S}{T}\ (T \neq 0)$$

Alternate Solution The probability of selecting a king is $\frac{4}{52}$ or $\frac{1}{13}$. Therefore, the probability that the card selected is *not* a king is

$$P(not\ K) = 1 - \frac{1}{13} = \frac{13}{13} - \frac{1}{13} = \frac{12}{13}$$

NW ***Now Work Problems 1d, e, and f.***

Empirical Probability

Throughout this section, our discussion has focused primarily on *theoretical probability*. As noted in the corresponding Math Connections, theoretical probability involves equally likely outcomes. It also implies that the probability is calculated without actually conducting an experiment. Suppose, however, that the outcomes of an experiment are not equally likely to occur, or that to determine these outcomes

Math Connections

Theoretical versus Empirical Probability

In our discussion of probability, we defined probability from the perspective of a ratio that compares the number of times an event can occur (i.e., the number of successful outcomes) to the total number of opportunities for the event to occur (i.e., the total number of all equally likely possible outcomes). Frequently, though, we speak of the probability of an event occurring in the future based on past occurrences. As an example, consider the following: What is the probability that a satellite will fall from space and land in a populated area? What is the probability that it will snow in Miami, Florida? What is the probability that a child growing up today will one day become president of the United States? What is the probability that the NASDAQ will break 10,000? Note that in each of these examples, we are seeking the probability of a (possibly) future event. To address these types of probability questions, it is useful to distinguish between theoretical probability and empirical probability.

Theoretical probabilities are independent of prior experience or actual observation. Instead, they are grounded in logic. For example, we state that the probability of getting heads on a flip of a coin is $\frac{1}{2}$ because heads is one of only two possible outcomes that are equally likely to occur when we flip a coin. In a similar manner, we state that the probability of getting 5 on a single roll of a die is $\frac{1}{6}$ because 5 is one of six possible outcomes that are equally likely to occur when we roll a die. When we state a theoretical probability, we are essentially providing a *guess of the relative frequency* of an event occurring "in the long run" *without actually observing this event occur.*

Empirical probabilities, on the other hand, do not have a logical basis. Instead, they are an *estimate of the relative frequency* of an event occurring *based on past observations*. For example, based on the millions of automobile accidents that have been reported and documented, we can determine the probability of a person being involved in an automobile accident. Similarly, based on the millions of recorded births throughout the world, we can determine the probability of giving birth to twins. In these two examples, each probability is based on the relative frequency of the occurrence of a related past event. In the first example, the probability is based on the relative frequency of drivers being involved in past automobile accidents, and in the second example, the probability is based on the relative frequency of past twins' births.

Our perspective throughout this chapter will be to view probability as a theoretical concept. By doing so, we will be able to use the knowledge acquired from our study to help us better understand and interpret probability based on empirical observations.

TABLE 4.1

Year	Male	Female	Total
1981	6000	2300	8300
1986	5300	2200	7500
1991	3900	1900	5800
1996	3800	2000	5800
2001	3700	1900	5600

Source: U.S. Department of Transportation (all numbers rounded).

requires a tremendous amount of work. As an illustration, consider Table 4.1, which summarizes the number of teenagers ages 13–19 who died in motor vehicle crashes in 1981, 1986, 1991, 1996, and 2001.

Note that for each given year, the probabilities involving teenage motor vehicle fatalities are slightly different. For example, in 1981, given the total number of teenagers who died in a motor vehicle accident, the probability that the teenager was male is $\frac{6000}{8300} = \frac{60}{83}$. Yet, in 1986, 1991, 1996, and 2001, these probabilities were respectively, $\frac{53}{75}$, $\frac{39}{58}$, $\frac{38}{58}$, and $\frac{37}{56}$. These differences can be seen more readily if we convert the probabilities to equivalent decimal values by dividing the denominator into the numerator. Thus, the respective decimal values for the five given years are 0.72, 0.71, 0.67, 0.66, and 0.66. Why are these probabilities not exactly the same? They are not the same because each year represents a different "experiment," which leads to different outcomes and different probabilities. This is an example of *empirical probability*. Unlike theoretical probability, empirical probability is only an approximate value. It is based on the *relative frequency* of the occurrence of a past event. In the present example, we are basing our probabilities on the frequency of teen motor vehicle deaths relative to the five given years shown in Table 4.1. Thus, when using survey data to calculate probabilities, we are actually calculating empirical probabilities and hence our results are only approximate. This is because each survey represents a probability experiment and the results of these experiments will be different each time the survey is conducted.

EXAMPLE 9 ***Using Relative Frequencies to Approximate Probability*** In a survey of 1500 office workers, people were asked how often they spend evenings or weekends with their coworkers. The results of this survey are given in Table 4.2. Given these results, if we were to select an office worker at random, what is the approximate probability that this person spends a few times per month with a coworker on evenings or weekends?

TABLE 4.2

Frequency	Number Responding
Every week	25
Every weekend	75
A few times per month	450
A few times per year	650
Never	300

Solution There were a total of 1500 respondents. Of these, 450 indicated that they spend a few times per month with their co-workers during the evening or on weekends. Thus, the estimated (or empirical) probability is $\frac{450}{1500} = \frac{45}{150} = \frac{3}{10}$.

EXAMPLE 10 ***Comparing Theoretical and Approximate Probabilities*** In a lesson on probability, Mrs. Ollis tossed a die 20 times and recorded the results as shown in Table 4.3.

a. Calculate the empirical probability of obtaining a number greater than 4.

b. Calculate the theoretical probability of obtaining a number greater than 4.

c. Compare the results from parts **a** and **b** and explain why they are different.

Solution

a. From Table 4.3, experiment numbers 6 and 18 resulted in a number greater than 4. Thus, based on Mrs. Ollis's demonstration, the probability of getting a number greater than 4 on a single toss of a die is estimated to be $\frac{2}{20} = \frac{1}{10}$.

TABLE 4.3

Experiment Number	Result	Experiment Number	Result
1	3	11	3
2	4	12	3
3	3	13	4
4	1	14	1
5	3	15	3
6	5	16	4
7	3	17	3
8	2	18	6
9	4	19	4
10	2	20	3

b. In Example 2, we found the probability of getting a number greater than 4 on a single toss of a die was $\frac{2}{6} = \frac{1}{3}$.

c. In part (a), we can expect to get a number greater than 4 (i.e., a 5 or 6) on a single toss of a die approximately once for every ten tosses, which is 10% of the time. In part (b), however, the likelihood of this occurring is once for every three tosses, which is more than three times as much. The reason these results are different is because the first probability is only an approximate value. It was based on 20 tosses. As the number of tosses increases, though, the empirical probability will get closer and closer to the theoretical probability.

Exercises for Section 4.2

1. A bag of M & M's usually contains 20 pieces of candy. Pamela purchased a bag that contained seven orange pieces, six green pieces, four yellow pieces, and three red pieces. Pam selects one piece of candy from the bag at random. Find the probability that she will select

NW **a.** a red piece of candy
NW **b.** a green piece of candy
NW **c.** a red or yellow piece of candy
NW **d.** neither a red nor a yellow piece of candy
NW **e.** anything but an orange piece of candy
NW **f.** a purple piece of candy

2. A box of assorted fruit-flavored gumdrops usually contains 20 gumdrops. Anthony purchased a box that contained three cherry-flavored gumdrops, two lime-flavored gumdrops, two lemon-flavored gumdrops, seven strawberry-flavored gumdrops, and six orange-flavored gumdrops. If Anthony chooses one gumdrop at random, find the probability that he chooses

a. a lemon-flavored gumdrop
b. a cherry-flavored gumdrop
c. a lemon- or a lime-flavored gumdrop
d. neither a strawberry- nor an orange-flavored gumdrop
e. anything but an orange-flavored gumdrop
f. a grape-flavored gumdrop

3. On the single toss of one die, find the probability of obtaining

a. a 4
b. an odd number
c. an even number
d. a number less than 4
e. a number greater than 4
f. an odd or an even number

4. On the single toss of one die, find the probability of obtaining

a. a number divisible by 3 (for example, 6 is divisible by 3 because 3 divides 6 evenly; that is, the remainder is zero)
b. a number divisible by 5
c. a number divisible by 2
d. a number divisible by 1
e. a number less than 1
f. a number less than 7

5. On a single draw from a shuffled standard deck of 52 cards, find the probability of obtaining

a. the ace of spades
b. the two of hearts
c. a deuce (2)
d. a red card
e. a diamond
f. a red jack

6. On a single draw from a shuffled standard deck of 52 cards, find the probability of obtaining
   a. a spade
   b. a club
   c. a spade or a club
   d. a spade and a club
   e. a three or a heart
   f. a three and a heart (Be careful on this one.)

7. On a single draw from a shuffled standard deck of 52 cards, find the probability of obtaining
   a. a picture card (jack, queen, king)
   b. a picture card or a heart
   c. a jack or a heart
   d. a jack and a heart (Be careful on this one.)
   e. a one-eyed jack (jack of spades or jack of hearts)
   f. a king with an axe (king of diamonds)

8. Harry Hose keeps all of his socks in the top drawer of his bureau. In the drawer there are four blue socks, six black socks, seven brown socks (he lost one in the laundry), and four red socks. Harry reaches in and pulls a sock out at random. Find the probability that the sock chosen is
   a. brown
   b. blue
   c. red
   d. black
   e. not brown
   f. neither brown nor blue

9. Gloria Glove keeps all of her mittens on the top shelf of her hall closet. On the shelf are four blue mittens, six brown mittens, and four green mittens. Gloria reaches up and pulls a mitten out at random. Find the probability that the mitten chosen is
   a. blue or brown
   b. blue or green
   c. not red
   d. green or red
   e. neither blue nor green

10. Refer to Exercise 9. If it is dark in the hall and the light doesn't work, what is the greatest number of mittens Gloria will have to pull out to make sure that she has two mittens of the same color?

11. Veronica has separated the 12 pages from last year's calendar, put them in a paper bag, and shaken them around. She will draw a page at random. Find the probability of obtaining
   a. a page with a month beginning with *A*
   b. a page with a month beginning with *M*
   c. a page with a month beginning with *B*
   d. a page with a month ending with *y*
   e. a page with a month containing 31 days
   f. a page with a month containing 30 days

12. Charlie's wallet contains a one-dollar bill, a five-dollar bill, a ten-dollar bill, and a twenty-dollar bill. If Charlie chooses one bill at random, find the probability that the value of the bill chosen is
   a. even
   b. odd
   c. greater than one dollar
   d. less than five dollars
   e. less than twenty dollars
   f. divisible by 3

13. A package of flower seeds contains 20 seeds, 12 seeds for red flowers and eight seeds for yellow flowers. A seed is randomly selected. Find the probability that it will produce
   a. a red flower
   b. a yellow flower
   c. a blue flower
   d. a red or a blue flower
   e. a yellow or a blue flower
   f. a red or a yellow flower

14. Each individual letter of the word *probability* is typed on a separate identical slip of paper; all 11 pieces of paper are placed in a bowl and the pieces of paper are mixed. If one letter is chosen at random, find the probability that the letter chosen is
   a. a *t*
   b. a vowel
   c. not a vowel
   d. a *b*
   e. a *b* or an *i*
   f. neither a *b* nor an *i*

15. A cookie jar contains four chocolate chip cookies, six sugar cookies, five peanut butter cookies, and three molasses cookies. Addie reaches into the jar and picks one cookie at random. Find the probability that she will pick
   a. a sugar cookie
   b. a peanut butter cookie
   c. a chocolate chip or sugar cookie
   d. neither a sugar cookie nor a chocolate chip cookie
   e. an oatmeal cookie
   f. anything but a chocolate chip cookie

16. A flower bucket at the supermarket contains six carnations, four roses, and eight zinnias. A customer chooses

one flower at random. Find the probability that the customer chooses

a. a rose

b. a carnation

c. a zinnia

d. a rose or a carnation

e. neither a rose nor a zinnia

f. a daisy

Writing Mathematics

17. What is a sample space?

18. What are equally likely outcomes?

19. Describe the difference between an impossible event and a certain event in probability. Be sure to provide examples.

Challenge Exercises

20. In our discussion of theoretical versus empirical probability in the Math Connections, empirical probability was presented from an *objective* nature. Empirical probability can also be viewed from a *subjective* nature, however. Define what you think a subjective empirical probability is and give at least two examples. (*Hint:* Think of the concepts of "educated guesses" or "hunches.") What type of probability do you think insurance companies base their life insurance policies and premiums on? Explain your answer.

Group/Research Activities

21. The topic of probability evolved when mathematicians began to solve problems that arose from gambling tables. Research the history of probability and some of the games that were played. Prepare a report to be presented, by your group, to your classmates.

22. Determine the approximate probability of winning your state's weekly lottery, or that of a neighboring state. This information is usually provided on a lottery ticket or in the related literature. Next, research the probability of a person being struck by lightning. Compare the respective probabilities and form a conclusion. Report this information to your classmates.

23. One of the classic introductions to the study of probability is the 1968 book, *An Introduction to Probability Theory and Its Applications*, by William Feller (John Wiley & Sons, New York). Get a copy of this book and review Feller's introduction as well as his chapters on sample space, conditional probability, and the laws of large numbers. Prepare a report on this material and present it to your classmates.

Real-World Data

24. Border patrols Table 4.4 shows the total number of illegal immigrants who were apprehended by immigration agents along the Southwest border of the United States in 2002.

a. Find the probability that an illegal immigrant caught along the Southwest border was caught in San Diego.

TABLE 4.4

Location Apprehended	Number
Tucson, AZ	335,000
El Centro, CA	110,000
San Diego, CA	100,000
El Paso, TX	95,000
McAllen, TX	90,000
Laredo, TX	85,000
Del Rio, TX	70,000
Yuma, AZ	44,000
Marfa, TX	11,000

Source: Bureau of Citizenship and Immigration Services (all numbers rounded).

b. Of the total number of immigrants apprehended along the Southwest border in 2002, if we were to select one of these immigrants at random and asked the agent to identify the state this person was apprehended in, what state do you think has the greatest likelihood of being named? Why?

25. Birth rates Let's assume that the number of births in the United States in 2003 was approximately 4,100,000. Let's further assume that the male–female birth ratio was 1.05 to 1, which indicates that there were 1.05 male births to every 1 female birth. Based on the 2003 birth rate, this means that 2,100,000 baby boys were born.

a. What is the probability that a child born in the United States in 2003 was female?

b. In theory, what is the probability that a child born is female?

c. Compare the results from parts **a** and **b**.

26. Alzheimer's disease The bar graph in Figure 4.3 shows the percentage of people 65 or older who have been diagnosed with probable Alzheimer's disease. For example, there is a 3% chance that a person who is between the ages of 65 and 74 has Alzheimer's disease. Thus, the probability is $\frac{3}{100}$ (see Chapter 12 for information on converting a percent to a fraction).

a. If we randomly select an 80-year old person, what is the probability that this person will have Alzheimer's disease?

b. If we randomly select a person who is at least 85 years old, what is the probability that this person does not have Alzheimer's disease?

c. Are the probabilities shown in Figure 4.3 theoretical or empirical? Why?

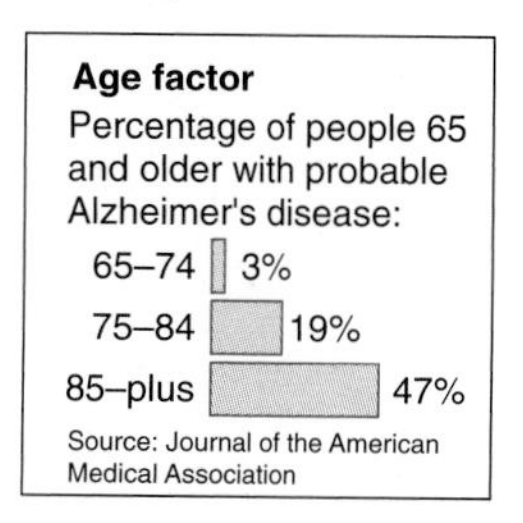

Figure 4.3

Just for fun

What is the answer to each problem?

a. $\frac{0}{1} = ?$ **b.** $\frac{1}{0} = ?$ **c.** $\frac{0}{0} = ?$

4.3 SAMPLE SPACES

Many times when we want to compute the probability of some event, the total number of possible outcomes of the experiment is not easy to determine. Consider the following examples:

EXAMPLE 1 ***Flipping a Coin and Tossing a Die*** A quarter is flipped and a die is tossed. What is the probability of obtaining a head or an odd number?

Solution A person might reason that there are two outcomes with the quarter and six outcomes with the die, so altogether there are eight possible outcomes. This same person might also reason that there is one successful outcome with the quarter (heads) and three successful outcomes with the die, $\{1, 3, 5\}$, so there are four successful outcomes altogether. Therefore, the probability of getting heads or an odd number is $\frac{4}{8}$. This is *wrong!*

In fact, there are 12 possible outcomes. The quarter may turn up two ways, heads or tails. If it lands heads up, then the heads can be matched with six different outcomes of the die:

$$H\text{–}1,\quad H\text{–}2,\quad H\text{–}3,\quad H\text{–}4,\quad H\text{–}5,\quad H\text{–}6$$

The same thing can happen with tails. If tails come up, then we could have

$$T\text{–}1,\quad T\text{–}2,\quad T\text{–}3,\quad T\text{–}4,\quad T\text{–}5,\quad T\text{–}6$$

Altogether, we have 12 *different* outcomes. Making a list of these outcomes, we have

$$\begin{array}{cc} H\text{–}1 & T\text{–}1 \\ H\text{–}2 & T\text{–}2 \\ H\text{–}3 & T\text{–}3 \\ H\text{–}4 & T\text{–}4 \\ H\text{–}5 & T\text{–}5 \\ H\text{–}6 & T\text{–}6 \end{array}$$

How many of these possible outcomes are successful? Remember, we wanted heads or an odd number. Six of the total outcomes contain a head, so they have to be considered successful. Three of the remaining possible outcomes are also successful: T–1, T–3, T–5. Hence, there are nine successful outcomes, and the probability of obtaining heads or an odd number is $\frac{9}{12}$, or $\frac{3}{4}$.

Recall from Section 4.2 that a list of all possible outcomes for an experiment is called a *sample space*. A method to determine the number of outcomes for two experiments together is to multiply the number of ways one can occur by the number of ways the other can occur. In Example 1, a quarter can come up two ways (assuming it can't land on its edge), and a die can turn up six ways. Therefore, we have $2 \times 6 = 12$ outcomes for both experiments together.

In general, if one experiment has m different outcomes and a second experiment has n different outcomes, then the first and second experiments performed together have $m \times n$ different outcomes. This idea may be extended if there are other experiments to follow: We would have $m \times n \times r \times \cdots \times t$ outcomes. This is often called the **fundamental counting principle**.

EXAMPLE 2 ***Sample Space for a Pair of Dice*** List a sample space showing all possible outcomes when a pair of dice is tossed.

Solution First of all, how many outcomes will we have in the sample space? There are two dice and each die has six possible outcomes. Using the counting principle, we have $6 \times 6 = 36$ total possible outcomes.

Let one die be blue and the other white, so that we can distinguish the two dice. If a 1 comes up on the white die, it can be paired with each of the six numbers on the blue die; that is,

$$(1, 1), \quad (1, 2), \quad (1, 3), \quad (1, 4), \quad (1, 5), \quad \text{and} \quad (1, 6)$$

We can do this in turn for each of the numbers that comes up on the white die, pairing it with all the numbers on the blue die. This pairing is illustrated in Figure 4.4.

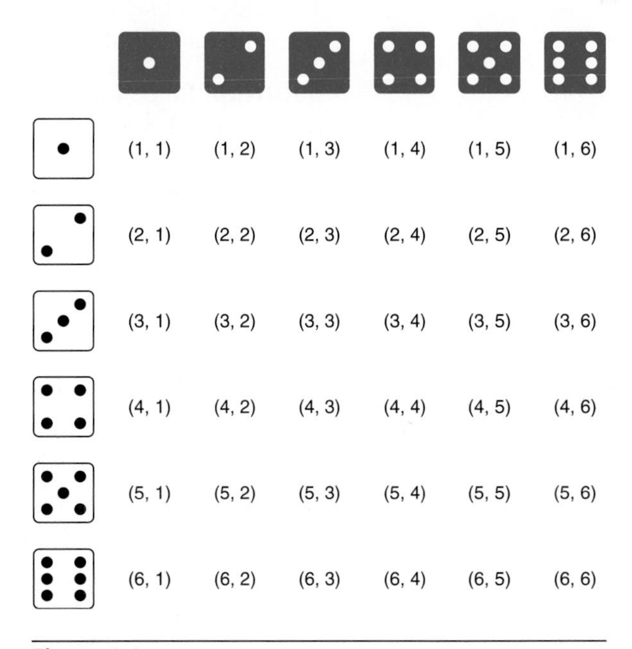

Figure 4.4

In games where a pair of dice is used, the sum of the two numbers is the primary concern. Note that there is only one outcome where the sum of the dice is 2 (the vernacular term is "snake eyes"); this is also the case for 12 ("boxcars"). The sums that may be obtained by rolling a pair of dice are 2, 3, 4, 5, 6, 7, 8, 9, 10, 11, and 12. These are the only possible totals with two dice.

Examining the sample space illustrated in Figure 4.4, we see that we have the following number of ways to obtain each of the sums:

Sum:	2	3	4	5	6	7	8	9	10	11	12
Number of ways:	1	2	3	4	5	6	5	4	3	2	1

Note that the sum 7 can occur in six ways and all of the other sums can occur in fewer ways. It is interesting (and helpful) to note that 6 and 8 can occur in five ways, the 5 and 9 can occur in four ways, and so on. This is a symmetric pattern, and it should help you to remember the total number of ways each sum can occur.

 **EXAMPLE 3** ***Tossing a Pair of Dice*** Use the sample space for the total number of possible outcomes when a pair of dice is tossed to find the probability of obtaining a sum of 7 on a single toss of a pair of dice.

Solution There are 36 total possible outcomes and six of these outcomes are successful, because there are six ways of obtaining a sum of 7. Hence, the probability of obtaining a 7 is $\frac{6}{36}$ or $\frac{1}{6}$:

$$P(7) = \frac{6}{36} = \frac{1}{6}$$

Note that a 7 is the most likely outcome when a pair of dice is tossed. It is believed that this is one of the reasons why many people consider 7 their lucky number. It keeps coming up for them more often than other numbers.

NW ***Now Work Problem 1.***

 **EXAMPLE 4** ***Over 7/Under 7*** A popular dice game in gambling establishments is "over and under." This is a game where the bettor may bet that the dice (when flipped, rolled, or tossed) will total over 7 or under 7. Larry decides to play this game, and he wants to bet on "under" because he feels that "under" should win. Is he right?

Solution Since Larry bet on "under," he will win if any of the following sums comes up: 2, 3, 4, 5, 6. According to the sample space, there are five ways for a 6 to occur, four ways for a 5 to occur, three ways for a 4 to occur, two ways for a 3 to occur, and one way for a 2 to occur. Hence, there are 15 outcomes that can be considered successful out of a possible 36 outcomes. Therefore, the probability of the dice turning up under 7 is $\frac{15}{36}$ or $\frac{5}{12}$:

$$P(\text{under } 7) = \frac{15}{36} = \frac{5}{12}$$

Is this a good bet? Out of a total of 36 possible outcomes, only 15 are considered successful. The other 21 outcomes can turn Larry into a loser. The probability that the dice will *not* be under 7 is $\frac{21}{36}$; since the sum of the probability in favor of an event and the probability against an event is 1, if $P(\text{under } 7) = \frac{15}{36}$, then $P(\textit{not} \text{ under } 7)$ is $1 - \frac{15}{36}$, or

$$\frac{36}{36} - \frac{15}{36} = \frac{21}{36}$$

Hence, he is more likely to lose than to win.

What happens if Larry changes his mind and decides to bet "over"? Has he got a better chance of winning? No! If we compute the probability of "over 7," we get $\frac{15}{36}$, and again Larry is more likely to lose because the probability of "not over 7" is $\frac{21}{36}$.

What happens if Larry decides to bet both "over" and "under"? The best he can do is get his own money back because if he wins on one side, he loses on the other side. Note also that he could be a double loser if 7 comes up because 7 is neither "over" nor "under"!

As we learned in Chapter 2, one of the uses of Venn diagrams is to show and separate numerical data that have been compiled. Venn diagrams can also be used to solve probability problems. Consider Examples 5 and 6.

 **EXAMPLE 5** ***Using Venn Diagrams (2 Sets)*** At a recent college registration, 100 students were interviewed. Eighty of the students stated that they

had registered for a mathematics course, 14 of the students stated that they had registered for a history course, and 5 of the students stated that they had registered for a mathematics course and a history course. What is the probability that a student in this survey registered only for history?

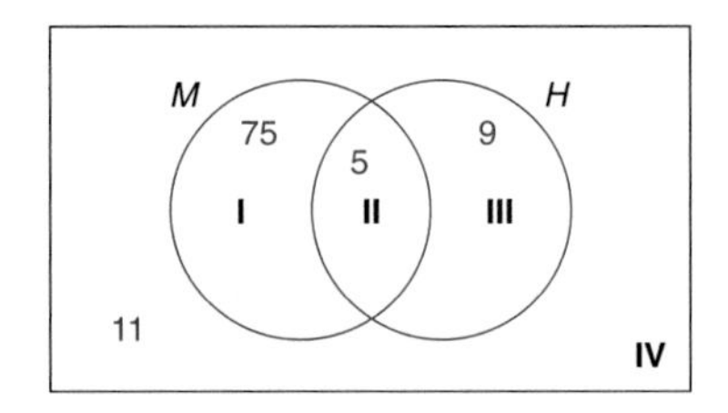

Figure 4.5
M = mathematics, and H = history.

Solution A Venn diagram using the given information is shown in Figure 4.5. With it, we can summarize the information that 11 of the students did not register for either a mathematics course or a history course, 9 students registered only for history, and 75 students registered only for mathematics.

This information can also be used to solve the probability problem. We note that the total number of students is 100, whereas 9 of them are registered only for history. Hence, the answer is $\frac{9}{100}$.

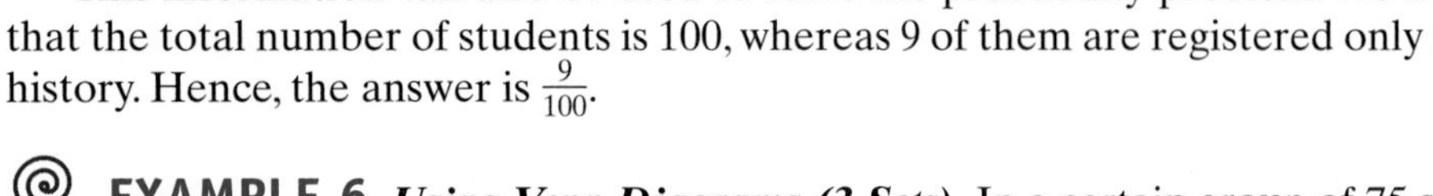

EXAMPLE 6 ***Using Venn Diagrams (3 Sets)*** In a certain group of 75 students, it has been determined that 16 students are taking statistics, chemistry, and psychology; 24 students are taking statistics and chemistry; 30 students are taking statistics and psychology; 22 students are taking chemistry and psychology; 6 students are taking only statistics; 9 students are taking only chemistry; and 5 students are taking only psychology.

a. What is the probability that a student is not taking any of the three subjects?

b. What is the probability that a student is taking chemistry?

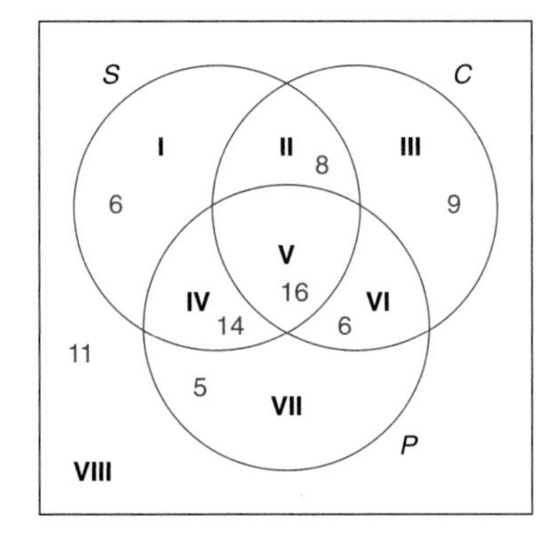

Figure 4.6
S = statistics, C = chemistry, and P = psychology.

Solution We first complete the necessary Venn diagram (see Figure 4.6). After completing the diagram, we can answer the questions.

a. The probability that a student is not taking any of the three subjects is $\frac{11}{75}$.

b. The probability that a student is taking chemistry is $\frac{39}{75}$.

The answer to question **b** is obtained by adding the number of students in each partition of the chemistry circle. Hence, there are 39 students taking chemistry.

NW ***Now Work Problem 11.***

Exercises for Section 4.3

NW **1.** A quarter is flipped and a die is tossed (see Example 1, this section).

a. How many outcomes are possible?

b. Construct a sample space showing the possible outcomes.

c. What is the probability of obtaining heads and a 6?

d. What is the probability of obtaining heads and a 7?

e. What is the probability of obtaining heads or a 7?

f. What is the probability of obtaining tails and an even number?

2. A nickel is flipped and a dime is flipped.

a. How many outcomes are possible?

b. Construct a sample space showing the possible outcomes.

c. What is the probability of obtaining two heads?

d. What is the probability of obtaining two tails?

e. What is the probability of obtaining exactly one head?

f. What is the probability of obtaining no heads?

3. A box contains a one-dollar bill, a five-dollar bill, a ten-dollar bill, a twenty-dollar bill, and a fifty-dollar bill. Two bills are chosen at random in succession. The first bill is not replaced before the second is drawn.

a. How many outcomes are possible?

b. Construct a sample space for this experiment.

c. What is the probability that the value of the first bill is even?

d. What is the probability that the value of the second bill is even?

e. What is the probability that the values of both bills are even?

f. What is the probability that the value of neither bill is even?

4. A bag contains five balls numbered 0 through 4. Two balls are chosen at random in succession. The first ball is replaced before the second is chosen.

a. How many outcomes are possible?

b. Construct a sample space showing the possible outcomes.

c. What is the probability that the first ball has an odd number?

d. What is the probability that the second ball has an odd number?

e. What is the probability that both balls have odd numbers?

f. What is the probability that neither ball has an odd number?

5. A quarter is flipped and a die is tossed (see Example 1, this section). Find the probability of obtaining

a. tails and a number less than 4

b. tails and a number less than 1

c. tails or a number less than 10

d. heads and a number divisible by 2

e. heads or a number divisible by 2

f. tails or a number divisible by 7

6. A quiz show uses a fish bowl that contains nine balls, numbered one through nine, for selecting participants. A ball is selected at random. Find the probability that the number chosen is

a. 6

b. greater than 6

c. an odd number

d. less than 6

e. an even number

f. an odd or even number

7. Using the sample space for the possible outcomes when a pair of dice is tossed (see Example 2, this section), find the probability that

a. the sum of the numbers is 7

b. the sum of the numbers is 11

c. the sum of the numbers is 1

d. the sum of the numbers is even

e. the sum of the numbers is odd or even

f. the sum of the numbers is divisible by 5

8. Thirty-four people are being considered as possible jurors in a court case. The accompanying table shows the distribution of people: male, female, workers, and retirees.

	Workers	Retirees
Male	10	9
Female	7	8

A juror is chosen at random. Find the probability of choosing

a. a worker

b. a retiree

c. a female

d. a male

e. a female retiree

f. a male worker

9. The accompanying table shows the number of students in a swimming class. To complete the course, each student must pass a skills test.

	Freshmen	Sophomores
Female	6	2
Male	5	7

The test is begun by selecting a student at random. Find the probability of choosing

a. a female

b. a male

c. a freshman

d. a sophomore

e. a male sophomore

f. a female freshman

10. In a survey of 30 contestants at a racquetball tournament, the following information was obtained: 10 contestants entered both the singles and doubles tournaments, 22 contestants were entered in the singles tournament, 16 entered the doubles tournament.

a. Display the results of the survey in a Venn diagram.

b. Find the probability that a contestant is entered in the singles tournament only.

c. Find the probability that a contestant is entered in the doubles tournament only.

11. NW At a meeting of 50 car salespeople, the following information was obtained: 12 salespeople sold Toyotas, 15 salespeople sold Fords, and 16 salespeople sold Chevrolets. Four salespeople sold Toyotas and Fords, six salespeople sold Fords and Chevrolets, and five salespeople sold both Toyotas and Chevrolets. Two salespeople sold all three kind of cars.

Using a Venn diagram, find the probability that a salesperson at this meeting sold

a. Toyotas only

b. Fords only

c. Chevrolets only

d. Toyotas and Fords, but not Chevrolets

e. Fords and Chevrolets, but not Toyotas

f. Fords or Toyotas

12. In a survey of 75 students who registered for courses, the following data were collected: 27 students were taking statistics, 26 were taking history, and 41 were taking English. Twelve students were taking statistics and history, 13 students were taking statistics and English, and 17 students were taking history and English. Four students were taking all three courses.

Using a Venn diagram, find the probability that a student participating in this survey is taking

a. only statistics
b. only English
c. statistics and English, but not history
d. statistics or English
e. none of the three subjects
f. exactly one of these three subjects

Writing Mathematics

13. State the **fundamental counting principle**, and provide examples using it to determine the number of outcomes when two or more events are performed together.

14. Write a probability problem for your classmates to solve. Be sure to pick an area or topic that will be familiar to most people.

Challenge Exercises

15. Almost all students have a Social Security number. It often serves as an ID number for students. How many possible Social Security numbers are there? (Assume repetition of digits. A Social Security number contains nine digits.)

16. There are eight horses entered in the featured race at Churchill Downs. A cash prize of $50,000 is awarded to any bettor that picks the eight horses in their exact order of finish. If a person wanted to ensure that all possibilities are covered, how many tickets would that bettor have to purchase? If each ticket costs $2, is it worthwhile to do so? Why or why not?

Group/Research Activities

17. Many probability problems involve the use of dice. Research the history of dice and their use over the centuries. Prepare a report on this topic. Be sure to include typical dice games that people played through out the time periods covered.

18. On the television game show *The Price Is Right,* contestants are provided with various challenges, which, if successful, result in winning a prize. Watch five episodes of this show and identify at least five different challenges and their corresponding probabilities of winning. Prepare a report that describes these challenges and how you calculated their probabilities.

Just for fun

Given the digits 1, 2, 3, 4, 5, 6, and 7, construct an addition problem whose sum is 100. You may use each digit only once.

4.4 TREE DIAGRAMS

A sample space consists of all possible outcomes for a particular experiment. A technique that shows us the sample space for two or more experiments that are performed together is a **tree diagram**.

A tree diagram consists of a number of "branches" that illustrate the possible outcomes for the experiments. We may read the possibilities directly from the branches. The following examples illustrate the use of a tree diagram to obtain a sample space.

EXAMPLE 1 ***Using a Tree Diagram to Construct a Sample Space*** When a coin is flipped, it may turn up heads or tails. How many different outcomes are possible when two coins are tossed, and what are the possible outcomes?

Solution There are two experiments and each experiment has two possible outcomes (head or tail). Using the counting principle, we have $2 \times 2 = 4$ possible outcomes.

Note of Interest

1 1
1 2 1
1 3 3 1
1 4 6 4 1
1 5 10 10 5 1

This array of numbers is known as "Pascal's triangle." We can construct such an array by the following method. In each row, the first and last numbers are always 1. Any number in the array is the sum of the two numbers it lies between on the line above. Five rows are shown (it may be extended).

The sum of the numbers in any row yields the total number of combinations possible within that group. For example, to find the probability of any boy–girl combination in a family of four children, we examine the fourth row. Note that the sum of the numbers is 16, the number of different outcomes. The numbers at the ends of the row represent the chances of the least likely events—that is, all boys or all girls: 1 in 16. The next number from the end represents the next type of combination, three boys and one girl or vice versa: 4 in 16. Finally, the chances of having two boys and two girls are 6 in 16.

Pascal did not invent the triangle that bears his name; some Chinese works of the early 1300s contain such an array of numbers. (See p. 374.) However, because he discovered and applied many of its properties, this triangular array has become known as Pascal's triangle.

In constructing the tree diagram, remember that the first experiment has two possible outcomes and each of these may be matched with the two possible outcomes of the second experiment. Hence, we have

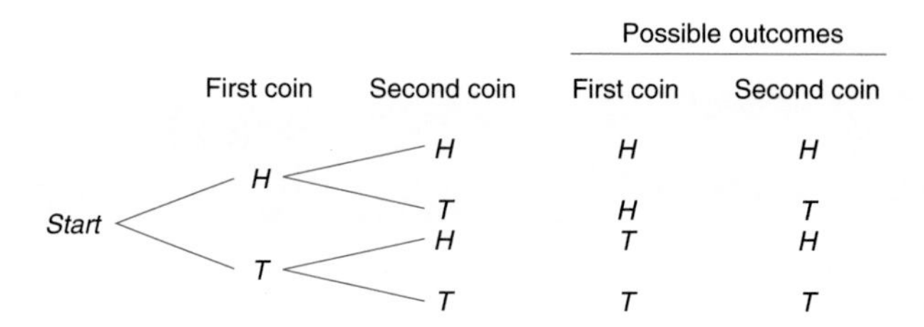

From the tree diagram we may obtain our sample space. The first branch gives us H–H, the second branch H–T, and so on, until we have the complete sample space: H–H, H–T, T–H, and T–T.

 **EXAMPLE 2** ***Using a Tree Diagram to Construct a Sample Space*** Mr. Examination is preparing a quickie quiz for his mathematics class to see if the students did their assignment. The quiz is to consist of three true–false questions. How many different arrangements of the answers are possible? What are the possible outcomes?

Solution We have three questions and each question has two possible outcomes (true or false). Using the counting principle, we compute $2 \times 2 \times 2 = 8$ total possible outcomes.

We can determine the various outcomes by means of a tree diagram. Remember that the quickie quiz consists of three questions and the answer to each question is either true or false.

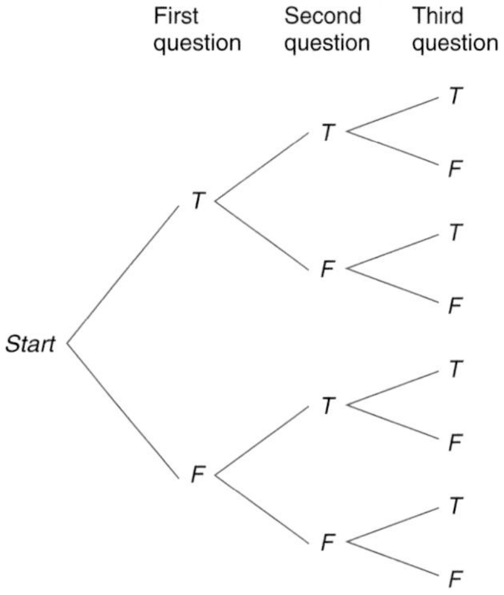

				Possible outcomes	
First question	Second question	Third question	First question	Second question	Third question
	T	T	T	T	T
		F	T	T	F
T		T	T	F	T
	F	F	T	F	F
Start		T	F	T	T
	T	F	F	T	F
F		T	F	F	T
	F	F	F	F	F

The sample space is listed beside the tree diagram.

NW ***Now Work Problem 1.***

EXAMPLE 3 ***Selecting Successful Outcomes from a Sample Space*** David, a student in Mr. Examination's class, did not do the assignment. David took the quiz and guessed at the answers. David answered the questions F–T–F. What is the probability that he answered all three questions correctly?

Solution The sample space in Example 2 shows each of the eight possible outcomes. Only one of these contains all three correct answers. Hence, the probability that F–T–F is the correct response is $\frac{1}{8}$.

NW ***Now Work Problem 7.***

Exercises for Section 4.4

NW **1.** Use a tree diagram to find the sample space showing the possible arrangements of boys and girls in a family with exactly two children.

- **a.** What is the probability that both children are boys?
- **b.** What is the probability that both children are girls?
- **c.** What is the probability that the first child is a boy and the second child is a girl?
- **d.** What is the probability that at least one of the children is a girl?
- **e.** What is the probability that neither of the children is a girl?

2. A box contains a one-dollar bill, a five-dollar bill, a ten-dollar bill, and a twenty-dollar bill. Two bills are chosen in succession without replacement. Use a tree diagram to list the sample space for this experiment.

- **a.** What is the probability that the values of both bills are even?
- **b.** What is the probability that the value of neither bill is even?
- **c.** What is the probability that the value of exactly one of the bills is even?
- **d.** What is the probability that the value of at least one of the bills is even?
- **e.** What is the probability that the total value of the bills chosen is $30?

3. Use a tree diagram to list the sample space showing the possible arrangements of boys and girls in a family with exactly three children.

- **a.** What is the probability that all three children are boys?
- **b.** What is the probability that two children are boys and one is a girl?
- **c.** What is the probability that at least one of the children is a girl?

d. What is the probability that none of the children is a boy?

e. What is the probability that all three children are of the same sex?

f. If a family has three children, all boys, what is the probability that a fourth child would be a girl?

4. Use a diagram to list the sample space showing the possible arrangements of heads and tails when four coins are tossed. Then use the sample space to find the probability that

a. all four coins come up heads

b. exactly two coins come up heads

c. exactly three coins come up tails

d. at most two coins come up heads

e. at least two coins come up heads

f. no more than three coins come up tails

5. An urn contains one blue and three orange balls. An experiment consists of selecting two balls in succession, without replacement. Use a tree diagram to list the sample space for this experiment. [*Hint:* Balls of the same color are indistinguishable, so you may find it helpful to use the symbols O_1, O_2, and O_3 to distinguish the three orange balls.] Then use the sample space to find the probability that

a. both balls are orange

b. at least one ball is orange

c. the first ball is blue

d. the second ball is blue

e. at least one ball is blue

f. one ball is blue and the other is orange

6. The Knicks won the first game against the Lakers in a basketball playoff. If the first team to win four games is the winner, construct a tree diagram for this playoff. [*Hint:* Some of the branches in the tree diagram will end sooner than others, as one of the teams will have won four games.]

NW 7. Jennie is at the dog kennel to select a puppy. The choices are in three categories.

Group A	Group B	Group C
Spaniel	Male	Brown
Terrier	Female	Black
Hound		White

a. Construct a tree diagram to list the sample space for the possible choices.

b. What is the probability that Jennie selects a spaniel?

c. What is the probability that Jennie selects a male puppy?

d. What is the probability that Jennie selects a brown male terrier?

e. What is the probability that Jennie selects a female hound that is brown or black?

f. What is the probability that Jennie selects a spaniel or hound that is black or white?

8. The Carriage Hill apartment complex contains a variety of different apartments. A person seeking an apartment has the following options.

A	B	C
one bedroom	one bathroom	first floor
two bedrooms	two bathrooms	second floor
three bedrooms		third floor

a. Construct a tree diagram to list the sample space for the possible apartment choices.

b. What is the probability that a person selects a one-bedroom apartment?

c. What is the probability that a person selects a one-bathroom apartment?

d. What is the probability that a person selects a one-bedroom, one-bathroom apartment on the second floor?

e. What is the probability that the apartment selected is a two-bedroom, two-bathroom above the first floor?

f. What is the probability that the choice is a one- or two-bedroom apartment on the third floor with one bathroom?

Writing Mathematics

9. What is a tree diagram and how do you construct one?

10. In an experiment two items are chosen in succession: one time *with replacement* and then another time *without replacement*. What does this mean, and what is the difference in the two trials?

Challenge Exercises

11. The Comets and Barons are two evenly matched teams. Each has an equally likely chance of winning a game. Suppose the two teams are in a playoff series that is won by winning two out of three games.

a. Make a tree diagram to list the sample space for the possible outcomes of the series.

b. What is the probability that the Barons will win the series?
c. What is the probability that the Barons will win the first two games?
d. What is the probability that the Comets will win two straight games?
e. What is the probability that the Comets will win at least one game?
f. What is the probability that the Comets will win the last two games?

12. A quarter is flipped and the outcome is noted. If a head results, then a die is rolled. If a tail results, then the quarter is flipped a second time.

a. How many different outcomes are possible?
b. Make a tree diagram to list the sample space for the possible outcomes.
c. What is the probability of obtaining a head?
d. What is the probability of obtaining an odd number?
e. What is the probability of obtaining no heads?
f. What is the probability of obtaining at least one tail?

13. Rachel can travel from Syracuse, New York to New York City by bus, train, car, or airplane. She can travel from New York City to London, England by airplane or ship.

a. How many choices for the trip from Syracuse to London are there?
b. Make a tree diagram to list the sample space for the possible outcomes.
c. What is the probability that Rachel will select a bus?
d. What is the probability that Rachel will select a ship?
e. What is the probability that Rachel will select an airplane?
f. If Rachel must return from London to New York City by airplane, and from New York City to Syracuse by airplane, how many ways may she make the round trip?

Group/Research Activity

14. Blaise Pascal and Girolamo Cardano were early contributors to the study of probability. Do research and write a report on the life of one of these men. Be sure to include their contributions to mathematics and other areas that might be of interest. References may include history of mathematics books and the Internet.

Just for fun

There are only four words in the English language that end in "dous." One of them is tremendous. Can you name the other three?

4.5 ODDS AND EXPECTATION

Odds

What odds will you give me?

The Browns are odds-on favorites to win the title.

The odds on Dead Last are 100 to 1.

The odds on getting a royal flush in poker are 649,739 to 1.

All of these expressions mention the word *odds*. We all have a nodding acquaintance with odds and probably have come in contact with them at some time or another. Most of us know that if the odds on a bet are 3 to 1 (3 : 1) and we bet a dollar and win, then we will win three dollars. But there is more to it than that.

Odds are usually expressed as a ratio (for example, 3 : 2). But they may also be expressed as 3 to 2, or even as a fraction $\frac{3}{2}$.

Odds can occur in two ways. We may discuss the *odds in favor of* an event or the *odds against* an event.

Odds are computed by finding a ratio. The **odds in favor of an event** A are found by taking the probability that an event A will occur and dividing it by the probability that an event A will not occur. We may state this as

$$\text{Odds in favor of } A = \frac{\text{Probability that } A \text{ will occur}}{\text{Probability that } A \text{ will not occur}}$$

EXAMPLE 1 ***Finding Odds in Favor*** What are the odds in favor of obtaining a sum of 7 when a pair of dice is tossed once?

Solution Using our definition, we first find the probability of obtaining a sum of 7 when a pair of dice is tossed. It is $\frac{6}{36}$. Next, we find the probability of not getting a 7. Recall that if $P(A) = S/T$, then $P(not\ A) = 1 - S/T$. Hence, we have

$$P(7) = \frac{6}{36} \quad \text{and} \quad P(not\ 7) = 1 - \frac{6}{36} = \frac{36}{36} - \frac{6}{36} = \frac{30}{36}$$

Alternatively, if there are 6 ways in 36 that we may obtain a 7, then the remaining 30 ways in 36 is the probability of not obtaining a 7.

Now we construct our ratio according to the definition:

$$\text{Odds in favor of } A = \frac{\text{Probability that } A \text{ will occur}}{\text{Probability that } A \text{ will not occur}}$$

$$\text{Odds in favor of a 7} = \frac{\frac{6}{36}}{\frac{30}{36}} = \frac{6}{36} \times \frac{36}{30} = \frac{6}{30} = \frac{1}{5}$$

The odds in favor of obtaining a 7 are $\frac{1}{5}$. But remember that the odds may also be stated as 1 to 5 or 1 : 5.

What does this mean? If we roll a pair of dice, over the long run we should obtain a 7 once out of six times. The other five times we would lose.

EXAMPLE 2 ***Finding Odds in Favor*** Find the odds in favor of obtaining two heads in a single toss of two coins.

Solution The probability in favor of obtaining two heads in a single toss of two coins is $\frac{1}{4}$. We can obtain this from the sample space

$$\{H\text{–}H,\quad H\text{–}T,\quad T\text{–}H,\quad T\text{–}T\}$$

The probability against two heads is

$$1 - \frac{1}{4} = \frac{4}{4} - \frac{1}{4} = \frac{3}{4}$$

Therefore, we have

$$\frac{\frac{1}{4}}{\frac{3}{4}} = \frac{1}{4} \times \frac{4}{3} = \frac{1}{3}$$

The odds in favor of two heads are 1 to 3, or 1 : 3.

NW ***Now Work Problems 1a, b, and c.***

If we change the ratio in Example 2 around (3 to 1, or 3 : 1), we will have the odds *against* obtaining two heads in a single toss of two coins. We must always make certain that we find the odds in favor of an event, if that is what we are asked to find. If we are asked to find the **odds against an event** A, then we should use the following ratio:

$$\text{Odds against } A = \frac{\text{Probability that } A \text{ will not occur}}{\text{Probability that } A \text{ will occur}}$$

We can also find the odds in favor of A and then reverse the ratio to obtain the odds against A.

EXAMPLE 3 ***Finding Odds Against*** What are the odds against obtaining an 11 when a pair of dice is tossed once?

Solution Using the definition, we still find the probability in favor of getting an 11 and the probability against getting an 11 when a pair of dice is tossed. The probability of getting an 11 is $\frac{2}{36}$, and the probability of not getting an 11 is

$$1 - \frac{2}{36} = \frac{36}{36} - \frac{2}{36} = \frac{34}{36}$$

Now we construct our ratio according to the definition:

$$\text{Odds against an 11} = \frac{\frac{34}{36}}{\frac{2}{36}} = \frac{34}{36} \times \frac{36}{2} = \frac{34}{2} = \frac{17}{1}$$

The odds against obtaining an 11 are $\frac{17}{1}$, which we would usually write as 17 to 1, or 17 : 1. Therefore, from the odds (17 : 1), we can see that we probably are not going to get an 11 when we roll the dice. Over the long run, we should get an 11 once out of every 18 tries. This is not a very good percentage of successful outcomes.

NW ***Now Work Problems 1d, e, and f.***

Odds enable us to play and bet fairly on games. We have just computed the odds against obtaining an 11 on the toss of a pair of dice, and it was 17 to 1. Suppose that Nevada Nellie is rolling a pair of dice and she bets she will roll an 11. If she bets a dollar, then according to the odds, she should receive $17 if she does toss an 11. This would make the game a "fair" one. Nevada Nellie should get an 11 once every 18 tries, over the long run. Hence, in 18 tries she should lose $17, $1 on each of 17 failures, and she should win once, and that one time should pay $17.

Two things are usually against a shooter like Nevada Nellie. One, the owner of the game (the house) normally cuts down the odds a little when they pay a winner, to 15 to 1 or 14 to 1. Two, Nevada Nellie has to play over the long run in order to regain her losses, which means she has to play for a long time. People like Nevada Nellie usually lose all their money first, or just get tired and bored, and then quit. The house can keep its game going because the casinos are open 24 hours a day.

B. C. by permission of Johnny Hart and Creators Syndicate, Inc.

EXAMPLE 4 ***Finding Probability Given Odds in Favor*** If the odds in favor of obtaining a 7 when a pair of dice is tossed are 1 : 5, what is the probability of obtaining a 7?

Solution We already know that $P(7) = \frac{6}{36}$ or $\frac{1}{6}$, but how do we obtain the answer when we are given the odds?

If the odds in favor of event A are $B : C$, then the probability of A is B divided by $B + C$; that is,

$$P(A) = \frac{B}{B + C}$$

Hence, for this particular problem, we have $B = 1, C = 5$, and

$$P(7) = \frac{1}{1 + 5} = \frac{1}{6}$$

EXAMPLE 5 ***Finding Probability Given Odds Against*** If the odds against the Dodgers winning the pennant are 8 : 5, what is the probability that the Dodgers will win the pennant?

Solution We can use the formula given in Example 4, but first, because the odds are given as 8 : 5 against, we restate them as 5 : 8 in favor. Now we can apply our formula.

If $B : C$ equals odds in favor of A, then

$$P(A) = B/(B + C), \quad B = 5, \quad C = 8,$$

and the probability that the Dodgers will win the pennant is

$$\frac{5}{5 + 8} = \frac{5}{13}$$

NW ***Now Work Problem 7.***

Mathematical Expectation

Mathematical probability and odds lead to another related topic: **mathematical expectation**. We can describe expectation (or expected value) using an example, as follows: Suppose that Lucky Louie is betting on a certain dice game, and his probability of winning is P, and if Louie wins then he receives M dollars. We would say that Lucky Louie's *mathematical expectation* is $P \times M$. You may not understand mathematical expectation yet, but at least we have a method for finding it. First, let's describe the way we find mathematical expectation a little more formally. Let

M = the amount that will be won if an event occurs

P = the probability that the event will occur

Then

$$Expectation = P \times M$$

Expectation tells us the expected value or "fair" price to pay to play a game, if the game can be described by the probability of equally likely outcomes. It also gives us the *average* amount of winnings we can expect for each game if we play a great many games.

Biography: Girolamo Cardano

Girolamo Cardano (1501–1576), also known as Jerome Cardan, is probably one of the most extraordinary people in the history of mathematics. He was born in Pavia, Italy, the illegitimate son of a local jurist.

He was a man of many contrasts. He began his professional life as a physician and during this time he also studied, taught, and wrote mathematics. His book, *Liber de Ludo Aleae* (*Book on Games of Chance*), is a gambler's handbook and is considered the first book on probability. In this book Cardan discussed the number of successes versus the number of outcomes, mathematical expectation, the different outcomes for two dice, and the different outcomes for three dice.

Later in his career, Cardan held important chairs at the universities of Pavia and Bologna. Eventually he became a distinguished astrologer, and at one time he was imprisoned for publishing a horoscope of Christ's life. Once, in a fit of rage, he cut off the ears of his younger son.

Cardan's life came to a dramatic end. Years before, he had made an astrological prediction of the date of his death. When the day arrived and he was still alive, he committed suicide to make his prediction come true.

Suppose Lucky Louie is playing a dice game in an Atlantic City casino and he is betting on 7 coming up. If a 7 comes up, then Louie receives a payoff of \$18. According to the formula, Lucky Louie's expectation is

$$E = P(7) \times \$18$$

$$E = \frac{6}{36} \times \$18 = \$3$$

That is, Lucky Louie should be willing to bet \$3 for the privilege of betting on 7. But unfortunately, the smallest denomination chip used at the casino is \$5. Hence, Lucky Louie must bet at least \$5 each time he plays the game. We have already computed his expectation, and it is \$3. If he has to bet \$5 each time he plays, then Lucky Louie can expect to lose an average of \$2 on each \$5 bet over the long run.

EXAMPLE 6 ***Mathematical Expectation and Raffle Tickets*** The Brighton Fire Department is running a raffle in which the prize for the lucky ticket is \$1,000. If 5,000 tickets are sold at \$2 each, what is the expectation of Hugh, who buys one ticket?

Solution The probability of having the winning ticket for this person is $\frac{1}{5000}$, and Hugh's expectation is

$$E = \frac{1}{5000} \times \$1000.00$$

$$E = \frac{\$1.00}{5} = \$0.20$$

Twenty cents represents a fair price to pay for a ticket. It appears that \$2.00 is too much to pay for a ticket, but this is how organizations make money.

EXAMPLE 7 ***Effect of Buying More Tickets on Mathematical Expectation*** Suppose Hugh changes his mind and decides to buy five tickets in the raffle for the $1,000 prize. Does this change his expectation?

Solution Hugh has increased only his chances of winning. Now the probability of winning is $\frac{5}{5{,}000}$, and his expectation is

$$E = \frac{5}{5{,}000} \times \$1{,}000.00 = \frac{\$5{,}000}{5{,}000} = \$1.00$$

Hugh's mathematical expectation is $1.00 for five tickets; it is still $0.20 per ticket.

EXAMPLE 8 ***Multiple Expectation*** A bag contains two red, five white, and three green balls. A prize of $10 is given if a red ball is drawn and a prize of $1 is given if a green ball is drawn. Nothing is won if a white ball is drawn. What is the expectation?

Solution First we must compute the probability of winning. The probability of drawing a red ball is $\frac{2}{10}$, and the probability of drawing a green ball is $\frac{3}{10}$. The expectation for the red ball is

$$E(\text{red}) = \frac{2}{10} \times \$10.00 = \frac{\$20}{10} = \$2.00$$

The expectation for the green ball is

$$E(\text{green}) = \frac{3}{10} \times \$1.00 = \frac{\$3}{10} = \$0.30$$

The total expectation is the sum of the two previous expectations. Therefore, we would have

$$E = E(\text{red}) + E(\text{green}) = \$2.00 + \$0.30 = \$2.30$$

Remember that the $2.30 represents a "fair" price to pay for the privilege of playing the game. It also represents the average amount you should expect to win when you play a great many games—that is, if play is over the long run—since these are equally likely events. Suppose you played the game 10 times. You should expect to win two times on red, which equals $20.00 in prizes; you should win three times on green, which equals $3.00 in prizes; and the other five times you should receive nothing. So, in 10 games you should receive $23.00. In a fair game, you should also bet $23.00, which is $2.30 per try.

NW ***Now Work Problem 11.***

In addition to helping us determine the "fair" price to pay to play a game, mathematical expectation also has applications in various industries. For example, life insurance companies can use mathematical expectation to determine the cost (called the *premium*) of an insurance policy. (*Note*: See Chapter 12 for additional information about insurance premiums.) This is done by calculating the probability that a person of a certain gender or age lives in a one-year period and then multiplying this probability by the cost of the life insurance policy. This product, which is the *expected value*, represents an estimate of the average claims a company must pay over the long run. Once this expected value has been determined, insurance companies can then decide the price of the premium.

EXAMPLE 9 ***Insurance Premiums*** A life insurance company has established a new $100,000 life insurance policy that will be marketed to 50-year-old males. Use mathematical expectation to find the annual premiums the company should charge for this policy.

Solution To use mathematical expectation, we need to know (1) what the probability is that a 50-year old man will die within a one-year period, and (2) the cost to the company if the policy holder dies. Note that we already know item (2)—it's the cost of the policy, namely, $100,000. To find item (1) we consult an "expected deaths" table that provides death rates based on age, gender, and years. Such tables are available from the Census Bureau's *Statistical Abstract of the United States*, which is accessible via the Internet at **http://www.census.gov/statab/www.** According to the data in one of these tables, the expected deaths per 1000 population for 50-year old males is 5.09. Rounding up, this implies for every 1000 50-year old males, we can expect 6 will die in the year they are 50. Expressed as a probability, this is $\frac{6}{1000}$. Thus, the probability that a 50-year old man dies during a one-year period is $\frac{6}{1000}$, and the expected value is

$$E = \frac{6}{1000} \times \$100{,}000$$

$$E = 6 \times \$100 = \$600$$

Interpreting this answer, if the insurance company sells a lot of $100,000 life insurance policies to 50-year old men, then the estimated average payout will be about $600 per policy. So, if the insurance company wants to cover any administrative costs as well as make a profit, then it should set the annual premium price for this policy greater than $600. A reasonable amount might be $800. In any event, though, it should not sell the policy for less than $600.

One final note: When we calculated the expected value, we did not consider the probability that the policy holder lives during the year. This probability is $1 - \frac{6}{1000} = \frac{1000}{1000} - \frac{6}{1000} = \frac{940}{1000} = \frac{94}{100}$. The corresponding expected value is $0 because the insurance company does not have to pay the premium. This can be confirmed mathematically by the expectation formula: $\frac{94}{100} \times \$0 = \$0$.

Exercises for Section 4.5

1. In a single toss of a pair of dice, find the odds (the number referred to is the sum of the numbers on the dice)

NW **a.** in favor of obtaining a 7

NW **b.** in favor of obtaining an 11

NW **c.** in favor of obtaining a 12

NW **d.** against obtaining a 12

NW **e.** against obtaining a 6

NW **f.** against obtaining a 10

2. In a single toss of a pair of dice, find the odds (the number referred to is the sum of the numbers on the dice)

a. in favor of obtaining a 4

b. in favor of obtaining a 6

c. against obtaining a 9

d. against obtaining a 3

e. in favor of obtaining a 3 or 11

f. against obtaining a number other than 9

3. On a single draw from a shuffled standard deck of 52 cards, find the odds

a. in favor of drawing an ace

b. in favor of drawing a club

c. in favor of drawing a red card

d. against drawing a deuce (2)

e. in favor of drawing a picture card (jack, queen, or king)

f. against drawing a picture card (jack, queen, or king) or a diamond

4. On a single draw from a shuffled standard deck of 52 cards, find the odds

a. in favor of drawing a queen

b. against drawing a queen

c. in favor of drawing a black card

d. against drawing a red card

e. in favor of drawing a heart

f. against drawing a heart or a diamond

5. In a single toss of two coins, make a sample space and find the odds

a. in favor of obtaining two tails

b. against obtaining two heads

c. in favor of obtaining exactly one tail

d. against obtaining no tails
e. in favor of obtaining at least one tail
f. in favor of obtaining a head and a tail

6. Find the probability that event A will happen if the odds are
   a. $5:3$ in favor of A
   b. $2:1$ in favor of A
   c. $5:1$ against A
   d. $4:7$ against A

NW 7. Find the probability that event B will happen if the odds are
   a. $8:3$ in favor of B
   b. $2:5$ in favor of B
   c. $7:5$ against B
   d. $3:5$ against B

8. The odds against the Yankees winning the pennant are $7:2$. What is the probability that the Yankees will win the pennant?

9. The odds against Laura winning a certain raffle are $99:1$. What is the probability that Laura will win?

10. In order to win a game, Nelson must throw a 7 in a single toss of a pair of dice. Larry bets that Nelson will; Benny bets that Nelson won't. What are the odds for and against this event? Do the odds favor Larry or Benny?

NW 11. A player will win $18 if he or she throws a double on the first toss of a pair of dice. What are the odds in favor of this player winning? What is a fair price to pay to play this game?

12. In a wheel game show four prizes are hidden on the wheel which contains 20 spaces. One prize is worth $500, two prizes are worth $1000 each, and the other prize is worth $3000. The remaining spaces contain no prizes. If a contestant picks a space at random, what is the mathematical expectation?

13. The Association for Conservation is awarding a $1000 cash prize to the winner of a raffle. A total of 5000 tickets are sold for $2 each. What is the mathematical expectation for a person who buys five tickets?

14. Many states conduct daily lotteries of one type or another. "Numbers" is a popular type. A player selects any three-digit number from 000 to 999 (any number that seems lucky to the player). In New York a player may select a number for $1. If the player wins, the payoff is $500. How many different three-digit numbers exist in this lottery? What is a player's chance of winning? What is a fair price to pay for a ticket? What can you conclude about the cost of playing this lottery?

15. A state lottery offers a weekly prize of $50,000. A person bought one ticket which cost $0.50. The winning ticket can be any six-digit number from 000,000 to 999,999. How many different six-digit numbers exist in this lottery? What were the person's chances of winning? What was a fair price to pay for it?

16. A special New York State Lottery offers a grand prize of $100,000, three second prizes of $10,000, and ten third prizes of $1000. The winning ticket can be any six-digit number from 000,000 to 999,999. If a person buys one ticket, what is a fair price to pay for the ticket?

17. Four envelopes are placed in a bag. Each envelope contains a single bill: a one-dollar bill, a five-dollar bill, a ten-dollar bill, and a twenty-dollar bill, respectively. The envelopes are mixed, and if you pay a certain amount, you may reach in and select one envelope at random. Would you be willing to pay $7 to select an envelope? Why?

18. One thousand raffle tickets are sold for a first prize of $300, a second prize of $150, and a third prize of $50. What is a fair price to pay for a raffle ticket? If Mary bought a raffle ticket for $1, what can you conclude about the cost of entering this raffle?

19. Five hundred raffle tickets were sold by a club for a drawing on a compact disc player valued at $600. What is a fair price to pay for a raffle ticket? If you buy a raffle ticket for $3, what can you conclude about the cost of the raffle ticket?

Writing Mathematics

20. In everyday language, odds and probability are often used interchangeably. Explain the difference between *the probability in favor of an event* and *the odds in favor of an event.*

21. What is the purpose of mathematical expectation? How is it typically used?

22. Explain how to find the probability of an event E occurring when given the odds in favor of event E are $4:7$.

Challenge Exercises

23. A magazine subscription company runs a contest where the first prize is $100,000, the second prize is $50,000, and the third prize is $25,000. If the company receives 1,000,000 entries, what is the expectation for a person who enters the contest once? If the entrant must use a first-class stamp to send in the entry, what can you conclude about the cost of entering the contest?

24. You are to toss two coins. If both land heads, you win $4. If one coin shows heads and the other tails, you lose $2. If both are tails, you neither win nor lose. What is your mathematical expectation?

Group/Research Activities

25. Do research and prepare a group report on state lotteries. Be sure to include information on the following: Which states do not have lotteries? Do different states have different types of lotteries? (Be sure to give examples.) What kind of annual profit do the states make on the lotteries? What do they do with the money? Be sure to include other information that might be of interest.

26. In this and previous sections of the chapter, we learned that games of "pure chance," such as tossing dice, card games, and lotteries, can be analyzed by probability theory. An adjunct to this is something called *game theory*, in which players have the opportunity to make choices and use rational strategies. One of the first publications in this field was by John Von Neumann (see related Biography in Section 4.8) called *The Theory of Games and Economic*

Behavior (Princeton University Press, Princeton). Review this publication and prepare a report on the concept of game theory as described by Von Neumann.

Real-World Data

27. Insurance premiums A life insurance company wants to determine the average payout per policy for $100,000 life insurance policies for 60-year-old women. If the expected deaths rate of 60-year-old women is 7.65 per 1000 population, determine the following:

a. The probability that a 60-year-old women dies during a one-year period.

b. The probability that a 60-year-old women does not die during a one-year period.

c. The expected value of the payouts for these policies. Interpret the results of this answer.

28. Insurance premiums Let's suppose that in Exercise 27, the insurance company is going to include an accidental death clause with its policy where the company will pay two times the policy amount if the death was ruled accidental. If the accidental death rate for 60-year-old women is 34 per 100,000, find the expected value of the payouts for these policies and interpret the results.

29. Theoretical versus empirical probability In Example 9 and Exercises 27 and 28, we consulted an expected deaths table to determine the probability that a person of a certain age or gender will die during a given year. Is this probability considered theoretical or empirical? Explain your answer.

30. Retail sales Kelly Yonkoski is manager of sales at a leading retail department store. In an attempt to improve her overall sales revenue, she asks her staff to assess how much additional money they think that can generate from their current customer base and the probability that they can actually do this. Let's assume that Rachel believes there is a 25% chance she can generate an additional $10,000 in sales from an existing client whose current sales total $25,000. What is the average amount of increased sales revenue that Rachel can expect from this client?

Just for fun

How many cubes are shown in the figure?

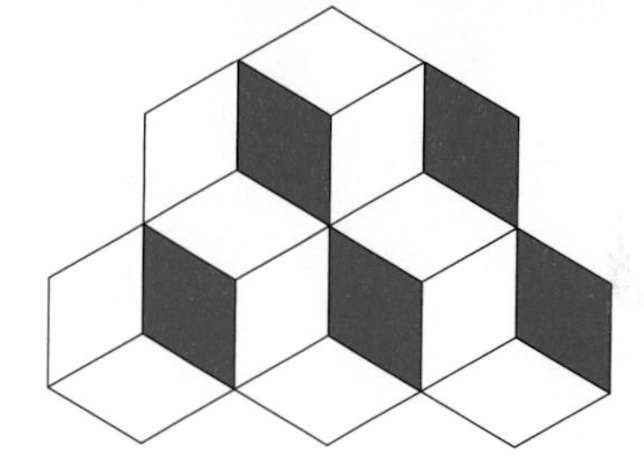

4.6 COMPOUND PROBABILITY

"AND" Probability with Independent and Dependent Events

Many events are **compound events**, made up of two or more simpler events. If we draw a card from a shuffled deck of cards, look at it, replace it, shuffle the cards again, and then draw another card, we have made two single simple drawings; but together they constitute a compound event. The problem in this example could be stated as, "What is the probability of drawing two red cards from a shuffled deck of 52 cards, if the first card is replaced"? We can also think of this as "What is the probability of drawing a red card *and then another* red card from a shuffled deck of 52 cards, if the first card is replaced?"

This probability problem leads us to two important questions:

1. How do we compute compound probabilities?
2. How does replacement influence the probability?

First, let us figure out how to compute compound probabilities. Recall the *counting principle*; it says that if one experiment has m different outcomes and a

Note of Interest

The term *probability* as applied to precipitation forecasting (for example, a 40% chance of precipitation) is the percentage chance that at least one-hundredth inch of precipitation (rain or the liquid equivalent of snow or other frozen precipitation) will fall at any selected point in the area and time period covered by the forecast.

A 60% probability indicates a 6 in 10 chance of precipitation, and a 4 in 10 chance of no precipitation, at any location in the forecast area. Forecasts usually cover a 12-hour period and a moderate-sized metropolitan area. Usually no differentiation is made for points within the forecast area.

If there is a 10% chance of precipitation, the qualifying forecast usually has no mention of precipitation. If there is a 20% chance of precipitation, the qualifying forecast calls for "slight or small chance of" precipitation. When there is a 30%, 40%, or 50% chance of precipitation, the qualifying forecast calls for a "chance of" precipitation. If there is a 60% or 70% chance of precipitation, the qualifying forecast states it is "likely that" precipitation will occur. If there is an 80% or 90% chance of precipitation, then there is no qualifying forecast term; precipitation is virtually assured.

second experiment has n different outcomes, then the first and second experiments performed together have $m \times n$ different outcomes. Thus, if the outcome of the first experiment does not influence the outcome of the second, we can use this idea to compute the probability of events occurring in succession; that is, we find the product of the probabilities of the two events. In general, we have

$$P(A \text{ and } B) = P(A) \times P(B)$$

Note that this rule may be extended for more than two events.

Second, let us decide how replacement influences the probability. In the preceding problem, the probability that the second card is red is the same as the probability that the first card is red, because we replaced the first card before drawing the second. The occurrence of the second event is **independent** of the first event; that is, the occurrence of the first event does not influence the probability of the second event.

Let us see why this is the case. The probability that the first card is red is $\frac{26}{52}$, or $\frac{1}{2}$, because there are 26 red cards (hearts and diamonds) out of a total of 52. The probability that the second card is red is also $\frac{26}{52}$, or $\frac{1}{2}$, because we replaced the first card. Therefore, the probability of drawing two red cards in succession, with replacement, is

$$\frac{26}{52} \times \frac{26}{52}, \quad \text{or} \quad \frac{1}{2} \times \frac{1}{2} = \frac{1}{4}, \quad \text{or} \quad 0.25$$

Suppose the first card had not been replaced. The probability that the second card is red then becomes $\frac{25}{51}$. Why? We know that there is one less card in the deck; hence the denominator of our fraction must be 51. But the 25 in the numerator confuses some students. Why not 26? How do we know that the first card is red? We don't necessarily know that it is red, but we must assume that it is, because the probability of success (that is, of drawing two red cards in succession) depends on the first card being red. The probability of the second card being red is dependent on whether the first card was red. The first event *did* influence the second event. Two events are **dependent** if the occurrence of one affects the probability of occurrence of the other.

Therefore, the probability of drawing two red cards in succession, without replacement, is

$$\frac{26}{52} \times \frac{25}{51} = \frac{1}{2} \times \frac{25}{51} = \frac{25}{102}, \quad \text{or} \quad 0.245$$

EXAMPLE 1 ***Compound Probability and Dependent Events*** What is the probability of being dealt two hearts in succession, without replacement, from a shuffled standard deck?

Solution This implies that we should use the formula

$$P(A \text{ and } B) = P(A) \times P(B)$$

These are dependent events. The probability that the first card is a heart is $\frac{13}{52}$, but the probability that the second card is a heart is $\frac{12}{51}$. Hence, the probability that both cards are hearts is

$$\frac{13}{52} \times \frac{12}{51} = \frac{156}{2{,}652} = \frac{1}{17}$$

Note: We must assume that event A has occurred before calculating the probability of event B (even when "assuming that event A has occurred" has not been stated).

EXAMPLE 2 ***Compound Probability and Independent Events*** A quarter is flipped three times. What is the probability of obtaining three tails in succession?

Solution The flipping of a coin is an independent event. What happens the first time does not affect what will happen the next time. The probability that a flip will yield a tail is $\frac{1}{2}$. To compute the probability of obtaining three tails in a row we have

$$P(T \text{ and } T \text{ and } T) = P(T) \times P(T) \times P(T)$$
$$P(3 \text{ tails}) = \frac{1}{2} \times \frac{1}{2} \times \frac{1}{2} = \frac{1}{8}$$

EXAMPLE 3 ***Compound Probability and Independent Events*** On a certain Sunday in October, the probability that the Rams will win their football game is 0.6 and the probability that the Giants will win their football game is 0.4. Assuming that they are not playing each other, what is the probability that both teams will win their games on this given Sunday?

Solution The probability that the Rams will win does not affect the probability that the Giants will win. Hence, we may consider these events to be independent, and we may compute the probability that the Rams *and* Giants will both win by multiplying their respective probabilities. Therefore, we have

$$P(\text{Rams and Giants will win}) = 0.6 \times 0.4 = 0.24$$

NW ***Now Work Problems 5a and c.***

"OR" Probability and Mutually Exclusive Events

The next question is how to compute the probability that *A or B* will occur.

For example, if a quarter is flipped and a die is tossed, what is the probability of obtaining a head *or* an odd number? To solve this problem previously, we generated a sample space (see Example 1, Section 4.3) and selected those outcomes that were considered to be successful. We could solve this problem by means of a sample space because we had to consider only 12 outcomes, but when the total possible number of outcomes becomes rather large we shall see that it is much more efficient to use a formula. Before computing the probability of *A or B* by means of a formula, we must first familiarize ourselves with a different kind of set of events called *mutually exclusive events*.

Events that are **mutually exclusive** are events that *cannot* happen at the same time: Only one of the events can occur at any one time. For example, when we flip a coin, we can get a head or a tail. Only one of these—not both—can occur for any one flip. If we roll a die, we can get an odd or an even number with one roll, but not both. These are mutually exclusive events. If two or more events cannot happen at the same time, they are mutually exclusive.

How do we compute the probability of A *or* B? Recall our discussion of the union of two sets A and B in Chapter 2. The *union* of sets A and B is the set of all elements that are elements of A *or* B, or elements of both. When we count the elements of the union of two sets, we find the elements of the two sets, but if an element is in both sets, we count it only once.

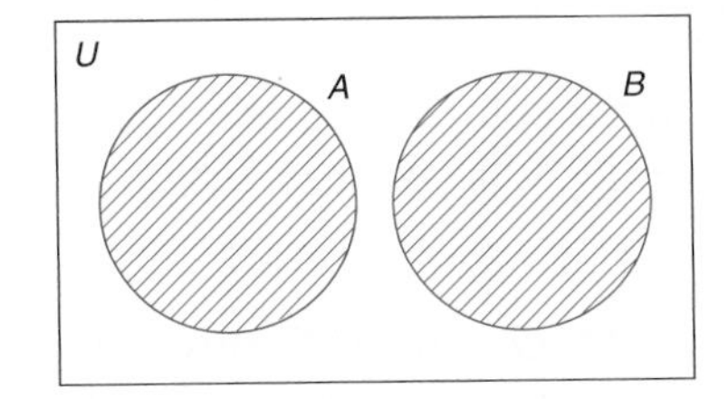

Figure 4.7

If two events are mutually exclusive, then they are disjoint (have no common elements). The Venn diagram for this situation is shown in Figure 4.7.

The probability of A *or* B, which we will denote by $P(A \text{ or } B)$, is the same as the probability of A union B, denoted by $P(A \cup B)$. From our diagram and the previous discussion, we have

$$P(A \cup B) = P(A) + P(B) = P(A \text{ or } B)$$

That is, in order to compute $P(A \text{ or } B)$, where A and B are mutually exclusive events, we add $P(A)$ and $P(B)$.

Suppose A and B are not mutually exclusive events; then what is $P(A \text{ or } B)$? In this case, A and B are not mutually exclusive events and therefore have a nonempty intersection. The Venn diagram for this situation is shown in Figure 4.8.

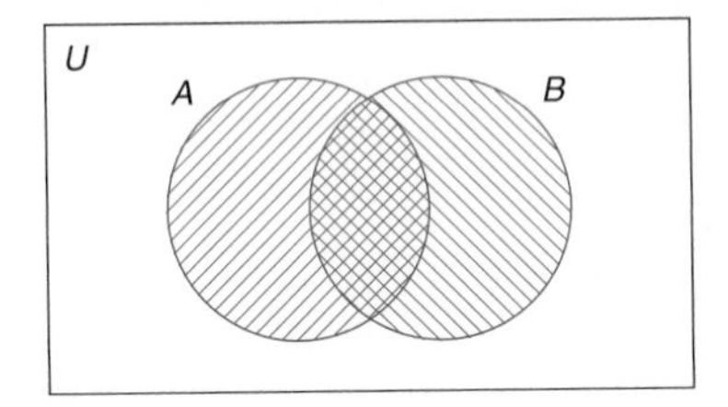

Figure 4.8

Here again, $P(A \text{ or } B)$ is the same as $P(A \cup B)$, but remember that this time the events are not mutually exclusive. Hence, we would have

$$P(A \cup B) = P(A) + P(B) - P(A \cap B)$$

We subtract $P(A \cap B)$, the probability of the intersection of A and B, from $P(A) + P(B)$ because $P(A \cap B)$ is included twice in $P(A) + P(B)$. Recall that the *intersection* of a set A and a set B is the set of elements common to both A *and* B. So $P(A \cap B)$ is the same as $P(A \text{ and } B)$, which is the probability of the outcomes that are common to both events A and B. Therefore, if A and B are not mutually exclusive events,

$$P(A \text{ or } B) = P(A) + P(B) - P(A \text{ and } B)$$

It should be noted that this equation is true even when A and B are mutually exclusive events; then $P(A \text{ and } B) = 0$, because A and B are disjoint, and so the intersection is empty. Then we have

$$P(A \text{ or } B) = P(A) + P(B) - 0.$$

EXAMPLE 4 ***Compound Probability and Mutually Exclusive Events*** In a single toss of a pair of dice, find the probability of obtaining a 7 or 11.

Solution A 7 and an 11 are mutually exclusive events. Hence, we have

$$P(\{7\} \cup \{11\}) = P(7 \text{ or } 11) = P(7) + P(11)$$
$$= \frac{6}{36} + \frac{2}{36} = \frac{8}{36} = \frac{2}{9}$$

NW ***Now Work Problems 5d and f.***

EXAMPLE 5 ***Compound Probability and Non–Mutually Exclusive Events*** One card is drawn from a shuffled standard deck of 52 cards. Find the probability that the card selected is a heart or a jack.

Solution We must first determine whether the events are mutually exclusive. Since a card may be selected that is both a heart and a jack—namely, the jack of hearts—the events are not mutually exclusive. Therefore, we should use the formula

$$P(A\ or\ B) = P(A) + P(B) - P(A\ and\ B)$$

and we have

$$P(\text{heart } or \text{ jack}) = P(H) + P(J) - P(H\ and\ J)$$
$$= \frac{13}{52} + \frac{4}{52} - \frac{1}{52} = \frac{16}{52} = \frac{4}{13}$$

EXAMPLE 6 ***Compound Probability and Non–Mutually Exclusive Events*** A quarter is flipped and a die is tossed. What is the probability of obtaining a head or a 6?

Solution These events are not mutually exclusive. We can get a head on the quarter and a 6 on the die at the same time.

We can compute probability problems of this type in a variety of ways. One technique is to rely on a sample space and read our answer directly from it. For this problem we would have

(*H* – 1)	(*H* – 4)	*T* – 1	*T* – 4
(*H* – 2)	(*H* – 5)	*T* – 2	*T* – 5
(*H* – 3)	(*H* – 6)	*T* – 3	(*T* – 6)

The outcomes that are circled in the sample space are the successful outcomes, so $P(H\ or\ 6) = \frac{7}{12}$.

Another way to solve this problem is to use the formula for events that are not mutually exclusive—namely,

$$P(A\ or\ B) = P(A) + P(B) - P(A\ and\ B)$$

Then

$$P(H\ or\ 6) = P(H) + P(6) - P(H\ and\ 6)$$
$$P(H\ or\ 6) = \frac{1}{2} + \frac{1}{6} - \frac{1}{12}$$
$$P(H\ or\ 6) = \frac{6}{12} + \frac{2}{12} - \frac{1}{12}$$
$$P(H\ or\ 6) = \frac{7}{12}$$

NW ***Now Work Problems 15d and e.***

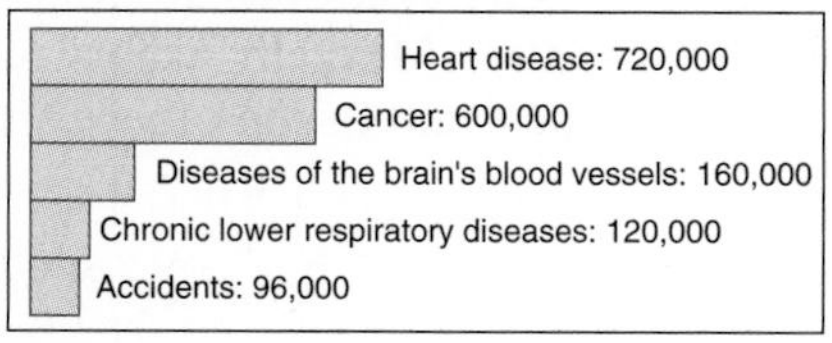

EXAMPLE 7 ***Causes of Death*** In 2000, there were approximately 2,400,000 deaths in the United States. The top five causes of death, which are considered mutually exclusive, and the number of people who died from these causes are given in Figure 4.9 (all numbers are rounded). Using these data, find

a. *P*(a person died from heart disease)

b. *P*(a person died from cancer *or* an accident)

Heart disease: 720,000
Cancer: 600,000
Diseases of the brain's blood vessels: 160,000
Chronic lower respiratory diseases: 120,000
Accidents: 96,000

Figure 4.9
Source: Center for Disease Control

c. P(a person died from a disease of the brain's blood vessels *and* from chronic lower respiratory disease)

d. P(a person died from one of these causes *except* diseases of the brain's blood vessels)

Solution

a. $P(\text{heart disease}) = \dfrac{720000}{2400000} = \dfrac{72}{240} = \dfrac{3}{10}$

Thus, there was a 3 in 10 chance (30%) that a person who died in 2000 died from heart disease.

b. P(a person died from cancer or an accident) represents two mutually exclusive events. As a result, we add their respective probabilities.

$$\begin{aligned}
P(\text{cancer or accident}) &= P(\text{cancer}) + P(\text{accident}) \\
&= \frac{600000}{2400000} + \frac{96000}{2400000} \\
&= \frac{6}{24} + \frac{96}{2400} \\
&= \frac{1}{4} + \frac{1}{25} \\
&= \frac{25}{100} + \frac{4}{100} \\
&= \frac{29}{100}
\end{aligned}$$

c. Disease of the brain's blood vessels and chronic lower respiratory represent two mutually exclusive events and hence cannot occur at the same time. Thus, the probability that a person died from both of these causes is zero.

d. If we let A, B, C, D, and E, respectively, represent the five leading causes, then the probability that a person died from one of five leading causes *except* diseases of the brain's blood vessels is $P(A) + P(B) + P(D) + P(E)$.

$$\begin{aligned}
P(A) + P(B) + P(D) + P(E) &= \left[\frac{3}{10} + \frac{1}{4} + \frac{1}{20} + \frac{1}{25}\right] \\
&= \left[\frac{30}{100} + \frac{25}{100} + \frac{5}{100} + \frac{4}{100}\right] \\
&= \frac{64}{100} = \frac{16}{25}
\end{aligned}$$

In summary, two or more events are **independent** if the occurrence of one event has no effect upon the occurrence or nonoccurrence of the other event(s). Two or more events are **dependent** if the occurrence of one event has an effect upon the occurrence or nonoccurrence of the other event(s). **Mutually exclusive** events cannot occur together at the same time. That is, one event or the other can occur, but both cannot occur simultaneously.

Math Connections

Genetics

One application of probability is in the field of genetics. For example, every expectant mother thinks about what her child will look like when it is born. Will it have blue eyes and dark hair, or will it have brown eyes and light hair? These are questions of probability (e.g., What is the probability that my child will have blue eyes?). To answer these type of questions, we first need to introduce some basic biological concepts.

The genetic constitution of an individual is called **genotype**. An individual's genotype for a specific trait such as eye color is determined by the genes of the person's parents. From the moment of conception (i.e., when the sperm and egg unite), the genotype of the baby is determined. If the sperm carries a brown-eyed gene (B) and the egg possesses a blue-eyed gene (b), then the baby's genotype consists of one brown-eyed gene and one blue-eyed gene (Bb). In the case of eye color, we say that brown is the **dominant characteristic** and blue is the **recessive characteristic**, which implies that when both genes are present, the baby's eye color will be the dominant one, namely, brown. The physical appearance of an individual trait is known as the **phenotype**. For our simple eye color example, three possible genotypes and two possible phenotypes are arranged as follows.

Genotype	Phenotype	Biological Term
BB	Brown	Pure dominant (*homozygous* brown)
Bb	Brown (because of the dominant B)	Hybrid (*heterozygous*)
bb	Blue (note the absence of the brown gene)	Pure recessive (*homozygous* blue)

Once we know the parents' genotypes, we can then place this information in a 2×2 grid (called a **Punnett square**) to determine the baby's possible genotypes. For example, if the father has blue eyes and the mother is heterozygous, then the father's genotype is bb and the mother's genotype is Bb. This is placed in the Punnett square as follows.

Father \ **Mother**	B	b
b	Bb	bb
b	Bb	bb

Note that there are four possibilities for the baby's genotype, two of which are bb. Thus, for parents who possess these genotypes, the probability that their baby will have blue eyes is $\frac{2}{4} = \frac{1}{2}$. The Bb entries in the Punnett square mean that the baby is a brown-eyed hybrid (heterozygous).

What do you think the probability is of a baby being born with blue eyes if both parents have brown eyes? To examine this probability, we need to know if the parents genotype is pure dominant (i.e., BB) or hybrid (i.e., Bb). From the preceding table, note that there are three different arrangements in which this can occur:

- Both parents are pure dominant (BB and BB).
- Both parents are hybrid (Bb and Bb).
- One parent is pure dominant and the other parent is hybrid (BB and Bb).

Thus, a Punnett square will have to be constructed for each arrangement. Doing so, leads to the following probabilities:

- If both parents are pure dominant, then $P(\text{blue-eyed baby}) = 0$.
- If both parents are hybrid, the $P(\text{blue-eyed baby}) = \frac{1}{4}$.
- If one parent is pure dominant and the other is hybrid, then $P(\text{blue-eyed baby}) = 0$.

In genetics, there is also the law of **independent assortment**, which states that genes (on different chromosomes) representing different traits unite independently. So the probability of two independent traits (e.g., gender and eye color) occurring together is found by multiplying the probabilities of the individual traits occurring separately. That is, $P(A \text{ and } B) = P(A) \times P(B)$. For example, the probability of a blue-eyed father (bb) and a brown-eyed mother (Bb) having a blue-eyed son is

$$\begin{aligned} P(\text{blue-eyed son}) &= P(\text{blue-eyed male}) \\ &= P(\text{blue-eyed}) \times P(\text{male}) \\ &= \frac{1}{2} \times \frac{1}{2} \\ &= \frac{1}{4} \end{aligned}$$

This concept can be extended to include additional traits. It can also be applied to most life forms in the animal kingdom, including plants. George Mendel, for example, who is considered the father of genetics, did his original experiments with garden peas using methods similar to those presented here.

Exercises for Section 4.6

1. A jar contains five red, one green, and two blue pencils. A pencil is selected and then replaced, after which a second pencil is selected. Find the probability that
   a. the first is red and the second is blue
   b. the first is green and the second is red
   c. the first is blue and the second is green
   d. both are blue
   e. both are red
   f. both are green

2. Each of the numbers 1 through 5 is painted on a separate ball. The five balls are placed in a bag and it is shaken to mix up the balls. A ball is drawn and then replaced, after which a second ball is drawn. Find the probability that
   a. the first ball has an odd number and the second ball has an even number
   b. the first ball has a number less than 4 and the second ball has a number greater than 2
   c. the first ball has a number less than 4 and so does the second
   d. both balls contain even numbers
   e. both balls contain odd numbers
   f. the sum of the numbers on both balls will be 10

3. In a certain political science class there are 20 students. The accompanying table shows the distribution of students: male, female, freshmen, and sophomores.

	Freshmen	Sophomores
Male	4	7
Female	3	6

   Mr. Harris, the instructor, wants to have a student panel discussion regarding recent events. If the panel is to consist of three students randomly chosen from the class, find the probability that the panel will consist of
   a. all freshmen
   b. all sophomores
   c. all females
   d. all males
   e. a freshman chosen first and then two sophomores
   f. a sophomore chosen first and then two freshmen

4. The accompanying table shows the number of students in a mathematics contest. Ms. Moderator will select contestants at random.

	Freshmen	Sophomores	Juniors	Seniors
Female	3	4	6	7
Male	4	5	5	6

   Find the probability that
   a. the first two contestants are female
   b. the first two contestants are sophomores
   c. the first two contestants are juniors
   d. the first contestant selected is a female and the second contestant selected is a male
   e. the first four contestants selected are female seniors
   f. the first two contestants selected are freshmen and the next two contestants selected are seniors

5. One card is randomly selected from a shuffled standard deck of 52 cards and then a quarter is flipped. Find the probability of obtaining
   NW a. a red card and a head
   b. a red card or a head
   NW c. a picture card and a tail
   NW d. a picture card or a tail
   e. a red picture card and a head
   NW f. a red picture card or a tail

6. One card is randomly selected from a shuffled standard deck of 52 cards and then a die is rolled. Find the probability of obtaining
   a. a red card and a 1
   b. a red card or a 1
   c. a picture card and a 1
   d. a picture card or a 1
   e. a red picture card and a 1
   f. a red picture card or a 1

7. A quarter is flipped and a die is tossed. Find the probability of obtaining
   a. a head and a 1
   b. a tail and a 1
   c. a head or a 1
   d. a tail or a 1
   e. a head and an even number
   f. a tail or an even number

8. A bag contains five slips of paper, each with one of the numbers 1 through 5. A box contains a red, a blue, and a green ball. One item is drawn from the bag and one from the box. Find the probability of obtaining
   a. an even number and a red ball
   b. an odd number and a blue ball
   c. an even number or a red ball
   d. an odd number or a blue ball
   e. a 2 and a green ball
   f. a 2 or a green ball

9. Two cards are selected at random from a shuffled standard deck of 52 cards, without replacement. Find the probability that
   a. both cards are picture cards
   b. the first card is an ace and the second card is a picture card
   c. the first card is the ace of spades and the second card is a picture card

d. both cards are kings

e. both cards are diamonds

f. both cards are of the same denomination

10. Lee's golf bag contains 30 golf tees. Ten are white, six are red, five are blue, and nine are purple.

a. Find the probability that a tee drawn at random is red.

b. Find the probability of drawing four blue tees in succession if there is no replacement after each draw.

c. Find the probability of drawing five blue tees in succession, if there is replacement after each draw.

d. If two tees are drawn in succession (without replacement), find the probability that both tees are white.

e. If two tees are drawn in succession (without replacement), what is the probability that the first tee is purple and the second is red?

11. An ice chest contains five cans of ginger ale, three cans of orange soda, and four cans of root beer. If three cans are drawn at random from the chest without replacement, find the probability of getting

a. three cans of root beer

b. three cans of orange

c. three cans of ginger ale

d. a can of orange soda, then a can of root beer, then a can of ginger ale

e. a can of orange soda, then a can of ginger ale, then a can of root beer

f. no root beer

12. One die is rolled, and then another die is rolled. Find the probability of obtaining

a. a 5 on the first die and a 4 on the second die

b. a 5 on the first die or a 4 on the second die

c. a 6 on both dice

d. an even number on the first die and an odd number on the second die

e. an even number on the first die or an odd number on the second die

f. an even number on the first die and a 3 on the second die

13. A track coach must decide who is going to run on the school's relay team. He must choose four runners from a list of six equally fast runners: Nenno, Nelson, Gilligan, Neanderthal, Clar, and Connelly. He decides that the easiest way to make the decision is to draw names from a hat in succession. (Note that there is no replacement since the coach needs four different runners.) Find the probability that

a. he chooses Neanderthal first

b. he chooses Neanderthal first and Nenno second

c. he chooses Nelson or Gilligan first

d. his third choice is a person whose name begins with N, given that Connelly and Clar were already chosen

e. Nenno is chosen first, Neanderthal is chosen second, Nelson is chosen third, and Gilligan is chosen to run the anchor leg (fourth)

14. Each of the numbers 0 through 9 is painted on a separate ball. The 10 balls are put in a can and the can is shaken to mix up the balls. Find the probability of

a. drawing an even-numbered ball

b. drawing a ball numbered 4 and then another numbered 4 on two successive draws with replacement

c. drawing a ball numbered 4 and then another numbered 4 on two successive draws without replacement

d. drawing a ball numbered 4 on the first draw, but not on the second draw, with replacement

e. drawing an even number on the first draw and an odd number on the second draw, without replacement

15. A bingo caller has a machine that contains 75 balls. The balls are marked in the following manner:

$$\begin{array}{ccccc} B1, & B2, & B3, & \ldots, & B15 \\ I16, & I17, & I18, & \ldots, & I30 \\ N31, & N32, & N33, & \ldots, & N45 \\ G46, & G47, & G48, & \ldots, & G60 \\ O61, & O62, & O63, & \ldots, & O75 \end{array}$$

The balls are mixed and each ball is drawn at random.

a. What is the probability of drawing a ball that has a B on it?

b. What is the probability of drawing three B balls in succession, if there is no replacement?

c. What is the probability of drawing a ball that has a double number (11, 22, ...) on it?

NW **d.** What is the probability of drawing a ball that has a B or a G on it?

NW **e.** What is the probability of drawing two balls, without replacement, so that the first one has a B on it and the second one has an I or N on it?

f. What is the probability of drawing a ball that has a number on it that is not divisible by 5?

16. A package of flower seeds contains ten seeds, six seeds for red flowers and four seeds for yellow flowers. A seed is randomly selected and planted and then another seed is randomly selected and planted. Find the probability that

a. both seeds will result in red flowers

b. both seeds will result in yellow flowers

c. one seed of each color flower is selected

17. Given that events A and B are independent, and $P(A) = 0.4$ and $P(B) = 0.5$, find

a. $P(A \text{ and } B)$ **b.** $P(A \text{ or } B)$
c. $P(\text{not } A)$ **d.** $P(\text{not } B)$
e. $P(A \text{ and not } B)$ **f.** $P(\text{not } A \text{ or } B)$

18. Given that events C and D are mutually exclusive, and $P(C) = 0.6$ and $P(D) = 0.3$, find

a. $P(C \text{ and } D)$ **b.** $P(C \text{ or } D)$
c. $P(\text{not } C)$ **d.** $P(\text{not } D)$
e. $P(C \text{ or not } D)$ **f.** $P(\text{not } C \text{ and not } D)$

19. Given $P(A) = \frac{5}{12}$, $P(B) = \frac{1}{4}$, and $P(A \text{ or } B) = \frac{7}{12}$, are events A and B mutually exclusive? Why?

20. Given $P(C) = \frac{7}{10}$, $P(D) = \frac{6}{10}$, and $P(C \text{ and } D) = \frac{40}{100}$, are events C and D independent? Why?

Writing Mathematics

21. What are independent events and how are they different than dependent events? Give at least two different examples.

22. What are mutually exclusive events? Give an example of mutually exclusive events. Give an example of two events that are not mutually exclusive.

23. Explain how to find probabilities of events that are not mutually exclusive. Give an example.

Challenge Exercises

24. A coin is tossed and a die is rolled. Let

event A = a tail appears
event B = an odd number appears
event C = a one or a head appears (but not both)

Find

a. $P(A \text{ or } B)$ **b.** $P(B \text{ or } C)$ **c.** $P(A \text{ or } C)$

25. What is the probability of obtaining at least one head on a toss of five coins?

26. What is the probability of obtaining at most two heads on a toss of five coins?

In Exercises 27 and 28, refer to the Math Connections on genetics.

27. Suppose two parents, both Bb, plan to have three children. What is the probability that all will be blue-eyed males?

28. If we let D = dark hair (dominant), d = blond hair, B = brown eyes (dominant), and b = blue eyes, then find the following probabilities given a mother whose genotype is $DdBb$ and a father whose genotype is $ddbb$:

a. P(baby will have blue eyes).

b. P(baby will be a blond-haired, blue-eyed girl).

c. If the parents plan on having seven children, P(all will have blond hair and blue eyes).

d. If the parents have seven children, P(all will be blond-haired, blue-eyed girls).

Group/Research Activities

29. Another type of probability that was not discussed in this section is *conditional probability,* which involves finding the probability of an event that is affected by a specific condition or stipulation. For example, suppose two dice are tossed. What is the probability that a 2 appears, *knowing that the sum of the dice is 7?* Conditional probability leads to an altered sample space. Research the concept of conditional probability and prepare a report on it. As part of your report, include the notation used to express a conditional probability, the general formula used for calculating conditional probabilities, and examples and solutions of conditional probability problems.

30. Computing conditional probabilities (as discussed in Exercise 29) can be simplified by using *Bayes's theorem*. Research this theorem and prepare a report that describes it along with examples of its use. Include in your report how the general formula for conditional probability from Exercise 29 can be used to verify Bayes's theorem.

Real-World Data

31. **Children and weight** Figure 4.10 shows the probabilities of children who are at the highest risk among their peers in terms of weight to be overweight when they are 35 (95th percentile). For example, the probability that a 3-year old boy will be overweight when he is 35 is 70%, which is $\frac{70}{100}$ (see Chapter 12 for information on converting a percent to a decimal). Use this figure to answer the following questions.

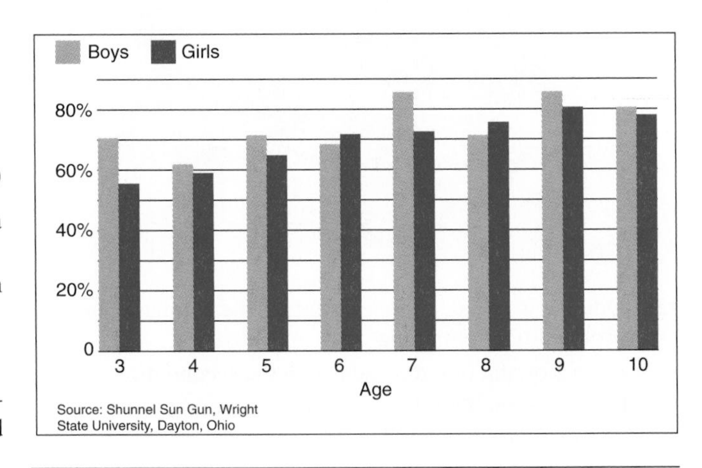

Figure 4.10

a. If we were to randomly select a 6-year old girl and a 6-year old boy who are in the 95th percentile among their peers in terms of weight, what is the probability that both children will be overweight when they are 35?

b. If we were to randomly select a 10-year old boy and a 10-year old girl who are in the 95th percentile among their peers in terms of weight, what is the probability that both children will be overweight when they are 35?

c. If we were to randomly select a 6-year old girl and a 9-year old girl who are in the 95th percentile among their peers in terms of weight, what is the probability that both girls will be overweight when they are 35?

d. If we were to randomly select a 5-year old boy and a 10-year old boy who are in the 95th percentile among their peers in terms of weight, what is the probability that both boys will be overweight when they are 35?

32. Children and diabetes risk Figure 4.11 shows the lifetime risk (or probability) of developing diabetes for children born in 2000. For example, the overall probability that a girl born in 2000 develops diabetes in her lifetime is 39%, which is $\frac{39}{100}$ (see Chapter 12 for information on converting a percent to a decimal). Use this figure to answer the following questions.

a. If we randomly select one boy and one girl born in 2000, what is the probability that both children will develop diabetes in their lifetime?

b. If we randomly select one Hispanic boy and one Hispanic girl born in 2000, what is the probability that both children develop diabetes in their lifetime?

c. If we randomly select one white boy and one black boy born in 2000, or two Hispanic girls born in 2000, what is the probability these children develop diabetes in their lifetime?

d. If we randomly select one Hispanic girl and one black boy born in 2000, or two white girls born in 2000, what is the probability that all four children will develop diabetes in their lifetime?

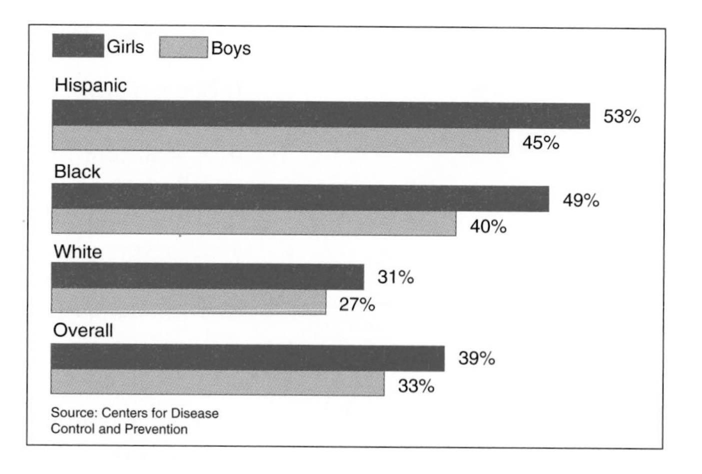

Figure 4.11

Just for fun

Try the following multiple-choice test and answer the questions after the test.

1. In 1956, Mickey Mantle led the major leagues in home runs with a total of
   a. 42 b. 48 c. 52 d. 49
2. The 1992 Olympic summer games were held in
   a. Atlanta b. Barcelona c. Seoul d. Sydney
3. Franklin D. Roosevelt was born in
   a. 1884 b. 1890 c. 1887 d. 1882
4. One of the world's longest railway tunnels is the English Channel Tunnel. Its length is approximately
   a. 11 miles b. 21 miles c. 31 miles d. 41 miles
5. Kansas, the Sunflower State, entered the union in
   a. 1861 b. 1846 c. 1816 d. 1836

Questions Pertaining to the Test

1. What is the probability of guessing correctly on any one given question?
2. What is the probability of guessing correctly on all of the questions?
3. What is the probability of guessing incorrectly on any one given question?
4. What is the expected number of correct answers?

Answers to Multiple-Choice Test

1. c 2. b 3. d 4. c 5. a

4.7 COUNTING, ORDERED ARRANGEMENTS, AND PERMUTATIONS (OPTIONAL)

What is counting? When we count something, we want to know "how many." Often we have to count all of the possible outcomes for probability problems. The basic definition of the probability that an event A will occur is the *number* of successful outcomes of the experiment divided by the total *number* of possible outcomes. In Section 4.3, we introduced the *counting principle* to enable us to efficiently count the possible outcomes for an experiment.

Counting Principle

If one experiment has m different outcomes, and a second experiment has n different outcomes, then the first and second experiments performed together have $m \times n$ different outcomes. This may be extended, if there are other experiments to follow, to $m \times n \times r \times \cdots \times t$.

How many different telephone numbers are possible in your local calling area? Every phone number consists of seven digits (10, if you also count the area code). As a counting example, we shall examine just the last four digits. There are four numbers of outcomes to multiply, one for each digit:

$$m \times n \times r \times s$$

In how many ways can each digit be chosen? Each digit may be any one of the numerals 0, 1, 2, 3, 4, 5, 6, 7, 8, 9, and therefore, there are 10 possibilities for each digit. (We are assuming that each numeral may be repeated in more than one digit.) Hence, each digit may be chosen in any one of 10 different ways, assuming no restrictions on the use of 0 or 1, so we have

$$10 \times 10 \times 10 \times 10 = 10{,}000$$

different telephone numbers in a local calling area or exchange. If we want to find out the number of telephone numbers in one general area that shares the same area code, we can expand this problem to seven digits.

EXAMPLE 1 ***Counting Principle with Replacement*** How many different license plates can be made if each license plate is to consist of three letters followed by two digits? (Assume replacement.)

Solution Here order is important. There are five slots to fill. The first three slots are to be filled by letters of the alphabet; because we have replacement, there are 26 choices for each of the first three slots. Each of the last two slots may be filled by any one of the 10 numerals, 0, 1, 2, 3, 4, 5, 6, 7, 8, 9. Therefore, the solution is

$$26 \times 26 \times 26 \times 10 \times 10 = 1{,}757{,}600$$

EXAMPLE 2 ***Counting Principle without Replacement*** How many different license plates can be made if each license plate is to consist of three letters followed by two digits, and no letters or digits may be repeated?

Solution This problem is similar to Example 1, but now we have a restriction that no repetition is allowed. We still have 26 choices for the first slot in the license plate, but there are only 25 choices for the second slot because we have already used a letter in the first slot. Similarly, there are 24 choices for the third slot, because we used one letter of the 25 in the second slot. The same thing happens when we fill in the slots for the digits. Hence, the solution is

$$26 \times 25 \times 24 \times 10 \times 9 = 1{,}404{,}000$$

NW ***Now Work Problem 3.***

EXAMPLE 3 ***Counting Principle Applied to Scheduling Courses*** Marie is planning her schedule for next semester. She must take the following five courses: English, history, geology, psychology, and mathematics.

a. In how many different ways can Marie arrange her schedule of courses?

b. How many of these schedules have mathematics listed first?

Solution

a. Since Marie has to take five different courses, there are five time slots to consider:

$$m \times n \times r \times s \times t$$

There are five choices for the first time slot, four for the second, and so on. The solution is

$$5 \times 4 \times 3 \times 2 \times 1 = 120$$

b. If mathematics is to be listed first, then this is a restriction, and there is only one choice for the first time slot. Mathematics is used to fill the first slot, so we have four choices for the second slot, three for the third slot, and so on. The solution is

$$1 \times 4 \times 3 \times 2 \times 1 = 24$$

EXAMPLE 4 ***Forming Numbers from Digits*** Given the set of digits $\{1, 3, 4, 5, 6\}$,

a. How many three-digit numbers can be formed?

b. How many three-digit numbers can be formed if the number must be even?

c. How many three-digit numbers can be formed if the number must be even and no repetition of digits is allowed?

Solution

a. There are three places, or slots, to consider. Each slot can be filled by any one of the five digits. Therefore, the solution is

$$5 \times 5 \times 5 = 125$$

b. The three-digit number must be even, so it must end either in a 4 or 6, the even digits in our given set. This is a restriction, so we should attend to it first: we have two choices, 4 or 6, for the last place. The counting principle gives us

$$\_\_ \times \_\_ \times 2$$

three-digit numbers. We have five choices for the first place and five choices for the second. Hence, we have

$$5 \times 5 \times 2 = 50$$

three-digit numbers.

c. There are two restrictions here: The number must be even, and no digit may be repeated. We still fill in the last place with two choices:

$$\_\_ \times \_\_ \times 2$$

But now there are only four choices for the first place, because one of the digits has been used in the last place. There are then three choices for the second place, and we have

$$4 \times 3 \times 2 = 24$$

three-digit numbers.

NW ***Now Work Problem 5.***

If a woman has individual photos of each of her three children, Mary, Scott, and Joe, how many ways can she arrange these photos in a row on her desk? From our previous discussion, we know that there are $3 \times 2 \times 1 = 6$ different possible outcomes, and they are

Mary–Scott–Joe
Mary–Joe–Scott
Scott–Joe–Mary
Scott–Mary–Joe
Joe–Scott–Mary
Joe–Mary–Scott

Each of these arrangements is different. The pictures are in a definite order, and the order in which the pictures are arranged is important. When we have a group of things arranged in a definite order, we have a *permutation*. A **permutation** is a particular ordering of the elements of a given set.

Permutations may use all the elements from a given set (as we did in our example), or only a certain number of them. If we are given the digits 1, 2, 3, 4, and 5, how many two-digit numbers can we form? This is a permutation (arrangement) of five things taken two at a time. In this case, we would have $5 \times 4 = 20$ different arrangements. If we have n items and want to count the possible arrangements using r of them at a time, we say that we want the number of **permutations of** n things taken r at a time. The notation for this is

${}_nP_r$ the number of permutations of n things taken r at a time

The notation ${}_nP_r$ says that we want to fill r places or slots from a total group of n things. For example ${}_5P_2$ is a permutation of five things taken two at a time, which is $5 \times 4 = 20$. Similarly, ${}_6P_4$ is a permutation of six things taken four at a time, and

$${}_6P_4 = 6 \times 5 \times 4 \times 3 = 360$$

If we have ${}_nP_r$, a permutation of n things taken r at a time, then we have r slots and must fill them from n things. The first slot may be filled in n different ways, the second slot may be filled in $n - 1$ different ways, the third slot may be filled in $n - 2$ different ways, and so on. The computation of the number of permutations of n things taken r at a time is as follows, along with two examples that we discussed previously:

$${}_5P_2 = 5 \times 4 = 20$$

$${}_6P_4 = 6 \times 5 \times 4 \times 3 = 360$$

$${}_nP_r = \underbrace{\overbrace{n}^{\textit{1st slot}} \times \overbrace{(n-1)}^{\textit{2nd slot}} \times \overbrace{(n-2)}^{\textit{3rd slot}} \times \overbrace{(n-3)}^{\textit{4th slot}} \times \cdots \times \overbrace{(n-r+1)}^{\textit{rth (last) slot}}}_{\textit{Number of ways each slot may be filled}}$$

This leads to the formula

$${}_nP_r = n \times (n-1) \times (n-2) \times (n-3) \times \cdots \times (n-r+1)$$

or

$${}_nP_r = n(n-1)(n-2)(n-3)(n-4) \times \cdots \times (n-r+1)$$

From this formula we now know that

$$\underset{n \text{ things} \quad r \text{ slots}}{{}_8P_3} = \underset{n}{8} \times \underset{(n-1)}{(8-1)} \times \underset{(n-r+1)}{(8-3+1)} = 8 \times 7 \times 6$$

What happens when $n = r$? If we examine ${}_nP_r$ when $n = r$, we have

$${}_nP_r = n(n-1)(n-2)\cdots(n-r+1)$$

and the factor $(n - r + 1)$ becomes $(n - n + 1)$, or just 1. Thus when $n = r$, we have ${}_nP_n = n(n-1)(n-2)(n-3)\cdots(3)(2)(1)$. The product of all the numbers from n down to and including 1 is called ***n* factorial**. The symbol for n factorial is $n!$. A table of factorials (Table 1) appears in the Appendix. Some examples are

$$(n \text{ factorial})\ n! = n(n-1)(n-2)(n-3)\cdots(3)(2)(1)$$

$$(3 \text{ factorial})\ 3! = 3 \times 2 \times 1 \quad \text{or} \quad (3)(2)(1) \quad \text{or} \quad 3\cdot2\cdot1 = 6$$

$$(5 \text{ factorial})\ 5! = 5 \times 4 \times 3 \times 2 \times 1 \quad \text{or} \quad (5)(4)(3)(2)(1)$$

$$\text{or} \quad 5\cdot4\cdot3\cdot2\cdot1 = 120$$

Zero factorial, 0!, cannot be defined by the rule used in the preceding examples. We agree to let 0! equal 1 as a matter of convenience. This special definition enables us to produce a simpler formula for ${}_nP_r$.

Examine the following examples:

$${}_6P_2 = 6\cdot5 = 30 \quad \text{or} \quad {}_6P_2 = \frac{6!}{4!} = \frac{6\cdot5\cdot4\cdot3\cdot2\cdot1}{4\cdot3\cdot2\cdot1} = \frac{720}{24} = 30$$

$${}_5P_3 = 5\cdot4\cdot3 = 60 \quad \text{or} \quad {}_5P_3 = \frac{5!}{2!} = \frac{5\cdot4\cdot3\cdot2\cdot1}{2\cdot1} = \frac{120}{2} = 60$$

In general, we have

$${}_nP_r = \frac{n!}{(n-r)!} \quad (r \le n)$$

For example,

$${}_5P_3 = \frac{5!}{(5-3)!} = \frac{5!}{2!} = \frac{5\cdot4\cdot3\cdot2\cdot1}{2\cdot1} = 60$$

$${}_6P_2 = \frac{6!}{(6-2)!} = \frac{6!}{4!} = \frac{6\cdot5\cdot4\cdot3\cdot2\cdot1}{4\cdot3\cdot2\cdot1} = 30$$

$${}_4P_4 = \frac{4!}{(4-4)!} = \frac{4!}{0!} = \frac{4\cdot3\cdot2\cdot1}{1} = 24$$

Note that 0! appears in the denominator of the last example, ${}_4P_4$. In order for our formula to work for this case, we must define $0! = 1$.

EXAMPLE 5 ***Evaluating Factorials*** Evaluate each of the following.

a. 4! **b.** 8!

c. ${}_7P_4$ **d.** ${}_8P_3$

e. ${}_{10}P_{10}$ **f.** ${}_5P_0$

Solution

a. $4! = 4 \cdot 3 \cdot 2 \cdot 1 = 24$

b. $8! = 8 \cdot 7 \cdot 6 \cdot 5 \cdot 4 \cdot 3 \cdot 2 \cdot 1 = 40{,}320$

c. $${}_7P_4 = \frac{7!}{(7-4)!} = \frac{7!}{3!} = \frac{7 \cdot 6 \cdot 5 \cdot 4 \cdot \not{3} \cdot \not{2} \cdot 1}{\not{3} \cdot \not{2} \cdot 1} = 840$$

d. $${}_8P_3 = \frac{8!}{(8-3)!} = \frac{8!}{5!} = \frac{8 \cdot 7 \cdot 6 \cdot \not{5} \cdot \not{4} \cdot \not{3} \cdot \not{2} \cdot 1}{\not{5} \cdot \not{4} \cdot \not{3} \cdot \not{2} \cdot 1} = 336$$

e. $${}_{10}P_{10} = \frac{10!}{(10-10)!} = \frac{10 \cdot 9 \cdot 8 \cdot 7 \cdot 6 \cdot 5 \cdot 4 \cdot 3 \cdot 2 \cdot 1}{1} = 3{,}628{,}800$$

f. $${}_5P_0 = \frac{5!}{(5-0)!} = \frac{5!}{5!} = \frac{\not{5} \cdot \not{4} \cdot \not{3} \cdot \not{2} \cdot 1}{\not{5} \cdot \not{4} \cdot \not{3} \cdot \not{2} \cdot 1} = 1$$

NW ***Now Work Problem 15.***

EXAMPLE 6 ***Selecting A Seat*** If seven people board an airplane, and there are nine aisle seats, in how many ways can they be seated if they all choose aisle seats?

Solution This is a permutation of nine things taken seven at a time. So we have

$$\begin{aligned} {}_9P_7 &= \frac{9!}{(9-7)!} \\ &= \frac{9!}{2!} \\ &= \frac{9 \cdot 8 \cdot 7 \cdot 6 \cdot 5 \cdot 4 \cdot 3 \cdot 2 \cdot 1}{2 \cdot 1} \\ &= 181{,}440 \end{aligned}$$

EXAMPLE 7 ***Arranging Music*** A disc jockey can play eight records in a 30-minute segment of her show. For a particular 30-minute segment, she has 12 records to select from. In how many ways can she arrange her program for the particular segment?

Solution This is a permutation of 12 things taken 8 at a time. Therefore, we have

$$\begin{aligned} {}_{12}P_8 &= \frac{12!}{(12-8)!} = \frac{12!}{4!} \\ &= \frac{12 \cdot 11 \cdot 10 \cdot 9 \cdot 8 \cdot 7 \cdot 6 \cdot 5 \cdot \not{4} \cdot \not{3} \cdot \not{2} \cdot 1}{\not{4} \cdot \not{3} \cdot \not{2} \cdot 1} \\ &= 19{,}958{,}400 \end{aligned}$$

NW ***Now Work Problem 19.***

We now know that ${}_3P_3$ is equal to 6. This means that there are six distinct arrangements of three things taken three at a time. Let's now look at a slightly different problem. Consider the word ALL and the number of arrangements that can be made from the letters in it. There are three letters, and therefore the number of arrangements is 3!. Let's call the first L in the word L_1 ("L sub one") and the second L in the word L_2 ("L sub two"). Now we can distinguish between the two L's. The arrangements of the three letters are

1. AL_1L_2	2. AL_2L_1
3. L_1AL_2	4. L_2AL_1
5. L_1L_2A	6. L_2L_1A

If we use the subscript notation as before, we can see the different arrangements. When we interchange the L's, we technically have a different arrangement. But without the subscripts, it is impossible to tell if the L's have been moved, and we have

1. $A\ L\ L$	2. $A\ L\ L$
2. $L\ A\ L$	4. $L\ A\ L$
3. $L\ L\ A$	6. $L\ L\ A$

Since we cannot tell if the L's have been interchanged, the arrangements are not all distinct. Arrangements 1 and 2 are the same; so are 3 and 4, and 5 and 6. Hence, there are only three distinct arrangements, namely, ALL, LAL, and LLA. The two L's can be arranged in 2! ways. In order to obtain the number of distinct arrangements, we must divide ${}_3P_3 = 3!$ by 2!. That is, there are

$$\frac{3!}{2!} = \frac{3 \cdot 2 \cdot 1}{2 \cdot 1} = 3$$

distinct arrangements from the letters in ALL.

If we want to find the number of **distinct arrangements of** n things when p of the things are alike, then we divide $n!$ by $p!$.

EXAMPLE 8 ***Distinct Arrangements*** How many distinct arrangements can be formed from all the letters of *GENESEE?*

Solution There are seven letters. If they were all different, we would have 7! arrangements, but there are 4 E's, which can be arranged in $4! = 24$ ways. So we divide 7! by 4! and the answer is

$$\frac{7!}{4!} = \frac{7 \cdot 6 \cdot 5 \cdot 4 \cdot 3 \cdot 2 \cdot 1}{4 \cdot 3 \cdot 2 \cdot 1} = 210$$

Suppose, in addition to p things of the same kind, we also have q things alike, and r things alike, and so on. If this occurs, then we just extend our computation to

$$\frac{n!}{p!q!\cdots}$$

the number of distinct arrangements of n things when p things are alike, q things are alike, and so on.

EXAMPLE 9 ***Distinct Arrangements*** In how many distinct ways can the letters of *OSMOSIS* be arranged?

Solution There are seven letters in the word *OSMOSIS*, and if they were all different we would have 7! arrangements. Since there are 2 O's, we must divide by 2!, and since there are 3 S's, we must divide by 3!. Hence, we have

$$\frac{7!}{2!3!} = \frac{7 \cdot 6 \cdot 5 \cdot 4 \cdot 3 \cdot 2 \cdot 1}{2 \cdot 1 \cdot 3 \cdot 2 \cdot 1} = 420$$

NW ***Now Work Problem 23a.***

Exercises for Section 4.7

1. A club of 20 people is going to elect a chairperson and a secretary. In how many different ways can this be done?

2. Mike has 8 sweatshirts and 6 pairs of pants. How many outfits does he have to choose from?

NW 3. A primary zip code is a five-digit numeral. How many of these zip codes can be formed if no digit can be repeated? How many of these zip codes can be formed if repetition is allowed?

4. In Riverhead there are 5 roads leading to a traffic circle. In how many ways can a driver enter the traffic circle by one road and leave by another?

NW 5. Given the set of digits $\{4, 5, 6, 7, 8, 9\}$, how many three-digit numbers can be formed if no digit can be repeated? How many three-digit numbers can be formed if repetition is allowed?

6. If there are 50 contestants in a beauty pageant, in how many ways can the judges award first and second prizes?

7. The Southampton Sports Car Club has 30 members. A slate of officers consists of a president, a vice president, a secretary, and a treasurer. If a person can hold only one office, in how many ways can a set of officers be formed?

8. A baseball manager has eight pitchers and three catchers on his squad. In how many ways can the manager select a starting battery (pitcher and catcher) for a game?

9. A conference room has four doors. In how many ways can a person enter and leave the conference room by a different door?

10. Given the set of digits $\{5, 6, 7, 8, 9\}$, how many four-digit numbers can be formed if no digit can be repeated? How many of these will be odd? How many of these will be divisible by 5? How many of these will be over 6,000? How many will be over 5,000?

11. How many license plates can be made if each plate must consist of two letters followed by three digits if we assume no repetition? How many are possible if the first digit cannot be zero? How many are possible if the first letter cannot be O and the first digit cannot be zero?

12. How many four-letter words can be formed from the set of letters $\{m, o, n, e, y\}$? Assume that any arrangement of letters is a word. How many four-letter words can be formed if the first letter must be y and the last letter must be m? (Assume no repetition.)

13. The Rochester Tennis Club is having a mixed-doubles tournament. If eight women and their husbands sign up for the tournament, how many mixed-doubles teams are possible? How many can be formed if no woman is paired with her husband?

14. At Finger Lakes Race Track, there are eight horses in each race. The daily double consists of picking the winning horses in the first and second races. If a bettor wanted to purchase all possible daily double tickets, how many would he have to purchase?

15. Evaluate the following:
   - NW **a.** $3!$ NW **b.** $5!$ NW **c.** $1!$
   - NW **d.** $\frac{7!}{4!}$ NW **e.** ${}_5P_2$ NW **f.** ${}_4P_4$

16. Evaluate the following:
   - **a.** $4!$ **b.** $6!$ **c.** $0!$
   - **d.** $\frac{100!}{99!}$ **e.** ${}_{50}P_2$ **f.** ${}_6P_0$

17. On a naval vessel, a signal can be formed by running up two flags on a flag pole, one above the other. If there are 10 different flags to choose from, how many different signals can be formed?

18. How many different ways can seven students be seated in seven seats on a subway car?

19. NW A baseball manager must hand in the lineup card before the game begins. The manager has decided which nine players will play, but not the batting order. How many different batting orders could be chosen?

20. Given the set of digits $\{3, 4, 5, 6, 7, 8, 9\}$, how many different numbers between 3,000 and 6,000 can be written using these digits if repetition of digits is allowed?

21. Working with a squad of 12 players, a basketball coach selects a guard and then a center. How many different ways can a coach do this?

22. A dictionary, an almanac, a catalog, and a diary are to be placed on a shelf. In how many ways can they be arranged?

23. In how many distinct ways can the letters of each word be arranged?
   - NW **a.** *ALGEBRA* **b.** *STATISTICS*
   - **c.** *CALCULUS* **d.** *SCIENCE*

24. In how many distinct ways can the letters of each word be arranged?
   - **a.** *DALLAS* **b.** *TORONTO*
   - **c.** *ATLANTA* **d.** *CINCINNATI*

25. Given the set of digits $\{1, 2, 3, 4, 5\}$, how many different numbers consisting of four digits can be formed when repetition is allowed?

26. A traveling book salesperson has five copies of a certain statistics book, four copies of a certain geometry book, and three copies of a certain calculus book. If these books are to be stored on a shelf in the salesperson's van, how many distinct arrangements are possible?

27. Telephone numbers consist of seven digits: three digits for the exchange, followed by four more digits. In order to call long distance, you must also use an area code, which consists of three more digits. How many long-distance telephone numbers are there if the first digit cannot be 0 or 1, and the fourth digit cannot be 0 or 1? (Assume repetition of digits.)

28. In the Olympics there are typically three awards for an event: a gold, a silver, and a bronze medal. How many ways exist for awarding the medals, if there are eleven entrants for an event?

Writing Mathematics

29. What is a permutation?

30. Explain the concept of *distinct arrangements*. Give an example.

31. What is $n!$? What does it mean? How do you evaluate $n!$? Without the use of a calculator, evaluate $\frac{100!}{98!}$.

Challenge Exercises

32. There are eight children in Mrs. Setek's pre-K class, five girls and three boys. The students are going to pose for a class picture. In how many ways can they line up for the photograph if all the boys want to stand together and all the girls want to stand together?

33. The State License Plate Company manufactures a certain type of license plate consisting of two letters followed by either two or three or four digits. How many different license plates of this type are possible? (Assume repetition.)

34. A librarian has twelve books to place on a shelf. Five are history books, three are mathematics books, and four are Spanish books. How many different shelf arrangements can be made if all books of the same subject are to stay together?

35. A computer password consists of five letters (A through Z) followed by a single digit. Assume that passwords are case sensitive (uppercase letters are different from lowercase letters and no repetition is permitted).

    a. How many different passwords are possible?

    b. How many different passwords begin with G?

    c. How many different passwords do not begin with either g or G?

Group/Research Activities

36. It is true that $0! = 1$ and $1! = 1$. Research why this is so and make a presentation to the class explaining why this is the case.

37. The increasing use of pagers, cellular phones, fax machines, and computers (World Wide Web) has affected the number of telephone numbers available. In fact, in certain area codes the supply of phone numbers has almost been depleted. Research the topic of phone numbers in this country. Include area codes, toll-free numbers, and the fact that some phone companies are developing phone numbers that use more than seven digits.

PEANUTS reprinted by permission of United Feature Syndicate, Inc.

Just for fun

Name the southernmost state in the United States.

4.8 COMBINATIONS (OPTIONAL)

In Section 4.7, we were concerned with the counting of ordered arrangements, or permutations. The techniques developed can be applied to situations such as the following: If a tennis squad has five players on it, the coach can select the first singles player in five ways and the second singles player in four ways. Hence, there are $5 \times 4 = 20$ different ways that the coach can select the first and second singles players. This type of choice is a permutation of five things taken two at a time.

Now, let's consider a slightly different type of choice. Suppose the five members of the squad are Stan, Bill, Mike, Alice, and Pam, and we want to know in how many ways the coach can select a doubles team from this group of five players. This is different because when the coach selects a doubles team, it does not matter which person is chosen first and which person is chosen second. That is, the order of choice is not important. The 20 different ordered arrangements are

Stan, Bill	Bill, Stan	Mike, Stan	Alice, Stan	Pam, Stan
Stan, Mike	Bill, Mike	Mike, Bill	Alice, Bill	Pam, Bill
Stan, Alice	Bill, Alice	Mike, Alice	Alice, Mike	Pam, Mike
Stan, Pam	Bill, Pam	Mike, Pam	Alice, Pam	Pam, Alice

Note that the doubles team of Stan and Bill is the same doubles team as that of Bill and Stan. This is also the case for the teams of Alice and Mike and of Mike and Alice. If we eliminate the duplicate doubles teams, we have

Stan, Bill	Bill, Mike
Stan, Mike	Bill, Alice
Stan, Alice	Bill, Pam
Stan, Pam	Mike, Alice
Alice, Pam	Mike, Pam

There are 10 different doubles teams that the coach can select. The 20 distinct ordered arrangements of two people are reduced to 10 groups of 2 people when we disregard order.

The problem we have just discussed can be classified as a *combination* problem. A **combination** is a distinct group of objects without regard to their arrangement. Committees are good examples of combinations. We are concerned only with who is on a committee, not with who is first, who is second, and so on.

Biography: John Von Neumann

John Von Neumann was born in Budapest, Hungary, in 1903. By the time he was 6 years old he was able to perform amazing mental calculations. For example, he was able to divide an eight-digit number by another eight-digit number in his head, such as 87,251,465 ÷ 38,376,325. By the time he was 8 years old, he had mastered calculus. Another thing he did to amaze people was to memorize on sight, the names, addresses, and telephone numbers in a column of a telephone book. At the age of 23 he authored a book, *Mathematical Foundations of Quantum Mechanics,* that was used in developing atomic energy.

He became a professor of mathematical physics at Princeton University in 1930. During his tenure at Princeton he became interested in computers and constructed one of the first modern electronic brains, called MANIAC (Mathematical Analyzer, Numerical Integrator and Computer). During World War II, he served as an advisor to the U.S. government and had a major influence in the design of nuclear weapons and missiles. In 1952, he completed work on the EDVAC (Electronic Discrete Variable Automatic Computer), which was the first computer developed in the United States that could store both program instructions and data.

John Von Neumann had many interests, but he derived the greatest satisfaction from solving problems. He was fascinated by games of chance and strategy. As a result of this work he helped develop a new field of mathematics called *game theory*. By using the probabilities associated with certain games of chance and the strategies that yield winners in decision-making games, *game theory* can be used to solve problems in science, economics, and military strategy. John Von Neumann died in 1957 after making many amazing contributions to the field of mathematics.

Consider the set of letters $\{x, y, z\}$. How many different three-letter arrangements can be formed from this set? The answer is 3!, or 6, and they are

$x, y, z \qquad y, z, x \qquad z, x, y$

$x, z, y \qquad y, x, z \qquad z, y, x$

How many three-letter combinations (distinct groups) can be formed from this set? We see that all six arrangements constitute the same group of elements. Therefore, there is only one combination. We can say that the number of combinations of three things taken three at a time is 1. Symbolically, we can write

$$_3C_3 = 1$$

In the doubles problem, we had combinations of five things taken two at a time. The number of these combinations was equal to 10; hence,

$$_5C_2 = 10$$

Note that

$$_5C_2 = \frac{_5P_2}{2!} = \frac{5!}{(5-2)!2!} = \frac{5!}{3!2!} = \frac{5 \cdot 4 \cdot \not{3} \cdot \not{2} \cdot 1}{\not{3} \cdot 2 \cdot 1 \cdot \not{2} \cdot 1} = 10$$

and

$$_3C_3 = \frac{_3P_3}{3!} = \frac{3!}{(3-3)!3!} = \frac{3!}{0!3!} = 1$$

In order to obtain the number of distinct combinations, we must eliminate the different ordered arrangements within each group, and we do this by division. From the set of five tennis players, we have five choices for the first selection and four for the second, but those two selections can order themselves in 2! ways, so we must divide by 2!.

The general formula for the number of combinations of n things taken r at a time is

$$_nC_r = \frac{n!}{(n-r)!r!}$$

Recall the formula for the number of permutations of n things taken r at a time:

$$_nP_r = \frac{n!}{(n-r)!}$$

If you know this formula, then the formula for $_nC_r$ is easy to remember because it is similar to $_nP_r$. If

$$_nP_r = \frac{n!}{(n-r)!} \quad \text{and} \quad _nC_r = \frac{n!}{(n-r)!r!}$$

then

$$_nP_r = {_nC_r} \cdot r! \quad \text{or} \quad _nC_r = \frac{_nP_r}{r!} = \frac{n!}{(n-r)!r!}$$

You should be thoroughly familiar with both formulas:

$$_nP_r = \frac{n!}{(n-r)!}$$ *the number of permutations of n things taken r at a time*

$$_nC_r = \frac{n!}{(n-r)!r!}$$ *the number of combinations of n things taken r at a time*

Note: Some books use the notation $\binom{n}{r}$ to represent the number of combinations of n things taken r at a time. We shall not use this notation, but be aware that $\binom{n}{r} = n!/(n-r)!r!$

EXAMPLE 1 ***Selecting Co-Captains of a Basketball Team*** Two co-captains are to be selected from the starting five for a basketball team. In how many ways can this be done?

Solution This is a combination problem since order is not important. We have a combination of five things taken two at a time. Therefore, we have

$$_5C_2 = \frac{5!}{(5-2)!2!} = \frac{5!}{3!2!} = \frac{5 \cdot 4 \cdot \cancel{3} \cdot \cancel{2} \cdot 1}{\cancel{3} \cdot 2 \cdot 1 \cdot \cancel{2} \cdot 1} = 10$$

EXAMPLE 2 ***Selecting Candidates to Serve on a Council*** The student association each year selects a council consisting of seven members. If there are 10 candidates for the seven-member council, how many different councils may be elected?

Solution Since order is not important, we treat this as a combination of 10 things taken 7 at a time. Therefore, we have

$$_{10}C_7 = \frac{10!}{(10-7)!7!} = \frac{10!}{3!7!} = \frac{10 \cdot 9 \cdot 8 \cdot \cancel{7} \cdot \cancel{6} \cdot \cancel{5} \cdot \cancel{4} \cdot \cancel{3} \cdot \cancel{2} \cdot 1}{3 \cdot 2 \cdot 1 \cdot \cancel{7} \cdot \cancel{6} \cdot \cancel{5} \cdot \cancel{4} \cdot \cancel{3} \cdot \cancel{2} \cdot 1} = 120$$

NW ***Now Work Problem 3.***

EXAMPLE 3 ***Number of Different Poker Hands*** How many different poker hands can be dealt from a standard deck of 52 cards? (Here we assume that a poker hand consists of five cards.)

Solution Order is not important, since a poker hand consisting of king of hearts, king of clubs, queen of clubs, jack of hearts, and king of spades is the same as one consisting of king of hearts, queen of clubs, king of spades, jack of hearts, and king of clubs. We have a combination of 52 things taken 5 at a time; that is,

$$_{52}C_5 = \frac{52!}{(52-5)!5!} = \frac{52!}{47!5!} = \frac{52!}{5!47!}$$
$$= \frac{52 \cdot 51 \cdot 50 \cdot 49 \cdot 48 \cdot \cancel{47!}}{5 \cdot 4 \cdot 3 \cdot 2 \cdot 1 \cdot \cancel{47!}} = 2{,}598{,}960$$

EXAMPLE 4 ***Selecting Committee Members*** How many committees can be selected from four teachers and 100 students if each committee must have two teachers and three students?

Solution This combination problem is somewhat different from the others we have considered. Here we must choose two teachers from four and three students from 100 to form our committee. Hence, we must use the counting principle. If we can choose the teachers in t ways and the students in s ways, then together the choosing can be done in $t \times s$ ways. Therefore, the number of committees consisting of two teachers and three students may be found by multiplying $_4C_2$ (four teachers taken two at a time) by $_{100}C_3$ (100 students taken three at a time):

$$_4C_2 = \frac{4!}{(4-2)!2!} = \frac{4!}{2!2!} = \frac{4 \cdot 3 \cdot \cancel{2} \cdot 1}{2 \cdot 1 \cdot \cancel{2} \cdot 1} = 6$$

$$_{100}C_3 = \frac{100!}{(100-3)!3!} = \frac{100!}{97!3!} = \frac{100 \cdot 99 \cdot 98 \cdot \cancel{97!}}{3 \cdot 2 \cdot 1 \cdot \cancel{97!}} = 161{,}700$$

Therefore,

$$_4C_2 \cdot {}_{100}C_3 = 6 \cdot 161{,}700 = 970{,}200$$

In doing problems of this nature, we must carefully examine the problem to determine whether we are doing a permutation problem or a combination problem. If we want distinct arrangements, then we are doing permutations. If the distinct arrangements are not to be counted, only the different groups, then we are doing combinations. Always be careful of special conditions in the problem. Try to take care of these first; then proceed with the rest of the problem.

NW ***Now Work Problem 17.***

Exercises for Section 4.8

1. Evaluate the following:
   a. $_5C_3$ b. $_5C_2$ c. $_7C_4$
   d. $_7C_3$ e. $_{10}C_{10}$ f. $_{10}C_0$
2. Evaluate the following:
   a. $_7C_5$ b. $_7C_2$ c. $_{50}C_2$
   d. $_{51}C_{50}$ e. $_{52}C_{48}$ f. $_{51}C_2$
3. (NW) If the Xerox Corporation has to transfer four of its ten junior executives to a new location, in how many ways can the four executives be chosen?
4. A newspaper boy discovers in delivering his papers that he is three papers short. He has eight houses left to deliver to, but only five papers left. In how many ways can he deliver the remaining newspapers?
5. Alice has a penny, a nickel, a dime, a quarter, and a half-dollar. She may spend any three coins. In how many ways can Alice do this? What is the most money she can spend using just three coins?
6. Joe has to take a math exam that consists of ten questions. He must answer only seven of the ten questions. In how many ways can Joe choose the seven questions? If he must answer the first and last questions and still answer a total of only seven, in how many ways can he do this?
7. In a mathematics class of 15 students, 10 students must do problems at the board on a given day. In how many ways can the 10 students be chosen?
8. A football coach has 40 candidates out for the squad. In how many ways can a starting 11 be selected without regard to the position that a candidate will play? (Indicate your answer; do not evaluate.)
9. At registration, a student needs two more courses to complete her schedule. If there are seven possible courses left to pick from, in how many ways can she choose the two courses?
10. A committee of 11 people, six women and five men, is forming a subcommittee that is to be made up of two women and three men. In how many ways can the subcommittee be formed?

11. A baseball squad consists of eight outfielders and seven infielders. If the baseball coach must choose three outfielders and four infielders, in how many ways can this be done?
12. An urn contains six blue balls and four orange balls.
    a. In how many ways can we select a group of three balls?
    b. In how many ways can we select two blue balls and one orange ball?
    c. In how many ways can we select two orange balls and one blue ball?
13. From a group of 12 sprinters and 10 distance runners, a medley relay team is to be formed. The relay team must consist of two sprinters and two distance runners. How many possible medley relay teams are there?
14. Don has to take a history exam that consists of 15 multiple-choice questions and five essay questions. If Don has to answer 10 multiple-choice questions and two essay questions, in how many ways can he choose them?
15. A student belongs to a music club. This month she has to purchase two CDs and three tapes. If there are 10 CDs and 10 tapes to choose from, in how many ways can she choose her purchases?
16. If six points are drawn on a plane, no three of which are on the same straight line, how many straight lines can be formed? (Two points determine a line.)
17. (NW) How many different committees, each composed of two Democrats, two Republicans, two Liberals, and one Conservative, can be formed from 12 Democrats, 11 Republicans, five Liberals, and three Conservatives?

18. The Speaker of the House wants to appoint a committee consisting of three representatives from New York, three representatives from California, two representatives from Ohio, and three representatives from Illinois. How many different committees can be formed if eight representatives from New York, ten representatives from California, five representatives from Ohio, and six representatives from Illinois are eligible?

Writing Mathematics

19. What is a combination?

20. Explain how to distinguish between a combination and a permutation. Give at least one example of each.

21. Give the formula for ${}_nC_r$. What does it mean?

22. Write a word problem that can be solved by evaluating ${}_{49}C_6$.

Challenge Exercises

23. A football team has 35 members. The coach must choose 11 members of the team to play offense and a different 11 members to play defense. How many choices are possible? (Assume that each player can play either offense or defense.) Indicate your answer. Do not evaluate.

24. From a group of nine people, how many committees made up of either three or four people may be formed?

25. A board of directors consists of 10 members. In how many ways can a committee of the board of directors be selected? Solve this problem using combinations and then see if you can develop an alternative formula that can be used to get the same results.

Group/Research Activities

26. A typical **combination lock** has a "combination" that consists of dialing a three-number sequence and each number can be selected from $0, 1, 2, 3, \ldots, 39$. Prepare a report explaining why this type of lock is actually a **permutation lock**. Be sure to include in your report whether repetition of numbers is permitted and how many such "combinations" exist.

27. Research at least three different state lotteries. Compare the chances of winning the grand prize in each lottery. Compare the number of combinations possible for each different lottery. For example, in one state a player must choose 6 of 50 numbers, while in another state it may be 6 of 52 numbers. Be sure to compute the number of combinations for each state considered.

Just for fun

A card and an envelope together cost 99 cents. The card costs 90 cents more than the envelope. How much does the envelope cost?

4.9 MORE PROBABILITY (OPTIONAL)

A poker hand consisting of the ten, jack, queen, king, and ace of any one suit is called a royal flush. The probability of being dealt a royal flush is

$$\frac{4}{{}_{52}C_5} \text{ or } \frac{4}{2{,}598{,}960} = \frac{1}{649{,}740}$$

We can utilize the counting principle, permutations, and combinations in solving many probability problems. Some of them are similar to those that we discussed previously, whereas others are a little more involved.

Consider the discussion at the beginning of Section 4.6. We want to find the probability of being dealt two red cards in succession, without replacement, from a shuffled standard deck of 52 cards. We found that the probability was

$$\frac{26}{52} \cdot \frac{25}{51} = \frac{650}{2{,}652} = \frac{25}{102}$$

There are 26 red cards out of 52 for the first card, and then 25 red cards out of 51 remaining cards for the second card. We multiplied the two probabilities together to find the probability of getting a red card *and* a red card.

Another way to solve this problem is by combinations. We are not concerned with the order in which the cards appear, just as long as they are red. In how many ways can two cards be chosen from a deck of 52? We have

$${}_{52}C_2 = \frac{52!}{(52-2)!2!} = \frac{52!}{50!2!} = \frac{52 \cdot 51 \cdot \cancel{50!}}{\cancel{50!} \cdot 2 \cdot 1} = \frac{52 \cdot 51}{2 \cdot 1} = 1{,}326$$

In how many ways can two red cards be chosen? This is the number of successful outcomes. There are 26 red cards in the deck, and a successful outcome is any combination of two of these cards. Therefore, we compute

$$_{26}C_2 = \frac{26!}{(26-2)!2!} = \frac{26!}{24!2!} = \frac{26 \cdot 25 \cdot \cancel{24!}}{\cancel{24!} \cdot 2 \cdot 1} = \frac{26 \cdot 25}{2 \cdot 1} = 325$$

Hence, the probability of being dealt two red cards in succession is

$$\frac{_{26}C_2}{_{52}C_2} = \frac{325}{1{,}326} = \frac{25}{102}$$

Let us look at some other examples.

EXAMPLE 1 ***Selecting Courses*** A student has to complete registration for the next semester by choosing three more courses. The courses left to choose from are five humanities courses and four science courses. If the three courses are chosen at random, what is the probability that they will all be humanities courses?

Solution There are nine courses to choose from, and the student must choose three of them. There are $_9C_3$ ways of choosing three courses. The three humanities courses may be chosen from the five offered in $_5C_3$ ways. Hence, we have

$$_5C_3 = \frac{5!}{(5-3)!3!} = \frac{5!}{2!3!} = \frac{5 \cdot 4 \cdot \cancel{3!}}{2 \cdot \cancel{3!}} = 10$$

$$_9C_3 = \frac{9!}{(9-3)!3!} = \frac{9!}{6!3!} = \frac{9 \cdot 8 \cdot 7 \cdot \cancel{6!}}{\cancel{6!} \cdot 3 \cdot 2 \cdot 1} = 84$$

$$P(3 \text{ humanities}) = \frac{_5C_3}{_9C_3} = \frac{10}{84}$$

EXAMPLE 2 ***Getting a Flush in Poker*** Find the probability of being dealt a hand in five-card poker that is all spades. (A *flush* is a hand where all the cards are of the same suit—all hearts, all spades, all diamonds, or all clubs.)

Solution There are 52 cards in a deck, and there are $_{52}C_5$ possible ways of being dealt five cards. There are 13 spades in a deck, and a successful outcome is being dealt any five of these spades. There are $_{13}C_5$ ways of being dealt five spades. Therefore,

$$_{13}C_5 = \frac{13!}{(13-5)!5!} = \frac{13!}{8!5!} = \frac{13 \cdot 12 \cdot 11 \cdot 10 \cdot 9 \cdot \cancel{8!}}{\cancel{8!} \cdot 5 \cdot 4 \cdot 3 \cdot 2 \cdot 1} = 1{,}287$$

$$_{52}C_5 = \frac{52!}{(52-5)!5!} = \frac{52!}{47!5!}$$

$$= \frac{52 \cdot 51 \cdot 50 \cdot 49 \cdot 48 \cdot \cancel{47!}}{\cancel{47!} \cdot 5 \cdot 4 \cdot 3 \cdot 2 \cdot 1} = 2{,}598{,}960$$

$$P(5 \text{ spades}) = \frac{_{13}C_5}{_{52}C_5} = \frac{1{,}287}{2{,}598{,}960} = \frac{33}{66{,}640}$$

NW ***Now Work Problem 3.***

Note of Interest

The odds against dealing a bridge hand consisting of 13 cards of one suit are 158,753,389,899 to 1. The odds against each of the four players receiving a complete suit (a perfect deal) are 2,235,197,406,895,366,368,301,559,999 to 1.

EXAMPLE 3 ***Getting a Full House in Poker*** Find the probability of being dealt a *full house* (three cards of one denomination and two of another) consisting of kings over deuces (three kings and two deuces).

Solution There are 52 cards in a deck and there are ${}_{52}C_5$ possible ways of being dealt five cards. A successful outcome consists of getting three kings and two deuces. There are four of each kind of card in a deck and so the number of ways of getting three kings is ${}_4C_3$. The number of ways of getting two deuces is ${}_4C_2$. Therefore, the total number of ways of obtaining three kings and two deuces is ${}_4C_3 \cdot {}_4C_2$. So we have

$$
\begin{aligned}
{}_4C_2 &= \frac{4!}{(4-2)!2!} = \frac{4!}{2!2!} = \frac{4\cdot 3\cdot \cancel{2}\cdot 1}{2\cdot 1\cdot \cancel{2}\cdot 1} = 6 \\
{}_4C_3 &= \frac{4!}{(4-3)!3!} = \frac{4!}{1!3!} = \frac{4\cdot \cancel{3}\cdot \cancel{2}\cdot 1}{1\cdot \cancel{3}\cdot \cancel{2}\cdot 1} = 4 \\
{}_{52}C_5 &= 2{,}598{,}960 \text{ (see Example 2)}
\end{aligned}
$$

and the probability of obtaining a full house consisting of three kings and two deuces is

$$P(3 \text{ kings and 2 twos}) = \frac{{}_4C_3 \cdot {}_4C_2}{{}_{52}C_5} = \frac{24}{2{,}598{,}960} = \frac{1}{108{,}290}$$

NW ***Now Work Problem 9.***

Table 4.5 provides some interesting information about five-card poker. It shows the number of various kinds of hands possible when a 52-card deck is used and nothing is wild. Note that the total number of different hands is 2,598,960, which is ${}_{52}C_5$.

TABLE 4.5

Straight flush	40
Four of a kind	624
Full house	3,744
Flush	5,108
Straight	10,200
Three of a kind	54,912
Two pairs	123,552
One pair	1,098,240
No pair	1,302,540
Total	2,598,960

Exercises for Section 4.9

1. Find the probability of being dealt two queens when you are dealt two cards from a shuffled deck of 52 cards.
2. Find the probability of being dealt three kings when you are dealt three cards from a shuffled deck of 52 cards.
3. (NW) You are dealt three cards from a shuffled deck of 52 cards. Find the probability that all three cards are hearts.
4. You are dealt five cards from a shuffled deck of 52 cards. Find the probability that all five cards are picture cards (king, queen, or jack). What is the probability that none of the cards is a picture card? (Indicate your answers; do not evaluate.)
5. On the track team of the York Athletic Club, there are 8 sprinters and 10 distance runners. A relay team consisting of four people must be chosen at random (without regard to who runs first, second, etc.). Find the probability that the relay team will
   - **a.** consist of sprinters only
   - **b.** consist of distance runners only
   - **c.** consist of two sprinters and two distance runners
   - **d.** consist of three sprinters and one distance runner
   - **e.** contain at least one sprinter
6. You are dealt five cards from a shuffled deck of 52 cards. Find the probability that
   - **a.** all five cards are aces
   - **b.** three cards are aces and two cards are picture cards
   - **c.** four cards are aces and one card is a picture card
7. A five-person committee is to be formed at random from seven Democrats and nine Republicans. Find the probability that the committee will consist of
   - **a.** all Democrats
   - **b.** all Republicans
   - **c.** two Democrats and three Republicans
   - **d.** two Republicans and three Democrats
   - **e.** four Democrats and one Republican
   - **f.** four Republicans and one Democrat

8. A football coach has five guards and seven tackles trying out for the squad. As a final cut, five of these players will be cut at random. Find the probability that the group of players cut will
   a. consist of guards only
   b. consist of tackles only
   c. consist of three guards and two tackles
   d. consist of four tackles and one guard
   e. consist of three tackles and two guards
   f. consist of four guards and one tackle

NW 9. In how many ways can 13 cards (a bridge hand) be selected from a deck of 52 cards? Find the probability of being dealt a bridge hand that consists of all spades. (Indicate your answer, but do not evaluate.)

10. A case of soda pop contains six bottles of root beer, six bottles of orange soda, seven bottles of cola, and five bottles of ginger ale. If you select three bottles at random, find the probability that
    a. all three bottles are cola
    b. all three bottles are root beer
    c. two bottles are ginger ale and one bottle is cola
    d. one bottle is orange soda and two bottles are root beer
    e. one bottle is cola, one bottle is orange soda, and one bottle is ginger ale
    f. one bottle is cola, one bottle is orange soda, and one bottle is root beer

11. A committee consisting of four people is to be selected at random from a group of seven people that includes Bob and Carol and Ted and Alice. Find the probability that the committee will consist of Bob, Carol, Ted, and Alice.

12. A jury of 12 people is to be randomly selected from a group consisting of nine men and eleven women. Find the probability that the jury will
    a. consist of women only
    b. consist of men only
    c. consist of six women and six men
    d. consist of seven women and five men
    e. consist of seven men and five women
    f. contain at least one man

13. An urn contains seven orange balls, five blue balls, and eight red balls. If three balls are drawn at random, indicate (but do not evaluate) the probability of selecting
    a. three orange balls
    b. three blue balls
    c. three red balls
    d. two orange balls and one blue ball
    e. two blue balls and one red ball
    f. two red balls and one orange ball

14. Pocket billiards (pool) is played with 15 balls numbered 1 through 15. The balls with numbers greater than eight are striped, whereas the rest are solid colors. Fast Eddie, a pool hustler, sinks two balls with a single shot. Find the probability that the balls sunk by Fast Eddie are
    a. both striped
    b. both solids
    c. 1 striped and 1 solid
    d. the number 1 ball and the number 15 ball

15. Find the probability of being dealt a flush (five cards of the same suit) in hearts in five-card poker.

16. What is the probability of being dealt a flush in hearts or spades in five-card poker?

17. Find the probability of being dealt two queens and three jacks in five-card poker.

18. Find the probability of being dealt two queens, two kings, and one ace in five-card poker.

19. Find the probability of being dealt a full house (three of a kind, together with two of a kind) in five-card poker.

Writing Mathematics

20. In five-card poker a full house is worth more than a flush. Justify this on the basis of probability.

21. If people understood the mathematics involving probabilities and gambling, do you think that they would spend money on such things? Explain your answer. Be sure to cite examples.

Challenge Exercises

22. A 10-question true–false quiz is given to a class. A score of 10 correct gets an A, a score of 9 correct = B, and a score of 8 correct = C. Mary Lou has not studied for the quiz and guesses at each answer. What is the probability that she will get a C or better?

23. A pinochle deck contains 48 cards. There are four suits, and each suit contains two each of the following: 9, 10, jack, queen, king, ace. Two cards are drawn at random. What is the probability of obtaining a pair?

Group/Research Activity

24. There is an old adage, "lightning never strikes twice in the same place," which some people believe to be true. What do you think? Research the topic of multiple lightning strikes and prepare a report, either supporting this adage or rebutting it, according to your research. Be sure to include research involving the probability of lightning strikes.

Just for fun

Take a standard deck of 52 cards and shuffle them as much as you want. Start with the top card and turn each card over one at a time. Is it a good bet that you will get two cards in a row that are of the same denomination and the same color (for example, 3 of clubs and 3 of spades, 7 of hearts and 7 of diamonds)?

CHAPTER REVIEW MATERIALS

Summary/Chapter 4

4.2 Definition of Probability

- A *sample space* is the set of all possible outcomes to a given experiment.
- An *event* is any subset of a sample space.
- If an experiment has a total of T *equally likely* outcomes, and if exactly S of these outcomes are considered successful—that is, they are the members of the event A—then the probability that event A will occur is

$$P(A) = \frac{\text{Number of successful outcomes}}{\text{Total number of all possible outcomes}} = \frac{S}{T}$$

- If an event, A, can never occur, then $P(A) = 0$; if an event, A, is certain to occur, then $P(A) = 1$.
- *Empirical probability* represent an estimate of the relative frequency of an event occurring based on past observations.
- *Theoretical probability* is an estimate of the relative frequency of an event occurring "in the long run" without actually observing this event occur.

4.3 Sample Spaces/4.4 Tree Diagrams

- When two or more experiments are performed together, the *counting principle* can sometimes be used to determine the total number of possible outcomes in the sample space.
- In general, if one experiment has m different outcomes and a second experiment has n different outcomes, then if performed together, these two experiments have $m \times n$ different outcomes.
- A *tree diagram* may be used to determine a sample space for a particular problem. A tree diagram contains "branches" that illustrate the possible outcomes for the experiments.

4.5 Odds and Expectations

- The *odds in favor of an event* A represent a ratio of the probability that event A will occur and the probability that the event will not occur. That is,

$$\text{Odds in favor of } A = \frac{\text{Probability that } A \text{ will occur}}{\text{Probability that } A \text{ will not occur}}$$

- The *odds against an event* A represent a ratio of the probability that event A will not occur and the probability that the event will occur. That is,

$$\text{Odds against } A = \frac{\text{Probability that } A \text{ will not occur}}{\text{Probability that } A \text{ will occur}}$$

- In addition to being expressed as a fraction, odds can also be expressed in words or as a ratio. For example, if the odds in favor of A are $\frac{3}{2}$, then we may express the odds as "3 to 2" or as 3 : 2.
- If the odds in favor of A are $B{:}C$, then $P(A) = \frac{B}{B+C}$.
- *Mathematical expectation* (also called *expected value*) is an estimate of the average cost of doing something in the long run. When applied to games of chance, mathematical expectation gives us the "fair" price to pay to play a game.
- Mathematical expectation, E, is the product of P, the probability that an event will occur, and M, the amount of money or cost involved. That is, $E = P \times M$.

4.6 Compound Probability

- *Compound probability* involves *compound events*, which comprise two or more simpler events.

- If the occurrence of one event does not affect the occurrence of a second event, then the events are *independent events*. If the occurrence of one event does have an effect on the occurrence of a second event, then the events are *dependent events*.
- To determine the probability of two or more events occurring together, we multiply the various probabilities together. That is, $P(A \text{ and } B) = P(A) \times P(B)$. When calculating this probability, we must assume that event A has occurred before calculating the probability of event B (even when "assuming that event A has occurred" has not been stated).
- If two or more events *cannot* occur at the same time, then they are *mutually exclusive events*.
- To compute the probability of mutually exclusive events, we add their respective probabilities. That is, $P(A \text{ or } B) = P(A) + P(B)$.
- If A and B are *not mutually exclusive* events, then we must subtract redundant instances when we calculate their probabilities. That is, $P(A \text{ or } B) = P(A) \times P(B) - P(A \text{ and } B)$.

4.7 Counting, Ordered Arrangements, and Permutations (Optional)

- *Permutations* are ordered arrangements of objects.
- The number of permutations of n objects taken r at a time is denoted ${}_nP_r$, where ${}_nP_r = \frac{n!}{(n-r)!}$.
- To find the number of *distinct arrangements of n* objects when p of the objects are alike, then we divide $n!$ by $p!$ to remove the redundancy.

4.8 Combinations (Optional)/4.9 More Probability (Optional)

- *Combinations* are distinct groups of objects without regard to their arrangement.
- The number of combinations of n objects taken r at a time is denoted ${}_nC_r$, where ${}_nC_r = \frac{n!}{(n-r)!r!}$.
- Permutations and combinations are useful in solving many probability problems, particularly those that involve a great many possible outcomes, such as the various probabilities for different poker hands.

Key Terms and Concepts

*combinations, *242*
dependent events, *224*
equally likely events, *198*
expectation, *218*
*factorial, *237*
fundamental counting principle, *207*
independent events, *224*
mutually exclusive events, *226*
odds, *215*
*permutations, *236*
probability, *198*
sample space, *198*
tree diagram, *211*

Note: * indicates optional material.

Review Exercises

4.2 Definition of Probability

1. On a single draw from a bag containing four red, six blue, and three green balls, find the probability of obtaining
   a. a red ball
   b. a blue ball
   c. a red or a green ball
   d. a ball that is not red
   e. a ball that is not green
   f. a ball that is neither red nor blue
2. A pair of dice is tossed. Find the probability that
   a. the sum of the numbers is 6
   b. the sum of the numbers is not 6
   c. the same number appears on both dice
   d. the sum of the numbers is 6 or 10
   e. the sum of the numbers is greater than 9
   f. the sum of the numbers is less than 2
3. **Best-selling books** Table 4.6 on the next page contains a list of the top ten best-selling books for the last week of June 2003. If one book is selected at random, what is the probability that the book was a *Harry Potter* book?

4.3 Sample Spaces/4.4 Tree Diagrams

4. Mary is choosing a pair of running shoes. In her size, the running shoes come in four different styles and six different colors. How many different pairs of running shoes can Mary choose from?
5. By means of a tree diagram, list the sample space showing the possible arrangements when three coins are tossed.
   a. Find the probability that all three coins are heads.
   b. Find the probability that at least one coin is heads.
   c. Find the probability that no coin is heads.

4.5 Odds and Expectations

6. On a single draw from a shuffled deck of 52 cards, find the odds
   a. in favor of drawing a king
   b. against drawing a king
   c. in favor of drawing a picture card (jack, queen, king)
   d. in favor of drawing a club or jack
   e. in favor of drawing a club and a jack
   f. against drawing the ace of spades

TABLE 4.6

Rank	Book/Author
1	*Harry Potter and the Order of the Phoenix*/J. K. Rowling
2	*Living History*/Hillary Rodham Clinton
3	*East of Eden*/John Steinbeck
4	*New Diet Revolution*/Robert C. Atkins
5	*The Da Vinci Code*/Dan Brown
6	*Harry Potter and the Goblet of Fire*/J. K. Rowling
7	*The Lake House*/James Patterson
8	*Harry Potter and the Prisoner of Azkaban*/J. K. Rowling
9	*The Secret of Bees*/Sue Monk Kidd
10	*Harry Potter and the Sorcerer's Stone*/J. K. Rowling

Source: USA Today.

7. If the odds are 8 to 1 against the Giants winning the Super Bowl, what is the probability that the Giants will win the Super Bowl?

8. The probability that the Bruins will win the Stanley Cup is $\frac{3}{11}$. Find the odds in favor of the Bruins winning the Stanley Cup.

9. Five thousand tickets are sold for a drawing on a boat valued at $100,000. If a woman buys one ticket, what is her mathematical expectation?

10. A fraternity sold 500 raffle tickets at $2 each on a color television set valued at $400. If Joe Kool buys five tickets, what is his mathematical expectation?

11. The New York State Lottery runs a daily lottery where a player chooses a three-digit number (any 3 digits from 000 through 999). A lottery ticket costs $1 and if a player chooses the winning number, the payoff (prize) is $500.

a. How many outcomes are possible?

b. What is a fair price to pay for a ticket?

c. What can you conclude about the cost of a lottery ticket?

12. Describe the difference between odds and probability.

4.6 Compound Probability

13. A bag contains five balls numbered 1 through 5. Two balls are chosen in succession. The first ball is replaced before the second is drawn. Are these events *dependent* or *independent?* Why?

14. The object of a game is to obtain a 7 or 11 with a single toss of a pair of dice. Are these events *mutually exclusive?* Why?

15. Two cards are randomly selected in succession from a shuffled deck of 52 cards, without replacement. Find the probability that

a. both cards are red

b. both cards are 3's

c. the first card is red and the second card is black

d. the first card is a heart and the second card is a club

e. the first card is a picture card and the second card is not a picture card

f. the first card is an ace and the second card is a king

16. One card is randomly selected from a shuffled deck of 52 cards and then a die is rolled. Find the probability of obtaining

a. a black card and a 1

b. a black card or a 1

c. a queen and a 1

d. a queen or a 1

e. a red ace and a 1

f. a red ace or a 1

17. Older people and disabilities Figure 4.12 shows the future concerns of disabled people ages 50 and older.

a. What is the probability that a person is concerned about loss of independence and loss of mobility (assume these events are independent)?

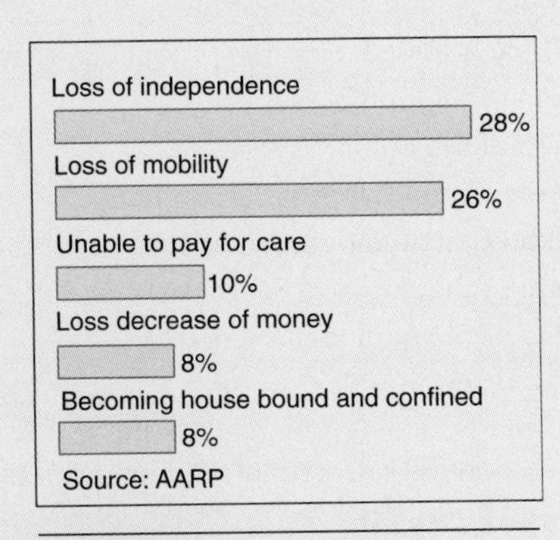

Figure 4.12

b. What is the probability that a person is concerned about being unable to pay for care or becoming house bound/confined (assume these events are mutually exclusive)?

4.7 Counting, Ordered Arrangements, and Permutations (Optional)

*18. There is a game called *poker dice*, which is based on rolling five dice. The resulting outcome is then treated as a poker hand.

a. How many different outcomes are possible?

b. What is the probability of getting five of a kind?

*19. How many license plates can be made if each plate must consist of two letters followed by three digits? (Assume no repetition.) How many are possible if the first letter cannot be *O* and the first digit cannot be zero? How many are possible if the letter *Q* cannot be used and zero cannot be used?

*20. How many different four-letter "words" (that is, arrangements of four letters) can be formed from the letters of the alphabet if each letter can be used only once and none of the vowels *a, e, i, o, u* may be used?

*21. In a certain collegiate basketball conference, there are 10 teams and each team plays all of the other teams in the conference twice. How many league games are played in a season?

*22. How many distinct arrangements are possible in using all the letters of the word *RECYCLE*?

*23. Evaluate the following:

a. $6!$ b. $3!$ c. $\frac{5!}{4!2!}$

d. ${}_4P_2$ e. ${}_{50}P_4$ f. ${}_5P_0$

4.8 Combinations (Optional)/4.9 More Probability (Optional)

*24. Evaluate the following:

a. ${}_5C_2$ b. ${}_6C_2$ c. ${}_7C_4$

d. ${}_7C_3$ e. ${}_nC_0$ f. ${}_nC_n$

*25. From five teachers and 50 students, how many committees can be selected if each committee is to have two teachers and three students?

*26. From a group of five freshmen, six sophomores, four juniors, and three seniors, a staff of three freshmen, three sophomores, two juniors, and two seniors is to be chosen for the school's radio station. In how many ways can this be done?

*27. Three balls are drawn simultaneously at random from a bag containing four red, four blue, and two yellow balls. Find the probability that

a. all three balls are blue

b. all three balls are red

c. two balls are yellow and one is red

d. two balls are red and one is blue

e. two balls are red and one is yellow

*28. You are dealt five cards from a shuffled deck of 52 cards. Indicate (but do not evaluate) the probability that

a. all five cards are red

b. none of the five cards is red

c. all five cards are picture cards

d. none of the five cards is a picture card

e. three of the cards are picture cards and two are not

*29. In the game of five-card poker, find the probability of being dealt a hand that is

a. four of a kind

b. a full house

c. a flush

*30. Describe the difference between a combination and a permutation.

*Note:** indicates optional material.

Biography: Blaise Pascal

Blaise Pascal (1623–1662) was born in France. He displayed signs of mathematical genius at an early age. By the time he was 12, he had discovered many of the theorems of elementary geometry. At 14, he participated in weekly meetings of French mathematicians from which the French Academy ultimately arose in 1666. Pascal provided the first proofs of theorems in projective geometry.

By the time he was 20, Pascal had built the first mechanical adding machine and, furthermore, sold over 50 of these machines to businessmen in the community. Some of these machines are still preserved in a museum in Paris. Pascal, together with Pierre de Fermat, another French mathematician, developed the basic mathematical theory of probability by solving a problem given to him by a gambler friend, the Chevalier de Méré.

In approximately 1650, Pascal decided to abandon his work in mathematics and science and devote himself to religious contemplation. Most of his life was spent with physical pain as he suffered from acute dyspepsia and at times was paralyzed. It is interesting to note that Pascal has also been credited with the invention of the one-wheeled wheelbarrow as we know it today.

Chapter Quiz

1. Find the probability of drawing an ace or a heart from a shuffled standard deck of 52 cards.
2. Find the probability of obtaining 7 or 11 when a pair of dice is tossed.
3. Find the odds in favor of obtaining a 7 when a pair of dice is tossed.
4. A quarter is flipped and a die is tossed. Find the probability of obtaining a head or a 6.
5. Find the probability that at least one head appears when two coins are tossed.
6. A die is rolled, and then one card is randomly selected from a shuffled deck of 52 cards. Find the probability of obtaining a 1 and an ace.
7. Using the information in question 6, find the probability of obtaining a 1 or an ace.
8. The odds against Angie winning a certain raffle are 999 : 1. Find the probability that Angie will win the raffle.
9. A special lottery offers a prize of $1000. The winning ticket can be any four-digit number from 0000 to 9999. What is a fair price to pay for a ticket?
10. A coin is flipped and a die is tossed. Find the probability of obtaining a tail or an even number.

Indicate whether each statement (11–25) is true or false.

11. If an event is certain to occur, then its probability is 1.
12. If an event can never occur, then its probability is 0.
13. If an event may occur or may not occur, then its probability is $\frac{1}{2}$.
14. The set of all possible outcomes of an experiment is called an *event.*
15. The odds against obtaining an 11 when a pair of dice is tossed are 17 : 1.
16. The more tickets you purchase for a raffle, the more chances you have of winning.
17. The more tickets you purchase for a raffle, the greater the mathematical expectation of a ticket.
18. Two events, A and B, are *independent* events if the occurrence of one event does not affect the probability of the other.
19. Drawing an ace from a standard deck of playing cards and then drawing a king from the same deck without replacing the first card are considered independent events.
20. If two or more events can happen at the same time, they are *mutually exclusive.*

*21. Given the set of digits $\{1, 3, 4, 5, 6\}$, 50 three-digit numbers can be formed that are even.

*22. $3! = 2! + 1!$

*23. The letters in *TOMATO* can be arranged in 90 distinct ways.

*24. Two co-chairpersons are to be selected from a group of five eligible people. This can be done in 20 ways.

*25. ${}_7C_3 = {}_7C_4$

Note: *indicates optional material.

Just for fun

Do you think that two people in your class have the same birthday (month and day)? Try it; you might be surprised. If there are 25 people in your class, the probability that two people have the same birthday is greater than 0.5. See also, the Math Connections at the end of the chapter.

Math Connections

The Birthday Problem

One of the classic problems in any discussion of probability is the *birthday problem*, which is usually expressed as "How many people do we need in a group before there is at least a 50% probability that two of them were born on the same day of the year?" The graph in the figure below shows the birthday probabilities for various group sizes. Notice that only 23 people are needed before we get a probability greater than 0.5. This phenomenon is sometimes called the **birthday paradox** because the number of people needed in a group before there is at least a 50% probability that two of them were born on the same day of the year is less than the number of days in the year. Note also that only 42 people are needed to get a probability greater than 0.9.

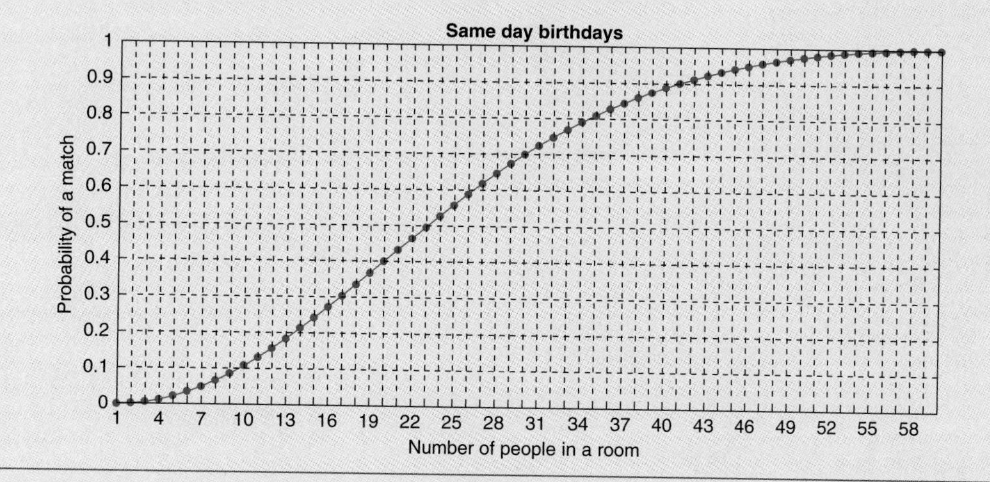

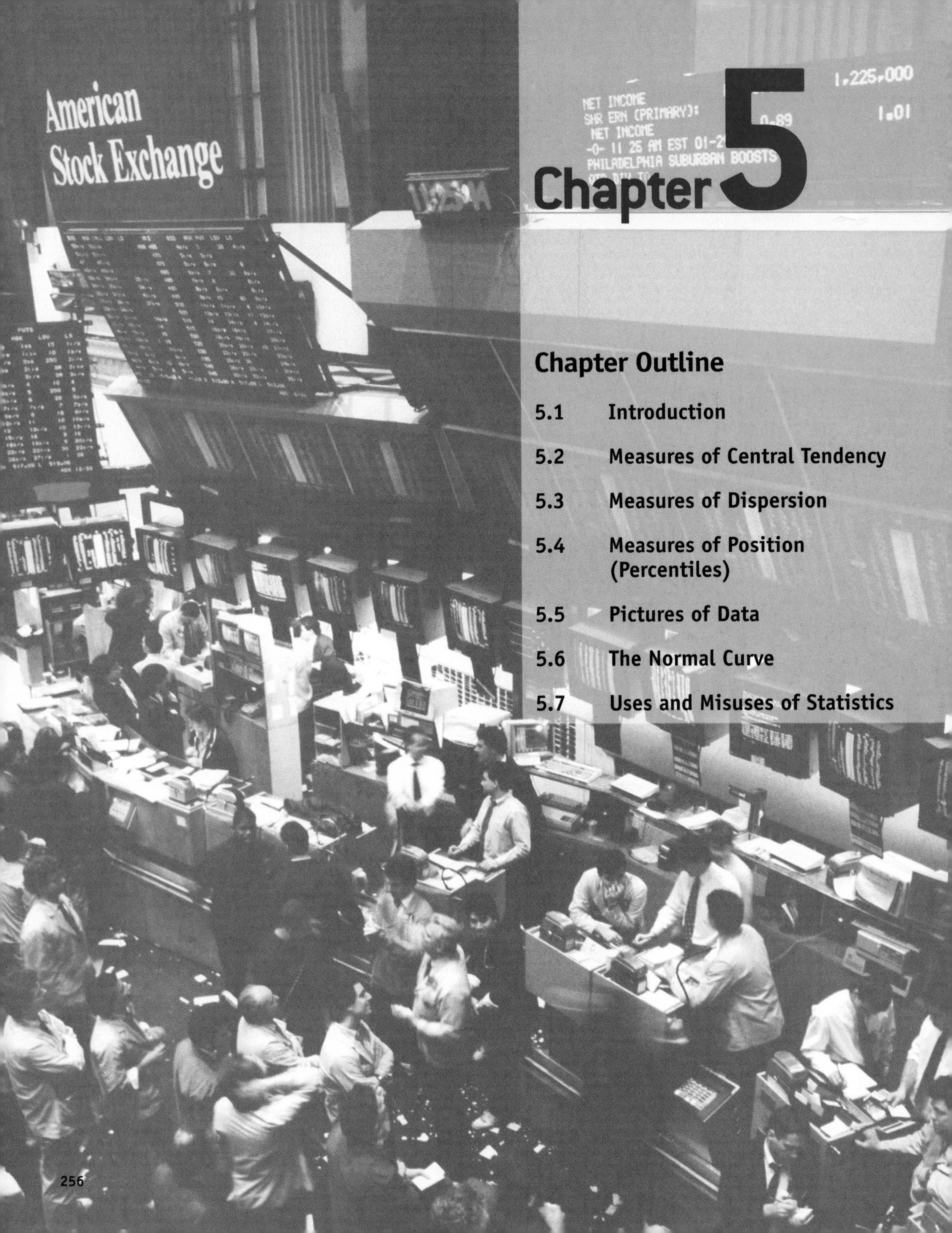

Chapter 5

Chapter Outline

Statistics

After Studying This Chapter, You Will Be Able to Do the Following

1. Compute the **mean, median, mode**, and **midrange** for a given set of data, and distinguish among them.
2. Compute the **range** and **standard deviation** for a given set of data, and distinguish between them.
3. Find the **percentile** or **quartile** of a single datum in a given set of data.
4. Construct a **vertical bar graph**, a **horizontal bar graph**, a **comparative bar graph**, a **pictogram**, and a **circle graph** from a given set of data.
5. Construct a **frequency distribution table** and **histogram** or **frequency polygon** from a given set of data.
6. Determine what percentage of **normally distributed data** is within a given number of **standard deviations** from the mean.

Note: *indicates optional material.

Notations Frequently Used in This Chapter

$\bar{x}$	mean (read "*x* bar")
a^2	*a* squared (for example, $3^2 = 3 \times 3 = 9$)
$\sqrt{a}$	the positive square root of *a* (for example, $\sqrt{9} = 3$)
σ	standard deviation (lowercase Greek letter sigma)
$\approx$	is approximately equal to

Visit the PH Companion Web site at www.prenhall.com/setek

OVERVIEW

Statisticians collect numerical data from different populations to determine information about the population. What do people watch on television? Which candidate do you favor in the next election? What is the most important problem facing America today?

Whenever we watch television, listen to the radio, read newspapers, magazines or books, we encounter statistics. We can find statistics in articles on business, the state of the economy, politics, science, education, sports, and many other subjects. In order to understand the information that is presented, we must possess some understanding of statistics.

In this chapter you will gain an understanding of some statistical terms and concepts. For example, are you aware that there exist several ways to compute an *average*? What is a *standard deviation*? We will cover topics such as *mean, median, mode, midrange, range, and standard deviation*. Additional topics include *measures of position, pictures of data, the normal curve*, and *uses and misuses of statistics*.

5.1 INTRODUCTION

Most people first encounter statistics in elementary school when they take standardized achievement tests: Schools want to know what the "average" student has learned or how a student compares to other students who have taken the standardized tests.

A **statistic** is a number derived from a set of data that (in some way) is used to represent or describe the data. Unfortunately, statistics are often misused or abused. Some people maintain that a statistician can prove whatever he or she wants to prove. A more familiar quote is "Figures don't lie, but liars figure." Adults and children alike are inundated by statistics. All too often they have no real understanding of the facts presented.

Advertising is one area where statistics are often abused. Statistics may be distorted by the manner in which the facts are gathered or by the manner in which the facts are presented. Statistics can be distorted in diagrams by emphasizing the wrong facts, exaggerating comparisons, or simply not showing all of the data. When we apply statistical methods to a specific area of interest, we are using statistics to help us gain a greater understanding of that subject.

It has often been said that if a person is going to be an intelligent member of today's society, he or she must possess some understanding of statistics. As H. G. Wells stated, "Statistical thinking will one day be as necessary for efficient citizenship as the ability to read and write." In this chapter, we discuss several important statistical concepts to help prepare you to better understand and interpret statistics when you encounter them, and to help you become a more effective and informed citizen.

5.2 MEASURES OF CENTRAL TENDENCY

One of the first concepts of statistics that people encounter, and one that is familiar to everyone, is the concept of "average." Whether or not a person can calculate an average, he or she has an intuitive idea of what it is. We are familiar with phrases such as "average miles per gallon," "average precipitation for the month," the "batting average" of a baseball player, and a "student's average" for a set of test scores.

The term *average* is used in different ways by different people. An average is a *measure of central tendency*. A **measure of central tendency** describes a set of data by locating the middle region of the set. The common measures of central tendency are the *arithmetic mean*, the *median*, the *mode*, and the *midrange*. The

Note of Interest

A *sample* or a *census* is often used to provide data for a statistical project. A *sample* is a study of only a portion of the items or people to be used in the analysis. To find the average height of the students in your college, you would probably use a sample of the students. The sample should be representative of all the students attending your college. The basketball team would probably not be a representative sample for finding the average heights of all students.

A *census* is a study or count of every item or person that is to be used in the analysis. For example, if you wanted to know the brand of calculator that is most popular with the students in your mathematics class, you would ask each student.

arithmetic mean is usually referred to simply as the *mean*. Each of the four measures of central tendency—mean, median, mode, and midrange—is an average because each describes the middle region of the data. Each one has its advantages and disadvantages; in a given situation, it may be more desirable to use one as opposed to the other three.

Mean

The **mean** (arithmetic mean) is the most familiar type of average that we encounter. You have probably found the mean for a set of data before, but instead of calling it a mean, you probably called it an average.

The mean for a set of data, scores, or facts is found by determining the sum of the data and dividing this sum by the total number of elements in the set.

The scores for five students on a quiz are 40, 20, 30, 25, and 15. To find the mean score for this group of students, we first find the sum of the scores:

$$40 + 20 + 30 + 25 + 15 = 130$$

We then divide the sum by 5, the number of scores:

$$\frac{130}{5} = 26$$

The mean for a set of data is often represented by the notation $\bar{x}$ (read "x bar"). Therefore, we can write $\bar{x} = 26$ for our given set of data. Even though the mean score is 26, note that none of the scores is 26. There are two scores, 40 and 30, above the mean and three scores, 25, 20, and 15, below the mean. This single measure of central tendency is used to represent the whole group of five students. It gives us information as to how the group of students performed, but it does not tell us anything about a particular student in the group.

EXAMPLE 1 ***Mean*** Find the mean for the set of test scores

$$71, \quad 75, \quad 60, \quad 84, \quad 71, \quad 63, \quad 66$$

Solution The sum of the scores is

$$71 + 75 + 60 + 84 + 71 + 63 + 66 = 490$$

and we have seven scores; hence, we divide 490 by 7:

$$\bar{x} = \frac{490}{7} = 70$$

EXAMPLE 2 ***Mean*** At the Surf and Sand Restaurant, a waitress earns $28 a night in tips for the five nights during the week. But on weekends (Saturday and Sunday) she earns $50 a night in tips. What is her average daily earnings in tips?

Solution We must find the total income from tips and then divide this sum by the number of elements in our set of data. We have

$$5 \times \$28 = \$140, \qquad 2 \times \$50 = \$100$$

and total tips are

$$\$140 + \$100 = \$240$$

Now we divide $240 by 7:

$$\bar{x} = \frac{\$240}{7} \approx \$34.29 \qquad \text{(daily average earnings in tips)}$$

In Example 1, the mean score of 70 tells us that the average score for the set is 70, but it does not necessarily represent any particular score. Note that the mean of 70 is somewhat centrally located in the set of data. However, a mean can be misleading in describing data because it can be affected by extreme values in the data. Example 3 illustrates this.

EXAMPLE 3 ***Mean*** Find the mean for the set of scores

82, 81, 80, 87, 20

Solution The sum of the scores is $82 + 81 + 80 + 87 + 20 = 350$; hence, the mean is

$$\bar{x} = \frac{350}{5} = 70$$

The mean is only 70, although most of the scores are in the 80s. 

Note of Interest

A *random sample* is one that you pick by chance. That is, an item is selected *at random* from a group of items if the selection process is such that each item in the group has an equal chance to be selected. For example, if you put the name of each person in your class on the same size card, put all the cards in a box, shake, shuffle, or toss them until they are all mixed up, and then select three cards without looking, you would have a random sample of three names from your class. If you wanted to know the opinions of students in your college, you would draw a random sample from the names of all the students registered at your college. You would not pick a sample of the students in your mathematics class, in the bookstore, or at a sporting event. Each of these places would give you a sample of a certain type of student, not students in general, or at random.

The method of selecting the sample might introduce a *bias*. This in turn can lead to the misuse of statistics found in the samples that are used as the basis for statistical research.

Median

Although the average, or mean, of the scores in Example 3 is 70, it is obvious that this number does not give a very accurate idea of the typical score. Suppose, however, that we arrange the scores in Example 3 in order, from lowest to highest:

20, 80, 81, 82, 87

The middle number in this arrangement is 81. This number, called the *median* of the data, gives a more accurate idea of the typical score in cases where the data include extreme values. The **median** can be described as the middle value of a set of data when the data are listed in order.

The median for a set of data is found by arranging the data (numbers) in sequential order and finding the middle number.

The scores for five students on a quiz are 40, 20, 30, 24, and 15. To find the median, we must first arrange the scores in sequential order: 15, 20, 24, 30, 40. The median (middle number) for these five scores is 24.

When we have an even number of scores, it is customary to use the number halfway between the two middle scores (i.e., the mean of the two middle numbers) as the median. For example, consider the following set of scores:

$$\{40, 20, 30, 24, 28, 15\}$$

What is the median? Arranging the scores in order gives us the following: 15, 20, 24, 28, 30, 40. Because there are six pieces of data (an even number), we must find the mean of the two middle numbers. The two middle numbers are 24 and 28. Therefore, the median is

$$\frac{24 + 28}{2} = \frac{52}{2} = 26$$

The median is the measure of central tendency that determines the middle of a given set of data—that is, the number, value, or score such that the number of scores below the median is the same as the number of scores above the median.

EXAMPLE 4 ***Median*** Find the median for the given set of data:

13, 16, 14, 12, 20, 19, 10, 18

Solution Arranging the numbers in sequential order, we have

10, 12, 13, 14, 16, 18, 19, 20

Note that there are eight pieces of data (an even number), so we must find the mean of the two middle numbers, 14 and 16:

$$\frac{14 + 16}{2} = \frac{30}{2} = 15$$

Therefore, the median is 15. Four of the pieces of data are less than 15, and four of the pieces of data are greater than 15.

EXAMPLE 5 ***Median*** The prices of nine stocks on the New York Stock Exchange are as follows:

$102, $85, $30, $92, $35, $70, $20, $18, $68

Find the median price.

Solution When we arrange the data in sequential order, we have

$18, $20, $30, $35, $68, $70, $85, $92, $102

There are nine pieces of data, so the median is the fifth one, $68. There are four pieces of data before it, and four pieces of data after it.

EXAMPLE 6 ***Mean and Median*** A group of students, Tom, Carlos, Janie, Irene, and Stacey, reported that they had earned the following amounts during summer vacation:

Tom	$800
Carlos	$900
Janie	$300
Irene	$900
Stacey	$600

Find the mean and median summer income for this group of students.

Solution

a. To find the mean income, we must find the total income and then divide this sum by 5 (the number of students):

$$\$800 + \$900 + \$300 + \$900 + \$600 = \$3{,}500$$

$$\frac{\$3{,}500}{5} = \$700 \quad \text{(mean income)}$$

b. To find the median income, we must arrange the amounts in sequential order and find the middle amount (the number of students is odd):

$300, $600, $800, $900, $900

The middle amount, $800, is the median income.

Mode

In Example 6 we calculated both the mean and the median for a given set of data. Notice, however, that although the mean was $700 and the median was $800, more people in the group earned $900 than earned any other amount. This figure, $900, is called the **mode** of the data in Example 6.

The mode for a given set of data is that number, item, or value that occurs most frequently.

The scores for five students on a quiz are 40, 24, 30, 24, and 15. The mode for this set of scores is 24 because it occurs twice, whereas each of the other scores occurs only once.

There are times when the mode is the most meaningful of the three measures of central tendency we have discussed thus far. Consider the case of Harry, who operates Harry's Hamburger Haven. During a typical 8-hour shift, Harry sells approximately 2400 hamburgers. This is a mean of 300 hamburgers per hour. But Harry's customers do not come into his Hamburger Haven at the same rate each hour. He is much busier some hours than others. In planning how many hamburgers to have ready each hour, Harry is interested in what hour(s) he sells the most hamburgers.

You should be aware that a set of data always has a mean and a median, but does not necessarily have a mode. Consider the set of data 1, 2, 3, 4, 5. The mean is 3 and the median is 3, but there is no mode, because there is no number that occurs most frequently.

It is also the case that a set of data can have more than one mode. Consider the set of data 1, 2, 3, 3, 4, 5, 5, 6. We see that the number 3 occurs twice and the number 5 occurs twice. Hence, there are two modes, 3 and 5. Since this given set of data has two modes, we refer to it as **bimodal**.

EXAMPLE 7 ***Mode*** Find the mode for the following data:

12, 10, 11, 13, 11, 14, 13, 11, 17

Solution The mode is 11. It occurs three times in the set of data.

EXAMPLE 8 ***Mode*** Find the mode for the following data:

18, 16, 13, 14, 12, 10, 11, 15, 17, 19

Solution There is no mode for this set of data because each value occurs only once.

EXAMPLE 9 ***Mean, Median, Mode*** Following is a list of scores that Terrie received on a series of agility tests:

52, 77, 74, 82, 74, 104, 83

Find **(a)** the mean, **(b)** the median, and **(c)** the mode for this set of data.

Solution

a. The sum of the scores is 546, and the mean is

$$\frac{546}{7} = 78$$

b. Arranging the scores in order, we have

52, 74, 74, 77, 82, 83, 104

The median is 77.

c. The mode is 74.

Midrange

In Example 9, we calculated the mean, median, and mode for a given set of data. Another measure that may be used to give an approximation for a measure of central tendency is the *midrange*. We can think of the **midrange** as the value midway between the end points of the set of data. Note that this is not the same as the median, although it may sometimes have the same value.

The midrange is found by adding the least value in the given set of data to the greatest value and dividing the sum by 2:

$$\text{Midrange} = \frac{L + G}{2}$$

In Example 9, the least value that occurs is 52 and the greatest value is 104. Therefore, the midrange for Example 9 is

$$\frac{52 + 104}{2} = \frac{156}{2} = 78$$

This is also the value of the mean in Example 9, but this is a coincidence. The midrange is used to estimate the central tendency of the data. Weather forecasters use the midrange when computing the average temperature for a given day. That is, they add the lowest temperature reading to the highest temperature reading, divide the sum by 2, and report the result as the average temperature.

EXAMPLE 10 ***Mean, Median, Mode, Midrange*** The following table lists the maximum speeds achieved by various animals (in mph).

Animal	Speed	Animal	Speed
Cheetah	70	Quarter horse	47
Antelope	61	Elk	45
Wildebeest	50	Cape dog	45
Lion	50	Coyote	43
Gazelle	50	Gray fox	42

Find the mean, median, mode, and midrange for this set of data.

Solution

$$\begin{aligned}\text{Mean} &= \frac{70 + 61 + 50 + 50 + 50 + 47 + 45 + 45 + 43 + 42}{10} \\ &= \frac{503}{10} = 50.3\end{aligned}$$

There are 10 pieces of data (an even number), so we must find the mean of the two middle numbers, 50 and 47:

$$\frac{50 + 47}{2} = \frac{97}{2} = 48.5 \quad \text{(median)}$$

Hence, the median is 48.5. Five pieces of data are less than 48.5 and five pieces of data are greater than 48.5. The *mode* is 50. It occurs three times in the set of data. Finally,

$$\text{Midrange} = \frac{42 + 70}{2} = \frac{112}{2} = 56$$

NW ***Now Work Problems 1, 3, and 7.***

You should be aware that each of these measures is a useful measuring device. There are instances when one of them (mean, median, mode, midrange) is more representative than the others. The mean is not a true indication of the "average" if there are values in the given set of data that are extreme values at one end or the other. A typical example that illustrates this phenomenon is one that involves wages. Consider the annual salaries of the employees of the Custom Moving Company. The manager earns \$75,000 per year, one driver earns \$40,000 per year, and another driver earns \$36,000 per year. One helper earns \$27,000 per year, while three other helpers each earn \$20,000 per year. The mean salary for these seven people is computed as follows:

$$\$75{,}000 + \$40{,}000 + \$36{,}000 + \$27{,}000 + \$20{,}000 + \$20{,}000 + \$20{,}000$$
$$= \$238{,}000$$
$$\frac{\$238{,}000}{7} = \$34{,}000$$

Carrying the discussion a little further, suppose the owner of the Custom Moving Company earns \$100,000. Now the mean salary for all eight people is

$$\$100{,}000 + \$75{,}000 + \$40{,}000 + \$36{,}000 + \$27{,}000 + \$20{,}000 + \$20{,}000$$
$$+ \$20{,}000 = \$338{,}000$$
$$\frac{\$338{,}000}{8} = \$42{,}250$$

The mean salary is \$42,250. This is certainly not a sensible representation of the average salary. This is a case where the mean should not be used to describe the situation for the given set of data. In contract bargaining, the management would probably like to use the mean (\$42,250) as a basis for negotiation, whereas the union would probably like to use the mode (\$20,000). Remember that the mean is not a true indication of the average if the given set of data contains extreme values at one end.

The median is the middle value (number). The number of data below the median is the same as the number of data above the median. The median is not affected by extreme values in the set of data. But it may not be a true representation of the average if the data occur in distinct, separate groups.

The mode is that item that occurs most frequently in a set of data. There can be situations where a set of data has no mode—that is, where each value occurs an equal number of times. In other situations, a set of data can have more than one mode—that is, two or more different values can occur the same number of times. The mode can be misleading at times; it does not take into account the other numbers in the set of data, as do the mean and median.

The midrange is the value midway between the end points of the set of data. It too can be misleading at times; it does not take into account the other values or how many pieces of data are contained in a set of data.

Many times information is presented in some form of a table. Consider the following table, which depicts the salary distribution for a certain company, DANA Enterprises:

Annual Income	Number Receiving This Income
\$100,000	1
50,000	2
20,000	8
15,000	10
10,000	4

Note how the data are reported in this table. Instead of listing the individual salaries for each employee, the *frequency* of each salary is reported. For example, *one* employee earns \$100,000, *two* employees each earn \$50,000, *eight* employees each earn \$20,000, *ten* employees each earn \$15,000, and *four* employees each earn \$10,000. This type of table is called a *frequency distribution table*. (Note: Frequency distribution tables are discussed in more detail in Section 5.5.)

The mean, median, mode, and midrange of the data listed in a frequency distribution table can be determined from the information given in the table. However, we must be careful how we compute them. For example, when calculating the mean from a frequency distribution table, it is very common for students to simply add the numbers in the data column while ignoring the frequency column. In the table of salaries given, to find the mean income we divide the total salary by the total number of employees. But what is the total salary? To find the total salary we multiply the amount of income by the number of people receiving that income (i.e., its frequency). We then sum these totals. That is,

$$\begin{aligned}
\$100{,}000 \times 1 &= \$100{,}000 \\
50{,}000 \times 2 &= 100{,}000 \\
20{,}000 \times 8 &= 160{,}000 \\
15{,}000 \times 10 &= 150{,}000 \\
10{,}000 \times 4 &= \underline{40{,}000} \\
&\ \$550{,}000 \text{ Total annual income}
\end{aligned}$$

Note: Total number ofemployees = 25

Therefore, the mean income is

$$\frac{\$550{,}000}{25} = \$22{,}000$$

The median income is determined by the number of pieces of data: 25. Since we have an odd number, the median is the middle piece of data, the 13th. Note that the salaries are already ranked in the table and the 13th employee's salary is the median income. We may begin from the top or bottom. Regardless of how we begin, we see that the 13th ranking income is \$15,000. Hence, the median income is \$15,000.

The mode is the item that occurs most frequently in a set of data. The table indicates that 10 people earned \$15,000, which is a greater number of people than any of those earning the other salaries. Therefore, the mode income is \$15,000.

The midrange of the incomes is computed as

$$\text{Midrange} = \frac{\$10{,}000 + \$100{,}000}{2} = \frac{\$110{,}000}{2} = \$55{,}000$$

Care must be exercised when deriving statistical data from tables. At first glance, it would appear that the median income is \$20,000 and the mean income is

$$\begin{aligned}
&(\$100{,}000 + \$50{,}000 + \$20{,}000 + \$15{,}000 + \$10{,}000) \div 5 \\
&= \$195{,}000 \div 5 = \$39{,}000
\end{aligned}$$

This is not correct!

NW ***Now Work Problem 23.***

Exercises for Section 5.2

In Exercises 1–10, find the mean, median, mode, and midrange for each set of data. (Round off any decimal answer to the nearest tenth.)

NW **1.** 1, 2, 3, 4, 4, 5, 9, 12

2. 12, 18, 16, 12, 13

NW **3.** 1, 2, 3, 4, 5, 6, 7, 8, 9, 10

4. 3, 3, 4, 4, 6, 11, 13, 12, 11

5. 2, 4, 8, 10, 6, 12

6. 4, 6, 12, 2, 1, 8, 10, 15

NW **7.** 11, 99, 77, 88, 66, 44, 55, 22, 33

8. 85, 21, 79, 37, 49, 25, 59, 42, 80

9. 1492, 1776, 1941, 1812

10. 1980, 2000, 1970, 1950, 2000

11. Baseball Listed in the following tables are the home run leaders of the American and National Leagues for the years 1993–2003. Find the **(a)** mean, **(b)** median, **(c)** mode, and **(d)** midrange for the number of home runs of each league. **(e)** Do you think the National League's mean is an adequate representation of the average number of home runs hit for this 11-year period? Why or why not?

American League Home Run Leaders

Year	Player	Number of Home Runs
1993	Juan Gonzalez	46
1994	Ken Griffey, Jr.	40
1995	Albert Belle	50
1996	Mark McGwire	52
1997	Ken Griffey, Jr.	56
1998	Ken Griffey, Jr.	56
1999	Ken Griffey, Jr.	48
2000	Troy Glaus	47
2001	Alex Rodriguez	52
2002	Alex Rodriguez	57
2003	Alex Rodriquez	47

National League Home Run Leaders

Year	Player	Number of Home Runs
1993	Barry Bonds	46
1994	Matt Williams	43
1995	Dante Bichette	40
1996	Andres Galarraga	47
1997	Larry Walker	49
1998	Mark McGwire	70
1999	Mark McGwire	65
2000	Sammy Sosa	50
2001	Barry Bonds	73
2002	Sammy Sosa	49
2003	Jim Thome	47

12. Great Lakes The Great Lakes form the largest body of fresh water in the world and their connecting waterways are the largest inland water transportation system. Following are the lakes and their length in miles:

Lake	Length (miles)
Superior	350
Michigan	307
Huron	206
Erie	241
Ontario	193

Find the **(a)** mean, **(b)** median, **(c)** mode, and **(d)** midrange for the length of the lakes.

13. Dams The following table gives the location and height of the 10 tallest dams in the United States:

Name	Location	Height (feet)
Oroville	California	770
Hoover	Nevada	726
Dworshak	Idaho	717
Glen Canyon	Arizona	710
New Bullards	California	637
New Melones	California	625
Swift	Washington	610
Mossyrock	Washington	607
Shasta	California	602
New Don Pedro	California	585

Find the **(a)** mean, **(b)** median, **(c)** mode, and **(d)** midrange for the heights.

Based on this photograph, do you think the median of the heights of the members of this class would be equal to the mean of their heights?

14. **Grades** An art teacher grades all student projects 1, 2, or 3, as follows:

    1—excellent
    2—acceptable
    3—unacceptable (must be redone)

    Given a set of data consisting of all the grades assigned to student projects during the term, which measure of central tendency would you use to determine the grade received by the greatest number of projects?

15. **Salaries** Women employees of a certain firm have complained of sex discrimination in the company's pay scale. If the women win their case, the firm will have to raise the salaries of all underpaid employees. The employees and their salaries are as follows:

Mr. Clar $35,000	Ms. Chung $27,000
Mr. Hart $30,000	Ms. Wolski $26,000
Mr. Hobb $40,000	Ms. Dugan $22,000

    a. What is the mean salary for all employees?
    b. What is the mean salary for women? For men?
    c. What is the median salary for women? For men?
    d. What is the midrange salary for women? For men?
    e. If you were a lawyer acting for the firm, which measure of central tendency would you use to describe the average salaries of all employees? Which measure of central tendency would you use if you were a lawyer arguing on behalf of the women employees?

16. In Sam's physics course, a mean score of 80 on 10 tests is necessary for a grade of B. Sam's mean score for the 10 tests was 79 and his instructor, Ms. Molecule, gave him a grade of C. Sam protested that because he was so close, his instructor should give him the "one lousy point" and hence a B. Did Sam need only one point for a B? Explain your answer.

17. **Baseball** In a certain week, the New York Mets played eight baseball games. The number of runs they scored in the respective games were 3, 2, 0, 6, 9, 4, 5, and 3. Find the mean, median, and mode for this set of data. Did the Mets win all eight games?

18. The mean score on a set of 10 scores is 71. What is the sum of the 10 test scores?

19. The mean score on a set of 13 scores is 77. What is the sum of the 13 test scores?

20. The mean score on four of a set of five scores is 75. The fifth score is 90. What is the sum of the five scores? What is the mean of the five scores?

21. Two sets of data are given. The first set of data has 10 scores with a mean of 70, and the second set of data has 20 scores with a mean of 80. What is the mean for both sets of data combined?

22. **Grade point average** Janet and Larry took the same courses last semester: calculus (4 credits), geology (4 credits), English (3 credits), history (3 credits), and physical education (1 credit). Janet received A, A, B, C, C, respectively, and Larry received C, C, B, A, A, respectively. Janet and Larry bet a dinner as to who would have the higher average. Janet maintains she won, whereas Larry maintains that they are even because they both got two A's, one B, and two C's. Who is right? [*Hint:* A grade-point average is found as follows: Allow 4 points for an A, 3 points for a B, and 2 points for a C. Multiply the number of points equivalent to the letter grade received in each course by the number of credits for the course to arrive at the total points earned in each course. Divide the sum of the points by the total number of credit hours. The answer is the grade point average.]

23. **Salaries** The accompanying table gives the annual salary distribution for the Ronolog Corporation. Using the information provided in the table, find the following:
NW
    a. the mean annual income
    b. the median annual income
    c. the mode annual income
    d. the midrange annual income

Annual Income	Number Receiving This Income
$110,000	1
60,000	1
25,000	7
22,000	10
19,000	8
16,000	13

24. **Age** The accompanying table indicates the ages of a group of 30 people who entered a video store on a Saturday

morning. Using the information provided in the table, find the following:

a. the mean age
b. the median age
c. the mode age
d. the midrange age

Age	Number of People
17	4
18	5
19	7
20	5
21	4
22	2
23	3

25. Commuting distance On the first day of class last semester, 30 students were asked for the one-way distance from home to college (to the nearest mile). The accompanying table provides the resulting data. Using the information provided in the table, find the following:

a. the mean distance
b. the median distance
c. the mode distance
d. the midrange distance

Distance (miles)	Number of People
30	2
26	1
22	2
18	5
14	6
10	7
6	5
2	2

26. Fastest trains The fastest scheduled runs for different passenger trains in the world are listed in the accompanying table. Using the information provided in the table, find the following:

a. the mean speed
b. the median speed
c. the mode speed
d. the midrange speed

Country	From	To	Speed (mph)
France	Paris	Macon	134.0
Japan	Nagoya	Yokohama	112.3
Great Britain	Peterborough	Stevenage	101.0
USA	Wilmington	Baltimore	100.1
Germany	Hamm	Bielefeld	96.2
USA	Rensselaer	Hudson	88.4
USA	Newark	Philadelphia	87.8
Italy	Rome	Chiusi	83.5

***27.** The following figure graphically represents a distribution of scores that is *positively skewed*. (Note: In a positively skewed distribution the highest frequency of scores is on the left-hand side, and the extreme scores are on the right-hand side. This is also called *skewed right*.) Study this graph carefully and identify the points on the graph—**a**, **b**, and **c**—that best correspond to the mean, median, and mode. Justify your answers.

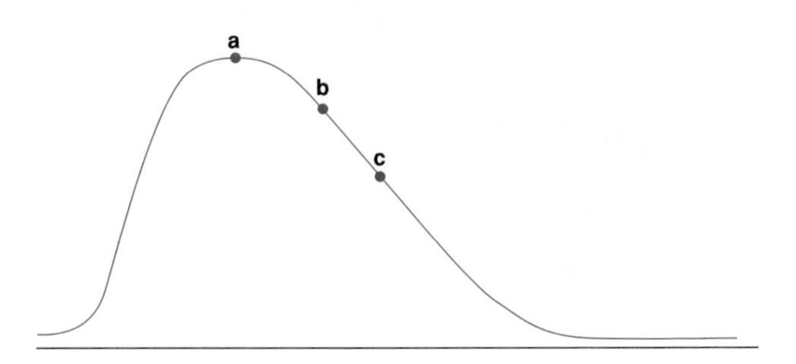

***28.** The following figure graphically represents a distribution of scores that is *negatively skewed*. (*Note*: In a negatively skewed distribution the highest frequency of scores is on the right-hand side, and the extreme scores are on the left-hand side. This is also called *skewed left*.) Study this graph carefully and identify the points on the graph—**a**, **b**, and **c**—that best correspond to the mean, median, and mode. Justify your answers.

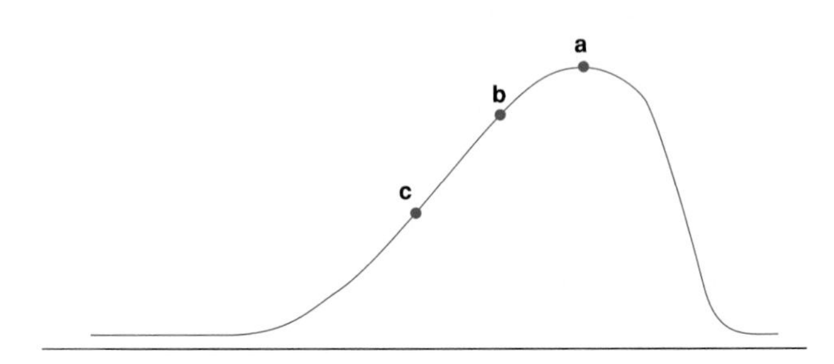

***29.** The following figure graphically represents a *symmetrical* distribution of scores. Study this graph carefully and determine where the mean, median, and mode would be placed. Justify your answers.

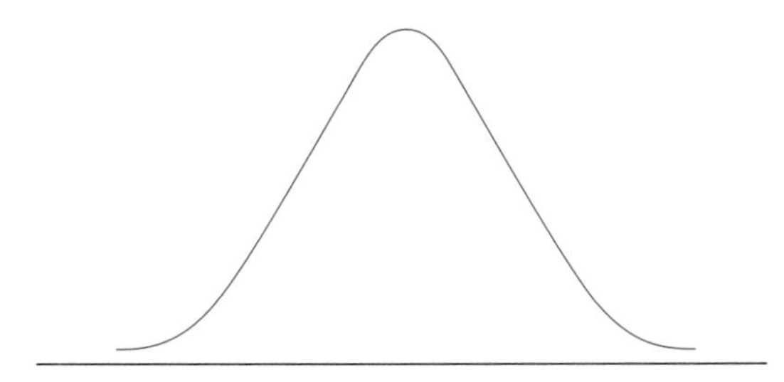

***30.** The following figure graphically represents a *uniform* (*rectangular*) distribution of scores. Study this graph carefully and determine where the mean, median, and mode would be placed. Justify your answers. (see next page)

Writing Mathematics

31. Explain the difference(s) among the four primary measures of central tendency.
32. Explain what it means to say that the mean gets "pulled" in the direction of the extreme scores. Give an example.
33. Can the mean be a negative number? Explain your answer and give an example.
34. If a survey of 100 Internet users revealed that 74 people answered yes to the question, "Do you have a broadband Internet connection such as a cable modem or DSL?", which measure of central tendency is most appropriate to report? Why?

Challenge Exercises

35. In addition to the arithmetic mean, there is something called the *harmonic mean*. This mean is the reciprocal of the arithmetic mean of the reciprocals of the data. For example, given the data 5, 10, 15, we first find the arithmetic mean of the reciprocals of the data; that is,

$$\frac{\frac{1}{5}+\frac{1}{10}+\frac{1}{15}}{3}=\frac{\frac{11}{30}}{3}=\frac{11}{90}$$

We then take the reciprocal of this result,

$$\frac{90}{11} \approx 8.2 \text{ (the harmonic mean)}$$

Given the data 2, 3, 3, 4, 15, calculate the harmonic mean and compare it to the corresponding arithmetic mean. Which mean do you think is more representative of the data? Why? When do you think the harmonic mean is most appropriate to report?

36. What do you think a *weighted mean* is? Give an example.

Research/Group Activities

37. Research the concept of the *golden mean*. [*Hint:* This mean is associated with the Fibonacci sequence, which is discussed in Chapter 1.]
38. Consult a statistics book and report on the specific properties associated with the mean.

Just for fun

If you are given 10 pennies and three coffee cups, can you place all of the pennies in the three coffee cups so that there is an odd number of coins in each coffee cup? Each cup has to contain at least one penny.

5.3 MEASURES OF DISPERSION

Range

In the previous section, we discussed measures of central tendency: mean, median, mode, and midrange. They are called measures of central tendency because each of them tells us something about the average of the data—that is, where the data tend to center or cluster. Measures of dispersion tell us how much the data tend to disperse or scatter—that is, the spread of the data. One measure of dispersion is the *range*.

The range for a set of data is found by subtracting the smallest value from the largest value in the given set of data.

Consider the respective quiz scores in a history class for Cathy and Juanita:

Juanita: 72, 74, 74, 77, 80, 83, 86
Cathy: 58, 74, 74, 77, 81, 82, 100

Computing the mean score for Juanita, we have

$$\frac{72+74+74+77+80+83+86}{7}=\frac{546}{7}=78$$

The mean score for Cathy is

$$\frac{58 + 74 + 74 + 77 + 81 + 82 + 100}{7} = \frac{546}{7} = 78$$

The mean score for each is 78. Comparing further, we see that the median for each is 77, the mode for each is 74, and the midrange for each is 79. Cathy and Juanita have the same measures of central tendency for their quiz scores. The one significantly different thing about their scores is the range. The range for Juanita's score is $86 - 72 = 14$, whereas the range for Cathy's scores is $100 - 58 = 42$.

Cathy's scores had the greater range. Because Cathy and Juanita cannot be compared using the measures of central tendency, we might use the range. But we must be careful. We might argue that Juanita is the more consistent of the two because her score range is only 14, but we might also argue that Cathy showed the most improvement because her score range is 42.

EXAMPLE 1 ***Mean and Range*** Find the mean and the range for the two sets of test scores:

Test A: 75, 75, 70, 70, 70, 65, 65, 65, 55, 40
Test B: 70, 70, 67, 66, 65, 65, 65, 62, 60, 60

Solution For Test A,

$$\text{Mean} = \frac{75 + 75 + 70 + 70 + 70 + 65 + 65 + 65 + 55 + 40}{10} = \frac{650}{10} = 65$$

$$\text{Range} = 75 - 40 = 35$$

For Test B,

$$\text{Mean} = \frac{70 + 70 + 67 + 66 + 65 + 65 + 65 + 62 + 60 + 60}{10} = \frac{650}{10} = 65$$

$$\text{Range} = 70 - 60 = 10$$

Note that both test A and test B had a mean score of 65, but the range for test A was 35, whereas the range for test B was only 10. There were both higher and lower grades on test A. There is less dispersion for the scores on test B.

Standard Deviation

The range, though easily computed, is not considered to be the best measure of dispersion because it involves only the two extreme values. It does not tell us anything about the remaining values in our set of data.

One measure of dispersion (variation) that considers all scores in a given set of data is the **standard deviation**.

In order to find the standard deviation for a given set of data, we must perform a number of tasks. First we must find the **deviation from the mean** for each value in the given set of data. As an example of this process, consider the set of scores $\{60, 65, 70, 70, 70\}$. The mean for this set of data is 67. The deviation from the mean is found by subtracting the mean from each value in the set of data:

The deviation for 70 is $70 - 67 = 3$.
The deviation for 65 is $65 - 67 = -2$.
The deviation for 60 is $60 - 67 = -7$.

This is summarized in Table 5.1.

TABLE 5.1

Value	Deviation from the Mean	
70	$70 - 67 = 3$	
70	$70 - 67 = 3$	$\text{Mean} = \frac{335}{5} = 67$
70	$70 - 67 = 3$	
65	$65 - 67 = -2$	
60	$60 - 67 = -7$	
Sum 335	0	

After we have found the deviation from the mean for each score, we square each of the deviations (see Table 5.2). We need to do this in order to find a value that is useful in measuring dispersion or deviation about the mean. Note that the sum of the deviations is 0. This is true for any set of data. Therefore, you can use this information to check the accuracy of your work, particularly the accuracy of your calculation of the mean.

TABLE 5.2

Value	Deviation	$(\text{Deviation})^2$
70	3	9
70	3	9
70	3	9
65	-2	4
60	-7	49
Sum 335	0	80

Now that we have squared each deviation, we have a set of positive values. The sum of these values in Table 5.2 is 80. We next find the mean of these squared deviations, called the **variance**. The mean of the squared values is $80 \div 5 = 16$. The variance also indicates dispersion, but we only mention it in passing; we need the variance in order to find the standard deviation. We find the standard deviation by finding the principal square root of the variance. Hence for our example,

$$\text{Standard deviation} = \sqrt{16} = 4$$

The standard deviation for a set of data is usually represented by the lowercase Greek letter, σ, called *sigma*. Therefore, we can write $\sigma = 4$ for our given set of data.

Admittedly, there is a lot of work involved in finding the standard deviation, but, if it is done in an orderly and precise manner, it is not difficult.

To find the *standard deviation* for a given set of data, we must follow these steps:

1. Find the *mean* for the given set of data.
2. Find the *deviation from the mean* for each value in the set of data.
3. Square each deviation.
4. Find the mean of the squared deviations.
5. Find the principal square root of this number.

EXAMPLE 2 ***Standard Deviation*** Find the standard deviation for the following data:

$$\{5, 7, 9, 13, 16\}$$

Solution We follow the steps outlined previously.

1. The mean is

$$\frac{5 + 7 + 9 + 13 + 16}{5} = \frac{50}{5} = 10$$

2. The deviation from the mean for each value is

$$5 - 10 = -5, \quad 7 - 10 = -3, \quad 9 - 10 = -1,$$
$$13 - 10 = 3, \quad 16 - 10 = 6$$

3. Now we square each deviation:

$$(-5)^2 = 25, \quad (-3)^2 = 9, \quad (-1)^2 = 1,$$
$$(3)^2 = 9, \quad (6)^2 = 36$$

4. To find the mean of the squared deviations, we sum the squares and divide by 5 (the number of values):

$$\frac{25 + 9 + 1 + 9 + 36}{5} = \frac{80}{5} = 16$$

5. We now find the principal square root of this number:

$$\sqrt{16} = 4$$

Hence,

$$\text{Standard deviation} = \sigma = 4$$

The fact that we obtained a standard deviation of 4, as we did in the discussion prior to this example, is pure coincidence.

EXAMPLE 3 ***Standard Deviation*** Find the standard deviation for the given set of data.

$$\{20, 22, 26, 26, 28, 34\}$$

Solution We combine the necessary operations in the accompanying table.

Value	Deviation from the Mean	$(\text{Deviation})^2$
34	$34 - 26 = 8$	64
28	$28 - 26 = 2$	4
26	$26 - 26 = 0$	0
26	$26 - 26 = 0$	0
22	$22 - 26 = -4$	16
20	$20 - 26 = -6$	36
Sum 156	0	120

$$\text{Mean} = \frac{156}{6} = 26$$

$$\text{Mean of squared deviations} = \frac{120}{6} = 20 \quad \text{(variance)}$$

$$\text{Standard deviation} = \sigma = \sqrt{20} \approx 4.5$$

NW ***Now work Problem 3.***

Biography: David H. Blackwell

David H. Blackwell (b. 1919) entered college at the University of Illinois at the age of sixteen with the ambition of becoming an elementary school teacher. By 1941 he had received a Ph.D. in mathematics, making him the sixth African-American to receive a doctorate in mathematics in this country (Elbert F. Cox [b. 1895, d. 1969] being the first, from Cornell University in 1925.) He had a distinguished career as a scholar and teacher, creating major contributions in the fields of statistics, probability, game theory, set theory, dynamic programming, and information theory. He has published over 90 papers and books. The book *Theory of Games and Statistical Decisions* (1954), coauthored with Meyer Girshick, is considered a classic in the area. Despite his influence in scholarly work, during an interview he was quoted as saying, "I'm not interested in doing research and never have been. I'm interested in understanding, which is quite a different thing." Dr. Blackwell was one of the first African-Americans to be elected to the National Academy of Science (1965) and was president of the Institute of Mathematical Statistics and the Bernoulli Society. He has been a recipient of the John von Neumann Theory Prize (1979) and the R. A. Fisher Award. Since 1964 he has been professor of statistics at the University of California at Berkeley. Finally, what does Dr. Blackwell have to say about mathematics? "There is beauty in mathematics at all levels, all levels of sophistication and all levels of abstraction."

The standard deviation for the set of data given in Example 3 is approximately 4.5. We approximated $\sqrt{20}$ by means of Table 2 in the Appendix, or by using a calculator with a square root key.

EXAMPLE 4 ***Standard Deviation*** Nancy took seven tests in her math class. Following is her set of test scores:

$$\{72, 74, 74, 77, 82, 83, 84\}$$

Find the standard deviation for her set of test scores.

Solution We combine the necessary operations in the following table:

Score	Deviation	(Deviation)^2
84	6	36
83	5	25
82	4	16
77	−1	1
74	−4	16
74	−4	16
72	−6	36
Sum 546	0	146

$$\text{Mean} = \frac{546}{7} = 78$$

$$\text{Mean of squared deviations} = \frac{146}{7} \approx 20.86 \quad \text{(variance)}$$

$$\text{Standard deviation} = \sigma = \sqrt{20.86} \approx 4.6$$

Note that in Example 4 the mean of the squared deviations did not come out evenly; that is, $\frac{146}{7}$ was rounded off to 20.86 and then we approximated the square root of 20.86 by using Table 2 in the Appendix, or by using a calculator.

NW ***Now Work Problem 5.***

A large standard deviation indicates that the data are scattered widely about the mean, whereas a small standard deviation indicates that the data are closely grouped about the mean. As an example, consider the following two sets of data:

$A = \{4, 5, 6, 7, 8\}$	$B = \{2, 4, 6, 8, 10\}$
The mean for set A is 6 and the standard deviation is $\sqrt{2} \approx 1.41$	The mean for set B is 6 and the standard deviation is $\sqrt{8} \approx 2.83$

Now that we are able to calculate the standard deviation, what can we do with it? Suppose that Tony scored 78 on a mathematics exam. The mean score for this exam was 73 and the standard deviation was 5. What does Tony know? First, he scored above the mean—that is, he obtained a better than average score on his exam. Second, he scored one standard deviation above the mean. On the next exam in his mathematics class, Tony scored 47. Did he do better or worse on the second test? In order to answer this, we must know the mean and the standard deviation. Suppose that the second exam has a mean of 41 and a standard deviation of 3. Therefore, Tony again scored above the mean and, in fact, he scored 2 standard deviations above the mean $[41 + (2 \times 3) = 47]$, which indicates that he performed better on the second exam than he did on the first, relative to the rest of his class.

EXAMPLE 5 ***Best Grade*** Sue and Jim took their midterm exams in their respective statistics courses. Sue is in Mr. Data's class, where she scored 79. In Mr. Data's class the mean was 75 and the standard deviation was 4. Jim is in Ms. Mode's class, where he also scored 79. But in Ms. Mode's class the mean was 73 and the standard deviation was 3. Who performed better in his or her respective class?

Solution Sue's score was actually one standard deviation above the mean, $(75 + 4 = 79)$, while Jim's score was two standard deviations above the mean for his class $[73 + (2 \times 3) = 79]$. Jim performed better compared to his classmates in Ms. Mode's class than did Sue compared to her classmates in Mr. Data's class.

EXAMPLE 6 ***Improved Grade*** Frank, Louie, Janie, and Irene are all in the same history class. Their scores for the first two exams in their history class are listed in the following table.

	Exam 1	Exam 2
Frank	78	66
Louie	66	66
Janie	90	70
Irene	72	70

The first exam has a mean of 78 and a standard deviation of 6, whereas the second exam had a mean of 66 and a standard deviation of 4. Which of the four people improved on the second exam, which did worse, and which performed the same, relative to the rest of the class?

Solution On each of the exams, Frank scored the mean score (78 on exam 1 and 66 on exam 2), so he performed the same on both exams. Louie scored 66 on the first exam, two standard deviations below the mean $[78 - (2 \times 6) = 66]$. On the second exam, he also scored 66, but that was the mean for exam 2; hence, Louie improved. Janie scored two standard deviations above the mean on the first exam $[78 + (2 \times 6) = 90]$, whereas on the second exam she scored only one standard deviation above the mean $[66 + (1 \times 4) = 70]$. Hence, Janie did worse on the second exam compared to the first. On the first exam, Irene scored one standard deviation below the mean $[78 - (1 \times 6) = 72]$, whereas on the second exam, she scored one standard deviation above the mean $[66 + (1 \times 4) = 70]$. Therefore, Irene's performance improved on the second exam.

NW ***Now Work Problem 17.***

Exercises for Section 5.3

In Exercises 1–10, find the standard deviation for each set of data. (Round off any decimal answer to the nearest tenth.)

1. $\{8, 10, 10, 16\}$

2. $\{1, 2, 3, 4, 5\}$

NW **3.** $\{16, 13, 13, 12, 9, 9\}$

4. $\{15, 10, 10, 8, 8, 9\}$

NW **5.** $\{38, 44, 46, 48, 48, 48, 50\}$

6. $\{18, 20, 21, 21, 23, 23, 28\}$

7. $\{1, 2, 7, 10, 15\}$

8. $\{10, 14, 16, 20\}$

9. $\{68, 68, 70, 72, 66, 73, 78, 65, 80, 70\}$

10. $\{82, 80, 81, 87, 86, 86, 88, 84, 82, 84\}$

11. **Bowling** A sample of 10 bowlers was taken in a tournament. The number of strikes recorded for each bowler in his or her first game was as follows: $\{2, 3, 4, 5, 5, 6, 3, 2, 7, 3\}$. For this set of data, find

a. mean **b.** median
c. mode **d.** range
e. midrange **f.** standard deviation

12. **Quiz** On a computer-lab quiz, the following scores were made in a class of 10 students: 72, 83, 86, 97, 90, 70, 65, 71, 80, 86. For this set of scores, find

a. mean **b.** median
c. mode **d.** range
e. midrange **f.** standard deviation

13. **Life span** The maximum recorded life spans (to the nearest year) of 10 different animals are listed here:

Baboon	36	Beaver	21
Polar bear	35	Dog (domestic)	20
Cat (domestic)	28	Horse	46
Elephant (African)	60	Lion	25
Gorilla	39	Cow	30

For this set of life spans, find the following:

a. mean **b.** median
c. mode **d.** range
e. midrange **f.** standard deviation

14. **Snowfall** The accompanying table gives the typical annual snowfall (in inches) for some of the "snowiest" places in the United States.

Place	Annual Snowfall (inches)
Anchorage, AK	68.5
Flagstaff, AZ	96.4
Denver, CO	59.8
South Bend, IN	72.2
Portland, ME	70.7
Marquette, MI	126.0
International Falls, MN	61.5
Buffalo, NY	91.7
Syracuse, NY	110.5
Burlington, VT	77.1
Casper, WY	80.1

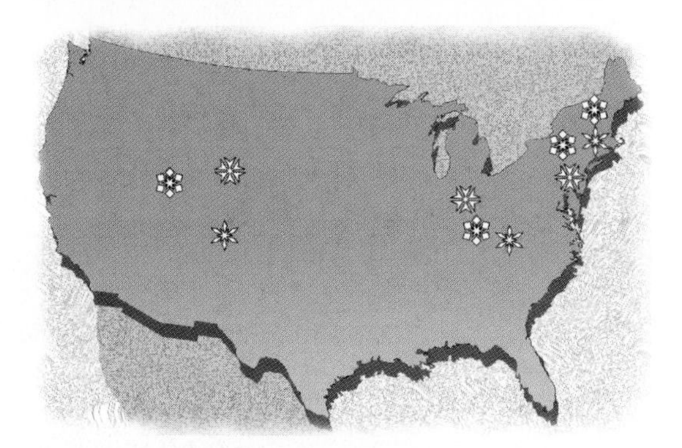

For this set of data, find the following (round decimal answers to the nearest tenth of an inch):

a. mean **b.** median
c. mode **d.** midrange
e. range **f.** standard deviation

*15. **Golf** The back nine holes of the Southampton Golf Club have the following lengths in yards: 150, 370, 310, 340, 200, 450, 490, 420, 375. For this set of measurements, find the following:

a. mean **b.** median
c. mode **d.** range
e. midrange **f.** standard deviation

16. **Temperature** The U.S. Weather Bureau reported the following high temperatures for 10 major cities in the United States: 75, 65, 50, 90, 80, 60, 70, 85, 80, 85. For this set of data, find

a. mean **b.** median
c. mode **d.** range
e. midrange **f.** standard deviation

NW 17. **Exam grade** Hugh, Norma, Ruth, Joe, and Doris are all in the same statistics class. Their scores for the first two exams in the class are listed in the accompanying table. The first exam had a mean of 84 and a standard deviation of 6, whereas the second exam had a mean of 78 and a standard deviation of 4.

	Exam 1	Exam 2
Hugh	84	78
Norma	90	74
Ruth	66	78
Joe	78	70
Doris	84	78

a. Who improved on the second exam?
b. Who did the poorest on the second exam?
c. Who performed the same on both exams?

18. **Exam grade** Two history exams were given. On the first exam, the mean was 74 and the standard deviation was 8. On the second exam, the mean was 48 and the standard deviation was 10. The scores of six students who took the exams are listed in the accompanying table.

a. Who improved on the second exam?
b. Who improved the most on the second exam?
c. Who did not improve on the second exam?

	Exam 1	Exam 2
Eric	82	48
Maria	78	53
Rudy	70	58
Maureen	58	58
Jeff	74	53
Mark	90	78

d. Considering both exams, which student did the poorest?
e. Who performed the same on both exams?
f. Considering both exams, which student has the best grade so far?

Writing Mathematics

19. Explain the difference between range and standard deviation.

20. Which of the two measures of dispersion discussed in this section do you think is more meaningful to the general public: range or standard deviation? Explain why.

21. Describe two purposes for calculating the sample standard deviation, σ.

Challenge

22. Consider for a moment the range of a set of scores. What effect do you think increasing the sample size will have on the range? Why?

Research/Group Activities

23. We briefly mentioned the concept of *variance*. Consult a statistics book and prepare a report on this concept.

24. As part of our procedures for calculating standard deviation, we found the *sum of squared deviations*. This sum is formally called a *sum of squares* and it is part of a statistical procedure called *analysis of variance* (ANOVA). Consult a statistics book and prepare a report on ANOVA and how sum of squares is used in ANOVA.

Just for fun

How many triangles are there in this figure?

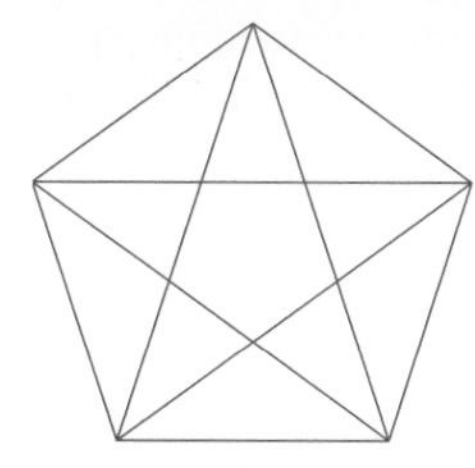

5.4 MEASURES OF POSITION (PERCENTILES)

Consider the case of Sue and Dale, who are applying for a job. Both people have to fill out an application form. On the form, one question asks for the applicant's high school class rank. Sue's rank in her class was 30th, whereas Dale's class rank was 10th. That is, Sue was 30th in her class, whereas Dale was 10th in his class. At first glance it seems that Dale was the better student and probably should get the job. But this is not necessarily the case, because Sue and Dale went to different high schools. We do not have enough information for a fair comparison.

Percentiles

Suppose Sue's rank was 30th in a class of 300, whereas Dale's rank was 10th in a class of 50. This points out that rank is useless as a measure of position unless we know how many are in the total group. It would appear that Sue's rank is similar to Dale's. In order to describe their relative positions better, we can find the **percentile rank**, or **percentile**, for Sue and Dale.

In Sue's class, there were $300 - 30 = 270$ people ranked below her. Therefore,

$$\frac{270}{300} = \frac{27}{30} = \frac{9}{10} = 0.90 = 90\%$$

of Sue's classmates were ranked lower than she was. We can say that Sue was in the 90th percentile. Again, this means that 90% of her classmates were ranked below her.

Computing Dale's percentile rank, we find that $50 - 10 = 40$ people ranked below him. Hence,

$$\frac{40}{50} = \frac{4}{5} = 0.80 = 80\%$$

of Dale's classmates were ranked lower than he was. Dale was in the 80th percentile in his class.

Percentiles divide sets of data into 100 equal parts; hence, 100% is the basis of measure. Practically every student has encountered a percentile rank at one time or another. Every graduating senior has a percentile rank in his or her graduating class. Standardized achievement tests, intelligence tests, and perception tests give results in terms of percentiles. When a student is told that he scored at the 77th percentile on an exam, he then knows that he scored better than 77% of those that took the exam or that 77% of those taking the exam scored lower than he did.

Note: In each of these examples, it is assumed that each measure is distinct. That is, no two of the measures are the same.

EXAMPLE 1 ***Percentile Rank*** In a class of 100 students, Bob has the rank of 12th. What is Bob's percentile rank in the class?

Solution There are $100 - 12 = 88$ students ranked below Bob. Hence, we compare 88 to 100:

$$\frac{88}{100} = 0.88 = 88\%$$

Bob's percentile rank is 88; he is in the 88th percentile.

Biography: Karl Friedrich Gauss

Karl Friedrich Gauss (1777–1855) was one of the eminent mathematicians of the nineteenth century. Along with Archimedes and Isaac Newton, he is considered one of the three greatest mathematicians of all time.

Gauss had many interests, including astronomy, physics, and mathematics. He discovered a technique for computing the orbits of the asteroids and helped develop electromagnetic theory. In mathematics, his major contributions were in the areas of number theory, theory of functions, probability, and statistics.

The "prince of mathematicians," as he was called by his contemporaries, loved to work long, difficult problems. As a result of this penchant, he was able to predict, after weeks of calculations, the correct orbit of the asteroid known as Ceres.

EXAMPLE 2 ***Percentile Rank*** In a class of 120 students, Sam has the rank of 32nd. What is Sam's percentile rank?

Solution The number of students ranked below Sam is $120 - 32 = 88$; his percentile is computed as

$$\frac{88}{120} = \frac{22}{30} = \frac{11}{15} \approx 0.73 = 73\%$$

Sam's percentile rank is 73.

NW ***Now Work Problems 1 and 3.***

EXAMPLE 3 ***Amount Below a Percentile*** In a statistics class of 40 students, an exam was given and Pam scored at the 75th percentile. How many students scored lower than Pam?

Solution Since Pam scored at the 75th percentile, 75% of the students scored less than she did.

$$75\% = 0.75, \qquad 0.75 \times 40 = 30$$

Therefore, 30 students scored lower than Pam on the exam.

NW ***Now Work Problem 5.***

EXAMPLE 4 ***Numerical Rank*** In a class of 30 students, Larry has a percentile rank of 70. What is Larry's rank in the class?

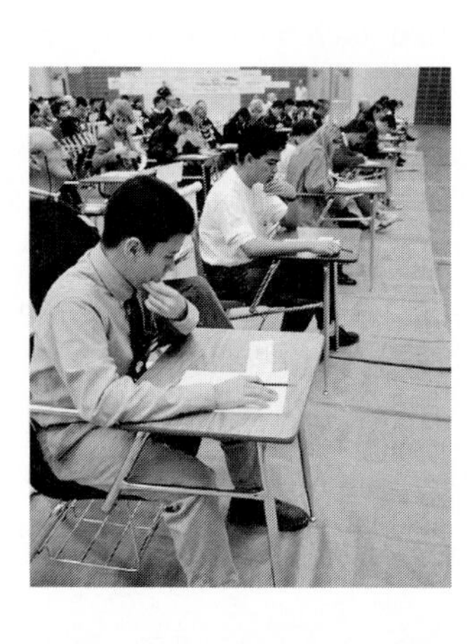

Solution Larry's percentile rank indicates that 70% of the students have scores lower than Larry's on the tests given to the class.

$$70\% \text{ of } 30 = 0.70 \times 30 = 21$$

Twenty-one students have scored lower than Larry. Hence, Larry's rank in the class is ninth, because $30 - 21 = 9$.

To determine at what percentile a student scored on an exam, we need to find the number of students who scored below the individual and divide that number by the total number of students in the class.

EXAMPLE 5 ***Percentile Rank*** The scores of 10 students on a math quiz were

4, 10, 12, 14, 16, 18, 20, 20, 22, 22

Tracy's score on the quiz was 16. What is her percentile rank?

Solution Since Tracy's score was 16, there were four scores lower than hers. We divide 4 by 10, the total number of students who took the quiz: $4 \div 10 = 0.40 = 40\%$. Therefore, Tracy's percentile rank is 40; she is in the 40th percentile.

EXAMPLE 6 ***Numerical and Percentile Rank*** The scores of 10 students on a history quiz were

8, 11, 13, 15, 17, 19, 20, 21, 24, 25

Mike's score on the quiz was 21.

a. What is the rank (from the top) for Mike's score?
b. What is Mike's percentile rank?

Solution

a. There were two scores higher than Mike's, 24 and 25. His score of 21 was third.
b. Since Mike's score was third, there were seven scores lower than his score of 21. We divide 7 by 10; $7 \div 10 = 0.70 = 70\%$. Hence, Mike is in the 70th percentile.

Quartiles

The 25th percentile, 50th percentile, and 75th percentile are probably the most commonly used percentiles in educational testing. They are unique in that they divide sets of data into fourths, or quarters; hence they are referred to as **quartiles**.

Each set of data has three quartiles. The procedure for determining the value of the quartiles is the same as for percentiles. The **first quartile** for a set of data is the value that has 25%, or one-fourth of the data (scores) below it. The **second quartile** is the value that has 50%, or one-half of the data (scores) below it. The **third quartile** is the value that has 75%, or three-fourths of the data (scores) below it.

EXAMPLE 7 ***Quartiles*** Given the following set of scores:

$\{60, 70, 72, 73, 73, 80, 82, 84, 84, 85, 87, 88\}$ find

a. the first quartile
b. the median
c. the third quartile

Solution

a. The first quartile is the value that has 25% of the scores below it. There are 12 scores; 25% (one-fourth) of 12 is 3. So the value for the first quartile is the fourth score (three scores will lie below it), which is 73.
b. The median is the measure that determines the middle of a given set of data. There are 12 scores (an even number), so we must find the mean of the sixth and seventh scores:

$$\frac{80 + 82}{2} = 81$$

The median is 81.

c. The third quartile is the value that has 75% of the scores below it; 75% (three-fourths) of 12 is 9. The value for the third quartile is the tenth score (nine scores lie below it), which is 85.

NW ***Now Work Problem 11.***

EXAMPLE 8 ***Quartiles*** Given the following set of data:

$$\{30, 35, 36, 37, 38, 39, 40, 41\} \quad \text{find}$$

a. the first quartile
b. the third quartile

Solution

a. There are eight pieces of data; one-fourth of 8 is 2. Hence, the third piece of data from the bottom, 36, is the first quartile, or 25th percentile.
b. Three-fourths of 8 is 6. Therefore, the seventh piece of data from the bottom, 40, is the third quartile, or 75th percentile.

You should be aware that a datum or score has no significance or meaning by itself. If Helen reported that she scored 92 on her last exam in history, many people would jump to the conclusion that she did well on the exam. But that might not be the case. If Helen had 92 points out of a possible 200, then the score has more meaning, and we would conclude that Helen did not do well on the exam. On the other hand, if Helen had 92 points out of a possible 100, then we could conclude that she did do well on the exam. We could also better judge Helen's performance on the exam if we knew her score's location relative to the other students' scores on the exam. If, for instance, Helen reported that her score was at the 90th percentile, then we know that 90% of the scores are below Helen's.

B.C. by permission of Johnny Hart and Creators Syndicate, Inc.

Exercises for Section 5.4

NW 1. In a class of 200 students, Julia has a rank of 12th. What is Julia's percentile rank in the class?

2. In a senior class of 300 students, Erin has a rank of 30th. What is Erin's percentile rank in the class?

NW 3. In a 10-kilometer "fun-run" Dave finished 14th. If 200 contestants entered the race, what was Dave's percentile rank?

4. Latoya scored at the 90th percentile on a certain skills test in her gymnastics class of 20 students. What is her rank in the class?

NW 5. In a statistics class of 30 students, an exam was given and Eddie scored at the 80th percentile. How many students scored lower than Eddie?

6. In Hugh's mathematics class, there are 32 students including Hugh. On the last exam, Hugh's score was at the third quartile. How many students scored lower than Hugh?
7. In a class of 40 students, Don has a percentile rank of 80. What is Don's rank in the class?
8. Doris is ranked at the first quartile in her senior class of 300 students. What is her rank in the class?
9. Jessie is ranked at the third quartile in her economics class of 60 students. What is her rank in the class?
10. The following data represent the heights in inches of the starting five for the Dunkem basketball team: 73, 78, 80, 82, 85.
    a. What is the rank (from the top) of the height of 82 inches?
    b. What is the percentile rank of the height of 82 inches?

NW 11. Given the 20 scores 62, 63, 64, 65, 68, 72, 73, 74, 75, 76, 77, 82, 83, 86, 87, 88, 89, 90, 91, and 92.
    a. What is the rank (from the top) of a score of 90?
    b. What is the percentile rank of a score of 90?
    c. What is the percentile rank of a score of 83?
    d. What score is at the first quartile?
    e. What score is at the third quartile?
    f. What score is at the 70th percentile?

12. Given the 20 scores 62, 64, 66, 69, 70, 72, 74, 75, 80, 82, 83, 86, 87, 88, 90, 91, 92, 94, 97, and 98.
    a. What is the rank (from the top) of a score of 87?
    b. What is the percentile rank of a score of 87?
    c. What score is at the first quartile?
    d. What score is at the third quartile?
    e. What score is at the 90th percentile?
    f. What score is at the 80th percentile?

13. Roberto is ranked 75th in his senior class of 200 students. Larry is in the same senior class and Larry has a percentile rank of 75. Of the two seniors, who has the higher standing in the class?
14. Sarah is ranked eighth in her mathematics class of 35 students. Helen is in the same mathematics class, and Helen has a percentile rank of 85. Who ranks higher, Sarah or Helen?
15. When asked how he did on a midterm exam, Ricci replied that he scored at the "100th percentile." What is your reaction to this statement?
16. Daniel is ranked at the third quartile in his English class of 60 students. Peter is in the same English class, and he is ranked 13th. Who ranks higher, Daniel or Peter?
17. In a class of 50 students, Jay has a rank of 50th. What is his percentile rank?
18. In a senior class of 300 students, Addie has a rank of 51st. What is her percentile rank in the class?
19. Barbara is ranked at the second quartile in her mathematics class of 40 students. What is her rank in the class?
20. In a large group lecture class of 200 psychology students, Bob has the rank of 54th. What is Bob's percentile rank in the class?
21. The weights (in pounds) of 40 elementary school students are given as follows:

62	43	62	90	84	78	46	53
65	73	61	66	76	53	58	87
94	87	83	71	96	64	58	77
85	74	68	63	47	68	86	75
90	42	84	84	53	58	84	62

    a. What is the rank (from the top) of a weight of 94?
    b. What is the percentile rank of 94?
    c. What is the rank of a weight of 78?
    d. What is the percentile rank of 78?
    e. What weight is at the first quartile?
    f. What weight is at the 80th percentile?

22. The heights of 40 students in a mathematics class were recorded (in inches). The resulting data were

65	67	71	73	68	72	59	76
59	65	71	60	68	69	61	74
67	66	75	70	70	64	62	66
70	70	70	61	70	63	62	65
64	67	60	66	67	58	62	64

    a. What is the rank (from the top) of a height of 69?
    b. What is the percentile rank of 69?
    c. What is the rank of a height of 58?
    d. What is the percentile rank of 58?
    e. What height is at the 90th percentile?
    f. What height is at the first quartile?

Writing Mathematics

23. Explain the concept of a percentile.
24. Explain the difference(s) between percentiles and quartiles.

25. Defend or refute the following statement and give an example to illustrate your reasoning: The 50th percentile, the second quartile, and the median are equivalent measures.

Challenge Exercises

26. Assume that a sample data set contains the three scores 20, 45, 87. What is the percentile rank of 87? Would you consider this percentile meaningful? Defend your answer. What conclusion can you make about sample size and percentiles?

27. Consider once again the three-item data set of Exercise 26. Do you think it is correct to assign the score 87 a percentile rank between 66 and 99? Defend your answer.

Research/Group Activities

28. Although a *percentile rank* is often referred to as a *percentile*, the two terms are not exactly synonymous. Consult an appropriate reference and write a short report that explains the difference between the two terms. Include an example as part of your explanation.

29. Two other measures of position that were not discussed in this section are *deciles* and *stanines*. Consult an appropriate reference and prepare a report that describes these two concepts. Include an example as part of your descriptions.

Just for fun

Almost every word that begins with the letter *q* has a *u* as the second letter. Can you name two *q* words that do not have *u* as the second letter? Can you name one?

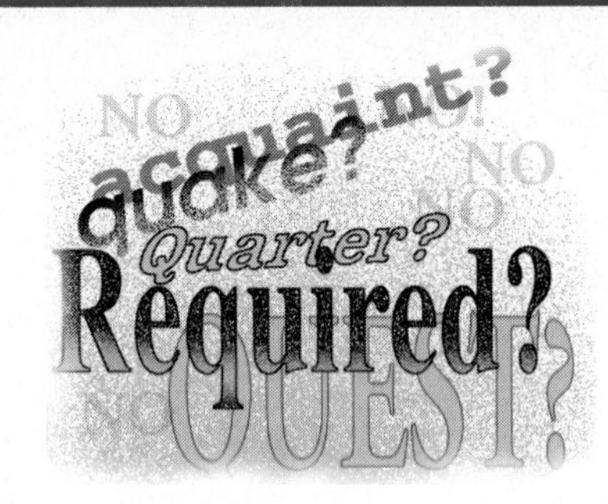

5.5 PICTURES OF DATA

It has been said that a picture is worth a thousand words. It is true that most of us can gain information more easily from pictures than from written material. Architects and contractors use drawings (blueprints) to exchange information. Large corporations use various types of graphs in their reports to stockholders to show how the company has performed. Newspapers and magazines use graphs to show the reader some types of information—usually the presentation of data.

Graphs are used in mathematics to show relationships between sets of numbers. Graphs are useful in the field of statistics because they can show the relationships in a set of data. In this section we shall examine the most common types of graphs used in statistics: the *bar graph*, the *circle* or *pie graph*, the *histogram*, and the *frequency polygon* or *line graph*.

Bar Graphs

A **bar graph** consists of a series of bars of uniform width, with some form of measure on a vertical or horizontal axis. Figure 5.1 is an example of a typical bar graph.

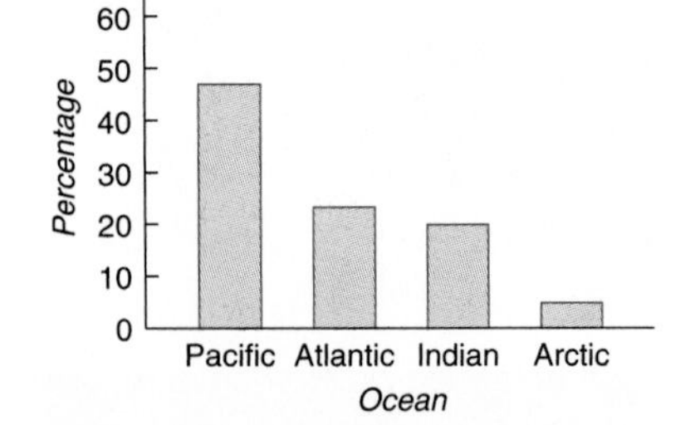

Figure 5.1
Percentage of world's water.

EXAMPLE 1 ***Reading a Graph*** Using Figure 5.1, answer the following:

a. What percentage of the world's water does the Indian Ocean contain?

b. Approximately what percentage of the world's water does the Pacific Ocean contain?

Solution

a. We locate the top of the bar representing the Indian Ocean and then read across to the corresponding percentage. The Indian Ocean contains 20% of the world's water.

b. The Pacific Ocean contains approximately 47% of the world's water.

NW Now Work Problem 7.

EXAMPLE 2 ***Vertical Bar Graph*** The total number of cars sold by Al's Automobile Agency during the years 2000–2004 are as follows:

2000—100 cars, 2001—150 cars, 2002—225 cars,
2003—200 cars, 2004—275 cars

Represent this information on a vertical bar graph.

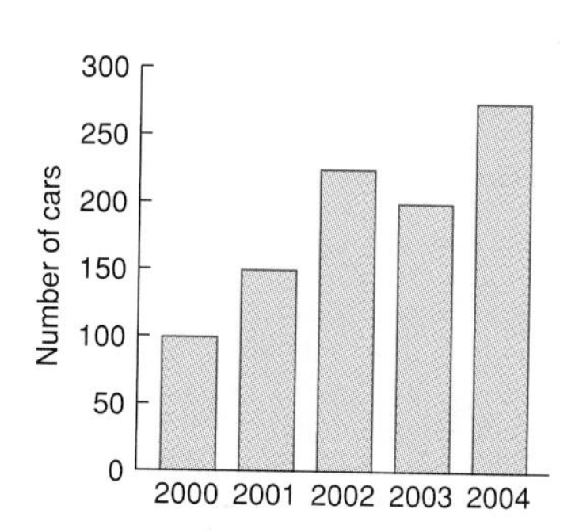

Figure 5.2

Number of cars sold by Al's Automobile Agency, 2000–2004.

Solution A scale should be chosen before the graph is drawn. When the data are given for a period of weeks, months, or years, the time factor is usually indicated on the horizontal axis. Bar graphs can be constructed by beginning with a set of axes and constructing a set of bars (the same width) parallel to each other. [*Note:* The bars are vertical for a vertical bar graph and horizontal for a horizontal bar graph.] See Figure 5.2.

NW Now Work Problem 1.

EXAMPLE 3 ***Horizontal Bar Graph*** The four longest suspension bridges in the United States and their lengths are as follows:

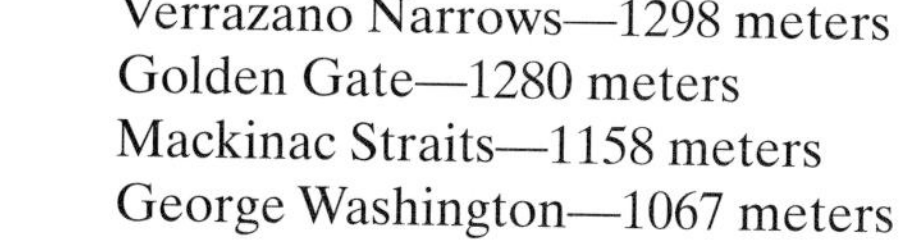

Verrazano Narrows—1298 meters
Golden Gate—1280 meters
Mackinac Straits—1158 meters
George Washington—1067 meters

Represent this information on a horizontal bar graph.

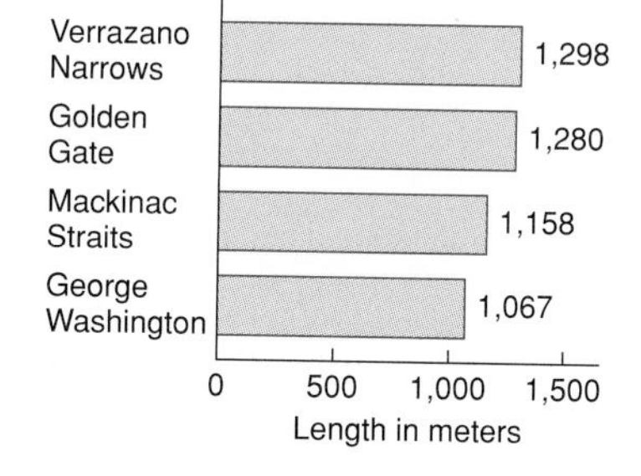

Figure 5.3

The four longest suspension bridges and their lengths.

Solution Figure 5.3 shows the graph. Note that the ends of the bars are labeled to indicate the exact length of each bridge.

NW Now Work Problem 5.

A **comparative bar graph** represents more than one set of related data within one bar graph. For example, suppose we obtain sales figures for two salespeople who sell sneakers at an athletic supply store. Consider the following table:

Day	Number of Pairs of Sneakers Sold by Tom	Number of Pairs of Sneakers Sold by Lisa
Monday	12	14
Tuesday	20	15
Wednesday	22	18
Thursday	16	20
Friday	28	20
Saturday	26	22

Figure 5.4 is a comparative vertical bar graph that represents this information. The graph compares the sales figures for Lisa and Tom, and also compares the sales figures for each day.

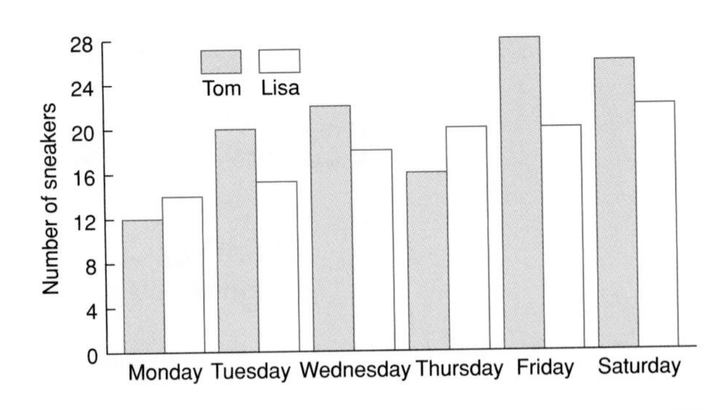

Figure 5.4
Number of pairs of sneakers sold.

Pictograms

A variation of the bar graph is the **picture graph** or **pictogram**. It consists of a series of identical symbols or pictures, each of which represents a specified quantity, and a key explaining what one symbol represents. The number of times each picture or symbol is drawn represents the required total for each quantity.

EXAMPLE 4 ***Pictogram*** Figure 5.5 shows the number of automobile registrations for the year 2004 for four different states.

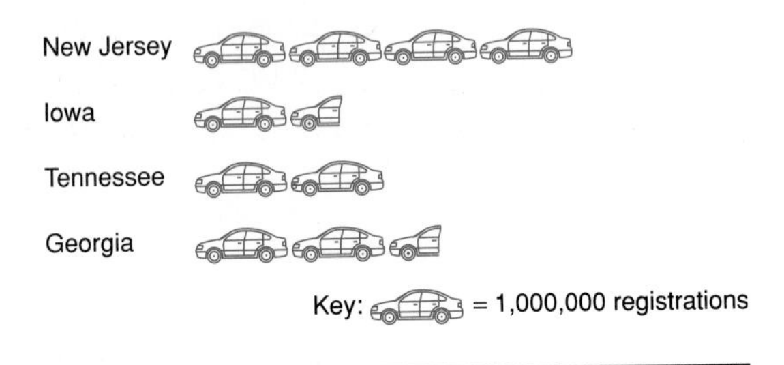

Figure 5.5
Automobile registrations for 2004 in four states.

a. How many automobiles were registered in Tennessee?
b. How many automobiles were registered in Iowa?

Solution

a. Since each picture represents 1,000,000 auto registrations, there were 2,000,000 automobiles registered in Tennessee.
b. There were approximately one and one-half million, or 1,500,000, automobiles registered in Iowa.

NW ***Now Work Problem 15.***

A pictogram is usually easy to read because of its simplicity, but sometimes this is a deterrent to accuracy. It is difficult to determine a precise number when a partial symbol is used.

EXAMPLE 5 ***Pictogram*** The numbers of cars sold by Al's Automobile Agency during the first six months of last year were as follows:

January—80	April—80
February—60	May—85
March—40	June—95

Represent this information by means of a pictogram. Let each symbol represent 10 cars sold.

Solution Figure 5.6 shows the completed pictogram.

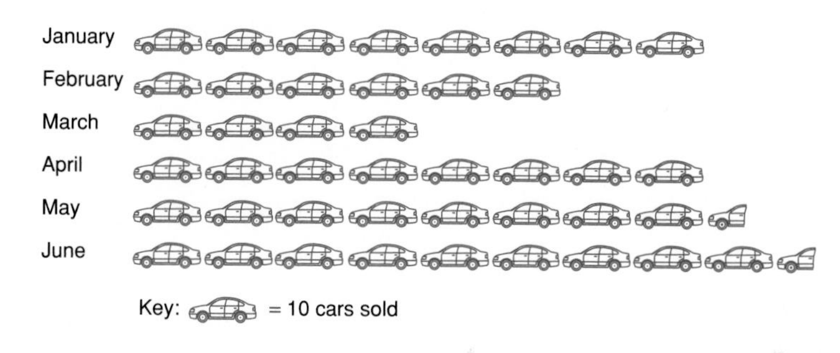

Figure 5.6

The number of cars sold by Al's Automobile Agency during the first six months of last year.

NW ***Now Work Problem 11.***

Circle Graph

The **circle** or **pie graph** is another type of graph that is used quite often. It is particularly useful in illustrating how a whole quantity is divided into parts. Budgets are often illustrated in this manner. Circle graphs used by government agencies often show how the tax dollar is spent or how the tax dollar is collected. Figure 5.7 is an example of a typical circle graph.

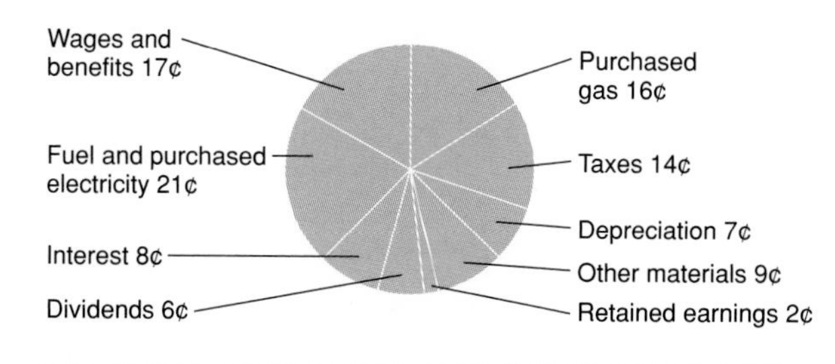

Figure 5.7

Where your utility dollar goes.

EXAMPLE 6 ***Reading a Graph*** Using Figure 5.7, answer the following:

a. How much of every utility dollar is used for wages and benefits?

b. How much of every utility dollar is used for taxes?

Solution

a. 17¢

b. 14¢

When we want to show how a whole quantity is divided into parts, we can use a circle graph. Consider the case of Sam Student. On any given weekday, Sam spends 8 hours sleeping, 6 hours in school, 4 hours doing homework, and 3 hours eating and doing odd jobs. He uses the remaining 3 hours for recreation, leisure, and miscellaneous. We can illustrate how Sam spends his time for a 24-hour period by means of a circle graph.

First recall that a circle contains 360 degrees. In order to illustrate the given data, we must find the percentage or ratio of a 24-hour day spent in each activity and then multiply the percentages by 360 degrees, so that we can divide the circle into proportional parts:

Sleeping:	8 hours	$\frac{8}{24} = \frac{1}{3}$	$\frac{1}{3}$ of $360° = 120°$
School:	6 hours	$\frac{6}{24} = \frac{1}{4}$	$\frac{1}{4}$ of $360° = 90°$
Homework:	4 hours	$\frac{4}{24} = \frac{1}{6}$	$\frac{1}{6}$ of $360° = 60°$
Meals, odd jobs:	3 hours	$\frac{3}{24} = \frac{1}{8}$	$\frac{1}{8}$ of $360° = 45°$
Miscellaneous:	3 hours	$\frac{3}{24} = \frac{1}{8}$	$\frac{1}{8}$ of $360° = 45°$
	24 hours		$360°$

We then use a protractor to construct the corresponding angles in a circle. Figure 5.8 shows the completed circle graph. It does not matter how large or how small the circle is, since every circle represents 100% of the data and can be divided into proportional parts.

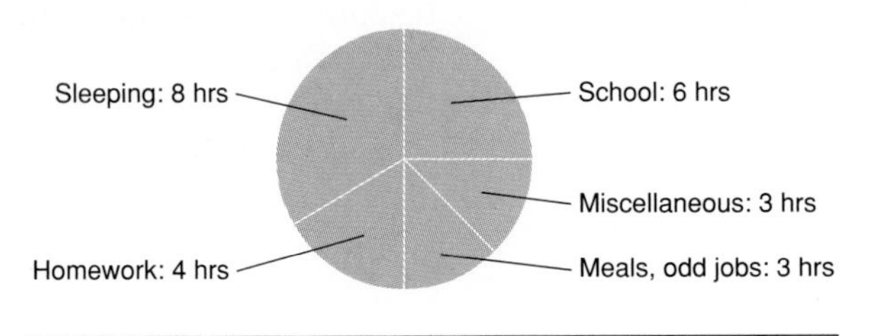

Figure 5.8

Circle graphs are used to show how a whole quantity is divided into parts; however, they are not as useful when we wish to picture other types of data. Consider the following set of 35 scores for an exam in a statistics class:

42,	48,	43,	47,	46,	45,	42
49,	50,	42,	44,	48,	47,	42
42,	48,	50,	45,	46,	42,	49
45,	44,	48,	49,	45,	47,	46
46,	46,	44,	46,	49,	50,	45

Examining these scores does not give us much information because the scores have been presented in the order of their occurrence. We probably could find the highest and lowest score without too much trouble, but we can get a better understanding of the scores if we do the following: First, list the scores in order from highest to lowest, and then count how many of each score we have.

TABLE 5.3

Score	Tally	Frequency
50	|||	3
49	||||	4
48	||||	4
47	|||	3
46	~~||||~~ |	6
45	~~||||~~	5
44	|||	3
43	|	1
42	~~||||~~ |	6

TABLE 5.4

Score	Tally	Frequency
50–51	|||	3
48–49	~~||||~~ |||	8
46–47	~~||||~~ ||||	9
44–45	~~||||~~ |||	8
42–43	~~||||~~ ||	7

Frequency Distribution

The number of times a score occurs is the **frequency** of the score, and Table 5.3 is called a **frequency distribution**. In this particular case, we have an ungrouped frequency distribution because each score value in the distribution is listed separately.

Note that Table 5.3 contains nine categories or classes. We can reduce this number by creating classes that contain more than one score. For example, Table 5.4 has five classes for the same data we used in Table 5.3. Note that each class interval is the same width, and the classes do not overlap. Also, each piece of data belongs in exactly one class. Table 5.4 is called a **grouped frequency distribution**. Grouped frequency distributions representing the same set of data may be different. That is, the classes or intervals constructed by one person may be different from those done by someone else.

But regardless of the preference of the individual, three guidelines should be followed:

1. A range of 5 to 12 classes is desirable to avoid difficulty in interpretation of the data.
2. Classes should not overlap; each piece of data belongs to only one class.
3. Each class should be the same width.

Histogram

The frequency distribution or frequency table in Table 5.3 lists all the different scores and the frequency with which each score occurs. This frequency distribution can be illustrated by means of a **histogram**. A histogram consists of a series of bars that are drawn all with the same width on the horizontal axis, and uniform units on the vertical axis. The frequencies are shown on the vertical axis (see Figure 5.9).

It is important that each bar in a histogram should be drawn in proportion to the frequency of values that occur. Note that both the horizontal and vertical scales have been labeled completely.

Figure 5.10 is a histogram for the grouped frequency distribution in Table 5.4. Note that the bars in a histogram meet one another, because these bars represent classes that are consecutive. That is, one class is 42–43, whereas the next is 44–45, and so on. The frequency is shown by the height of the vertical bar—the higher the bar, the greater the frequency.

Frequency Polygon

A **frequency polygon**, sometimes called a **line graph**, can also be used to graph a frequency distribution. It is constructed in much the same manner as a histogram, using the same kind of vertical and horizontal scales. If we connect the midpoints of the top of each bar in a histogram, the resulting line graph, shown in Figure 5.11, is a frequency polygon. Figure 5.11 uses the same data as Figure 5.9.

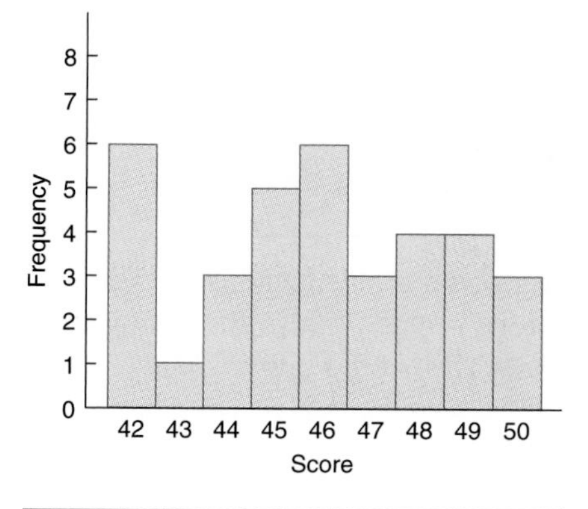

Figure 5.9

Figure 5.10

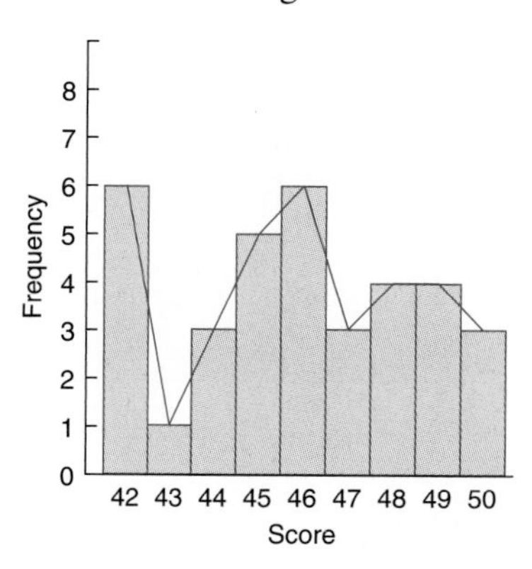

Figure 5.11

EXAMPLE 7 ***Histogram*** Following are the final numerical grades for 30 students in an elementary statistics class. Construct a histogram for these data.

92, 93, 90, 91, 84, 85, 86, 87, 89, 88,
93, 93, 84, 85, 85, 92, 92, 91, 88, 87,
86, 92, 92, 86, 90, 86, 88, 88, 87, 86

Solution First we construct a frequency distribution for the given data:

Grade	Tally	Frequency
93	///	3
92	𝍸	5
91	//	2
90	//	2
89	/	1
88	////	4
87	///	3
86	𝍸	5
85	///	3
84	//	2

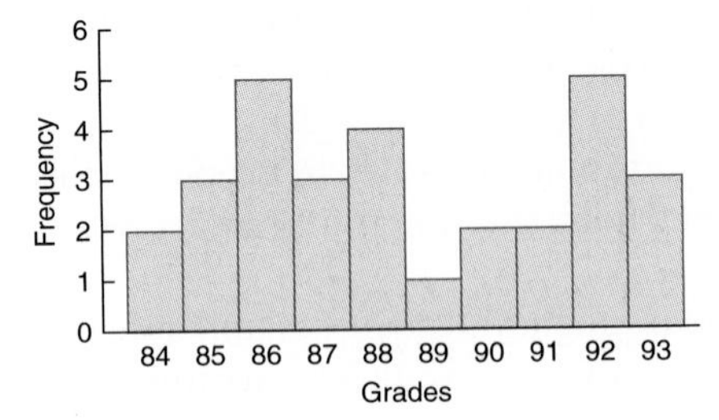

Figure 5.12

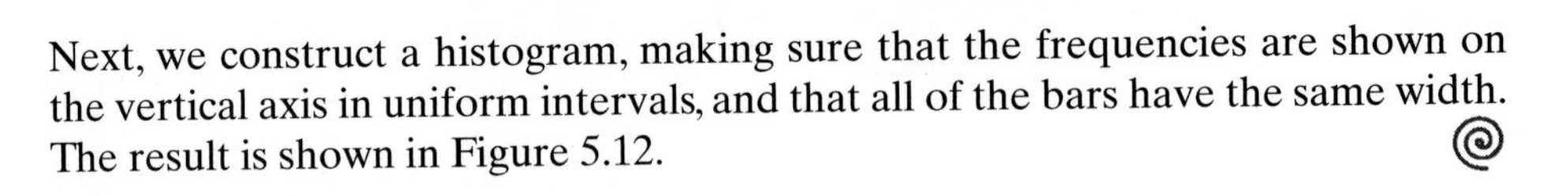

Next, we construct a histogram, making sure that the frequencies are shown on the vertical axis in uniform intervals, and that all of the bars have the same width. The result is shown in Figure 5.12.

NW ***Now Work Problem 23.***

EXAMPLE 8 ***Frequency Polygon*** A student rolled a die 30 times and obtained the following results. Construct a frequency polygon for these data.

6, 5, 4, 4, 5, 6, 1, 2, 1, 6, 4, 3, 3, 3, 4,
2, 2, 5, 6, 4, 1, 2, 4, 3, 5, 5, 3, 3, 4, 2

Solution First we construct a frequency distribution for the given data:

Number	Tally	Frequency
6	....	4
5	𝍸	5
4	𝍸 //	7
3	𝍸 /	6
2	𝍸	5
1	///	3

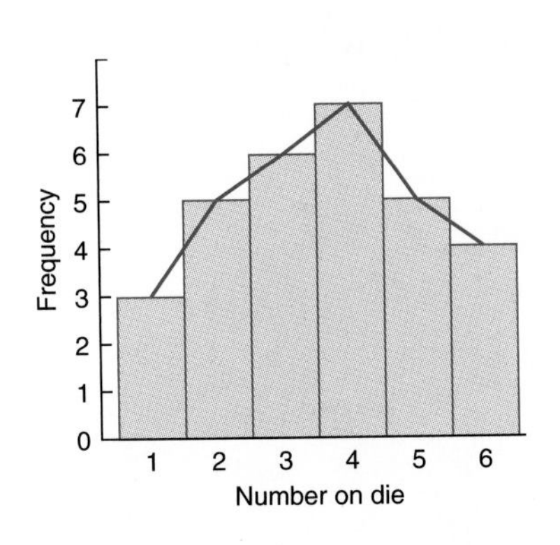

Figure 5.13

To aid us in drawing the frequency polygon, we construct a histogram for the data and then connect the midpoints of the tops of each bar in the histogram. The resulting line graph is the frequency polygon shown in Figure 5.13.

NW ***Now Work Problem 29.***

Stem-and-Leaf Display

As an alternative to frequency distribution tables, data may also be represented using **stem-and-leaf displays**. These are a combination of a graphic technique and a sorting technique. In a stem-and-leaf display, each numerical datum is divided into two parts: the leading digit(s) becomes the **stem**, and the trailing digit(s) becomes the **leaf**. For example, a data value of 12 would be separated into a stem of 1 and a leaf of 2, while 259 would have 25 as the stem and 9 as the leaf. To construct a stem-and-leaf display, we typically use a two-column table with the stems written in the left column and the leaves written in the right column. Each leaf is written in the row that corresponds to the stem of the number from which it was separated.

To illustrate a stem-and-leaf display, consider the following set of 24 scores:

67	35	82	63	71	64	72	74
68	51	58	95	88	80	48	93
95	42	55	85	53	64	61	95

We see there are scores in the 30s, 40s, 50s, 60s, 70s, 80s, and 90s. Use the first digit of each score as the stem and the second digit as the leaf. Draw a vertical line and place the stems, in order, to the left of the line.

Stem	Leaf
3	
4	
5	
6	
7	
8	
9	

Next we place each leaf on its stem. This is done by putting the trailing digit on the right side of the vertical line opposite its corresponding "stem" or first digit. Our first score, 67, has a stem of 6 and a leaf of 7. Thus, we place a 7 opposite the stem 6, 6|7. The next score in the first column is 68, hence we place the 8 after the 7 in the same display, like so: 6|78.

The next score in the first column is 95, so a leaf of 5 is placed on the 9 stem.

6	78
7	
8	
9	5

We continue until each of the other 24 leaves is placed on the display and we get the following table:

Stem	Leaf
3	5
4	28
5	1853
6	783441
7	124
8	2580
9	5535

In the table, note that we have three 5s listed in the leaf column for the stem 9. This is because the score 95 appears three times in our data set. Similarly, the stem 6 has six leaves.

In contrast to frequency distributions, stem-and-leaf displays are relatively simple to construct, they allow you to see every score, and they give you a visual picture of the entire distribution of scores. For example, if we rotate the stem-and-leaf display 90 degrees counterclockwise, we automatically get the histogram of its corresponding grouped frequency distribution.

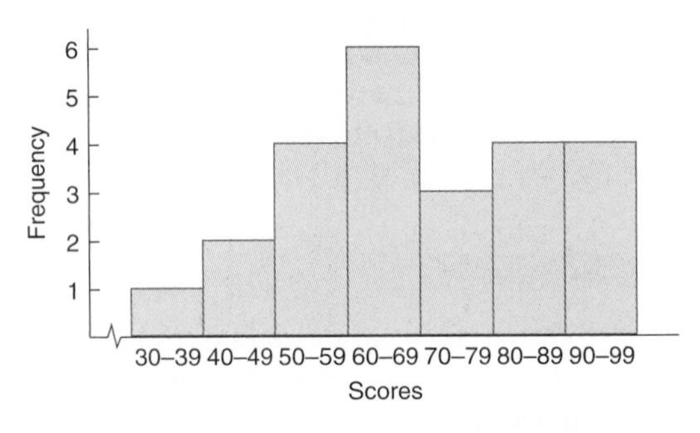

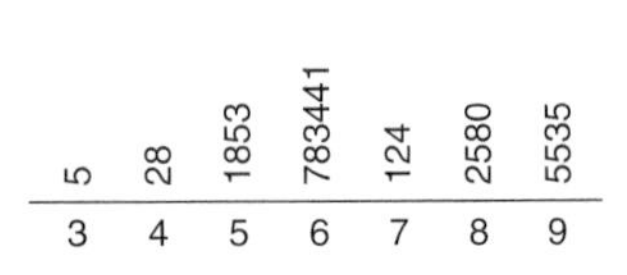

Stem-and-leaf displays are not practical for large data sets. However, they can be used for preliminary data analysis to help statisticians examine a distribution of scores in a relatively simple and easy manner.

Exercises for Section 5.5

NW **1. Life span** The maximum recorded life spans (to the nearest year) of six different animals are listed here:

Elephant (African)	60	Baboon	36
Horse	46	Polar bear	35
Gorilla	39	Cow	30

Construct a vertical bar graph representing this information.

2. Speed of animals The maximum recorded speeds (to the nearest mph) of six different animals are listed here:

Cheetah	70	Lion	50
Pronghorn antelope	61	Quarter horse	47
Wildebeest	50	Elk	45

Construct a vertical bar graph representing this information.

3. Canals The accompanying table gives the location and length (in miles) of six notable canals.

Name	Location	Length (miles)
Amsterdam–Rhine	The Netherlands	45.0
Beaumont–Port Arthur	United States	40.0
Houston	United States	43.0
Panama	Canal Zone	50.7
Kiel	Germany	61.3
Welland	Canada	27.5

Construct a vertical bar graph representing this information.

4. Tallest buildings The five highest buildings in the U.S. and their approximate heights (in stories) are as follows:

Sears Tower (Chicago)	110
Empire State (N.Y.C.)	102
John Hancock Center (Chicago)	100
Aon Centre (Chicago)	80
Chrysler (N.Y.C.)	77

Construct a horizontal bar graph representing this information.

NW **5. Tunnels** In New York City, four notable tunnels are used by motor vehicles. The accompanying table gives the name and length (in kilometers) of each.

Name	Length
Lincoln	4.0
Brooklyn-Battery	3.4
Holland	2.7
Queens-Midtown	2.1

Construct a horizontal bar graph representing this information.

6. **Types of milk** Four types of milk typically available in a supermarket are whole milk, 2%, skim, and buttermilk. The table lists the amount of cholesterol (in milligrams) found in 1 cup of each type.

Variety	Cholesterol (mg)
Whole	33
2%	18
Skim	5
Buttermilk	9

Construct a horizontal bar graph representing this information.

NW 7. **Annual sales** The graph shown in Figure 5.14 represents the annual sales of the Ronolog Corporation for the years 2000–2004.

a. Which year had the lowest sales amount?
b. Approximately how much were sales in 2004?
c. In what years did sales decrease?
d. What was the approximate total sales for the five years?

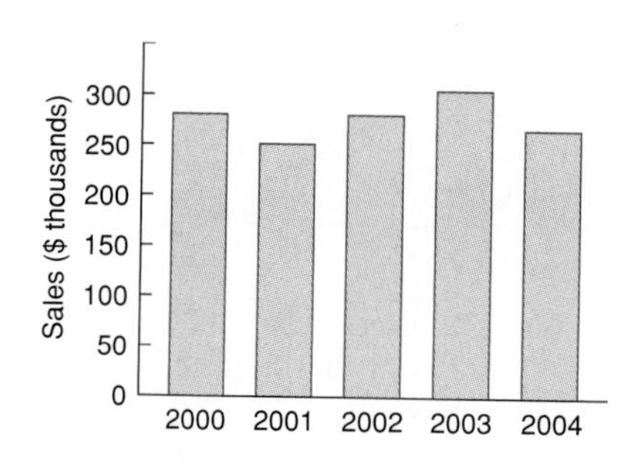

Figure 5.14

Annual sales of the Ronolog Corporation for the years 2000–2004.

8. The graph shown in Figure 5.15 represents the attendance at the annual Jazz Concert at Monroe Community College for 2001–2004.

a. Which year had the lowest attendance?
b. How many people attended the concert in 2003?
c. In what years did attendance increase compared with the previous year?
d. What was the total attendance for the four years?

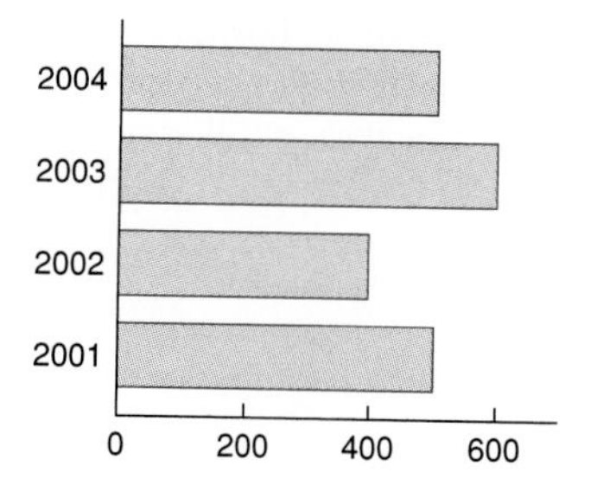

Figure 5.15

Number of people at Jazz Concert.

9. Bill, Mary, Francis, and Floyd are all in the same mathematics class. Their scores for the first two exams in the class are listed in the following table:

	Exam 1	Exam 2
Bill	72	76
Mary	70	74
Francis	78	70
Floyd	80	76

Construct a comparative vertical bar graph representing this information.

10. The heights (in inches) of five students were recorded during the past two years. The results were as follows:

	Last Year	Two Years Ago
Hugh	72	70
Norma	62	60
Ruth	66	64
Julia	64	64
Martin	72	68

Construct a comparative horizontal bar graph representing this information.

NW 11. **Pizza** The numbers of pizzas sold by Phil's Pizza Palace during the last seven days were as follows:

Sunday	90	Thursday	70
Monday	closed	Friday	105
Tuesday	70	Saturday	85
Wednesday	50		

Represent this information by means of a pictogram. Let each symbol ○ represent 10 pizzas sold.

12. **Net income** The net income of the Ketes Company for the years 1999–2004 is as follows:

1999	$80,000	2002	$110,000
2000	$90,000	2003	$130,000
2001	$100,000	2004	$120,000

Represent this information by means of a pictogram. Let each symbol $ represent $20,000.

13. **Bowling** A sample of six bowlers was taken at a tournament. The number of strikes recorded for each bowler in his or her first game was as follows:

Ann	4	Sophie	5
Tom	3	Chris	6
Floyd	2	Martina	4

Represent this information by means of a pictogram. Let each symbol represent one strike.

14. **Tires** The numbers of tires sold by Joe's Tire Mart during the last week were as follows:

Monday	75	Thursday	55
Tuesday	55	Friday	90
Wednesday	35	Saturday	70

Represent this information by means of a pictogram. Let each symbol represent 10 tires sold.

NW 15. **Net income** The graph in Figure 5.16 shows the net income of the Josco Company for the years 2000–2004.

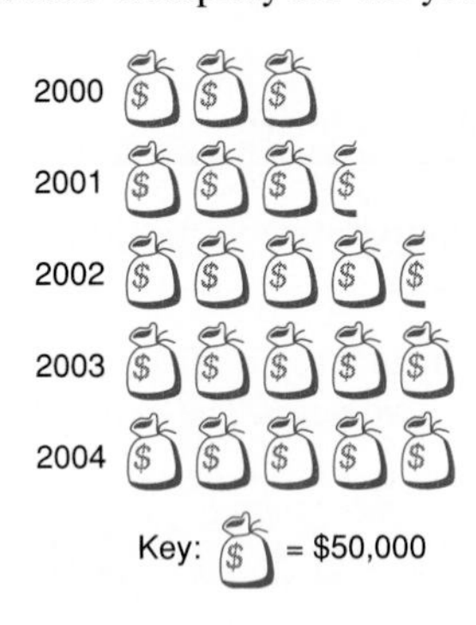

Figure 5.16

Josco Company net income.

a. What was the net income for 2001?

b. What was the net income for 2002?

c. What was the total net income for 2002 and 2003?

d. What was the total net income for 2000–2004?

16. **Parcels delivered** The graph in Figure 5.17 shows the number of parcels delivered by the Divided Parcel Service for the years 2001–2004.

Figure 5.17

Divided Parcel Service deliveries.

a. How many parcels were delivered in 2001?

b. How many parcels were delivered in 2002?

c. What is the total number of parcels delivered in 2002 and 2003?

d. What is the total number of parcels delivered for the years 2001–2004?

17. **Types of transportation** In a survey of 180 students, it was determined that 90 students came to school by bus, 60 students came by car, and 30 students walked. Construct a circle graph representing this information.

18. **Budget** The Garcia family has a budget. Each month they use their income in the following manner: 30% for food, 25% for household expenses, 20% for transportation, 10% for savings, 5% for entertainment, and 10% for unexpected expenses. Construct a circle graph representing this information.

19. **Taxes** The Land of Taxes obtains its money from taxes. Each dollar that the Land of Taxes collects is obtained in the following manner: 25¢ comes from personal income taxes, 25¢ comes from corporate income taxes, 15¢ comes from excise taxes, 20¢ comes from sales taxes, 10¢ comes from highway taxes, and 5¢ comes from miscellaneous taxes. Construct a circle graph representing this information.

20. **Continental land area** The accompanying table indicates the approximate land area of the seven continents. Construct a circle graph (each angle to the nearest degree) representing this information.

Continent	Area in Square Miles	Percentage of World's Land
Asia	16,988,000	29.5
Africa	11,506,000	20.0
North America	9,390,000	16.3
South America	6,795,000	11.8
Europe	3,745,000	6.5
Australia	2,975,000	5.2
Antarctica	5,500,000	9.6
Miscellaneous	687,000	1.1

21. **Revenue** For the fiscal year 2004 the government of Gotham City received the indicated amount of revenue from the following sources:

Federal aid	$70,000,000
Licenses	$10,000,000
State aid	$30,000,000
Sales tax	$20,000,000
Property tax	$60,000,000
Other	$10,000,000

Construct a circle graph representing this information.

22. **Rolling a die** A student rolled a die 40 times; the results are as follows. Construct a frequency distribution and a histogram to represent these data.

6, 4, 3, 5, 6, 2, 4, 4, 5, 2, 1, 5, 2, 4, 1, 6, 6, 6, 1, 5,
2, 6, 3, 1, 6, 1, 6, 2, 1, 2, 2, 4, 4, 6, 4, 1, 3, 4, 5, 5

NW **23. Student heights** The heights of 28 students in a mathematics class were recorded (in inches), as shown. Construct a frequency distribution and a histogram to represent these data.

65, 69, 64, 65, 70, 71, 75, 60, 60, 61, 68, 70, 67, 69,
67, 70, 67, 66, 67, 71, 70, 72, 70, 66, 68, 70, 72, 64

24. Radar A police radar unit measured the speed of 25 cars on a certain street. The resulting data were

23, 38, 24, 26, 18,
23, 52, 30, 45, 27,
28, 25, 28, 37, 29,
33, 27, 34, 36, 32,
23, 18, 23, 38, 21

a. Construct a grouped frequency distribution of the data by using 15–19 as the first class.

b. Using the frequency distribution from part **a**, construct a frequency polygon that represents the data.

25. Coin toss A student tossed three coins together 32 times and after each toss recorded the number of heads. The table presents the results. Construct a frequency polygon that represents these data.

Number of Heads	Frequency
0	4
1	12
2	12
3	4

26. Study time On final-exam day in a statistics class last semester, 40 students were asked to state the number of hours they spent studying for the exam. The resulting data were

3, 6, 6, 6, 14, 8, 10, 9, 4, 1
5, 6, 2, 7, 12, 8, 10, 12, 5, 6
5, 3, 7, 9, 12, 9, 10, 11, 4, 6
9, 4, 8, 8, 11, 13, 10, 11, 6, 4

a. Construct a grouped frequency distribution of the data by using 1–3 as the first class.

b. Using the frequency distribution from part **a**, construct a histogram that represents the data.

27. Test scores The following test scores were received by 33 students in a statistics class:

56, 91, 85, 66, 72, 81, 60, 90, 70, 71, 77,
84, 75, 58, 89, 67, 98, 96, 70, 87, 74, 64,
64, 59, 87, 73, 91, 63, 86, 81, 72, 72, 73

a. Construct a grouped frequency distribution for these scores using the intervals 95–99, 90–94, 85–89, and so on.

b. Use the frequency distribution from part **a** to construct a histogram that represents the data.

28. Snack A survey of 40 students was made in the cafeteria to determine the cost (in cents) of each student's snack, as follows. Construct an ungrouped frequency distribution and a frequency polygon for these data.

92, 93, 86, 93, 93, 92, 90, 84, 92, 91,
85, 86, 84, 85, 90, 86, 92, 86, 87, 92,
87, 92, 88, 89, 91, 88, 87, 88, 88, 86,
93, 84, 86, 84, 85, 86, 85, 92, 92, 88

29. Children's weight The weights (in pounds) of 50 elementary school students are given as follows: NW

62, 43, 62, 90, 84, 78, 46, 53, 44, 92,
65, 73, 61, 66, 76, 53, 58, 87, 83, 71,
94, 87, 83, 71, 96, 64, 58, 77, 76, 58,
85, 74, 68, 63, 47, 68, 86, 75, 77, 71,
90, 42, 84, 84, 53, 58, 84, 62, 68, 74

a. Construct a grouped frequency distribution for these weights using the intervals 95–99, 90–94, 85–89, and so on.

b. Using the frequency distribution from part **a**, construct a frequency polygon that represents the data.

30. Given the following distribution:

Scores	Frequency
56–60	2
61–65	4
66–70	6
71–75	7
76–80	6
81–85	10
86–90	6
91–95	5
96–100	4

a. Construct a histogram and a frequency polygon for these data.

b. What do you think a "good guess" would be for the mode?

c. What do you think a "good guess" would be for the median?

31. The circle graph in Figure 5.18 is titled "Where Your Tax Dollar Goes." Does the graph give enough information? Why or why not?

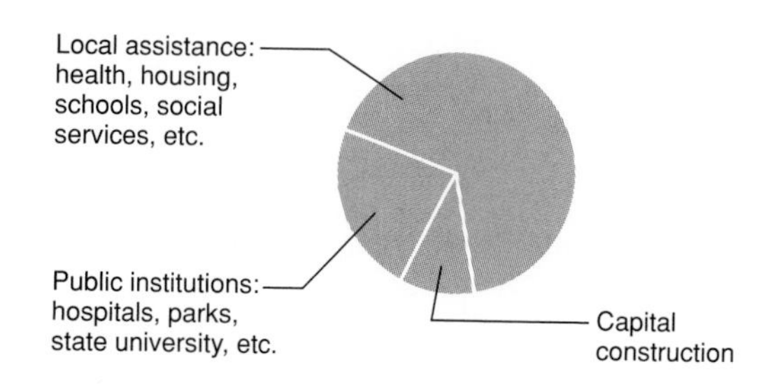

Figure 5.18
Where Your Tax Dollar Goes

32. What can you deduce from the following statement: "Statistics show that more men than women are involved in auto accidents"?

33. Toss a die 30 times and tabulate the results.

a. Construct a frequency distribution showing the results.

b. Construct a histogram that represents the results.

c. Construct a frequency polygon that represents the results.

34. Toss three coins together 32 times and tabulate the number of heads appearing on each toss—that is, 0, 1, 2, or 3 heads.

a. Construct a frequency distribution showing the results.

b. Construct a frequency polygon that represents the results.

c. Construct a percentage distribution for part **a**.

d. Compare the percentages in part **c** with the theoretical percentages:

0 heads—12.5%
1 head—37.5%
2 heads—37.5%
3 heads—12.5%

e. Toss the three coins 32 more times and tabulate the results with those of part **a**. Construct a frequency polygon for the 64 tosses.

Writing Mathematics

35. Briefly describe each of the common types of graphs used in statistics that were discussed in this section.

36. Explain the difference between an ungrouped frequency distribution and a grouped frequency distribution.

37. Defend or refute the following statement: A bar graph and a histogram are the same type of graphs. Include an example of when one type is more appropriate than the other.

38. Describe the advantages and disadvantages of a stem-and-leaf display.

Challenge Exercises

39. When preparing the classes or intervals of a grouped frequency distribution, describe the effect of using intervals that are either too large or too narrow. Give an example as part of your description.

40. Grouped frequency distributions generally use *closed* intervals (e.g., 40–44, 45–49). Under what situation do you think there is a need to use *open* intervals (e.g., <40 or 50+)? Give a sample data set in which an open interval might be used. What is the effect of using an open interval in a grouped frequency distribution?

Research/Group Activities

41. When describing a set of data using a frequency polygon or histogram, the overall appearance of the plot or histogram provides us with information about how the data set is distributed. Several different distributions are possible. Some of the most common ones include *uniform, J-shaped, reversed J-shaped, U-shaped, bimodal, skewed left, skewed right*, and *normal*. Consult a statistics textbook and prepare a report that includes a written and visual description of each of these distributions. As part of your report, include an example of when each type of distribution might result.

42. In addition to ungrouped and grouped frequency distributions, there are *cumulative* and *relative* frequency distributions. Consult a statistics textbook and prepare a report that describes each of these distributions. As part of your report, include advantages and disadvantages of each and when using these types of frequency distributions is appropriate.

43. Select one of your classes in which your teacher gives a multiple choice exam. After the test is graded and returned to the class, collect the raw data about the test results from your teacher or directly from your classmates. The data you collect should include (among others) the number of students who got each question correct as well as the total number of correct responses. Using these data, prepare a graph that describes the test results in a clear, informative, and easy-to-read and understandable manner.

Just for fun

Can you make 10 by using just nine matches?

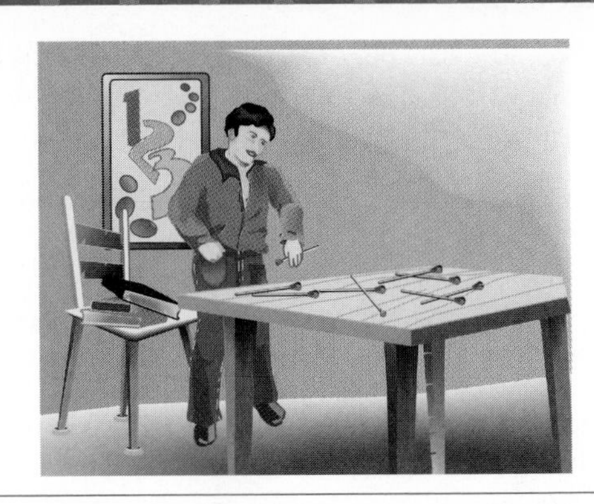

5.6 THE NORMAL CURVE

Two students, Scott and Joe, decided to toss four coins 100 times, record the number of heads appearing for each toss, and then construct a frequency polygon representing the outcomes. Their results were as follows:

1, 2, 2, 4, 2, 2, 1, 1, 0, 3, 1, 0, 2, 2, 3, 1, 4, 2, 2, 3,
2, 1, 1, 4, 2, 2, 2, 1, 2, 4, 2, 3, 2, 2, 2, 1, 2, 2, 3, 2,
2, 0, 3, 3, 2, 1, 1, 3, 2, 1, 4, 0, 2, 3, 2, 1, 2, 1, 3, 1,
1, 3, 3, 3, 3, 3, 2, 1, 2, 2, 1, 1, 2, 1, 0, 3, 2, 3, 2, 3,
3, 3, 2, 2, 1, 0, 2, 1, 2, 1, 4, 1, 4, 0, 0, 2, 1, 3, 2, 2

After recording the number of heads appearing on each toss, Scott and Joe next constructed the frequency distribution in Table 5.5. After constructing the frequency distribution, they constructed the frequency polygon in Figure 5.19.

Figure 5.19

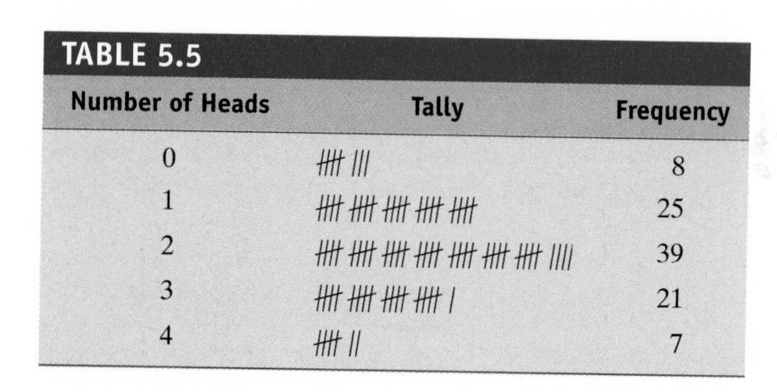

TABLE 5.5

Number of Heads	Tally	Frequency
0	𝍸 |||	8
1	𝍸 𝍸 𝍸 𝍸 𝍸	25
2	𝍸 𝍸 𝍸 𝍸 𝍸 𝍸 𝍸 ||||	39
3	𝍸 𝍸 𝍸 𝍸 |	21
4	𝍸 ||	7

Computing the mean for this experiment, we find it is 1.94. The mode is 2, as is the median. If Scott and Joe had increased the number of tosses, their outcomes would have more closely approached the outcomes for tossing four coins according to the laws of probability. For example, they tabulated 39 outcomes out of 100 where two heads appeared. Recall from Exercise 4b in Section 4.4 that if four coins are tossed, the probability of obtaining two heads is $\frac{6}{16}$, or 0.375, which is the same as 375 out of 1,000. This ratio is quite close to the results that Scott and Joe had $(\frac{39}{100})$. Note that the frequency polygon in Figure 5.19 is similar in shape to the bell-shaped curve in Figure 5.20.

Figure 5.20

Normal Curve

One particular bell-shaped curve has been studied extensively by statisticians and is named the **normal curve**. The normal curve has some unique properties. If we have a **normal distribution**, then the mean, median, and mode all have the same value, and all occur exactly at the center of the distribution. For normally distributed data, it can also be shown that *approximately* 68% of the data will be included within an interval of one standard deviation about the mean—that is, from one standard deviation above the mean to one standard deviation below the mean. For two standard deviations about the mean—that is, from two standard deviations above the mean to two standard deviations below the mean—*approximately* 95% of the data will be included. Practically all of the data, *about* 99.7%, will be included in the interval from three standard deviations above the mean to three standard deviations below the mean. The given percentages for a normal curve are illustrated in Figure 5.21. The area under the curve represents all of the frequencies for a normal distribution.

Note that the tails of the curve do not touch the horizontal axis and they will not, no matter how far they may be extended. Data that lie more than three standard deviations from the mean are rare.

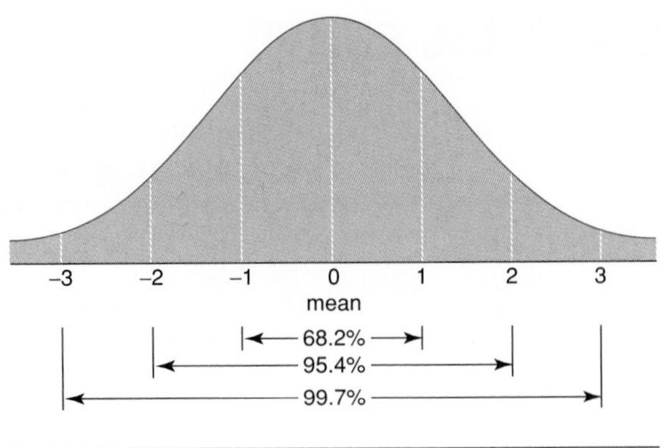

Figure 5.21

An interesting occurrence in natural phenomena is that the data in large samples are often distributed so that the frequency polygon approximates the normal curve. A classic example is the distribution of intelligence quotient (IQ) scores for the entire population of a city, state, or country.

EXAMPLE 1 ***Normal Distribution*** The weights of the first-graders in an elementary school are found to be approximately normally distributed with a mean of 60 pounds and a standard deviation of 5 pounds.

a. What percentage of the students in this group weighs between 55 and 65 pounds?

b. What percentage of these students weighs between 55 and 70 pounds?

Solution Given that the weights are approximately normally distributed with a mean of 60 and a standard deviation of 5, we can draw a normal curve for the weights of this group of first-graders, as shown in Figure 5.22.

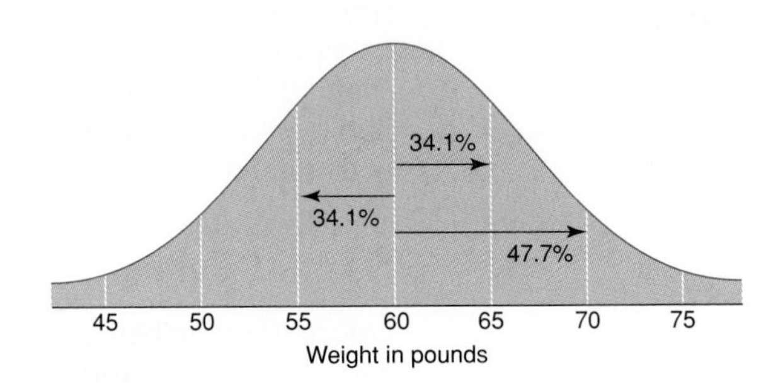

Figure 5.22

Table 5.6 on page 297 indicates what percentage of the data is contained within certain distances from the mean (in one direction only), where distance is measured as a multiple of the standard deviation.

a. Using Figure 5.22, we see that 34.1% + 34.1% = 68.2% of the students weigh between 55 and 65 pounds.

b. 34.1% + 47.7% = 81.8% of the students weigh between 55 and 70 pounds.

NW ***Now Work Problem 1.***

TABLE 5.6

Number of Standard Deviations from Mean (in one direction only)	Percentage of Data	Number of Standard Deviations from Mean (in one direction only)	Percentage of Data
0.1	4.0	1.6	44.5
0.2	7.9	1.7	45.5
0.3	11.8	1.8	46.4
0.4	15.5	1.9	47.1
0.5	19.2	2.0	47.7
0.6	22.6	2.1	48.2
0.7	25.8	2.2	48.6
0.8	28.8	2.3	48.9
0.9	31.6	2.4	49.2
1.0	34.1	2.5	49.4
1.1	36.4	2.6	49.5
1.2	38.5	2.7	49.65
1.3	40.3	2.8	49.7
1.4	41.9	2.9	49.8
1.5	43.3	3.0	49.87

EXAMPLE 2 ***Normal Distribution*** The results of an exam in a statistics class are approximately normally distributed, with a mean of 70 and a standard deviation of 10. The instructor decides that a student will receive an A on the exam if the student scores more than two standard deviations above the mean. What is the lowest grade a student can get and still receive an A?

Solution Because the scores are approximately normally distributed with a mean of 70 and a standard deviation of 10, we can draw a normal curve for this class (See Figure 5.23).

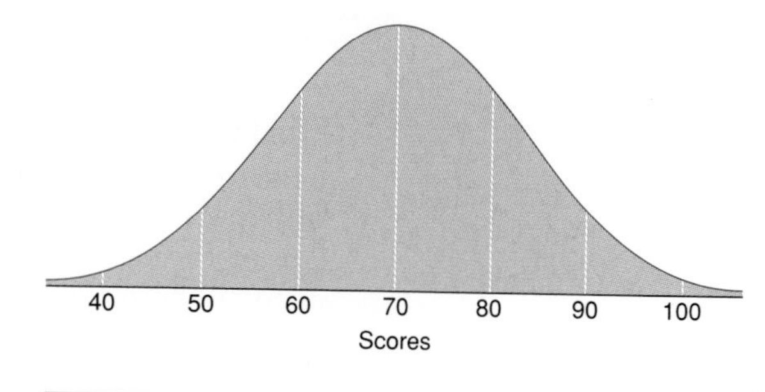

Figure 5.23

A score of 90 is exactly two standard deviations above the mean of 70. But the instructor said that only those students scoring *more* than two standard deviations above the mean would receive an A. Hence, a student must have a score greater than 90 in order to receive an A.

EXAMPLE 3 ***Reading a Normal Curve*** A statistician recorded the mileage on each of 1,000 cars in a parking lot. A frequency polygon of the mileages *approximated* the normal curve in Figure 5.24.

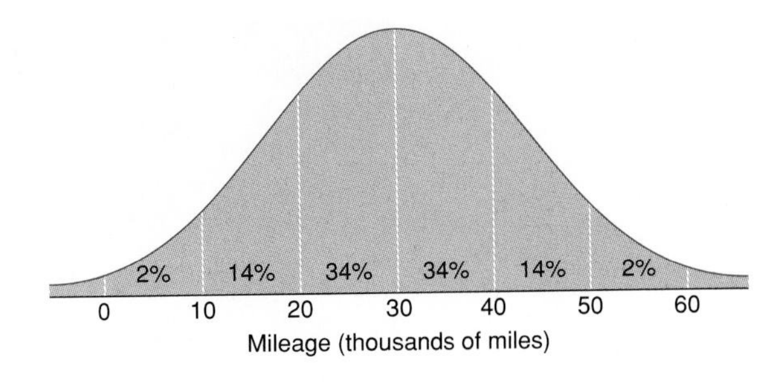

Figure 5.24

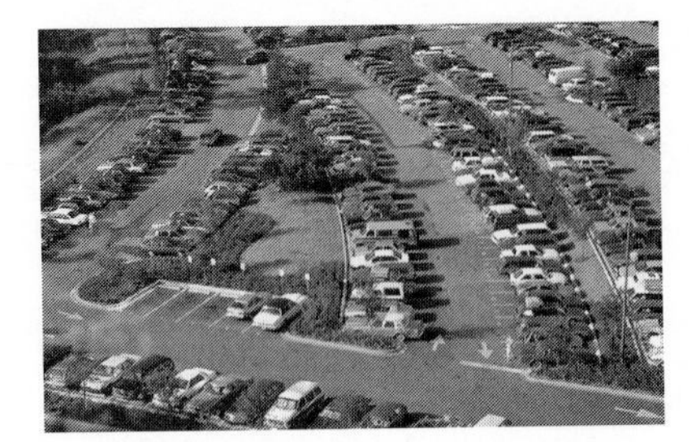

a. What is the mean mileage for this set of cars?

b. What is the standard deviation?

c. What percentage of the cars had less than 20,000 miles on the odometer?

d. How many cars had more than 50,000 miles on the odometer?

Solution

a. Since the data are approximated by the normal curve in Figure 5.24, we see that the mean is 30,000 miles.

b. Vertical lines have been drawn on the graph to indicate the standard deviations from the mean. The standard deviation is 10,000 miles.

c. Since the mean is 30,000 miles and the standard deviation is 10,000 miles, those cars with less than 20,000 miles on the odometer must be more than one standard deviation below the mean. The data approximate the normal curve, so 50% of the data lie to the left of the mean. Figure 5.24 shows that 34% of the data are between 20,000 and 30,000. Hence, $50\% - 34\% = 16\%$ of the cars have been driven less than 20,000 miles.

d. Cars with more than 50,000 miles on the odometer are more than two standard deviations above the mean. Note that 50% of the data lie to the right of the mean, and 48% of the data lie between 30,000 and 50,000. Hence, $50\% - 48\% = 2\%$ of the cars have more than 50,000 miles. Now, since there are 1000 cars, we have 2% of 1000 or $0.02 \times 1000 = 20$ cars.

NW ***Now Work Problem 3.***

EXAMPLE 4 ***Interpreting a Normal Curve*** Testing indicates that the lifetimes of 1000 Dimlite light bulbs are approximately normally distributed with a mean of 100 hours and a standard deviation of 8. A frequency polygon of the lifetimes approximated the normal curve in Figure 5.25. How many of these 1000 Dimlite bulbs will last

a. more than 116 hours? **b.** less than 92 hours?

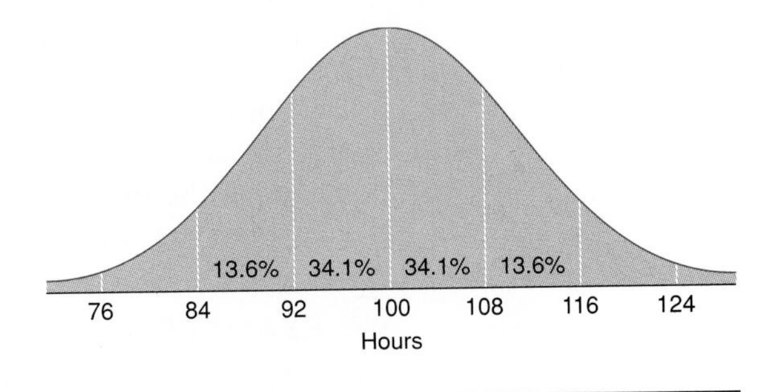

Figure 5.25

Historical Note

Many of the symbols that we now use in mathematics were adopted through repeated use by different people, so their exact origins are not known. For example, the sign for plus (+) is believed to have been adapted from the symbol for "and," the ampersand, &. After repeated writing, the symbol for "and" began to appear as + instead of &. The plus sign did not originate with mathematicians—it was first used by merchants in the 1400s to indicate excess. For example, if a bolt of cloth was supposed to contain 10 yards of cloth but it actually contained 13 yards, then a merchant had an excess of 3 yards and would mark "10 + 3" on the package.

The exact origin of the symbol for equality (=) is known. It first appeared in an English algebra book, Whetstone of Witte (1557), written by Robert Recorde. The author stated that he chose a pair of horizontal line segments to represent equality because "noe 2 thynges can be moare equable." Recorde's symbol for equality was greeted cautiously and not universally accepted until the early 1700s.

Solution Since the lives of the light bulbs are approximately normally distributed with a mean of 100 and a standard deviation of 8, the normal curve shown in Figure 5.25 reflects the lifetimes of this set of bulbs.

a. The light bulbs lasting more than 116 hours are those that lie more than two standard deviations above the mean. Figure 5.25 shows that 34.1% + 13.6% = 47.7% of the data are between 100 and 116 hours. Since the data are approximately normally distributed, 50% of the data are greater than (or above) 100 hours. Hence, 50% − 47.7% = 2.3% of the 1000 Dimlite bulbs will last more than 116 hours; 2.3% of 1000 is $0.023 \times 1000 = 23$ bulbs.

b. The light bulbs lasting less than 92 hours are those that lie more than one standard deviation below the mean. That is, 50% − 34.1% = 15.9% of the 1000 bulbs will last less than 92 hours; 15.9% of 1000 is $0.159 \times 1000 = 159$ bulbs.

NW ***Now Work Problem 5.***

Z-Score

In an earlier discussion, it was pointed out that approximately 68% of normally distributed data are included within an interval of plus or minus one standard deviation about the mean, or 34.1% of the data are between the mean and one standard deviation above the mean. Similarly, approximately 95% of normally distributed data are included within an interval of plus or minus two standard deviations about the mean, and 99.7% of the data are included within three deviations about the mean. Sometimes we may wish to consider a portion of the data that is a fractional part of a standard deviation from the mean. Table 5.6 indicates what percentages of the data are contained between the mean and various multiples of a standard deviation from the mean.

Many problems require fractional parts of standard deviations. Consider the following example. In a statistics class, the scores on the last exam were approximately normally distributed. The instructor decided to assign a grade of C for any score within 1.2 standard deviations of the mean. What percentage of the class will receive a grade of C? Table 5.6 indicates that 38.5% of the data are within 1.2 standard deviations of the mean on only one side of the mean. Since the normal curve is symmetric, we have the same percentage on the other side, and we therefore double the percentage: 2 × 38.5% = 77% of the class will receive a grade of C.

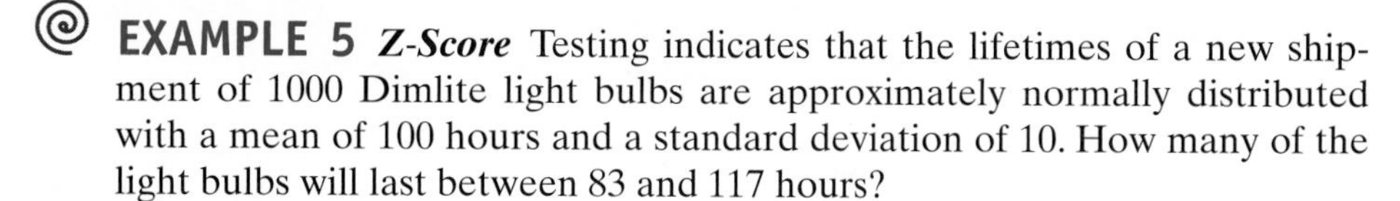

EXAMPLE 5 ***Z-Score*** Testing indicates that the lifetimes of a new shipment of 1000 Dimlite light bulbs are approximately normally distributed with a mean of 100 hours and a standard deviation of 10. How many of the light bulbs will last between 83 and 117 hours?

Solution This problem is similar to Example 4, but the boundaries involve fractional parts of the standard deviation from the mean. The mean is 100, and the standard deviation is 10. Then

$$100 - 83 = 17, \quad \frac{17}{10} = 1.7$$

and

$$117 - 100 = 17, \quad \frac{17}{10} = 1.7$$

Therefore, both boundaries are 1.7 standard deviations from the mean. Table 5.6 indicates that 45.5% of the data lie within 1.7 standard deviations on one side of the mean. We have 1.7 standard deviations on each side of the mean; hence $2 \times 45.5\% = 91\%$, or 910 light bulbs, will last between 83 and 117 hours.

It should be noted at this point that the position of a piece of data in terms of the number of standard deviations it is located from the mean is called the *Z-score* or *standard score*. The ***Z*-score** is found by the formula

$$Z = \frac{\text{Piece of data} - \text{Mean}}{\text{Standard deviation}}$$

Therefore, for Example 5 we have

$$Z = \frac{117 - 100}{10} = \frac{17}{10} = 1.7$$

and

$$Z = \frac{83 - 100}{10} = \frac{-17}{10} = -1.7$$

Data that lie below the mean will always have negative Z-scores. This is an indication of the position of the data with respect to the mean. Similarly, data that lie above the mean will always have positive Z-scores. We will still use Table 5.6 to find the percentages of data contained within various intervals, whether we call them Z-scores or the number of standard deviations from the mean.

NW ***Now Work Problem 11.***

EXAMPLE 6 ***Interpreting Z-scores*** The final-exam results in a statistics class were approximately normally distributed. Grades were assigned in the following manner:

A for a score more than 1.6 standard deviations above the mean
B for a score between 1.1 and 1.6 standard deviations above the mean
C for a score between 1.1 standard deviations above the mean and 1.1 standard deviations below the mean
D for a score between 1.1 and 1.6 standard deviations below the mean
F for a score more than 1.6 standard deviations below the mean

What percentage of the class received each grade?

Note of Interest

Table 5.7 gives a percentile-rank interpretation for a normal distribution. For example, consider a national competency exam in mathematics. If the scores are approximately normally distributed with a mean of 100 and a standard deviation of 10, then a score of 110 on the exam would yield a percentile rank of 84. That is, 84% of the people taking the exam would rank below a person having a score of 110.

TABLE 5.7

Number of Standard Deviations from the Mean	Percentile Rank in a Normal Distribution
+2.0	98
+1.5	93
+1.0	84
+0.5	69
0	50
−0.5	31
−1.0	16
−1.5	7
−2.0	2

Solution

a. Table 5.6 indicates that 44.5% of the population lies between the mean and 1.6 standard deviations above the mean. Hence, 50% − 44.5% = 5.5% of the class was above 1.6 standard deviations from the mean and received an A.

b. Since 44.5% of the population lies within 1.6 standard deviations and 36.4% lies within 1.1 standard deviations, we subtract 44.5% − 36.4% to determine that 8.1% of the class received a grade of B.

c. Since 36.4% of the class lies within 1.1 standard deviations on one side of the mean, and we want those both above and below the mean, we multiply 36.4% by 2 to determine that 72.8% of the class received a grade of C.

d. This is similar to part **b**, but we are concerned with standard deviations below the mean. Hence, we have 44.5% − 36.4% = 8.1% of the class received a grade of D.

e. Calculations like those in part **a** show that 5.5% of the class received a grade of F.

NW ***Now Work Problem 9.***

EXAMPLE 7 ***Z-Score*** A standardized examination is given to all entering students at a certain college. The scores are approximately normally distributed with a mean of 180 and a standard deviation of 30.

a. What percentage of the students have scores above 258?

b. What percentage of the students have scores between 120 and 258?

c. What percentage of the students have scores less than 195?

Solution Given that the scores are approximately normally distributed with a mean of 180 and a standard deviation of 30, we can draw a normal curve for the scores of this group of students, as shown in Figure 5.26.

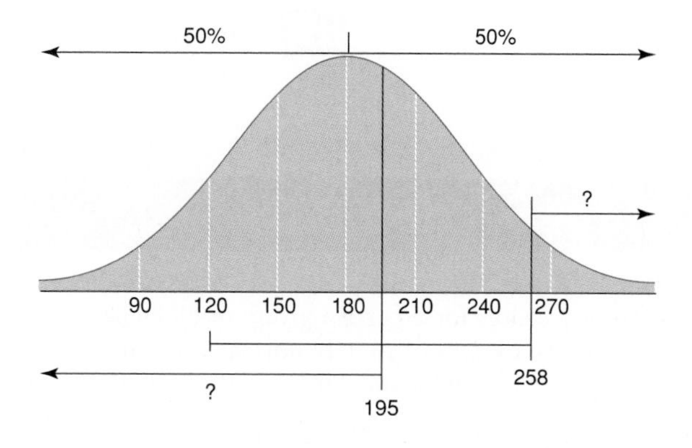

Figure 5.26

a. A score of 258 does not fall on a boundary line (a full standard deviation); it is between 240 and 270. We approximate its position and draw an arrow to the right of it because we want to find the percentage of scores above 258. Whenever we have an arrow like this in our diagram (we can call it a "tail test"), it is a reminder that we will probably need to use the fact that the mean of a normal distribution divides the curve so that there is 50% of the population to the right and 50% of the population to the left of the mean. Hence, we also note this fact in our diagram.

Since 258 does not lie on a boundary line, we must use the Z-score formula to find the number of standard deviations it is from the mean:

$$Z = \frac{258 - 180}{30} = \frac{78}{30} = 2.6$$

Table 5.6 tells us that 49.5% of the students have scores between the mean (180) and 258, but we want the percentage with scores above 258. Hence, we subtract 49.5% from 50%: 50% − 49.5% = 0.5% of the students score above 258.

b. We have already determined that 49.5% of the students score between 180 and 258. A score of 120 is two standard deviations below the mean (See Figure 5.26). The percentage of students with scores between the mean and 2 standard deviations on one side of the mean is 47.7% (see Table 5.6.) Now we add these percentages because we want to go from two standard deviations below the mean to 2.6 standard deviations above the mean [*Note:* We are working on opposite sides of the mean so we typically add.]:

$$47.7\% + 49.5\% = 97.2\%$$

c. A score of 195 does not fall on a boundary line. We approximate its position and draw an arrow to the left because we want to find the percentage of scores below 195. Note that we cross the mean and have another "tail test" so we will probably need to use the fact that 50% of the scores are to the left of or below the mean. To find the percentage of scores between the mean and 195 we must use the Z-score formula:

$$Z = \frac{195 - 180}{30} = \frac{15}{30} = 0.5$$

Table 5.6 tells us that 19.2% of the students have scores between the mean (180) and 195. We want to find the percentage with scores below 195, so we add 19.2% to 50%:

$$50\% + 19.2\% = 69.2\% \quad \text{scored less than 195.}$$

NW ***Now Work Problem 13.***

Exercises for Section 5.6

NW **1. IQ scores** The IQ scores for a certain group of elementary school students are approximately normally distributed with a mean of 100 and a standard deviation of 10.

a. What percentage of the students have IQ scores between 90 and 110?

b. What percentage of the students have IQ scores between 80 and 120?

c. What percentage of the students have IQ scores between 70 and 130?

2. Test results An examination is given to all entering students at a certain college. The scores are approximately normally distributed with a mean of 100 and a standard deviation of 15.

a. What percentage of the students have scores between 85 and 115?

b. What percentage of the students have scores between 70 and 130?

c. What percentage of the students have scores between 55 and 145?

d. What percentage of the students have scores above 115?

NW **3. Odometer readings** A group of students in a statistics class recorded the mileages of 2,000 cars in the student parking lot at their college. A frequency polygon of the mileages approximated the normal curve in Figure 5.27.

a. What is the mean mileage for this set of cars?

b. What is the standard deviation?

c. What percentage of cars had less than 62,000 miles on the odometer?

d. How many cars had more than 62,000 miles on the odometer?

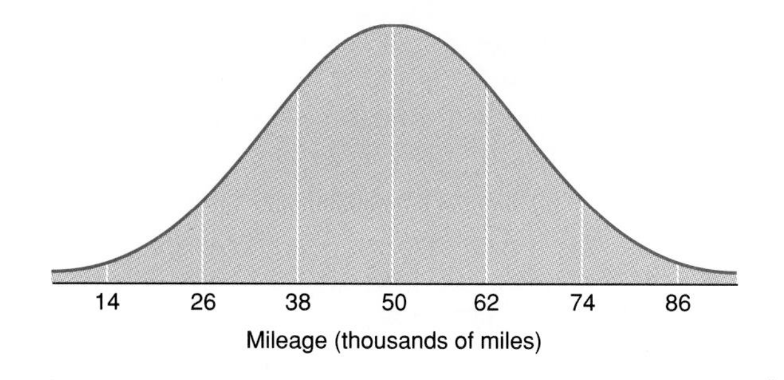

Figure 5.27

4. **Student heights** A statistician recorded the heights of 1000 students at a particular college. A frequency polygon of the heights approximated the normal curve in Figure 5.28.

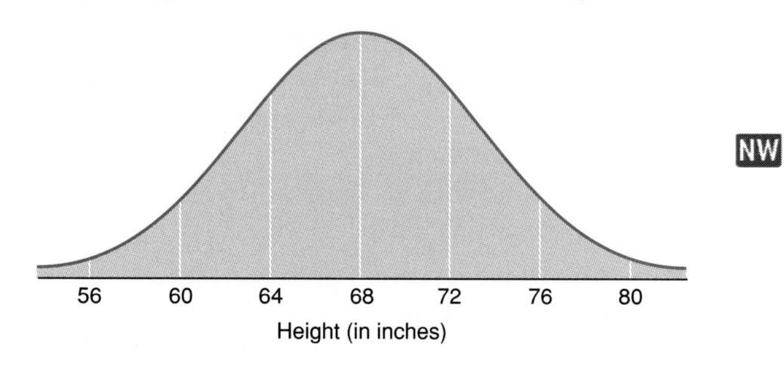

Figure 5.28

 a. What is the mean height for this set of students?
 b. What is the standard deviation?
 c. What percentage of the students was more than 5 feet tall and less than 6 feet tall?
 d. How many students were less than 5 feet tall?

NW 5. **Butane lighters** Testing indicates that the lifetimes of a shipment of disposable butane lighters are approximately normally distributed with a mean of 1000 lights and a standard deviation of 100. A shipment contains 5000 of these lighters.
 a. Approximately how many of the lighters will light more than 1100 times?
 b. Approximately how many of the lighters will light more than 1200 times?
 c. Approximately how many of the lighters will light less than 700 times?
 d. Approximately how many of the lighters will light between 800 and 1100 times?

6. **Test results** The scores on a physics exam for a class of 50 students were approximately normally distributed with a mean of 40 and a standard deviation of 5.
 a. Approximately how many of the students scored above 45?
 b. Approximately how many of the students scored above 50?
 c. Approximately how many of the students scored below 35?
 d. Approximately how many of the students scored between 30 and 45?
 e. Find the approximate number of students that scored between 25 and 50.

7. **Tire life** It has been shown that the lifetimes for a certain type of belted radial tire are approximately normally distributed with a mean of 40,000 miles and a standard deviation of 5000 miles.
 a. What percentage of these tires will last more than 50,000 miles?
 b. What percentage of these tires will last less than 35,000 miles?
 c. What percentage of these tires will last between 35,000 and 50,000 miles?
 d. What percentage of these tires will last between 30,000 and 55,000 miles?

8. **TV Warranty** The manufacturer of Zap television sets guarantees its picture tube for 2 years. Testing indicates that the lifetimes of Zap picture tubes are approximately normally distributed with a mean of 3 years and a standard deviation of 6 months ($\frac{1}{2}$ year). What percentage of Zap television sets will have to be replaced under the guarantee?

NW 9. **Test Results** The scores on a mathematics exam were approximately normally distributed, and the instructor assigned grades in the following manner:
 A for a score more than 1.7 standard deviations above the mean
 B for a score between 1.1 and 1.7 standard deviations above the mean
 C for a score between 1.1 standard deviations above the mean and 1.2 standard deviations below the mean
 D for a score between 1.2 and 1.8 standard deviations below the mean
 F for a score more than 1.8 standard deviations below the mean

 What percentage of the class received each grade?

10. Given that the scores on a certain exam are approximately normally distributed with a mean of 50 and a standard deviation of 10, use Table 5.6 in this section to find
 a. the percentage of scores greater than 55
 b. the percentage of scores greater than 65
 c. the percentage of scores greater than 71
 d. the percentage of scores between 45 and 55
 e. the percentage of scores between 33 and 44
 f. the percentage of scores less than 44

NW 11. **Student heights** The heights of the students at a local school are approximately normally distributed with a mean height of 64 inches and a standard deviation of 4 inches. Use Table 5.6 in this section to find
 a. the percentage of students between 62 and 66 inches tall
 b. the percentage of students between 58 and 66 inches tall
 c. the percentage of students over 6 feet tall
 d. the percentage of students shorter than 5 feet 6 inches
 e. the percentage of students between 5 feet 6 inches and 6 feet tall

12. **IQ scores** The IQs for a class of mathematics students are approximately normally distributed with a mean of 100 and a standard deviation of 10. Use Table 5.6 in this section to find

a. the percentage of students with an IQ between 95 and 115
b. the percentage of students with an IQ between 93 and 119
c. the percentage of students with an IQ between 115 and 119
d. the percentage of students with an IQ greater than 122
e. the percentage of students with an IQ less than 117

NW 13. **Test results** The scores on a placement test in mathematics administered to college freshmen are approximately normally distributed with a mean of 100 and a standard deviation of 20. Use Table 5.6 in this section to find

a. the percentage of students who scored between 90 and 120
b. the percentage of students who scored above 114
c. the percentage of students who scored less than 134
d. the percentage of students who scored between 82 and 118
e. the percentage of students who scored between 118 and 138

14. **Weight of cereal boxes** A certain brand of cereal is supposed to contain 16 ounces of cereal in each box. The amount of cereal in each box is approximately normally distributed with a mean of 16 ounces and a standard deviation of 0.5 ounce. Use Table 5.6 in this section to find

a. the percentage of boxes containing more than 16.2 ounces of cereal
b. the percentage of boxes containing less than 16.4 ounces of cereal
c. the percentage of boxes containing between 16.2 and 16.4 ounces of cereal
d. the percentage of boxes containing between 15.8 and 16.2 ounces of cereal
e. the percentage of boxes containing more than 14.8 ounces of cereal

15. **Study time** A survey was taken among students in the college cafeteria to find out how many hours a week they spent studying for courses for which they were currently enrolled. The data were approximately normally distributed with a mean of 24 hours and a standard deviation of 5 hours. Use Table 5.6 in this section to find

a. the percentage of students who studied more than 28 hours per week
b. the percentage of students who studied more than 33 hours per week
c. the percentage of students who studied less than 16 hours per week
d. the percentage of students who studied between 22 and 30 hours per week
e. the percentage of students who studied between 28 and 33 hours per week

16. **Coffee machine** The coffee machines in a certain snack bar are supposed to dispense 5 ounces of coffee in each cup. The amount of coffee in each cup is approximately normally distributed with a mean of 5.0 ounces and a standard deviation of 0.5 ounce.

a. What percentage of cups contain between 4 and 5.5 ounces of coffee?
b. What percentage of cups contain between 5 and 5.8 ounces of coffee?
c. What percentage of cups contain between 4.8 and 5.8 ounces of coffee?
d. What percentage of cups will overflow if 6-ounce cups are used in the machine?

17. **Postage machine** A survey indicates that a postmarking machine in a certain post office cancels an average (mean) of 65 letters per minute, with a standard deviation of 10 letters per minute. Assume that the number of letters cancelled per minute by the machine is approximately normally distributed.

a. What percentage of the time does the machine cancel more than 90 letters per minute?
b. What percentage of the time does the machine cancel less than 50 letters per minute?

18. **Phone call duration** The average (mean) length of a phone call from a certain pay phone is 5 minutes, with a standard deviation of 1 minute. Assume that the lengths of the phone calls are approximately normally distributed.

a. What percentage of the calls last more than $6\frac{1}{2}$ minutes?
b. What percentage of the phone calls last between $5\frac{1}{2}$ and $6\frac{1}{2}$ minutes?
c. What percentage of the calls cost extra, if a caller has to pay extra for any call over 3 minutes in length?

19. A manufacturer has been awarded a government contract to manufacture bleeps, which must be between 1.45 and 1.55 inches in length. Testing indicates that the manufacturer's machines produce bleeps whose lengths are normally distributed with a mean length of 1.5 inches and a standard deviation of 0.1 inch.

a. What percentage of bleeps meet the government specifications?
b. What percentage of bleeps do not meet the government specifications?

20. **Test results** The results of an exam in a statistics class are approximately normally distributed with a mean of 40 and a standard deviation of 8.

a. What percentage of the scores are above 52?
b. What percentage of the scores are above 60?

c. If 100 students took the exam and all students with grades from 52 to 60 received a B, how many students received a B?

21. **Light bulb life** Testing indicates that the lifetimes of a certain lot of light bulbs are approximately normally distributed with a mean of 100 hours and a standard deviation of 10.
    a. What is the probability that a bulb selected at random will last more than 100 hours?
    b. What is the probability that a bulb selected at random will last more than 105 hours?
    c. What is the probability that a bulb selected at random will last between 95 and 108 hours?

22. **Test results** The results of an exam in a history class are normally distributed with a mean of 60 and a standard deviation of 6.
    a. What is the probability that a student who took the exam will have a score between 57 and 63?
    b. What is the probability that a student who took the exam will have a score between 45 and 75?

Writing Mathematics

23. Provide a written description of a normal curve.
24. What is a Z-score?
25. Explain why a Z-score is more informative than a raw score.
26. Assume that the mean and standard deviation of a set of data collected from a national survey of 5000 respondents were 50 and 5, respectively. If the mean of a locally collected sample was 45, explain how you would determine the number of respondents from the national survey who scored higher than the mean of the local sample.

Challenge Exercises

27. If you review each of the examples in this section you will find that all of the examples have three common elements. Determine what these are and why it is necessary to include them in problems of this nature.
28. Suppose that 40% of the data from a given data set was found to be between a Z-score of 0.5 and the mean. What conclusion(s) can you make?

Research/Group Activities

29. The normal distribution discussed in this section is actually a *mathematical distribution based on a theoretical model* of the relative frequencies of different scores in a population. This means that the normal distribution is not a distribution of real scores but instead is a hypothetical distribution that is based on an infinite number of observations. Other types of theoretical distributions include the t-distribution, the F-distribution, the *chi-square* distribution, and the *binomial* distribution. Consult an appropriate reference and prepare a report that describes these distributions.
30. A very important theorem in statistics is called the ***central limit theorem***. Prepare a report that describes this theorem and its application when selecting a sample from a population.
31. Interview a teacher at your school who "grades on the curve." As part of your interview, ask this teacher his or her reasons for doing so and what the advantages and disadvantages are for the students. Based on the results of this interview, combined with the knowledge of statistics you have acquired thus far from this chapter, what is your opinion about grading on the curve?

Just for fun

Can you connect all nine dots with only four connecting line segments and without raising your pencil from the paper?

• • •

• • •

• • •

5.7 USES AND MISUSES OF STATISTICS

As noted in the introduction to this chapter, the application of statistics can be found in nearly everything we encounter. From newspaper articles to the latest medical research findings to television ratings to education and sports, we live in a world filled with statistical information. When we apply statistical methods to a specific area of interest, we are using statistics to help us gain a greater understanding of that subject. For example, medical researchers use statistics in a variety of situations to help them answer specific questions such as, What effect does eating a low-fat diet have on heart disease? What is the relationship between second-hand smoke and cancer? Are there any physiological differences between individuals who are sedentary versus those who walk at least three miles per week? Social scientists might use statistics to study the possible effects single-parent homes have on juvenile deliquency, or the degree to which a person's religious convictions might affect self-esteem. Classroom teachers also have a need to use statistics. For example, teachers might be interested in examining the effects a specific method of instruction has on student achievement, or whether children who are home-schooled are equivalent in academic ability to those taught in a formal school setting.

In order to answer these questions it is necessary to conduct an *experiment.* This involves identifying a specific group, called the *population*, which is the focus of the study. The population represents the group to which a researcher wants to apply the results of an experiment. For example, a medical researcher studying the relationship between second-hand smoke and cancer might focus on nonsmoking waiters and waitresses who work in the smoking section of a restaurant, nonsmoking spouses who are married to and live with spouses who smoke, or nonsmoking children who live with parents who smoke. All of these are illustrations of possible populations. It is the researcher's responsibility to identify the population in which he or she is interested.

If you are thinking that a population represents a large set, you are right. It is usually impossible or impractical to work with every member of a specific population. As a result, researchers must select a subgroup (i.e., subset) from the population. This subgroup, called a *sample,* represents those individuals or scores that are accessible or available to the researcher, and the process of selecting a sample is called *sampling*. Although rare, it is possible for a sample and its corresponding population to be identical. In such cases the sample is called a *census.* The results obtained from a census can be extended unconditionally to the population. Thus census results describe the corresponding population exactly.

Using a sample instead of a population is very cost-effective and productive. However, care must be exercised when selecting a sample from a population. It is critical that the sample be *representative* of the population. If not, then any results we obtain from the experiment will not be applicable to the population. There are many ways in which a representative sample can be selected from a population. One way is to use as large a sample as possible because the larger the sample, the more likely the sample is representative of the population. The best method, though, is to select members *randomly*. This means that every member of a population has an equal and independent chance of being selected. Special random number tables are available to help in this process. A random sample produces a *statistically unbiased sample*. To contrast between a biased sample and an unbiased sample, consider the following situation:

> *You have been asked to conduct a survey to determine the opinions of voters about a school board referendum that, if approved, will increase all Brevard County, Florida, homeowner's school taxes by an average of $20 per year for three years. The population for this survey is all registered voters of Brevard County, Florida. Following are several strategies for selecting a sample: A. Send a survey to all registered voters and have voters voluntarily mail in their surveys; B. Select a random sample of voters from the list of registered voters who work for Harris Corporation, the county's largest employer; C. Select a random sample from the list of all persons who have children attending school in the county; and D. Select a random sample from a compiled list of all registered voters in Brevard County, Florida, and survey these people.*

In this example, note that sample A is a biased sample because only those individuals who choose to mail in their survey will make up the sample. Sample B is biased toward employees who work for Harris Corporation. Sample C also is biased; the opinions given will only reflect the perspective of people who have children attending school in the county. Of the four possible samples, only sample D is unbiased. Every registered voter in Brevard County has an equal and independent chance of participating, and it is not biased toward any one group.

Once a sample is selected, the researcher implements the experiment and collects data. Depending on the research objective, these data are then analyzed and reported in one of two ways. We can either describe the data, or we can infer the results to the overall population (or both). Describing data represents the general category of statistics known as *descriptive statistics*. Most of the concepts we discussed in this chapter—mean, median, mode, standard deviation, frequency distributions, histograms, and so forth—are examples of descriptive statistics. Each of these concepts enables us to summarize a set of data in a brief but clear and informative manner. Thus the goal of descriptive statistics is to summarize and describe a group of scores or a set of data.

The other general category of statistics involves drawing inferences that extend beyond the sample data. This is known as *inferential statistics,* and involves using sample data to make conjectures about a population. Inferential statistics employs a specific statistical procedure called *hypothesis testing,* which is based on probability. The concept of hypothesis testing is to use sample data to evaluate the credibility of a specific hypothesis about a population. For example, a researcher might hypothesize that second-hand smoke leads to cancer. To test this hypothesis, the researcher identifies the population, selects a random sample from the population, collects data from the sample, and then compares these sample data with the hypothesis. If the sample data are consistent with the population hypothesis, then the researcher has evidence to support the hypothesis and extends (i.e., *generalizes*) the results of the sample to the population. If the data are not consistent with the hypothesis, that is, there is a large discrepancy between the sample data and the hypothesis, then there is evidence for the researcher to reject the hypothesis. Thus the goal of inferential statistics is to use limited information from samples to draw conclusions about populations. Once again, note how critical it is for the sample to be representative of the population if we intend to obtain credible results.

Descriptive and inferential statistics are both meaningful and useful when applied correctly and properly. Unfortunately, statistics also can be presented in a manner that misleads you. By accident or on purpose, essentially accurate information can be used to present distorted or biased information. In the remainder of this section, we present several examples of specific situations in which statistics are misused or misinterpreted.

Contradictory Conclusions

Who do you believe when you hear reports that seem to contradict each other? For example, one medical study might find that eating a low-fat diet reduces your risk of cancer. Another study might contradict the results of the first study and conclude that there is no connection between how much fat you eat and cancer. A third study might find that too little fat in a low-fat diet is not good for you. What do you do? Which report should you believe? This is a classic illustration of one problem of inferential statistics. Each study might have merit, but in order to completely understand the results, we need to place them in their proper perspective. Remember, inferential statistics uses sample data to draw conclusions about populations. So when reviewing the results of any statistical findings, you should always consider what the population was, how the sample was selected, the size of the sample, the manner in which data were collected, and how the results are being interpreted before you draw any conclusions.

Different Horizontal or Vertical Axes

Statistics are often given by means of graphs or charts. These too can be misleading or deceptive. For example, look at the graphs shown in Figure 5.29. The graphs show the total daily sales for two salespeople, Finnie and Goldfarb, for a one-week

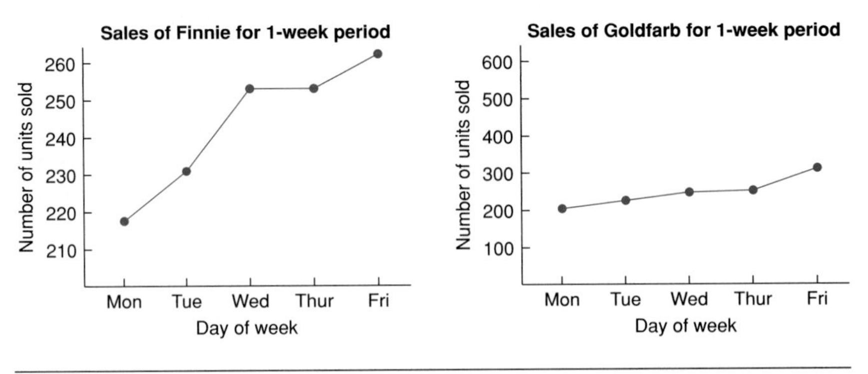

Figure 5.29

period in a store. Which person appears to be more successful? Note that both graphs present similar daily sales figures; the difference, though, is that the vertical scale of Finnie is exaggerated. Its vertical scale ranges from 200 to 260 in increments of 10 units, whereas the vertical scale of Goldfarb ranges from 0 to 600 in increments of 100 units. Consequently, the graph of Finnie gives the impression that her sales increased substantially, where those of Goldfarb were flat. Which of these graphs is correct? Actually, neither is very good because one graph exaggerates the number of sales and the other conceals them. Perhaps a better way to display this information is in chart form. So, when reviewing graphs, carefully examine both vertical and horizontal scales. Also see if a scale begins at zero; if it does not, then relative changes might appear to be more significant than they actually are.

Unethical Practices

Frequently we hear reports that claim that some large percentage of people polled prefer a particular brand. For example, an automobile dealer might claim that 48 out of 50 people who purchase new cars from this dealership are satisfied with their deal. In another example, a grocery store chain might claim that in a survey of 500 shoppers, 90% believed that this store had lower prices than the other major grocery store. Although these claims might be accurate, the manner in which the survey is conducted might be unethical. For example, the car dealership might continuously poll a random sample of 50 new car buyers until it gets a sample in which at least 48 people claim they were pleased with their deal, and then disregard all of the data collected from the other samples. This is unethical practice.

Implied Cause–Effect Relationships

In statistics there is a technique that enables us to measure and describe the relationship between two entities. This technique is called *correlation*. For example, if a researcher is interested in investigating the relationship between birth order and success in school, a correlation study is conducted. One of the most common errors in interpreting the result of a correlation study is to assume that the relationship is a *cause–effect* relationship. For example, if a researcher discovered that there is indeed a relationship between birth order and success in school (e.g., second-born children are more successful in school than first-born children), one might conclude that birth order *causes* school success. This is incorrect. Correlation does not imply a cause–effect relationship. It tells us if a relationship exists. Thus relationships such as those between drinking alcohol and birth defects, eating carrots and good vision, or taking an aspirin daily and heart attacks might be highly correlated (i.e., strong relationship). However, this does not mean that drinking alcohol causes birth defects, or eating carrots causes good vision, or taking an aspirin daily causes a reduction in risk of heart attacks.

Misinterpretation of Averages

As we discussed in Section 5.2, there are several ways in which to compute an average (mean, median, mode, midrange). Depending on the setting, one method is more appropriate than another. For example, the mean is not a true indication of the average if other scores in a given set of data are extreme values. This is because the mean gets "pulled" in the direction of the extreme scores. Consequently, we must examine the context in which an average is given to ensure that an appropriate "average" is reported. It also helpful to know the range of scores when a mean is reported so that we can determine the extent to which the mean is getting "pulled." Finally, a "good" mean is not necessarily good. For example, a school district might learn that its mean on a standardized mathematics exam taken by all its 10th-grade students was higher than the state and national averages. On the surface this is terrific news. However, what does this mean tell us? Not much. In some settings (particularly in education) it is important to know what is influencing the mean. There is a technique called *data disaggregation* that partitions the results of a test into separate components. In our school district example, data disaggregation might reveal that the high mean the district received was attributed to Asian-American students, and that all other students actually scored below the state and national averages. So a "good" mean does not always mean you did "good."

These examples should serve as illustrations of how statistics can be misused or misinterpreted. There are many more. If you are interested in this subject or would like to read about other examples, the book *How to Lie with Statistics* by Darrell Huff (New York: W. W. Norton, 1993) is an excellent resource to consider. In the final analysis, statistics provide valuable information when dealt with correctly. The misuse or misinterpretation of them by some should not detract from their value when used properly.

Exercises for Section 5.7

1. **Enrollment** The president of a 4-year college reports to the board of trustees that enrollment is up by 10%. However, personnel from the office of student affairs claim that enrollment is down 2%. What would you give as possible reasons for this conflicting information?
2. **Research grants** Faculty at a research university acquire research grants from various funding institutions. Many of these grants are multiyear in length. For example, a $500,000, three-year grant implies that the researcher receives a total of $500,000 over a three-year period (not $500,000 each year for three years). Identify one way in which the director of sponsored research at this university can use statistics to show how well the university is doing as a whole in acquiring research grants.
3. **Handgun control** Handgun Control Incorporated (**http://www.handguncontrol.org**) makes the following quote: "Thinking of buying a handgun to protect your home? You may want to remember that guns in the home for self-protection are 43 times more likely to kill a family member or friend than to kill in self-defense." Discuss how statistics are being misused in this quote.
4. **Network crimes** In 1991, an FBI study revealed that over 80% of network crimes were caused by insiders at a company. In 1997, through a joint study by the FBI and the Computer Security Institute, the number was adjusted to about 75%. In 2000, this number approached 50%. Do you think this trend is accurate? Why or why not? What are some possible reasons for this trend?
5. **Commuting** A survey was given by demographic area to a group of people who commute to and from work. The questions asked of the respondents were in regard to distance and duration of the person's commute. The question regarding distance, though, did not specify if it was one way or round trip. In what way can the results of this survey be misleading?
6. **Drug testing** Most reliable sources state that the accuracy of drug tests is approximately 95%. What do you think this means? Given this information, what do you think the chances are that, if tested, your result will be a "false-positive"—that is, that the test will show you are taking drugs even though you are not?
7. **Computer–student ratio** To prepare a report on the average computer-student ratio, a school district divides the total number of students enrolled by all of the computers in the district. It then reports that there is approximately a 1 to 3 ratio; that is, there is one computer available for every three students. In what way do you think this figure is misleading?
8. **Class size** In an effort to show that the current administration has reduced class sizes, school leaders of a school

district are asked to tally the total number of educators in the district who have a valid state-approved teaching certificate. It then divides the total number of students enrolled in the district by the total number of certified teachers and reports that the student-teacher ratio is 20 to 1. Thus, average class size is 20. In what way(s) can this information be misleading?

Writing Mathematics

9. Briefly describe the primary ways in which statistics can be misused as discussed in this section.
10. Explain how reporting data using line graphs that have different horizontal or vertical axes can lead to misinterpretations.
11. Give one example of an unethical statistical practice.
12. What does the term *data disaggregation* mean?

Challenge

13. Give an illustration of each of the following as an example of misusing statistics.
    a. Vague language
    b. Misleading claims
    c. Selective reporting
14. When interpreting the results of a statistical analysis, it is sometimes important to make a distinction between the concepts "statistical significance" and "practical significance." Explain the difference between these terms and give an example as part of your explanation.

Group/Research Activities

15. Prepare a report on how statistics are used in sports, education, or business.
16. Identify three advertisements or television commercials that misuse statistics and explain how each is misleading.
17. Select three graphs from *U.S.A. Today* and modify them so that the information reported can be misinterpreted.
18. Consult the book *How to Lie with Statistics,* referenced at the end of this section, and identify at least two other ways statistics can be misused or misinterpreted.

Just for fun

How much is a billion dollars? To give you some idea, suppose that you could spend $1000 a day, every day. How long would it take you to spend $1,000,000,000? Assume that each year contains 365 days.

CHAPTER REVIEW MATERIALS

Summary/Chapter 5

5.2 Measures of Central Tendency

- The *mean* for a set of data is found by determining the sum of the data and dividing this sum by the total number of elements in the set.
- The *median* for a set of data is found by arranging the data in sequential order and finding the middle number. When we have an even number of pieces of data, it is customary to use the number halfway between the two middle numbers.
- The *mode* for a given set of data is the item or value that occurs most frequently. There can be times when a set of data has no mode, when each value occurs an equal number of times. A set of data can also have more than one mode.
- The *midrange* is found by adding the least value in the given set of data to the greatest value and dividing the sum by 2.

5.3 Measures of Dispersion

- The *range* for a set of data is found by subtracting the smallest value in the set from the largest value in the set.
- To find the *standard deviation* for a given set of data, we must follow these steps:
    1. Find the *mean* of the given set of data.
    2. Find the *deviation from the mean* for each value in the set of data.
    3. Square each deviation.
    4. Find the mean of the squared deviations (*variance*).
    5. Find the principal square root of this number.

5.4 Measures of Position

- *Percentiles* and *quartiles* are *measures of location* of data. Percentiles divide sets of data into 100 equal parts; hence, 100% is the basis of the measure. The *first quartile* for a set of data is the value that has 25% of the data below it. The *second quartile* is the value that has 50% of the data below it, and the *third quartile* is the value that has 75% of the data below it.

5.5 Pictures of Data

- Pictures of data are often used to transmit or summarize statistical information. Graphs covered in this chapter include: *bar graphs* (vertical, horizontal, and comparative), *picture graphs* or *pictograms*, and *circle* or *pie graphs*.
- A *histogram* consists of a series of bars that are all the same width. Uniform units are used on the vertical axis to show the frequency of the data. A *frequency distribution table* is used to tabulate the given data. A *frequency polygon* (*line graph*) can also be used to graph the information in a frequency distribution table.
- *Stem-and-leaf displays* are used as an alternative to frequency distributions to describe and analyze a set of data.

5.6 The Normal Curve

- When we have a *normal distribution*, the area under the curve represents the entire set of data, or population. If the data are approximately normally distributed, then the mean, median, and mode all have approximately the same value. Practically all of the data (99.7%) will be included in the interval from three standard deviations below the mean to three standard deviations above the mean.
- The *Z-score*, or *standard score*, is the position of a piece of data in terms of the number of standard deviations it is located from the mean. The *Z*-score is found by the formula

$$Z = \frac{\text{Piece of data} - \text{mean}}{\text{Standard deviation}}$$

5.7 Uses and Misuses of Statistics

- In any research setting in which statistics are used, a *population* is first defined. From this population a *random sample* is selected, and data are then collected from this sample and analyzed. This procedure of using sample data to describe a population is called *inferential statistics*.

Key Terms and Concepts

bar graph, *282*
bimodal, *262*
central tendency, *258*
comparative bar graph, *283*
correlation, *308*
descriptive statistics, *307*
frequency distribution, *287*
frequency polygon, *287*
histogram, *287*
horizontal bar graph, *283*
hypothesis testing, *307*
inferential statistics, *307*
mean, *259*
median, *260*
midrange, *263*
mode, *261*
normal distribution, *295*
percentile, *277*
pictogram, *284*
picture graph, *284*
pie graph, *285*
population, *306*
quartile, *279*
random sample, *260, 306*
range, *269*
sample, *259, 306*
standard deviation, *270*
stem-and-leaf display, *289*
unbiased sample, *306*
vertical bar graph, *283*
Z-score, *299*

Review Exercises

5.2 Measures of Central Tendency

1. a. Name four measures of central tendency.
b. Which measure of central tendency gives the value that occurs most frequently?
c. Which measure of central tendency gives the middle value of the given data?
d. Which measure of central tendency gives the average value of the given data?

In Exercises 2–5, find the mean, median, mode, and midrange for each set of data. (Round off any decimal answer to the nearest tenth.)

2. 12, 18, 16, 12, 19
3. 2, 3, 4, 5, 5, 6, 10, 13
4. 2, 3, 5, 7, 11, 13, 17, 19
5. 99, 11, 88, 22, 33, 77
6. In order to receive a grade of C in her statistics class, Susan needs a mean score of 72 on five tests. If Susan had scores of 62, 78, 80, and 68 on her first four tests, what is the lowest score that Susan can get on her last test and still receive a grade of C in the course?
7. The mean of four of a set of five scores is 74. The fifth score is 88. What is the sum of the five scores? What is the mean of the five scores?

5.3 Measures of Dispersion

8. Find the range, midrange, and mean for the following set of test scores: {70, 70, 67, 66, 65, 65, 65, 62, 60, 60}.
9. Find the range, midrange, and mean for the following set of data: {58, 76, 76, 79, 84, 85, 100}.
10. Find the standard deviation for the following set of data: {10, 11, 13, 13, 14, 17}. (Round your answer to the nearest tenth.)
11. Find the standard deviation for the following set of data: {10, 14, 16, 26, 14, 16}. (Round your answer to the nearest tenth.)

5.4 Measures of Position

12. In a senior class of 300 students, Joe has the rank of 45th. What is Joe's percentile rank in the class?
13. Andy is ranked at the third quartile in his statistics class of 32 students. Julie is in the same statistics class, and she is ranked sixth. Who ranks higher, Andy or Julie?

5.5 Pictures of Data

14. **Temperatures** The average high temperature (°F) for selected cities during January is as follows:

London	44°	Berlin	35°
Moscow	21°	Vienna	34°
Tokyo	47°	Madrid	47°

Construct a vertical bar graph representing this information.

15. **Gasoline tax** The gasoline tax (per gallon) for selected states is as follows:

Arizona	17¢	Minnesota	20¢
Idaho	18¢	Tennessee	20¢
Iowa	20¢	California	9¢

Construct a horizontal bar graph representing this information.

16. **Dividends** The annual dividend paid to its shareholders by the Peters Corporation for the years 1999–2004 is as follows:

1999	$1.50	2002	$2.25
2000	$1.75	2003	$2.50
2001	$2.00	2004	$2.50

Represent this information by means of a pictogram.
Let each symbol ¢ represent $0.50.

17. **Television** In a survey of 200 students regarding their favorite television network, it was determined that 50 students chose ESPN, 40 students chose CBS, 80 students chose ABC, and 30 students chose NBC. Construct a circle graph representing this information.

18. A student tossed a die 50 times. His results are as follows. Construct **(a)** a frequency distribution, **(b)** a histogram, and **(c)** a frequency polygon that represent these data.

2, 5, 4, 4, 1, 5, 3, 5, 6 6,
2, 5, 2, 5, 5, 1, 1, 5, 4 6,
2, 4, 4, 6, 6, 3, 2, 6, 1, 1,
6, 1, 6, 4, 5, 4, 1, 6, 3, 2,
3, 4, 4, 5, 6, 4, 6, 1, 2, 2

19. **Student heights** The heights (in inches) of 30 students in a physical education class were recorded as follows:

65, 69, 64, 65, 70, 71, 75, 69, 70, 64,
72, 70, 66, 68, 70, 72, 64, 68, 70, 69,
67, 70, 67, 66, 67, 71, 70, 75, 69, 70

a. Construct a frequency distribution for these data.
b. Construct a histogram for these data.
c. Construct a frequency polygon for these data.
d. Does the frequency polygon obtained in part **c** approximate a normal curve? Why?

5.6 The Normal Curve

20. **Test scores** One thousand test scores are approximately normally distributed with a mean of 60 and a standard deviation of 8.

a. What percentage of the scores are above 68?
b. What percentage of the scores are above 76?
c. What percentage of the scores are below 52?
d. What percentage of the scores are between 52 and 76?
e. If a grade of C is assigned to the scores between 52 and 68, how many scores will receive a grade of C?

21. **Sleep** On the first day of class last semester, a group of students was asked for the number of hours of sleep they obtained the previous night. The data were approximately normally distributed with a mean of 5.0 and a standard deviation of 0.5. Find

a. the percentage of students who slept more than 5.8 hours
b. the percentage of students who slept less than 4.8 hours
c. the percentage of students who slept between 4.8 and 5.8 hours
d. the percentage of students who slept more than 5.4 hours
e. the percentage of students who slept between 5.4 and 5.8 hours

22. Two exams were given to a certain mathematics class. On the first exam, the mean was 64 and the standard deviation was 8. On the second exam, the mean was 48 and the standard deviation was 12. The scores of four students who took both exams were

	Exam 1	Exam 2
Bill	64	63
Louie	70	60
Jack	60	54
Steve	80	66

a. Who improved on the second exam?
b. Who did not do as well on the second exam as on the first?
c. Who performed the same on both exams?
d. Who has the best grade for the two exams combined?

Various Sections

23. **Babe Ruth** Babe Ruth was one of the first inductees into the Baseball Hall of Fame. He played in the major leagues from 1914 to 1935. The following is a list of the number of home runs he hit in each season: 0, 4, 3, 2, 11, 29, 54, 59, 35, 41, 46, 25, 47, 60, 54, 46, 49, 46, 41, 34, 22, 6. For this set of data, find the following:

a. mean **b.** median **c.** mode
d. midrange **e.** range

24. Use the set of data

$$\{18, 20, 22, 24, 26, 26, 28, 36\}$$

to find the following:

a. mean **b.** median **c.** mode
d. range **e.** midrange
f. standard deviation (to the nearest tenth)
g. the score at the first quartile
h. the score at the third quartile

25. Given the following set of 12 scores: {29, 22, 24, 27, 22, 20, 20, 20, 15, 24, 22, 34,}

a. What is the rank (from the top) of the score of 27?
b. What is the percentile rank of the score of 27?

c. What score is at the third quartile?

d. What is the median?

e. What is the mode?

26. **Part-time employment** A survey in a statistics class of 35 students revealed that most students worked part time. The accompanying table shows the resulting data indicating how long the students worked (to the nearest hour) in a typical week.

Time (hrs)	Number of People
15	3
12	3
10	4
8	6
6	9
3	3
2	2
0	5

Using the information provided, find the following (to the nearest tenth):

a. the mean time **b.** the median time

c. the mode time **d.** the midrange time

27. Obtain a set of 50 exam scores from your instructor.

a. Rank the data from smallest to largest.

b. Compute the mean, median, mode, midrange, range, and standard deviation for this set of data.

c. Construct a frequency distribution, histogram, and frequency polygon for the set of scores.

28. As we complete the statistics unit, you are to hand in examples of the following types of graphs: circle graph, line graph, horizontal bar graph, vertical bar graph, histogram, and pictograph. You need two of each for a total of 12. You are to use actual graphs from newspapers, magazines, and other publications. If you are not able to find any of these graphs, you may draw your own. But, they must be neat, orderly, and labeled.

The graphs should be affixed to standard size notebook paper, or its equivalent; and clearly labeled and credited. You may use only one type of graph from a given article.

Chapter Quiz

Determine whether each statement (1–20) is true or false.

1. Each of the measures of central tendency—mean, median, mode, and midrange—is an "average."
2. The *median* for a set of data is found by arranging the data in sequential order and finding the middle value.
3. A given set of data may have more than one *mode*.
4. The *range* for a set of data is also a measure of central tendency.
5. The *midrange* is a measure of dispersion.
6. If Angie is first in her class, then she is at the 100th percentile.
7. The mean of the squared deviations will always be a nonnegative value.
8. When a student is told that he scored at the 77th percentile on an exam, he then knows that he received a score of 77% on the exam.
9. The position of a piece of data, in terms of standard deviations from the mean, is called the *Z-score*.
10. The first quartile for a set of data is the value that has 75% of the data below it.
11. If Sam scored at the zero percentile, then he scored the highest in his class.
12. Grouped frequency distributions representing the same set of data may be different.
13. The tails of a normal curve must touch the horizontal axis.
14. Circle graphs are used in showing how a whole quantity is divided into parts.
15. A histogram consists of a series of bars that are drawn all with the same widths on the vertical axis, and uniform units on the horizontal axis.
16. A frequency polygon is a type of histogram.
17. A normal bell-shaped distribution will have a range that is approximately equal to six standard deviations.
18. For normally distributed data, approximately 68% of the data will be included within an interval of one standard deviation about the mean.
19. If we have data that are approximately normally distributed, then the mean, median, and mode all have approximately the same value.
20. All normal curves appear the same.
21. Given the set of data $\{70, 70, 70, 65, 60\}$, find the standard deviation.
22. In a class of 30 students, Mike has a percentile rank of 70. What is his numerical rank?
23. In a class of 40 students Irene has a rank of 10th. What is her percentile rank?
24. Given the set of data $\{30, 35, 36, 37, 38, 39, 40, 41\}$, what score is at the first quartile?
25. Sandra is ranked 50th in her senior class of 200 students. Eileen is in the same senior class and Eileen has a percentile rank of 72. Who has the higher standing in the class?

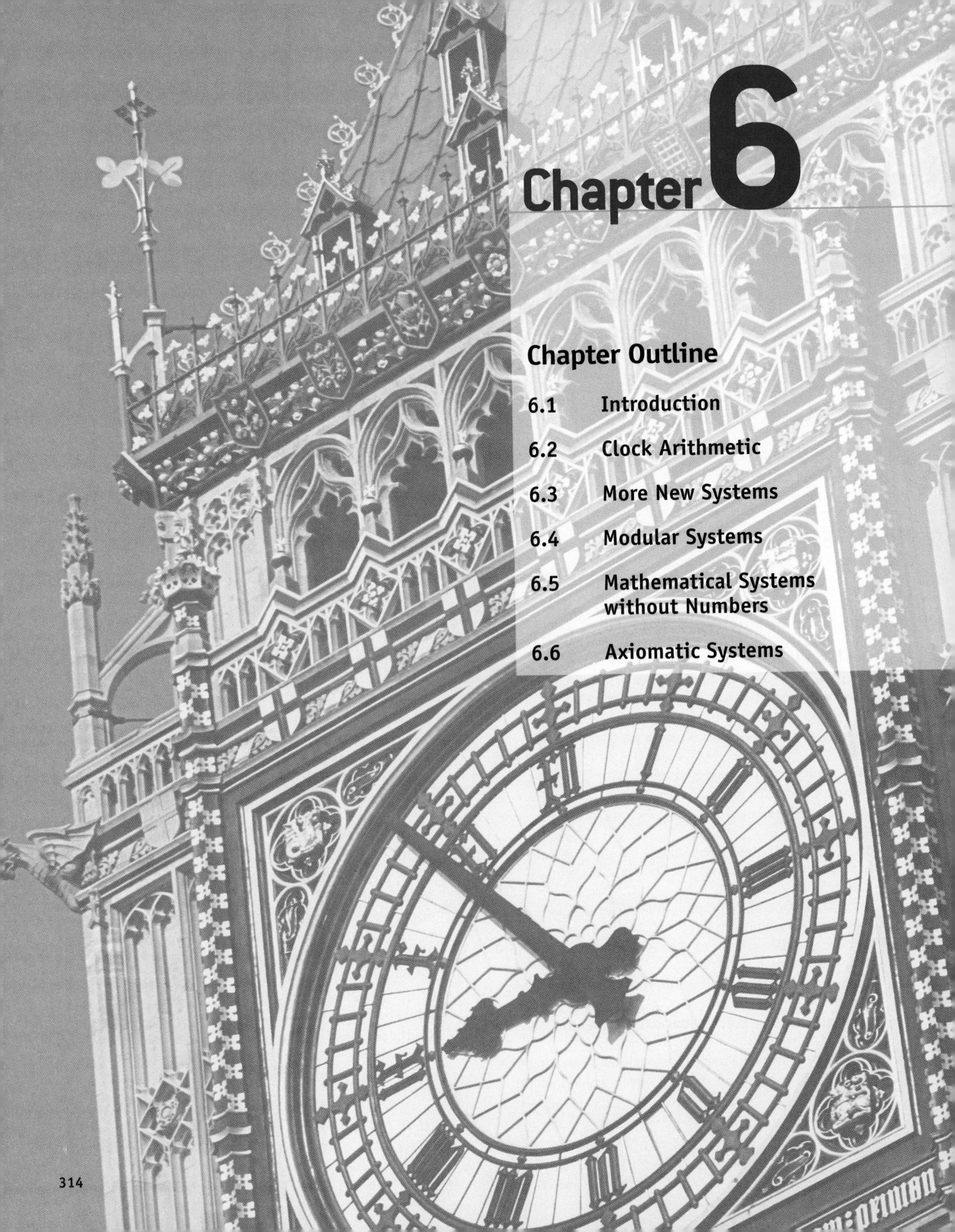

Chapter 6

Chapter Outline

Mathematical Systems

After Studying This Chapter, You Will Be Able to Do the Following:

1. Add, subtract, and multiply in the **12-hour clock system**.
2. Identify the basic parts of a **mathematical system**.
3. Given a mathematical system, determine if the system is—with respect to its corresponding operation—**closed**, if an **identity element** exists, if each element has an **inverse**, if the operation is **associative**, if the operation is **commutative**, and if the system forms a **group**.
4. Add, subtract, and multiply in other **modular systems**.
5. Given a table that defines an operation for the elements in an **abstract system**, evaluate problems corresponding to the system and determine whether the properties of a group are satisfied by the system.
6. Identify the basic parts of an **axiomatic system**, construct a **diagram** (i.e., model) for which all the axioms of a system are satisfied, and prove a **theorem** given the undefined terms, defined terms, and axioms for the system.

Note: *indicates optional material.

Visit the PH Companion Web site at www.prenhall.com/setek

OVERVIEW

In this chapter, we shall examine mathematical systems and their properties. That is, we shall study the nature and structure of mathematical systems. Regardless of what area of mathematics is examined (sets, logic, etc.), there are certain basic common characteristics that these topics possess. For the present time, we shall think of a **mathematical system** as a set of elements together with one or more operations (rules) for combining elements of the set. We shall expand upon this idea later in the chapter, but for now this concept of a system is sufficient.

6.1 INTRODUCTION

***A mathematical system** consists of a set of elements and at least one binary operation. A binary operation is an operation that is applied to two objects. For example, any two numbers can be added.*

One of the first mathematical systems to which we are exposed in school is the set of counting numbers $\{1, 2, 3, \ldots\}$ together with the operation of addition. This system is considered to be an infinite system, because there is an infinite number of elements in the set. We shall begin our study of mathematical systems by examining a finite system, one that has some unusual properties.

6.2 CLOCK ARITHMETIC

Clock arithmetic is an example of a finite mathematical system that will enable us to understand the nature and structure of mathematical systems.

Consider the following addition problems:

$$1 + 2 = 3 \qquad 5 + 7 = 12 \qquad 9 + 10 = 7$$
$$3 + 4 = 7 \qquad 5 + 8 = 1 \qquad 11 + 11 = 10$$
$$5 + 6 = 11 \qquad 6 + 12 = 6 \qquad 9 + 9 = 6$$

Each addition problem listed is correct; there are no mistakes. The reason all of these examples are correct is that they come from a system called clock arithmetic. The first four examples in the list look exactly like examples from ordinary arithmetic, but $5 + 8 = 1$ and $9 + 9 = 6$ do not. In clock arithmetic, $5 + 8 = 1$ because 1:00 comes 8 hours after 5:00. Similarly, 6:00 comes 9 hours after 9:00, so $9 + 9 = 6$. It also follows that if it is 6:00 now, then 12 hours from now it will be 6:00, so $6 + 12 = 6$. Hence, we see that our mathematical system, clock arithmetic, has a set of elements, $\{1, 2, 3, 4, 5, 6, 7, 8, 9, 10, 11, 12\}$, the numerals 1 through 12 on the face of a clock, and it also has an operation (addition), which consists of counting hours in a clockwise direction.

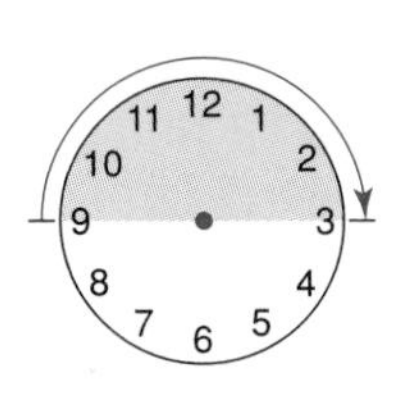

Figure 6.1

Using the clock face in Figure 6.1, we can see that $9 + 6 = 3$. We start at 9 and count 6 units (hours) in a clockwise direction. We complete the counting at 3. Therefore, $9 + 6 = 3$ in clock arithmetic. Any of the examples listed earlier can be figured out in this manner.

By using this technique, we can also verify that $6 + 12 = 6, 9 + 10 = 7$, $11 + 11 = 10$, and $9 + 9 = 6$. In fact, we can construct a table of addition facts for a 12-hour clock using the set of elements $\{1, 2, 3, 4, 5, 6, 7, 8, 9, 10, 11, 12\}$ and the operation of addition. (See Table 6.1.)

The Closure Property

Table 6.1 gives us the answer when we add any two numbers on a 12-hour clock. All answers are included, since we have combined each element in the set with every other element in the set. This underscores the fact that we are working with a finite

TABLE 6.1

+	1	2	3	4	5	6	7	8	9	10	11	12
1	2	3	4	5	6	7	8	9	10	11	12	1
2	3	4	5	6	7	8	9	10	11	12	1	2
3	4	5	6	7	8	9	10	11	12	1	2	3
4	5	6	7	8	9	10	11	12	1	2	3	4
5	6	7	8	9	10	11	12	1	2	3	4	5
6	7	8	9	10	11	12	1	2	3	4	5	6
7	8	9	10	11	12	1	2	3	4	5	6	7
8	9	10	11	12	1	2	3	4	5	6	7	8
9	10	11	12	1	2	3	4	5	6	7	8	9
10	11	12	1	2	3	4	5	6	7	8	9	10
11	12	1	2	3	4	5	6	7	8	9	10	11
12	1	2	3	4	5	6	7	8	9	10	11	12

mathematical system. Every answer in the table is a member of the original set $\{1, 2, 3, 4, 5, 6, 7, 8, 9, 10, 11, 12\}$, so there are no new elements in the set of answers. This is a special characteristic for some systems and is called **closure**.

A system is said to be *closed* with respect to an operation (in this case, addition) if, when we operate on any two elements in the system, the result is also an element in the system. In the case of addition of clock numbers, when we add any two clock numbers, the sum is also a clock number.

More formally, we can say that

> ***A system consisting of a set of elements $\{a, b, c, \ldots\}$ and an operation $*$ is* closed *if for any two elements a and b in the set, $a * b$ (read "a operation b") is also a member of the set.***

EXAMPLE 1 ***Adding Numbers in the 12-Hour Clock System*** Using 12-hour clock addition, evaluate each of the following:

a. $4 + 7$ **b.** $7 + 7$ **c.** $10 + 9$

d. $5 + 8$ **e.** $9 + 7$ **f.** $8 + 12$

Solution We use the table of addition facts for a 12-hour clock (Table 6.1) to find the answer to each problem.

a. $4 + 7 = 11$ **b.** $7 + 7 = 2$ **c.** $10 + 9 = 7$

d. $5 + 8 = 1$ **e.** $9 + 7 = 4$ **f.** $8 + 12 = 8$

NW ***Now Work Problem 1.***

It may have occurred to you that the technique of using a table to solve the problems in Example 1 is not the only way that these problems could be done. There is a more efficient way that we shall now explore.

The armed forces and many factories operate on a 24-hour clock. These clocks begin the same as the 12-hour clock, but once it becomes noon, the 12-hour system starts over, whereas the 24-hour system continues. For example, 1:00 P.M. becomes 1300 hours, or 13:00. Similarly, 2:00 P.M. is the same as 14:00, 3:00 P.M. is the same as 15:00, and so on. This idea can be used to express any number as one of the numbers on a 12-hour clock, that is, as one of the numbers in the set $\{1, 2, 3, 4, 5, 6, 7, 8, 9, 10, 11, 12\}$. For example, 13 can be expressed as 1, since 13 hours is the same as 1 rotation around the clock (12 hours) plus 1 additional hour, that is, $13 = 12 + 1$. The number 15 can be expressed as 3, since $15 = 12 + 3$.

The Identity Element

The number 12 has a special property in the 12-hour clock system: Whenever we add 12 to a number, we obtain that number as a solution ($8 + 12 = 8$, $2 + 12 = 2$, and so on). The number 12 is the *identity element* in this system.

> ***A system consisting of a set of elements {a, b, c, . . .} and an operation * has an*** **identity element** ***(we will call it e) if for every element a in the system,***
>
> $$a * e = a \quad \textbf{and} \quad e * a = a$$

The identity element does not change any element when it is operated on together with that element. In ordinary arithmetic, the identity element for the operation of addition is 0 (zero), because $4 + 0 = 4$ and $0 + 4 = 4$. The sum of any number and zero is that number. This is known as the addition property of zero. The identity element for the operation of multiplication is 1 (one), because $5 \times 1 = 5$ and $1 \times 5 = 5$. The product of any number and one is that number. This is known as the multiplication property of one.

We have already changed some numbers to one of the numbers in the system, but suppose we want to change a number such as 55. It can be expressed as 7 in the 12-hour system as follows: Starting at 12, we complete four rotations around the clock (48 hours), plus 7 more hours, to get 55 hours. That is,

$$55 = 12 + 12 + 12 + 12 + 7$$

Hence, the number 55 can be expressed as 7 in our system, because 12 is the identity element and does not change the identity of the number 7. Therefore, 55 and 7 are in the same position on a 12-hour clock.

Another way to show that 55 can be expressed as 7 is by means of division. If we divide 55 by 12, the remainder is 7:

$$\begin{array}{r} 4 \\ 12\overline{)55} \\ \underline{48} \\ 7 \end{array}$$

This division indicates that there are four 12s in 55. The 12s do not affect the value of the number in the 12-hour clock system; therefore, the remainder, 7, is our answer.

In order to convert any number into a number in the 12-hour clock system, we divide it by 12 and record the remainder. The number 116, for example, can be expressed as 8:

$$\begin{array}{r} 9 \\ 12\overline{)116} \\ \underline{108} \\ 8 \end{array}$$

The nine 12s contained in 116 do not affect the value of the number in the 12-hour system, so the remainder, 8, is our answer.

EXAMPLE 2 ***Using the Identity Element to Find Numerical Equivalence***

Find the equivalent of each of the following on a 12-hour clock:

a. 124 **b.** 258 **c.** 2,000 **d.** 300

Solution

a. We divide 124 by 12 and record the remainder.

$$\begin{array}{r} 10 \\ 12\overline{)124} \\ \underline{12} \\ 04 \\ \underline{0} \\ 4 \end{array}$$

Hence, 124 is equivalent to 4 on a 12-hour clock.

b. Using the same technique we used in part **a**, we divide 258 by 12. The remainder is 6.

c. The number 2000 is equivalent to 8 on a 12-hour clock, because the remainder when 2000 is divided by 12 is 8.

d. The number 300 is equivalent to 12 on a 12-hour clock. Recall that 12 is the identity element in our system under the operation of addition; it has the same property as zero for the operation of addition in ordinary arithmetic. Thus 300 hours would take 25 complete rotations around the clock and would stop at the same place it began, at 12. Hence, after 300 hours, the clock is again at the beginning position.

NW ***Now Work Problem 3.***

EXAMPLE 3 ***Additions That Equal the Identity Element*** Evaluate each of the following on a 12-hour clock:

a. $8 + 4$ **b.** $9 + 3$ **c.** $6 + 6$

Solution

a. $8 + 4 = 12$ **b.** $9 + 3 = 12$ **c.** $6 + 6 = 12$

The Inverse Property

The problems presented in Example 3 all have the answer 12. Recall that 12 is the identity element in our system, and that when 12 is added to a number, the identity or position of the number will not be changed. That is, $2 + 12 = 2$, $3 + 12 = 3$, and so on. In the problems in Example 3, we have added one number to another number and obtained the identity element as the result. These problems illustrate another property found in mathematical systems. In the problem $8 + 4 = 12$, 4 is called the *inverse* or *additive inverse* of 8 because when we add 4 to 8 we obtain the identity element 12. Given any clock number in the 12-hour system $\{1, 2, 3, 4, 5, 6, 7, 8, 9, 10, 11, 12\}$, we can find another clock number such that the sum of the two numbers is the identity element.

Each element in a system consisting of a set of elements {a, b, c, . . .} and an operation * has an* inverse *if for every element a in the system there exists an element b (also in the system) such that

$$a * b = e \quad \text{and} \quad b * a = e$$

where e is the identity element of the system.

Note that if a system has no identity element, then the inverses of elements cannot occur. In the 12-hour clock system, every element has an inverse. For example, 11 is the inverse of 1, because $1 + 11 = 12$ and $11 + 1 = 12$.

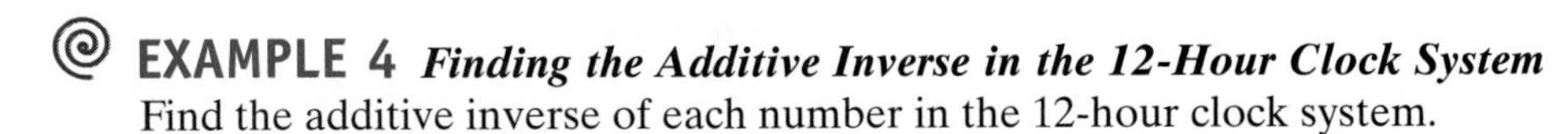

EXAMPLE 4 *Finding the Additive Inverse in the 12-Hour Clock System*

Find the additive inverse of each number in the 12-hour clock system.

a. 7 **b.** 2 **c.** 12

Solution

a. The additive inverse of 7 is 5. We have $5 + 7 = 12$ and $7 + 5 = 12$.

b. The additive inverse of 2 is 10: $2 + 10 = 12$ and $10 + 2 = 12$.

c. The additive inverse of 12 is 12 because in a 12-hour clock system, $12 + 12 = 12$. Thus 12 is its own inverse.

Figure 6.2

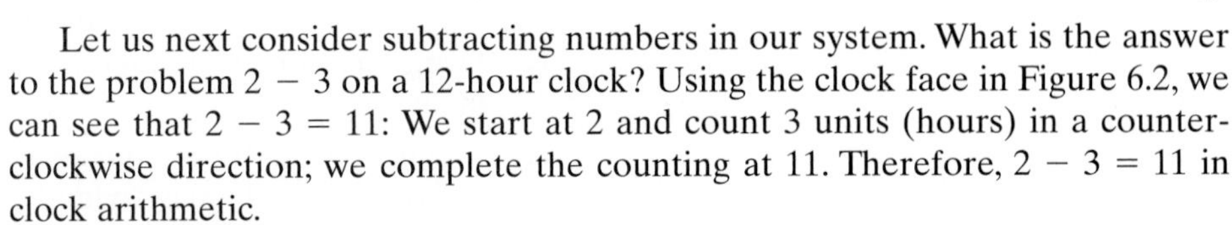

Let us next consider subtracting numbers in our system. What is the answer to the problem $2 - 3$ on a 12-hour clock? Using the clock face in Figure 6.2, we can see that $2 - 3 = 11$: We start at 2 and count 3 units (hours) in a counterclockwise direction; we complete the counting at 11. Therefore, $2 - 3 = 11$ in clock arithmetic.

This problem may also be solved in another manner. Recall that 12 is the identity element in our system, so if we add 12 to a number we will not change its value in the system. Therefore, when we add 12 to a number we will not change its position on the clock face. Hence, 2:00 is the same as 14:00, and we can think of the problem $2 - 3$ as $14 - 3$. Therefore, $2 - 3 = 14 - 3 = 11$ in the 12-hour clock system.

EXAMPLE 5 *Subtracting Numbers in the 12-Hour Clock System*

Evaluate each of the following on a 12-hour clock:

a. $4 - 7$ **b.** $3 - 8$ **c.** $4 - 12$

Solution

a. In the 12-hour clock system, we can add 12 to a number and not change its identity. Hence $4 - 7 = 16 - 7 = 9$.

b. $3 - 8 = 15 - 8 = 7$.

c. $4 - 12 = 16 - 12 = 4$ is one way of solving this problem. An alternate method is to recall that 12 may also be thought of as zero in our system. Therefore, we have $4 - 12 = 4 - 0 = 4$.

NW ***Now Work Problem 5.***

Historical Note

Évariste Galois was a brilliant French mathematician who was killed in a duel (in 1832) when he was 20 years old. The night before the duel he worked furiously on a set of notes and proofs that solved a long-standing mathematical problem. His calculations laid the basis for what has become the theory of groups.

This type of mathematical abstraction is considered to be a powerful tool because it has provided structural links among arithmetic, algebra, geometry, coding theory, crystallography, quantum mechanics, and elementary-particle physics. Some physicists believe that they will obtain a most intimate insight into the basic structures of the universe because of group theory.

The Associative Property

Thus far in our discussion of clock arithmetic, we have encountered three properties of a mathematical system: The clock system has the *closure property* with respect to the operation of addition, it has an *identity element* for addition, and each element has an *inverse* with respect to addition. This system also has a property that has not been mentioned yet—the ***associative property for addition***.

An operation is associative if the location of parentheses in a problem does not affect the answer. To be more specific, consider the following addition problem in ordinary arithmetic:

$$4 + 6 + 9$$

We can find the sum by adding 4 and 6 first, obtaining 10, and then adding 9 to get an answer of 19; that is,

$$(4 + 6) + 9 = 10 + 9 = 19$$

Or we might add 6 and 9 first, obtaining 15, and then add 4 to get an answer of 19; that is,

$$4 + (6 + 9) = 4 + 15 = 19$$

Regardless of which two numbers are added first, the answer is the same. Hence, we can say that

$$(4 + 6) + 9 = 4 + (6 + 9)$$

If a system consists of a set of elements {a, b, c, . . .} and an operation *, we say that the operation is* associative *(or has the associative property) if, for all of the elements in the system,

$$(a * b) * c = a * (b * c)$$

The associative property does not hold for all operations. We have seen that $(4 + 6) + 9 = 4 + (6 + 9)$ in ordinary arithmetic. But consider the operation of subtraction in ordinary arithmetic. Let us see if

$$(7 - 4) - 3 = 7 - (4 - 3)$$

is true. In computing the answer, we always operate inside the parentheses first. Therefore $(7 - 4) - 3 = 3 - 3 = 0$, while $7 - (4 - 3) = 7 - 1 = 6$, and we have shown that

$$(7 - 4) - 3 \neq 7 - (4 - 3)$$

Hence, the associative property does not hold for the operation of subtraction in ordinary arithmetic, because one example showing that a property does not hold is sufficient to illustrate that a property does not work for all elements in the system. Such an example is sometimes called a **counterexample**.

On a 12-hour clock, we have the set of elements

$$\{1, 2, 3, 4, 5, 6, 7, 8, 9, 10, 11, 12\}$$

We can illustrate the associative property for addition by considering the following example:

$$(7 + 8) + 10 \stackrel{?}{=} 7 + (8 + 10)$$

Working with the left side of the equation first, we have

$$(7 + 8) + 10 = 15 + 10$$

But on a 12-hour clock, 15 is the same as 3; hence, $15 + 10 = 3 + 10 = 13$, which is the same as 1 on a 12-hour clock. Hence, $(7 + 8) + 10 = 1$ on a 12-hour clock. Now working with the right side, we have $7 + (8 + 10) = 7 + 18$. But 18 on a 12-hour clock is the same as 6, so $7 + 18 = 7 + 6 = 13$, which is the same as 1 on a 12-hour clock. Hence, $7 + (8 + 10) = 1$ on a 12-hour clock. We have verified that the associative property holds for the example

$$(7 + 8) + 10 = 7 + (8 + 10).$$

Space does not permit the verification of the associative property for addition for all of the elements in the 12-hour clock system, but the associative property for addition does hold for this system.

Groups

Thus far, the 12-hour clock system, consisting of the set of elements {1, 2, 3, 4, 5, 6, 7, 8, 9, 10, 11, 12} and the operation of addition

1. is *closed* with respect to addition;
2. contains an *identity element* with respect to addition;
3. contains an *inverse* for each of its elements with respect to addition;
4. is *associative* with respect to addition.

When a set of elements and an operation satisfy these properties, we say that the elements form a **group** under the operation (in this case, addition). The group operation must be a **binary operation**—that is, one that combines two elements to produce a third element. Addition is a binary operation because it acts on two numbers to produce a third. Squaring a number is not a binary operation, because it acts on only one number.

EXAMPLE 6 ***Determining If a Mathematical System Forms a Group***
Consider the set of counting numbers $\{1, 2, 3, 4, \dots\}$, and the operation of addition. Does this system form a group? Why or why not?

Solution In order to form a group, a system must satisfy the closure property, identity property, inverse property, and associative property. The set of counting numbers $\{1, 2, 3, 4, \dots\}$ does not have an identity element under the operation of addition, because zero is not included in the given set of elements. In addition, the elements have no additive inverses. Therefore, this system *does not* form a group.

EXAMPLE 7 ***Determining If a Mathematical System Forms a Group***
Consider the set of counting numbers, $\{1, 2, 3, 4, \dots\}$, and the operation of multiplication. Does this system form a group? Why or why not?

Solution In order to form a group under the operation of multiplication, the system must satisfy the four properties: closure, identity, inverse, and associative. We shall check to see if these properties hold under the operation of multiplication.

a. The system is closed under multiplication. Whenever we multiply two counting numbers, the product is a counting number.

b. The system does contain an identity element under multiplication, the number 1. The number 1 does not change the identity of a number when the two are multiplied together. That is,

$$2 \times 1 = 1 \times 2 = 2, \quad 3 \times 1 = 1 \times 3 = 3$$

and so on.

c. The system does *not* contain an inverse element under multiplication for each element. The number 1 is the only element in the set that has an inverse, and it is its own inverse: $1 \times 1 = 1$. The number 2 does not have an inverse in the set of elements: There is no number b in the set such that $2 \times b = 1$. This counterexample shows that the set of counting numbers $\{1, 2, 3, 4, \dots\}$ under the operation of multiplication *does not* form a group.

NW ***Now Work Problem 15.***

The Commutative Property and Commutative Groups

When you find the sum of 4 and 5, whether you add them as $4 + 5$, or as $5 + 4$, the answer is 9. That is, $4 + 5 = 5 + 4$. Similarly, if you multiply 4 and 5 together, you can multiply 4×5 or 5×4, and the answer is 20. So $4 \times 5 = 5 \times 4$. It does not matter in which order you do the operation; the answer is the same. In other words, we can switch the elements around; we can *commute* them. When we can do this for all elements in a system using a given operation, we say that the operation is *commutative*.

Given a system consisting of a set of elements $\{a, b, c, \dots\}$ and an operation $*$, we say that the operation is **commutative** ***if for all elements a and b in the system***

$$a * b = b * a$$

Not all operations are commutative. We have seen that addition and multiplication of counting numbers are commutative operations. But consider the operation of subtraction: Does $7 - 6 = 6 - 7$? The answer is no, because $7 - 6 = 1$, whereas $6 - 7 = -1$. This verifies that $7 - 6 \neq 6 - 7$, and that subtraction is not commutative. A group with operation $*$ is called a **commutative group** if $a*b = b*a$ for all elements a and b in the group.

Thus far in this section, we have performed the operations of addition and subtraction on the numbers in a 12-hour clock system. Now let's examine the operation of multiplication. Consider the problem 3×9 on a 12-hour clock. We can consider this as an addition problem, because $3 \times 9 = 9 + 9 + 9$. On a 12-hour clock, $9 + 9 = 18 = 6$, so $(9 + 9) + 9$ becomes $6 + 9$, and $6 + 9 = 3$. Therefore, on a 12-hour clock, $3 \times 9 = 3$. This process is quite tedious. We can multiply two numbers on a 12-hour clock in a more efficient manner: First multiply the numbers as you would in ordinary arithmetic, and then convert your answer to its equivalent on the 12-hour clock. Using this technique for 3×9, we have $3 \times 9 = 27$, which is equivalent to 3 on the 12-hour clock. Hence $3 \times 9 = 3$ on a 12-hour clock.

EXAMPLE 8 ***Multiplying Numbers in the 12-hour Clock System*** Evaluate each of the following on a 12-hour clock:

a. 4×5 **b.** 6×7 **c.** 8×10

d. 11×11 **e.** 10×12

Solution

a. $4 \times 5 = 20$ and 20 is equivalent to 8; hence, $4 \times 5 = 8$.
b. $6 \times 7 = 42$ and 42 is equivalent to 6; hence, $6 \times 7 = 6$.
c. $8 \times 10 = 80$ and 80 is equivalent to 8; hence, $8 \times 10 = 8$.
d. $11 \times 11 = 121$ and 121 is equivalent to 1; hence, $11 \times 11 = 1$.
e. $10 \times 12 = 120$ and 120 is equivalent to 12; hence, $10 \times 12 = 12$.

NW ***Now Work Problem 7.***

Note of Interest

Clocks and watches that use Roman numerals typically use IX to represent nine. But on these same timepieces, four is almost always represented by IIII instead of IV. Why?

Some historians claim that when mechanical clocks were first displayed in public places, such as on a cathedral, the clocks had no hour or minute hand. The clocks simply chimed once on the hour. Eventually, in Italy, Roman numerals were affixed to these clocks. But the common people (peasants) could not read Roman numerals and they could not subtract. They counted on their fingers. Hence, four slash marks (IIII) was much easier for them to deal with than taking one away from five (IV). Therefore, many clock makers always showed four and nine with slash marks instead of IV or IX.

Today Roman numerals are still used on some clocks and watches because of tradition. Many consumers also like the "classic" look of Roman numerals. Ironically, the tradition of IIII instead of IV is also still being carried on.

We shall not explore in detail the operation of division in a 12-hour clock system because of the problems that arise in doing division problems in this system. Consider the problem $8 \div 4$. In ordinary arithmetic, the answer is 2; we can verify this because $2 \times 4 = 8$. But in a 12-hour clock system, $8 \div 4 = 2$, $8 \div 4 = 5$, $8 \div 4 = 8$, and $8 \div 4 = 11$, because $4 \times 2 = 8$, $4 \times 5 = 8$, $4 \times 8 = 8$, and $4 \times 11 = 8$ on a 12-hour clock. Thus some problems have more than one answer, whereas other problems, such as $8 \div 3$, have no answer. In ordinary arithmetic, the answer to this is $2\frac{2}{3}$, but in a 12-hour clock system there is no answer. In order for 3 to divide 8 in a 12-hour clock system, the answer must be one of the elements in the set $\{1, 2, 3, 4, 5, 6, 7, 8, 9, 10, 11, 12\}$. But not one of these numbers will satisfy the statement $3 \times a = 8$. Hence, there is no number in this system that when multiplied by 3 will yield 8 as an answer. In other words, $8 \div 3$ has no answer in a 12-hour clock system.

Exercises for Section 6.2

NW **1.** Evaluate each sum on a 12-hour clock.

a. $2 + 4$ **b.** $7 + 6$
c. $9 + 2$ **d.** $11 + 11$
e. $10 + 11$ **f.** $9 + 10$

2. Evaluate each sum on a 12-hour clock.

a. $9 + 7$ **b.** $7 + 10$
c. $9 + 9$ **d.** $10 + (8 + 10)$
e. $(9 + 8) + 5$ **f.** $11 + (7 + 6)$

NW **3.** Find the equivalent of each number on a 12-hour clock.

a. 33 **b.** 44
c. 55 **d.** 66
e. 277 **f.** 188

4. Find the equivalent of each number on a 12-hour clock.

a. 438 **b.** 261
c. 328 **d.** 1,996
e. 2,565 **f.** -29

NW **5.** Evaluate each difference on a 12-hour clock.

a. $5 - 7$ **b.** $6 - 8$
c. $7 - 11$ **d.** $9 - 10$
e. $2 - 5$ **f.** $3 - 7$

6. Evaluate each difference on a 12-hour clock.

a. $2 - 7$ **b.** $5 - 10$
c. $4 - 12$ **d.** $1 - 11$
e. $3 - (4 - 8)$ **f.** $3 - (7 - 10)$

NW **7.** Evaluate each product on a 12-hour clock.

a. 6×8 **b.** 4×6
c. 3×7 **d.** 4×10
e. 5×11 **f.** 9×9

8. Evaluate each product on a 12-hour clock.

a. 7×11 **b.** 4×13
c. 10×10 **d.** $3 \times (4 + 7)$
e. $2 \times (9 + 10)$ **f.** $5 \times (6 - 10)$

9. Evaluate each product on a 12-hour clock.

a. $2 \times (5 + 8)$ **b.** $3 \times (9 - 11)$
c. $8 \times (7 + 6)$ **d.** $9 \times (11 + 10)$
e. $10 \times (2 - 11)$ **f.** $11 \times (1 - 11)$

10. State the property of the 12-hour clock system that is illustrated by each of the following:

a. $(4 + 3) + 5 = 4 + (3 + 5)$
b. $7 + 6 = 1$, a number in the 12-hour clock system
c. $9 + 12 = 9$
d. $8 + 9 = 9 + 8$
e. $7 \times 7 = 1$
f. $4 \times 5 = 8$, a number in the 12-hour clock system

11. State the property of the 12-hour clock system that is illustrated by each of the following:

a. $7 + 6 = 1$
b. $7 \times 5 = 5 \times 7$
c. $(7 + 6) + 3 = (6 + 7) + 3$
d. $(2 \times 4) \times 3 = 2 \times (4 \times 3)$
e. $5 \times 5 = 1$
f. $12 + 7 = 7$

12. Construct a complete table of multiplication facts for the 12-hour clock system. Use your table to answer each question.

a. What is the identity element for multiplication in this system?
b. Does the closure property hold for this system?
c. Does the commutative property hold for this system?

d. Verify that the associative property holds for one specific instance in this system.

e. Which elements in this system have an inverse?

13. Determine whether each statement is true or false. Given the set of counting numbers $\{1, 2, 3, 4, \ldots\}$, it

a. is closed with respect to addition.

b. is closed with respect to subtraction.

c. is associative with respect to addition.

d. is commutative with respect to division.

e. contains an identity element for addition.

f. contains an identity element for multiplication.

14. Determine whether each statement is true or false for the set of all integers, $\{\ldots, -2, -1, 0, 1, 2, \ldots\}$.

a. The set is closed with respect to addition.

b. The set is closed with respect to subtraction.

c. The set is closed with respect to division.

d. The set contains an identity element for addition.

e. The set contains an identity element for multiplication.

f. The set contains an additive inverse element for each element in the set.

NW **15.** Consider the set of all integers $\{\ldots, -2, -1, 0, 1, 2, \ldots\}$ and the operation of addition. Does this system form a group? Why or why not?

16. Consider the set of whole numbers $\{0, 1, 2, 3, \ldots\}$ and the operation of addition. Does this system form a group? Why or why not?

17. Replace each question mark with a number from the 12-hour clock system to give a true statement.

a. $6 + ? = 1$ b. $? - 4 = 9$

c. $5 \times ? = 1$ d. $5 \times (8 + 7) = ?$

e. $? \times 7 = 9$ f. $? \times (2 + 11) = 2$

18. Replace each question mark with a number from the 12-hour clock system to give a true statement.

a. $3 + ? = 1$ b. $4 - ? = 9$

c. $4 - 7 = ?$ d. $3 \times (4 + ?) = 9$

e. $8 + ? = 2$ f. $4 + ? = 8 - ?$

19. Is the following statement ever true in the 12-hour clock system? If so, give an example. For any numbers a, b, and c, $(a - b) - c = a - (b - c)$.

20. Is the following statement ever true in the 12-hour clock system? If so, give an example. For any numbers a and b, $a - b = b - a$.

Writing Mathematics

21. Describe (in your own words) what *closure* means in a mathematical system.

22. What is a binary operation? Give at least two examples.

23. What is an identity element? What are the additive and multiplicative identity elements for the set of whole numbers?

"Daddy, 28 plus (36 plus 49) equals (28 plus 36) plus what--using the associative principle?"

Reprinted with special permission of King Features Syndicate.

24. List all the properties of a group, and give an example of each property.

Challenge Exercises

25. Consider the following sets: $A = \{1, 3, 5, 7, \ldots\}$, $B = \{2, 4, 6, 8, \ldots\}$, $C = \{5, 10, 15, 20, \ldots\}$, $D = \{1, 3, 9, 27, \ldots\}$. In order for a set to have a binary operation, the set must be closed under the operation.

a. In which of the given sets is addition a binary operation?

b. In which of the given sets is multiplication a binary operation?

26. A set that satisfies the associative property under a binary operation, but is not a group, is called a *semigroup*. Find an example of a semigroup. Verify your example.

27. Construct a 12-hour clock "face" similar to the ones shown in Figures 6.1 and 6.2 that uses *negative numbers*. Associate each negative number with its corresponding positive number. [*Hint:* Negative numbers in the 12-hour clock system imply a *counterclockwise* rotation.]

28. At the end of this section, we indicated that the problem with division in the 12-hour clock system is that some division questions have no answers and some division questions have more than one answer. Given the elements of the 12-hour clock system, determine which numbers have no common factors with 12.

Group/Research Activity

29. **Abelian group** A group that satisfies the commutative property is called a commutative group or *Abelian group*, named after Niels Henrik Abel (1802–1829). Do research and prepare a report on Abel. Be sure to include his contributions to the development of mathematics and other areas that might be of interest.

Just for fun

Are the horizontal lines straight?

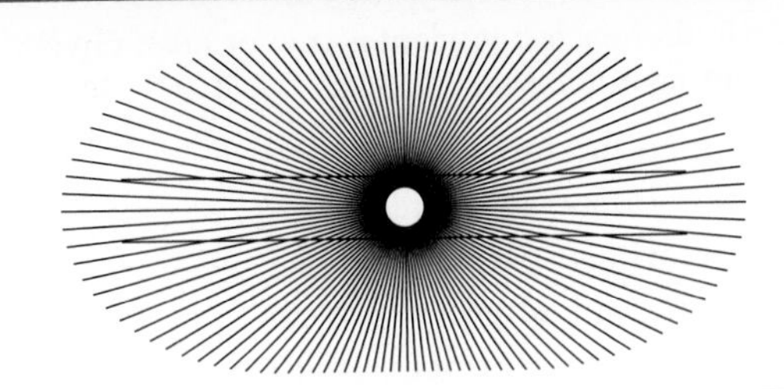

6.3 MORE NEW SYSTEMS

The Four Seasons as a Mathematical System

In the preceding section, we examined the nature of clock arithmetic. Using the set of numbers

$$\{1, 2, 3, 4, 5, 6, 7, 8, 9, 10, 11, 12\}$$

and the properties of a 12-hour clock, we constructed a new system of arithmetic.

There are many other systems of arithmetic that can be created. For another example, let

$$1 = \text{spring} \qquad 2 = \text{summer} \qquad 3 = \text{fall} \qquad 4 = \text{winter}$$

If it is now spring, then three seasons from now it will be winter. That is, $1 + 3 = 4$, or spring + fall = winter. Suppose it is now winter. What season will it be five seasons from now? Starting with winter and counting off five seasons, we wind up at spring. This suggests that $4 + 5 = 1$ in our new system. That is, we have a cycle of four seasons, and so once we reach 4 in our counting, we start over again. For the previous problem we have $4 + 5 = 9$ in ordinary arithmetic, but $9 = 4 + 4 + 1$, so in our new system 9 is equivalent to 1. To help us in our calculations, we could think of this new system in terms of a 4-hour clock. Tables 6.2 and 6.3 show the addition and multiplication facts for the seasons system.

TABLE 6.2

+	1	2	3	4
1	2	3	4	1
2	3	4	1	2
3	4	1	2	3
4	1	2	3	4

TABLE 6.3

×	1	2	3	4
1	1	2	3	4
2	2	4	2	4
3	3	2	1	4
4	4	4	4	4

Biography: Julia Robinson

The *Ladies' Home Journal* has included Julia Robinson (1919–1985) in its list of the 100 most outstanding women in America. So who is Julia Robinson? She was a mathematician who specialized in the field of where logic and number meet. Her work contributed to the solution of Hilbert's tenth problem (can there be a general method for determining whether any diophantine equation has a solution in integers?), which was solved by Yuri Matijasevič (in the negative, 1970). She got her Ph.D. (on definability and decision problems in arithmetic) in 1948 under the leadership of Alfred Tarski, a leading logician at Berkeley. This was the same year she started working on Hilbert's tenth problem. She was the first woman president of the American Mathematical Society (1983–1984) and the first woman to be elected to the National Academy of Sciences (1975). She preferred to be remembered not as the first woman this or that, but as a mathematician should be remembered, for the problems that she solved and the theorems that she proved.

EXAMPLE 1 ***Adding and Multiplying in the Four Seasons System*** Let spring = 1, summer = 2, fall = 3, and winter = 4. Evaluate the following:

a. summer + summer
b. fall + summer
c. summer × fall
d. fall × fall

Solution

a. Since summer = 2, we have 2 + 2 = 4, and 4 = winter, so summer + summer = winter.

b. Fall = 3 and summer = 2; therefore, we have 3 + 2 = 1, according to Table 6.2. Since 1 = spring, fall + summer = spring.

As an alternate solution, 3 + 2 = 5, which is equivalent to 1 in this system, and 1 = spring.

c. Summer = 2 and fall = 3; therefore, we have 2 × 3 = 2, according to Table 6.3, because 2 = summer, summer × fall = summer.

Alternatively, we have 2 × 3 = 6, which is equivalent to 2 in this system, and 2 = summer.

d. Fall = 3; therefore, we have 3 × 3 = 1, according to Table 6.3, because 1 = spring, fall × fall = spring.

As an alternate solution, we have 3 × 3 = 9, which is equivalent to 1 in this system, and 1 = spring.

NW ***Now Work Problems 1, 3, and 7.***

Recall that Table 6.2 illustrates the addition facts for the seasons system. If we substitute the names of the seasons for the numbers in Table 6.2, we have Table 6.4, which may appear strange at first glance. But remember, the only thing that we have done is substitute the names of the seasons in place of the numbers in Table 6.2. This system may seem more abstract, but it is still a system.

TABLE 6.4

+	Spring	Summer	Fall	Winter
Spring	summer	fall	winter	spring
Summer	fall	winter	spring	summer
Fall	winter	spring	summer	fall
Winter	spring	summer	fall	winter

Note that we have closure in this system, because whenever we combine any two seasons the result is a member of the set {spring, summer, fall, winter}. What is the identity element in this system? Recall that the identity element does not change any element when it is operated on together with that element. Upon examination of Table 6.4 we see that winter is the identity element in our seasons system for the operation of addition. We can verify this by checking each season's addition with winter: for example, spring + winter = spring, and winter + spring = spring.

EXAMPLE 2 ***Finding the Additive Inverse in the Four Seasons System*** Using Table 6.4, find the additive inverse of each of the following:

a. spring
b. summer
c. fall
d. winter

Solution Recall that each element a in a system has an inverse element if there exists an element b such that $a * b = e$ and $b * a = e$, where e is the identity element of the system and * is the operation of the system. Since winter is the identity element in the seasons system, we examine the table to see which season added to a given season yields winter as the answer.

a. The additive inverse of spring is fall.
b. The additive inverse of summer is summer.
c. The additive inverse of fall is spring.
d. The additive inverse of winter is winter. 

NW ***Now Work Problem 15.***

The Days of the Week as a Mathematical System

Suppose that today is Wednesday, the fourth day of the month. What day of the week is the 27th? One way we could determine this would be to examine a calendar. Another interesting technique is the following: Because Wednesday corresponds to 4 (it is the fourth day of the month), then Thursday corresponds to 5, Friday corresponds to 6, and so on. Twenty-one days from now it will be Wednesday again (the days of the week are in a cycle of 7), and the date will be the 25th; hence, the 27th will fall on a Friday. We can also solve this problem in another way: Since every seven days brings us back to the same day of the week, 27 is equivalent to 6 in a week system, because $27 = 7 + 7 + 7 + 6$. Recall that Wednesday corresponded to 4; hence, Friday must correspond to 6, and the 27th must be a Friday. This discussion suggests that we can use the days of the week to create another new system of arithmetic. Let

1 = Sunday	5 = Thursday
2 = Monday	6 = Friday
3 = Tuesday	7 = Saturday
4 = Wednesday	

If today is Sunday, then three days from now it will be Wednesday; that is, $1 + 3 = 4$. If today is Tuesday, then four days from now it will be Saturday; that is, $3 + 4 = 7$. Suppose that today is Monday; what day will it be eight days from now? We know that eight days from now it will be Tuesday, because seven days from now it will be Monday again, and one more day will bring us to Tuesday. But we could also say Monday = 2, and $2 + 8 = 10$. Ten is equivalent to 3 in the week system, because 7 acts as an identity element in this system for the operation of addition and $10 = 3 + 7$.

Table 6.5 shows the addition facts for the week system. Note that whenever 7 is added to another number in the system, the result is always the original number. We also have closure in this new system, because the results of adding elements in

TABLE 6.5

+	1	2	3	4	5	6	7
1	2	3	4	5	6	7	1
2	3	4	5	6	7	1	2
3	4	5	6	7	1	2	3
4	5	6	7	1	2	3	4
5	6	7	1	2	3	4	5
6	7	1	2	3	4	5	6
7	1	2	3	4	5	6	7

the system are always in the system. There is an identity element (7), so we can determine the additive inverse for an element in this system. For example, the additive inverse of 2 is 5, because $2 + 5 = 7$ and $5 + 2 = 7$.

EXAMPLE 3 *Adding and Multiplying in the Days of the Week System*

Given that Sunday = 1, Monday = 2, Tuesday = 3, Wednesday = 4, Thursday = 5, Friday = 6, and Saturday = 7, evaluate each of the following:

a. Sunday + Tuesday
b. Monday + Wednesday
c. Friday + Tuesday
d. Wednesday + Friday

Solution

a. Because Sunday = 1 and Tuesday = 3, we have $1 + 3 = 4$, and 4 = Wednesday; Sunday + Tuesday = Wednesday.

b. Monday = 2 and Wednesday = 4; therefore we have $2 + 4 = 6$, and 6 = Friday; Monday + Wednesday = Friday.

c. Friday = 6 and Tuesday = 3; therefore we have $6 + 3 = 2$ according to the table, and 2 = Monday. Hence, Friday + Tuesday = Monday.

As an alternative solution, $6 + 3 = 9$, which is equivalent to 2 in this system, and 2 = Monday.

d. Wednesday = 4 and Friday = 6; hence, $4 + 6 = 3$ according to the table, and 3 = Tuesday. Therefore, Wednesday + Friday = Tuesday.

We could also note that $4 + 6 = 10$, which is equivalent to 3 in this system, and 3 = Tuesday.

NW ***Now Work Problems 19 and 21.***

EXAMPLE 4 *Evaluating Expressions in the Days of the Week System*

Using the week system, evaluate the following:

a. Wednesday − Thursday
b. Monday × Tuesday
c. Wednesday × Thursday
d. (Friday × Monday) − Wednesday

Solution

a. Wednesday = 4 and Thursday = 5; therefore, we have $4 - 5$. In order to subtract 5 from 4, we must change 4 to an equivalent number: $4 + 7 = 11$, so 4 is equivalent to 11. Hence, in our system, $4 - 5 = 11 - 5 = 6$, and 6 = Friday. Hence, Wednesday − Thursday = Friday.

b. Monday = 2 and Tuesday = 3; $2 \times 3 = 6$ = Friday. Hence, Monday × Tuesday = Friday.

c. Wednesday = 4 and Thursday = 5; $4 \times 5 = 20$, which is equivalent to 6, because in the week system, $20 = 7 + 7 + 6$, and 6 = Friday. Hence, Wednesday × Thursday = Friday.

d. Friday = 6, Monday = 2, and Wednesday = 4. Working inside the parentheses first, we have (6×2), which equals 12, and $12 - 4 = 8$, which is equivalent to 1 in the week system. Hence, (Friday × Monday) − Wednesday = Sunday.

Exercises for Section 6.3

In Exercises 1–12, you can assume that 1 = spring, 2 = summer, 3 = fall, and 4 = winter. Evaluate each of the following, giving your answer in terms of a season.

NW 1. summer + fall
2. spring + summer
NW 3. winter + summer
4. spring − fall
5. spring − summer
6. winter − spring
NW 7. summer × summer
8. fall × summer
9. fall × winter
10. spring × (fall + winter)
11. fall × (winter − spring)
12. spring × winter
13. What is the identity element in the seasons system for the operation of multiplication?
14. What is the multiplicative inverse of summer?
NW 15. What is the multiplicative inverse of fall?
16. Does the seasons system form a group under the operation of addition?
17. Does the seasons system form a group under the operation of multiplication?

In Exercises 18–34, assume that 1 = Sunday, 2 = Monday, 3 = Tuesday, 4 =Wednesday, 5 = Thursday, 6 = Friday, and 7 = Saturday. Evaluate each of the following, giving your answer in terms of a day of the week.

18. Monday + Thursday
NW 19. Monday + Saturday
20. Wednesday + Sunday
NW 21. Thursday + Friday
22. Saturday + Thursday
23. Sunday + Sunday
24. Thursday − Tuesday
25. Thursday − Friday
26. Friday − Saturday
27. Thursday − Saturday
28. Tuesday − Friday
29. Tuesday − Thursday
30. Tuesday × Thursday
31. Wednesday × Saturday
32. Saturday × Saturday
33. Tuesday × (Monday + Friday)
34. Friday × (Friday − Saturday)
35. What is the identity element in the week system for the operation of addition?
36. What is the identity element in the week system for the operation of multiplication?
37. What is the additive inverse of
    a. Monday
    b. Tuesday
    c. Wednesday
38. What is the multiplicative inverse of
    a. Monday
    b. Friday
39. Verify that Sunday × (Tuesday × Friday) = (Sunday × Tuesday) × Friday.
40. Does the week system form a group under the operation of multiplication?
41. Does the week system form a group under the operation of addition?

In Exercises 42–58, assume that 1 = up, 2 = down, 3 = left, and 4 = right. Evaluate each of the following, giving your answer in terms of a direction.

42. left + up
43. right + down
44. left − down
45. up − right
46. down × up
47. left × left
48. (right + down) × left
49. up × (left + down)
50. right × (down − left)
51. (up − right) × left
52. What is the identity element in the directions system for the operation of multiplication?
53. What is the identity element in the directions system for the operation of addition?
54. What is the multiplicative inverse of
    a. left
    b. up
    c. right
55. What is the additive inverse of
    a. right
    b. left
    c. up
56. Verify that left × (right × up) = (left × right) × up.
57. Does the directions system form a group under the operation of addition?
58. Does the directions system form a group under the operation of multiplication?

Writing Mathematics

59. What are the components of a mathematical system?
60. If a mathematical system forms a commutative group, then what properties must be satisfied?

Challenge Exercises

61. Given the set of elements {w, x, y, z} and the operation &, we have the following table:

&	w	x	y	z
w	x	y	z	w
x	y	z	w	x
y	z	w	x	y
z	w	x	y	z

Using the table, evaluate the following:
    a. x & y
    b. y & w
    c. w & (w & z)
    d. (y & x) & (z & w)
62. Using the table in Exercise 61, answer the following:
    a. Is the set closed under the operation of &?
    b. What is the identity element, if any?
    c. Does each element of the set have an inverse? If so, name them.
    d. Does the set form a group? Why or why not?

63. Create a *subtraction table* for the seasons system to complement the addition and multiplication tables (Tables 6.2 and 6.3).

64. Review the Math Connections (on page 353) titled "Math Patterns" and construct a design based on the multiplication table for the seasons system (Table 6.3).

65. **Ceiling fans** Consider a four-way switch that operates a ceiling fan. In most cases this switch is controlled by a chain that is part of the ceiling fan housing. If the fan is in the "off" position, then pulling down (and releasing) the chain places the fan in "low." A second tug on the chain places the fan in "medium," and a third pull places the fan in "high." Let 0 = off, 1 = low, 2 = medium, and 3 = high. Also let ♦ represent the operation of the fan's speed switch. Construct a table that shows the fan's final speed depending on the fan switch's initial position and the number of "pulls" you make on the chain.

Group/Research Activities

66. As mentioned in Section 6.2, Evariste Galois was a French mathematician whose calculations laid the basis for the theory of groups. Do research and write a report on Galois. Be sure to include his contributions to the development of mathematics and other areas that might be of interest.

67. Read the article "Using Mathematical Structures to Generate Artistic Designs," which appeared in the May 1984 (pp. 393–398) edition of *The Mathematics Teacher*. Prepare a summary of this article, along with some of the author's designs as well as your own designs, and give a report to your class.

Just for fun

Take any size piece of paper and fold it in half, then fold it in half again, and keep doing this as many times as possible. You probably cannot do it eight times.

6.4 MODULAR SYSTEMS

The Modulo 5 and Modulo 7 Systems

Thus far in this chapter, we have examined various new systems of arithmetic. One of the systems discussed was the 12-hour clock system, since everyone is familiar with the 12-hour clock. We will next consider a mathematical system that is also based on a clock, but this time we will consider a 5-hour clock.

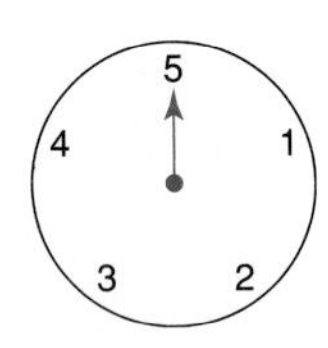

Figure 6.3

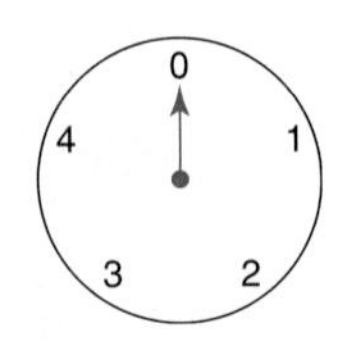

Figure 6.4

A 5-hour clock might be like the one shown in Figure 6.3, but to make our new clock a little easier to understand, we will make it look like a timer. You have probably seen some sort of a timer before—for example, an egg timer or a stopwatch. These timers have zero at the top of the face, rather than some other number. Figure 6.4 shows our new 5-hour clock. This 5-hour clock system contains the elements 0, 1, 2, 3, 4. If we begin to count in this system (clockwise), we have 1, 2, 3, 4, 0, and then the system starts to repeat itself; that is, 1, 2, 3, 4, 0, 1, 2, 3, 4, 0, 1, 2, ...

If it is now 1:00 on our 5-hour clock, then 3 hours from now it will be 4:00. Thus $1 + 3 = 4$. Suppose it is now 2:00 on the 5-hour clock. What time will it be 4 hours from now? To find out, we begin at 2 and count 4 units in a clockwise direction, ending up at 1. Therefore, in this new system, $2 + 4 = 1$.

Similarly, we can verify that $4 + 3 = 2$ using the 5-hour clock. Continuing with this technique, we can construct a table of addition facts for a 5-hour clock using the set of elements $\{0, 1, 2, 3, 4\}$ (see Table 6.6).

TABLE 6.6

+	0	1	2	3	4
0	0	1	2	3	4
1	1	2	3	4	0
2	2	3	4	0	1
3	3	4	0	1	2
4	4	0	1	2	3

When we have a system such as the 5-hour or 12-hour clock that repeats itself in a cycle, we call it a **modular system**. The 5-hour clock is called the *modulo 5 system*. We abbreviate this to *mod 5 system*.

Recall that in order to convert a given number into a number in the 12-hour clock system, we divided it by 12 and recorded the remainder. Twelve did not affect the answer because it was the identity element. In order to convert a given number into a number in the 5-hour clock system (modulo 5), we divide it by 5 and record the remainder. For example, from Table 6.6 we know that $4 + 4 = 3$ in the modulo 5 system, but we can also do this by finding the sum of 4 and 4 in ordinary arithmetic, 8, and converting it to a number in the modulo 5 system:

$$\begin{array}{r} 1 \\ 5\overline{)8} \\ \underline{5} \\ 3 \end{array}$$

When we divide 8 by 5, the remainder is 3, our desired result. Therefore, we can write

$$4 + 4 \equiv 3 \pmod 5$$

This is read "4 + 4 is equivalent to 3, mod 5."

Similarly, we can write $6 \equiv 1 \pmod 5$, $21 \equiv 1 \pmod 5$, and $36 \equiv 1 \pmod 5$. In each case, this indicates that 1 is the remainder when the numbers 6, 21, and 36 are divided by 5. In general terms,

$$a \equiv b \pmod m$$

means that a and b both have the same remainder when they are divided by m. We say that *a is equivalent to b mod m*.

EXAMPLE 1 ***Adding Numbers in the Mod 5 System*** Evaluate the following in the modulo 5 system:

a. $4 + 4$ **b.** $2 + 4$ **c.** $3 + 2$

Solution

a. Using Table 6.6, we note that the sum of 4 and 4 is 3; that is, $4 + 4 \equiv 3 \pmod 5$.

b. According to the table, the sum of 2 and 4 is 1, or $2 + 4 \equiv 1 \pmod{5}$.
c. Since $3 + 2$ is equivalent to 0, $3 + 2 \equiv 0 \pmod{5}$.

Alternate Solution

a. $4 + 4 = 8$; divide 5 into 8, and the remainder is 3. Hence, $4 + 4 \equiv 3 \pmod{5}$.
b. $2 + 4 = 6$; divide 5 into 6, and the remainder is 1. Hence, $2 + 4 \equiv 1 \pmod{5}$.
c. $3 + 2 = 5$; divide 5 into 5, and the remainder is 0. Hence, $3 + 2 \equiv 0 \pmod{5}$.

NW ***Now Work Problems 1a, b, and c.***

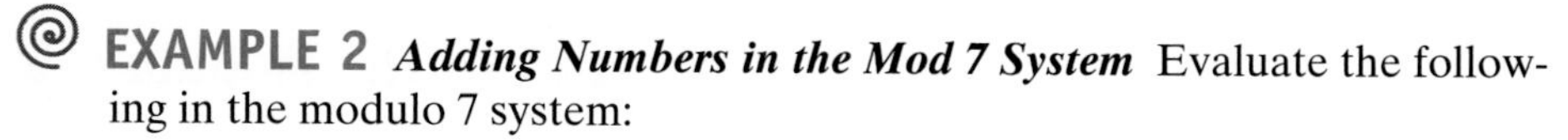

EXAMPLE 2 ***Adding Numbers in the Mod 7 System*** Evaluate the following in the modulo 7 system:

a. $4 + 5$ **b.** $4 + 4$ **c.** $6 + 5$ **d.** $4 + 3$

Solution

a. $4 + 5 = 9$; divide 7 into 9, and the remainder is 2. Therefore, $4 + 5 \equiv 2 \pmod{7}$.
b. $4 + 4 = 8$; divide 7 into 8, and the remainder is 1. Therefore, $4 + 4 \equiv 1 \pmod{7}$.
c. $6 + 5 = 11$; divide 7 into 11, and the remainder is 4. Therefore, $6 + 5 \equiv 4 \pmod{7}$.
d. $4 + 3 = 7$; divide 7 into 7, and the remainder is 0. Therefore, $4 + 3 \equiv 0 \pmod{7}$.

NW ***Now Work Problem 11a.***

Note of Interest

One of the more popular toys and puzzles of the 1980s was Rubik's Cube. It was developed in 1974 by Erno Rubik, a Hungarian design professor who wanted to give his students more experience in understanding three-dimensional solids.

Each side of a Rubik's Cube is made up of nine squares. In the beginning, all nine squares on each face are of one color; that is, there are six faces and six colors. The cube has a system of axles such that it can rotate about its center so that each corner's small cube can rotate in any of the three dimensions. Also, with as few as four twists of the cube the solid faces of the cube will become completely scrambled. The challenge is to manipulate the cube back to its original state. Most people cannot do it, even with a solutions manual.

One of the reasons for the difficulty in solving it is that the cube has approximately 43,252,003,274,489,856,000 (43 quintillion) positions.

In addition to the fascination of trying to solve Rubik's Cube, it should be noted that it satisfies the properties of a group. That is, the moves are closed: No matter how you twist it, the result is a member of the original set. Each move has an inverse, and identity moves exist for each one as well. Similarly, the moves are associative.

EXAMPLE 3 ***Subtracting Numbers in the Mod 5 System*** Evaluate the following in the modulo 5 system:

a. $2 - 4$ **b.** $1 - 3$ **c.** $3 - 4$ **d.** $4 - 0$

Solution

a. In the mod 5 system, we can add 5 to a number and not change its identity. Therefore, $2 - 4 = 7 - 4 = 3$, so $2 - 4 \equiv 3 \pmod 5$.

b. $1 - 3 = 6 - 3 = 3$, so $1 - 3 \equiv 3 \pmod 5$

c. $3 - 4 = 8 - 4 = 4$, so $3 - 4 \equiv 4 \pmod 5$

d. $4 - 0 \equiv 4 \pmod 5$

NW ***Now Work Problems 5a, b, c, and d.***

Properties of the Modulo 5 System

Let's examine the properties of the modulo 5 mathematical system with the operation of addition. This system consists of the set of elements $\{0, 1, 2, 3, 4\}$. The addition operation on this set is shown in Table 6.6 in this section.

1. **Closure property:** All of the entries in the table are elements of the given set; no new elements appear in the table. Therefore, the system satisfies the closure property.
2. **Identity property:** There is an element, 0, in the set of elements that does not change any element when it is added to that element:

$$0 + 0 = 0, \quad 1 + 0 = 1, \quad 2 + 0 = 2,$$
$$3 + 0 = 3, \quad 4 + 0 = 4$$

3. **Inverse property:** Each element in this system has an inverse, an element which when added to the given element results in the identity element, 0. Note that

$$0 + 0 = 0, \quad 1 + 4 = 0, \quad 2 + 3 = 0.$$
$$3 + 2 = 0, \quad 4 + 1 = 0$$

   Therefore, the inverse of 0 is 0, the inverse of 1 is 4, the inverse of 2 is 3, the inverse of 3 is 2, and the inverse of 4 is 1.
4. **Associative property:** The addition operation is associative if $(a + b) + c = a + (b + c)$. Trying an example, we have $(1 + 2) + 3 = 3 + 3 = 1$ and $1 + (2 + 3) = 1 + 0 = 1$, so that

$$(1 + 2) + 3 = 1 + (2 + 3)$$

   Any other example for this system also works; hence, the system satisfies the associative property.
5. **Commutative property:** Addition is commutative if $a + b = b + a$. Trying an example, we have $2 + 4 = 1$ and $4 + 2 = 1$; therefore, $2 + 4 = 4 + 2$. Any other example for this system also works; hence, the system satisfies the commutative property.

The mathematical system (modulo 5) that consists of the set of elements $\{0, 1, 2, 3, 4\}$ and the operation of addition satisfies all of the properties listed. Since a group is composed of a set of elements together with a binary operation that satisfies the closure property, identity property, inverse property, and associative property, the modulo 5 system with the operation of addition forms a group. The fact that the system also satisfies the commutative property means that we have a commutative group under the operation of addition.

TABLE 6.7

×	0	1	2	3	4
0	0	0	0	0	0
1	0	1	2	3	4
2	0	2	4	1	3
3	0	3	1	4	2
4	0	4	3	2	1

Now let's consider the modulo 5 system with the operation of multiplication. This system consists of the set of elements $\{0, 1, 2, 3, 4\}$ and the operation of multiplication. We first construct a multiplication table for mod 5 (see Table 6.7). We next determine which properties are satisfied under this system. The system is closed with respect to multiplication. It has an identity element, 1; that is, $a \times 1 = a$ for any a in the system. But not every element has an inverse. In particular, there is no number in the system that when multiplied by zero will yield the identity element 1. Zero times a number is zero for every number in the system. Hence, the modulo 5 system with the operation of multiplication does not form a group because the system does not satisfy all the properties of a group.

EXAMPLE 4 ***Finding Equivalence in the Mod 5 System*** Using the elements of the mod 5 system, find replacements for the question marks, so that each of the following is true.

a. $4 + 3 \equiv ?(\text{mod } 5)$ **b.** $? + 4 \equiv 1(\text{mod } 5)$

c. $3 - ? \equiv 4(\text{mod } 5)$ **d.** $3 \times ? \equiv 1(\text{mod } 5)$

Solution

a. $4 + 3 = 7$; divide 5 into 7, and the remainder is 2. Hence, $4 + 3 \equiv 2$ (mod 5).

b. Because the mod 5 system contains the set of numbers $\{0, 1, 2, 3, 4\}$, we can try each one of these in place of the question mark. Starting with 0, we have: $0 + 4 = 4$; $1 + 4 = 0$; $2 + 4 = 6$, but $6 \equiv 1(\text{mod } 5)$. Therefore, our answer is 2.

c. Using the same technique, we have $3 - 0 = 3$, $3 - 1 = 2$, $3 - 3 = 0$, and $3 - 4 = 8 - 4 = 4$. Therefore, the solution is 4.

d. We see that $3 \times 0 = 0$, $3 \times 1 = 3$, and $3 \times 2 = 6$. Because $6 \equiv 1$ (mod 5), the solution is 2.

NW ***Now Work Problems 17b and d.***

EXAMPLE 5 ***An Application of Modular Systems*** Joe, an avid sports fan, collects sports cards. Each day he studies his collection of special cards. On Monday, he divides his collection into piles of 5 with 2 left over; on Tuesday, he divides his set of cards into piles of 4 with 2 left over; and on Wednesday, he divides his set into piles of 7 with 0 left over. If Joe's collection of special sports cards consists of fewer than 50 cards, how many does he have?

Solution There is no information given as to exactly how many cards Joe has, but we do know how many are in each pile on each day. Let x represent the number of cards. On Monday each pile contained 5 cards with 2 left over. Therefore, if 5 is divided into x, there is a remainder of 2; that is, $x \equiv 2(\text{mod } 5)$. On Tuesday, each pile contained 4 cards with 2 left over. Therefore, if 4 is divided into x, there is a remainder of 2; so $x \equiv 2(\text{mod } 4)$. On Wednesday, each pile contained 7 cards with 0 left over. Hence, if 7 is divided into x, there is a remainder of 0; so $x \equiv 0(\text{mod } 7)$.

Now we must find a value that satisfies all three statements:

$$x \equiv 2(\text{mod } 5), \quad x \equiv 2(\text{mod } 4), \quad x \equiv 0(\text{mod } 7)$$

For the first statement, $x \equiv 2(\text{mod } 5)$, the set of possible replacements for x is $\{7, 12, 17, 22, 27, 32, 37, 42, 47\}$. We stop at 47 because Joe has fewer than 50 cards. Now which of these numbers also satisfies the second statement, $x \equiv 2(\text{mod } 4)$?

That is, which of these numbers has a remainder of 2 when divided by 4? These numbers are 22 and 42. Now which of these two numbers also satisfies the third statement, $x \equiv 0 \pmod 7$? The remainder is zero when 42 is divided by 7, and therefore Joe has 42 cards in his collection.

NW ***Now Work Problem 19.***

Exercises for Section 6.4

1. Evaluate each sum on a 5-hour clock—that is, in the modulo 5 system.
   NW **a.** $1 + 3$ NW **b.** $2 + 3$
   NW **c.** $4 + 2$ **d.** $3 + 4$
   e. $(3 + 2) + 4$ **f.** $(4 + 3) + 4$

2. Evaluate each sum in the modulo 5 system.
   a. $1 + 0$ **b.** $2 + 1$
   c. $4 + 0$ **d.** $3 + 3$
   e. $2 + (4 + 3)$ **f.** $(4 + 1) + 1$

3. Find the equivalent of each number in the modulo 5 system.
   a. 33 **b.** 44 **c.** 55
   d. 342 **e.** 780 **f.** -8

4. Find the equivalent of each number in the modulo 5 system.
   a. 47 **b.** 56 **c.** 48
   d. 377 **e.** 1,001 **f.** -27

5. Evaluate each difference in the modulo 5 system.
   NW **a.** $2 - 4$ NW **b.** $1 - 4$
   NW **c.** $1 - 3$ NW **d.** $3 - 4$
   e. $3 - (2 - 4)$ **f.** $2 - (3 - 4)$

6. Evaluate each difference in the modulo 5 system.
   a. $2 - 3$ **b.** $1 - 2$
   c. $4 - 4$ **d.** $(2 - 3) - 4$
   e. $(3 - 4) - 1$ **f.** $1 - (2 - 4)$

7. Evaluate each product in the modulo 5 system.
   a. 4×3 **b.** 2×4
   c. 3×2 **d.** $2 \times (4 \times 3)$
   e. $3 \times (4 + 2)$ **f.** $2 \times (3 - 4)$

8. Evaluate each product in the modulo 5 system.
   a. 6×6 **b.** 9×6
   c. 6×3 **d.** $2 \times (2 \times 3)$
   e. $(4 + 2) \times 2$ **f.** $(3 - 4) \times 3$

9. Given the set of elements $\{0, 1, 2, 3, 4, 5, 6\}$, construct a complete table of addition facts for the modulo 7 system.
   a. Does the closure property hold for this system?
   b. What is the identity element (if any) for the operation of addition in this system?
   c. Does the commutative property hold for this system?
   d. Verify that the associative property holds for one specific instance in this system.
   e. List the elements in this system that have an additive inverse, and list their inverses.
   f. Does this system form a group under the operation of addition?

10. Given the set of elements $\{0, 1, 2, 3, 4, 5, 6\}$, construct a complete table of multiplication facts for the modulo 7 system.
    a. Does the closure property hold for this system?
    b. What is the identity element (if any) for the operation of multiplication?
    c. Does the commutative property hold for this system?
    d. List the elements in this system that have a multiplicative inverse, and list their inverses.
    e. Does this system form a group under the operation of multiplication?

NW 11. Evaluate each of the following in the modulo 7 system.
    a. $5 + 6$ **b.** $3 - 6$
    c. $2 \times (5 + 3)$ **d.** $3 \times (1 - 6)$
    e. $2 - (3 - 4)$ **f.** $4 \times (3 - 5)$

12. Evaluate each of the following in the modulo 7 system.
    a. $7 + 5$ **b.** $2 - 4$
    c. 4×8 **d.** $4 - (1 - 3)$
    e. $2 \times (5 + 4)$ **f.** $3 \times (4 - 5)$

13. Determine whether each statement is true or false.
    a. $18 \equiv 3 \pmod 5$ **b.** $22 \equiv 1 \pmod 5$
    c. $144 \equiv 4 \pmod 5$ **d.** $33 \equiv 5 \pmod 7$
    e. $49 \equiv 0 \pmod 7$ **f.** $99 \equiv 2 \pmod 7$

14. Determine whether each statement is true or false.
    a. $44 \equiv 9 \pmod 5$ **b.** $27 \equiv 2 \pmod 5$
    c. $17 \equiv 3 \pmod 5$ **d.** $140 \equiv 88 \pmod 7$
    e. $213 \equiv 12 \pmod 7$ **f.** $1{,}000 \equiv 55 \pmod 5$

15. Determine whether each statement is true or false.
    a. $10 \equiv 3 \pmod 7$ **b.** $122 \equiv 1 \pmod 3$
    c. $1{,}234 \equiv 27 \pmod 5$ **d.** $0 \equiv 171 \pmod 2$
    e. $184 \equiv 32 \pmod 8$ **f.** $121 \equiv 22 \pmod{11}$

16. Using the elements of the indicated modular system, find a replacement for each question mark so that each statement is true.
    a. $4 + ? \equiv 1 \pmod 5$ **b.** $3 + ? \equiv 2 \pmod 5$
    c. $? + 4 \equiv 3 \pmod 5$ **d.** $? + 3 \equiv 2 \pmod 7$
    e. $2 - ? \equiv 4 \pmod 6$ **f.** $3 - ? \equiv 4 \pmod 7$

17. Using the elements of the indicated modular system, find a replacement for each question mark so that each statement is true.

a. $2 \times ? \equiv 3(\text{mod } 7)$ NW **b.** $? \times 4 \equiv 1(\text{mod } 5)$
c. $? - 6 \equiv 2(\text{mod } 7)$ NW **d.** $1 - ? \equiv 4(\text{mod } 5)$
e. $2 \times ? \equiv 3(\text{mod } 9)$ **f.** $2 - ? \equiv 3(\text{mod } 12)$

18. Grocery stock Stan, a stock boy in a local supermarket, had to change the price on a set of soup cans on a particular shelf. In order to make the job easier, Stan decided to arrange the cans in stacks of 10, but when he did this he had 1 left over. Trying again, he arranged them in stacks of 7, but then he had 4 left over. Finally, he arranged them in stacks of 3, and there were none left over. If the shelf can hold only 100 cans, on how many cans did Stan have to change the price?

NW **19. Restaurant checks** Irene is a cashier in a restaurant. After the rush hour, she began to tabulate the customers' checks. First, she arranged the checks in stacks of 5, and there was 1 left over. Next, Irene arranged the checks in stacks of 7, and there were 2 left over. Finally, she arranged them in stacks of 4, and there were 3 left over. If Irene had fewer than 100 checks, how many did she have?

20. Bicycle accident A girl was carrying a basket of tomatoes, and a man on a bicycle hit the basket and smashed all the tomatoes. Wishing to pay for the damages, he asked the girl how many tomatoes she had. The girl replied that she didn't know, but she remembered that when she counted them by 2s, there was 1 tomato left over; when she counted them by 3s, there was also 1 tomato left over; when she counted them by 4s, there was also 1 left over; but when she counted them by 5s, there were no tomatoes left over. How many tomatoes do you think the girl had in her basket? Why?

Writing Mathematics

21. What does a modular system consist of? Describe its characteristics.

22. What does $a \equiv b(\text{mod } m)$ mean? Describe its meaning. Give a specific example.

23. Describe two ways of finding $10 + 10$ in a modulo 12 system.

Challenge Exercises

24. How many elements does a modulo m system have, and what are they?

25. The seasons of a year can be considered a modulo 4 system. Find two other systems that occur in cycles. What modulo system may be used for each of your examples?

26. Consider the days of the week as a modulo 7 system. In a certain year, Valentine's Day, February 14, is the 45th day of the year and falls on a Monday. In that same year, on what day of the week does July 4, the 185th day of the year, fall? On what day does Halloween, the 304th day of the year, fall?

27. In a 24-hour clock system, the first two digits represent hours and the last two digits represent minutes. For example, 0930 means 9:30 A.M. while 2045 means 8:45 P.M. $(20 - 12 = 8)$. For the following, evaluate each problem in the 24-hour clock system:

a. $0700 + 1330$ **b.** $1345 + 0935$
c. $1450 + 1720$ **d.** $1530 + 1845$

Group/Research Activities

28. Fields Consider the *distributive property*, which states that

$$a \times (b + c) = a \times b + a \times c$$

The distributive property involves both addition and multiplication. A mathematical system for which there are two operations (not necessarily addition and multiplication) and whose elements satisfy the five properties for a commutative group for both operations, plus the distributive property (a total of 11 properties), forms a special system known as a *field*.

Prepare a group report on fields. Provide examples and also give uses of a field. References may include the Internet and other mathematics texts.

29. Check digits The concept of modular arithmetic has important applications in many industries in the form of *identification numbers*. For example, transportation companies such as UPS and FedEx use tracking numbers to identify their packages uniquely; in the banking industry, there is a string of numbers at the bottom of all checks that uniquely identifies each check; and in the publishing industry, books are labeled with an International Standard Book Number (ISBN) that is used to identify each book uniquely. Common among all of these identification numbers is something called a *check digit*, which is used to detect errors when these numbers are entered into a computer; check digits also provide a measure of security. A check digit is determined by mathematically manipulating the identification number (less the check digit, of course). The result of this manipulation is then expressed in a particular mod, which becomes the check digit. As a group project, investigate how check digits are determined for (a) FedEx and UPS tracking numbers, (b) bank checks, and (c) ISBN labels. Prepare a report of your findings and present your report to your class. Include in your report a demonstration of how each check digit is determined and the mod system used for each application.

Math Connections

Application of Modular Systems

The concept of modular systems has considerable practical applications in the field of computer science. One such application is the *RSA algorithm*, which is used to encode and decode messages. This algorithm, which is named after its designers—Ronald Rivest, Adi Shamir, and Len Adleman—provides a relatively high degree of security for transmitting messages electronically.

RSA is a widely accepted and implemented method known as *public-key encryption* that requires two "keys" for each user, one public and the other private. Each user's public key is available to anyone, whereas the private key is kept secret and known only to the user. A message coded with the public key is decoded with the private key and vice versa. This system provides three distinct methods of sending coded messages. The first method involves sending a coded message to a person from someone whose identity is not verifiable. To illustrate,

> *If we wish to send you a coded message, we code our message using your public key. This way you can decode it using your private key.*

The problem with this method is that the receiver of the message can never be absolutely certain who the sender is because it is the receiver's public key that is being used to code the message.

The second method involves sending a coded message from someone whose identity is verifiable, but the message can be decoded by anyone. To illustrate,

> *If we wish to send you a coded message and we want you to be absolutely certain that it is from us, then we will code our message using our private key. You in turn will decode our message using our public key. Since it is our public key that actually decodes our message, you are certain that the message came from us.*

Authentication of a sender's identity, as demonstrated by this second method, is very important in certain transactions, such as in electronic funds transfers.

Now, to be absolutely clandestine, if we want to send you a coded message and we want you to know it is from us, a third method is used:

> *We first code our message using your public key. (This makes the message secret.) We then encode this code once more using our private key. (This guarantees that the message is from us.) Upon receiving the message, you must decode it twice, first using our public key and then using your private key. Now, only you can read the message and it could only have been sent by us.*

In each of the preceding examples, note how a person's public and private keys work in tandem. If a message is coded using our public key, then the message can only be decoded using our private key. Similarly, if we code a message using our private key, then this message can only be decoded using our public key. With this information under our belt, let's now see how a message actually gets coded and decoded using the RSA algorithm.

The **encryption stage** of the RSA algorithm involves applying a person's public key to the equation $E(s) = s^e (\text{mod } n)$, where s is a given message and e and n represent a previously derived pair of integers that specifies a person's public key. As an illustration, let's assume that our public-key integer pair is $e = 3$ and $n = 55$. Let's further assume that our "message" consists of the single character "D." We first convert the message to a corresponding digit by replacing D with 4. Thus, our message, s, is equal to 4. (*Note:* In practice, messages are not sent one letter at a time but instead an entire message is converted to decimal form with A = 01, B = 02, and so forth, and a blank space = 00.) Applying these values to the encryption function, we get

$$\begin{aligned} E(s) &= s^e (\text{mod } n) \\ &= 4^3 (\text{mod } 55) \\ &= 64 (\text{mod } 55) \\ &= 9 (\text{mod } 55) \end{aligned}$$

Thus, our encrypted message, $E(s) = 9$. This is what gets transmitted.

The **decryption stage** of the algorithm involves applying the person's private key to the equation $s = [E(s)]^d (\text{mod } n)$, where $E(s)$ is the encrypted message that we received and d and n represent a previously derived pair of integers that specifies a person's private key. In both the encryption and decryption stages, the modular system being used, n, is the same and is the product of two extremely large (e.g., 300 digits each) *prime numbers* (see Chapter 8). To decrypt a message, we apply the private key to the decryption function.

Just for fun

Read the following sentence once, slowly. Now count the number of F's. Please count them only once.

FINISHED FILES ARE THE RESULT OF YEARS OF SCIENTIFIC STUDY COMBINED WITH THE EXPERIENCE OF MANY, MANY YEARS.

6.5 MATHEMATICAL SYSTEMS WITHOUT NUMBERS

TABLE 6.8

*	A	B	C	D
A	A	B	C	D
B	B	C	D	A
C	C	D	A	B
D	D	A	B	C

Thus far in this chapter, we have examined mathematical systems such as clock arithmetic, the system of seasons, the week system, and various modular systems. Recall that a mathematical system consists of a set of elements together with one or more operation (rules) for combining elements of the set. The systems that we have considered so far have been based on the use of numbers, but now we want to consider systems that are more abstract.

Consider the set of elements $\{A, B, C, D\}$, together with the operation $*$. We define this operation by means of Table 6.8. The operation $*$ is a binary operation because we combine two elements of the given set to obtain each result. For example, to find the answer to $B*C$, we find the first element, B in the vertical column under $*$ and the second element, C, in the top horizontal row following $*$. The answer is found where the row containing B and the column containing C intersect, at D. Hence, $B*C = D$. Similarly, $C*D = B$ and $D*C = B$.

Let's examine the mathematical properties of this particular system, the set of elements $\{A, B, C, D\}$ and the operation $*$:

1. **Closure property:** All of the entries in the table are elements in the given set; no new elements appear in the table. Therefore, the system satisfies the closure property.
2. **Identity property:** There is an element, A, in the set that does not change any element when it is operated on together with that element. Note that $A*A = A, B*A = B, C*A = C, D*A = D$.
3. **Inverse property:** Each element in this system has an inverse, an element, which when operated on with the given element results in the identity element A. Note that $A*A = A, B*D = A, C*C = A, D*B = A$. From these equations, we can see that the inverse of A is A, the inverse of B is D, the inverse of C is C, and the inverse of D is B.
4. **Associative property:** An operation is associative if the location of parentheses does not affect the answer. Trying an example, we have $(B*C)*D = D*D = C$, and $B*(C*D) = B*B = C$; hence, $(B*C)*D = B*(C*D)$. Any other example for this system also works, and therefore the system satisfies the associative property.
5. **Commutative property:** An operation is commutative if it does not matter in which order you perform the operation. For an example in this system, we find that $B*C = D$ and $C*B = D$; therefore, $B*C = C*B$. Any other example for this system also works, so the system satisfies the commutative property.

Note: A quick check for the commutative property may be performed by examining the table itself. Look at the "diagonal" from the upper left-hand corner to the lower right-hand corner. If there is "symmetry" along the diagonal, then the system is commutative. In Table 6.8 we examine the diagonal consisting of A–C–A–C. We see that there exists a B on each side of the diagonal; similarly, we have a C on each side, D–D on each side, A on each side, and another B on each side. This is known as symmetry or a mirror image. Hence, the system satisfies the commutative property.

The abstract mathematical system consisting of the set of elements $\{A, B, C, D\}$ and the binary operation $*$ satisfies all of the properties listed. Therefore, this system forms a group, and because it satisfies the commutative property, it forms a commutative group.

Historical Note

A group is Abelian or commutative if, in addition to satisfying the four basic properties of a group, it also satisfies the commutative property. That is, $a*b = b*a$, where a and b are any two members of the group.

The Abelian group is named after a famous Norwegian mathematician, Niels Henrik Abel (1802–1829). As a child he was considered a mathematical genius. He thoroughly studied the works of many great mathematicians and began producing original work of his own while still in his teens.

He received little recognition during his lifetime, lived in poverty, and died at the age of 27 from a variety of ills.

EXAMPLE 1 ***The Odd/Even Math System*** The following table defines the operation $\times$ for the set of elements $\{odd, even\}$:

$\times$	***odd***	***even***
odd	*odd*	*even*
even	*even*	*even*

Find each product using this table.

a. $odd \times even$

b. $odd \times odd$

c. $even \times even$

Solution

a. $odd \times even = even$

b. $odd \times odd = odd$

c. $even \times even = even$

NW ***Now Work Problems 1, 3, and 5.***

EXAMPLE 2 ***Examining Properties of the Odd/Even Math System*** Using the table in Example 1, answer the following:

a. Is the set closed with respect to the operation $\times$? Why or why not?

b. What is the identity element (if any)?

c. Which elements of the set have an inverse? Name the inverse of each of these elements.

d. Does the commutative property hold for this system? Verify your answer.

e. Does this set form a group under the operation $\times$?

Solution

a. Yes, the set is closed with respect to the operation $\times$. There are no new elements appearing in the table.

b. The identity element is *odd*. It does not change any of the elements in the given set when it is operated on with them: $odd \times odd = odd$, $even \times odd = even$.

c. The element *even* does not have an inverse. However, $odd \times odd = odd$, so *odd* has an inverse, itself.

d. Yes, the commutative property does hold for this system. Checking the elements, we have $odd \times even = even$ and $even \times odd = even$. Therefore,

$$odd \times even = even \times odd$$

Note that $odd \times odd = odd \times odd$ and $even \times even = even \times even$.

e. No, the set of elements $\{odd, even\}$ with the operation $\times$ does not form a group. Not every element has an inverse, because *even* does not have an inverse.

NW ***Now Work Problem 11.***

EXAMPLE 3 ***Military Drill Commands As a Mathematical System***
Sergeant Gig, a drill instructor, drills his drill team daily. During the drills, he issues four commands: *right face, left face, about face,* and *as you were*. Let

$$r = \textit{right face} \quad a = \textit{about face}$$
$$l = \textit{left face} \quad y = \textit{as you were}$$

Sometimes Sergeant Gig gives two commands in succession, such as *right face* followed by *left face*. If a person followed these commands he would wind up in the original position, the position of *as you were*. If we let $\otimes$ stand for the operation *followed by*, then $r \otimes l = y$. From this information, evaluate the following:

a. $r \otimes r$
b. $a \otimes r$
c. $l \otimes a$

Solution

a. $r \otimes r$ means *right face* followed by *right face*, which would be the same as *about face*. Therefore, $r \otimes r = a$.

b. $a \otimes r$ means *about face* followed by *right face*, which would be the same as *left face*. Therefore, $a \otimes r = l$.

c. $l \otimes a$ means *left face* followed by *about face*, which would be the same as *right face*. Therefore, $l \otimes a = r$.

TABLE 6.9

$\otimes$	r	l	a	y
r	a	y	l	r
l	y	a	r	l
a	l	r	y	a
y	r	l	a	y

Using the idea of drill commands in Example 3, we can create a table to illustrate the results when any two commands are given. Using the notation $r = \textit{right face}$, $l = \textit{left face}$, $a = \textit{about face}$, and $y = \textit{as you were}$, and letting $\otimes$ stand for the operation *followed by*, we have a system consisting of a set of elements $\{r, l, a, y\}$ and an operation $\otimes$. Table 6.9 illustrates the operation $\otimes$ in this system.

From Table 6.9, we can verify the results in Example 3: $r \otimes r = a$, $a \otimes r = l$, and $l \otimes a = r$. Examining the entries in the table, we see that the set is closed with respect to the operation $\otimes$. The set also contains an identity element for the operation $\otimes$, the element y. Since y is the identity element, we can find the inverse of each element. The inverse of r is l, the inverse of l is r, a is its own inverse, and y is its own inverse. It can also be shown that the set satisfies the associative property for the operation $\otimes$. Hence, the set of elements $\{r, l, a, y\}$ with the operation $\otimes$ satisfies the closure property, identity property, inverse property, and associative property, and therefore forms a group.

It is interesting to note that many of the examples discussed in this chapter are different, yet similar. We have discussed examples that contain different sets of elements that are combined using different operations, but they still satisfy the same properties of closure, identity, and so on. This is one of the unique characteristics of mathematical systems.

Exercises for Section 6.5

For Exercises 1–14, use Table 6.10, which defines an operation for the set of elements $\{P, Q, R, S\}$.

TABLE 6.10

:	P	Q	R	S
P	P	Q	R	S
Q	Q	R	S	P
R	R	S	P	Q
S	S	P	Q	R

Evaluate the following:

NW **1.** $P : Q$
2. $S : R$
NW **3.** $S : S$
4. $P : P$
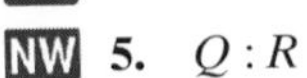 **5.** $Q : R$
6. $Q : Q$
7. $R : (Q : S)$
8. $(Q : Q) : P$
9. $(P : Q) : R$
10. $P : (P : S)$

NW **11.** Is the set closed with respect to the operation :? Why or why not?

12. What is the identity element (if any)?

13. Which elements of the set have an inverse? Name the inverse of each of these elements.

14. Does this set form a group under the operation :?

For Exercises 15–29, use Table 6.11, which defines an operation $\odot$ *for the set of elements* {$, ?, ¢}.

TABLE 6.11

$\odot$	$	?	¢
$	$	!	?
?	?	¢	!
¢	¢	?	$

Evaluate the following:

15. $ ⊙ ?
16. ? ⊙ $
17. ? ⊙ ¢
18. ¢ ⊙ ¢
19. ¢ ⊙ ?
20. $ ⊙ $
21. $ ⊙ (¢ ⊙ ?)
22. ($ ⊙ ¢) ⊙ ?
23. ¢ ⊙ (? ⊙ $)
24. (¢ ⊙ ?) ⊙ $

25. Is the set closed with respect to the operation ⊙? Why or why not?

26. What is the identity element (if any)?

27. Which elements of the set have an inverse? Name the inverse of each of these elements.

28. Is the commutative property satisfied for this system? Why or why not?

29. Does the set form a group under the operation ⊙?

For Exercises 30–46, use Table 6.12, which defines an operation $*$ *for the set of elements* $\{a, b, c, d, e\}$.

TABLE 6.12

*	a	b	c	d	e
a	a	b	c	d	e
b	b	c	d	e	a
c	c	d	e	a	b
d	d	e	a	b	c
e	e	a	b	c	d

Evaluate the following:

30. $c * e$
31. $c * d$
32. $b * c$
33. $d * a$
34. $c * b$
35. $d * d$
36. $b * b$
37. $c * c$
38. $(b * b) * c$
39. $(c * d) * e$
40. $d * (c * e)$
41. $b * (a * d)$

42. Is the set closed with respect to the operation *? Why or why not?

43. What is the identity element (if any)?

44. Which elements of the set have an inverse? Name the inverse of each of these elements.

45. Is the commutative property satisfied for this system? Why or why not?

46. Does this set form a group under the operation *?

For Exercises 47–64, use Table 6.13, which defines an operation # for the set of elements $\{w, x, y, z\}$.

TABLE 6.13

#	w	x	y	z
w	z	y	x	w
x	y	w	z	x
y	x	z	w	y
z	w	x	y	z

Evaluate the following:

47. $z \# x$ **48.** $y \# w$ **49.** $y \# z$

50. $x \# w$ **51.** $z \# z$ **52.** $y \# x$

53. $x \# (y \# z)$ **54.** $(z \# y) \# y$ **55.** $(w \# w) \# y$

56. $(w \# x) \# x$ **57.** $((x \# y) \# z) \# w$ **58.** $((w \# y) \# z) \# x$

59. Is the set closed with respect to the operation #? Why or why not?

60. What is the identity element (if any)?

61. Which elements of the set have an inverse? Name the inverse of each of these elements.

62. Is the commutative property satisfied for this system? Why or why not?

63. Does this set form a group under the operation #?

64. Give one example where the associative property is true for this system.

65. Given the set of counting numbers $\{1, 2, 3, 4, \dots\}$ and the operation of addition, answer the following:

- **a.** Is this set closed with respect to the operation of addition?
- **b.** What is the identity element (if any)?
- **c.** Which elements of the set have an inverse with respect to addition?
- **d.** Does this set form a group under the operation of addition?

66. Given the set of counting numbers $\{1, 2, 3, 4, \dots\}$ and the operation of multiplication, answer the following:

- **a.** Is the set closed with respect to the operation of multiplication?
- **b.** What is the identity element (if any)?
- **c.** Which elements of the set have an inverse with respect to the operation of multiplication?
- **d.** Does this set form a group under the operation of multiplication?

67. Geometric rotations Given the equilateral triangle ABC (all sides equal), with a point in the center of the triangle so that the triangle may be rotated, as in Figure 6.5a. If we rotate the triangle 120° clockwise (call it rotation a), the triangle changes position: A goes to B, B goes to C, and C goes to A (see Figure 6.5b). If we rotate the triangle 240° clockwise (call it rotation b), the triangle again changes position: A goes to C, B goes to A, and C goes to B (see Figure 6.5c). If we rotate the triangle 360° clockwise (call it rotation c), the triangle returns to its original position: A goes to A, B goes to B, C goes to C (see Figure 6.5a).

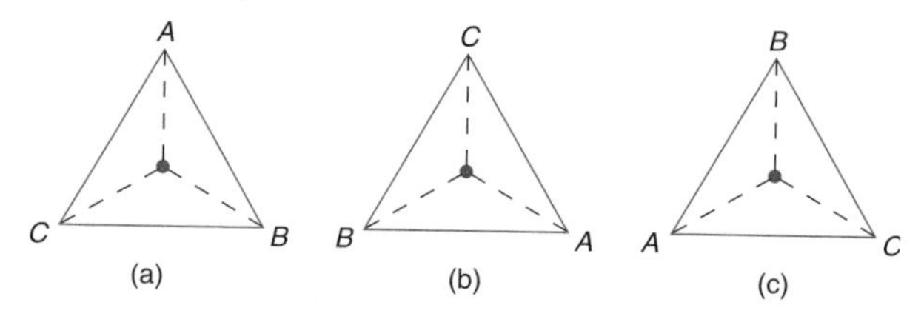

Figure 6.5

Let's define an operation $*$ to mean "followed by." Therefore, $a * b$ would mean rotation a followed by rotation b, which is the same as a rotation of 360°, or rotation c. Hence, for this system, $a * b = c$.

- **a.** Complete the following table for this system:

*	a	b	c
a		c	
b			b
c	a		

- **b.** Is this set closed with respect to the operation $*$?
- **c.** What is the identity element (if any)?
- **d.** Which elements of the set have an inverse? Name the inverse of each of these elements.
- **e.** Does this set form a group under the operation $*$?

68. Geometric rotations Given the square $ABCD$ (all sides equal), with a point in the center of the square so that the square can be rotated (see Figure 6.6). Define the following rotations:

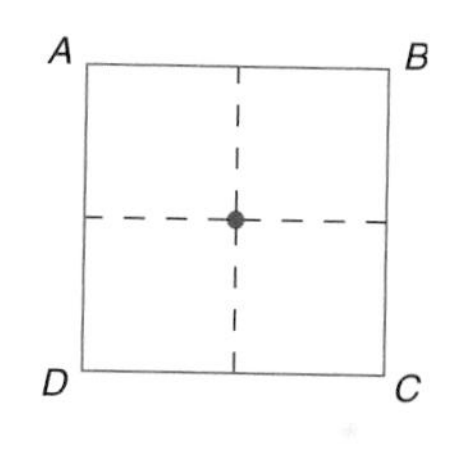

Figure 6.6

a = a clockwise rotation of 90°

b = a clockwise rotation of 180°

c = a clockwise rotation of 270°

d = a clockwise rotation of 360°

Let the operation $*$ mean "followed by."

- **a.** Complete the following table for this system:

*	a	b	c	d
a		c		
b		d		
c			b	
d	a			

- **b.** Is this set closed with respect to the operation $*$?
- **c.** What is the identity element (if any)?
- **d.** Which elements of the set have an inverse? Name the inverse of each of these elements.
- **e.** Does the set form a group under the operation $*$?

Writing Mathematics

69. Explain one method of determining whether a system defined by a table is commutative or not under the given operation.

70. Given the set of counting numbers $\{1, 2, 3, 4, \dots\}$ and the operation of addition, give two different examples to illustrate the associative property of addition.

Challenge Exercises

For each of the following exercises, let ♣ mean "select the second of the two numbers" and let ♦ mean "select the larger of the two numbers." Evaluate each of the following:

71. 5 ♣ 4 **72.** 8 ♣ 13 **73.** 5♦4

74. 8♦13 **75.** 7♣(11♦13) **76.** 9♦(17♣19)

77. Verify that 7♣(11♦13) = (7♣11)♦(7♣13).

78. Verify that 9♦(12♣17) = (9♦12)♣(9♦17).

79. Exercises 77 and 78 illustrate what property?

Group/Research Activities

80. Karl Friedrich Gauss (1777–1855) was one of the eminent mathematicians of the nineteenth century. He made many contributions to the development of mathematics. Among these contributions was the development of modular systems. Do research and write a report on Gauss. Be sure to include his contributions to mathematics and also his contributions to other areas, such as astronomy and physics.

81. **Klein-four group** A special group known as the *Klein-four group* is characterized by the fact that each element is its own inverse, and when the identity element is excluded, the product of any two of the remaining three elements is the third element. Research the concept of Klein-four groups and provide three examples of such groups.

Just for fun

If a certain kind of egg sells for \$1 per dozen, which costs more, $\frac{1}{2}$ *dozen dozen* eggs, or 6 *dozen dozen* eggs?

6.6 AXIOMATIC SYSTEMS

It is not everything that can be proved, otherwise the chain of proof would be endless. You must begin somewhere, and you start with things admitted but undemonstrable. There are first principles common to all sciences which are called axioms or common opinions.

Aristotle

Throughout this chapter we have examined the basic structure of mathematical systems. We have considered a mathematical system to be a set of elements together with one or more operations (rules) for combining any two elements of the set. In this section we will examine more closely the structure of mathematics itself.

Regardless of the branch of mathematics we choose to examine, the various branches are similar in the way that they are constructed. We can compare the basic characteristics of mathematics to the basic characteristics of a game.

Most games have a vocabulary of special terms, some defined and some undefined. After a player acquires this vocabulary, he learns the rules of the games—that is, what moves he can make, and what moves he cannot make. Normally, he accepts these rules without question. For instance, in the game of baseball, one rule says that a runner must run the bases in a counterclockwise direction. When a youngster is first learning how to play baseball, she sometimes wants to run bases in a clockwise direction. But when she is told that the rules state that a runner must run the bases the other way (counterclockwise), she accepts this. Similarly,

Historical Note

Two of the most important ideas left behind by the ancient Greeks were the notion of axiom and proof. Around the nineteenth century mathematicians were looking for a list of axioms from which all of mathematics could be deduced. That is, given a finite list of axioms, build up the whole of mathematics known to date. In the early part of this century two mathematicians, Bertrand Russell and Alfred North Whitehead, came very close to achieving this in their work titled *Principia Mathematica* (after some 362 pages they were finally able to prove that $1 + 1 = 2$). Other mathematicians were also trying to find "the list of axioms" to build all of mathematics. But in 1930–1931 Kurt Gödel, a logician, proved that any list of axioms rich enough (that is, with enough axioms to support the axioms of arithmetic) would never be able to prove all the theorems of the system. That is, there would be statements in the system of axioms that could not be proven true or shown false using only the axioms of the system. This was a severe blow to those mathematicians trying to find "the list of axioms" to build all of mathematics. This result is a hallmark of twentieth century mathematics and is known as Gödel's *First Incompleteness Theorem*.

Gödel (b. April 28, 1906, Brünn, Austria-Hungary—d. Jan. 14, 1978, Princeton, N.J., U.S.).

we all accept the fact that a queen ranks higher than a jack in card games. Why? Because the rules say so!

In mathematics, the rules are called *axioms*. An **axiom** is a statement that is accepted as true without proof. Each mathematical system must have axioms that are consistent. They must not contradict one another, just as the rules of a game should not contradict one another.

After learning the undefined terms, defined terms, and rules of a game, we are ready to play. Once we have mastered the elementary moves of the game, we usually try more complicated moves using the rules (as in the game of chess). In mathematics, these new results that have evolved from the undefined terms, defined terms, and axioms (rules) are called *theorems*. **Theorems** are logical deductions that are made from undefined terms, defined terms, and axioms. Some theorems are even logical deductions from other theorems.

In essence, an *axiomatic system* consists of four main parts:

1. Undefined terms
2. Defined terms (definitions)
3. Axioms (also called postulates)
4. Theorems

Undefined terms are necessary in an axiomatic system, as they are used to form a fundamental vocabulary with which other terms can be defined. Even though a term may be undefined, that does not mean that we do not know what it is. In a high school geometry course, the terms *point* and *line* are not defined terms, yet we know what they are. In Chapter 2 of this text, *set* is an undefined term, but that does not prevent us from having an intuitive idea of what *set* means.

Definitions of defined terms may use undefined terms or terms that have been previously defined. Definitions should be concise, consistent, and not circular.

As we noted earlier, axioms are statements that are accepted as true without proof. In a given system, no axiom should contradict another; that is, the axioms must be consistent. Axioms are necessary in a system because not everything can be proved. Axioms are needed to derive other statements. The derived statements are the theorems.

Let us now examine a proved theorem for a system that uses undefined terms, defined terms, and axioms. The following proof is found in a high school geometry course. The statement (theorem) to be proved is "If two straight lines intersect, then the vertical angles thus formed are equal."

We first construct a diagram representing this situation (see Figure 6.7).

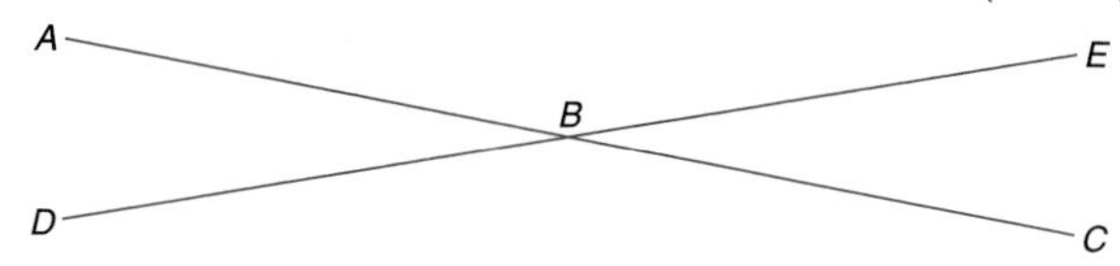

Figure 6.7

This diagram shows straight lines AC and DE intersecting at B and forming vertical angles ABD and EBC. We want to prove that the measure of angle ABD equals the measure of angle EBC—that is, $m\angle ABD = m\angle EBC$. The following is a formal proof.

Statements	Reasons
1. AC and DE are straight lines	1. Given
2. Angles ABC and DBE are straight angles	2. Definition of a straight angle
3. $m\angle ABC = m\angle DBE$	3. All straight angles are equal.
4. $m\angle ABE = m\angle ABE$	4. Identity
5. $m\angle ABD = m\angle EBC$	5. If the same quantity is subtracted from two equal quantities, the remainders are equal.

This proof uses undefined terms (straight lines), definitions (the definition of a straight angle), and axioms. The axioms appear in reason 3 (All straight angles are equal) and reason 5 (If the same quantity is subtracted from two equal quantities, the remainders are equal). The axioms used are consistent; that is, they do not contradict each other. The diagram in Figure 6.7 could be considered a *model*—that is, a physical interpretation of the undefined terms which satisfies the axioms.

Let's examine another example, but one that is less familiar. We start with the following axioms, and the undefined terms *road*, *town*, and *stop sign*.

1. There is at least one road in the town.
2. Every stop sign is on exactly two roads.
3. Every road has exactly two stop signs on it.

We wish to prove that there is at least one stop sign in the town. It is helpful to construct a diagram (model) that satisfies all of the axioms (see Figure 6.8).

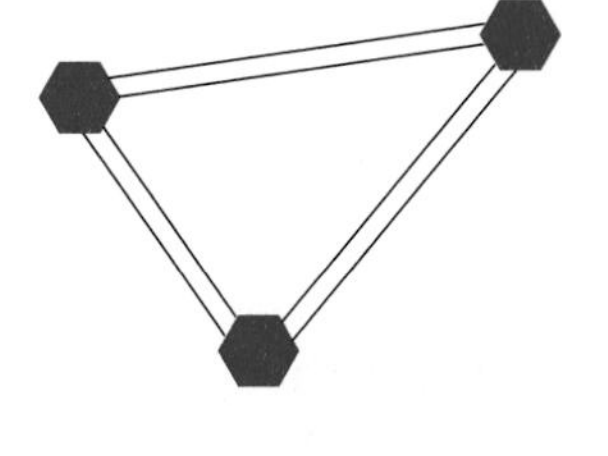

Figure 6.8

Axiom 1 states that there must be at least one road in the town, and axiom 3 says that every road has exactly two stop signs on it. Hence, there must be at least one stop sign in the town.

Note that from this set of axioms we could have derived other conclusions, but we derived only the desired conclusion.

EXAMPLE 1 ***Working with an Axiomatic System*** Given

Axiom 1: There is exactly one road between any two traffic lights.
Axiom 2: For every road there exists a traffic light not on that road.
Axiom 3: There exist at least two traffic lights.
Prove: There exist at least three roads.

Math Connections

Basis of Axiomatic Systems

As indicated in the text, the basis of a mathematical system is to begin with undefined terms. Many people often criticize this beginning as being rather dubious: How can the basic, underlying foundation of a system be a set of undefined terms? The need for undefined terms is necessary, though, because we cannot always define *all* elements of a system without eventually encountering *circular definitions*. That is, at some point we will begin to define an element in terms of itself. As an example, consider the definition of the word *disease*, which is extracted from the tenth edition of *Merriam Webster's Collegiate Dictionary* (1996).

- *Disease*: A condition of the living animal or plant body or one of its parts that impairs normal functioning; *sickness; malady*.
- *Sickness:* ill health; illness; a disordered, weakened, or unsound condition; a specific **disease**; nausea, queasiness.
- *Malady*: A **disease** or disorder of the animal body.

It doesn't take too long before *disease* is being defined in terms of itself.

Interestingly, Euclid, who developed plane geometry (which is one of the earliest mathematical systems), did not see the need for undefined terms. He believed that it was possible to define all terms. For example, Euclid defined his basic elements, *point, line,* and *plane*, as follows:

- A **point** is that which has no part.
- A **line** is a breadthless length.
- A **plane** surface is a surface that lies evenly with the straight line on itself.

Unfortunately, these definitions are deficient because they include terms that have not been previously defined. For example, to understand what a *point* is, we must first know what the word *part* means. Similarly, what does a "breadthless length" mean in the definition of *line?* To answer this, we must first define *breadthless* and *length*. Euclid's *Elements*, which is one of the most famous geometry books of the ancient world, begins with 23 definitions, many of which involve terms that have not been previously defined. As a result, these definitions really do not define anything. Modern mathematicians have realized and accepted the need for undefined terms, and hence today we consider *point* and *line* to be undefined terms. Furthermore, to help understand the meaning of undefined terms, we rely on synonyms for our terms, and we usually provide an intuitive picture or model of these undefined terms.

(continued)

(continued)

Once we have stated our undefined terms, we then expand our system by defining terms as needed. A definition must be meaningful and consistent. Once a definition is accepted, we do not challenge it or try to prove it as a theorem. A definition needs no proof; it is accepted as it stands.

An axiomatic system is further developed by incorporating axioms or postulates. Again, as indicated in the text, postulates are basic facts that we accept as true without proof. Postulates also possess three properties. First, the list of postulates of an axiomatic system should be enough to develop the system. This is known as the property of *completeness*. A second property of postulates is that of *independence*. That is, a postulate cannot be a logical consequence of other postulates. The third property of postulates is that of *consistency*. A group of postulates is said to be consistent if the group does not develop contradictions within itself. Testing the postulates of an axiomatic system for these three properties can be a long and exhaustive process. For example, testing Euclid's postulates of plane geometry was an obsession for many mathematicians for centuries.

Collectively, the set of undefined elements, defined terms, and postulates represents the basis of an axiomatic system. Once determined and accepted, the system can then be organized and developed by formulating theorems and by utilizing the basis of the system logically to construct proofs of the theorems. Moreover, the system can grow like a plant by introducing new definitions, which expands the system's basis, as well as by following up with additional theorems and their proofs. This is exactly how the Euclidean system of geometry has evolved.

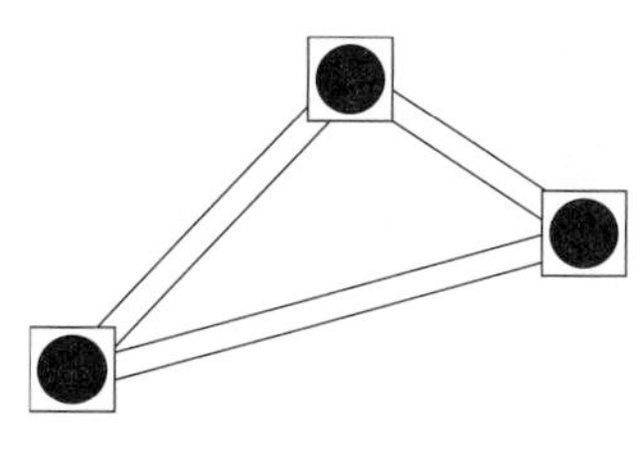

Figure 6.9

Solution First we construct a model that satisfies all of the given axioms. Let ● represent a traffic light and ═══ represent a road. Our model appears in Figure 6.9; this helps us understand the axioms.

Axiom 3 tells us that there are at least two traffic lights, and axiom 1 states that there is exactly one road between any two traffic lights. So far we have one road. Now axiom 2 tells us that there has to be another traffic light not on the given road. But if there is another traffic light, then there must be two more roads: Axiom 1 tells us that there is exactly one road between any two traffic lights, and since there is a third traffic light, it must be connected to the other two lights by means of two roads. Therefore, we have at least three roads.

Exercises for Section 6.6

NW **1.** Consider the following set of axioms:

I: There are at least two buildings on campus.

II: There is exactly one sidewalk between any two buildings.

III: Not all of the buildings are on the same sidewalk.

If the capital letters *A*, *B*, *C*, and so on represent buildings and the lines represent sidewalks, which of the following models represent the given axiomatic system?

a.

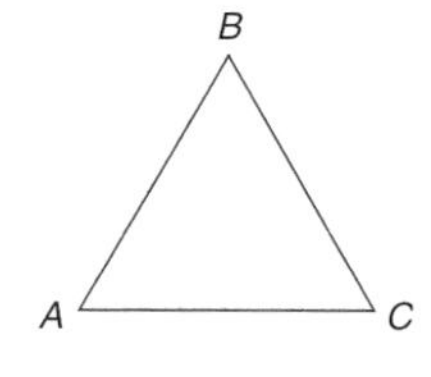

b.

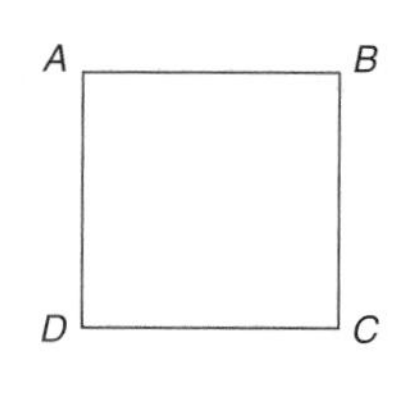

c.

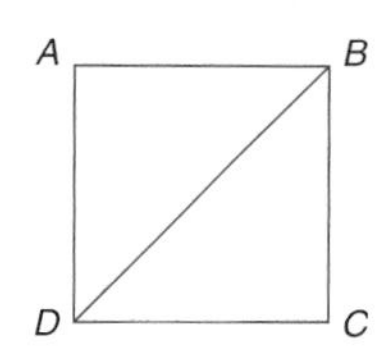

d.

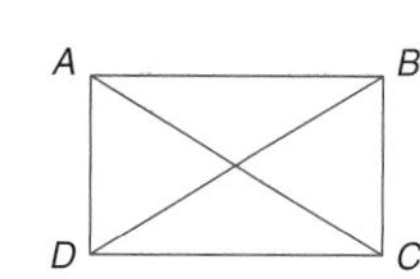

e.

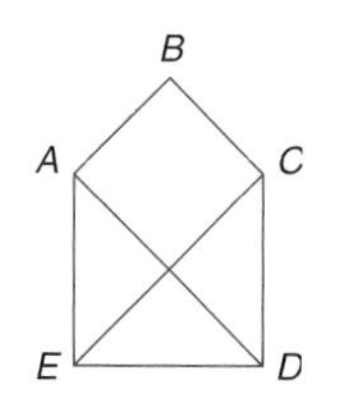

f.

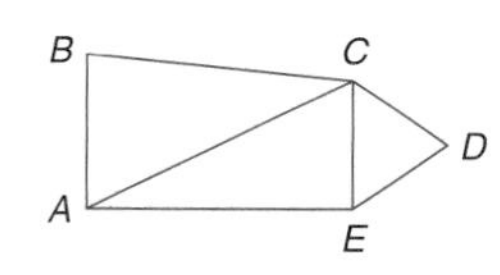

2. Consider the following set of axioms:

I: There are exactly four bleeps.
II: Each bleep is on a cleep.
III: No bleep is on a cleep by itself.

If the capital letters, *A*, *B*, *C*, and so on, represent bleeps and the lines represent cleeps, which of the following models represent the given axiomatic system?

a.

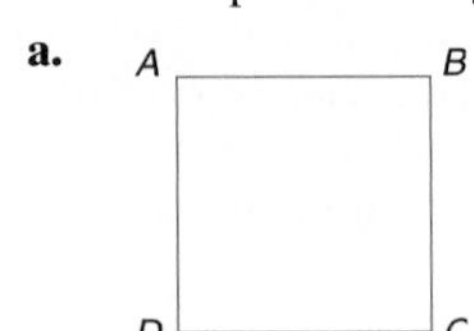

b.

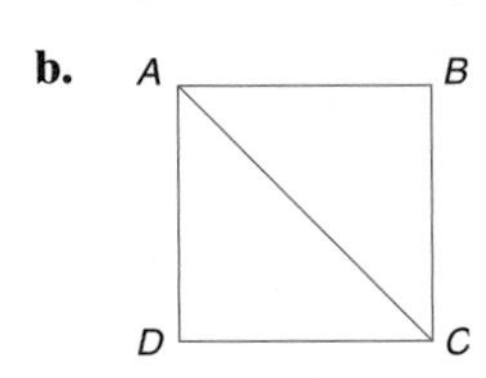

c.

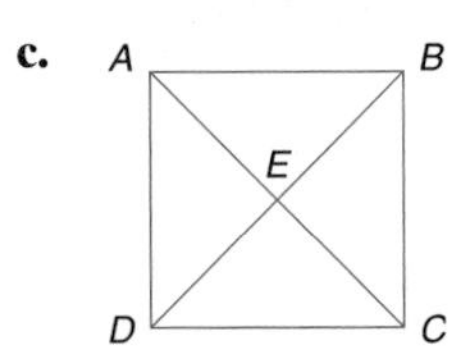

d.

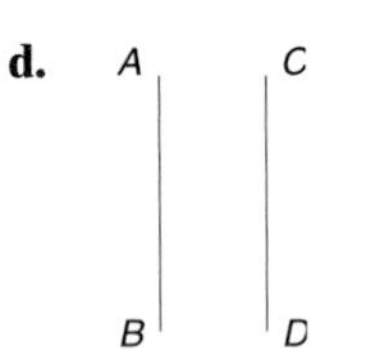

e.

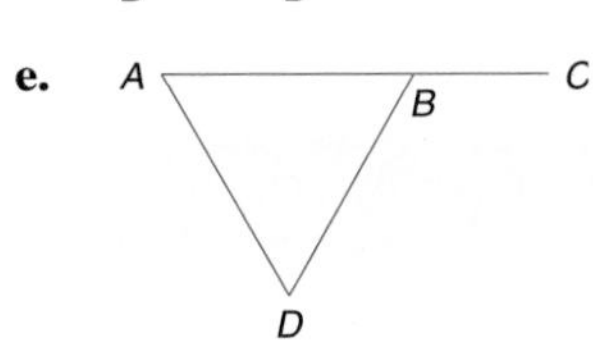

f.

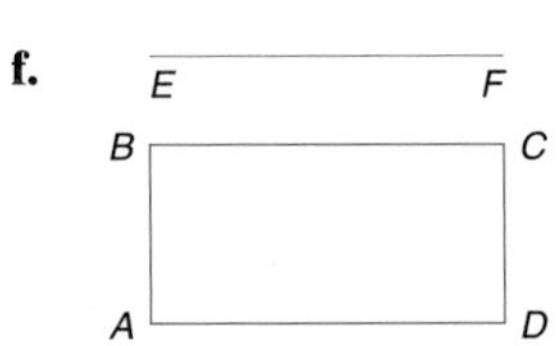

3. What is the basic structure of an axiomatic system?

4. Given the following axioms,

I: There are at least two buildings on campus.
II: There is exactly one sidewalk between any two buildings.
III: Not all of the buildings are on the same sidewalk.

Prove: There exist at least three buildings on campus.

5. Given the following axioms,

I: There are exactly two points on each line.
II: There is at least one line.
III: For each pair of points, there is one and only one line containing them.
IV: Corresponding to each line there is exactly one other line which has no point in common with it.

Prove: There are at least four points.

6. Given the following axioms,

I: Not all cars are in the same garage.
II: There exist at least two cars.
III: There is exactly one garage for any two cars.

Prove: There exist at least two garages.

7. Given the following axioms,

I: If equal quantities are added to equal quantities, then the sums are equal.
II: If equal quantities are subtracted from equal quantities, then the differences are equal.
III: If equal quantities are multiplied by equal quantities, then the products are equal.
IV: If equal quantities are divided by equal quantities (except zero), then the quotients are equal.
V: A quantity may be substituted for its equal in any process.

State which axiom is used in each step in solving the following equations:

a. Given $6x + 4 = 40$

1. $6x = 36$
2. $x = 6$

b. Given $3x - 2 = 25$

1. $3x = 27$
2. $x = 9$

c. Given $\frac{x}{3} - 6 = 3$

1. $\frac{x}{3} = 9$
2. $x = 27$

d. Given $\begin{cases} x + 5y = 9 \\ x + y = 5 \end{cases}$

1. $4y = 4$
2. $y = 1$
3. $x + 1 = 5$
4. $x = 4$

e. Given $\begin{cases} 4x + y = 10 \\ x - y = 5 \end{cases}$

1. $5x = 15$
2. $x = 3$
3. $3 - y = 5$
4. $3 = 5 + y$
5. $-2 = y$

8. Using the axioms given in Exercise 7, prove the desired conclusion for each of the following:

a. Given: $\overline{AC} = \overline{BC}$
$\overline{DC} = \overline{EC}$
Prove: $\overline{AE} = \overline{BD}$

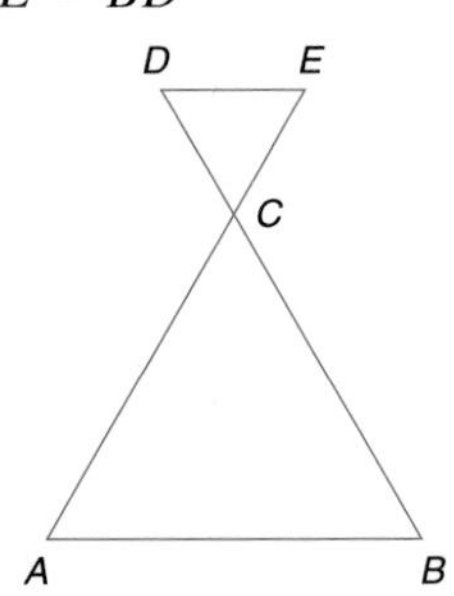

b. Given: $m\angle EBC = m\angle DBA$
Prove: $m\angle 1 = m\angle 2$

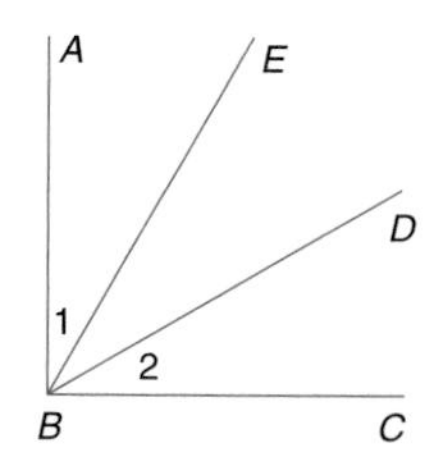

c. Given: $m\angle ABC = m\angle EFG$
$m\angle 1 = m\angle 3$
Prove: $m\angle 2 = m\angle 4$

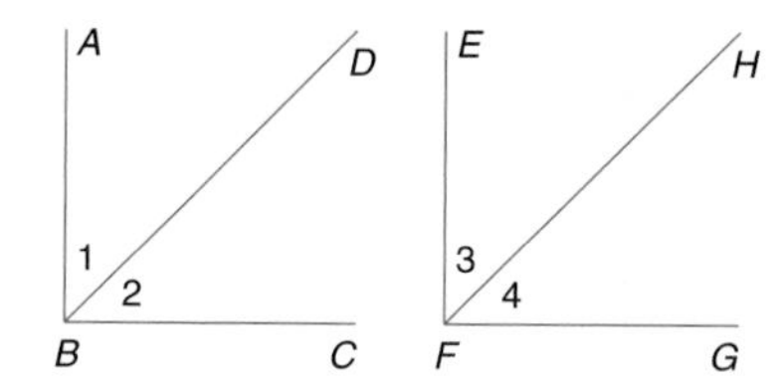

d. Given: $m\angle 1 = m\angle 2$
$m\angle 1 = m\angle 3$
$m\angle 3 = m\angle 4$
Prove: $m\angle 4 = m\angle 2$

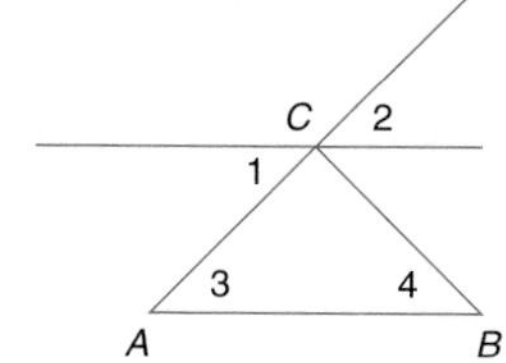

e. Given: $\overline{AB}$, $\overline{CD}$, and $\overline{EF}$ are straight lines.
$m\angle b = m\angle c$
$m\angle a = m\angle b$
Prove: $m\angle a = m\angle c$

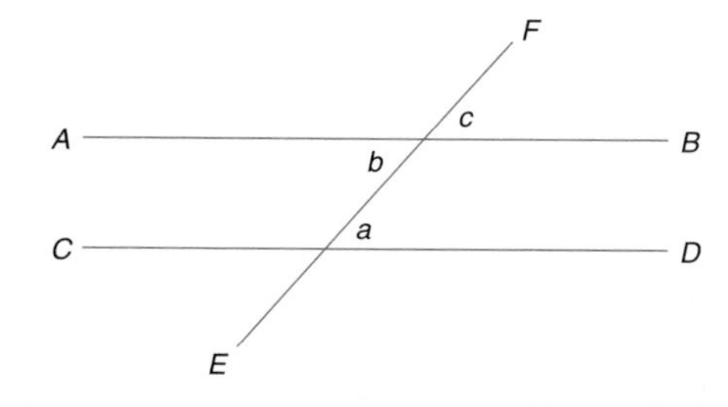

Writing Mathematics

9. What is an axiom?

10. What is a theorem?

11. What is the difference between a theorem and an axiom?

12. Explain what is meant by an *undefined term.*

Challenge Exercises

13. Consider the following set of axioms:

I. There are exactly three sets.
II. Every two sets have exactly one element in common.
III. Every element belongs to exactly two sets.

a. What is the total number of elements?
b. Prove your answer to part **a**.

Group/Research Activities

14. **Peano's postulates** Giuseppe Peano (1858–1932) developed a set of axioms known as *Peano's postulates* (the word *postulate* is often used in place of *axiom*). *Peano's postulates* are considered to be an important set of axioms because an extremely large number of theorems can be proved from them. Do research and prepare a report on Peano and his postulates.

15. In addition to undefined terms, defined terms, axioms, and theorems, an axiomatic system can also include *hypotheses, lemmas,* and *corollaries*. Research each of these terms and report on how they fit into an axiomatic system.

16. Review Euclid's *Elements* and prepare a report that summarizes the salient aspects of this publication relative to axiomatic systems.

17. **Euclid's parallel postulate** One of Euclid's postulates, called *Euclid's parallel postulate*, was the subject of much attention by many mathematicians throughout history. Research this postulate and report on what was special about it that garnered so much attention. Include in your report what the postulate stated, the issues surrounding it, and the various developments that came from it.

Just for fun

Betsy wanted to plant some tomato plants in her garden. Altogether she had 12 tomato plants. She wanted to plant these in six rows, but with four tomato plants in each row. How did Betsy do it?

CHAPTER REVIEW MATERIALS

Summary/Chapter 6

6.2 Clock Arithmetic

- The *12-hour clock system* is an example of a finite mathematical system that consists of the set of elements {1, 2, 3, 4, 5, 6, 7, 8, 9, 10, 11, 12}.
- When performing *clock arithmetic*, the 12-hour clock system is *closed* with respect to addition and multiplication because when we add or multiply two elements of the given set the result is also a member of the set.
- The number 12 in the 12-hour clock system is the *identity element* of the system because when 12 is added to or multiplied by any element in the set, the answer is the element itself. In general, if an element e in a given set does not change any element when it is operated on together with the element, then e is the identity element.
- If two elements of a given set are operated on and the result of this operation is the identity element, then the two elements are *inverses* of each other. Thus, 8 and 4 are *additive inverses* in the 12-hour clock system because $8 + 4 = 12$. Similarly, 3 and 4 are *multiplicative inverses* in the 12-hour clock system because $3 \times 4 = 12$.
- An operation is *associative* if the order in which the elements are operated on does not affect the answer. Thus, $7 + 8 + 10$ is an example of the *associative property for addition* in the 12-hour clock system because $(7 + 8) + 10 = 7 + (8 + 10)$. In general, if the associative property under a specific operation holds for all the elements of a given set, then the mathematical system is associative.
- When a set of elements and a corresponding operation satisfy the closure, identity, inverse, and associative properties, then the set of elements forms a *group* under the given operation.
- When performing *binary operations* (i.e., two elements are operated on to produce a third element), if the order in which the elements are operated on is switched but the results are the same, then the operation is *commutative*. If the set of elements of a mathematical system forms a group and all the elements are commutative under the corresponding operation, then the group is called a *commutative group*.

6.3 More New Systems

- Similar to the 12-hour clock system, we can construct a *four season system* and a *days of the week system*.
- In the season system, the system is closed under addition, and winter is the identity element.
- In the week system, the system is closed under addition and Saturday is the identity element.

6.4 Modular Systems

- A mathematical system that repeats itself or is cyclical is called a *modular system*.
- The notation $a \equiv b \pmod{m}$ is read *a is equivalent to b mod m* and indicates that a and b both have the same remainder when they are divided by m.
- The 12-hour clock system is an example of a modular system. More specifically, it is a *mod 12 system*.

6.5 Mathematical Systems without Numbers

- By extending our basic notion of a mathematical system, we can design *nonnumeric systems*, which are abstract mathematical systems without numbers.
- The elements in the set of a nonnumeric system can be anything (although they are usually letters), and the operation can be defined in any way.
- An example of a nonnumeric mathematical system is the *odd-even* system that contains the set of elements {odd, even} and the operation of multiplication. A second example of such a system is the *military drill system* that contains the set of elements {left face, right face, about face, as you were} and the operation $\otimes$.
- An abstract system can form a group provided it satisfies the properties of a group.

6.6 Axiomatic Systems

- An *axiomatic system* consists of *undefined terms, defined terms, axioms*, and *theorems*.
- A *model* of an axiomatic system is a physical interpretation or diagram of the undefined terms; it must be constructed so that all the axioms of the system are satisfied.
- The axioms of a system must be consistent; that is, they must not contradict each other.
- All branches of mathematics are similar in that they are based on axiomatic systems.

Key Terms and Concepts

associative, *320*
axiom, *345*
axiomatic system, *344*
closure, *317*
commutative, *322*
defined terms, *345*
group, *322*
identity element, *318*
inverse element, *319*
mathematical system, *316*
model, *346*
modulo, *332*
theorem, *345*
undefined terms, *345*

Review Exercises

6.2 Clock Arithmetic

1. Evaluate each of the following on a 12-hour clock:
   a. 6 + 8 b. 8 + 5
   c. 9 + 6 d. 13 + 10
   e. 11 + 9 f. 10 + 7
2. Evaluate each of the following on a 12-hour clock:
   a. 4 − 6 b. 5 − 9
   c. 4 − 8 d. 1 − 11
   e. 8 − 12 f. 9 − 10
3. Evaluate each of the following on a 12-hour clock:
   a. 6 × 6 b. 4 × 5
   c. 9 × 6 d. 5 × 3
   e. 7 × 6 f. 8 × 3
4. Evaluate each of the following on a 12-hour clock:
   a. 7 × (8 + 6) b. 4 × (3 − 6)
   c. 4 − (6 − 8) d. (4 − 6) − 8
   e. 2 × (6 − 8) f. (2 × 6) − 8
5. Describe in your own words what a *mathematical system* is.
6. Given the set of whole numbers {0, 1, 2, 3, . . . }, tell whether each statement is true or false.
   a. The set is closed with respect to addition.
   b. The set is closed with respect to subtraction.
   c. The set is closed with respect to multiplication.
   d. The set is closed with respect to division.
7. True or false?
   a. The set of whole numbers {0, 1, 2, 3, . . . } contains an identity element for the operation of addition.
   b. The set of whole numbers contains an additive inverse element for each element in the set.
   c. The 12-hour clock system contains an identity element for the operation of multiplication.
   d. The 12-hour clock system contains a multiplicative inverse element for each element.
8. True or false? The set of whole numbers
   a. is associative with respect to addition.
   b. is commutative with respect to addition.
9. True or false? The set of whole numbers
   a. is a group with respect to addition.
   b. is a group with respect to multiplication.

6.3 More New Systems

For Exercises 10 and 11, evaluate the following, given that 1 = spring, 2 = summer, 3 = fall, and 4 = winter. (Your answer should be in terms of a season.)

10. a. summer + summer
    b. winter + summer
    c. fall + winter
    d. summer − fall
    e. fall − winter
    f. spring − summer
11. a. fall × fall
    b. summer × fall
    c. fall × (fall + winter)
    d. fall × (spring − winter)
    e. winter × (fall − winter)
    f. summer × (winter + spring)

6.4 Modular Systems

12. Find the equivalent to each of the following in the modulo 5 system:
    a. 42 b. 61 c. 89
    d. −32 e. −108 f. 2003
13. Using the elements of the indicated modular systems, find a replacement for each question mark so that the statement is true.
    a. 4 + ? ≡ 2 (mod 5)
    b. 2 − ? ≡ 4 (mod 7)
    c. 3 × ? ≡ 2 (mod 7)
    d. ? × 2 ≡ 3 (mod 7)
    e. 4 × (3 + ?) ≡ 1 (mod 5)
    f. 2 × (3 − ?) ≡ 4 (mod 7)
14. **Coin collection** Carl the coin collector obtained two rolls of pennies from the bank. After examining the coins, he divided the pennies into two groups: those with mint marks, and those without mint marks. Working with the coins that had mint marks, Carl arranged these pennies in stacks of 5, with 2 left over. Next he arranged these coins in stacks of 7, with 1 left over. Finally, he arranged them in stacks of 3, with 0 left over. How many of the pennies had a mint mark?
15. **Automobile odometer** The speedometer on a car indicates how fast a car travels; the odometer indicates how many miles a car travels. The odometer is an example of a modular system. Why? What modular system is used on the odometer of an ordinary car?

6.5 Mathematical Systems without Numbers

16. Evaluate the following and answer the questions by using Table 6.14, which defines an operation $*$ for the elements of the set $\{\$, ¢, \&, ?\}$:

TABLE 6.14

*	$	¢	&	?
$	¢	$	?	&
¢	$	¢	&	?
&	?	&	π	$
?	&	?	$	π

a. $ * ¢ **b.** ¢ * ? **c.** & * &

d. ? * ? **e.** ¢ * & **f.** & * ?

g. ¢ * (& * $) **h.** & * (? * $)

i. Is this set closed with respect to the operation *? Why or why not?

j. What is the identity element (if any)?

k. Which elements of the set have an inverse?

l. Does this set form a group under the operation *?

6.6 Axiomatic Systems

17. Name the basic parts of an axiomatic system.

18. Given the following axioms,

I: There are at least three squirrels.

II: Each squirrel is in exactly one tree.

III: No squirrel is in a tree by itself.

IV: For every tree, there is a squirrel that is not in that tree.

Prove: There exist at least four squirrels.

19. The sides of a triangle are represented by $3x + 4$, $2x + 8$, and $5x - 4$. If the perimeter of the triangle is 48, prove that the triangle is equilateral—that is, all three sides are equal.

20. Given: $\overline{EOA}$ is a straight line.

Prove: $\overline{OC}$ bisects $\angle BOD$.

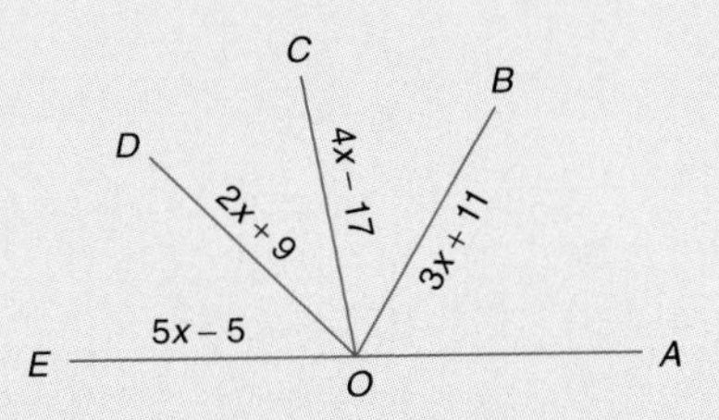

Chapter Quiz

Determine whether each statement (1–15) is true or false.

1. The set of whole numbers is closed for the operation of addition.

2. The set of whole numbers contains an identity element for the operation of multiplication.

3. For every element in a group, there must be an element such that combining the two produces the identity element.

4. $(a*b)*c = (b*a)*c$ is an example of the associative property.

5. If $a \equiv b \pmod{m}$, then a and b both have the same remainder when they are divided by m.

6. If a system consisting of a set of elements $\{a, b, c, \ldots\}$ and an operation * forms a group, then we may conclude that the operation is commutative.

7. The set of integers is closed with respect to division.

8. $99 \equiv 44 \pmod{7}$

9. $3 \times (4 - 2) \equiv 2 \times (5 - 6) \pmod{7}$

10. The set of counting numbers forms a group under the operation of addition.

11. The identity element in the "week" system for the operation of addition is Sunday.

12. A mathematical system consists of a set of elements together with one or more rules for combining elements of the set.

13. An axiomatic system consists of undefined terms, definitions, axioms, and theorems.

14. The 12-hour clock system under the operation of addition contains an inverse for each of its elements.

15. Theorems are logical deductions that are made from undefined terms, defined terms, and axioms.

In Exercises 16–20, use the elements of the indicated modular system to find a replacement for each question mark so that each statement is true.

16. $5 + ? \equiv 2 \pmod{8}$

17. $? + 6 \equiv 5 \pmod{9}$

18. $4 - ? \equiv 5 \pmod{7}$

19. $? \times 7 \equiv 1 \pmod{8}$

20. $2 \times ? \equiv 5 \pmod{11}$

Use the following table to answer Exercises 21–25.

×	a	b	c
a	d	c	b
b	a	b	c
c	b	c	d

21. $(a \times b) \times c$

22. $(c \times b) \times (a \times c)$

23. Is this system closed with respect to the operation ×?

24. What is the identity element (if any) for this system?

25. Does this system form a group under the operation ×? Why or why not?

Just for fun

Most proofs are done by means of deduction; that is, we proceed from the premises, step by step, to the conclusion. As we go from one step to the next, we must have a reason for each step to show that it follows logically. The following is an example of a proof that does not obey the rules. Even though the derivation appears to be correct, it is not. Can you find the error?

Statements	Reasons
1. $a = b$	Given
2. $a^2 = ab$	Multiplying both sides by a
3. $a^2 - b^2 = ab - b^2$	Subtracting b^2 from both sides
4. $(a + b)(a - b) = b(a - b)$	Factoring both sides
5. $\dfrac{(a + b)(a - b)}{(a - b)} = \dfrac{b(a - b)}{(a - b)}$	Dividing both sides by $(a - b)$
6. $(a + b) = b$	Result of step 5
7. $b + b = b$	Recall that $a = b$ (step 1), so we may substitute b for a
8. $2b = b$	Combining $b + b$
9. $\dfrac{2b}{b} = \dfrac{b}{b}$	Dividing both sides by b
10. Therefore, $2 = 1$	Result of step 9 (this is the conclusion)

Math Connections

Math Patterns

The modular addition and subtraction tables discussed in this chapter can be used to create artistic designs. Although there are many ways to create basic design patterns from these tables, the traditional way is to assign a *patterned square* to each numbered cell in the interior of a table. Thus, instead of using numbers, we use a pattern. As an example, consider the addition table for the seasons system (Table 6.2).

+	1	2	3	4
1	2	3	4	1
2	3	4	1	2
3	4	1	2	3
4	1	2	3	4

We begin by assigning patterned squares to the numbers appearing in the interior of the table. The pattern we choose is as follows and is arbitrary. You can choose any pattern you want to produce a desired effect.

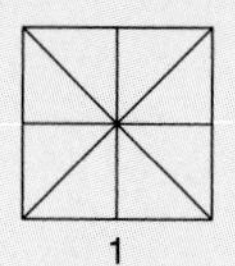

1

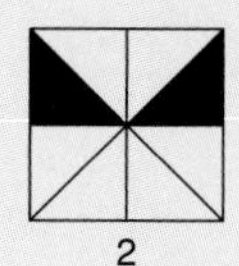

2

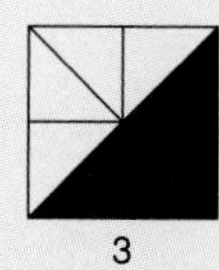

3

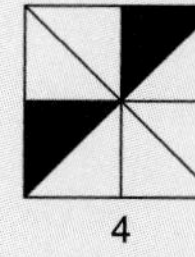

4

We next fill each interior cell of the addition table with the corresponding numbered pattern. Thus, wherever there is a 1 in the table, we replace it with the pattern labeled 1; wherever there is a 2 in the table, we replace it with the pattern labeled 2; and so forth.

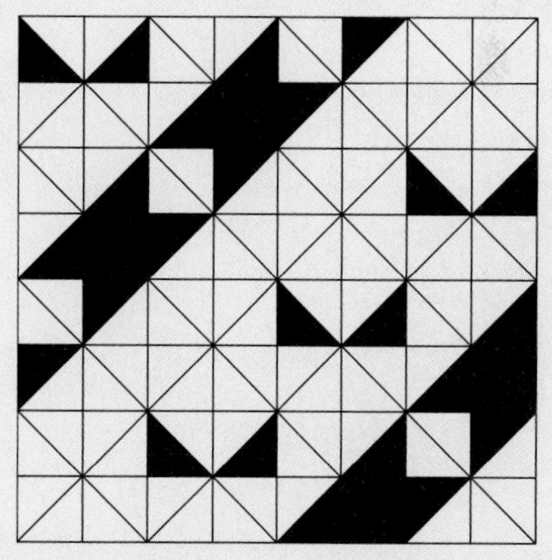

To be even more creative, we can take this 4×4 square and combine it with itself using transformational geometry properties of translation, reflection, and rotation to form larger (e.g., 8×8, 16×16, or 32×32) and possibly more artistic designs.

Chapter 7

Chapter Outline

Systems of Numeration

After Studying This Chapter, You Will Be Able to Do the Following:

1. Express a counting number as a **Roman numeral**, express a Roman numeral as a base 10 numeral, and add and subtract Roman numerals.
2. Use an **abacus** to add two counting numbers.
3. Use **Napier's rods** to multiply two counting numbers.
4. Express a counting number as an **Egyptian numeral**, express an Egyptian numeral as a base 10 numeral, and add and subtract Egyptian numerals.
5. List the distinguishing characteristics of and identify systems of numeration that use a **simple grouping**, a **multiplicative grouping**, or a **place-value system**.
6. Convert a numeral in any base other than 10 to a base 10 numeral using **expanded notation**, and convert a base 10 numeral to any desired base using the **division algorithm**.
7. Perform arithmetic operations in base 2, base 5, and base 16.

Note: *indicates optional material.

Visit the PH Companion Web site at www.prenhall.com/setek

OVERVIEW

Regardless of the human culture we examine, the people in that culture need a system of counting. People need numbers: They must be able to count. No matter what type of job a person has, he or she must be able to count in order to cope with situations encountered in everyday living, even in such common activities as paying bills and making change. Humans have needed some form of counting (numeration) system ever since they began to reason and to develop a civilization.

7.1 INTRODUCTION

Anthropologists maintain that many of the primitive tribes of prehistoric times had some system for counting. The most primitive counting systems went only as high as three or four, with anything greater being described as "many"; however, many so-called primitive peoples had much more sophisticated systems.

Probably the first form of counting was done by matching the things to be counted (such as animals) with something else, such as fingers, toes, stones, or sticks. A pile of pebbles may have been matched, one-to-one, to the number of animals in a certain herd. A herdsman might have placed the pebbles in a pile as he was letting the animals out to graze, one pebble corresponding to each animal. When the animals returned, the herdsman would match a pebble from the pile with each animal. If, after all the animals had passed by, there were some pebbles still remaining that had not been matched, then the herdsman would know that there were some animals not accounted for—perhaps one, two, three, or even "many."

In this chapter, we examine several different methods of counting, computing, and calculation that have been developed throughout time. Formally known as *systems of numeration*, these methods range from early counting systems such as the abacus to our current base 10 (i.e., decimal) number system to methods of computation used by computers.

7.2 EARLY COUNTING AND CALCULATION SYSTEMS

Early Primitive Systems

Humans have invented many different ways to record numbers. For example, the ancient Chinese recorded numbers by tying knots on a string, as did the Incas. Primitive people drew pictures or made slashes (i.e., marks) in the dirt; each mark represented an animal that a person owned. Other civilizations recorded marks on stones or pieces of clay, or made notches in a stick. Even today we still use tally marks to record numbers. For example, we use the tally mark ||| to represent the number three, whereas |||||| represents the number six. Tally marks form an excellent system of numeration. Unfortunately, its use is somewhat cumbersome when we try to use it to represent large numbers. For example, it is much easier for us to distinguish quickly between the numbers fifteen and sixteen by using the numerals 15 and 16 than it is by using the tally marks ||||||||||||||| and ||||||||||||||||.

Over the ages, several improvements were made to the tally system of numeration. One improvement was the use of geometric patterns. For example, instead of using nine tally marks to represent the number nine, a 3 × 3 "square" of dots was used: $\begin{smallmatrix}\bullet&\bullet&\bullet\\ \bullet&\bullet&\bullet\\ \bullet&\bullet&\bullet\end{smallmatrix}$. Note that this idea of patterns is still in use today. Consider, for example, the markings used on dice. Once again, note how much easier it is for us to distinguish between ||||| and |||||| by using $\begin{smallmatrix}\bullet&&\bullet\\ &\bullet&\\ \bullet&&\bullet\end{smallmatrix}$ and $\begin{smallmatrix}\bullet&\bullet\\ \bullet&\bullet\\ \bullet&\bullet\end{smallmatrix}$.

The Roman Numeral System of Numeration

Another system of numeration that evolved from using marks and notches to represent numbers was **Roman numerals**. Consider, for a moment, the symbols I, II, and III. Even today, we find the use of Roman numerals on clocks, room numbers, corner stones, and at the end of motion picture films or television shows. Although there are earlier systems of numeration than Roman numerals (e.g., the Egyptians use of hieroglyphics), we will examine Roman numerals first because most of us are somewhat familiar with them. Other systems will then be discussed in later sections of the chapter.

The Romans were one of the first to recognize that a symbol did not have to look like the number it represented. For example, the Romans used V to represent five and X to represent ten. In all, the Romans used seven capital letters to denote numbers. The letters and their corresponding values are listed in Table 7.1.

TABLE 7.1 The Roman Numerals

Roman Numeral	Numerical Value
I	One (1)
V	Five (5)
X	Ten (10)
L	Fifty (50)
C	One Hundred (100)
D	Five Hundred (500)
M	One Thousand (1000)

Unlike some of the numeration systems that preceded it, the Roman system did not have a one-to-one correspondence between each symbol and the number it represented. The Romans instead relied on only these seven letters to represent numbers. Since more than seven numbers are possible, we must have some rules for combining these letters together to represent other numbers. These rules are summarized as follows:

Rules for Evaluating Roman Numerals

1. If we repeat a letter, we repeat its numerical value. For example,

- XX = 2 *tens*, or 20
- II = 2 *ones*, or 2
- CCC = 3 *hundreds*, or 300

2. If a letter is positioned immediately **after** another letter of greater numerical value, we *add* its value to that of the letter of greater value that precedes it. For example,

- XI = X + I = 10 + 1 = 11
- LX = L + X = 50 + 10 = 60
- CL = C + L = 100 + 50 = 150
- XVI = X + V + I = 10 + 5 + 1 = 16

3. If a letter is positioned immediately **before** another letter of greater numerical value, we *subtract* its value from the letter of greater value that follows it. For example,

- IV = V − I = 5 − 1 = 4
- IX = X − I = 10 − 1 = 9
- CM = M − C = 1000 − 100 = 900

Math Connections

Number versus Numeral

When discussing systems of numeration, students are often confronted with and even confused by the terms *number* and *numeral*. Exactly what is the difference between these terms? From a purely technical perspective, the word *number* is used to refer to anything that can be counted (e.g., the number of students in a mathematics class or the number of calories in a chocolate bar). On the other hand, the word *numeral* refers to any character, symbol, or letter that is used to denote a number. For example, the number three can be represented by the numerals 3, III, iii, as well as by the "symbols" $1 + 2$, $5 - 2$, 3×1, and $6 \div 2$. Thus, many different numerals can be used to denote the same number. Conversely, the same numerals can be used to denote different numbers. For example, consider the numbers 27 and 72. These numbers are certainly different (they both specify a different count or amount), but they also contain the same numerals, namely, 2 and 7.

The distinction between a number and a numeral can also be compared to the distinction between the letter "oh" (O) and the number zero (0). There is a distinct difference of meaning between the two, but we all know what people mean when they tell us that their telephone area code is "three-oh-five" (i.e., 305), or that the highway they drive on is route "one-oh-four" (i.e., 104). Although from a technical perspective we should distinguish between number and numeral (as we should for "oh" and zero), in everyday language or nontechnical usage it is completely acceptable to use the word *number* for numeral. Throughout this chapter we shall emphasize the distinction between these two words only when it is helpful to do so.

4. If a letter is positioned **between** two letters, each of greater numerical value, its value is *subtracted* from the sum of the values of those letters of greater value. For example,

   - XIX = (X + X) − I = (10 + 10) − 1 = 20 − 1 = 19
   - LIX = (L + X) − I = (50 + 10) − 1 = 60 − 1 = 59
   - XLIV = (L − X) + V − I = (50 − 10) + 5 − 1 = 44
   - MCMLXXVI = 1976

Note the importance of "position" in the Roman numeral system. The rules for combining the numerical values of the letters are dependent on each letter's position relative to other letters. Note further, though, that the position of a letter does not actually affect the numerical value of a letter (i.e., its *denomination*). It only determines whether that letter's numerical value is to be added or subtracted. Thus, the Roman system really was not a *place-value* system in which the numerical value of a letter is dependent on the position it occupies within a numeral. (*Note:* Place-value systems are discussed in Section 7.4.)

EXAMPLE 1 ***Expressing Roman Numerals as Counting Numbers*** Evaluate the following Roman numerals.

a. XXX **b.** XL **c.** XIV **d.** DC **e.** MCMLXXXIV

Solution

a. This is an example of the first rule in which X is being repeated three times. If we repeat a letter, then we repeat its value. Thus, XXX = three *tens*, or more simply, 30.

b. This is an example of the third rule in which a numerically lower-valued letter (X) precedes a numerically higher-valued letter (L). Thus, we must subtract. XL = L − X = 50 − 10 = 40.

c. This is an example of the fourth rule in which a letter (I) is placed between two letters of greater numerical value than itself (X and V). We first add the two higher-valued letters and then subtract the numerical value of I from this sum. Thus, XIV = (X + V) − I = (10 + 5) − 1 = 15 − 1 = 14.

d. This is an example of the second rule in which a numerically lower-valued letter (C) follows a numerically higher-valued letter (D). Thus, we add: DC = D + C = 500 + 100 = 600.

e. MCMLXXXIV is evaluated by using all of the given rules. Working from left to right, note that the first three letters have a letter positioned between two letters of greater value (rule 4); hence, MCM = 1900. Next we have L (50) followed by XXXIV (34); that is, 50 + 34 = 84. Therefore, MCMLXXXIV = 1984.

NW ***Now Work Problem 17.***

EXAMPLE 2 ***Expressing Counting Numbers as Roman Numerals*** Express the following as Roman numerals.

a. 1000 **b.** 2000 **c.** 25 **d.** 14

Solution

a. 1000 is represented by one of the seven letters, namely, M.
b. 2000 is equal to *two* 1000s (i.e., 1000 + 1000), which is MM.
c. 25 is equal to 20 + 5, which is XX + V = XXV.
d. 14 is equal to 10 + 4, which is X + IV, or XIV.

NW ***Now Work Problem 7.***

Adding and Subtracting Roman Numerals

It is relatively easy to add and subtract numbers expressed as Roman numerals. For example, to add numbers in this system, we group the same symbols together and then rewrite (i.e., simplify) the expression. To do this, however, you must remember the following facts, which are needed to simplify an expression:

VV = X   IIIII = V
LL = C   XXXXX = L
DD = M   CCCCC = D

As an illustration, consider the following addition problem:

DCLXXII + CCCLXIII

To add these two numbers, we first group the same symbols together. This yields

DCLXXII + CCCLXIII = DCCCCLLXXXIIIII
= DCCCCCXXXV
= DDXXV
= MXXXV

To subtract two numbers that are expressed as Roman numerals, we simply remove (i.e., take away) those letters that are contained in both numbers. In some cases it might be necessary to rewrite some numerals in terms of other letters (e.g., L = XXXXX). As an illustration, consider the following subtraction problem:

DCLXXII − CCCLXIII

To subtract these two numbers, we first rewrite the *minuend* (i.e., the number you subtract from):

DCLXXII − CCCLXIII
= CCCCCCLXVIIIIII − CCCLXIII   *Rewrite D as CCCCC; X as VIIIII*

(continued on p. 360.)

= CCCCCCLXVIIII − CCCLX	*Subtract III*
= CCCCCCVIIII − CCC	*Subtract LX*
= CCCVIIII	*Subtract CCC*
= CCCIX	*Rewrite VIIII as IX*

EXAMPLE 3 ***Adding Roman Numerals*** Add the following Roman numerals.

a. VI + IV **b.** IX + XIX **c.** XLI + XXXIV

Solution

a. VI + IV = VI + IIII *Rewrite IV as IIII*
= VIIIII *Combine Is*
= VV *Rewrite IIIII*
= X *Rewrite VV*

b. IX + XIX = VIIII + XVIIII *Rewrite IX*
= XVVIIIIIIII *Combine Vs and Is*
= XVVVIII *Rewrite IIIII*
= XXVIII *Rewrite VV*

c. XLI + XXXIV
= XXXXI + XXXIIII *Recall XL = L − X = XXXX and IV = V − I = IIII*
= XXXXXXXIIIII *Combine similar symbols*
= LXXV *Rewrite XXXXX and IIIII*

EXAMPLE 4 ***Subtracting Roman Numerals*** Subtract the following Roman numerals.

a. XVII − XIII **b.** XII − VII **c.** LXXV − XXXII

Solution

a. XVII − XIII
= XIIIIIII − XIII *Rewrite V as IIIII*
= IIII *Subtract X and III*
= IV

b. XII − VII
= X − V *Subtract II*
= VV − V *Rewrite X as VV*
= V *Subtract V*

c. LXXV − XXXII
= XXXXXXXIIIII − XXXII *Rewrite L and V*
= XXXXIII *Subtract XXX and II*
= XLIII *Rewrite XXXX*

NW ***Now Work Problems 29 and 45.***

It might seem rather awkward to add and subtract using Roman numerals, but the technique does not require a great deal of skill. The only facts that have to be memorized are the numerical values of the seven letters and how to write some numerals in terms of others. This was a relatively simple task for the Romans, compared to our addition and subtraction "facts," which take three or four years of nearly daily drilling to learn (assuming you are not using a calculator).

Math Connections

Additional Information about Roman Numerals

Recall the third rule of evaluating Roman numerals: If a letter is positioned before another letter of greater numerical value, we subtract its value from that of the letter of greater value that follows it. Without further clarification, this rule can lead to ambiguity. As an example, consider trying to evaluate XLC. Notice in this expression that L is positioned before a letter with a greater numerical value, namely, C. Thus, following rule 3 we have

$$\begin{aligned}&(C - L) - X\\ &= (100 - 50) - 10\\ &= 50 - 10\\ &= 40\end{aligned}$$

However, we also have the situation where X is positioned before a letter of greater numerical value, namely, L, which itself is placed before a letter of greater value, namely, C. Thus, according to rule 3, we should do the following:

$$\begin{aligned}&C - (L - X)\\ &= 100 - (50 - 10)\\ &= 100 - 40\\ &= 60\end{aligned}$$

To avoid this type of situation in which there might be three or more letters positioned in ascending order, the Romans used the following rules:

1. I precedes only V or X.
2. X precedes only L or C.
3. C precedes only D or M.

These rules limit the arrangement of letters in pairs to "four" or "nine" combinations: IV (4), IX (9), XL (40), XC (90), CD (400), and CM (900).

Note further that the Roman numerals I, X, C, and M are equivalent to our current decimal place-value system, which is based on *powers of ten*. Specifically, these four symbols respectively represent 1, 10, 100, and 1000. Using only these four letters, the Romans could represent any number from 1 to 9999. However, to express numbers greater than 9999, the Romans would have to invent a new symbol for each succeeding power of ten such as 10,000, 100,000, and 1,000,000. (Note that this is not necessarily true because the Romans could always use I to represent any number similar to the way tally marks are used. Doing so, though, is a step backward in trying to improve our system of numeration.) To address this limitation, the Romans used another rule: *A bar placed over a letter multiplies the numerical value of that letter by 1000.* For example,

$$\overline{V} = 1000 \times 5 = 5000, \qquad \overline{X} = 1000 \times 10 = 10{,}000,$$

$$\overline{C} = 1000 \times 100 = 100{,}000, \text{ and}$$

$$\overline{M} = 1000 \times 1000 = 1{,}000{,}000.$$

Counting Boards and the Abacus

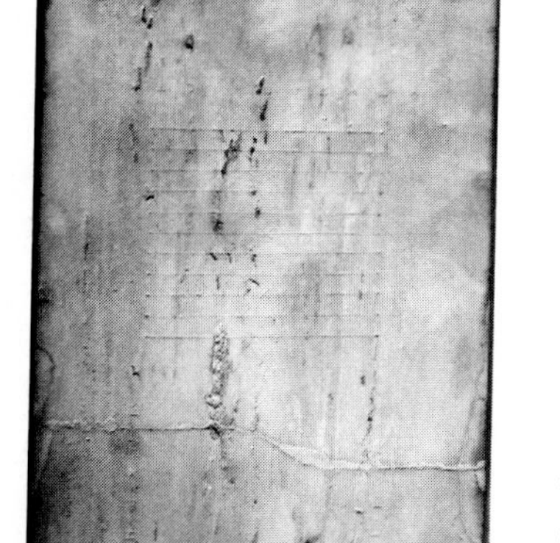

Figure 7.1
The Salamis Tablet.

The **abacus** was one of the first mechanical aids to computation. In its simplest form, the abacus was known as a *counting board* that consisted of lines and pebbles. The earliest counting board probably involved drawing lines in the sand (the space between two lines would represent the units 10s, 100s, etc.) and placing small pebbles within those lines. With the need for something more durable, wooden counting boards were created and these in turn led to the construction of boards made of metal or stone such as marble. Furthermore, instead of pebbles, beads were used. The oldest surviving counting board is the **Salamis tablet** (Figure 7.1), which was used by the Babylonians around 300 B.C. The tablet was discovered in 1899 on the island of Salamis, which is located in the Saronic Isles off the Greek coast. It consists of a slab of marble that is marked with two sets of 11 vertical lines that yield ten columns per set. There is also a blank space between the set of lines as well as a horizontal line crossing each. Greek symbols are located along the top and bottom.

The evolution of the abacus, from early counting boards to its present-day form, is quite extensive. From 500 B.C. to 500 A.D., counting boards evolved from lines drawn in the sand to Greek and Roman times, in which they were constructed from stone and metal. Pebbles and beads were also placed vertically along the lines. During the Middle Ages (500 A.D.–1400 A.D.), counting boards were primarily made of wood and the beads were placed horizontally along the lines. The abacus eventually disappeared completely from Europe when written numbers were used for computation. Nevertheless, it is widely speculated that early Christians brought the abacus to the Orient, where it began to appear first in

China around 1200 A.D., and then in Japan and Korea around 1600 A.D. In Chinese, the abacus is called *suan-pan*; in Japanese it is called *soroban*. Figure 7.2 shows sketches of these early devices.

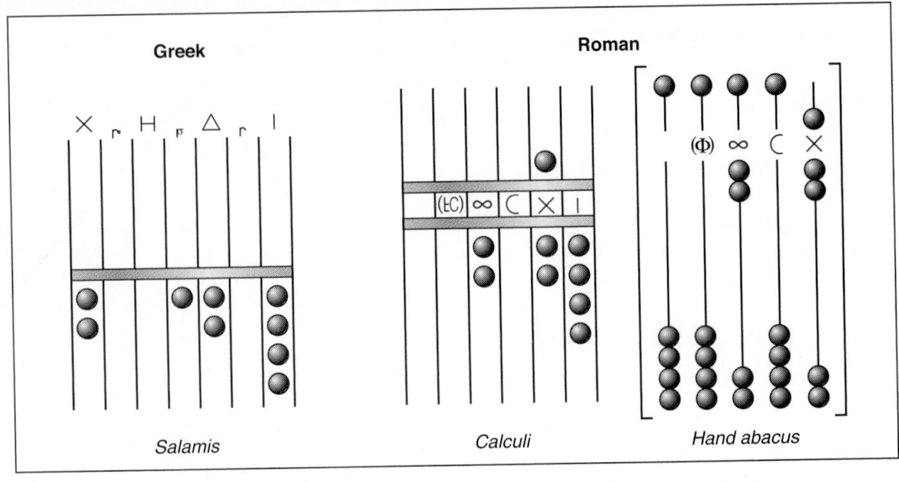

Ancient Times: circa 500 B.C.–c. 500 A.D.

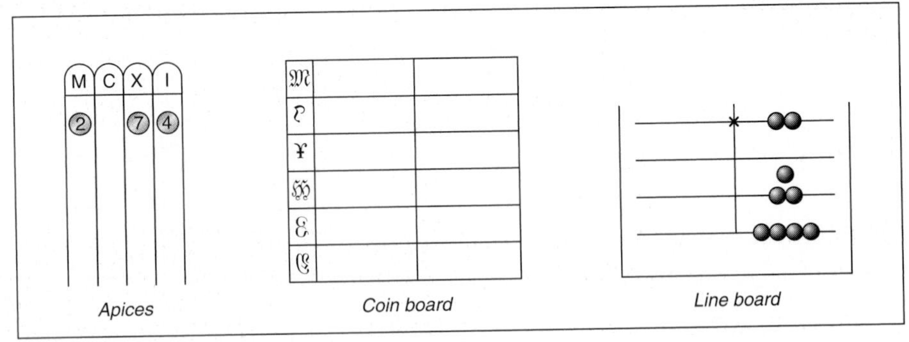

Middle Ages: c. 500 A.D.–c. 1400 A.D.

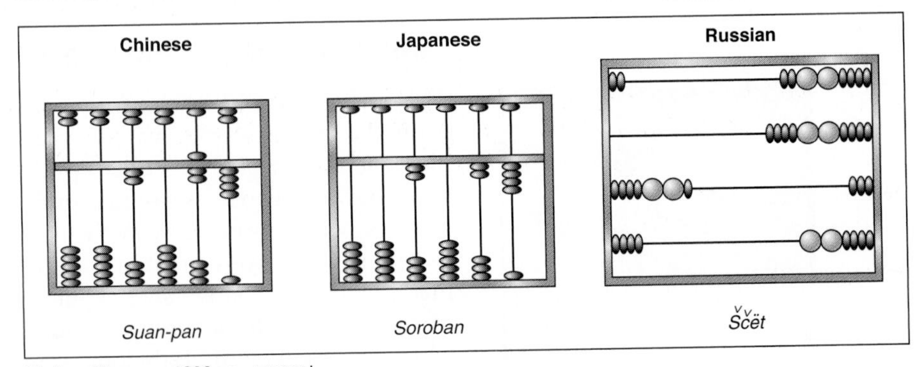

Modern Times: c. 1200 A.D.–present

Figure 7.2
The evolution of the abacus.

The abacus represented one of the earliest *place-value systems*. The lines represented a specific *place value* and the pebbles or beads were placed on these lines to represent one unit. Although each pebble or bead was equal to one unit, its numerical value was based on which line the pebble was placed. This is the basic concept of place value; that is, the numerical value of a symbol is dependent on the location or position of that symbol. Thus, the abacus represented a place-value system because the *position* of a pebble or bead was critical to interpreting its numerical value. Note that this was unlike the Roman system of numeration. Although the Romans also relied on the concept of position, the position of a letter in a Roman

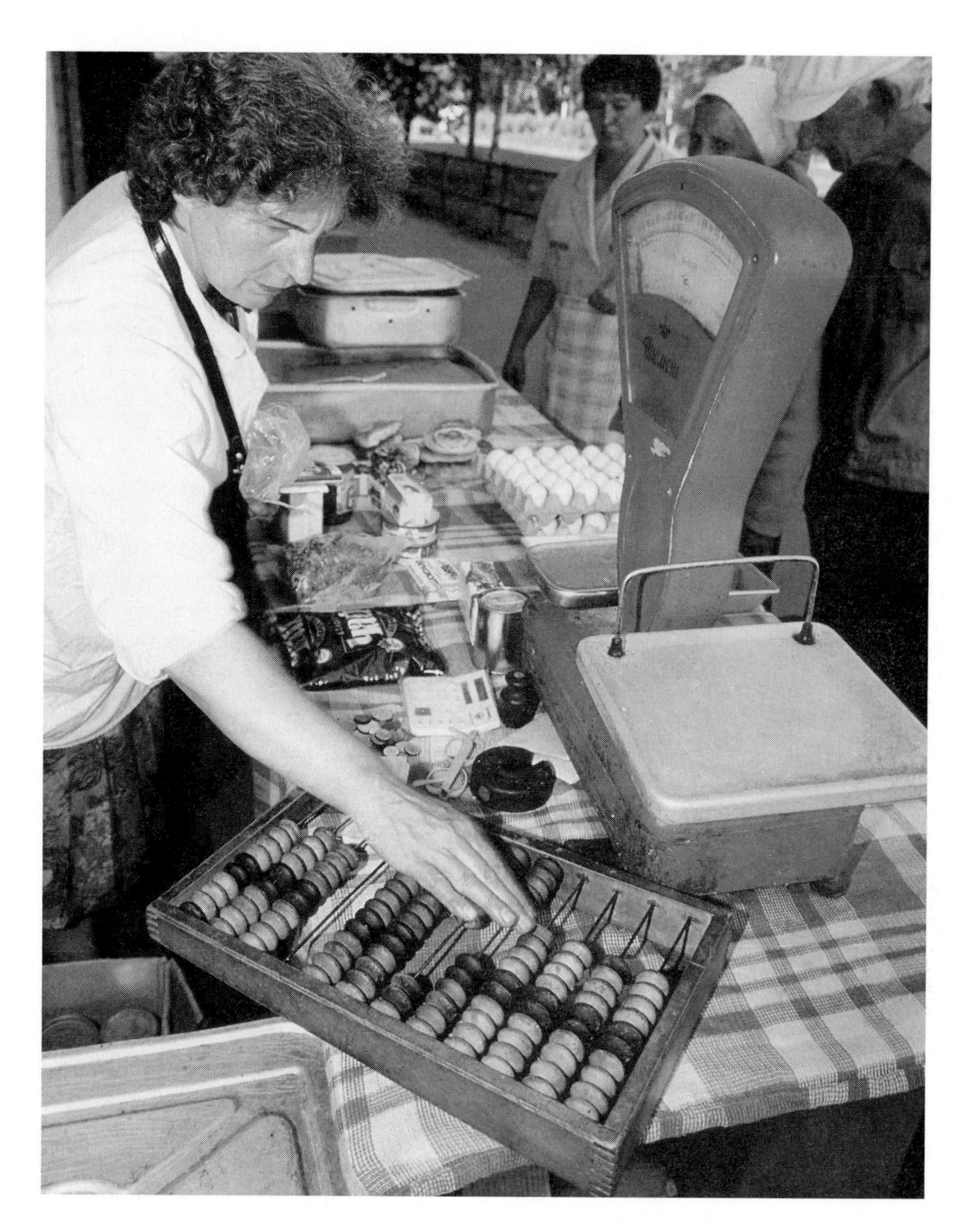

An abacus.

Note of Interest

Scott F. Hall, an art professor at the University of Central Florida, created the world's longest and most powerful abacus: A handmade art object entitled "Supercomputer." The fully functional wooden computer is 17.5 feet long and can perform calculations up to the 279th decimal place. Like the world's tiniest abacus which was recently built by engineers at IBM, Hall's huge instrument employs freely moveable ball bearings as counters—1395 of them!

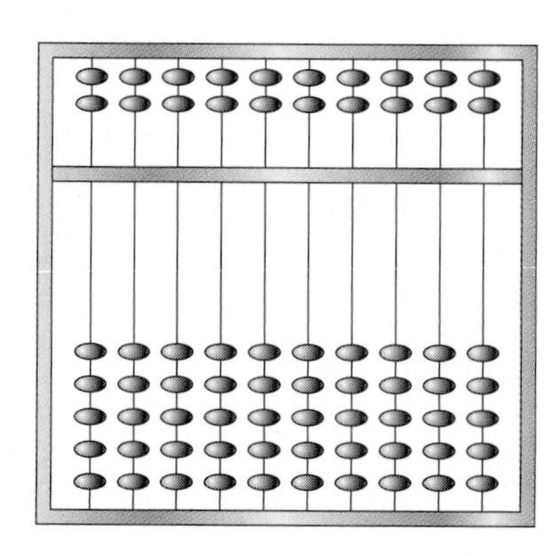

Figure 7.3

numeral does not actually affect the numerical value of a letter. It only determines whether that letter's numerical value is to be added or subtracted. We discuss place-value systems in more detail in Section 7.4.

The conventional abacus consists of a rectangular frame with vertical rods and a horizontal center bar that partitions the frame into two unequal sized decks (see Figure 7.3). The upper deck is smaller in area than the lower deck and contains two beads per rod; the lower deck contains five beads per rod. Each rod represents a decimal value and order similar to our number system. Thus, working from right to left, the rods respectively represent 1s, 10s, 100s, 1000s, and so forth. The position of the beads on a particular rod denotes a digit in that particular decimal position. The beads in the upper deck have a numerical value of 5; the beads in the lower deck have a numerical value of 1. Using the abacus for arithmetic operations involves moving beads toward the center bar. Hence, beads are moved either up from the lower deck or down from the upper deck. Results are then interpreted from the beads along the center bar. For example, in Figure 7.4, the beads denote the number 204,658. (See p. 364)

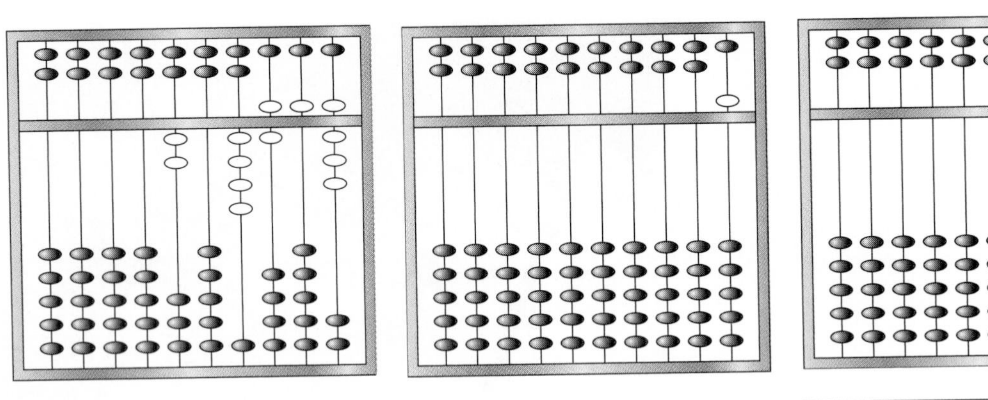

Figure 7.4 Figure 7.5a Figure 7.5b

Addition on the abacus involves moving the beads toward the center bar to denote the numbers being added. If done correctly, the sum of the numbers is read directly from the abacus. For example, to add $5 + 3$, we first move one bead down from the upper deck of the first rod; this denotes 5 (see Figure 7.5a). We then move three beads up from the lower deck of the same rod; this denotes 3 (see Figure 7.5b). The answer is found by reading the resulting bead positions along the center bar. Thus, $5 + 3 = 8$.

EXAMPLE 5 ***Using an Abacus to Add Numbers*** Use an abacus to perform the following additions.

a. $4 + 2$ **b.** $9 + 8$ **c.** $7 + 6$

Solution

a. We first move four beads up along the first rod (see Figure 7.6a). We then try to move two beads up along the first rod but cannot do so because there is only one bead available. Our only alternative is to move one bead down from the upper deck (see Figure 7.6b). However, this single bead denotes 5, which is three more beads than what we need. To correct this "overrun," we *remove* three beads from the lower deck. This is done by moving three beads away from the center bar (see Figure 7.6c). Reading the beads along the center bar in Figure 7.6c, we see that $4 + 2 = 6$.

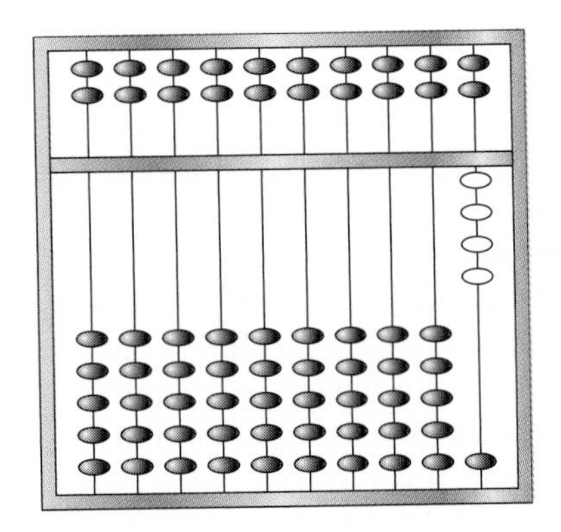

Figure 7.6a

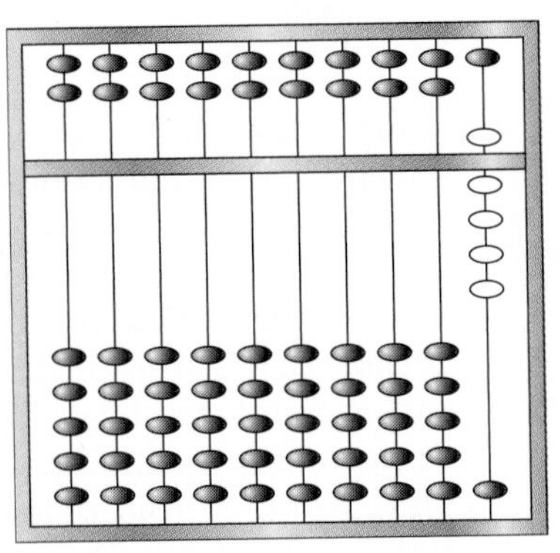

Figure 7.6b

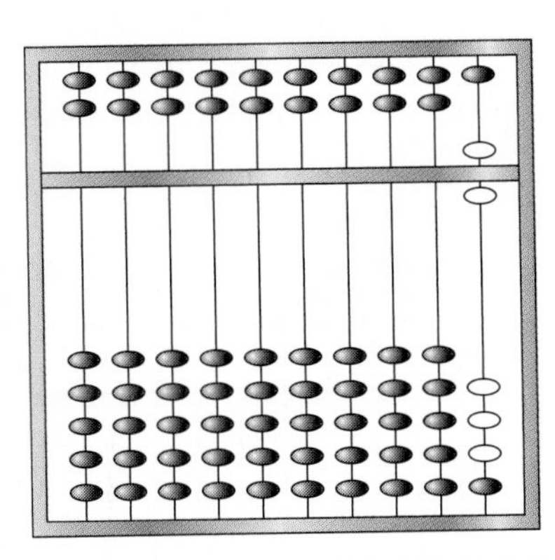

Figure 7.6c

b. We first record 9 on the abacus by moving one bead down from the upper deck and four beads up from the lower deck (see Figure 7.7a). Now, to record 8, we observe that 8 is 2 less than 10; that is, $10 - 2 = 8$. This is recorded on the abacus by moving one bead up from the lower deck on the *next rod* (i.e., the rod immediately to the left of the current rod) and removing two beads from the lower deck on the first rod (see Figure 7.7b). Remember: Working from right to left, the rods represent 1s, 10s, 100s, 1000s, and so forth. Looking at Figure 7.7b, we see that the beads along the center bar denote $10 + 5 + 2 = 17$. Thus, $9 + 8 = 17$.

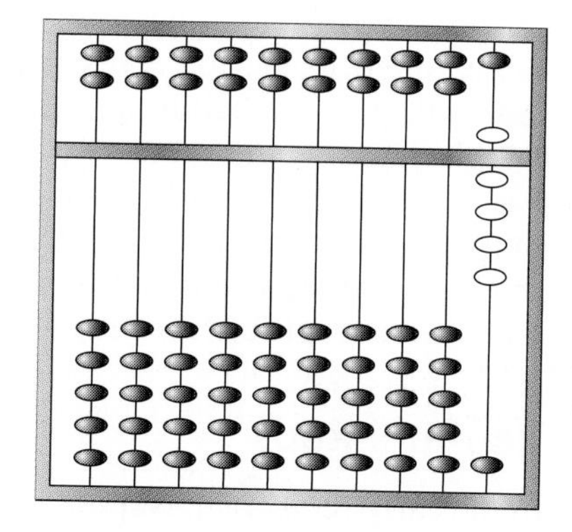

Figure 7.7a

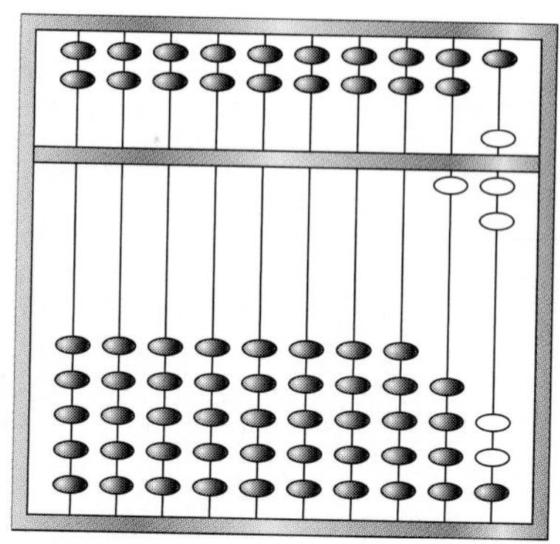

Figure 7.7b

c. We first record 7 on the abacus by moving one bead down from the upper deck and two beads up from the lower deck (see Figure 7.8a). Now, to record 6, we would normally move one bead down from the upper deck and one bead up from the lower deck. Doing so, however, yields 10 in the upper deck of the first column. This is not acceptable because 10 is represented by a single bead from the lower deck of the next rod. So, instead of moving a second bead down from the upper deck of the first rod, we *remove* the initial bead from the upper deck in Figure 7.8a (i.e., we take away 5) and move one bead up from the lower deck along the next rod (i.e., we add 10). This is demonstrated in Figure 7.8b. We are still not done, though, because these two actions yield 5, that is, $10 - 5 = 5$. Our objective was to represent 6. Hence, we must move one bead up from the lower deck of the first rod (see Figure 7.8c). Note how the combined result of these last three actions enabled us to represent 6, namely, $6 = 10 - 5 + 1$. Reading the center bar of Figure 7.8c, we see that $7 + 6 = 13$.

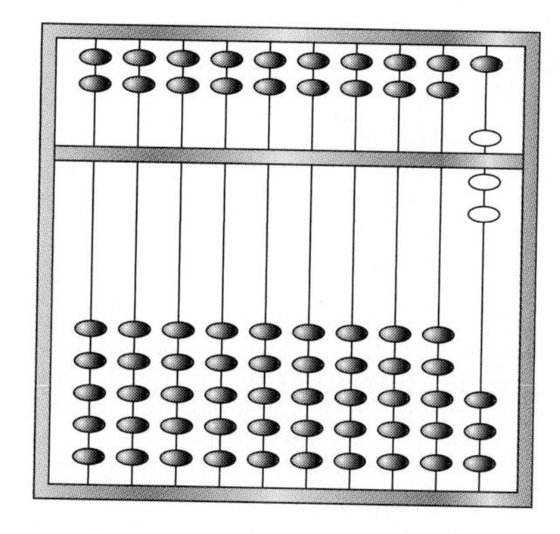

Figure 7.8a

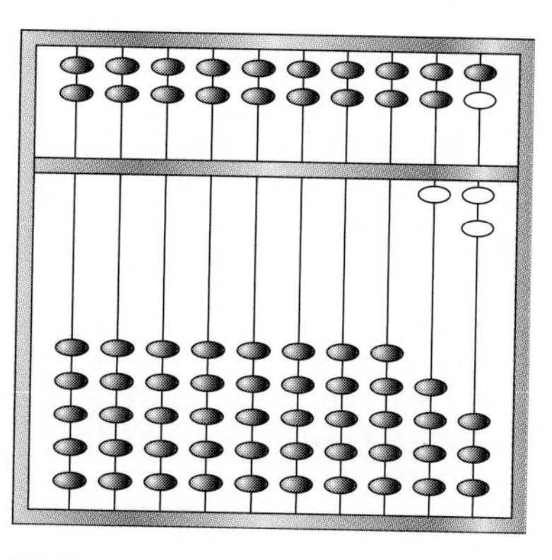

Figure 7.8b

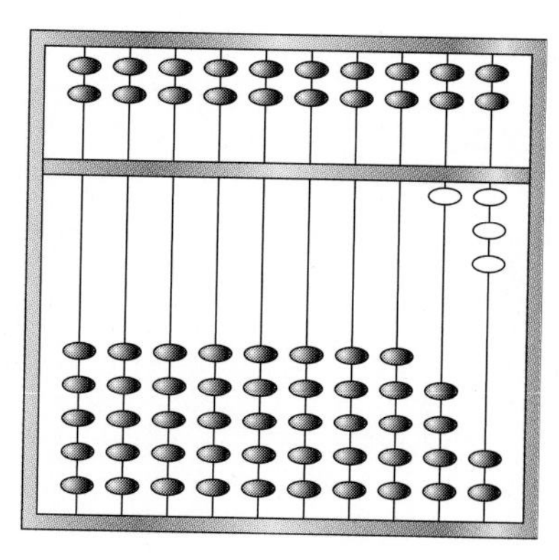

Figure 7.8c

NW ***Now Work Problem 57.***

Subtraction is performed by recording the *minuend* (i.e., the first number of the subtraction problem) along the center bar. Then, starting from the left, we begin removing beads from the lower or upper decks. The final bead positions represent the answer. This operation is left as a challenge exercise for the reader.

From Example 5, you might suspect that the abacus is not a true computation device such as a calculator. This is true. The abacus was used primarily as a mechanical aid for counting. The user mentally performed the actual calculations and used the abacus to keep track of any "carrying" in addition or "borrowing" in subtraction. In the hands of a skilled user, however, the abacus rivals a mechanical calculator in both speed and accuracy. In fact, on November 12, 1946, a contest was held in Tokyo between the Japanese abacus and the electric calculating machine. The abacus won decisively. The abacus is still used today by many Chinese and Japanese merchants, and it is also used to help teach elementary school children the concepts of "carrying" in addition and "borrowing" in subtraction.

Napier's Rods

In the early 1600s John Napier, a Scottish mathematician (1550–1617), developed a calculating device called **Napier's rods** or *Napier's bones*, which were used to perform multiplication. This device was supposedly made of lengths of bones. The numerals 0 through 9 were inscribed at the top of each "bone" (rod) and multiples of the numerals were listed below it in the manner shown in Figure 7.9.

Index	0	1	2	3	4	5	6	7	8	9
1	0/0	0/1	0/2	0/3	0/4	0/5	0/6	0/7	0/8	0/9
2	0/0	0/2	0/4	0/6	0/8	1/0	1/2	1/4	1/6	1/8
3	0/0	0/3	0/6	0/9	1/2	1/5	1/8	2/1	2/4	2/7
4	0/0	0/4	0/8	1/2	1/6	2/0	2/4	2/8	3/2	3/6
5	0/0	0/5	1/0	1/5	2/0	2/5	3/0	3/5	4/0	4/5
6	0/0	0/6	1/2	1/8	2/4	3/0	3/6	4/2	4/8	5/4
7	0/0	0/7	1/4	2/1	2/8	3/5	4/2	4/9	5/6	6/3
8	0/0	0/8	1/6	2/4	3/2	4/0	4/8	5/6	6/4	7/2
9	0/0	0/9	1/8	2/7	3/6	4/5	5/4	6/3	7/2	8/1

Figure 7.9

To multiply 34 by 5 in a manner similar to how the "bones" work, first note that

$$\begin{array}{r} 34 \\ \times\ 5 \\ \hline 20 \\ 15\ \ \\ \hline 170 \end{array}$$

Now consider the row of numerals alongside the "index" 5 and below the 3 and 4 columns. Hence, we have

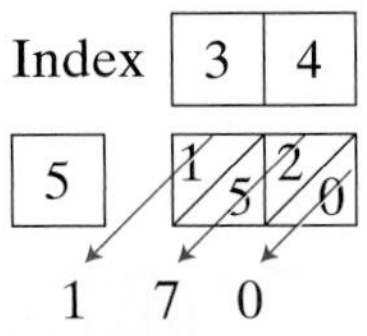

Index	3	4
5	1/5	2/0

1 7 0

If we add along the paths or diagonals, we obtain 170. Therefore, $5 \times 34 = 170$ using Napier's rods. Let's try another example. What is 9×76? First, we find 9 on the "index" and examine the numerals in the row alongside it and below the 7 and 6 columns. We find

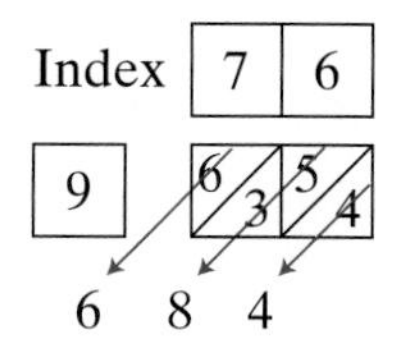

Adding along the diagonals, we obtain 684. Therefore, $9 \times 76 = 684$ by means of Napier's rods.

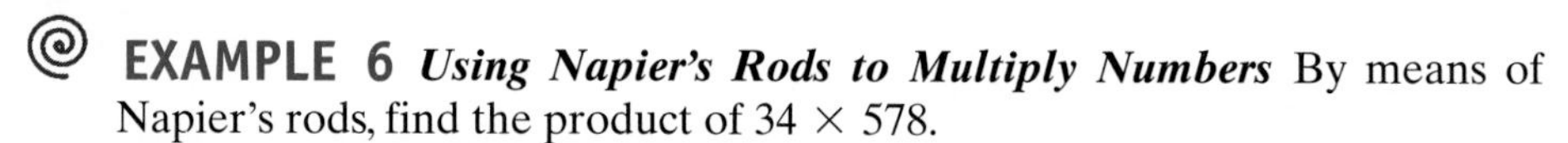

EXAMPLE 6 ***Using Napier's Rods to Multiply Numbers*** By means of Napier's rods, find the product of 34×578.

Solution We are able to find only one-digit products from Napier's rods. Hence, we must perform the calculations in the following manner:

$$\begin{array}{r} 578 \\ \underline{\times\ 34} \end{array} \quad \text{is the same as } 4 \times 578 + 30 \times 578$$

For 4×578 we have

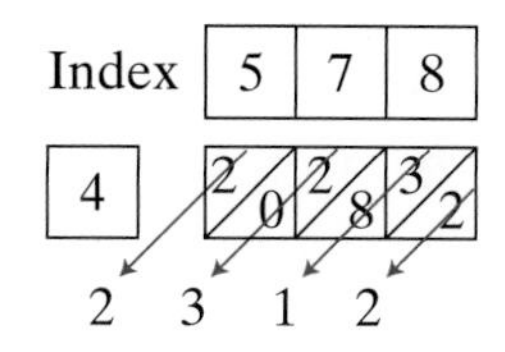

which equals 2,312.

For 30×578 we have

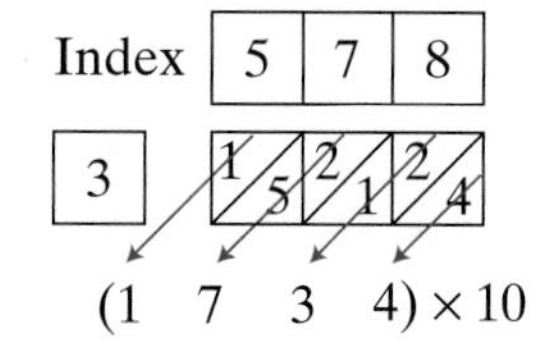

which equals 17,340.

Thus,

$$\begin{array}{r} 4 \times 578 = 2{,}312 \\ \underline{30 \times 578} = \underline{17{,}340} \\ 34 \times 578 = 19{,}652 \end{array}$$

NW ***Now Work Problem 67.***

The development of different systems of numeration and the history of calculation and computation devices is quite interesting. In the remaining sections of this chapter we will examine many different number systems, including early Egyptian and Greek systems, Chinese and Japanese systems, and the Babylonian system. We will also discuss our own base 10 number system as well as number systems in bases other than 10, including base 2 and base 16, which are the number systems used in the computer industry.

Exercises for Section 7.2

In Exercises 1–12, express the given number in Roman numerals.

1. 8 **2.** 13 **3.** 16 **4.** 14

5. 21 **6.** 191 NW **7.** 276 **8.** 341

9. 474 **10.** 676 **11.** 1988 **12.** 2004

In Exercises 13–25, evaluate the given Roman numerals.

13. XVI **14.** XXIV

15. CLI **16.** CXIX

NW **17.** CDIV **18.** CMLIV

19. MDXII **20.** DCVIII

21. DCCLIV **22.** MMDX

23. MCCIV **24.** DCLXIII **25.** $\overline{\text{X}}$

In Exercises 26–40, add the given Roman numerals and express answers as counting numbers.

26. XVI + IIII **27.** XV + VI

28. IV + VI NW **29.** XIX + XXIX

30. LX + XI **31.** LXXV + XXV

32. CCCVI + CCCXV **33.** DCC + CCCI

34. CCC + CCI **35.** DLX + DXL

36. CXV + XCV **37.** MCCL + CCL

38. MCCLV + CCL **39.** MMCC + CCCL

40. DCCV + DCCV

In Exercises 41–52, subtract the given Roman numerals and express answers as counting numbers.

41. XVI − III **42.** XV − VI

43. VI − IV **44.** XXIX − XIX

NW **45.** LX − XL **46.** XXIV − XIX

47. XXXIX − XXV **48.** CCV − LXII

49. CV − XL **50.** DLXV − DXLII

51. MMDC − DCL **52.** MCC − DL

In Exercises 53–55, write the numeral depicted by the given abacus.

53.

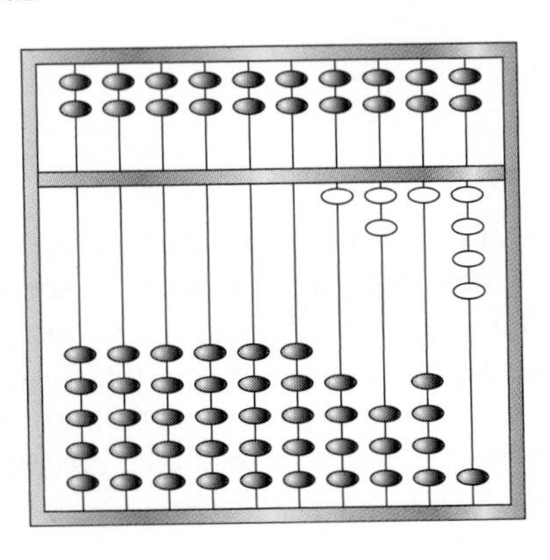

54.

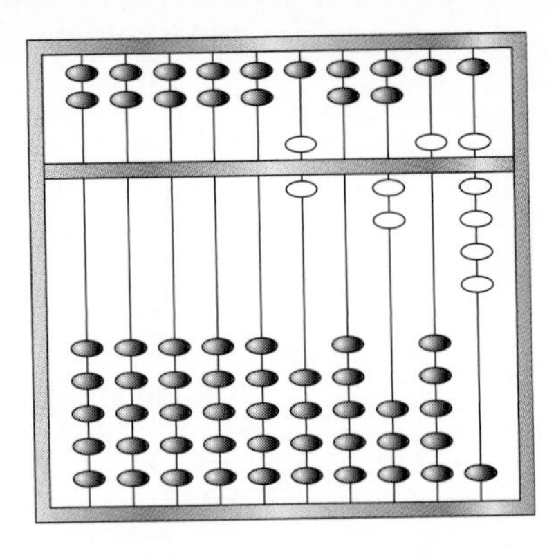

55.

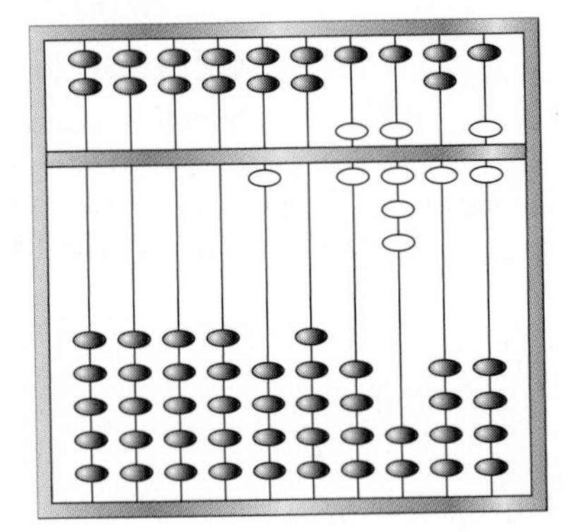

In Exercises 56–65, perform the indicated additions using an abacus. You should sketch a picture of an abacus and show the bead movements similar to the way we did in Example 5.

56. 1 + 3 NW **57.** 3 + 6

58. 4 + 4 **59.** 1 + 4

60. 8 + 2 **61.** 4 + 6

62. 6 + 9 **63.** 8 + 7

64. 5 + 7 **65.** 5 + 9

Each diagram in Exercises 66–70 illustrates a use of Napier's rods. Determine the indicated multiplication problem and find the product. (Use Figure 7.9.)

66.

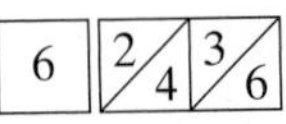

NW **67.**

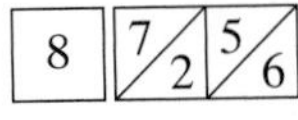

68. 2 | 1/4 | 0/8

69. 4 | 3/6 | 2/8 | 2/0

70. 3 | 1/5 | 1/2 | 2/7

Writing Mathematics

71. In what way(s) was the Roman numeral system an improvement over earlier numeration systems?

72. Explain the significance of "position" (i.e., where a letter is placed) in the Roman numeral system.

73. Explain why the Roman numeral system was not a place-value system.

74. Assume that you have a blind classmate who is trying to visualize an abacus. How would you describe the abacus to him or her?

Challenge Exercises

75. Consider the numbers eight and seven. When these numbers are divided by five, the first division yields a remainder of three and the second a remainder of two. Further note that the sum of the two original numbers is divisible by five (i.e., it produces a zero remainder). Use tally marks to explain why this is true. Do you think this phenomenon is true in all such cases? Why?

76. Review the Math Connections on number versus numeral. Now consider "none" and 0 (zero). Explain these two entities from a number-versus-numeral perspective.

In Exercises 77–79, use an abacus to perform the given subtraction. Sketch a picture of an abacus and show the bead movements similar to the way we did in Example 5.

77. $8 - 2$ **78.** $6 - 2$ **79.** $12 - 6$

80. Complete the following chart to describe how to add the given numbers using an abacus:

If the First Number is	To Add	Move the Beads in the Lower Deck	Move the Beads in the Upper Deck
1, 2, 3, or 4	4		
2, 3, or 4	3		
3 or 4	2		
4	1		

Group/Research Activities

81. Research Napier and prepare a biographical sketch of him.

82. Research the history behind writing the copyright of motion pictures and television shows in Roman numerals.

83. Interview a merchant who still uses an abacus. As part of your interview try to discover where and how the merchant learned to use the abacus, why he or she continues to use it today, and what the merchant's opinion is about calculators. If acceptable, ask the merchant for a demonstration and perhaps invite him or her to participate in a calculation speed contest with you using a calculator and he or she the abacus.

Just for fun

Fill in the blanks correctly to obtain the sum shown.

1. Fahrenheit _____ ( )
2. House of _____ Gables ( )
3. Life begins at _____ ( )
4. The Indianapolis _____ ( )
5. Ali Baba and the _____ Thieves ( )
6. _____ Horsemen of the Apocalypse ( )
7. _____ Downing Street ( )
8. The boiling point of water is _____ Celsius ( )
9. _____ Leagues under the Sea ( )
10. Into the valley of death rode the _____ (_____)

21,752

7.3 GROUPING SYSTEMS

Simple Grouping Systems

One of the oldest known systems of numeration is that of the Egyptians. The Egyptian culture was a fairly advanced one, and consequently it had an advanced system of numeration. The pyramids are proof of the technical ingenuity of the Egyptians. Pictures called *hieroglyphics* were painted on the walls of these tombs to represent numbers.

The Egyptians used hieroglyphics as early as 3400 B.C. A single line or stroke represented items up to 10. After reaching 10, a new symbol is used to indicate a set of 10 things. This is an example of *simple grouping.*

A simple grouping system is a system in which the position of a symbol does not affect the number represented, and in which a different symbol is used to indicate a certain number or group of things.

In the Egyptian system, special symbols were used to represent tens, hundreds, thousands, ten-thousands, hundred-thousands, and millions. That is, the Egyptians used a different symbol for each power of 10. A **power of 10** is 10 raised to a number. For example, $10^1 = 10$ is the first power of 10, $10^2 = 10 \times 10 = 100$ is the second power of 10, and $10^3 = 10 \times 10 \times 10 = 1000$ is the third power of 10. Because the Egyptians used a different symbol for each power of 10, we say that their system had a **base** of 10. In the expression 10^3, 10 is the base and 3 is the exponent, and 1000, or 10^3, is the third power of 10.

Some of the Egyptian hieroglyphic numerals and their corresponding values are given in Table 7.2. It should be noted that the order and shape of the figures may vary from one reference to another, because different examples were obtained from different sources (inscriptions) and are a reflection of the individuals who made the original inscriptions.

TABLE 7.2 Egyptian Numerals and Their Values

Egyptian Numerals	Name	Value	Power of 10
𓏺	Stroke	1	10^0
𓎆	Heelbone	10	10^1
𓍢	Coiled rope	100	10^2
𓆼	Lotus flower	1,000	10^3
𓂭	Pointed finger	10,000	10^4
𓆐	Polywog	100,000	10^5
𓁨	Astonished man	1,000,000	10^6

The only fractions that the ancient Egyptians considered were those they called *unit fractions*—that is, the numeral 1 over an ordinary number. These unit fractions were denoted by placing the symbol 𓂋 over the number for the denominator. For example,

𓂋 over 𓏺𓏺𓏺 $= \frac{1}{3}$ and 𓂋 over 𓎆𓎆 $= \frac{1}{20}$

Fractions that were not unit fractions were expressed as sums of unit fractions. In the Egyptian system $\frac{3}{4}$ would be expressed as the sum of $\frac{1}{2}$ and $\frac{1}{4}$. That is, $\frac{3}{4}$ would be written as

$\frac{1}{2} + \frac{1}{4} =$ 𓂋 over 𓏺𓏺 + 𓂋 over 𓏺𓏺𓏺𓏺

Because the Egyptians used a simple grouping system, the position of a symbol does not affect the number represented. That is, the numeral 𓎆𓏺𓏺 represents 12, as do 𓏺𓏺𓎆 and 𓏺𓎆𓏺. This is certainly not the case for other systems of numeration. For example, 23 does not equal 32 and 128 does not equal 821 in our system of numeration.

A simple grouping system can also be thought of as an "additive" system, because we must add the values of the symbols rather than concern ourselves with the position of the symbols. Sometimes the Egyptians wrote their numerals in left-to-right order, from largest to smallest; at other times the numerals were written in a right-to-left order. Regardless of the arrangement of the symbols, the value of the number represented is not affected. Both of the following pictures represent 2304:

𓆼𓆼𓍢𓍢𓍢𓏺𓏺𓏺𓏺 𓆼𓍢𓍢𓆼𓏺𓏺𓏺𓏺𓍢

In order to evaluate the Egyptian numeral represented, we just add the values of the hieroglyphics, regardless of their position.

To add numbers that are expressed as Egyptian numerals, we group the same symbols together, and then simplify (rewrite) the expression. Consider the following addition problem:

```
  @∩∩∩∩∩∩|||||||
+ @∩∩∩||||
```

In order to add these two numbers, we first group the same symbols together. Then to obtain our final answer, we rewrite 10 of the strokes as a heelbone. This results in a string of 10 heelbones, which we can rewrite as a coiled rope.

```
  @@∩∩∩∩∩∩∩∩∩|||||||||||
= @@∩∩∩∩∩∩∩∩∩∩|
= @@@|
```

In order to subtract two numbers that are expressed as Egyptian numerals, we simply take away those symbols that are contained in both numbers. It may be necessary to rewrite some numerals in terms of other symbols, such as ∩ = ||||||||||. Consider the following subtraction problem:

```
  @∩∩∩∩∩∩|||
- @∩∩∩|||||
```

Note of Interest

One of the Great Pyramids in Egypt that is considered to have a technically perfect architectural design is Cheops' pyramid. The four sides of the pyramid are almost perfectly aligned with true north, south, east, and west. The base of the pyramid covers an area that is over 13 acres! Yet each side of the pyramid joins the other at almost perfect right angles. The blocks that cover the exterior of the pyramid are fitted together with joints that measure approximately $\frac{1}{50}$ of an inch. Truly, the ancient Egyptians were masters of the mathematical tools they possessed.

Following is an illustrative example of multiplication using a technique employed by the ancient Egyptians. Consider 12×13. Using the Egyptians' method, prepare two columns. In the first column successively double the number 1; in the second column successively double the multiplicand (13). Stop when you have obtained numbers in the first column whose sum equals the multiplier (12). For example,

1	13
2	26
*4	→ 52
*8	→ 104
STOP (4 + 8 = 12)	156

Add the corresponding "doubles" in the second column and the result is the desired answer. Therefore,

$$12 \times 13 = 156$$

As we cannot subtract five strokes from three strokes, we rewrite one of the heel-bones as ten strokes, and we have

```
  𓍢∩∩∩∩∩|||||||||||||
– 𓍢∩∩∩     |||||
---------------------
      ∩∩     ||||||||
```

Remember that in the Egyptian system of computation, it is necessary to group the symbols together, and it is also sometimes necessary to rewrite some numerals in terms of other symbols.

EXAMPLE 1 ***Expressing Counting Numbers as Egyptian Numerals*** Express the following as Egyptian numerals:

a. 13 **b.** 231 **c.** 13,423

Solution

a. 13 is one 10 and three 1s; that is, ∩|||

b. 231 is two 100s, three 10s, and one 1; that is, 𓍢𓍢∩∩∩|

c. 13,423 is one 10,000, three 1000s, four 100s, two 10s, and three 1s; that is,

𓂭𓆼𓆼𓆼𓍢𓍢𓍢𓍢∩∩|||

NW ***Now Work Problems 1a and 1d.***

EXAMPLE 2 ***Expressing Egyptian Numerals as Counting Numbers*** Evaluate the following Egyptian numerals:

a. 𓍢𓍢∩∩∩∩|||| **b.** 𓆼𓆼𓍢∩∩∩| **c.** 𓁨𓂭𓍢∩||

Solution

a. 𓍢𓍢∩∩∩∩|||| = two 100s, four 10s, four 1s; that is,

$$(2 \times 100) + (4 \times 10) + (4 \times 1) = 200 + 40 + 4 = 244.$$

b. 𓆼𓆼𓍢∩∩∩| = two 1000s, one 100, three 10s, and one 1; that is,

$$(2 \times 1000) + (1 \times 100) + (3 \times 10) + (1 \times 1)$$
$$= 2000 + 100 + 30 + 1 = 2131$$

c. 𓁨𓂭𓍢∩|| = one 1,000,000, one 10,000, one 100, one 10, and two 1s; that is,

$$(1 \times 1{,}000{,}000) + (1 \times 10{,}000) + (1 \times 100) + (1 \times 10) + (2 \times 1)$$
$$= 1{,}000{,}000 + 10{,}000 + 100 + 10 + 2 = 1{,}010{,}112.$$

NW ***Now Work Problems 5a and 5d.***

EXAMPLE 3 ***Adding Egyptian Numerals*** Add

a. ∩∩∩|| + ∩∩∩∩|||

b. 𓆼∩∩∩∩∩∩|||||| + 𓆼∩∩∩|||||

Solution

a. In order to add these two numbers, we simply combine the symbols:

∩∩∩|| + ∩∩∩∩||| = ∩∩∩∩∩∩∩|||||

b. In order to add these two numbers, we first group the same symbols together and then simplify:

𓆼∩∩∩∩∩∩|||||| + 𓆼∩∩∩|||||
= 𓆼𓆼∩∩∩∩∩∩∩∩∩|||||||||||
= 𓆼𓆼∩∩∩∩∩∩∩∩∩∩|
= 𓆼𓆼𓍢|

EXAMPLE 4 ***Subtracting Egyptian Numerals*** Subtract.

a. ∩∩∩∩||| − ∩∩∩||

b. 𓍢∩∩∩|| − ∩∩∩∩|||||||

Solution

a. In order to subtract two numbers that are expressed as Egyptian numerals, we subtract (or take away) those symbols that are contained in both numerals.

∩∩∩∩||| − ∩∩∩|| = ∩

b. This subtraction cannot be performed immediately; we must first rewrite the first number, since it does not contain 7 ones or 4 tens as such.

𓍢∩∩∩|| − ∩∩∩∩|||||||
= ∩∩∩∩∩∩∩∩∩∩∩∩|||||||||||| − ∩∩∩∩|||||||
= ∩∩∩∩∩∩∩∩|||||

NW ***Now Work Problems 9b and 11c.***

Multiplicative Grouping Systems

The Greek System of Numeration

One of the numeration systems developed by the Greeks used letters to represent numbers; that is, letters were used as numerals. Listed in Table 7.3 are the Greek numerals together with their corresponding values.

One thing that the Greeks did differently from the other systems of numeration is to create symbols for multiples of 5. Since the Greeks had no symbol for 50, they thought of 50 as five 10s and wrote it as ΓΔ. Similarly, 500 was thought of as five 100s and written as ΓH. It follows that ΓX = 5000 and ΓM = 50,000.

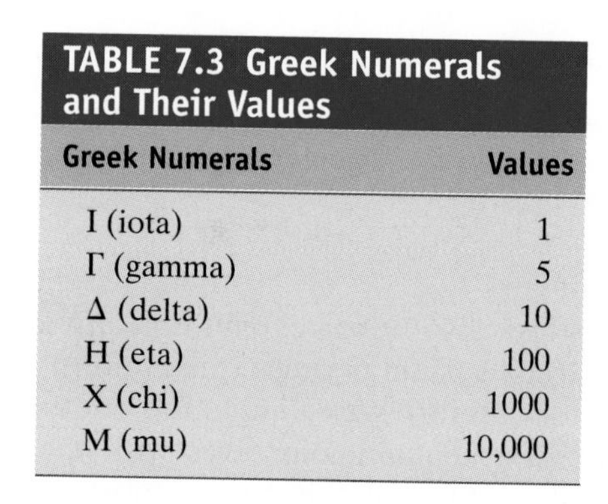

TABLE 7.3 Greek Numerals and Their Values

Greek Numerals	Values
I (iota)	1
Γ (gamma)	5
Δ (delta)	10
H (eta)	100
X (chi)	1000
M (mu)	10,000

In order to express 1984 in Greek numerals, we first think of 1984 as 1000 + 900 + 80 + 4. But 900 = 500 + 400 and 80 = 50 + 30. Therefore, 1984 = 1000 + 500 + 400 + 50 + 30 + 4, or

1984 = XΓHHHHHΓΔΔΔΔIIII

Note that the number is represented by using multiples of 5; this enabled the Greeks to use fewer symbols to express a number. The use of multiples of 5 is an example of *multiplicative grouping*.

> **A multiplicative grouping system** ***is a system that uses certain symbols for numbers in a basic group, together with a second symbol or notation to represent numbers that are multiples of the basic group.***

In the Greek system, 5 is the basic group. The symbol for 5, Γ, is used together with other symbols to represent numbers that are multiples of 5.

EXAMPLE 5 ***Expressing Counting Numbers as Greek Numerals*** Express each number with a Greek numeral.

a. 12 **b.** 56 **c.** 88 **d.** 167 **e.** 1776

Solution

a. 12 is one 10 and two 1s: ΔII

b. 56 is five 10s, one 5, and one 1: ΓΔΓI

c. 88 is one 50, three 10s, one 5, and three 1s: ΓΔΔΔΔΓIII

d. 167 is one 100, one 50, one 10, one 5, and two 1s: HΓΔΔΓII

e. 1776 is one 1000, one 500, two 100s, one 50, two 10s, one 5, and one 1: XΓHHHΓΔΔΔΓI

NW ***Now Work Problems 13a and 13d.***

EXAMPLE 6 ***Expressing Greek Numerals as Counting Numbers*** Evaluate each Greek numeral.

a. ΓΔΓII **b.** ΓHHHΓΔI **c.** ΓXXXΓHHΓΔΔII

Solution

a. ΓΔ is five 10s, or 50, Γ is 5, and II is 2; hence,

$$\text{ΓΔΓII} = 50 + 5 + 2 = 57$$

b. ΓH = 500, HH = 200, ΓΔ = 50, and I = 1; hence,

$$\text{ΓHHHΓΔ I} = 500 + 200 + 50 + 1 = 751$$

c. ΓX = 5000, XX = 2000, ΓH = 500, H = 100, ΓΔ = 50, Δ = 10, and II = 2; hence,

$$\text{ΓXXXΓHHΓΔΔII} = 5000 + 2000 + 500 + 100 + 50 + 10 + 2 = 7662$$

NW ***Now Work Problems 17a and 17d.***

The Greek system of numeration uses six symbols. The system is repetitive and it uses multiples of 5. The use of multiples of 5 in the Greek numeration system is an example of multiplicative grouping.

The Chinese–Japanese System of Numeration

The Chinese–Japanese system of numeration also involves multiplicative grouping. But this system differs from the Greek system because it uses multiples of 10, 100, and 1000 in its grouping system. Before we examine the multiplicative grouping of this system, we must first acquaint ourselves with the characteristics of the Chinese–Japanese system. One of the most important things to remember is that the system uses vertical instead of horizontal writing. Listed in Table 7.4 are the Chinese–Japanese numerals together with their corresponding values.

Because the Chinese–Japanese system of numeration is a system that uses multiplicative grouping, a number such as 2347 is thought of as two 1000s, three 100s, four 10s, and 7. Two 1000s is written as 二千, three 100s is written as 三百, and four 10s is written as 四十. Hence, 2347 is written as shown in the margin.

二千 two 1000s
三百 three 100s
四十 four 10s
七 7

This excerpt from a fourteenth-century Chinese mathematical treatise shows an array of numbers known in the West as Pascal's triangle.

TABLE 7.4 Chinese–Japanese Numerals and Their Values

Chinese–Japanese Numerals	Values	Chinese–Japanese Numerals	Values
一	1	七	7
二	2	八	8
三	3	九	9
四	4	十	10
五	5	百	100
六	6	千	1,000

EXAMPLE 7 ***Expressing Counting Numbers as Chinese–Japanese Numerals*** Express each number with a Chinese–Japanese numeral.

a. 12 **b.** 56 **c.** 88 **d.** 167 **e.** 2776

Solution

a. 十二 **b.** 五十六 **c.** 八十八 **d.** 百六十七 **e.** 二千七百七十六

NW ***Now Work Problems 19a and 19d.***

EXAMPLE 8 ***Expressing Chinese-Japanese Numerals as Counting Numbers*** Evaluate each Chinese–Japanese numeral.

a. 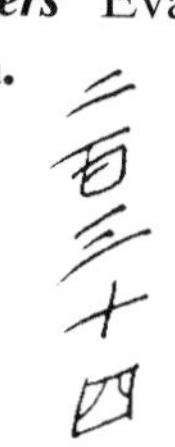

b. 百六十七

c. 五千百五十四

Solution

a. Because the Chinese–Japanese system of numeration uses multiplicative grouping, we have two 100s, three 10s, and 4:

$$(2 \times 100) + (3 \times 10) + 4 = 234$$

b. We have one 100, six 10s, and 7:

$$100 + (6 \times 10) + 7 = 167$$

c. This numeral contains five 1000s, one 100, five 10s, and 4:

$$(5 \times 1000) + 100 + (5 \times 10) + 4 = 5154$$

NW ***Now Work Problems 23a and 23d.***

Exercises for Section 7.3

1. Express each number with an Egyptian numeral.
NW **a.** 18 **b.** 23 **c.** 34
NW **d.** 102 **e.** 201 **f.** 1132

2. Express each number with an Egyptian numeral.
a. 15 **b.** 24 **c.** 104
d. 1321 **e.** 10,212 **f.** 21,111

3. Express each number with an Egyptian numeral.
a. 14 **b.** 21 **c.** 132
d. 1776 **e.** 1984 **f.** 10,001

4. Evaluate each Egyptian numeral.
a. ∩|∩ **b.** 𓍢𓍢| **c.** 𓍢∩|
d. 𓆼𓍢∩ **e.** 𓆼𓆼∩ **f.** 𓂭𓆼𓍢

5. Evaluate each Egyptian numeral.
NW **a.** ∩∩|| **b.** 𓆼𓍢𓍢∩|||
c. 𓆼𓍢| NW **d.** 𓂭𓆼𓆼𓍢𓍢∩||
e. ∩∩𓆼𓍢𓍢|| **f.** |∩|𓆼𓍢𓆼𓍢

6. Evaluate each Egyptian numeral.
a. ∩∩||∩ **b.** 𓍢𓍢∩|∩
c. 𓆼||∩ **d.** 𓆼𓆼𓍢𓍢||||
e. 𓂭𓆼|𓍢𓆼 **f.** |∩𓍢𓆼𓂭

7. Add the following:
a. ∩| + |∩|
b. 𓍢∩∩∩ + ∩∩∩∩∩
c. |||||| + ||||
d. ∩∩∩∩∩∩∩ + ∩∩∩

8. Add the following:
a. ∩∩∩| + ∩∩|
b. 𓍢𓍢∩∩∩||||| + 𓍢∩|||||
c. 𓍢𓍢∩∩∩|| + 𓍢∩∩∩∩||
d. 𓆼𓆼𓍢𓍢𓍢 + 𓍢𓍢𓍢𓍢𓍢𓍢𓍢

9. Add the following:
a. |||||| + |||||
NW **b.** 𓍢∩∩∩∩∩|||| + |||||∩
c. ∩∩∩∩ + ∩∩∩∩∩∩∩
d. 𓆼𓍢𓍢𓍢𓍢𓍢𓍢 + 𓆼𓍢𓍢𓍢𓍢

10. Perform each indicated subtraction.
a. ∩∩∩| − ∩∩||||||
b. 𓍢∩∩|| − ∩∩∩|||
c. 𓍢𓍢∩∩∩∩|| − 𓍢∩∩|
d. 𓆼𓆼𓍢𓍢 − 𓍢∩∩|||

11. Perform each indicated subtraction.
a. ∩| − |||||
b. ∩∩ − ∩|||||||
NW **c.** 𓍢∩| − ∩∩∩∩∩∩∩||
d. 𓆼𓍢 − 𓍢𓍢∩∩∩∩|||||

12. Perform each indicated subtraction.
a. ∩∩ − ||||||
b. 𓍢 − ∩∩
c. ∩∩| − ∩||||
d. 𓍢𓍢 − ∩∩∩|||

13. Express the following as Greek numerals:
NW **a.** 18 **b.** 23 **c.** 34
NW **d.** 44 **e.** 187 **f.** 598

14. Express the following as Greek numerals:
a. 26 **b.** 51 **c.** 127
d. 1391 **e.** 2132 **f.** 6575

15. Express the following as Greek numerals:
a. 21 **b.** 57 **c.** 137
d. 1776 **e.** 1984 **f.** 2001

16. Evaluate each Greek numeral.
a. ΓΙ **b.** ΔΙΙΙΙ **c.** ΔΓΙ
d. ΗΔΔΔΓΙ **e.** Η𐅄ΓΙΙΙ **f.** 𐅅Η𐅄ΓΙΙ

17. Evaluate each Greek numeral.
NW **a.** ΔΓΙΙ **b.** ΗΔΓΙΙ **c.** ΜΧΔ
NW **d.** 𐅅Η𐅄ΓΙ **e.** 𐅆Χ𐅅ΗΙ **f.** 𐅆𐅅𐅄Γ

18. Evaluate each Greek numeral.
a. ΔΙΙ **b.** ΔΙΙΙ **c.** ΔΓΙΙΙ
d. ΗΔΓΙ **e.** Η𐅄Ι **f.** 𐅅ΔΙ

19. Express the following as Chinese–Japanese numerals:
NW **a.** 21 **b.** 57 **c.** 137
NW **d.** 1776 **e.** 1984 **f.** 2001

20. Express the following as Chinese–Japanese numerals:
a. 18 **b.** 23 **c.** 34
d. 46 **e.** 234 **f.** 477

21. Express the following as Chinese–Japanese numerals:
a. 16 **b.** 54 **c.** 147
d. 897 **e.** 3473 **f.** 4176

22. Evaluate each Chinese–Japanese numeral.

a. 二
十

b. 七
百
五

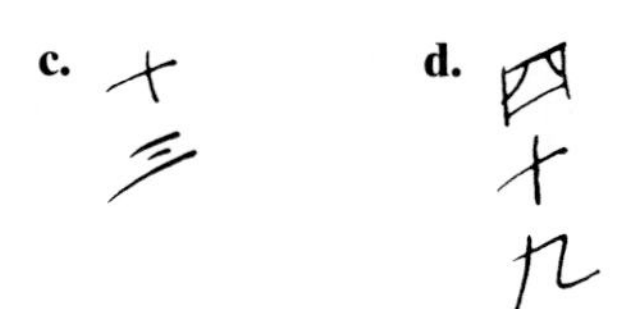

23. Evaluate each Chinese–Japanese numeral.

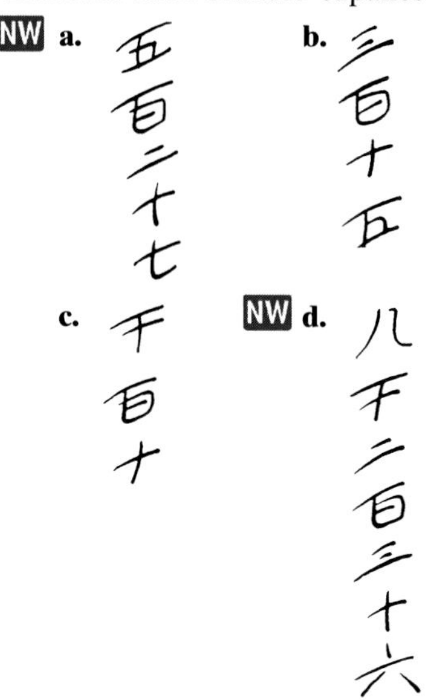

24. Evaluate each Chinese–Japanese numeral.

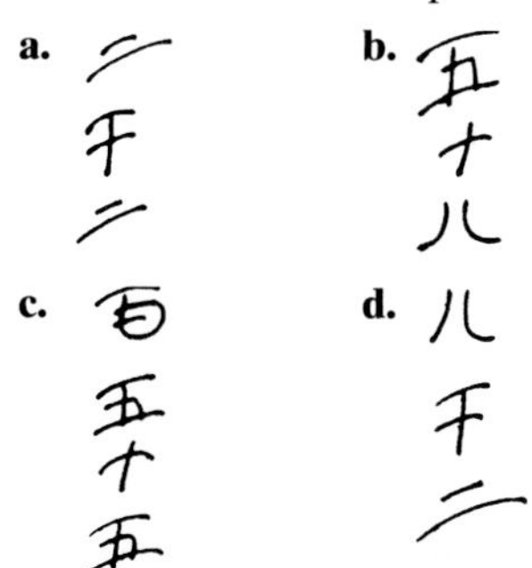

Writing Mathematics

25. Describe the difference(s) between a simple grouping system and a multiplicative grouping system. Give an example of each as part of your description.

26. Explain the difference between the Egyptian system of numeration and our decimal system of numeration.

27. Compare and contrast the Roman system of numeration (Section 7.2) and the Greek system of numeration.

28. When using tally marks, we usually express five as 𝍸. Explain how this is similar to the Greek system of numeration.

29. Which numeration system is more similar to our decimal system: the Greek system or the Chinese–Japanese system? Why?

Challenge Exercises

30. Explain why the Egyptian system does not need a symbol for zero.

31. As indicated in the text, the Egyptians used the symbol ⬭ (called a "mouth," which meant "part") to represent fractions and only considered *unit fractions* (i.e., fractions of the form "1 over a nonzero number"). Given this premise, evaluate the following Egyptian numeral:

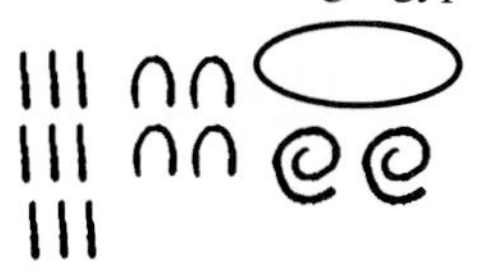

32. Do you think that there is a maximum or largest number that can be represented by Egyptian numerals? If so, what is it? If not, why?

Group/Research Activities

33. The hieroglyphics presented in this section did not remain constant throughout ancient Egyptian civilization. In fact, ancient Egyptian civilization is often partitioned into three distinct periods. Conduct research to determine each of these periods. As part of your research include the different numeral hieroglyphics that were used in these different periods.

34. After the Egyptians began writing on papyrus, they used another number system composed of *hieratic numerals*. Conduct research to find out what these symbols were and what they represented. Compare them to the Egyptian system discussed in this section.

35. Much of our information about Egyptian mathematics has been derived from the Rhind or Ahmes Papyrus. Conduct research to discover what this is and the kind of information that was recorded on it.

36. In the text, we indicated that numerals in the Chinese–Japanese system are written vertically instead of horizontally. Interview both young and old Chinese- or Japanese-Americans to see if they perform mathematics using Chinese–Japanese numerals in this tradition. As part of your interview and research, try to uncover the rationale for this tradition.

Historical Note

In ancient times people wrote in the dirt and on walls, pieces of stone, bricks, pottery, and clay tablets. The Egyptians improved upon these methods with the discovery of a large water plant with peculiar properties. By trial and error the Egyptians learned to cut this plant into thin strips, lay the strips together, and place a similar layer of strips crosswise over the first. Next they pressed the layers together and allowed them to dry, after which they had something to write on. The coarse, brown writing material was called *papyrus* because it was obtained from the papyrus reed. It is believed that the word *paper* is derived from the term for this ancient writing material.

Just for fun

The following is an addition problem in which each letter represents a number. Two different letters cannot represent the same number. What numbers do the letters represent?

```
  SEND
+ MORE
 MONEY
```

7.4 PLACE-VALUE SYSTEMS

Thus far in our discussion of systems of numeration, we have examined a simple grouping system of numeration—the Egyptian system—and two multiplicative grouping systems—the Greek and the Chinese–Japanese. We shall now examine two systems of numeration that use a *place-value system*—the Babylonian system and the Hindu–Arabic system.

> ***A* place-value system *is a system in which the position of a symbol matters; that is, the value that any symbol represents depends on the position it occupies within the numeral.***

The Babylonian System of Numeration

The Babylonian system of numeration uses only two symbols to represent numbers, ▼ to represent 1 and ◀ to represent 10. Because the signs looked like little wedges, they were called *cuneiform* signs, a word meaning "wedge shaped." The Babylonians pressed the end of a stick into a clay tablet in order to write their numbers. They used these two symbols, ▼ = 1, and ◀ = 10, to write any number up to 60. That is, ▼ = 1, ▼▼ = 2, ▼▼▼ = 3, and so on. They used the principle of addition to write numbers. For example, to record the number 7, the Babylonians would write seven as

7 = ▼▼▼▼
▼▼▼

Since ◀ = 10, we can express 11 as one 10 and one 1 : 11 = ◀▼, 12 = ◀▼▼, 13 = ◀▼▼▼, and so on. Forty-three can be expressed as four 10s and three 1s:

43 = ◀◀▼▼▼
◀◀

The Babylonians used a place-value system, in which the position of a symbol is important. The symbols for 10 were always placed to the left of the symbols for 1. Therefore, to represent the number 57 in the Babylonian system, we use five 10s and seven 1s arranged as follows:

57 = ◀◀◀◀◀ ▼▼▼▼
▼▼▼

In order to represent a number greater than 60, such as 85, the Babylonians used a *sexagesimal system*, a system based on 60. The number 85 was thought of as one 60 and 25—that is, as $(1 \times 60) + 25$. To indicate this, the Babylonians placed a symbol for one, ▼, to the left of the numeral for 25.

85 = ▼ ◀◀ ▼▼▼▼▼

one 60 two 10s 5

In our system, a number such as 1984 is read "one thousand, nine hundred, eighty-four." In other words, 1984 is composed of one 1000, nine 100s, eight 10s, and four 1s. Symbolically, we can write this as $(1 \times 10^3) + (9 \times 10^2) + (8 \times 10^1) + (4 \times 1)$. Our system is based on powers of 10; hence, it is called a **decimal system** of numeration. (The word *decimal* is derived from the Latin word *decem,* which means "ten.") The Babylonian system is based on powers of 60; hence, it is called a **sexagesimal system**. Therefore, a Babylonian numeral such as

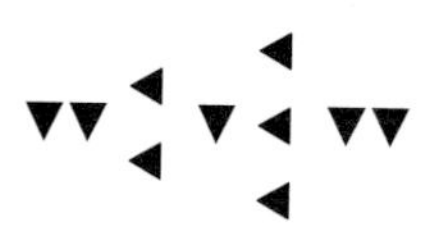

is interpreted as $(2 \times 60^2) + (21 \times 60^1) + 32$, or $(2 \times 3600) + (21 \times 60) + 32 = 7200 + 1260 + 32 = 8492$.

EXAMPLE 1 *Expressing Counting Numbers as Babylonian Numerals*

Express the following as Babylonian numerals:

a. 12 **b.** 42 **c.** 56 **d.** 88 **e.** 147

Solution

a. 12 is one 10 and two 1s; hence,

12 = ◀▼▼

b. 42 is four 10s and two 1s; hence,

42 = ◀◀◀◀▼▼

c. 56 is five 10s and six 1s; hence,

56 = ◀◀◀◀◀▼▼▼▼▼▼

d. 88 is one 60 and 28, or one 60, two 10s, and eight 1s; hence,

88 = ▼ ◀◀▼▼▼▼▼▼▼▼

e. 147 is two 60s, two 10s, and seven 1s; hence,

147 = ▼▼ ◀◀▼▼▼▼▼▼▼

NW ***Now Work Problems 1a and 1d.***

EXAMPLE 2 *Expressing Babylonian Numerals as Counting Numbers*

Evaluate each Babylonian numeral.

a. ◀▼▼▼ **b.** 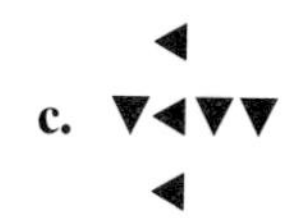 **c.** ▼ ◀◀◀▼▼

Solution

a. In this expression, we have one 10 and three 1s; $10 + 3 = 13$.
b. Here we have four 10s and four 1s; $(4 \times 10) + 4 = 44$.
c. In this expression, we have one 60; three 10s, and two 1s;

$$(1 \times 60) + (3 \times 10) + 2 = 60 + 30 + 2 = 92.$$

NW ***Now Work Problems 5a and 5d.***

The Babylonian system of numeration uses only two symbols, and the symbols can be repeated for numbers up to 60. After that, the numbers are expressed in powers of 60. The numeral

represents one 60, two 10s, and two 1s; that is

$$(1 \times 60) + (2 \times 10) + 2 = 82$$

The Hindu–Arabic System of Numeration

The system of numerals we use today is called the **Hindu–Arabic system**. That is, we use Hindu–Arabic numerals to express numbers. The symbols we use are 0, 1, 2, 3, 4, 5, 6, 7, 8, 9. These symbols had their beginning in India; it was the Arabs who were responsible for making their existence known in Europe. The Arabs did their calculations with these new numerals and then exposed the Europeans to their new techniques.

One of the most significant contributions of the Hindu numerals was zero. Zero evolved from a need for a placeholder, because it was important to distinguish between numerals such as 501 and 51. The transition to Hindu–Arabic numerals was a slow process, and it was not until the end of the sixteenth century that the changeover was fairly complete.

Recall that the Egyptian system of numeration uses a different symbol for each power of 10, but it has no place value. That is, the positions of the symbols do not affect the value of the number. Our system of numeration also uses powers of 10, but it does have **place value**: The position that a symbol has within a numeral is important.

The Hindu–Arabic system of numeration uses multiplicative grouping based on powers of 10. For example, in the numeral 1978, the 8 represents eight 1s, the 7 represents seven 10s, the 9 represents nine 100s, and the 1 represents one 1000. One, 10, 100, and 1000 are all powers of 10:

$$\begin{aligned} 1 &= 10^0 \\ 10 &= 10^1 \\ 100 &= 10 \times 10 = 10^2 \\ 1000 &= 10 \times 10 \times 10 = 10^3 \end{aligned}$$

The number 1978 is the result of combining multiples of these powers of 10:

$$\begin{aligned} 8 &= 8 \times 10^0 \\ 70 &= 7 \times 10^1 \\ 900 &= 9 \times 10^2 \\ \underline{1000} &\underline{= 1 \times 10^3} \\ 1978 & \end{aligned}$$

Notice that the positions of the numerals are important. The only way that we know that the 8 in 1978 represents eight 1s is by its position; in 1987, the 8 represents eight 10s. Thus we say that the Hindu–Arabic system of numeration is a place-value system.

When we write a number in terms of powers of 10, we are writing it in **expanded notation**. In expanded notation,

$$1978 = (1 \times 10^3) + (9 \times 10^2) + (7 \times 10^1) + (8 \times 10^0)$$

When expressing a number in expanded notation, it is convenient to start with the 1s place and proceed from right to left.

EXAMPLE 3 ***Writing Counting Numbers in Expanded Notation*** Write each number in expanded notation.

a. 123 **b.** 2347 **c.** 2003

Solution

a. The Hindu–Arabic numeral 123 is composed of one 100, two 10s, and three 1s. Therefore, we have $(1 \times 100) + (2 \times 10) + (3 \times 1)$. Rewriting this using powers of 10, we have $(1 \times 10^2) + (2 \times 10^1) + (3 \times 10^0)$. (Do recall that $10^0 = 1$. Any number raised to the zero power, except zero, equals 1.)

b. $2347 = (2 \times 1000) + (3 \times 100) + (4 \times 10) + (7 \times 1) = (2 \times 10^3) + (3 \times 10^2) + (4 \times 10^1) + (7 \times 10^0)$.

c. $2003 = (2 \times 10^3) + (0 \times 10^2) + (0 \times 10^1) + (3 \times 10^0)$.

Note that there are no 10s or 100s, but we still must indicate this in expanded notation, as the zeros are placeholders, and 2003 is not the same as 23.`

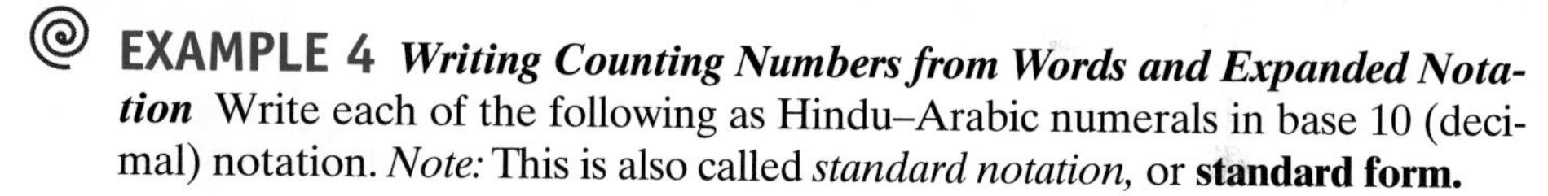

EXAMPLE 4 ***Writing Counting Numbers from Words and Expanded Notation*** Write each of the following as Hindu–Arabic numerals in base 10 (decimal) notation. *Note:* This is also called *standard notation,* or **standard form.**

a. One hundred eighty-seven

b. Two thousand three hundred forty-one

c. One thousand two

d. $(3 \times 10^3) + (2 \times 10^2) + (1 \times 10^1) + (0 \times 10^0)$

e. $(4 \times 10^3) + (2 \times 10^0)$

Solution

a. 187

b. 2341

c. 1002 (Note that we did not write "one thousand *and* two." In mathematics, the word *and* is used to indicate the position of the decimal point, as in one hundred three and two-tenths, which is 103.2.)

d. $(3 \times 10^3) + (2 \times 10^2) + (1 \times 10^1) + (0 \times 10^0)$
$= (3 \times 1000) + (2 \times 100) + (1 \times 10) + (0 \times 1)$
$= 3000 + 200 + 10 + 0 = 3210$

e. $(4 \times 10^3) + (2 \times 10^0) = (4 \times 1000) + (2 \times 1) = 4000 + 2 = 4002$.

(Note that the 100s and 10s places were omitted in the expanded notation, but we were able to obtain the correct result by proceeding in an orderly manner.)

NW ***Now Work Problems 9a and 11d.***

Exercises for Section 7.4

1. Express the following as Babylonian numerals:

   NW **a.** 18 **b.** 23 **c.** 82

   NW **d.** 102 **e.** 349 **f.** 864

2. Express the following as Babylonian numerals:

   a. 34 **b.** 44 **c.** 93

   d. 201 **e.** 423 ***f.** 3674

3. Express the following as Babylonian numerals:

   a. 42 **b.** 58 **c.** 65

   d. 132 **e.** 193 ***f.** 3682

4. Evaluate each Babylonian numeral.

   a. ◀▼ **b.** ◀◀▼▼▼

   c. ◀◀◀▼▼ **d.** ▼◀▼▼

5. Evaluate each Babylonian numeral.

   NW **a.** ▼◀◀◀▼ **b.** ▼▼◀◀▼▼

   c. ▼◀▼ NW **d.** ▼◀◀▼

6. Evaluate each Babylonian numeral.

   a. ◀▼▼▼▼ **b.** ◀◀▼▼▼▼

   c. ◀◀◀▼▼▼ **d.** ▼◀▼▼▼

7. Write each number in expanded notation.

   a. 243 **b.** 378 **c.** 1234

   d. two thousand fifty-one

   e. ten thousand four hundred one

8. Write each number in expanded notation.

   a. 345 **b.** 1776 **c.** 19,876

   d. three thousand four hundred fifty-six

   e. twelve thousand nine hundred three

9. Write each number in expanded notation.

   NW **a.** 402 **b.** 1476 **c.** 20,182

   d. two thousand three hundred eighty-one

   e. ten thousand fifty

10. Write each of the following in base 10 (decimal) notation:

    a. two thousand one

    b. one billion three

    c. forty-five thousand twenty-five

    d. $(3 \times 10^3) + (2 \times 10^1) + (4 \times 10^0)$

    e. $(1 \times 10^4) + (4 \times 10^2) + (2 \times 10^0)$

    f. $(4 \times 10^5) + (2 \times 10^4) + (6 \times 10^3) + (5 \times 10^0)$

11. Write each of the following in base 10 (decimal) notation:

    a. two hundred forty

    b. two thousand three hundred eleven

    c. one thousand seven hundred seventy-six

    NW **d.** $(4 \times 10^3) + (2 \times 10^2) + (1 \times 10^1) + (3 \times 10^0)$

    e. $(2 \times 10^2) + (0 \times 10^1) + (4 \times 10^0)$

    f. $(4 \times 10^4) + (3 \times 10^2) + (1 \times 10^0)$

12. Write each of the following in base 10 (decimal) notation:

    a. three hundred forty-five

    b. forty thousand two

    c. one million one

    d. $(5 \times 10^3) + (3 \times 10^2) + (2 \times 10^1) + (1 \times 10^0)$

    e. $(2 \times 10^4) + (3 \times 10^2) + (4 \times 10^1) + (5 \times 10^0)$

    f. $(7 \times 10^5) + (8 \times 10^3) + (9 \times 10^1)$

Writing Mathematics

13. Define in your own words the concept of place value.

14. Explain the primary difference between the Egyptian and Babylonian systems of numeration. Give an example of each as part of your explanation.

15. Explain the similarities and differences between the Babylonian system of numeration and the decimal system.

16. Explain the similarities and differences between the number 222 from our decimal system and the Babylonian number ▼▼ ▼▼ ▼▼.

Challenge Exercises

17. Name two areas in which the Babylonian system is still in use today.

18. The base 60 appears to have been consciously adopted by the Babylonians because of the number of times the number can be evenly partitioned. Determine the total number of subdivisions of 60 and list them.

19. Explain how the Babylonian number ▼▼ ▼▼ could be interpreted as either 122 or 7202.

Group/Research Activities

20. It was at first thought that the Babylonians did not have any clear way to represent a zero symbol, although they sometimes left a gap where a zero was intended. (This led to some ambiguity, as illustrated in challenge exercise 19.) However, a special zero symbol was indeed invented. Research what this symbol was and how it was written. As part of your research, include the manner in which this symbol was used and why its use still led to ambiguity.

21. In the previous section we indicated that the Egyptians used an oval symbol to represent fractions and that only unit fractions were considered. Research how the Babylonians expressed fractions and contrast their method to that of the Egyptians.

Just for fun

Look at the equation

XI + I = X

Can you make it a correct statement without adding to, crossing out, or changing anything in the equation?

7.5 NUMERATION IN BASES OTHER THAN 10

A dozen roses.

Working in Base 12

It is common practice to group items by 10s. For example, a decade is a period of 10 years. A dime is equal to 10 pennies and 10 dimes make one dollar. A decathlon is a famous Olympic athletic contest in which each contestant participates in 10 events. But it is not uncommon to group items in some other manner. We group things such as socks and mittens by 2s. Another common grouping is by 12s. How do you purchase doughnuts and eggs? We buy items like these by the dozen—that is, in groups of 12. Three dozen doughnuts is three 12s, or 36 doughnuts.

Other common groupings of 12 include 12 inches in a foot, 12 hours in one complete cycle of the clock, and 12 months in a year. Consider the instructor who is ordering supplies and orders a *gross* of chalk. One *gross* is a dozen dozen, or twelve 12s. Therefore, the instructor ordered 144 pieces of chalk. There is also another type of *gross* grouping, called a *great gross:* 1 great gross = 12 gross. Hence, a great gross is a dozen gross or 12×144 or 1728 units. Gross and great gross are common units when ordering items in bulk; usually wholesalers order items in this manner.

Suppose we have $2\frac{1}{2}$ dozen doughnuts. How many doughnuts do we have? We have 2 dozen and $\frac{1}{2}$ of a dozen, or $(2 \times 12) + 6 = 24 + 6 = 30$. Because we are grouping by dozens, we could have written this as two dozen + 6. But we also could have said 2 dozen + six 1s, and this is the same as $(2 \times 12) + (6 \times 1)$. Since any number (except zero) raised to the zero power is 1, we can write this as $(2 \times 12^1) + (6 \times 12^0)$ This expanded notation indicates that we are grouping by 12s. In decimal notation, we grouped by 10s and hence were working in base 10. Now we are grouping by 12s and therefore we can say that we are working in base 12. Hence, we have

$$(2 \times 12^1) + (6 \times 12^0) = 26_{\text{twelve}}$$

When we write numerals in some base other than base 10, we must indicate what base we are working with. We do this by using a subscript. The numeral

$$47_{\text{twelve}}$$

indicates that we are grouping by 12s, and for this particular example we have four 12s and seven 1s—that is,

$$(4 \times 12^1) + (7 \times 12^0) \quad \text{or} \quad (4 \times 12) + (7 \times 1) = 55$$

EXAMPLE 1 ***Expressing Base 12 Numbers as Base 10 Numbers*** Change to base 10 notation.

a. 42_{twelve} **b.** 30_{twelve} **c.** 234_{twelve}

Solution

a. In order to change a numeral to base 10 notation, we first write the numeral in expanded notation:

$$42_{\text{twelve}} = (4 \times 12^1) + (2 \times 12^0) = (4 \times 12) + (2 \times 1)$$
$$= 48 + 2 = 50$$

b. $30_{\text{twelve}} = (3 \times 12^1) + (0 \times 12^0) = (3 \times 12) + (0 \times 1)$

$$= 36 + 0 = 36$$

c. To write numerals in expanded notation, we start from the right (note that this is the units place) and proceed to the left in successive higher powers of the indicated base. Therefore,

$$234_{\text{twelve}} = (2 \times 12^2) + (3 \times 12^1) + (4 \times 12^0)$$
$$= (2 \times 144) + (3 \times 12) + (4 \times 1)$$
$$= 288 + 36 + 4 = 328$$

NW ***Now Work Problem 11b.***

Working in Base 5

Many people believe that the reason we normally group items by 10s is that humans have 10 fingers. But we could just as easily group items by 5s because we have five fingers on one hand. The *base 5 system* is a system of numeration that groups items by 5s. Consider the numeral 13: In base 10, this is one 10 and three 1s, but in base 5, it is two 5s and three 1s. Therefore, $13 = 23_{\text{five}}$. If we are given a numeral in base 5 notation, we can convert it to base 10 by writing the base 5 numeral in expanded notation. Recall that in order to write numerals in expanded notation we start from the right, which is the units place. This represents the zero power of the indicated base. Then we proceed to the left in successive higher powers of the indicated base. Therefore, we have

$$23_{\text{five}} = (2 \times 5^1) + (3 \times 5^0) = (2 \times 5) + (3 \times 1) = 10 + 3 = 13$$

Next, let us convert 342_{five} to base 10 notation:

$$342_{\text{five}} = (3 \times 5^2) + (4 \times 5^1) + (2 \times 5^0)$$
$$= (3 \times 25) + (4 \times 5) + (2 \times 1) = 75 + 20 + 2 = 97$$

Remember that when we want to convert from a given base to base 10, we use expanded notation.

EXAMPLE 2 ***Expressing Base 5 Numbers as Base 10 Numbers*** Write each number in base 10 notation.

a. 40_{five} **b.** 121_{five} **c.** 203_{five}

Solution

a. We first write the numeral in expanded notation, then simplify.

$$40_{\text{five}} = (4 \times 5^1) + (0 \times 5^0) = (4 \times 5) + (0 \times 1)$$
$$= 20 + 0 = 20$$

b. $121_{\text{five}} = (1 \times 5^2) + (2 \times 5^1) + (1 \times 5^0)$

$$= (1 \times 25) + (2 \times 5) + (1 \times 1)$$
$$= 25 + 10 + 1 = 36$$

c. $$\begin{aligned} 203_{\text{five}} &= (2 \times 5^2) + (0 \times 5^1) + (3 \times 5^0) \\ &= (2 \times 25) + (0 \times 5) + (3 \times 1) \\ &= 50 + 0 + 3 = 53 \end{aligned}$$

NW ***Now Work Problem 5b.***

Converting Base 10 Numbers to Other Bases

Thus far we have only considered changing numerals in a given base to base 10 notation. We should also be able to express base 10 numerals in terms of other bases. Suppose we want to express 13 as a numeral in base 5. Because base 5 groups items by 5s, we determine how many 5s are contained in 13. There are two 5s and three 1s. Therefore,

$$13 = 23_{\text{five}}$$

There is a convenient rule that enables us to convert from base 10 to any given base. We can convert any number from base 10 to another base by recording the remainders of successive divisions. We stop dividing when we obtain a quotient of zero. Using the last example, we can illustrate the procedure involved. We wish to convert 13 to base 5, so we divide 13 by 5 and record the remainders.

$$\begin{array}{r|l l} 5 & 13 & \\ 5 & 2 & 3 \\ & 0 & 2 \end{array} \quad \text{Remainders} \uparrow$$

Stop (a zero quotient)

The answer is determined by reading the remainders *from bottom to top*. Therefore, $13 = 23_{\text{five}}$.

Consider the number 97; let's express it with a base 5 numeral. Performing the successive divisions, we have

$$\begin{array}{r|l l} 5 & 97 & \\ 5 & 19 & 2 \\ 5 & 3 & 4 \\ & 0 & 3 \end{array} \uparrow$$

Stop

Reading the remainders from bottom to top, we have

$$97 = 342_{\text{five}}$$

We can check the answer by converting the base 5 numeral to base 10:

$$\begin{aligned} 342_{\text{five}} &= (3 \times 5^2) + (4 \times 5^1) + (2 \times 5^0) \\ &= (3 \times 25) + (4 \times 5) + (2 \times 1) \\ &= 75 + 20 + 2 \\ &= 97 \end{aligned}$$

The answer checks.

EXAMPLE 3 ***Expressing Base 10 Numbers in Base 5*** Express each number with base 5 notation.

a. 43 **b.** 147 **c.** 520

Solution In each case, we perform successive divisions by 5, the new base, and record the remainders for each division. We determine the answer by reading the remainders from bottom to top.

a.

```
5|43
 5|8   3 ↑
  5|1  3 |
    0  1 |
   ↗
Stop
```

$43 = 133_{\text{five}}$

b.

```
5|147
 5|29  2 ↑
  5|5  4 |
  5|1  0 |
    0  1 |
   ↗
Stop
```

$147 = 1042_{\text{five}}$

c.

```
5|520
5|104  0 ↑
 5|20  4 |
  5|4  0 |
    0  4 |
   ↗
Stop
```

$520 = 4040_{\text{five}}$

NW ***Now Work Problem 7b.***

EXAMPLE 4 ***Expressing a Base 10 Number in Base 6 and Base 8*** Change 34 to the indicated base.

a. base 6 **b.** base 8

Solution

a. To express 34 as a numeral in base 6, we divide 34 by 6:

```
6|34
 6|5  4 ↑
   0  5 |
```

Therefore, $34 = 54_{\text{six}}$.

b. To express 34 as a numeral in base 8, we divide 34 by 8:

```
8|34
 8|4  2 ↑
   0  4 |
```

Therefore, $34 = 42_{\text{eight}}$.

EXAMPLE 5 ***Expressing a Base 10 Number in Base 12*** Change the following to base 12 notation:

a. 43 **b.** 100 **c.** 520

Solution In each case, we shall apply our handy rule and perform successive divisions. Remember that the answer is determined by reading the remainders from bottom to top. Since we are converting to base 12, we shall divide by 12.

a.

```
12|43
 12|3  7 ↑
    0  3 |
   ↗
Stop
```

$43 = 37_{\text{twelve}}$

b.

```
12|100
 12|8  4 ↑
    0  8 |
   ↗
Stop
```

$100 = 84_{\text{twelve}}$

c.

```
12|520
12|43  4 ↑
 12|3  7 |
    0  3 |
   ↗
Stop
```

$520 = 374_{\text{twelve}}$

NW ***Now Work Problem 13b.***

The Base 12 System of Numeration

Suppose you buy 22 of something (eggs, doughnuts, or bagels, for example). You have purchased 1 dozen plus 10 more. How can we express this in base 12 notation? We cannot say $22 = 110_{\text{twelve}}$. Why? If we evaluate 110_{twelve}, we have $(1 \times 12^2) + (1 \times 12^1) + (0 \times 12^0) = (1 \times 144) + (1 \times 12) + (0 \times 1) = 144 + 12 + 0 = 156$, which is not 22!

In our decimal system of numeration we use 10 symbols: 0, 1, 2, 3, 4, 5, 6, 7, 8, 9. Consequently we should use 12 symbols in the base 12 numeration system. Because we already have 10 symbols we can borrow from the base 10 system, let us agree to use the additional symbols T and E in the base 12 system of numeration. Let T stand for 10 and E stand for 11. Then the 12 symbols that we shall use in base 12 are 0, 1, 2, 3, 4, 5, 6, 7, 8, 9, T, E. Keep in mind that, in the base 10 system of numeration, 11 represents one 10 and one 1, or eleven. But in the base 12 system of numeration, T represents 10, and 11_{twelve} represents one 12 and one 1, or 13.

EXAMPLE 6 ***Expressing a Base 12 Number in Base 10 Notation*** Change to base 10 notation.

a. 40_{twelve} **b.** $4T_{\text{twelve}}$ **c.** $ET2_{\text{twelve}}$

Solution

a.
$$\begin{aligned} 40_{\text{twelve}} &= (4 \times 12^1) + (0 \times 12^0) = (4 \times 12) + (0 \times 1) \\ &= 48 + 0 = 48 \end{aligned}$$

b.
$$\begin{aligned} 4T_{\text{twelve}} &= (4 \times 12^1) + (T \times 12^0) = (4 \times 12) + (10 \times 1) \\ &= 48 + 10 = 58 \end{aligned}$$

c.
$$\begin{aligned} ET2_{\text{twelve}} &= (E \times 12^2) + (T \times 12^1) + (2 \times 12^0) \\ &= (E \times 144) + (T \times 12) + (2 \times 1) \\ &= (11 \times 144) + (10 \times 12) + (2 \times 1) \\ &= 1584 + 120 + 2 \\ &= 1706 \end{aligned}$$

NW ***Now Work Problem 17b.***

Converting Numbers to/from Bases Other Than 10

The base 12 system of numeration is a fairly common system. It is also called the **duodecimal system** of numeration, which indicates that it is a system of numeration with 12 as its base, as opposed to the decimal system of numeration, which has 10 as its base.

So far all of our conversions have involved changing a base 10 numeral to an equivalent numeral in a base other than 10 and vice versa. Let's now examine how to convert numerals from one base to another where neither base is base 10. To do so, it is best to convert first to base 10 and then reconvert this result to the desired base. For example, suppose we want to convert 45_{six} to base 8. First we convert 45_{six} to base 10 using expanded notation:

$$\begin{aligned} 45_{\text{six}} &= (4 \times 6^1) + (5 \times 6^0) \\ &= (4 \times 6) + (5 \times 1) \\ &= 24 + 5 \\ &= 29 \end{aligned}$$

Now we convert 29 to base 8 by performing successive divisions:

$$\begin{array}{r|l} 8 & 29 \\ \hline 8 & 3 \quad 5 \uparrow \\ \hline & 0 \quad 3 \end{array}$$

Therefore, $45_{\text{six}} = 35_{\text{eight}}$.

EXAMPLE 7 ***Converting between Bases 5 and 6*** Convert 354_6 to base 5.

Note: 354_6 is the same as 354_{six}.

Solution First we convert 354_6 to base 10:

$$\begin{aligned} 354_6 &= (3 \times 6^2) + (5 \times 6^1) + (4 \times 6^0) \\ &= (3 \times 36) + (5 \times 6) + (4 \times 1) \\ &= 108 + 30 + 4 = 142 \end{aligned}$$

Now we have $354_6 = 142$, and we can convert 142 to a base 5 numeral:

$$\begin{array}{r|l l} 5 & 142 & \\ 5 & 28 & 2 \uparrow \\ 5 & 5 & 3 \\ 5 & 1 & 0 \\ & 0 & 1 \end{array}$$

Therefore, $354_6 = 1032_5$.

Check:

$$\begin{aligned} 354_6 &= (3 \times 6^2) + (5 \times 6^1) + (4 \times 6^0) \\ &= (3 \times 36) + (5 \times 6) + (4 \times 1) \\ &= 108 + 30 + 4 = 142 \\ 1032_5 &= (1 \times 5^3) + (0 \times 5^2) + (3 \times 5^1) + (2 \times 5^0) \\ &= (1 \times 125) + (0 \times 25) + (3 \times 5) + (2 \times 1) \\ &= 125 + 0 + 15 + 2 \\ &= 142 \end{aligned}$$

NW *Now Work Problem 25b.*

EXAMPLE 8 ***Converting between Bases 4 and 6*** Which is greater, 123_4 or 45_6?

Solution We will convert each of the numerals to base 10 and compare the results:

$$\begin{aligned} 123_4 &= (1 \times 4^2) + (2 \times 4^1) + (3 \times 4^0) \\ &= (1 \times 16) + (2 \times 4) + (3 \times 1) \\ &= 16 + 8 + 3 = 27 \end{aligned}$$

$$\begin{aligned} 45_6 &= (4 \times 6^1) + (5 \times 6^0) \\ &= (4 \times 6) + (5 \times 1) \\ &= 24 + 5 = 29 \end{aligned}$$

Because $45_6 = 29$ and $123_4 = 27$, $45_6 > 123_4$.

Exercises for Section 7.5

1. In each example, items are grouped by 12. Perform the indicated operations.

a.
```
  4 years 7 months
+ 2 years 9 months
```

b.
```
  7 years 3 months
- 4 years 10 months
```

c.
```
  13 feet 11 inches
+ 11 feet 10 inches
```

d.
```
  14 feet 6 inches
- 10 feet 7 inches
```

e.
```
  2 gross 3 dozen 8 units
+ 4 gross 11 dozen 6 units
```

f.
5 gross 9 dozen 3 units
− 2 gross 11 dozen 7 units

2. In each example, items are grouped by 12. Perform the indicated operations. (*Note:* 1 great gross = 12 gross.)

a.
3 years 9 months
+ 2 years 7 months

b.
8 years 2 months
− 3 years 9 months

c.
3 gross 3 dozen 9 units
+ 5 gross 8 dozen 4 units

d.
7 gross 2 dozen 5 units
− 3 gross 9 dozen 7 units

e.
3 great gross 2 gross 4 dozen
+ 4 great gross 10 gross 8 dozen

f.
3 great gross 2 gross 4 dozen
− 1 great gross 3 gross 7 dozen

3. In each example, items are grouped by 12. Perform the indicated operations. (*Note:* 1 great gross = 12 gross.)

a.
4 feet 7 inches
+ 5 feet 8 inches

b.
6 feet 2 inches
− 3 feet 9 inches

c.
2 years 4 months
+ 3 years 11 months

d.
8 years 5 months
− 7 years 7 months

e.
4 gross 10 dozen 8 units
+ 9 gross 5 dozen 7 units

f.
4 great gross
− 2 gross 3 dozen 2 units

4. Change to base 10 notation.
a. 22_{five} **b.** 11_{five} **c.** 24_{five}
d. 103_{five} **e.** 143_{five} **f.** 2041_{five}

5. Change to base 10 notation.
a. 13_{five} NW **b.** 44_{five} **c.** 231_{five}
d. 304_{five} **e.** 100_{five} **f.** 4021_{five}

6. Change to base 10 notation.
a. 32_{five} **b.** 33_{five} **c.** 142_{five}
d. 201_{five} **e.** 4001_{five} **f.** 2302_{five}

7. Change to base 5 notation.
a. 6 NW **b.** 19 **c.** 38
d. 3 **e.** 121 **f.** 497

8. Change to base 5 notation.
a. 9 **b.** 27 **c.** 4
d. 243 ***e.** 2003 ***f.** 3421

9. Change to base 5 notation.
a. 60 **b.** 2 **c.** 11
d. 48 **e.** 101 ***f.** 1001

10. Change to base 10 notation.
a. 36_{twelve} **b.** 24_{twelve} **c.** 61_{twelve}
d. 126_{twelve} **e.** 343_{twelve} **f.** 1005_{twelve}

11. Change to base 10 notation.
a. 40_{twelve} NW **b.** 54_{twelve} **c.** 99_{twelve}
d. 137_{twelve} **e.** 243_{twelve} **f.** 2001_{twelve}

12. Change to base 10 notation.
a. 13_{twelve} **b.** 27_{twelve} **c.** 44_{twelve}
d. 105_{twelve} **e.** 133_{twelve} **f.** 1011_{twelve}

13. Change to base 12 notation.
a. 42 NW **b.** 53 **c.** 60
d. 137 **e.** 234 **f.** 876

14. Change to base 12 notation.
a. 49 **b.** 66 **c.** 118
d. 341 ***e.** 3421 ***f.** 5736

15. Change to base 12 notation.
a. 43 **b.** 61 **c.** 112
d. 201 **e.** 432 ***f.** 4101

16. Change to base 10 notation.
a. $2E_{twelve}$ **b.** TE_{twelve}
c. $T2E_{twelve}$ **d.** $E2T_{twelve}$

17. Change to base 10 notation.
a. $T3_{twelve}$ NW **b.** ET_{twelve}
c. $3E4_{twelve}$ **d.** $TE5_{twelve}$

18. Change to base 10 notation.
a. $3E_{twelve}$ **b.** $E3_{twelve}$
c. $1ET_{twelve}$ **d.** $1TE_{twelve}$

19. Change to base 10 notation.
a. 47_{nine} **b.** 52_{nine} **c.** 34_{six}
d. 52_{six} **e.** 25_{seven} **f.** 462_{seven}

20. Change to base 10 notation.
a. 23_{four} **b.** 201_{four} **c.** 53_{eight}
d. 612_{eight} **e.** 110_{three} **f.** 102_{three}

21. Change to base 9 notation.
a. 34 **b.** 60 **c.** 102
d. 135 **e.** 234 **f.** 716

22. Change to base 4 notation.
a. 21 **b.** 65 **c.** 83
d. 137 **e.** 260 **f.** 301

23. An instructor ordered the following classroom supplies: a gross of pencils, a great gross of chalk (1 great gross is 12 gross), 3 dozen notebooks, $\frac{1}{2}$ dozen pens, and one grade book. What is the total number of items ordered?

24. Jake, the bagel maker, prepared the following order: 6 dozen salt bagels, 3 dozen poppy bagels, 6 dozen dozen rye bagels, a gross of plain bagels, and a gross of onion bagels. How many bagels did Jake make?

25. Convert the given numeral to base 5.
    a. 47_{twelve} NW b. 51_{six} c. 246_{seven}
    d. 302_{four} e. 567_{eight} f. 835_{nine}

26. Determine which numeral is the larger for each given pair.
    a. 38_{nine} and 42_{eight} b. 26_{seven} and 25_{eight}
    c. 122_{three} and 17_{eight} d. 555_{six} and 443_{seven}
    e. 91_{eleven} and 1023_{four} f. 1001_{twelve} and 12000_{six}

Writing Mathematics

27. Interpret in your own words what the notation 10_{twelve} means.

28. Explain why you think number bases other than base 10 were invented and used throughout history.

29. Explain why it is necessary to specify two new symbols to write numerals in base 12.

30. Describe two ways in which to convert a base x numeral to a base y numeral, where neither x nor y is base 10.

Challenge Exercises

31. In the text, we used the letters T and E to denote 10 and 11, respectively, in base 12. Other notations besides T and E have also been suggested. For example, we could let X denote 10 and H denote 11 in base 12. On what basis do you think X and H are appropriate and informative?

32. Construct addition and multiplication tables for base 12.

33. Which numeration system lends itself more easily to division: base 10 or base 12? Why?

34. Explain the following reasoning process: Given 232_{four} and 314_{five}. Since $232_{\text{four}} < 1000_{\text{four}} = 64_{\text{ten}}$ and since $314_{\text{five}} > 300_{\text{five}} = 75_{\text{ten}}$, we now have $232_{\text{four}} < 64_{\text{ten}} < 75_{\text{ten}} < 314_{\text{five}}$. Thus, $232_{\text{four}} < 314_{\text{five}}$.

Group/Research Activities

35. Another system of numeration not discussed in the book is the *Mayan system*, which was a base 20 system. Research this number system and prepare a report on its history. As part of your report include a brief overview of the Mayan culture, the speculation why 20 was selected as the grouping number, the symbols the Mayans used to represent numbers, and the role zero played in this system of numeration.

36. The issue of selecting symbols to denote 10 and 11 in base 12 is not new. For example, in the 1800s. Sir Isaac Pitman suggested that an upside down 2 (i.e., ↊) and a backward 3 (i.e., ↋) be used to denote 10 and 11, respectively. Research the rationale for selecting these symbols. As part of your research include a brief biography of Sir Isaac Pitman, who was well known in the secretarial field. (*Suggestion:* A good starting point, if the link is still available, is to go to the Web site **http://www.dsgb.orbix.co.uk/basics.html.**)

37. The duodecimal system has a zealous following. There is even a formal society called *The Dozenal Society of Great Britain*, which actively promotes a change from base 10 to base 12. Go to the Web site **http://www.dsgb.orbix.co.uk** to learn more about this movement. Prepare a report and discuss this with your classmates.

Just for fun

Following are some units of measure that are probably not familiar to you. See if you can find equivalent measures (for example, 1 yard = 3 feet).

1 fathom = ? (A fathom is used in measuring depths at sea.)

1 hand = ? (A hand is used in measuring the height of a horse.)

3 barleycorns = ? (A barleycorn is used by shoe manufacturers in measuring the length of a foot.)

7.6 BASE 5 ARITHMETIC

Historical Note

The earliest known use of coins occurred around 550 B.C. Before this time, merchants traded their wares for other items and services they needed. Next, they began to sell their merchandise for silver and copper coins. The significance of this is that the invention of "money" made it necessary for people to know more about counting and working with numbers.

Thus far we have converted base 10 numerals to equivalent numerals in bases other than 10, including base 12 and base 5. We have also reversed this procedure. That is, we have converted non–base 10 numerals to base 10. As part of these conversion processes, we also learned to convert numerals from one base to another where neither numeral is in base 10. In a manner of speaking, we have learned to count in some bases other than base 10.

The next step is to perform some arithmetic operations in a base system other than base 10. A convenient base to work with is base 5. Using conventional Hindu–Arabic numerals, the base 5 system involves the numerals 0, 1, 2, 3, 4. Recall that the numeral 14_{five} is composed of one 5 and four 1s; hence, $14_{\text{five}} = 9$ in base 10. For the sake of convenience, we shall now write all numerals in base five with the numeral 5 as a subscript. Therefore, 14_{five} is the same as 14_5. Recall that if no subscript is written—that is, if no base is indicated—then it is understood that the numeral is written in base 10 notation.

Base 5 Addition

Now suppose we want to perform the following addition: $14_5 + 24_5$. At first glance, you might want to say that the answer is 38_5. But 38_5 is an impossible answer! Why? For one reason, the base 5 system of numeration uses only digits 0, 1, 2, 3, 4, and therefore we cannot have an 8 in our answer.

Let us try again. One way to find an answer to the problem $14_5 + 24_5$ is to convert each of the base 5 numerals to base 10, add the base 10 numerals, and then convert the answer back to base 5.

$$14_5 = 9 \quad \text{and} \quad 24_5 = 14$$
$$9 + 14 = 23 \quad \text{and} \quad 23 = 43_5$$

Therefore,

$$14_5 + 24_5 = 43_5$$

This may seem like the best method, but it isn't—especially if we want to add three- and four-digit numerals such as 1223_5 and 4223_5. One thing that would help us to add numerals in base 5 is a table of addition facts. Table 7.5 is a table of addition facts for base 5.

TABLE 7.5 Base 5 Addition Table

+	0_5	1_5	2_5	3_5	4_5
0_5	0_5	1_5	2_5	3_5	4_5
1_5	1_5	2_5	3_5	4_5	10_5
2_5	2_5	3_5	4_5	10_5	11_5
3_5	3_5	4_5	10_5	11_5	12_5
4_5	4_5	10_5	11_5	12_5	13_5

The entries in this table may seem odd at first. Let us check a few of them so we can see that they do make sense. How does $3_5 + 4_5 = 12_5$? If we add 3 + 4 in base 10, we get 7. Since 7 is composed of one 5 and two 1s, $7 = 12_5$, and so $3_5 + 4_5 = 12_5$. Similarly, $4_5 + 4_5 = 13_5$, because 4 + 4 = 8 in base 10, and 8 is composed of one 5 and three 1s; hence, $8 = 13_5$. Let us solve an addition problem:

$$\begin{array}{r} 23_5 \\ +\ 34_5 \\ \hline \end{array}$$

We start with the units place: $3_5 + 4_5 = 12_5$. Writing the 2 and carrying the 1, we have

$$\begin{array}{r} {}^{1} \\ 23_5 \\ +\ 34_5 \\ \hline 2_5 \end{array}$$

Next we add the 5s: $1_5 + 2_5 + 3_5 = 3_5 + 3_5 = 11_5$. Therefore, the completed problem is

$$\begin{array}{r} 23_5 \\ +\ 34_5 \\ \hline 112_5 \end{array}$$

We can check our work in base 10:

$$\begin{aligned} 23_5 &= (2 \times 5^1) + (3 \times 5^0) = (2 \times 5) + (3 \times 1) = 10 + 3 = 13 \\ +\ 34_5 &= (3 \times 5^1) + (4 \times 5^0) = (3 \times 5) + (4 \times 1) = 15 + 4 = \underline{19} \\ 112_5 &= (1 \times 5^2) + (1 \times 5^1) + (2 \times 5^0) \\ &= (1 \times 25) + (1 \times 5) + (2 \times 1) = 25 + 5 + 2 \qquad = 32 \end{aligned}$$

Let us try the preceding problem without using the addition table. In order to add 23_5 and 34_5, we can combine $3 + 4$ as we normally would; that is, $3 + 4 = 7$. Remember that the numerals 0, 1, 2, 3, 4 have the same meaning in base 5 as they do in base 10. Our only problem is the 7, but 7 is composed of one 5 and two 1s, and therefore $3_5 + 4_5 = 12_5$. We write the 2 and carry the 1 as before:

$$\begin{array}{r} {}^{1} \\ 23_5 \\ +\ 34_5 \\ \hline 2_5 \end{array}$$

We now add $1 + 2 + 3 = 6$, but in base 5 notation $6 = 11_5$ (one 5 and one 1). Hence, $23_5 + 34_5 = 112_5$.

EXAMPLE 1 ***Adding in Base 5*** Find the sum of 234_5 and 341_5.

Solution Starting with the units place, $4 + 1 = 5$, but in base 5, $5 = 10_5$ (one 5 and no 1s). Placing the 0 and carrying the 1, we have $1 + 3 + 4 = 8$; but in base 5, $8 = 13_5$ (one 5 and three 1s). Placing the 3 and carrying the 1, we now have $1 + 2 + 3 = 6$, and in base 5, $6 = 11_5$ (one 5 and one 1). Our addition is complete:

$$\begin{array}{r} {}^{11} \\ 234_5 \\ +\ 341_5 \\ \hline 1130_5 \end{array}$$

EXAMPLE 2 ***Adding in Base 5*** Find the sum of 133_5 and 341_5.

Solution Starting with the units place, $3 + 1 = 4$. This is the same for base 5 as for base 10, because the numerals 0, 1, 2, 3, 4 have the same meaning in base 5 as they do in base 10. Next we proceed to the 5s place: $3 + 4 = 7$, but in base 5, $7 = 12_5$ (one 5 and two 1s). Placing the 2 and carrying the 1, we now have $1 + 1 + 3 = 5$, and in base 5, $5 = 10_5$ (one 5 and no 1s). Our addition is complete:

$$\begin{array}{r} {}^{1} \\ 133_5 \\ +\ 341_5 \\ \hline 1024_5 \end{array}$$

EXAMPLE 3 ***Adding in Base 5*** Find the sum of 342_5 and 324_5.

Solution

$$\begin{array}{r} {\scriptstyle 1\,1} \\ 342_5 \\ +\ 324_5 \\ \hline 1221_5 \end{array}$$

Check:

$$\begin{aligned} 342_5 &= (3 \times 5^2) + (4 \times 5^1) + (2 \times 5^0) \\ &= (3 \times 25) + (4 \times 5) + (2 \times 1) = 75 + 20 + 2 = 97 \\ +\ 324_5 &= (3 \times 5^2) + (2 \times 5^1) + (4 \times 5^0) \\ &= (3 \times 25) + (2 \times 5) + (4 \times 1) = 75 + 10 + 4 = \underline{89} \\ \overline{1221_5} &= (1 \times 5^3) + (2 \times 5^2) + (2 \times 5^1) + (1 \times 5^0) \\ &= (1 \times 125) + (2 \times 25) + (2 \times 5) + (1 \times 1) \\ &= 125 + 50 + 10 + 1 \qquad = 186 \end{aligned}$$

NW ***Now Work Problems 1a and 1d.***

Base 5 Subtraction

Another arithmetic operation that goes hand in hand with addition is the operation of subtraction. Before we try a subtraction problem in base 5, let's review subtraction in base 10. Suppose we wish to subtract 248 from 735—that is, 735 − 248. We set the problem up as shown. Note that the parts of the problem have been labeled.

$$\begin{array}{rl} 735 & \textit{minuend} \\ -\ 248 & \textit{subtrahend} \\ \hline \end{array}$$

The answer is usually called the *difference*, but it is also sometimes called the *remainder*. Performing the subtraction, we have

$$\begin{array}{rrrr} & {\scriptstyle 6} & {\scriptstyle 12} & {\scriptstyle 15} \\ & 7 & 3 & 5 \\ - & 2 & 4 & 8 \\ \hline & 4 & 8 & 7 \end{array}$$

Notice that we must rename three 10s, five 1s as two 10s, fifteen 1s. Seven 100s, two 10s are then renamed as six 100s, twelve 10s. The numbers in small type over the minuend are shown only to indicate the new arrangement of the number in order to facilitate the subtraction.

Now let us try a subtraction problem in base 5. Consider the following:

$$\begin{array}{r} 42_5 \\ -\ 13_5 \\ \hline \end{array}$$

We note that we cannot subtract 3 from 2, so we must borrow from the 4. But what do we borrow? Since we are in base 5, we borrow one 5. So we now have one 5 and two 1s in the units place—that is, 12_5. But this is the same as 7, and 3 from 7 is 4. Now we have only 3 in the 5s place, and 1 from 3 is 2. This process is illustrated next.

$$\begin{array}{rrr} & {\scriptstyle 3} & {\scriptstyle 1} \\ & \not{4} & 2_5 \\ - & 1 & 3_5 \\ \hline & 2 & 4_5 \end{array}$$

Remember to indicate the base with which you are working in your answer. If no subscript is written, then it is understood that the numeral is written in base 10 notation.

Let us check our answer to the preceding problem by translating it into base 10:

$$\begin{array}{rl} 42_5 = (4 \times 5^1) + (2 \times 5^0) = (4 \times 5) + (2 \times 1) = 20 + 2 = & 22 \\ -\ 13_5 = (1 \times 5^1) + (3 \times 5^0) = (1 \times 5) + (3 \times 1) = 5 + 3 = & -\ 8 \\ \hline 24_5 = (2 \times 5^1) + (4 \times 5^0) = (2 \times 5) + (4 \times 1) = 10 + 4 = & 14 \end{array}$$

Since $22 - 8 = 14$, the answer checks.

Now let us try another problem. Consider

$$\begin{array}{r} 431_5 \\ -\ 132_5 \\ \hline \end{array}$$

Since we cannot subtract 2 from 1, we borrow 1 from the 3 in the 5s place, leaving a 2 in the 5s place and giving us 11_5 in the units place. We know 11_5 is the same as 6, and 2 from 6 is 4. So far we have

$$\begin{array}{rrr} & \overset{2}{\cancel{3}} & \overset{1}{1_5} \\ 4 & & \\ -\ 1 & 3 & 2_5 \\ \hline & & 4_5 \end{array} \qquad 11_5 = 6 \quad \text{so} \quad 11_5 - 2_5 = 4_5$$

We cannot subtract 3 from 2 in the 5s place, so again we borrow. This time we borrow 1 from the 4 in the next place (the 25s). This means we have borrowed one 25, or five 5s, which gives us a total of seven 5s, and three 5s from seven 5s is four 5s. We can also think of this as subtracting 3_5 from 12_5, or 3 from 7, which gives us a 4 in the 5s place. Now we have only 3 in the 25s place, and 1 from 3 is 2. The completed subtraction process is

$$\begin{array}{rrr} \overset{3}{\cancel{4}} & \overset{12}{\cancel{3}} & \overset{1}{1_5} \\ -\ 1 & 3 & 2_5 \\ \hline 2 & 4 & 4_5 \end{array} \qquad 12_5 = 7 \quad \text{so} \quad 12_5 - 3_5 = 4_5$$

Table 7.5, the base 5 addition table, may be helpful to you in subtraction problems. Remember that when you borrow, you are borrowing a number in the indicated base, not a 10. In these examples we are borrowing 5s and 25s. For problems in other bases, you may be borrowing 7s, 4s, and so on.

EXAMPLE 4 ***Subtracting in Base 5*** Subtract 24_5 from 33_5.

Solution Since we cannot subtract 4 from 3, we borrow 1 from the 5s place, which gives us one 5 and three 1s in the units place, or $13_5 = 8$; hence, 4_5 from 13_5 is 4_5. We are left with a 2 in the 5s place, and 2 from 2 is zero:

$$\begin{array}{rr} \overset{2}{\cancel{3}} & \overset{1}{3_5} \\ -\ 2 & 4_5 \\ \hline & 4_5 \end{array} \quad \text{or} \quad \begin{array}{r} 33_5 \\ -\ 24_5 \\ \hline 4_5 \end{array}$$

EXAMPLE 5 ***Subtracting in Base 5*** Subtract 234_5 from 433_5.

Solution We cannot subtract 4 from 3, so we borrow 1 from the 5s place. This gives one 5 and three 1s in the units place, or $13_5 = 8$, and 4 from 8 is 4. In the 5s place, we cannot subtract 3 from 2, so we borrow 1 from the 4 in the 25s place.

Historical Note

John Nash

One afternoon at Princeton University, John von Neumann (see p. 250) was visiting the common room in the mathematics department. While there he noticed two students humped over a game board (see Figure 7.10); he asked a colleague what they were playing and the colleague replied, "Nash."

Originally, Piet Hein invented this game a few years earlier and John Nash created it independently of Hein (Parker Bros. now holds the rights to it and it is called Hex). The players take turns placing stones, white or black, to make a connected path from black to black if you have the black stones or white to white for the white stones. The first player to make a connected path wins. Nash proved that on an n by n Hex board, the player with the first move can always win.

In 1950, at age 22, John Nash was awarded his Ph.D. in mathematics from Princeton University for a 27-page thesis titled *Non-Cooperative Games*. Several years later, this rising star of mathematics began to suffer from schizophrenia, which impaired his mathematical research for three decades. Remarkably, his mental illness went into remission, and in 1994 he (along with Reinhard Selten and John Harsanyi) won the Nobel Prize in economics for his 1950 dissertation. In that dissertation, Nash laid the foundation for what is now called Nash equilibrium and Nash bargaining. These concepts are used is situations like owner–player negotiations in major league sports. You can read more about John Nash in the book *A Beautiful Mind*, by Sylvia Nasar, or watch the movie by the same title.

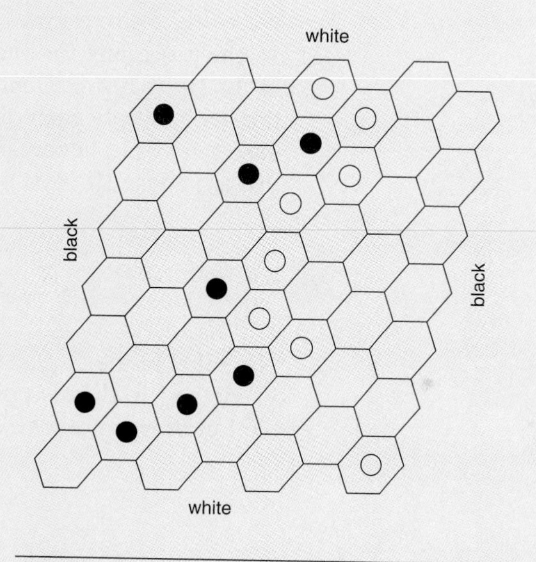

Figure 7.10

One 25 and two 5s gives us seven 5s, and three 5s from seven 5s is four 5s. We are left with a 3 in the 25s place, and 2 from 3 is 1:

$$\begin{array}{r} \overset{3}{\not4}\ \overset{12}{\not3}\ \overset{1}{3}_5 \\ -\ 2\ \ 3\ \ 4_5 \\ \hline 1\ \ 4\ \ 4_5 \end{array} \quad \text{or} \quad \begin{array}{r} 433_5 \\ -\ 234_5 \\ \hline 144_5 \end{array}$$

NW ***Now Work Problems 5a and 5d.***

Base 5 Multiplication

We shall also examine the operation of multiplication in base 5. The procedure for multiplication in base 5 (or in any other base) is the same as that in base 10. Table 7.6 is a table of multiplication facts for multiplying numerals in base 5.

We see that $3_5 \times 3_5 = 14_5$ because $3 \times 3 = 9$ and 9 is equal to one 5 and four 1s. Similarly, $4_5 \times 3_5 = 22_5$; $4 \times 3 = 12$, but 12 is two 5s and two 1s.

We must master single-digit multiplication before we can proceed to other examples. To help in an example such as 4_5 times 4_5, just remember that $4 \times 4 = 16$, and 16 equals three 5s and one 1; therefore $4_5 \times 4_5 = 31_5$.

TABLE 7.6 Base 5 Multiplication Table

×	0_5	1_5	2_5	3_5	4_5
0_5	0_5	0_5	0_5	0_5	0_5
1_5	0_5	1_5	2_5	3_5	4_5
2_5	0_5	2_5	4_5	11_5	13_5
3_5	0_5	3_5	11_5	14_5	22_5
4_5	0_5	4_5	13_5	22_5	31_5

Consider the following multiplication problem:

$$\begin{array}{r} 21_5 \\ \times \quad 31_5 \\ \hline 21_5 \\ 113_5 \\ \hline 1201_5 \end{array}$$

As the procedure for multiplication is the same for any base, we first multiply 21_5 by 1_5: $21_5 \times 1_5 = 21_5$. Now we multiply by the next digit, that is, $21_5 \times 3_5$. To do this, we multiply each digit of 21_5 by 3_5: $3 \times 1 = 3$ and $3 \times 2 = 6$. However, in base 5, $6 = 11_5$; hence, $21_5 \times 3_5 = 113_5$. Note that the partial product is indented just as in base 10. Next we find the sum of the partial products:

$$\begin{array}{r} 21_5 \\ +\ 113_5 \\ \hline 1201_5 \end{array}$$

Note that in the 5s place $2 + 3 = 5$, but in base 5, $5 = 10_5$, so we write the 0 and carry the 1 to the next place, adding it to the 1 already there.

Let us try another example. Consider

$$\begin{array}{r} 433_5 \\ \times \quad 2_5 \\ \hline \end{array}$$

One way to do this problem is shown next. Note that partial products are used here, and each one is indented to indicate the powers of 5 involved:

$$\begin{array}{r} 433_5 \\ \times \quad 2_5 \\ \hline 11_5 \\ 11_5 \\ 13_5 \\ \hline 1421_5 \end{array}$$

We can do this problem in another way, but we will have to do some work mentally. The solution to the problem could have appeared as

$$\begin{array}{r} 433_5 \\ \times \quad 2_5 \\ \hline 1421_5 \end{array}$$

To do the multiplication in the shortened form, we first multiply the 3 in the units place by 2: $2 \times 3 = 6 = 11_5$. Therefore, we write 1 and carry 1. In the 5s place, we again have 2×3, which equals 11_5, but we carried 1 from the units place, so we have $11_5 + 1_5 = 12_5$. Hence, we write the 2 and carry the 1. In the 25s place, we have 2×4, which equals 13_5, and we carried a 1 from the 5s place, and so $13_5 + 1_5 = 14_5$ and our multiplication is complete.

EXAMPLE 6 ***Multiplying in Base 5*** Find the product

$$\begin{array}{r} 342_5 \\ \times \quad 23_5 \\ \hline \end{array}$$

Solution First, we multiply 342_5 by 3_5: $3 \times 2 = 6$, but $6 = 11_5$, so we write 1 and carry 1. We have $3 \times 4 = 12 = 22_5$, and because we carried a 1, we have $22_5 + 1_5 = 23_5$. We write the 3 and carry the 2. Now, $3 \times 3 = 9 = 14_5$, and because we carried a 2, we have $14_5 + 2_5 = 21_5$. Therefore, our partial product $(342_5 \times 3_5)$ appears in the problem as

$$\begin{array}{r} 342_5 \\ \times \quad 23_5 \\ \hline 2131_5 \end{array}$$

Now we are ready to multiply by the next digit, 2: $2 \times 2 = 4 = 4_5$; we write the 4 and proceed. We have $2 \times 4 = 8$, but $8 = 13_5$, so we write the 3 and carry the 1: $2 \times 3 = 6 = 11_5$ and because we carried a 1, we have $11_5 + 1_5 = 12_5$. Now the problem looks like

$$\begin{array}{r} 342_5 \\ \times \quad 23_5 \\ \hline 2131_5 \\ 1234_5 \\ \hline \end{array}$$

Note that the partial product is indented to indicate the powers of 5 involved. Next, we find the sum of the partial products, and the completed problem is

$$\begin{array}{r} 342_5 \\ \times \quad 23_5 \\ \hline 2131_5 \\ 1234_5 \\ \hline 20021_5 \end{array}$$

EXAMPLE 7 ***Multiplying in Base 5*** Multiply 44_5 by 23_5.

Solution According to Table 7.6, $3_5 \times 4_5 = 22_5$; we write 2 and carry 2. Again, we have $3_5 \times 4_5 = 22_5$ and because we carried a 2, we have $22_5 + 2_5 = 24_5$. Therefore, the partial product is 242_5. We find the next partial product as follows: $2_5 \times 4_5 = 13_5$, so we write the 3 and carry the 1; again, we have $2_5 \times 4_5 = 13_5$, and because we carried a 1, we have $13_5 + 1_5 = 14_5$. This partial product is 143_5. We next find the sum of the partial products and obtain the final result, 2222_5:

$$\begin{array}{r} 44_5 \\ \times \quad 23_5 \\ \hline 242_5 \\ 143_5 \\ \hline 2222_5 \end{array}$$

NW ***Now Work Problems 7a and 7d.***

Base 5 Division

A true test of understanding the process of computation in base 5 is the operation of division. We divide in base 5 in the same way that we divide in base 10. In fact, we can think in base 10, but we must write our computation and answer in base 5. Consider the problem 123_5 divided by 2_5. We first set up this problem just as we would in base 10, but remember that we are working in base 5. Therefore, we have

$$2_5\overline{)123_5}$$

At first glance, we might want to say that 2 divides 12 six times; but this is not the case. We are dividing 2_5 into 12_5. Besides, we cannot have a 6 for an answer because base 5 uses only the numerals 0, 1, 2, 3, 4. Since we are dividing 2_5 into 12_5, we can think of this as dividing 2 into $7 = 12_5$. Two divides 7 three times, and $2_5 \times 3_5 = 11_5$. Thus, we have

$$\begin{array}{r} 3 \\ 2_5\overline{)123_5} \\ \underline{11_5} \end{array}$$

Next, we subtract and bring down the next digit. The problem now appears as

$$\begin{array}{r} 3 \\ 2_5\overline{)123_5} \\ \underline{11_5} \\ 13_5 \end{array}$$

Now we must divide 2_5 into 13_5. We can think of this as dividing 2 into 8, which equals 4, and $2_5 \times 4_5 = 13_5$. We have completed the division, and there is no remainder. The completed problem and check appear as follows:

$$\begin{array}{r} 34_5 \leftarrow \text{Quotient} \\ \text{Divisor} \rightarrow 2_5\overline{)123_5} \leftarrow \text{Dividend} \\ \underline{11_5} \\ 13_5 \\ \underline{13_5} \\ 0_5 \end{array}$$

Check:

$$\begin{array}{rl} 34_5 & \longleftarrow \text{Quotient} \\ \underline{\times \quad 2_5} & \longleftarrow \text{Divisor} \\ 123_5 & \longleftarrow \text{Dividend} \end{array}$$

EXAMPLE 8 ***Dividing in Base 5*** Divide 141_5 by 2_5.

Solution First we divide 2_5 into 14_5 (think of this as 2 into 9), which equals 4. As we do in base 10, we now multiply $4_5 \times 2_5 = 13_5$; we then subtract this from 14_5 in the dividend. We bring down the next digit, 1. Now we must divide 2_5 into 11_5 (think of this as 2 into 6), which equals 3, and $3_5 \times 2_5 = 11_5$. We have completed the division, and there is no remainder:

$$\begin{array}{r} 43_5 \\ 2_5\overline{)141_5} \\ \underline{13_5} \\ 11_5 \\ \underline{11_5} \\ 0_5 \end{array}$$

$$\text{Recall that} \begin{cases} 1_5 \times 2_5 = 2_5 \\ 2_5 \times 2_5 = 4_5 \\ 3_5 \times 2_5 = 11_5 \\ 4_5 \times 2_5 = 13_5 \end{cases}$$

EXAMPLE 9 ***Dividing in Base 5*** Divide 234_5 by 4_5.

Solution Dividing 4_5 into 23_5 (think of this as 4 into 13), we get 3. Multiplying $3_5 \times 4_5$, we obtain 22_5, and subtract this from 23_5 in the dividend. After bringing down the next digit, we divide 4_5 into 14_5 (think of this as 4 into 9), which equals 2, and $2_5 \times 4_5 = 13_5$. Subtracting this from 14_5, we obtain a remainder of 1_5, and the division is complete:

$$\begin{array}{r} 32_5 \\ 4_5\overline{)234_5} \\ \underline{22_5} \\ 14_5 \\ \underline{13_5} \\ 1_5 \end{array} \quad \text{remainder: } 1_5$$

NW ***Now Work Problems 11a and 11e.***

Remember that we can always check our work in division. In order to check, we multiply the quotient by the divisor and add the remainder, if any, to the resulting product. If the result is equal to the dividend, then the work is correct.

Exercises for Section 7.6

1. Perform the following additions in base 5. Check your work by converting to base 10.

NW **a.** $\begin{array}{r} 13_5 \\ +\ 23_5 \\ \hline \end{array}$ **b.** $\begin{array}{r} 14_5 \\ +\ 22_5 \\ \hline \end{array}$ **c.** $\begin{array}{r} 23_5 \\ +\ 32_5 \\ \hline \end{array}$

NW **d.** $\begin{array}{r} 123_5 \\ +\ 124_5 \\ \hline \end{array}$ **e.** $\begin{array}{r} 231_5 \\ +\ 222_5 \\ \hline \end{array}$ **f.** $\begin{array}{r} 343_5 \\ +\ 112_5 \\ \hline \end{array}$

2. Perform the following additions in base 5. Check your work by converting to base 10.

a. $\begin{array}{r} 23_5 \\ +\ 23_5 \\ \hline \end{array}$ **b.** $\begin{array}{r} 32_5 \\ +\ 24_5 \\ \hline \end{array}$ **c.** $\begin{array}{r} 12_5 \\ +\ 34_5 \\ \hline \end{array}$

d. $\begin{array}{r} 1123_5 \\ +\ 2132_5 \\ \hline \end{array}$ **e.** $\begin{array}{r} 4312_5 \\ +\ 1144_5 \\ \hline \end{array}$ **f.** $\begin{array}{r} 3243_5 \\ +\ 2122_5 \\ \hline \end{array}$

3. Perform the following additions in base 5. Check your work by converting to base 10.

a. $\begin{array}{r} 31_5 \\ +\ 32_5 \\ \hline \end{array}$ **b.** $\begin{array}{r} 41_5 \\ +\ 44_5 \\ \hline \end{array}$ **c.** $\begin{array}{r} 32_5 \\ +\ 14_5 \\ \hline \end{array}$

d. $\begin{array}{r} 134_5 \\ +\ 122_5 \\ \hline \end{array}$ **e.** $\begin{array}{r} 342_5 \\ +\ 234_5 \\ \hline \end{array}$ **f.** $\begin{array}{r} 1234_5 \\ +\ 1441_5 \\ \hline \end{array}$

4. Perform the following subtractions in base 5. Check your work by converting to base 10.

a. $\begin{array}{r} 23_5 \\ -\ 14_5 \\ \hline \end{array}$ **b.** $\begin{array}{r} 22_5 \\ -\ 13_5 \\ \hline \end{array}$ **c.** $\begin{array}{r} 12_5 \\ -\ 4_5 \\ \hline \end{array}$

d. $\begin{array}{r} 321_5 \\ -\ 231_5 \\ \hline \end{array}$ **e.** $\begin{array}{r} 231_5 \\ -\ 132_5 \\ \hline \end{array}$ **f.** $\begin{array}{r} 411_5 \\ -\ 122_5 \\ \hline \end{array}$

5. Perform the following subtractions in base 5. Check your work by converting to base 10.

NW **a.** $\begin{array}{r} 32_5 \\ -\ 23_5 \\ \hline \end{array}$ **b.** $\begin{array}{r} 22_5 \\ -\ 14_5 \\ \hline \end{array}$ **c.** $\begin{array}{r} 11_5 \\ -\ 3_5 \\ \hline \end{array}$

NW **d.** $\begin{array}{r} 434_5 \\ -\ 332_5 \\ \hline \end{array}$ **e.** $\begin{array}{r} 4211_5 \\ -\ 1232_5 \\ \hline \end{array}$ **f.** $\begin{array}{r} 3212_5 \\ -\ 2233_5 \\ \hline \end{array}$

6. Perform the following subtractions in base 5. Check your work by converting to base 10.

a. $\begin{array}{r} 33_5 \\ -\ 24_5 \\ \hline \end{array}$ **b.** $\begin{array}{r} 41_5 \\ -\ 12_5 \\ \hline \end{array}$ **c.** $\begin{array}{r} 13_5 \\ -\ 4_5 \\ \hline \end{array}$

d. $\begin{array}{r} 341_5 \\ -\ 43_5 \\ \hline \end{array}$ **e.** $\begin{array}{r} 421_5 \\ -\ 122_5 \\ \hline \end{array}$ **f.** $\begin{array}{r} 4121_5 \\ -\ 1213_5 \\ \hline \end{array}$

7. Perform the following multiplications in base 5. Check your work by converting to base 10.

NW **a.** $\begin{array}{r} 231_5 \\ \times\ 3_5 \\ \hline \end{array}$ **b.** $\begin{array}{r} 432_5 \\ \times\ 2_5 \\ \hline \end{array}$ **c.** $\begin{array}{r} 432_5 \\ \times\ 4_5 \\ \hline \end{array}$

NW **d.** $\begin{array}{r} 231_5 \\ \times\ 21_5 \\ \hline \end{array}$ **e.** $\begin{array}{r} 324_5 \\ \times\ 23_5 \\ \hline \end{array}$ **f.** $\begin{array}{r} 432_5 \\ \times\ 34_5 \\ \hline \end{array}$

8. Perform the following multiplications in base 5. Check your work by converting to base 10.

a. $\begin{array}{r} 242_5 \\ \times\ 3_5 \\ \hline \end{array}$ **b.** $\begin{array}{r} 423_5 \\ \times\ 2_5 \\ \hline \end{array}$ **c.** $\begin{array}{r} 221_5 \\ \times\ 21_5 \\ \hline \end{array}$

d. $\begin{array}{r} 232_5 \\ \times\ 41_5 \\ \hline \end{array}$ **e.** $\begin{array}{r} 434_5 \\ \times\ 24_5 \\ \hline \end{array}$ **f.** $\begin{array}{r} 434_5 \\ \times\ 342_5 \\ \hline \end{array}$

9. Perform the following multiplications in base 5. Check your work by converting to base 10.

a. $\begin{array}{r} 342_5 \\ \times\ 2_5 \\ \hline \end{array}$ **b.** $\begin{array}{r} 123_5 \\ \times\ 3_5 \\ \hline \end{array}$ **c.** $\begin{array}{r} 121_5 \\ \times\ 31_5 \\ \hline \end{array}$

d. $\begin{array}{r} 312_5 \\ \times\ 13_5 \\ \hline \end{array}$ **e.** $\begin{array}{r} 434_5 \\ \times\ 23_5 \\ \hline \end{array}$ **f.** $\begin{array}{r} 132_5 \\ \times\ 213_5 \\ \hline \end{array}$

10. Perform the following divisions in base 5. Check your work by converting to base 10.

a. $2_5\overline{)11_5}$ b. $3_5\overline{)22_5}$ c. $4_5\overline{)31_5}$

d. $2_5\overline{)32_5}$ e. $3_5\overline{)212_5}$ f. $4_5\overline{)103_5}$

11. Perform the following divisions in base 5. Check your work by converting to base 10.

NW a. $2_5\overline{)124_5}$ b. $3_5\overline{)343_5}$ c. $4_5\overline{)342_5}$

d. $3_5\overline{)1234_5}$ NW e. $11_5\overline{)243_5}$ f. $13_5\overline{)341_5}$

12. Perform the following divisions in base 5. Check your work by converting to base 10.

a. $2_5\overline{)13_5}$ b. $3_5\overline{)14_5}$ c. $4_5\overline{)103_5}$

d. $3_5\overline{)113_5}$ e. $12_5\overline{)223_5}$ f. $14_5\overline{)311_5}$

Writing Mathematics

13. Given the addition problem $23_5 + 14_5$, explain how "carrying" is performed in base 5.

14. Given the subtraction problem $23_5 - 14_5$, explain how "borrowing" is performed in base 5.

15. The product of 4_5 and $3_5 = 22_5$. Explain how this result is obtained and what it means.

16. Defend or refute the following statement: When performing base 5 arithmetic, instead of working exclusively in base 5, it is much simpler to convert everything to base 10 first, do the arithmetic, and then reconvert the answer back to base 5.

Challenge

17. Perform each operation in the indicated base.

a. $23_6 + 34_6$ b. $352_7 + 405_7$

c. $41_7 - 23_7$ d. $241_6 - 42_6$

18. Perform each operation in the indicated base.

a. $21_4 + 33_4$ b. $35_6 + 42_6$

c. $32_4 - 13_4$ d. $44_6 - 25_6$

19. Perform each operation in the indicated base.

a. $32_7 \times 43_7$ b. $24_6 \times 43_6$

c. $46_8 \times 23_8$ d. $56_9 \times 34_9$

20. Perform each operation in the indicated base.

a. $121_3 + 122_3$ b. $234_5 + 141_5$

c. $441_6 - 212_6$ d. $325_7 - 146_7$

21. Perform each operation in the indicated base.

a. $212_3 \times 22_3$ b. $424_6 \times 32_6$

c. $423_5 \times 22_5$ d. $325_7 \times 46_7$

Group/Research Activity

22. To introduce the base 5 number system to a child, try experimenting with quarters (Q), nickels (N), and pennies (P). First create a four-column table. The first three columns (working from left to right) are labeled Q, N, P, and the last column (i.e., rightmost) is labeled Cents, which will be the sum of the three columns for each row. Next, give the child 4 quarters, 4 nickels, and 4 pennies and have him or her represent varying amounts up to 124 cents. Each amount should be recorded as a separate row in the table. For example, 55 cents is equivalent to $2_Q + 1_N + 0_P$.

Just for fun

What three words in the English language are pronounced the same as the numeral 4?

7.7 THE LANGUAGE OF COMPUTERS: BINARY AND HEXADECIMAL BASES

First Computer

For over 30 years many people had thought that the first computer ever built was the specialized one (for cryptanalysis) called Colossus built in England around 1943 and, in the United States, the ENIAC (electronic numerical integrator and computer) built in 1946, the world's first general-purpose electronic computer. However, the world's first electronic digital computer was invented in 1939 by John V. Atanasoff. He wanted to essentially test two of his ideas: (1) Can capacitors store data in binary form, and (2) can you create electronic logic circuits to perform addition and subtraction? He was able to succeed in his objectives and his computer was the first machine to manipulate binary numbers through electronic means.

John V. Atanasoff was born in Hamilton, New York on October 4, 1903. He earned his Ph.D. in theoretical physics and was a professor of mathematics and physics at Iowa State University when he and his graduate assistant, Clifford E. Berry, set to work on the electronic computer. However, with the advent of World War II, Atanasoff abandoned his research and went to work at the Naval Ordnance Laboratory, Washington, D.C. In 1981 he received the Computer Pioneer Medal and in 1990 he received the National Medal of Technology. More information can be found out about Atanasoff in two books, *The First Electronic Computer: The Atanasoff Story*, and *Atanasoff: Forgotten Father of the Computer*. He died at the age of 91 on June 15, 1995 in Frederick, Maryland.

In previous sections of this chapter, we examined several different place-value systems, including the decimal number system (i.e., base 10) and systems in bases other than 10. In each case we learned that when working in a particular base, we group items relative to that base. For example, in base 10, we group items by 10s; in base 5, we group items by 5s; and in group 12, we group items by 12s. We also learned that regardless of the place-value system under discussion, we can always change a base 10 numeral to a numeral in a different base by successive divisions, and we can convert a non–base 10 numeral to base 10 by expanded notation. In this section we extend these concepts to base 2 and base 16 systems of numeration. The base 2 system is called the *binary number system* and numbers represented in the binary system are said to be in **binary notation.** The base 16 system is called the *hexadecimal* (or *hex* for short) *number system* and numbers represented in the hexadecimal system are said to be in **hexadecimal notation**. The binary and hex systems are important because, as we will see, they represent the language of computers.

The Binary System: Base 2

The binary number system uses only two symbols, 0 and 1, and groups items by 2s. From a place-value perspective this means that when working from right to left each succeeding position is two times greater in numerical value than the previous position. For example, the binary numeral 10_2 represents 1 *two* and 0 *ones*. Writing this in expanded notation, we have

$$10_2 = (1 \times 2^1) + (0 \times 2^0) = (1 \times 2) + (0 \times 1) = 2 + 0 = 2$$

Therefore, $2 = 10_2$. How would we express 3 in base 2? Three is composed of 1 *two* and 1 *one*; hence, $3 = 11_2$.

As another illustration, consider the binary numeral 1011_2. It represents 1 *eight*, 0 *fours*, 1 *two*, and 1 *one*. That is,

Eights	*Fours*	*Twos*	*Ones*
1	0	1	1

Once again, writing this in expanded notation, we have

$$\begin{aligned} 1011_2 &= (1 \times 2^3) + (0 \times 2^2) + (1 \times 2^1) + (1 \times 2^0) \\ &= (1 \times 8) + (0 \times 4) + (1 \times 2) + (1 \times 1) \\ &= 8 + 0 + 2 + 1 \\ &= 11 \end{aligned}$$

Therefore, $11 = 1011_2$. Be careful not to confuse these numerals. The numeral 11 without a base subscript implies a base 10 number and is read *eleven*. The numeral 1011_2 denotes a base 2 number and is read *one zero one one*.

To convert a base 10 numeral to base 2, we perform successive divisions by 2, the new base, and record the remainders for each separate division. The answer is determined by reading the remainders from the bottom to the top.

To express 7 as a base 2 numeral, we perform the division in the following manner:

$$\begin{array}{r|l l} 2 & 7 & \\ 2 & 3 & 1 \uparrow \\ 2 & 1 & 1 \\ & 0 & 1 \end{array} \qquad 7 = 111_2$$

We can check our answer by converting 111_2 back to base 10. We can do this by writing 111_2 in expanded notation:

$$\begin{aligned} 111_2 &= (1 \times 2^2) + (1 \times 2^1) + (1 \times 2^0) \\ &= (1 \times 4) + (1 \times 2) + (1 \times 1) \\ &= 4 + 2 + 1 \\ &= 7 \end{aligned}$$

Once again, note that because we are now working in base 2, each place value is a power of 2:

$$2^0, \quad 2^1, \quad 2^2, \quad 2^3, \quad \text{and so on}$$

Computers perform all of their internal calculations in base 2. The results are converted to base 10 for display on the monitor screen.

The binary system is unique in that it uses only two symbols to represent any number. This is important for computers because they are made of electrical circuits. Each of these circuits, like a light switch, has only two possible positions or "states": The circuit is either on or off. If a circuit is on (i.e., electrical current is flowing), then it represents the numeral 1; if it is off (i.e., no current is flowing), it represents the numeral 0. The nature of the binary system effectively doubles the number of possible circuit states whenever a new circuit is incorporated into a computer's design. For example,

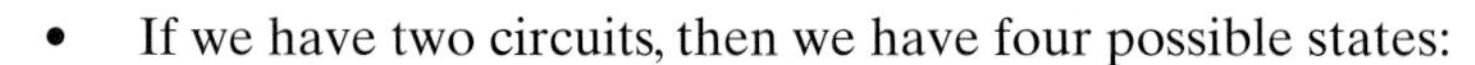

- If we have two circuits, then we have four possible states:

  off-off on-off
  off-on on-on

- If we have three circuits, then we have eight possible states:

  off-off-off on-off-off
  off-off-on on-off-on
  off-on-off on-on-off
  off-on-on on-on-on

Four circuits provide us with 16 possible states; five circuits provide us with 32 possible states; six circuits provide 64 possible states; and so forth. This is an extremely powerful concept. Using only these two symbols, a computer can perform billions of calculations per second.

Note of Interest

Computers are used for inventory and price control. The Universal Product Code was adopted by the supermarket industry in 1973. It is a system that identifies each item sold in the stores with a unique 12-digit code. A scanner reads the bar code and the computer registers the price and notes that the item has been purchased. The code for a book such as this one gives country and publisher (the first six digits) and the identifying number for the particular title.

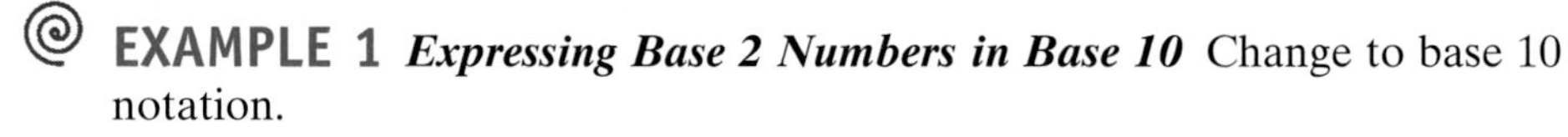

EXAMPLE 1 ***Expressing Base 2 Numbers in Base 10*** Change to base 10 notation.

a. 101_2 **b.** 1101_2

Solution

a. In order to change a numeral to base 10 notation, we must write the numeral in expanded notation:

$$\begin{aligned} 101_2 &= (1 \times 2^2) + (0 \times 2^1) + (1 \times 2^0) \\ &= (1 \times 4) + (0 \times 2) + (1 \times 1) \\ &= 4 + 0 + 1 \\ &= 5 \end{aligned}$$

b.
$$\begin{aligned} 1101_2 &= (1 \times 2^3) + (1 \times 2^2) + (0 \times 2^1) + (1 \times 2^0) \\ &= (1 \times 8) + (1 \times 4) + (0 \times 2) + (1 \times 1) \\ &= 8 + 4 + 0 + 1 \\ &= 13 \end{aligned}$$

 **EXAMPLE 2** ***Expressing Base 10 Numbers in Base 2*** Change to base 2 notation.

a. 9 **b.** 15

Solution We perform successive divisions by 2 and record the remainders for each division. The answer is determined by reading the remainders from bottom to top.

a.

	Quotient	Remainder
2	15	
2	7	1 ↑
2	3	1
2	1	1
	0	1

b.

	Quotient	Remainder
2	9	
2	4	1 ↑
2	2	0
2	1	0
	0	1

$9 = 1001_2 \qquad 15 = 1111_2$ 

NW ***Now Work Problems 1e and 3d.***

Base 2 Addition and Subtraction

TABLE 7.7 Base 2 Addition Table

+	0_2	1_2
0_2	0_2	1_2
1_2	1_2	10_2

In order to perform addition in base 2, we must use the four addition facts listed in Table 7.7. We must also remember the "carrying" process. Consider the following addition problem:

$$\begin{array}{r} 111_2 \\ +\ 110_2 \\ \hline \end{array}$$

We start with the units place: $0 + 1 = 1$, and this is the answer for both base 2 and base 10, because the numerals 0 and 1 have the same meaning in base 2 and base 10. The sum of 0 and 1 is 1, and there is nothing to carry. We proceed to the 2s place: $1 + 1 = 2$, but in base 2, $2 = 10_2$. We write the 0 and carry the 1. Next we have $1 + 1 + 1$, which equals 3; in base 2, $3 = 11_2$ (one 2 and one 1). The addition is complete:

$$\begin{array}{r} {\scriptstyle 1} \\ 111_2 \\ +\ 110_2 \\ \hline 1101_2 \end{array}$$

 **EXAMPLE 3** ***Adding in Base 2*** Find the sum of 1010_2 and 1011_2.

Solution

$$\begin{array}{r} {\scriptstyle 1} \\ 1010_2 \\ +\ 1011_2 \\ \hline 10101_2 \end{array}$$

Check:

$$\begin{aligned} 1010_2 &= (1 \times 2^3) + (0 \times 2^2) + (1 \times 2^1) + (0 \times 2^0) \\ &= (1 \times 8) + (0 \times 4) + (1 \times 2) + (0 \times 1) \\ &= 8 + 0 + 2 + 0 && = 10 \\ +\ 1011_2 &= (1 \times 2^3) + (0 \times 2^2) + (1 \times 2^1) + (1 \times 2^0) \\ &= (1 \times 8) + (0 \times 4) + (1 \times 2) + (1 \times 1) \\ &= 8 + 0 + 2 + 1 && = 11 \\ \hline 10101_2 &= (1 \times 2^4) + (0 \times 2^3) + (1 \times 2^2) + (0 \times 2^1) + (1 \times 2^0) \\ &= (1 \times 16) + (0 \times 8) + (1 \times 4) + (0 \times 2) + (1 \times 1) \\ &= 16 + 0 + 4 + 0 + 1 && = 21 \end{aligned}$$

EXAMPLE 4 ***Subtracting in Base 2*** Subtract 111_2 from 1011_2.

Solution In the units place and 2s place, we subtract 1 from 1 and obtain 0. But in the 4s place, we cannot subtract 1 from 0. Therefore, we borrow 1 from the 8s place. One 8 gives us two 4s, and one 4 from two 4s is one 4. The subtraction is complete:

$$\begin{array}{r} \not{1}011_2 \\ -\ 111_2 \\ \hline 100_2 \end{array}$$

NW ***Now Work Problems 7d and 11c.***

Base 2 Multiplication

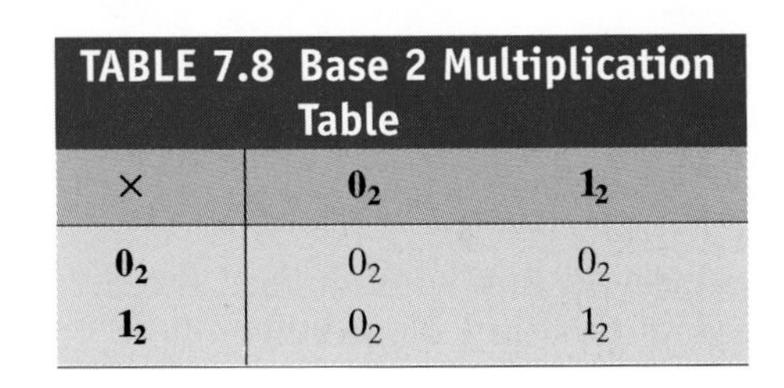

TABLE 7.8 Base 2 Multiplication Table

×	0_2	1_2
0_2	0_2	0_2
1_2	0_2	1_2

Multiplication in base 2 does not present much of a problem, provided we can add the partial products, because we have to remember only the four multiplication facts listed in Table 7.8.

$$0 \times 0 = 0, \quad 0 \times 1 = 0, \quad 1 \times 0 = 0, \quad \text{and} \quad 1 \times 1 = 1$$

Consider the following multiplication problem:

$$\begin{array}{r} 101_2 \\ \times\ 11_2 \\ \hline 101_2 \\ 101_2 \\ \hline 1111_2 \end{array}$$

Because the procedure for multiplication is the same for any base, we first multiply 101_2 by 1_2, which equals 101_2. Now we multiply by the next digit, thus giving us $101_2 \times 1_2$, which equals 101_2. Note that the partial product is indented, as in base 10. Next we find the sum of the partial products. For this example, we do not have to carry anything.

EXAMPLE 5 ***Multiplying in Base 2*** Multiply 110_2 by 11_2.

Solution

$$\begin{array}{r} 110_2 \\ \times\ 11_2 \\ \hline 110_2 \\ 110_2 \\ \hline 10010_2 \end{array}$$

The only problem that occurs here is in the adding of the partial products:

$$\begin{array}{r} 110_2 \\ +\ 110_2 \\ \hline 10010_2 \end{array}$$

Note that in the 4s place we have $1 + 1$, which equals 10_2, so we place the 0 and carry the 1. This again gives $1 + 1$, which is 10_2. Since our addition is completed, we write down this sum to give us the final answer, 10010_2.

EXAMPLE 6 ***Multiplying in Base 2*** Multiply 101_2 by 101_2.

Solution The only difference between the following two solutions is that the first solution indicates the multiplication by 0 in the second partial product, whereas the second solution actually shows the multiplication by 0:

$$\begin{array}{r} 101_2 \\ \times\ 101_2 \\ \hline 101_2 \\ 1010_2 \\ \hline 11001_2 \end{array} \quad \text{or} \quad \begin{array}{r} 101_2 \\ \times\ 101_2 \\ \hline 101_2 \\ 000_2 \\ 101_2 \\ \hline 11001_2 \end{array}$$

NW ***Now Work Problem 13c.***

The Hexadecimal System: Base 16

Although binary notation is appropriate for computer use, numbers expressed in binary form are extremely cumbersome to work with from a human perspective. In other words, binary numerals are not very human friendly. For example, consider the 32-digit binary numeral $01011000111110011011011100101100_2$. Can you look at this numeral and copy it down from memory without making a mistake? Suppose you need to compare this to the numeral $00100011011101100001010100010101_2$. Can you easily distinguish between the two of them and identify which positions are different?

To avoid making errors when working with binary numerals, computer programmers usually use the *hexadecimal* system. Since the hex system is a base 16 system of numeration, this implies that it contains 16 symbols. However, because we only have ten symbols (0–9) in the base 10 system, we must agree on six new symbols to represent 10, 11, 12, 13, 14, and 15. These new symbols are, respectively, A, B, C, D, E, and F. To convert between base 2 and base 16, we could do what

Math Connections

Binary Magic

In the base 2 system of numeration (also called the binary number system), only the numerals 0 and 1 are used. To express a decimal number (i.e., a base 10 number) as a binary number, we can use a base 2 place-value system in which each place-value position is two times greater than the position on its right. For example,

128 64 32 16 8 4 2 1

Using this place-value system, we can easily convert numbers from base 10 to base 2. To do this, we select a pattern of 0s and 1s such that when each digit is multiplied by its respective place value, the sum of these products is equal to the base 10 number. For example, the base 10 number 67 is equivalent to the binary number 1000011 because when we multiply each binary digit by its respective base 2 place value, the sum of the products is equal to 67.

___	1	0	0	0	0	1	1	← Base 2
128	64	32	16	8	4	2	1	← Base 10

Following are the binary representations of the first 15 counting numbers in base two.

1: 1_2 **2:** 10_2 **3:** 11_2
4: 100_2 **5:** 101_2 **6:** 110_2
7: 111_2 **8:** 1000_2 **9:** 1001_2
10: 1010_2 **11:** 1011_2 **12:** 1100_2
13: 1101_2 **14:** 1110_2 **15:** 1111_2

We can now use these representations in a mathematical puzzle. Following are four rows of numbers ranging from 1 to 15. The first row consists of only those numbers whose binary representation has a 1 as its rightmost digit. The second row consists of only those numbers whose binary representation has a 1 as its second rightmost digit. The third row consists of only those numbers whose binary representation has a 1 as its third rightmost digit. The fourth row consists of only those numbers whose binary representation has a 1 as its fourth rightmost digit.

Row 1: 1, 3, 5, 7, 9, 11, 13, 15
Row 2: 2, 3, 6, 7, 10, 11, 14, 15
Row 3: 4, 5, 6, 7, 12, 13, 14, 15
Row 4: 8, 9, 10, 11, 12, 13, 14, 15

Ask someone to pick a number from 1 to 15 and tell you only the rows that contain the number. Add the first number from each of these rows and this is the person's number. For example, if we choose 7, then we would identify that our number is located in rows 1, 2, and 3. If you now add the first digits in each of these rows, you get $1 + 2 + 4 = 7$.

we did in Section 7.5—namely, first convert one numeral to base 10 and then reconvert this result to the other base. Instead of doing this, though, let's see how we can convert directly between the two bases.

To change a base 2 numeral directly to base 16, we observe that the relationship between base 2 and base 16 is $2^4 = 16$. This implies that four binary digits are needed to represent a single base 16 symbol. Thus, converting from base 2 to base 16 involves assigning a group of four binary numerals to a corresponding hexadecimal symbol. This correspondence is shown in Table 7.9. Note that the base 10 equivalences shown in Table 7.9 can be confirmed by changing each binary grouping to base 10 by expanded notation.

TABLE 7.9

Base 10 Equivalence	Binary Grouping	Hex Symbol
0	0000	0
1	0001	1
2	0010	2
3	0011	3
4	0100	4
5	0101	5
6	0110	6
7	0111	7
8	1000	8
9	1001	9
10	1010	A
11	1011	B
12	1100	C
13	1101	D
14	1110	E
15	1111	F

We can now change $01011000111110011011011100101100_2$ directly to base 16 by first marking off the binary digits in groups of four (working from right to left) and then using Table 7.9 to assign each grouping a corresponding hex symbol.

$$\begin{array}{cccccccc} 0101, & 1000, & 1111, & 1001, & 1011, & 0111, & 0010, & 1100 \\ \mathbf{5} & \mathbf{8} & \mathbf{F} & \mathbf{9} & \mathbf{B} & \mathbf{7} & \mathbf{2} & \mathbf{C} \end{array}$$

Thus, $01011000111110011011011100101100_2 = 58F9B72C_{16}$. This representation is much easier to work with than binary notation.

EXAMPLE 7 ***Converting Numbers from Base 2 to Base 16*** Convert each base 2 numeral to base 16 directly.

a. 001001011101_2 **b.** 11111001010_2

Solution

a. We first mark off groups of four binary digits working from right to left and then we assign each group a corresponding hexadecimal symbol.

$$\begin{array}{ccc} 0010, & 0101, & 1101_2 \\ 2 & 5 & D \end{array}$$

Therefore, $001001011101_2 = 25D_{16}$.

b. In this example, note that when we mark off our base 2 numeral in groups of four, the last group only contains three binary digits. To make this last group a "four-group" we simply "pad" the left side of the group with a 0. This does not affect the numerical value of the group or the overall base 2 numeral.

"Padded" 0

↓

$0111, \quad 1100, \quad 1010_2$

$7 \quad\quad C \quad\quad A$

Therefore, $11111001010_2 = 7CA_{16}$.

EXAMPLE 8 ***Converting Numbers from Base 16 to Base 2*** Convert each base 16 numeral to base 2 directly.

a. ABC_{16} **b.** $45D_{16}$

Solution

a. Change each base 16 symbol to its corresponding base 2 form using Table 7.9.

$$\begin{array}{ccc} A & B & C \\ = 1010 & 1011 & 1100 \end{array}$$

Therefore, $ABC_{16} = 101010111100_2$.

b. Using Table 7.9, we see that $45D_{16} = 010001011101_2$.

Base 16 Addition and Multiplication

Adding and multiplying base 16 numerals is similar to adding or multiplying in any other base. We first need to construct hexadecimal addition and multiplication tables. These are shown in Tables 7.10 and 7.11.

EXAMPLE 9 ***Adding in Base 16*** Do the indicated additions.

a. $3C2_{16} + B6_{16}$ **b.** $A19F_{16} + BEA_{16}$

TABLE 7.10 Base 16 Addition Table

+	0_{16}	1_{16}	2_{16}	3_{16}	4_{16}	5_{16}	6_{16}	7_{16}	8_{16}	9_{16}	A_{16}	B_{16}	C_{16}	D_{16}	E_{16}	F_{16}
0_{16}	0_{16}	1_{16}	2_{16}	3_{16}	4_{16}	5_{16}	6_{16}	7_{16}	8_{16}	9_{16}	A_{16}	B_{16}	C_{16}	D_{16}	E_{16}	F_{16}
1_{16}	1_{16}	2_{16}	3_{16}	4_{16}	5_{16}	6_{16}	7_{16}	8_{16}	9_{16}	A_{16}	B_{16}	C_{16}	D_{16}	E_{16}	F_{16}	10_{16}
2_{16}	2_{16}	3_{16}	4_{16}	5_{16}	6_{16}	7_{16}	8_{16}	9_{16}	A_{16}	B_{16}	C_{16}	D_{16}	E_{16}	F_{16}	10_{16}	11_{16}
3_{16}	3_{16}	4_{16}	5_{16}	6_{16}	7_{16}	8_{16}	9_{16}	A_{16}	B_{16}	C_{16}	D_{16}	E_{16}	F_{16}	10_{16}	11_{16}	12_{16}
4_{16}	4_{16}	5_{16}	6_{16}	7_{16}	8_{16}	9_{16}	A_{16}	B_{16}	C_{16}	D_{16}	E_{16}	F_{16}	10_{16}	11_{16}	12_{16}	13_{16}
5_{16}	5_{16}	6_{16}	7_{16}	8_{16}	9_{16}	A_{16}	B_{16}	C_{16}	D_{16}	E_{16}	F_{16}	10_{16}	11_{16}	12_{16}	13_{16}	14_{16}
6_{16}	6_{16}	7_{16}	8_{16}	9_{16}	A_{16}	B_{16}	C_{16}	D_{16}	E_{16}	F_{16}	10_{16}	11_{16}	12_{16}	13_{16}	14_{16}	15_{16}
7_{16}	7_{16}	8_{16}	9_{16}	A_{16}	B_{16}	C_{16}	D_{16}	E_{16}	F_{16}	10_{16}	11_{16}	12_{16}	13_{16}	14_{16}	15_{16}	16_{16}
8_{16}	8_{16}	9_{16}	A_{16}	B_{16}	C_{16}	D_{16}	E_{16}	F_{16}	10_{16}	11_{16}	12_{16}	13_{16}	14_{16}	15_{16}	16_{16}	17_{16}
9_{16}	9_{16}	A_{16}	B_{16}	C_{16}	D_{16}	E_{16}	F_{16}	10_{16}	11_{16}	12_{16}	13_{16}	14_{16}	15_{16}	16_{16}	17_{16}	18_{16}
A_{16}	A_{16}	B_{16}	C_{16}	D_{16}	E_{16}	F_{16}	10_{16}	11_{16}	12_{16}	13_{16}	14_{16}	15_{16}	16_{16}	17_{16}	18_{16}	19_{16}
B_{16}	B_{16}	C_{16}	D_{16}	E_{16}	F_{16}	10_{16}	11_{16}	12_{16}	13_{16}	14_{16}	15_{16}	16_{16}	17_{16}	18_{16}	19_{16}	$1A_{16}$
C_{16}	C_{16}	D_{16}	E_{16}	F_{16}	10_{16}	11_{16}	12_{16}	13_{16}	14_{16}	15_{16}	16_{16}	17_{16}	18_{16}	19_{16}	$1A_{16}$	$1B_{16}$
D_{16}	D_{16}	E_{16}	F_{16}	10_{16}	11_{16}	12_{16}	13_{16}	14_{16}	15_{16}	16_{16}	17_{16}	18_{16}	19_{16}	$1A_{16}$	$1B_{16}$	$1C_{16}$
E_{16}	E_{16}	F_{16}	10_{16}	11_{16}	12_{16}	13_{16}	14_{16}	15_{16}	16_{16}	17_{16}	18_{16}	19_{16}	$1A_{16}$	$1B_{16}$	$1C_{16}$	$1D_{16}$
F_{16}	F_{16}	10_{16}	11_{16}	12_{16}	13_{16}	14_{16}	15_{16}	16_{16}	17_{16}	18_{16}	19_{16}	$1A_{16}$	$1B_{16}$	$1C_{16}$	$1D_{16}$	$1E_{16}$

TABLE 7.11 Base 16 Multiplication Table

×	0_{16}	1_{16}	2_{16}	3_{16}	4_{16}	5_{16}	6_{16}	7_{16}	8_{16}	9_{16}	A_{16}	B_{16}	C_{16}	D_{16}	E_{16}	F_{16}
0_{16}	0_{16}	0_{16}	0_{16}	0_{16}	0_{16}	0_{16}	0_{16}	0_{16}	0_{16}	0_{16}	0_{16}	0_{16}	0_{16}	0_{16}	0_{16}	0_{16}
1_{16}	0_{16}	1_{16}	2_{16}	3_{16}	4_{16}	5_{16}	6_{16}	7_{16}	8_{16}	9_{16}	A_{16}	B_{16}	C_{16}	D_{16}	E_{16}	F_{16}
2_{16}	0_{16}	2_{16}	4_{16}	6_{16}	8_{16}	A_{16}	C_{16}	E_{16}	10_{16}	12_{16}	14_{16}	16_{16}	18_{16}	$1A_{16}$	$1C_{16}$	$1E_{16}$
3_{16}	0_{16}	3_{16}	6_{16}	9_{16}	C_{16}	F_{16}	12_{16}	15_{16}	18_{16}	$1B_{16}$	$1E_{16}$	21_{16}	24_{16}	27_{16}	$2A_{16}$	$2D_{16}$
4_{16}	0_{16}	4_{16}	8_{16}	C_{16}	10_{16}	14_{16}	18_{16}	$1C_{16}$	20_{16}	24_{16}	28_{16}	$2C_{16}$	30_{16}	34_{16}	38_{16}	$3C_{16}$
5_{16}	0_{16}	5_{16}	A_{16}	F_{16}	14_{16}	19_{16}	$1E_{16}$	23_{16}	28_{16}	$2D_{16}$	32_{16}	37_{16}	$3C_{16}$	41_{16}	46_{16}	$4B_{16}$
6_{16}	0_{16}	6_{16}	C_{16}	12_{16}	18_{16}	$1E_{16}$	24_{16}	$2A_{16}$	30_{16}	36_{16}	$3C_{16}$	42_{16}	48_{16}	$4E_{16}$	54_{16}	$5A_{16}$
7_{16}	0_{16}	7_{16}	E_{16}	15_{16}	$1C_{16}$	23_{16}	$2A_{16}$	31_{16}	38_{16}	$3F_{16}$	46_{16}	$4D_{16}$	54_{16}	$5B_{16}$	62_{16}	69_{16}
8_{16}	0_{16}	8_{16}	10_{16}	18_{16}	20_{16}	28_{16}	30_{16}	38_{16}	40_{16}	48_{16}	50_{16}	58_{16}	60_{16}	68_{16}	70_{16}	78_{16}
9_{16}	0_{16}	9_{16}	12_{16}	$1B_{16}$	24_{16}	$2D_{16}$	36_{16}	$3F_{16}$	48_{16}	51_{16}	$5A_{16}$	63_{16}	$6C_{16}$	75_{16}	$7E_{16}$	87_{16}
A_{16}	0_{16}	A_{16}	14_{16}	$1E_{16}$	28_{16}	32_{16}	$3C_{16}$	46_{16}	50_{16}	$5A_{16}$	64_{16}	$6E_{16}$	78_{16}	82_{16}	$8C_{16}$	96_{16}
B_{16}	0_{16}	B_{16}	16_{16}	21_{16}	$2C_{16}$	37_{16}	42_{16}	$4D_{16}$	58_{16}	63_{16}	$6E_{16}$	79_{16}	84_{16}	$8F_{16}$	$9A_{16}$	$A5_{16}$
C_{16}	0_{16}	C_{16}	18_{16}	24_{16}	30_{16}	$3C_{16}$	48_{16}	54_{16}	60_{16}	$6C_{16}$	78_{16}	84_{16}	90_{16}	$9C_{16}$	$A8_{16}$	$B4_{16}$
D_{16}	0_{16}	D_{16}	$1A_{16}$	27_{16}	34_{16}	41_{16}	$4E_{16}$	$5B_{16}$	68_{16}	75_{16}	82_{16}	$8F_{16}$	$9C_{16}$	$A9_{16}$	$B6_{16}$	$C3_{16}$
E_{16}	0_{16}	E_{16}	$1C_{16}$	$2A_{16}$	38_{16}	46_{16}	54_{16}	62_{16}	70_{16}	$7E_{16}$	$8C_{16}$	$9A_{16}$	$A8_{16}$	$B6_{16}$	$C4_{16}$	$D2_{16}$
F_{16}	0_{16}	F_{16}	$1E_{16}$	$2D_{16}$	$3C_{16}$	$4B_{16}$	$5A_{16}$	69_{16}	78_{16}	87_{16}	96_{16}	$A5_{16}$	$B4_{16}$	$C3_{16}$	$D2_{16}$	$E1_{16}$

Solution

a. Working from right to left and using Table 7.10,

- $2_{16} + 6_{16} = 8_{16}$
- $C_{16} + B_{16} = 17_{16}$; write 7 and carry 1
- $1_{16} + 3_{16} = 4_{16}$

Therefore, $3C2_{16} + B6_{16} = 478_{16}$.

$$\begin{array}{r} {\scriptstyle 1} \\ 3C2_{16} \\ +\ B6_{16} \\ \hline 478_{16} \end{array}$$

b. Working from right to left and using Table 7.10:

- $F_{16} + A_{16} = 19_{16}$; write 9 and carry the 1
- $1_{16} + 9_{16} + E_{16} = A_{16} + E_{16} = 18_{16}$; write 8 and carry 1
- $1_{16} + 1_{16} + B_{16} = 2_{16} + B_{16} = D_{16}$
- Bring down A_{16}.

Therefore, $A19F_{16} + BEA_{16} = AD89_{16}$.

$$\begin{array}{r} {\scriptstyle 1\,1} \\ A\,1\,9\,F_{16} \\ +\ \ BEA_{16} \\ \hline AD\,8\,9_{16} \end{array}$$

EXAMPLE 10 ***Multiplying in Base 16*** Do the indicated multiplications.

a. $2B3_{16} \times D_{16}$ **b.** $6C8_{16} \times B2_{16}$

Solution

a. Working from right to left and using Tables 7.10 and 7.11,

- $D_{16} \times 3_{16} = 27_{16}$; write 7 and carry 2
- $D_{16} \times B_{16} = 8F_{16}$ and $8F_{16} + 2_{16} = 91_{16}$; write 1 and carry 9
- $D_{16} \times 2_{16} = 1A_{16}$ and $1A_{16} + 9_{16} = 23_{16}$.

Therefore, $2B3_{16} \times D_{16} = 2317_{16}$.

$$\begin{array}{r} {\scriptstyle 9\,2} \\ 2B3_{16} \\ \times\ \ D_{16} \\ \hline 2317_{16} \end{array}$$

b. Working from right to left and using Tables 7.10 and 7.11,

Multiply $6C8_{16}$ *by* 2_{16}

- $2_{16} \times 8_{16} = 10_{16}$; write 0 and carry 1
- $2_{16} \times C_{16} = 18_{16}$ and $18_{16} + 1_{16} = 19_{16}$; write 9 and carry 1
- $2_{16} \times 6_{16} = C_{16}$ and $C_{16} + 1_{16} = D_{16}$.

$$\begin{array}{r} {\scriptstyle 8\,5} \\ {\scriptstyle 1\,1} \\ 6C8_{16} \\ \times\ \ B2_{16} \\ \hline D90_{16} \end{array}$$

Multiply $6C8_{16}$ *by* B_{16}

- $B_{16} \times 8_{16} = 58_{16}$; write 8 and carry 5.
- $B_{16} \times C_{16} = 84_{16}$ and $84_{16} + 5_{16} = 89_{16}$; write 9 and carry 8
- $B_{16} \times 6_{16} = 42_{16}$ and $42_{16} + 8_{16} = 4A_{16}$; write 4A

$4A98_{16}$

Add the Partial Products

- $D90_{16} + 4A980_{16} = 4B710_{16}$.

$$\begin{array}{r} D90_{16} \\ +\ 4A98_{16} \\ \hline 4B710_{16} \end{array}$$

Therefore, $6C8_{16} \times B2_{16} = 4B710_{16}$.

NW Now Work Problems 23b and 25c.

Math Connections

Binary and Hexadecimal Representation Using the Abacus

Although the abacus is essentially a base 10 device, we can still use it to represent binary and hexadecimal numbers. (See Section 7.2 for more information about the abacus.) To represent **binary numbers**, we use only the upper deck of the abacus and interpret the rods using base 2 notation. Thus, working from right to left the rods respectively denote 1, 2, 4, 8, 16, 32, and so forth. For example, Figure 7.11 denotes the binary number 11011. Note that the base 10 equivalence of 11011 can be found directly from the abacus by multiplying each binary digit by its corresponding rod's numerical decimal value.

To represent **hexadecimal numbers**, we manipulate the beads of the first (i.e., rightmost) rod in the lower and upper decks to denote the numbers A, B, C, D, E, and F (Figures 7.12 a–f). Working from right to left, the rods will now represent powers of 16, namely, 1, 16, 256, 4096, and so forth. For example, Figure 7.13 denotes the hexadecimal number 4B2A. Once again note that the base 10 equivalence can be found directly from the abacus by multiplying each hex digit by its corresponding rod's numerical decimal value.

Decimal equivalence of each rod

512 256 128 64 32 16 8 4 2 1

1 1 0 1 1

Binary interpretation

Base 10 Equivalence

$$(1 \times 16) + (1 \times 8) + (0 \times 4) + (1 \times 2) + (1 \times 1)$$
$$= 16 + 8 + 0 + 2 + 1$$
$$= 27_{10}$$

Figure 7.11

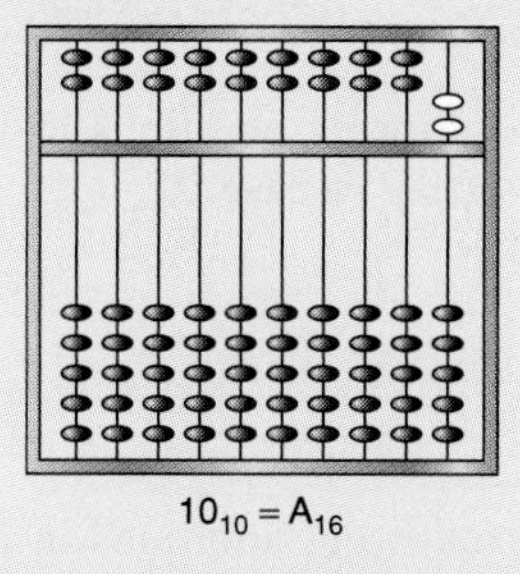

$10_{10} = A_{16}$

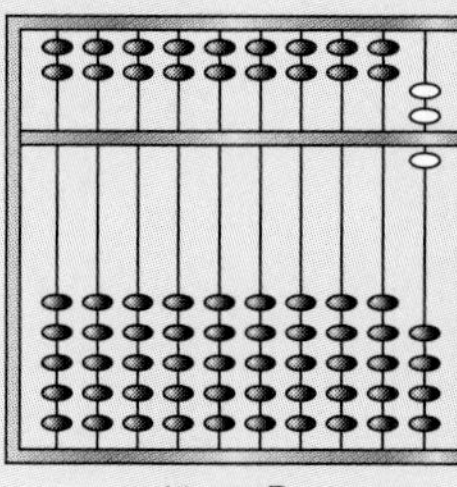

$11_{10} = B_{16}$

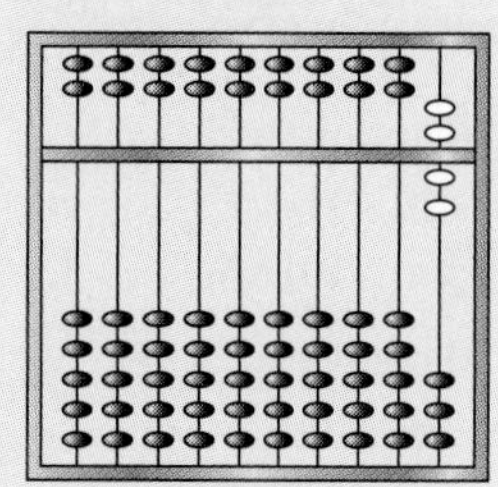

$12_{10} = C_{16}$

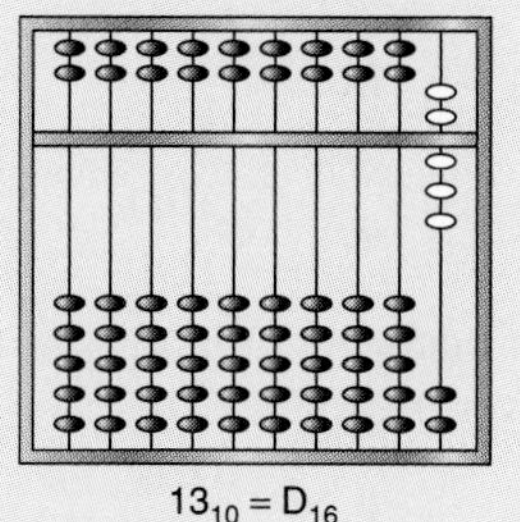

$13_{10} = D_{16}$

$14_{10} = E_{16}$

$15_{10} = F_{16}$

Figure 7.12

(continued)

(continued)

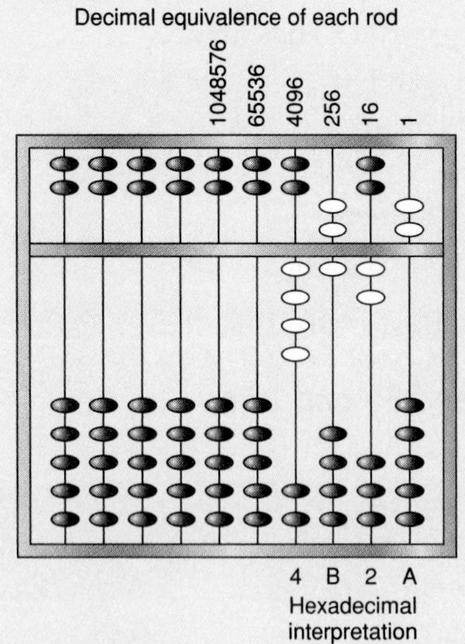

Base 10 Equivalence

$$\begin{aligned} & (4 \times 4096) + (B \times 256) + (2 \times 16) + (A \times 1) \\ &= (4 \times 4096) + (11 \times 256) + (2 \times 16) + (10 \times 1) \\ &= 16384 + 2816 + 32 + 10 \\ &= 19{,}242_{10} \end{aligned}$$

Figure 7.13

Conchi by: James Childress © By Permission News American Syndicate.

Exercises for Section 7.7

1. Change to base 10 notation.
   a. 10_2 **b.** 11_2 **c.** 101_2
   d. 111_2 NW **e.** 1011_2 **f.** 11011_2

2. Change to base 10 notation.
   a. 100_2 **b.** 110_2 **c.** 1101_2
   d. 1010_2 **e.** 10110_2 **f.** 11111_2

3. Change to base 10 notation.
   a. 1000_2 **b.** 1001_2 **c.** 1100_2
   NW **d.** 1111_2 **e.** 10000_2 **f.** 10001_2

4. Change to binary notation.
   a. 3 **b.** 11 **c.** 14
   d. 15 **e.** 18 **f.** 24

5. Change to binary notation.
   a. 6 **b.** 9 **c.** 13
   d. 21 **e.** 25 **f.** 33

6. Change to binary notation.
   a. 2 **b.** 19 **c.** 20
   d. 26 **e.** 28 **f.** 35

7. Perform the following additions in base 2. Check your work by converting to base 10.
   a. $10_2 + 11_2$ **b.** $11_2 + 11_2$ **c.** $10_2 + 10_2$
   NW **d.** $110_2 + 10_2$ **e.** $100_2 + 101_2$ **f.** $110_2 + 110_2$

8. Perform the following additions in base 2. Check your work by converting to base 10.
   a. $101_2 + 11_2$ **b.** $111_2 + 10_2$ **c.** $110_2 + 101_2$
   d. $111_2 + 111_2$ **e.** $1011_2 + 1110_2$ **f.** $1111_2 + 1110_2$

9. Perform the following additions in base 2. Check your work by converting to base 10.
   a. $1000_2 + 101_2$ **b.** $1001_2 + 111_2$ **c.** $1110_2 + 111_2$
   d. $1110_2 + 1110_2$ **e.** $1001_2 + 1011_2$ **f.** $1111_2 + 1111_2$

10. Perform the following subtractions in base 2. Check your work by converting to base 10.

a. $10_2 - 1_2$ b. $11_2 - 10_2$ c. $110_2 - 10_2$
d. $101_2 - 11_2$ e. $110_2 - 101_2$ f. $1011_2 - 101_2$

11. Perform the following subtractions in base 2. Check your work by converting to base 10.

a. $11_2 - 1_2$ b. $111_2 - 10_2$ NW c. $111_2 - 101_2$
d. $1010_2 - 101_2$ *e. $1001_2 - 11_2$ *f. $1001_2 - 110_2$

12. Perform the following subtractions in base 2. Check your work by converting to base 10.

a. $111_2 - 110_2$ b. $1011_2 - 100_2$ c. $1011_2 - 110_2$
d. $1100_2 - 10_2$ *e. $1100_2 - 110_2$ *f. $1100_2 - 111_2$

13. Perform the following multiplications in base 2. Check your work by converting to base 10.

a. $11_2 \times 11_2$ b. $11_2 \times 10_2$ NW c. $101_2 \times 11_2$
d. $110_2 \times 11_2$ e. $101_2 \times 111_2$ f. $101_2 \times 101_2$

14. Perform the following multiplications in base 2. Check your work by converting to base 10.

a. $110_2 \times 10_2$ b. $111_2 \times 10_2$ c. $111_2 \times 110_2$
d. $1010_2 \times 11_2$ e. $1100_2 \times 111_2$ f. $110_2 \times 100_2$

15. Perform the following multiplications in base 2. Check your work by converting to base 10.

a. $111_2 \times 11_2$ b. $100_2 \times 10_2$ c. $100_2 \times 11_2$
d. $100_2 \times 101_2$ e. $1100_2 \times 101_2$ f. $1001_2 \times 101_2$

16. Change to base 10 notation.

a. 23_{16} b. 34_{16} c. 71_{16}
d. 123_{16} e. 703_{16} f. 958_{16}

17. Change to base 10 notation.

a. $2A_{16}$ b. $3C_{16}$ c. $D5_{16}$
d. AB_{16} e. $2DB_{16}$ f. $68F_{16}$

18. Change to base 10 notation.

a. $6A5E_{16}$ b. $D47A_{16}$ c. $0AC2_{16}$
d. $AC6B_{16}$ e. $30FF_{16}$ f. $3A0D_{16}$

19. Change the given base 10 numeral to base 16 notation.

a. 52 b. 47 c. 99
d. 107 e. 169 f. 202

20. Change the given base 10 numeral to base 16 notation.

a. 333 b. 468 c. 673
d. 1234 e. 3579 f. 4567

21. Change the given base 10 numeral to base 16 notation.

a. 2009 b. 9999 c. 3697
d. 2045 e. 4077 f. 3750

22. Perform the following additions in base 16. Check your work by converting to base 10.

a. $15_{16} + 45_{16}$ b. $27_{16} + 34_{16}$ c. $88_{16} + 17_{16}$
d. $73_{16} + 45_{16}$ e. $65_{16} + 74_{16}$ f. $92_{16} + 43_{16}$

23. Perform the following additions in base 16. Check your work by converting to base 10.

a. $5A_{16} + 26_{16}$ NW b. $B8_{16} + 19_{16}$ c. $6C_{16} + 95_{16}$
d. $BC_{16} + 36_{16}$ e. $AF_{16} + 48_{16}$ f. $CE_{16} + 95_{16}$

24. Perform the following additions in base 16. Check your work by converting to base 10.

a. $BA_{16} + C5_{16}$ b. $EE_{16} + 3A_{16}$ c. $9C_{16} + AB_{16}$
d. $FE_{16} + AB_{16}$ e. $DC_{16} + FA_{16}$ f. $ABC_{16} + DEF_{16}$

25. Perform the following multiplications in base 16. Check your work by converting to base 10.

a. $29_{16} \times 13_{16}$ b. $83_{16} \times 51_{16}$ NW c. $74_{16} \times 63_{16}$
d. $5A_{16} \times 73_{16}$ e. $B8_{16} \times 65_{16}$ f. $F5_{16} \times 39_{16}$

26. Perform the following multiplications in base 16. Check your work by converting to base 10.

a. $A29_{16} \times 84_{16}$ b. $2B5_{16} \times 39_{16}$ c. $6C3_{16} \times 78_{16}$
d. $B5E_{16} \times A6_{16}$ e. $3AD_{16} \times D4_{16}$ f. $B9E_{16} \times 5E_{16}$

27. Perform the following multiplications in base 16. Check your work by converting to base 10.

a. $88E_{16} \times AB_{16}$ b. $6B7_{16} \times DE_{16}$ c. $CF9_{16} \times BE_{16}$
d. $CDE_{16} \times FAD_{16}$ e. $AEF_{16} \times CCA_{16}$ f. $FCE_{16} \times DBA_{16}$

Writing Mathematics

28. Explain why 11_{10} and 11_2 are not equivalent and why they are not read the same.

29. Explain the significance of studying the base 2 number system.

30. Describe the benefit of studying the hexadecimal number system.

31. Explain why it is necessary to define six new symbols for the hexadecimal system.

Challenge Exercises

32. In the field of computer networks, many computers are assigned *Ethernet addresses*, which consist of 48 binary digits represented in hexadecimal notation. An example of an Ethernet address is 08:00:20:01:D6:2A. Note that the hex numerals are given in pairs separated by a colon. Given the following Ethernet addresses, convert each one to its binary and decimal equivalent forms.

a. 10:00:D4:AC:61:3B

b. 08:00:20:01:D6:2A

c. 08:00:69:1A:79:AE

33. Several people like to express their age in hex because it will make them sound younger than they really are. As an example, what is a 50-year-old man's age in hex? What is your age in hex? Why do you think the hexadecimal value is less than the decimal value? Will this always be true? Why?

34. Represent the seasons in binary notation.

35. Represent the days of the week in binary notation.

36. In addition to hexadecimal notation, computer professionals also use octal notation. What number system do you think this represents? Why? Construct a translation table between binary notation and octal notation similar to Table 7.9.

37. Do the following divisions in the binary number system.

a. $10_2\overline{)10101_2}$ **b.** $11_2\overline{)11011_2}$

c. $101_2\overline{)10101101_2}$

38. Decode the following message by converting the base 10 numerals into base 16.

3501 T10K14S 10 3243 TO TH14 51966 10N13
IS 4077 48879

Group/Research Activities

39. As a group activity, use the hexadecimal system to construct coded messages similar to that given in Exercise 38.

40. Internet Web pages are constructed using a language called *hypertext markup language*, or HTML. In HTML, six hexadecimal digits are used to create a color. Working from left to right, the first two digits represent red, the second two digits represent green, and the third two digits represent blue. Furthermore, the least amount of color is assigned 0 (zero) and the strongest amount of color is assigned F. As a group project consult various Web pages and identify what six-digit hex numerals are being used to generate the color scheme displayed. For example, what hex numerals do you think correspond respectively to the colors white, black, blue, yellow, gray, and orange?

Just for fun

Three Yankees fans and three Mets fans have to ride an elevator up to the top floor (there are no stops between the ground floor and the top), but the elevator will hold no more than two people. The Mets fans always start an argument if they are left in a situation where they outnumber the Yankees fans, but they are fine if they are left alone or if they are with the same or a greater number of Yankees fans. How do they all get to the top floor, using the elevator, without any arguments?

CHAPTER REVIEW MATERIALS

Summary/Chapter 7

7.2 Early Counting and Calculation Systems

- Early counting and calculation systems included drawing pictures, making slashes (or marks) in the dirt, and recording marks on stones. As counting methods evolved, improvements were made to the early tally systems of numeration. These improvements included using patterns and symbols to represent numbers.
- One of the first numeration systems that used symbols to represent numbers was the *Roman numeral system*. Roman numerals used seven symbols, I, V, X, L, C, D, and M, to represent

the numbers 1, 5, 10, 50, 100, 500, and 1000, respectively. Numbers were represented and manipulated in Roman numerals by following specific rules that were dependent on a symbol's "position."

- The *abacus* was one of the first mechanical aids to computation and represented one of the earliest *place-value systems*. Unlike Roman numerals where the numerical value of a symbol was retained regardless of its position, the numerical value of a bead on an abacus was dependent on where the bead was located.
- An early multiplication device was *Napier's rods*.

7.3 Grouping Systems

- A *simple grouping system* is one in which the position of a symbol does not affect the number represented. It also involves using a different symbol to represent a certain number or group of objects. An example of a simple grouping system is the *Egyptian system of numeration*.
- The Egyptians used a simple grouping system that involved grouping items by 10s. In this system, special symbols called hieroglyphic numerals were used to represent powers of 10 (see Table 7.2).
- A *multiplicative grouping system* is one in which a symbol is used to represent a basic group and a second symbol is used to represent multiples of this basic group. Examples of multiplicative grouping systems include the *Greek* and *Chinese–Japanese systems of numeration*.
- In the Greek system, 5 is the basic group and numbers are represented using multiples of 5. For example, because there is no symbol for 50, the Greeks considered 50 as five 10s; similarly, 500 was represented as five 100s. The Greek numerals are given in Table 7.3.
- In the Chinese–Japanese system, numbers are arranged vertically instead of horizontally. This system also uses positional notation and groups items in powers of 10.

7.4 Place-Value Systems

- A *place-value system* is one in which the numerical value of a symbol is dependent on the position the symbol occupies within the numeral. Examples of place-value systems include the *Babylonian* and *Hindu–Arabic systems of numeration*.
- The Babylonian system uses only two symbols to represent numerals and is based on powers of 60. As such, it is called a *sexagesimal* system.
- The Hindu–Arabic system uses 10 symbols to represent numerals and is a multiplicative grouping system based on powers of 10. As such, it is called a *decimal* system. Our current system of numeration uses Hindu–Arabic numerals to express numbers.
- Hindu-Arabic numbers may be expressed in *expanded notation* or *standard notation*. Expanded notation involves writing the number using powers of 10. For example, $945 = (9 \times 10^2) + (4 \times 10^1) + (5 \times 10^0)$. Standard notation implies writing the number in conventional form.

7.5 Numeration in Bases Other Than 10

- Although we use the decimal system of numeration, we commonly group items in some other manner such as pairs (2s) or dozens (12s). When we group items by a number other than 10, we say that we are working in a *base* different from 10. When we write numerals in a base other than 10, we denote the base in which we are working by using a subscript. For example, 23_5 indicates that we are grouping by 5s.
- To change a numeral in some other base to base 10 notation, we write the numeral in expanded notation. For example, $23_5 = (2 \times 5^1) + (3 \times 5^0) = (2 \times 5) + (3 \times 1) = 10 + 3 = 13$. If a numeral is expressed without a subscript, then it is assumed to represent a base 10 number.
- To convert any numeral from base 10 to another base, we perform successive divisions by the new base and record the remainders from each division; we stop dividing when we get a quotient of zero. The answer in the new base is found by reading the remainders from bottom to top.

7.6 Base 5 Arithmetic

- Numbers expressed in base 5 notation use the symbols 0, 1, 2, 3, and 4.
- When we perform an arithmetic operation in base 5, it is helpful to think in terms of base 10 because the arithmetic operations are performed in exactly the same manner. The difference, though, is that we must consciously be aware that we are working in base 5. For example, when adding $4_5 + 2_5$, we add $4 + 2$ as we would in base 10. However, the result 6 must be represented in base 5 as one 5 and one 1, that is, 11_5.

7.7 The Language of Computers: Binary and Hexadecimal Bases

- The *binary number system* (base 2) involves only two symbols, 0 and 1, and groups items by 2s.
- The *hexadecimal number system* (base 16) involves 16 symbols, 0, 1, 2, 3, 4, 5, 6, 7, 8, 9, A, B, C, D, E, F, and groups items by 16s. The symbols A–F, respectively, represent the base 10 numbers 10–15.
- Arithmetic operations involving base 2 and base 16 numbers are performed in a similar manner as operations involving base 5.
- Conversions between bases 2 and 16 can be done directly without first converting numbers to base 10. This is done by assigning a group of four binary numerals to a corresponding hexadecimal symbol. This correspondence is shown in Table 7.9.

Key Terms and Concepts

Review Exercises

7.2 Early Counting and Calculation Systems

1. Express the given number in Roman numerals.

a. 78 **b.** 124 **c.** 920 **d.** 2006

2. Evaluate the given Roman numeral.

a. DXIX **b.** DCL **c.** DCM **d.** CDXIV

3. Evaluate the given Roman numerals.

a. CCLV + XXVII **b.** CXV + XCV

c. CCV − LXII **d.** DLXV − DXLII

4. Write the numeral depicted by the given abacus.

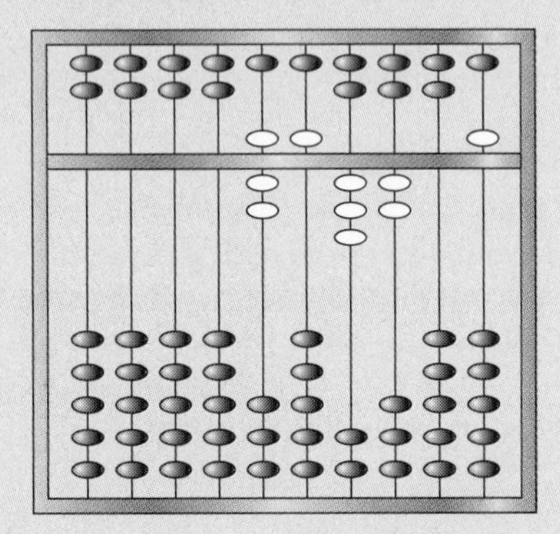

5. Sketch a picture of an abacus that shows the bead movements for adding 6 + 8.

6. Following is an illustration of Napier's rods. Determine the indicated multiplication problem and find the product (use Figure 7.9).

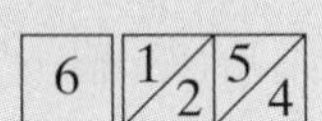

7.3 Grouping Systems

7. Express 78 as

a. an Egyptian numeral **b.** a Greek numeral

c. a Chinese–Japanese numeral

8. Express the following as Egyptian numerals:

a. 32 **b.** 211 **c.** 1111

9. Express the following Egyptian numerals as base 10 numerals:

a. 𓆼𓍢𓍢𓎆𓎆𓎆𓏺𓏺𓏺

b. 𓂭𓆼𓆼𓎆𓎆𓏺𓏺

10. Perform the indicated operations. Express your answers in Egyptian numerals.

a. 𓎆𓎆𓏺𓏺𓏺 + 𓎆𓎆𓎆𓏺𓏺

b. 𓍢𓎆𓎆𓎆𓏺𓏺𓏺 + 𓍢𓎆𓏺𓏺𓏺𓏺𓏺𓏺𓏺

c. 𓍢𓎆𓎆𓏺𓏺 − 𓎆𓎆𓎆𓏺𓏺𓏺

d. 𓆼 − 𓍢𓎆𓎆

11. List the distinguishing characteristics of

a. a simple grouping system

b. a multiplicative grouping system

12. Name a system of numeration that uses

a. simple grouping

b. multiplicative grouping

7.4 Place-Value Systems

13. Express 78 as a Babylonian numeral.

14. List the distinguishing characteristics of a place-value system.

15. Name a system of numeration that uses place value.

16. Write in expanded notation.

a. 345 **b.** 342_5

c. 10111_2 (see Section 7.7)

d. $E4_{16}$ (see Section 7.7)

7.5 Numeration in Bases Other Than 10

17. Convert to base 10.

a. 48_{12} **b.** 44_5 **c.** 242_5

d. 37_8 **e.** 10111_2 **f.** $4TE_{12}$

g. $D5A_{16}$ **h.** $FFFF_{16}$

18. Convert to the indicated base.

a. $41 = \_\_\_\_\__5$ **b.** $53 = \_\_\_\_\__5$

c. $31 = \_\_\_\_\__{12}$ **d.** $35 = \_\_\_\_\__2$

e. $62 = \_\_\_\_\__2$ **f.** $36 = \_\_\_\_\__{12}$

g. $194 = \_\_\_\_\__{16}$

19. Name at least two systems of numeration, other than base 10, to which we are exposed in everyday life, and give examples of how each is used.

20. Convert to the indicated base.

a. $321_5 = \_\_\_\_\__6$ **b.** $42_5 = \_\_\_\_\__3$

c. $10111_2 = \_\_\_\_\__5$ **d.** $77_8 = \_\_\_\_\__2$

e. $5F_{16} = \_\_\_\_\__2$

f. $11011001_2 = \_\_\_\_\__{16}$

21. Express each of the following as a numeral in base 5 and base 2:

a. 7 **b.** 13 **c.** 33

22. Convert to base 10.

a. 58_9 **b.** 34_7 **c.** 65_8

d. 402_9 **e.** 321_6 **f.** 465_7

23. Convert to the indicated base.

a. $33 = \_\_\_\_\__9$ **b.** $46 = \_\_\_\_\__7$

c. $65 = \_\_\_\_\__4$ **d.** $92 = \_\_\_\_\__3$

e. $75 = \_\_\_\_\__6$ **f.** $135 = \_\_\_\_\__8$

***24.** Perform each operation in the indicated base.

a. $\begin{array}{r} 432_6 \\ +\ 324_6 \\ \hline \end{array}$ **b.** $\begin{array}{r} 321_4 \\ +\ 233_4 \\ \hline \end{array}$

c. $\begin{array}{r} 521_7 \\ -\ 436_7 \\ \hline \end{array}$ **d.** $\begin{array}{r} 843_9 \\ -\ 627_9 \\ \hline \end{array}$

***25.** Perform each operation in the indicated base.

a. $\begin{array}{r} 314_6 \\ \times\ 42_6 \\ \hline \end{array}$ **b.** $\begin{array}{r} 122_3 \\ \times\ 12_3 \\ \hline \end{array}$

c. $\begin{array}{r} 346_7 \\ \times\ 34_7 \\ \hline \end{array}$ **d.** $\begin{array}{r} 843_9 \\ \times\ 27_9 \\ \hline \end{array}$

7.6 Base 5 Arithmetic

26. Perform the indicated operations in base 5.

a. $\begin{array}{r} 24_5 \\ +\ 21_5 \\ \hline \end{array}$ **b.** $\begin{array}{r} 324_5 \\ +\ 222_5 \\ \hline \end{array}$

c. $\begin{array}{r} 32_5 \\ -\ 13_5 \\ \hline \end{array}$ **d.** $\begin{array}{r} 333_5 \\ -\ 134_5 \\ \hline \end{array}$

27. Perform the indicated operations in base 5.

a. $\begin{array}{r} 231_5 \\ \times\ 3_5 \\ \hline \end{array}$ **b.** $\begin{array}{r} 123_5 \\ \times\ 4_5 \\ \hline \end{array}$

c. $\begin{array}{r} 231_5 \\ \times\ 34_5 \\ \hline \end{array}$ **d.** $2_5\overline{)124_5}$

7.7 The Language of Computers: Binary and Hexadecimal Bases

28. Perform the indicated operations in base 2.

a. $\begin{array}{r} 101_2 \\ +\ 110_2 \\ \hline \end{array}$ **b.** $\begin{array}{r} 101_2 \\ +\ 101_2 \\ \hline \end{array}$

c. $\begin{array}{r} 1101_2 \\ +\ 1101_2 \\ \hline \end{array}$ **d.** $\begin{array}{r} 101_2 \\ -\ 10_2 \\ \hline \end{array}$

29. Perform the indicated operations in base 2.

a. $\begin{array}{r} 111_2 \\ -\ 11_2 \\ \hline \end{array}$ **b.** $\begin{array}{r} 1011_2 \\ -\ 101_2 \\ \hline \end{array}$

c. $\begin{array}{r} 111_2 \\ \times\ 10_2 \\ \hline \end{array}$ **d.** $\begin{array}{r} 1011_2 \\ \times\ 101_2 \\ \hline \end{array}$

30. Perform the indicated operations in base 16.

a. $\begin{array}{r} 2A_{16} \\ +\ 5D_{16} \\ \hline \end{array}$ **b.** $\begin{array}{r} DA_{16} \\ +\ FB_{16} \\ \hline \end{array}$

c. $\begin{array}{r} 76_{16} \\ \times\ 39_{16} \\ \hline \end{array}$ **d.** $\begin{array}{r} A8_{16} \\ \times\ D6_{16} \\ \hline \end{array}$

Just for fun

Three books stand in order on a shelf. The pages of each book take up 2 inches on the shelf, and their front and back covers are each $\frac{1}{4}$-inch thick. What is the minimum distance from the first page of Volume 1 to the last page of Volume 3?

Chapter Quiz

Determine whether each statement (1–10) is true or false.

1. Any symbol for a number is called a numeral.
2. The Egyptians used a multiplicative grouping system.
3. In the Egyptian system of numeration

   ∩|| = |∩|

4. If the position of a symbol does not affect the number represented, then the system is a simple grouping system.
5. The Greek system of numeration is a simple grouping system.
6. One of the advantages of a multiplicative grouping system is that it enables us to use fewer symbols to express a number as compared with a simple grouping system.
7. The Chinese–Japanese system of numeration uses multiplicative grouping.
8. The Babylonian system of numeration uses only two symbols.
9. One of the most significant contributions of the Babylonian system was devising a numeral for zero.
10. The base 12 system of numeration uses 10 symbols to represent numbers.

In Exercises 11–21, convert each number to the indicated base.

11. $543_{12} = \underline{\qquad}_{10}$ **12.** $121_5 = \underline{\qquad}_{10}$

13. $ET1_{12} = \underline{\qquad}_{10}$ **14.** $78_9 = \underline{\qquad}_5$

15. $345_6 = \underline{\qquad}_5$ **16.** $82_9 = \underline{\qquad}_3$

17. $421_5 = \underline{\qquad}_2$ **18.** $43_{16} = \underline{\qquad}_2$

19. $BA_{16} = \underline{\qquad}_2$ **20.** $100110110_2 = \underline{\qquad}_{16}$

21. $11111011011_2 = \underline{\qquad}_{16}$

In Exercises 22–31, perform each operation in the indicated base.

22. $433_5 + 234_5$ **23.** $33_5 \times 44_5$

24. $1011_2 + 111_2$ **25.** $1001_2 - 111_2$

26. $111_2 \times 11_2$ ***27.** $134_7 + 265_7$

***28.** $432_6 \times 43_6$ ***29.** $68_{12} + 54_{12}$

30. $4D8_{16} + 7AB_{16}$ **31.** $FA_{16} \times 3D_{16}$

Chapter 8

Chapter Outline

Sets of Numbers and Their Structure

After Studying This Chapter, You Will Be Able to Do the Following:

1. Determine whether a natural number is **prime** or **composite**, and find the **prime factors** of a given composite number.
2. Find the **greatest common divisor** and the **least common multiple** for a given pair of numbers.
3. Add, subtract, multiply, and divide **integers**.
4. Add, subtract, multiply, and divide **rational numbers**, and express rational numbers as decimals.
5. Express a **terminating decimal** or a **repeating nonterminating decimal** as a quotient of integers.
6. Identify an **irrational number** as a nonterminating and nonrepeating decimal, and identify the set of **real numbers** as the union of the sets of rational and irrational numbers.
7. Use **scientific notation** to evaluate expressions that contain very large or very small numbers.

Note: *indicates optional material.

Visit the PH Companion Web site at www.prenhall.com/setek

OVERVIEW

Without a doubt, there is not a day that goes by where we do not use numbers in some fashion. For example, we use social security numbers, cell phone numbers, and credit card numbers for identification. We use numbers to represent the cost of goods, such as groceries or automobiles and we use numbers to represent data in charts and graphs. Some of us even have lucky numbers. Numbers are everywhere! In using them in different contexts, there are also many different types of numbers. A few different examples are: whole numbers, signed numbers, fractions and decimals. Mathematicians have always been fascinated by numbers, and have discovered surprising and interesting patterns among them. (See Chapter 1 for examples of such patterns.) In this chapter, we will discuss the many different types of numbers and some of their most important properties.

8.1 INTRODUCTION

In the previous chapter, we provided information about the development of different systems of numeration, including those used by the Babylonians, Chinese–Japanese, Egyptians, Greeks, and Romans. The development of these number systems was evolutionary in nature. As was the case with these early systems, our present-day decimal-based system of numeration, called the Hindu–Arabic system, also evolved over time and included the development of different types of numbers that were used for different purposes. For example, the first set of numbers that formed was the *counting numbers*, which consist of 1, 2, 3, 4, and so forth. When the counting numbers proved insufficient in certain applications such as partitioning land parcels into equal parts, the set of *fractions* (e.g., 2/3 and 8/5) was developed. Eventually, fractions led to the development of yet another set of numbers when the Greeks (particularly Pythagoras) discovered that some lengths could not be represented by fractions. This set was called the set of the *irrational numbers* (e.g., $\sqrt{2}$). Still another development occurred when it was necessary to represent the difference between two numbers where a larger number was to be subtracted from a smaller number. This led to the set of *negative numbers*. Finally, two additional sets of numbers were formed: the set of *imaginary numbers* and the set of *complex numbers*. Imaginary numbers were developed to handle cases where it was necessary, for example, to take the square root of a negative number (e.g., $\sqrt{-2}$). Complex numbers were developed to facilitate the arithmetic of imaginary numbers and the other sets of numbers (e.g., $5 + 7\sqrt{-11}$). In this chapter, we will discuss all of these different sets of numbers except imaginary and complex numbers.

8.2 NATURAL NUMBERS—PRIMES AND COMPOSITES

Cardinal, Ordinal, and Nominal Numbers

Numbers, like many other things, can be described or classified in a variety of ways depending on the context in which they are used. For example, in Chapter 2, we discussed *cardinal numbers*, which are used to tell us "how many." In Chapter 2, we used cardinal numbers to determine the number of elements in a set. Other examples of cardinal numbers include 4 cats, 15 surfboards, and 5 candy bars. In each instance, the numbers 4, 15, and 5 indicate "how many." Another number classification is *ordinal numbers*, which are used to indicate rank or position. In other words, ordinal numbers specify the order of things in a set; that is, first, second, third, and so forth. Examples include first in line, fifth fastest time, and the twentieth caller. Note that first, fifth, and twentieth all refer to order or rank. A third classification is *nominal numbers*, which are used to name or identify something. Examples include a telephone number, student number, and Social Security number. Note that in each case, these numbers do not show quantity or rank; they are used strictly for identification purposes.

Natural Numbers

In addition to cardinal, ordinal, and nominal numbers, another number classification is the set of natural numbers. In the introduction to Chapter 7, we stated that when human beings first started to count, the most primitive counting system went only as high as three or four, with anything greater being described as "many." As the sophistication of human society increased, there was a need to maintain an accurate count of larger quantities. This need gave rise to the set of *counting numbers*, which we today more formally call the set of *natural numbers*.

> ***The* counting numbers, *or* natural numbers, *are the numbers* 1, 2, 3, 4, 5, 6, 7, 8, 9, ... *Written in set notation, we have* $N = \{1, 2, 3, 4, 5, \ldots\}$, *where N stands for natural numbers.***

Any natural number can be expressed as the product of two or more natural numbers. For example,

$$6 = 3 \times 2, \quad 8 = 4 \times 2, \quad 3 = 3 \times 1, \quad \text{and} \quad 1 = 1 \times 1$$

The numbers that are multiplied together to form a number are the *factors* of the number. For example, because $6 = 3 \times 2$, 3 and 2 are factors of 6. Factors are also called *divisors*. A factor or divisor of a number divides the given number with a zero remainder. Thus, 2 and 3 can also be thought of as divisors of 6 because when we divide 6 by either 2 or 3 we get a zero remainder.

Prime and Composite Numbers

Given the set of natural numbers, the Greeks recognized that 1 was special and called it a *unit*. The special nature of 1 was that every natural number had 1 as a divisor. The Greeks also observed that for some natural numbers greater than 1, there were *only two natural number divisors*, namely, the number itself and 1. For example:

- The only natural number divisors of 2 are 2 and 1. That is, the only way to represent 2 as a product of natural numbers is 2×1.
- The only natural number divisors of 3 are 3 and 1. That is, the only way to represent 3 as a product of natural numbers is 3×1.
- The only natural number divisors of 5 are 5 and 1. That is, the only way to represent 5 as a product of natural numbers is 5×1.

On the other hand, other natural numbers greater than 1 had *more than two natural number divisors*. For example, although we can express the product of 4 as 4×1, we can also express the product of 4 as 2×2. Thus, the natural number divisors of 4 are 1, 2, and 4. Similarly, the natural number divisors of 6 are 1, 2, 3, and 6, and the natural number divisors of 8 are 1, 2, 4, and 8.

The Greeks called numbers such as 2, 3, 5, and 7, which had only themselves and 1 as natural number divisors, *prime numbers*; other numbers such as 4, 6, 8, and 9 were called *composite numbers*. Thus, the Greeks essentially partitioned the set of natural numbers into three distinct subsets: unit, primes, and composites.

> ***A* prime number *is any natural number greater than 1 that is divisible only by itself and 1.***

> ***A* composite number *is any natural number other than 1 that is not prime. That is, it is a natural number greater than 2 that has more than two factors.***

Note that by definition, 1 is not a prime number. Note also that the first prime number, 2, is unique in that it is the only even prime number. Any other even natural number, such as 10, 200, or 484, is divisible by 2 and therefore cannot be prime because a prime number is divisible only by itself and 1.

Determining If a Number Is Prime by Checking Divisors

How can we tell if a number is prime? There is no quick and easy solution; no formula exists for finding primes. Suppose we want to determine whether 29 is prime. How do we go about it? In order to determine whether a number is prime, we check divisors to see if they divide it. The first divisor to check is 2: 29 is not divisible by 2. Try 3: 29 is not divisible by 3. How about 4? We need not test 4 because it is a composite number ($4 = 2 \times 2$) and contains smaller divisors (2) that have already been tried. Next we try 5: 29 is not divisible by 5. We do not need to test 6. Why? Because $6 = 3 \times 2$ and neither 3 nor 2 divide 29. Try 7: 29 is not divisible by 7. How far do we keep testing? All the way to 29? No; in fact, we should have stopped at 6, because $6 \times 6 = 36$ and $36 > 29$. If 6 or a number greater than 6 divided 29, then the quotient would be less than 6 and would also be a factor of 29. But we have already tested all possible factors less than 6, and they all failed. Therefore, our conclusion is that 29 is prime.

> ***To determine whether a number is prime, we need to test only the prime divisors {2, 3, 5, 7, 11, . . . } up to the largest natural number whose square is less than or equal to the number we are testing.***

Remember that we do not have to check composite divisors, because a composite number can be expressed as a product of prime factors.

Fanciful Numbers and Number Patterns

The subject of number theory has generated various observations and patterns of numbers. Several of the more interesting ones involving natural numbers are summarized here.

Perfect Numbers

The number 6 has historically been associated with the creation of the universe and hence called a *perfect number* after it was observed that the sum of its divisors, 1, 2, and 3, is the number itself (e.g., $6 = 1 + 2 + 3$). Another perfect number is 28, which corresponds to the lunar month. The Greeks discovered five such numbers: 6, 28, 496, 8128, and 33, 550, 336.

Deficient Numbers

Following the idea of a perfect number, if the sum of a number's divisors is less than the number itself, then the number is called a *deficient number*. For example, 4, 8, 9, and 10 are deficient numbers.

Abundant Numbers

Following the idea of a deficient number, if the sum of a number's divisors is greater than the number itself, then the number is called an *abundant number*. For example, 12 is an abundant number.

Masculine and Feminine Numbers

In the days of Pythagoras, odd numbers were considered to be *masculine* and even numbers were thought of as being *feminine*. One rationale for these associations was that males generally exhibit a certain "odd" characteristic, whereas females are even tempered. It was also believed that the number 1 was the source of all numbers and hence was neither male nor female. Thus, 2 was the first female number, and 3 was the first male number. It was further believed that the sum of 2 and 3, namely, 5, was considered the union of males and females and hence represented marriage.

Amicable Numbers

Another number curiosity that dates back to the days of Pythagoras is the concept of *amicable numbers*, which involve pairs of numbers. Two numbers are amicable if the sum of their respective divisors (excluding the numbers themselves) is equal to the other number. For example, 220 and 284 are amicable because (a) the sum of the divisors of 220 (1, 2, 4, 5, 10, 11, 20, 22, 44, 55, and 110) is equal to 284, and (b) the sum of the divisors of 284 (1, 2, 4, 71, 142) is equal to 220. The Greeks believed that two people would remain friendly if the numbers they carried with them were amicable.

Triangular Numbers

Another type of number investigated by the Greek mathematicians was *triangular numbers*, which, when represented by dots, take on a triangular form. For example,

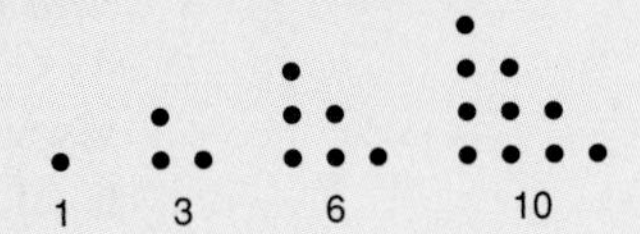

Magic Triangles

Along the same line of triangular numbers, natural numbers can also be placed in a special triangular arrangement such that the sum of the numbers on each side of the triangle is always the same. This sum is called the *magic sum*, and the triangle itself is called a *magic triangle*. For example, the first six natural numbers form the following magic triangle, which has a magic sum of 11. This triangle is also a *third-order* triangle because each side has three numbers.

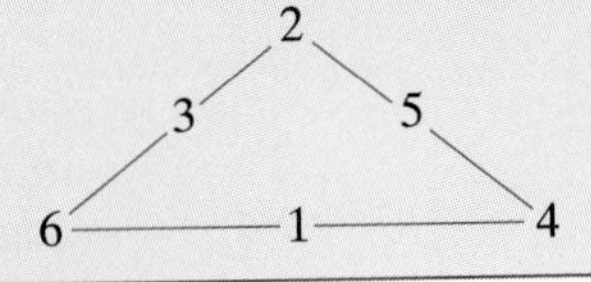

EXAMPLE 1 ***Determining If a Number Is Prime*** Is 43 prime?

Solution Yes, 43 is not divisible by 2, 3, 5, or 7. We need not check any other divisors, since $7^2 = 49$ and $49 > 43$.

EXAMPLE 2 ***Determining If a Number Is Prime*** Is 91 prime?

Solution No. We need not check past 10, since $10^2 = 100$ and $100 > 91$. In fact, we have to check only through 7, because 8, 9, and 10 are all composite numbers. Two does not divide 91, 3 does not divide 91, and 5 does not divide 91. But 7 does divide 91: $91 = 7 \times 13$. Hence, 91 is not prime.

EXAMPLE 3 ***Determining If a Number Is Prime*** Is 1001 prime?

Solution No. The primes less than 100 are 2, 3, 5, 7, 11, 13, 17, 19, 23, 29, 31, 37, 41, 43, 47, 53, 59, 61, 67, 71, 73, 79, 83, 89, and 97. We do not have to check past 37, since $37^2 = 1369$ and $1369 > 1001$. In fact, we have to check only through 31, since the other natural numbers up to 37 are composite. We see that 2 does not divide 1001, 3 does not divide 1001, and 5 does not divide 1001. But 7 does divide 1001: $1001 = 7 \times 143$. Therefore, 1001 is not prime.

EXAMPLE 4 ***Determining If a Number Is Prime*** Is 2003 prime?

Solution Yes. We need not check past 45, since $45^2 = 2025$ and $2025 > 2003$. None of the prime numbers 2, 3, 5, 7, 11, 13, 17, 19, 23, 29, 31, 37, 41, or 43 divides 2003. This is far enough to check, since 44 and 45 are composite numbers. Therefore, 2003 is prime.

NW ***Now Work Problems 3a and 3b.***

Divisibility Rules

As demonstrated in Examples 1–4, to determine if a number is prime requires dividing the given number by prime divisors. There are many situations in mathematics and in every day life where it is helpful to know whether one number divides another number evenly (i.e., with a zero remainder). For example, if an organization with 171 employees wants to run three shifts 24 hours a day with an equal number of employees on each shift, it needs to know if 171 is divisible by 3. This can be determined by dividing 3 into 171 using hand calculations or by using a calculator. Doing so yields an answer of 57. Thus, 3 divides into 171 exactly 57 times and hence 171 is divisible by 3. Sometimes, though, it is helpful not to have to resort to either mode of calculation but instead determine whether a number is divisible by another number using visual inspection. This can be done in a somewhat limited sense by applying the rules of divisibility found on the next page.

EXAMPLE 5 ***Using the Divisibility Rules*** Test each of the following numbers for divisibility by 2, 3, 4, 5, 8, 9, 10, and 11.

a. 330 **b.** 410 **c.** 369 **d.** 1331

Solution

a. 330 is divisible by 2, 3, 5, 10, and 11.
b. 410 is divisible by 2, 5, and 10.
c. 369 is divisible by 3 and 9.
d. 1331 is divisible by 11.

NW ***Now Work Problem 5a.***

Divisibility Rules for 2, 3, 4, 5, 6, 8, 9, 10, and 11	
A Natural Number Is Divisible by	**If:**
2	The number is an even number (i.e., it ends in 0, 2, 4, 6, or 8). *Example:* 2754 is divisible by 2 because 2754 ends in 4 and 4 is even.
3	The sum of the digits is divisible by 3. *Example:* 2754 is divisible by 3 because $2 + 7 + 5 + 4 = 18$, and 18 is divisible by 3.
4	The number formed by its last two digits is divisible by 4. *Example:* 2924 is divisible by 4 because the last two digits, 24, form a number divisible by 4.
5	The rightmost digit is 0 or 5 (i.e., the number ends in 0 or 5). *Example:* 1350 and 234,795 are divisible by 5 because they end in 0 or 5.
6	The number is divisible by 2 and 3. *Example:* 2160 is divisible by 6 because (a) it is divisible by 2 (2160 an even number) and (b) the sum of its digits (i.e., $2 + 1 + 6 + 0 = 9$) is divisible by 3.
8	The number formed by the last three digits is divisible by 8. *Example:* 4560 is divisible by 8 because the last three digits, 560, form a number divisible by 8.
9	The sum of the digits is divisible by 9. *Example:* 2754 is divisible by 9 because $2 + 7 + 5 + 4 = 18$, and 18 is divisible by 9.
10	The number ends in 0. *Example:* 1350 and 408,750 are divisible by 10 because they end in 0.
11	The difference between the sum of the digits in the odd places and the sum of the digits in the even places is 0 or divisible by 11. *Example:* 368,610 is divisible by 11 because (a) the sum of the digits in the odd places is $6 + 6 + 0 = 12$, (b) the sum of the digits in the even places is $3 + 8 + 1 = 12$, and (c) the difference between these sums is $12 - 12 = 0$.

Finding Prime Numbers: The Sieve of Eratosthenes

There exists an interesting and ancient technique for finding primes. It is called the **sieve of Eratosthenes** and was invented by a Greek scholar, Eratosthenes (276–195 B.C.), who was also head of the famous library in Alexandria. We shall use the numbers from 1 to 100 to illustrate Eratosthenes' method:

~~1~~	(2)	(3)	~~4~~	(5)	~~6~~	(7)	~~8~~	~~9~~	~~10~~
(11)	~~12~~	(13)	~~14~~	~~15~~	~~16~~	(17)	~~18~~	(19)	~~20~~
~~21~~	~~22~~	(23)	~~24~~	~~25~~	~~26~~	~~27~~	~~28~~	(29)	~~30~~
(31)	~~32~~	~~33~~	~~34~~	~~35~~	~~36~~	(37)	~~38~~	~~39~~	~~40~~
(41)	~~42~~	(43)	~~44~~	~~45~~	~~46~~	(47)	~~48~~	~~49~~	~~50~~
~~51~~	~~52~~	(53)	~~54~~	~~55~~	~~56~~	~~57~~	~~58~~	(59)	~~60~~
(61)	~~62~~	~~63~~	~~64~~	~~65~~	~~66~~	(67)	~~68~~	~~69~~	~~70~~
(71)	~~72~~	(73)	~~74~~	~~75~~	~~76~~	~~77~~	~~78~~	(79)	~~80~~
~~81~~	~~82~~	(83)	~~84~~	~~85~~	~~86~~	~~87~~	~~88~~	(89)	~~90~~
~~91~~	~~92~~	~~93~~	~~94~~	~~95~~	~~96~~	(97)	~~98~~	~~99~~	~~100~~

Note of Interest

Throughout the history of mathematics, many mathematicians have tried to find the largest prime number. In 1876, Anatole Lucas, a French mathematician, discovered a prime number that is 39 digits long. Since the invention of the computer, much larger prime numbers have been discovered. In 1952, a prime number with 386 digits was found with the aid of a computer. In the early 1980s, two college students with the assistance of a university computer discovered the largest known prime to be $2^{216,091} - 1$. This prime number contains 65,050 digits. The students spent approximately three years working on the project. In 1989, another group found $2^{216,193} - 1$ to be the largest known prime number. In 1993, D. Slowinski, using a Cray computer, found $2^{859,433} - 1$ to be the largest known prime. The decimal expression for this number contains 258,716 digits.

On June 1, 1999, Nayan Hajratwala found a 2,098,960-digit Mersenne prime: $2^{6972593} - 1$. A Mersenne prime is a prime number that can be expressed as a power of two, minus one. For example, the first few Mersenne primes are $2^2 - 1 = 3$, $2^3 - 1 = 7$, $2^5 - 1 = 31$, and $2^7 - 1 = 127$. The Hajratwala discovery was part of the Great Internet Mersenne Prime Search (GIMPS) challenge and won him a $50,000 award. This is the GIMPS fourth Mersenne prime in as many years and the 38th known Mersenne prime.

Hajratwala, a Price Waterhouse Coopers employee, used his home computer, a 350-MHz IBM Aptiva. It took him 111 days of idle time to find this prime. It is estimated that it could have been done in three weeks if his computer was used full time for this endeavor. Although Hajratwala is credited with the discovery, he actually had the assistance of two key individuals: George Woltman, who started GIMPS so that free software as well as a database to coordinate the efforts of those involved could be provided to find Mersenne primes; and Scott Kurowski, whose company, Entropia.com, provided the distributed computing technology and services for anyone to join in on the search.

On November 17, 2003 Michael Shafer's computer found the 40th known Mersenne prime $2^{20,996,011} - 1$. This number "weighs in" at a whopping 6,320,430 decimal digits! This is also the largest nown prime number, surpassing GIMPS' last discovery by over 2 million digits

For more information, visit the GIMPS Website at **http://www.mersenne.org/prime.htm**.

We exclude 1 because 1 is not prime. Eratosthenes determined that 2 is prime, and also that every second number (beginning with 2) is not prime because it has 2 as a factor. Therefore, we cross out 4, 6, 8, 10, ... Similarly, 3 is a prime number, and every third number (beginning with 3) is not prime because it has 3 as a factor. Therefore, we cross out 6, 9, 12, ... Note that some of these numbers, such as 6, 12, and 18, have already been crossed out. The next number to consider is 5; it is prime, so we cross out every fifth number.

We can continue this process, but for how long? We do not have to go past 11, since $11^2 = 121$ and $121 > 100$. In fact, after determining that 7 is prime and crossing out every seventh number, we have a list of all the prime numbers less than 100. This process is called the sieve of Eratosthenes because instead of crossing out the numbers he punched them out with a sharp stick to form a "sieve."

Prime Factorization and the Fundamental Theorem of Arithmetic

In our discussion of natural numbers, we have learned that every natural number greater than 1 is either prime or composite, but not both. Thus, if a number is not prime, then it is composite and vice versa. Another observation of the natural numbers is that any composite number can be expressed as a product of prime factors. For example, 6 can be written as 2×3, and 2 and 3 are both prime numbers. Now consider the composite number 12. We can write 12 as 2×6. Note that 2 is prime but 6 is not. However, as we just observed, 6 can be written as 2×3, where both 2 and 3 are prime. Therefore, 12 can be expressed as a product of prime factors as $2 \times 2 \times 3$. Similarly, $20 = 2 \times 2 \times 5$, and $50 = 2 \times 5 \times 5$. If we disregard the order of the factors (e.g., $6 = 2 \times 3$ and $6 = 3 \times 2$ are the same if we ignore order), then every composite number can be expressed as a product of prime factors in one and only one way. This concept of *unique factorization into primes* is formally known as the *fundamental theorem of arithmetic*.

FUNDAMENTAL THEOREM OF ARITHMETIC: *Every natural number greater than 1 is either prime or can be expressed as a product of prime factors. Except for the order of the factors, this can be done in one and only one way.*

Because every natural number greater than 1 is either a prime number or can be expressed as a product of primes, the concept of prime numbers is extremely important in mathematics. In fact, prime numbers are the fundamental building blocks of all the *positive integers*, which is another set of numbers we will discuss later in this chapter. In mathematics, prime numbers are sometimes referred to as the "atoms" of the natural numbers. This reference is historical in nature and comes from science in which the atom is a hypothetical ultimate particle of matter that is so small that it is incapable of being divided further.

There are several approaches we can take to determine the prime factors of a composite number. One approach is to first find any two factors of the given number that might appear "obvious." For example, to determine the prime factors of 60, we observe that $60 = 2 \times 30$. We next examine the factors and repeat this procedure for any remaining composite factor. Thus, in our illustration we next try to factor 30 and observe that it also has 2 as a factor, namely $30 = 2 \times 15$. We now have $60 = 2 \times 2 \times 15$. But we are not finished yet because 15 is not prime. Two factors of 15 are 3 and 5, which are both prime. Hence, $60 = 2 \times 2 \times 3 \times 5$ and is now expressed as a product of prime factors.

To help organize our work, we can use a *factor tree*. For example, the factor tree that corresponds to the factoring process we worked through for 60 is

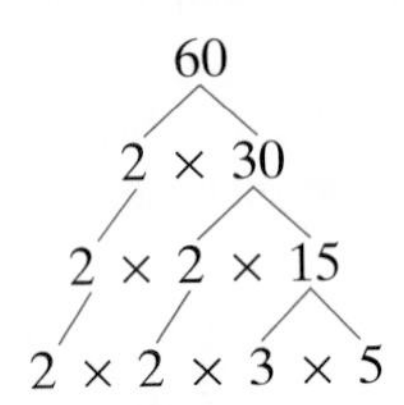

Write each composite number as a product of natural numbers. When a prime is determined, bring it down to the next line so it does not get "lost":

$$60 = 2 \times 2 \times 3 \times 5$$

Note also that although the final factorization is unique (as given by the fundamental theorem of arithmetic), the factorization *process* is not unique. For example, another factor tree for 60 is shown next, where we begin with 4 and 15 as factors of 60 instead of with 2 and 30.

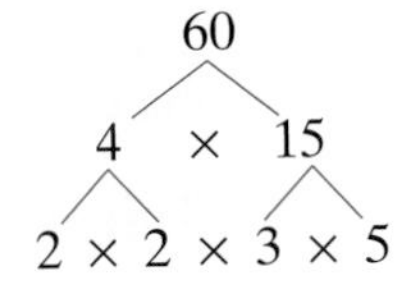

Another way to determine the prime factors of 60 is by successive divisions by prime divisors. This division technique is the same as we used in Chapter 7, but now we must always have remainders of zero. In order to use the technique of successive divisions, we first determine the smallest prime that will divide into 60, which is 2. We divide 2 into 60 and continue in this manner until we reach a quotient that is prime. This indicates that no more divisions are necessary. This technique is illustrated as follows:

$$\begin{array}{r|l} 2 & 60 \\ \hline 2 & 30 \\ \hline 3 & 15 \\ \hline & 5 \end{array}$$

Note of Interest

In the 1970s, three mathematicians, Ronald Rivest, Adi Shamir, and Leonard Adleman, developed a public-key cryptographic system known as RSA (the last-name initials of the three mathematicians). The success of their system relies on the difficulty of factoring large composite numbers (that is, numbers with more than three hundred digits). The essence of the system is first to find three large prime numbers; these are used to encode the message and eventually to decode it. To protect your message you multiply the two prime numbers together. You could publish the composite number if you wish without significantly risking security. In order for someone to decrypt your message, they will need knowledge of the prime numbers, but a three hundred digit number will take at least several thousand years to factor. So you can rest assured that your message will be kept safe. For homework, watch the movie *Sneakers*.

Because 5 is prime, no more divisions are necessary. Therefore,

$$60 = 2 \times 2 \times 3 \times 5$$

Note that we could also express this as $60 = 2 \times 3 \times 5 \times 2$. This is not a different factorization, but merely a rearrangement of the original one. A composite number can be expressed as a product of its prime factors in one and only one way, *disregarding order*: We still have two factors of 2, one factor of 3, and one factor of 5.

EXAMPLE 6 ***Finding Prime Factors*** Determine the prime factors of 345.

Solution 345 is not divisible by 2 since 345 is not even. But 345 is divisible by 3, since the sum of the digits $3 + 4 + 5 = 12$ is divisible by 3. So we have

$$\begin{array}{r|l} 3 & 345 \\ \hline & 115 \end{array}$$

Because 115 ends in 5, it is divisible by 5:

$$\begin{array}{r|l} 5 & 115 \\ \hline & 23 \quad \text{STOP: 23 is prime} \end{array}$$

Putting the two steps together, we have

$$\begin{array}{r|l} 3 & 345 \\ \hline 5 & 115 \\ \hline & 23 \end{array}$$

and $345 = 3 \times 5 \times 23$.

EXAMPLE 7 ***Finding Prime Factors*** Determine the prime factors of 4830.

Solution

$$\begin{array}{r|l} 2 & 4830 \\ \hline 3 & 2415 \\ \hline 5 & 805 \\ \hline 7 & 161 \\ \hline & 23 \quad \text{STOP: 23 is prime} \end{array}$$

$$4830 = 2 \times 3 \times 5 \times 7 \times 23$$

EXAMPLE 8 ***Finding Prime Factors*** Determine the prime factors of 900.

Solution

$$\begin{array}{r|l} 2 & 900 \\ \hline 2 & 450 \\ \hline 3 & 225 \\ \hline 3 & 75 \\ \hline 5 & 25 \\ \hline & 5 \quad \text{STOP: 5 is prime} \end{array}$$

$$900 = 2 \times 2 \times 3 \times 3 \times 5 \times 5$$

(*Note*: We also could have said $900 = 2 \times 3 \times 5 \times 3 \times 2 \times 5$ and still be correct, as this is still the same set of prime factors. We can also write this as $900 = 2^2 \times 3^2 \times 5^2$.)

NW ***Now Work Problems 9a and 9c.***

Math Connections

Laws of Natural Numbers

The set of natural numbers has been proven to follow specific mathematical "laws." Following is a description of these laws as applied to the set of natural numbers.

Commutative Law of Addition

This law states that the *order* in which we add two natural numbers is unimportant. For example, $3 + 2 = 2 + 3$. In general,

> *If a and b represent any two natural numbers, then* $a + b = b + a$.

(Note that there is no commutative law of subtraction for the set of natural numbers. Why?)

Associative Law of Addition

This law states that when adding three or more natural numbers, we can *group* these numbers in any way we want without affecting the sum. For example, $(2 + 5) + 9 = 2 + (5 + 9)$. Note that unlike the commutative law of addition, the order remains the same here; it is the *grouping* of the numbers that is different. In general,

> *If a, b, and c represent any three natural numbers, then* $(a + b) + c = a + (b + c)$.

Closure Law of Addition

This law states that if we add two natural numbers, the sum is also a natural number. In general,

> *If a and b represent two natural numbers, then the sum,* $a + b$*, is also a natural number.*

(Note that there is no closure law of subtraction for the set of natural numbers. Why?)

Commutative Law of Multiplication

This law states that the *order* in which we multiply two natural numbers is unimportant. For example, $3 \times 2 = 2 \times 3$. In general,

> *If a and b represent any two natural numbers, then* $a \times b = b \times a$.

(Note that there is no commutative law of division for the set of natural numbers. Why?)

Associative Law of Multiplication

This law states that when multiplying three or more natural numbers, we can *group* these numbers in any way we want without changing the product. For example, $(2 \times 5) \times 9 = 2 \times (5 \times 9)$. In general,

> *If a, b, and c represent any three natural numbers, then* $(a \times b) \times c = a \times (b \times c)$.

Closure Law of Multiplication

This law states that if we multiply two natural numbers, the product is also a natural number. In general,

> *If a and b represent two natural numbers, then the product,* $a \times b$*, is also a natural number.*

(Note that there is no closure law of division for the set of natural numbers. Why?)

Exercises for Section 8.2

1. Determine whether the number used in each of the following is used as a cardinal number, an ordinal number, or a nominal number.
   - **a.** He shot 82 for a round of golf.
   - **b.** Julie was third in line.
   - **c.** My Social Security number is 089-20-4944.
   - **d.** Ben's phone number is 727-3100.
2. Determine whether the number used in each of the following is used as a cardinal number, an ordinal number, or a nominal number.

   - **a.** In the marathon race, Matt came in eighth.
   - **b.** Concetta won two out of three racquetball games.
   - **c.** The telephone company installed 42 new phones last week.
   - **d.** Nema's student number is 127-72-1326.
3. Determine whether each number is prime or composite.
   - NW **a.** 97 NW **b.** 89 **c.** 1
   - **d.** 243 **e.** 741 **f.** 1955
4. Determine whether each number is prime or composite.
   - **a.** 107 **b.** 299 **c.** 207
   - **d.** 2907 **e.** 1003 **f.** 4,159,731
5. Test each of the following numbers for divisibility by 2, 3, 4, 5, 8, 9, 10, and 11:
   - NW **a.** 2688 **b.** 73,440 **c.** 3,290,154
6. Test each of the following numbers for divisibility by 2, 3, 4, 5, 8, 9, 10, and 11:
   - **a.** 7128 **b.** 67,255 **c.** 4,194,168
7. List the first ten prime numbers.
8. What is the largest prime number less than 100?
9. Determine the prime factors of each number.
   - NW **a.** 36 **b.** 72 NW **c.** 216
   - **d.** 475 **e.** 625 **f.** 147

10. Determine the prime factors of each number.

a. 234	**b.** 213	**c.** 891
d. 1331	**e.** 902	**f.** 7429

11. Two is the only even prime number; all of the other primes are odd. Hence, any two consecutive odd primes must differ by at least 2, as do 3 and 5, and 5 and 7. Prime numbers that differ by exactly 2 are called *twin primes*. Find three pairs of twin primes other than those already mentioned.

12. Numerals 37 and 73 are prime numbers with reversed digits. What are the other pairs of two-digit primes with reversed digits that are less than 100?

Writing Mathematics

13. Explain what it means to say that 3 is a factor of 15.

14. Explain the difference between a prime number and a composite number. Give an example as part of your explanation.

15. Describe what you do to determine if a number is prime. Give an example as part of your explanation.

16. Defend or refute the following statement: 625,830 is divisible by 2, 3, 4, 5, 6, and 10.

17. Express the fundamental theorem of arithmetic in your own words.

18. Can a prime number ever end in the digit 5 or 6? Explain why.

Challenge Exercises

19. How many prime number years do you think there were during the last half of the twentieth century (1951–2000)? What are they?

20. The fundamental theorem of arithmetic refers to the concept of *unique prime factorization*. It is also possible to express a number as the *sum of primes*. For example, one of Goldbach's conjectures states that every odd number greater than 5 can be expressed as the sum of three prime numbers (e.g., $7 = 2 + 2 + 3$). Show that this conjecture holds for the set of all odd numbers between 9 and 33. (See also the Just for Fun at the end of Section 8.3 for another Goldbach conjecture involving prime numbers.)

21. Consider the set of prime numbers between 5 and 19, inclusive. Add 1 to each of these primes and then divide by 6. Observe the pattern that emerges from the remainders of these divisions and then make a general statement that describes this observed relationship.

22. Find four consecutive natural numbers that are not prime.

For Exercises 23 and 24, refer to the Math Connections on fanciful numbers and number patterns.

23. Find the next triangular number.

24. Is it possible to find a magic triangle of order two? Why?

Group/Research Activities

25. Over 2000 years ago Euclid proved that there are infinitely many prime numbers. His method of proof was *proof by contradiction*. Consult some math resources, including doing a Web search on prime numbers, and examine Euclid's proof. Prepare a report on his proof and present it to your class.

26. A question that has been the focus of mathematicians for centuries is "Is there a formula that will produce only prime numbers?" One person to investigate this was Pierre Fermat (1601–1665). Research this question and prepare a report that summarizes the various attempts made to identify such a formula. Include in your report the formulas posited by Fermat and Euler as well as the most recent attempts made to answer this question.

27. As a group activity, research the current efforts and status of finding perfect numbers and amicable numbers.

Just for fun

Pick a three-digit natural number and then rewrite the digits again in exactly the same order to form a six-digit natural number. For example, if we start with 629, then we will have 629629. Now divide this result by 13. You will note that there is a zero remainder. Thus, 629629 is divisible by 13. Believe it or not, this works for *any* 3-digit natural number. Check it out for yourself. Can you explain why this happens?

8.3 GREATEST COMMON DIVISOR AND LEAST COMMON MULTIPLE

Now that we are able to determine the prime factors of a given natural number, we shall use this process to determine some other properties of natural numbers. Two important concepts that make use of prime factorization are the *greatest common divisor* and the *least common multiple*.

Greatest Common Divisor

In the previous section we introduced the notion of factors and indicated that factors of a number divide the number evenly (i.e., with a zero remainder). As such, factors are known as divisors, and hence, the terms factors and divisors may be used interchangeably. We also learned that prime numbers have only two factors, 1 and the number itself. Composite numbers, however, have more than two factors. When we are given two or more composite numbers, it is useful to identify their common factors. For example, factors of 12 are 1, 2, 3, 4, 6, 12, and factors of 16 are 1, 2, 4, 8, 16. Note that among these divisors, 1, 2, and 4 are common to both numbers. Thus, 1, 2, and 4 are common divisors of 12 and 16. In examining properties of numbers, we are often interested in determining the largest or greatest common divisor. In the case of 12 and 16, the greatest common divisor is 4. This leads to the following definition:

> ***The* greatest common divisor (GCD) *of two natural numbers is the greatest (largest) natural number that divides a given pair of natural numbers with remainders of zero.***

It should be noted here that the greatest common divisor is also called the *greatest common factor* (GCF) and either of these two names may be used.

Consider the two natural numbers 32 and 40. We list the set of divisors of each number:

$$\begin{aligned} 32&: \quad \{1, 2, 4, 8, 16, 32\} \\ 40&: \quad \{1, 2, 4, 5, 8, 10, 20, 40\} \end{aligned}$$

If we find the intersection of these two sets of divisors,

$$\{1, 2, 4, 8, 16, 32\} \cap \{1, 2, 4, 5, 8, 10, 20, 40\}$$

we have $\{1, 2, 4, 8\}$. As 8 is the greatest number in the intersection, it is the greatest common divisor of 32 and 40.

When we are given two prime numbers, their greatest common divisor is 1. But there are also pairs of composite numbers whose greatest common divisor is 1. Consider the two natural numbers 24 and 25. The sets of divisors of these numbers are

$$\begin{aligned} 24&: \quad \{1, 2, 3, 4, 6, 8, 12, 24\} \\ 25&: \quad \{1, 5, 25\} \end{aligned}$$

The intersection of these two sets of divisors is

$$\{1, 2, 3, 4, 6, 8, 12, 24\} \cap \{1, 5, 25\} = \{1\}$$

Therefore, 1 is the greatest common divisor of 24 and 25. Two numbers whose GCD is 1 are said to be **relatively prime**.

Do we always have to list the sets of divisors to determine the greatest common divisor? The answer is no. We can do it in a more efficient manner. Let's consider our original example, 32 and 40, and find the prime factors (divisors) of each number:

$$\begin{array}{r|l} 2 & 32 \\ \hline 2 & 16 \\ \hline 2 & 8 \\ \hline 2 & 4 \\ \hline & 2 \end{array} \qquad\qquad \begin{array}{r|l} 2 & 40 \\ \hline 2 & 20 \\ \hline 2 & 10 \\ \hline & 5 \end{array}$$

$$32 = 2 \times 2 \times 2 \times 2 \times 2 \qquad 40 = 2 \times 2 \times 2 \times 5$$

Now examine the two sets of prime factors and determine the factors common to both sets:

$$\begin{aligned} 32 &= \textcircled{2} \times \textcircled{2} \times \textcircled{2} \times 2 \times 2 \\ 40 &= \textcircled{2} \times \textcircled{2} \times \textcircled{2} \times 5 \end{aligned}$$

Note that both sets of prime divisors contain $2 \times 2 \times 2$. Therefore, the greatest common divisor of 32 and 40 is $2 \times 2 \times 2$, or 8.

Let's try another example. What is the greatest common divisor of 30 and 45? First we find the prime factors of each:

$$30 = 2 \times 3 \times 5$$
$$45 = 3 \times 3 \times 5$$

Examining the two sets of prime factors (divisors), we see that the intersection is 3×5. Therefore, the greatest common divisor (or greatest common factor) of 30 and 45 is 3×5, or 15.

EXAMPLE 1 ***Finding the GCD*** Find the greatest common divisor of 8 and 12.

Solution First we find the prime factors of 8 and 12:

$$\begin{array}{r|l} 2 & 8 \\ \hline 2 & 4 \\ \hline & 2 \end{array} \quad 8 = 2 \times 2 \times 2 \qquad \begin{array}{r|l} 2 & 12 \\ \hline 2 & 6 \\ \hline & 3 \end{array} \quad 12 = 2 \times 2 \times 3$$

Next, we examine the intersection of the two sets of prime factors and we note that 2×2 is common to both sets; hence, $2 \times 2 = 4$ is the GCD.

EXAMPLE 2 ***Finding the GCD*** Find the greatest common divisor of 8 and 15.

Solution First we find the prime factors of 8 and 15:

$$\begin{array}{r|l} 2 & 8 \\ \hline 2 & 4 \\ \hline & 2 \end{array} \quad 8 = 2 \times 2 \times 2 \qquad \begin{array}{r|l} 3 & 15 \\ \hline & 5 \end{array} \quad 15 = 3 \times 5$$

Note that the two sets of prime factors have no elements in common; their intersection is empty. When this occurs, the GCD for the two numbers is 1, so the numbers are relatively prime. Note that we did not list 1 as a prime factor for either number because 1 is not a prime number.

EXAMPLE 3 ***Finding the GCD*** Find the greatest common divisor of 342 and 380.

Solution Finding the prime factors of each number, we have

$$\begin{array}{r|l} 2 & 342 \\ \hline 3 & 171 \\ \hline 3 & 57 \\ \hline & 19 \end{array} \qquad \begin{array}{r|l} 2 & 380 \\ \hline 2 & 190 \\ \hline 5 & 95 \\ \hline & 19 \end{array}$$

$$342 = 2 \times 3 \times 3 \times 19 \qquad 380 = 2 \times 2 \times 5 \times 19$$

Examining the intersection of the two sets of prime factors, we see that 2×19 is common to both; hence, the GCD of 342 and 380 is $2 \times 19 = 38$.

NW ***Now Work Problems 1a and 1d.***

The GCD and Fractions

One application of the greatest common divisor is in the reduction of fractions. We can simplify, or reduce, fractions if we determine the greatest common divisor of both the numerator and denominator. Suppose we are asked to reduce the fraction

$$\frac{65}{91}$$

We can do this in the following manner. We find the greatest common divisor of both the numerator and denominator; that is, we find the GCD of 65 and 91:

$$\begin{array}{r|l} 5 & 65 \\ \hline & 13 \end{array} \quad 65 = 5 \times 13 \qquad \begin{array}{r|l} 7 & 91 \\ \hline & 13 \end{array} \quad 91 = 7 \times 13$$

The GCD of 65 and 91 is 13. We now rewrite the original fraction:

$$\frac{65}{91} = \frac{\cancel{13} \times 5}{\cancel{13} \times 7} = \frac{5}{7}$$

The numerator and denominator of the reduced fraction are relatively prime; that is, their GCD is 1.

EXAMPLE 4 ***Using GCD to Reduce Fractions*** Reduce $\frac{130}{455}$ to lowest terms.

Solution We first find the GCD of 130 and 455.

$$\begin{array}{r|l} 2 & 130 \\ \hline 5 & 65 \\ \hline & 13 \end{array} \quad 130 = 2 \times 5 \times 13 \qquad \begin{array}{r|l} 5 & 455 \\ \hline 7 & 91 \\ \hline & 13 \end{array} \quad 455 = 5 \times 7 \times 13$$

The GCD of 130 and 455 is 5 × 13. Now we rewrite the original fraction.

$$\frac{130}{455} = \frac{\cancel{(5 \times 13)} \times 2}{\cancel{(5 \times 13)} \times 7} = \frac{2}{7}$$

EXAMPLE 5 ***Using GCD to Reduce Fractions*** Reduce $\frac{310}{460}$ to lowest terms.

Solution We first find the GCD of 310 and 460:

$$\begin{array}{r|l} 2 & 310 \\ \hline 5 & 155 \\ \hline & 31 \end{array} \quad 310 = 2 \times 5 \times 31 \qquad \begin{array}{r|l} 2 & 460 \\ \hline 2 & 230 \\ \hline 5 & 115 \\ \hline & 23 \end{array} \quad 460 = 2 \times 2 \times 5 \times 23$$

The GCD of 310 and 460 is 2 × 5, and we can rewrite the original fraction.

$$\frac{310}{460} = \frac{\cancel{(2 \times 5)} \times 31}{\cancel{(2 \times 5)} \times 2 \times 23} = \frac{31}{2 \times 23} = \frac{31}{46}$$

NW ***Now Work Problems 5a and 5d.***

If we want to find the greatest common divisor for three or more numbers, we extend the process for finding the GCD for two numbers. We find the prime factors that are common to all the sets of prime factors.

Least Common Multiple

When we multiply two factors, the product is called a *multiple* of each number. For example, 6 is a multiple of 3 and 2 because 3 × 2 = 6. Similarly, 21 is a multiple of 3 and 7 because 3 × 7 = 21. Multiples of a given number are easy to generate—we simply "count" in terms of the given number. For example, to generate multiples of 2, we count by twos: 2, 4, 6, 8, 10, 12, 14, and so forth. To generate multiples of 3, we count by threes: 3, 6, 9, 12, 15, 18, 21, and so forth. Note that for any given natural number, there are an unlimited number of multiples. Note further that given any two natural numbers, there is also an unlimited number of *common multiples*. In examining properties of numbers, we are often interested in determining the smallest or least common multiple. This leads to the following definition:

> ***The* least common multiple (LCM) *of two natural numbers is the smallest (least) natural number that is a multiple of each of the two given numbers.***

The least common multiple can also be thought of as the smallest (least) natural number that is divisible by both of the given numbers. Four is a multiple of 2, as are 6, 8, 10, and so on, because each of these numbers has 2 as a factor. The multiples of 3 are $\{3, 6, 9, 12, 15, \ldots\}$. Listing the sets of multiples for 2 and 3, we have

$$2: \quad \{2, 4, 6, 8, 10, \ldots\}$$

$$3: \quad \{3, 6, 9, 12, 15, \ldots\}$$

Upon inspection, we note that 6 is the least common multiple (LCM) of 2 and 3. Also, it is the smallest number that is divisible by both of the given numbers, 2 and 3.

Let's consider another example, the LCM of 10 and 12. Listing the sets of multiples for 10 and 12, we have

$$10: \quad \{10, 20, 30, 40, 50, 60, \ldots\}$$

$$12: \quad \{12, 24, 36, 48, 60, 72, \ldots\}$$

Inspection of these two sets of multiples indicates that 60 is the least common multiple of 10 and 12. It is the smallest number that is divisible by both of the given numbers, 10 and 12.

Can we do this problem another way? The answer is yes. First find the greatest common divisor of 10 and 12. It is 2. Note the $10 \times 12 = 120$. If we divide 120 by the G.C.D. of 10 and 12, that is by 2, we have $120 \div 2 = 60$, and 60 is the least common multiple (L.C.M.) of 10 and 12.

In general, we can determine the *least common multiple* of two natural numbers by dividing the product of the two numbers by their greatest common divisor.

EXAMPLE 6 ***Finding the LCM*** Find the least common multiple of 8 and 12.

Solution

Multiples of 8: $\{8, 16, 24, 32, 40, 48, \ldots\}$
Multiples of 12: $\{12, 24, 36, 48, 60, \ldots\}$

The common multiples include 24 and 48. The least common multiple (LCM) of 8 and 12 is 24.

EXAMPLE 7 ***Finding the LCM*** Find the LCM of 48 and 72.

Solution

Multiples of 48: $\{48, 96, 144, 192, \ldots\}$
Multiples of 72: $\{72, 144, 216, \ldots\}$

The least common multiple of 48 and 72 is 144.

NW ***Now Work Problems 9a and 9d.***

The LCM and Fractions

When do we use the least common multiple? It is usually used in combining fractions. Suppose we want to add $\frac{2}{5}$ and $\frac{1}{6}$:

$$\frac{2}{5} + \frac{1}{6}$$

We cannot add these two fractions as they are represented here. Before we can add or subtract two fractions, they must have a common denominator; when they have a common denominator, we can add or subtract the numerators. Because $\frac{2}{5}$ and $\frac{1}{6}$ do not have a common denominator, we must rewrite them as equivalent fractions that have the same denominator. In doing this, we usually use the **least common denominator**, which is the least common multiple of the given denominators. The least common multiple of 5 and 6 is 30. Therefore, we now have

$$\frac{2}{5} + \frac{1}{6} = \frac{12}{30} + \frac{5}{30} = \frac{17}{30}$$

EXAMPLE 8 ***Using the LCM to Add Fractions*** Add $\frac{1}{4}$ and $\frac{2}{9}$.

Solution Before we can add these two fractions, they must have a common denominator. A common denominator is the LCM of 4 and 9. Because 4 and 9 are relatively prime, their GCD is 1. Therefore, the LCM is

$$\frac{4 \times 9}{1} = 36$$

Then

$$\frac{1}{4} + \frac{2}{9} = \frac{9}{36} + \frac{8}{36} = \frac{17}{36}$$

EXAMPLE 9 ***Using the LCM to Subtract Fractions*** Subtract $\frac{1}{6}$ from $\frac{2}{9}$.

Solution We must first rewrite the fractions so that they have a common denominator, the LCM of 9 and 6. First we find the GCD of 9 and 6, which is 3. Therefore, the LCM is

$$\frac{9 \times 6}{3} = \frac{54}{3} = 18$$

Now we can subtract:

$$\frac{2}{9} - \frac{1}{6} = \frac{4}{18} - \frac{3}{18} = \frac{1}{18}$$

NW Now Work Problems 13a and 13d.

EXAMPLE 10 ***Applying the Concept of GCD*** Rachel teaches three fifth-grade classes at a local elementary school. Her respective class sizes are 28, 18, and 24. Rachel wants to order some math manipulatives to help her students understand some basic math concepts as part of a lesson on fractions. Because of limited available funding, Rachel needs to order the manipulatives so that they can be used by equal-sized groups in each class. What is the largest-sized group in each class so that each group has the same number of students?

Solution This problem is an application of greatest common divisor. To find the largest-sized group in each class, Rachel needs to find the GCD of 28, 18, and 24. This is shown using factor trees:

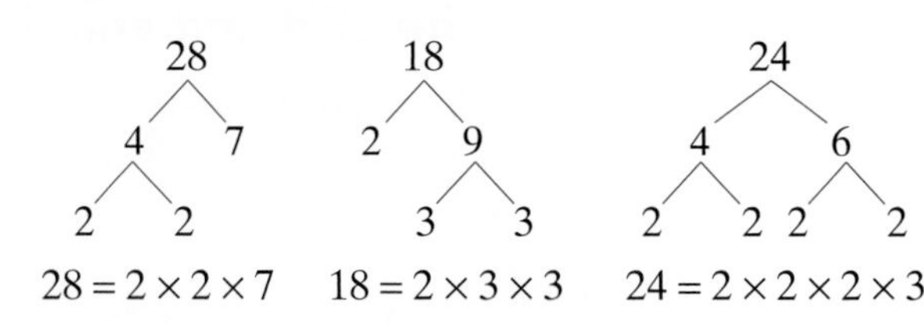

The GCD of 28, 18, and 24 is 2. Thus, the largest-sized group in each class that can be formed so that the manipulatives can be used by equal-sized groups in each class is 2.

NW Now Work Problem 19.

EXAMPLE 11 ***Applying the Concept of LCM*** Three bicyclists, Sue Ann, Chuck, and Janie, are doing roundtrip "loops" along River Road, where they ride along the road to the first major intersection, turn around, and then ride back to where they started. The roundtrip distance is 5 miles, and all three bicyclists want to do 50 miles. It takes Sue Ann 24 minutes to complete one roundtrip, Chuck can do one roundtrip in 16 minutes, and Janie requires 22 minutes. If all 3 begin riding at the same time, will they ever be together at any point along the road after they begin?

Solution This problem is an application of least common multiple. We first note that if all three bicyclists are going to ride 50 miles, then each will do 10 "loops." We further note that since Sue Ann is the slowest, her total riding time will be the maximum time the three riders will ride. Since it takes Sue Ann 24 minutes per "loop," no rider will ride for more than 240 minutes. Thus, if the three bicyclists are to ride together again after they begin, they will have to do so within this 240-minute time frame. We now find the LCM of each person's roundtrip time:

Multiples of 24: $\{24, 48, 72, 96, 120, 144, 168, 192, 216, 240, \ldots\}$
Multiples of 16: $\{16, 32, 48, 64, 80, 96, 112, 128, 144, 160, 176, 192, 208, 224, 240, \ldots\}$
Multiples of 22: $\{22, 44, 66, 88, 110, 132, 154, 176, 198, 220, 242, \ldots\}$

Note that we were unable to find a common multiple for all three numbers before the 240-minute threshold. Therefore, the answer to the question is no. Once Sue Ann, Chuck, and Janie begin riding, there will be no point during their ride where they will be riding together again.

NW Now Work Problem 25.

Exercises for Section 8.3

1. Find the greatest common divisor for each of the following:
   - NW **a.** 8 and 14
   - **b.** 14 and 28
   - **c.** 15 and 24
   - NW **d.** 52 and 78
   - **e.** 111 and 267
   - **f.** 24, 48, and 60
2. Find the greatest common divisor for each of the following:
   - **a.** 7 and 17
   - **b.** 11 and 23
   - **c.** 24 and 84
   - **d.** 224 and 430
   - **e.** 220 and 400
   - **f.** 30, 60, 225
3. Find the greatest common divisor for each of the following:
   - **a.** 12 and 20
   - **b.** 22 and 68
   - **c.** 40 and 23
   - **d.** 9 and 26
   - **e.** 45 and 63
   - **f.** 36 and 42
4. Find the greatest common divisor for each of the following:
   - **a.** 24 and 60
   - **b.** 18 and 54
   - **c.** 5 and 23
   - **d.** 32 and 80
   - **e.** 63 and 105
   - **f.** 80 and 240
5. Reduce each fraction to lowest terms.
   - NW **a.** $\frac{30}{36}$
   - **b.** $\frac{42}{54}$
   - **c.** $\frac{39}{65}$
   - NW **d.** $\frac{120}{180}$
   - **e.** $\frac{294}{304}$
   - **f.** $\frac{195}{390}$
6. Reduce each fraction to lowest terms.
   - **a.** $\frac{18}{30}$
   - **b.** $\frac{13}{40}$
   - **c.** $\frac{30}{76}$
   - **d.** $\frac{52}{275}$
   - **e.** $\frac{222}{460}$
   - **f.** $\frac{120}{460}$
7. Reduce each fraction to lowest terms.
   - **a.** $\frac{14}{42}$
   - **b.** $\frac{13}{52}$
   - **c.** $\frac{16}{96}$
   - **d.** $\frac{3}{57}$
   - **e.** $\frac{17}{51}$
   - **f.** $\frac{28}{280}$

8. Reduce each fraction to lowest terms.

a. $\frac{45}{900}$ b. $\frac{72}{360}$ c. $\frac{35}{210}$

d. $\frac{36}{324}$ e. $\frac{132}{165}$ f. $\frac{209}{323}$

9. Find the least common multiple for each of the following:

NW a. 8 and 14 b. 14 and 28

c. 15 and 24 NW d. 52 and 78

e. 66 and 90 *f. 111 and 267

10. Find the least common multiple for each of the following:

a. 6 and 17 b. 20 and 25

c. 42 and 70 d. 12 and 15

*e. 70 and 220 *f. 711 and 900

11. Find the least common multiple for each of the following:

a. 12 and 8 b. 3 and 10

c. 5 and 9 d. 8 and 10

e. 14 and 21 f. 15 and 35

12. Find the least common multiple for each of the following:

a. 12 and 30 b. 4 and 7

c. 12 and 16 d. 9 and 11

e. 45 and 18 f. 36 and 12

13. Perform the indicated operations and reduce your answers to lowest terms.

NW a. $\frac{3}{4}+\frac{2}{9}$ b. $\frac{5}{9}+\frac{1}{12}$ c. $\frac{1}{6}+\frac{3}{4}$

NW d. $\frac{7}{8}-\frac{1}{12}$ e. $\frac{10}{11}-\frac{4}{5}$ f. $\frac{13}{15}-\frac{3}{20}$

14. Perform the indicated operations and reduce your answers to lowest terms.

a. $\frac{9}{15}+\frac{1}{50}$ b. $\frac{3}{8}+\frac{1}{4}$ c. $\frac{3}{5}+\frac{4}{11}$

d. $\frac{5}{7}-\frac{1}{6}$ e. $\frac{5}{24}-\frac{1}{12}$ f. $\frac{9}{13}-\frac{3}{5}$

15. Perform the indicated operations and reduce your answers to lowest terms.

a. $\frac{3}{7}+\frac{4}{11}$ b. $\frac{5}{9}+\frac{7}{36}$ c. $\frac{8}{15}+\frac{3}{40}$

d. $\frac{8}{11}-\frac{3}{5}$ e. $\frac{7}{18}-\frac{3}{24}$ f. $\frac{8}{9}-\frac{3}{5}$

16. Perform the indicated operations and reduce your answers to lowest terms.

a. $\frac{5}{6}+\frac{1}{4}$ b. $\frac{7}{9}+\frac{5}{12}$ c. $\frac{4}{9}+\frac{3}{5}$

d. $\frac{13}{15}-\frac{7}{20}$ e. $\frac{10}{11}-\frac{2}{5}$ f. $\frac{7}{8}-\frac{5}{12}$

17. **Making bookshelves** Benny has two pieces of plywood. Both pieces have a height of 6 feet. But one piece is 54 inches wide whereas the other is 36 inches wide. If Benny is going to cut the pieces of plywood into 6-foot lengths to make book shelves, what is the widest shelf that can be cut from both pieces without any wood being left over?

18. **Adjusting shelves** Larry has adjustable shelves in his shoe store. He received a shipment of shoes and boots. The shoes come in boxes that are 9 inches high, whereas the boots come in boxes that are 12 inches high. Both types of boxes must fit exactly between the shelves. What height can Larry set the shelves at so that a stack of shoe boxes and a stack of boot boxes will fit on the same shelf?

NW 19. **Tiling a floor** Dee Dee is converting a spare bedroom into a home office and wants to replace the carpet in the bedroom with tile. The "tile man" suggested that she get the largest size tile possible so that there will be a minimum amount of grout lines. If Dee Dee's spare bedroom measures 12 feet by 15 feet, what is the largest size square tile that can be used without cutting any tiles?

20. **Tiling a floor** In Exercise 19, when Dee Dee informed the tile man what size tile she wanted, he indicated that tiles for residential homes are not available in such a large size and that the largest size tile available was 18″ square. If Dee Dee selects this size, is it still possible for no tiles to be cut? Why?

21. **Forming groups** Johnny is teaching two liberal arts mathematics classes. One class has 24 students and the second class has 36 students. Johnny wants to assign group projects to both classes but would like to have equal-sized groups in each class. In how many ways can this be done, and what are the possible group sizes?

22. **E-mail and word processing** Kristin has configured her e-mail program to check her company's e-mail server every 15 minutes for mail. She also has configured her word processing program to automatically save a backup copy of any open files every 10 minutes. If Kristin invokes her e-mail and word processing programs at exactly 9 A.M. and closes both programs at 5 P.M., how many times during this period will these two applications be "working" at the same time?

23. **Learning to count in multiples** Bill is teaching his 3-year-old granddaughter Jillian how to count in multiples of 3 and 5 by arranging small blocks of wood into two separate collections: One arrangement places the blocks in groups of 5, the other in groups of 3. What is the smallest number of wood blocks Bill needs so that each collection can be arranged without any blocks left over?

24. **Planning Passover dinner** Richard and Annie are planning to host a Passover dinner at their house this year. Because

Passover falls on a weekday, they do not know how many people will attend. After talking to family and friends about their status, Richard informs Annie that he thinks they can expect 6, 14, or 21 people. Assuming Richard is correct, how many matzoh balls should Annie order so that there are an equal number available for each guest without any left over?

NW **25.** **Walking dogs** Neighbors Mark and Andrew walk their dogs along the same circular path within the subdivision in which they live. It takes Mark 12 minutes to walk around the path whereas it takes Andrew 16 minutes. If both boys begin walking their dogs in front of Mark's house at the same time, how long will it take before they both arrive at Mark's house again at the same time?

26. **Work schedules** Joe and his wife Laurie both work for the local police department but have different work schedules. Joe's schedule is set up so that he gets every fourth day off but Laurie gets every sixth day off. If both have the day off for their wedding anniversary, which is February 28, when will they next have a day off together? (Assume the current year is not a leap year.)

Writing Mathematics

27. Explain the difference between a common factor and the greatest common divisor.

28. Explain the difference between a multiple and the least common multiple.

29. Describe how the process used to find the greatest common divisor differs from the process used to find the least common multiple.

30. Explain how the greatest common divisor concept is used to reduce the fraction $\frac{12}{18}$.

31. Explain how the least common multiple concept is used to add $\frac{1}{9} + \frac{5}{12}$.

Challenge Exercises

An alternative method for finding the greatest common divisor of two natural numbers is called the *Euclidean algorithm*. This algorithm is demonstrated below to find the greatest common divisor of 95 and 430. We use a chart to help organize our work.

Step 1: Make a two-column chart and label the first column q (for quotient) and the second column r (for remainder). Place the two numbers in the r column with the larger number on top, and place a hyphen at the top of the q column.

q	r
—	**430**
	95

Step 2: Divide the smaller number into the larger number ($430 \div 95 = 4$ with remainder 50). Place the quotient 4 in the q column and the remainder 50 in the r column.

q	r
—	430
4	95
	50

Step 3: Divide the current remainder 50 into the previous divisor 95 (i.e., $95 \div 50 = 1$ with remainder 45). Place the quotient 1 in the q column and the remainder 45 in the r column.

q	r
—	430
4	95
1	50
	45

Step 4: Divide the current remainder 45 into the previous remainder 50 (i.e., $50 \div 45 = 1$ with remainder 5). Place the quotient 1 in the q column and the remainder 5 in the r column.

q	r
—	430
4	95
1	50
1	45
	5

Step 5: Repeat Step 4 until you get a zero remainder.

- $45 \div 5 = 9$ with a remainder 0. STOP.

The GCD is the last nonzero remainder (i.e., the last nonzero entry in the r column). Therefore, the GCD of 430 and 95 is 5.

q	r
—	430
4	95
1	50
1	45
9	5
	0

In Exercises 32–37, find the GCD of the given number pairs using the Euclidean algorithm.

32. 68 and 94

33. 120 and 460

34. 152 and 464

35. 472 and 564

36. 1234 and 1356

37. 3280 and 5425

38. An alternative approach to the Euclidean algorithm is to recall that division is essentially *repeated subtraction*. For example, to find the GCD of 95 and 430, we observe that

- 95 can be subtracted from 430 four times with a remainder of 50:

$$430 - \mathbf{95} = 335 - \mathbf{95} = 240 - \mathbf{95} = 145 - \mathbf{95} = 50$$

- 50 can be subtracted from 95 once with a remainder of 45:

$$95 - \mathbf{50} = 45$$

- 45 can be subtracted from 50 once with a remainder of 5:

$$50 - \mathbf{45} = 5$$

- 5 can be subtracted from 45 nine times with a remainder of 0:

$$45 - \mathbf{5} = 40 - \mathbf{5} = 35 - \mathbf{5} = 30 - \mathbf{5} = 25 - \mathbf{5} = 20 - \mathbf{5} = 15 - \mathbf{5} = 10 - \mathbf{5} = 5 - \mathbf{5} = 0$$

Once again, the GCD is the last nonzero remainder. Hence, the GCD of 95 and 430 is 5. Repeat Exercises 32–37 and find the GCD by using repeated subtractions.

Group/Research Activities

39. The concept of relatively prime factors is an important component of the RSA encryption algorithm (see the Note of Interest in this section). Using the Internet, find out how this concept is used in the algorithm and why it makes messages coded by the algorithm difficult to crack.

40. An interesting relationship between the GCD and LCM of two numbers is that the product of the two numbers is equal to the product of their GCD and LCM. As an example, consider 15 and 20: the GCD is 5, the LCM is 60, $15 \times 20 = 300$, and GCD $\times$ LCM $= 300$. Examine this relationship to see if it is true for all natural numbers. As part of your research, try to discover why this relationship exists.

41. John Farey was a surveyor during the Napoleonic era who is credited with the discovery of a mathematical relation involving fractions. Research Farey to discover this relationship and prepare a report for your class.

Just for fun

It has been stated that "every even number greater than 2 can be expressed as the sum of two prime numbers." For example,

$$6 = 3 + 3, \quad 8 = 3 + 5, \quad 12 = 7 + 5, \quad \text{and} \quad 14 = 11 + 3$$

This statement is called Goldbach's conjecture. It is a conjecture because it has never been proved. Show that Goldbach's conjecture is true for all even numbers except 2, up to and including 30.

8.4 INTEGERS

In Section 8.2, we introduced the set of natural numbers

$$\{1, 2, 3, 4, \ldots\}$$

Many countries, including Colombia, salute the study of math by means of their postage.

The first natural number is 1. This is the first number that human beings used in counting. The use of zero did not come about until approximately 700 A.D. Zero was first used to indicate position, distinguishing between numbers such as 32 and 302. Later, zero began to be used as a starting point in counting. The set of natural numbers was expanded to include zero, forming the set $\{0, 1, 2, 3, 4, \ldots\}$. This set is called the set of **whole numbers** to indicate that it is different from the set of natural numbers, which does not include zero:

$$\{1, 2, 3, 4, \ldots\} = \text{natural numbers}$$

$$\{0, 1, 2, 3, 4, \ldots\} = \text{whole numbers}$$

Recall that 0 is the identity element for the operation of addition: It does not change the identity of a number when it is added to that number. That is, $1 + 0 = 1, 2 + 0 = 2, 3 + 0 = 3$, and so on. Recall also that -3 is the additive inverse of 3, because $-3 + 3 = 0$, and -1 is the additive inverse of 1, as $-1 + 1 = 0$. The set of numbers consisting of the set of whole numbers and their additive inverses is called the set of **integers**; that is,

$$\{\ldots, -4, -3, -2, -1, 0, 1, 2, 3, 4, \ldots\} = \text{integers}$$

It should be noted that a number such as 2 may be classified as a natural number, a whole number, or an integer, because it is an element of all three of these sets. But a number such as -2 can be classified only as an integer, and not as a natural number or a whole number. We can also classify -2 as a negative integer, whereas 2 is a positive integer. Note that 0 is neither positive nor negative, but it is an integer.

It is possible to picture the set of integers,

$$\{\ldots, -4, -3, -2, -1, 0, 1, 2, 3, 4, \ldots\}$$

on a **number line**, as shown in Figure 8.1. First, draw a line, pick a point on the line, and label it 0. Next, mark off equal units to the right and left of 0. Label the endpoints of the intervals to the right of 0 with the *positive integers*, and those to the left of 0 with the *negative integers*.

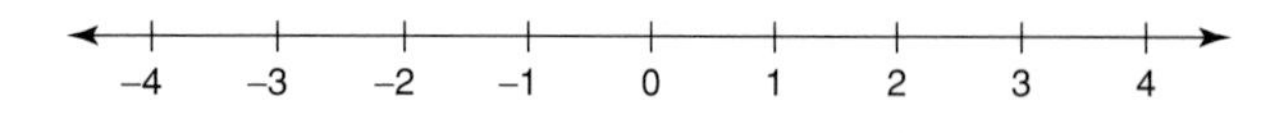

Figure 8.1

Numbers are placed on a number line in numerical order from left to right. Thus, moving from left to right the numbers become larger in value, whereas moving from right to left the numbers become smaller in value.

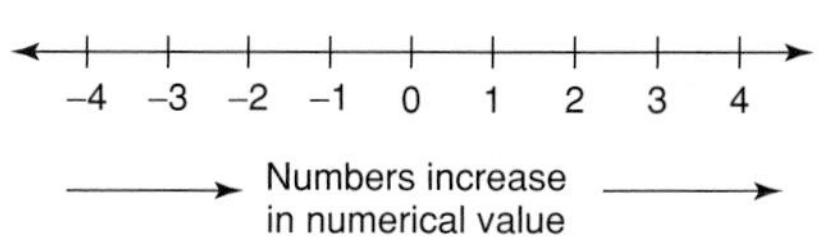

The number line may be extended indefinitely in both directions—that is, to the left and right of zero—as indicated by the arrowheads. Remember that zero is an integer, but it is neither positive nor negative.

Using the number line, we can determine the order of the integers. Three is greater than 2 ($3 > 2$) because 3 is to the right of 2 on the number line. Similarly, 4 is greater than 1 ($4 > 1$) because 4 is to the right of 1 on the number line. For the same reason, 0 is greater than -1 ($0 > -1$) and -1 is greater than -4 ($-1 > -4$). Other observations that we might make are

$$4 > 0, \quad 1 > 0, \quad 1 > -1, \quad 2 > -2, \quad -2 > -4$$

Instead of stating that 3 is greater than 2 ($3 > 2$), we could have said that 2 is less than 3 ($2 < 3$), because 2 is to the left of 3 on the number line. Similarly, $1 < 4$ (1 is less than 4) because 1 is to the left of 4 on the number line. We can also say that

$$0 < 4, \quad 0 < 1, \quad -1 < 1, \quad -2 < 2, \quad -4 < -2$$

Adding and Subtracting Integers

Prior to our discussion of the set of integers, we used the minus sign ($-$) only to indicate subtraction. Now we are also using it to label negative integers. The number 3 is three units to the right of 0 on the number line and its "opposite," negative 3 (-3), is three units to the left of 0 on the number line. The opposite of 4 is negative 4 (-4), and the opposite of 1 is negative 1 (-1). Note that the sum of any number and its opposite is 0:

$$3 + (-3) = 0, \quad 4 + (-4) = 0, \quad \text{and} \quad 1 + (-1) = 0$$

It should also be noted that the opposite of 0 is 0: $0 + 0 = 0$. You see that the opposite of any number is also the additive inverse of the given number.

How do we combine integers? Consider the sum of the positive integers 2 and 3. We know that $2 + 3 = 5$. However, let us work this problem on the number line. To find the solution to $2 + 3$ on the number line, we start at 0 and proceed to the right two units to 2. Because we want to add 3, we proceed three more units to the right; this brings us to 5. Hence, $2 + 3 = 5$. This process is illustrated in Figure 8.2.

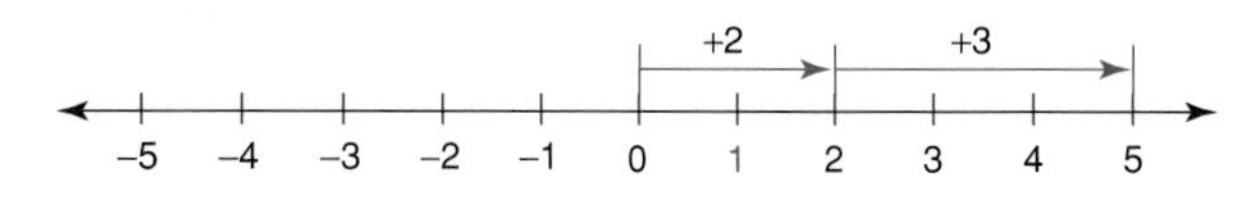

Figure 8.2 $2 + 3 = 5$.

Now let's combine a positive and a negative integer, such as $2 + (-3)$, on a number line. We start at 0 and proceed two units to the right to 2. Because we want to add -3 to 2, we move three units to the *left* from 2. We end up at -1.

Therefore,

$$2 + (-3) = -1$$

This is illustrated in Figure 8.3.

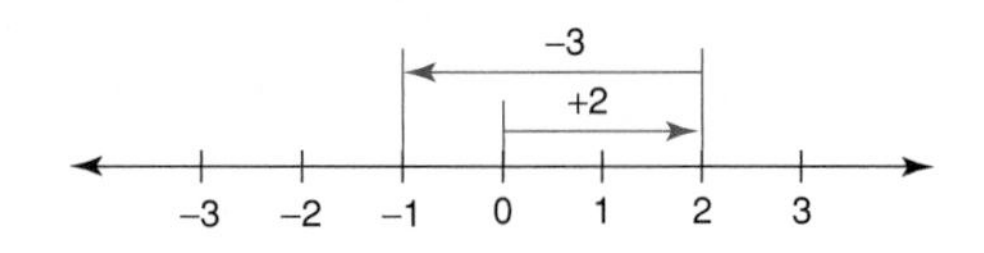

Figure 8.3 $2 + (-3) = -1$.

We could do this problem in another way: -3 can be expressed as $(-2) + (-1)$; then the original problem becomes

$$2 + (-3) = 2 + (-2) + (-1)$$

But (-2) is the opposite, or additive inverse, of 2 and thus

$$2 + (-2) + (-1) = 0 + (-1) = -1$$

Summarizing these steps, we have

$$2 + (-3) = [2 + (-2)] + (-1) = 0 + (-1) = -1$$

We are not advocating one technique over the other. The method that you best understand is the one to use.

EXAMPLE 1 ***Adding Integers*** Evaluate $1 + (-4)$.

Solution $1 + (-4) = -3$

On the number line in Figure 8.4, we start at 0 and proceed one unit to the right to 1. Next, we move four units in a negative direction—that is, to the left—from 1. We end up at -3.

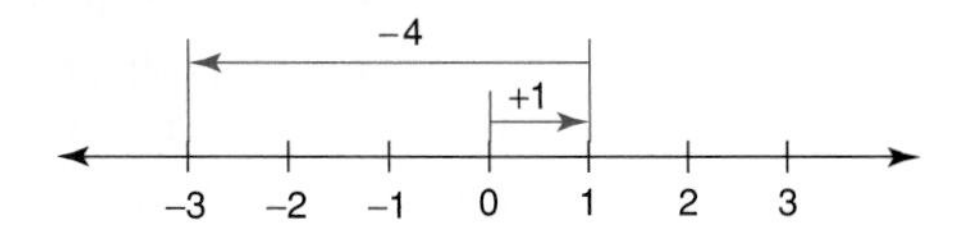

Figure 8.4 $1 + (-4) = -3$.

Alternate Solution

$$1 + (-4) = [1 + (-1)] + (-3) = 0 + (-3) = -3$$

EXAMPLE 2 ***Adding Integers*** Evaluate $-2 + 5$.

Solution $-2 + 5 = 3$

We start at 0 on the number line in Figure 8.5 and move two units to the left to -2. Next we move five units to the right (positive direction) from -2; we end up at 3.

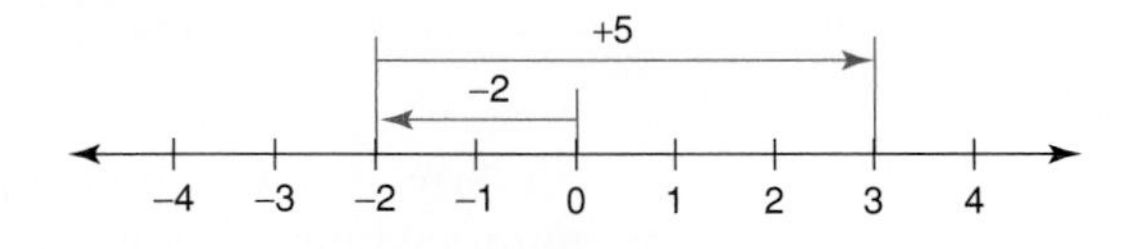

Figure 8.5 $-2 + 5 = 3$.

Alternate Solution

$$-2 + 5 = [-2 + 2] + 3 = 0 + 3 = 3$$

NW ***Now Work Problems 1a and 1d.***

If we add two positive integers, then their sum will be a positive integer. If we add two negative integers, then their sum will be a negative integer. If we add a positive integer and a negative integer, then their sum may be a positive integer, a negative integer, or zero.

The problem $8 - 5$ has the same answer as $8 + (-5)$. Similarly,

$$7 - 4 = 7 + (-4) \quad \text{and} \quad 3 - 1 = 3 + (-1)$$

This seems to indicate that for integers a subtraction problem can be thought of as an addition problem. That is, subtracting integers is the same as adding the opposite of the second integer to the first integer.

You probably do not need this rule for problems like $8 - 5$ and $7 - 4$. But what about $5 - 8$? According to the above rule, we can think of $5 - 8$ as $5 + (-8)$. Now we have a problem similar to the ones that we did in Examples 1 and 2. Using the number line, as in Figure 8.6, $5 + (-8) = -3$.

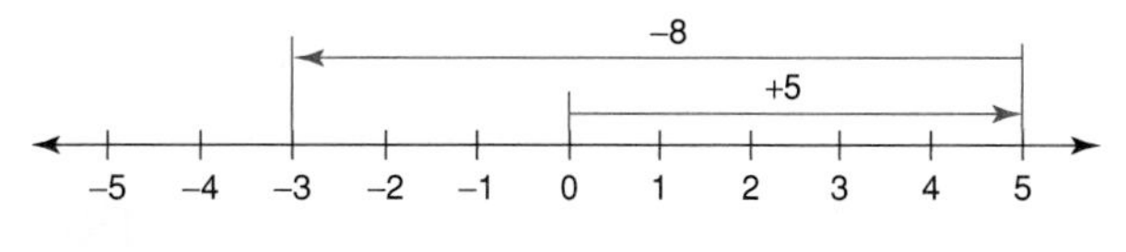

Figure 8.6 $5 + (-8) = -3$.

Alternate Solution

$$5 - 8 = 5 + (-8) = [5 + (-5)] + (-3) = 0 + (-3) = -3$$

EXAMPLE 3 ***Subtracting Integers*** Evaluate $3 - 4$.

Solution $3 - 4 = -1$

$3 - 4$ is the same as $3 + (-4)$. Using the number line as in Figure 8.7, we have $3 + (-4) = -1$.

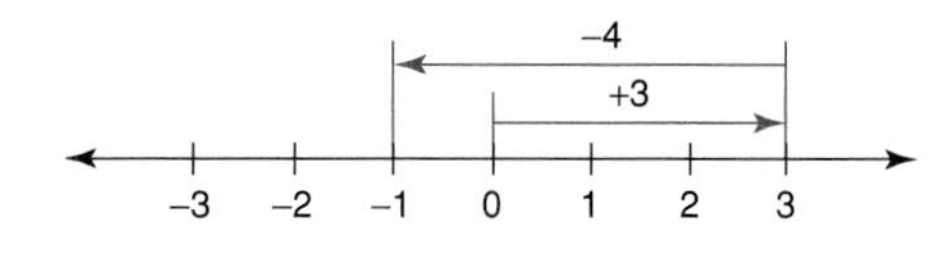

Figure 8.7 $3 + (-4) = -1$.

Alternate Solution

$$3 - 4 = 3 + (-4) = [3 + (-3)] + (-1) = 0 + (-1) = -1$$

EXAMPLE 4 ***Subtracting Integers*** Evaluate $-3 - 2$.

Solution $-3 - 2 = -5$

$-3 - 2$ is the same as $-3 + (-2)$. Using the number line, as in Figure 8.8, we have $-3 + (-2) = -5$. Note that when we add two negative integers their sum is a negative integer.

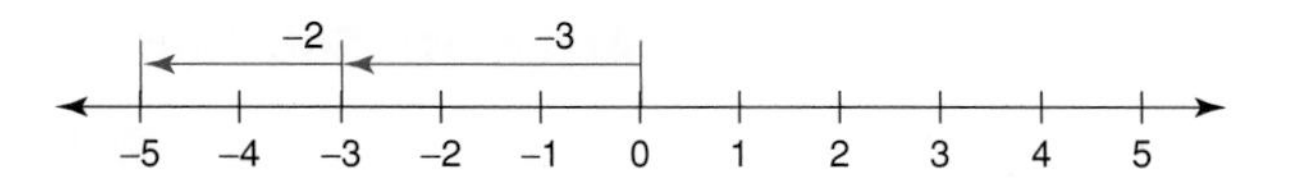

Figure 8.8 $-3 + (-2) = -5.$

Alternate Solution

$$-3 - 2 = -3 + (-2) = -(3 + 2) = -5$$

NW ***Now Work Problems 3a and 3c.***

EXAMPLE 5 ***Subtracting a Negative Integer*** Evaluate $-2 - (-3)$.

Solution $-2 - (-3) = 1$

Subtracting integers is the same as adding the opposite of the second integer to the first integer. Therefore, $-2 - (-3)$ is the same as $-2 + 3$. (Note that 3 is the opposite of -3.) Using the number line, we see that $-2 + 3 = 1$, as shown in Figure 8.9.

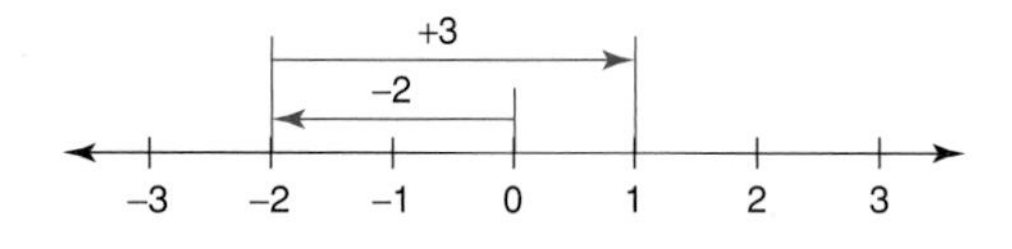

Figure 8.9 $-2 + 3 = 1.$

Alternate Solution

$$-2 - (-3) = -2 + 3 = [-2 + 2] + 1 = 0 + 1 = 1$$

EXAMPLE 6 ***Subtracting a Negative Integer*** Evaluate $-5 - (-3)$.

Solution $-5 - (-3) = -2$

This problem is similar to Example 5: $-5 - (-3)$ is the same as $-5 + 3$. By means of the number line in Figure 8.10, we see that $-5 + 3 = -2$.

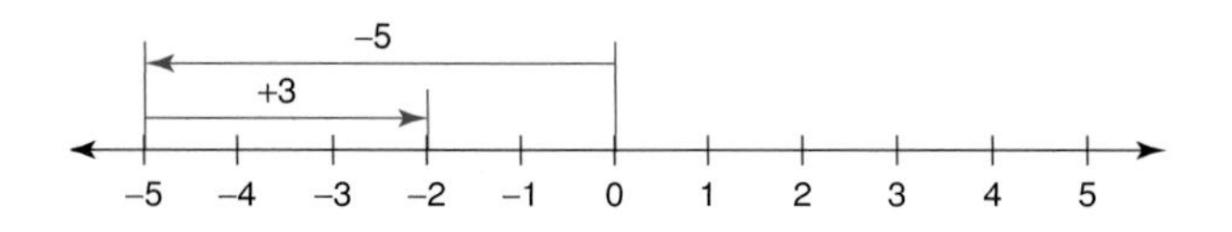

Figure 8.10 $-5 + 3 = -2.$

Alternate Solution

$$-5 - (-3) = -5 + 3 = -2 + [(-3) + 3] = -2 + 0 = -2$$

NW ***Now Work Problems 5a and 5b.***

To add two integers with opposite signs, consider the distance each integer is from zero. Subtract the shorter distance from the longer distance, and use the sign of the longer distance in the answer.

To add integers with the same sign, add without regard to the signs. The answer contains the same sign as the given numbers.

To subtract two integers, add the opposite of the second integer to the first integer.

Multiplying Integers

This desert area has a negative altitude because it is below sea level.

Once we have mastered addition and subtraction of integers, we can proceed to the operation of multiplication. We already know how to multiply the natural numbers: for example,

$$4 \times 2 = 8, \quad 5 \times 3 = 15, \quad \text{and} \quad 7 \times 5 = 35$$

These problems are also examples of multiplying positive integers. One observation that we can make from these examples is the following:

A positive integer multiplied by a positive integer always yields a positive integer.

But what happens when we multiply a positive integer by a negative integer? Consider the example $2 \times (-4)$. What is the answer? We can think of $2 \times (-4)$ as $(-4) + (-4)$, which equals -8. Therefore, $2 \times (-4) = -8$. Let's try another example, $3 \times (-5)$; this can also be expressed as $(-5) + (-5) + (-5)$, which equals -15. Therefore, $3 \times (-5) = -15$.

Whenever we multiply a positive integer by a negative integer, the answer is always a negative integer.

Before we consider the product of two negative integers, let us consider the *distributive property.* It states that

$$a \times (b + c) = a \times b + a \times c$$

for all integers a, b, and c. In terms of a specific example,

$$3 \times (2 + 5) = 3 \times 2 + 3 \times 5$$

More formally, we call this the **distributive property (or law) for multiplication over addition**. When we are given an expression such as $3 \times (2 + 5)$, we can evaluate it in two different ways:

$$3 \times (2 + 5) = 3 \times 7 = 21$$

or

$$3 \times (2 + 5) = 3 \times 2 + 3 \times 5$$
$$= 6 + 15 = 21$$

Similarly,

$$(-2) \times (5 + 6) = (-2) \times 11 = -22$$

or

$$(-2) \times (5 + 6) = (-2) \times 5 + (-2) \times 6$$
$$= -10 + (-12) = -22$$

Consider the example $2 \times (-4 + 4)$. We can also evaluate this in two ways:

$$2 \times (-4 + 4) = 2 \times (0) = 0$$

or

$$2 \times (-4 + 4) = 2 \times (-4) + 2 \times 4 = -8 + 8 = 0$$

Now consider the example $(-2) \times (-4 + 4)$. Evaluating this both ways, we have

$$(-2) \times (0) = 0 \quad \text{and} \quad (-2) \times (-4) + (-2) \times 4 = ?$$

Here we run into the problem of multiplying two negative integers. But now we can determine the answer. Because

$$a \times (b + c) = a \times b + a \times c$$

for all integers a, b, and c, $(-2) \times (-4 + 4)$ must have the same answer regardless of which way we evaluate it. We already know that

$$(-2) \times (-4 + 4) = (-2) \times (0) = 0$$

Consequently,

$$(-2) \times (-4 + 4) = (-2) \times (-4) + (-2) \times 4$$

must also equal 0. We already know that $(-2) \times 4 = -8$; we must determine the answer to $(-2) \times (-4)$. We also know that when the answer to $(-2) \times (-4)$ is added to -8, the final answer must be 0. What number added to -8 will give an answer of 0? The opposite, or additive inverse, of -8, namely, 8. Therefore, $(-2) \times (-4)$ must equal 8. We list the steps again:

$$\begin{aligned}(-2) \times (-4+4) &= (-2) \times (-4) + \underbrace{(-2) \times 4}_{} = 0 \\ &= \underbrace{(-2) \times (-4)}_{} + \quad (-8) \quad = 0 \\ &= \quad 8 \quad + \quad (-8) \quad = 0\end{aligned}$$

Whenever we multiply two negative integers, the product is always positive.

Table 8.1 provides a summary of the rules we have developed for multiplying integers.

TABLE 8.1

Rule	Notation	Example
1. The product of two positive integers is positive.	**1.** $(+) \times (+) = +$	**1.** $2 \times 3 = 6$
2. The product of a positive integer and a negative integer is negative.	**2.** $(+) \times (-) = -$	**2.** $2 \times (-3) = -6$ or $(-2) \times 3 = -6$
3. The product of two negative integers is positive.	**3.** $(-) \times (-) = +$	**3.** $(-2) \times (-3) = +6$
4. The product of any integer and zero is zero.	**4.** $(+) \times 0 = 0$ and $(-) \times 0 = 0$	**4.** $(-2) \times 0 = 0$, $100 \times 0 = 0$

EXAMPLE 7 ***Multiplying Integers*** Evaluate each of the following:

a. 4×3 **b.** $4 \times (-3)$

c. $(-4) \times (-3)$ **d.** 4×0

Solution

a. $4 \times 3 = 12$. The product of two positive integers is positive.

b. $4 \times (-3) = -12$. The product of a positive integer and a negative integer is negative.

c. $(-4) \times (-3) = 12$. The product of two negative integers is positive.

d. $4 \times 0 = 0$. The product of any integer and zero is zero.

EXAMPLE 8 ***Multiplying Integers*** Evaluate each of the following:

a. $7 \times (-8)$ **b.** $(-3) \times 9$

c. $(-9) \times (-4)$ **d.** $(-6) \times (-6)$

Solution

a. $7 \times (-8) = -56$ **b.** $(-3) \times 9 = -27$

c. $(-9) \times (-4) = 36$ **d.** $(-6) \times (-6) = 36$

NW ***Now Work Problems 7a and 7c.***

Math Connections

Multiplication Shortcuts

To add 9 + 6 most would know that the answer is 15 because of the addition facts. But consider that

$$9 + 6 = 9 + 1 + 6 - 1$$

Note that $1 - 1 = 0$, so the original expression $9 + 6$ is equal to itself. Now, consider the following:

$$\begin{aligned} 9 + 6 &= 9 + 1 + 6 - 1 \\ &= 10 + 5 = 15 \end{aligned}$$

For any number N,

$$9 + N = 10 + (N - 1)$$

A similar statement is also true for 99, 999, etc; that is,

$$99 + N = 100 + (N - 1)$$

and

$$999 + N = 1000 + (N - 1)$$

Ninety-nine can be added to any number by adding 100 and subtracting 1. Similarly, 999 can be added to any number by adding 1000 and subtracting 1. For example,

$$\begin{aligned} 99 + 44 &= 100 + (44 - 1) \\ &= 100 + 43 = 143 \end{aligned}$$

and

$$\begin{aligned} 999 + 462 &= 1000 + (462 - 1) \\ &= 1000 + 461 = 1461 \end{aligned}$$

Multiplication is repeated addition and it therefore follows that similar shortcuts exist for multiplication by 9, 99, and 999.

Consider the multiplication problem, 9×7. We can rewrite this as

$$\begin{aligned} 9 \times 7 &= (10 \times 7) - 7 \\ &= 70 - 7 = 63 \end{aligned}$$

From this example, we can see that 9 times a number is the same as 10 times the number if we subtract the number. That is,

$$9N = \underbrace{(10 - 1)}_{9}N = 10N - N$$

Similarly, multiplying by 99, we have

$$99N = \underbrace{(100 - 1)}_{99}N = 100N - N$$

Therefore,

$$\begin{aligned} 99 \times 75 &= (100 - 1)75 = 7500 - 75 \\ &= 7425 \\ 99 \times 32 &= (100 - 1)32 = 3200 - 32 \\ &= 3168 \end{aligned}$$

If we multiply by 999, we have

$$999N = \underbrace{(1000 - 1)}_{999}N = 1000N - N$$

and

$$\begin{aligned} 999 \times 432 &= (1000 - 1)432 \\ &= 432{,}000 - 432 \\ &= 431{,}568 \end{aligned}$$

Practice these shortcuts, and you will be amazed at how quickly you will be able to perform certain tedious calculations.

Dividing Integers

Earlier in this section, we noted that every integer has an opposite. For example, the opposite of 4 is -4 and the opposite of -3 is 3. In mathematics, an integer and its opposite are formally called *additive inverses* of each other. This concept of additive inverse (or opposite) is what enables us to treat subtraction problems involving integers as addition problems. For example, $8 - (-3)$ can be thought of as adding the opposite of -3 to 8 because the additive inverse of -3 is 3. Therefore, $8 - (-3) = 8 + (3) = 11$. This relationship implies that subtraction is the inverse (i.e., opposite) operation of addition.

In a similar manner, division can be thought of as being the opposite or inverse operation of multiplication. To illustrate this relationship, consider the division problem $8 \div 4 = 2$.

To divide 8 by 4 means to find a number (?) so that when it is multiplied by 4 we get a product of 8.

$$4\overline{\smash{)}8}^{\;?}$$

$$? \times 4 = 8$$

$$2 \times 4 = 8$$

Thus, $8 \div 4 = 2$ because $2 \times 4 = 8$.

As a result of this relationship between multiplication and division, we can extend and apply our rules for multiplying integers to dividing integers. For example, consider the following two division problems:

$$6 \div 2 \quad \text{and} \quad -6 \div -2$$

In the first problem, we know that $6 \div 2 = 3$ because $3 \times 2 = 6$. This division was pretty straightforward since both integers were positive. In the second problem, though, we are dividing two negative integers. Following the same approach as before, we re-express the problem as an equivalent multiplication problem. That is, $-6 \div -2 = ? \times -2 = -6$. In other words, we ask ourselves what do we have to multiply -2 by to get -6. Notice that the answer must be positive because we learned that whenever we multiply a positive integer by a negative integer the answer is always a negative integer. Thus, the answer is 3. As a result of these two problems, we make the following observation:

> ***When dividing two positive integers or two negative integers, the quotient will always be positive.***

Let's now examine what happens when we divide two integers that do not have the same sign. As an example, consider the following two problems:

$$-6 \div 2 \quad \text{and} \quad 6 \div -2$$

Following the same approach as before, we first re-express each division problem as a multiplication. The first problem, $-6 \div 2$, is rewritten as $? \times 2 = -6$. Recalling our rules for multiplying integers, we observe that the answer must be -3 because a negative integer multiplied by a positive integer yields a negative integer. In the second problem, $6 \div -2$ is rewritten as $? \times -2 = 6$, and we note that the answer must also be -3 because a negative integer multiplied by a negative integer yields a positive integer. As a result of these two problems, we make the following observation:

> ***When dividing a positive integer by a negative integer, or a negative integer by a positive integer, the quotient will always be negative.***

EXAMPLE 9 ***Dividing Integers*** Evaluate each of the following:

a. $14 \div -2$ **b.** $-8 \div 4$

c. $-9 \div -3$ **d.** $-21 \div -7$

Solution

a. $14 \div -2 = -7$ **b.** $-8 \div 4 = -2$

c. $-9 \div -3 = 3$ **d.** $-21 \div -7 = 3$

NW ***Now Work Problems 9a and 9f.***

Order of Operations

Frequently in mathematics, a numerical expression involves a mix of more than one operation. For example, $6 + (3 \times 4)$, $9 + [(3 + 4) \times 5]$, and $4 \times 4 \div 8 - 2 \times 9$ are three such expressions. To evaluate these expressions, we must agree on a specific *order of operations* because if not, then it is possible to arrive at different answers. A simple illustration is the problem $2 + 3 \times 4$. Without a prescribed order

Historical Note

Rules for multiplying positive and negative numbers were known to Diophantus of Alexandria, approximately A.D. 275. But his rules or ideas were not known in Europe until approximately 1600. Hindu mathematicians, as early as A.D. 700, considered positive numbers to represent possession and negative numbers to indicate debt.

A Hindu named Bhaskara (approximately A.D. 1150) is believed to be one of the first to indicate negative numbers as roots of equations. But European mathematicians did not allow or consider negative numbers as roots of equations until the fifteenth century. It is interesting to note that the negative numbers were long in gaining acceptance to represent loss or debt, and it took even longer for negative numbers to be allowed as solutions of equations. For example, a simple equation such as $x + 1 = 0$ has no solution if negative numbers are not allowed. Historians speculate that this was probably due to the fact that earlier mathematicians viewed numbers as existing before humankind, and therefore possessed some mystical properties.

The first appearance in print of our present + and − signs is in an arithmetic book published in the late 1400s by Johann Widman. The signs in this book were not used to indicate operations, but indicated excess and deficiency. It is thought that the plus sign originated from the Latin word et (and), which was used to indicate addition, and maybe the minus sign originated from the abbreviation $\overline{M}$ for minus.

A number of books on arithmetic were published in Europe during the late 1400s and early 1500s. They were largely devoted to explaining the writing of numbers, computation with them, and some applications to partnerships and barter. In 1491, a book by Filippo Calandri was published in Florence, Italy, that contained the first printed example of long division as we know it today. This means that for people of that era, such as Christopher Columbus, higher mathematics consisted of problems involving long division.

of operations to follow, this problem can be solved in two ways and has two different answers. This is shown next.

Solution 1	*Solution 2*
$2 + 3 \times 4$	$2 + 3 \times 4$
$= 5 \quad \times 4$	$= 2 + \quad 12$
$= 20$	$= 14$

To prevent situations such as this, mathematicians have agreed to follow a specific order of operations.

Order of Operations

1. Evaluate all expressions within grouping symbols first. If more than one set of grouping symbols is used, then always begin with the innermost set.
2. Working from left to right, perform all multiplications and divisions.
3. Working from left to right, perform all additions and subtractions.

As a result, the correct answer to the preceding problem is 14 because when working from left to right, and in the absence of any grouping symbols, multiplication is performed before addition.

To evaluate an expression that contains two or more grouping symbols, the expression in the innermost grouping symbol is evaluated first. Hence,

$$9 + [(3 + 4) \times 5] = 9 + [7 \times 5] = 9 + 35 = 44$$

If no grouping symbols appear in a numerical expression, the standard convention is to perform all multiplications and divisions in the order in which they occur (from left to right) and then do all the additions and subtractions in the order in which they occur (from left to right). Therefore,

$$\begin{aligned} 4 \times 3 + 5 - 2 \times 9 - 11 &= 12 + 5 - 18 - 11 \\ &= 17 - 18 - 11 \\ &= -1 - 11 \\ &= -12 \end{aligned}$$

EXAMPLE 10 ***Using the Order of Operations*** Evaluate each of the following:

a. $3 + (6 \times 4)$ **b.** $12 - (3 \times 4 + 5)$

c. $2 \times 4 + 5 - 6 \div 3$ **d.** $7 + [(3 \times 4) \div 6]$

Solution

a. $3 + (6 \times 4) = 3 + 24 = 27$

b. $12 - (3 \times 4 + 5) = 12 - (12 + 5) = 12 - (17) = -5$

c. $2 \times 4 + 5 - 6 \div 3 = 8 + 5 - 2 = 13 - 2 = 11$

d. $7 + [(3 \times 4) \div 6] = 7 + [12 \div 6] = 7 + (2) = 9$

NW ***Now Work Problems 11d and 11e.***

Even and Odd Integers

Every integer can be classified as either an even integer or an odd integer. An integer is even if it can be expressed as 2 times another integer. For example, 6 is an even integer because $6 = 2 \times 3$. Similarly, -2 is even because $-2 = 2 \times (-1)$, and 0 is even since $0 = 2 \times 0$. In general, any even integer may be expressed in the form $2 \times n$, $2 \cdot n$ or $2(n)$, where n represents some integer.

If an integer is not even, then it must be odd. If $2 \cdot n$ represents an even integer, then adding 1 more to it will make it odd. Therefore, $2 \cdot n + 1$ represents an odd integer. Because 6 is even, adding 1 more will yield an odd integer, 7:

$$7 = 2 \cdot 3 + 1$$

Another odd integer is -5:

$$-5 = 2 \cdot (-3) + 1$$

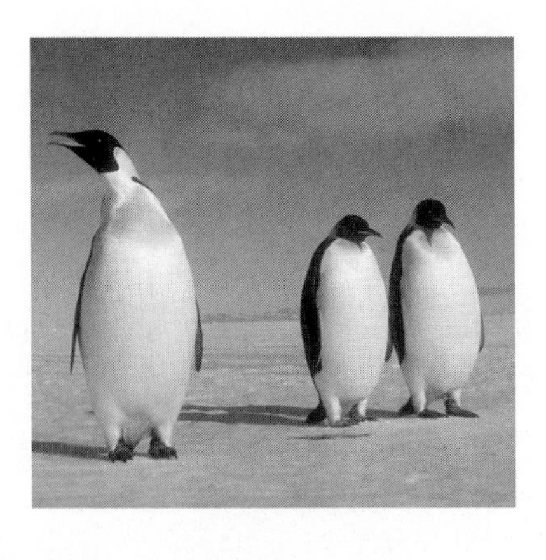

Some interesting observations can be made about odd and even integers. For example, what happens when you add two even integers? Considering a few examples, we see that $2 + 2 = 4$, $2 + 4 = 6$, $8 + 10 = 18$, and $12 + 6 = 18$. It appears that the sum of two even integers is also even. Are you sure? Considering some more examples, we see that $6 + 6 = 12$, $8 + 12 = 20$, and so on.

You are probably convinced that the sum of two even integers is even, but have we proved it? No, we have just observed what happens for a certain set of examples. We have not examined all of the possibilities, and therefore we cannot be sure that it is always true. If we can show that it is true for *any* two even integers, then we can be certain that the sum of any two specific even integers is even. Any even integer can be expressed as 2 times another integer, so let's consider the even integers $2 \times n$ and $2 \times k$, where n and k are also integers. Then we have

$$(2 \times n) + (2 \times k)$$

But

$$(2 \times n) + (2 \times k) = 2 \times (n + k)$$

by means of the distributive property. Note that $(n + k)$ is an integer because the sum of any two integers is an integer. Therefore, $2 \times (n + k)$ is an even integer, since 2 times any integer is an even integer, and we have shown that the sum of any two even integers is also an even integer.

What happens when you add two odd integers? From $3 + 5 = 8$, $1 + 3 = 4$, and $5 + 7 = 12$, it appears that the sum of two odd integers is an even integer. Let's examine what happens when we add *any* two odd integers. Consider the odd integers $2 \times n + 1$ and $2 \times k + 1$. Adding them, we have

$$\begin{aligned}(2 \times n + 1) + (2 \times k + 1) &= 2 \times n + 2 \times k + 2 \\ &= 2 \times (n + k + 1)\end{aligned}$$

by means of the distributive property. Note that $(n + k + 1)$ is an integer because $n + k$ is an integer and the sum of 1 and an integer is still an integer. Therefore, $2 \times (n + k + 1)$ is an even integer because 2 times any integer is an even integer.

EXAMPLE 11 ***Even and Odd Integers*** Prove that the product of any two even integers is an even integer.

Solution Any two even integers may be expressed as $2 \times n$ and $2 \times k$. Consequently, we have

$$(2 \times n) \times (2 \times k) = 2 \times (2 \times n \times k)$$

$(2 \times n \times k)$ represents an integer because the product of three integers is an integer. Therefore, $2 \times (2 \times n \times k)$ is an even integer. Hence, the product of any two even integers is an even integer.

NW ***Now Work Problem 17a.***

EXAMPLE 12 ***Income Gain/Loss*** Figure 8.11 shows a comparison of airline America West's income for the second quarter of each given year.

a. What was the difference in second-quarter income between 2001 and 2002?

b. What was the difference in second-quarter income between 2002 and 2003?

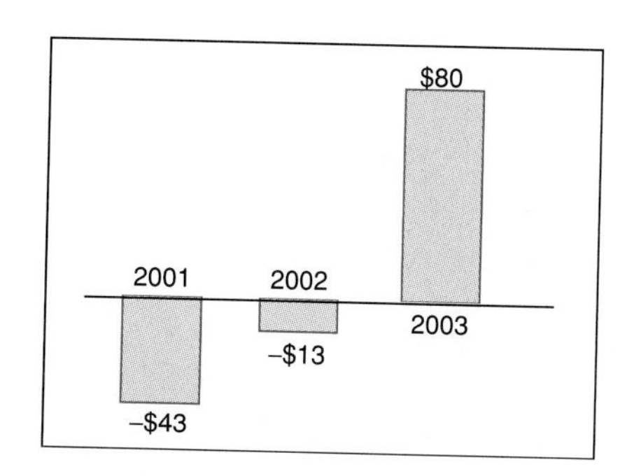

Figure 8.11

Second-quarter net income/loss (rounded to nearest million dollars).
Source: America West

Solution

a. The difference in second-quarter income between 2001 and 2002 was $-13 - (-43) = -13 + 43 = 30$. Thus, America West earned \$30 million more in the second quarter of 2002 than it did in the second quarter of 2001.

b. The difference in second-quarter income between 2002 and 2003 was $80 - (-13) = 80 + 13 = 93$. Thus, America West earned \$93 million more in the second quarter of 2003 than it did in the second quarter of 2002.

Math Connections

Absolute Value

An important concept in mathematics is that of *absolute value*. Although there are formal mathematical definitions, we can think of absolute value simply as the *distance from zero*. As an example, let's examine the following number line.

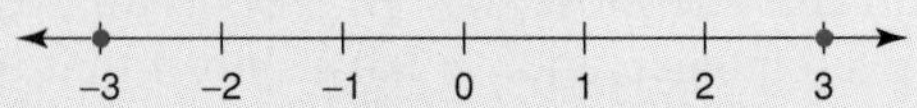

Notice that each point on the number line is exactly three units away from zero. As a result, we say that the absolute value of 3 is equal to 3 and that the absolute value of −3 is also equal to 3. We denote absolute value by placing a pair of vertical bars on either side of the number. Thus, in our example we have $|3| = 3$ and $|-3| = 3$.

Since absolute value can be regarded as the distance from zero, the absolute value of a number is never negative. This is because absolute value only asks the question, "how far?" and not "in which direction?" Once again, this is why $|3| = |-3| = 3$; both 3 and −3 are three units from zero, irrespective of their direction. Another way of looking at the concept of absolute value is to think of it as "magnitude," which in effect disregards (or ignores) the sign of a number. Following are examples of how absolute value expressions are evaluated:

$$|-5| = 5$$
$$|0 - 9| = |-9| = 9$$
$$|7 - 3| = |4| = 4$$
$$|3 - 7| = |-4| = 4$$
$$|0 \times (-8)| = |0| = 0$$
$$|2 + 3 \times (-4)| = |2 - 12| = |-10| = 10$$
$$5 - |3 - 9| = 5 - |-6| = 5 - 6 = -1$$

When comparing two absolute values, it is important to keep in mind that we are examining the respective magnitudes of each number and not their "face values." For example, $-7 < -4$ because −7 lies to the left of −4 on the number line. However, $|-7| > |-4|$ because $|-7|$ implies a distance of 7 units from zero and $|-4|$ implies a distance of 4 units from zero. Since $|-7|$ represents the larger distance, it is greater in value than $|-4|$.

Exercises for Section 8.4

1. Evaluate each of the following:
NW **a.** $2 + (-3)$ **b.** $4 + (-7)$
c. $5 + (-8)$ NW **d.** $-3 + 5$
e. $-7 + 8$ **f.** $-9 + 12$

2. Evaluate each of the following:
a. $-12 + 15 + (-9)$
b. $-2 + (-2) + (-2)$
c. $-4 + (-2) + 6$
d. $9 + 6 + 9$
e. $4 + (-8) + (-4)$
f. $7 + (-2) + (-5) + (-3)$

3. Evaluate each of the following:
NW **a.** $3 - 5$ **b.** $6 - 9$
NW **c.** $-8 - 6$ **d.** $6 - (-5)$
e. $-2 - (-1)$ **f.** $-7 - (-3)$

4. Evaluate each of the following:
a. $17 - (-1)$
b. $-18 - 2$
c. $14 - [-6 - (-1)]$
d. $-19 - [-3 - (-2)]$
e. $-12 - [-5 - (-8)]$
f. $[-16 - 9] - (-3)$

5. Evaluate each of the following:
NW **a.** $-7 - (-2)$ NW **b.** $-10 - (-10)$
c. $12 - (-13)$ **d.** $6 \times (-5 + 5)$
e. $7 \times (-5)$ **f.** $(-8 + 5) \times (-3)$

6. Evaluate each of the following:
a. $-8 + 8$ **b.** $-13 - (-12)$
c. $-6 - (8 + 3)$ **d.** $(-8) \times 9$
e. $(5 - 3) \times (-2)$ **f.** $(-2 - 1) \times (-3)$

7. Evaluate each of the following:
NW **a.** $(-4) \times (-5)$
b. $(-3) \times 2 \times 4$
NW **c.** $(-4) \times (-6) \times (-1)$
d. $(-2)(-2)(-2)(-2)$
e. $4 \times (-5)(-2)$
f. $7 \times (-3) \times (-3) \times 2$

8. Evaluate each of the following:
a. $12 \div -6$ **b.** $-9 \div 3$
c. $10 \div -5$ **d.** $-14 \div 2$
e. $20 \div -4$ **f.** $22 \div 2$

9. Evaluate each of the following:
NW **a.** $-9 \div -9$ **b.** $-21 \div -3$
c. $-25 \div -5$ **d.** $-100 \div -5$
e. $-300 \div -10$ NW **f.** $-567 \div 3$

10. Evaluate each of the following:
a. $0 \div -1$ **b.** $-5 \div 10$
c. $-25 \div -100$ **d.** $(5 - 9) \div 2$
e. $(-30 + 15) \div -5$ **f.** $-20 \div (24 - 28)$

11. Evaluate each of the following:
  a. $4(3 + 2) - 5$
  b. $(7 + 6 \times 4) - 12$
  c. $15 - (3 \times 2 - 1)$
  NW d. $8 \times 2 - 5 + 6 \div 3$
  NW e. $8 \div 2 \times 4 + 5 - 3$
  f. $7 \times [(3 + 15) \div 6]$

12. Evaluate each of the following:
  a. $3 \times (5 - 2) + 8$
  b. $(6 \times 4 + 3) \div 9$
  c. $16 - 2(3 + 4 \div 2)$
  d. $16 \div 4 - 2 + 6 \times 3$
  e. $12 \times 2 \div 4 + 8 - 2$
  f. $24 \div [(5 - 3) \times 3]$

13. Evaluate each of the following:
  a. $200 \div 10 + (-20) \times 4$
  b. $18 \div 9 \times (3 - 5) + (-16)$
  c. $5 \div 5 - 5 \times 5 + 4 \div 4 - 4$
  d. $-5 \div (6 - 5) \times 5 + 4 \div 4 - 4$
  e. $6 \div (3 - 5) + 5 \times [10 \div (9 - 4)]$
  f. $25 \div [(15 - 5) \div (6 - 8)] \times 4 - 4$

14. Replace each question mark with =, >, or < to make the sentence true.
  a. $13 ? 5$
  b. $3 ? 4$
  c. $3 ? -4$
  d. $-5 + 7 ? 7 - 5$
  e. $1 + (-5) ? 4 \times (-1)$
  f. $(-1)(4) ? (-1)(-4)$

15. Replace each question mark with =, >, or < to make the sentence true.
  a. $-1 ? -2$
  b. $-10 ? -3$
  c. $-1 - (-1) ? 0$
  d. $-3 + 2 ? 2 - (-3)$
  e. $-1 - (-2) ? -3 + 4$
  f. $(-1)(-3) ? -3 + (-1)$

16. Classify each integer as odd or even and express it in the form $2 \times n + 1$ or $2 \times n$. (*Example:* 7 is an odd integer; $7 = 2 \times 3 + 1$.)
  a. 14 b. 13
  c. 41 d. -7

17. Classify each integer as odd or even and express it in the form $2 \times n + 1$ or $2 \times n$. (*Example:* 6 is an even integer; $6 = 2 \times 3$.)
  NW a. 12 b. 17 c. -6 d. 101

18. One of Goldbach's conjectures states that any odd number greater than 7 can be expressed as the sum of three odd primes (for example, $9 = 3 + 3 + 3$). Show that this conjecture is true for all positive odd integers from 11 up to and including 29.

19. On a particular Monday, the Dow Jones Average opened at 10,842. At the close of trading on Monday, it had fallen 11 points. On Tuesday at the close of trading it had gained 5 points. On Wednesday at the close of trading it had gained 12 points. On Thursday at the close of trading it had fallen 15 points. What was the opening Dow Jones Average on Friday?

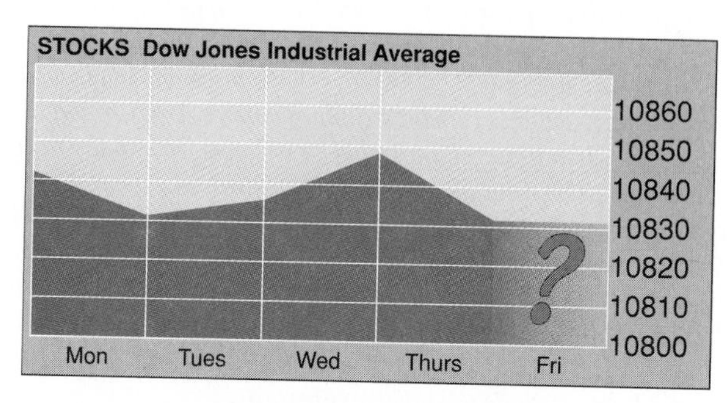

20. An elevator in the Empire State Building started at the 35th floor, rose 7 floors, descended 12 floors, rose 4 floors, descended 11 floors, descended 3 floors, and then rose 15 floors and stopped. At what floor did the elevator stop? Assuming that there are 15 feet between floors, how far did the elevator travel?

21. If Aristotle was born in 384 B.C. and Euclid was born in 300 B.C., who was born first?

22. Prove that the product of any two odd integers is an odd integer.

23. Prove that the sum of an odd integer and an even integer is an odd integer.

24. When you multiply an even integer by an odd integer, is the answer even or odd? Prove your answer.

Writing Mathematics

25. Explain the differences among the set of natural numbers, the set of whole numbers, and the set of integers.

26. What does it mean to say that two numbers are opposites (or additive inverses) of each other?

27. Rewrite the problem, $4 - (-1)$, as an equivalent addition problem and express in words what the equivalent relationship is between the given problem and the rewritten form.

28. What is the general rule for multiplying integers?

29. What is the general rule for dividing integers?

30. Describe the order of operations and explain why it is used.

31. Describe the distributive property for multiplication over addition and give an example of how it is used.

Challenge Exercises

32. Demonstrate how the number line can be used to conclude that the product of a negative integer and a negative integer is positive. (*Hint:* Observe the pattern that emerges when a negative integer such as -3 is multiplied by a positive integer that is consistently reduced by 1. For example, $-3 \times 3, -3 \times 2, -3 \times 1, -3 \times 0$, etc.).

33. Demonstrate how the number line can be used to conclude that the product of a positive integer and a negative integer is negative. (*Hint:* Observe the pattern that emerges when a positive integer such as 2 is multiplied by a positive integer that is consistently reduced by 1. For example, $2 \times 3, 2 \times 2, 2 \times 1, 2 \times 0$, etc.).

In our discussion of multiplying integers, we used the fact that multiplication is a form of repeated addition to help us develop a rule for finding the correct sign of a product. For example, $(-4) \times (2) = (-4) + (-4) = -8$. In a similar manner, we use exponents to express repeated multiplication. An exponent is written as the superscript of a number to indicate how many times the number is to be multiplied by itself. For example, the multiplication problem $2 \times 2 \times 2 \times 2$ can be re-expressed as 2^4, where 2 is called the base or factor and 4 is called the exponent. This expression is read as "2 to the fourth power" and means that the base 2 is to be multiplied by itself four times. Other examples include $5^2 = 5 \times 5 = 25$ and $3^4 = 3 \times 3 \times 3 \times 3 = 81$. In Exercises 34–44, evaluate the given expression.

34. 4^2 **35.** 3^2 **36.** 2^3

37. $2^2 \times 3^3$ **38.** $7^2 + 2^7$

39. $(-3)^2$ **40.** $(-5)^3$

41. $(-2)^3 + (-3)^2$ **42.** $(-4)^2 - (-5)^2$

43. -3^2 **44.** -5^3

Research/Group Activities

45. Dividing integers involving zero is not always a simple process. As an example, consider the three division problems, $0 \div 1$, $1 \div 0$, and $0 \div 0$. Research the concept of *division by zero* and prepare a report that explains the differences among these three division problems and then generalize your findings.

46. Many people conventionally think of "negative" as being bad. For example, it is not good to have a negative balance in your checking account, a negative outdoor temperature reading such as −15°F is unpleasant for most of us, and "gaining" negative yardage in football implies that you have lost yardage. There are, however, several situations in which "negative" is good. Working in a group, try to discover at least three instances where this is indeed the case and present your findings to your classmates.

Real-World Data

47. Wind chill Table 8.2 is a wind chill chart that shows how cold the air feels when wind speed is combined with air temperature.

a. What is the difference in wind chill between a thermometer reading of 20°F and wind speed of 15 miles per hour compared to the same temperature reading but with a wind speed of 25 miles per hour?

b. What is the difference in wind chill between a thermometer reading of 30°F and wind speed of 15 miles per hour compared to a thermometer reading of 10°F and wind speed of 10 miles per hour?

c. Which has the greater "danger" (as defined in Table 8.2) and what is the difference in temperature: A thermometer reading of 10°F above zero with a wind speed of 40 miles per hour or a thermometer reading of 20°F below zero and a wind speed of 5 miles per hour?

TABLE 8.2

	°Fahrenheit									
Wind Speed (miles/hour)	**50**	**40**	**30**	**20**	**10**	**0**	**−10**	**−20**	**−30**	**−40**
Calm	50	40	30	20	10	0	−10	−20	−30	−40
5	48	37	27	16	6	−5	−15	−26	−36	−47
10	40	28	16	4	−9	−21	−33	−46	−58	−70
15	36	22	9	−5	−18	−36	−45	−58	−72	−85
20	32	18	4	−10	−25	−39	−53	−67	−82	−96
25	30	16	0	−15	−29	−44	−59	−74	−88	−104
30	28	13	−2	−18	−33	−48	−62	−79	−94	−109
35	27	11	−4	−20	−35	−49	−67	−82	−98	−113
40	26	10	−6	−21	−37	−53	−69	−85	−100	−116

Little Danger More Danger Great Danger

48. Budget predictions Figure 8.12 shows the U.S. Congressional Budget Office's deficit projections released in January 1993 and this same office's surplus projections released in January 2001.

a. What was the total projected deficit for the 5-year period shown?

b. What was the total projected surplus for the 5-year period shown?

c. What is the difference between the total amounts calculated in parts **a** and **b**?

49. Bone density A bone density test measures a person's bone health. The result of the test is a *T*-score. If your *T*-

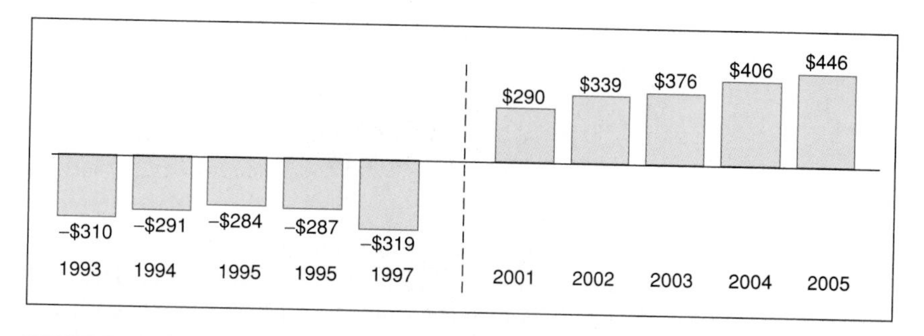

Figure 8.12

Congressional Budget Office deficit and surplus projections released in January 1993 and January 2001, respectively. (Figures are given in billions of dollars.) Source: Congressional Budget Office.

score is −1.0, then your bone mass is 10% below normal. A *T*-score of −2.0 means that your bone mass is 20% below normal, and you have osteoporosis (see Figure 8.13). In reporting this information, television and print ads frequently use the following statement: "The bigger the negative number, the lower your bone mass." What is wrong about the way this statement is worded and how should it be corrected?

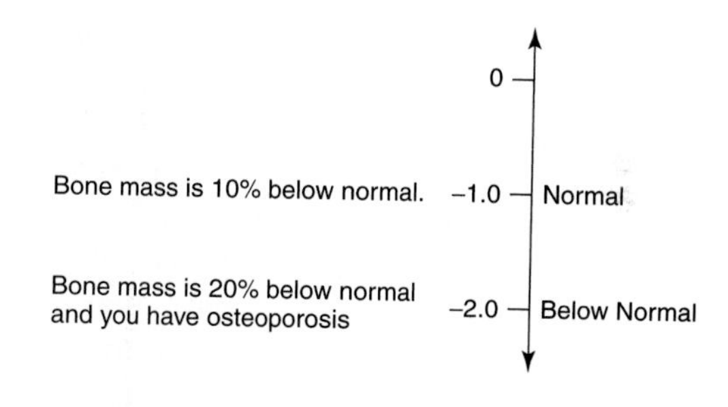

Figure 8.13 *What your T-score means.*

Just for fun

A numismatist (coin collector) was examining a collection of coins. In this collection he discovered a coin dated 384 B.C. What is your conclusion regarding this coin?

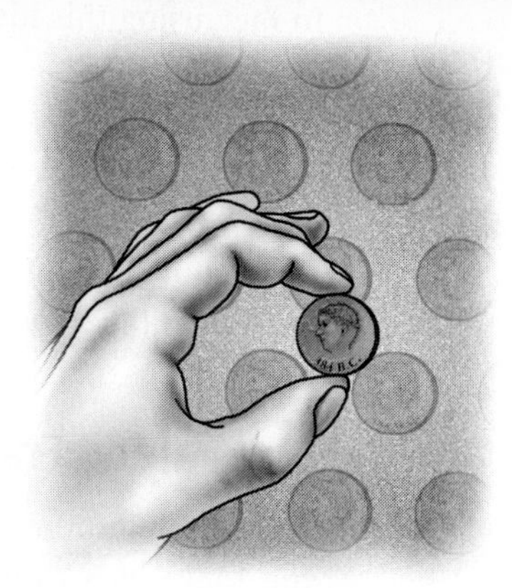

8.5 RATIONAL NUMBERS

Thus far we have examined the set of natural numbers, the set of whole numbers, and the set of integers. All of these can be shown on a number line like that in Figure 8.14.

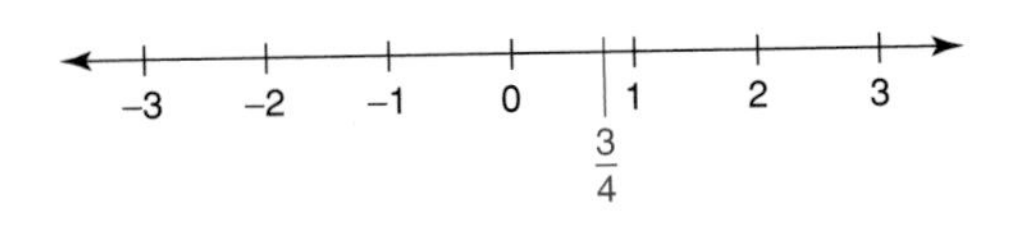

Figure 8.14

But what about the intervals between the numbers on the number line? Do any other numbers belong in these intervals? The answer is yes. Consider the number $\frac{3}{4}$: It is greater than 0 and less than 1, so it belongs in the interval between 0 and 1, as shown in Figure 8.14. What kind of number is it? It is not a natural number, it is not a whole number, and it is not an integer. It is a *rational number*.

A **rational number** is a number that can be expressed in the form a/b, where a and b are integers and $b \neq 0$ (we cannot divide by zero). In other words, a rational number is any number that can be expressed as the quotient of two integers. A rational number such as $\frac{3}{4}$ is commonly referred to as a fraction. Remember that both the numerator and denominator of the fraction must be integers, and that the denominator of the fraction cannot be 0.

*In this context, the term **rational** does not mean reasonable or in good mental health. It comes from the word ratio. Thus, rational numbers are so named because they can be written as the ratio of two integers.*

Note that a rational number also denotes a *ratio*, which is a quotient of two numbers. A ratio provides a comparison between two entities. For example, in a popular television advertisement, we are told that 4 out of 5 dentists recommend a particular product. The ratio in this advertisement is "4 out of 5 dentists" and is commonly expressed as 4 to 5, 4 : 5, or $\frac{4}{5}$. A ratio of two measurements that have differents units of measure is called *rate*. For example, if we travel 20 miles on 1 gallon of gasoline, then the ratio $\frac{20 \text{ miles}}{1 \text{ gallon}}$ is called a rate. In this example, the rate is 20 miles per gallon. We will study the concept of ratio in more detail in Chapter 12.

Is the number 4 a rational number? The answer is yes, because a rational number is any number that can be expressed as the quotient of two integers, and

$$4 = \frac{4}{1}$$

In fact, using this idea we can see that any integer can be expressed as a quotient of two integers:

$$7 = \frac{7}{1}, \quad 10 = \frac{10}{1}, \quad -4 = \frac{-4}{1}, \quad -8 = \frac{-8}{1}, \quad \text{and} \quad 0 = \frac{0}{1}$$

Every integer is also a rational number, but remember that not all rational numbers are integers. (The fraction $\frac{3}{4}$ is an example of a rational number that is not an integer.)

Equivalent Fractions

A rational number may be expressed in many different but equivalent ways. Recall that 4 is a rational number because $4 = \frac{4}{1}$. However, we can also express 4 using many different but equivalent rational numbers. For example,

$$4 = \frac{4}{1} = \frac{8}{2} = \frac{16}{4} = \frac{32}{8} = \dots$$

These equivalent rational numbers, which are commonly referred to as *equivalent fractions*, can be generated by multiplying a given number by 1. Recall that 1 is the identity element for multiplication, which means that the product of a number and 1 is the number itself (i.e., for any number a, $a \times 1 = a$). In the case of rational numbers, though, we generate equivalent fractions by multiplying by a "disguised" 1. For example, note that

$$\frac{4}{1} \times \frac{2}{2} = \frac{8}{2}, \quad \frac{8}{2} \times \frac{2}{2} = \frac{16}{4}, \text{ and } \frac{16}{4} \times \frac{2}{2} = \frac{32}{8}, \ldots$$

In each instance, new and equivalent rational numbers were formed by multiplying by the special fraction $\frac{2}{2}$, which is equal to 1. In general, for any fraction $\frac{a}{b}$, if k is any number other than zero, then

$$\frac{a \times k}{b \times k} = \frac{a}{b}$$

This rule is helpful in reducing, or simplifying, fractions. Given the fraction $\frac{15}{25}$, we can reduce it by factoring both the numerator and denominator into prime factors:

$$\frac{15}{25} = \frac{3 \times 5}{5 \times 5}$$

Applying the rule, we have

$$\frac{15}{25} = \frac{3 \times \not{5}}{5 \times \not{5}} = \frac{3}{5}$$

Note that we eliminated the factors that were common to both the numerator and denominator.

How do we know that our answer is correct? Does $\frac{15}{25} = \frac{3}{5}$? An expression such as this can be verified by cross-multiplying to see if the products are equal:

$$15 \times 5 \stackrel{?}{=} 25 \times 3$$
$$75 = 75$$

The product of the means and extremes of a proportion are found by cross-multiplying:

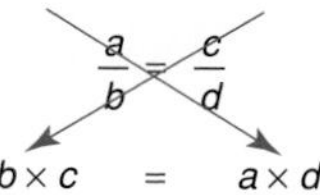

$b \times c \quad = \quad a \times d$

The multiplication process forms a cross and the left cross-product ($b \times c$) is equal to the right cross-product ($a \times d$).

Another way of looking at this is to recall that the expression $\frac{15}{25} = \frac{3}{5}$ denotes the equivalence of two ratios, which is called a *proportion*.

For any proportion to be true, the product of the *means* must equal the product of the *extremes.* More formally, we say that

$$\frac{a}{b} = \frac{c}{d} \text{ if and only if } \underbrace{a \times d}_{\text{Extremes}} = \underbrace{b \times c}_{\text{Means}}$$

We will examine proportions in more detail in Chapter 12.

EXAMPLE 1 ***Determining If Two Fractions Are Equivalent*** Does $\frac{3}{11} = \frac{9}{33}$?

Solution The answer is yes, because $3 \times 33 = 9 \times 11$; that is, $99 = 99$.

Alternate Solution We reduce $\frac{9}{33}$ and see if the result is $\frac{3}{11}$:

$$\frac{9}{33} = \frac{\not{3} \times 3}{\not{3} \times 11} = \frac{3}{11}$$

EXAMPLE 2 ***Reducing a Fraction*** Reduce $\frac{42}{54}$.

Solution

$$\frac{42}{54} = \frac{\not{2} \times \not{3} \times 7}{\not{2} \times \not{3} \times 3 \times 3} = \frac{7}{3 \times 3} = \frac{7}{9}$$

Check:

$$\frac{42}{54} = \frac{7}{9} \quad \text{if and only if} \quad 42 \times 9 = 7 \times 54$$
$$378 = 378$$

NW ***Now Work Problem 1a.***

Adding and Subtracting Rational Numbers

How do we combine rational numbers? Because 2 and 3 are rational numbers and $2 + 3 = 5$, we already know how to combine some rationals. But consider the problem of adding the rational numbers $\frac{1}{5} + \frac{2}{3}$. In Section 8.3 we discussed the process of adding and subtracting fractions. However, those problems were considered only with regard to the use of the least common multiple. Let's state a general rule for adding any two rational numbers.

If $\frac{a}{b}$ and $\frac{c}{d}$ are rational numbers, then $\frac{a}{b} + \frac{c}{d} = \frac{ad + bc}{bd}$.

Therefore, for the example $\frac{1}{5} + \frac{2}{3}$, we have

$$\frac{1}{5} + \frac{2}{3} = \frac{1 \times 3 + 5 \times 2}{5 \times 3} = \frac{3 + 10}{15} = \frac{13}{15}$$

Recall that in order to add two fractions, we rewrite them so that they have the same denominator, and then we add the numerators. The preceding procedure is just another way of doing this. Because the two given fractions do not have the same denominator, we could have found a common denominator by finding the least common multiple of 5 and 3, which is 15. This would result in the same answer as obtained previously:

$$\frac{1}{5} + \frac{2}{3} = \frac{3}{15} + \frac{10}{15} = \frac{13}{15}$$

EXAMPLE 3 ***Adding Fractions*** Add $\frac{1}{3} + \frac{2}{5}$.

Solution Using the rule for addition, we have

$$\frac{1}{3} + \frac{2}{5} = \frac{1 \times 5 + 3 \times 2}{3 \times 5} = \frac{5 + 6}{15} = \frac{11}{15}$$

EXAMPLE 4 ***Adding Fractions*** Add $\frac{2}{6} + \frac{3}{9}$.

Solution

$$\frac{2}{6} + \frac{3}{9} = \frac{2 \times 9 + 6 \times 3}{6 \times 9} = \frac{18 + 18}{54} = \frac{36}{54} = \frac{\cancel{2} \times 2 \times \cancel{3} \times \cancel{3}}{\cancel{2} \times 3 \times \cancel{3} \times \cancel{3}} = \frac{2}{3}$$

NW ***Now Work Problem 3a.***

How do we subtract two rational numbers? Consider the problem $\frac{2}{3} - \frac{1}{5}$. The expression $-\frac{1}{5}$ is equivalent to $\frac{1}{-5}$ and also to $\frac{-1}{5}$. Hence the problem $\frac{2}{3} - \frac{1}{5}$ is the equivalent to $\frac{2}{3} + \frac{-1}{5}$, which turns out to be an addition problem similar to those that we have been considering. Therefore,

$$\frac{2}{3} - \frac{1}{5} = \frac{2}{3} + \frac{-1}{5} = \frac{2 \times 5 + 3 \times (-1)}{3 \times 5} = \frac{10 + (-3)}{15} = \frac{7}{15}$$

Biography: Srinivasa Ramanujan

Srinivasa Ramanujan (1887–1920) was one of India's greatest mathematical geniuses. He was the first Indian to be elected to the Royal Society of London. At the age of 15 he started to teach himself mathematics. Soon he was developing his own ideas on properties of numbers. In 1903 he was awarded a scholarship, but a year later it was taken away. It seems that Ramanujan had a great interest in mathematics but not in the other required topics. He eventually settled for a clerk job in order to make a living. Around 1911 he published some of his works in the Journal of the Indian Mathematical Society. Eventually his genius was recognized and in 1914 he traveled to England to receive further education in mathematics. He collaborated with Godfrey Hardy, a famous British mathematician, and became very close friends with him. Much of his work in mathematics is considered brilliant. In 1917 Ramanujan contracted tuberculosis.

An interesting story told by Hardy is, "I remember once going to see him when he was lying ill at Putney. I had ridden in taxi-cab No. 1729, and remarked that the number seemed to me rather a dull one, and that I hoped it was not an unfavorable omen. 'No,' he replied, 'it is a very interesting number; it is the smallest number expressible as a sum of two cubes in two different ways.'" (Can you find the numbers?) Ramanujan returned to India in 1919 and died a year later. He was generally unknown to the world, but considered by mathematicians as a phenomenal genius.

EXAMPLE 5 ***Subtracting Fractions*** Subtract $\frac{4}{5} - \frac{1}{3}$.

Solution

$$\frac{4}{5} - \frac{1}{3} = \frac{4}{5} + \frac{-1}{3} = \frac{4 \times 3 + 5 \times (-1)}{5 \times 3} = \frac{12 + (-5)}{15} = \frac{7}{15}$$

NW ***Now Work Problem 3d.***

Multiplying Rational Numbers

Now that we have examined the operations of addition and subtraction for rational numbers, we next examine multiplication. Most students think that multiplication is the easiest operation to perform with fractions. In order to multiply two fractions, we simply multiply numerator times numerator and denominator times denominator.

If $\frac{a}{b}$ and $\frac{c}{d}$ are rational numbers, then $\frac{a}{b} \times \frac{c}{d} = \frac{a \times c}{b \times d}$.

If we want to find the product of $\frac{3}{5}$ and $\frac{2}{7}$, we simply multiply the numerators together to find the numerator of the product and multiply the denominators together to find the denominator of the product. Therefore,

$$\frac{3}{5} \times \frac{2}{7} = \frac{3 \times 2}{5 \times 7} = \frac{6}{35}$$

Consider the problem $\frac{5}{18} \times \frac{6}{25}$. We can do this problem in the same manner as the previous example; that is,

$$\frac{5}{18} \times \frac{6}{25} = \frac{5 \times 6}{18 \times 25} = \frac{30}{450} = \frac{\cancel{2} \times \cancel{3} \times \cancel{5}}{\cancel{2} \times 3 \times \cancel{3} \times \cancel{5} \times 5} = \frac{1}{15}$$

Or we can make the problem a little easier by simplifying it before performing the actual multiplication:

$$\frac{5}{18} \times \frac{6}{25} = \frac{5}{2 \times 3 \times 3} \times \frac{2 \times 3}{5 \times 5} = \frac{\cancel{5} \times \cancel{2} \times \cancel{3}}{\cancel{2} \times 3 \times \cancel{3} \times \cancel{5} \times 5} = \frac{1}{15}$$

Note that we eliminated the factors that are common to both the numerator and denominator.

EXAMPLE 6 ***Multiplying Fractions*** Multiply $\frac{4}{9} \times \frac{2}{5}$.

Solution

$$\frac{4}{9} \times \frac{2}{5} = \frac{4 \times 2}{9 \times 5} = \frac{8}{45}$$

EXAMPLE 7 ***Multiplying Fractions*** Multiply $\frac{7}{16} \times \frac{40}{42}$.

Solution

$$\frac{7}{16} \times \frac{40}{42} = \frac{7}{2 \times 2 \times 2 \times 2} \times \frac{2 \times 2 \times 2 \times 5}{2 \times 3 \times 7}$$
$$= \frac{\cancel{7} \times \cancel{2} \times \cancel{2} \times \cancel{2} \times 5}{\cancel{2} \times \cancel{2} \times \cancel{2} \times 2 \times 2 \times 3 \times \cancel{7}} = \frac{5}{2 \times 2 \times 3} = \frac{5}{12}$$

NW ***Now Work Problem 5a.***

Dividing Rational Numbers

Division of fractions can be defined in terms of multiplication:

$$\frac{a}{b} \div \frac{c}{d} = \frac{a}{b} \times \frac{d}{c}$$

Math Connections

Visualizing Multiplication of Fractions

Multiplication of simple fractions can be pictured using an array of dots. For example, in the discussion, we found $\frac{3}{5} \times \frac{2}{7} = \frac{3 \times 2}{5 \times 7} = \frac{6}{35}$. This problem can be pictured as follows:

1. Represent the product of the numerator (3×2) as an array consisting of 3 rows and 2 columns.

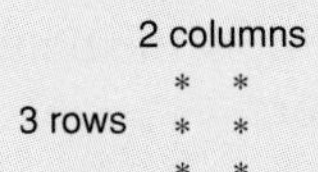

2. Represent the product of the denominator (5×7) as an array consisting of 5 rows and 7 columns by building on the previous array. Thus, we will need to add two more rows and five more columns.

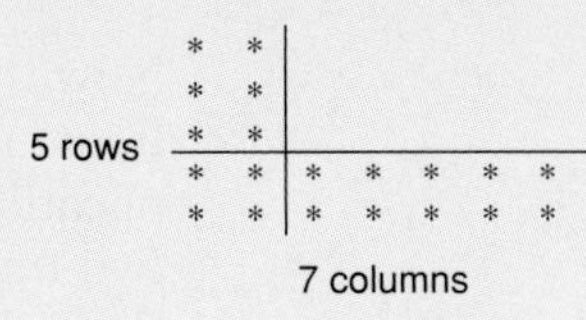

3. Complete the array so that we have a rectangular pattern. (Why?)

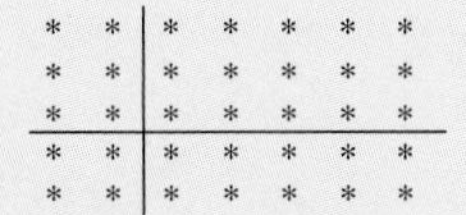

4. The product can now be read directly from the array. The numerator is in the top left corner and the denominator is the total number of "points." Thus, $\frac{3}{5} \times \frac{2}{7} = \frac{6}{35}$.

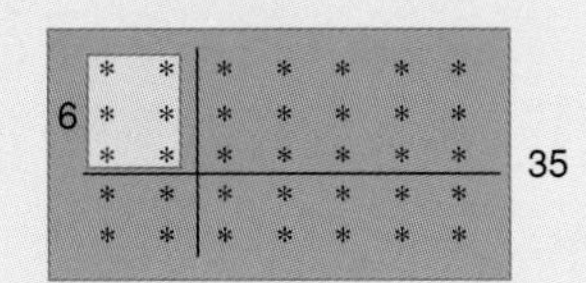

You may recall a rule that you learned previously: "In order to divide two fractions, invert the divisor and multiply." Why does this work? Consider the problem $\frac{2}{3} \div \frac{1}{2}$. According to the rule,

$$\frac{2}{3} \div \frac{1}{2} = \frac{2}{3} \times \frac{2}{1} = \frac{2 \times 2}{3 \times 1} = \frac{4}{3}$$

Another way to look at this problem is

$$\frac{\frac{2}{3}}{\frac{1}{2}}$$

This is a **complex fraction**, because the numerator or (inclusive *or*) denominator of the fraction is also a fraction. The complex fraction would no longer be complex if the denominator were 1 because dividing the numerator by one equals the numerator itself. In order to convert this denominator to 1, we must multiply $\frac{1}{2}$ by its *reciprocal*, $\frac{2}{1}$. But, if we multiply the denominator by $\frac{2}{1}$, we must multiply the numerator by $\frac{2}{1}$. Therefore,

$$\frac{\frac{2}{3}}{\frac{1}{2}} = \frac{\frac{2}{3} \times \frac{2}{1}}{\frac{1}{2} \times \frac{2}{1}} = \frac{\frac{2 \times 2}{3 \times 1}}{\frac{1 \times 2}{2 \times 1}} = \frac{\frac{4}{3}}{\frac{2}{2}} = \frac{\frac{4}{3}}{1} = \frac{4}{3}$$

In general terms, we have

$$\frac{\frac{a}{b}}{\frac{c}{d}} = \frac{\frac{a}{b} \times \frac{d}{c}}{\frac{c}{d} \times \frac{d}{c}} = \frac{\frac{a}{b} \times \frac{d}{c}}{1} = \frac{a}{b} \times \frac{d}{c}$$

From the illustrative example, we can see that in order to divide two rational numbers, we multiply the first rational number (the *dividend*) by the multiplicative inverse of the second rational number (the *divisor*).

For any rational numbers $\frac{a}{b}$ and $\frac{c}{d}$,

$$\boldsymbol{\frac{a}{b} \div \frac{c}{d} = \frac{a}{b} \times \frac{d}{c} \qquad b \neq 0, c \neq 0, d \neq 0}$$

EXAMPLE 8 ***Dividing Fractions*** Divide $\frac{9}{11} \div \frac{5}{4}$.

Solution

$$\frac{9}{11} \div \frac{5}{4} = \frac{9}{11} \times \frac{4}{5} = \frac{9 \times 4}{11 \times 5} = \frac{36}{55}$$

EXAMPLE 9 ***Dividing Fractions*** Divide $\frac{6}{7} \div \frac{9}{14}$.

Solution

$$\frac{6}{7} \div \frac{9}{14} = \frac{6}{7} \times \frac{14}{9} = \frac{2 \times 3}{7} \times \frac{2 \times 7}{3 \times 3} = \frac{2 \times \cancel{3} \times 2 \times \cancel{7}}{\cancel{7} \times 3 \times \cancel{3}} = \frac{4}{3}$$

Alternate Solution

$$\frac{6}{7} \div \frac{9}{14} = \frac{6}{7} \times \frac{14}{9} = \frac{84}{63} = \frac{2 \times 2 \times \cancel{3} \times \cancel{7}}{3 \times \cancel{3} \times \cancel{7}} = \frac{4}{3}$$

NW ***Now Work Problem 5d.***

Note that the only difference between the two solutions in Example 9 is that in the first solution the prime factors are determined first and those common to both the numerator and denominator are eliminated before the answer is determined.

EXAMPLE 10 ***Evaluating Complex Fractions*** Evaluate the following:

a. $\dfrac{\frac{2}{3}}{\frac{5}{7}}$ **b.** $\dfrac{4}{\frac{2}{3}}$ **c.** $\dfrac{\frac{3}{5}}{7}$

Solution All of these expressions are complex fractions. The major fraction line (division line) is made longer to avoid confusion about the numerator and denominator of the complex fraction. We can evaluate these by performing the indicated division.

a. $\dfrac{\frac{2}{3}}{\frac{5}{7}} = \frac{2}{3} \div \frac{5}{7} = \frac{2}{3} \times \frac{7}{5} = \frac{14}{15}$

b. $\dfrac{4}{\frac{2}{3}} = 4 \div \frac{2}{3} = \frac{4}{1} \times \frac{3}{2} = \frac{12}{2} = 6$

c. $\dfrac{\frac{3}{5}}{7} = \frac{3}{5} \div 7 = \frac{3}{5} \times \frac{1}{7} = \frac{3}{35}$

NW ***Now Work Problem 9a.***

Sometimes a complex fraction may contain more than one fraction or whole number in the numerator or denominator (or both). For example, consider

$$\frac{1 - \frac{2}{3}}{\frac{1}{2} + \frac{1}{3}}$$

To simplify this fraction we must first perform the indicated operations in the numerator and denominator, and then simplify the resulting complex fraction by performing the indicated division. Therefore,

$$\frac{1 - \frac{2}{3}}{\frac{1}{2} + \frac{1}{3}} = \frac{\frac{1}{3}}{\frac{5}{6}} = \frac{1}{3} \times \frac{6}{5} = \frac{6}{15} = \frac{2}{5}$$

Mixed Numbers and Improper Fractions

Throughout this section on rational numbers, we have limited our discussion to *proper fractions*, which are rational numbers in which the numerical value of the numerator is less than the numerical value of the denominator. For instance, $\frac{3}{4}, \frac{5}{9}, \frac{-7}{8}$, and $\frac{-13}{15}$ are all examples of proper fractions. We now extend our discussion to the concept of mixed numbers and improper fractions. A **mixed number** represents the sum of an integer and a rational number expressed without the plus $(+)$ sign. For example, $2\frac{3}{4}$ denotes $2 + \frac{3}{4}$. Observe that the integer 2 can be rewritten as the rational number $\frac{8}{4}$ and that $\frac{8}{4} + \frac{3}{4} = \frac{11}{4}$. Thus, $2\frac{3}{4} = \frac{11}{4}$. A fraction such as $\frac{11}{4}$, in which the numerical value of the numerator is greater than or equal to the numerical value of the denominator, is called an **improper fraction**.

A simple way to convert a mixed number to a rational number is to multiply the denominator of the fraction by the integer and then add this sum to the numerator. This final sum is then placed over the initial denominator. For example, $3\frac{5}{8}$ is rewritten as an equivalent rational number as follows: $3\frac{5}{8} = \frac{(8 \times 3) + 5}{8} = \frac{24 + 5}{8} = \frac{29}{8}$. Thus, $3\frac{5}{8} = \frac{29}{8}$.

We can also rewrite an improper fraction as a mixed number. To do this, we divide the denominator into the numerator and record the quotient and remainder. The integer part of the mixed number is equal to the quotient of the division, and the fractional part is obtained by writing the remainder as the numerator of the fraction. The denominator remains unchanged. For example, to rewrite $\frac{43}{5}$ as a mixed number, we observe that $43 \div 5 = 8$ with a remainder of 3. Therefore, $\frac{43}{5} = 8\frac{3}{5}$.

When adding, subtracting, multiplying, or dividing mixed numbers, it is helpful first to express mixed numbers as rational numbers and then perform the indicated operation. This is demonstrated in Example 11.

EXAMPLE 11 ***Working with Mixed Numbers*** Perform the indicated operations:

a. $2\frac{1}{4} + 3\frac{2}{3}$ **b.** $-2\frac{1}{2} + (-1\frac{2}{3})$

c. $3\frac{2}{5} \times \frac{-3}{2}$ **d.** $-1\frac{3}{4} \div (-2\frac{5}{8})$

Solution

a. We first rewrite each mixed number as a rational number:

$$2\frac{1}{4} = \frac{(4 \times 2) + 1}{4} = \frac{8 + 1}{4} = \frac{9}{4} \quad \text{and} \quad 3\frac{2}{3} = \frac{(3 \times 3) + 2}{3} = \frac{9 + 2}{3} = \frac{11}{3}$$

Therefore, $2\frac{1}{4} = \frac{9}{4}$ and $3\frac{2}{3} = \frac{11}{3}$.

We now add the two fractions using the rule for addition.

$$\frac{9}{4} + \frac{11}{3} = \frac{9 \times 3 + 4 \times 11}{4 \times 3} = \frac{27 + 44}{12} = \frac{71}{12}$$

An alternative solution is to combine integer and fractional components instead of converting to rational numbers:

$$2\frac{1}{4} + 3\frac{2}{3} = 2 + \frac{1}{4} + 3 + \frac{2}{3} = (2 + 3) + \left(\frac{1}{4} + \frac{2}{3}\right)$$

$$= 5 + \left(\frac{1 \times 3 + 4 \times 2}{4 \times 3}\right) = 5 + \left(\frac{3 + 8}{12}\right) = 5 + \frac{11}{12} = 5\frac{11}{12}$$

Note that $5\frac{11}{12} = \frac{71}{12}$.

b. First rewrite each mixed number as a rational number:

$$-2\frac{1}{2} = \frac{-5}{2} \quad \text{and} \quad \left(-1\frac{2}{3}\right) = \frac{-5}{3}$$

We now add the two fractions using the rule for addition.

$$\frac{-5}{2} + \frac{-5}{3} = \frac{(-5) \times 3 + 2 \times (-5)}{2 \times 3} = \frac{-15 + (-10)}{6} = \frac{-25}{6}$$

Therefore, $-2\frac{1}{2} + (-1\frac{2}{3}) = \frac{-25}{6}$. Note that this is equivalent to $-4\frac{1}{6}$.

c. First rewrite $3\frac{2}{5}$ as $\frac{17}{5}$. Now multiply the two fractions.

$$\frac{17}{5} \times \frac{-3}{2} = \frac{17 \times (-3)}{5 \times 2} = \frac{-51}{10}$$

d. First rewrite $-1\frac{3}{4}$ as $\frac{-7}{4}$ and $-2\frac{5}{8}$ as $\frac{-21}{8}$. Now divide the two fractions by multiplying the first fraction by the reciprocal of the second fraction.

$$\frac{-7}{4} \div \left(\frac{-21}{8}\right) = \frac{-7}{4} \times \left(\frac{8}{-21}\right) = \frac{-7 \times 8}{4 \times (-21)}$$

$$= \frac{-7 \times 2 \times 2 \times 2}{2 \times 2 \times 3 \times (-7)} = \frac{2}{3}$$

NW ***Now Work Problems 17a, 17e, and 17i.***

EXAMPLE 12 ***Stock Portfolio*** Figure 8.15 shows a circle graph that represents how some financial planners think investors who are 50 years or older should structure their investment portfolio. Using this graph, what fraction of a person's portfolio do financial planners recommend should be in small-to-midsized stocks?

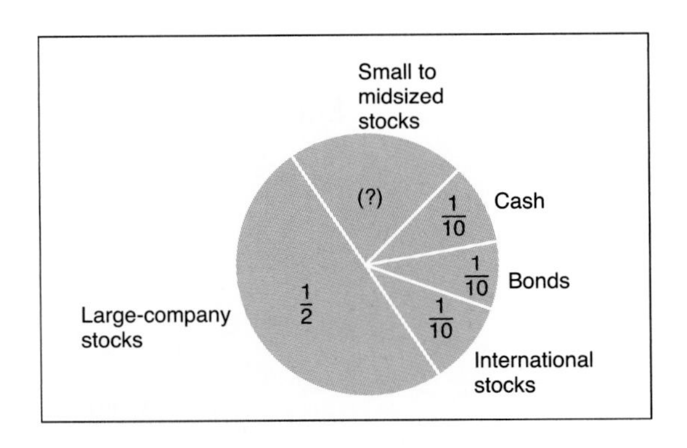

Figure 8.15

Recommended structure of an investment portfolio for people 50 years or older.

Solution To answer this question, we first add the given fractions and then subtract this sum from 1. We will first add the three common fractions (i.e., those with the same denominator).

$$\frac{1}{10} + \frac{1}{10} + \frac{1}{10} = \frac{3}{10}$$

We next add $\frac{3}{10}$ to $\frac{1}{2}$.

$$\frac{3}{10} + \frac{1}{2} = \frac{3 \times 2 + 10 \times 1}{10 \times 2} = \frac{6 + 10}{20} = \frac{16}{20}$$

Now we subtract this sum from 1.

$$1 - \frac{16}{20} = 1 + \frac{-16}{20} = \frac{1}{1} + \frac{-16}{20} = \frac{1 \times 20 + 1 \times (-16)}{1 \times 20} = \frac{20 - 16}{20} = \frac{4}{20}$$

Reducing $\frac{4}{20}$ we to lowest form, we get

$$\frac{4}{20} = \frac{2 \times 2}{2 \times 2 \times 5} = \frac{1}{5}$$

Thus, small-to-midsized stocks should represent $\frac{1}{5}$ of a person's portfolio.

Exercises for Section 8.5

1. Reduce each fraction to lowest terms.
   - NW a. $\frac{6}{16}$ b. $\frac{8}{72}$
   - c. $\frac{81}{129}$ d. $\frac{54}{448}$
2. Reduce each fraction to lowest terms.
   - a. $\frac{24}{72}$ b. $\frac{4}{9}$
   - c. $\frac{484}{576}$ d. $\frac{775}{1325}$
3. Perform the indicated operations.
   - NW a. $\frac{4}{5} + \frac{1}{7}$ b. $\frac{2}{3} + \frac{1}{4}$ c. $\frac{1}{9} + \frac{1}{8}$
   - NW d. $\frac{9}{11} - \frac{2}{3}$ e. $\frac{13}{16} - \frac{4}{5}$ f. $\frac{8}{9} - \frac{1}{3}$
4. Perform the indicated operations.
   - a. $\frac{5}{11} + \frac{3}{14}$ b. $\frac{7}{15} + \frac{3}{13}$ c. $\frac{5}{22} + \frac{2}{11}$
   - d. $\frac{5}{11} - \frac{1}{4}$ e. $\frac{11}{14} - \frac{2}{3}$ f. $\frac{7}{12} - \frac{1}{7}$
5. Perform the indicated operations.
   - NW a. $\frac{4}{5} \times \frac{2}{7}$ b. $\frac{3}{11} \times \frac{4}{5}$ c. $\frac{8}{13} \times \frac{4}{7}$
   - NW d. $\frac{4}{9} \div \frac{2}{7}$ e. $\frac{3}{11} \div \frac{4}{9}$ f. $\frac{8}{13} \div \frac{4}{7}$
6. Perform the indicated operations.
   - a. $\frac{5}{9} \times \frac{6}{11}$ b. $\frac{3}{9} \times \frac{2}{3}$ c. $\frac{6}{13} \times \frac{26}{15}$
   - d. $\frac{6}{25} \div \frac{3}{5}$ e. $\frac{8}{15} \div \frac{2}{5}$ f. $\frac{10}{55} \div \frac{2}{11}$
7. Simplify each of the following:
   - a. $\dfrac{\frac{3}{11}}{\frac{4}{9}}$ b. $\dfrac{\frac{2}{13}}{\frac{5}{7}}$ c. $\dfrac{\frac{2}{5}}{3}$
   - d. $\dfrac{1 + \frac{1}{2}}{2 - \frac{1}{3}}$ e. $\dfrac{2 + \frac{1}{4}}{3 - \frac{1}{2}}$ f. $\dfrac{\frac{1}{2} + \frac{1}{3}}{\frac{1}{4} + \frac{4}{5}}$
8. Simplify each of the following:
   - a. $\dfrac{\frac{3}{8}}{7}$ b. $\dfrac{9}{\frac{1}{3}}$ c. $\dfrac{\frac{2}{4}}{\frac{4}{5}}$
   - d. $\dfrac{3 - \frac{1}{3}}{4 - \frac{1}{2}}$ e. $\dfrac{1 - \frac{1}{4}}{3 + \frac{1}{3}}$ f. $\dfrac{\frac{3}{8} + \frac{1}{2}}{\frac{1}{5} - \frac{1}{7}}$
9. Simplify each of the following:
   - NW a. $\dfrac{\frac{3}{4}}{\frac{2}{5}}$ b. $\dfrac{8}{\frac{3}{5}}$ c. $\dfrac{\frac{4}{5}}{3}$
   - d. $\dfrac{\frac{3}{8}}{\frac{5}{4}}$ e. $\dfrac{2\frac{1}{5}}{3\frac{3}{4}}$ f. $\dfrac{1\frac{1}{2}}{2\frac{3}{4}}$
10. Simplify each of the following:
    - a. $\dfrac{1 + \frac{1}{5}}{2 - \frac{1}{2}}$ b. $\dfrac{1 + \frac{3}{4}}{2 - \frac{2}{3}}$ c. $\dfrac{\frac{2}{3} + \frac{3}{4}}{\frac{5}{8} - \frac{1}{4}}$

d. $\dfrac{3\frac{1}{2}}{2 - \frac{1}{4}}$ **e.** $\dfrac{2\frac{5}{6}}{3 + \frac{1}{3}}$ **f.** $\dfrac{3 - \frac{3}{4}}{2 + \frac{1}{2}}$

11. Simplify each of the following:

a. $\left(\frac{3}{4} + \frac{2}{3}\right) \times \left(\frac{7}{8} - \frac{1}{2}\right)$ **b.** $\frac{1}{2} \times \left(\frac{5}{6} + 2\right)$

c. $\left(1 - \frac{2}{3}\right) \times \left(\frac{3}{8} + \frac{1}{2}\right)$ **d.** $\left(\frac{3}{4} \times \frac{5}{6} + \frac{1}{2}\right) \times 2$

12. Simplify each of the following:

a. $\left(\frac{1}{2} + \frac{3}{4}\right) \div \left(\frac{7}{8} - \frac{1}{2}\right)$

b. $3 \div \left(\frac{1}{4} + \frac{2}{3} - \frac{1}{5}\right)$

c. $\left(\frac{1}{4} + \frac{1}{3}\right) \times \left(\frac{3}{4} - \frac{1}{3}\right)$

d. $\frac{3}{8} \times \left(\frac{4}{5} - \frac{1}{2} + \frac{1}{3}\right)$

13. Determine whether each statement is true or false. (*Hint:* Convert the fractions under consideration to fractions with the same denominator.)

a. $\frac{4}{7} > \frac{2}{3}$ **b.** $\frac{3}{4} < \frac{7}{8}$ **c.** $\frac{4}{11} > \frac{3}{7}$

d. $\frac{4}{9} < \frac{16}{36}$ **e.** $\frac{5}{11} > \frac{11}{5}$ **f.** $\frac{6}{7} < \frac{8}{8}$

14. Determine whether each statement is true or false (see Exercise 13).

a. $\frac{4}{3} > \frac{5}{4}$ **b.** $\frac{6}{4} < \frac{8}{9}$ **c.** $\frac{8}{9} > \frac{6}{5}$

d. $\frac{8}{33} < \frac{4}{11}$ **e.** $\frac{8}{9} < \frac{2}{3}$ **f.** $\frac{12}{33} = \frac{4}{11}$

15. Determine whether each statement is true or false.

a. Every rational number is an integer.
b. Every integer is a rational number.
c. Every rational number is a natural number.
d. Every natural number is a rational number.
e. Every rational number is a whole number.

16. Determine whether each statement is true or false.

a. Every whole number is a rational number.
b. Every whole number is an integer.
c. Every integer is a whole number.
d. The rationals are a subset of the integers.
e. The integers are a subset of the rationals.

17. Perform the indicated operations.

NW **a.** $2\frac{1}{4} - \frac{17}{4}$ **b.** $3\frac{2}{3} - \left(-\frac{20}{3}\right)$

c. $-3\frac{4}{5} - \left(-1\frac{1}{15}\right)$ **d.** $-2\frac{5}{9} + \left(-1\frac{2}{7}\right)$

NW **e.** $2\frac{3}{8} \times \frac{1}{4}$ **f.** $3\frac{3}{4} \times \left(\frac{-8}{5}\right)$

g. $-3\frac{2}{3} \times \left(-1\frac{1}{5}\right)$ **h.** $-2\frac{4}{5} \times \left(-1\frac{1}{4}\right)$

NW **i.** $-2\frac{1}{2} \div \frac{1}{4}$ **j.** $\frac{3}{7} \div \left(-1\frac{2}{7}\right)$

k. $-1\frac{2}{3} \div \left(-2\frac{5}{6}\right)$ **l.** $-2\frac{4}{5} \div \left(-1\frac{1}{4}\right)$

18. **Hiking** Lynn, Sue, Jane, and Mike decided to hike to Moose Lake, which was approximately $8\frac{2}{5}$ miles away from their hotel. After hiking for about an hour, the lake was still $5\frac{1}{3}$ miles away. How far did the group hike thus far?

19. **Fishing** George and Mildred took their grandchildren fishing out in the ocean one Sunday afternoon. After four hours of fishing, the group only caught two fish, respectively weighing $5\frac{1}{4}$ pounds and $3\frac{2}{3}$ pounds. What was the total weight of their catch?

20. **Beach cleanup** The senior class of Bayside High School engaged in a beach cleanup project encompassing $6\frac{1}{2}$ miles of shoreline. Mr. Clark, the senior class advisor, divided the class into four groups. How many miles of shoreline will each group cleanup?

21. **Bicycle riding** During his bicycle training rides, Ross does $3\frac{1}{2}$-minute miles. At this rate, how long will it take Ross to ride 10 miles?

22. **Bicycle riding** While riding a stationary bicycle, Patti burns on average 450 calories every $\frac{1}{2}$ hour. Based on this rate, how many calories will Patti burn if she rides the bike for $1\frac{3}{4}$ hours?

23. **Saving money** Kelly earns $42,000 per year as the assistant manager of a department store. If she saves $\frac{3}{7}$ of her salary, how much does Kelly save per year?

24. **Pay rate** If Kristin's pay rate is $\$8\frac{2}{5}$ per hour and at the end of one week her paycheck is for $420, how many hours did Kristin work that week?

25. **Movies** Jim and Kathy went to a movie that was $1\frac{3}{4}$ hours long. After watching $\frac{2}{3}$ of the movie, they both thought the movie was too absurd and promptly left. How long did they stay at the movie?

26. In a class of 30 students, $\frac{2}{5}$ of the girls have blond hair. If $\frac{2}{3}$ of the class is female, how many girls in the class have blond hair?

27. **Map legend** The legend of a map indicates that 1 inch = 24 miles. If the measured distance between two cities is $3\frac{5}{8}$ inches, how far apart are the two cities?

Writing Mathematics

28. Define a rational number. As part of your definition, include what distinguishes a rational number from an integer.

29. Explain how you can determine if the rational numbers $\frac{3}{8}$ and $\frac{5}{7}$ are equivalent. Are they? Why?

30. In what way(s) are adding and subtracting integers similar to adding and subtracting rational numbers? In what way(s) are they different?

31. Define the concept of a complex fraction. Give an example as part of your definition.

Challenge Exercises

In Exercises 32–35, use the number line shown below.

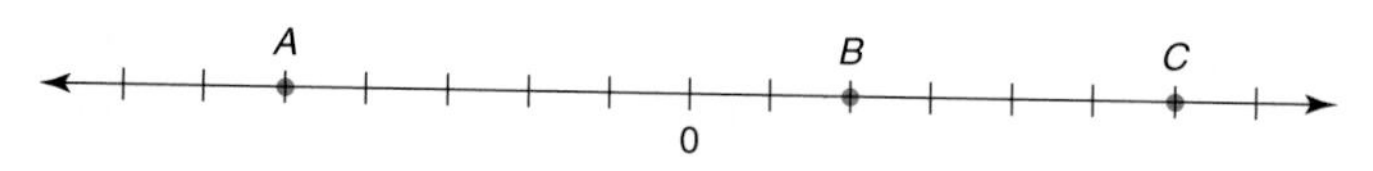

32. If the number line is scaled in thirds, give the fraction in simplest form for points A, B, and C.

33. If the number line is scaled in fifths, give the fraction in simplest form for points A, B, and C.

34. If the number line is scaled in ninths, give the fraction in simplest form for points A, B, and C.

35. If the number line is scaled in fifteenths, give the fraction in simplest form for points A, B, and C.

36. The circle graph in Figure 8.16 shows what Jill generally does on Saturdays. What fraction of the day does Jill spend napping? How many hours does this equal?

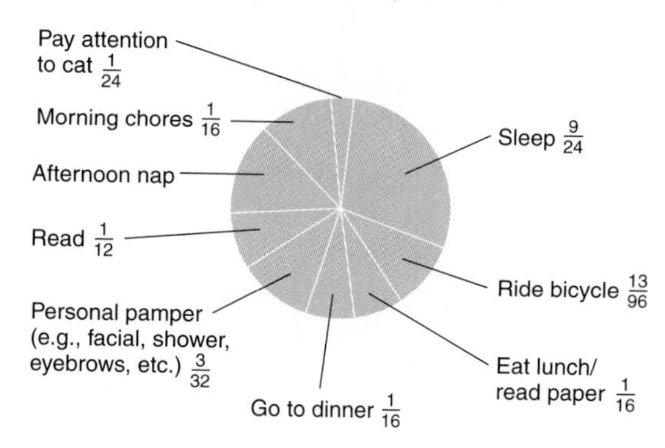

Figure 8.16

Research/Group Activities

37. Although many people rely on calculators to assist them with arithmetic calculations, most calculators are not capable of directly adding, subtracting, multiplying, or dividing fractions. See what your calculator's capability is relative to this and, if the capability exists, learn how to perform these operations. Also conduct a poll of your classmates and record whether or not their calculators can work with fractions directly. Report the results to your class.

38. Interview at least 10 people from different occupations (e.g., a butcher, a carpenter, an electrician, etc.) to see how they use rational numbers. Prepare a report from your interviews and share what you learn with your classmates.

Real-World Data

39. **Doctor visits** Figure 8.17 shows the results of a survey taken in 2000 that represent how often we visited the doctor's office or hospital emergency room in a 12-month period. (All results were rounded.) What fraction represents 10 or more visits?

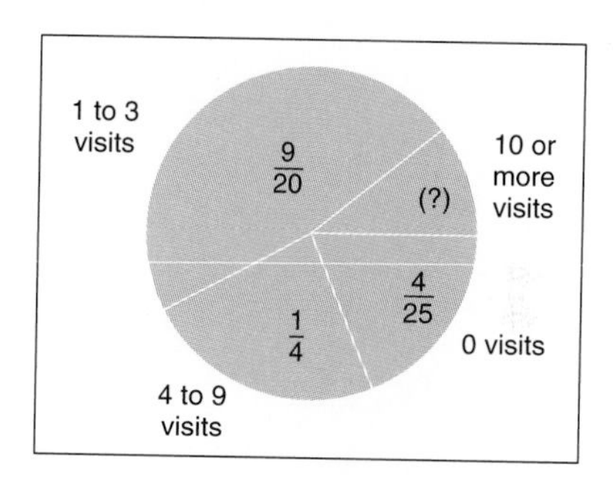

Figure 8.17

Visits to the doctor's office or hospital emergency room during 2000. Source: National Center for Health Statistics.

40. **Energy consumption** Figure 8.18 shows the primary sources of energy consumed in the United States. (All figures are rounded.) What fraction represents hydroelectric power or other?

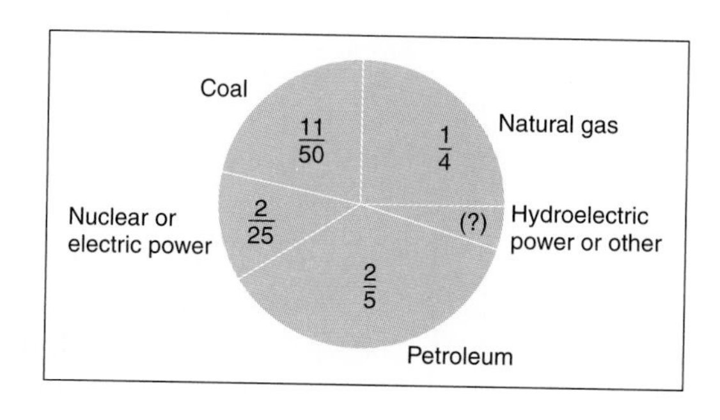

Figure 8.18

Sources of energy consumed in the United States.

Just for fun

If you double $\frac{1}{4}$ of a certain fraction and multiply it by that fraction, the answer is $\frac{1}{8}$. What is the fraction?

8.6 RATIONAL NUMBERS AND DECIMALS

Expressing Fractions in Decimal Form

Before we continue our discussion regarding what other numbers belong on the number line, let's review the topic of *decimals.* **Decimals** are fractions that have a power of 10, such as 10, 100, 1000, or 10,000, for their denominator. The word *decimal* comes from the Latin word *decem*, which means *ten*. The fraction $\frac{3}{10}$ is represented by the decimal 0.3.

The *decimal point* is a period (.) that appears just to the left of the tenths place in the decimal. Some other examples of fractions expressed as decimals are

$$\frac{4}{10} = 0.4, \quad \frac{31}{100} = 0.31, \quad \frac{471}{1000} = 0.471$$

The decimal 0.4 is read as "4 tenths," 0.31 is read as "31 hundredths," and 0.471 is read as "471 thousandths." The following example indicates the names of the places for decimals:

Thousands	Hundreds	Tens	Units		Tenths	Hundredths	Thousandths	Ten-thousandths
4	3	2	1	.	2	1	3	4
Integers					Decimals			

In order to change any fraction in the form *a/b* to a decimal, we divide the denominator into the numerator:

$$\begin{array}{r} 0.2 \\ 5\overline{)1.0} \\ \underline{1\,0} \\ 0 \end{array} \quad \frac{1}{5} = 0.2 \qquad\qquad \begin{array}{r} 0.375 \\ 8\overline{)3.000} \\ \underline{2\,4} \\ 60 \\ \underline{56} \\ 40 \\ \underline{40} \\ 0 \end{array} \quad \frac{3}{8} = 0.375$$

Decimals such as 0.2 and 0.375 are called **terminating decimals** because at some point in the division a remainder of zero is obtained; that is, when we change the fraction to a decimal, the division terminates.

Not all fractions can be expressed as terminating decimals. For example, consider the rational number $\frac{1}{3}$. Converting $\frac{1}{3}$ to a decimal, we have

$$\begin{array}{r} 0.3333\ldots \\ 3\overline{)1.0000} \\ \underline{9} \\ 10 \\ \underline{9} \\ 10 \\ \underline{9} \\ 10 \\ \underline{9} \\ 1 \end{array}$$

The division does not terminate: We will never obtain a remainder of zero, but the remainder of 1 will keep reappearing at regular intervals.

Instead of writing the decimal expression for $\frac{1}{3}$ as $0.3333\ldots$, we can express it in a more convenient and efficient way by placing a bar over the 3, which indicates that the 3 repeats endlessly—that is, $0.333\ldots = 0.\overline{3}$. In the same manner, $0.121212\ldots = 0.\overline{12}$. In this case the digits 12 repeat endlessly, so we place a bar over both the 1 and the 2. Decimals such as $0.\overline{3}$ and $0.\overline{12}$ are called **repeating nonterminating decimals**.

EXAMPLE 1 ***Expressing a Fraction as a Decimal*** Express $\frac{5}{8}$ as a decimal.

Solution

$$\begin{array}{r} 0.625 \\ 8\overline{)\,5.000} \\ \underline{4\,8} \\ 20 \\ \underline{16} \\ 40 \\ \underline{40} \\ 0 \end{array} \qquad \frac{5}{8} = 0.625$$

EXAMPLE 2 ***Expressing a Fraction as a Decimal*** Express $\frac{4}{9}$ as a decimal.

Solution

$$\begin{array}{r} 0.44\ldots \\ 9\overline{)4.00} \\ \underline{36} \\ 40 \\ \underline{36} \\ 4 \end{array} \qquad \frac{4}{9} = 0.\overline{4}$$

EXAMPLE 3 ***Expressing a Fraction as a Decimal*** Express $\frac{3}{7}$ as a decimal.

Solution

$$\begin{array}{r} 0.4285714 \\ 7\overline{)\,3.0000000} \\ \underline{2\,8} \\ 20 \\ \underline{14} \\ 60 \\ \underline{56} \\ 40 \\ \underline{35} \\ 50 \\ \underline{49} \\ 10 \\ \underline{7} \\ 30 \\ \underline{28} \\ 2 \end{array} \qquad \frac{3}{7} = 0.\overline{428571}$$

Note that we place a bar over the six digits $0.\overline{428571}$. The last remainder of 2 is a repeat of a remainder that we had previously. Therefore, the pattern of division will repeat, and the digits 428571 will repeat endlessly.

NW ***Now Work Problems 1a and 1d.***

Expressing Decimals in Fractional Form

In examining the set of rational numbers, we learned how to express a fraction as a decimal. For example,

$$\frac{5}{8} = 0.625, \quad \frac{4}{9} = 0.\overline{4} \quad \text{and} \quad \frac{3}{7} = 0.\overline{428571}$$

The next thing to consider is whether, given the decimal expression, we can find its equivalent fraction. We already know that $0.\overline{3} = \frac{1}{3}$ and $0.25 = \frac{1}{4}$, but what about other examples we might encounter?

To convert a terminating decimal to a fraction, we simply omit the decimal point and supply the proper denominator. For example,

$$0.3 = \frac{3}{10}, \quad 0.25 = \frac{25}{100} = \frac{1}{4}, \quad 0.125 = \frac{125}{1000} = \frac{1}{8}$$

This technique does not work for repeating decimals, so we will need to develop another method for repeating decimals. Consider the decimal $0.3\overline{3}$. Let $x = 0.3\overline{3}$. Then $10x = 3.3\overline{3}$. (Multiplying a number by 10 moves the decimal point one place to the right, multiplying by 100 moves the decimal point two places to the right, and so on.) Thus far we have

$$\begin{aligned} 10x &= 3.3\overline{3} \\ x &= 0.3\overline{3} \end{aligned}$$

The decimal points are lined up in the same position.

Now subtract x from $10x$ and $0.3\overline{3}$ from $3.3\overline{3}$:

$$\begin{aligned} 10x &= 3.3\overline{3} \\ x &= 0.3\overline{3} \\ \hline 9x &= 3 \end{aligned}$$

All of the repeating 3s are subtracted from repeating 3s

Next divide both sides of the resulting equation by 9:

$$\frac{9x}{9} = \frac{3}{9}$$

$$x = \frac{3}{9} = \frac{1}{3}$$

Therefore,

$$0.3\overline{3} = \frac{1}{3}$$

Let's try another example. Suppose we wish to convert the repeating decimal $0.\overline{13}$ to a fraction. Let $x = 0.\overline{13}$; because the digits repeat in cycles of two, we multiply both sides of the equation by 100. If $x = 0.\overline{13}$, then $100x = 13.\overline{13}$. We subtract, which gives

$$\begin{aligned} 100x &= 13.\overline{13} \\ x &= 0.\overline{13} \\ \hline 99x &= 13 \end{aligned}$$

Dividing both sides of the equation by 99, we have

$$\frac{99x}{99} = \frac{13}{99}$$

$$x = \frac{13}{99}$$

Therefore,

$$0.\overline{13} = \frac{13}{99}$$

EXAMPLE 4 ***Expressing a Decimal as a Fraction*** Express $0.\overline{25}$ as a quotient of integers (that is, convert $0.\overline{25}$ to a fraction).

Solution Let $x = 0.\overline{25}$ and multiply both sides of the equation by 100 (two digits repeating). Subtracting, we have

$$\begin{aligned} 100x &= 25.\overline{25} \\ x &= 0.\overline{25} \\ \hline 99x &= 25 \end{aligned}$$

Dividing both sides by 99,

$$\begin{aligned} \frac{99x}{99} &= \frac{25}{99} \\ x &= \frac{25}{99} \end{aligned}$$

Therefore,

$$0.\overline{25} = \frac{25}{99}$$

EXAMPLE 5 ***Expressing a Decimal as a Fraction*** Express $3.\overline{162}$ as a quotient of integers.

Solution Let $x = 3.\overline{162}$. Multiplying both sides of the equation by 1000 (three digits repeating) and subtracting, we have

$$\begin{aligned} 1000x &= 3162.\overline{162} \\ x &= 3.\overline{162} \\ \hline 999x &= 3159 \end{aligned}$$

Dividing both sides by 999,

$$\begin{aligned} \frac{999x}{999} &= \frac{3159}{999} \\ x &= \frac{3159}{999} \\ &= \frac{117}{37} \end{aligned}$$

Therefore,

$$3.\overline{162} = \frac{117}{37}$$

EXAMPLE 6 ***Expressing a Decimal as a Fraction*** Express $2.14\overline{27}$ as a quotient of integers.

Solution Let $x = 2.14\overline{27}$, and multiply both sides of the equation by 100 (two digits repeating). We place the decimal points in the same position and subtract:

$$\begin{aligned} 100x &= 214.27\overline{27} \\ x &= 2.14\overline{27} \\ \hline 99x &= 212.13 \end{aligned}$$

Dividing both sides by 99,

$$\frac{99x}{99} = \frac{212.13}{99}$$

$$x = \frac{212.13}{99}$$

But we are not finished. We were supposed to express $2.14\overline{27}$ as a quotient of two integers, and 212.13 is not an integer—it is a decimal.

In our earlier discussion of rational numbers (Section 8.5), we noted that

$$\frac{a}{b} = \frac{a \times k}{b \times k}$$

Using this idea, we multiply $\frac{212.13}{99}$ by $\frac{100}{100}$ (we use 100, as we have two decimal places). Therefore,

$$x = \frac{212.13}{99} \times \frac{100}{100} = \frac{21{,}213}{9900} = \frac{2357 \times \cancel{3} \times \cancel{3}}{1100 \times \cancel{3} \times \cancel{3}} = \frac{2357}{1100}$$

Therefore,

$$2.14\overline{27} = \frac{2357}{1100}$$

NW ***Now Work Problems 5a and 5d.***

Density Property of Rational Numbers

Thus far we have examined the set of natural numbers, the set of whole numbers, the set of integers, and the set of rational numbers. All of these can be shown on

Note of Interest

During 1999 Sarah Flannery (age 16), a high school student in Blarney, Ireland, entered a research project titled *Cryptography—A New Algorithm versus the RSA*, at the ISEF (International Science and Engineering Fair).

There she won the top prize, "Young Scientist of the Year." What makes this newsworthy is that Sarah's project was a new cryptographic algorithm that challenged RSA (see p. 424) by being up to 10 to 30 times faster in encrypting messages. This has great implications because of all of the business transactions that are done on the Internet (for example, it would help in keeping your credit card number safe while you shop online). She called this new algorithm the Cayley-Purser algorithm, after Arthur Cayley, who was a nineteenth-century British mathematician, and Michael Purser, is a cryptographer who inspired her at Baltimore Technologies (in Dublin, Ireland) where she was doing a two-week work experience program. Because of her project, she has won several awards and wrote a book detailing her story titled, *In Code: A Mathematical Journey*. However, Sarah doesn't see herself as a child prodigy; she says that she sees herself as a regular teenager with a wide range of hobbies and interests.

a number line. For example, Figure 8.19 is a number line showing the integers 0 and 1 and the rational numbers $\frac{1}{2}$, $\frac{14}{16}$, and $\frac{15}{16}$.

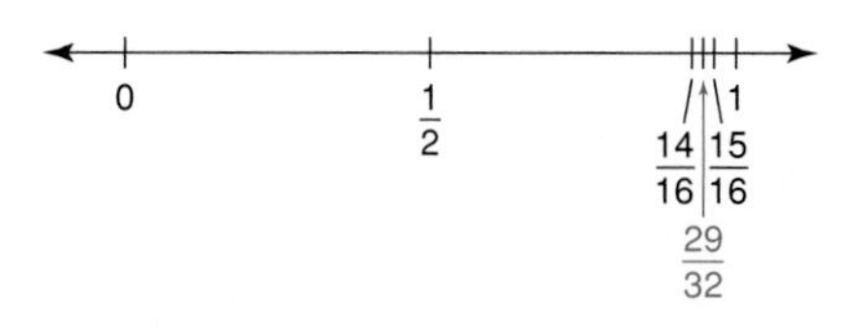

Figure 8.19

If two rational numbers are indicated on a number line, can we find other numbers that fit between them? The answer is yes. As an example, we will find a number that lies between $\frac{14}{16}$ and $\frac{15}{16}$ on the line in Figure 8.19. To do this, we convert $\frac{14}{16}$ to $\frac{28}{32}$ and $\frac{15}{16}$ to $\frac{30}{32}$. Now we can see that the number $\frac{29}{32}$ is between the two given fractions. This process can be continued indefinitely; it is possible to find a rational number between any two given rational numbers. This process may seem to "fill up" the number line, but such is not the case, because there is always room for one more number. In fact, between any two rational numbers, there is always another rational number. This particular property is called the **density property of rational numbers**. We can also say that the rational numbers are **dense**. Note that the density property does not hold for all kinds of numbers. The natural numbers are not dense, because there is no natural number between the natural numbers 3 and 4.

We have seen one example of how to find a rational number between two given rational numbers, but how do we do it for *any* two rational numbers? One way to do this is to find the arithmetic mean of the two given rational numbers. In general terms, if we let x equal the number we are seeking, and we let a/b and c/d equal the given rational numbers, then

$$x = \frac{1}{2} \times \left(\frac{a}{b} + \frac{c}{d}\right)$$

Using $\frac{14}{16}$ and $\frac{15}{16}$ from our previous discussion, we have

$$x = \frac{1}{2} \times \left(\frac{14}{16} + \frac{15}{16}\right)$$

$$x = \frac{1}{2} \times \left(\frac{29}{16}\right)$$

$$x = \frac{29}{32}$$

EXAMPLE 7 ***The Density Property*** Find a rational number between $\frac{5}{7}$ and $\frac{6}{7}$.

Solution Finding the arithmetic mean of the given numbers, we have

$$x = \frac{1}{2} \times \left(\frac{5}{7} + \frac{6}{7}\right)$$

$$x = \frac{1}{2} \times \left(\frac{11}{7}\right)$$

$$x = \frac{11}{14}$$

Check:

$$\frac{5}{7} = \frac{10}{14} \quad \text{and} \quad \frac{6}{7} = \frac{12}{14}$$

$$\frac{10}{14} < \frac{11}{14} \quad \text{and} \quad \frac{11}{14} < \frac{12}{14}$$

EXAMPLE 8 ***The Density Property*** Find a rational number between $\frac{1}{5}$ and $\frac{1}{9}$.

Solution

$$x = \frac{1}{2} \times \left(\frac{1}{5} + \frac{1}{9}\right)$$

$$x = \frac{1}{2} \times \left(\frac{9+5}{45}\right) = \frac{1}{\cancel{2}} \times \left(\frac{\overset{7}{\cancel{14}}}{45}\right)$$

$$x = \frac{7}{45}$$

Check:

$$\frac{1}{9} = \frac{5}{45} \quad \text{and} \quad \frac{1}{5} = \frac{9}{45}$$

$$\frac{5}{45} < \frac{7}{45} \quad \text{and} \quad \frac{7}{45} < \frac{9}{45}$$

Note that there are other rational numbers that are also between $\frac{1}{5}$ and $\frac{1}{9}$. However, using the formula

$$x = \frac{1}{2} \times \left[\left(\frac{a}{b}\right) + \left(\frac{c}{d}\right)\right]$$

we found the arithmetic mean, which is exactly halfway between the two given numbers.

NW ***Now Work Problems 9a and 9d.***

In summary, whenever we are given two rational numbers a/b and c/d, we can find another rational number (call it x) such that $a/b < x$ and $x < c/d$, or $a/b > x$ and $x > c/d$. This particular property is called the *density property of rational numbers.* The rational numbers are dense, but the integers are not. For example, there is no integer between the integers 1 and 2. The density of rational numbers can also be stated as follows:

Between every two rational numbers there is another rational number.

Exercises for Section 8.6

1. Express each fraction as a decimal.

NW **a.** $\frac{3}{8}$ **b.** $\frac{5}{16}$ **c.** $\frac{2}{3}$

NW **d.** $\frac{7}{33}$ **e.** $\frac{1}{11}$ **f.** $\frac{15}{37}$

2. Express each fraction as a decimal.

a. $\frac{4}{5}$ **b.** $\frac{19}{20}$ **c.** $\frac{6}{27}$

d. $\frac{1}{9}$ **e.** $\frac{15}{33}$ **f.** $\frac{5}{9}$

3. Express each fraction as a decimal.

 a. $\frac{7}{8}$ b. $\frac{4}{9}$ c. $\frac{9}{11}$

 d. $\frac{1}{16}$ e. $\frac{5}{7}$ f. $\frac{8}{17}$

4. Express each decimal as a quotient of integers, in simplest form.

 a. 0.65 b. 0.045 c. $0.\overline{6}$

 d. $0.\overline{1}$ e. $0.1\overline{22}$ f. $2.6\overline{26}$

5. Express each decimal as a quotient of integers, in simplest form.

 NW a. 0.125 b. 0.0025 c. $0.\overline{7}$

 NW d. $0.\overline{34}$ e. $6.2\overline{81}$ f. $0.\overline{9}$

6. Express each decimal as a quotient of integers, in simplest form.

 a. 0.012 b. 0.0005 c. 0.55

 d. $1.\overline{3}$ e. $0.\overline{85}$ f. $3.\overline{12}$

7. Express each decimal as a quotient of integers, in simplest form.

 a. 0.75 b. 0.875 c. $0.\overline{8}$

 d. $0.\overline{45}$ e. $0.\overline{235}$ f. $4.59\overline{6}$

8. Express each decimal as a quotient of integers, in simplest form.

 a. 0.43 b. 0.375 c. $0.\overline{4}$

 d. $2.\overline{147}$ e. $3.1\overline{45}$ f. $2.54\overline{9}$

9. Find a rational number between each pair of rational numbers.

 NW a. $\frac{1}{2}, \frac{1}{3}$ b. $\frac{1}{3}, \frac{1}{4}$ c. $\frac{1}{4}, \frac{1}{5}$

 NW d. $\frac{2}{3}, \frac{7}{8}$ e. $\frac{3}{4}, \frac{9}{11}$ f. $\frac{7}{11}, \frac{15}{16}$

10. Find a rational number between each pair of rational numbers.

 a. $\frac{1}{5}, \frac{2}{5}$ b. $\frac{2}{6}, \frac{5}{7}$ c. $\frac{3}{7}, \frac{4}{11}$

 d. $\frac{3}{5}, \frac{7}{9}$ e. $\frac{4}{9}, \frac{11}{12}$ f. $\frac{7}{13}, \frac{9}{17}$

11. Find a rational number between each pair of rational numbers.

 a. $\frac{2}{8}, \frac{3}{8}$ b. $\frac{4}{9}, \frac{5}{9}$ c. $\frac{1}{8}, 0$

 d. $\frac{9}{10}, 1$ e. $\frac{5}{7}, \frac{3}{8}$ *f. 1.999, 2

12. Find a number that is between 0 and 0.1.

13. Find a number that is between 0 and 0.01.

14. Find a number that is between 0 and 0.001.

Writing Mathematics

15. Explain the difference between a rational number and a decimal.

16. Define *terminating decimal* and give an example.

17. Define *repeating nonterminating decimal* and give an example.

18. Explain what it means to say that rational numbers are dense.

Challenge Exercises

19. While performing a division on his calculator, Dr. Shapiro was unexpectedly interrupted by his nurse. When he returned to his calculation 10 minutes later, he forgot what numbers he was dividing. If the display on the calculator showed 136.373737, what numbers might have Dr. Shapiro been dividing?

20. A calculator display reads 28.766666.

 a. If the 6 repeats forever, then what is one possible rational expression that could have produced this result?

 b. If the 6 does *not* repeat forever (in other words, the display shows the entire answer), then what rational expression could have produced this result?

21. You are informed that a hat contains no more than 15 marbles of different colors. Some are red and some are blue. You are also told that if you were to select a marble from the hat without looking, the probability of selecting a red marble is $0.\overline{36}$.

 a. Express this probability as a quotient of integers in simplest form.

 b. Use your answer from part **a** to determine how many marbles are in the hat.

Research/Group Activity

22. The New York Stock Exchange and NASDAQ recently switched from trading stock in fractions to decimals. Research the history of these exchanges relative to the trading "units." As part of your research, include what the basic fractional unit was and why it was selected. Also include the rationale for switching to decimals as well as the perceived advantages and disadvantages.

Just for fun

Write down four odd digits that will add up to 19.

8.7 IRRATIONAL NUMBERS AND THE SET OF REAL NUMBERS

Irrational Numbers

We have seen that any rational number can be expressed as a decimal, and that this decimal will be a terminating decimal or a repeating nonterminating decimal. For example, $\frac{5}{8}$ and $\frac{1}{11}$ are rational numbers and, expressing each as a decimal, we have

$$\frac{5}{8} = 0.625 \quad \text{(a terminating decimal)}$$

$$\frac{7}{11} = 0.\overline{63} \quad \text{(a repeating nonterminating decimal)}$$

Are all decimals either terminating or repeating nonterminating decimals? The answer is no. We can construct a decimal that does not terminate, yet does not repeat, as follows: Choose any digit and write it after the decimal point, then write a zero; repeat the digit, then write two zeros; repeat the digit, then write three zeros; repeat the digit, then write four zeros; repeat the digit, and so on. Using this idea, we construct decimals such as

$$0.1010010001000010000001\ldots$$
$$0.2020020002000020000002\ldots$$
$$0.9090090009000090000009\ldots$$

For each of these decimals, we have no repeating cycle as we did for $\frac{1}{3}$ and $\frac{3}{7}$. Regardless of how far we extend these decimals, there will be no repeating set of digits. These decimals have a pattern, but no repeating cycle. Decimals that are *nonterminating* and *nonrepeating* are called **irrational numbers**.

Probably the most famous irrational number is *pi* (π). The value of π for the first 50 decimal places is

$$\pi = 3.14159265358979323846264338327950288419716939937510$$

You will note that there is no repeating sequence of digits for this decimal.

The formula for finding the circumference of a circle says that to find the circumference of a circle, we multiply its diameter times *pi*; that is, $C = \pi d$. Therefore, π is the ratio of the circumference of a circle to its diameter. In other words, to find the value of π, you must divide the circumference of a circle by its diameter. You will obtain a nonrepeating nonterminating decimal. Computers have been used to find the value of π well beyond the 50 decimal places shown, but no one has ever reached the end, or ever will.

There are many other examples of irrational numbers—numbers that, when they are expressed as decimals, are nonterminating and nonrepeating. Some other examples are

$$2.7182818284\ldots$$
$$0.12122122212222122222\ldots$$
$$\sqrt{2},\quad \sqrt{3},\quad \sqrt{5},\quad \sqrt{6},\quad \sqrt{7}$$

Historical Note

Rather than describe a number as the square root of another number, the ancient Greeks referred to it as a side of a square number. For example, 2 is the side of a square number 4, 3 is the side of a square number 9, and so on. From this the Arabs developed the idea of a root of a number; that is, 2 is the root of 4, 3 is the root of 9, and so forth. Later, European mathematicians used the Latin word radix to indicate the root of a number. When abbreviations began to be used in algebra (about 100 A.D.), the word radix was abbreviated as R_x. Therefore, $R_x4 = 2$, $R_x9 = 3$. This notation was used for some time, until a small *r* began to be used in place of R_x. R_x4 became *r*4. Most historians believe that the radical sign, $\sqrt{\ }$, as we know it today, is a representation of the small letter *r* as it was written by mathematicians at the time (because everything had to be copied by hand).

A **perfect square** is a number that is the product of an integer times itself. For example, 4 is a perfect square because $4 = 2 \cdot 2$. Some other examples of perfect squares are 1, 9, 16, 25, and 36. The square root of any positive integer that is *not* a perfect square is an irrational number.

For example, $\sqrt{2}$ is an irrational number. If we were to express $\sqrt{2}$ as a decimal, it would be a nonterminating nonrepeating decimal. Some other examples of irrational numbers are $\sqrt{3}$, $\sqrt{5}$, $\sqrt{6}$, $\sqrt{7}$, and so on. These are all square roots of nonnegative numbers that are not perfect squares.

EXAMPLE 1 *Distinguishing between Rational and Irrational Numbers*

Classify each number as rational or irrational.

a. $0.\overline{13}$ **b.** $0.131131113\ldots$ **c.** $\sqrt{11}$ **d.** $\sqrt{49}$

Solution

a. $0.\overline{13}$ is a rational number. It is a nonterminating decimal, but it is repeating.

b. $0.131131113\ldots$ is an irrational number. It is a decimal that is nonterminating and nonrepeating.

c. $\sqrt{11}$ is an irrational number. The square root of a nonnegative number that is not a perfect square is an irrational number.

d. $\sqrt{49}$ is a rational number because 49 is a perfect square. In fact, $\sqrt{49} = 7$.

NW ***Now Work Problems 1a–1f.***

Historical Note

Irrational numbers were studied as early as 500 B.C. by Pythagoras, a famous Greek mathematician. He discovered that there is no rational number for the square root of 2; that is, there is no rational number whose square is 2.

Pythagoras and his colleagues also found that there is no common unit of length for the length of a side of a square and its diagonal at the same time. That is, no unit of length will ever measure exactly both the diagonal of a square and its side.

These mathematicians were supposedly so upset with this discovery that they vowed not to reveal it, and threatened to punish

any member of the group who divulged this "secret." They were afraid that this information would cause people to doubt their ability as mathematicians.

Consider the Pythagorean theorem, which in essence states that the square of the hypotenuse of a right triangle is equal to the sum of the squares of the other two sides, or given right triangle ABC, $c^2 = a^2 + b^2$.

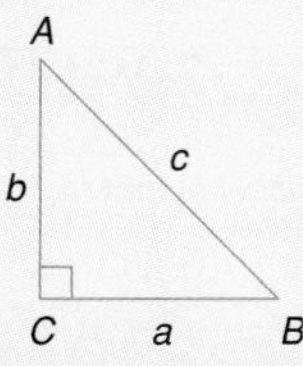

Hence, the length of the hypotenuse of a right triangle that has sides of length 1 is $\sqrt{2}$. This implies that we can easily construct a line of a certain length. However, it is not measurable, and this is what concerned Pythagoras.

Real Numbers

It is interesting to note that the set of rational numbers and the set of irrational numbers are disjoint sets; that is, their intersection is empty. If we take the union of these two sets, then we have all of the numbers on the number line. The union of the set of rational numbers and the set of irrational numbers yields a set of numbers that is called the set of **real numbers**. Therefore, any rational or irrational number is a real number. Note that a rational number may be expressed as either a terminating decimal or a repeating nonterminating decimal, and an irrational number may be expressed as a nonrepeating nonterminating decimal. Therefore, every decimal is a real number and every real number may be expressed as a decimal.

Recall that we began our discussion of the classification of numbers with the set of natural numbers. Including zero with the set of natural numbers produced the set of whole numbers. The set of integers was composed of the set of whole numbers and their additive inverses. The set of integers and the set of fractions yielded the set of rational numbers. Finally came the set of irrational numbers, and the union of the set of rationals and the set of irrationals yielded the set of real numbers. This process is illustrated by the following diagram:

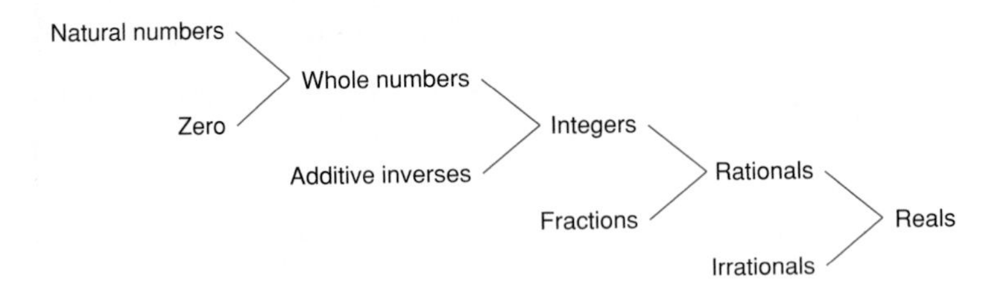

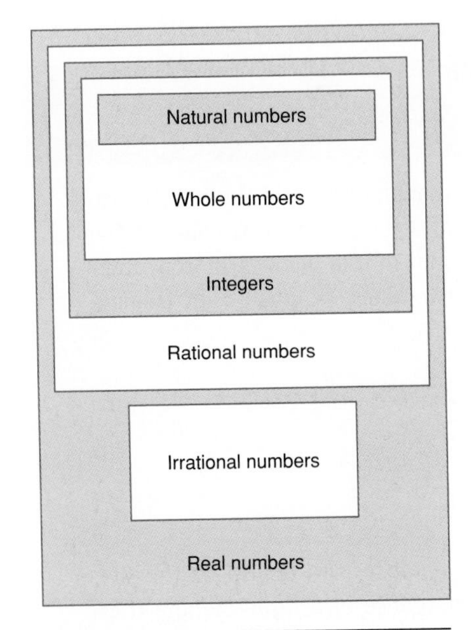

Figure 8.20

Figure 8.20 illustrates the relationship of these sets of numbers to the set of real numbers.

EXAMPLE 2 *Manipulating Different Sets of Numbers*

Let R={realnumbers}, I={integers, Y=rational numbers}, Z = {irrational numbers}. Find

a. $R \cap I$ **b.** $I \cap Y$ **c.** $Y \cap Z$ **d.** $Y \cup Z$

Solution We can find these sets by examining Figure 8.14.

a. $R \cap I = I$ The set of integers is common to both sets.

b. $I \cap Y = I$ The set of integers is common to both sets.

c. $Y \cap Z = \emptyset$ The set of rational numbers and the set of irrational numbers have no elements in common.

d. $Y \cup Z = R$ The set of rational numbers together with the set of irrational numbers forms the set of real numbers.

NW ***Now Work Problems 9a, 9b, and 9c.***

Math Connections

Laws of Real Numbers

Earlier in this chapter, we presented a Math Connections on the laws of natural numbers. We now extend these laws and apply them to the set of real numbers.

A. Commutative Laws of Addition and Multiplication

1. The *commutative law of addition* states that the order in which we add two real numbers is unimportant. In general, if a and b represent any two real numbers, then

$$a + b = b + a$$

2. The *commutative law of multiplication* states that the order in which we multiply two real numbers is unimportant. In general, if a and b represent any two real numbers, then

$$a \times b = b \times a$$

B. Associative Laws of Addition and Multiplication

1. The *associative law of addition* states that when adding three or more real numbers, we can group these numbers in any way we want without affecting the sum. In general, if a, b, and c represent any three real numbers, then

$$(a + b) + c = a + (b + c)$$

2. The *associative law of multiplication* states that when multiplying three or more real numbers, we can group these numbers in any way we want without affecting the product. In general, if a, b, and c represent any three real numbers, then

$$(a \times b) \times c = a \times (b \times c)$$

C. Identity Laws of Addition and Multiplication

1. The *identity law of addition* states that there is a unique zero (0)—called the *identity element of addition*—such that any real number a added to 0 yields the number itself. In general, for all real numbers a,

$$a + 0 = a$$

2. The *identity law of multiplication* states that there is a unique unit (1)—called the *identity element of multiplication*—such that any real number a multiplied by 1 yields the number itself. In general, for all real numbers a,

$$a \times 1 = a$$

D. Inverse Laws of Addition and Multiplication

1. The *inverse law of addition* states every real number a has a unique opposite real number $-a$ (called the *additive inverse*) such that when a is added to its opposite, the sum is the additive identity element, 0. In general, for every real number a:

$$a + (-a) = 0$$

2. The *inverse law of multiplication* states every nonzero real number a has a unique reciprocal $\frac{1}{a}$ (called the *multiplicative inverse*) such that when a is multiplied by its opposite, the product is the multiplicative identity element 1. In general, for any nonzero real number a,

$$a \times \frac{1}{a} = 1$$

E. Distributive Law

In the real number system *multiplication is distributive over addition.* That is,

$$a \times (b \times c) = a \times b + a \times c$$

F. Laws of Order

The real numbers are also *simply ordered* and obey the following laws of order:

1. Trichotomy Law
   Given two real numbers, a and b, exactly one of the following relations is true:
   (a) $a < b$, **(b)** $a = b$, or **(c)** $a > b$
2. Transitive Law
   If $a < b$ and $b < c$, then $a < c$.
3. Order and Addition
   If $a < b$, then $a + c < b + c$.
4. Order and Multiplication
   If $a < b$ and $c > 0$, then $a \times c < b \times c$.

Exercises for Section 8.7

NW **1.** Classify each number as rational or irrational.

a. $\frac{1}{3}$ **b.** -2
c. $0.\overline{3}$ **d.** $\sqrt{2}$
e. $\sqrt{3}$ **f.** $\sqrt{4}$

2. Classify each number as rational or irrational.

a. $\frac{5}{8}$ **b.** $\sqrt{7}$
c. $\sqrt{49}$ **d.** 2.4494897
e. 0.131131113... **f.** $0.\overline{6}$

3. Classify each number as rational or irrational.

a. $\frac{3}{4}$ **b.** $\frac{2}{3}$

c. $\sqrt{16}$ **d.** 3.14159

e. $0.\overline{21}$ **f.** $\sqrt{99}$

4. Determine whether each of the following can be represented by a terminating decimal, a repeating decimal, or a nonterminating nonrepeating decimal.

a. $\frac{2}{5}$ **b.** $\frac{1}{3}$

c. $\sqrt{7}$ **d.** $\sqrt{16}$

e. $\frac{5}{9}$ **f.** $\sqrt{77}$

5. Determine whether each of the following can be represented by a terminating decimal, a repeating decimal, or a nonterminating nonrepeating decimal:

a. $\frac{1}{8}$ **b.** $\frac{3}{7}$

c. $\sqrt{3}$ **d.** $\sqrt{\frac{9}{16}}$

e. $\sqrt{99}$ **f.** $\frac{1}{11}$

6. Determine whether each of the following can be represented by a terminating decimal, a repeating decimal, or a nonterminating nonrepeating decimal:

a. $\sqrt{5}$ **b.** $\sqrt{625}$

c. $\frac{3}{8}$ **d.** $\frac{5}{7}$

e. $\frac{22}{7}$ **f.** $\sqrt{225}$

7. Determine whether each statement is true or false.

a. A real number is either a rational or irrational number.

b. A real number is positive, negative, or zero.

c. A repeating nonterminating decimal is a rational number.

d. A nonrepeating nonterminating decimal is a rational number.

e. A terminating decimal is an irrational number.

8. Determine whether each statement is true or false.

a. The intersection of the set of rational numbers and the set of irrational numbers is not empty.

b. The union of the set of rational numbers and the set of irrational numbers is the set of real numbers.

c. All rational numbers are real numbers.

d. All real numbers are rational numbers.

e. Every real number can be expressed as a terminating decimal or a repeating decimal.

9. Let $R = \{\text{real numbers}\}$, $I = \{\text{integers}\}$, $Y = \{\text{rational numbers}\}$, and $Z = \{\text{irrational numbers}\}$. Determine whether each of the following is true or false:

NW **a.** $R \cap Y = I$ NW **b.** $I \cup Y = Z$

NW **c.** $Y \subset R$ **d.** $R \subset Z$

e. $Y \cap Z = R$ **f.** $Y \cup Z = R$

10. Answer yes or no to tell whether each of the following numbers is (I) a natural number, (II) a whole number, (III) an integer, (IV) a rational number, (V) an irrational number, and (VI) a real number:

a. 7 **b.** -3

c. 3.14 **d.** $\sqrt{2}$

e. 0 **f.** $2.\overline{123}$

Writing Mathematics

11. Explain the difference between a *rational number* and an *irrational number.*

12. Define a *real number.*

13. Define the concept of a *perfect square* and give an example of one that was not presented in the book.

Challenge Exercises

When representing numbers on a number line, we can use a closed circle, • *to represent a specific number, an open circle,* ∘, *to represent a number that is being* excluded, *and a shaded line segment,* —, *to indicate that every number is to be included. Given these three symbols, identify the set of numbers being represented by the number lines in Exercises 14–19. Express your set as being part of the natural numbers, whole numbers, integers, rationals, or real numbers. For example, we would describe the number line below as representing the set of all natural numbers less than or equal to 10.*

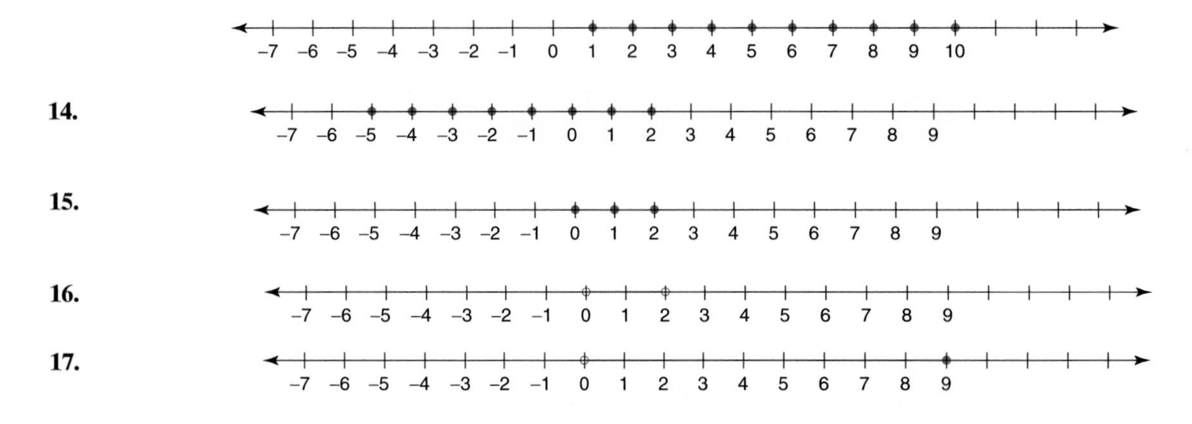

18.

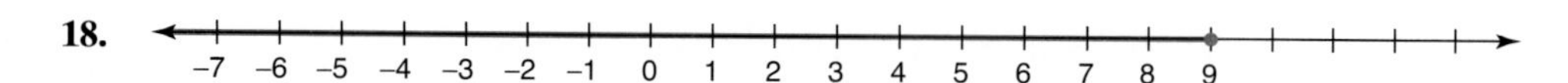

19.

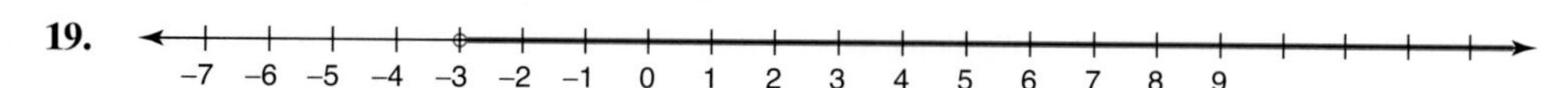

20. Archimedes, who was one of the greatest mathematicians of all time, knew that π was between $3\frac{1}{7}$ and $3\frac{10}{71}$. Calculate the decimal equivalents of these mixed numbers. Why do you think Archimedes did not have a decimal equivalent for π?

Research/Group Activities

21. There are several ongoing π-related research projects. Using the Internet, find three Web sites that are related to these projects. As part of your investigation, find out the current greatest number of decimal places to which π has been calculated. Also include a description of the research being conducted.

22. In addition to the set of real numbers, there is the set of *imaginary numbers*. Conduct research to discover what these numbers are. As part of your research, investigate the history of imaginary numbers, the symbol used to represent them, and applications of their use.

Just for fun

If *pi* is an irrational number (a number that is a nonrepeating nonterminating decimal), then why is it that we are usually *told* that pi(π) is $\frac{22}{7}$, when $\frac{22}{7} = 3.\overline{142857}$?

8.8 SCIENTIFIC NOTATION (OPTIONAL)

Scientific Notation for Large Numbers

It is estimated that the average human breath contains as many as ten sextillion atoms. This extremely large number, when expressed in *standard form*, appears as

10,000,000,000,000,000,000,000

To write this number in *scientific notation* we imagine a decimal point after the first nonzero digit, and count the number of decimal places from there to the actual decimal point.

Standard form	*Scientific notation*
10,000,000,000,000,000,000,000 =	1.0×10^{22}
22 places	Write the number as a product of 1.0 and 10^{22}.

Note: Multiplying by 10^{22} moves the decimal point 22 places to the *right*. To write a number in scientific notation, write it as a product so that

1. The first factor is greater than or equal to 1 and less than 10.
2. The second factor is a power of 10.

EXAMPLE 1 ***Writing Large Numbers in Scientific Notation*** Write each of the following in scientific notation:

a. 190,000,000 **b.** 67,200 **c.** 150

Solution

a. 190,000,000
8 places
1.9×10^8

b. 67,200
4 places
6.72×10^4

c. 150
2 places
1.5×10^2

NW ***Now Work Problems 3 and 5.***

Scientific Notation for Small Numbers

One millimicron is approximately 0.0000000394 inch. Scientific notation can also be used to write very small numbers more compactly. Remember that scientific notation consists of writing a number in the general form $N \times 10^m$, where N is greater than or equal to 1 and less than 10, and m is an integer. Imagine a decimal point after the first nonzero digit. Count the number of places from there to the actual decimal point. Note that we use a negative exponent with 10 when the original number is less than 1.

$$0.0000000394 = 3.94 \times 10^{-8}$$

8 places

Note: Multiplying by 10^{-8} moves the decimal point 8 places to the *left.*

EXAMPLE 2 ***Writing Small Numbers in Scientific Notation*** Write each of the following in scientific notation:

a. 0.00245 **b.** 0.00074 **c.** 0.0271

Solution

a. 0.00245
3 places
2.45×10^{-3}

b. 0.00074
4 places
7.4×10^{-4}

c. 0.0271
2 places
2.71×10^{-2}

NW ***Now Work Problems 11 and 15.***

EXAMPLE 3 ***Converting from Scientific Notation to Standard Form*** Express each of the following in standard form:

a. 5.21×10^5 **b.** 3.9×10^{-3} **c.** 1.05×10^{-6}

Note of Interest

In 1614, John Napier invented logarithms. The word *logarithm* means "ratio number." It was derived from the Greek words *logos*, which means "ratio," and *arithmos*, which means "number." Napier first published a table of logarithms for trigonometric functions. Later, Henry Briggs (1561–1630), an English mathematician, visited Napier to discuss the logarithms. As a result of this visit, Briggs introduced what are now known as common logarithms. The combined work of these two men led to the invention of the slide rule in England by Edmund Gunter in 1620. A slide rule can be used to find products, quotients, powers, and roots using the basic laws of logarithms. Before the advent of today's pocket calculators, a slide rule was a common calculating device for thousands of college students studying mathematics, science, and engineering.

Napier announced his discovery of logarithms in a book titled *Mirifici Logarithmorum Canonis Descriptio (A Description of the Admirable Table of Logarithms)*. In a prefatory paragraph, he wrote

> Seeing there is nothing, right well-beloved students of mathematics, that is so troublesome to mathematical practice, nor doth more molest and hinder calculators, than multiplication, division, square and cubical extractions of great numbers, which besides the tedious expense of time are for the most part subject to many slippery errors, I began therefore to consider in my mind by what certain and ready art I might remove those hindrances.

Solution

a. The exponent of 10 is positive. Hence, move the decimal point five places to the right:

$$5.21 \times 10^5 = 521{,}000$$

b. The exponent of 10 is negative. Therefore, move the decimal point three places to the left:

$$3.9 \times 10^{-3} = 0.0039$$

c. Move the decimal point six places to the left:

$$1.05 \times 10^{-6} = 0.00000105$$

NW ***Now Work Problems 21 and 23.***

Multiplying and Dividing Numbers in Scientific Notation

We can use scientific notation in computing answers to problems that contain very large or very small numbers. Since scientific notation makes use of integral exponents, we can apply the rules for exponents. If b is any real number and m and n are natural numbers, then $b^m \times b^n = b^{m+n}$. For example,

$$10^3 \times 10^2 = 10^{3+2} = 10^5 \quad \text{and} \quad 10^{-2} \times 10^{-4} = 10^{-2+(-4)} = 10^{-6}$$

Consider this multiplication problem:

$$12{,}000{,}000 \times 130{,}000{,}000$$

Many of the pocket calculators that are so popular today would not be able to handle this problem because of the number of digits involved. An electronic calculator that can handle this multiplication most likely uses scientific notation in performing the necessary computation.

To perform the indicated multiplication, we must first write the given numbers in scientific notation:

$$12{,}000{,}000 \times 130{,}000{,}000 = (1.2 \times 10^7)(1.3 \times 10^8)$$

Next, we make use of the commutative and associative properties of multiplication, and write an equivalent expression:

$$(1.2 \times 10^7)(1.3 \times 10^8) = (1.2 \times 1.3)(10^7 \times 10^8)$$

Performing these multiplications, we obtain

$$(1.2 \times 1.3)(10^7 \times 10^8) = 1.56 \times 10^{15}$$

Therefore,

$$12{,}000{,}000 \times 130{,}000{,}000 = 1.56 \times 10^{15} = 1{,}560{,}000{,}000{,}000{,}000$$

Note that we performed the operations separately. That is, we performed the multiplication of the numbers between 1 and 10, and separately we performed the multiplication of the powers of 10.

A similar procedure can be used for the operation of division. But, instead of multiplying powers of 10, we divide powers of 10.

If b is any real number ($b \neq 0$) and m and n are natural numbers, then

$$\frac{b^m}{b^n} = b^{m-n}$$

For example,

$$\frac{10^6}{10^2} = 10^{6-2} = 10^4$$

$$\frac{10^2}{10^4} = 10^{2-4} = 10^{-2}$$

$$\frac{10^3}{10^{-2}} = 10^{3-(-2)} = 10^{3+2} = 10^5$$

Powers of Ten

The following table contains the word name and number representation for various powers of ten. It is interesting to note that although many people feel compelled to use the word *zillion* to express an extremely large quantity, "zillion" is not a real number; that is, there is no equivalent power of ten or number representation. For a visual journey of the powers of ten that correspond to our universe—from the inside of a proton to the farthest reaches of the universe—see **http://www.wordwizz.com/pwrsof10.htm**. This site contains a series of 42 drawings and pictures that illustrate our known universe, which is currently estimated to be 13.7 thousand million years old, at least 2×10^{26} meters in diameter, and made up of quarks and electrons that are smaller than 10^{-16} meters. Each successive image shows a view ten times wider or narrower than its neighbor.

Power of 10	Word Name	Written as a Number	How Written
10^1	ten	10	1 followed by 1 zero
10^2	one hundred	100	1 followed by 2 zeros
10^3	one thousand	1000	1 followed by 3 zeros
10^4	ten thousand	10,000	1 followed by 4 zeros
10^5	hundred thousand	100,000	1 followed by 5 zeros
10^6	one million	1,000,000	1 followed by 6 zeros
10^9	one billion	1,000,000,000	1 followed by 9 zeros
10^{12}	one trillion	1,000,000,000,000	1 followed by 12 zeros
10^{15}	one quadrillion	1,000,000,000,000,000	1 followed by 15 zeros
10^{18}	one quintillion	1,000,000,000,000,000,000	1 followed by 18 zeros
10^{21}	one sextillion	1,000,000,000,000,000,000,000	1 followed by 21 zeros
10^{24}	one septillion	1,000,000,000,000,000,000,000,000	1 followed by 24 zeros
10^{27}	one octillion	1,000,000,000,000,000,000,000,000,000	1 followed by 27 zeros
10^{30}	one nonillion	1,000,000,000,000,000,000,000,000,000,000	1 followed by 30 zeros

This pattern continues with the names *decillion, undecillion, duodecillion, tredecillion, quattuordecillion, quindecillion, sexdecillion, septdecillion, octodecillion, novemdecillion*, and *vigintillion*, respectively, used to denote 10^{33}, 10^{36}, 10^{39}, 10^{42}, 10^{45}, 10^{48}, 10^{51}, 10^{54}, 10^{57}, 10^{60}, and 10^{63}. Other notable names include **googol**, which is 10^{100} (i.e., 1 followed by 100 zeros), and **googolplex**, which is 10^{google} (i.e., $10^{10^{100}}$). Finally, it is also interesting to note that the conventional European naming system for large numbers is different than ours beginning with what we regard as one billion (e.g., the Europeans denote 10^9 as "milliard"). For additional information, see **http://g42.org/MiscInfo/numbers.html**.

Consider the division problem

$$144{,}000{,}000 \div 7{,}200{,}000$$

We write this as $\frac{144{,}000{,}000}{7{,}200{,}000}$. Now we write the given numbers in scientific notation:

$$\frac{144{,}000{,}000}{7{,}200{,}000} = \frac{1.44 \times 10^8}{7.2 \times 10^6} = \frac{1.44}{7.2} \times \frac{10^8}{10^6}$$

Performing the indicated operations separately, we obtain

$$\frac{1.44}{7.2} \times \frac{10^8}{10^6} = 0.2 \times 10^{8-6} = 0.2 \times 10^2 = 20$$

EXAMPLE 4 ***Performing Calculations in Scientific Notation*** Use scientific notation to compute the following:

a. $0.0035 \times 2{,}400{,}000$ **b.** $0.000143 \div 0.0013$

Solution We first express all numbers in scientific notation and then separately perform the operations with the numbers between 1 and 10, and the powers of 10.

a.
$$\begin{aligned} 0.0035 \times 2{,}400{,}000 &= (3.5 \times 10^{-3})(2.4 \times 10^6) \\ &= (3.5 \times 2.4)(10^{-3} \times 10^6) \\ &= 8.4 \times 10^{-3+6} \\ &= 8.4 \times 10^3 = 8400 \end{aligned}$$

b.
$$\begin{aligned} 0.000143 \div 0.0013 &= \frac{0.000143}{0.0013} \\ &= \frac{1.43 \times 10^{-4}}{1.3 \times 10^{-3}} = \frac{1.43}{1.3} \times \frac{10^{-4}}{10^{-3}} \\ &= 1.1 \times 10^{-4-(-3)} = 1.1 \times 10^{-4+3} \\ &= 1.1 \times 10^{-1} \\ &= 0.11 \end{aligned}$$

NW ***Now Work Problems 29 and 37.***

In computations with scientific notation, it is best to perform separately the multiplication or division of the numbers between 1 and 10, and then separately perform the multiplication or division of the powers of 10.

Exercises for Section 8.8

For Exercises 1–16, write each number in scientific notation.

1. 240 **2.** 5300 NW **3.** 125,000

4. 147,100 NW **5.** 4,300,000 **6.** 68,200,000

7. 739.2 **8.** 432.12 **9.** 0.231

10. 0.0068 NW **11.** 0.0002345 NW **12.** 0.0000008

13. 0.0012 **14.** 0.00301 **15.** 0.00000009

16. 0.000209

17. The earth is approximately 4,000,000,000 years old. Write this number in scientific notation.

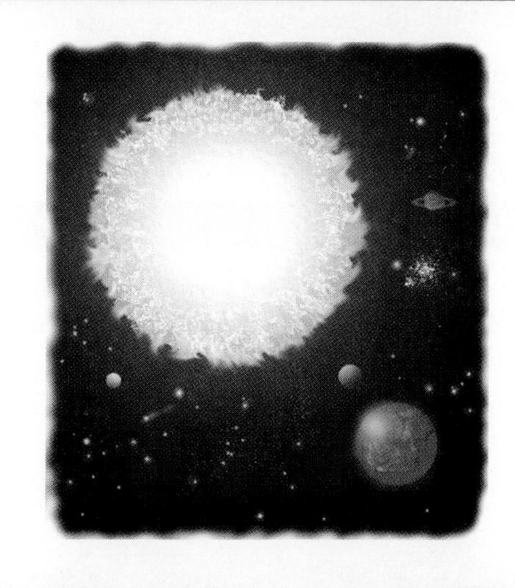

18. An electron is approximately 0.0000000000001 centimeter in diameter. Write this number in scientific notation.

19. An angstrom is a unit of length and it is 0.0000001 millimeter long. Write this number in scientific notation.

20. The sun is believed to be 4,500,000,000 years old. Write this number in scientific notation.

For Exercises 21–28, write the given expressions in standard form.

NW **21.** 4.1×10^{-3} **22.** 9.12×10^{4}

NW **23.** 3.142×10^{5} **24.** 3.17×10^{-5}

25. 4.859×10^{7} **26.** 5.4321×10^{-6}

27. 4.8×10^{-5} **28.** 9.8×10^{-7}

For Exercises 29–49, use scientific notation to evaluate each of the given expressions. Answers may be left in scientific notation.

NW **29.** $(3.4 \times 10^{4}) \times (2.6 \times 10^{3})$

30. $(1.5 \times 10^{-2}) \times (4.7 \times 10^{-5})$

31. 1800×180 **32.** 0.000301×0.003

33. $60{,}500 \times 0.0034$ **34.** $75{,}000 \times 1300$

35. 0.0058×750 **36.** $76{,}000 \times 120{,}000$

NW **37.** $\dfrac{4.2 \times 10^{6}}{1.4 \times 10^{3}}$ **38.** $\dfrac{6.25 \times 10^{7}}{2.5 \times 10^{2}}$

39. $\dfrac{450{,}000}{9000}$ **40.** $\dfrac{105{,}000{,}000}{21{,}000}$

41. $\dfrac{255{,}000{,}000}{170{,}000}$ **42.** $\dfrac{0.000143}{0.0013}$

43. $\dfrac{0.024}{0.00012}$ **44.** $\dfrac{10^{5} \times 10^{-6} \times 10^{2}}{10^{-3} \times 10^{-5}}$

45. $\dfrac{10^{1} \times 10^{4} \times 10^{-5}}{10^{3} \times 10^{-2} \times 10^{-3} \times 10^{4}}$ ***46.** $\dfrac{0.0018 \times 12{,}000}{0.000027 \times 0.002}$

***47.** $\dfrac{24{,}000 \times 140{,}000}{210{,}000 \times 0.0016}$ ***48.** $\dfrac{320{,}000 \times 18{,}000}{0.0012 \times 0.000048}$

***49.** $\dfrac{0.00036 \times 0.024}{0.00012 \times 0.0015}$

50. Loose snow weighs approximately 0.125 gram per cubic centimeter. Approximately how much does 1,000,000,000 cubic centimeters of loose snow weigh?

Writing Mathematics

51. What is scientific notation? As part of your response, include a description of the two parts that make up scientific notation expressions.

52. The star *Alpha Centauri* is approximately 26,000,000,000,000 miles away from earth. Describe how you would convert this number to scientific notation.

53. One atom of uranium weighs 0.000000000000000000000396 gram. Describe how you would convert this number to scientific notation.

54. Explain the product rule for exponents and include an example as part of your explanation.

55. Explain the quotient rule for exponents and include an example as part of your explanation.

Challenge Exercises

56. Mr. Setek, in an effort to demonstrate the "size" of one million, announced to his liberal arts math class at the beginning of the term that anyone who can manually write the first one million natural numbers prior to the end of the term will automatically receive an "A" for the course. With the exception of Carl, several students accepted Mr. Setek's challenge but quit after the first day of writing the numbers. Carl, on the other hand, persevered a little longer but eventually stopped when he got to 100,000. If all of the numbers Carl wrote were expressed as a horizontal string of digits, how many zeros do you think this single "number" has?

57. Calculators capable of expressing numbers in scientific notation have an EXP, EE, or equivalent key. Thus, to enter the number 1.2×10^{42}, you enter 1.2 EE 42 and the display shows 1.2 42. Similarly, to enter the number 3.05×10^{-7}, you enter 3.05 EE 7+/− and the display shows 3.05 −07. Given the key sequence 6.435 EE 9+/−, express the given number in *decimal* form.

58. Without doing any calculations, which number is larger: 1×10^{10} or 9×10^{9}? Why?

Research/Group Activities

59. Research answers to the following questions and then express your answers in both decimal and scientific notation forms. How long is a light-year? What is the diameter of the universe? What is the diameter of the smallest particle visible to the naked human eye? What is the current population of the earth? How many seconds are there in one year? What is the diameter of a red blood corpuscle? What is the approximate weight of the largest blue whale ever measured? What is the current U.S. national debt?

60. In 1811, the Italian chemist, Amadeo Avogadro, correctly conjectured that equal volumes of gases at the same temperature and pressure contain equal numbers of molecules. This conjecture has since been confirmed many times and is now known as Avogadro's law. Associated with this law is the *Avogadro number*. Find out what this number is and what it means. Report back to your class.

Just for fun

Four men, in different colored overcoats, sat on two facing bench seats in a train. Two sat next to the window and two next to the aisle. The Englishman sat on Mr. Bell's left. Alto wore a tan-colored coat. The man in olive was on the German's right. Mr. Cardinal was the only cigar smoker. Mr. Dunn was across from the American. The Russian was in khaki, and the Englishman stared out the window on his left. Who was the man in the rust-colored coat?

CHAPTER REVIEW MATERIALS

Summary/Chapter 8

8.2 Natural Numbers—Primes and Composites

- *Cardinal numbers* are used to tell us "how many." Examples include 3 dogs, 4 cars, and 7 days in a week.
- *Ordinal numbers* are used for rank or position. Examples include first place, third in line, and ninth on the list.
- *Natural numbers* are also known as the *counting numbers* and consist of the set $\{1, 2, 3, 4, \dots\}$.
- Every natural number is divisible (i.e., has a remainder of zero) by itself and 1.
- A *prime number* is any natural number greater than 1 that is divisible by only 1 and itself. Thus, the first five prime numbers are 2, 3, 5, 7, and 11.
- A *composite number* is any natural number greater than 1 that is not prime.
- The *fundamental theorem of arithmetic* states that every natural number greater than 1 is either prime or can be expressed as a unique product of prime factors.

8.3 Greatest Common Divisor and Least Common Multiple

- The *greatest common divisor* (GCD) of a pair of natural numbers is the greatest number that divides both of the given numbers evenly (i.e., with a zero remainder). Two numbers whose GCD is 1 are said to be *relatively prime*.
- The *least common multiple* (LCM) of two natural numbers is the least (i.e., smallest) natural number that is a multiple of each of the given numbers.

8.4 Integers

- Expanding the set of natural numbers to include zero gives us the set of *whole numbers* $\{0, 1, 2, 3, 4, \dots\}$.
- The set of numbers consisting of the set of whole numbers and their additive inverses is called the set of *integers*. Thus, the set of integers is $\{\dots, -3, -2, -1, 0, 1, 2, 3, \dots\}$. The set of natural numbers and the set of whole numbers are subsets of the set of integers.
- A *number line* can be used to determine the numerical order of the integers. Numbers are placed on the number line in increasing numerical order from left to right. Thus, $-1 > -3$ because -1 is to the right of -3 on the number line.
- When *combining integers*, the following observations are noted: (1) If we add two positive integers, then the sum is a positive integer; (2) if we add two negative integers, then the sum is a negative integer; and (3) if we add a positive integer and a negative integer, the sum is either positive, negative, or zero.
- When *multiplying or dividing integers*, the following observations are noted: (1) The product or quotient of two positive integers is positive; (2) the product or quotient of two negative integers is positive; and (3) the product or quotient of a positive integer and a negative integer is negative.
- When evaluating expressions involving integers, we follow a prescribed *order of operations*: (1) Working from left to right, evaluate all expressions within grouping symbols first; (2) working from left to right, perform all multiplications and divisions; and (3) working from left to right, perform all additions and subtractions.

- An integer is *even* if it can be expressed in the form $2 \times n$, where n is an integer. Similarly, an integer is *odd* if it can be expressed in the form $2 \times n + 1$, where n is an integer.

8.5 Rational Numbers

- A *rational number* is a number than can be expressed as a quotient of integers; that is, in the form $\frac{a}{b}$, where a and b are integers and $b \neq 0$.
- Two rational numbers are equal if their *cross-products* are equal.
- *Adding and subtracting rational numbers* follows the general rule $\frac{a}{b} + \frac{c}{d} = \frac{ad + bc}{b \times d}$, where $b \neq 0$ and $d \neq 0$.
- *Multiplying rational numbers* follows the general rule $\frac{a}{b} \times \frac{c}{d} = \frac{a \times c}{b \times d}$, where $b \neq 0$ and $d \neq 0$.
- *Dividing rational numbers* follows the general rule $\frac{a}{b} \div \frac{c}{d} = \frac{a}{b} \times \frac{d}{c} = \frac{a \times d}{b \times c}$, where $b \neq 0, c \neq 0$ and $d \neq 0$.
- A *complex fraction* is a fraction in which either the numerator, denominator, or both the numerator and denominator are fractions.
- A *mixed number* represents the sum of an integer and a fraction expressed without the plus (+) sign. Examples include $7\frac{3}{5}$ and $11\frac{9}{30}$.
- An *improper fraction* is a fraction in which the numerator is greater in numerical value than the denominator. Examples include $\frac{9}{2}$ and $\frac{17}{4}$. Improper fractions can be expressed as mixed numbers by dividing the denominator into the numerator. The quotient becomes the integer component of the mixed number; the fractional component of the mixed number is obtained by writing the remainder as the numerator of the fraction.

8.6 Rational Numbers and Decimals

- Rational numbers can be written as decimals by dividing the denominator into the numerator.
- Rational numbers can be classified as either *terminating decimals* or *repeating nonterminating decimals*.

8.7 Irrational Numbers and the Set of Real Numbers

- An *irrational number* is a nonterminating and nonrepeating decimal. Examples of irrational numbers include π, $\sqrt{2}$, $\sqrt{3}$, and 0.1010010001 . . . .
- The set of *real numbers* is the union of the set of rational numbers and the set of irrational numbers. This set represents all the numbers on a number line.

8.8 Scientific Notation (Optional)

- *Scientific notation* is used to write very large or very small numbers more compactly. It involves writing a number that is greater than or equal to 1 and less than 10, times an integral power of 10. That is, $N \times 10^m$, where $1 \leq N < 10$ and m is an integer.

Key Terms and Concepts

complex fraction, *457*
composite number, *419*
density property, *468*
distributive property, *441*
fundamental theorem of arithmetic, *423*
greatest common divisor, *427*
improper fraction, *459*
integer, *436*
irrational number, *472*
least common multiple, *430*
mixed number, *459*
natural number, *419*
order of operations, *444*
prime number, *419*
rational number, *451*
real number, *474*
sieve of Eratosthenes, *422*
whole number, *436*

Review Exercises

8.2 Natural Numbers—Primes and Composites

1. Determine whether each number is prime or composite.

a. 99 **b.** 97
c. 83 **d.** 431
e. 657 **f.** 10,101

2. Determine the prime factors of each natural number.

a. 42 **b.** 115
c. 525 **d.** 174
e. 705 **f.** 2222

8.3 Greatest Common Divisor and Least Common Multiple

3. Find the greatest common divisor for each pair of numbers.

a. 30 and 48
b. 42 and 55
c. 48 and 72
d. 66 and 90
e. 111 and 231
f. 342 and 612

4. Find the least common multiple for each pair of numbers.
   a. 21 and 36
   b. 32 and 44
   c. 40 and 50
   d. 72 and 150
   e. 30 and 100
   f. 75 and 220

8.4 Integers

5. Evaluate each of the following:
   a. $3 + (-7)$
   b. $4 + (-5)$
   c. $-8 + 7$
   d. $-7 - (-3)$
   e. $13 - (-3)$
   f. $-8 - 5$

6. Evaluate each of the following:
   a. $(-2) \times (-4)$
   b. $(-5) \times 3$
   c. $(-1) \times (-2) \times (-3)$
   d. $(-3 + 7) \times (-2)$
   e. $(2 - 4) \times 2$
   f. $(-3 - 5) \times (-3 - 4)$

7. Evaluate each of the following:
   a. $3\,(4 - 3) + 5$
   b. $(7 \times 6 - 2) \div 10$
   c. $4 \times 3 - 5 + 8 \div 4$
   d. $(8 \times 2 - 4) \div 6 \times 2$
   e. $48 \div [(10 - 6) \times 6]$
   f. $16 \div 4 \times 8 + 10 - 6$

8.5 Rational Numbers

8. Perform the indicated operation.
   a. $\frac{2}{6} + \frac{1}{4}$
   b. $\frac{3}{7} + \frac{4}{9}$
   c. $\frac{2}{11} + \frac{1}{13}$
   d. $\frac{4}{7} - \frac{1}{3}$
   e. $\frac{3}{5} - \frac{1}{2}$
   f. $\frac{7}{8} - \frac{2}{5}$

9. Perform the indicated operation.
   a. $\frac{1}{3} \times \frac{2}{5}$
   b. $\frac{3}{7} \times \frac{4}{9}$
   c. $\frac{2}{11} \times \frac{3}{5}$
   d. $\frac{2}{3} \div \frac{4}{7}$
   e. $\frac{3}{5} \div \frac{1}{2}$
   f. $\frac{6}{11} \div \frac{3}{5}$

10. Simplify each of the following:
   a. $\dfrac{\frac{3}{8}}{\frac{4}{9}}$
   b. $\dfrac{\frac{3}{4}}{5}$
   c. $\dfrac{5}{\frac{3}{4}}$
   d. $\dfrac{3 - \frac{1}{2}}{2 + \frac{4}{5}}$
   e. $\dfrac{\frac{1}{2} + \frac{3}{4}}{\frac{2}{3} + \frac{1}{6}}$
   f. $\dfrac{\frac{3}{8} - \frac{1}{4}}{\frac{2}{5} + \frac{1}{10}}$

8.6 Rational Numbers and Decimals

11. Express each of the following as a decimal:
   a. $\frac{7}{8}$
   b. $\frac{7}{16}$
   c. $\frac{2}{11}$
   d. $\frac{13}{37}$
   e. $\frac{5}{13}$
   f. $\frac{3}{7}$

12. Express each of the following as a quotient of integers:
   a. 0.75
   b. 0.213
   c. 3.14
   d. $0.\overline{46}$
   e. $2.\overline{49}$
   f. $4.1\overline{23}$

8.7 Irrational Numbers and the Set of Real Numbers

13. Which of the following numbers are rational and which are irrational?
   a. 2.1
   b. $2.\overline{1}$
   c. $\sqrt{5}$
   d. 3.141141114...
   e. 2.121121112...
   f. 3.1415926

14. Determine whether each statement is true or false.
   a. Every real number is an irrational number.
   b. Every real number is a rational number.
   c. Every real number is either a rational number or an irrational number.
   d. Every rational number is a real number.
   e. Every irrational number is a real number.
   f. The union of the sets of rational and irrational numbers is the set of real numbers.

15. Answer yes or no to tell whether each of the following numbers is (I) a natural number, (II) a whole number, (III) an integer, (IV) a rational number, (V) an irrational number, (VI) a real number.

a. 6 b. -1 c. 0

d. 2.89 e. $1.\overline{34}$ f. π

g. $\sqrt{16}$ h. $\sqrt{5}$ i. $\frac{2}{3}$

8.8 Scientific Notation (Optional)

16. Use scientific notation to evaluate each of the given expressions. Answers should be left in scientific notation.

a. $355{,}000 \times 8000$

b. 0.000065×0.00084

c. $\dfrac{24{,}000 \times 140{,}000}{210{,}000 \times 0.0016}$

d. $\dfrac{0.024 \times 0.00036}{0.0015 \times 0.00012}$

Just for fun

The value of π has been computed to hundreds of decimal places by modern computers. The value of π to seven decimal places is

$\pi = 3.1415926\ldots$

The following diagram shows a trick for remembering these digits:

3	1	4	1	5	9	2	6
May	I	have	a	large	container	of	coffee?

How does this trick work?

Chapter Quiz

Determine whether each statement (1–15) is true or false.

1. "Janet finished third" is an example of the use of a cardinal number.
2. An example of a prime number is 111.
3. The smallest prime number is 1.
4. An example of a composite number is 97.
5. The greatest common divisor of two natural numbers is the largest natural number that is a multiple of each of the two given numbers.
6. Two numbers whose greatest common divisor (or greatest common factor) is 1 are said to be relatively prime.
7. A prime number cannot be written as the product of natural numbers that are less than itself.
8. The least common multiple of two natural numbers is the smallest natural number that divides a given pair of natural numbers with remainders of zero.
9. Whenever we multiply two negative integers, the product is always positive.
10. If n and k are even integers, then $2 \times (n + k + 1)$ is an even integer.
11. A rational number is a number that can be expressed in the form a/b, where a and b are integers and $b \neq 0$.
12. The set of rational numbers between $\frac{5}{11}$ and $\frac{6}{11}$ is empty.
13. Decimals that are nonterminating are called irrational numbers.
14. The union of the set of rational numbers and the set of irrational numbers is called the set of real numbers.

15. Every irrational number is a real number.

For Exercises 16–21, perform the indicated operation and simplify your answer completely.

16. $\frac{3}{5} + \frac{5}{8}$

17. $\frac{5}{9} - \frac{3}{7}$

18. $\dfrac{1 + \frac{1}{2}}{2 - \frac{1}{3}}$

19. $\dfrac{\frac{1}{2} + \frac{3}{5}}{\frac{7}{8} - \frac{1}{2}}$

20. $\left(\frac{3}{4} \times \frac{5}{6} - \frac{1}{2}\right) \times 2$

21. $-4 - (-7)$

22. Find the least common multiple and greatest common divisor of 48 and 72.

23. Express $\frac{1}{7}$ as a decimal.

24. Express $4.\overline{123}$ as a quotient of integers, in simplest form.

For Exercises 25–27, classify each number as rational or irrational.

25. $\sqrt{11}$

26. $-\sqrt{16}$

27. $3.\overline{06}$

For Exercises 28–30, use scientific notation to evaluate each of the given expressions. Answers may be left in scientific notation.

*28. $(1.7 \times 10^3) \times (5.2 \times 10^7)$

*29. $\dfrac{1.21 \times 10^5}{1.1 \times 10^3}$

*30. $\dfrac{48{,}000 \times 300{,}000}{0.00012 \times 0.00015}$

Chapter 9

Chapter Outline

An Introduction to Algebra

After Studying This Chapter, You Will Be Able to Do the Following:

1. Understand that algebra is the area of mathematics that generalizes the facts of arithmetic by the use of variables.
2. Define a **variable** as a symbol (letter) that is used to represent an unknown member of a set.
3. Distinguish between **true sentences**, **false sentences**, and **open sentences.**
4. **Solve** and **graph** the solution set of simple **equations** and **inequalities** in one variable.
5. Translate word phrases into mathematical expressions (equations) and solve verbal problems involving one variable.
6. Find at least three solutions for a **linear equation** in two variables, and graph a linear equation in two variables on the **Cartesian plane**.
7. Graph an inequality involving two variables on the Cartesian plane.

Note: *indicates optional material.

Notations Frequently Used in This Chapter

$>$	greater than
$<$	less than
$\geq$	greater than or equal to
$\leq$	less than or equal to

Visit the PH Companion Web site at www.prenhall.com/setek

OVERVIEW

What is algebra? Basically, algebra is a branch of mathematics that generalizes the facts of arithmetic by using letters to represent basic arithmetic relations. Algebra is very similar to arithmetic. In fact, algebra deals with the same basic operations as arithmetic, including addition, subtraction, multiplication, and division. The difference between the two subjects, though, is that arithmetic cannot generalize mathematical relations. Arithmetic can only produce specific instances of these relations. For example, in arithmetic we encounter numerical expressions such as $4 + 2$, which simplifies to 6. In algebra, we encounter *algebraic expressions* such as $x + y$, which cannot be simplified until we know the values of x and y. If we replace x with 4 and y with 2, then we have the specific arithmetic expression of $4 + 2$. Similarly, if x is replaced with 29 and y with 13, then we have the arithmetic expression $29 + 13$, which simplifies to 42. So the algebraic expression $x + y$ is an expression that *generalizes* adding two numbers.

9.1 INTRODUCTION

In this chapter, we provide an introduction to the study of algebra. We begin in Section 9.2 by presenting some basic concepts of algebra, including the symbols used to represent numbers and relations, the sets of numbers used to replace algebraic symbols, the fundamental operations of arithmetic and their laws, and the evaluation of algebraic expressions. Many of these concepts were discussed in Chapter 8 but are presented here for completeness. In Section 9.3 we provide a first look at solving algebraic equations from an intuitive perspective, and then, in Section 9.4, we extend this discussion to solving equations using the rules of algebra. In Section 9.5 we present applications of algebra by examining how algebra is used to solve various word problems. Solving algebraic equations is further extended in Section 9.6 when we introduce equations involving two unknown quantities. In Section 9.7 we learn how to graph the solution to these equations in the Cartesian coordinate system. Finally, in Section 9.8, we apply the skills we learned from the previous two sections to solve algebraic inequalities involving two unknown quantities.

Historical Note

The study of algebra began in early Egypt. The first documented occurrence of an algebra-type problem is in a papyrus copied by the Egyptian priest Ahmes (approximately 1600 B.C.). His papyrus was a copy of a book written even earlier, approximately 2200 B.C. It is significant because it illustrates the first recorded use of a "symbol" to represent an unknown quantity. The Egyptians used the word *hau* to represent any unknown quantity in a mathematical problem.

Diophantus (approximately 200 A.D.), a Greek mathematician, was one of the first mathematicians to use symbols to represent unknown quantities. He supposedly used a combination of the first two letters of the Greek word *arithmos*, which means "number."

One of the most famous mathematicians of this era was Muhammed ibn Musa al-Khwarizmi (approximately 800 A.D.), who wrote a book titled *Al-jabr wa Al-muzabalah*, which contributed greatly to the adoption of Hindu–Arabic numerals by the scholars of that time. It is believed that the title word *Al-Jabr* gave rise to the word *algebra* as we know it today.

The title of al-Khwarizmi's book is best translated into English as "restoration and reduction." By restoration, we are moving terms to the other side of the equal sign; by reduction we are combining like terms. For example, $6 - x = 2x$ becomes $6 = 2x + x$ by *al-Jabr* (restoration); and by *al-Muzabalah* (reduction), we have $6 = 3x$ and as a result $2 = x$.

9.2 BASIC CONCEPTS OF ALGEBRA

The Symbols and Language of Algebra

Variables and Constants

One of the most important characteristics of algebra is that symbols are used to denote numbers or a set of numbers. The symbols commonly used are letters of the alphabet such as a, b, m, n, x, and y. Although we can choose any letter we want, we usually select a letter that is easily recognized for the quantity we want to represent. For example, it is more meaningful to represent motion by the letter m, time by t, area by a, and distance by d.

A letter that represents a number is called a **variable**. Thus, the symbols x and y in the algebraic expression $x + y$ are variables. A variable is so named because it represents an unknown quantity, which implies that it can take on different numerical values. That is, its numerical value can *vary* depending on a given situation. For example, in the expression $x + y$, we can replace x and y by any quantity, which will result in a different sum that corresponds to each replacement. If we replace x with 5 and y with 19, we get a sum of $5 + 19 = 24$. If we replace x with 10 and y with 7, we get a sum of 17. Notice that we are always adding two different quantities. In contrast to a variable is a **constant**, which specifies a fixed numerical value; that is, a constant's numerical value will always remain the same. For example, in the algebraic expression $x + 3$, regardless of what numerical value we replace x with, we will always add 3 to it.

EXAMPLE 1 ***Generalizing Patterns Using Variables and Constants***

Consider the following numerical pattern and then generalize this pattern using a variable and constant.

$$6 + 5 > 6$$
$$15 + 5 > 15$$
$$-8 + 5 > -8$$
$$\tfrac{1}{2} + 5 > \tfrac{1}{2}$$
$$\pi + 5 > \pi$$

Solution There are actually three patterns to note. First, observe that in each instance 5 is being added to the given number. This implies that 5 is a *constant*. Second, note that the numbers being added to 5 consist of integers $(6, 15, -8)$, rational numbers $\left(\frac{1}{2}\right)$, and irrational numbers (π). This implies that the pattern involves all real numbers. Finally, the overall pattern shows that when we add 5 to a real number, the sum will be greater than (>) the number itself. Thus, if we let $R =$ any real number, then the given pattern can be generalized as $R + 5 > R$.

Arithmetic Operations

You might notice that the algebraic expression $x + y$ denotes a sum as indicated by the + symbol. This is indeed the case. In fact, with the exception of multiplication, the symbols used to operate on variables in algebra are exactly the same as those used in arithmetic. For example, we use + for addition, − for subtraction, and ÷ for division. Furthermore, the vocabulary associated with these operations is maintained in algebra. Therefore,

- $m + n$ denotes the *sum* of two numbers m and n.
- $m - n$ denotes the *difference* between two numbers m and n.
- $m \div n$ denotes the *quotient* of two numbers m and n. (We may also use the fractional form $\frac{m}{n}$ to express a quotient.)

When it comes to multiplication, the symbol with which we are most familiar, namely, ×, is usually not used in algebra. Although this symbol maintains its meaning in algebra (i.e., it denotes multiplication), we avoid using this symbol in algebra because it can easily be confused with the variable x. As a result, to represent the product of two numbers such as 2 and y, we use any of the following:

$$(2)(y), \quad (2)y, \quad 2(y), \quad 2 \cdot y, \quad \text{or} \quad 2y$$

Each of the these representations means "2 times y" and hence denotes a *product*. Of the various notations available for expressing a product in algebra, the method most commonly used is the last one (i.e., $2y$). It is important to note that we use this method only when expressing a product involving variables. Thus, $5x$ means "5 times x," and $7z$ means "7 times z." However, the product of 2 and 3 must *not* be written as 23 because this is the numeral for the number twenty-three. We can, however, represent this product using any of the other methods. Therefore, $(2)(3)$, $2(3)$, $(2)3$, and $2 \cdot 3$ all are interpreted as "2 times 3."

Also recall from arithmetic that whenever we multiply two numbers such as 2 and 3, the numbers being multiplied are called **factors.** This vocabulary is also extended to algebra. For example, the expression $7y$ contains the factors 7 and y. Similarly, the algebraic expression xyz contains three factors, namely, x, y, and z.

Besides the operation symbols, we also use the same arithmetic symbols to express algebraic relationships. Thus, $=$ (equal), $\neq$ (not equal), $>$ (greater than), and $<$ (less than) are used to show relationships between quantities. Finally, we use parentheses ( ) to group quantities. As a result, with few exceptions, all of the symbols we learned to use in arithmetic are also used in exactly the same manner and have exactly the same meaning in algebra. We will expand on the use and interpretation of these symbols later in Section 9.3.

EXAMPLE 2 ***Writing Algebraic Expressions Using the Signs of Arithmetic*** Represent the following algebraically:

a. The sum of the numbers a and b

b. The difference between $7x$ and 8

c. The product of the factors $(a + d)$ and 9

d. The quotient of the numbers $(x + y)$ and 5

Solution

a. $a + b$

b. $7x - 8$

c. Preferably, $9(a + d)$. The following are also acceptable: $(9)(a + d)$ and $9 \cdot (a + d)$.

d. $(x + y) \div 5$ or $\dfrac{(x + y)}{5}$

NW ***Now Work Problems 1 and 3.***

Factors, Terms, and Coefficients

Thus far in our discussion, we have used the phrase "algebraic expression" in a somewhat intuitive manner. For example, when we defined variables we said "the symbols x and y in the algebraic expression $x + y$ are variables." But what exactly is an algebraic expression? An **algebraic expression** is a mathematical structure consisting of variables and constants, which are being combined by addition, subtraction, multiplication, or division. For instance, each of the following are examples of algebraic expressions:

$$18, \quad x + 6, \quad 2x - y, \quad 3ab - b, \quad 5(2x + 13), \quad \frac{a + b}{2c}$$

Note the absence of the symbols used to show a relationship between two entities such as an equality ($=$) or an inequality ($\neq, >, <$). This is because the inclusion of these symbols produces an *equation* or *inequality*. For example, $x + y = 3$ is an equation that equates the two expressions $x + y$ and 3, and $x + y > 3$ is an inequality that indicates the expression $x + y$ is greater than 3. We will discuss algebraic equations and inequalities later in this chapter.

All algebraic expressions are composed of **terms**, which are individual quantities separated by the signs of addition ($+$) and subtraction ($-$). For example,

- The expression $3x$ consists of a single term, namely, $3x$.
- The expression $5x + y + 7$ consists of three terms, namely, $5x$, y, and 7.
- The expression $ab + 3(x - y) + 6$ consists of the three terms ab, $3(x - y)$, and 6.
- The expression $\frac{m+9}{5} - (2n + 2)$ contains the two terms $\frac{m+9}{5}$ and $-(2n + 2)$.
- The expression $b^2 - 4ac$ contains the two terms b^2 and $-4ac$ (Recall that b^2 means b times b.)

In the last two illustrations, note that we included the $-$ sign as part of the second term. In general, a term includes the sign that immediately precedes it. However, if the sign is $+$, we do not need to list it as part of the term.

It is important to understand the difference between terms and factors. Terms are individual quantities of an expression separated by $+$ or $-$ signs. Factors, on the other hand, represent products and hence denote multiplication. For example, as noted previously, the expression $5x + y + 7$ consists of the three terms $5x$, y, and 7. Note that the first term, $5x$, represents a product, which consists of the factors 5 and x. If a term consists of a product involving numerical and variable factors (e.g., $10x$ or $-9xyz$), then we refer to the numerical factor as a **numerical coefficient**, or **coefficient** for short. If a term consists of only variable factors (e.g., a or xyz), then the term's coefficient is understood to be 1. Thus, in the expression $5x + y + 7$, the coefficient of the first term is 5 and the coefficient of the second term is 1. The last term is a constant.

Math Connections

Polynomials

In our discussion, we indicated that algebraic expressions are composed of *terms*, which are individual quantities separated by the signs of addition ($+$) or subtraction ($-$). Algebraic expressions are usually named by the number of terms they contain. The names and definitions of the three most common expressions follow:

Monomial—An algebraic expression that consists of only one term and does not include any division by variables is called a *monomial*. The following are examples of monomials:

$$6x, \quad -3a^2, \quad \text{and} \quad \frac{2xy}{7}$$

Binomial—If an algebraic expression consists of two monomials, then it is called a *binomial*. The following are examples of binomials:

$$x + 7, \quad 3x^2 + 2x, \quad -5ab^2 - 4ab, \quad \text{and} \quad 4x^3 - 2x^2$$

Trinomial—An algebraic expression that consists of three monomials is called a *trinomial*. The following are examples of trinomials:

$$x + y + z, \quad ax^2 + bx + c, \quad 3x^2 - 5xy + y^2,$$

$$\text{and} \quad 7x^2 - 2x - 8$$

In general, any algebraic expression that consists of two or more monomials is called a **polynomial**. Thus, we may use the term *polynomial* to specify binomial or trinomial.

EXAMPLE 3 ***Identifying Factors, Terms, Constants, Coefficients*** Using the vocabulary developed thus far, identify all the parts of the following algebraic expressions.

a. $-6x$

b. $5x + 3$

c. $2x^2 + 4x - 9$

Solution

a. $-6x$ is a single *term*; -6 and x are *factors* of the term; and -6 is the *coefficient* of the term.

b. The *terms* of the expression are $5x$ and 3. The first term consists of the *factors* 5 and x and the term's *coefficient* is 5; the second term, 3, is a *constant*.

c. The *terms* of the expression are $2x^2$, $4x$, and -9. The first term consists of the *factors* 2 and x^2 and its *coefficient* is 2; the second term consists of the *factors* 4 and x and its *coefficient* is 4; and the third term is the constant -9. (Note that the first term, $2x^2$, is really $2xx$. Thus, the factors can also be expressed as 2, x, and x.)

NW ***Now Work Problem 13.***

The Numbers of Algebra

When working with algebraic expressions or equations, we will need to replace variables with specific numbers. For example, most of us know that to find the area of a rectangle we multiply the length of the rectangle by its width. In other words, using the language of algebra, we would denote the area of any rectangle as $l \times w$. If we now wanted to determine the area of a specific rectangle that measures 5 feet long and 2 feet wide, we need to replace the variables l and w with 5 and 2, respectively. The numbers we use to replace variables are from the various sets of numbers discussed in Chapter 8. These are summarized here.

- Natural numbers are the *counting numbers*.

$$N = \{1, 2, 3, 4, \dots\}$$

Math Connections

Numerical versus Literal Coefficient

In our discussion, we stated that if a term consists of a product involving numerical and variable factors (e.g., $10x$ or $-9xyz$), then we refer to the numerical factor as a *numerical coefficient*. It is also possible for a term to have a *variable coefficient*, which is formally called a **literal coefficient**. For example, given the term x^2y, we may say that x^2 is the coefficient of y and that y is the coefficient of x^2. Since neither of these factors is numerical, we call them literal coefficients to distinguish them from numerical coefficients. In general, ***if a term consists of several factors, then any single factor or combination of factors is the coefficient of the rest of the term.*** For example, given the term $5x^2yz$, then the following are all true:

- $5x^2y$ is the coefficient of z.
- $5x^2z$ is the coefficient of y.
- x^2y is the coefficient of $5z$.
- y is the coefficient of $5x^2z$.
- z is the coefficient of $5x^2y$.
- 5 is the coefficient of x^2yz.

- Whole numbers are the set of natural numbers and 0.

$$W = \{0, 1, 2, 3, \dots\}$$

- Integers are the negative natural numbers, 0, and the positive natural numbers.

$$I = \{\dots, -3, -2, -1, 0, 1, 2, 3, \dots\}$$

- Rational numbers are fractions of the form p/q, where both p and q are integers and $q \neq 0$.

$$Y = \left\{ \frac{p}{q} \middle| p \in I, q \in I, q \neq 0 \right\}$$

- Irrational numbers are any numbers that are not rational.

$$Z = \{x | x \notin Y\}$$

- Real numbers consist of the set of rational and irrational numbers.

$$R = \{x | x \in Y \text{ or } x \in Z\}$$

Fundamental Operations and Their Laws

As noted earlier, the fundamental operations of algebra include addition, subtraction, multiplication, and division. In one of the Math Connections of Chapter 8 (Laws of Real Numbers), we presented the various laws that govern these operations relative to real numbers. These laws (or *properties*) are applicable to algebra as well. A summary of some of these properties follows:

Properties of Addition

The addition of real numbers is consistent with the following two properties:

- *Commutative Property of Addition*
  If a and b represent any two real numbers, then $a + b = b + a$. That is, the addition of two real numbers is not affected by the order in which the two numbers are added. For example,

$$9 + 8 = 8 + 9$$

- *Associative Property of Addition*
  If a, b, and c represent any three real numbers, then $(a + b) + c = a + (b + c)$. That is, the addition of three real numbers is not affected by the way in which the numbers are grouped. For example,

$$(2 + 3) + 4 = 2 + (3 + 4)$$
$$5 + 4 = 2 + 7$$
$$9 = 9$$

Properties of Multiplication

The multiplication of real numbers is consistent with the following two properties:

- *Commutative Property of Multiplication*
  If a and b represent any two real numbers, then $ab = ba$. That is, the multiplication of two real numbers is not affected by the order in which the two numbers are multiplied. For example,

$$9 \times 8 = 8 \times 9$$

- *Associative Property of Multiplication*
  If a, b, and c represent any three real numbers, then $(ab)c = a(bc)$. That is, the multiplication of three real numbers is not affected by the way in which the numbers are grouped. For example,

$$\begin{aligned}(2 \times 3) \times 4 &= 2 \times (3 \times 4)\\ 6 \times 4 &= 2 \times 12\\ 24 &= 24\end{aligned}$$

The Distributive Property

If a, b, and c represent any three real numbers, then $a(b + c) = ab + ac$. That is, multiplication is distributive with respect to addition. For example,

$$5 \times (6 + 8) = (5 \times 6) + (5 \times 8) = 30 + 40 = 70$$

Inverse Properties

- *Inverse Property of Addition*
  If a is any real number, then $a + (-a) = 0$. That is, every real number a has a unique opposite real number $-a$ such that when a is added to its opposite, the sum is 0. This opposite is called the *additive inverse*. For example,

$$3 + (-3) = 0$$

  Furthermore, addition and subtraction are considered inverse operations because subtraction will undo addition and addition will undo subtraction. For example,

$$5 + 2 = 7 \quad \text{and} \quad 7 - 2 = 5$$

- *Inverse Property of Multiplication*
  If a is any nonzero real number, then $a \times \frac{1}{a} = 1$. That is, every nonzero real number a has a unique opposite real number $\frac{1}{a}$ such that when a is multiplied by its opposite, the product is 1. This opposite is called the *multiplicative inverse*. For example,

$$5 \times \frac{1}{5} = 1$$

  Furthermore, multiplication and division are considered inverse operations because division will undo multiplication and multiplication will undo division. For example,

$$5 \times 6 = 30 \quad \text{and} \quad 30 \div 6 = 5$$

Properties of Identity

- *Identity Property of Addition*
  If a is any real number, then $a + 0 = a$. That is, there is a unique zero (0) such that any real number a added to 0 yields the number itself. Zero is called the *identity element of addition*. For example,

$$5 + 0 = 5$$

- *Identity Property of Multiplication*
  If a is any real number, then $a \times 1 = a$. That is, there is a unit (1) such that any real number a multiplied by 1 yields the number itself. One is

called the *identity element of multiplication*. For example,

$$5 \times 1 = 5$$

Finally, it is acknowledged that when operating with real numbers, *division by zero is excluded*. We will apply these properties when we begin solving equations later in the chapter.

EXAMPLE 4 ***Identifying Properties of Real Numbers*** Determine the appropriate property of real numbers that is being illustrated in the following:

a. $3 + 4 = 4 + 3$

b. $4(x + 2) = 4x + 4 \cdot 2$

c. $3 + (4 + 5) = (3 + 4) + 5$

d. $1 \cdot y = y \cdot 1 = y$

e. $x(yz) = (xy)z$

f. $x + 0 = 0 + x = x$

Solution

a. $3 + 4 = 4 + 3$ illustrates the *commutative property of addition*.

b. $4(x + 2) = 4x + 4 \cdot 2$ illustrates the *distributive property of multiplication over addition*.

c. $3 + (4 + 5) = (3 + 4) + 5$ illustrates the *associative property of addition*.

d. $1 \cdot y = y \cdot 1 = y$ illustrates the *identity property of multiplication*.

e. $x(yz) = (xy)z$ illustrates the *associative property of multiplication*.

f. $x + 0 = 0 + x = x$ illustrates the *identity property of addition*.

NW ***Now Work Problems 19 and 21.***

Evaluating Algebraic Expressions

As we indicated earlier, an algebraic expression such as $2x + 9$ represents an unknown quantity until the variable, x, is replaced by a specific number. If we replace x with 3, then the expression equals 15 because $2(3) + 9$ is equal to $6 + 9$, which is 15. This process of substituting a variable with a specific number and then evaluating the resulting expression is called *evaluating algebraic expressions*. When evaluating algebraic expressions, we apply the same rules and order of operations for adding, subtracting, multiplying, and dividing integers that we discussed in Chapter 8.

A summary of the order of operations is provided in Table 9.1, and a summary of the rules we developed for adding, subtracting, multiplying, and dividing integers is provided in Table 9.2.

TABLE 9.1

Summary of the Order of Operations

Order of Operations
1. Evaluate all expressions within grouping symbols first. If more than one set of grouping symbols is used, then always begin with the innermost set.
2. Working from left to right evaluate all powers.
3. Working from left to right, perform all multiplications and divisions.
4. Working from left to right, perform all additions and subtractions.

TABLE 9.2

Summary of the Operations with Integers

Addition of Integers

1. To add integers with the same sign, add without regard to the signs. The answer contains the same sign as the given numbers.

Examples:

a. $5 + 9 = 14$ **b.** $(-5) + (-9) = -14$

2. To add two integers with opposite signs, first consider the distance each integer is from zero using a number line. Then subtract the shorter distance from the longer distance and use the sign of the longer distance in the answer.

Examples:

a. $5 + (-9) = -4$ **b.** $(-5) + 9 = 4$

Subtraction of Integers

To subtract two integers, add the opposite of the second integer to the first integer. In other words, add the additive inverse of the second integer to the first integer using the rules for addition of integers.

Examples:

a. $15 - 8 = 15 + (-8) = 7$ **b.** $12 - (-3) = 12 + 3 = 15$
c. $-23 - 7 = -23 + (-7) = -30$ **d.** $-9 - (-11) = -9 + 11 = 2$

Multiplication of Integers

1. The product of two positive integers or two negative integers is always positive.

Examples:

a. $(8)(4) = 32$ **b.** $(-5)(-9) = 45$

2. The product of two integers that are opposite in sign is always negative.

Examples:

a. $(-8)(4) = -32$ **b.** $(5)(-9) = -45$

Division of Integers

1. The quotient of two positive integers or two negative integers is always positive.

Examples:

a. $20 \div 5 = 4$ **b.** $(-24) \div (-8) = 3$

2. The quotient of two integers that are opposite in sign is always negative.

Examples:

a. $(-8) \div (4) = -2$ **b.** $35 \div (-7) = -5$

EXAMPLE 5 ***Evaluating Expressions*** In **a** through **c**, evaluate the given expression when $x = 3$, $y = -2$, and $z = 5$.

a. $3x + y$ **b.** $3(x + y)$ **c.** $5xy + 2(y - z)$

Solution

a. $3x + y = 3(3) + (-2) = 9 - 2 = 7$

b. $3(x + y) = 3((3) + (-2)) = 3(3 - 2) = 3(1) = 3$

c.
$$\begin{aligned} 5xy + 2(y - z) &= 5(3)(-2) + 2((-2) - (5)) \\ &= 15(-2) + 2(-2 - 5) \\ &= -30 + 2(-7) \\ &= -30 - 14 \\ &= -44 \end{aligned}$$

EXAMPLE 6 ***Evaluating Expressions*** Evaluate each of the following algebraic expressions for the indicated numerical replacement.

a. $4z - 6$ when $z = 2$ **b.** $3x^2 + 7x$ when $x = 3$

c. πr^2 when $r = 4$ **d.** $ay^3 + by$ when $y = 2$

e. $-x^2$ when $x = 3$ **f.** x^2 when $x = -3$

g. $4a - [b - (5x + 2y)]$ when $a = -1, b = 2, x = -4$, and $y = 8$

h. $-b - (b^2 - 4ac)$ when $a = 5, b = -4$, and $c = -2$

Solution

a.
$$\begin{aligned} 4z - 6 &= 4(2) - 6 \\ &= 8 - 6 \\ &= 2 \end{aligned}$$

b.
$$\begin{aligned} 3x^2 + 7x &= 3(3)^2 + 7(3) \\ &= 3(9) + 7(3) \\ &= 27 + 21 \\ &= 48 \end{aligned}$$

c.
$$\begin{aligned} \pi r^2 &= \pi(4)^2 \\ &= \pi(16) \\ &= 16\pi \end{aligned}$$

d.
$$\begin{aligned} ay^3 + by &= a(2)^3 + b(2) \\ &= a(8) + 2b \\ &= 8a + 2b \end{aligned}$$

e.
$$\begin{aligned} -x^2 &= -(3)^2 \\ &= -9 \end{aligned}$$

f.
$$\begin{aligned} x^2 &= (-3)^2 \\ &= 9 \end{aligned}$$

Note in part **e** that the negative sign precedes the expression x^2. We are asked to square 3, not -3. Similarly, in part **f** we are asked to square -3 and $(-3)(-3) = 9$.

g.
$$\begin{aligned} 4a - [b - (5x + 2y)] &= 4(-1) - [2 - (5(-4) + 2(8)] \\ &= -4 - [2 - (-20 + 16)] \\ &= -4 - [2 - (-4)] \\ &= -4 - [2 + 4] \\ &= -4 - [6] \\ &= -4 - 6 \\ &= -10 \end{aligned}$$

h.
$$\begin{aligned} -b - (b^2 - 4ac) &= -(-4) - [(-4)^2 - 4(5)(-2)] \\ &= 4 - [16 - 20(-2)] \\ &= 4 - [16 - (-40)] \\ &= 4 - [16 + 40] \\ &= 4 - 56 \\ &= -52 \end{aligned}$$

NW ***Now Work Problems 33 and 45.***

EXAMPLE 7 ***Evaluating Formulas*** The formula $I = PRT$ is the simple interest formula used in business, where $I =$ interest, $P =$ principal, $R =$ rate, and $T =$ time. Find I when $P = \$1000$, $R = 6\%$, and $T = 2$ years, 6 months.

Solution Substitute the specific values of P, R, and T into the formula and evaluate the resulting expression. Note that the rate of 6% must first be expressed as the equivalent decimal 0.06, and the time of 2 years, 6 months is equivalent to 2.5 years.

$$\begin{aligned} I &= PRT \\ &= (1000)(0.06)(2.5) \\ &= (60)(2.5) \\ &= 150 \end{aligned}$$

Therefore, the amount of simple interest earned on a principal of $1000 invested for two and one-half years at a rate of 6% is equal to $150.

NW ***Now Work Problem 59.***

Exercises for Section 9.2

In Exercises 1–10, represent the given statement as an algebraic expression.

NW **1.** The sum of b and 3

2. 8 times the number a

NW **3.** The quotient of the numbers $(x - y)$ and a

4. The difference between 19 and the number $(x + y)$

5. The number $(a - b)$ divided by $(a + b)$

6. The sum of 3 times a number x and 5 times a number y

7. The difference between 5 times x and y

8. The product of x, y, and z

9. The difference between the product of x and y, and 4 times a number b

10. The sum of a number $(a - b)$ and the square of x

In Exercises 11–18, use the vocabulary developed in this section to identify all the parts of the following algebraic expressions.

11. $4xy$

12. $3x + 6$

NW **13.** $4s + 7t$

14. $2x - 3y - 7$

15. $5x^2 - 3x + 6$

16. $-x^2 + 7$

17. $8x^3y^2z + 2x^2y$

18. $4y^3 + 3y^2 + 2(y - 1)$

In Exercises 19–28, identify the correct property of real numbers that is being illustrated.

NW **19.** $48 + 72 = 72 + 48$

20. $12 \cdot 13 = 13 \cdot 12$

NW **21.** $4(3 + 2) = 4 \cdot 3 + 4 \cdot 2$

22. $1 + (2 + 3) = (1 + 2) + 3$

23. $2(3 \cdot 4) = (2 \cdot 3)4$

24. $4(2 - 3) = 4 \cdot 2 - 4 \cdot 3$

25. $8 + 0 = 0 + 8 = 8$

26. $17 + 13 = 13 + 17$

27. $4 \cdot 1 = 1 \cdot 4 = 4$

28. $4 \cdot 3 + 4 \cdot 5 = 4(3 + 5)$

In Exercises 29–32, use the commutative or distributive property to find equivalent expressions for each of the following.

29. $7 \cdot 4 + 7 \cdot 2$

30. $3 \cdot 2 - 4 \cdot 2$

31. $\pi x + \pi y$

32. $(x + y)x + (x + y)y$

In Exercises 33–42, evaluate each expression using the given values for the variables.

NW **33.** $3y - 8x \quad (x = -2, y = -3)$

34. $2x - 3y + z \quad (x = -5, y = 2, z = 3)$

35. $2a + 2b - ab \quad (a = -4, b = -3)$

36. $(x + y)(x - y) \quad (x = 4, y = -2)$

37. $-b - (4ab + c) \quad (a = 3, b = -2, c = 1)$

38. $3x + (4 + y) \quad (x = 9, y = 6)$

39. $5x - (x - y) \quad (x = -2, y = 4)$

40. $x(x - y) \quad (x = 3, y = -3)$

41. $b + (4a - 2x) \quad (a = 3, b = 9, x = -2)$

42. $2x + (x - (a + b)) \quad (a = -1, b = -1, x = 3)$

In Exercises 43–52, evaluate each expression using the given values for the variables.

43. $3x^3$ when $x = 3$

44. $4y^2 + 2y + 3$ when $y = 0$

NW **45.** $2z^2 + 6z - 6$ when $z = 2$

46. $2\pi r$ when $r = 4$

47. $x^2y - 2xy^3$ when $x = 1$ and $y = -2$

48. $2x^3y + 3xy^2$ when $x = 2$ and $y = 3$

49. $3x^2y - 2xy + 8$ when $x = 1$ and $y = 2$

50. $3x^3y^2z^2 + 2x^3y^2z^2$ when $x = -2$, $y = 2$, and $z = 2$

51. $ax^2 + bx + c$ when $x = 3$

52. $ay^2 + by + c$ when $y = -2$

In Exercises 53–58, evaluate each expression using the given values for the variables.

53. $\frac{a+b}{c}$ when $a = -2$, $b = 3$, and $c = -4$

54. $\frac{b^2 - 4ac - 1}{2a}$ when $a = -1$, $b = -3$, and $c = 4$

55. $p(1 + r)^n$ when $p = 5$, $r = 10$, and $n = 3$

56. $\frac{a(1 - r^n)}{1 - r}$ when $a = 2$, $r = 3$, and $n = 2$

57. ar^{n-1} when $a = 2$, $r = 2$, $n = 4$

58. $\frac{n}{2}[2a + (n - 1)d]^a$ when $a = 3$, $n = 6$, and $d = -1$

NW 59. **Electrical circuits** To find voltage, amperage, or resistance in electrical circuits, we use Ohm's law: $I = E/R$, where I = amperes, E = voltage, and R = resistance (ohms). **a.** How many amperes are needed for a circuit that contains 12 volts with a resistance of 30 ohms? **b.** How much current will flow during the starting process of an automobile that contains a 12-volt battery and starting motor with a resistance of 0.05 ohms?

60. **Compound interest** The formula $A = P(1 + rt)$, where P = principal, r = rate of interest, and t = time, is used to determine the amount of money accumulated, A, over a period of time. If Yoshi invests \$10,000 at a rate of 6%, how much money will she accumulate at the end of 18 months?

61. **IQ** The formula IQ = (MA/CA) × 100 is used to determine a person's intelligent quotient (IQ) based on that individual's mental age (MA) and chronological age (CA). **a.** If Kristin is 28 years old and has a mental age of 40, what is her IQ? **b.** As Kristin gets older, her mental age remains the same. What effect will this have on her IQ?

www

62. **Temperature conversion** To convert from degrees Celsius to degrees Fahrenheit, the formula $F = (9/5)C + 32$ is used. What is the Fahrenheit equivalent of 20 degrees Celsius?

63. **Nozzle pressure** The formula $h = 0.5p + 26$ is used to determine the length, h, a stream of water will travel (in feet) from a hose given a nozzle pressure, p. If the nozzle pressure for a 3/4-inch nozzle is 18 psi, what is the range of the water?

64. **Height of a Hit Baseball** If a ball is hit upward with an initial velocity of 120 feet per second, its height, h (in feet), after t seconds is given by the formula $h = 120t - 16t^2$. If a baseball player hits a ball with an initial velocity of 120 feet per second, how high will the ball travel if it takes the fielder exactly 6 seconds to catch it from the time the ball is hit?

65. **Constructing a Brick Wall** Some bricklayers use the formula $N = 7lh$ to determine the number of bricks (N) needed to build a wall that is l feet long and h feet high. **a.** How many bricks are needed to build a wall 10 feet high and 35 feet long? **b.** Are 1000 bricks enough to build a wall that is 12 feet high 14.5 feet long?

Writing Mathematics

66. Explain the difference between a variable and a constant. Give an example of each.

67. Explain the difference between a factor and a term. Give an example of each.

68. Define an algebraic expression and give three examples.

69. What does it mean to evaluate an expression?

Challenge Exercises

70. Explain the difference between a coefficient and a constant. (Think about this one very carefully.)

71. Represent the cost of one rose as an algebraic expression if a dozen roses cost $(2x + 3)$ dollars.

72. Represent the cost of n shirts as an algebraic expression if each shirt costs $(x - y)$ dollars.

73. Represent the total number of books in a bookcase as an algebraic expression if the bookcase has x shelves and there are $(2x + 9)$ books on each shelf.

74. Represent the number of seats in a classroom as an algebraic expression if there are $(r + 6)$ rows and $2s$ seats per row.

75. Given the term $2\pi r$, identify each factor as a coefficient, constant, or variable.

76. Given the term πr^2, identify each factor as a coefficient, constant, or variable.

77. Try evaluating the expression $\sqrt{b^2 - 4ac}$ for $a = 1$, $b = -2$, and $c = 2$. What seems to be the problem? Write a general statement about what $b^2 - 4ac$ must equal so that when the given expression $\sqrt{b^2 - 4ac}$ is evaluated using specific values of a, b, and c, the answer is a real number.

78. Try evaluating the expression $\frac{2x}{x - y}$ for $x = -1$ and $y = -1$. What seems to be the problem? Write a general statement about what $x - y$ must equal so that when the given expression $\frac{2x}{x - y}$ is evaluated using specific values of x and y, the answer is a real number.

Group/Research Activities

79. In 1591, a French mathematician, François Viète, first used variables to describe patterns. For this reason, Viète is sometimes considered the "father of algebra." Research Viète's life and prepare a report on his influence in the development of algebra.

80. Interview various trades or crafts people to discover the formulas they use in their professions. Provide examples on how these formulas are applied to specific situations.

Just for fun

Can you name 100 different words that do not contain the letters *A, B, C,* or *Q*? (*Hint:* What is one of the first things you do with numbers?) Time limit: 3 minutes.

9.3 OPEN SENTENCES AND THEIR GRAPHS

Solving Open Sentences Based on a Replacement Set

You may recall that in Chapter 3 we encountered statements such as the following:

February has 30 *days.*

$3 + 2 = 1$

Bill Clinton was President of the United States.

$5 + 4 + 3 + 2 + 1 = 14$

Each of these statements is either true or false (but not both). We can tell whether one of these statements is true or false upon reading the statement.

Consider the sentence

$$x + 1 = 3$$

Is this sentence true or false? We cannot answer that question yet. The sentence is neither true nor false until we replace x with a number. Sentences such as $x + 1 = 3$, which cannot be classified as true or false until we replace x by a number, are called **open sentences**. Note that any real number can be substituted for x in the open sentence $x + 1 = 3$. The sentence will be true or false depending on the value we substitute for x. Because x represents an unknown number, we can call x a variable.

If we replace x by 4 in the open sentence $x + 1 = 3$, we have $4 + 1 = 3$, which is a false statement. If we replace x by 2, we have $2 + 1 = 3$, which is a true statement. Therefore, 2 is a solution to the open sentence $x + 1 = 3$. Because it can be shown that 2 is the only solution to $x + 1 = 3$, we call $\{2\}$ the *solution set* for the open sentence. The **solution set** for an open sentence is the set of numbers that make the sentence true when they are substituted for x.

If we replace the equal sign in the open sentence $x + 1 = 3$ with the symbol $>$, we have

$$x + 1 > 3$$

an open sentence of *inequality*. The sentence $x + 1 > 3$ is read as "$x + 1$ is greater than 3." Note that if we replace x by 2 in this statement, we have $2 + 1 > 3$, which is not a true statement. Suppose we restrict our replacements for x to the set of integers. What will the solution set be? The integers 3, 4, 5, and so on will make the sentence true. Therefore, the solution set is $\{3, 4, 5, \ldots\}$.

Some other examples of open sentences of inequality are

$x > 2$ x is greater than 2

$x < 2$ x is less than 2

$x \geq 2$ x is greater than or equal to 2

$x \leq 2$ x is less than or equal to 2

It is important to realize that an open sentence may have different solution sets, depending on the restrictions placed on the set of permissible replacements, the **replacement set**. For instance, the equation $x + 1 = 0$ has no solution set (the solution is the empty set, $\emptyset$) if we restrict the set of permissible replacements to the set of natural numbers. But if the replacement set is the set of integers, then the solution set for $x + 1 = 0$ is $\{-1\}$.

Consider the following sentence:

$$x + 1 < 5$$

If the replacement set is the set of natural numbers, then the solution set is $\{1, 2, 3\}$. If the replacement set is the set of whole numbers, then the solution set is $\{0, 1, 2, 3\}$. If the replacement set is the set of integers, then the replacement set is $\{\ldots, -2, -1, 0, 1, 2, 3\}$. If the replacement set is the set of real numbers, then the solution set is {all real numbers less than 4}.

EXAMPLE 1 ***Finding the Solution Set of an Open Sentence*** Find the solution set for $x + 5 = 9$, where x is any real number.

Solution There is only one real number that will make this sentence true. Because $4 + 5 = 9$, the solution set is $\{4\}$.

EXAMPLE 2 ***Finding the Solution Set of an Open Sentence of Inequality*** Find the solution set for $x + 2 \geq 7$, where x is an integer.

Solution The sentence is read as "$x + 2$ is greater than or equal to 7," and the integers that satisfy this sentence are 5, 6, 7, and so on. Hence, the solution set is $\{5, 6, 7, \ldots\}$. Note that 5 is a member of this set because the sentence reads "greater than *or* equal to."

EXAMPLE 3 ***Finding the Solution Set of an Open Sentence of Inequality*** Find the solution set for $x + 2 < 3$, where x is a natural number.

Solution We want the sum $(x + 2)$ of two natural numbers to be less than 3. There are no natural numbers that we may substitute for x to make this statement true; hence, the solution set is the empty set, $\emptyset$.

EXAMPLE 4 ***Finding the Solution Set of an Open Sentence of Inequality*** Find the solution set of $x + 2 < 3$, where x is an integer.

Solution Now we are considering a different replacement set than in Example 3. There are integers that can be added to 2 to yield a sum less than 3—namely, $0, -1, -2$, and so on. Therefore, the solution set is $\{\ldots, -2, -1, 0\}$.

NW ***Now Work Problems 1, 5, and 13.***

Graphing Solution Sets on the Number Line

Once we find the solution set for an open sentence, we can make a "picture" of the solution set. That is, we can **graph** the solution set. In Chapter 8 we introduced the concept of a *number line*. First, we drew a horizontal line, picked a point on the line, and labeled it zero. Next, we marked off equal units to the right and left of zero. We labeled the endpoints of the intervals to the right of zero with the positive integers, and those to the left of zero with the negative integers. The number line can be extended indefinitely in each direction (see Figure 9.1).

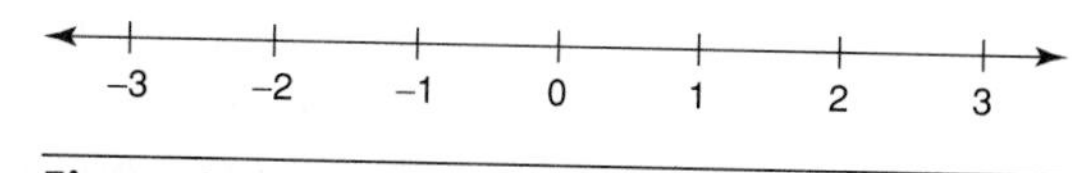

Figure 9.1

Recall that each integer corresponds to a particular point on the number line. Also, each point on the number line corresponds to a real number. The set of real numbers is composed of the set of rational numbers and the set of irrational numbers. It is not physically possible to label every point on the number line, but we do know that there exists a point $\frac{3}{4}$ unit from 0, a point -1.5 units from 0, and points $\sqrt{3}$ units and π units from 0. There is a point on the number line that corresponds to every rational or irrational number, and there is a rational or irrational number that corresponds to every point on the number line. Hence, we can use the number line to represent the set of real numbers.

How can we picture these numbers on the real number line? We can represent individual numbers on the number line by marking a heavy solid dot on the line at the point that corresponds to that number. For example, Figure 9.2 is a graph of the number 1. Figure 9.3 shows the graph of the set of integers $\{-1, 0, 1\}$.

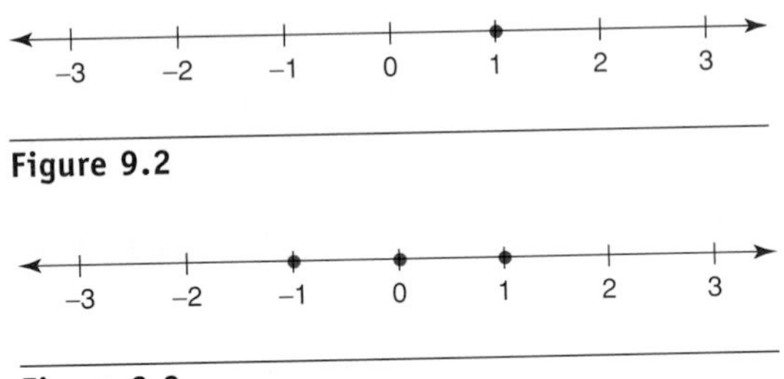

Figure 9.2

Figure 9.3

Note that we can also call Figure 9.3 the graph of the integers between -2 and 2. Figure 9.3 is the correct graph because "between -2 and 2" means that we do not include the integers -2 and 2.

Suppose we want a graph that represents all of the real numbers between -2 and 2. This graph would have to include all of the integers, rational numbers, and irrational numbers between -2 and 2. We indicate this by using open circles or dots at -2 and 2 and drawing a solid bar between the open circles, as shown in Figure 9.4.

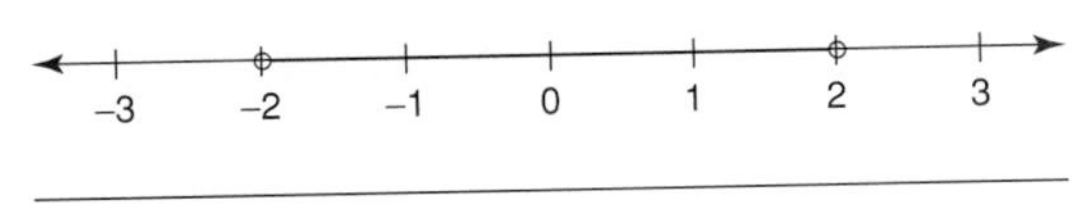

Figure 9.4

The fact that the dots at points -2 and 2 are not solid indicates that these points are not included in our solution set. Speaking in terms of algebra and the set of real numbers, Figure 9.4 is a graph of all real numbers x such that $-2 < x$ and $x < 2$. We can shorten this expression to $-2 < x < 2$.

Now consider the algebraic sentence $-2 \leq x \leq 2$, read "-2 is less than or equal to x, and x is less than or equal to 2." If we are to represent the solution set of this sentence on the number line, where x is any real number, then we must include the endpoints. Consequently, they would be colored in, as shown in Figure 9.5.

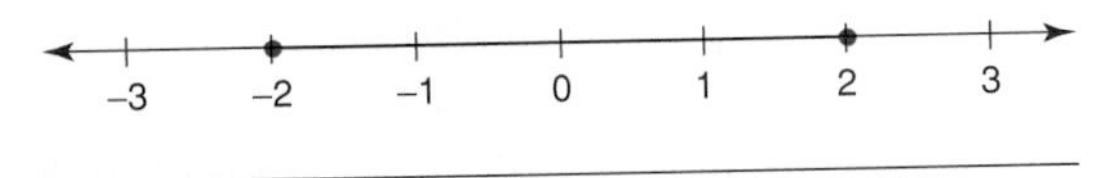

Figure 9.5

Next, let's consider the open sentence $x + 2 < 3$, where x is a real number. This statement is true for any real number that is less than 1—that is, $x < 1$. We can represent this on the number line as shown in Figure 9.6.

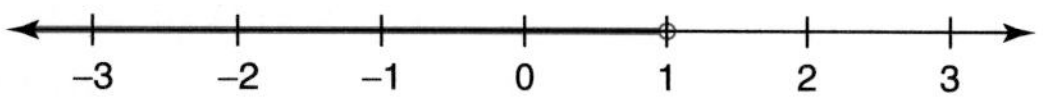

Figure 9.6

The solid arrow indicates that the solution set extends to the left indefinitely. Note that we have $x < 1$ (x is less than 1) and therefore we do not include 1 in the solution. If we had the statement $x \le 1$ (x is less than or equal to 1), then we would have included 1 in the solution, and 1 would have been marked with a solid dot.

The graph in Figure 9.6 is sometimes called a **half-line**. If the point at 1 were included, then the graph would be called a **ray**. Figure 9.4 depicts an **open line segment**, whereas Figure 9.5 depicts a **line segment**.

EXAMPLE 5 ***Graphing the Solution Set of an Open Sentence*** Graph the solution of $x - 2 > 1$, where x is a real number.

Solution In order for the open sentence $x - 2 > 1$ to be true, x must be a number greater than 3—that is, $x > 3$. The solution is the half-line shown in Figure 9.7.

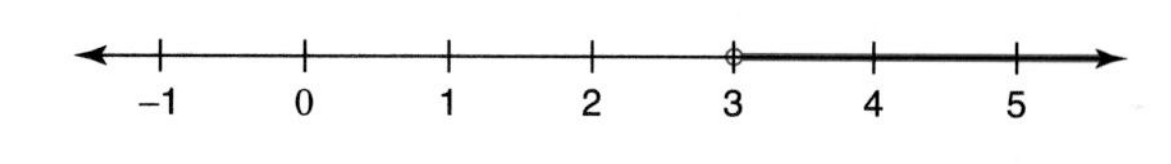

Figure 9.7

EXAMPLE 6 ***Graphing the Solution Set*** Graph the solution of $x + 2 \le 1$, where x is a real number.

Solution In order for the open sentence $x + 2 \le 1$ to be true, x must be a number less than or equal to negative 1—that is, $x \le -1$. The solution is the ray shown in Figure 9.8.

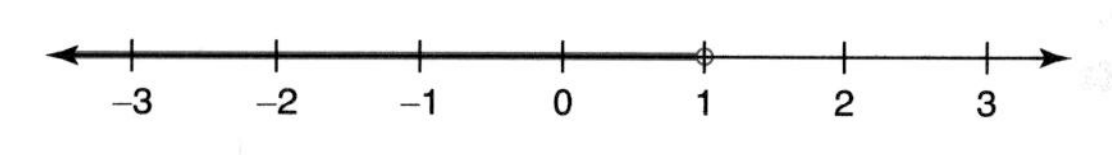

Figure 9.8

EXAMPLE 7 ***Graphing the Solution Set*** Graph the solution of $x + 2 > x$, where x is a real number.

Solution Notice that there are no restrictions on x in order for the open sentence $x + 2 > x$ to be true. Regardless of the number with which we replace x, we have a true statement. For example,

$$1 + 2 > 1, \quad 0 + 2 > 0, \quad \text{and} \quad -2 + 2 > -2$$

Therefore, the solution set is the set of all real numbers. The graph of the set of all real numbers is a **line**, as shown in Figure 9.9.

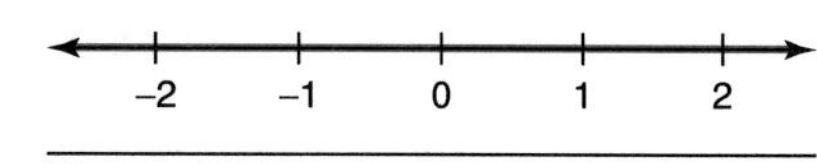

Figure 9.9

EXAMPLE 8 ***Graphing the Solution Set*** Graph the solution of the open sentence $-1 < x < 3$, where x is a real number.

Solution The sentence $-1 < x < 3$ is read as "-1 is less than x, and x is less than 3." This means that a member of the solution set must satisfy both conditions at the same time. That is, if 0 is a member of the solution set, then -1 must be less than it and at the same time it must be less than 3.

The solution set for this sentence is the set of real numbers between -1 and 3. The graph of the solution set is the open line segment shown in Figure 9.10.

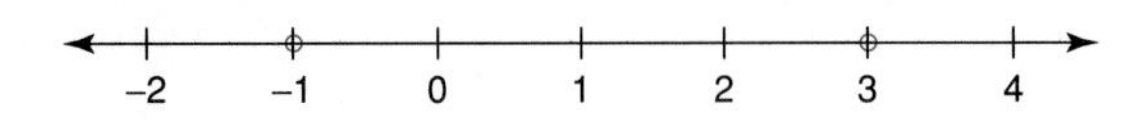

Figure 9.10

NW ***Now Work Problems 23 and 29.***

Exercises for Section 9.3

For Exercises 1–20, find the solution set for each open sentence. Unless otherwise noted, the replacement set is the set of real numbers.

NW **1.** $x + 2 = 5$ **2.** $x + 3 = 7$

3. $0 = 1 + x$ **4.** $6 = x + 5$

NW **5.** $5 - 2 = 4 + x$ (x is a whole number)

6. $x - 4 = 5 + 1$ (x is a whole number)

7. $x - 17 = 23$ **8.** $17 = x - 7$

9. $13 = x + 5$ **10.** $18 = 6 + x$

11. $x + 2 < 7$ (x is a natural number)

12. $x + 4 \leq 6$ (x is a natural number)

NW **13.** $x + 4 > 5$

14. $x - 4 > -6$

15. $0 < x < 3$ (x is an integer)

16. $-1 \leq x \leq 3$ (x is an integer)

17. $0 \leq x < 5$ (x is an integer)

18. $0 < x \leq 5$ (x is an integer)

19. $-2 < x \leq 3$ (x is an integer)

20. $-2 \leq x \leq 2$ (x is a whole number)

For Exercises 21–36, graph the solution set for each open sentence on the number line. In each case, the replacement set is the set of real numbers.

21. $x + 3 = 4$ **22.** $x - 1 = 0$

NW **23.** $4 = x - 5$ **24.** $6 = 7 - x$

25. $5 - x = 2$ **26.** $3 = 7 - x$

27. $x < 4$ **28.** $x \geq 2$

NW **29.** $x + 2 \geq 0$ **30.** $x - 2 > 1$

31. $2 < x < 4$ **32.** $-1 \leq x \leq 3$

33. $-2 < x \leq 1$ **34.** $-1 \leq x < 1$

***35.** $x + 1 > x$ ***36.** $x + 1 < x$

In Exercises 37 and 38, write an inequality that describes the given situation and then graph the solution set of each inequality on the number line using the replacement set of real numbers.

37. The speed limit on Interstate 95 includes a minimum speed of 45 mph and a maximum speed of 70 mph. Marion is driving d miles per hour and obeying the speed limit.

38. Lori's annual income is d dollars, which is less than $40,000.

Writing Mathematics

39. Explain the concept of an open sentence and give an example.

40. Given the open sentence $x + 1 = 0$, explain why the solution set is dependent on the set of numbers under which we are operating. Include illustrations as part of your explanation by considering the sets of natural, whole, integer, rational, and real numbers as replacement sets.

41. Explain how to graph the solution set $-2 \leq x < 2$ and then graph it (x is a real number).

42. Explain under what circumstances a solution set is the entire set of real numbers.

Challenge Exercises

In Exercises 43–46, determine the solution set for each open sentence and then graph it on the number line. In each case the replacement set is the set of real numbers.

43. $\frac{x}{3} > 0$ **44.** $\frac{3}{x} > 0$

45. $\frac{x}{3} > 1$ **46.** $\frac{3}{x} \geq 1$

47. Write an open sentence that contains $x < 3$ as its solution set if the replacement set is real numbers.

48. Consider the two open sentences of inequality, $x + 2 > 0$ and $x - 3 < 0$. Graph the solution set of each inequality on separate number lines and then place the first graph directly under the second one. (Be sure to align the number lines properly.) Now determine the solution set that satisfies *both* inequalities by examining the common points between the two lines. Draw a third number line that shows this intersection. This should now be the graph of the solution set that satisfies both inequalities.

49. Repeat Exercise 48 for $x - 3 \geq 0$ and $x + 2 < 0$.

Group/Research Activity

50. Review Chapter 2 on sets and examine the relationship of set-builder notation and set roster form from the perspective of open sentences and solution sets. Write a report that explains this relationship and include specific examples that demonstrate the relationship.

Just for fun

How many triangles are there in the figure?

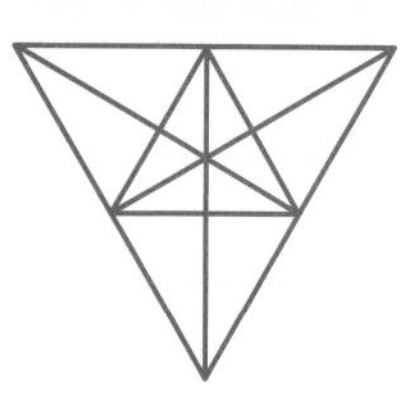

9.4 MORE OPEN SENTENCES—SOLVING LINEAR EQUATIONS

Thus far in our discussion of open sentences, we have only considered sentences such as $x - 2 > 1$ and $x + 2 = 3$. It is not too difficult to find replacements for x such that the given sentences will be true, as we can usually do this by observation or by trial and error. That is, we can replace x by a number and see if the resulting statement is true. But how about a sentence such as the following?

$$4x - 2 = x + 7$$

This open sentence is an equation because it is a statement of equality. We could probably find the solution to the equation $4x - 2 = x + 7$ by trial and error, but there does exist a more efficient method. By using certain techniques, we can systematically solve many equations in one unknown.

Solving algebraic equations involves simplifying the expressions on each side of the equal sign and then applying specific axioms, or rules. Thus, before we generate specific techniques for solving equations, we need to first discuss the concept of simplifying expressions and then review the rules we will use to solve equations.

Simplifying Algebraic Expressions

If we have an expression such as $2y + 3y$, we can simplify this expression by writing $2y + 3y = 5y$. This can be illustrated in the following manner:

$$2y + 3y = (2 + 3)y = 5y$$

In the second step, the y has been factored out of $2y + 3y$ using the distributive property; we then add $(2 + 3)$ and obtain 5. We can also think of this as combining two like quantities: If we have 2 yaks plus 3 yaks, we will obtain 5 yaks; that is, 2 yaks + 3 yaks = 5 yaks.

An expression such as $5z - 2z$ can also be simplified.

$$5z - 2z = (5 - 2)z = 3z$$

We can also think of this as 5 zebras minus 2 zebras, resulting in 3 zebras; that is, 5 zebras − 2 zebras = 3 zebras.

It should be noted that when we wish to write $1x$, we usually do this by writing x by itself. Similarly, $1y$ is written as y. Therefore, an expression such as

$$3x + x$$

is the indicated sum of $3x$ and $1x$, so that $3x + x = 4x$. Also, $4y - y = 3y$ and $x + x = 2x$.

Summarizing what we have done thus far,

If the terms of an expression contain the same variable or quantity, then we can simplify the expression into a single term by adding or subtracting the coefficients of each term. Thus, $\mathbf{2x + 5x = 7x}$ *and* $\mathbf{9y - 5y = 4y}$*. This process is known as* ***combining like terms***.

EXAMPLE 1 ***Simplifying Expressions by Combining Like Terms*** Simplify each of the following expressions.

a. $5h + 8h$
b. $14x - 23x$
c. $3a + 8 + 5a - 6$

Solution

a. $5h + 8h = 13h$
b. $14x - 23x = (14 - 23)x = -9x$
c. Prior to simplifying this expression, it is helpful to first *collect* like terms before we combine them. By collecting like terms, we mean writing like terms next to each other to facilitate combining them.

$$\begin{aligned} 3a + 8 + 5a - 6 &= 3a + 5a + 8 - 6 \\ &= 8a + 2 \end{aligned}$$

NW ***Now Work Problems 17 and 21.***

We can also simplify expressions that involve multiplication. For example, if we have $3x$ and we wish to double the amount, we can multiply $3x$ by 2; that is,

$$2(3x) = 3x + 3x = 6x$$

We may use the distributive property to perform indicated multiplications, such as $5(2x + 6)$. Basically, the distributive property says that for any numbers a, b, and $c, a(b + c) = ab + ac$. Hence,

$$5(2x + 6) = 5(2x) + 5(6)$$
$$5(2x + 6) = 10x + 30$$

The following are some other examples of indicated multiplications and the resulting products:

$$3(4x + 2) = 12x + 6, \qquad 2(5x - 3) = 10x - 6,$$
$$3(4x) = 12x \qquad 2(6y) = 12y,$$
$$9(z) = 9(1z) = 9z, \qquad \tfrac{1}{2}(4y) = 2y,$$

The last multiplication example also illustrates another operation. If we multiply a quantity by $\frac{1}{2}$, then this is the same as dividing the quantity by 2. In each of the previous examples, we are working with only the **coefficients, not the variables**. For example, $2x + 3x = 5x, 3y - y = 2y$, and $3(2z) = 6z$. The same holds true for the operation of division. If we have $9x$ and divide that quantity by a number such as 3, the quotient is $3x$:

$$9x \div 3 = \frac{9x}{3} = 3x$$

The following are some other examples of indicated divisions and the resulting quotients:

$$\frac{5y}{5} = y, \quad \frac{10z}{5} = 2z, \quad \frac{6x}{2} = 3x, \quad \frac{3y}{1} = 3y, \quad \frac{8x}{2} = 4x$$

EXAMPLE 2 ***Simplifying Expressions Involving Products*** Simplify each of the following expressions.

a. $4(3x + 8)$

b. $9x + 2(4x - 7)$

Solution

a. To simplify this expression, we apply the distributive property by multiplying each term within parentheses by 4. Thus,

$$\begin{aligned} 4(3x + 8) &= 4(3x) + 4(8) \\ &= 12x + 32 \end{aligned}$$

Note that we cannot simplify this expression any further because $12x$ and 32 are not like terms.

b. To simplify this expression we first apply the distributive property and then combine all like terms.

$$\begin{aligned} 9x + 2(4x - 7) &= 9x + 2(4x) + 2(-7) \\ &= 9x + 8x - 14 \\ &= 17x - 14 \end{aligned}$$

NW ***Now Work Problems 35 and 43.***

Axioms for Solving Equations

Now that we know how to simplify expressions, we next discuss the axioms, or rules, for solving equations.

AXIOM 1: If the same quantity or two equal quantities are added to two equal quantities, then the sums are equal.

As an example of axiom 1, we have

$$\begin{array}{rl} x - 1 = 2 & \textit{given} \\ +1 = +1 & \textit{adding 1 to both sides} \\ \hline x - 1 + 1 = 2 + 1 & \textit{sums are equal} \\ x = 3 & \end{array}$$

AXIOM 2: If the same quantity or two equal quantities are subtracted from two equal quantities, then the remainders (differences) are equal.

The following example illustrates axiom 2:

$$\begin{array}{rl} x + 4 = 5 & \textit{given} \\ -4 = -4 & \textit{subtracting 4 from both sides} \\ \hline x + 4 - 4 = 5 - 4 & \textit{remainders are equal} \\ x = 1 & \end{array}$$

AXIOM 3: If equal quantities are multiplied by equal quantities, then the products are equal.

As an example of axiom 3, we have

$$
\begin{array}{rl}
\frac{x}{2} = 4 & \textit{given} \\
\underline{\times 2 = \times 2} & \textit{multiplying both sides by 2} \\
\frac{x}{2} \cdot 2 = 4 \cdot 2 & \textit{products are equal} \\
x = 8 &
\end{array}
$$

AXIOM 4: If equal quantities are divided by equal quantities (other than zero), then the quotients are equal.

The idea behind solving equations is to find a sentence of the form

$x =$ _______ or _______ $= x$

that is equivalent to the original equation.

To illustrate axiom 4, we have

$$
\begin{array}{rl}
3x = 9 & \textit{given} \\
\underline{\div 3 = \div 3} & \textit{dividing both sides by 3} \\
\frac{3x}{3} = \frac{9}{3} & \textit{quotients are equal} \\
x = 3 &
\end{array}
$$

One thing to keep in mind in solving any equation is that an equation is like a balance scale, and in order to keep the equation in balance, whatever we do to one side of the equation, we must do to the other side of the equation. Our ultimate goal is to get x by itself on one side of the equal sign. It does not matter whether we wind up with x on the left side or the right side of the equal sign, as long as it is by itself.

Now let's solve the equation $4x - 2 = x + 7$ using the axioms whenever necessary. We must get x by itself on one side of the equation. We will first eliminate the -2 from the left side. We can do this by adding 2 to the left side; but if we add 2 to the left side, then, by axiom 1, we must also add 2 to the right side. Hence, we have

$$4x - 2 + 2 = x + 7 + 2$$

which simplifies to

$$4x = x + 9$$

Remember that we must get all of the x's on one side of the equal sign. We can eliminate x from the right side, $x + 9$, by subtracting x from it. But if we subtract x from the right side, we must also subtract x from the left side in order to maintain the balance (axiom 2). Therefore, we have

$$4x - x = x - x + 9$$

$$\text{or} \quad 3x = 9$$

Because we want to solve for x, we must divide the expression $3x$ (3 times x) by 3. Therefore, we divide both sides by 3 (axiom 4):

$$\frac{3x}{3} = \frac{9}{3}$$

$$x = 3$$

We can check our answer by replacing x with 3 in the original equation to see if the left side of the equation equals the right side. Replacing x with 3, we have

$$4(3) - 2 \stackrel{?}{=} 3 + 7$$

$$12 - 2 \stackrel{?}{=} 10$$

$$10 = 10$$

The solution checks.

The following is the solution of the equation $4x - 2 = x + 7$ without the discussion. We list the reasons for each step of the solution.

$4x - 2 = x + 7$	*given*
$4x - 2 + 2 = x + 7 + 2$	*axiom 1: adding 2 to both sides*
$4x = x + 9$	*combining like terms*
$4x - x = x - x + 9$	*axiom 2: subtracting x from both sides*
$3x = 9$	*combining like terms*
$\frac{3x}{3} = \frac{9}{3}$	*axiom 4: dividing both sides by 3*
$x = 3$	*simplifying*

EXAMPLE 3 ***Solving an Equation*** Solve $4x - 7 = x + 5$ for x.

Solution

$4x - 7 = x + 5$	*given*
$4x - 7 + 7 = x + 5 + 7$	*axiom 1: adding 7 to both sides*
$4x = x + 12$	*combining like terms*
$4x - x = x - x + 12$	*axiom 2: subtracting x from both sides*
$3x = 12$	*combining like terms*
$\frac{3x}{3} = \frac{12}{3}$	*axiom 4: dividing both sides by 3*
$x = 4$	*simplifying*

EXAMPLE 4 ***Solving an Equation*** Solve $2y + 6 = 6y - 10$ for y.

Solution

$2y + 6 = 6y - 10$	*given*
$2y + 6 + 10 = 6y - 10 + 10$	*axiom 1: adding 10 to both sides*
$2y + 16 = 6y$	*combining like terms*
$2y - 2y + 16 = 6y - 2y$	*axiom 2: subtracting 2y from both sides*
$16 = 4y$	*combining like terms*
$\frac{16}{4} = \frac{4y}{4}$	*axiom 4: dividing both sides by 4*
$4 = y$	*simplifying*

NW ***Now Work Problem 63.***

EXAMPLE 5 ***Solving an Equation*** Solve $\frac{3z}{4} + 2 = 8$ for z.

Solution

$\frac{3z}{4} + 2 = 8$	*given*
$\frac{3z}{4} + 2 - 2 = 8 - 2$	*axiom 2: subtracting 2 from both sides*
$\frac{3z}{4} = 6$	*combining like terms*

$$\frac{3z}{4}\cdot 4 = 6\cdot 4 \qquad \textit{axiom 3: multiplying both sides by 4}$$

$$3z = 24 \qquad \textit{simplifying}$$

$$\frac{3z}{3} = \frac{24}{3} \qquad \textit{axiom 4: dividing both sides by 3}$$

$$z = 8 \qquad \textit{simplifying}$$

EXAMPLE 6 ***Solving an Equation Involving the Distributive Property*** Solve $2(3x + 2) = 10$ for x.

Solution

$$2(3x + 2) = 10 \qquad \textit{given}$$

$$2\cdot 3x + 2\cdot 2 = 10 \qquad \textit{distributive property}$$

$$6x + 4 = 10 \qquad \textit{simplifying}$$

$$6x + 4 - 4 = 10 - 4 \qquad \textit{axiom 2: subtracting 4 from both sides}$$

$$6x = 6 \qquad \textit{combining like terms}$$

$$\frac{6x}{6} = \frac{6}{6} \qquad \textit{axiom 4: dividing both sides by 6}$$

$$x = 1 \qquad \textit{simplifying}$$

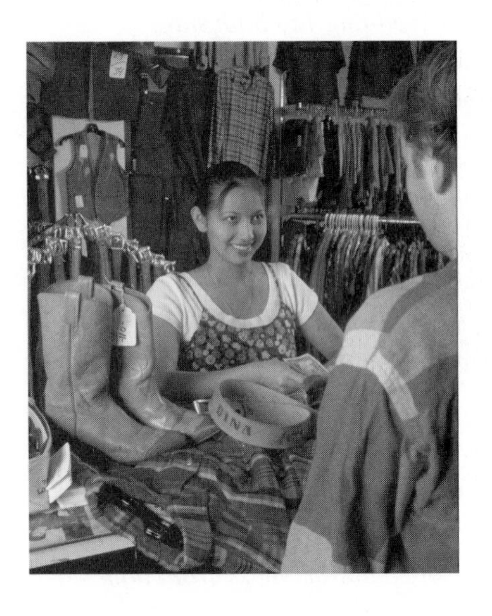

NW ***Now Work Problems 75 and 83.***

EXAMPLE 7 ***Solving Formulas*** In business, the profit earned on an item is determined by subtracting the cost of producing the item from its selling price. This is represented mathematically by the formula $p = s - c$, where p = profit, s = selling price, and c = production cost. Write an equivalent formula that can be used to determine what the selling price of an item should be.

Solution To find an equivalent formula for the selling price, we must solve the formula for s.

$$p = s - c$$

$$p + c = s - c + c \qquad \textit{add c to both sides}$$

$$p + c = s$$

Therefore, $s = p + c$.

NW ***Now Work Problem 105.***

Exercises for Section 9.4

For Exercises 1–48, simplify each of the algebraic expressions.

1. $2x + 3x$

2. $6y + 3y$

3. $2x + x$

4. $5x - 3x$

5. $3y - y$

6. $4z - 2z$

7. $2(4x)$

8. $2(3z)$

9. $2(7y)$

10. $6z \div 3$

11. $10x \div 2$

12. $8y \div 4$

13. $\frac{4x}{2}$

14. $\frac{12z}{3}$

15. $\frac{15x}{3}$

16. $5 + 4x + x$

NW 17. $2y + 2 + 3y$

18. $6z - 4z + 4$

19. $8y + 2y + 4$

20. $4x - x + 2$

NW 21. $2z - z + 2$

22. $2x + 3 - x - 2$

23. $4y - 2 - y + 3$

24. $6z + 2 - z - 3$

25. $4(3x) + 2$

26. $6(2x + 3)$

27. $3(x - 2)$

28. $2(4y - 3) + 2y$

29. $5(2z) + 3 - z$

30. $y + 3(4 - y)$

31. $0(3x - 4) + 1$

32. $3(4y) - 4(2y)$

33. $4x + 3x + 5 - 3$

34. $2y - y + 7 + 2$

NW 35. $3(5z) + 2$

36. $(16x + 4) \div 4$

37. $(3x - x) \div 2$

38. $4(2y) + 3(2y)$

39. $4z + 3z - 7 + 5$

40. $2x - x + 7 + 4x$

41. $2(4x) + 3(5x)$

42. $8y + 3y + 2(2y + 4)$

NW 43. $4x - 2 - x + 3(x + 1)$

44. $4(x - 2) + 3$

45. $(3y + 6) \div (2 + 1)$

46. $(8y - 2) \div (5 - 3)$

47. $y + 4(5 - 2y)$

48. $(3x - x)(5 + 3)$

Solve each of the following equations in the system of real numbers:

49. $x + 4 = 6$

50. $x - 7 = 5$

51. $4 = x - 8$

52. $7 = x + 3$

53. $3y = 18$

54. $5z = 35$

55. $3y - 4 = 8$

56. $5x + 2 = 12$

57. $2x = x + 3$

58. $3z - 1 = 2z$

59. $7y = 5y + 2$

60. $3y - 6 = 2y$

61. $4x + 3 = 3x + 5$

62. $5y - 9 = 4y + 2$

NW 63. $6x + 4 = 4x - 8$

64. $7x - 3 = 4x - 6$

65. $9y - 2 = 7y + 10$

66. $3y + 5 = 6y - 4$

67. $\frac{x}{3} = 5$

68. $\frac{y}{4} = 2$

69. $\frac{x}{5} + 3 = 5$

70. $\frac{z}{3} - 2 = 7$

71. $\frac{x}{5} = x + 8$

72. $\frac{y}{2} + y = 6$

73. $\frac{z}{3} + z = 4$

74. $\frac{x}{5} = 6 - x$

NW 75. $\frac{x}{9} - 3 = x - 3$

76. $\frac{z}{3} + 2 = 6 - z$

77. $\frac{x}{4} + 6 = x - 3$

78. $\frac{y}{3} + 1 = 2y - 4$

79. $3(2x + 1) = 9$

80. $2(5x + 4) = 28$

81. $5(x - 1) = 20$

82. $4(x - 6) = 20$

NW 83. $3(2x + 5) = 27$

84. $5(3x - 2) = 20$

85. $3x + 3(4 + 2x) = 30$

86. $5x + 3(9 - 5x) = 47$

87. $12 = 4(2x - 3)$

88. $2(2x + 3) = 2(x - 1)$

89. $3(x - 2) = 2(x + 5)$

90. $4(5x - 3) = 3(5x + 1)$

Writing Mathematics

91. Explain what it means to simplify an algebraic expression by combining like terms. Include an example as part of your explanation.

92. Explain and demonstrate how the distributive property is used to combine the terms $3x + 5x$.

93. Explain why $3x$ and $4t$ are not like terms.

94. Explain the steps needed to solve the equation $2x + 3 = 5x - 4$.

Challenge Exercises

In Exercises 95–100, solve the equations in the system of real numbers.

95. $6 + 3(5x + 6) = -3(4 - 3x)$

96. $5x - 3 + 3(-2x - 7) = 5x - (12 + 3x)$

97. $7x + 2(6x - 4) - (x + 5) = 2x + 4(3x - 6) - 1$

98. $14 + 2x - 3(x + 5) = 3(2x - 9) + 5 - 6(4 + 2x)$

99. $5x + 2 - [2(3x + 2) - 8] + 4x = 2(x + 6) - 3x + 2$

100. $2 - [5x + 4(3 - x) - 6 - 2x - (5 - 2x)]$
$= 4[2(3x - 4) - 6(x - 1)]$

101. A classmate simplified $5x - x$ to 4 instead of the correct answer, which is $4x$. Explain how the distributive property can be used to demonstrate to your classmate why he is incorrect in his reasoning.

102. In solving the equation $3(x + 3) = 18$, Kelly got an answer of 5, Kristin got an answer of 3, and Melanie got an answer of 12. Checking their answers, we find that Kristin is correct. See if you can determine the error(s) made by Kelly and Melanie:

Kelly's Solution	Melanie's Solution
$3(x + 3) = 18$	$3(x + 3) = 18$
$3x + 3 = 18$	$x + 3 = 15$
$3x = 15$	$x = 12$
$x = 5$	

103. **Long distance** A telephone company provides two long distance plans. Plan A is $0.99 for the first 20 minutes and then 10 cents per minute for each minute after 20 minutes. There is no monthly service charge for this plan. Plan B is a flat rate of 7 cents per minute but carries a $4.95 monthly service charge.

a. Write an algebraic expression that describes each plan.

b. At what point will both plans be equal? That is, how many long distance minutes must a person on each plan use before they both pay the same amount?

Group/Research Activity

104. Using the concept of a balance bar to solve equations, try playing the following game, which involves a moderator and two competing groups. The moderator's job is to construct an equation that has an integer solution and the groups' job is to guess the solution using trial and error. Without showing the equation to either group, the moderator asks the first group (selected by a coin toss) to guess the solution. The moderator then indicates which side of the equation is larger. The group that is able to identify the correct solution first wins the contest.

Real-World Data

NW **105. Physics** In physics, the formula $F = MA$ represents the relationship among force (F), mass (M), and acceleration (A) in terms of force (i.e., force is equal to mass times acceleration). Write an equivalent formula that can be used to express this relationship in terms of mass.

106. Circumference. The circumference (C) of a circle can be found by the formula $C = 2\pi r$, where π(pi) is the ratio of a circle's circumference to its diameter ($\pi \approx 3.14$) and r is equal to the circle's radius. Write an equivalent formula that expresses this relationship in terms of a circle's radius.

107. Temperature conversion The formula $F = \frac{9}{5}C + 32$ is used to convert from temperatures expressed in degrees Celsius (C) to temperatures in degrees Fahrenheit (F). Write an equivalent formula that can be used to convert from degrees Fahrenheit to degrees Celsius.

108. Volume In geometry, the volume (V) of a pyramid can be found by using the formula $V = \frac{1}{3}Bh$, where B represents the base of the pyramid and h is equal to its height. Write an equivalent formula that expresses this relationship in terms of a pyramid's height.

109. Interest The formula $A = P(1 + rt)$ is used to calculate the amount of money accumulated (A) given the principal (P), rate of interest (r), and time period (t) in years. Write an equivalent formula that expresses this formula in terms of time.

110. Automobile design Some automobiles are designed with a tapered front end. A formula that can be used to determine the taper is $T = \frac{12(w - x)}{L}$, where T is the taper, w is the length of the longer height, x is the length of the shorter height, and L is the overall length. Write an equivalent formula that expresses this relationship in terms of w.

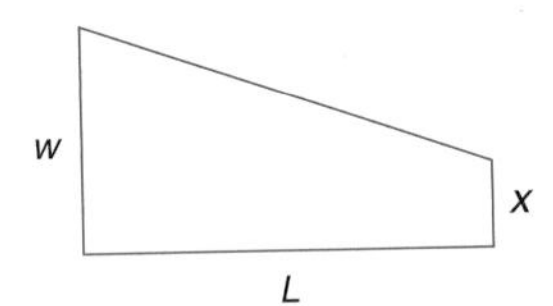

Just for fun

Ask a friend to perform the steps in the left column while you do the steps in the right column.

1. Pick a number.	1. Let $x =$ the number
2. Subtract 2.	2. $x - 2$
3. Multiply by 4.	3. $4(x - 2) = 4x - 8$
4. Add 10.	4. $4x - 8 + 10 = 4x + 2$
5. Divide by 2.	5. $(4x + 2) \div 2 = 2x + 1$
6. Add 9.	6. $2x + 1 + 9 = 2x + 10$
7. Divide by 2.	7. $(2x + 10) \div 2 = x + 5$
8. Give the result.	8. Give the original number. The result will always be 5 more than the starting number.

9.5 APPLICATIONS OF LINEAR EQUATIONS (PROBLEM SOLVING)

One of the oldest applications of algebra is solving word problems. As early as 2000 B.C., the Egyptians worked on word problems. You can probably solve some word problems by means of simple arithmetic, but most word problems require the use of algebra in order to find the solution in a systematic manner, as opposed to trial and error.

Consider the following sentence:

What number when decreased by 5 equals 15?

This sentence is an example of a word problem. What is the number? We can find the number if we can translate the sentence into an equation. In translating word problems into algebra, we must pay strict attention to what the words say. For instance, in the given sentence the phrase *what number* is the key phrase; it tells us that we are looking for a certain number. Therefore, we let x equal the number. (We could have just as easily let some other variable such as n or y equal the number.) Next we have the phrase *decreased by 5*. What does this mean? Exactly what it says; that is, *decreased by 5* is the same as *minus 5*, or -5. The last part of the sentence, *equals 15*, can be expressed as "$=15$." Therefore, the equation for the sentence

What number when decreased by 5 equals 15?

is

$$x - 5 = 15$$

Now we solve the equation:

$$\begin{aligned} x - 5 &= 15 \\ x - 5 + 5 &= 15 + 5 \\ x &= 20 \end{aligned}$$

As a check, observe that $20 - 5 = 15$.

Whenever we attempt to solve a word problem, it is necessary to translate the given sentence into an equation. The first thing we must do is let the unknown be represented by a variable such as x, y, z, m, or n. Next we must translate the given relationship with the unknown into an equation. The next step is to solve the equation in order to find the value of the unknown.

Mathematical translations of common English phrases are listed in Table 9.3. The italicized phrase is represented in the corresponding mathematical phrase.

TABLE 9.3

English Phrase	Mathematical Phrase
Addition	
4 *more than* a number	$x + 4$
a number *increased by* 5	$y + 5$
the *sum* of x and y	$x + y$
a number *added* to 3	$3 + n$
x *plus* y	$x + y$
Subtraction	
a number *decreased* by 5	$y - 5$
5 *less than* a number	$n - 5$
the *difference* between x and y	$x - y$
x *minus* y	$x - y$
Multiplication	
the *product* of a and b	$a \cdot b$
x *multiplied* by a	$a \cdot x$
twice a number	$2 \cdot n$
$\frac{1}{2}$ of y	$\frac{1}{2} \cdot y$
32 *percent* of z	$(0.32)(z)$
x *times* y	$x \cdot y$
Division	
the *quotient* of x and y	$\frac{x}{y}$
the *quotient* of y and x	$\frac{y}{x}$
the *ratio* of x and y	$\frac{x}{y}$
x *divided* by y	$\frac{x}{y}$

Biography: Diophantus

Diophantus of Alexandria was an ancient mathematician who contributed greatly to the development of algebra. He produced three different known works in the area of mathematics. The titles of these books are *Arithmetica, Porisms,* and *Polygonal Numbers*. It is thought that all these works dealt with the properties of rational or integral numbers. His most famous work, *Arithmetica,* deals with algebraic equations and certain problems that require positive rational numbers as solutions. In his presentation, Diophantus used special signs and abbreviations, such as the first letter of a word. Therefore, a great deal of his work had standardized notation that made it easier to understand. This was one of the first steps toward a formalization of algebra and the creation of mathematical notation.

Not much is known about Diophantus other than his mathematical contributions. Historians disagree as to when he lived: perhaps 75 A.D. or 250 A.D. They even disagree on the spelling of his name: Diophantas or Diophantus. Nevertheless, he was immortalized because of an epitaph that appeared in a Greek anthology. It gives the following information:

Diophantus spent one-sixth of his life as a child, one-twelfth as a youth, one-seventh more as a bachelor. In the fifth year of his marriage he had a son. The son died 4 years prior to the death of Diophantus, which made the life span of the son half that of Diophantus. This information gives us

$$\frac{x}{6} + \frac{x}{12} + \frac{x}{7} + 5 + \frac{x}{2} + 4 = x$$

where x equals the age of Diophantus.

Note that we can choose a variety of letters to represent the unknown. Normally x is used, but it does not have to be the chosen letter.

The word *is* or some form of the verb *to be* is often used to indicate equality (=). For example, the English sentence, "The sum of a number and 10 is 32," is translated into the mathematical sentence, $x + 10 = 32$. Table 9.4 gives some examples of English sentences translated into mathematical sentences.

TABLE 9.4

English Sentence	Mathematical Sentence
A number increased by 2 is 6.	$x + 2 = 6$
The product of a number and 3 is 18.	$3y = 18$
The difference between a number and 3 is 10.	$n - 3 = 10$
Two less than four times a number is 34.	$4z - 2 = 34$
One-half of a number and 6 more is 14.	$\frac{1}{2}n + 6 = 14$
Three more than twice a number is 9.	$2x + 3 = 9$
If a number is increased by 4, twice the sum is 50.	$2(y + 4) = 50$

EXAMPLE 1 ***Translating a Sentence into an Equation*** Write an equation to illustrate "What number increased by 2 is 5?"

Solution Let x equal the unknown. The phrase *increased by 2* means "plus 2," or "+2." Another word for *is* is *equals*. Therefore, we have

$$x + 2 = 5$$

EXAMPLE 2 ***Translating a Sentence into an Equation*** Write an equation to illustrate "The sum of what number and twice that number equals 9?"

Solution Let n equal the unknown. Twice a number can be expressed as $2n$. Therefore, we have

$$n + 2n = 9$$

EXAMPLE 3 ***Translating a Sentence into an Equation*** Write an equation to illustrate "Two less than four times what number is 34?"

Solution Let x equal the unknown. The phrase *two less* indicates that we subtract 2, but from what? We subtract 2 from "four times what number"—that is, $4x$. Therefore, we have

$$4x - 2 = 34$$

EXAMPLE 4 ***Translating a Sentence into an Equation*** Write an equation to illustrate "One-half of what number and 5 more is 12?"

Solution Let n equal the unknown. One-half of a number can be expressed as $\frac{1}{2}n$ or $n/2$. The phrase *and 5 more* indicates that we add 5 to $n/2$. Therefore, we have

$$\frac{n}{2} + 5 = 12$$

EXAMPLE 5 ***Writing an Equation to Represent Dimensions of a Pool*** A rectangular swimming pool is 6 feet longer than twice its width w. If the perimeter of the pool is 120 feet, write an equation to find its dimensions.

Solution This is stated in a little more formal language, but we treat it the same as the other examples. We are already given that the width is w, so the length l can be expressed as $2w + 6$. Using the fact that the perimeter of a rectangle is $2w + 2l$, we express the perimeter as

$$2w + 2(2w + 6) = 120$$

EXAMPLE 6 ***Writing an Equation to Represent a Sum of Integers*** The sum of two consecutive odd integers is 40. Write an equation for this problem.

Solution Let x equal the first odd integer. The next (consecutive) odd integer is $x + 2$. (It cannot be $x + 1$, because if x is odd, then $x + 1$ would be even.) Therefore, the sum of the two consecutive odd integers is $x + (x + 2)$. Hence, we have

$$x + (x + 2) = 40$$

Following is a list of suggestions that will aid us in solving word or verbal problems.

1. Read the problem carefully.
2. Reread the problem carefully.
3. If possible, draw a diagram to assist in interpreting the given information.

4. Translate the English phrases into mathematical phrases and choose a variable for the unknown quantity.
5. Write the equation using all of the preceding information.
6. Solve the equation.
7. Check the solution to determine whether it satisfies the original problem (not the equation).

Now that we have had some practice in translating verbal problems into algebraic expressions, let us complete the process and solve the problem. Consider the following problem:

The sum of two consecutive integers is 99. Find the value of the smaller integer.

The first thing we must do is represent the unknown in terms of a variable. Let x equal the first integer; then $x + 1$ will equal the next consecutive integer. Translating the given relationship with the unknown into an equation, we have

$$x + (x + 1) = 99$$

Next we solve the equation:

$$\begin{aligned} 2x + 1 &= 99 && \textit{combining like terms} \\ 2x + 1 - 1 &= 99 - 1 && \textit{subtracting 1 from both sides} \\ 2x &= 98 && \textit{combining like terms} \\ \frac{2x}{2} &= \frac{98}{2} && \textit{dividing both sides by 2} \\ x &= 49 && \textit{the solution} \end{aligned}$$

We can check our solution to see if it is correct. If the first integer is 49, then the next consecutive integer is $x + 1$—that is, $49 + 1$, or 50. Is the sum of 49 and 50 equal to 99? Yes $49 + 50$ does equal 99. Our answer checks, so the solution is correct.

EXAMPLE 7 ***Solving a Word Problem Involving Ages*** Scott is 2 years older than Joe and the sum of their ages is 46. What are the ages of Scott and Joe?

Solution Let x = Joe's age; then $x + 2$ = Scott's age. The sum of their ages is 46. Therefore, we have the equation

$$x + x + 2 = 46$$

Solving the equation, we have

$$\begin{aligned} 2x + 2 &= 46 && \textit{combining like terms} \\ 2x + 2 - 2 &= 46 - 2 && \textit{subtracting 2 from both sides} \\ 2x &= 44 && \textit{combining like terms} \\ \frac{2x}{2} &= \frac{44}{2} && \textit{dividing both sides by 2} \\ x &= 22 && \textit{Joe's age} \\ x + 2 = 22 + 2 &= 24 && \textit{Scott's age} \end{aligned}$$

NW ***Now Work Problem 7.***

EXAMPLE 8 ***Solving a Word Problem to Find a Number*** Two less than four times what number is 34?

Solution Let x = the number. "Two less than four times x is 34" gives us the equation

$$4x - 2 = 34$$

Next, we solve the equation:

$$4x - 2 + 2 = 34 + 2 \quad \textit{adding 2 to both sides}$$
$$4x = 36 \quad \textit{combining like terms}$$
$$\frac{4x}{4} = \frac{36}{4} \quad \textit{dividing both sides by 4}$$
$$x = 9 \quad \textit{the solution}$$

NW ***Now Work Problem 9.***

EXAMPLE 9 ***Solving a Word Problem to Find the Dimensions of a Rectangle*** The width of a rectangle is 2 feet less than its length. If the perimeter is 32 feet, find the dimensions of the rectangle.

Solution Let x = the length of the rectangle; then $x - 2$ = the width of the rectangle. Because the perimeter is equal to the sum of the lengths of all the sides, we have the equation

$$x + x - 2 + x + x - 2 = 32$$

Solving the equation,

$$4x - 4 = 32 \quad \textit{combining like terms}$$
$$4x - 4 + 4 = 32 + 4 \quad \textit{adding 4 to both sides}$$
$$4x = 36 \quad \textit{combining like terms}$$
$$\frac{4x}{4} = \frac{36}{4} \quad \textit{dividing both sides by 4}$$
$$x = 9 \quad \textit{length of the rectangle}$$
$$x - 2 = 9 - 2 = 7 \quad \textit{width of the rectangle}$$

NW ***Now Work Problem 13.***

EXAMPLE 10 ***Solving a Word Problem to Find the Angles of a Triangle*** The sum of two angles of a triangle is 90°. If one angle is 10° more than three times the smaller angle, find the angles.

Solution Let y = the smaller angle; then $3y + 10$ = the larger angle. Because the sum of the two angles is 90°, we have the equation

$$y + 3y + 10 = 90$$

Solving the equation,

$$4y + 10 = 90 \quad \textit{combining like terms}$$
$$4y + 10 - 10 = 90 - 10 \quad \textit{subtracting 10 from both sides}$$
$$4y = 80 \quad \textit{combining like terms}$$
$$\frac{4y}{4} = \frac{80}{4} \quad \textit{dividing both sides by 4}$$
$$y = 20 \quad \textit{the smaller angle}$$
$$3y + 10 = 3 \cdot 20 + 10 = 60 + 10 = 70 \quad \textit{the larger angle}$$

EXAMPLE 11 ***Solving a Coin Problem*** Pam has 85 cents in her change purse. If there are only nickels and dimes in her change purse, and she has 13 coins altogether, how many nickels and dimes does she have?

Solution Let x = number of dimes; then $13 - x$ = the number of nickels. Now if Pam has x dimes in her purse, then the value of these dimes in cents is $10x$. For

Historical Note

We mentioned earlier that the first algebra-type problems occurred in a papyrus copied by Ahmes. This document also contained one of the oldest verbal problems in the world. It was stated as "Hau, its whole, its seventh, it makes 19," which roughly translates to "Find a number such that if the whole number is added to one-seventh of it, then the result is 19." This problem was considered quite difficult by the Egyptians and its solution was complicated. If we translate the English phrases into mathematical phrases, we can obtain an equation and determine the solution to the problem. The desired equation is

$$y + \frac{y}{7} = 19$$

instance, if she had three dimes, then the value is $10 \cdot 3$, or 30 cents. Similarly, the value of the nickels in Pam's purse is $5(13 - x)$. The total value of the coins in her purse is 85 cents, and hence we have the equation

$$10x + 5(13 - x) = 85$$

Solving the equation,

$$\begin{aligned} 10x + 65 - 5x &= 85 && \textit{distributive property} \\ 5x + 65 &= 85 && \textit{combining like terms} \\ 5x + 65 - 65 &= 85 - 65 && \textit{subtracting 65 from both sides} \\ 5x &= 20 && \textit{combining like terms} \\ \frac{5x}{5} &= \frac{20}{5} && \textit{dividing both sides by 5} \\ x &= 4 && \textit{number of dimes} \\ 13 - x &= 13 - 4 = 9 && \textit{number of nickels} \end{aligned}$$

Check:

$x = 4$, and $13 - x = 9$, so it checks that Pam possesses 13 coins. But we must also check the value of the coins; that is, $4 \cdot 10 = 40$ (the value of the dimes), $9 \cdot 5 = 45$ (the value of the nickels), and $40 + 45 = 85$ cents (the total value of the coins). Our solution is verified.

NW *Now Work Problem 19.*

EXAMPLE 12 ***Solving a Consecutive Integer Problem*** The sum of three consecutive even integers is 30. Find the integers.

Solution Let $z =$ the first even integer; then the next consecutive even integer is $z + 2$. The third consecutive even integer is 2 more than the second one—that is, $z + 2 + 2$, or $z + 4$. Hence, we have

$$\begin{aligned} z &= \text{first even integer} \\ z + 2 &= \text{second consecutive even integer} \\ z + 4 &= \text{third consecutive even integer} \end{aligned}$$

As the sum of these integers is 30, we have the equation

$$z + (z + 2) + (z + 4) = 30$$

Solving the equation,

$$\begin{aligned} 3z + 6 &= 30 && \textit{combining like terms} \\ 3z + 6 - 6 &= 30 - 6 && \textit{subtracting 6 from both sides} \\ 3z &= 24 && \textit{combining like terms} \\ \frac{3z}{3} &= \frac{24}{3} && \textit{dividing both sides by 3} \\ z &= 8 && \textit{the first even integer} \\ z + 2 &= 10 && \textit{the second consecutive even integer} \\ z + 4 &= 12 && \textit{the third consecutive even integer} \end{aligned}$$

Check:

The three consecutive even integers are 8, 10, and 12, and their sum is $8 + 10 + 12$, or 30. This satisfies the original problem and shows that the solution is correct.

NW *Now Work Problem 17.*

Exercises for Section 9.5

Solve each of the following problems. Only an algebraic solution will be accepted.

1. Ten more than a certain number is 12. Find the number.
2. Three less than a certain number is 9. Find the number.
3. The sum of a certain number and 9 is 16. Find the number.
4. Four more than twice a certain number is 8. Find the number.
5. The sum of two consecutive integers is 21. Find the numbers.
6. Chan is 4 years older than Soo and the sum of their ages is 46. How old is each?
7. NW Angela is 9 years younger than Richard and the sum of their ages is 83. How old is each?
8. Five less than three times a number is 40. Find the number.
9. NW The sum of three times a number and 6 is 48. Find the number.
10. The sum of one-half of a number and 6 is 14. Find the number.
11. The width of a rectangle is 5 feet less than its length. If the perimeter of the rectangle is 66 feet, find the dimensions of the rectangle.
12. The length of a rectangular garden is 5 meters longer than its width. If the perimeter of the garden is 50 meters, find the dimensions of the garden.
13. NW The length of a rectangle is 6 meters longer than twice its width. If its perimeter is 228 meters, find the dimensions of the rectangle.
14. The sum of two consecutive integers is 101. Find the numbers.
15. The sum of two consecutive odd integers is 28. Find the numbers.
16. The sum of two consecutive even integers is 38. Find the numbers.
17. NW The sum of two consecutive even integers is 18. Find the numbers.
18. The sum of three consecutive odd integers is 123. Find the numbers.
19. NW Lorraine has \$7.70 in dimes and quarters in her car coin carrier for tolls. If the number of quarters is two more than twice the number of dimes, how many of each type does she have?
20. Daniel, a newspaper carrier, has \$2.90 in nickels, dimes, and quarters. If he has three more nickels than dimes and twice as many dimes as quarters, how many of each type of coin does he have?
21. In a collection of 16 coins the number of quarters is one-half the number of dimes and the number of nickels is 5 less than twice the number of dimes. How many of each type of coin (quarters, dimes, and nickels) are in the collection?
22. A rectangular garden is enclosed by 460 feet of fencing. If the length of the garden is 10 feet less than three times the width, find the dimensions of the garden.

23. **Concert Tickets** Tickets for a Rolling Stones concert sold for \$51, \$63, and \$75. The number of \$63 tickets sold was three times the number of \$51 tickets sold, and the number of \$75 tickets sold was twice that of \$51 tickets. The total number of tickets sold was 36,000, and the total amount of money collected was \$2,340,000. How many of each ticket were sold?
24. The width of a rectangle is 3 inches more than one-half of its length. If the perimeter is 60 inches, find the length and width of the rectangle.
25. The measure of one base angle of a triangle is twice that of the "top" angle, and the measure of the second base angle is 5 degrees more than the first base angle. Find the number of degrees in each angle of the triangle. (*Hint:* The sum of the angles of a triangle is 180°.)
26. One number is 20 more than another. If the greater number is increased by 4, the result is five times the smaller. Find the two numbers.
27. A vending machine contains \$118.60 worth of coins. If there are 24 more quarters than dimes and 36 more dimes than nickels, how many of each coin are in the machine?
28. The perimeter of a triangle is 17 centimeters. One side is 3 centimeters longer than the shortest side, whereas the third side is 2 centimeters shorter than twice the shortest side. Find the lengths of the three sides.
29. A square and an equilateral triangle have the same perimeter. Each side of the triangle is 12 meters. Find the length of each side of the square.
30. Seventy-seven mathematics students are separated into two groups. The first group is 4 less than twice the second group. How many students are in each group?

Writing Mathematics

In Exercises 31–33, write an English statement that corresponds to the given equation.

31. $2x - 4 = 10$
32. $\frac{x}{2} + 4 = 5$
33. $27 + 5x = 42$
34. Explain why x and $x + 1$ represent two consecutive integers.

35. If x is an even integer, explain why $x + 2$ must be the next consecutive even integer.

Challenge Exercises

36. **Long distance** The following table represents the cost of making a long distance call to a particular international location.
    a. Write an equation that represents the data in this table. Your equation should be of the form C = something in terms of m, where C = the cost of a call and m = the number of minutes of the call.
    b. Use this equation to determine the cost of an 18-minute call.

Cost (C)	Number of Minutes (m)
$ 3.45	1 or less
5.60	2
7.75	3
9.90	4
12.05	5

37. **Tour bus trips** Every Christmas Eve, bonfires are lit along the Mississippi River outside of New Orleans as part of Cajun tradition to help Papa Noel find his way. A tour bus company that sponsors trips to the bonfires uses buses that can hold a total of 48 people per bus.
    a. Write an equation that the bus company can use to determine the total number of buses needed for this event.
    b. If tickets cost $45 per person and the company sells $38,700 worth of tickets, how many buses will the company need to use?

38. **Sales commission** Dee works part time in the cosmetics department at Burdine's Department Store. Her salary is $150 per week plus a 7.5% commission of her total weekly sales and she gets paid every two weeks.
    a. Write an equation that can be used to show Dee's salary structure, using S to represent her salary and t to represent her total weekly sales.
    b. If Dee's most recent paycheck is for $815, determine the average amount of Dee's sales during the period of time reflected by her paycheck.

39. **Automobile expenses** Kallie earns $82,750 per year as a sales representative for an Internet company. In addition to her annual salary, Kallie also receives an annual reimbursement for the use of her personal vehicle for business-related travel. Her business-related automobile expenses are reimbursed at the rate of 30 cents per mile and her reimbursement is based on the percentage of miles used for business.
    a. Write an equation that can be used to model Kallie's annual income. Let S represent her total salary, t represent the total of miles she travels, and b represent the total business-related miles she travels.
    b. If Kallie traveled 38,520 miles last year, of which 14,740 miles were for personal use, how much income did Kallie receive last year? How much of this income was derived from her auto reimbursement expense?

Group/Research Activities

40. **Book publishing** Some book publishing companies require authors to submit a manuscript that contains a specific number of words. For example, a short story usually contains less than 10,000 words and some textbooks have a word count in the hundreds of thousands. Research the number of pages editors usually assign to published works such as novels. As part of your research, determine the mathematical model (i.e., equation) that is most commonly used to estimate the number of pages contained in a book and the accuracy of this model. Use this model to estimate the number of words in this book.

41. Find an example of a mathematical model (i.e., equation) in a newspaper or magazine article.

42. **Taxes** Review the current year's tax chart from the IRS and see if you can develop an equation that can be used to determine the amount of taxes a person must pay based on his or her annual salary.

Just for fun

Riddles and puzzles have also helped to develop the study of word problems. Many ancient texts or manuscripts contained various types of riddles. The following is a very old children's rhyme from England:

As I was going to St. Ives,
I met a man with seven wives,
Every wife had seven sacks,
Every sack had seven cats,
Every cat had seven kits,
Kits, cats, sacks, and wives—
How many were going to St. Ives?

9.6 LINEAR EQUATIONS IN TWO VARIABLES

Solving Linear Equations in Two Variables

Thus far in our discussion of open sentences and equations, we have dealt with sentences that contained only one unknown, or variable. But this is not the only type of open sentence. Consider the equation

$$x + 2y = 10$$

This is an equation in two unknowns, or two variables. If we replace x by 2, that is,

$$2 + 2y = 10$$

we still have an open sentence, but now we can solve the equation $2 + 2y = 10$ for y. When $x = 2$ and $y = 4$, the open sentence $x + 2y = 10$ is true. This means that $x = 2$ and $y = 4$ is a *solution* to the equation $x + 2y = 10$.

Rather than write $x = 2$ and $y = 4$, we can denote this in another manner—namely, by means of the ordered pair (2, 4). We say *ordered pair* because the order in which the numbers appear is important. By convention, the first number in an ordered pair is always a value for x, and the second number is always a value for y. We know that the ordered pair (2, 4) is a solution to the equation $x + 2y = 10$: When we replace x by 2 and y by 4, the resulting statement is true; that is, $2 + 2(4) = 10$, or $2 + 8 = 10$. Note that the ordered pair (4, 2) is not a solution to the equation $x + 2y = 10$; that is, $4 + 2(2) \neq 10$.

An ordered pair is always of the form (x, y); that is, the x value is listed first and the y value is listed second. The x value is formally called the **abscissa** and the y value is called the **ordinate,** but in our discussion we shall refer to them as the x and y values.

It should be noted that the ordered pair (2, 4) is not the only ordered pair that will make the open sentence $x + 2y = 10$ a true statement; it is not the only solution for the given equation. In fact, there are infinitely many ordered pairs in the solution set of the equation $x + 2y = 10$.

Biography: Mary Frances Winston Newson

Mary Frances Winston Newson (1869–1959) was the first American woman to earn a Ph.D. in mathematics from a German university. She received her bachelor's degree from the University of Wisconsin and did one year of graduate work at Bryn Mawr. Dr. Newson was one of the first five people accepting fellowships to study mathematics at the University of Chicago when it opened in 1892. A year later, with encouragement from two faculty members in Chicago, she left America for Germany without being certain of financial aid, and in spite of the fact that German universities did not admit women. After some time she did receive a fellowship to study mathematics at Göttingen, Germany. She graduated magna cum laude and earned her doctorate under Felix Klein.

After three years she returned to America and taught high school for one year. She was then appointed chair of the mathematics department at Kansas State Agricultural College. She resigned her position in 1900 to marry and raise three children. Her husband died unexpectedly, and she returned to teaching at Washington College in Topeka, Kansas. She later taught at Eureka College in Illinois and retired after she reached the age of 70.

Any equation of the form

$$Ax + By = C$$

where A, B, and C are real numbers, is called a **linear equation.** Thus $x + 2y = 10$ is a linear equation. It is called a linear equation because when we graph such an equation in the Cartesian plane, we get a straight line.

We mentioned before that there are infinitely many ordered pairs in the solution set of the equation $x + 2y = 10$. We already have the ordered pair $(2, 4)$; now let's obtain some more ordered pairs that are in the solution set. One way to do this is to let x be any value we choose and then solve for y. Suppose $x = 0$. Then we have

$$\begin{aligned} x + 2y &= 10 \\ \text{let } x = 0\text{:} \quad 0 + 2y &= 10 \\ \frac{2y}{2} &= \frac{10}{2} && \textit{dividing both sides by 2} \\ y &= 5 \end{aligned}$$

Hence, when $x = 0$ and $y = 5$, we have another solution to the equation $x + 2y = 10$. We can, therefore, say that the ordered pair $(0, 5)$ is a solution.

To find another solution for the equation $x + 2y = 10$, let $x = 4$. Then we have

$$\begin{aligned} x + 2y &= 10 \\ \text{let } x = 4\text{:} \quad 4 + 2y &= 10 \\ 4 + 2y - 4 &= 10 - 4 && \textit{subtracting 4 from both sides} \\ 2y &= 6 && \textit{combining like terms} \\ \frac{2y}{2} &= \frac{6}{2} && \textit{dividing both sides by 2} \\ y &= 3 \end{aligned}$$

Hence, $(4, 3)$ is a solution.

Thus far, we have three ordered pairs that satisfy the linear equation $x + 2y = 10$—namely, $(2, 4)$, $(0, 5)$, and $(4, 3)$.

In order to find the solutions for any linear equation, select a value for x and substitute it for x in the equation; then solve the resulting equation for y. We can also find a solution for a linear equation by selecting a value for y, substituting it for y in the equation, and then solving the resulting equation for x.

EXAMPLE 1 ***Solving a Linear Equation*** Find three solutions for $3x + 4y = 15$.

Solution

a. We let $x = 1$. Then we have

$$\begin{aligned} 3x + 4y &= 15 \\ \text{let } x = 1\text{:} \quad 3(1) + 4y &= 15 \\ 3 + 4y &= 15 \\ 3 + 4y - 3 &= 15 - 3 && \textit{subtracting 3 from both sides} \\ 4y &= 12 && \textit{combining like terms} \\ \frac{4y}{4} &= \frac{12}{4} && \textit{dividing both sides by 4} \\ y &= 3 \end{aligned}$$

Therefore, $(1, 3)$ is a solution.

b. Let $x = 5$; then

$$
\begin{aligned}
3(5) + 4y &= 15 \\
15 + 4y &= 15 \\
15 + 4y - 15 &= 15 - 15 && \textit{subtracting 15 from both sides} \\
4y &= 0 && \textit{combining like terms} \\
\frac{4y}{4} &= \frac{0}{4} && \textit{dividing both sides by 4} \\
y &= 0
\end{aligned}
$$

Therefore, $(5, 0)$ is a solution.

c. Let $x = 0$; then

$$
\begin{aligned}
3(0) + 4y &= 15 \\
0 + 4y &= 15 \\
4y &= 15 \\
\frac{4y}{4} &= \frac{15}{4} && \textit{dividing both sides by 4} \\
y &= \frac{15}{4}
\end{aligned}
$$

Therefore, $(0, \frac{15}{4})$ is a solution.

NW ***Now Work Problem 5.***

The third solution obtained in Example 1, $(0, \frac{15}{4})$, contains a fraction. However, as we shall see in the next section, it is advantageous to obtain integral solutions—that is, solutions that contain only integers. We can avoid obtaining a fractional solution if we are careful about the values that we choose for x. For instance, in Example 1, if $x = -3$, then $y = 6$, and $(-3, 6)$ is a solution for $3x + 4y = 15$.

EXAMPLE 2 ***Solving a Linear Equation*** Find three solutions for $2x - 3y = 12$.

Solution

a. Selecting a value for x, we let $x = 0$. Hence, we have

$$
\begin{aligned}
2x - 3y &= 12 \\
\text{let } x = 0\text{:} \quad 2(0) - 3y &= 12 \\
0 - 3y &= 12 \\
-3y &= 12 \\
\frac{-3y}{-3} &= \frac{12}{-3} && \textit{dividing both sides by } -3 \\
y &= -4
\end{aligned}
$$

Therefore, $(0, -4)$ is a solution.

b. Let $x = 3$; then

$$
\begin{aligned}
2(3) - 3y &= 12 \\
6 - 3y &= 12
\end{aligned}
$$

$$6 - 3y - 6 = 12 - 6 \quad \textit{subtracting 6 from both sides}$$
$$-3y = 6 \quad \textit{combining like terms}$$
$$\frac{-3y}{-3} = \frac{6}{-3} \quad \textit{dividing both sides by } -3$$
$$y = -2$$

Therefore, $(3, -2)$ is a solution.

c. Let us try a different approach, selecting a value for y. Let $y = 0$; then

$$2x - 3(0) = 12$$
$$2x - 0 = 12$$
$$2x = 12$$
$$\frac{2x}{2} = \frac{12}{2} \quad \textit{dividing both sides by 2}$$
$$x = 6$$

Therefore, $(6, 0)$ is a solution. Note that even though we selected a value for y first, we list the x value first and the y value second when we write the solution as an ordered pair.

NW *Now Work Problem 9.*

Instead of listing the solutions to the equation $2x - 3y = 12$ in Example 2 as the ordered pairs $(0, -4)$, $(3, -2)$, and $(6, 0)$, we can list the solutions as a **table of values**, as follows.

$$2x - 3y = 12$$

x	y
0	−4
3	−2
6	0

Mathematical Modeling (Optional)

Linear equations in two variables are commonly used to model data. To illustrate this concept, consider Table 9.5, which contains data collected from the records of a local police department over a one-week period. As shown by the table headings, the data represent the relationship between the number of motor vehicle citations issued and drivers' ages. For example, 25 citations were issued to drivers who were 16 years old, 30 citations were issued to drivers who were 18 years old, and so forth.

Because we have two variables, we can use a linear equation in two variables to represent the relationship between the variables. In our earlier work in this section, we were given an equation and asked to find solutions for the equation. In the current illustration, however, notice that we have the reverse situation. That is, the data given in Table 9.5 may be thought of as solutions to an equation and it is our job to find the equation. The procedure used to do this is called *linear regression*, and the equation that results from this process is referred to as a *mathematical model.*

Although we do not discuss the process of linear regression in this book, the concept of mathematical modeling is important because models enable us to predict the result of future events. For example, mathematical models are used by weather forecasters to predict the potential path of a hurricane. Models are also used by medical researchers to predict the likelihood of a person developing a particular illness. In each instance, specific data are collected and then analyzed using regression methods to derive a model (i.e., an equation). In the given example, the mathematical model derived from the data in Table 9.5 is $y = 2x - 7$.

TABLE 9.5

Y = Number of Motor Vehicle Citations	X = Age of Individual (Years)
25	16
30	18
34	19
39	21
43	25
48	28
55	32
60	33
60	35
64	36
66	38
69	40
77	43
78	45
90	50
120	55
100	58
100	60
105	62
110	65
110	70
133	74
136	78
200	80
150	81

Because a mathematical model is derived from a set of data, it is important to note that the model will not be perfect. What we mean by this is that the data will not necessarily "fit" the derived equation exactly; the model will only represent an approximation of the data. To illustrate this, consider the first two ages given in Table 9.5. If we substitute 16 for x into our derived model, we get $y = 2(16) - 7 = 32 - 7 = 25$, which matches the corresponding y value. However, when we substitute 21 into the model for x, the corresponding y value is 35, which is "close" to the actual value of 39 but is not exactly equal to it. The accuracy of the model is dependent on many factors. Nevertheless, if we assume that a model accurately represents the data, then we can use it to predict values on one variable from the knowledge of values on the other variable(s). For example, using our model, we predict that 53 citations will be issued to 30-year old drivers in any given one-week period; that is: $y = 2(30) - 7 = 60 - 7 = 53$.

TABLE 9.6

Y = Year	X = Percentage of All U.S. Households
1980	5.7%
1982	11.0%
1984	11.9%
1986	20.0%
1988	24.4%
1990	25.0%
1992	27.0%
1994	30.7%
1996	37.2%
1998	44.0%
2000	49.0%
2002	49.6%

Source: Investment Company Institute.

EXAMPLE 3 ***Mutual Funds*** Table 9.6 shows the percentage of U.S. households that owned mutual funds from 1980 to 2002, in two-year increments. The mathematical model derived from these data is $Y = \left(\frac{1}{2}\right)p + 1977$, where Y is the year and p is the percentage of households. Assuming this model is accurate, in what year will the percentage of U.S. households that own mutual funds be

a. 60%

b. 90%

TABLE 9.7

Cost ($)	Speed (MPH)
52	39
61	42
70	45
79	48
85	50
115	60
160	75
175	80

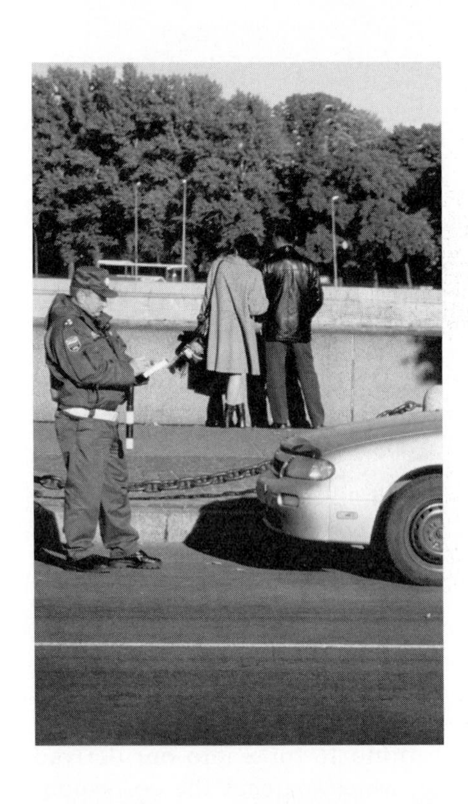

Solution

a. If $p = 60\%$, $Y = \left(\frac{1}{2}\right)p + 1977 = \left(\frac{1}{2}\right)(60) + 1977 = 30 + 1977 = 2007$. Thus, we predict that in 2007, 60% of all U.S. households will own mutual funds.

b. If $p = 90\%$, $Y = \left(\frac{1}{2}\right)p + 1977 = \left(\frac{1}{2}\right)(90) + 1977 = 45 + 1977 = 2022$. Thus, we predict that in 2022, 90% of all U.S. households will own mutual funds.

EXAMPLE 4 ***Speeding Fines*** Table 9.7 contains data collected from a local beachside police department and shows the cost of a speeding ticket based on a car's speed. The mathematical model derived from these data is $C = 3s - 65$, where C = the cost of the ticket in dollars and s = the speed of a car in miles per hour (mph).

a. Assuming this model is accurate, how much will a speeding ticket cost if someone is driving 55 mph?

b. Suppose you received a fine for \$250. How fast were you going?

Solution

a. If $s = 55$, $C = 3s - 65 = 3(55) - 65 = 165 - 65 = 100$. Thus, if you are caught going 55 mph in this beachside community, you will be given a \$100 speeding fine.

b. To determine the speed of your car based on a \$250 speeding ticket, we substitute 250 for C and solve for s.

$$C = 3s - 65$$
$$250 = 3s - 65$$
$$250 + 65 = 3s - 65 + 65 \quad \textit{adding 65 to both sides}$$
$$315 = 3s \quad \textit{combining like terms}$$
$$\frac{315}{3} = s \quad \textit{dividing both sides by 3}$$
$$105 = s$$

Therefore, your car must have been traveling 105 mph.

Exercises For Section 9.6

Find three solutions for each of the following equations:

1. $x + y = 5$
2. $x + y = 6$
3. $x - y = 3$
4. $x - y = 5$
5. NW $2x + y = -6$
6. $3x - y = 6$
7. $3x - y = -10$
8. $3x + 4y = -12$
9. NW $3x - 2y = 8$
10. $5x + 3y = 15$
11. $3x + 5y = -15$
12. $x + y = 0$
13. $x - y = 0$
14. $x - 2y = 0$
15. $x + 2y = 0$
16. $x + 3y = 13$
17. $2x - 3y = -11$
18. $3x - 5y = -9$
19. $-2x - 3y = 7$
20. $-3x - 2y = -3$
21. $-5x - 4y = -12$

In Exercises 22–25, determine if the ordered pair $(-3, -2)$ is a solution to the following equations. Answer yes or no.

22. $x - y + 1 = 0$
23. $-x + 2y - 1 = 0$
24. $-3x - 2y + 13 = 0$
25. $2x + 3y = 0$

In Exercises 26–29, use the equation $p = 22s + 15b$, which represents the purchase price, p (in dollars), for buying s shirts and b belts to answer each question. Assume that the ordered pair (s, b) is used to represent a specific purchase.

26. What is the purchase price of one shirt and one belt? Express this purchase as an ordered pair.
27. Does the ordered pair (3, 1) correctly represent a purchase of \$81? Why?
28. Suppose a sales clerk mistakenly wrote (1, 3) to represent the purchase of Exercise 27. What is the effect of doing this?

29. What set(s) of numbers (e.g., natural, whole, integer, rational, real) is the solution to this equation appropriate? Why?

Writing Mathematics

30. Explain the concept of ordered pairs relative to linear equations in two variables.
31. When discussing the relationship between two variables, what advantage does an equation provide?
32. When discussing the relationship between two variables, what advantage does a table of values provide?

Challenge Exercises

33. **Postal rates** The following table provides the 2003 U.S. postal rates for mailing a first class letter based on weight. The relationship between the two variables weight, w (in ounces), and cost, C (in cents), is an *arithmetic progression*. To find the equation that models this relationship, we use the formula $C = c_1 + d(w - 1)$, where c_1 is equal to the initial cost based on a weight of 1 ounce or less, and d is the common difference among the various costs.
    a. Find the equation that represents these data.
    b. Use this equation to determine the cost of mailing a first class letter that weighs between 8 and 9 ounces; 10 and 11 ounces.

Weight (W) in Ounces	Cost (C) in Cents
1	37
2	60
3	83
4	106
5	129
6	152

34. Write a linear equation in two variables that is satisfied by the ordered pair $(3, 4)$.
35. Write a linear equation in two variables that is satisfied by the ordered pairs $(0, 2)$ and $(2, 0)$.

Group/Research Activities

36. The concept of an ordered pair has various applications outside of mathematics. For example, life insurance companies might use ordered pairs to represent life expectancy based on a person's current age. Thus, in this context, Male: $(45, 75)$ implies that the current life expectancy for a 45-year-old male is 75 years. Discover other contexts in which ordered pairs may be used and provide examples of such use.
37. Find examples of linear equations in two variables by reviewing articles in various newspapers and graphs. Note that many such articles will provide this information in a table instead of an actual equation. If this is indeed the case, try representing the data as a linear equation.

Real-World Data

38. Brad is interested in studying the relationship between students' mathematics ability and their music ability. To study this relationship, Brad selected 11 high school students and administered to them the *Jimi Hendrix Musical Aptitude Test* (JHMAT) and the *Euler Mathematical Aptitude Test* (EMAT). The scores for each test are given in the following table. The mathematical model derived from these data is $y = x + 3$, where y is a person's musical aptitude (measured by the JHMAT) and x is a person's math aptitude (measured by the EMAT). Use this model to predict someone's JHMAT score given an EMAT score of 16.

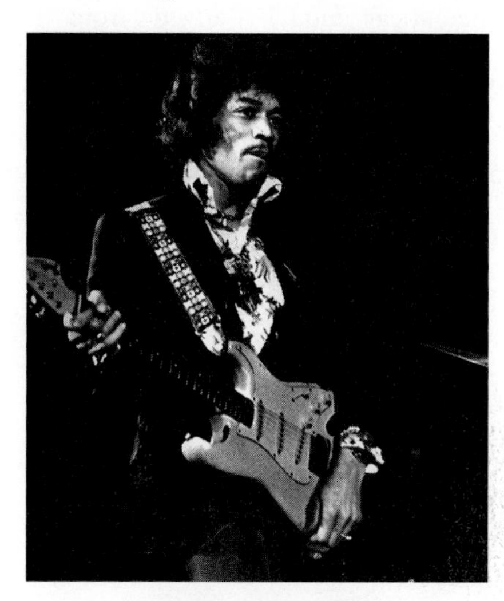

JHMAT	EMAT
5	6
7	4
12	9
5	2
8	7
9	4
20	11
5	3
9	5
5	7
8	1

39. Art Guttman examined the relationship between birth order and IQ using data acquired from 10 adult employees who work at a local grocery store. The results are given in the following table. The mathematical model derived from these data is $Q = 3.5b + 100$, where Q is a person's IQ score and b is a person's birth order. Use this model to predict the IQ of a person whose birth order is six.

IQ	Birth Order
85	2
102	4
114	1
103	3
123	4
106	2
100	1
126	3
112	2
115	3

40. Cecilia is interested in examining the relationship between students' study time and their achievement in her statistics class. To study this relationship, Cecilia selected 10 students from her introductory statistics class and compared these students' final exam grades to their self-reported estimates of how long they studied for the final. The results are given in the following table. The mathematical model derived from these data is $E = -\left(\frac{1}{4}\right)t + 102$, where E is a person's final exam score and t is a person's study time (in minutes).

a. Use this model to predict what a person's final exam score will be if he or she studies for 32 minutes.

b. Based on this model, how long would a person have to study to score 100 on the final exam?

Final Exam Score	Study Time (minutes)
96	120
88	50
67	110
76	115
84	60
52	150
91	90
89	40
77	90
68	120

41. Lesley examined the relationship between students' attitudes toward the environment and the number of specific environmentally friendly acts they perform (e.g., recycle paper, plastic, and glass; install low-flow shower heads; not use pesticides on the lawn; etc.). Attitude scores were acquired using the *Garner Eco-Attitude Scale* (the higher the score the more positive the attitude), which was administered to a group of students in her Life Science class. The data are shown in the following table. The mathematical model derived from these data is $A = -2.5e + 119$, where A is a person's attitude and e is the number of environmentally friendly acts performed.

a. Use this model to predict a person's attitude toward the environment given 10 environmentally friendly acts.

b. Based on this model, how many eco-friendly acts do you predict will be performed by someone who has an attitude score of 114?

Attitude Score	# of Eco-Friendly Acts
98	5
123	7
112	1
115	1
111	4
80	8
110	5
118	4
106	7
95	6

Just for fun

In the "old, old days" students had to know the following kinds of measure: the tierce, hogshead, pipe, butt, and tun. These are similar kinds of measure. What do they measure?

9.7 GRAPHING EQUATIONS

The Cartesian Plane

Now that we are able to find solution sets for linear equations, our next goal is to illustrate these solution sets by means of a graph. We can do this on a grid called the **Cartesian plane**, named after the French mathematician-philosopher René Descartes.

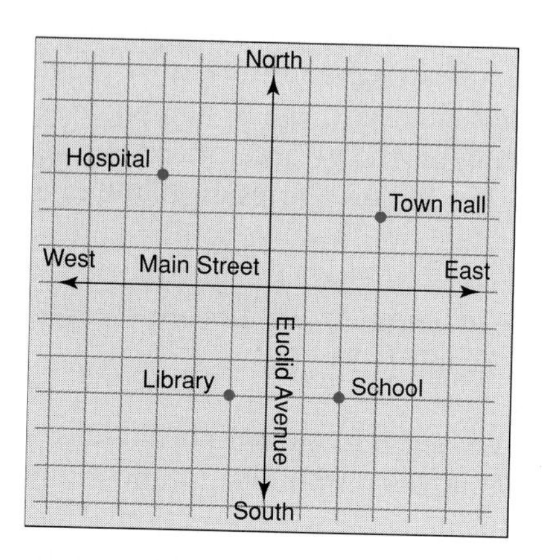

Figure 9.11

In order to develop an understanding of the Cartesian plane and how to graph ordered pairs on it, consider the map of Anytown in Figure 9.11. Assume that we are at the intersection of Main Street (east–west) and Euclid Avenue (north–south). In order to get to the town hall, we must go three blocks east and two blocks north; to get to the school, we must go two blocks east and three blocks south. Similarly, to get to the hospital, we must go three blocks west and three blocks north, and to get to the library we must go one block west and three blocks south.

We can use this same idea to graph, or **plot**, ordered pairs. In Chapter 8, we used a horizontal number line like the line representing the east–west Main Street in Figure 9.11. Recall that zero was in the middle of the line, the positive numbers were to the right (east) of zero, and the negative numbers were to the left (west) of zero. We shall use such a horizontal line and call it the x-number line, or the x-axis. Now we construct another number line perpendicular to the x-axis (like the north–south street Euclid Avenue in Figure 9.11) and passing through zero. We mark off the numbers on this new line as we did for the x-number line. The numbers above the x-axis (north) are positive, and those below the x-axis (south) are negative. This vertical number line is called the y-number line, or y-axis. The Cartesian plane is based on the x- and y-axes. They intersect at the common point where $x = 0$ and $y = 0$. This particular point is represented by the ordered pair $(0, 0)$ and is called the **origin** (see Figure 9.12).

Instead of giving directions to the town hall in Figure 9.11 by saying "go three blocks east and then two blocks north," we can now say "go 3 units in the positive x-direction and then 2 units in the positive y-direction." We can shorten this even more by writing the ordered pair $(3, 2)$. If we are asked to plot the ordered pair $(3, 2)$ on the Cartesian plane, we start at the origin and move 3 units in the positive x-direction and then 2 units in the positive y-direction (see Figure 9.13).

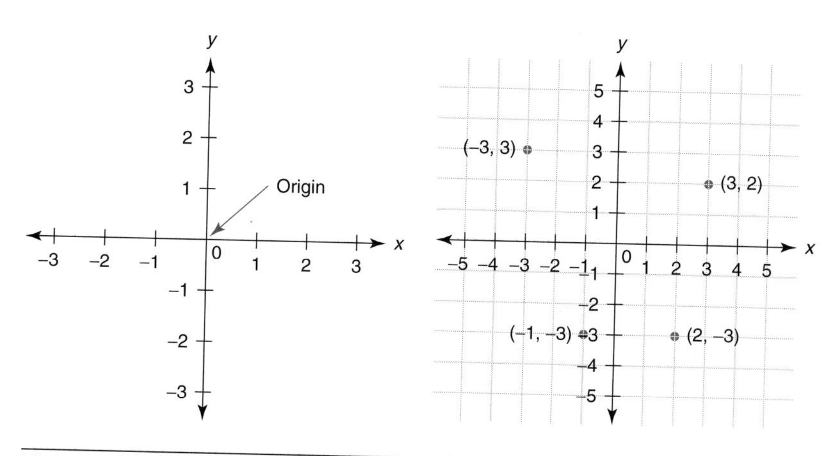

Figure 9.12

Figure 9.13

Using this idea, we can also graph the ordered pairs $(2, -3)$, $(-3, 3)$, and $(-1, -3)$ on the Cartesian plane shown in Figure 9.13. In order to graph an ordered pair, we always start at the origin $(0, 0)$ and proceed from there. The point represented by $(2, -3)$ is obtained by moving 2 units in a positive x-direction and then 3 units in a negative y-direction. The point represented by $(-3, 3)$ is obtained by starting at the origin and moving 3 units in a negative x-direction and then 3 units in a positive y-direction. To plot the point represented by $(-1, -3)$, we start at the origin and move 1 unit in a negative x-direction and then 3 units in a negative y-direction.

In order to obtain any point on the plane, we start at the origin and then move in a positive or negative x-direction (east–west, right–left); next we move in a positive or negative y-direction (north–south, up–down). The first number in an ordered pair is the x-value, and the second number is the y-value. These two numbers are called the **coordinates** of the point; hence, the Cartesian plane is also called the **Cartesian coordinate system**.

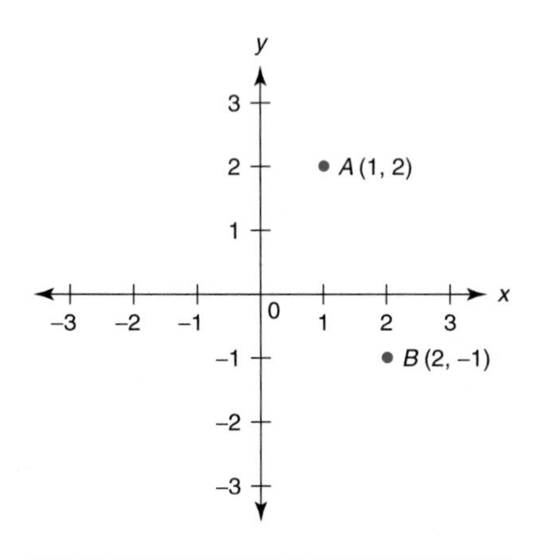

Figure 9.14

EXAMPLE 1 ***Locating Points in the Cartesian Plane*** Locate the points corresponding to the ordered pairs $A(1, 2)$ and $B(2, -1)$.

Solution The given ordered pairs are labeled A and B to aid us in our discussion. Point A, corresponding to the ordered pair $(1, 2)$, is found by moving 1 unit in a positive x-direction from the origin and then 2 units in a positive y-direction. Point B, corresponding to $(2, -1)$, is found by moving 2 units in a positive x-direction from the origin and then 1 unit in a negative y-direction. Points A and B are shown in Figure 9.14.

EXAMPLE 2 ***Locating Points in the Cartesian Plane*** Locate the points corresponding to the ordered pairs $C(-3, 2)$ and $D(-2, -3)$.

Solution Point C, $(-3, 2)$, is found by moving 3 units in a negative x-direction from the origin and then 2 units in a positive y-direction. Point D, $(-2, -3)$, is

Math Connections

Latitude and Longitude

One very important application of the Cartesian coordinate system is the mapping of the earth into coordinates called *latitude* and *longitude*. Latitude represents angular distance north or south from the earth's equator measured through 90 , and longitude represents distance east or west on the earth's surface measured as an arc of the equator up to 180 . Thus, longitude corresponds to the x-coordinate because movement is left or right, and latitude corresponds to the y-coordinate because movement is up or down. When reporting a specific location on earth, the coordinates are always given in the order of latitude followed by longitude. For example, one of the authors lives at approximately 28 north latitude, 81 longitude. Using the following map, we can see that this location is along the east central Florida coast.

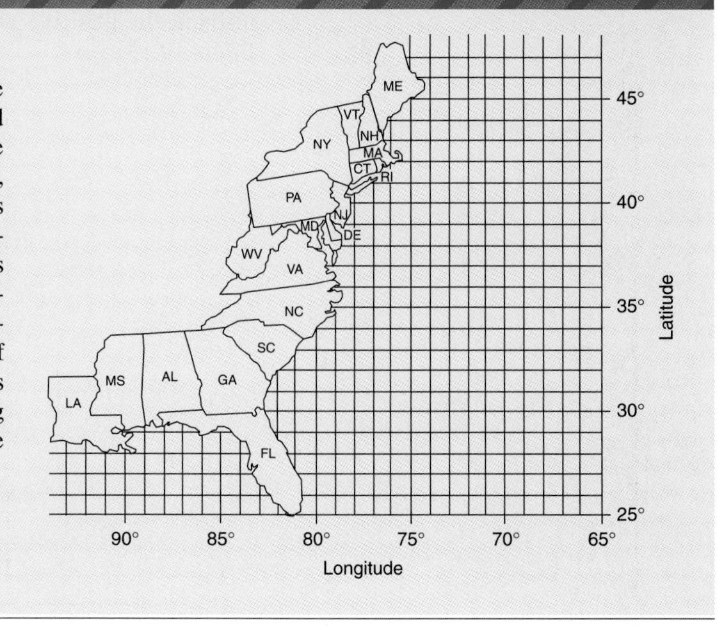

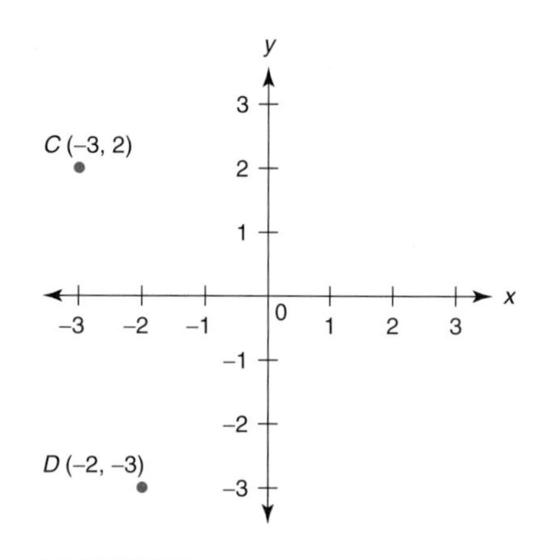

Figure 9.15

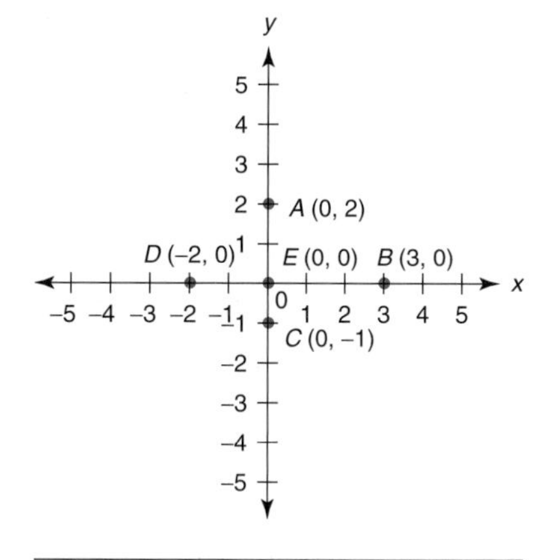

Figure 9.16

found by moving 2 units in a negative x-direction from the origin and then 3 units in a negative y-direction (see Figure 9.15).

EXAMPLE 3 ***Locating Points in the Cartesian Plane*** Locate the points corresponding to the ordered pairs

$$A(0, 2),\quad B(3, 0),\quad C(0, -1),\quad D(-2, 0),\quad \text{and}\quad E(0, 0)$$

Solution Each of the given ordered pairs contains 0 as one of its values. Point $A(0, 2)$ is found by moving 0 units in the x-direction from the origin and then 2 units in a positive y-direction. Point $B(3, 0)$ is found by moving 3 units in a positive x-direction from the origin and 0 units in the y-direction. Point $C(0, -1)$ is found by moving 0 units in the x-direction from the origin and then 1 unit in a negative y-direction. Point $D(-2, 0)$ is found by moving 2 units in the negative x-direction from the origin and then 0 units in the y-direction. Point $E(0, 0)$ is the origin.

Points A, B, C, D, and E are shown in Figure 9.16.

NW ***Now Work Problems 1a, 1b, 1c, and 1d.***

Graphing Linear Equations

Because each ordered pair represents a point on the plane, let us now locate some ordered pairs that are solutions to the linear equations that we discussed in the previous section.

Consider the equation

$$x + 2y = 10$$

Three ordered pairs that satisfy this equation are (4, 3), (2, 4), and (0, 5). We can plot all of these ordered pairs in the Cartesian plane, as shown in Figure 9.17. Note that these points tend to form a pattern; that is, they appear to lie on the same path. If we connect these points, we see that all of these points do in fact lie on the same straight line. This line, shown in Figure 9.18, is a graph of the equation $x + 2y = 10$.

The fact that the graph of this type of equation—that is, an equation of the form $Ax + By = C$, where A, B, and C are real numbers—is a straight line is the reason that such equations are called **linear equations**.

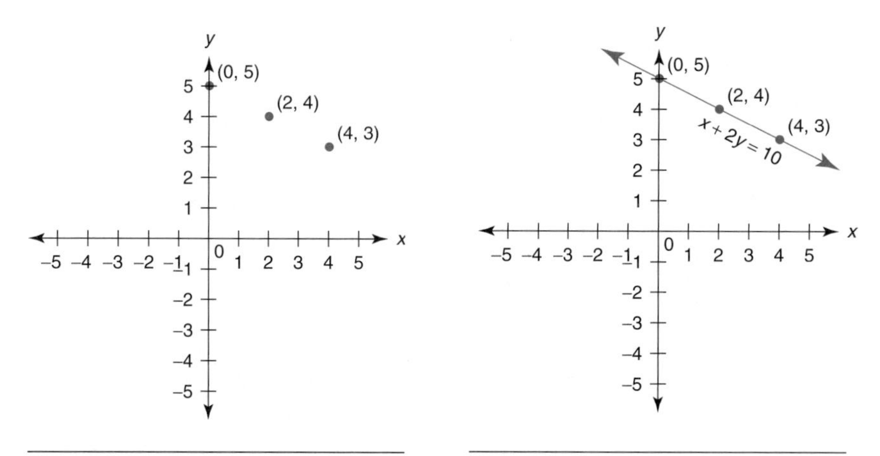

Figure 9.17

Figure 9.18

If the graph of a linear equation is always a straight line, then why did we use three points to determine the graph of $x + 2y = 10$? It is true that two points determine a line; that is, through two given points one and only one straight line can be drawn. We use three points because the third point is a check. We locate the third point to check that all three points lie on the same straight line. If one of the points is not on the line, then we must go back and check for an error in determining the solutions to the linear equation.

EXAMPLE 4 ***Graphing a Linear Equation*** Graph $x + y = 5$.

Solution In order to graph $x + y = 5$, we must first find three ordered pairs that are solutions to the equation.

a. Selecting a value for x, we let $x = 1$. Hence, we have

$$\begin{aligned} & x + y = 5 \\ \text{let } x = 1: \quad & 1 + y = 5 \\ & y = 4 \end{aligned}$$

and $(1, 4)$ is a solution.

b. Let $x = 4$; then

$$\begin{aligned} 4 + y &= 5 \\ y &= 1 \end{aligned}$$

and $(4, 1)$ is a solution.

c. Let $x = 2$; then

$$\begin{aligned} 2 + y &= 5 \\ y &= 3 \end{aligned}$$

and $(2, 3)$ is a solution.

Now that we have three different ordered pairs, $(1, 4)$, $(4, 1)$, and $(2, 3)$, that are solutions to the equation $x + y = 5$, we plot these points and draw the line that contains them, as shown in Figure 9.19.

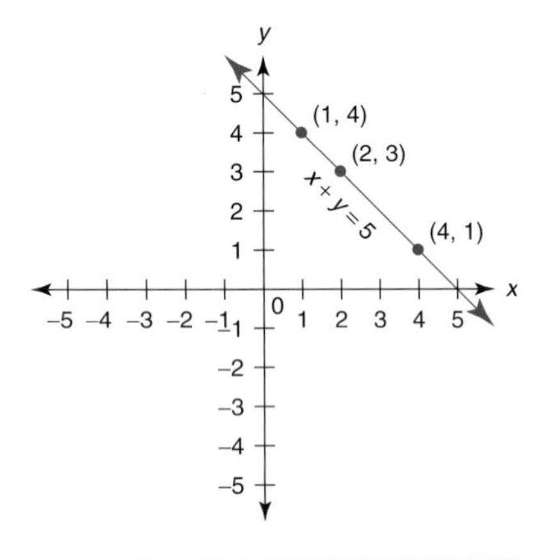

Figure 9.19

NW ***Now Work Problem 5.***

When we graph a linear equation, it is often convenient to find the points where the graph crosses each axis. Note that $y = 0$ at the point where a line crosses the x-axis, and $x = 0$ at the point where a line crosses the y-axis. The point at which a line crosses the x-axis is called the ***x*-intercept**, and similarly, the point at which it crosses the y-axis is called the ***y*-intercept**. In order to find the intercepts of the equation in Example 4, $x + y = 5$, we proceed as follows:

$$\begin{aligned} & x + y = 5 \\ \text{let } x = 0: \quad & 0 + y = 5 \\ & y = 5, \quad (0, 5) \text{ is a solution} \end{aligned}$$

The y-intercept is $(0, 5)$. Now let $y = 0$; then

$$\begin{aligned} x + 0 &= 5 \\ x &= 5 \quad (5, 0) \text{ is a solution} \end{aligned}$$

The x-intercept is $(5, 0)$.

If we want to graph $x + y = 5$ by using the intercepts, we should find a third point as a check:

$$\begin{aligned} \text{let } x = 2: \quad & 2 + y = 5 \\ & y = 3 \quad (2, 3) \text{ is a solution} \end{aligned}$$

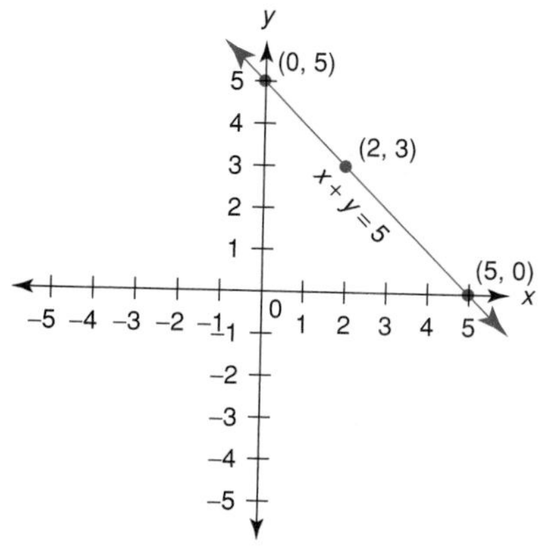

Figure 9.20

Using this point and the intercepts, we graph $x + y = 5$, as in Figure 9.20. Note that we get the same graph as in Figure 9.19, although we used different ordered pairs.

Remember that there is an infinite number of solutions to a linear equation and that each solution (ordered pair) is a point on the line. All the solutions of the equation are points on the line, and all the points on the line are solutions of the equation. In graphing an equation it is sometimes convenient to find the intercepts because they contain zeros, which simplifies computation.

But if letting x or y be zero produces an intercept with a fractional value, then it is better to find solutions to the equation that contain only integers. For example, consider the equation $3x + 4y = 15$. If $x = 0$, then $y = \frac{15}{4}$, and $(0, \frac{15}{4})$ is the y-intercept. It will be more difficult to locate this point accurately on the Cartesian plane than a pair of integers, such as $(-3, 6)$. Therefore it is better to use $(-3, 6)$ than $(0, \frac{15}{4})$ as one of the ordered pairs for graphing $3x + 4y = 15$.

EXAMPLE 5 ***Graphing a Linear Equation Using Intercepts*** Graph $2x - y = 4$.

Solution First we shall find the intercepts. Therefore, the value we select for x is 0:

$$2x - y = 4$$
$$\text{let } x = 0: \quad 2(0) - y = 4$$
$$0 - y = 4$$
$$-y = 4$$
$$-1(-y) = -1(4) \quad \textit{multiplying both sides by } -1 \textit{ to make the y-term positive}$$
$$y = -4$$

The ordered pair $(0, -4)$ is a solution; $(0, -4)$ is also the y-intercept. Next, let $y = 0$; then

$$2x - 0 = 4$$
$$2x = 4$$
$$x = 2$$

The ordered pair $(2, 0)$ is a solution; $(2, 0)$ is also the x-intercept.

Selecting a third point as a check, we let $x = 3$; then

$$2(3) - y = 4$$
$$6 - y = 4$$
$$-y = -2$$
$$-1(-y) = -1(-2) \quad \textit{multiplying both sides by } -1 \textit{ to make the y-term positive}$$
$$y = 2$$

The pair $(3, 2)$ is a solution.

Using these points, we graph $2x - y = 4$ in Figure 9.21.

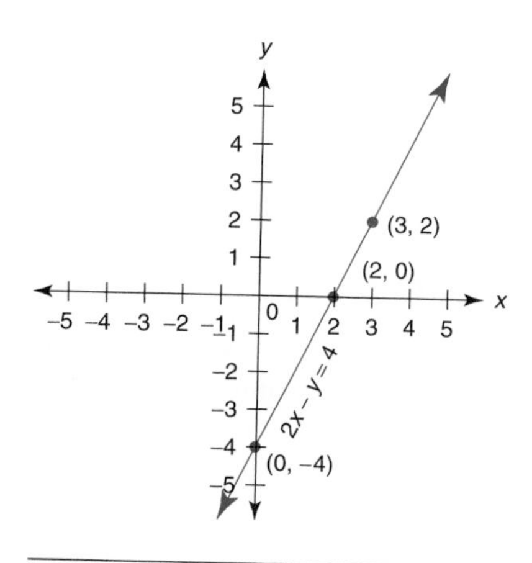

Figure 9.21

EXAMPLE 6 ***Graphing a Linear Equation Using Intercepts*** Graph $2x + 3y = 12$.

Solution We find the intercepts first:

$$2x + 3y = 12$$
$$\text{let } x = 0: \quad 2(0) + 3y = 12$$
$$0 + 3y = 12$$
$$3y = 12$$
$$y = 4$$

The y-intercept is $(0, 4)$.

$$\text{let } y = 0: \quad 2x + 3(0) = 12$$
$$2x + 0 = 12$$
$$2x = 12$$
$$x = 6$$

The x-intercept is $(6, 0)$.

We then select a third point:

$$\text{let } x = 3: \quad 2(3) + 3y = 12$$
$$6 + 3y = 12$$
$$3y = 6$$
$$y = 2$$

The point $(3, 2)$ is a solution.

Using these points, we graph $2x + 3y = 12$ in Figure 9.22. Note that we can also list the solutions to the equation $2x + 3y = 12$ in a table of values such as the one shown next to Figure 9.22.

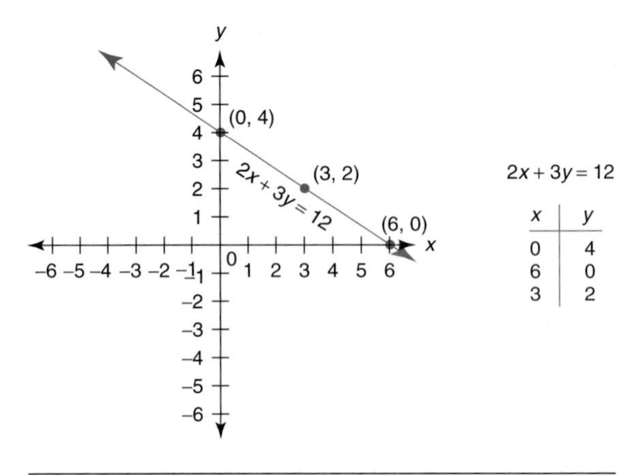

2x + 3y = 12

x	y
0	4
6	0
3	2

Figure 9.22

NW ***Now Work Problem 13.***

Consistent, Inconsistent, and Dependent Systems

Sometimes it is useful to examine more than one graph on the same set of axes. For example, Figure 9.23 shows a comparison between customers who switched

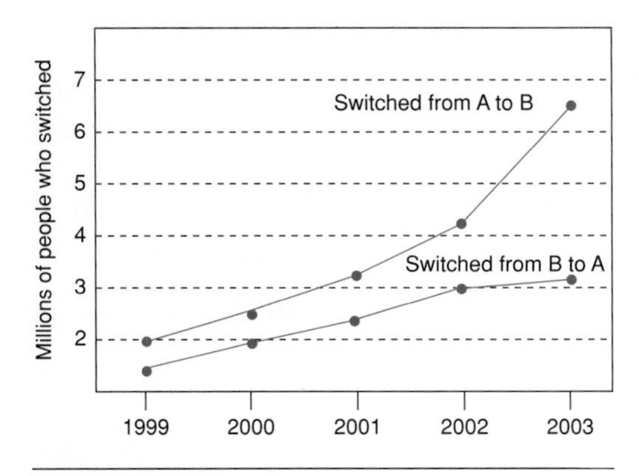

Figure 9.23

Customers who switched from one long-distance carrier to another.

from long distance service A to long distance service B, and vice versa. Although we could have used two separate graphs to display this information, note how much easier it is to compare the data when they are graphed on the same axes. Applying this concept to our discussion on graphing linear equations, whenever we have two (or more) equations that are to be examined together, we refer to them as a **system of linear equations**, or a **system of simultaneous equations**.

To solve a system of equations by graphing, we graph each equation on the same set of axes. Next, we read the coordinates of the point of intersection from the graph. If the graphs of the two equations intersect at a point, the equations are **consistent**, and the coordinates of this point represent the solution for the given system. The coordinates of the point of intersection must satisfy both equations simultaneously.

EXAMPLE 7 ***Graphing a Consistent System*** Graph $-3x + y = -9$ and $4x + y = 5$ on the same set of axes.

Solution We determine a table of values for each equation as we did in previous examples. Then we graph each equation (see Figure 9.24).

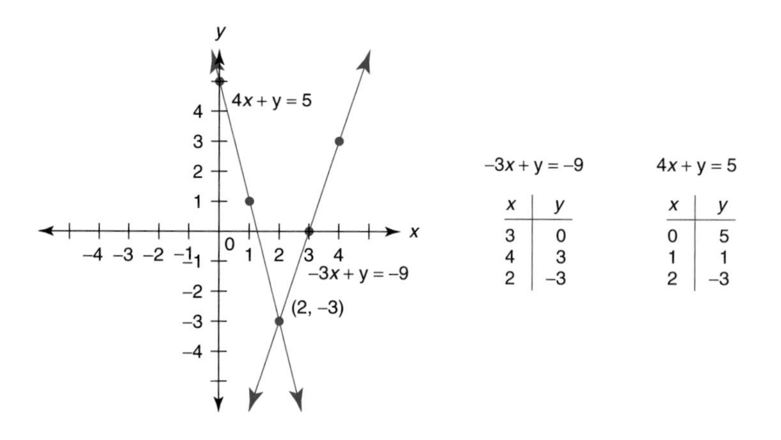

$-3x + y = -9$

x	y
3	0
4	3
2	−3

$4x + y = 5$

x	y
0	5
1	1
2	−3

Figure 9.24

NW ***Now Work Problem 23.***

Note that in Example 7, the graphs of the two lines intersect in a common point, $(2, -3)$. This point is common to both lines and the coordinates

$$x = 2, \quad y = -3$$

satisfy both equations, $-3x + y = -9$ and $4x + y = 5$. Therefore, we can say that $(2, -3)$ is a solution to the system of equations:

$$-3x + y = -9 \quad \text{and} \quad 4x + y = 5$$

It should be noted that two lines may have no points in common (if they are parallel) and therefore never intersect. When this occurs, we say that the given system of equations is **inconsistent**.

For example, if we graph $x + y = 5$ and $x + y = 2$ on the same set of axes, the graphs are parallel lines and therefore have no common point of intersection. The equations are said to be **inconsistent**. Another possibility that can exist when determining the solution for a system of linear equations is that the two graphs may be the same line. When this occurs, any solution of one equation is also a solution of the other. In this case, the equations are called **dependent**.

Exercises for Section 9.7

1. Locate the points corresponding to the given ordered pairs on a Cartesian plane.

NW **a.** $(-3, 2)$ NW **b.** $(-2, -3)$
NW **c.** $(4, 0)$ NW **d.** $(3, -4)$
e. $(3, 6)$ **f.** $(-6, 1)$
g. $(-5, 0)$ **h.** $(0, 7)$
i. $(0, -3)$ **j.** $(-6, -5)$
k. $(0, 0)$

2. Use the following graph to name the coordinates of each point.

a. A
b. B
c. C
d. D
e. E
f. F
g. G
h. H
i. I
j. J

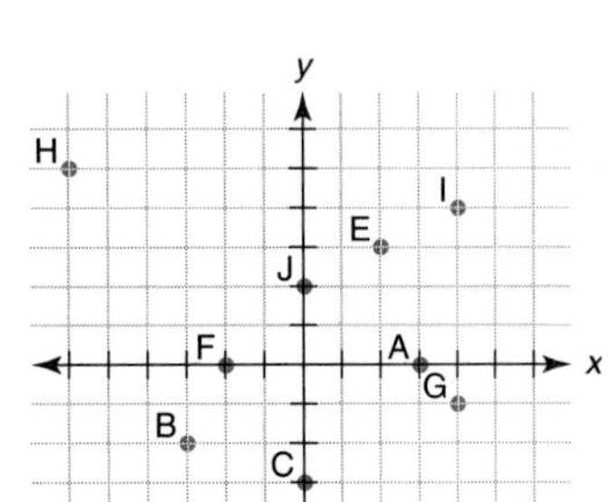

For Exercises 3–22, graph the given equation on a Cartesian plane.

3. $x - y = 0$ **4.** $x - y = 5$
NW **5.** $x + y = 3$ **6.** $2x + y = 8$
7. $3x + y = 5$ **8.** $2x - y = -10$
9. $2x + y = -7$ **10.** $x - y = 7$
11. $x - 3y = 4$ **12.** $x - 2y = -6$
NW **13.** $3x + 2y = 18$ **14.** $x - 3y = 0$
15. $2x - y = -16$ **16.** $x + 2y = 0$
17. $-2x - 3y = 16$ **18.** $3x - 2y = 7$
19. $x = -6$ **20.** $-2x - 3y = 6$
21. $3y = -6$ **22.** $-2x + y = -8$

For Exercises 23–30, solve each system of equations graphically.

NW **23.** $3x + y = 1$, $4x + 2y = 0$ **24.** $2x + 3y = -2$, $-x + 4y = 1$
25. $2x - y = 6$, $3x + 6y = 9$ **26.** $-x + y = -6$, $x + 2y = 3$
27. $3x - 2y = 4$, $5x + 6y = -12$ **28.** $2x + 3y = 0$, $x - y = 5$
29. $4x + 3y = 7$, $3x - y = -11$ **30.** $2x - 3y = -6$, $-4x + 6y = 18$

Writing Mathematics

31. Define the Cartesian plane. As part of your definition, give an example and identify all the parts of the plane.

32. Explain the relationship between an ordered pair, which we discussed in the previous section, and the Cartesian plane.

33. Explain how the ordered pair $(-3, 1)$ is graphed in the Cartesian plane.

34. Explain the difference between an x-intercept and a y-intercept. At what point in the plane are these two intercepts equal to each other?

35. Explain the difference among consistent, dependent, and inconsistent systems of linear equations. Include an example of each as part of your explanation.

Challenge Exercises

36. Consider the graph of the line in which the x- and y-coordinates are equal—for example, $(-5, -5)$, $(2, 2)$, and $(0.6, 0.6)$. Without plotting any points, what do you think the graph of this line looks like? Write the equation of this line in the form $Ax + By = C$.

Miniature golf *In Exercises 37–39, use the graph below which is the outline of a miniature golf hole.*

37. What are the coordinates of the tee?

38. What are the coordinates of the hole?

39. Scott believes that he can make a hole-in-one by hitting his tee shot to the midpoint of the arc, where it will then bounce off the edge and roll into the hole. What are the coordinates of the arc's midpoint? Is this shot possible from the tee? Why or why not?

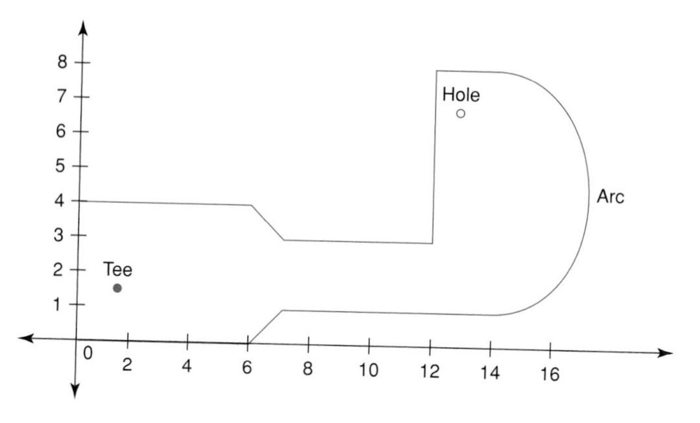

Group/Research Activities

40. Basically there are three distinct stages in the historical development of algebraic notation: *rhetorical algebra, syncopated algebra*, and *symbolic algebra*. Research these three stages, and write a paper on how these stages played an important role in the development of algebra.

41. In an earlier exercise, we "mapped" out a miniature golf hole. Identify other objects that can be mapped in the Cartesian plane and then prepare drawings that show these graphs.

42. In addition to the *x-y* coordinate system, objects can also be graphed in the *x-y-z* system, which provides a three-dimensional aspect of an object. Consult some other references, such as graphic design books or calculus books, and prepare a report on the *x-y-z* system.

43. Working as a team, identify the latitude and longitude of at least five European cities and plot these values as coordinates in the Cartesian plane. If done correctly, the resulting graph will accurately depict these cities' locations relative to each other.

Just for fun

A bicycle dealer was asked how many bikes he had in stock. He answered, "If one-half, one-third, and one-quarter of the number of bikes were added together, they would make 13." How many bikes did he have in stock?

9.8 INEQUALITIES IN TWO VARIABLES

In Section 9.3, we graphed the solution sets of open sentences in one variable. The open sentences that we considered were either equations or inequalities, such as $x + 3 = 4$ and $x > 4$. In Section 9.7, we graphed the solution sets of equations in two unknowns, such as $2x - 2y = 10$ and $x + 2y = 11$. In this section, we shall graph the solution sets of inequalities involving two variables.

When the equation $x + y = 3$ is graphed on the Cartesian plane, we note that the points $(3, 0)$ and $(0, 3)$ are solutions of the given equation. In fact, there is an infinite number of solutions to the linear equation $x + y = 3$. Each solution (ordered pair) is a point on the line. All the solutions of the equation are points on the line, and all the points on the line are solutions of the equation. But what about the points (ordered pairs) that are not on the line? What about points such as $(0, 0)$ and $(4, 0)$? Because they are not on the line, they are not solutions of the equation $x + y = 3$. But how do they compare with the solutions? Let's find out.

We can evaluate the equation $x + y = 3$ for the point $(0, 0)$. Let $x = 0$ and $y = 0$; then we have

$$0 + 0 = 3$$
$$0 = 3 \quad \text{(not true)}$$

Substituting 0 for x and 0 for y in the equation $x + y = 3$ gives us a false statement of equality. The value 0 on the left side of the equal sign is *less than* the value 3 on the right side:

$$0 < 3 \quad \text{(true)}$$

Let's try the other point, $(4, 0)$. Let $x = 4$ and $y = 0$; then we have

$$4 + 0 = 3$$
$$4 = 3 \quad \text{(not true)}$$

Again we have a false statement of equality, but note that this time the value 4 on the left side of the equal sign is *greater than* the value 3 on the right side:

$$4 > 3 \quad \text{(true)}$$

Notice that $(0, 0)$ is a solution of the inequality $x + y < 3$ and $(4, 0)$ is a solution of $x + y > 3$. We have found three different types of points: those like (3, 0) and (0, 3), for which $x + y = 3$; those like (0, 0), for which $x + y < 3$; and those like $(4, 0)$, for which $x + y > 3$.

In fact, the equation $x + y = 3$ separates the Cartesian plane into three sets. One of these sets is the graph of $x + y = 3$. Because this graph is a line, we call it "the line $x + y = 3$." The other two sets are the **half-planes** that lie above and below this line. One of these half-planes contains $(0, 0)$ and is the graph of the inequality $x + y < 3$; the other contains (4, 0) and is the graph of the inequality $x + y > 3$. These two half-planes are called "the half-plane $x + y < 3$" and "the half-plane $x + y > 3$," respectively.

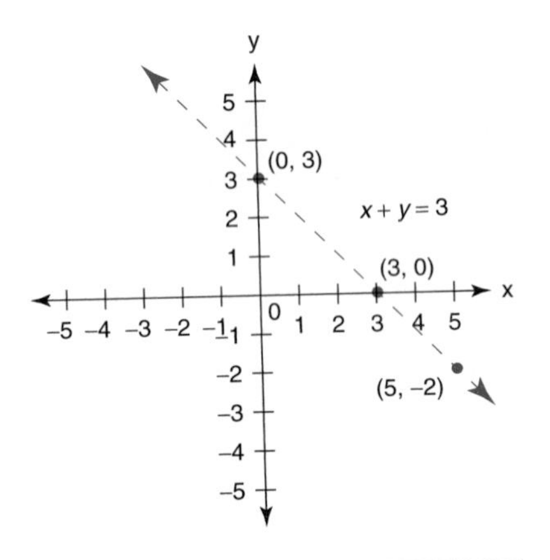

Figure 9.25

How do we go about graphing an inequality such as $x + y < 3$? Recall that the line $x + y = 3$ divides the plane into two half-planes. One half-plane contains those points (x, y) such that $x + y < 3$, whereas the other half-plane contains those points (x, y) such that $x + y > 3$. The set of points that satisfy the equation $x + y = 3$ is those points that lie on the line. In order to graph $x + y < 3$, we locate the boundary of this half-plane, which is the line $x + y = 3$ (see Figure 9.25). Note that we draw the graph of $x + y = 3$ as a dashed line because the points on the line do not satisfy the inequality $x + y < 3$.

Which half-plane is the correct half-plane? The one whose points satisfy the inequality $x + y < 3$. Let's test a point on either side of the line to see which one satisfies the given inequality. One convenient point to try is the origin, $(0, 0)$. If we let $x = 0$ and $y = 0$; then we have

$$0 + 0 < 3$$
$$0 < 3 \quad \text{(true)}$$

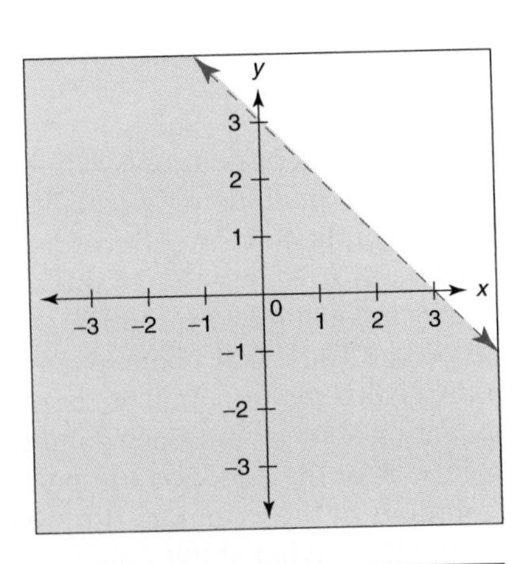

Figure 9.26 $x + y < 3$.

Now, if a point satisfies the given inequality, then all of the points in that half-plane must also satisfy the same inequality. It should also be noted that if a point does not satisfy the given inequality, then none of the points in that half-plane satisfies the given inequality and the desired half-plane must be on the other side of the line.

Because the point (0, 0) does satisfy the given inequality, $x + y < 3$, we shade the half-plane that contains this point. In this case we shade the half-plane that is below the line $x + y = 3$ to indicate the solution (see Figure 9.26).

If we were asked to graph $x + y \leq 3$ instead of $x + y < 3$, then we would have the same picture with one exception. The exception would be to draw the line $x + y = 3$ as a solid line instead of a dashed line. We do this because

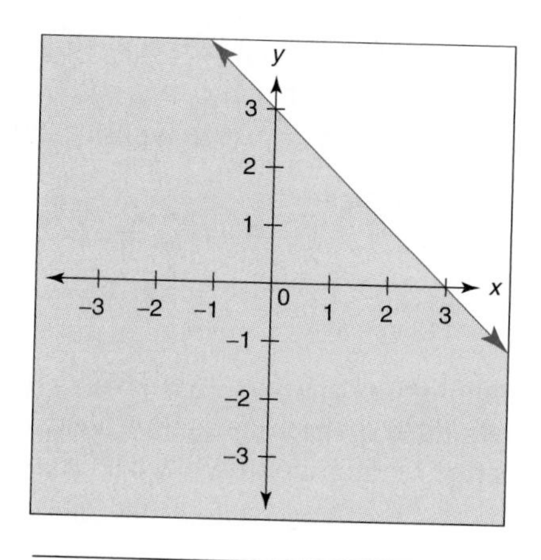

Figure 9.27 $x + y \le 3$.

$x + y \le 3$ means that we want the set of points that satisfy the inequality $x + y < 3$ *or* the equation $x + y = 3$. The sentence $x + y \le 3$ is read as "$x + y$ *is less than or equal to* 3." Recall that *or* is the same as set union, and therefore we unite the half-plane $x + y < 3$ and the line $x + y = 3$. Figure 9.27 is the graph of $x + y \le 3$.

In both Figures 9.26 and 9.27, the line $x + y = 3$ is the boundary for the solution set, a half-plane. In Figure 9.26 the boundary is not included in the solution, whereas in Figure 9.27 the boundary is included in the solution.

In order to graph inequalities in two variables, we must do the following:

1. Find the boundary of the half-plane. We do this by graphing the equation derived from the inequality.
2. The boundary should be a dashed line for *greater than* ($>$) or *less than* ($<$). For $\ge$ or $\le$, the boundary should be a solid line.
3. Indicate the half-plane that is the solution by shading. We do this by testing a point. If a point satisfies the given inequality, then all of the points in that half-plane must also satisfy the same inequality.

EXAMPLE 1 ***Graphing an Inequality in Two Variables*** Graph $x + 2y > 6$.

Solution First we determine the boundary by graphing the equation $x + 2y = 6$. Note that this boundary will be a dashed line because we have a strict inequality in $x + 2y > 6$. Next we test a point to see if it satisfies the given inequality. Let's try the origin, $(0, 0)$. Let $x = 0$ and $y = 0$; then we have

$$\begin{aligned} 0 + 2(0) &> 6 \\ 0 + 0 &> 6 \\ 0 &> 6 \quad \text{(not true)} \end{aligned}$$

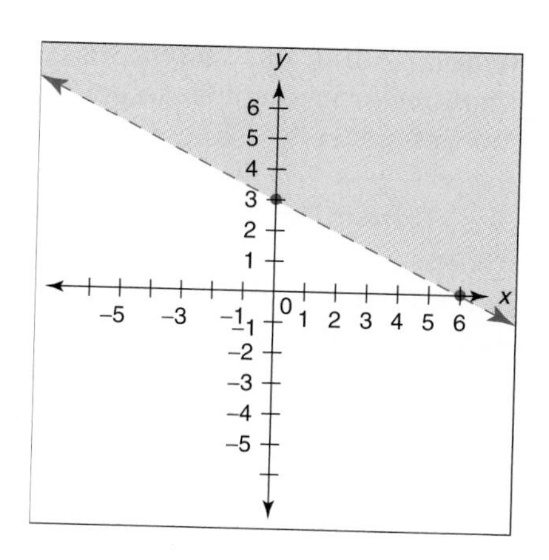

Figure 9.28 $x + 2y > 6$.

The point $(0, 0)$ does not satisfy the statement $x + 2y > 6$. Thus, we can try a point on the other side of the line $x + 2y = 6$, $(3, 3)$. Let $x = 3$ and $y = 3$; then we have

$$\begin{aligned} 3 + 2(3) &> 6 \\ 3 + 6 &> 6 \\ 9 &> 6 \quad \text{(true)} \end{aligned}$$

If a point satisfies a given inequality, then all of the points in that half-plane must also satisfy the same inequality. Therefore, we shade the half-plane that contains the point $(3, 3)$. In this case, we shade the half-plane that is above the line $x + 2y = 6$ to indicate the solution (see Figure 9.28).

NW ***Now Work Problem 3.***

EXAMPLE 2 ***Graphing an Inequality in Two Variables*** Graph $2x - y \le 6$.

Solution First we locate the boundary by graphing $2x - y = 6$. Note that this boundary will be a solid line because of the relationship $\le$. Next we test a point to see if it satisfies the given inequality. Let's try the origin $(0, 0)$. Let $x = 0$ and $y = 0$; then we have

$$\begin{aligned} 2(0) - 0 &\le 6 \\ 0 - 0 &\le 6 \\ 0 &\le 6 \quad \text{(true)} \end{aligned}$$

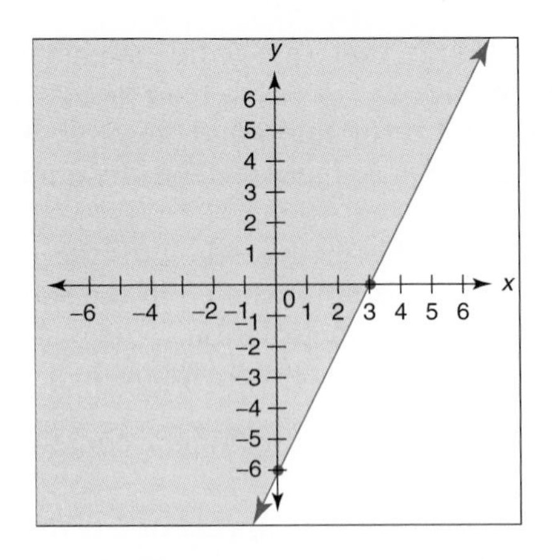

Figure 9.29 $2x - y \leq 6$.

If a point satisfies a given inequality, then all of the points in that half-plane must also satisfy the same inequality. Therefore, we shade the half-plane that contains the point (0, 0)—that is, the half-plane above the line $2x - y = 6$ (see Figure 9.29). Note that the boundary is a solid line.

EXAMPLE 3 ***Graphing an Inequality in Two Variables*** Graph $-3x + 4y < 12$.

Solution First we locate the boundary by graphing the equation $-3x + 4y = 12$. This boundary will be a dashed line because we have a strict inequality. Next we test a point to see if it satisfies the given inequality. Testing the origin, we let $x = 0$ and $y = 0$; then we have

$$\begin{aligned} -3(0) + 4(0) &< 12 \\ 0 + 0 &< 12 \\ 0 &< 12 \quad \text{(true)} \end{aligned}$$

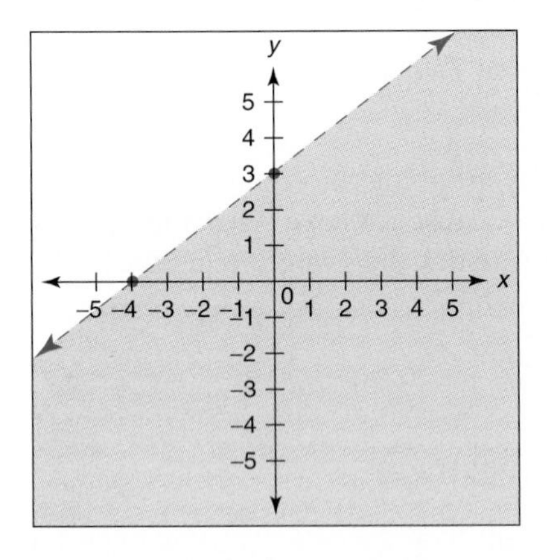

Figure 9.30 $-3x + 4y < 12$.

Since (0, 0) satisfies the given inequality, all of the points in that half-plane will also satisfy the given inequality. Therefore, we shade the half-plane that contains the point (0, 0), which is the half-plane below the line $-3x + 4y = 12$ (see Figure 9.30). Note that the boundary is a dashed line.

NW ***Now Work Problem 11.***

In the preceding section we graphed equations, and in this section we have graphed inequalities. We can also combine equations or inequalities to get compound sentences—for example, the conjunction of $x - y \leq 2$ and $x + y < 1$. The graph of the conjunction of $x - y \leq 2$ and $x + y < 1$ is the intersection of the individual graphs of $x - y \leq 2$ and $x + y < 1$. In order to obtain this intersection, we draw both graphs on the same Cartesian plane. The solution is the region where the two half-planes intersect. The coordinates of all the points in the intersection will satisfy both of the given sentences and hence form the solution set of their conjunction. The graphs of $x - y \leq 2$ and $x + y < 1$ are shown separately in Figures 9.31a and 9.31b. Figure 9.31c is the graph of $x - y \leq 2$ and $x + y < 1$. Note that Figure 9.31c is the intersection of the graphs in Figures 9.31a and 9.31b.

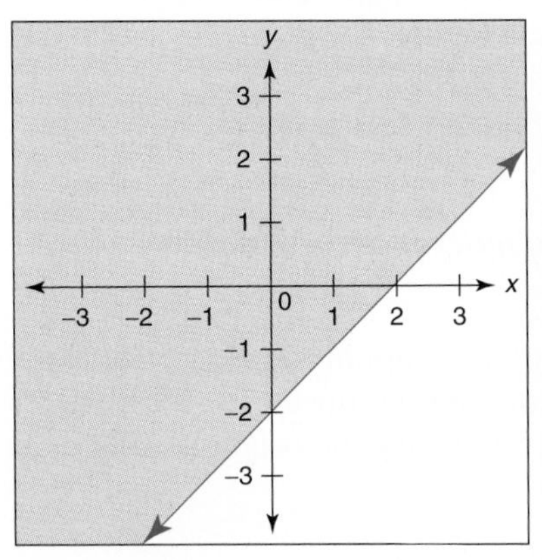

Figure 9.31a $x - y \leq 2$.

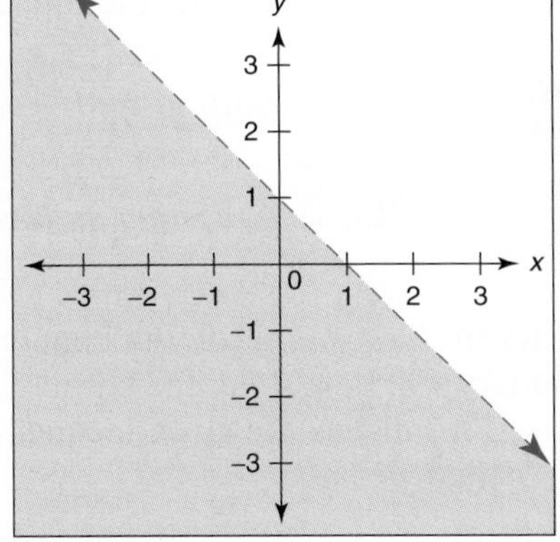

Figure 9.31b $x + y < 1$.

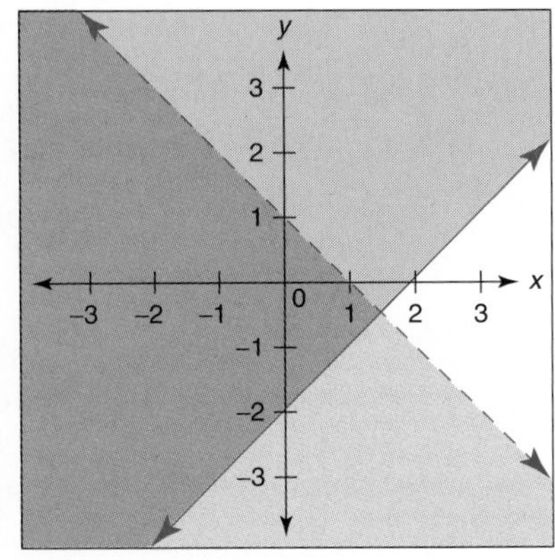

Figure 9.31c $x - y \leq 2$ and $x + y < 1$.

Exercises for Section 9.8

Graph each of the following inequalities on the Cartesian plane:

1. $x - y \geq -2$
2. $x + y < 6$
3. NW $x - y \geq 6$
4. $x - y > 3$
5. $x - y < -6$
6. $2x - y \geq 6$
7. $3x - y < 4$
8. $x - y \geq 0$
9. $3x - y > 0$
10. $x - 3y < 0$
11. NW $2x + y \leq -5$
12. $3x - 5y \leq 15$
13. $-3x + 2y < 6$
14. $-3x - 5y < -15$
15. $5x - 4y \geq 20$
16. $-2x - 3y > -6$
17. $-3x + 4y > -24$
18. $x + y \geq 2$
19. $2x + 6y \leq 0$ and $x + 2y > 3$
20. $x + y \leq 3$ and $x - y > 3$
21. www $3x + y > -3$ and $-2x + y \leq 4$

Writing Mathematics

22. Explain the similarities and differences between a linear equation and a linear inequality, including their respective graphs.
23. Define the phrase *boundary of the half-plane* relative to graphing linear inequalities.
24. Why are some boundaries represented as a solid line and other boundaries represented as a dashed line?
25. Explain how we determine the region that corresponds to the solution set when graphing a system of linear inequalities.

Challenge Exercises

26. Can a system of linear inequalities be consistent, dependent, or inconsistent as is the case with a system of linear equations? Explain why or why not.
27. Suppose you were given the two inequalities $x > 0$ and $y > 0$. Where in the x-y coordinate system will the solution set be located? Describe a situation in which these restrictions on x and y are necessary.
28. Graph the solution to the following system of inequalities:

$$20 \leq x \leq 70$$
$$-50 \leq y \leq 75$$

Group/Research Activity

29. **Automobile lease agreements** Automobile lease agreements vary from dealer to dealer. For example, some leases provide an upper limit on mileage and then begin charging so much per mile after you reach this limit. Other lease agreements provide unlimited mileage. Consult an auto dealership and compare two lease agreements, one based on restricted mileage, the other with unlimited mileage. Apply your knowledge of inequalities to determine which lease agreement is less expensive.

Just for fun

If one peacock lays three eggs in one day, then how many eggs will 33 peacocks lay in 11 days? (Be careful!)

CHAPTER REVIEW MATERIALS

Summary/Chapter 9

9.2 Basic Concepts of Algebra

- Algebra is the area of mathematics that generalizes the facts of arithmetic.
- In algebra, letters are used to represent an unknown quantity. These letters are called *variables* because their numerical value can vary from one application to another. For example, in the algebraic expression $x + y$, the variables are x and y and the expression represents a sum of any two numbers.
- In contrast to a variable, a *constant* represents a fixed numerical value. Unlike a variable, a constant's numerical value will always remain the same. For example, in the algebraic expression $x + 3$, the constant is 3.
- All algebraic expressions consist of *terms*, which are individual quantities separated by + or − signs. For example, the algebraic expression, $3x + 2y - 5$, consists of three terms: $3x$, $2y$, and -5. Terms always include the sign that immediately precede them. This is why the third term of the given expression is -5.
- If an algebraic term represents a product, then the components of that term are called *factors*. For example, in the expression $3x + 2y - 5$, the first term ($3x$) contains the factors 3 and x; similarly, the second term ($2y$) contains the factors 2 and y.
- The *laws of real numbers* and the *order of operations*, both of which were discussed in Chapter 8 and reviewed in this chapter, are applicable to algebra.

9.3 Open Sentences and Their Graphs

- Sentences such as $x + 1 = 3$, which cannot be classified as true or false until we replace x by a number, are called *open sentences*. There are open sentences of *equality* (e.g., $x + 1 = 3$) and open sentences of *inequality* (e.g., $x + 1 > 3$).
- We can find the *solution sets* for these sentences using trial and error; and we can also graph these solution sets on the number line.
- Because open sentences may have different solution sets, it is important that we identify what restrictions (if any) are being placed on the replacement set. For example, the solution set for $x + 5 = 0$ is the empty set if the replacement set is restricted to the set of natural numbers. However, if the replacement set is the set of integers, then the solution set is $\{-5\}$.

9.4 More Open Sentences—Solving Linear Equations

- Whenever the terms of an algebraic expression contain exactly the same variable(s), then the terms are regarded as *like terms*. For example, $5y$ and y, $4xy$ and $10xy$, and $6(a + b)$ and $3(a + b)$ are examples of pairs of like terms. In contrast to this, $2a$ and $3x$, and $5a$ and $2ab$ are examples of pairs of *unlike terms*.
- If an algebraic expression contains *like* terms, then we may combine these like terms into a single term.
- When an algebraic expression is set equal to a quantity (which itself may be another algebraic expression), the resulting statement is called an *equation*. Thus, an equation is a mathematical statement that shows an equality. Examples of *linear equations* are $x + 6 = 14$, $12x - 4x = 16$, and $6x - 20 = 2x - 8$.
- When given an equation, our goal is to solve it. This involves modifying the equation to the point where we end up with an open sentence that is equivalent to the given equation but in the form of x = something or something = x.
- In solving most equations, we can make use of the following axioms:

  1. If we *add* the same number to both sides of an equation, then the sums remain equal.
  2. If we *subtract* the same number from both sides of an equation, then the differences remain equal.
  3. If we *multiply* both sides of an equation by the same nonzero number, then the products remain equal.
  4. If we *divide* both sides of an equation by the same nonzero number, then the quotients remain equal.

- Once we have found a solution to an equation, we can check our answer by replacing the variable with the solution in the original equation. If the left side of the equation is equal to the right side of the equation, then we know that our solution is correct.

9.5 Applications of Linear Equations (Problem Solving)

- One of the oldest applications of algebra is solving *word problems*. Examples include number problems, consecutive integer problems, age problems, and coin problems.
- Most word problems require the use of algebra to find the solution in a systematic manner, as opposed to trial and error. This involves assigning a variable to represent an unknown quantity, setting up an equation that contains this variable and represents the given problem, and then solving the equation for the variable.

9.6 Linear Equations in Two Variables

- Linear equations in two unknowns are equations of the form $Ax + By = C$, where A, B, and C are real numbers. They are called linear equations because the graphs of such equations are straight lines.
- Because a linear equation such as $x + y = 1$ contains two variables, namely, x and y, each of its solutions is an *ordered pair (x, y)*. There are infinitely many ordered pairs that will satisfy a given linear equation.
- To solve a linear equation in two variables, we first select a value for x and then substitute this value for x into the given equation. We now solve the resulting equation for y. Finally, we check to see that the two solutions of the equation, that is, the ordered pair (x, y), satisfies the given equation.

- Linear equations in two variables can be used for *mathematical modeling*. In this context, an equation is derived from a set of data and the equation (i.e., model) is used for prediction purposes.

9.7 Graphing Equations

- The solutions to a linear equation in two variables may be graphed on the *Cartesian plane*. This plane is a rectangular coordinate system that consists of the intersection of two number lines—one horizontal and one vertical. The horizontal line is called the x-axis, the vertical line is called the y-axis, and the point of intersection is called the *origin*.
- Each ordered pair represents a point on the Cartesian plane. Points are plotted by first moving x units from the origin (move right if x is positive, move left if x is negative), and then from this location, moving y units (move up if y is positive, move down if y is negative).
- Because each ordered pair represents a point on the plane, we can locate (i.e., plot) several solutions for a linear equation on the plane and then connect these points to obtain a graph of the equation. When we graph a linear equation, it is usually convenient to find the points where the line crosses each axis—that is, the *x-intercept* and the *y-intercept*. We should also find a third solution to the equation to check that all three points lie on the same straight line.
- Whenever we have two (or more) equations that are to be examined together, we refer to them as a *system of linear equations*, or a *system of simultaneous equations*.
- The solution set for a system of equations involving two linear equations can be obtained by graphing both equations on the same set of axes. If the graphs of the two equations intersect at a point, then the equations are *consistent* and the coordinates of this point represent the solution for the given system.
- If the solution set for a system of equations is empty (i.e., the graphs of the lines do not intersect), then the system is called *inconsistent*. If, on the other hand, every solution of one equation is also a solution of the other equation, then the system is called *dependent*.

9.8 Inequalities in Two Variables

- *Linear inequalities* in two variables are similar to linear equations in two variables except the $=$ sign is replaced with an inequality symbol (i.e., $<$, $>$, $\leq$, or $\geq$).
- To graph the solution set of a linear inequality in two variables, we note that the graph of the corresponding linear equation divides the plane into three components: all of the points that lie on the line itself, the points that lie above the line, and the points that lie below the line. In this context, the line is called the *boundary line*, and the two regions above and below the line are called *half-planes*.
- To graph an inequality in two variables, we do the following:
  1. Find the boundary of the half-plane by graphing the linear equation that corresponds to the given inequality.
  2. Draw the boundary line as a dashed line for strict inequalities (i.e., $<$ or $>$), but use a solid line for $\leq$ and $\geq$.
  3. Use shading to indicate the half-plane that is the solution. We do this by choosing a *test point* that is either above or below the boundary line. If the test point satisfies the given inequality, then all of the points in that half-plane must also satisfy the given inequality. Thus, this is the region that gets shaded. If, on the other hand, the test point yields a false solution, then we shade the "other" region (i.e., the half-plane that does not contain the test point).

Key Terms and Concepts

Review Exercises

9.2 Basic Concepts of Algebra

In Exercises 1–4, identify each of the following as a variable, constant, factor, term, or coefficient.

1. The x in $14x + 12$
2. $12x$
3. The 5 in $x + 5$
4. The 8 in $3x + 8y$
5. Given the expression $8x^2 - 3x + 2$, do the following:
   a. Determine the number of terms.
   b. Write the terms.
   c. Write the coefficient of each term.
   d. Write the factors of each term.

In Exercises 6–8, identify the correct property of real numbers that is being illustrated.

6. $(6)(0) = (0)(6)$
7. $5(6 + 8) = 5(6) + 5(8)$
8. $6(ab) = (6a)b$

In Exercises 9–12, evaluate each expression using the given values for the variables.

9. $4a - 3b$ when $a = 6$ and $b = 3$

10. $3x - 2(y + z)$ when $x = 3$, $y = 4$, and $z = 8$

11. $4x^2y^3 - 3xy$ when $x = 3$ and $y = -2$

12. $N = \dfrac{D}{S - m}$ when $D = 4000$, $S = 500$, and $m = 100$

13. A New York City hot dog street vendor claims that his monthly gross sales, G (in dollars), is calculated using the formula, $G = 2P + C$, where P is the number of people who work within two miles of his stand, and C is the number of cars that pass his stand each day. What is the monthly gross sales if $P = 6000$ and $C = 8000$?

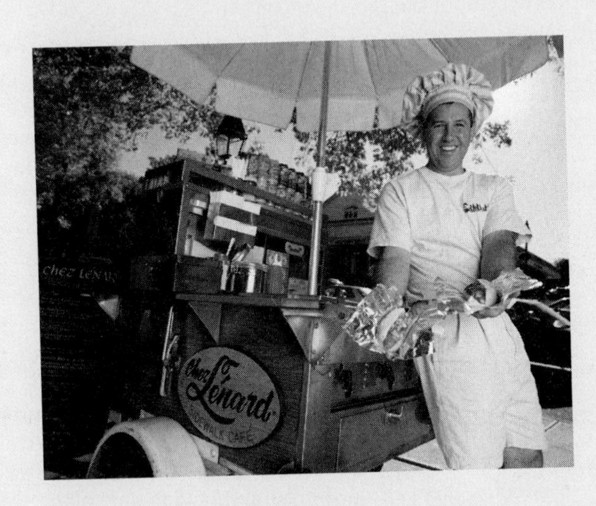

9.3 Open Sentences and Their Graphs

14. Find the solution set for each of the following open sentences. Unless otherwise noted, the replacement set is the set of real numbers.

   a. $x - 4 = 0$

   b. $x + 5 = -1$

   c. $x - 3 = 5 - 2$

   d. $x + 3 \leq 6$ (x is a natural number)

   e. $x - 3 < 5$ (x is a whole number)

   f. $-3 < x < 3$ (x is an integer)

   g. $-4 \leq x \leq 3$ (x is a natural number)

15. Graph the solution set (on a number line) for each of the following open sentences. In each case, the replacement set is the set of real numbers.

   a. $x < 3$ b. $x - 1 \geq 1$

   c. $-1 < x < 2$ d. $-2 \leq x \leq 1$

   e. $0 < x \leq 3$ f. $x - 2 < x$

9.4 More Open Sentences—Solving Linear Equations

16. Solve each of the following equations. The replacement set is the set of real numbers.

   a. $2x - 2 = 10$ b. $2y + 4 = y + 6$

   c. $5z - 3 = 2z + 3$ d. $2y + 6 = 5y - 3$

   e. $\dfrac{x}{2} - 2 = 3$ f. $\dfrac{z}{3} = 2 - z$

9.5 Applications of Linear Equations (Problem Solving)

17. The sum of two consecutive odd integers is 48. Find the numbers.

18. Diego is 3 years older than Andy and the sum of their ages is 49. How old is each?

19. Six less than four times a number is 34. Find the number.

20. The width of a rectangle is 5 meters less than its length. If the perimeter of the rectangle is 30 meters, find the dimensions of the rectangle.

21. Frank has \$3.00 in dimes and quarters in his pocket. If he has twice as many quarters as dimes, how many of each type of coin does he have?

22. Ten less than four times a number is equal to twice that number. Find the number.

23. A piece of wire 64 meters long is bent into the shape of a rectangle whose length is three times its width. Find the dimensions of the rectangle.

24. In a collection of 50 coins, the number of quarters is five more than twice the number of nickels and the number of dimes is five less than twice the number of nickels. How many of each type of coin are in the collection?

25. Tickets for a particular concert cost \$5 each if purchased at the advance sale and \$7 each if bought at the box office on the day of the concert. For this particular concert, 1200 tickets were sold and the receipts were \$6700. How many tickets were bought at the box office on the day of the concert?

9.6 Linear Equations in Two Variables

26. Find three solutions for each of the following equations, and then use the solutions to graph the equation on a Cartesian plane.

   a. $x - y = 2$ b. $2x + y = 4$

   c. $2x - y = 4$ d. $-2x + y = 6$

   e. $3x - 5y = 15$ f. $y = x$

9.7 Graphing Equations

27. Solve each system of equations graphically.

a. $x + y = 1$
$2x + y = 3$

b. $x - y = 1$
$x + y = -5$

c. $2x - y = -3$
$-x + 2y = 0$

9.8 Inequalities in Two Variables

28. Graph each of the following inequalities on the Cartesian plane.

a. $x + y > 7$

b. $x - y \leq 2$

c. $-x + y > 2$

d. $3x + 5y > -15$

e. $x - 2y \leq 0$

f. $-3x - 2y < -6$

Just for fun

A train 1 mile long travels through a tunnel 1 mile long at a rate of 1 mile per hour. How long will it take the train to pass completely through the tunnel?

Chapter Quiz

Determine whether each statement (1–15) is true or false.

1. The graph of $x < 1$ is a ray.
2. The graph of $-2 \leq x \leq 3$ is a line segment.
3. If the replacement set is the set of whole numbers, then the equation $x + 4 = 5 - 2$ has no solution.
4. If $3x + 4y = 15$, then $(1, 3)$, $(-3, 6)$, and $(0, 5)$ are all solutions of the equation.
5. The graph of the equation $2x - 3y = 12$ passes through the point $(-2, 3)$.
6. The origin is located at the point where the x- and y-axes intersect on the Cartesian plane.
7. The point at which a line crosses the x-axis is called the y-intercept.
8. The solution to the system of equations $4x + y = 5$ and $-3x + y = 9$ is $(2, -3)$.
9. The system of equations $-2x + y = 1$ and $2x - y = 5$ is inconsistent.
10. The system of equations $x + y = 5$ and $x + y = 2$ is dependent.
11. The boundary of the graph $2x - y \leq 6$ is a solid line.
12. The graph of $y = x - 2$ is a vertical straight line.
13. $(4x + 2) \div 2 = 2x + 2$
14. $4(5x + 3 - 2x + 1 - 3x) = 16$
15. If $2x + 10 = 6x - 6$, then $x = 3$.

For Exercises 16–23, find the solution set for each equation or inequality. In each case, the replacement set is the set of real numbers.

16. $3x + 4 = 10$
17. $5y + 4 = 2y + 10$
18. $\frac{z}{4} + 4 = 2z - 3$
19. $3x - 2 < x$
20. $6x - 12 = 30$
21. $3(2x - 4) - 2x = x + 5 + x + 3$
22. $3x + 5 = 8x - 15$
23. $\frac{y}{3} + \frac{y}{2} = 5$
24. Mary is five years older than Monique and the sum of their ages is 27. How old is each?
25. The width of a rectangle is 3 feet less than its length. If the perimeter is 30 feet, find the dimensions of the rectangle.
26. Addie has \$2.15 in her change purse, consisting of nickels, dimes, and quarters. If she has twice as many nickels as quarters and one less dime than quarters, how many of each type of coin does she have?

For Exercises 27 and 28, graph the solution on a number line for the given inequality.

27. $3x + 9 \leq 3$
28. $4 < x \leq 7$

Chapter 10

Chapter Outline

Selected Topics in Algebra

After Studying This Chapter, You Will Be Able to Do the Following:

1. Find the **slope** of a line, given the coordinates of two different points on the line, by means of the formula $m = \frac{y_2 - y_1}{x_2 - x_1}$.
2. Determine the **equation** of a line, given the coordinates of two different points on the line, by means of the **point-slope formula** $y - y_1 = m(x - x_1)$ and by the **slope-intercept formula** $y = mx + b$.
3. Graph a line by means of the slope-intercept formula.
4. Solve a **linear programming problem** by representing the given information as a mathematical model, **graphing** the inequalities of the model on the same set of axes, and using the **corner-point evaluation** technique to determine the values of the variables that will provide the desired maximum or minimum values for a given objective.
5. Solve any quadratic equation by applying the **quadratic formula** $x = \frac{-b \pm \sqrt{b^2 - 4ac}}{2a}$.
6. Graph a **parabola** of the form $y = ax^2 + bx + c$ on the Cartesian plane.

Note: *indicates optional material.

Notations Frequently Used in This Chapter

x_1, y_1, x_2, y_2	read x sub one, y sub one, x sub two, y sub two; the coordinates of points on a line
m	the slope of a line
$m\overleftrightarrow{AB}$	the slope of line AB
$\overline{AB} \parallel \overline{CD}$	side AB is parallel to side CD
$\overline{AB} \perp \overline{CD}$	side AB is perpendicular to side CD

Visit the PH Companion Web site at www.prenhall.com/setek

OVERVIEW

Algebra is one of the cornerstones of mathematics and provides methods for solving a multitude of problems.

It appears that algebra had its basic beginnings in Egypt in approximately 1700 B.C. The Greeks did some work with algebra, as they used it to solve equations; but it was René Descartes (1596–1650) who introduced the concept of using a letter of the alphabet to represent the unknown quantity of an equation.

In this chapter we extend our study of algebra by examining various additional topics.

10.1 INTRODUCTION

In the previous chapter, we learned that algebra generalizes the facts of arithmetic by using variables to represent basic arithmetic relations. Building on this concept, we learned how to solve algebraic equations and inequalities, and how to represent these equations and inequalities visually by graphing them. We also learned how to solve verbal problems by first writing an equation to represent the problem and then solving the equation. In this chapter, we extend our study of algebra by examining topics such as the slope of a line, the equation of a line, linear programming problems, solving quadratic equations, and graphing quadratic functions.

10.2 THE SLOPE OF A LINE

We have graphed various types of lines. Some were vertical, others were horizontal. Those that were neither horizontal nor vertical can be called *oblique lines.* Now we shall consider the steepness or slope of these lines.

Consider the ramp pictured in Figure 10.1. Such a ramp may be found at the entrance to a building, or a stadium, or a parking garage. This particular ramp has a length of 50 feet and an elevation (height) of 5 feet. For every 10 feet of horizontal distance, it rises 1 foot in vertical distance. Its steepness or slope is the ratio of the change of elevation to the change in horizontal distance. The ramp rises 1 foot for every change of 10 feet in horizontal distance. That is, its steepness or slope is $\frac{5}{50} = \frac{1}{10}$. We can use this same technique to measure the steepness or slope of a line. The slope of a line can be found by determining the ratio of the change in vertical distance to the change in horizontal distance that the line makes; that is, we can compare the "rise" of the line to the "run" of the line.

Figure 10.1

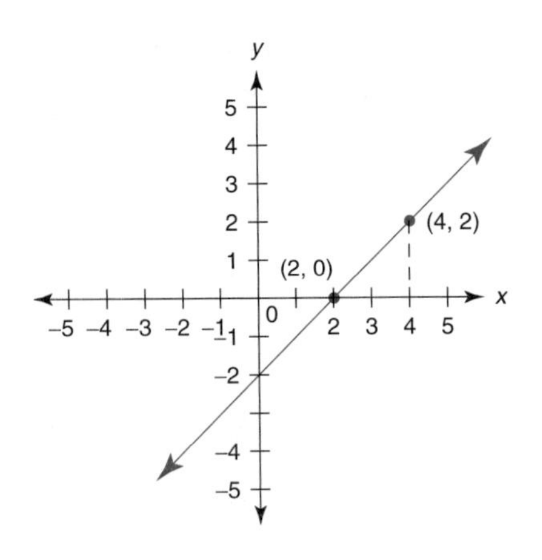

Figure 10.2

Consider the line pictured in Figure 10.2. It passes through the points (2, 0) and (4, 2). We shall use these as reference points. Note that as we proceed "up" the incline of the line (from left to right) the vertical distance (rise) changes one

Bicycling is always easier going down a slope.

unit for every one unit that the horizontal distance (run) changes. Since the slope of a line is the ratio of the change in vertical distance to the change in horizontal distance (rise over run), we have

$$\text{slope} = \frac{\text{rise}}{\text{run}} = \frac{1}{1} = 1$$

Note that the slope of the line pictured in Figure 10.2 is still 1 regardless of how we measure the change in vertical and horizontal distances. If we start at the point $(2, 0)$ and proceed to the point $(4, 2)$, the change in the vertical distance is 2 and the change in the horizontal distance is 2. Hence, the ratio is $\frac{2}{2}$, which also equals 1. This indicates that the slope of a line is the same or constant, regardless of which two points on the line are used to determine the change in the vertical and horizontal distances.

Rather than count the unit changes in the vertical and horizontal distances of a line as we proceed from point to point, we can determine the changes in a more efficient manner. Consider the line shown in Figure 10.3. The points $(-1, 2)$ and $(1, 6)$ have been selected to form a "ramp" and the coordinates of the end of the base of the ramp are $(1, 2)$.

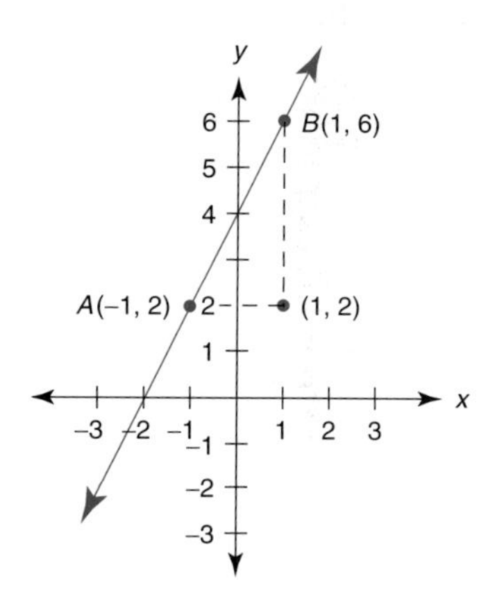

Figure 10.3

From this picture and these points, we can determine that the change in the vertical distance is 4 and the change in the horizontal distance is 2. Therefore, the slope of the line in Figure 10.3 is

$$\text{slope} = \frac{\text{rise}}{\text{run}} = \frac{4}{2} = \frac{2}{1} = 2$$

Note that the change in the vertical distance is the same as the change in the y coordinates and the change in the horizontal distance is the same as the change in the x coordinates. The change in the vertical distance is 4: The y coordinate of B is 6 and the y coordinate of A is 2; $6 - 2 = 4$. Similarly, the change in the horizontal distance is 2: The x coordinate of B is 1, and the x coordinate of A is -1; $1 - (-1) = 1 + 1 = 2$.

To find the slope of a line containing two points, we can find the rise by subtracting the y values, and we can find the run by subtracting the x values; that is,

If a line contains the points (x_1, y_1) and (x_2, y_2) then the slope of the line is defined to be the rise of the line divided by the run of the line:

$$\textbf{slope} = \frac{\textbf{rise}}{\textbf{run}} = \frac{y_2 - y_1}{x_2 - x_1}$$

***Note:* x_1 is read "x sub one."**

Consider the line shown in Figure 10.4. It passes through $C(3, 4)$ and $D(5, 7)$. To find the slope of the line, we designate one point as a first point and the other as

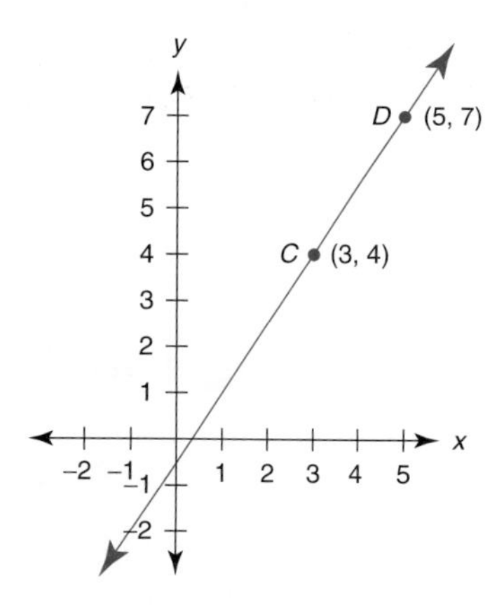

Figure 10.4

a second point (it does not matter which). Let's assume that C is the first point and D is the second point. Hence, $x_1 = 3$, $y_1 = 4$ and $x_2 = 5$, $y_2 = 7$. The slope of a line is often indicated by the letter m. Therefore, the slope of the line in Figure 10.4 can be called m, and we have

$$\text{slope} = m = \frac{\text{rise}}{\text{run}} = \frac{y_2 - y_1}{x_2 - x_1} = \frac{7 - 4}{5 - 3} = \frac{3}{2}$$

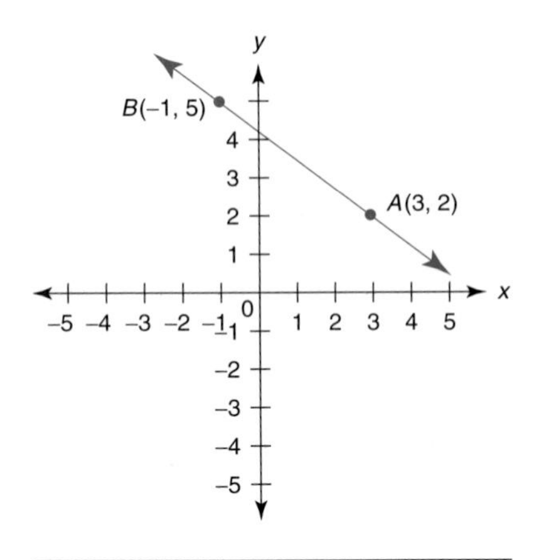

Figure 10.5

The slope of the line is the same if we choose D to be the first point and C to be the second point. In this case $x_1 = 5$, $y_1 = 7$ and $x_2 = 3$, $y_2 = 4$, so we have

$$m = \frac{y_2 - y_1}{x_2 - x_1} = \frac{4 - 7}{3 - 5} = \frac{-3}{-2} = \frac{3}{2}$$

If a line slants upward to the right, then its slope is positive. If a line slants downward to the right, then its slope is negative. Consider the line in Figure 10.5. It passes through $B(-1, 5)$ and $A(3, 2)$. If B is the first point and A is the second point, then $x_1 = -1$, $y_1 = 5$ and $x_2 = 3$, $y_2 = 2$. The slope of the line in Figure 10.5 is

$$m = \frac{y_2 - y_1}{x_2 - x_1} = \frac{2 - 5}{3 - (-1)} = \frac{-3}{3 + 1} = \frac{-3}{4} \quad \text{or} \quad -\frac{3}{4}$$

EXAMPLE 1 ***Finding the Slope of a Line Given Two Points*** Find the slope of the line containing the points $A(2, 3)$ and $B(5, 6)$.

Solution The slope of a line is the change in the y values divided by the change in the x values for any two different points on the line. Let A be the first point and B be the second point. $x_1 = 2$, $y_1 = 3$ and $x_2 = 5$, $y_2 = 6$. The slope of the line is

$$m = \frac{y_2 - y_1}{x_2 - x_1} = \frac{6 - 3}{5 - 2} = \frac{3}{3} = 1$$

EXAMPLE 2 ***Finding the Slope of a Line Given Two Points*** Find the slope of the line containing the points $C(-2, 3)$ and $D(-3, -2)$.

Solution Let C be the first point and D be the second point. Then $x_1 = -2$, $y_1 = 3$ and $x_2 = -3$, $y_2 = -2$. The slope of the line is

$$m = \frac{y_2 - y_1}{x_2 - x_1} = \frac{-2 - 3}{-3 - (-2)} = \frac{-5}{-3 + 2} = \frac{-5}{-1} = 5$$

Note: Extreme care must be exercised when combining negative values such as $-3 - (-2)$. Also, remember that the slope will be the same regardless of which point is chosen first and which point is chosen second. In this example we can let $x_1 = -3$, $y_1 = -2$, $x_2 = -2$, and $y_2 = 3$; then

$$m = \frac{y_2 - y_1}{x_2 - x_1} = \frac{3 - (-2)}{-2 - (-3)} = \frac{3 + 2}{-2 + 3} = \frac{5}{1} = 5$$

which is the same answer we obtained previously.

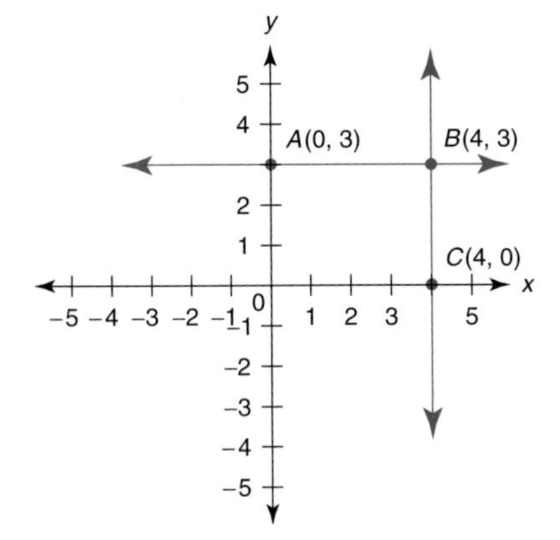

Figure 10.6

NW ***Now Work Problem 1.***

The Slope of Horizontal and Vertical Lines

The slopes of horizontal and vertical lines deserve special consideration. Consider the lines shown in Figure 10.6. The line containing the points $A(0,3)$ and $B(4, 3)$ is a horizontal line parallel to the x-axis. If we let A be the first point and B be the second point, then $x_1 = 0$, $y_1 = 3$ and $x_2 = 4$, $y_2 = 3$. The slope of the horizontal line containing these points is

$$m = \frac{y_2 - y_1}{x_2 - x_1} = \frac{3 - 3}{4 - 0} = \frac{0}{4} = 0$$

The slope of any horizontal line (parallel to the x-axis) is 0.

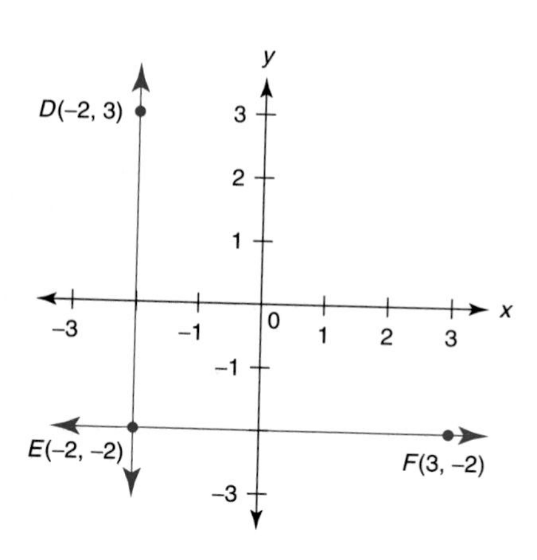

Figure 10.7

The line containing the points $B(4, 3)$ and $C(4, 0)$ is a vertical line parallel to the y-axis. If we let B be the first point and C be the second point, then $x_1 = 4$, $y_1 = 3$, and $x_2 = 4$, $y_2 = 0$. The slope of the vertical line containing these points is

$$m = \frac{y_2 - y_1}{x_2 - x_1} = \frac{0 - 3}{4 - 4} = \frac{-3}{0}, \quad \text{which is undefined.}$$

Since division by 0 is undefined, this line has no slope.

The slope of a vertical line (parallel to the y-axis) is undefined.

EXAMPLE 3 ***Finding the Slope of a Line Given Its Graph*** Find the slopes of the lines pictured in Figure 10.7.

Solution To find the slope of the line containing D and E, let D be the first point and E be the second point. Then $x_1 = -2$, $y_1 = 3$ and $x_2 = -2$, $y_2 = -2$. The slope of the line is

$$m = \frac{y_2 - y_1}{x_2 - x_1} = \frac{-2 - 3}{-2 - (-2)} = \frac{-5}{-2 + 2} = \frac{-5}{0}, \quad \text{undefined.}$$

The line containing D and E is a vertical line and the slope of a vertical line is undefined.

To find the slope of the line containing E and F, let E be the first point and F be the second point. Then $x_1 = -2$, $y_1 = -2$ and $x_2 = 3$, $y_2 = -2$. The slope of the line is

$$m = \frac{y_2 - y_1}{x_2 - x_1} = \frac{-2 - (-2)}{3 - (-2)} = \frac{-2 + 2}{3 + 2} = \frac{0}{5} = 0$$

The line containing E and F is a horizontal line and the slope of a horizontal line is zero.

NW ***Now Work Problems 7 and 11.***

The Slope of Parallel and Perpendicular Lines

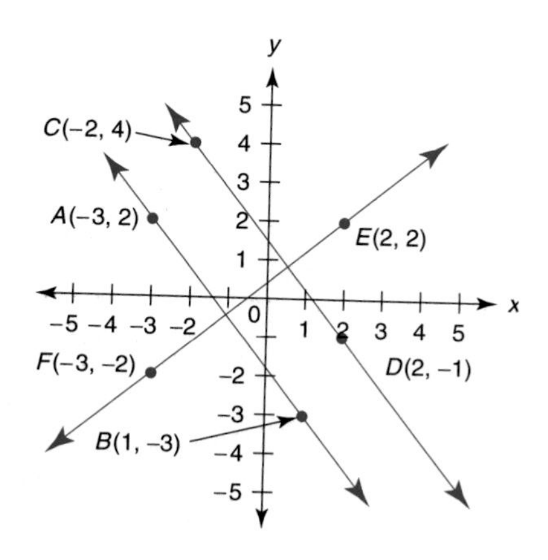

Figure 10.8

Now we shall consider the slopes of sets of lines. In particular, we shall examine the slopes of parallel and perpendicular lines. Consider the lines in Figure 10.8. We are given that the line containing A and B is parallel to the line containing C and D. Also, the line containing E and F is perpendicular to the line containing C and D. How do the slopes of these lines compare? Let's find out. To find the slope of the line containing A and B, let $x_1 = -3$, $y_1 = 2$, $x_2 = 1$, $y_2 = -3$. The slope of the line containing A and B, or the slope of the line AB (denoted by $m\overleftrightarrow{AB}$) is

$$m\overleftrightarrow{AB} = \frac{y_2 - y_1}{x_2 - x_1} = \frac{-3 - 2}{1 - (-3)} = \frac{-5}{1 + 3} = \frac{-5}{4}$$

To find the slope of $\overleftrightarrow{CD}$, let $x_1 = -2$, $y_1 = 4$ and $x_2 = 2$, $y_2 = -1$. The slope of $\overleftrightarrow{CD}$ is

$$m\overleftrightarrow{CD} = \frac{y_2 - y_1}{x_2 - x_1} = \frac{-1 - 4}{2 - (-2)} = \frac{-5}{2 + 2} = \frac{-5}{4}$$

We have shown that for these two particular parallel lines, their slopes are equal. It shall not be proved here, but it can be shown that, in general, *if lines are parallel, then they must have the same or equal slopes.* Also, lines that have the same slope are parallel (or they may be coincident).

To find the slope of $\overleftrightarrow{EF}$, let $x_1 = 2, y_1 = 2$ and $x_2 = -3, y_2 = -2$. Then the slope of $\overleftrightarrow{EF}$ is

$$m\overleftrightarrow{EF} = \frac{y_2 - y_1}{x_2 - x_1} = \frac{-2 - 2}{-3 - 2} = \frac{-4}{-5} = \frac{4}{5}$$

Recall that the slope of $\overleftrightarrow{CD}$ is $-\frac{5}{4}$. How does this compare with the slope of $\overleftrightarrow{EF}$? They are negative reciprocals of each other. We have shown for these two particular perpendicular lines that the slope of one line is the negative reciprocal of the slope of the other line. It can be shown that, in general, *if lines are perpendicular, then their slopes are negative reciprocals of each other.*

Also, if lines have slopes that are negative reciprocals of each other, then the lines are perpendicular. The exceptions to this occur when one line is vertical and the other is horizontal. The slope of a horizontal line is 0, and the slope of a vertical line is undefined. Hence, we cannot apply the preceding rule to horizontal and vertical lines, even though they are perpendicular.

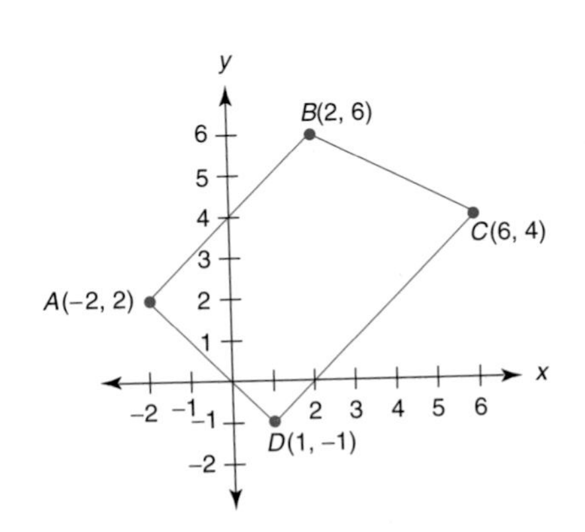

Figure 10.9

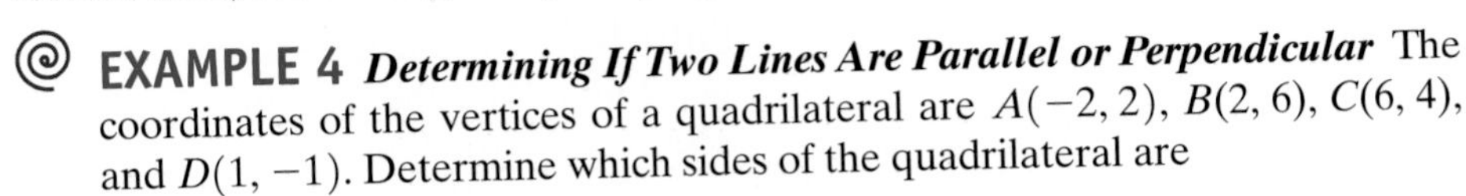

EXAMPLE 4 ***Determining If Two Lines Are Parallel or Perpendicular*** The coordinates of the vertices of a quadrilateral are $A(-2, 2)$, $B(2, 6)$, $C(6, 4)$, and $D(1, -1)$. Determine which sides of the quadrilateral are

a. parallel (if any) **b.** perpendicular (if any)

Solution First we plot the points on the Cartesian plane and connect them to obtain the appropriate diagram. See Figure 10.9.

Next, we find the slope of each side:

$$m\overline{AB} = \frac{y_2 - y_1}{x_2 - x_1} = \frac{6 - 2}{2 - (-2)} = \frac{4}{2 + 2} = \frac{4}{4} = 1$$

$$m\overline{BC} = \frac{y_2 - y_1}{x_2 - x_1} = \frac{4 - 6}{6 - 2} = \frac{-2}{4} = -\frac{1}{2}$$

$$m\overline{CD} = \frac{y_2 - y_1}{x_2 - x_1} = \frac{-1 - 4}{1 - 6} = \frac{-5}{-5} = 1$$

$$m\overline{AD} = \frac{y_2 - y_1}{x_2 - x_1} = \frac{2 - (-1)}{-2 - 1} = \frac{2 + 1}{-3} = \frac{3}{-3} = -1$$

a. Sides $\overline{AB}$ and $\overline{CD}$ have the same slope, and lines that have the same or equal slopes are parallel. Therefore, $\overline{AB}$ is parallel to $\overline{CD}$. (*Note:* We may denote this by $\overline{AB} \parallel \overline{CD}$.)

b. The slope of side $\overline{AD}$ is -1 and the slope of side $\overline{AB}$ is 1. If lines have slopes that are negative reciprocals of each other, then the lines are perpendicular. Therefore, $\overline{AB}$ is perpendicular to $\overline{AD}$. Also, a straight line perpendicular to one of two parallel lines is perpendicular to the other. Since $\overline{AB} \parallel \overline{CD}$ we can also state that $\overline{AD}$ is perpendicular to $\overline{CD}$. (*Note:* We may denote this by $\overline{AD} \perp \overline{AB}$ and $\overline{AD} \perp \overline{CD}$.)

NW ***Now Work Problems 17 and 19.***

EXAMPLE 5 ***Identifying, Calculating, and Interpreting Slope Based on Graphical Data*** The following figure shows the average tuition at public and private four-year colleges during the past 10 years (from 1993–1994 to 2002–2003). All costs were adjusted for inflation in 2002 dollars and tuitions were rounded.

a. Determine if the slope of each line is positive, negative, zero, or undefined and explain why.

b. Find the slope of each line using the points (1993, 3100) and (2003, 4100) for public college tuition, and (1993, 13500) and (2003, 18500) for private college tuition. Compare the slopes and interpret this comparison in the context of the given illustration.

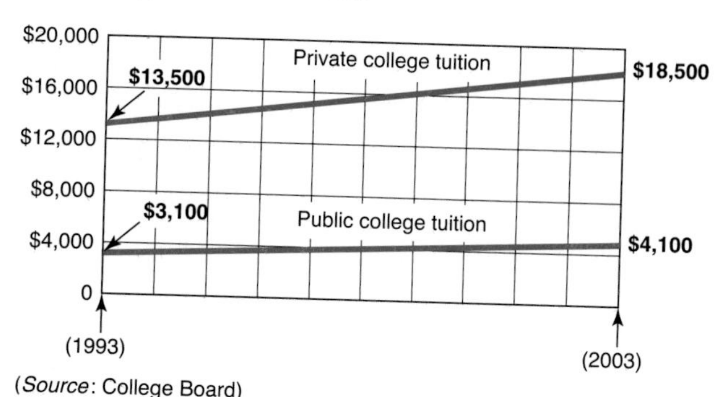

Solution

a. The slope of the line representing private college tuition is positive because the costs are rising throughout the given time period. The slope of the line representing public college tuition is also positive because the costs are rising throughout the given time period. Of the two, however, private college tuition has increased much more dramatically over the past 10 years than public college tuition and hence has a greater slope.

b. Slope of public college tuition is

$$\frac{y_2 - y_1}{x_2 - x_1} = \frac{4100 - 3100}{2003 - 1993} = \frac{1000}{10} = \frac{100}{1}$$

This slope means that there was a \$100 increase in public college tuition each year during the 10-year period. The slope of private college tuition is

$$\frac{y_2 - y_1}{x_2 - x_1} \frac{18500 - 13500}{2003 - 1993} = \frac{5000}{10} = \frac{500}{1}$$

This slope means that there was a \$500 increase in private college tuition each year during the 10-year period. Comparing slopes, the increase in private college tuition during the past 10 years was five times greater than that of public college tuition.

NW ***Now Work Problem 35.***

Summarizing, the slope of a line may be classified as being positive, negative, zero, or undefined.

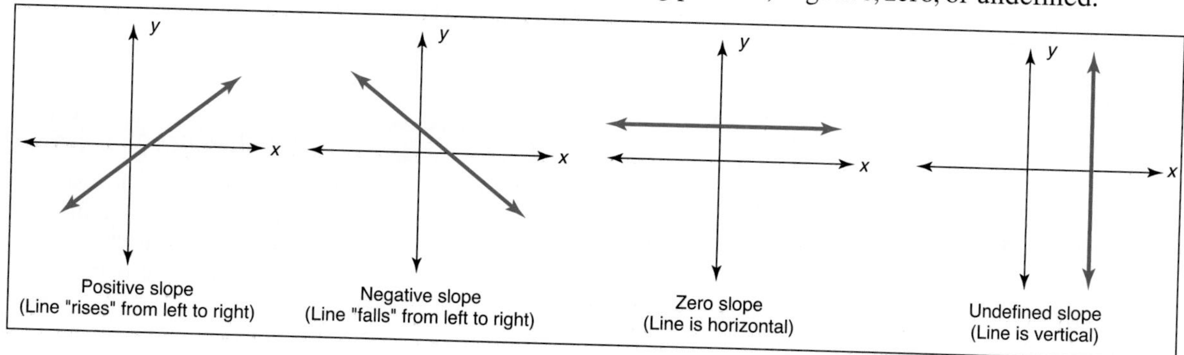

Exercises for Section 10.2

For Exercises 1–12, find the slope of the line containing the indicated pair of points.

NW **1.** $A(2,3)$ and $B(1,6)$

2. $C(-2,3)$ and $D(4,6)$

3. $E(-1,4)$ and $F(3,8)$

4. $G(3,-4)$ and $H(5,-3)$

5. $I(2,3)$ and $J(-2,4)$

6. $K(2,0)$ and $L(-3,4)$

NW **7.** $M(-3,2)$ and $N(-3,0)$

8. $O(6,-1)$ and $P(-3,-1)$

9. $Q(2,-6)$ and $R(-2,0)$

10. $S(-4,5)$ and $T(3,-4)$

NW **11.** $U(5,-5)$ and $V(6,-5)$

12. $W(5,-3)$ and $X(5,7)$

13. Find the slope of a line that is parallel to a line whose slope is

a. 6 **b.** -3 **c.** $\frac{2}{3}$ **d.** 0

14. Find the slope of a line that is parallel to a line whose slope is

a. -2 **b.** 5 **c.** $\frac{5}{4}$ **d.** $\frac{3}{2}$

15. Find the slope of a line that is perpendicular to a line whose slope is

a. 6 **b.** -3 **c.** $\frac{2}{3}$ **d.** 0

16. Find the slope of a line that is perpendicular to a line whose slope is

a. -2 **b.** $-\frac{3}{4}$ **c.** $\frac{4}{3}$ **d.** 1

NW **17.** Show that the line that passes through the points $A(3,-1)$ and $B(4,2)$ is parallel to the line that passes through the points $C(0,4)$ and $D(-2,-2)$.

18. The points $A(-2,-4)$, $B(-4,2)$, $C(2,-2)$, and $D(4,6)$ are vertices of a trapezoid. Which two sides are parallel?

NW **19.** Show that the line that passes through the points $M(-4,-1)$ and $N(2,1)$ is perpendicular to the line that passes through $O(3,-1)$ and $P(4,-4)$.

20. The vertices of a quadrilateral are $L(3,1)$, $M(5,6)$, $N(7,6)$, and $O(10,2)$. Show that the diagonals of the quadrilateral are perpendicular to each other.

***21.** Line AB is parallel to line CD. The slope of line AB is $\frac{-2}{3}$. The slope of line CD is $\frac{x}{12}$. Find x.

***22.** Line EF is parallel to line GH. The slope of line EF is $\frac{-2}{3}$. The slope of line GH is $\frac{8}{x-6}$. Find x.

***23.** Line AB is perpendicular to line MN. The slope of line AB is $\frac{2}{5}$. The slope of line MN is $\frac{x}{8}$. Find x.

***24.** Line PQ is perpendicular to line ST. The slope of line PQ is $\frac{x-1}{4}$. The slope of line ST is $\frac{8}{3}$. Find x.

***25.** The vertices of quadrilateral $ABCD$ are $A(x,y)$, $B(x,2y)$, $C(3x,y)$ and $D(-x,-y)$.

a. Is side AD parallel to side CD? Why or why not?

b. Is side BC perpendicular to side CD? Why or why not?

c. If $x=0$ and $y=1$, is side AB parallel to side AD? Why or why not?

d. If $x=1$ and $y=2$, is side BC perpendicular to side CD? Why or why not?

***26.** The coordinates of the vertices of quadrilateral $ABCD$ are $A(0,5)$, $B(3,4)$, $C(0,-5)$, and $D(-3,-4)$.

a. Is side AB parallel to side CD? Why or why not?

b. Is side BC parallel to side AD? Why or why not?

Writing Mathematics

27. Describe what is meant by the *slope* of a line.

28. Explain what it means when the slope of a line is *undefined*. How can this be?

29. When is the slope of a line *zero*? How can this be?

Challenge Exercises

30. The vertices of quadrilateral $PQRS$ are $P(0,0)$, $Q(6a,3b)$, $R(3a,4b)$, $S(a,3b)$.

a. Find the slopes of sides PQ and RS.

b. What is your conclusion regarding these two sides?

c. Is side PS parallel to side QR? Why?

31. Sides AB and DC of quadrilateral $ABCD$ are parallel to each other. The coordinates of the vertices are $A(7,-3)$, $B(2x,2)$, $C(x,12)$, and $D(3,2)$.

a. Find the slopes of sides AB and DC (express your answer in terms of x).

b. Write an equation that can be used to solve for x, and solve the equation for x.

32. The coordinates of the vertices of triangle ABC are $A(0,-5)$, $B(3,4)$, and $C(-3,-4)$. After determining the slopes of the sides of the triangles, show that triangle ABC is a right triangle.

Group/Research Activities

33. René Descartes (1596–1650), a French mathematician-philosopher, made significant contributions to the development of algebra. Write a paper on his life and contributions to algebra.

34. Algebra is one of the cornerstones of mathematics. It provides methods for solving a multitude of problems. Write a paper on the history and development of algebra.

Real-World Data

35. NW **Homeowner's insurance** The graph below shows the annual premiums (i.e., the cost) for different amounts of homeowner's insurance from two insurance companies. Find the slope of each line using the points (400, 60000) and (800, 100000) for Cook Insurance, and (500, 40000) and

(1000, 80000) for TriState Insurance. Compare the slopes and interpret this comparison in the context of the given illustration.

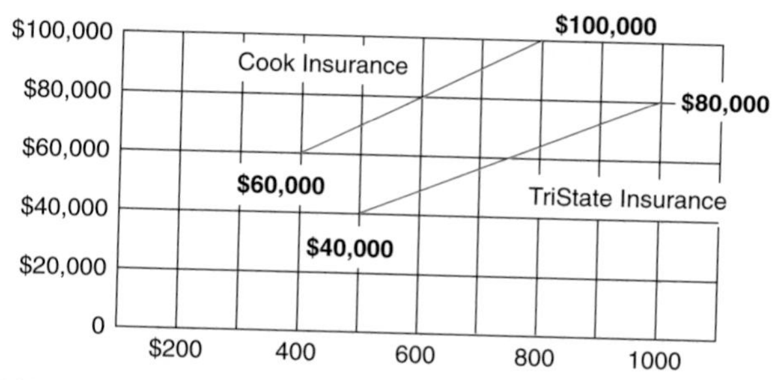

36. **Life expectancy** The graph below shows the life expectancy of males and females for selected years (all ages were rounded). Find the slope of each line using the points (1950, 71) and (2000, 80) for females, and (1950, 66) and (2000, 74) for males. Compare the slopes and interpret this comparison in the context of the given illustration.

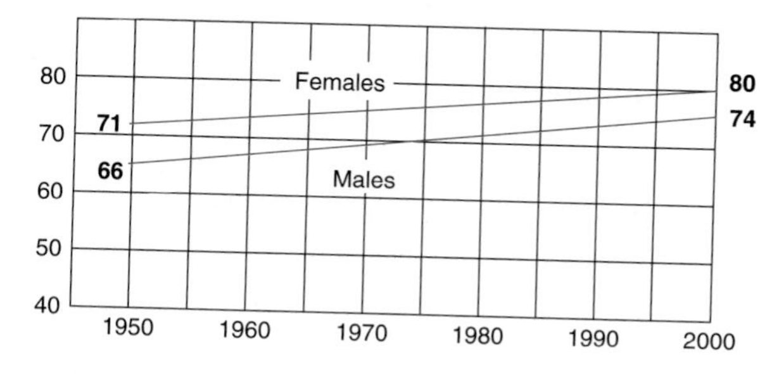

Source: National Center for Health Statistics

37. **Camera sales** The graph below shows the number of U.S. camera sales in millions between traditional and digital cameras between 2000 and 2003 (numbers were rounded). Find the slope of each line using the points (2000, 4) and (2003, 13) for Digital, and (2000, 20) and (2003, 12) for Traditional. Compare the slopes and interpret this comparison in the context of the given illustration.

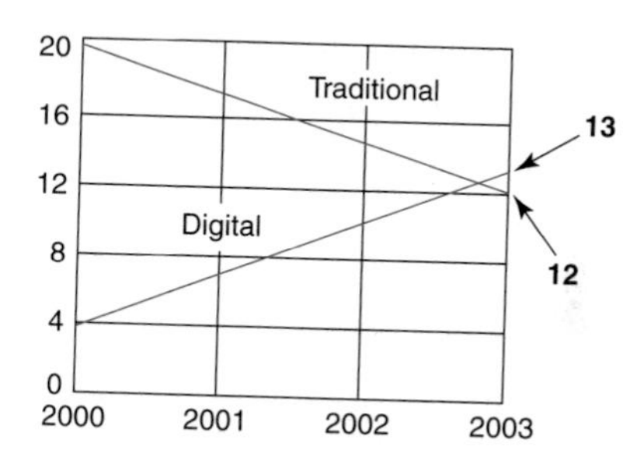

Source: Photo Marketing Association.

Note of Interest

Alfred B. Nobel (1833–1896), the inventor of dynamite, bequeathed $9,000,000 and the interest from this money to be distributed yearly to those judged to have had most benefited humankind in chemistry, physics, medicine-physiology, literature, and the promotion of peace. These prizes were first awarded in 1901. In 2000 each winner was given a large solid gold medal and a cash award of almost $900,000.

Many people have wondered why Alfred Nobel did not establish such an award for mathematicians. No one knows for sure, but there is a lot of speculation why this is the case. Some think that Nobel did not consider mathematics a practical science that would produce benefits in the same manner as physics or chemistry. Others believe that due to the existence of a Scandinavian prize for mathematics, Nobel did not want to add a competing prize.

There is a prestigious award for mathematicians; it is known as the **Fields Medal.** John Fields (1863–1932), a Canadian mathematician, endowed an award that recognizes outstanding achievement in completed mathematical works and the potential future work of young mathematicians. In 1932, the International Congress of Mathematicians adopted the award, to be given every four years to mathematicians under the age of 40. Four medals have been awarded at each congress since 1966, along with a cash prize. In addition, since 1983, the International Congress of Mathematicians has made an award for exemplary work in the field of theoretical computer science; it is called the Nevanlinna Prize.

Just for fun

Can you arrange three 7s so they will equal 20? You may use only three 7s!

10.3 THE EQUATION OF A LINE

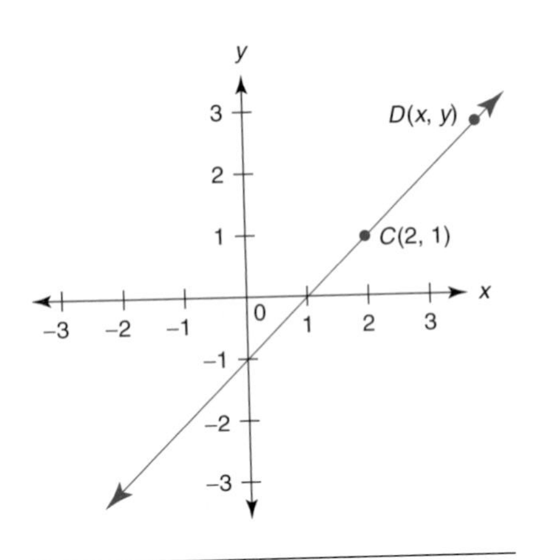

Figure 10.10

Given any two points on a line, we can find the slope of that line. For example, if a line contains the points $A(2, 1)$ and $B(4, 4)$, then the slope of that line is

$$m = \frac{y_2 - y_1}{x_2 - x_1} = \frac{3}{2}$$

But consider the line pictured in Figure 10.10. Note that it contains $C(2, 1)$ and $D(x, y)$. The values of x and y in point D are not given. We can still use the formula for the slope of a line to find this slope, but it will be expressed in terms of x and y. That is,

$$m = \frac{y - 1}{x - 2}$$

If we eliminate the fractions from this equation, we have $m(x - 2) = y - 1$, or we can rewrite the equation as $y - 1 = m(x - 2)$. This equation also represents the equation of the line pictured in Figure 10.10. Instead of a particular point such as (2, 1), we could have chosen some other point (different from C), such as $P(x_1, y_1)$ and we would have the general equation

$$y - y_1 = m(x - x_1).$$

The Point-Slope Formula

This equation is true whenever the points (x, y) and (x_1, y_1) lie on the same straight line. What is the significance of the equation $y - y_1 = m(x - x_1)$? It is a general equation of the line. This formula can be used when we are given a point on the line and the slope of the line. $y - y_1 = m(x - x_1)$ is known as the **point-slope formula** for the equation of a straight line.

We shall use the point-slope formula to find the equation of a line. Consider the line whose slope is 3, which passes through the point (5, 2). This is all the information we need to find the equation of the line. Using the point-slope formula,

Biography: Benjamin Banneker

Benjamin Banneker (1731–1806), an African-American born of a free mother and a slave father, was eventually able to buy his own freedom. He had very little formal education, so he was for the most part self-taught in mathematics, astronomy, geology, and physics. From 1792 through 1799 he published *Benjamin Banneker's Almanac and Ephemeris*. One feature of the almanac was his reporting and calculation of the seventeen-year cycle for the locust plagues. His almanac was the first scientific work published by an African-American. Through his scientific work, Mr. Banneker gained a respected reputation that won him a position on the survey commission of six men to assess the land that was to become the District of Columbia. When the head of the commission, Major L'Enfant, resigned and returned to France with all the plans, Mr. Banneker's excellent memory was able to reconstruct the plans in their entirety.

$y - y_1 = m(x - x_1,)$ with $m = 3$ and $x_1 = 5$, $y_1 = 2$, we have

$$\begin{aligned} y - y_1 &= m(x - x_1) && \textit{point-slope formula} \\ y - 2 &= 3(x - 5) && \textit{substituting for } m, x_1, y_1 \\ y - 2 &= 3x - 15 && \textit{distributive property} \\ y &= 3x - 13 && \textit{adding 2 to both sides} \end{aligned}$$

The desired equation is $y = 3x - 13$.

The point-slope formula for the equation of a line can be used to find the equation of a line whenever we have the slope of the line and a point on the line.

EXAMPLE 1 ***Finding the Equation of a Line Using the Point-Slope Formula*** Find the equation of the line that passes through the point $(-3, 2)$ and has a slope of 2.

Solution Using the point-slope formula $y - y_1 = m(x - x_1)$ we have $m = 2$ and $x_1 = -3$, $y_1 = 2$. Therefore,

$$\begin{aligned} y - y_1 &= m(x - x_1) && \\ y - 2 &= 2(x - (-3)) && \textit{substituting} \\ y - 2 &= 2(x + 3) && \textit{simplifying} \\ y - 2 &= 2x + 6 && \textit{distributive property} \\ y &= 2x + 8 && \textit{adding 2 to both sides} \end{aligned}$$

EXAMPLE 2 ***Finding the Equation of a Line Using the Point-Slope Formula*** Find the equation of the line that passes through the point $(-2, -4)$ and has a slope of $-\frac{1}{2}$.

Solution We have $m = -\frac{1}{2}$, $x_1 = -2$, and $y_1 = -4$. The point-slope formula is applied:

$$\begin{aligned} y - y_1 &= m(x - x_1) && \\ y - (-4) &= -\tfrac{1}{2}(x - (-2)) && \textit{substituting} \\ y + 4 &= -\tfrac{1}{2}(x + 2) && \textit{simplifying} \\ y + 4 &= -\tfrac{1}{2}\cdot x - \tfrac{1}{2}\cdot 2 && \textit{distributive property} \\ y + 4 &= -\tfrac{1}{2}x - 1 && \textit{simplifying} \\ y &= -\tfrac{1}{2}x - 5 && \textit{subtracting 4 from both sides} \end{aligned}$$

NW ***Now Work Problem 1.***

We can use the point-slope formula to find the equation of a line when we are given only two points that lie on the line, and not the slope. For example, if the points $A(3, 2)$ and $B(5, 6)$ lie on the same straight line, we can find the equation of the line. First we must find the slope of the line. Using the formula for slope, we have

$$m = \frac{y_2 - y_1}{x_2 - x_1} = \frac{6 - 2}{5 - 3} = \frac{4}{2} = 2$$

The slope of the line passing through A and B is 2, and now we can use either point A or point B, and the point-slope formula to find the equation of the line. If we choose A, then $m = 2$, $x_1 = 3$, $y_1 = 2$, and we have

$$\begin{aligned} y - y_1 &= m(x - x_1) \\ y - 2 &= 2(x - 3) && \textit{substituting} \\ y - 2 &= 2x - 6 && \textit{distributive property} \\ y &= 2x - 4 && \textit{adding 2 to both sides} \end{aligned}$$

Since B also lies on the same line, we will get the same equation if we choose point B. In this case, $m = 2$, $x_1 = 5$, $y_1 = 6$, and we have

$$\begin{aligned} y - y_1 &= m(x - x_1) \\ y - 6 &= 2(x - 5) && \textit{substituting} \\ y - 6 &= 2x - 10 && \textit{distributive property} \\ y &= 2x - 4 && \textit{adding 6 to both sides} \end{aligned}$$

Note that we obtain the same equation that resulted when we used point A. This must be the case because both points lie on the same line and therefore must satisfy the equation. Which point to use in determining the equation is a matter of personal preference.

EXAMPLE 3 ***Finding the Equation of a Line Given Two Points of the Line*** Find the equation of the line that passes through the points $C(3,6)$ and $D(7,2)$.

Solution Since we are not given the slope of the line, we must determine it. (*Note:* It does not matter in which order the points are chosen.)

$$m = \frac{y_2 - y_1}{x_2 - x_1} = \frac{6 - 2}{3 - 7} = \frac{4}{-4} = -1$$

Next we use the point-slope formula. Choosing point C, we have $m = -1$, $x_1 = 3$, $y_1 = 6$.

$$\begin{aligned} y - y_1 &= m(x - x_1) \\ y - 6 &= -1(x - 3) && \textit{substituting} \\ y - 6 &= -1x + 3 && \textit{distributive property} \\ y &= -x + 9 && \textit{adding 6 to both sides} \end{aligned}$$

EXAMPLE 4 ***Finding the Equation of a Line Given Two Points of the Line*** Find the equation of the line that passes through the points $E(-2, -2)$ and $F(-4, -1)$.

Solution The slope of the line is

$$m = \frac{y_2 - y_1}{x_2 - x_1} = \frac{-2 - (-1)}{-2 - (-4)} = \frac{-2 + 1}{-2 + 4} = \frac{-1}{2} = -\frac{1}{2}$$

Using the point-slope formula and choosing point E, we have $m = -\frac{1}{2}$, $x_1 = -2$, and $y_1 = -2$:

$$\begin{aligned} y - y_1 &= m(x - x_1) \\ y - (-2) &= -\tfrac{1}{2}(x - (-2)) && \textit{substituting} \\ y + 2 &= -\tfrac{1}{2}(x + 2) && \textit{simplifying} \\ y + 2 &= -\tfrac{1}{2}x - 1 && \textit{distributive property} \\ y &= -\tfrac{1}{2}x - 3 && \textit{subtracting 2 from both sides} \end{aligned}$$

Alternate Solution Using the point-slope formula and choosing point F, we have $m = -\frac{1}{2}$, $x_1 = -4$, and $y_1 = -1$. Then the point-slope formula giv

$y - y_1 = m(x - x_1)$	
$y - (-1) = -\frac{1}{2}(x - (-4))$	*substituting*
$y + 1 = -\frac{1}{2}(x + 4)$	*simplifying*
$y + 1 = -\frac{1}{2}x - 2$	*distributive property*
$y = -\frac{1}{2}x - 3$	*subtracting 1 from both sides*

which is the same equation obtained with point E.

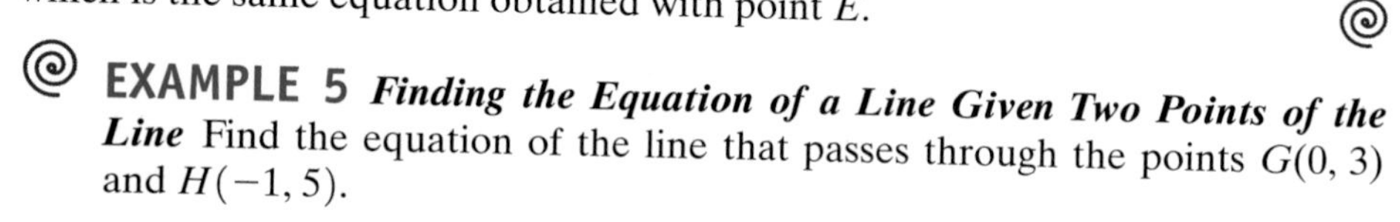

EXAMPLE 5 ***Finding the Equation of a Line Given Two Points of the Line*** Find the equation of the line that passes through the points $G(0, 3)$ and $H(-1, 5)$.

Solution The slope of the line is

$$m = \frac{y_2 - y_1}{x_2 - x_1} = \frac{5 - 3}{-1 - 0} = \frac{2}{-1} = -2$$

Using the point-slope formula and choosing point G, we have $m = -2$, $x_1 = 0$, and $y_1 = 3$:

$y - y_1 = m(x - x_1)$	
$y - 3 = -2(x - 0)$	*substituting*
$y - 3 = -2x + 0$	*distributive property*
$y = -2x + 3$	*adding 3 to both sides*

NW ***Now Work Problems 13 and 15.***

Slope-Intercept Formula

In Example 5, we note that $G(0, 3)$ is the y-intercept, the slope of the line is -2, and the equation of the line is $y = -2x + 3$. The coefficient of x, -2, is the slope of the line and the constant term, $+3$, is the y-intercept. This is not a coincidence.

Consider a line that crosses the y-axis at $(0, b)$ and has a slope m. Using the point-slope formula, we have

$y - y_1 = m(x - x_1)$	
$y - b = m(x - 0)$	*substituting*
$y - b = mx - 0$	*distributive property*
$y = mx + b$	*adding b to both sides*

The equation of the line is $y = mx + b$. The coefficient of x (m) is the slope of the line and the constant term ($+b$) is the y-intercept.

This equation, $y = mx + b$, is known as the slope-intercept form for the equation of a line.

$$y = mx + b$$

coefficient of x = slope of line   constant term = y-intercept

EXAMPLE 6 ***Finding the Slope and Y-Intercept of a Given Line*** Find the slope and y-intercept of the line whose equation is

a. $y = 3x + 2$ **b.** $y = -2x - 8$

c. $3x - 5y = 15$ **d.** $4x - 2y = 0$

Solution

a. The equation $y = 3x + 2$ is of the general form $y = mx + b$. The coefficient of x is the slope of the line and the constant term is the y-intercept. Therefore, $m = 3$ and the y-intercept is $(0, 2)$.

b. $y = -2x - 8$: Slope $= m = -2$; y-intercept is $(0, -8)$.

c. The equation $3x - 5y = 15$ is not of the general form $y = mx + b$. Hence, we must transform the given equation into the general form. We do this by solving for y:

$$3x - 5y = 15$$

$$-5y = -3x + 15 \quad \textit{subtracting 3x from both sides}$$

$$\frac{-5y}{-5} = \frac{-3}{-5}x + \frac{15}{-5} \quad \textit{dividing both sides by } -5$$

$$y = \frac{3}{5}x - 3 \quad \textit{simplifying}$$

The equation is now of the general form, with slope $= m = \frac{3}{5}$ and y-intercept $(0, -3)$.

d. The equation $4x - 2y = 0$ is not of the form $y = mx + b$. Solving it for y, we have

$$4x - 2y = 0$$

$$-2y = -4x + 0 \quad \textit{subtracting 4x from both sides}$$

$$\frac{-2y}{-2} = \frac{-4x}{-2} + \frac{0}{-2} \quad \textit{dividing both sides by } -2$$

$$y = 2x + 0 \quad \textit{simplifying}$$

$$y = 2x$$

Note that $y = 2x$ is the same as $y = 2x + 0$. The slope is 2 and the y-intercept is $(0, 0)$. The y-intercept of $(0, 0)$ also tells us that the line passes through the origin.

Graphing Lines Using the Slope-Intercept Formula

We can also use the slope-intercept formula for the equation of a line to graph the line. Consider the line whose equation is $y = \frac{2}{3}x + 2$. Note that the slope is $\frac{2}{3}$ and the y-intercept is $(0, 2)$. To graph the line, locate the y-intercept and then sketch a triangle for the slope. The slope is $\frac{2}{3}$ and we can think of this slope as rise over run:

$$\frac{2}{3} = \frac{\text{rise}}{\text{run}}$$

We start at the y-intercept $(0, 2)$ and move in a positive x-direction three units (for the run) and then up in a positive y-direction (for the rise), which brings us to the point $(3, 4)$. Note that $(3, 4)$ also satisfies the equation $y = \frac{2}{3}x + 2$. Since two points determine a line, we can graph the line as shown in Figure 10.11.

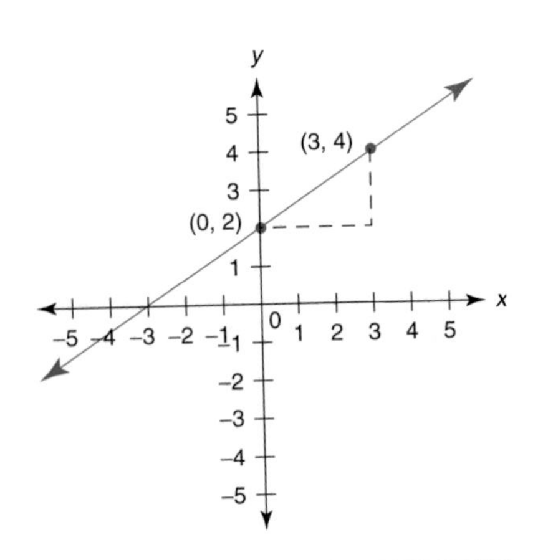

Figure 10.11

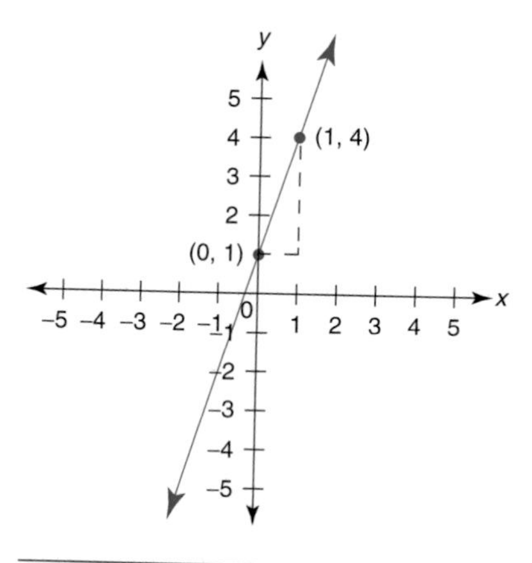

Figure 10.12a

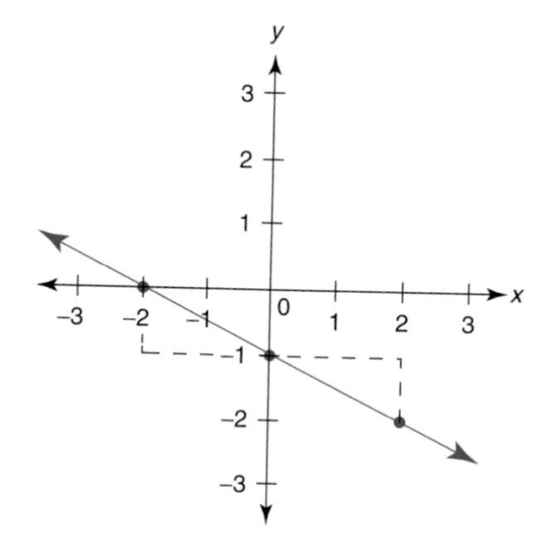
Figure 10.12b

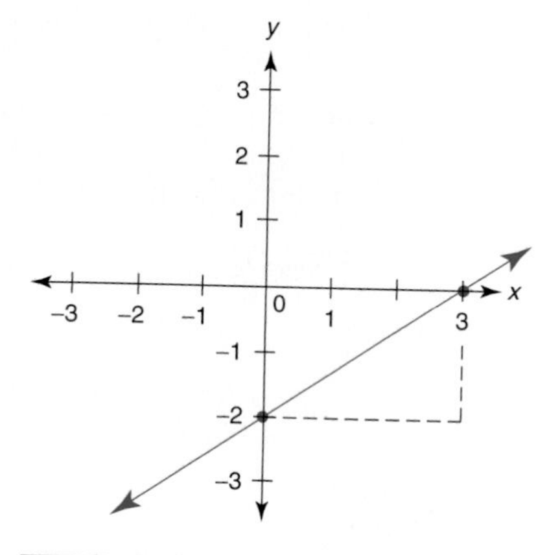
Figure 10.12c

EXAMPLE 7 ***Graphing a Line Expressed in Slope-Intercept Form*** Draw graphs of each of the following equations:

a. $y = 3x + 1$ **b.** $y = -\frac{1}{2}x - 1$ **c.** $2x - 3y = 6$

Solution

a. The given equation, $y = 3x + 1$, is in the slope-intercept form. The slope is 3 or $\frac{3}{1}$ and the y-intercept is $(0, 1)$. We start at $(0, 1)$ and move in a positive x-direction one unit (run) and then up in a positive y-direction three units (rise), which brings us to the point $(1, 4)$, enabling us to graph the line; see Figure 10.12a. Note that a line with a positive slope slants upward to the right.

b. The same procedure is used to graph $y = -\frac{1}{2}x - 1$. The y-intercept is $(0, -1)$. Note that the slope is $-\frac{1}{2} = \frac{-1}{2}$. Hence, we move in a positive x-direction two units from $(0, -1)$ and then down $(-)$ in a negative y-direction one unit. Since a slope of $-\frac{1}{2}$ is the same as $\frac{1}{-2}$ we can also determine a point by moving in negative x-direction two units from $(0, -1)$ and then up in a positive y-direction one unit. See Figure 10.12b. Note that a line with a negative slope slants downward to the right.

c. We must first transform $2x - 3y = 6$ into the general form $y = mx + b$ to find the slope and y-intercept:

$$2x - 3y = 6$$

$$-3y = -2x + 6 \qquad \textit{subtracting } 2x \textit{ from both sides}$$

$$\frac{-3y}{-3} = \frac{-2x}{-3} + \frac{6}{-3} \qquad \textit{dividing both sides by } -3$$

$$y = \frac{2}{3}x - 2 \qquad \textit{simplifying}$$

The slope is $\frac{2}{3}$ and the y-intercept is $(0, -2)$. From the point $(0, -2)$ we move in a positive x-direction three units (run) and then up in a positive y-direction two units (rise), which enables us to graph the line. See Figure 10.12c.

NW ***Now Work Problem 23.***

We shall not use the slope-intercept formula to graph horizontal or vertical lines. *The graph of any equation of the form $x = c$ is a vertical line parallel to the y-axis and passing through the point $(c, 0)$. The graph of any equation of the form $y = c$ is a horizontal line parallel to the x-axis and passing through the point $(0, c)$, which is the y-intercept.*

Comparing Point-Slope and Slope-Intercept Formulas

When we are given any two points that lie on a line, we can find the equation of the line by first finding the slope of the line and then using the coordinates of either point in the *point-slope formula*. We can also use the *slope-intercept formula* to find the equation of a line when we are given two points. We shall find the equation of a line using both techniques.

Consider the points $A(2, 4)$ and $B(6, 2)$. We want to find the equation of the line that passes through A and B. We must first find the slope:

$$m = \frac{y_2 - y_1}{x_2 - x_1} = \frac{2 - 4}{6 - 2} = \frac{-2}{4} = -\frac{1}{2}$$

Using the point-slope formula and choosing point A, we have $m = -\frac{1}{2}$, $x_1 = 2$, and $y_1 = 4$:

$y - y_1 = m(x - x_1)$	
$y - 4 = -\frac{1}{2}(x - 2)$	*substituting*
$y - 4 = -\frac{1}{2} \cdot x + (-\frac{1}{2})(-2)$	*distributive property*
$y - 4 = -\frac{1}{2}x + 1$	*simplifying*
$y = -\frac{1}{2}x + 5$	*adding 4 to both sides*

Now we shall use the slope-intercept formula, $y = mx + b$, to obtain the same equation. We first determine the slope in the same manner as before. Therefore, $m = -\frac{1}{2}$ and $y = mx + b$ becomes $y = -\frac{1}{2}x + b$. Now we must find a value for b and the equation will be complete. Since points $A(2, 4)$ and $B(6, 2)$ both lie on the line, their coordinates must satisfy the equation of the line. Hence, we can substitute the coordinates of A or B in the equation $y = -\frac{1}{2}x + b$ and solve it for b. Choosing point A, we have $x = 2$ and $y = 4$. Therefore,

$y = -\frac{1}{2}x + b$	
$4 = -\frac{1}{2} \cdot 2 + b$	*substituting 4 for y and 2 for x*
$4 = -1 + b$	*multiplying*
$5 = b$	*adding 1 to both sides*

We have found the value of b and we can substitute it in the equation $y = -\frac{1}{2}x + b$. The equation of the line is

$$y = -\frac{1}{2}x + 5$$

which is the same equation we obtained previously. One technique is not preferred over the other. You should be aware that there is more than one method for finding the equation of a line. In some cases you may choose to use the point-slope formula, whereas in other cases you may want to use the slope-intercept formula.

EXAMPLE 8 ***Using the Slope-Intercept Formula and Two Points to Find the Equation of a Line*** Using the slope-intercept formula, find the equation of the line that passes through the points $A(3, 6)$ and $B(7, 2)$.

Solution We first find the slope of the line:

$$m = \frac{y_2 - y_1}{x_2 - x_1} = \frac{2 - 6}{7 - 3} = \frac{-4}{4} = -1$$

Using the slope-intercept formula, $y = mx + b$, with $m = -1$, we have

$$y = -1x + b$$

Next, solve for b by substituting for x and y, using point A or point B. If we choose point B, we have $x = 7$ and $y = 2$:

$y = -1x + b$	
$2 = -1 \cdot 7 + b$	*substituting*
$2 = -7 + b$	*multiplying*
$9 = b$	*adding 7 to both sides*

Therefore, the equation of the line is $y = -1x + 9$. Note that we can rewrite the equation as $1x + y - 9 = 0$ or $x + y - 9 = 0$. This is known as the general form for the equation of a straight line. That is, the general form is $Ax + By + C = 0$.

EXAMPLE 9 ***Using the Slope-Intercept Formula and Two Points to Find the Equation of a Line*** Using the slope-intercept formula, find the equation of the line that passes through the points $C(-2, -2)$ and $D(-4, -1)$.

Solution We first find the slope of the line:

$$m = \frac{y_2 - y_1}{x_2 - x_1} = \frac{-1 - (-2)}{-4 - (-2)} = \frac{-1 + 2}{-4 + 2} = \frac{1}{-2} = -\frac{1}{2}$$

Hence, $y = mx + b$ becomes $y = -\frac{1}{2}x + b$.

Choosing point $D(-4, -1)$ and substituting for x and y, we have

$$y = -\tfrac{1}{2}x + b$$

$$-1 = -\tfrac{1}{2}(-4) + b \qquad \textit{substituting}$$

$$-1 = 2 + b \qquad \textit{multiplying}$$

$$-3 = b \qquad \textit{subtracting 2 from both sides}$$

Therefore, the equation of the line is $y = -\frac{1}{2}x - 3$.

Math Connections

Linear Models

The concepts of algebra we have discussed thus far can be used to model various real-world settings. The following is such an example.

Linear Depreciation

Most capital assets, such as farm machinery, automobiles, and business equipment, wear out or become obsolete over time and hence lose their value. This amount of value an asset loses each year during its useful life is called *depreciation*. One of the most widely used methods for calculating depreciation approved by the IRS is the *straight-line* or *linear depreciation* method, which is based on a constant depreciation rate for the entire working life of an asset. Let's now consider the following problem: The value, v, of an automobile declines linearly from its new purchase price of \$42,000 to \$6000 when it is 9 years old. To represent this relationship as a linear model, we observe the following:

- We must have an equation of the form $y = mx + b$, where m is the slope and b is the intercept. Because the value, v, of the car depends on its age, a, our linear model is of the form $v = ma + b$. Thus, v becomes our "y" and a becomes our "x."
- The given data can be equated to ordered pairs of the form (a, v). Thus, the ordered pairs are (0, 42000) and (9, 6000,)
- To find the slope, m, we calculate the change in y and divide this change by the change in x. In the context of this problem, we find the change in value, v, and divide this change by the change in age, a.

$$m = \frac{\text{change in value}}{\text{change in age}} = \frac{42{,}000 - 6000}{0 - 9} = \frac{36{,}000}{-9} = -4000$$

This results in the equation $v = -4000\,a + b$.

- To find b, we note that the ordered pair (0, 42,000) is the y-intercept. Hence, $b = 42{,}000$.

Therefore, our linear model is $v = -4000a + 42000$. This equation can now be used to determine the tax value of the given automobile based on the number of years since its purchase. A graphical representation of this model is shown below.

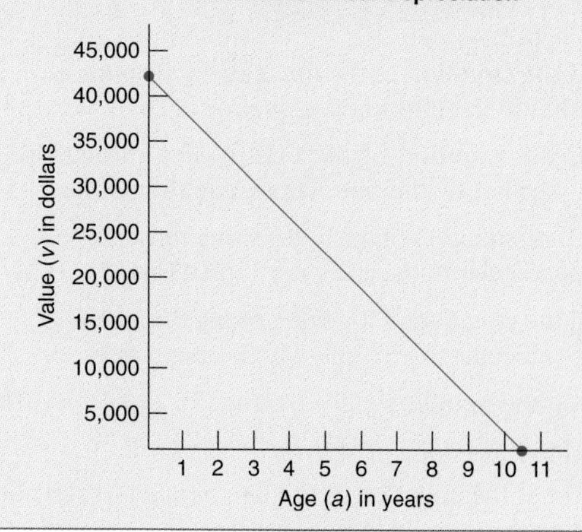

Exercises for Section 10.3

For Exercises 1–12, write the equation of the line that passes through the indicated point and has the indicated slope.

NW **1.** $A(3, 4), m = 1$

2. $B(3, 4), m = -3$

3. $C(-2, 0), m = -6$

4. $D(-4, 6), m = \frac{1}{2}$

5. $E(0, -4), m = \frac{3}{5}$

6. $F(0, -3), m = \frac{2}{3}$

7. $G(-6, 2), m = \frac{-3}{4}$

8. $H(-4, -2), m = -\frac{3}{2}$

9. $I(-1, 0), m = \frac{1}{7}$

10. $J(3, 3), m = \frac{1}{3}$

11. $K(-2, 3), m = 0$

12. $L(4, 0), m = \frac{3}{2}$

For Exercises 13–22, write the equation of the line that passes through the indicated points.

NW **13.** $A(3, 2)$ and $B(0, 1)$

14. $C(-2, 3)$ and $D(3, -2)$

NW **15.** $E(-2, 4)$ and $F(1, -2)$

16. $G(-2, 0)$ and $H(0, -3)$

17. $I(0, -4)$ and $J(-3, 0)$

18. $K(5, -2)$ and $L(6, 7)$

19. $M(1, -6)$ and $N(-2, -3)$

20. $O(2, 3)$ and $P(-1, 1)$

21. $Q(-5, -2)$ and $R(-3, -1)$

22. $S(-2, -3)$ and $T(-7, 0)$

For Exercises 23–32, find the slope and y-intercept of the line whose equation is given and then graph the line.

NW **23.** $y = 3x + 4$

24. $y = -2x + 3$

25. $y = -x - 6$

26. $y = -x + 1$

27. $x - 2y = -4$

28. $3x - 2y = 6$

29. $3x - 5y + 10 = 0$

30. $-2x - 3y = 6$

31. $2x - y = 0$

32. $x - y = 0$

33. Find the equation of the line passing through $(-1, 2)$ and parallel to the line whose equation is $2x - y = -3$.

34. Find the equation of the line passing through $(-2, -3)$ and parallel to the line whose equation is $2x - 5y = 6$.

35. Find the equation of the line passing through $(-3, -1)$ and perpendicular to the line whose equation is $2x + 3y = -1$.

36. Find the equation of the line passing through $(-1, -1)$ and perpendicular to the line whose equation is $x + 2y = 6$.

37. Given the points $A(-2, -3)$, $B(2, 5)$, and $C(-1, 0)$:

a. Find the slope of $\overline{AB}$.

b. Find the equation of the line parallel to $\overline{AB}$ and passing through the point $(3, 10)$.

c. Find the equation of the line that passes through A and C.

d. Find the equation of the line that passes through B and C.

38. Given the points $P(1, -1)$, $Q(4, 2)$, and $R(6, 4)$,

a. Find the slopes of line segments $\overline{PQ}$ and $\overline{PR}$.

b. From your answer to part **a**, what can you conclude regarding the points P, Q, and R?

c. Find the equation of the line that passes through P and R.

d. Does the point $S(10, 8)$ lie on the line PR? Why or why not?

39. Given that quadrilateral $ABCD$ is a rectangle whose vertices are $A(-4, 4)$, $B(-4, 2)$, $C(4, 4)$, and $D(4, 2)$,

a. Find the equation of side AC.

b. Find the equation of side CD.

c. Find the equation of diagonal AD.

d. Find the equation of diagonal BC.

Writing Mathematics

40. Write the point-slope formula for the equation of a straight line, and explain how to use it to obtain the equation of a straight line.

41. Write the slope-intercept formula for the equation of a straight line, and explain how to use it to obtain the equation of a straight line.

Challenge Exercises

42. The vertices of parallelogram $ABCD$ are $A(-2, 4)$, $B(2, 6)$, $C(7, 2)$, and $D(k, 0)$.

a. Find the slope of side AB.

b. Express the slope of side CD in terms of k.

c. Using the results found in parts **a** and **b**, find the value of k.

d. Find the equation of side AD.

43. Given the points $A(3, 2)$, $B(-2, -3)$, and $C(3, 0)$,

a. Determine the equation of the line that is parallel to the x-axis and passes through A.

b. Determine the equation of the line that is parallel to the y-axis and passes through B.

c. Determine the equation of the line that is parallel to line AB and passes through C.

d. Determine the equation of the line that is perpendicular to line AB and passes through C.

e. Determine the equation of the line that is perpendicular to the x-axis and passes through C.

44. Given points $A(-2, 3)$, $B(2, 3)$, and $C(0, 0)$,

a. What is the slope of AB?

b. Write an equation of the straight line that is parallel to AB and passes through the point $(4, 1)$.

c. Write the equation of the line that represents all points that are equidistant from A and B.

d. Write an equation of the line passing through B and C.

e. Write an equation of the line passing through A and C.

Research/Group Activities

45. One of the most influential English textbook authors of the sixteenth century was Robert Recorde (1510–1558). Recorde wrote several texts, most notably an algebra text, *The Whetstone of Witte*, published in 1557. Write a paper on his life and contributions to algebra.

46. Michael Stifel (1487–1567) was a notable German algebraist of the sixteenth century. He is described as one of the oddest personalities in the history of mathematics. Write a paper on his life and contributions to algebra.

Real-World Data

47. **Cost of a Long-Distance Call** A long-distance telephone service offers residential customers a choice of two different calling plans: Plan A, a flat-rate of 7¢ per minute, and Plan B, 39¢ to connect plus 3¢ per minute. The graphs of the lines that correspond to these price plans are shown below. Write the equation of each line where x = the number of minutes and C = the cost of the call based on x minutes. How much will a 20-minute long-distance call cost under each plan?

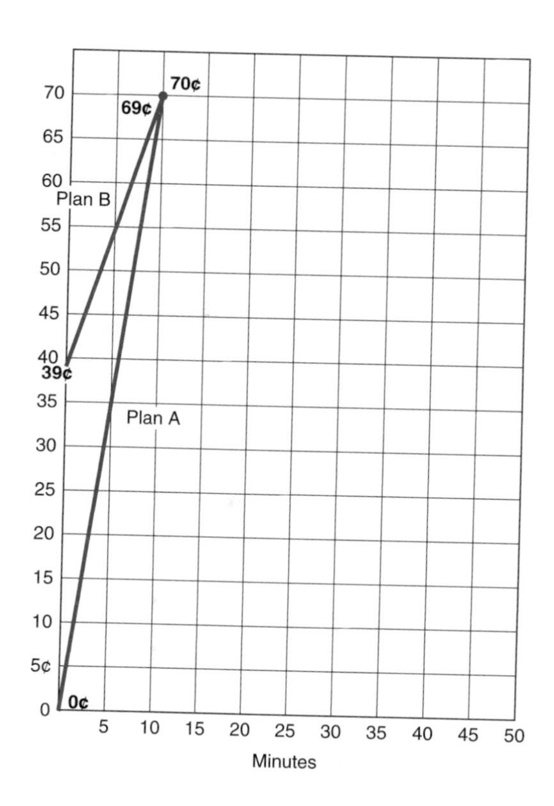

48. **Cost of a trip to Italy or Scotland** John and Mimi are considering a trip to either Italy or Scotland. They estimated that a trip to Italy will cost them $800 for round-trip air fare and $100 per day for each day that they stay there. The trip to Scotland, however, will cost them $1000 for round-trip air fare and $200 per day for each day that they stay there. The graphs of the lines that correspond to these trips are given below. Write the equation of each line where x = the number of days they stay in Italy or Scotland and C = the cost of the trip. How much will a two-week trip to either destination cost John and Mimi?

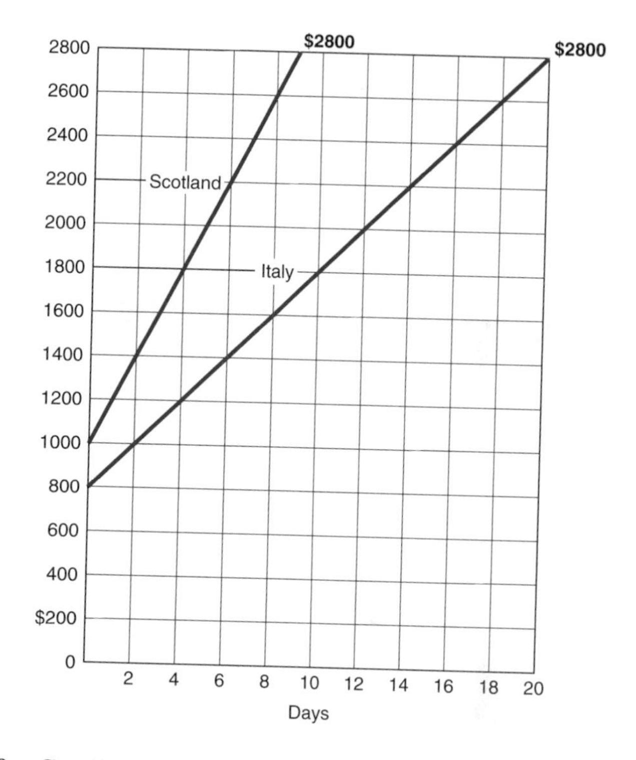

49. **Gasoline mileage** A study was done to determine the relationship between a car's weight and its gasoline mileage between cars manufactured in 1966 and those manufactured in 2002. The results are shown in the graph below. Write the equation of each line where W = weight of a vehicle in pounds and Y = gasoline mileage in miles per gallon (mpg). Compare the gasoline mileage for a 2500 pound vehicle manufactured in each of the given years.

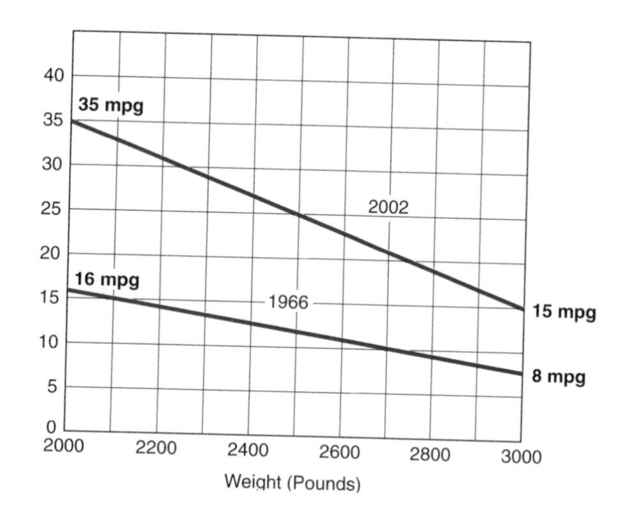

Just for fun

Mr. Apple and Mr. Jack have an 8-gallon container full of cider. They also have two empty containers whose capacities are 5 gallons and 3 gallons, respectively. How can cider be poured, using all three containers, so that Mr. Apple and Mr. Jack each have 4 gallons of cider?

10.4 LINEAR PROGRAMMING

Overview

In Chapter 9, we learned how to solve and graph linear equations and inequalities in two variables. As part of this discussion, we also learned that linear equations can be used to model data. In this section, we apply these concepts to solve a special type of problem called a ***linear programming problem***. To understand what a linear programming problem involves, let's consider the following situation.

> *The E-Z Furniture Company manufactures tables and chairs for family rooms and patios. All of the company's furniture is constructed in two buildings. Tables require 3 hours of work in Building I and 2 hours of work in Building II. Chairs require 2 hours of work in Building I and 4 hours of work in Building II. The profit from each table is $6, and the profit from each chair is $5. Due to union regulations, the two buildings can operate for at most 8 hours per day. Given these restrictions, how many tables and how many chairs should the company produce each day in order to maximize its profits?*

In this illustration, the ultimate goal is to maximize profits based on a set of restrictions. This is a common feature of linear programming problems. In fact, problems in which we try to maximize or minimize a situation based on a set of restrictions are called ***optimization problems***, and linear programming is a strategy or tool for solving optimization problems. Although the theory of linear programming has been expanded to solve extremely complex problems, our goal in this section is to introduce the concept of linear programming and hence we will restrict our discussion to relatively simple problems similar to the E-Z Furniture Company example.

To solve optimization problems similar to that of the E-Z Furniture Company, it is helpful to first represent the problem using a *mathematical model*, which is nothing more than a mathematical representation of an actual situation. By representing a situation mathematically, we can gain a better understanding of the given situation, which in turn can help us make better and informed decisions. Mathematical models used for linear programming problems all involve the following components:

- **Decision variables**, which describe the decisions to be made. For example, in the E-Z Furniture Company illustration, the company must decide how many tables and chairs to construct each day.

- **Objective function**, which represents the ultimate goal relative to the decision variables. In linear programming problems, the goal is to maximize something (usually revenue or profit) or minimize something (usually costs). In the E-Z Furniture Company illustration, the objective function is to maximize profit relative to the number of tables and chairs the company can construct.
- **Constraints**, which describe the set of restrictions that limit the objective function. In the E-Z Furniture Company illustration, the buildings that are used to make the tables and chairs cannot operate for more than 8 hours per day, which limit the number of tables and chairs that can be constructed daily.
- **Sign restrictions**, which indicate if the decision variables can assume only nonnegative values (i.e., greater than or equal to zero), or if the variables are unrestricted in sign and can take on positive, negative, and zero values. For the E-Z Furniture Company example, the decision variables are only meaningful if they are nonnegative.

In summary, linear programming is a strategy for solving an optimization problem, which involves maximizing or minimizing a situation. To do this, we represent the problem as a mathematical model. This model consists of decision variables that describe the decisions we want to make, an objective function that indicates what we want to maximize or minimize, a set of constraints that restrict the value of the decision variables, and a sign restriction that specifies whether the decision variables are nonnegative or unrestricted in sign. We now turn our attention to solving a linear programming problem.

Solving a Linear Programming Problem

To demonstrate how to solve a linear programming problem, we will solve the E-Z Furniture Company problem. As noted above, to solve a linear programming problem, we must represent the problem as a mathematical model. This involves defining the decision variables, constraints, objective function, and sign restrictions. Before we begin, though, it is sometimes helpful to first summarize all of the given information in a table. The summary table for the E-Z Furniture Company example is shown below.

	Time Needed (hours)		
	Building I	Building II	Profit for Each
Table	3	2	\$6
Chair	2	4	\$5
Time limit	8	8	

Let's now develop the model that represents this information.

Decision variables. The decision variables of the E-Z model may be defined as follows:

$x =$ the number of tables to be constructed
$y =$ the number of chairs to be constructed

Objective function. The objective of the E-Z Furniture Company is to maximize profits, P. Based on the given information, a profit of \$6 is made on each table. Therefore, two tables would yield a profit of $6 \times 2 = 12$ dollars. Similarly, x tables would yield a profit of $6x$ dollars. A profit of \$5 is made on each chair. Hence, y chairs would yield a profit of $5y$ dollars. An objective function that represents the goal of maximizing profits relative to the decision variables, then, is

$$P = 6x + 5y$$

Constraints. The constraints of the problem can be constructed directly from the information given in the table. For example, if a table requires 3 hours in Building I, then 2 tables require $2 \times 3 = 6$ hours in Building I. Therefore, x tables require $3x$ hours in Building I. Similarly, y chairs would require $2y$ hours in Building I. Now for the constraints. What do we know about the time spent in Building I? It must be less than or equal to 8 hours, due to union rules. This leads to the following inequality:

$$3x + 2y \leq 8 \text{ (Building I constraint)}$$

Similarly, x tables require $2x$ hours in Building II, and y chairs require $4y$ hours in Building II. Building II can also be used for a maximum of 8 hours. Therefore, a second constraint is

$$2x + 4y \leq 8 \text{ (Building II constraint)}$$

Sign restrictions. Because x and y represent the respective number of tables and chairs to be manufactured, x and y cannot be negative quantities. Therefore, the sign restrictions on the variables are $x \geq 0$ and $y \geq 0$.

Putting all of this information together, the E-Z Furniture Company model may be represented as follows:

1. $3x + 2y \leq 8$	*Constraint 1*
2. $2x + 4y \leq 8$	*Constraint 2*
3. $P = 6x + 5y$	*Profit*
4. $x \geq 0$	*Sign restriction on x*
5. $y \geq 0$	*Sign restriction on y*

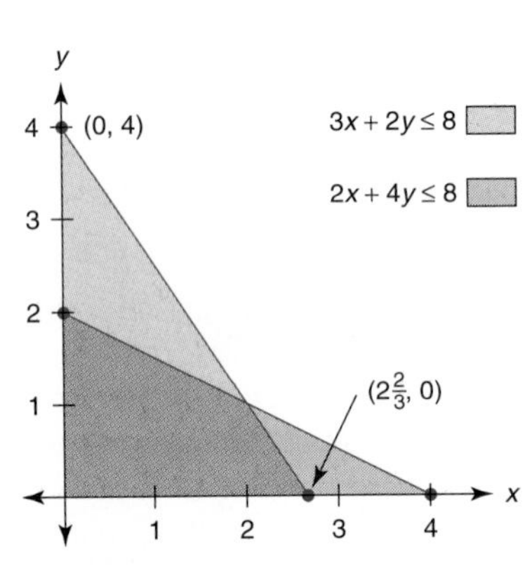

Figure 10.13

$3x + 2y \leq 8$ and $2x + 4y \leq 8$.

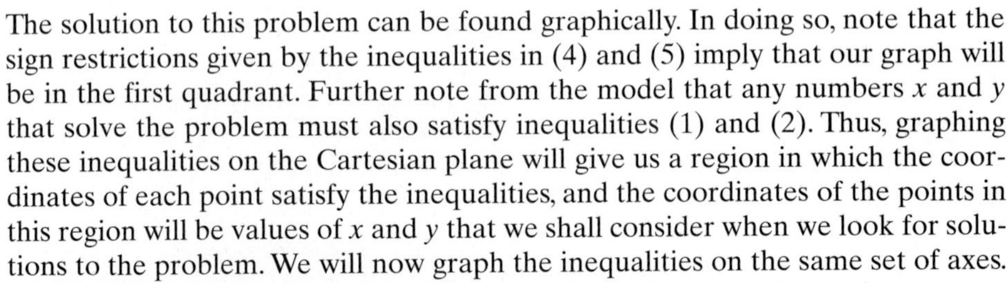

The solution to this problem can be found graphically. In doing so, note that the sign restrictions given by the inequalities in (4) and (5) imply that our graph will be in the first quadrant. Further note from the model that any numbers x and y that solve the problem must also satisfy inequalities (1) and (2). Thus, graphing these inequalities on the Cartesian plane will give us a region in which the coordinates of each point satisfy the inequalities, and the coordinates of the points in this region will be values of x and y that we shall consider when we look for solutions to the problem. We will now graph the inequalities on the same set of axes.

The graphs of the two inequalitites are shown in Figure 10.13. We are concerned only with the portion of this figure where the two half-planes intersect. This is because the coordinates of each point in the intersection satisfy both of the inequalities, and therefore the intersection forms the solution set of the conjunction of the two inequalities.

Let us examine this region more closely: Figure 10.14 shows this region with its four corners labeled. The corner points are $A(0, 0)$, $B(2\frac{2}{3}, 0)$, $C(2, 1)$, and $D(0, 2)$.

It can be proved, using more advanced mathematics, that in a linear programming problem, any maximum or minimum values of the objective function always occur at a vertex (corner). If an objective function has an optimal solution, then that solution will be at a vertex of the set of feasible solutions. Hence, our profit expression P will have its maximum or minimum value at a corner of the region in Figure 10.14. We can test these values by substituting them in the profit expression, $P = 6x + 5y$:

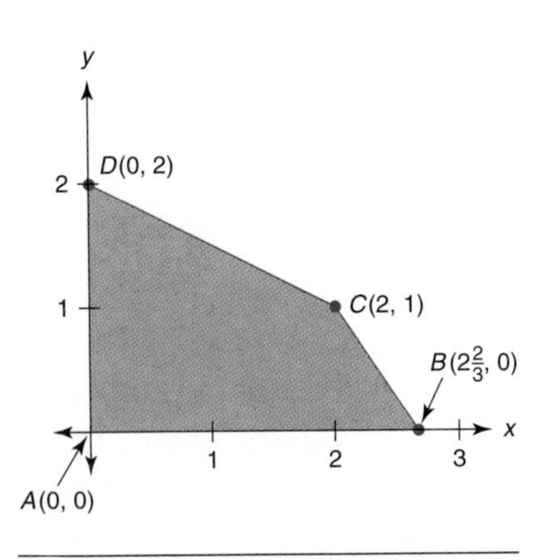

Figure 10.14

At $A(0, 0)$, we have	$P = 6 \cdot 0 + 5 \cdot 0 = 0 + 0 = 0.$
At $B(2\frac{2}{3}, 0)$, we have	$P = 6 \cdot \frac{8}{3} + 5 \cdot 0 = \frac{48}{3} + 0 = 16.$
At $C(2, 1)$, we have	$P = 6 \cdot 2 + 5 \cdot 1 = 12 + 5 = 17.$
At $D(0, 2)$, we have	$P = 6 \cdot 0 + 5 \cdot 2 = 0 + 10 = 10.$

Biography: George Bernard Dantzig

George Bernard Dantzig (Father of Linear Programming) (1914–) was named after the famous writer George Bernard Shaw. His father was hoping that his son would become a writer, but instead he became a mathematician. In 1947, Dantzig invented linear programming and the simplex method (the technique used to solve linear programming problems). During his first year as a graduate student at Berkeley he arrived late to class one day and noticed two problems written on the board. He copied them down thinking they were homework problems due the next day. He went to his professor a few days later apologizing for their late return, stating, "the problems seemed to be a little harder to do than usual." Six weeks later his professor came to his house congratulating him for solving two famous unsolved problems in statistics.

Dantzig has worked for the Bureau of Labor Statistics, the Pentagon, the RAND Corporation, and at Stanford University. In 1975, he received from President Ford the Medal of Science. In the same year Leonid Kantoronich and Tjalling Koopmans received the Nobel Prize in economics for their work on the relationship between linear programming and the theory of optimum allocation of resources. Koopmans has expressed regret that Dantzig was not also honored by the Nobel Committee.

The maximum profit occurs when $x = 2$ and $y = 1$, and it is \$17. The E-Z Furniture Company should produce two tables and one chair to obtain a maximum profit. Note that the minimum profit, \$0, occurs at $A(0, 0)$.

Keep in mind that this is an illustrative example of a linear programming problem and one method for solving such problems. More complex problems involve more variables and must be solved in a different manner.

In solving this example, we evaluated the objective function at the corner points of the graph of the constraint inequalities. As frequently happens, one of these corner points—$B(2\frac{2}{3}, 0)$—does not have integer coordinates. What if the profit expression P had its maximum value at $B(2\frac{2}{3}, 0)$ instead of at $C(2, 1)$? (This could easily happen if the problem were slightly different.) In terms of the original problem, this would mean that the E-Z Furniture Company should manufacture $2\frac{2}{3}$ tables and 0 chairs. But it does not make much sense to manufacture $\frac{2}{3}$ of a table. In such a case, linear programming yields only an approximate answer to the problem. A more sophisticated mathematical technique, *integer programming*, must be used to find exact answers to problems that require whole numbers as answers.

EXAMPLE 1 ***Solving a Linear Programming Problem*** The Blivit Electronic Company manufactures bleeps and peeps. Manufacturing a bleep requires 2 hours on machine A and 1 hour on machine B. Manufacturing a peep requires 1 hour on machine A and 1 hour on machine B. Machine A cannot be used more than 7 hours a day, and machine B cannot be used more than 5 hours a day. If the profit from a bleep is \$5 and that from a peep is \$4, how many of each should be produced to maximize profit?

Solution We begin by letting

x = the number of bleeps to be manufactured

y = the number of peeps to be manufactured

Next we list all of the given information in a table.

	Time Needed (hours)		
	Machine A	Machine B	Profit for Each
Bleep	2	1	\$5
Peep	1	1	\$4
Time limit	7	5	

We know that x bleeps require $2 \cdot x$, or $2x$, hours on machine A, and y peeps require $1y$, or y, hours on machine A. The time limit on machine A is 7 hours. Therefore, we have

$$2x + y \le 7$$

Similarly, x bleeps require $1x$, or x, hours on machine B, and y peeps require $1y$, or y, hours on machine B. The time limit on machine B is 5 hours. Therefore, we have

$$x + y \le 5$$

The profit for x bleeps is $5x$ and the profit for y peeps is $4y$. The total profit, P, is $5x + 4y$. That is,

$$P = 5x + 4y$$

We want to maximize P.

We now have three mathematical statements from the given information in the table:

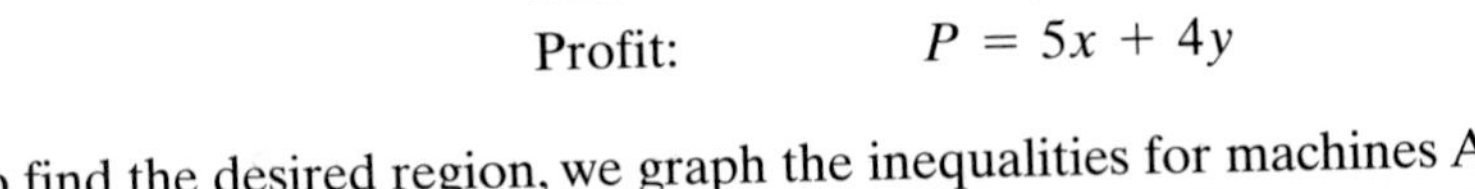

Machine A: $2x + y \le 7$
Machine B: $x + y \le 5$
Profit: $P = 5x + 4y$

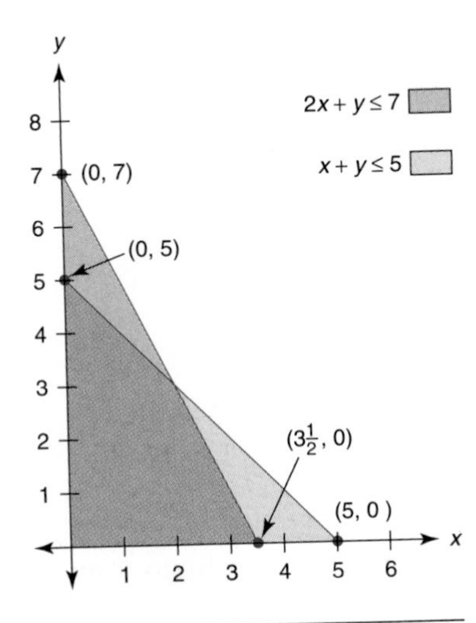

Figure 10.15
$2x + y \le 7$ and $x + y \le 5$.

To find the desired region, we graph the inequalities for machines A and B on the same set of axes. We do this only where $x \ge 0$ and $y \ge 0$, as shown in Figure 10.15.

The feasible region whose vertices will provide a maximum or minimum is that region where the half-planes intersect. The region and its corners are shown in Figure 10.16.

We now test the profit expression $P = 5x + 4y$ at each of these vertices:

At $A(0, 0)$, $P = 5 \cdot 0 + 4 \cdot 0 = 0 + 0 = 0.$
At $B(3\frac{1}{2}, 0)$, $P = 5 \cdot \frac{7}{2} + 4 \cdot 0 = \frac{35}{2} + 0 = 17.5.$
At $C(2, 3)$, $P = 5 \cdot 2 + 4 \cdot 3 = 10 + 12 = 22.$
At $D(0, 5)$, $P = 5 \cdot 0 + 4 \cdot 5 = 0 + 20 = 20.$

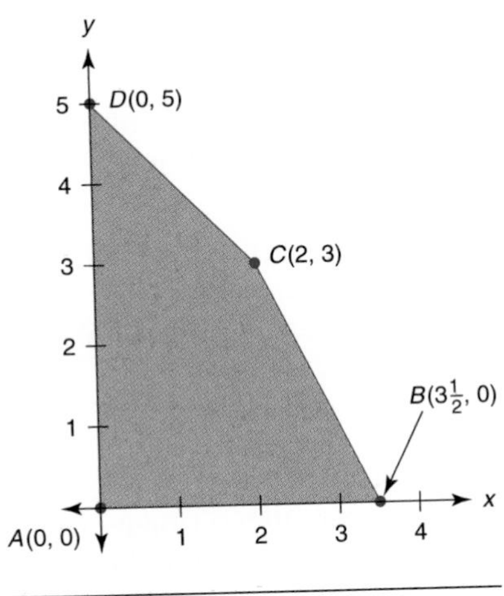

Figure 10.16

The maximum profit occurs when $x = 2$ and $y = 3$, and it is \$22. The Blivit Electronic Company should produce two bleeps and three peeps to obtain a maximum profit.

NW *Now Work Problems 1 and 11.*

When linear programming problems involve only two variables, the technique used in Example 1 (and in the E-Z Furniture Company problem) is convenient, and it is the only technique we will consider here. For problems containing many variables, more complicated techniques, usually executed by a computer, must be used. The most common of these techniques is called the *simplex method*.

Exercises for Section 10.4

NW 1. Find the maximum value of $P = 4x + 2y$ under the following conditions:

$$x \geq 0, \quad y \geq 0, \quad x + y \leq 5, \quad x - y \leq 1$$

2. Find the maximum value of $P = 3x + 8y$ under the following conditions:

$$x \geq 0, \quad y \geq 0, \quad x + y \leq 7, \quad 3x + y \leq 15$$

3. Find the maximum value of $P = 4x + 3y$ under the following conditions:

$$x \geq 0, \quad y \geq 0, \quad 2x + y \leq 12, \quad x + y \leq 7$$

4. Find the maximum value of $P = x + 4y$ under the following conditions:

$$x \geq 0, \quad y \geq 0, \quad x + y \leq 10, \quad 3x - y \leq 6$$

5. Find the maximum value of $P = 2x + y$ under the following conditions:

$$x \geq 0, \quad y \geq 0, \quad x + y \leq 16, \quad x + y \leq 10$$

6. Find the maximum value of $P = 2x + y$ under the following conditions:

$$x \geq 0, \quad y \geq 0, \quad x + y \leq 16, \quad x + y \geq 10$$

7. Find the maximum value of $P = 5x + 4y$ under the following conditions:

$$x \geq 0, \quad y \geq 0, \quad 2x - 3y \leq 12,$$
$$x - 3y + 9 \geq 0, \quad x \leq 6$$

8. Find the maximum value of $P = 3x - 2y$ under the following conditions:

$$x \geq 0, \quad y \geq 0, \quad 3x - 5y + 15 \geq 0,$$
$$x \leq 5, \quad y \leq 8$$

9. Find the maximum value of $P = 3x - 2y$ under the following conditions:

$$x \geq 0, \quad y \geq 0, \quad 3x - 5y + 15 \leq 0,$$
$$x \leq 5, \quad y \leq 8$$

10. A manufacturer makes bikes and wagons. To produce a bike requires 2 hours on machine A and 4 hours on machine B. To produce a wagon requires 3 hours on machine A and 2 hours on machine B. Machine A can operate at most 12 hours per day and machine B can operate at most 16 hours per day. If the manufacturer makes a profit of \$12 on a bike and \$10 on a wagon, how many of each should be produced in order to maximize profit?

NW 11. A manufacturer makes lawn mowers and snow blowers. To produce a lawn mower requires 3 hours on machine A and 2 hours on machine B. To produce a snow blower requires 2 hours on machine A and 4 hours on machine B. Machine A can operate at most 18 hours per day and machine B can operate at most 20 hours per day. If the manufacturer makes a profit of \$20 on a lawn mower and \$30 on a snow blower, how many of each should be produced in order to maximize profit?

12. Frank Sloane raises pheasants and partridges and has room for at most 100 birds. It costs him \$2 to raise a pheasant and \$3 to raise a partridge, and he has \$240 to cover these costs. If he can make a profit of \$7 on each pheasant and \$8 on each partridge, how many of each bird should he raise in order to maximize his profit?

13. Betty Juarez has a 100-acre farm where she raises two crops, potatoes and cauliflower. It costs her \$20 to raise an acre of potatoes and \$40 to raise an acre of cauliflower, and she has \$2600 to cover the costs. If she can make a profit of \$35 on each acre of potatoes and \$60 on each acre of cauliflower, how many acres of each crop should she plant in order to maximize her profit?

14. The Long Island Shellfish Company processes (cleans, sorts, opens, and freezes) oysters and clams. In a given week, the company can process 600 bushels of shellfish, of which 100 bushels of oysters and 200 bushels of clams are required by regular customers (restaurants). The profit on a bushel of oysters is \$8 and on a bushel of clams is \$10. How many bushels of oysters and of clams should the company process in order to maximize its profit?

15. A service station manager stocks two brands of motor oil, X and Y. The manager has room for no more than 60 cans of oil in the station. He also knows that at least twice as much brand X is sold as brand Y. If he makes a profit of 10 cents a can on brand X and 12 cents a can on brand Y, how many cans of each kind should he stock in his station to make the maximum profit?

Writing Mathematics

16. What are *constraints* in a linear programming problem?

17. What is the *objective function* in a linear programming problem?

18. In your own words, describe *linear programming* to a classmate.

Challenge Exercises

19. Minimize the objective function $C = 7x + 4y$ with the constraints

$$3x + y \geq 6, \quad x + y \geq 4, \quad x + 3y \geq 6, \quad x \geq 0, y \geq 0$$

20. The Jilrac Kid Company operates two factories, each producing three items: infant car seats, strollers, and high chairs. The following table shows the daily output of each factory:

	Infant Car Seats	Strollers	High Chairs
Factory A	100	200	400
Factory B	100	500	200

The daily cost of operation for Factory A is $10,000. The daily cost of operation for Factory B is $20,000. To fill orders, the Jilrac Kid Company must produce at least 5000 infant car seats, 15,000 strollers, and 16,000 high chairs next quarter. How many days should each factory be in production to minimize costs if, in addition, due to union regulations, the sum of the days both factories can be open next quarter cannot exceed 100 days?

Research/Group Activities

21. Linear programming has been used successfully to solve a wide range of problems, from military logistics to problems in business and the social sciences. Research the history of linear programming and its use in solving practical problems. Prepare a report to be presented in class.

22. Three people associated with the development of linear programming are George Bernard Dantzig, L. G. Khachion, and Narendra Karmarkar. Choose one of these people and write a paper on his life and contributions to the development of linear programming.

Just for fun

Two people have the same parents. They were born on the same day, and at the same place, but they are not twins. How are they related?

10.5 SOLVING QUADRATIC EQUATIONS

In our previous work, we solved equations of the form $ax + b = 0, a \neq 0$. Equations of this form are called **linear equations.** In this section, we will solve another type of equation called a **quadratic equation.** The **standard form** of a quadratic equation is

$$ax^2 + bx + c = 0, \quad a \neq 0$$

Note that a is the coefficient of x^2, b is the coefficient of the x term, and c is a constant. The following are all examples of a quadratic equation expressed in standard form:

$$\begin{array}{ll} (1)\ x^2 + 4x - 6 = 0 & (a = 1, b = 4, \text{and } c = -6) \\ (2)\ x^2 + 10 = 0 & (a = 1, b = 0, \text{and } c = 10) \\ (3)\ 2x^2 - 7x = 0 & (a = 2, b = -7, \text{and } c = 0) \\ (4)\ 2x^2 = 0 & (a = 2, b = 0, \text{and } c = 0) \end{array}$$

In the equations above, equations (2), (3), and (4) are *incomplete quadratic equations* because they are "missing" either the b term or the c term (or both). That is, in equation (2) $b = 0$, in equation (3) $c = 0$, and in equation (4) $b = 0$ and $c = 0$.

Although there are many different methods for solving quadratic equations, we will discuss only one. We will solve quadratic equations by applying the **quadratic formula.** To solve quadratic equations using this formula, the equation must be in standard form. That is, if the equation $ax^2 + bx + c = 0$ has solutions in the real number system, then the solutions can be found by applying the formula

$$x = \frac{-b \pm \sqrt{b^2 - 4ac}}{2a}$$

Biography: Emmy Amalie Noether

Born in Germany, and the daughter of a mathematician, Emmy Amalie Noether (1882–1935) grew up in a world of mathematics. She received a degree from the University of Erlangen, and in 1916 Noether began further studies at the University of Göttingen, where she worked with two well-known and respected mathematicians, David Hilbert and Felix Klein. Together they developed the mathematical aspects of the theory of relativity.

At that time women were not allowed to hold faculty rank at universities; however, at the insistence of the mathematics faculty she was given a nonfaculty teaching position.

She taught classes of enthusiastic students at Göttingen until 1933, when the Nazis came to power and people of Jewish descent were denied the right to teach. Emmy Noether then came to the United States and taught at Bryn Mawr College until her death. Her greatest contributions were in the area of abstract algebra, particularly with the theory of rings.

As indicated above, this formula can be used to solve any quadratic equation that is expressed in standard form and which has real number solutions. You should learn it and be able to state it as easily as your phone number or e-mail address. In the numerator of this formula, note that we have the symbol $\pm$. This symbol is read as "plus $(+)$ or minus $(-)$" and indicates that there are two solutions of a quadratic equation. More specifically, the formula can be reexpressed as

$$\text{Solution 1} = x = \frac{-b + \sqrt{b^2 - 4ac}}{2a}$$

and

$$\text{Solution 2} = x = \frac{-b - \sqrt{b^2 - 4ac}}{2a}$$

This will be demonstrated in the examples that follow. The development of the quadratic formula is based on a method called *completing the square*, which can be found in most algebra textbooks. (See also Problem 35.)

We shall now look at some examples to see how we can apply the formula. Consider the equation

$$x^2 + 6x + 8 = 0$$

We shall use the quadratic formula to find the roots of the equation. Since the equation is already in standard form, our next task is to determine the values of a, b, and c. Note that $x^2 + 6x + 8 = 0$ is the same as $1x^2 + 6x + 8 = 0$, and the standard form is $ax^2 + bx + c = 0$. Therefore, we have

$$a = 1, \quad b = 6, \quad c = 8$$

Next, we substitute these values in the quadratic formula:

$$x = \frac{-b \pm \sqrt{b^2 - 4ac}}{2a}$$

$$x = \frac{-6 \pm \sqrt{6^2 - 4 \cdot 1 \cdot 8}}{2 \cdot 1}$$

Simplifying, we have

$$x = \frac{-6 \pm \sqrt{36 - 32}}{2}$$

$$x = \frac{-6 \pm \sqrt{4}}{2}$$

$$x = \frac{-6 \pm 2}{2}$$

$$x = \frac{-6 + 2}{2} \quad \text{or} \quad x = \frac{-6 - 2}{2}$$

$$x = \frac{-4}{2} \qquad\qquad x = \frac{-8}{2}$$

$$x = -2 \qquad\qquad x = -4$$

The solution set is $\{-2, -4\}$.

Check:

If $x = -2$,

$$(-2)^2 + 6(-2) + 8 = 0$$
$$4 - 12 + 8 = 0$$
$$0 = 0$$

If $x = -4$,

$$(-4)^2 + 6(-4) + 8 = 0$$
$$16 - 24 + 8 = 0$$
$$0 = 0$$

The solution checks.

To solve a quadratic equation by means of the quadratic formula, we must be sure that the equation is in standard form before we determine the values for a, b, and c. Also, in determining these values, we must be sure to include the correct signs for a, b, and c.

EXAMPLE 1 ***Solving a Quadratic Equation*** Solve for x by means of the quadratic formula:

$$2x^2 - 3x + 1 = 0$$

Solution The equation is in standard form. Therefore, $a = 2$, $b = -3$, and $c = 1$.

$$x = \frac{-b \pm \sqrt{b^2 - 4ac}}{2a}$$

$$x = \frac{-(-3) \pm \sqrt{(-3)^2 - 4(2)(1)}}{2(2)}$$

$$x = \frac{3 \pm \sqrt{9 - 8}}{4}$$

$$x = \frac{3 \pm \sqrt{1}}{4}$$

$$x = \frac{3 \pm 1}{4}$$

$$x = \frac{3 + 1}{4} \quad \text{or} \quad x = \frac{3 - 1}{4}$$

$$x = \frac{4}{4} \qquad\qquad x = \frac{2}{4}$$

$$x = 1 \qquad\qquad x = \frac{1}{2}$$

The solution set is $\{1, \frac{1}{2}\}$.

Check:

If $x = 1$,

$$2 \cdot 1^2 - 3 \cdot 1 + 1 = 0$$
$$2 \cdot 1 - 3 + 1 = 0$$
$$2 - 3 + 1 = 0$$
$$0 = 0$$

If $x = \frac{1}{2}$,

$$2 \cdot \left(\frac{1}{2}\right)^2 - 3 \cdot \frac{1}{2} + 1 = 0$$
$$2 \cdot \frac{1}{4} - \frac{3}{2} + 1 = 0$$
$$\frac{1}{2} - \frac{3}{2} + 1 = 0$$
$$0 = 0$$

The solution checks.

EXAMPLE 2 ***Solving a Quadratic Equation*** Solve for x by means of the quadratic formula:

$$6x^2 + 7x = 3$$

Solution First write the equation in standard form:

$$6x^2 + 7x - 3 = 0$$

Now that the equation is in standard form, we have $a = 6, b = 7, c = -3$. Note that the value of c is -3, not 3.

$$x = \frac{-b \pm \sqrt{b^2 - 4ac}}{2a}$$
$$x = \frac{-7 \pm \sqrt{7^2 - 4(6)(-3)}}{2 \cdot 6}$$
$$x = \frac{-7 \pm \sqrt{49 + 72}}{12}$$
$$x = \frac{-7 \pm \sqrt{121}}{12}$$
$$x = \frac{-7 + 11}{12} \quad \text{or} \quad x = \frac{-7 - 11}{12}$$
$$x = \frac{4}{12} \qquad x = \frac{-18}{12}$$
$$x = \frac{1}{3} \qquad x = -\frac{3}{2}$$

The solution set is $\{\frac{1}{3}, -\frac{3}{2}\}$.

Check:

If $x = \frac{1}{3}$,

$$6\left(\frac{1}{3}\right)^2 + 7\left(\frac{1}{3}\right) = 3$$
$$6 \cdot \frac{1}{9} + \frac{7}{3} = 3$$
$$\frac{2}{3} + \frac{7}{3} = 3$$
$$\frac{9}{3} = 3$$
$$3 = 3$$

If $x = -\frac{3}{2}$,

$$6\left(-\frac{3}{2}\right)^2 + 7\left(-\frac{3}{2}\right) = 3$$
$$6 \cdot \frac{9}{4} - \frac{21}{2} = 3$$
$$\frac{27}{2} - \frac{21}{2} = 3$$
$$\frac{6}{2} = 3$$
$$3 = 3$$

The solution checks.

EXAMPLE 3 ***Solving a Quadratic Equation*** Solve for x by means of the quadratic formula:

$$x^2 + 2x - 24 = 0$$

Solution The equation is in standard form. Therefore, $a = 1, b = 2, c = -24$. Note that the value of c is -24, not 24. Next we substitute these values in the quadratic formula to get

$$x = \frac{-2 \pm \sqrt{2^2 - 4(1)(-24)}}{2(1)}$$

Simplifying the expression, we obtain

$$x = \frac{-2 \pm \sqrt{4 + 96}}{2}$$

$$x = \frac{-2 \pm \sqrt{100}}{2}$$

$$x = \frac{-2 \pm 10}{2}$$

$$x = \frac{-2 + 10}{2} \quad \text{or} \quad x = \frac{-2 - 10}{2}$$

$$x = \frac{8}{2} \qquad\qquad x = \frac{-12}{2}$$

$$x = 4 \qquad\qquad x = -6$$

The solution set is $\{4, -6\}$.

Check:

If $x = 4$,

$$4^2 + 2(4) - 24 = 0$$
$$16 + 8 - 24 = 0$$
$$24 - 24 = 0$$
$$0 = 0$$

If $x = -6$,

$$(-6)^2 + 2(-6) - 24 = 0$$
$$36 - 12 - 24 = 0$$
$$36 - 36 = 0$$
$$0 = 0$$

The solution checks.

NW *Now Work Problems 1 and 5.*

In this section we have worked with a formula that can be used to solve any quadratic equation. The solution set of most quadratic equations is obtained by means of this formula. Therefore, for the sake of convenience, this formula should be memorized just as you would memorize your student identification number.

Exercises for Section 10.5

Solve the following equations (1–28) by means of the quadratic formula:

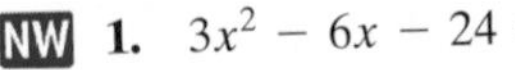

NW **1.** $3x^2 - 6x - 24 = 0$

2. $2x^2 + 8x + 6 = 0$

3. $2y^2 - 2y - 24 = 0$

4. $2y^2 - 8y - 10 = 0$

NW **5.** $6x^2 + 24x + 24 = 0$

6. $x^2 - 6x + 8 = 0$

7. $y^2 + 2y - 15 = 0$

8. $y^2 + y - 12 = 0$

9. $x^2 + 5x - 36 = 0$

10. $2x^2 - 11x + 5 = 0$

11. $2x^2 + 5x - 12 = 0$

12. $10x^2 - 17x + 3 = 0$

13. $2y^2 - y - 15 = 0$

14. $15y^2 - y - 2 = 0$

15. $6x^2 = 11x + 10$

16. $3y^2 = 4y + 4$

17. $3y^2 - 22y = 16$

18. $6x^2 + 1 = 7x$

19. $4y^2 = -8y - 3$

20. $8x^2 = 6x - 1$

*21. $x^2 - 2x = 2$

*22. $x^2 + 3x = 1$

*23. $y^2 - 2 = 10y$

*24. $y^2 = 3y + 2$

*25. $x^2 + x = 1$

*26. $16y^2 - 5 = 0$

*27. $2y^2 - 5 = 0$

*28. $2x^2 - 1 = 0$

Writing Mathematics

29. Write the quadratic formula.
30. Explain the difference between a quadratic equation and a linear equation.
31. What does it mean for a quadratic equation to be in standard form?

Challenge Exercises

32. The sum of the roots of a quadratic equation equals $2\frac{1}{6}$ and the product of the roots equals 1; find the roots of the equation.
33. The sum of the roots of a quadratic equation equals $1\frac{5}{6}$ and the product of the roots equals $\frac{1}{2}$; find the roots of the equation.
34. Find to the nearest tenth, the roots of the equation $2x^2 = 5x + 1$.

Research/Group Activities

35. The quadratic formula may be derived from the general quadratic equation $ax^2 + bx + c = 0, a \neq 0$, by a method known as *completing the square*. Research this method (it can be found in most algebra texts), demonstrate it to your classmates, and derive the quadratic formula.
36. Most graphing calculators are equipped to solve quadratic equations. Borrow such a graphing calculator from your instructor and determine how to solve quadratic equations with it. Finally, using an attachment and an overhead projector, demonstrate how such a graphing calculator can solve quadratic equations.

Just for fun

Where are the following located?

a. Tunnel of Corti
b. Island of Reil
c. McBurney's Point

10.6 GRAPHING QUADRATIC FUNCTIONS

In the previous section, we learned to solve quadratic equations by using the quadratic formula. We now turn our attention to graphing quadratic equations; that is, we will examine graphs of equations of the form $y = ax^2 + bx + c$. Before doing so, though, we need to first introduce one of the most important ideas in all of mathematics: the concept of *function*.

Concept of Function

In our earlier work in graphing linear equations in two variables (Chapter 9), we plotted ordered pairs in the Cartesian plane and then drew the line that contained these points. The ordered pairs that we used were generated by first selecting specific values for x and then substituting these x values into the given equation and solving for y. In each instance, x was considered the independent variable because we were free to select any x value we wanted, and y was considered the dependent variable because its value was dependent on the x value we used. This relationship between an independent variable and dependent variable can be represented in words as follows:

A dependent variable is a function of an independent variable.

Note that the phrase **is a function of** really means that the dependent variable **depends on** the independent variable. Thus, whenever the value of one variable *depends on* the values of a second variable, then the first variable *is a function of* the second variable. To consider this idea further, consider the following examples, which are all illustrations of *functional relationships*.

1. Hotel owners typically increase their room rates during weekends because that is when there is a greater demand for rooms. Thus, the price of a room is function of the day of the week. In this case, the independent variable is day of week and the dependent variable is price of room.
2. The attendance at a concert generally depends on who is performing. Thus, attendance is a function of the performer. Here, the independent variable is performer and the dependent variable is attendance.
3. Monthly electric bills depend on how much energy we use per month. This is a function of several things including how high or low we set our thermostat, how much hot water we use, and the number of lights we have on. In this example, there are several independent variables (e.g., thermostat setting, hot water use, light use) and the dependent variable is electricity cost.

In each of these examples, we verbally described a functional relationship; that is, we used a verbal description to express the relationship between an independent variable and a dependent variable. In addition to verbal descriptions, functional relationships can also be expressed using formulas. To illustrate this, consider the following description:

The distance we travel in a car depends on how fast the car is going and how long we are traveling.

Based on this description, if we are traveling at a constant rate of 70 mph for 2 hours, then the distance we travel is equal to the product of rate and time; that is, $70 \times 2 = 140$ miles. Thus, distance is dependent on (i.e., is a function of) rate and time, which are the independent variables. If we let D = distance, R = rate, and T = time, then this relationship can be expressed by the formula $D = R \times T$.

Functional relationships can be expressed in many other ways, including a table of values and a mathematical equation. For example, the relationship expressed in (1) above between the price of a hotel room and the day of the week could be expressed in table form as follows:

Day of week	Monday	Tuesday	Wednesday	Thursday	Friday	Saturday	Sunday
Room price	\$70	\$70	\$70	\$70	\$120	\$150	\$150

This idea of functional relationship, which denotes the relationship between an independent variable and dependent variable, is the heart of a very important concept in mathematics known as **function**. Specifically,

If we are given two variables x and y, then y is a function of x if for each permissible value of x there corresponds only one value of y.

The key to understanding this concept is to understand the phrase "for each permissible value of x there corresponds only one value of y." What this means is that each value of x must be unique. From a graphical perspective, this implies that given a set of ordered pairs, no two ordered pairs will have the same first coordinate and different second coordinates. To illustrate this, consider the following two quadratic equations and the corresponding set of points that satisfy these equations:

$y = x^2$	$x = y^2$
(0, 0)	(0, 0)
(1, 1)	(1, 1)
(−1, 1)	(1, −1)
(2, 4)	(4, 2)
(−2, 4)	(4, −2)

Note that in the first equation, the independent variable is x and the dependent variable is y. Furthermore, each x value is unique in the set of ordered pairs that satisfy this equation. In the second equation, x is now dependent on y. Furthermore, the x values of the ordered pairs that satisfy this equation are no longer unique. There are two ordered pairs where $x = 1$ and two ordered pairs where $x = 4$. As a result, the first equation is considered a function, but the second equation is not. In this section, we will only consider the graphs of **quadratic functions**. Additional information about the concept of function is provided in the corresponding Math Connections.

Quadratic functions are found in many different applications. For example, the distance an object drops from a specific height (e.g., from the top of a building) is represented by the quadratic function $S = 16t^2$, where S = distance and t = time. Note that the distance it takes to stop a car is found by the quadratic function $D = kV^2$, where D = stopping distance, k = a constant relative to a given situation, and V = the velocity or speed the car is traveling. In the health sciences, the equation $C = 42s^2 - 177s + 396$ is used to determine the number of calories (C) consumed by a person walking at a speed of s miles per hour (where s is between 2 and 5).

Note of Interest

First Programmers (Women in Mathematics)

In 1996 the computing world was celebrating the fiftieth anniversary of the first general-purpose electronic computer, known as the ENIAC (electronic numerical integrator and computer). The inventors, John W. Mauchly and John P. Eckert, were the center of focus. But there was something missing, the programmers! It's fine to invent a computer, but if no one tells it what to do, it does nothing. What makes this an interesting story is that the world's first programmers were women. In fact, there were six of them: Kay Mauchley Antonelli, Jean Bartik, Betty Holberton, Marlyn Meltzer, Frances Spence, and Ruth Teitelbaum (actually two were given recognition at the fiftieth anniversary, but not the rest). Because these ladies were classified as "subprofessionals," they never got the credit for inventing the field of programming.

The ENIAC was invented to calculate ballistic trajectories (parabolic paths) during World War II. Usually, there was a group of 80 women mathematicians doing the calculations by hand, as you can imagine, this would take quite a bit of time. Necessity being the mother of invention, the computer was invented. So six of the female mathematicians were selected to learn the ENIAC and figure out by themselves how to program the math problems into the computer (working six days a week). Mind you, this was during a time when there weren't such books a "ENIAC Programming for Dummies." The war ended, Mauchly and Eckert went on to become famous computer scientists, but what of the six ladies who invented programming?

The association Women in Technology International (WITI) inducted the six ladies in their Hall of Fame. A documentary is in the works and the WITI hopes that it will serve as an inspiration for girls and to dislodge the "math is for boys" mentality.

Graphing Quadratic Functions

To begin our discussion on graphing quadratic functions, let's examine $y = x^2$. This is a quadratic equation of the form $y = ax^2 + bx + c$; in this case, $a = 1, b = 0$, and $c = 0$. To graph $y = x^2$, we first find some ordered pairs that satisfy the given equation. Recall that the ordered pairs that satisfy $y = x^2$ are a function. Therefore, for every value of x, there is a unique value of y for which $y = x^2$. For example, if $x = 0$, then $y = (0)^2 = 0$. Similarly, if $x = 1$, then $y = (1)^2 = 1$, and if $x = -1$, then $y = (-1)^2 = 1$. If $x = 2, y = 4$ and if $x = -2, y = 4$. These points can be listed in a table of values as follows:

x	−2	−1	0	1	2
y	4	1	0	1	4

If we plot these points and connect them by means of a smooth curve, we obtain the curve shown in Figure 10.17. Note that the graph cannot be a straight line, because $y = x^2$ is not a linear equation.

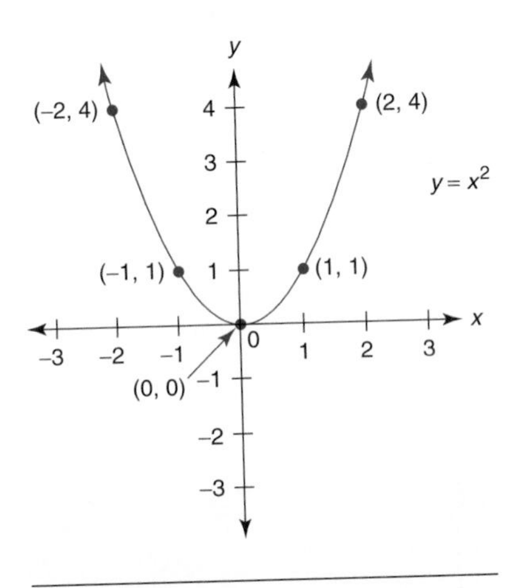

Figure 10.17

These freeways form a pattern of intersecting parabolas and straight lines.

The smooth curve in Figure 10.17 is called a **parabola**. The path of most projectiles is shaped like a parabola and the cables on suspension bridges hang in the shape of a parabola.

In Figure 10.17, note that the point (0, 0) is a turning point of the parabola $y = x^2$. That is, as we proceed from left to right along the x-axis, the values of y decrease until we reach (0, 0); then the values of y begin to increase. The turning point of a parabola is called the **vertex** of the parabola.

For any parabola with equation of the form $y = ax^2 + bx + c$, the x value of the vertex is

$$\frac{-b}{2a}$$

We can use this information to help graph a given parabola. For example, $y = x^2$ is of the form $y = ax^2 + bx + c$, with $a = 1$ and $b = 0$. Hence, the x value of the vertex is

$$\frac{-b}{2a} = \frac{-0}{2 \cdot 1} = -\frac{0}{2} = 0$$

The y value is found by substituting zero for x in the original equation. If $x = 0, y = 0^2 = 0$. Therefore, the vertex of $y = x^2$ is $(0, 0)$.

It should be noted that the value of a in $y = x^2$ is $+1$. If a is positive in any parabola with an equation of the form $y = ax^2 + bx + c$, then the parabola opens upward. If a is negative, then the parabola opens downward. These two facts are useful when graphing parabolas.

EXAMPLE 1 ***Graphing a Quadratic Function with a > 0*** Graph $y = x^2 - 4$.

Solution The general form is $y = ax^2 + bx + c$. In this case, $a = 1$, $b = 0$, and $c = -4$. Since $a = 1$ is positive, the parabola opens upward. The x value of the vertex is

$$\frac{-b}{2a} = \frac{-0}{2 \cdot 1} = -\frac{0}{2} = 0$$

If $x = 0$, $y = 0^2 - 4 = -4$. Therefore, the vertex is $(0, -4)$.

To obtain a better idea of the shape of the parabola, we choose two or three values for x on each side of the vertex and find the corresponding values of y. To the left of the vertex, we choose $x = -1$, $x = -2$, and $x = -3$, and to the right of the vertex, we choose $x = 1$, $x = 2$, and $x = 3$. The corresponding values of y are listed in the following table:

x	−3	−2	−1	0	1	2	3
y	5	0	−3	−4	−3	0	5

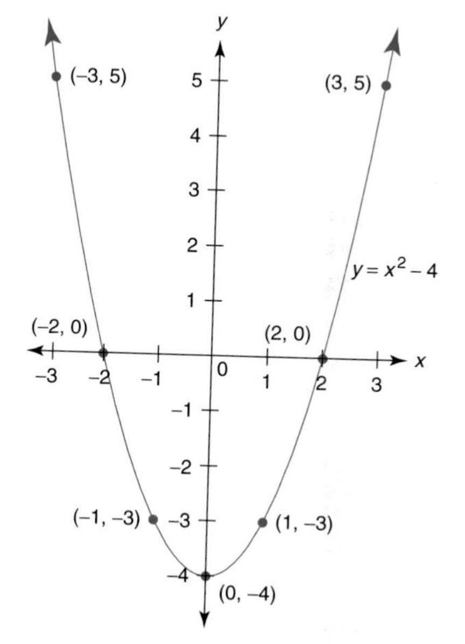

Figure 10.18

Using these points, we graph $y = x^2 - 4$ in Figure 10.18.

NW ***Now Work Problem 1.***

EXAMPLE 2 ***Graphing a Quadratic Function with a < 0*** Graph $y = -x^2 + 4x$.

Solution In this case, $a = -1$, $b = 4$, and $c = 0$. Because $a = -1$ is negative, the parabola opens downward. The x value of the vertex is

$$\frac{-b}{2a} = \frac{-(4)}{2(-1)} = \frac{-4}{-2} = 2$$

If $x = 2$, $y = -2^2 + 4 \cdot 2 = -4 + 8 = 4$. Therefore, the vertex is $(2, 4)$. To graph the parabola, we choose $x = 1$, $x = 0$, and $x = -1$ to the left of the vertex, and $x = 3$, $x = 4$, and $x = 5$ to the right of the vertex. The corresponding values of y are listed in the following table:

x	−1	0	1	2	3	4	5
y	−5	0	3	4	3	0	−5

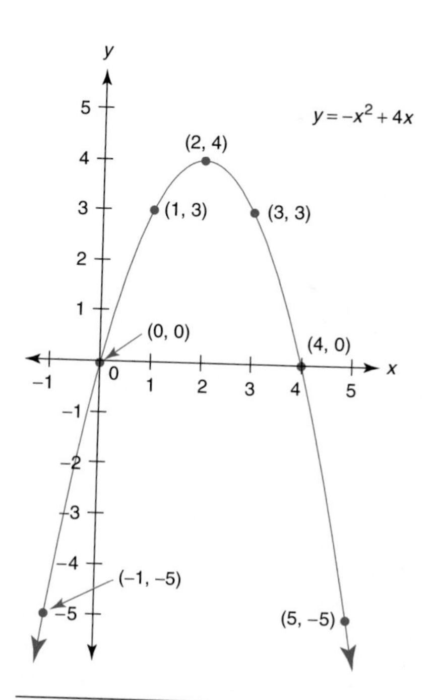

Figure 10.19

Using these points, we graph $y = -x^2 + 4x$ in Figure 10.19.

NW ***Now Work Problem 3.***

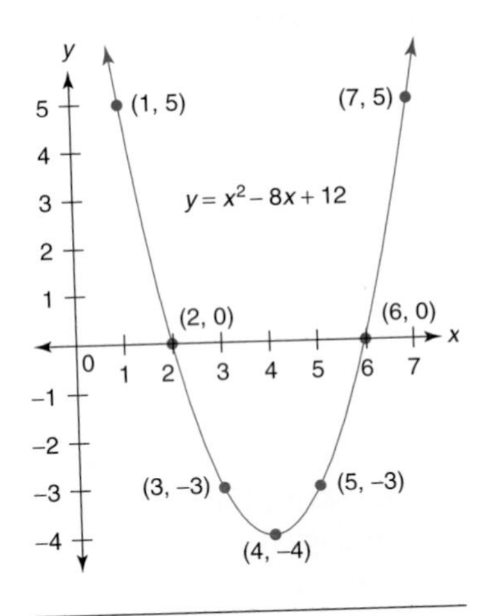

Figure 10.20

EXAMPLE 3 ***Graphing a Quadratic Function*** Graph $y = x^2 - 8x + 12$.

Solution In this case, $a = 1, b = -8$, and $c = 12$. Since $a = 1$ is positive, the parabola opens upward. The x value of the vertex is

$$\frac{-b}{2a} = \frac{-(-8)}{2 \cdot 1} = \frac{8}{2} = 4$$

If $x = 4$,

$$y = 4^2 - 8 \cdot 4 + 12 = 16 - 32 + 12 = -4$$

Therefore, the vertex is $(4, -4)$. To graph the parabola, we choose $x = 3, x = 2$, and $x = 1$ to the left of the vertex, and $x = 5, x = 6$, and $x = 7$ to the right of the vertex. The corresponding values of y are listed as follows:

x	1	2	3	4	5	6	7
y	5	0	−3	−4	−3	0	5

Using these points, we graph $y = x^2 - 8x + 12$ in Figure 10.20.

NW ***Now Work Problem 7.***

EXAMPLE 4 ***Driver Age and Automobile Fatalities*** The following graph shows the relationship between age and passenger vehicle driver deaths per 100,000 licensed drivers in 2001.

a. Does the graph depict a quadratic function? Why?

b. If you were to write the equation that represents this graph, would the a term be positive or negative? Why?

c. Determine the independent and dependent variables and write a verbal description of the functional relationship between these variables.

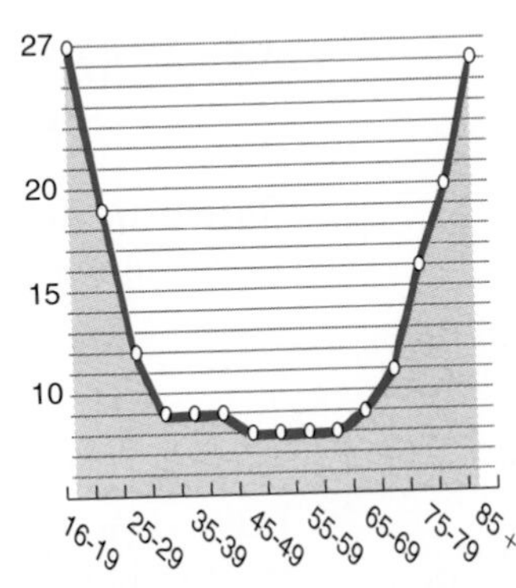

Source: Insurance Institute for Highway Safety

Solution

a. The graph approximates a quadratic function because it is a parabola and for each ordered pair that is graphed, the x coordinate is unique.

b. The a term would be positive because the parabola opens "up."

c. The independent variable is age and the dependent variable is passenger vehicle driver deaths. The functional relationship that exists between these variables is as follows: The number of passenger vehicle driver deaths is a function of driver age. More specifically, younger (less than 25) and older (greater than 70) drivers have higher passenger vehicle driver deaths than drivers not in these age groups.

Math Connections

Functions

The concept of *function* is one of the most important ideas in mathematics. A function represents a special correspondence between the elements of two sets and represents a unifying thread among the various mathematical curricula. For example,

- In arithmetic, a function denotes the operations of addition, subtraction, multiplication, and division. A pair of numbers from one set corresponds to a number from a second set relative to a specific operation. Thus, $8 + 2 = 10$ implies that the sum 10 is a function of adding 8 and 2.
- In algebra, a function denotes the relationship between variables that represent numbers. For example, if you ride your bicycle at an average rate of 15 miles per hour, then the distance you travel is determined by how long you travel (i.e., time). This relationship can be represented algebraically as the equation $d = 15t$, where d = distance and t = time (in hours). Once again, note the correspondence: For each specific time value there corresponds exactly one distance value. In this example, distance is defined as a function of time, or distance depends on time.

The function concept is also inherent in geometry (e.g., the correspondence between a set of points and the image the points represent), probability (e.g., the correspondence between a given event such as getting a tail on a flip of a coin and the probability of that event occurring), and every other area of mathematics. Functions can also be expressed via table form. For example, Table 10.1 provides the 2004 U.S. postal rates for mailing a first-class letter based on weight. The data in the table represent the relationship between the weight (in ounces) and cost (in cents). In this situation, the cost of mailing a letter is a function of how much it weighs. Finally, the concept of function also has meaning in nonmathematical settings. For example, we might say that a person's grade on an exam depends on (i.e., is a function of) how much he or she studies, or we might say that a person's overall health is a function of his or her diet.

TABLE 10.1

Weight (W) in Ounces	Cost (C) in Cents
1	37
2	60
3	83
4	106
5	129
6	152

An important consideration in learning mathematics is understanding the relationship between algebraic and graphical representations of functions. For example, we now know that equations of the form $y = mx + b$, or the more general form, $Ax + By = C$, are graphically represented as straight lines, and that equations of the form $y = ax^2 + bx + c$ are quadratic equations and are graphically represented as parabolas. These relationships are summarized below.

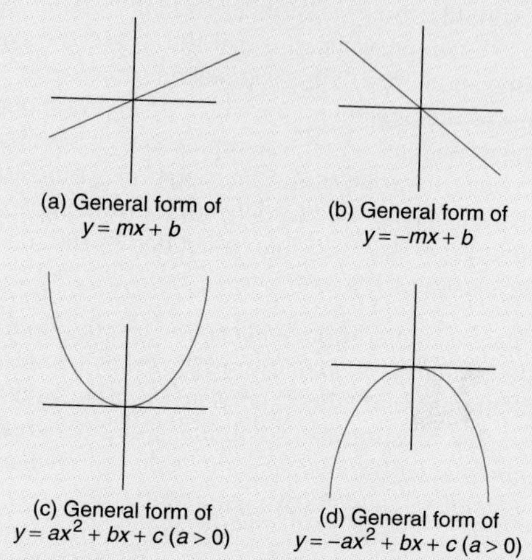

(a) General form of $y = mx + b$

(b) General form of $y = -mx + b$

(c) General form of $y = ax^2 + bx + c\ (a > 0)$

(d) General form of $y = -ax^2 + bx + c\ (a > 0)$

Thus, if we are given the equation $Ax + 3y = -6$ and asked to find A, the coefficient of x, we should be able to explain how we can find this coefficient by relating the algebraic and graphical representations of the given equation. Other important functions that we generally encounter in mathematics include the *cubic function*, the *exponential function*, and the *logarithmic function*. The general equations that correspond to these functions and their related graphical representations are shown below.

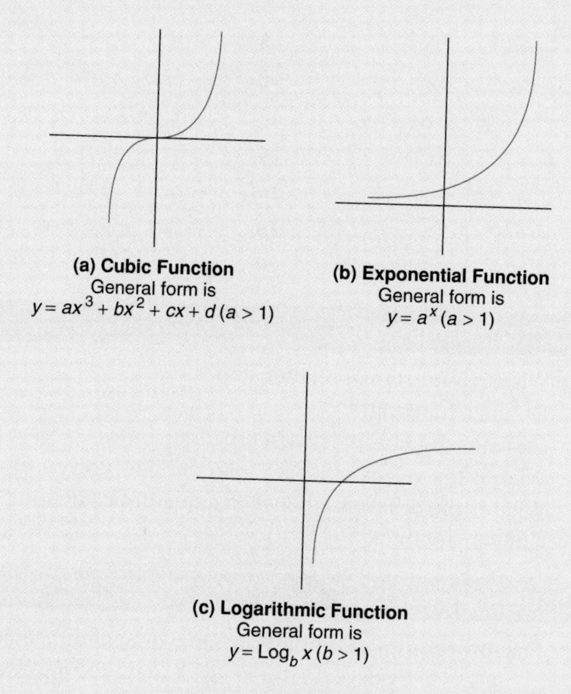

(a) Cubic Function General form is $y = ax^3 + bx^2 + cx + d\ (a > 1)$

(b) Exponential Function General form is $y = a^x\ (a > 1)$

(c) Logarithmic Function General form is $y = \text{Log}_b\, x\ (b > 1)$

(continued)

(continued)

Another aspect of functions is the concept of domain and range. When expressed algebraically, functions describe a relationship between two variables. More specifically, the dependent variable is written as a function of the independent variable. For example, the area of a circle depends on the radius of the circle and hence is expressed algebraically as $A = \pi r^2$. Here, A is the dependent variable and r is the independent variable. Similarly, in our bicycle example, we stated that the distance traveled depends on time and hence expressed this relationship as $d = 15t$. Once again, d is the dependent variable and t is the independent variable. The **domain** of a function represents all possible values of the independent variable, and the **range** of a function represents all possible values of the dependent variable that correspond to the independent variable as dictated by the relationship between the variables. For example,

- In our bicycle illustration, where $d = 15t$, the domain is all real numbers greater than or equal to zero because the independent variable, time, cannot be negative. Similarly, the range of this function is also all real numbers greater than or equal to zero because for any value of t where $t \geq 0$, when we multiply t by 15, we get $d \geq 0$.
- Given the function that describes the area of a circle, $A = \pi r^2$, the domain is $r > 0$ because it is meaningless to talk about a circle with a zero or negative radius. Given this restriction on r, we observe that the values of A will also be greater than zero. Thus, the range is $A > 0$.
- Given a quadratic function $y = x^2$, note that there is no restriction on the independent variable, and, hence, the domain for x is any real number. However, given the relationship that exists between y and x—namely, y is equal to the square of x—the dependent variable will never be negative. Thus, the range is $y \geq 0$.

A distinguishing feature of functions is that they represent a special relationship between two variables as follows: If we let set A represent the set of all domain elements and set B represent the set of all range elements, then each element in A corresponds to exactly one element in B. This implies that a function is a set of ordered pairs (x, y), where $x \in A$ and $y \in B$, and that no two ordered pairs have the same first coordinate and different second coordinates. From a graphical perspective, this suggests that if a vertical line cuts a graph in more than one point, then the graph is not the graph of a function. This is known as the **vertical line test**.

Exercises for Section 10.6

For Exercises 1–20, graph the given equation on a Cartesian plane.

NW **1.** $y = x^2 - 1$

2. $y = x^2 + 1$

NW **3.** $y = -x^2 + 2$

4. $y = -x^2 + 1$

5. $y = -x^2 + 6x$

6. $y = -x^2 - 3x$

NW **7.** $y = x^2 + 2x + 1$

8. $y = x^2 + 2x - 2$

9. $y = x^2 - 4x + 3$

10. $y = -2x^2 - 4x - 2$

11. $y = -x^2 + 6x - 9$

12. $y = x^2 + 2x - 8$

13. $y = x^2 - 6x + 5$

14. $y = x^2 - 4x + 2$

15. $y = -x^2 + 4$

16. $y = x^2 - 2x + 1$

17. $y = x^2 + 2x$

18. $y = 2x^2 + 4x - 6$

19. $y = x^2 + 3x + 2$

20. $y = x^2 + 5x + 6$

Writing Mathematics

21. What is a functional relationship?

22. Define a function.

23. Explain why $x = y^2$ is not a function.

24. What is the vertex of a parabola, and how do you find it?

25. What is the difference between a quadratic equation and a linear equation?

Challenge Exercises

26. Find the coordinates of the maximum or minimum point of

a. $y = x^2 - 8x + 15$ **b.** $y = 12 - 4x - x^2$

27. Find an equation of the form $y = ax^2 + bx + c$ whose graph passes through the points $(0, -12)$, $(2, 2)$, and $(5, -7)$.

28. Solve the following system of equations graphically:

$$y = 3x - 11$$
$$y = x^2 - 4x - 5$$

29. If a ball is thrown upward with an initial velocity of 120 feet per second, its height after t seconds is given by the formula $h = 120t - 16t^2$. Find the maximum height the ball will reach. (*Hint:* Think of t and h as x and y.)

30. What do you think the graphs of the equations $x = y^2$ and $x = -y^2$ look like? Explain why these equations are not considered functions.

Group/Research Activities

31. Graphing calculators can be used to display graphs of the parabolas discussed in this section. Borrow a graphing calculator from your instructor and determine how to graph parabolas and other functions on it. Finally, using an attachment and an overhead projector, demonstrate the graphing calculator to your class.

32. The reflecting mirror of a telescope is designed in the shape of a parabola. Car headlights and flashlights also make use of the parabola in their design. Research the design of these items and how they use a parabola. Write a paper explaining how a parabolic design enables these items to work as well as they do.

33. Johannes Kepler (1571–1630), a German astronomer/mathematician, used a pencil, a piece of string, and a crude version of a T-square to construct parabolas. Research the life and contributions of Kepler. Prepare a report on the life and contributions of Kepler. Be sure to include an explanation of how he drew his parabolas.

Just for fun

Which is worth more, a box full of \$10 gold pieces or an identical box half full of \$20 gold pieces?

Math Connections

The Roots of a Quadratic Function

In Section 10.5, we solved quadratic equations by using the quadratic formula. For example, given the quadratic equation, $x^2 - 2x - 3 = 0$, the solutions to this equation can be found by substituting $a = 1, b = -2$, and $c = -3$ into the quadratic formula and solving for x.

$$\begin{aligned} x &= \frac{-b \pm \sqrt{b^2 - 4ac}}{2a} \\ &= \frac{-(-2) \pm \sqrt{(-2)^2 - 4(1)(-3)}}{2(1)} \\ &= \frac{2 \pm \sqrt{4 + 12}}{2} \\ &= \frac{2 \pm \sqrt{16}}{2} \\ &= \frac{2 \pm 4}{2} \end{aligned}$$

Thus, $x = \frac{2 + 4}{2} = \frac{6}{2} = 3$ and $x = \frac{2 - 4}{2} = \frac{-2}{2} = -1$.

Let's now graph the function $y = x^2 - 2x - 3$. A table of values that contains the vertex $(1, -4)$ and three points on each side of the vertex is given below along with the corresponding graph.

Note that the graph of this function intersects the x axis at two points, $(-1, 0)$ and $(3, 0)$. We highlighted this by placing a circle around the points. Note further that the x-values of these two points are the same as the solutions to the corresponding equation $x^2 - 2x - 3 = 0$, which was solved using the quadratic formula. Thus, the solutions to a quadratic equation, which are called its *roots*, can also be obtained graphically. More specifically, the roots of the quadratic equation $x^2 - 2x - 3 = 0$ can be read directly from the graph where the parabola cuts the x-axis. Here the roots are $x = 3$ and $x = -1$. In general, quadratic equations that have real roots (i.e., their solutions are real numbers) can be solved graphically by reading from the graph where the parabola cuts the x-axis. In the case of integer solutions, we can interpret the roots exactly. In all other cases, though, the roots can only be determined approximately. If the parabola does not cut the x-axis, then the equation has no real roots.

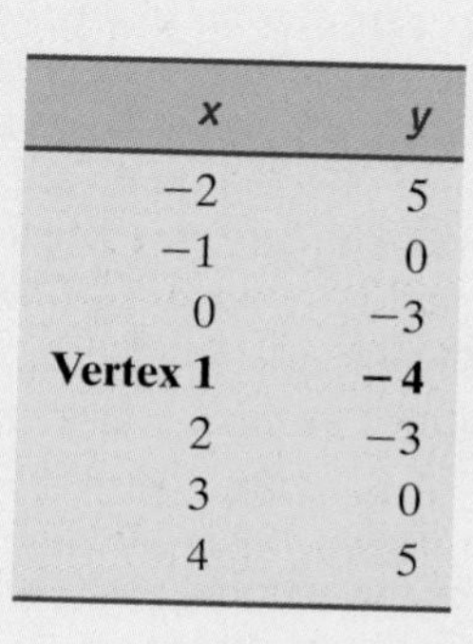

x	y
−2	5
−1	0
0	−3
Vertex 1	**−4**
2	−3
3	0
4	5

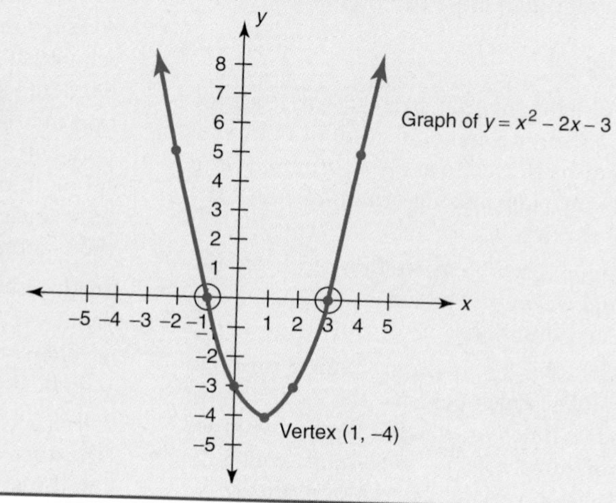

CHAPTER REVIEW MATERIALS

Summary/Chapter 10

10.2 The Slope of a Line

- The *slope* of a line is defined as the "rise" of the line divided by the "run" of the line. The "rise" is the change in the vertical distance and the "run" is the change in the horizontal distance. Thus, if a line contains any two points (x_1, y_1) and (x_2, y_2), then

$$\text{slope} = \frac{\text{rise}}{\text{run}} = \frac{y_2 - y_1}{x_2 - x_1}$$

One way in which to remember this is to think about how you walk up or down a staircase: You have to lift your foot up (or down) first followed by a horizontal movement.

- If a line slants upward to the right, then its slope is positive; if a line slants downward to the right, then its slope is negative.
- The slope of a horizontal line is 0 and the slope of a vertical line is undefined.

10.3 The Equation of a Line

- We can determine the equation of a line by the *point-slope formula*, which states that $y - y_1 = m(x - x_1)$. In this equation, m is the slope of the line and (x_1, y_1) represents the coordinates of any point on the line.
- We can also determine the equation of a line by the *slope-intercept formula*, which states that $y = mx + b$. In this equation, m is the slope of the line and b is the y intercept of the line.
- If we are given the equation of a line, we can draw its graph by first solving the equation for y so that it is in the form $y = mx + b$. Once the equation is in this form, we can plot the y intercept $(0, b)$ and then move "slope units" from this point to a second point on the graph. For example, if the intercept is $(0, 3)$ and the slope is $\frac{1}{2}$, then a second point may be found "1 unit up" (change in y) and "2 units to the right" (change in x) of $(0, 3)$. Thus, a second point will be located at $(2, 4)$. We now sketch the line that passes through these two points.

10.4 Linear Programming

- *Linear programming* is a mathematical strategy (or tool) for solving *optimization problems*. A common feature of optimization problems, which are also referred to as *linear programming problems*, is that they involve maximizing or minimizing a situation based on a set of restrictions.
- To solve a linear programming problem, we first represent the problem using a *mathematical model*, which is a mathematical representation of an actual situation.
- The mathematical model of a linear programming problem consists of *decision variables*, which describe the decisions we want to make; an *objective function*, which indicates what we want to maximize or minimize; a *set of constraints*, which restrict the value of the decision variables; and a *sign restriction*, which specifies whether the decision variables are nonnegative or unrestricted in sign.
- For the type of linear programming problems we solved in this chapter, we used two decision variables, x and y; the objective function was a linear equation in terms of x and y; the constraints were represented as linear inequalities in x and y; and the sign restrictions on the variables were $x \geq 0$ and $y \geq 0$.
- To solve a linear programming problem of the type discussed in this chapter, we graph the constraint and sign restriction inequalities on the same set of axes in the Cartesian plane. This graph results in a polynomial region. We then substitute into the objective function the x and y coordinates of the ordered pairs that correspond to each "corner point," or vertex, of the region. The ordered pair that results in the largest (if we are maximizing a situation) or smallest (if we are minimizing a situation) value is the solution to the problem.

10.5 Solving Quadratic Equations

- A *quadratic equation* is an equation of the form $ax^2 + bx + c = 0$. This is also called the *standard form* of a quadratic equation. In this equation, a is the coefficient of x^2 and $a \neq 0$, b is the coefficient of the x term, and c is a constant.
- An *incomplete quadratic equation* is a quadratic equation in which $b = 0$ or $c = 0$, or both $b = 0$ and $c = 0$.
- A quadratic equation may be solved by applying the quadratic formula

$$x = \frac{-b \pm \sqrt{b^2 - 4ac}}{2a}$$

- To solve quadratic equations using the quadratic formula, the equation must be in standard form. We then substitute the numerical values of a, b, and c into the formula and solve for x.

10.6 Graphing Quadratic Functions

- A *function* is a mathematical concept that denotes a special relationship between an independent variable, x, and a dependent variable, y. This relationship indicates that y is a function of x if for each permissible value of x there corresponds only one value of y. In other words, if we were to consider the set of all ordered pairs that satisfies a functional relationship between x and y, then we will find that for each ordered pair the x-coordinates are unique.
- In this chapter, we only considered the graphs of *quadratic functions*. In general, the graph of a quadratic function $y = ax^2 + bx + c$ is a *parabola*, which is a u-shaped curve. If $a > 0$, then the parabola opens upward; if $a < 0$, then the parabola opens downward.
- The turning point of a parabola is called the *vertex* of the parabola. For any parabola of the form $y = ax^2 + bx + c$, the x-coordinate of the vertex is found by substituting the numerical values of a and b into the expression $-b/(2a)$. The

corresponding y-coordinate is found by substituting this numerical value for x into the original equation and solving for y.

- To graph a quadratic function, we first find the coordinates of the parabola's vertex and plot this point. We then assign two or three convenient values to x, obtain the corresponding y values by substituting the x-values into the equation and solving for y, and then plot the resulting ordered pairs. The graph is the result of tracing a smooth curve that passes through the plotted points.

Key Terms and Concepts

function, *580*
linear programming, *568*
objective function, *569*
parabola, *582*
point-slope formula, *558*
quadratic equation, *574*
slope, *551*
slope-intercept formula, *561*
standard form, *574*
vertex, *582*

Review Exercises

10.2 The Slope of a Line

1. Find the slope of the line containing the indicated pair of points.

a. $A(-2, 4)$ and $B(5, 7)$

b. $C(3, 2)$ and $D(0, 5)$

10.3 The Equation of a Line

2. Write the equation of the straight line that passes through the indicated points:

a. $A(-2, 3)$ and $B(3, -2)$

b. $C(-2, 0)$ and $D(0, -3)$

c. $E(5, -2)$ and $F(6, 7)$

d. $G(2, 3)$ and $H(-1, 1)$

3. Find the slope and y-intercept of the line whose equation is given.

a. $y = -2x + 5$

b. $x + y = 3$

c. $-2x - 3y = 6$

d. $x - 3y + 3 = 0$

4. Use the slope-intercept formula to graph each of the following:

a. $y = x + 2$

b. $y = -2x - 1$

c. $x + y = 0$

d. $-2x - 3y = 6$

5. Given the points $P(1, -1)$, $Q(4, 2)$, and $R(6, 4)$.

a. Find the slopes of line segments $\overline{PQ}$ and $\overline{PR}$.

b. What can you conclude about points P, Q, and R?

c. Find the equation of the line that passes through P and R.

d. Find the coordinates of the point where line PR crosses the x-axis.

6. Given the points $A(3, 1)$, $B(0, -1)$, and $C(-3, -3)$.

a. Write an equation of the line that passes through point A and is parallel to the y-axis.

b. Write an equation of the line that passes through point B and has the slope of 1.

c. Show that A, B, and C lie on the same straight line.

d. Write an equation of the line that is parallel to line AB and passes through the origin.

10.4 Linear Programming

7. Find the maximum value of $P = 2x + 3y$ under the following conditions:

$$x \geq 0, \quad y \geq 0, \quad 2x + y \leq 10, \quad x + 3y \leq 15$$

8. A manufacturer makes couches and recliners. To produce a couch requires 3 hours in the frameshop and 2 hours in the upholstery shop. To produce a recliner requires 4 hours in the frame shop and 4 hours in the upholstery shop. The frame shop can operate at most 24 hours per day, and the upholstery shop can operate at most 20 hours per day. If the manufacturer makes a profit of \$30 on a couch and \$45 on a recliner, how many of each should be produced in order to maximize profit?

9. The Safety-First Corporation manufactures two types of smoke alarms, a standard model and a deluxe model. To produce a

standard smoke alarm requires 1 hour on machine A and 1 hour on machine B. To produce a deluxe smoke alarm requires 1 hour on machine A and 2 hours on machine B. Due to costs and safety regulations, machine A can operate at most 35 hours per week and machine B can operate at most 40 hours per week. If the manufacturer makes a profit of $7 on each standard model and $10 profit on each deluxe model, how many of each should the company produce in order to maximize profits?

10.5 Solving Quadratic Equations

Solve the following equations by means of the quadratic formula:

10. $3y^2 - 4y - 4 = 0$

11. $y^2 - y = 2$

12. $4x^2 - 4 = 0$

13. $x^2 - 4x = 0$

14. $8x^2 = 6x - 1$

15. $x^2 + 2x = 2$

16. $2y^2 - y - 1 = 0$

17. $2z^2 + 6z - 20 = 0$

10.6 Graphing Quadratic Functions

18. Graph the given equations on a Cartesian plane.

a. $y = x^2$

b. $y = x^2 - 4$

c. $y = -x^2 + 2$

d. $y = x^2 - 6x$

e. $y = x^2 + 2x - 3$

f. $y = -x^2 + 6x - 5$

Just for fun

Following is an algebraic expression for an indicated product.

$(x - a)(x - b)(x - c) \cdots (x - z) = ?$

Can you determine the product? Time limit for this question is 10 seconds.

Chapter Quiz

Determine whether each statement (1–10) is true or false.

1. The set of ordered pairs $\{(-2, 1), (1, -2), (3, -2), (4, 1)\}$ is a function.

2. The slope of the line passing through $G(2, 3)$ and $H(-1, 1)$ is $\frac{2}{3}$.

3. Perpendicular lines have equal slopes.

4. The slope of any horizontal line parallel to the x-axis is 0.

5. The equation of the line parallel to the y-axis and passing through the point $(5, 3)$ is $y = 3$.

6. The solution set of the quadratic equation $4y^2 - 4 = 0$ is $\{1, -1\}$.

7. The set of ordered pairs $\{(1, 3), (2, 5), (3, 6), (1, 0)\}$ is a function.

8. The objective function in a linear programming problem describes the relationship between the items produced and the profit that can be realized from selling them.

9. Any maximum or minimum values of the objective function always occur at a vertex.

10. The graph of the parabola $y = -x^2 + 6x$ opens downward and its vertex is located at the origin.

11. Find the slope of the line that passes through the points $A(2, -3)$ and $B(5, 7)$.

12. Find the equation of the line that passes through $(-4, -1)$ and has a slope of 3.

13. Find the equation of the line that passes through the points $C(-2, 0)$ and $D(4, 3)$.

14. Find the equation of the line passing through $(-2, -3)$ and parallel to the line whose equation is $x + 2y = 6$.

For Exercises 15–18, use the points $P(-1, -3)$, $Q(3, 1)$, and $R(7, 5)$.

15. Find the slopes of line segments $\overline{PQ}$ and $\overline{PR}$.

16. What can you conclude regarding points P, Q, and R?

17. Determine the equation of the line that passes through P and R.

18. Find the coordinates of the point where line PR crosses the x-axis.

19. Graph $y = -x^2 + 4x$ on the Cartesian plane.

20. Find the coordinates of the vertex of the parabola $y = x^2 - 4x + 6$.

21. Find the maximum value of $P = 3x + 2y$ under the following conditions:

$$x \geq 0, \quad y \geq 0, \quad x + y \leq 5, \quad x - y \leq 1$$

22. Find the maximum value of $P = 5x + 8y$ under the following conditions:

$$x \geq 0, \quad y \geq 0, \quad x + y \leq 7, \quad 3x + y \leq 15$$

For Exercises 23–30, solve the given quadratic equations.

23. $x^2 - 3x - 4 = 0$

24. $6x^2 + x - 2 = 0$

25. $6x^2 + 7x - 3 = 0$

26. $x^2 - 2x - 15 = 0$

27. $2y^2 + 6y - 20 = 0$

28. $3x^2 + 8x = 3$

29. $x^2 - 4x = 0$

30. $3x^2 = 2 - x$

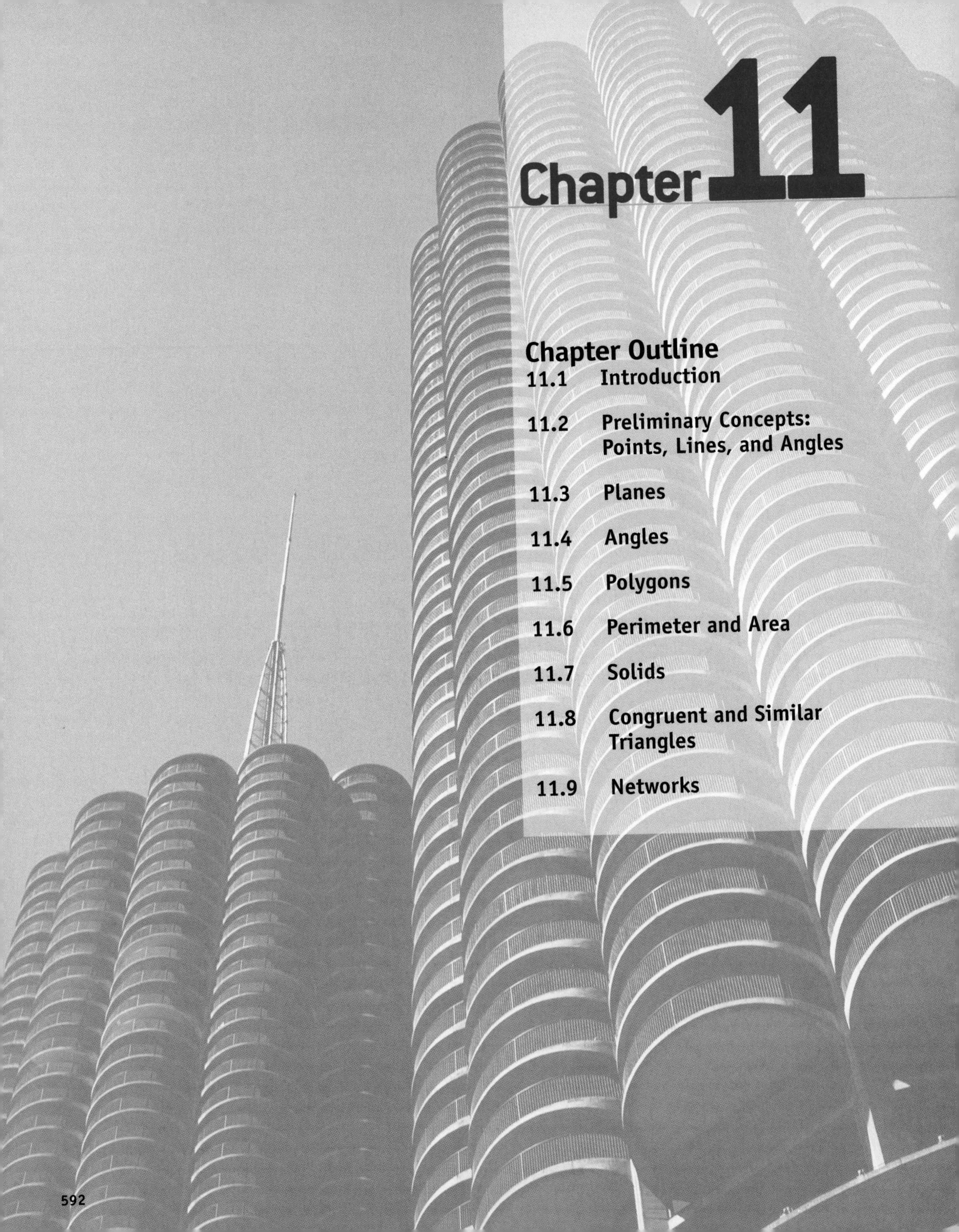

Chapter 11

Chapter Outline

An Introduction to Geometry

After Studying This Chapter, You Will Be Able to Do the Following:

1. Identify and find the intersection and union of **points, lines, half-lines, planes, half-planes, rays**, and **line segments.**
2. Identify and find the measure of **acute, obtuse, right, straight, vertical, complementary**, and **supplementary angles**, and **angles formed by two parallel lines cut by a transversal.**
3. Use the **Pythagorean theorem** to find the length of a side of a right triangle given the measures of the other two sides.
4. Identify, from a figure or stated properties, and find the values of certain parts of **equilateral, isosceles, right, acute, obtuse, equiangular, similar**, and **congruent triangles.**
5. Identify **trapezoids, parallelograms, rhombuses, rectangles, squares**, and **pentagons**, from a figure or stated properties.
6. Find the **perimeter** of any given **polygon**, the **circumference** of any given **circle**, and the **area** of **rectangles, squares, parallelograms, triangles, trapezoids**, and **circles.**
7. Find the **surface area** and **volume** of geometric **solids** such as cylinders, cones, spheres, prisms, and pyramids.
8. Determine whether a **network is traversable.**

Note: *indicates optional material.

Notations Frequently Used in This Chapter

Notation	Meaning
$\cdot C$	point C
$\overleftrightarrow{AB}$	line AB
$\overrightarrow{AB}$	ray AB
$\overline{AB}$	line segment AB
$\overset{\circ\!\rightarrow}{AB}$	half-line AB
$\angle RST$	angle RST
$m\angle RST$	measure of $\angle RST$
$\overset{\circ\!-\!\circ}{AB}$	open line segment AB
$m(\overline{AB})$	measure of line segment AB
$^\circ$	degree (of an angle)
$'$	minute (unit of a degree)
$''$	second (unit of a minute)
$\overleftrightarrow{MN} \perp \overleftrightarrow{BT}$	line MN is perpendicular to line BT
$\overleftrightarrow{MN} \parallel \overleftrightarrow{BT}$	line MN is parallel to line BT
ΔABC	triangle ABC
⊾	right angle
$\sim$	(is) similar to
$\cong$	(is) congruent to
$\square ABCD$	parallelogram $ABCD$
π	pi; equal to 3.14, rounded to nearest hundredth; the constant ratio of the circumference of a circle to its diameter

Visit the PH Companion Web site at www.prenhall.com/setek

OVERVIEW

What is geometry? To begin with, the word *geometry* is derived from two Greek words. The first part of the word *geometry* is taken from the Greek word *ge*, which means "earth," and the second part is taken from the Greek word *metron*, which means "measure." Therefore, we can safely assume that early geometry concerned itself with the measure of the earth, or earth measurement.

11.1 INTRODUCTION

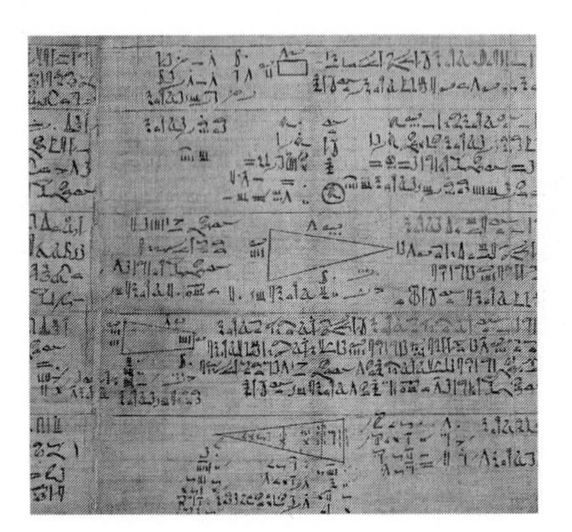

The Egyptians were among the first people to use geometry. One of the ways that the Egyptians used geometry was to survey land, or "measure earth." The Egyptians had to pay taxes for their land. The kings sent workers to measure the people's land, and taxes were levied accordingly. The Egyptians used this form of taxation as early as 1300 B.C.

From this beginning, the study of geometry was carried on by the Greeks. Thales was one of the first Greeks to concern himself with proving that mathematical statements were true. Pythagoras was the next major contributor to the study of geometry (approximately 550 B.C.). You may be familiar with the Pythagorean theorem, which states that the sum of the squares of the legs of a right triangle is equal to the square of the hypotenuse.

The next great Greek scholar in the field of geometry was Plato. He founded a school in the city of Athens. Plato considered geometry so important that he had the slogan "Let no one ignorant of geometry enter my doors" posted at the entrance to his school.

Euclid, another Greek, is known as the "father of geometry." In approximately 300 B.C., Euclid gathered all of the mathematical works known to exist at that time and produced a book called *The Elements*.

11.2 PRELIMINARY CONCEPTS: POINTS, LINES, AND ANGLES

Points

The most basic terms of geometry are *point, line*, and *plane*. Each of these terms presents a problem for us since we cannot define it. Granted, you can find descriptions for these words in the dictionary—but can you define them? For instance, what is a **point**? You might want to describe a point as a dot on a piece of paper, but what size dot? Other concepts of a point are of a point on the number line or a point on the Cartesian plane. The important thing to remember is that a geometric point has no dimension; that is, it has no length, breadth, or thickness. The one thing that a geometric point does possess is position. We can represent a point by a dot, just as we use the symbol "2" to represent the number two. Points are labeled or named by using capital letters. Therefore, if we wish to represent a point C, we make a dot and label it with a capital C as shown:

$$\cdot C$$

We now have an intuitive idea of what a point is and how we shall represent it. We shall not define a point, just as we did not define a set in Chapter 2. (Recall that one of the ingredients of an axiomatic system is a collection of undefined terms that are used to define other terms.)

A **line** is formed by the intersection of two flat surfaces. For example, the intersection of two walls in one corner of a room can be thought of as a line. A point in geometry has only one position, whereas a line in geometry has only one dimension, length. A geometric line has no width. Note that when we refer to a *line* we are talking about a *straight line*, unless otherwise noted.

We can think of a line as a set of points. The points are *on* the line and the line passes *through* the points. It is interesting to note that a line may be extended infinitely in either direction. What this means is that a line has no endpoints. Normally we work with only pieces or parts of lines; these are described later in this section.

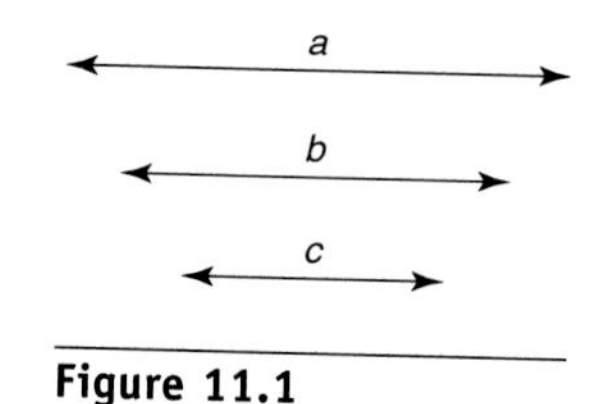

Figure 11.1

We can name a line in different ways. We can, for example, name a line by writing a lowercase letter such as *a, b, c*, or *d* near the line. Figure 11.1 shows three lines—line *a*, line *b*, and line *c*. It is more common to name a line by using two points on the line. The line shown in Figure 11.2 can be denoted by $\overleftrightarrow{AB}$, $\overleftrightarrow{AC}$, $\overleftrightarrow{BC}$, $\overleftrightarrow{CD}$, and so on. Since all the points *A, B, C*, and *D* lie on the same line, we say that they are *collinear*. **Collinear points** are points that lie on the same line.

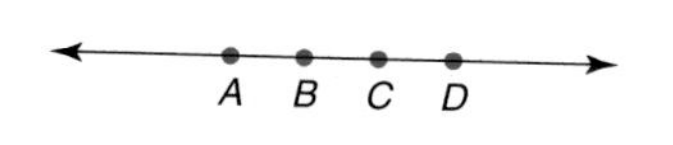

Figure 11.2

We can name a line by using only two letters, such as $\overleftrightarrow{AB}$, because one and only one straight line can be drawn through any two points. This is another way of saying that any two different points determine a unique line; or we could say that there is exactly one line containing any two different points.

Lines and Rays

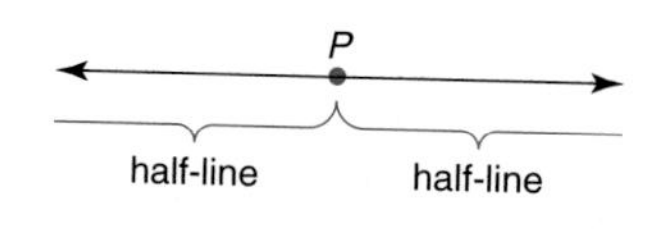

Figure 11.3

A point on a line separates one part of the line from another part. In fact, when we place a point on a line, as in Figure 11.3, we separate the line into three sets: the given point and two *half-lines*. The point *P* is not a point on either of the half-lines. A **half-line** is a set of points; if we include point *P* with the set of the points that constitute a half-line, we get what is known as a *ray*. A **ray** has only one end point and may be extended indefinitely in only one direction from that end point. Figure 11.4 illustrates ray *AB*. The notation used to denote ray *AB* is $\overrightarrow{AB}$.

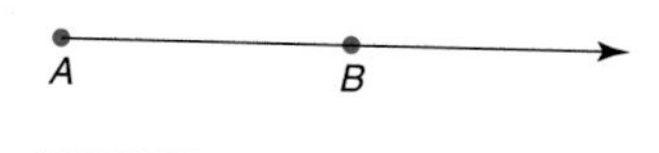

Figure 11.4

A **line segment** has two end points. The end points are used to name the segment, as in Figure 11.5. Line segment *AB* is denoted by $\overline{AB}$. Remember that a line may be extended indefinitely in either direction, a ray can be extended indefinitely in only one direction, and a line segment cannot be extended at all.

Figure 11.5

Thus far in our discussion we have covered the concepts of point, line, half-line, ray, and line segment. These terms and their corresponding notations are summarized below.

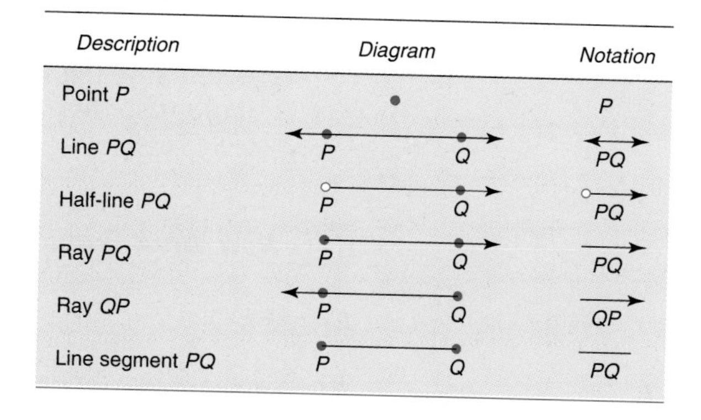

Description	*Diagram*	*Notation*
Point *P*	•	*P*
Line *PQ*	*P* *Q*	$\overleftrightarrow{PQ}$
Half-line *PQ*	*P* *Q*	$\overset{\circ\!\!\longrightarrow}{PQ}$
Ray *PQ*	*P* *Q*	$\overrightarrow{PQ}$
Ray *QP*	*P* *Q*	$\overrightarrow{QP}$
Line segment *PQ*	*P* *Q*	$\overline{PQ}$

Ray *PQ* is denoted by $\overrightarrow{PQ}$ which means that the end point of the ray is *P* and the ray is directed toward the point *Q*. Ray *QP* is denoted by $\overrightarrow{QP}$, which means that the end point of the ray is *Q* and the ray is directed toward the point *P*. (Refer to the chart.) Ray *PQ* ($\overrightarrow{PQ}$) and ray *QP* ($\overrightarrow{QP}$) are distinct; they involve different sets of points.

Figure 11.6

By now you have probably discovered that lines, rays, half-lines, and line segments that pass through a given pair of points have some points not in common. We can illustrate this by means of the two set operations, intersection and union.

Consider the line PQ in Figure 11.6. Using this line, what is $\overrightarrow{PQ} \cap \overrightarrow{QP}$? $\overrightarrow{PQ} \cap \overrightarrow{QP}$ is the intersection of ray PQ and ray QP. Ray PQ consists of the set of points that has the end point P and is directed toward the right through Q. Ray QP consists of the set of points that has the end point Q and is directed toward the left through P. Their intersection is the set of points common to both rays—that is, line segment PQ:

$$\overrightarrow{PQ} \cap \overrightarrow{QP} = \overline{PQ}$$

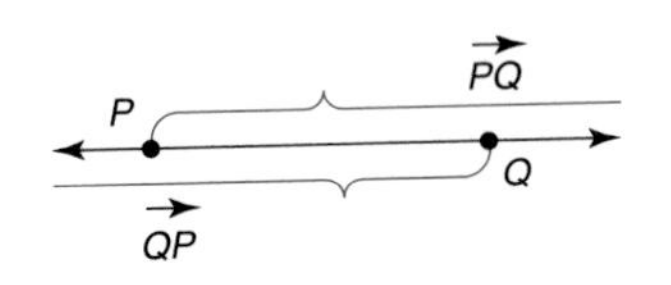

Figure 11.7

See Figure 11.7.

What is $\overrightarrow{PQ} \cup \overrightarrow{QP}$? We again refer to Figures 11.6 and 11.7, but instead of the intersection we want the union of the two sets of points. That gives us all of the points on the line PQ. Therefore, we have

$$\overrightarrow{PQ} \cup \overrightarrow{QP} = \overleftrightarrow{PQ}$$

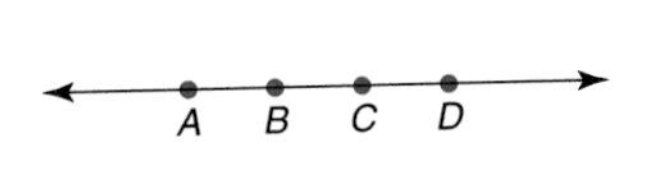

Figure 11.8

EXAMPLE 1 ***Finding Intersections and Unions of Rays and Line Segments*** Use the line in Figure 11.8 with the indicated points to find each of the following:

a. $\overrightarrow{AB} \cap \overrightarrow{CA}$ **b.** $\overrightarrow{AB} \cap \overrightarrow{BC}$

c. $\overrightarrow{BA} \cup \overrightarrow{BC}$ **d.** $\overline{AB} \cap \overline{CD}$

Solution

a. $\overrightarrow{AB} \cap \overrightarrow{CA}$ is the intersection of ray AB and ray CA. The set of points common to both of these rays is line segment AC. Therefore, $\overrightarrow{AB} \cap \overrightarrow{CA} = \overline{AC}$.

b. $\overrightarrow{AB} \cap \overrightarrow{BC}$ is the intersection of ray AB and ray BC. These two rays are both directed to the right and have all of the points in ray BC in common. Therefore, $\overrightarrow{AB} \cap \overrightarrow{BC} = \overrightarrow{BC}$.

c. $\overrightarrow{BA} \cup \overrightarrow{BC}$ is the union of the two rays, and, because they have the same end point and are directed in opposite directions, their union will result in all of the points on the line. Therefore, $\overrightarrow{BA} \cup \overrightarrow{BC} = \overleftrightarrow{AC}$. Note that it would also be correct to denote the answer as $\overleftrightarrow{AB}$, $\overleftrightarrow{BC}$, $\overleftrightarrow{CD}$, $\overleftrightarrow{BD}$, and so on.

d. $\overline{AB} \cap \overline{CD}$ is the intersection of line segment AB and line segment CD. Examining the diagram, we see that there are no points common to these two line segments. Their intersection is empty. Therefore, $\overline{AB} \cap \overline{CD} = \emptyset$.

NW ***Now Work Problems 5 and 7.***

Angles

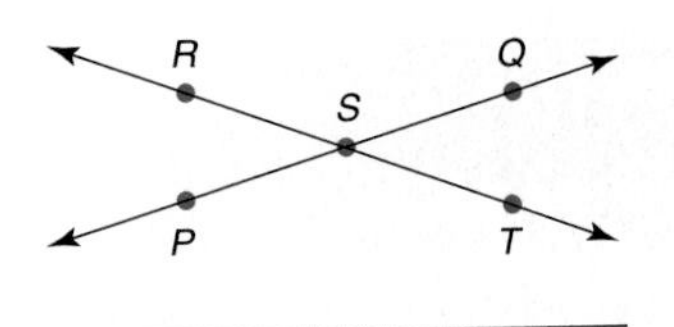

Figure 11.9

When we have two different lines that contain the same point, these lines are said to **intersect** at that point. In Figure 11.9, lines PQ and RT intersect at point S; that is, $\overleftrightarrow{PQ} \cap \overleftrightarrow{RT} = \{S\}$.

The set of points formed by two intersecting lines has many interesting subsets. For example, consider $\overrightarrow{SQ} \cup \overrightarrow{ST}$. What is the result when we unite the sets of points in ray SQ and ray ST? The result is a geometric figure formed by two rays drawn from the same point, as shown in Figure 11.10. This figure is called an

Figure 11.10

angle. An **angle** ($\angle$) is the union of two rays that have a common end point. The rays are called the **sides** of the angle and the common end point is called the **vertex** of the angle. Therefore,

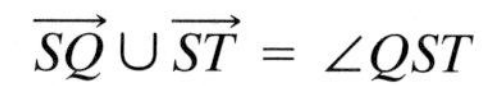

$$\overrightarrow{SQ} \cup \overrightarrow{ST} = \angle QST$$

We use capital letters to label an angle, and the name of the angle is written with the vertex letter in the middle. The first and third letters are used to designate the sides of the angle. In Figure 11.9, we can see that

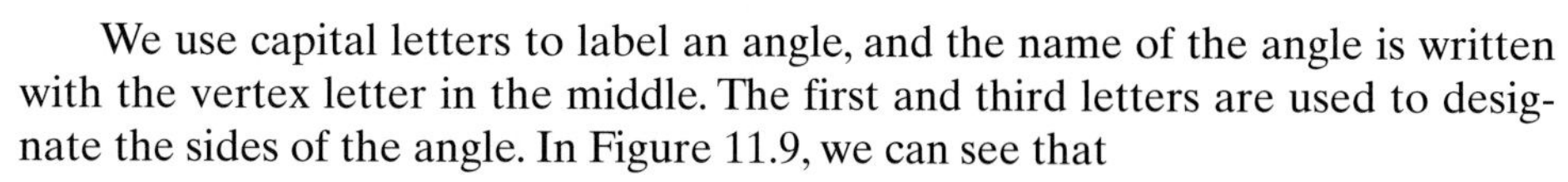

$$\overrightarrow{SR} \cup \overrightarrow{SP} = \angle RSP \qquad \overrightarrow{SR} \cup \overrightarrow{SQ} = \angle RSQ \qquad \overrightarrow{SP} \cup \overrightarrow{ST} = \angle PST$$

Angles RSP and RSQ are **adjacent angles** because they have the same vertex and a common side between them. Angles RSQ and QST are also adjacent angles, but angles RSP and QST are *not* adjacent angles. They do have the same vertex, but they do not have a common side. Angles RSP and QST are angles where the sides of one angle extend through the vertex and form the sides of the other. Angles of this type are called **vertical angles**. In Figure 11.9, angles RSQ and PST are also vertical angles.

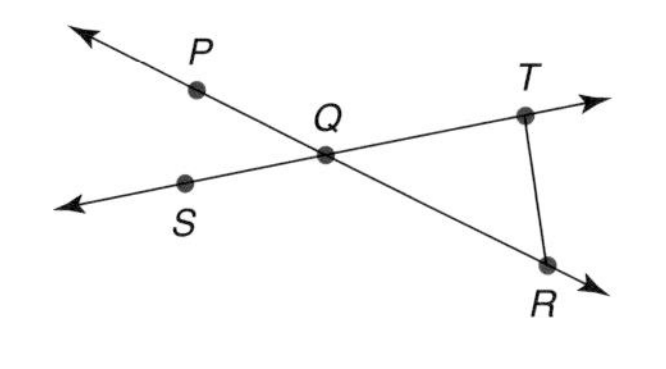

Figure 11.11

Consider the union of rays SR and ST in Figure 11.9. We see that $\overrightarrow{SR} \cup \overrightarrow{ST} = \overleftrightarrow{RT}$. But because an angle is the union of two rays that have a common end point, we can also say that $\overrightarrow{SR} \cup \overrightarrow{ST} = \angle RST$. Angle RST is a special kind of angle because its sides form a straight line. Angle RST is referred to as a **straight angle**. Therefore, a line such as PR in Figure 11.11 can also be thought of as $\angle PQR$. (Remember that the vertex letter of an angle is always written in the middle when we label the angle.)

EXAMPLE 2 ***Finding Intersections and Unions of Lines and Rays*** Use Figure 11.11 with the indicated points to find each of the following:

a. $\overleftrightarrow{PR} \cap \overleftrightarrow{ST}$ **b.** $\overrightarrow{QP} \cup \overrightarrow{QT}$

c. $\overrightarrow{PR} \cap \overrightarrow{RP}$ **d.** $\overrightarrow{QP} \cup \overrightarrow{QR}$

Solution

a. $\overleftrightarrow{PR} \cap \overleftrightarrow{ST}$ is the intersection of lines PR and ST, and the two lines intersect at point Q. Therefore, $\overleftrightarrow{PR} \cap \overleftrightarrow{ST} = \{Q\}$.

b. $\overrightarrow{QP} \cup \overrightarrow{QT}$ is the union of rays QP and QT. The union of two rays that have a common end point is an angle. Therefore, we see that $\overrightarrow{QP} \cup \overrightarrow{QT} = \angle PQT$.

c. $\overrightarrow{PR} \cap \overrightarrow{RP}$ is the intersection of two rays having opposite directions. The set of points that they have in common is line segment PR. Therefore, $\overrightarrow{PR} \cap \overrightarrow{RP} = \overline{PR}$.

d. $\overrightarrow{QP} \cup \overrightarrow{QR}$ is the union of two rays of opposite direction and having the same end point, so their union will result in all of the points on the line. Therefore, $\overrightarrow{QP} \cup \overrightarrow{QR} = \overleftrightarrow{PR}$ (*Note:* Recall that we can also say that $\overrightarrow{QP} \cup \overrightarrow{QR} = \angle PQR$.)

NW ***Now Work Problems 15 and 17.***

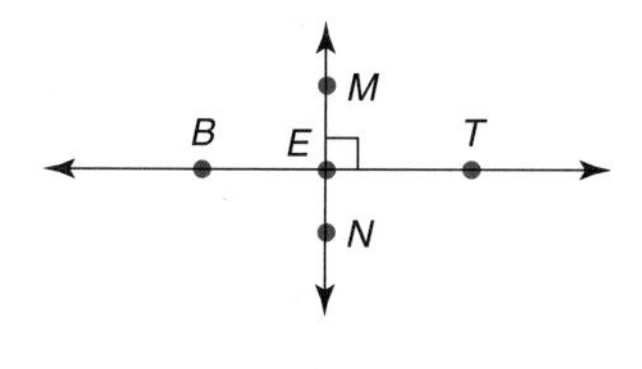

Figure 11.12

There are many other kinds of angles that can be defined. However, before we can discuss these angles, we must define *perpendicular lines*. Two lines that intersect so as to form a pair of congruent adjacent angles are called **perpendicular lines**. Each line is said to be perpendicular to the other. In Figure 11.12, $\overleftrightarrow{MN}$ is perpendicular to $\overleftrightarrow{BT}$; we denote this by $\overleftrightarrow{MN} \perp \overleftrightarrow{BT}$. Note also that $\overleftrightarrow{EM} \perp \overleftrightarrow{BT}$.

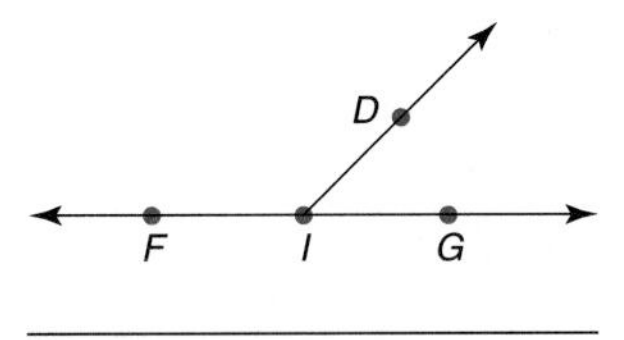

Figure 11.13

A **right angle** is an angle whose sides are perpendicular. In Figure 11.12, $\angle MET$ is a right angle, as are $\angle BEM$, $\angle BEN$, and $\angle NET$. An angle that is wider than a right angle but narrower than a straight angle is called an **obtuse angle**. An angle that is narrower than a right angle is called an **acute angle**. In Figure 11.13, $\angle FID$ is an obtuse angle, and $\angle DIG$ is an acute angle.

The measurement of these angles will be discussed in Section 11.4.

EXAMPLE 3 ***Finding Intersections and Unions of Rays, Line Segments, and Angles*** Use Figure 11.13 with the indicated points to find each of the following:

a. $\overrightarrow{FG} \cap \overrightarrow{ID}$ **b.** $\overrightarrow{IF} \cup \overrightarrow{ID}$

c. $\overline{FI} \cap \overline{IG}$ **d.** $\angle FID \cap \angle DIG$

Solution

a. $\overrightarrow{FG} \cap \overrightarrow{ID}$ is the intersection of two rays; the only thing they have in common is point I. Therefore, $\overrightarrow{FG} \cap \overrightarrow{ID} = \{I\}$.

b. $\overrightarrow{IF} \cup \overrightarrow{ID}$ is the union of two rays with a common end point. Therefore, $\overrightarrow{IF} \cup \overrightarrow{ID} = \angle FID$.

c. $\overline{FI} \cap \overline{IG}$ is the intersection of two line segments with point I in common. Therefore, $\overline{FI} \cap \overline{IG} = \{I\}$.

d. $\angle FID \cap \angle DIG$ is the intersection of two angles. They are adjacent angles because they have the same vertex and a common side. The intersection of these two angles is the common side—namely, $\overrightarrow{ID}$. Therefore, $\angle FID \cap \angle DIG = \overrightarrow{ID}$.

NW ***Now Work Problems 27 and 31.***

Remember that any dot we place on a piece of paper will have some measurement, such as height, width, and even thickness. The same can be said for a line segment that we may draw on paper; that is, any line that we draw will have some width and thickness in addition to length. But in geometry, points and lines do not have such characteristics. The points and lines that we draw are diagrams that represent the points and lines of geometry.

Exercises for Section 11.2

For Exercises 1–14, use Figure 11.14 to find each of the following:

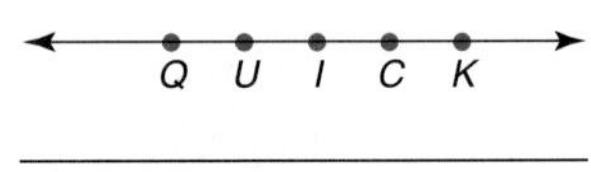

Figure 11.14

1. $\overline{QI} \cap \overline{IC}$
2. $\overline{QU} \cap \overline{CK}$
3. $\overline{QC} \cup \overline{CK}$
4. $\overline{QU} \cap \overline{IC}$
5. NW $\overrightarrow{QC} \cap \overline{QI}$
6. $\overrightarrow{QI} \cap \overrightarrow{KI}$
7. NW $\overrightarrow{UQ} \cup \overrightarrow{UK}$
8. $\overleftrightarrow{QU} \cap \overline{IC}$
9. $\overrightarrow{IK} \cap \overline{IU}$
10. $\overrightarrow{IK} \cup \overline{IU}$
11. $\overrightarrow{IK} \cap \overrightarrow{IQ}$
12. $\overrightarrow{IK} \cap \overrightarrow{IC}$
13. $\overrightarrow{CI} \cap \overline{QU}$
14. $\overline{QU} \cap \overrightarrow{IC}$

For Exercises 15–28, use Figure 11.15 to find each of the following:

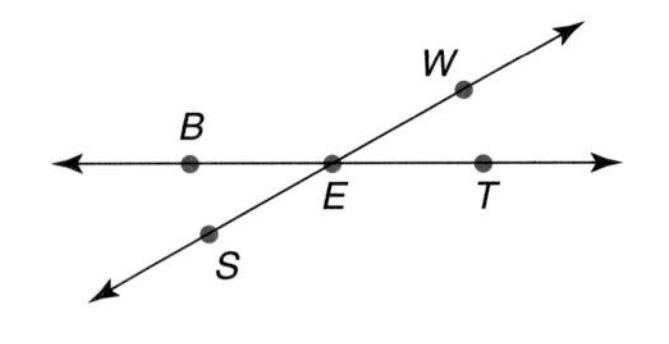

Figure 11.15

15. NW $\overleftrightarrow{BT} \cap \overleftrightarrow{SW}$
16. $\overrightarrow{BT} \cap \overrightarrow{SW}$
17. NW $\overrightarrow{TB} \cap \overrightarrow{WS}$
18. $\overline{BE} \cup \overline{ET}$
19. $\overline{SE} \cap \overline{ET}$
20. $\overline{BE} \cap \overline{ET}$
21. $\overrightarrow{ET} \cup \overrightarrow{EW}$
22. $\overrightarrow{EB} \cup \overrightarrow{ES}$

23. $\overrightarrow{ET} \cup \overrightarrow{EB}$

24. $\overrightarrow{BE} \cap \overrightarrow{ET}$

25. $\overrightarrow{BE} \cup \overrightarrow{ET}$

26. $\overrightarrow{EW} \cup \overrightarrow{EB}$

NW 27. $\angle BEW \cap \angle WET$

28. $\angle SET \cap \angle WET$

For Exercises 29–41, use Figure 11.16 to find each of the following:

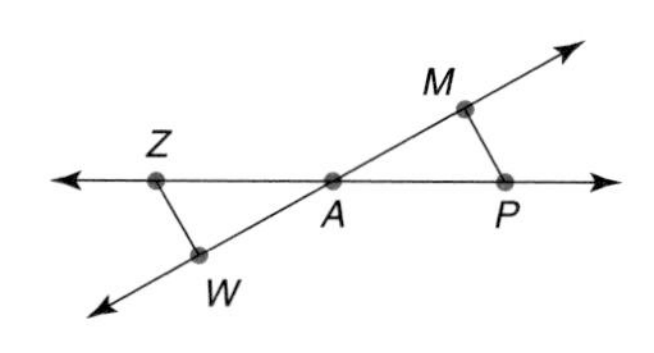

Figure 11.16

29. $\overleftrightarrow{WM} \cap \overleftrightarrow{ZP}$

30. $\overrightarrow{ZA} \cap \overrightarrow{PA}$

NW 31. $\overrightarrow{AP} \cup \overrightarrow{AZ}$

32. $\overrightarrow{AP} \cap \overrightarrow{AZ}$

33. $\overline{ZW} \cap \overline{MP}$

34. $\overrightarrow{AP} \cup \overrightarrow{AM}$

35. $\overrightarrow{AZ} \cup \overrightarrow{AW}$

36. $\overrightarrow{PA} \cap \overline{ZW}$

37. $\angle PAM \cap \angle ZAM$

38. $\angle PAW \cap \angle ZAW$

39. $\angle PAM \cap \angle ZAW$

40. $\angle ZAP \cap \angle WAM$

*41. $\overline{ZA} \cup \overline{AW} \cup \overline{ZW}$

Writing Mathematics

42. What are collinear points?

43. Define a ray.

44. What is the difference between a line segment and a line?

45. Define an angle.

Challenge Exercises

46. In this section we discovered that one and only one straight line can be drawn through any two given points. We can also say that two points determine a unique line. How many points determine a unique circle? What is the minimum number of points required?

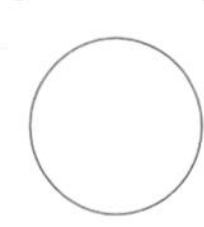

47. What are the differences among an acute angle, a right angle, an obtuse angle, a straight angle, a reflex angle, and a round angle?

48. In your own words, define perpendicular lines.

Group/Research Activities

49. There were many Greek scholars who contributed to the development of geometry. One man who was keenly interested in geometry was Plato. Do research and write a paper on the life of Plato and his contributions to the development of geometry.

50. Euclid is known as the father of geometry. Prepare a report on his life and his contributions to the development of geometry.

Just for fun

Can you draw a straight line that intersects all three sides of triangle *ABC*?

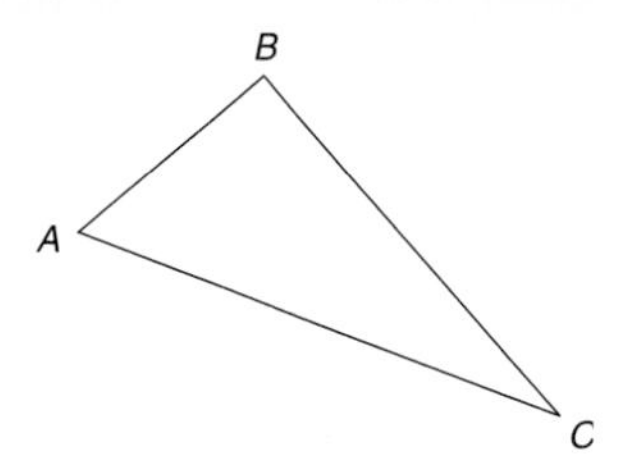

11.3 PLANES

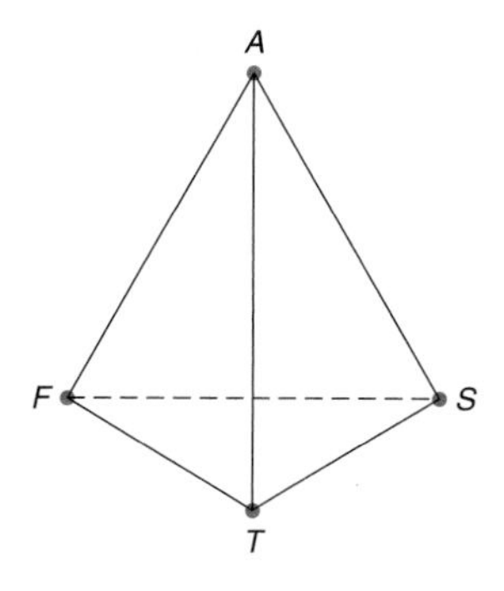

Figure 11.17

A **plane** can be thought of as a flat surface such as the floor of a room, or a table top, or a desk top. A plane or surface divides, or separates, one portion of space from another. A floor separates the space above the floor from the space below the floor. A wall is a plane that separates the space in one room from the space in the adjoining room. The wall of a building is a plane that separates the space inside from the space outside. In Figure 11.17, each of the faces of the pyramid is a flat, or plane, surface.

A plane has two dimensions, length and width. In geometry, a plane does not have any thickness. *Note that a flat surface suggests a plane, and a true plane is endless or infinite.*

Just as *point* and *line* are undefined terms in geometry, so is *plane*. Although our concept of a plane is intuitive, we can still discuss some of the properties of a plane. For example, we can think of a line as a set of points, and we can do the same for a plane. A plane is a set of points. The points are on the plane, and the plane contains the points. **Coplanar** points are points that are on the same plane, just as collinear points are points that are on the same line.

A unique plane is determined by any three noncollinear points. In other words, if we are given three distinct points that are not all on the same line, then there is one and only one plane that can contain all three points. For example, in Figure 11.17 points F, S, and T are not on the same line, and they determine a unique plane, namely, plane FST. Note that none of the other planes in Figure 11.17 contains all three of these points. The other planes do contain two of the points; by using a different third point, a different plane is determined.

Have you ever noticed that things such as easels, telescopes, cameras, and Christmas trees are mounted on stands with three legs? These stands, or tripods, have only three legs because three legs will always rest on some plane. A stand, table, or chair with four legs tends to "wobble" unless all four legs are exactly the same length; this is not the case for an object with three legs.

Figure 11.18

Since lines and planes are both composed of points, we can make the following observation: If two different points of a line are on a plane, then all the points on the line are also on the plane. In other words, if two points of a line are on a plane, then the line must also be on the plane, since two points determine a line. For an example, see line AB in Figure 11.18.

Figure 11.18 illustrates another important concept regarding the relationship between lines and planes. Any line on a given plane divides that plane into two half-planes. Note that in Figure 11.18 $\overleftrightarrow{AB}$ separates the points on the plane. There are those points to the right of $\overleftrightarrow{AB}$, and those points to the left of $\overleftrightarrow{AB}$. The points that are on $\overleftrightarrow{AB}$ are not points on either half-plane. This concept is similar to that of a point dividing a line into two half-lines. A line on a plane divides the plane into two half-planes, but the result is three sets of points: the points on the line, the points on one half-plane, and the points on the other half-plane.

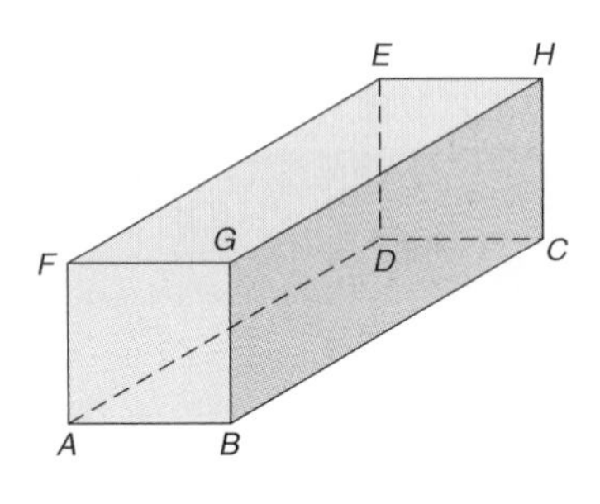

Figure 11.19

Either two different planes intersect or they do not. That is, two distinct planes must meet in a line or they must be parallel. The plane of a wall intersects the plane of the floor; that is, they meet in a line. But the floor and ceiling planes will never meet in a line, no matter how far they are extended. These two planes are **parallel planes**. In Figure 11.19, planes $ABCD$ and $EFGH$ are parallel planes, but planes $ABCD$ and $BGHC$ intersect in a line.

It should also be noted that lines BC and AD in Figure 11.19 are parallel lines. We can denote this by $\overleftrightarrow{BC} \parallel \overleftrightarrow{AD}$. Two lines that are on the same plane and do not intersect, however far they are extended, are called **parallel lines**. In Figure 11.19, lines AB and GH will not intersect, no matter how far we extend them, but they are not parallel lines. Why not? They are not parallel lines because they are not in the same plane. If two lines do not lie in the same plane, they are *skew lines*.

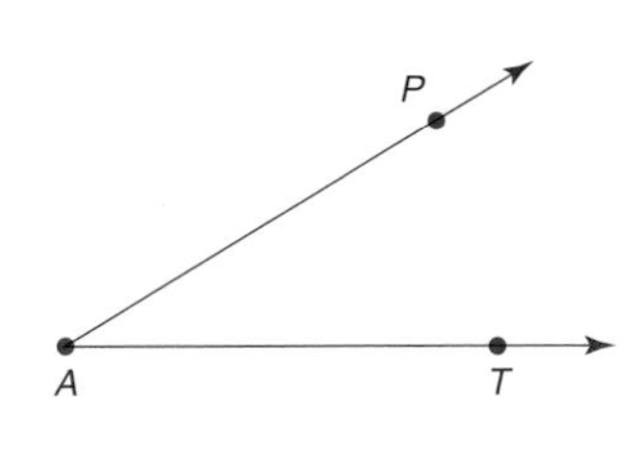

Figure 11.20

In the preceding section, we noted that an angle is the union of two rays that have a common end point, such as angle PAT in Figure 11.20. An angle on a plane divides that plane into two half-planes, and that angle produces three sets of points: the points on the angle, the points on one half-plane, and the points on the other half-plane. Consider this page as a plane. It is a set of points, and angle PAT divides those points into three parts. There are those points on angle PAT, such as point P. There are also points inside angle PAT; that is, they are in the interior of angle PAT. Finally, there are points outside angle PAT—that is, in the exterior of angle PAT. In Figure 11.21, point O is an exterior point, and point K is an interior point. Note that points P, A, and T are neither interior nor exterior points, because they are on the angle.

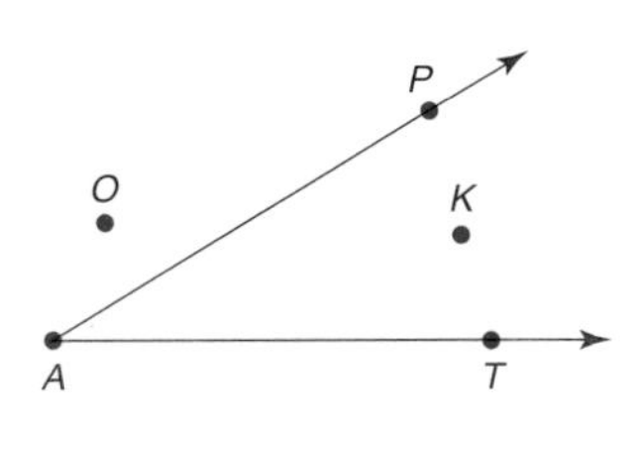

Figure 11.21

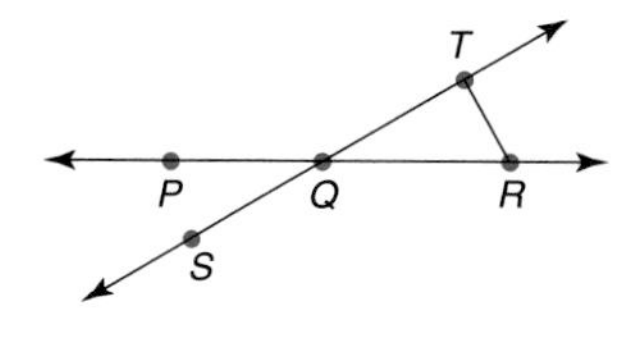

Figure 11.22

Let's combine these new concepts of interior and exterior points with set intersection. Consider Figure 11.22 with the indicated points. What is the intersection of angle TQR and $\overline{TR}$? That is, what is $\angle TQR \cap \overline{TR}$? The given figure is on a plane—namely, this page. An angle is the union of two rays with a common end point. When we are discussing $\angle TQR$, we are in effect discussing those points on the angle. Therefore, the points that $\angle TQR$ and $\overline{TR}$ have in common are points T and R; that is,

$$\angle TQR \cap \overline{TR} = \{T, R\}$$

Now consider the intersection of $\overline{TR}$ and the interior of $\angle TQR$—that is,

$$\overline{TR} \cap (\text{interior } \angle TQR)$$

The points that these two sets of points have in common are those points that are in the interior of $\angle TQR$ and are also on line segment TR. Is the answer $\overline{TR}$? No. We cannot list our answer as line segment TR because that would mean that points T and R are also members of the solution, and from our previous discussion we know that points T and R are on $\angle TQR$; therefore, they cannot be inside $\angle TQR$. Our solution consists of all those points on $\overline{TR}$ except T and R. This is an open line segment, and we denote it by $\overset{\circ\!-\!\circ}{TR}$. Therefore,

$$\overline{TR} \cap (\text{interior } \angle TQR) = \overset{\circ\!-\!\circ}{TR}$$

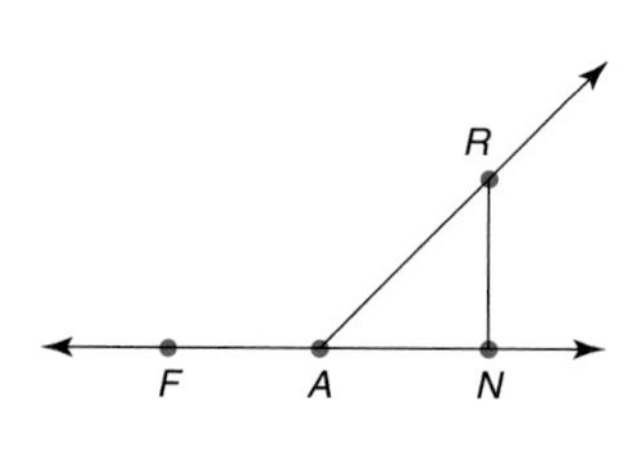

Figure 11.23

What is the intersection of ray QP and the exterior of angle TQR? That is, what is $\overrightarrow{QP} \cap (\text{exterior } \angle TQR)$? The exterior of $\angle TQR$ consists of all those points that are not on $\angle TQR$ and not inside $\angle TQR$. Now all of the points on ray QP lie outside of $\angle TQR$, except one point—namely Q. Therefore, the solution is the half-line QP (excluding Q), and we denote this by $\overset{\circ\!\rightarrow}{QP}$. Therefore,

$$\overrightarrow{QP} \cap (\text{exterior } \angle TQR) = \overset{\circ\!\rightarrow}{QP}$$

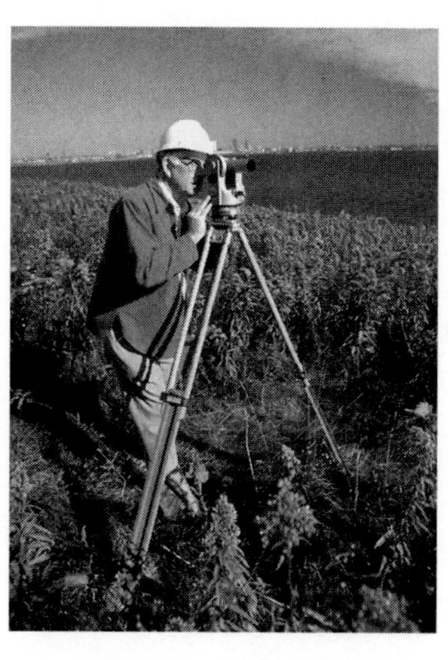

This surveyer's instrument is mounted on a three-legged stand for stability on any kind of terrain.

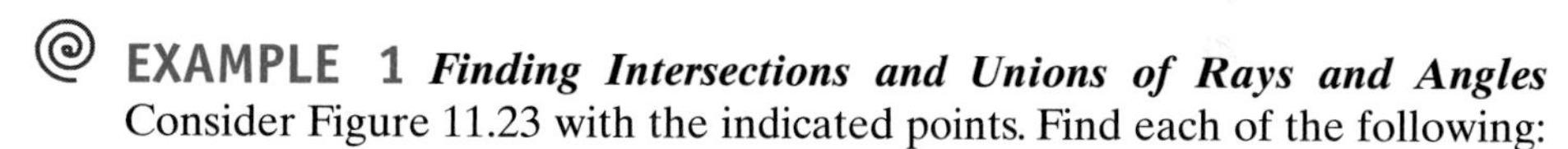

EXAMPLE 1 ***Finding Intersections and Unions of Rays and Angles***
Consider Figure 11.23 with the indicated points. Find each of the following:

a. $\overrightarrow{AR} \cup \overrightarrow{AN}$

b. $\overrightarrow{AF} \cap (\text{exterior } \angle RAN)$

c. $\overline{RN} \cap (\text{interior } \angle RAN)$

d. $(\text{interior } \angle RAN) \cap (\text{exterior } \angle RAN)$

Solution

a. $\overrightarrow{AR} \cup \overrightarrow{AN}$ is the union of two rays with a common end point—that is, an angle. Therefore,

$$\overrightarrow{AR} \cup \overrightarrow{AN} = \angle RAN$$

b. $\overrightarrow{AF} \cap (\text{exterior } \angle RAN)$ is the intersection of those points on $\overrightarrow{AF}$ with those points that are in the exterior of $\angle RAN$. This is all the points on $\overrightarrow{AF}$ except A. Therefore,

$$\overrightarrow{AF} \cap (\text{exterior } \angle RAN) = \overset{\circ\!\rightarrow}{AF}$$

c. $\overline{RN} \cap (\text{interior } \angle RAN)$ is the intersection of those points that are on $\overline{RN}$ and also in the interior of $\angle RAN$. This is all the points on $\overline{RN}$ except R and N. Therefore,

$$\overline{RN} \cap (\text{interior } \angle RAN) = \overset{\circ\!-\!\circ}{RN}$$

d. The set of points that are in the interior of $\angle RAN$ and the points that are in the exterior of $\angle RAN$ do not intersect. They are separated by those points that are on $\angle RAN$. Therefore,

$$(\text{interior } \angle RAN) \cap (\text{exterior } \angle RAN) = \emptyset$$

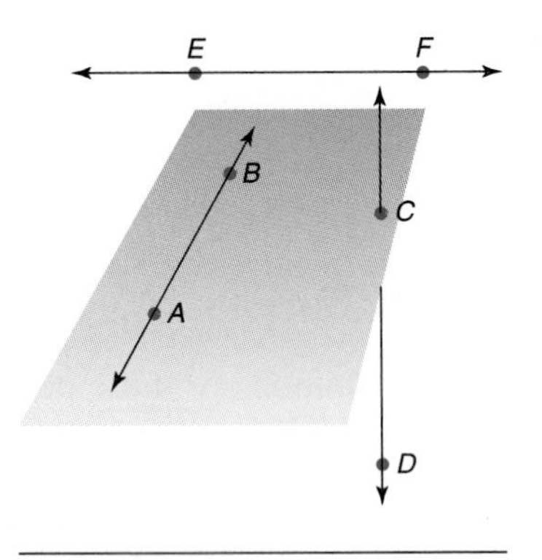

Figure 11.24

***NW* Now Work Problems 3 and 5.**

Remember that a plane is a flat surface, and that it separates one part of space from an adjoining part of space. It is also a set of points. A line on a plane divides that plane into half-planes, but there are other possible relationships between lines and planes.

Line AB in Figure 11.24 is an example of a line on a plane; since two points of $\overleftrightarrow{AB}$ are on the plane, all of the points on $\overleftrightarrow{AB}$ are on the plane. Line CD intersects the plane at only one point, while line EF does not intersect the plane at all. Line EF is parallel to the given plane and skew to line AB. A line parallel to the horizon is called a horizontal line, while a line that is perpendicular to a horizontal line is called a vertical line.

Exercises for Section 11.3

For Exercises 1–8, use Figure 11.25 with the indicated points to find each of the following:

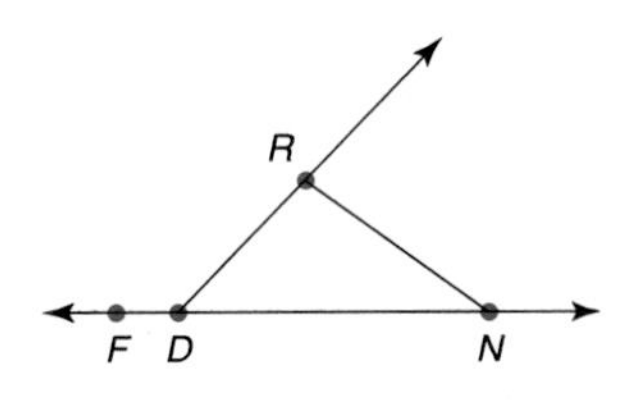

Figure 11.25

1. $\overline{RD} \cap \overrightarrow{DN}$
2. $\overrightarrow{DR} \cap \overline{DN}$
3. NW $\overline{RN} \cap (\text{interior } \angle RDN)$
4. $\overline{RN} \cap (\text{exterior } \angle RDN)$
5. NW $\overrightarrow{DR} \cup \overrightarrow{DF}$
6. $\overrightarrow{DN} \cap (\text{interior } \angle FDR)$
7. $\overrightarrow{DN} \cap (\text{exterior } \angle FDR)$
8. $(\text{interior } \angle FDR) \cap (\text{exterior } \angle RDN)$

For Exercises 9–18, use Figure 11.26 with the indicated points to find each of the following:

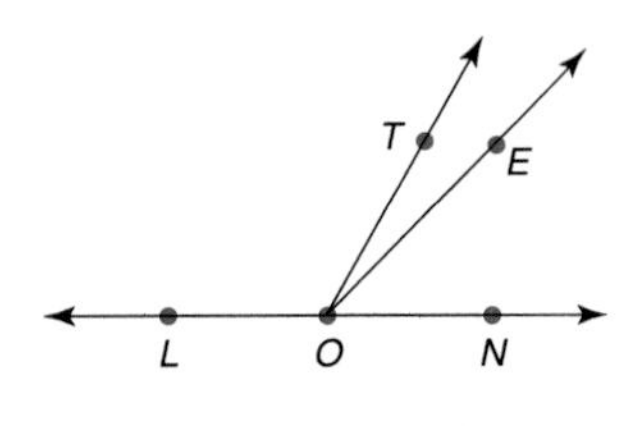

Figure 11.26

9. $\overrightarrow{OT} \cup \overrightarrow{ON}$
10. $\overrightarrow{OT} \cap \overline{ON}$
11. $\overrightarrow{OE} \cap (\text{interior } \angle TON)$
12. $(\text{interior } \angle TOE) \cap (\text{exterior } \angle EON)$
13. $(\text{exterior } \angle TOL) \cap (\text{interior } \angle EON)$
14. $(\text{interior } \angle EON) \cup (\text{interior } \angle TOE)$
15. $(\text{interior } \angle TOE) \cap (\text{interior } \angle EON)$
16. $(\text{exterior } \angle EON) \cap \overrightarrow{OL}$
17. $(\text{exterior } \angle TOE) \cap (\text{interior } \angle EON)$
18. $(\text{interior } \angle TOE) \cap (\text{exterior } \angle LOT)$

For Exercises 19–30, use Figure 11.27 with the indicated points to find each of the following:

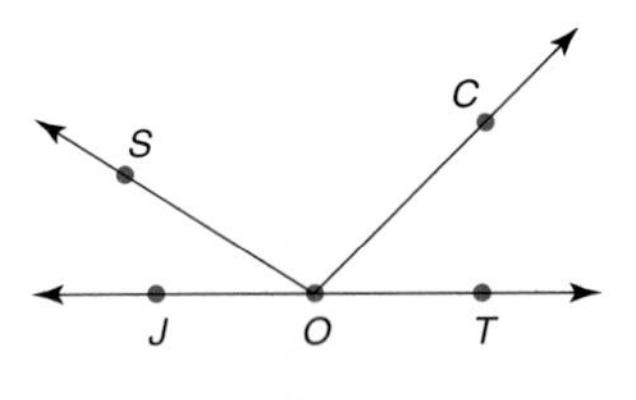

Figure 11.27

19. $\overrightarrow{OT} \cup \overrightarrow{OC}$
20. $\overrightarrow{OC} \cup \overrightarrow{OS}$
21. $\overrightarrow{OJ} \cup \overrightarrow{OS}$
22. $\overrightarrow{OS} \cap \overrightarrow{OT}$
23. $(\text{interior } \angle SOC) \cap (\text{interior } \angle SOT)$
24. $(\text{exterior } \angle COT) \cap (\text{interior } \angle JOS)$
25. $\overrightarrow{OT} \cap (\text{exterior } \angle JOS)$
26. $\angle COT \cap \angle SOC$
27. $\angle JOS \cap \angle COT$
28. $(\text{exterior } \angle COT) \cap (\text{exterior } \angle COS)$
29. $(\text{exterior } \angle SOJ) \cap (\text{exterior } \angle SOC)$
30. $(\text{interior } \angle JOS) \cap (\text{interior } \angle JOC)$

Determine whether each statement (31–38) is true or false.

31. Two parallel planes will never intersect.
32. The intersection of two planes that are not parallel is a line.
33. The intersection of a line and a plane is never a point.
34. Two lines on the same plane must intersect at some point.
35. Skew lines intersect at some point.
36. Parallel lines intersect at some point.
37. An angle is formed by the union of two rays with a common end point.
38. The interior of an angle is the intersection of two half-planes.

Writing Mathematics

39. What two dimensions does a *plane* have?
40. In your own words, describe a plane in geometry.
41. What is the difference between parallel planes and parallel lines?
42. Explain what *skew lines* are and sketch an example.

Challenge Exercises

43. Three intersecting planes can have **(a)** no common intersection, **(b)** intersection in a common line, or **(c)** intersection in one common point. Conditions **(a)** and **(b)** are shown in Figure 11.28. Can you draw three planes so that their intersection is one common point?

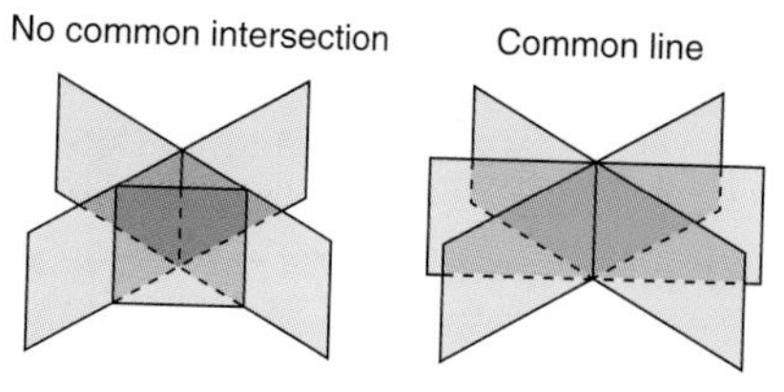

Figure 11.28

44. Sketch an example of each of the following:
    a. A plane and a line whose intersection is a point.
    b. A plane and a line whose intersection is a line.
    c. A plane and a line that do not intersect.

Research/Group Activities

45. The geometry that we use in this book, and in daily living, is called Euclidean geometry. *Hyperbolic geometry,* also known as Lobachevskian geometry (Nicolai Lobachevsky, 1793–1856), is a non-Euclidean geometry that has properties different from Euclidean geometry. Research this topic, noting the differences between hyperbolic geometry and Euclidean geometry. Prepare a report for class.
46. Another non-Euclidean geometry (see Exercise 45) is *Elliptical geometry,* also known as Riemannian geometry (Georg Riemann, 1826–1866). Research this topic, noting the differences between elliptical geometry and Euclidean geometry. Prepare a report for class.

Just for fun

Irene and Tom were each hired to work at a camp for 20 days. They had two choices regarding their salary: (A) they could receive $25 per day, or (B) receive 1 cent the first day, 2 cents the second day, 4 cents the third day, and so on. That is, the first day's pay was to be $0.01 and each succeeding day's pay would be twice the previous day's pay.

Irene promptly chose plan B, whereas Tom thought a while and then chose plan A. The person in charge then announced that Irene would be chief counselor and Tom assistant counselor. Why?

11.4 ANGLES

In the two previous sections we discussed to some extent the concept of **angles**. For example, in Section 11.2, we learned that an angle is the union of two rays that have a common endpoint. The rays are the *sides* of the angle and the common endpoint is the *vertex* of the angle. Using this information, an angle is named using three letters: The first and third letters represent the sides of the angle, and the middle letter represents its vertex. Thus, if given angle CAB, which is denoted as $\angle CAB$, we know that its vertex is at point A and that its sides are rays AC and AB.

We also introduced several different types of angles that are named by the manner in which they are formed. Four in particular were adjacent, vertical, straight, and right angles. **Adjacent angles** have the same vertex and a common side between them. **Vertical angles** are two angles where the sides of one angle extend through its vertex and form the sides of the second angle. A **straight angle** is an angle where its sides form a straight line; that is, its sides lie on the same line and extend in opposite directions from the vertex. A **right angle** is an angle whose sides are perpendicular. Finally, for completeness, we identified two other angles based on their "width" relative to right or straight angles. We said that an **obtuse angle** is wider than a right angle but narrower than a straight angle, and an **acute angle** is narrower than a right angle.

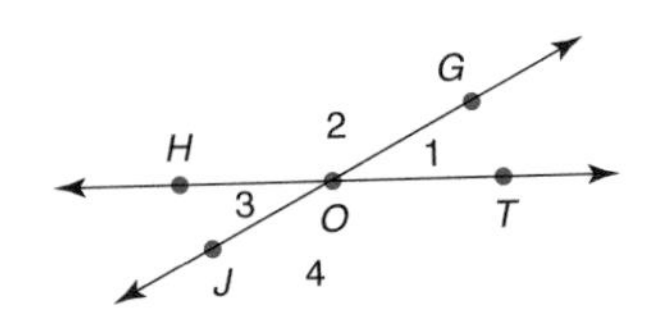

Figure 11.29

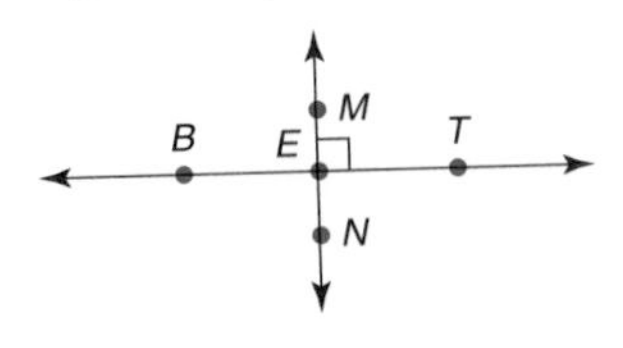

Figure 11.30

We can use Figures 11.29 and 11.30 to illustrate some of these angles. For example, in Figure 11.29, angle GOT and angle TOJ are adjacent angles; they have the same vertex and a common side between them. Angle HOJ and angle GOT are vertical angles; the sides of one angle are extended through the vertex and form the sides of the other angle. Angle JOG is an example of a straight angle; its sides lie on the same straight line and extend in opposite directions from the vertex. Note that angle GOT can also be named $\angle 1$ and, similarly, that

$$\angle GOH = \angle 2, \quad \angle HOJ = \angle 3, \quad \text{and} \quad \angle JOT = \angle 4$$

In Figure 11.30, $\overleftrightarrow{MN}$ is perpendicular to $\overleftrightarrow{BT}$; we denote this by $\overleftrightarrow{MN} \perp \overleftrightarrow{BT}$. Note also that by $\overleftrightarrow{EM} \perp \overleftrightarrow{BT}$. Therefore, $\angle MET$ is a right angle, as are $\angle BEM$, $\angle BEN$, and $\angle NET$. In this section, we extend our presentation of angles to include a discussion of their measures. Before doing so though, let's first prepare ourselves by identifying angles from a given figure.

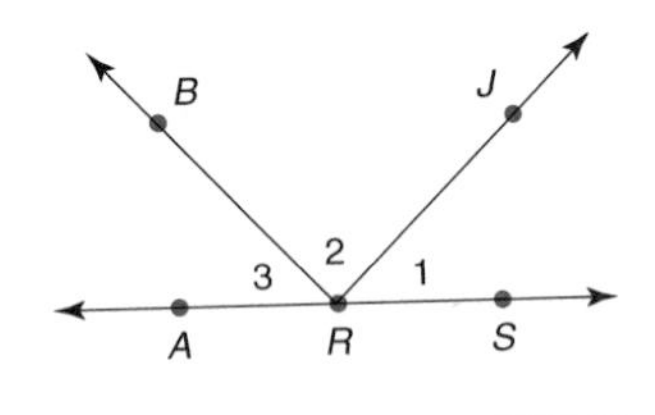

Figure 11.31

EXAMPLE 1 ***Identifying Angles*** Name six different angles in Figure 11.31. Line AS is a horizontal line.

Solution At first glance, there may appear to be only three angles in the given figure. But remember that an angle is formed by the union of two rays that have a common endpoint. We shall number the angles as we list them, but this does not mean that they must be listed in this order.

1. $\angle JRS$ or $\angle 1$
2. $\angle JRB$ or $\angle 2$
3. $\angle ARB$ or $\angle 3$
4. $\angle SRB$
5. $\angle JRA$
6. $\angle ARS$

NW ***Now Work Problem 1.***

Measures of Angles

Angles are usually measured in **degrees**. If the measure of an angle is 90 degrees, this is denoted by 90°. Just as feet can be divided into inches, and meters can be divided into centimeters, we can divide a degree into smaller units. This is usually done when better accuracy in measuring an angle is desired. Each degree is divided into 60 smaller units called **minutes**. A minute is denoted by $'$. Therefore, $1° = 60'$. Similarly, a minute is divided into 60 smaller units called **seconds**. A second is denoted by $''$. Therefore, $1' = 60''$. The fact that a degree is divided into 60 equal minutes and a minute is divided into 60 equal seconds is interesting, especially if we recall from Chapter 7 that the Babylonians used a sexagesimal system of numeration; that is, their system was based on 60.

The measure of a straight angle is 180°, and the measure of a right angle is 90°. Given the measure of an angle, we can find the measure of its *supplement* and its *complement*. Two angles whose measures sum to that of a straight angle (180°) are called **supplementary angles**, and two angles whose measures sum to that of a right angle (90°) are called **complementary angles**. Therefore, if we want to find the measure of the supplement of an angle that measures 60°, we must subtract 60° from 180°:

$$180^\circ - 60^\circ = 120^\circ$$

In order to find the measure of the complement of an angle measuring 60°, we must subtract 60° from 90°:

$$90^\circ - 60^\circ = 30^\circ$$

Math Connections

Transformational Geometry

Consider the following scenario: You are eating lunch with some friends when you pull out a new CD from your book bag. You look at the front cover and then you "swing" it around to read the words written vertically along the left side of the cover. While looking at it, one of your friends who is at the other end of the table asks if she can see the CD. You oblige by sliding it to her across the table. These two actions—swing and slide—can be more formally thought of as a *rotation* and *translation*, respectively. When you "swung" the CD around, you essentially rotated it 90° in a clockwise direction. When you slid the CD to your friend, you transferred it to a new location by sliding it in a certain direction and at a certain angle.

These two operations, rotation and translation, are formally known as *transformations*. A transformation is an operation that results in altering the position, size, or shape of a figure based on a given rule. The study of geometric transformations is called **transformational geometry**.

In addition to rotation and translation, there are two other basic transformations: *reflection* and *glide reflection*. A reflection represents the mirrored image of a figure. For example, when you look into a mirror you see a reflection of yourself. A reflection basically "flips" all the points in the plane over a line, which is called the *mirror*. Thus, a reflection results in an image that is "opposite" in form to the original image. In our CD illustration, a reflection involves flipping the CD from front to back. In doing so, we are actually reflecting the CD over an imaginary mirror, formally called the line of reflection.

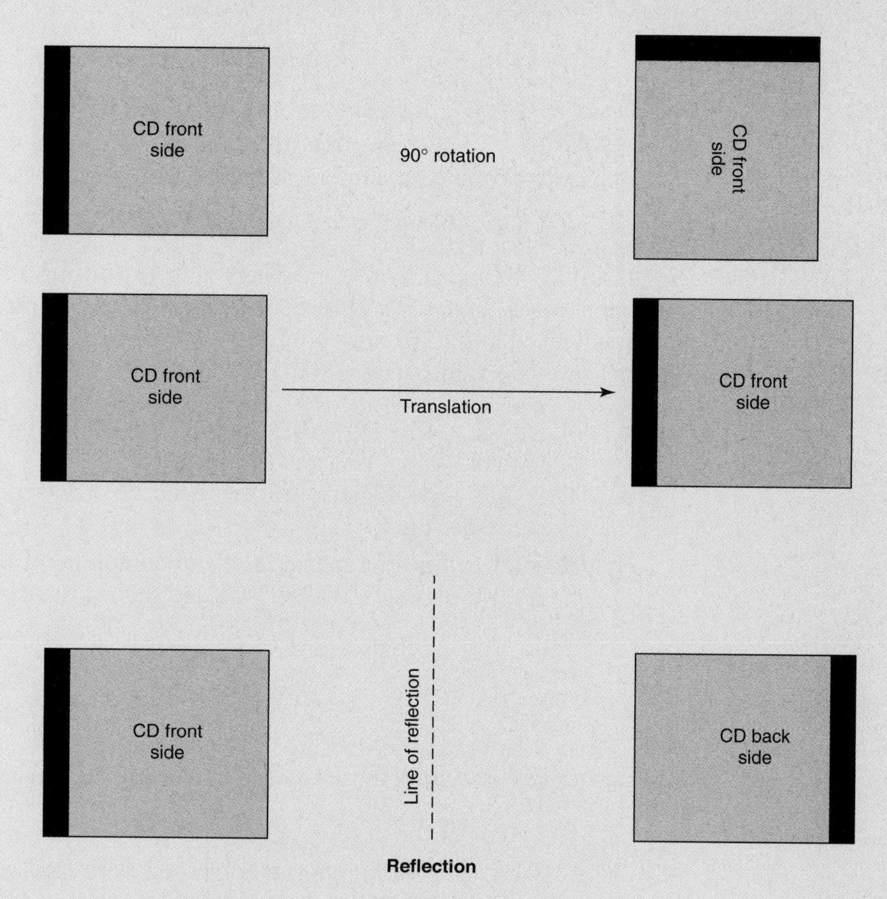

(continued)

(continued)

A glide reflection is the combination of a translation and a reflection. It involves the translation and reflection across a mirror that is parallel to the direction of the translation. A glide reflection is the only transformation that involves two "moves."

Collectively, translation, rotation, reflection, and glide reflection are the four basic transformations of transformational geometry. Furthermore, in any geometric transformation, the original object is called the *preimage* and the new object formed as a result of a transformation is called the *image*. If the preimage and the image are congruent to each other, then the transformation is called an *isometry* or *rigid transformation*.

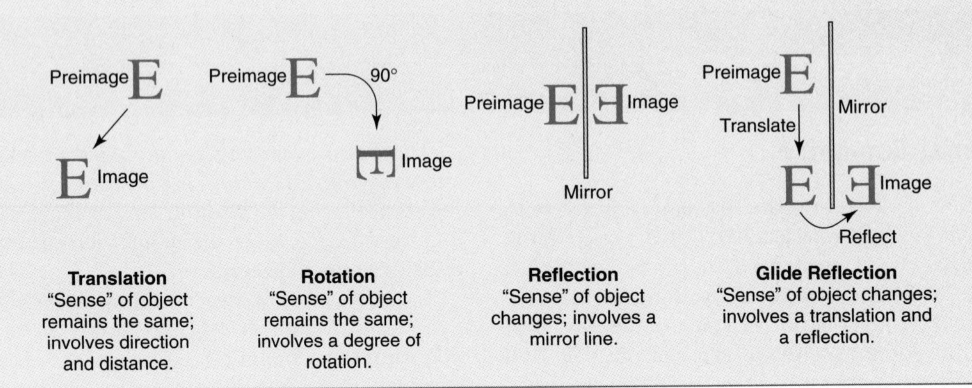

Suppose we want to find the measure of the complement of an angle measuring $42°45'$. Since we cannot subtract $42°45'$ from $90°$ we write $90°$ as $89°60'$, because a degree is divided into 60 minutes. Therefore, we have

$$\begin{array}{r} 89°60' \\ -42°45' \\ \hline 47°15' \end{array}$$

EXAMPLE 2 ***Finding the Measure of Complementary Angles*** Find the measure of the complement of angles with each of the following measures:

a. $62°$ **b.** $45°$ **c.** $22°15'$

Solution Two angles whose measures sum to that of a right angle are called complementary angles. Therefore, in order to find the measure of the complement of a given angle, we must subtract the measure of that angle from $90°$, since the measure of a right angle is $90°$.

a. $90° - 62° = 28°$. Therefore, the measure of the complement of an angle measuring $62°$ is $28°$.

b. $90° - 45° = 45°$. Therefore, the measure of the complement of an angle measuring $45°$ is $45°$.

c. In order to find the measure of the complement of an angle measuring $22°15'$, we rewrite $90°$ as $89°60'$, and then subtract $22°15'$ from it:

$$\begin{array}{r} 89°60' \\ -22°15' \\ \hline 67°45' \end{array}$$

Therefore, the measure of the complement of an angle measuring $22°15'$ is $67°45'$.

EXAMPLE 3 ***Finding the Measure of Supplementary Angles*** Find the measure of the supplement of angles with each of the following measures:

a. $62°$ **b.** $22°15'$ **c.** $128°42'16''$

Solution Two angles whose measures sum to that of a straight angle are called supplementary angles. Therefore, in order to find the measure of the supplement of a given angle, we must subtract the measure of that angle from 180°, since a straight angle contains 180°.

a. $180° - 62° = 118°$. Therefore, the measure of the supplement of an angle measuring 62° is 118°.

b. In order to find the measure of the supplement of an angle measuring 22°15′, we must rewrite 180° as 179°60′, and then subtract 22°15′ from it:

$$\begin{array}{r} 179°60' \\ -22°15' \\ \hline 157°45' \end{array}$$

Therefore, the measure of the supplement of an angle measuring 22°15′ is 157°45′.

c. In order to find the measure of the supplement of an angle measuring 128°42′16″, we must rewrite 180° as 179°59′60″, and then subtract 128°42′16″ from it:

$$\begin{array}{r} 179°59'60'' \\ -128°42'16'' \\ \hline 51°17'44'' \end{array}$$

Therefore, the measure of the supplement of an angle measuring 128°42′16″ is 51°17′44″.

NW ***Now Work Problems 5a and 5d.***

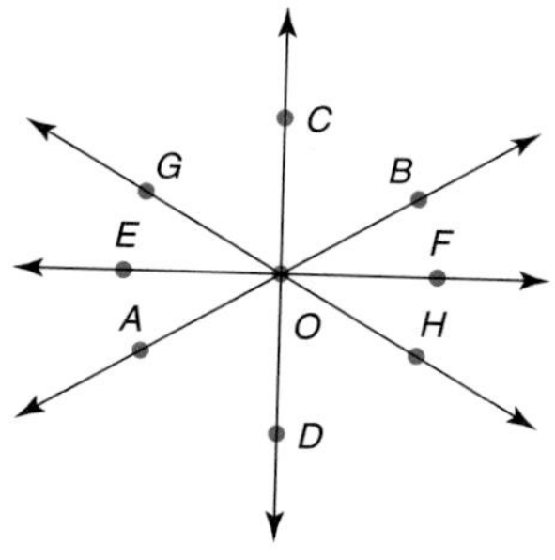

Figure 11.32

In Figure 11.32, we have four lines intersecting at point O, and $\overleftrightarrow{CD} \perp \overleftrightarrow{EF}$. An **acute angle** is an angle whose measure is greater than 0° and less than 90°. In Figure 11.32, $\angle COF$ is a right angle and $\angle BOF$ is an acute angle. An **obtuse angle** is an angle whose measure is greater than 90° and less than 180°. In Figure 11.32, $\angle EOF$ is a straight angle and $\angle GOF$ is an obtuse angle.

Two angles whose measures sum to that of a right angle are called complementary angles. In Figure 11.32, $\angle COB$ and $\angle BOF$ are complementary angles. $\angle COB$ is the complement of $\angle BOF$, and $\angle BOF$ is the complement of $\angle COB$. Note that there exist other complementary angles in Figure 11.32, such as $\angle EOG$ and $\angle GOC$.

Two angles whose measures sum to that of a straight angle are called supplementary angles. In Figure 11.32, $\angle EOC$ and $\angle COF$ are supplementary angles, as are $\angle EOG$ and $\angle GOF$. $\angle EOG$ is the supplement of $\angle GOF$, and $\angle GOF$ is the supplement of $\angle EOG$. Note that $\angle HOF$ is also the supplement of $\angle GOF$. Therefore, $m\angle EOG = m\angle HOF$, where m means "the measure of." (Supplements of the same angle must be equal in measure.) Therefore, since $\angle HOF$ and $\angle EOG$ are vertical angles, we may conclude that the measures of vertical angles are equal.

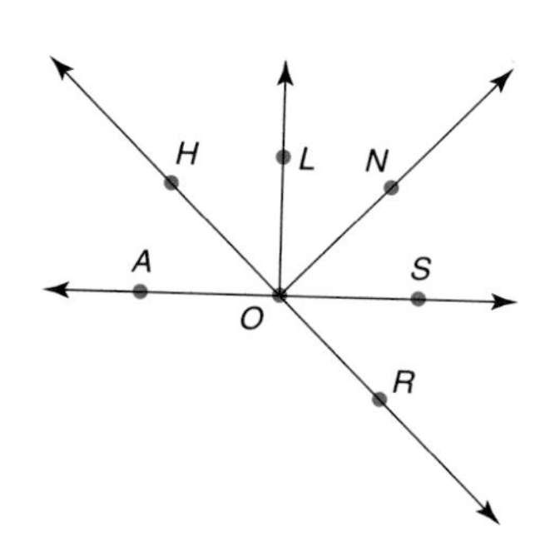

Figure 11.33

EXAMPLE 4 ***Identifying Angles*** In Figure 11.33, straight lines $\overleftrightarrow{AS}$ and $\overleftrightarrow{HR}$ intersect at point O and $\overrightarrow{ON} \perp \overleftrightarrow{HR}$. Using this information, find each of the following:

a. an acute angle
b. an obtuse angle
c. a right angle
d. a straight angle
e. adjacent angles
f. vertical angles
g. complementary angles
h. supplementary angles

Solution $\overleftrightarrow{AS}$ and $\overleftrightarrow{HR}$ are straight lines and therefore form straight angles. $\overrightarrow{ON} \perp \overleftrightarrow{HR}$ and therefore $\angle NOR$ and $\angle NOH$ are right angles.

- **a.** An acute angle is an angle whose measure is greater than 0° and less than 90°. Therefore, $\angle NOS$ is an acute angle. Note that there are others (such as $\angle SOR$), but we are asked to find only one.
- **b.** An obtuse angle is an angle whose measure is greater than 90° and less than 180°. Therefore, $\angle ROL$ is an obtuse angle.
- **c.** Angle NOR is a right angle.
- **d.** A straight angle is an angle whose sides lie on the same straight line and extend in opposite directions from the vertex. Therefore, $\angle AOS$ is a straight angle.
- **e.** Adjacent angles are angles that have the same vertex and a common side between them. Therefore, $\angle NOS$ and $\angle SOR$ are adjacent angles.
- **f.** Vertical angles are two angles where the sides of one angle extend through the vertex to form the sides of the other angle. Therefore, $\angle HOA$ and $\angle SOR$ are vertical angles.
- **g.** Two angles whose measures sum to that of a right angle are complementary angles. Therefore, $\angle NOS$ and $\angle SOR$ are complementary angles.
- **h.** Two angles whose measures sum to that of a straight angle are supplementary angles. Therefore, $\angle HOL$ and $\angle LOR$ are supplementary angles.

NW ***Now Work Problem 3.***

Finding the Measure of an Angle Using a Protractor

A **protractor** is the instrument used to measure the number of degrees contained in an angle. One type of protractor is shown in Figure 11.34. A protractor can also be used to draw an angle of any size. Note that 0° is marked on the right side of the protractor and 180° is marked on the left side. This is because an angle is *generated* in a counterclockwise direction. An angle is generated in the following manner: We are given two rays, $\overrightarrow{AB}$ and $\overrightarrow{AC}$, that lie in the same straight line and extend in the same direction, as shown in Figure 11.35a. Now we rotate $\overrightarrow{AC}$ in a counterclockwise direction for a certain number of degrees and then stop as shown in Figure 11.35b.

Figure 11.35a

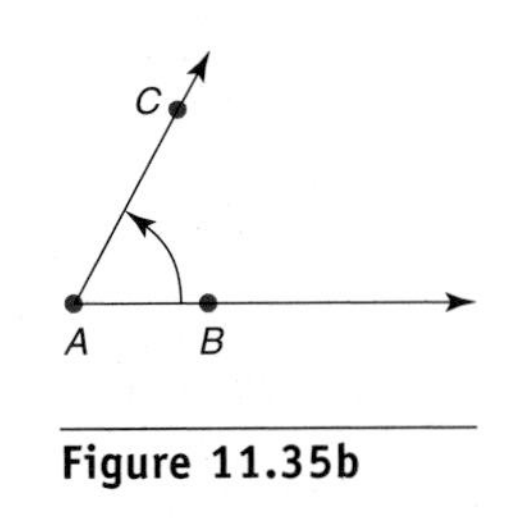

Figure 11.35b

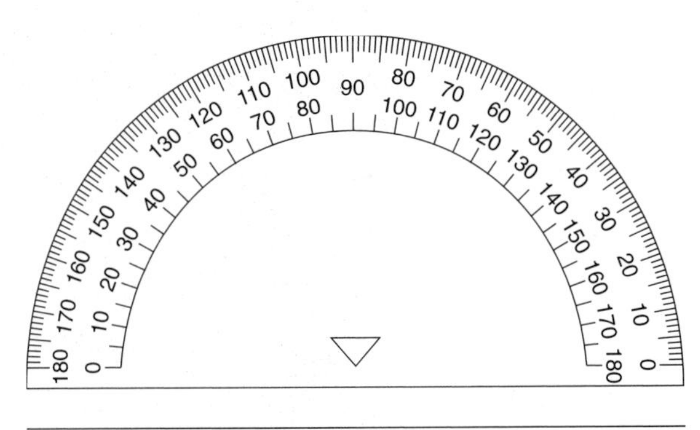

Figure 11.34

In generating $\angle CAB$, $\overrightarrow{AB}$ is called the **initial side**, since it is the ray where the angle begins, and $\overrightarrow{AC}$ is called the **terminal side**, since it is the ray where the angle ends, or terminates.

In order to measure the number of degrees in an angle, we place the marked point of the protractor at the vertex of the angle, and the 0° line along the initial side of the angle. The number of degrees in the angle is found by reading the position of the terminal side of the angle on the edge of the protractor. Using this technique, as shown in Figure 11.36, we see that angle CAB contains 60°. We denote this by writing $m\angle CAB = 60°$. For example, $m\angle BOE = 88°$ means that the measure of $\angle BOE$ is 88°.

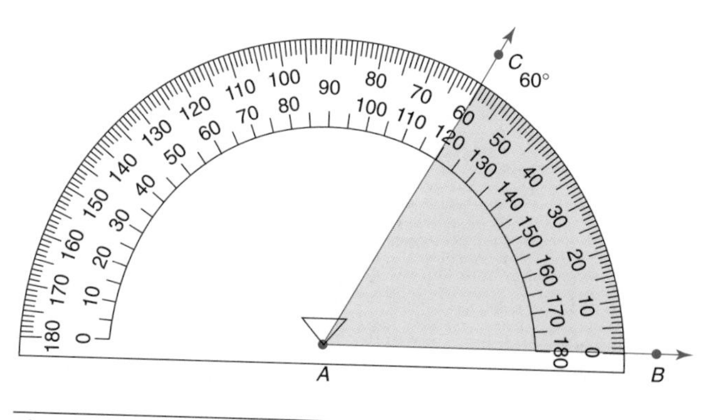

Figure 11.36

Notice that a protractor can be used to measure any angle up to and including 180°. But there are angles of greater magnitude. If we move the hands of a clock back one hour, as we do when we change from daylight saving time to standard time, then the minute hand is rotated one complete revolution, or 360°. The hands of the clock have generated an angle greater than 180°. When the terminal side of an angle has generated an angle of 360°, the angle is called a **round angle** or **perigon**. When the terminal side of an angle has generated an angle greater than 180° but less than 360°, the angle is called a **reflex angle**. Two angles whose measures sum to that of a round angle are called **conjugate angles**.

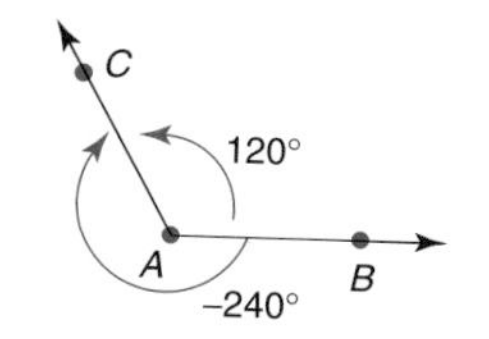

Figure 11.37

Angles are normally generated so that the terminal side of the angle has been moved in a counterclockwise direction. The measure of these angles is considered to be positive. But angles can be generated so that the terminal side of the angle has been moved in a clockwise direction. The measure of angles generated in a clockwise direction is considered to be negative. If the measure of $\angle BAC$ in Figure 11.37 is 120°, we could also say that its measure is −240°.

Finding the Measure of an Angle Algebraically

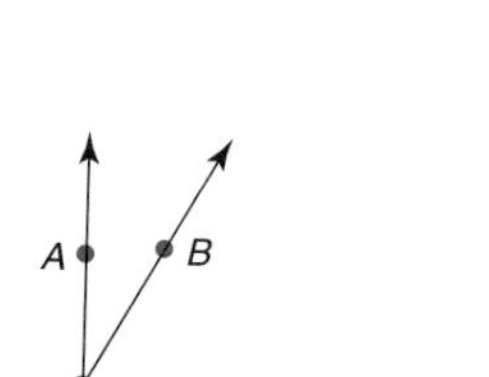

Figure 11.38

We can also use algebra to find the number of degrees in an angle. Consider $\angle ARS$ in Figure 11.38. Suppose we are given that $\angle ARS$ is a right angle, and that $m\angle BRS$ is 30° larger than $m\angle ARB$. How many degrees are in each angle? Since we are given that $\angle ARS$ is a right angle, we know that its measure is 90°. Also, $\angle ARB$ and $\angle BRS$ are complementary angles because the sum of their measures is 90°. Let $x = m\angle ARB$; then $m\angle BRS = x + 30$ because $m\angle BRS$ is 30° larger than $m\angle ARB$. We can now write the equation $x + x + 30 = 90$. Next, we solve the equation for x and find the number of degrees in each angle:

$$\begin{aligned} x + x + 30 &= 90 \\ 2x + 30 &= 90 \\ 2x &= 60 \\ x &= 30° \qquad m\angle ARB \\ x + 30 &= 60° \qquad m\angle BRS \end{aligned}$$

EXAMPLE 5 ***Using Algebra to Find the Measure of Complementary Angles*** Two angles are complementary and the measure of one angle is 20° larger than the measure of the other. How many degrees are there in each angle?

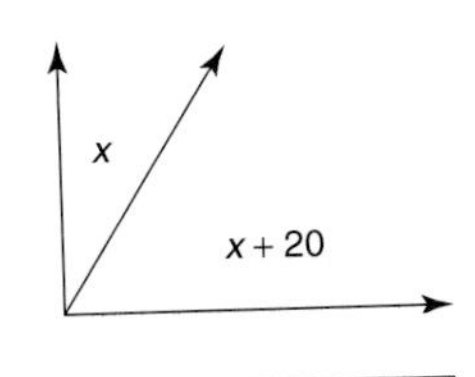

Figure 11.39

Solution We can draw a diagram (Figure 11.39) to aid us in solving this problem. The two angles are complementary, so we know their measures sum to that of a right angle, and one angle is 20° larger than the other. The resulting equation is

$$x + x + 20 = 90$$

Solving the equation for x,

$$\begin{aligned} 2x + 20 &= 90 \\ 2x &= 70 \\ x &= 35^\circ \quad \textit{first angle} \\ x + 20 &= 55^\circ \quad \textit{second angle} \end{aligned}$$

NW ***Now Work Problem 9.***

EXAMPLE 6 ***Using Algebra to Find the Measure of Supplementary Angles*** Two angles are supplementary and the measure of one angle is 60° less than the measure of the other. How many degrees are there in each angle?

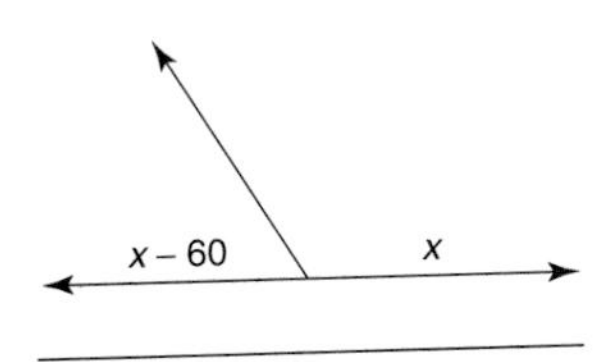

Figure 11.40

Solution Drawing a diagram to aid us in solving this problem, we have Figure 11.40.

The two angles are supplementary; therefore, their measures sum to that of a straight angle. Also, one angle is 60° less than the other, so if the measure of one angle is x, then the measure of the other is $x - 60°$. The resulting equation is

$$x + x - 60 = 180$$

Solving the equation for x,

$$\begin{aligned} 2x - 60 &= 180 \\ 2x &= 240 \\ x &= 120^\circ \quad \textit{first angle} \\ x - 60 &= 60^\circ \quad \textit{second angle} \end{aligned}$$

NW ***Now Work Problem 7.***

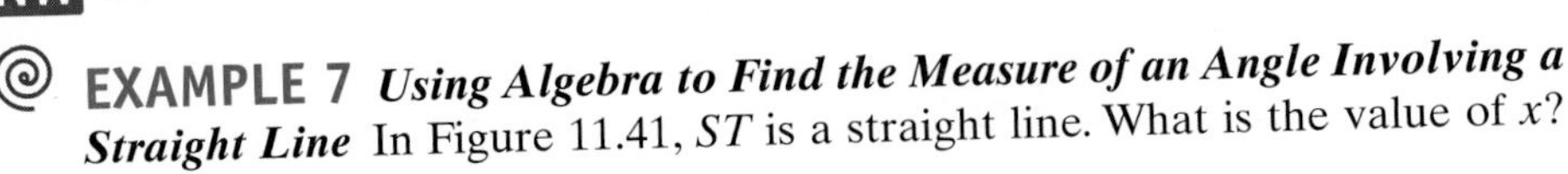

EXAMPLE 7 ***Using Algebra to Find the Measure of an Angle Involving a Straight Line*** In Figure 11.41, ST is a straight line. What is the value of x?

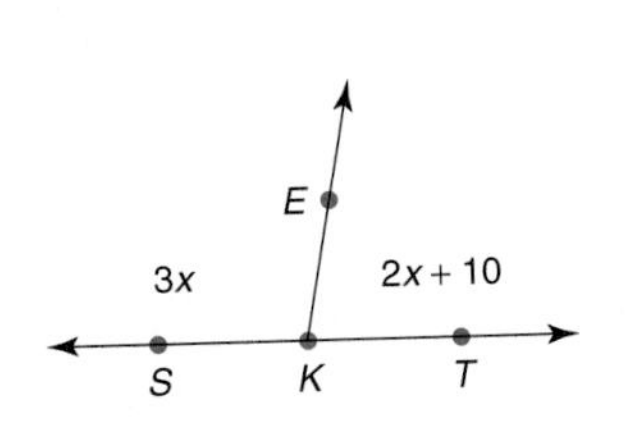

Figure 11.41

Solution If ST is a straight line, then $\angle SKT$ is a straight angle and its measure is 180°. Therefore, the sum of $m\angle SKE$ and $m\angle EKT$ is 180°, and the resulting equation is

$$3x + 2x + 10 = 180$$

Solving the equation for x,

$$\begin{aligned} 5x + 10 &= 180 \\ 5x &= 170 \\ x &= 34 \end{aligned}$$

Note that we are not asked to find the measures of the two angles, but only to find the value of x.

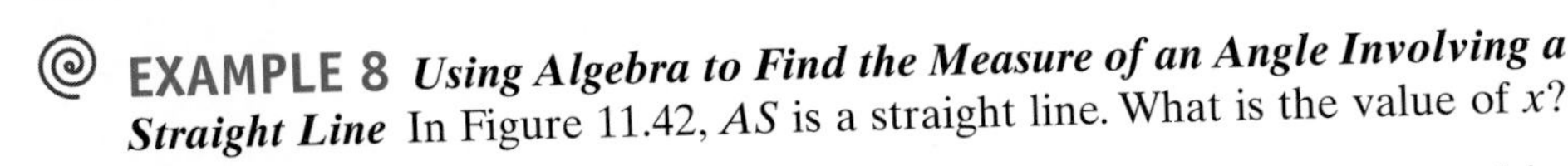

EXAMPLE 8 ***Using Algebra to Find the Measure of an Angle Involving a Straight Line*** In Figure 11.42, AS is a straight line. What is the value of x?

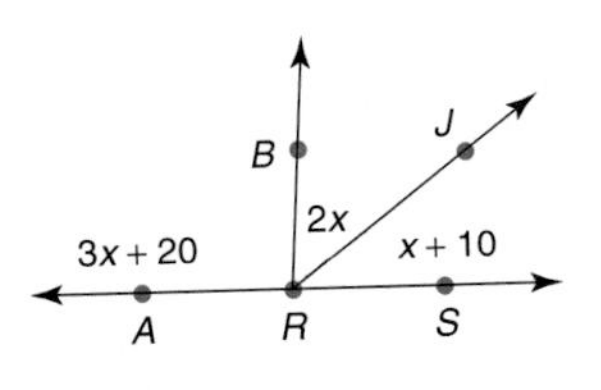

Figure 11.42

Solution If AS is a straight line, then $\angle ARS$ must be a straight angle and its measure is 180°. Therefore, the sum of $m\angle ARB$, $m\angle BRJ$, and $m\angle JRS$ is 180°, and the resulting equation is

$$3x + 20 + 2x + x + 10 = 180$$

Solving the equation for x,

$$\begin{aligned} 6x + 30 &= 180 \\ 6x &= 150 \\ x &= 25 \end{aligned}$$

NW ***Now Work Problem 11.***

Note that the size of an angle is a function of how many degrees it contains, not of how big we draw it. For example, both $\angle ARS$ and $\angle WMS$ in Figure 11.43 contain 30° and are therefore equal in size: $m\angle ARS = m\angle WMS$.

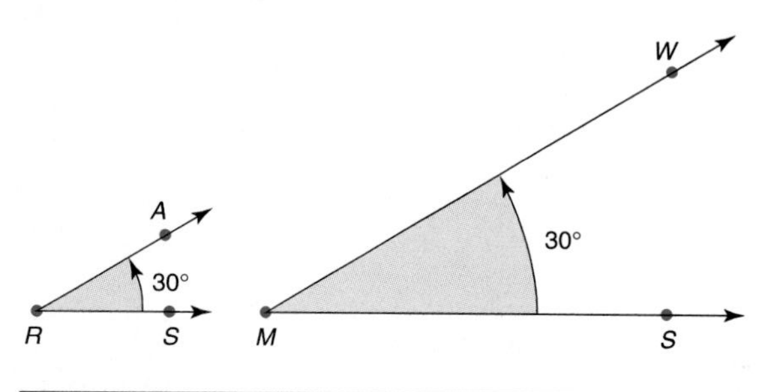

Figure 11.43

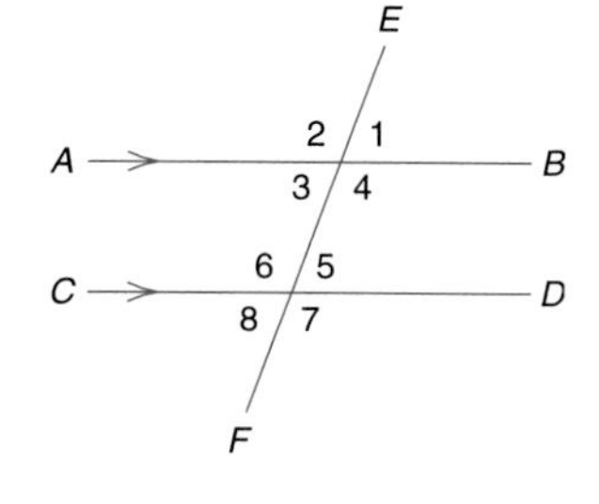

Figure 11.44

In Figure 11.44, lines AB and CD are parallel and they are cut by a transversal EF. A **transversal** is a line that intersects (cuts) two or more lines. (*Note:* To mark two given parallel lines, use arrowheads as shown in Figure 11.44.) In addition to vertical angles, there are other types of angles in the given figure. *Interior angles* are angles formed by a transversal and a line and which are inside the lines cut by a transversal. Hence, angles 3, 6, 4 and 5, are interior angles. *Alternate interior angles* are a pair of nonadjacent interior angles on opposite sides of the transversal. Therefore, in the given figure, angles 3 and 5 are alternate interior angles, as are angles 4 and 6. *Corresponding angles* have the same position with respect to their lines and the transversal. In Figure 11.44, $\angle 1$ and $\angle 5$, or $\angle 4$ and $\angle 7$, or $\angle 2$ and $\angle 6$, or $\angle 3$ and $\angle 8$, are a pair of corresponding angles.

It can be shown that if two parallel lines are cut by a transversal, then

1. Alternate interior angles are **congruent**.
2. Corresponding angles are **congruent**.
3. Interior angles on the same side of the transversal are **supplementary**.

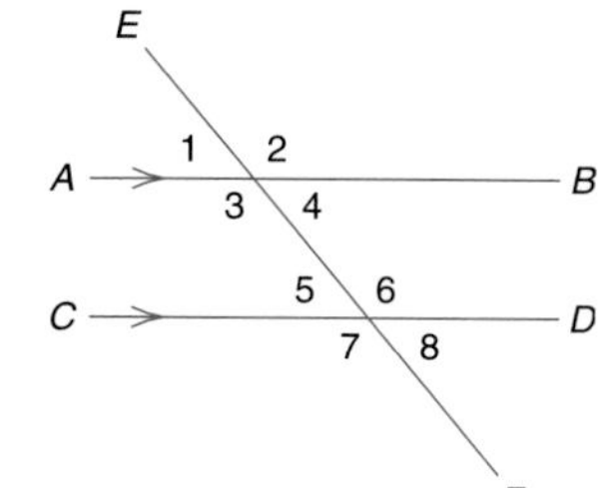

Figure 11.45

EXAMPLE 9 ***Finding the Measure of Angles Involving a Transversal*** In Figure 11.45, $\overline{AB} \parallel \overline{CD}$ cut by a transversal EF. If $m\angle 1 = 50°$, find the number of degrees in each of the other seven angles.

Solution

a. $m\angle 2 = 130°$, supplementary angles, $\angle 1$ and $\angle 2$.

b. $m\angle 3 = 130°$, vertical angles, $\angle 2$ and $\angle 3$.

c. $m\angle 4 = 50°$, vertical angles, $\angle 1$ and $\angle 4$, or supplementary angles, $\angle 3$ and $\angle 4$.

d. $m\angle 5 = 50°$, alternate interior angles, $\angle 5$ and $\angle 4$.

e. $m\angle 6 = 130°$, interior angles on the same side of the transversal are supplementary, $\angle 4$ and $\angle 6$, or corresponding angles, $\angle 2$ and $\angle 6$.

f. $m\angle 7 = 130°$, corresponding angles, $\angle 3$ and $\angle 7$.

g. $m\angle 8 = 50°$, supplementary angles, $\angle 7$ and $\angle 8$.

NW ***Now Work Problem 23.***

Exercises for Section 11.4

NW **1.** In Figure 11.46, how many different angles are there? Name them.

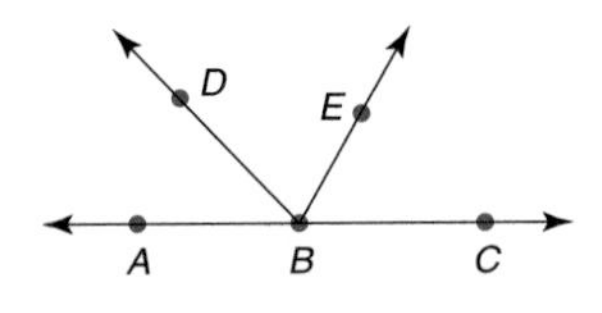

Figure 11.46

2. In Figure 11.47, how many different angles are there? Name them.

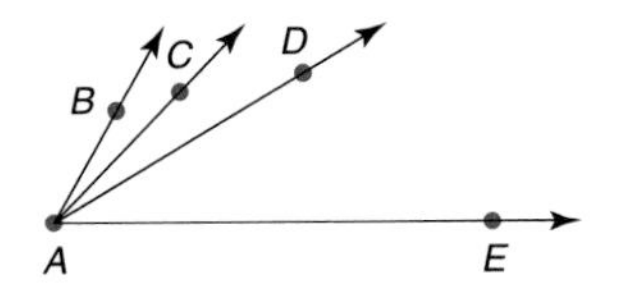

Figure 11.47

NW **3.** In Figure 11.48, straight lines $\overleftrightarrow{NS}$, $\overleftrightarrow{WE}$, $\overleftrightarrow{LR}$, and $\overleftrightarrow{BF}$ intersect at O. Also, $\overleftrightarrow{NS} \perp \overleftrightarrow{WE}$ and $\overleftrightarrow{LR} \perp \overleftrightarrow{BF}$. Using this information, find each of the following:

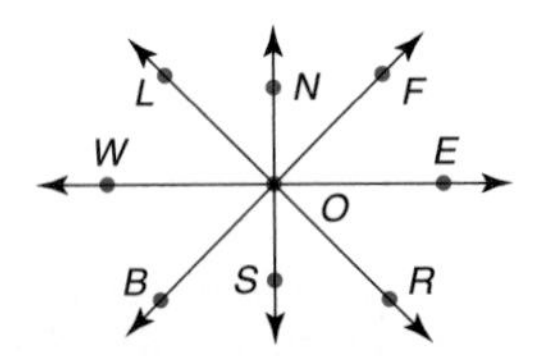

Figure 11.48

a. two acute angles **b.** two obtuse angles
c. two right angles **d.** two straight angles
e. two pairs of adjacent angles
f. two pairs of vertical angles
g. two pairs of complementary angles
h. two pairs of supplementary angles

4. Find the measure of the complement of the angle with each of the following measures:

a. 40° **b.** 65° **c.** 76°
d. 24°24′ **e.** 38°40′ **f.** 48°32′15″

5. Find the measure of the supplement of the angle with each of the following measures:

NW **a.** 80° **b.** 100° **c.** 70°
NW **d.** 120°20′ **e.** 38°30′ **f.** 100°50′25″

6. Two angles are complementary and one angle measures 10° less than the other. How many degrees are there in each angle?

NW **7.** Two angles are supplementary and one angle measures 60° more than the other. How many degrees are there in each angle?

8. Two angles are supplementary and one angle measures 30° more than two times the other. How many degrees are there in each angle?

NW **9.** Two angles are complementary and one angle measures 30° less than three times the other. How many degrees are there in each angle?

10. In Figure 11.49, $\overleftrightarrow{AB}$ is a straight line. How many degrees are in $\angle AOC$ and $\angle COB$?

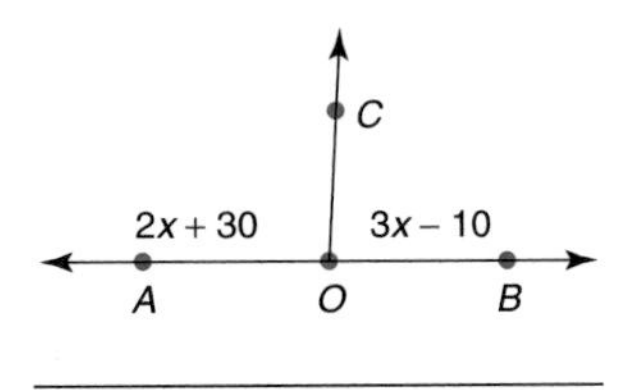

Figure 11.49

NW **11.** In Figure 11.50, what value of x will make AB a straight line?

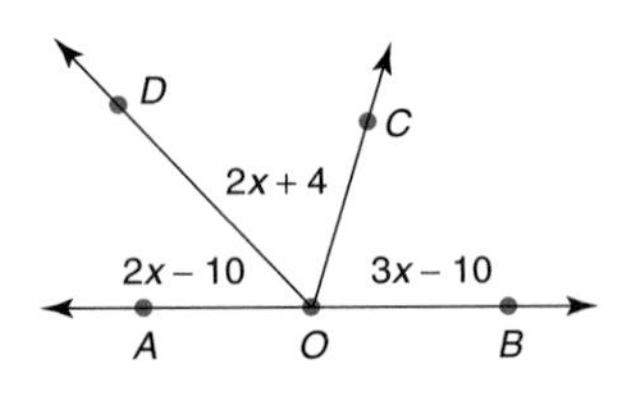

Figure 11.50

12. The sum of the measures of the complement and the supplement of a certain angle is 100°. Find the number of degrees in the angle.

13. The number of degrees in angle ACB is equal to one-third the number of degrees in its supplement. Find the number of degrees in angle ACB.

Determine whether each statement (14–21) is true or false.

14. The sides of a right angle are perpendicular to each other.

15. The complement of an acute angle is acute.

16. The supplement of an acute angle is acute.

17. If an acute angle is doubled, then the result must be an acute angle.

18. If an obtuse angle is doubled, then the result must be a reflex angle.

19. If the sum of two angles is a straight angle, then one of the angles must be acute.

20. $m\angle BAC + m\angle ABC + m\angle ACB = 180°$. Therefore, these angles are supplementary angles.

21. $m\angle BAC + m\angle ABC = 90°$. Therefore, these angles are complementary angles.

22. Find the measure of all the angles in Figure 11.51 if $\overline{AB} \parallel \overline{CD}$ and $m\angle 1 = 40°$.

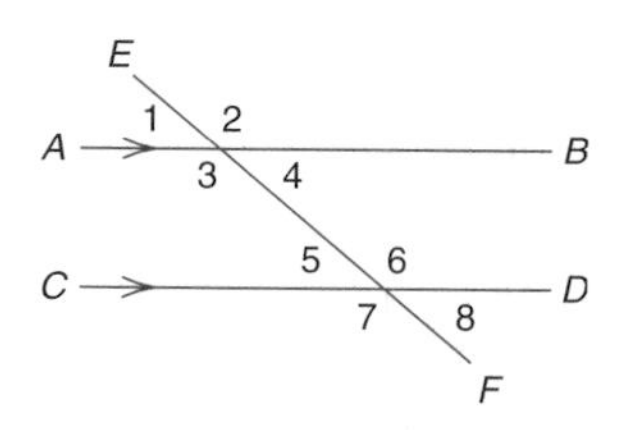

Figure 11.51

NW 23. Find the measure of all the angles in Figure 11.52 if $\overline{CD} \parallel \overline{EF}$ and $m\angle 3 = 50°$.

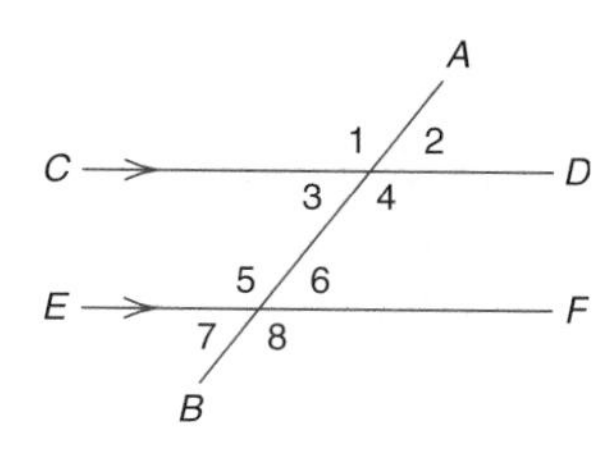

Figure 11.52

24. Find the measure of all the angles in Figure 11.53 if $\overline{MN} \parallel \overline{OP}$ and $m\angle 7 = 135°$.

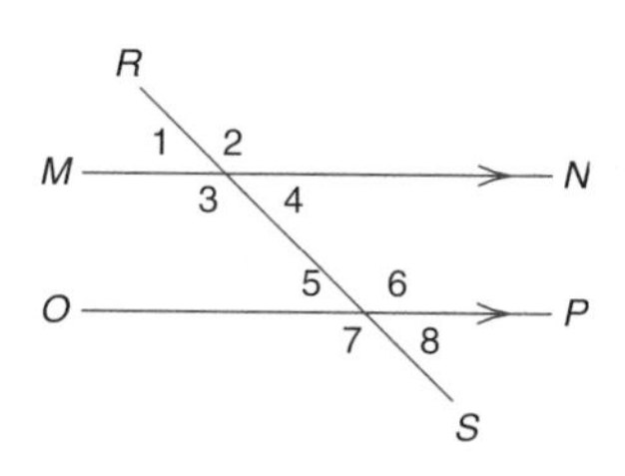

Figure 11.53

25. Find the measure of all the angles in Figure 11.54 if $\overline{AB} \parallel \overline{CD}$ and $m\angle 5 = 155°$.

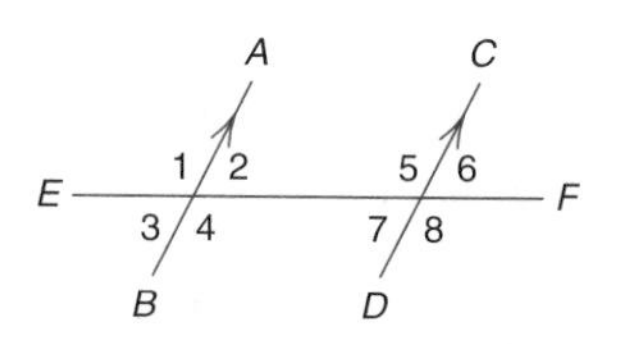

Figure 11.54

For Exercises 26–30, refer to Figure 11.55 in which $\overline{AB} \parallel \overline{CD}$.

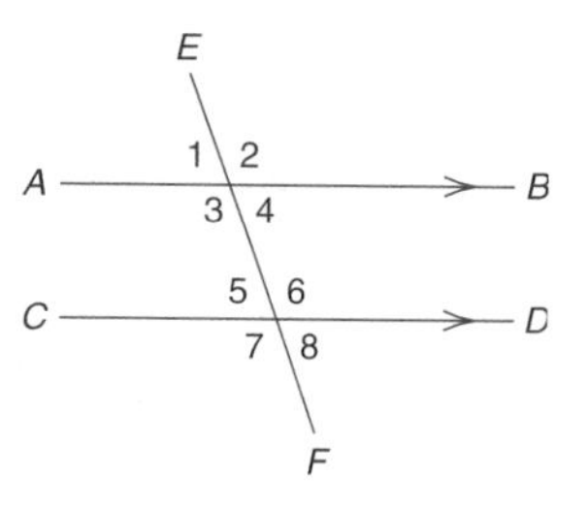

Figure 11.55

26. Find $m\angle 1$ and $m\angle 5$, if $m\angle 1 = (130 - x)°$ and $m\angle 5 = 9x°$.
27. Find $m\angle 3$ and $m\angle 5$, if $m\angle 3 = (x + 20)°$ and $m\angle 5 = x°$.
28. Find $m\angle 4$ and $m\angle 5$, if $m\angle 4 = 2x°$ and $m\angle 5 = (x + 40)°$.
29. Find $m\angle 4$ and $m\angle 8$, if $m\angle 4 = 2x°$ and $m\angle 8 = (x + 60)°$.
30. Find $m\angle 5$ and $m\angle 7$, if $m\angle 1 = (x + 20)°$ and $m\angle 6 = (x + 60)°$.

Writing Mathematics

31. What are *vertical angles?*
32. What are *adjacent angles?*
33. Explain the difference between *supplementary angles, complementary angles*, and *conjugate angles*.
34. Explain the difference between *degrees, minutes*, and *seconds* as used in this section.

Challenge Exercises

35. In parallelogram $ABCD$, the number of degrees in angle D is 30 less than twice the number of degrees in angle A. Find the number of degrees in angle A.
36. In parallelogram $ABCD$, the number of degrees in angle D is 30 more than twice the number of degrees in angle A. Find the number of degrees in angle C.

Research/Group Activities

37. The *degree* had its origin in ancient times. It had to do with noting the sun's position each day. Research the history and development of the degree. Prepare a report to be presented in class.
38. A *protractor* is the instrument used to measure the number of degrees contained in an angle. Write a paper on the history and use of the protractor.

Just for fun

What has four equal sides, but is drawn with six lines?

11.5 POLYGONS

Introduction

A **broken line** is a set of connected line segments—that is, a set of line segments that have been placed end to end. The four figures in Figure 11.56 are examples of broken lines.

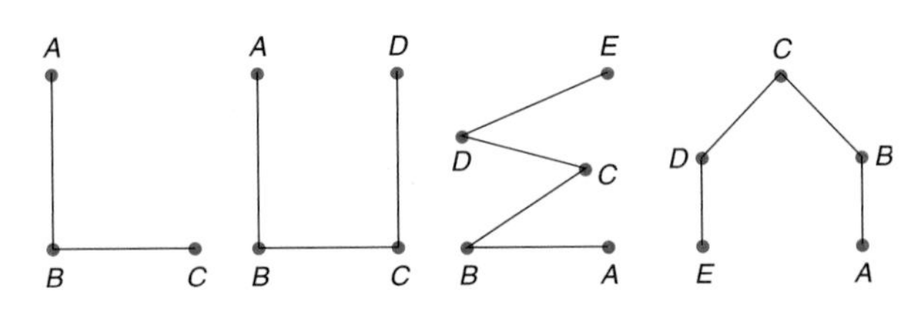

Figure 11.56

Now let us take each of the broken lines in Figure 11.56 and "close" it so that it appears as shown in Figure 11.57. A **closed broken line** begins and ends at the same point. You will note that each of the figures in Figure 11.57 is a closed broken line.

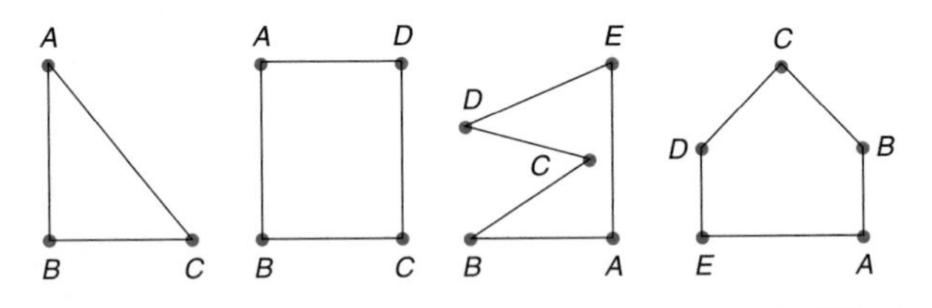

Figure 11.57

However, in addition, each is also a *simple closed broken line*. A **simple closed broken line** is one that does not intersect itself. None of the closed broken line figures shown in Figure 11.58 are simple because in each case the broken line intersects itself.

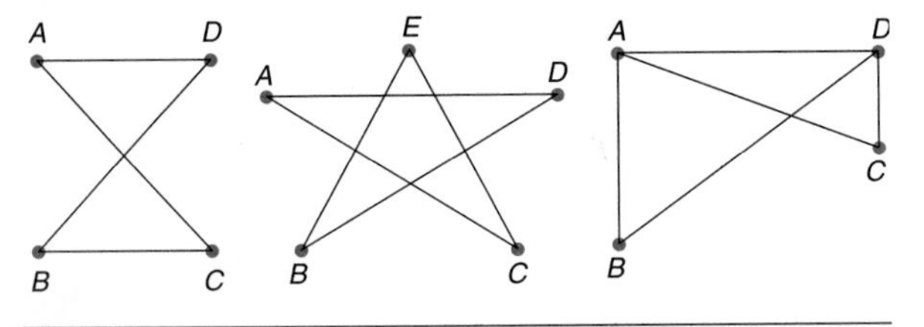

Figure 11.58

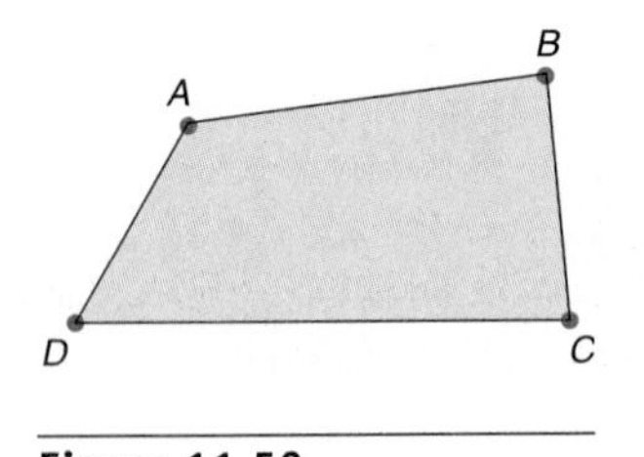

Figure 11.59

A simple closed broken line is called a **polygon**. The connected line segments are the **sides** of the polygon, and the points at which the line segments are connected are the **vertices** of the polygon. Note that any two consecutive sides of a polygon form an angle of the polygon. In Figure 11.59, polygon $ABCD$ is a simple closed broken line. Its sides are $\overline{AB}$, $\overline{BC}$, $\overline{CD}$, and $\overline{DA}$; its vertices are A, B, C, and D; and its angles are $\angle DAB$, $\angle ABC$, $\angle BCD$, and $\angle CDA$.

Polygons are classified according to the number of sides they have. The following is a partial list of some types of polygons and the number of sides of each.

Polygon	Number of Sides	Polygon	Number of Sides
Triangle	3	Octagon	8
Quadrilateral	4	Nonagon	9
Pentagon	5	Decagon	10
Hexagon	6	Dodecagon	12
Heptagon	7	Icosagon	20

Classifying Triangles by Their Sides

In order to form a simple closed broken line, we need at least three line segments. If we form such a closed broken line, then we have a polygon with three sides—that is, a **triangle**. There are many different kinds of triangles, and they can be classified according to the characteristics of their sides or their angles. A triangle in which the measures of all three sides are equal is called an **equilateral triangle**. A triangle in which two sides are of equal length is called an **isosceles triangle**. A triangle in which no two sides are of equal length is called a **scalene triangle**. An example of each type of triangle is shown in Figure 11.60.

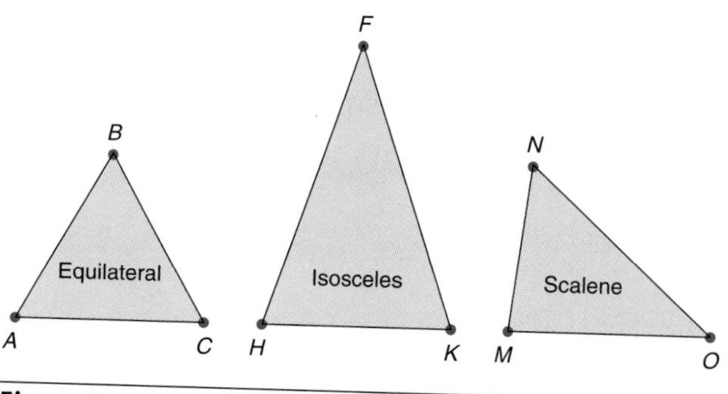

Figure 11.60

Can we form a triangle when given any three segments? The answer is no. There is a certain requirement that must be fulfilled in order to construct a triangle from three segments. For example, suppose we are given three line segments that measure 10 centimeters, 5 centimeters, and 2.5 centimeters. Can we form a triangle with these three line segments?

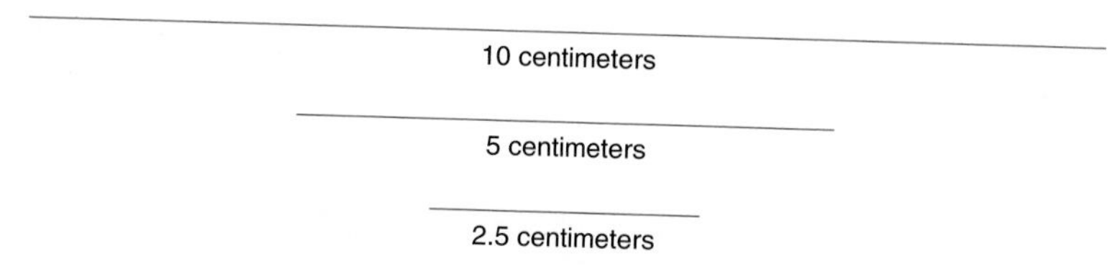

If we attempt to construct a triangle with these segments, we have the situation illustrated in Figure 11.61.

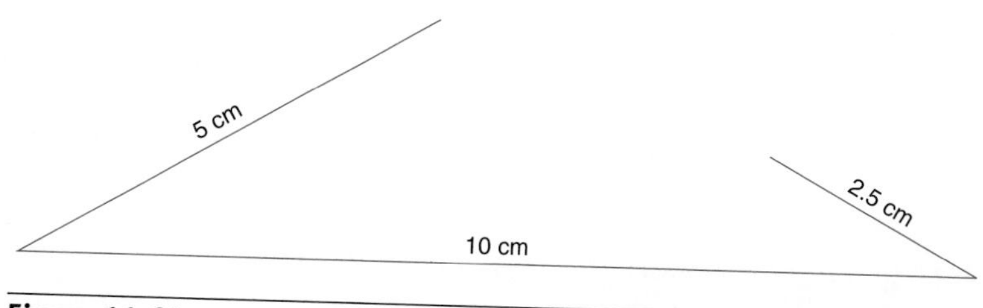

Figure 11.61

As you can see, we are not able to form a simple *closed* broken line using these three line segments. In fact, in order to construct any triangle, the sum of the measures of any two of the line segments must be greater than the measure of the third line segment. If we already have a triangle, then we know that the sum of the lengths of any two of the sides is greater than the length of the third side.

Is it possible to construct a triangle whose sides measure 4, 4, and 8? The answer is no, since $4 + 4$ is not greater than 8. Is it possible to construct a triangle whose sides measure 4, 5, 6? Yes, because

$$4 + 5 > 6, \quad 4 + 6 > 5, \quad \text{and} \quad 5 + 6 > 4$$

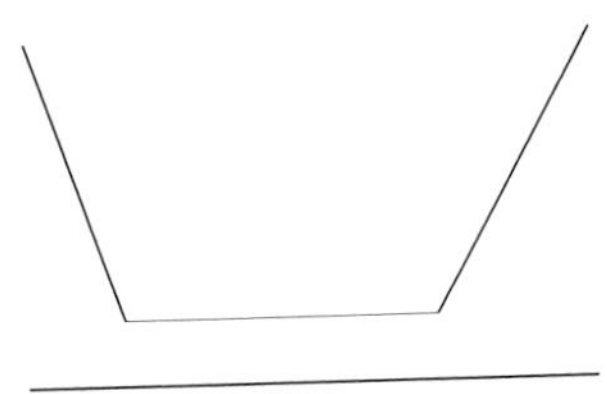

Figure 11.62

Classifying Triangles by Their Angles

We can also classify triangles according to the types of angles that are in the triangle. If all of the angles in the triangle are acute angles, then the triangle is called an **acute triangle**. If a triangle has an obtuse angle in it, then it is called an **obtuse triangle**. Can a triangle have more than one obtuse angle in it? We cannot form such a triangle because the figure formed is not a closed broken line, as shown in Figure 11.62.

If the measures of all of the angles in a triangle are equal, then the triangle is called an **equiangular triangle**. It can be shown that the sum of the measures of the interior angles of a triangle is equal to 180°. Therefore, if a triangle is equiangular, each angle will measure exactly 60° and it is also an acute triangle.

If a triangle contains a right angle, then it is called a **right triangle**.

Examples of an acute triangle, an obtuse triangle, a right triangle, and an equiangular triangle are shown in Figure 11.63.

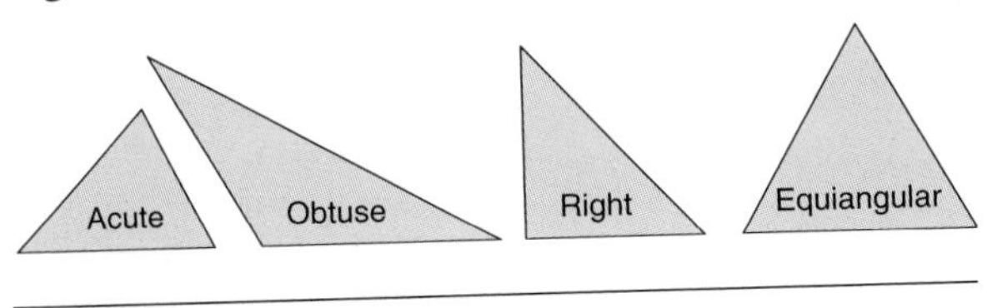

Figure 11.63

The Pythagorean Theorem

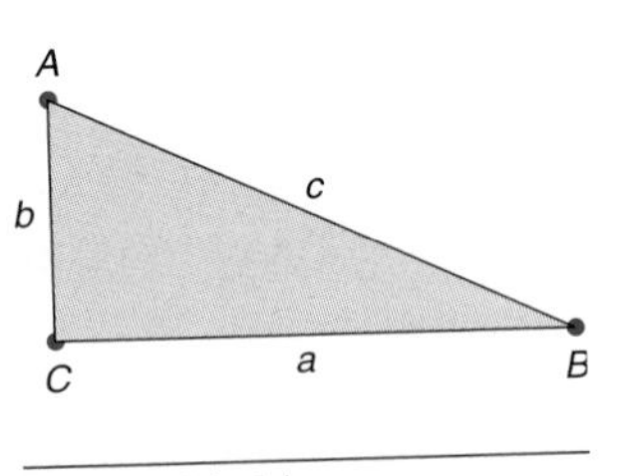

Figure 11.64

Consider right triangle ABC in Figure 11.64. The right angle is $\angle BCA$. The sides of this triangle that form the right angle are called the **legs** of the right triangle (a and b in Figure 11.64) and the side opposite the right angle is called the **hypotenuse** (c in Figure 11.64).

Probably one of the most famous theorems in geometry deals with the right triangle. It is called the **Pythagorean theorem** because its proof was supposedly discovered by Pythagoras. The Pythagorean theorem states that

> ***The square of the hypotenuse of a right triangle is equal to the sum of the squares of the legs.***

Biography: Pythagoras

Unfortunately, much of the documented history of Greek geometry was either lost or destroyed through the centuries. However, scholarly research has created an approximate history of early Greek mathematics.

It is believed that Pythagoras was born in approximately 570 B.C. on the Aegean island called Samos. After becoming an educated young man, Pythagoras spent several years traveling to different lands to learn all that he could about mathematics. After this, he settled in the Greek colonial seaport of Crotona, which is located in southern Italy. It was here that he founded the famous Pythagorean academy (in approximately 540 B.C.). In addition to mathematics, he taught his students (disciples) to worship numbers, to believe in reincarnation, and to sign the name of the Pythagorean brotherhood to any writing or discovery. As a result, it is now difficult to know which mathematical findings should be credited to Pythagoras himself, and which to other members of the brotherhood.

The best known of the Pythagorean teachings is the theorem that states, in any right triangle, the square of the hypotenuse is equal to the sum of the squares of the two other sides.

An interesting theory promoted by Pythagoras was that the entire universe could be expressed by whole number relationships. He even classified numbers as "amicable" and "perfect." He labeled the even whole numbers as feminine and the odd whole numbers as masculine, except 1, which he considered to be the generator of all the numbers. The number 5 represented marriage, the sum of the first feminine number (2) and the first masculine number (3).

Using right triangle ABC in Figure 11.64 as a reference, we can also state the Pythagorean theorem as

$$c^2 = a^2 + b^2$$

Suppose in Figure 11.64 we are given that $m(\overline{AC}) = 6$ and $m(\overline{BC}) = 8$, where $m(\overline{AC})$ is the measure of $\overline{AC}$ and $m(\overline{BC})$ is the measure of $\overline{BC}$. What is the length of $\overline{AB}$?

We can use the Pythagorean theorem to find the answer. Using the formula $c^2 = a^2 + b^2$, we are given $a = 8$ and $b = 6$. Therefore, we can substitute the appropriate values in the formula and solve it for c:

$$c^2 = a^2 + b^2$$

$$c^2 = 8^2 + 6^2 \qquad \textit{substituting } a = 8, b = 6$$

$$c^2 = 64 + 36$$

$$c^2 = 100$$

$$c = \sqrt{100}$$

$$c = 10 \qquad \text{length of } \overline{AB}$$

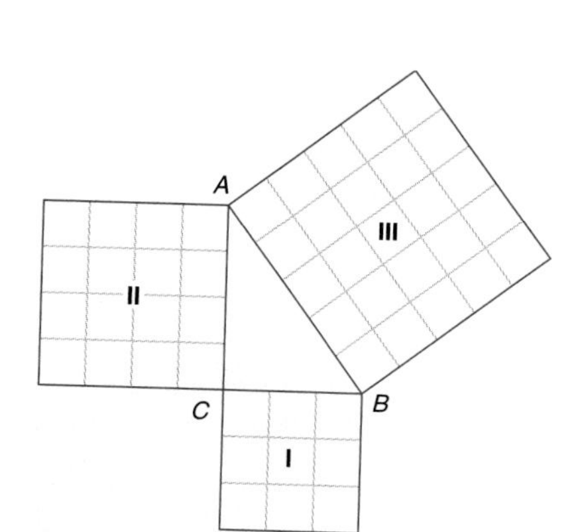

Figure 11.65

It should be noted that Pythagoras and his followers were studying geometry long before the development of algebraic notation. Therefore, they expressed the theorem in terms of its geometric representation. They probably stated the theorem in a manner similar to the following:

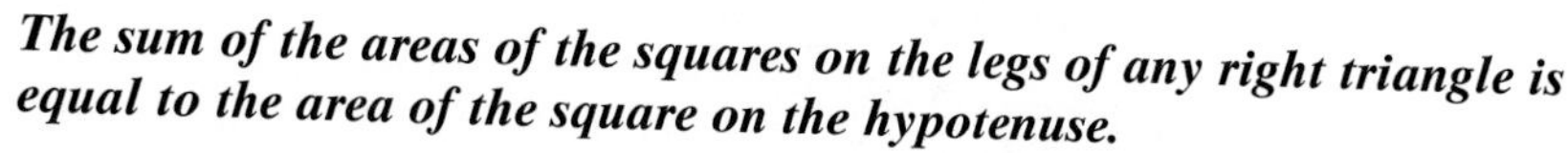
The sum of the areas of the squares on the legs of any right triangle is equal to the area of the square on the hypotenuse.

Figure 11.65 illustrates this statement for right triangle ABC whose sides are 3 and 4 and whose hypotenuse is 5.

$$\text{Area of square I} = 3 \times 3 = 9$$

$$\text{Area of square II} = 4 \times 4 = 16$$

$$\text{Area of square III} = 5 \times 5 = 25$$

$$\begin{array}{ccccc} 9 & + & 16 & = & 25 \\ \text{Area of square I} & + & \text{Area of square II} & = & \text{Area of square III} \end{array}$$

This can be shown for all right triangles.

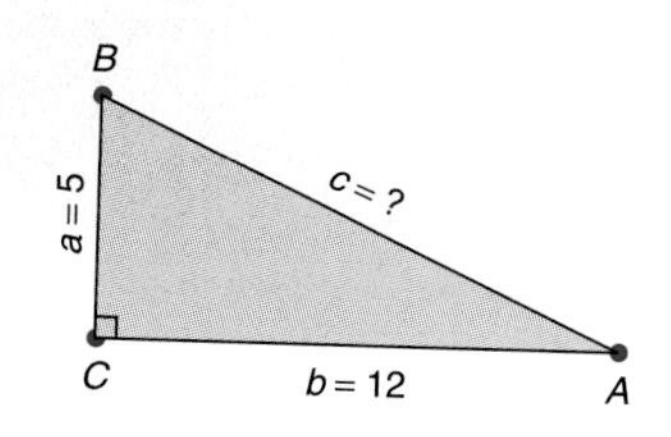

Figure 11.66

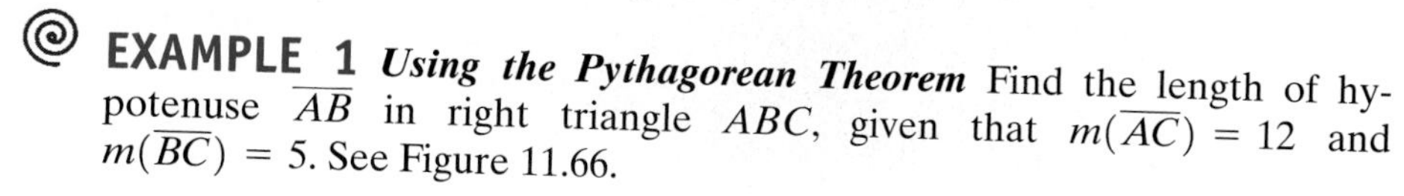
EXAMPLE 1 ***Using the Pythagorean Theorem*** Find the length of hypotenuse $\overline{AB}$ in right triangle ABC, given that $m(\overline{AC}) = 12$ and $m(\overline{BC}) = 5$. See Figure 11.66.

Solution Using the formula $c^2 = a^2 + b^2$, we have $a = 5$ and $b = 12$. Therefore,

$$c^2 = a^2 + b^2$$

$$c^2 = 5^2 + 12^2$$

$$c^2 = 25 + 144$$

$$c^2 = 169$$

$$c = \sqrt{169}$$

$$c = 13 \qquad \text{length of } \overline{AB}$$

NW ***Now Work Problem 7.***

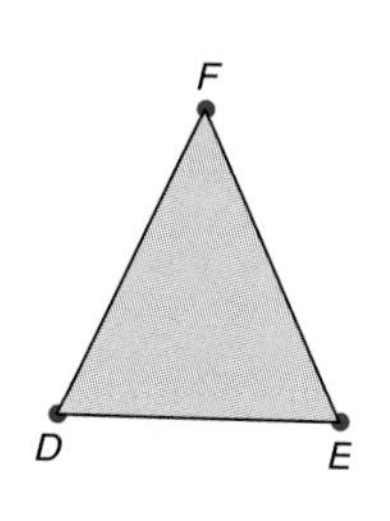

Figure 11.67

If a triangle is isosceles, then it has two sides of equal length. It can also be shown that the measures of the two angles opposite the sides of equal length are equal. For example, given isosceles triangle DEF in Figure 11.67, with

$m(\overline{DF}) = m(\overline{FE})$, then the angles opposite these sides are also equal. That is, $m\angle FDE = m\angle FED$. The measure of these two angles depends on the size of the third angle.

Quadrilaterals

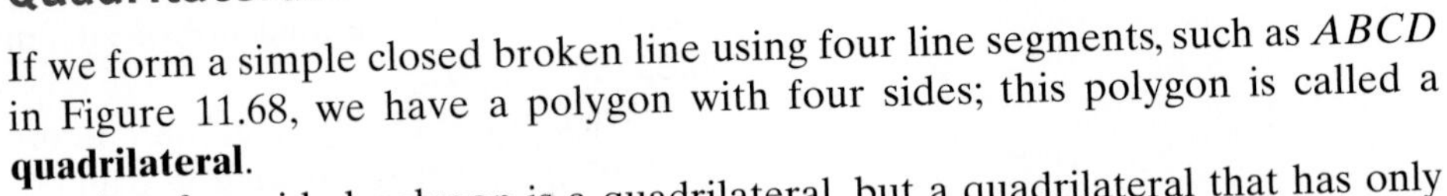

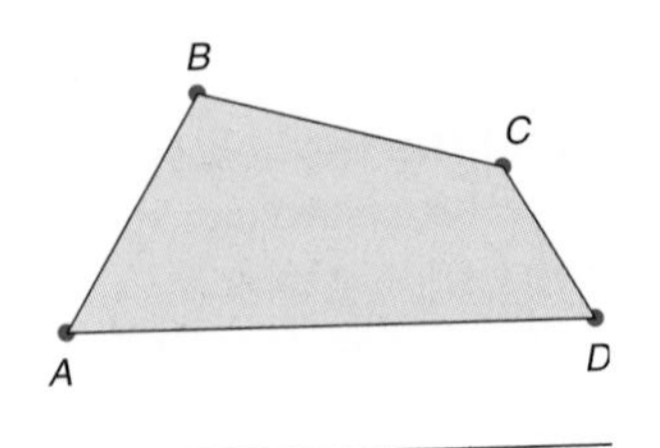

Figure 11.68

If we form a simple closed broken line using four line segments, such as $ABCD$ in Figure 11.68, we have a polygon with four sides; this polygon is called a **quadrilateral**.

Any four-sided polygon is a quadrilateral, but a quadrilateral that has only two parallel sides is called a **trapezoid**. The two parallel sides are called the **bases** of the trapezoid, and if the two nonparallel sides are equal in length, then the trapezoid is an **isosceles trapezoid**. A trapezoid and an isosceles trapezoid are shown in Figure 11.69.

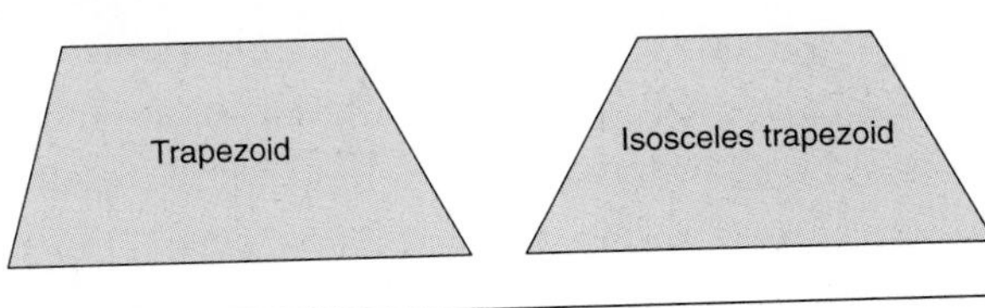

Figure 11.69

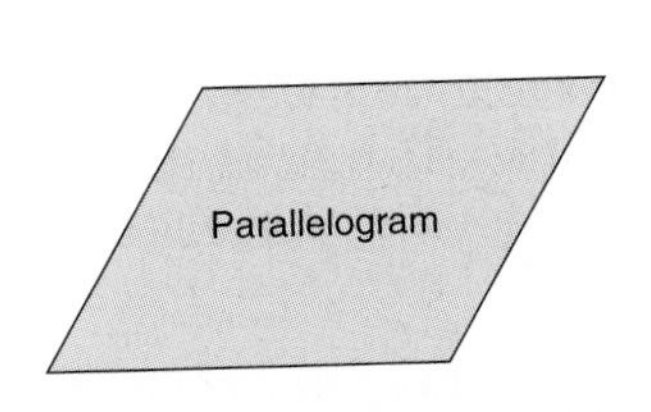

Figure 11.70

If a quadrilateral has both pairs of opposite sides parallel, then it is called a **parallelogram**. Note that we did not say anything about the angles of a parallelogram; they can be any type, as long as both pairs of opposite sides are parallel. Figure 11.70 shows a parallelogram.

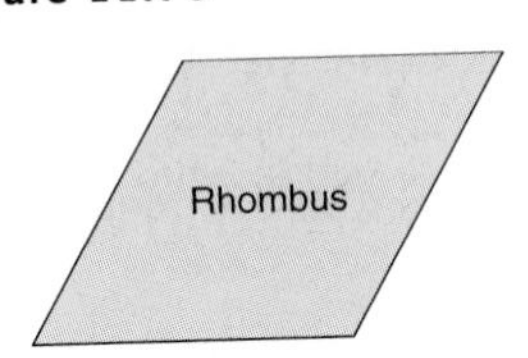

Figure 11.71

A parallelogram that has adjacent sides of equal length is called a **rhombus**. Note that if the adjacent sides of a parallelogram are of equal length, then all of its sides are of equal length. Therefore, we can also describe a rhombus as a parallelogram with four equal sides (see Figure 11.71).

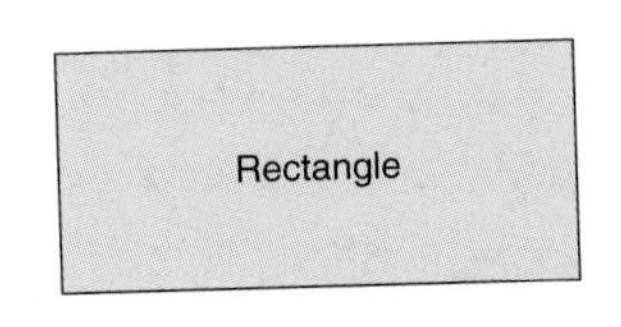

Figure 11.72

A parallelogram that contains a right angle is called a **rectangle**. If a parallelogram contains one right angle, then it must contain four right angles. Also, any quadrilateral with four right angles is a rectangle, as shown in Figure 11.72.

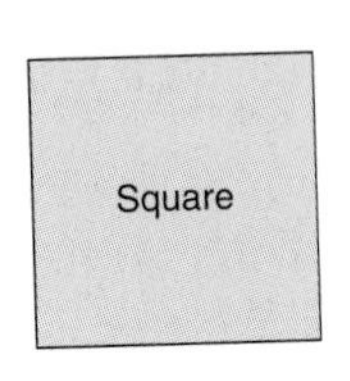

Figure 11.73

A **square** can be described in a number of ways. For example, a square is a rectangle with two adjacent sides of equal length (see Figure 11.73). We can also say that a square is a quadrilateral with four sides of equal length and four angles of equal measure. Note that the square can be considered to be a special case of the rhombus, because a rhombus is a parallelogram with four sides of equal length.

Following is a diagram that illustrates the relationships among the various quadrilaterals:

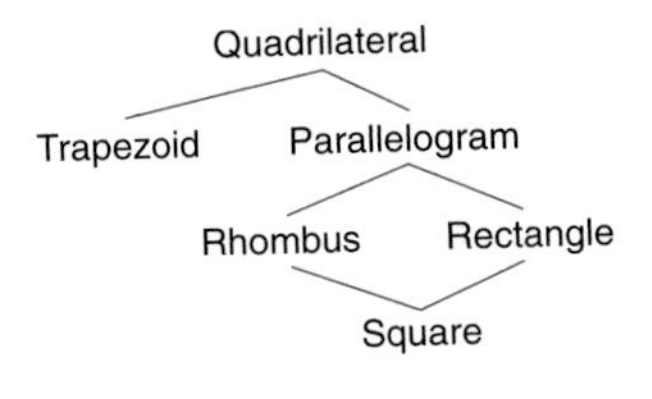

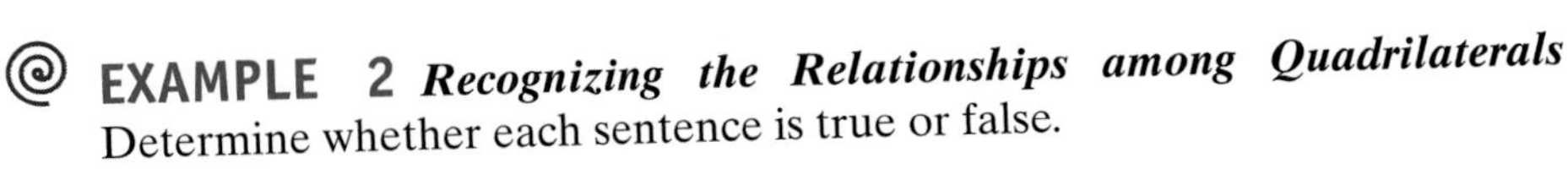

EXAMPLE 2 *Recognizing the Relationships among Quadrilaterals*

Determine whether each sentence is true or false.

a. Every square is a rectangle.
b. Every rectangle is a square.
c. Every square is a rhombus.
d. Every rhombus is a square.
e. Every parallelogram is a rectangle.
f. Every trapezoid is a parallelogram.

Solution

- **a.** True. A square is a rectangle with two adjacent sides equal.
- **b.** False. A rectangle is a quadrilateral with four right angles; the adjacent sides are not necessarily equal.
- **c.** True. A square is a special case of the rhombus.
- **d.** False. A rhombus is a parallelogram with four equal sides; the four angles are not necessarily equal.
- **e.** False. A parallelogram is a quadrilateral with both pairs of opposite sides parallel. The angles of a parallelogram are not necessarily right angles.
- **f.** False. A trapezoid is a quadrilateral that has only two sides parallel.

NW ***Now Work Problem 17.***

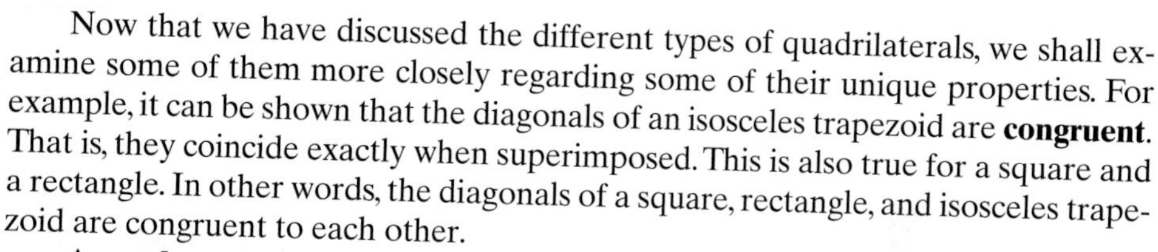

Now that we have discussed the different types of quadrilaterals, we shall examine some of them more closely regarding some of their unique properties. For example, it can be shown that the diagonals of an isosceles trapezoid are **congruent**. That is, they coincide exactly when superimposed. This is also true for a square and a rectangle. In other words, the diagonals of a square, rectangle, and isosceles trapezoid are congruent to each other.

A **regular polygon** is one that is equilateral and equiangular. That is, its sides are equal in measure, and its angles are equal in measure. Hence, a square is a **regular quadrilateral**. In addition to its diagonals being congruent, they are perpendicular to each other and also bisect each other (or divide each other into two equal parts).

Two other quadrilaterals whose diagonals bisect each other are the parallelogram and the rhombus. It should be noted that the diagonals of a rhombus are also perpendicular to each other. But, in neither case are the diagonals congruent. Hence, if we are given a quadrilateral whose diagonals bisect each other, we cannot conclude what the quadrilateral is. This is a property of parallelograms, rhombuses, rectangles, and squares. We need more information to determine the nature of the quadrilateral.

EXAMPLE 3 ***Identifying Quadrilaterals*** For what kind(s) of quadrilateral is each of the following true?

- **a.** The diagonals bisect each other.
- **b.** The opposite sides are congruent and parallel.
- **c.** Only two sides are parallel.
- **d.** Two sides are parallel and the other two are congruent but not parallel.

Solution

- **a.** The diagonals of a parallelogram, rhombus, rectangle, and square bisect each other.
- **b.** The opposite sides are congruent and parallel for a parallelogram, rhombus, rectangle, and square.
- **c.** A trapezoid has only two sides parallel.
- **d.** An isosceles trapezoid has two sides parallel and the other two congruent but not parallel.

Exercises for Section 11.5

1. Which of the following are broken lines?

a. 

b. 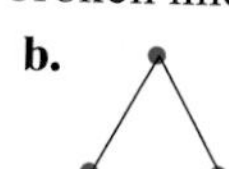

c. 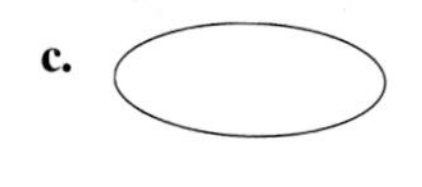

d. **e.** 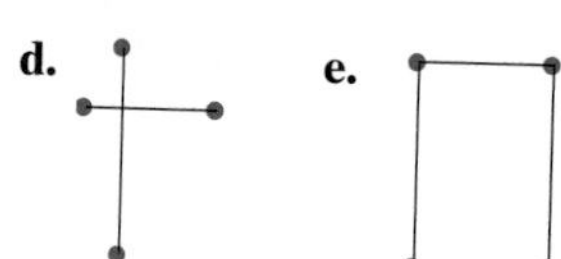

2. Which of the following are simple closed broken lines?

a.

b.

c.

d.

e.

3. Which of the following are polygons?

a.

b.

c.

d.

e.

4. Identify each of the following polygons:

a.

b.

c.

d.

e.

f.

g.

h.

5. Indicate whether each statement is true or false.

a. An equilateral triangle has all three sides equal.

b. A scalene triangle has two sides equal.

c. A quadrilateral that has both pairs of opposite sides parallel is a parallelogram.

d. An isosceles triangle has all three sides equal.

e. The sides of the triangle that form the right angle are called the legs of the right triangle.

6. Determine whether each statement is true or false.

a. The set of numbers $\{2, 7, 10\}$ may represent the lengths of the sides of a triangle.

b. The lengths of two sides of a triangle are 5 and 6. The third side may have length 11.

c. The set of numbers $\{2, 3, 4\}$ may represent the lengths of the sides of a right triangle.

d. The set of numbers $\{3, 4, 5\}$ may represent the lengths of the sides of a right triangle.

e. In any triangle, the square of one side is equal to the sum of the squares of the other two sides.

NW 7. Find the length of hypotenuse $\overline{AB}$ in right triangle ABC, given that $m(\overline{AC}) = 12$ and $m(\overline{BC}) = 5$.

8. Find the length of hypotenuse $\overline{AB}$ in right triangle ABC, given that $m(\overline{AC}) = 8$ and $m(\overline{BC}) = 6$.

9. Find the length of hypotenuse $\overline{AB}$ in right triangle ABC, given that $m(\overline{AC}) = 15$ and $m(\overline{BC}) = 36$.

10. A rectangular field is 75 meters wide and 100 meters long. What is the length of a diagonal path connecting two opposite corners?

11. Mary rode her moped 3 miles south and then 4 miles east. How far was she from her starting point?

12. A 13-foot ladder is leaning against the side of a building. The bottom of the ladder is 5 feet from the base of the building. At what height does it touch the building?

13. A 25-foot ramp covers 24 feet of ground. How high does it rise?

14. Find the length of leg $\overline{AC}$ in right triangle ABC, given that $m(\overline{BC}) = 12$, $m(\overline{AB}) = 13$, and $\overline{AB}$ is the hypotenuse.

15. Find the length of leg $\overline{BC}$ in right triangle ABC, given that $m(\overline{AC}) = 4$, $m(\overline{AB}) = 5$, and $\overline{AB}$ is the hypotenuse.

16. Identify each of the following quadrilaterals:

a.

b.

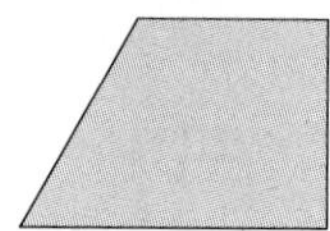

c.

d.

e.

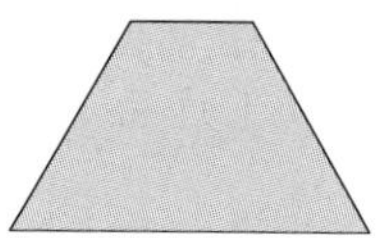

f.

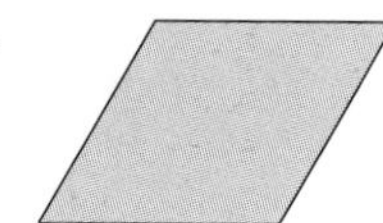

NW **17.** Indicate whether each statement is true or false.

a. A square is a rhombus.

b. A parallelogram is a rectangle with a right angle.

c. A parallelogram is a polygon whose opposite sides are parallel.

d. A rectangle is a parallelogram with four right angles.

e. A trapezoid is a quadrilateral whose opposite sides are parallel.

18. Indicate whether each statement is true or false.

a. An isosceles triangle is a triangle that has exactly two sides that are equal in length.

b. An acute triangle has only one acute angle.

c. A triangle can be both isosceles and obtuse.

d. An equilateral triangle can be an obtuse triangle.

e. A square is a rectangle.

19. Arrange the following terms in the order in which the definitions of each should be given: triangle, hypotenuse, polygon, right triangle.

20. Arrange the following terms in the order in which the definitions of each should be given: square, quadrilateral, parallelogram, polygon, rectangle.

21. For what kind(s) of parallelogram is each statement true?

a. The diagonals are equal.

b. The diagonals are perpendicular to each other.

22. Which figures *do not* always have congruent diagonals?

a. rectangle **b.** square

c. rhombus **d.** isosceles trapezoid

23. In quadrilateral $ABCD$ if $m\angle A = 40°$, $m\angle B = 140°$, quadrilateral $ABCD$ must (choose one)

a. be a rhombus.

b. have at least one pair of sides parallel.

c. be an isosceles trapezoid.

d. have at least one right angle.

24. In parallelogram $ABCD$, $m(\overline{AB}) = 5x - 4$ and $m(\overline{CD}) = 2x + 14$. Find the value of x.

25. In parallelogram $ABCD$, $m\angle A = 3x$ and $m\angle B = x + 40$. Find the value of x.

26. Which statement about the diagonals of an isosceles trapezoid is *always* true?

a. They bisect each other.

b. They are congruent.

c. They are perpendicular to each other.

d. They divide the trapezoid into four congruent triangles.

27. In isosceles trapezoid $DEFG$, $m\angle D$ is three times $m\angle F$. Find $m\angle F$.

28. A rectangle has a diagonal of length 10 and one side of length 6. What is the length of the other side?

29. Which statement is true?

a. All parallelograms are quadrilaterals.

b. All parallelograms are rectangles.

c. All quadrilaterals are trapezoids.

d. All trapezoids are parallelograms.

30. In rhombus $ABCD$, $m(\overline{AB}) = 4x - 12$ and $m(\overline{BC}) = 3x + 3$. Find the value of x.

For Exercises 31 and 32, choose the best answer.

31. The opposite angles of a parallelogram are

a. complementary **b.** congruent

c. supplementary **d.** right

32. Two opposite angles of an isosceles trapezoid are

a. equal **b.** complementary

c. supplementary

Writing Mathematics

33. What is a *broken line?*

34. What is a *closed broken line?*

35. What is a *simple closed broken line?*

36. In words, state the *Pythagorean theorem*.

Challenge Exercises

37. In rhombus $ABCD$, $m(\overline{AB}) = 4x - 2$ and $m(\overline{BC}) = 3x + 3$. Find the value of x.

38. In an isosceles right triangle, the measure of an acute angle is represented by $2x + 5$. Find the value of x.

39. In parallelogram $ABCD$, $m(\overline{AB}) = 5x - 4$ and $m(\overline{CD}) = 2x + 14$. Find the length of $\overline{AB}$.

40. The length of a side of a square is $\sqrt{3}$. What is the length of a diagonal of the square?

41. Two roads intersect at right angles. A sign on a pole is 30 meters from one road and also 40 meters from the other road. How far is the sign on the pole from the point where the roads intersect?

42. In the adjoining figure below, find the length of line segment $\overline{AG}$. Express your answer to the nearest tenth.

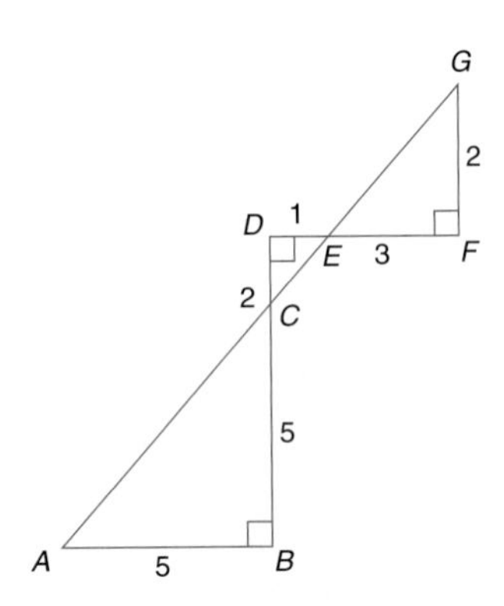

Group/Research Activities

43. The Pythagorean theorem may be expressed algebraically by the equation

$$x^2 + y^2 = z^2$$

Problems arise when we attempt to find only those solutions where x, y, and z are positive integers. For example, 3, 4, 5 is called a *Pythagorean triple* because 3, 4, 5 are positive integers that satisfy the equation

$$\begin{aligned} x^2 + y^2 &= z^2 \\ 3^2 + 4^2 &= 5^2 \\ 9 + 16 &= 25 \\ 25 &= 25 \end{aligned}$$

Note that the only common divisor of these numbers is one, which means that a triple like 6, 8, 10 is just another form of the triple 3, 4, 5.

Find at least four other Pythagorean triples such that the only common divisor of the numbers is one.

44. In Chapter 8, we located natural numbers, whole numbers, integers, and rational numbers on the number line. We can also locate irrational numbers on the number line.

For example, suppose we want to locate the point that corresponds to $\sqrt{2}$. We can do this by drawing the hypotenuse of a right triangle whose legs are 1 unit in length along the number line with one vertex at 0 (see the accompanying figure).

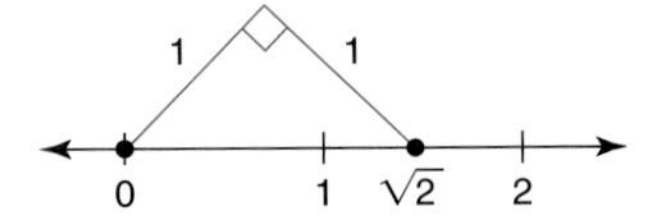

Recall that the Pythagorean theorem tells us that the square of the hypotenuse of a right triangle equals the sum of the squares of the legs; that is, if a and b are the lengths of the legs, and c is the length of the hypotenuse, then $c^2 = a^2 + b^2$.

If $a = 1$ and $b = 1$, then

$$c^2 = 1^2 + 1^2 = 1 + 1 = 2$$

and therefore,

$$c = \sqrt{2}$$

The length corresponds to a point on the number line in the figure.

Using the fact that $1^2 + 2^2 = 5$, see if you can graph $\sqrt{5}$ on the number line.

Just for fun

How many planets are in our solar system? Ranging from nearest to farthest from the sun, what are the names of these planets?

B. C. by permission of Johnny Hart and Creators Syndicate, Inc.

11.6 PERIMETER AND AREA

Perimeter

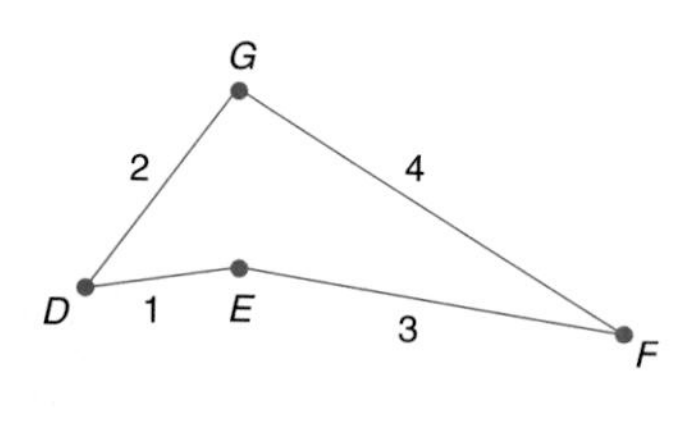

Figure 11.74

Geometry in its beginnings dealt with the measure of surfaces, and this has remained one of its most practical applications. We can measure a polygon in different ways. The **perimeter** of a polygon is the sum of the lengths of its sides. The perimeter of quadrilateral $DEFG$ in Figure 11.74 is determined by finding the sum of the lengths of its sides. Therefore, the perimeter of quadrilateral $DEFG$ is $1 + 3 + 4 + 2 = 10$.

Formulas can be used to find the perimeters of certain polygons. For example, the perimeter of a square whose side is s units in length is $4s$; that is, $P = 4s$. The perimeter of a rectangle with length l and width w is twice the length plus twice the width; that is $P = 2l + 2w$. The perimeter of a parallelogram with sides a and b is $P = 2a + 2b$. These formulas are shown in Figure 11.75.

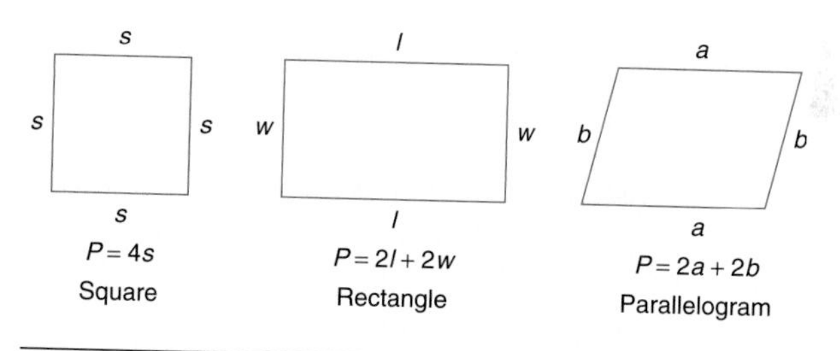

Figure 11.75

EXAMPLE 1 ***Finding Perimeter*** Find the perimeter of a square whose side measures 1.2 centimeters.

Solution

$$P = 4s$$
$$P = 4(1.2)$$
$$P = 4.8 \text{ cm}$$

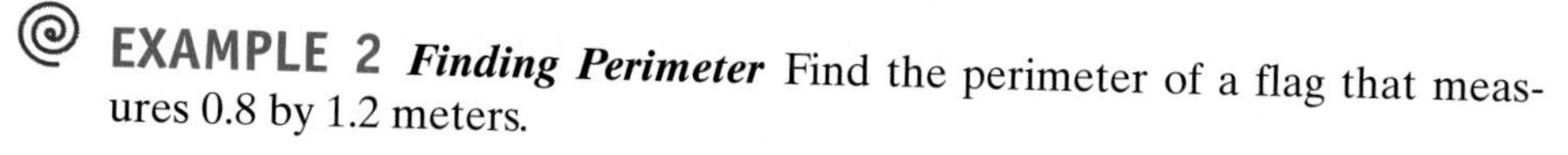

EXAMPLE 2 ***Finding Perimeter*** Find the perimeter of a flag that measures 0.8 by 1.2 meters.

Solution The longer side, 1.2, is the length and the shorter side, 0.8, is the width. The shape of the flag is a rectangle. Hence,

$$P = 2l + 2w$$
$$P = 2(1.2) + 2(0.8)$$
$$P = 2.4 + 1.6$$
$$P = 4.0 \text{ m}$$

NW ***Now Work Problems 1 and 3.***

Area of Rectangles

Another measure of a polygon is *area*. The **area** of a polygon tells us how many square units (inches, feet, yards, meters, centimeters, and so on) that polygon contains.

Suppose we want to put carpet tiles on the floor of a room. Suppose that the room is of a fairly standard size, 10 feet by 8 feet, and the carpet tiles are 1

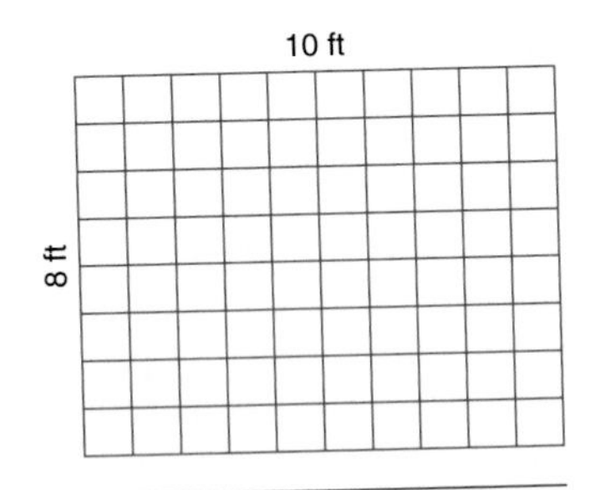

Figure 11.76

foot by 1 foot. How many carpet tiles do we need to cover the floor of the room? One way to do this would be to start laying the tiles down on the floor, placing one next to the other until the entire floor is covered. The result would look like Figure 11.76. Because the carpet tiles measure 1 foot by 1 foot, they are squares, and we can say that their measure is 1 square foot. Counting the number of carpet tiles in Figure 11.76, we see that there are 80 of them; that is, there are 80 square feet of carpet tiles on the floor. We can also say that the *area* of the floor is 80 square feet. Recall that the area of a polygon tells how many square units of a certain kind that polygon contains. Note that we could have also obtained the answer 80 square feet by multiplying 10 times 8; that is, $10 \times 8 = 80$.

Figure 11.76 is a rectangle whose length is 10 feet and width is 8 feet, and the rectangle contains 80 square feet. Therefore, we can conclude that

The area of a rectangle is equal to the product of its length and width; that is, area of a rectangle = length × width.

Symbolically, we have

$$A = l \times w \quad \text{or} \quad A = l \cdot w \quad \text{or} \quad A = lw$$

We can also say that the area of a rectangle is equal to the product of its base and its height. The term **base** is another name for the length of a rectangle, and the term **height** is another name for a width of a rectangle. Therefore,

$$A = b \times h$$

EXAMPLE 3 ***Finding the Area of a Rectangle*** Find the area of a rectangle whose base is 7 centimeters and whose height is 4 centimeters.

Solution The area of a rectangle is equal to the product of its base and height. Therefore, the area of the given rectangle is $7 \times 4 = 28$ square centimeters. Remember that area is measured in terms of square units.

EXAMPLE 4 ***Finding the Area of a Rectangle*** The area of a rectangle is 40 square meters, and the length is 8 meters. What is the width?

Solution Using the formula $A = lw$, we have $A = 40$ and $l = 8$. Substituting these values in the formula, we have

$$40 = 8w$$

Now we solve for w:

$$\frac{40}{8} = \frac{8w}{8} \quad \textit{dividing both sides by 8}$$

$$5 = w$$

The width is 5 meters.

NW ***Now Work Problem 13.***

EXAMPLE 5 ***Finding the Area of a Rectangle*** Find the area of a rectangle, given that its length is 15 inches and the length of its diagonal is 17 inches.

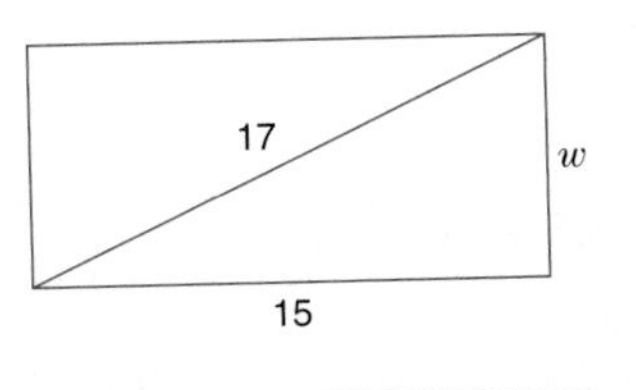

Figure 11.77

Solution This problem is a little more involved than the previous examples. In order to find the area, we must first find the width of the rectangle. We are given its length and diagonal. A diagonal of a polygon is a line segment that connects two vertices that are not consecutive. Therefore, the given rectangle could appear as in Figure 11.77.

Note that the triangle in the rectangle is a right triangle. We can find w by means of the Pythagorean theorem, which states that the square of the hypotenuse of a right triangle is equal to the sum of the squares of the legs. Therefore, we have

$$\begin{aligned} 17^2 &= 15^2 + w^2 \\ 289 &= 225 + w^2 \\ 64 &= w^2 \\ \sqrt{64} &= w \\ 8 &= w \end{aligned}$$

The width is 8 inches.

Now, we can find the area using the formula:

$$\begin{aligned} A &= lw \\ A &= 15 \times 8 \\ A &= 120 \text{ square inches} \end{aligned}$$

NW ***Now Work Problem 15.***

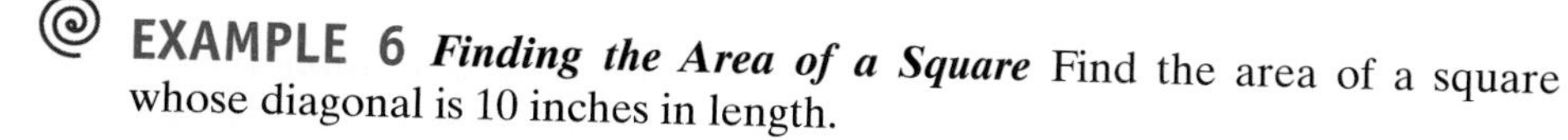

EXAMPLE 6 ***Finding the Area of a Square*** Find the area of a square whose diagonal is 10 inches in length.

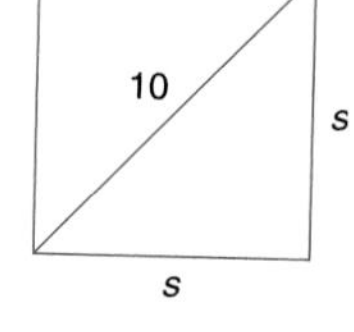

Figure 11.78

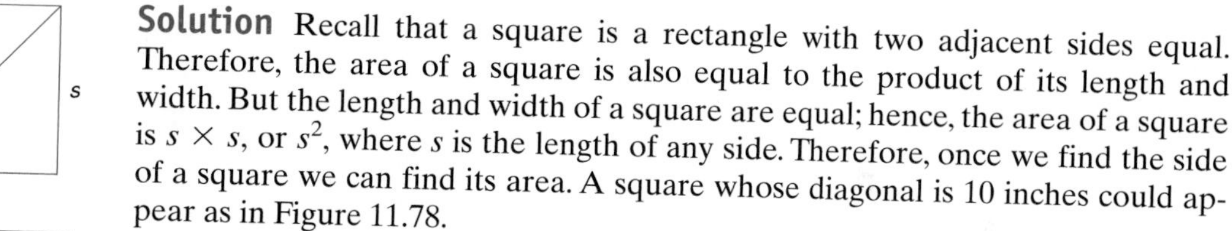

Solution Recall that a square is a rectangle with two adjacent sides equal. Therefore, the area of a square is also equal to the product of its length and width. But the length and width of a square are equal; hence, the area of a square is $s \times s$, or s^2, where s is the length of any side. Therefore, once we find the side of a square we can find its area. A square whose diagonal is 10 inches could appear as in Figure 11.78.

Note that we again have a right triangle (as in Example 5), and we can find s by means of the Pythagorean theorem. Therefore, we have

$$\begin{aligned} 10^2 &= s^2 + s^2 \\ 100 &= 2s^2 \\ 50 &= s^2 \end{aligned}$$

This is the answer; we do not need to do any more computation. Recall that the area of a square is equal to the product of two of its sides—that is, $s \times s$, or s^2, which is what we have. Therefore, the area of a square whose diagonal is 10 inches is 50 square inches.

NW ***Now Work Problem 17.***

Area of Parallelograms

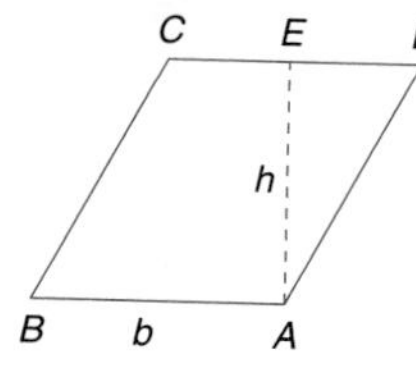

Figure 11.79a

Now that we are able to find the area of a rectangle and a square, let's consider the area of a parallelogram. We can state formally that

> ***The area of a parallelogram is equal to the product of its base and height; that is, $A = bh$.***

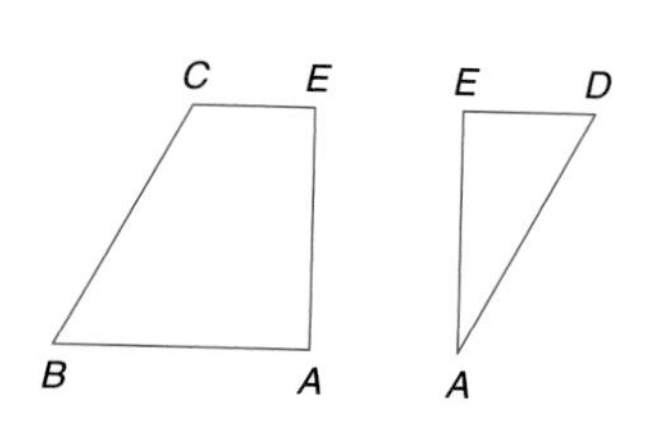

Figure 11.79b

Let's see why this formula works. Consider parallelogram $ABCD$ in Figure 11.79a, with the base AB denoted by b and its height AE denoted by h.

The height of a parallelogram is a line segment drawn perpendicular to the side from a point in the opposite side. We have chosen point A in order to form triangle ADE. Now, if we were to cut off triangle ADE from parallelogram $ABCD$, we would have Figure 11.79b.

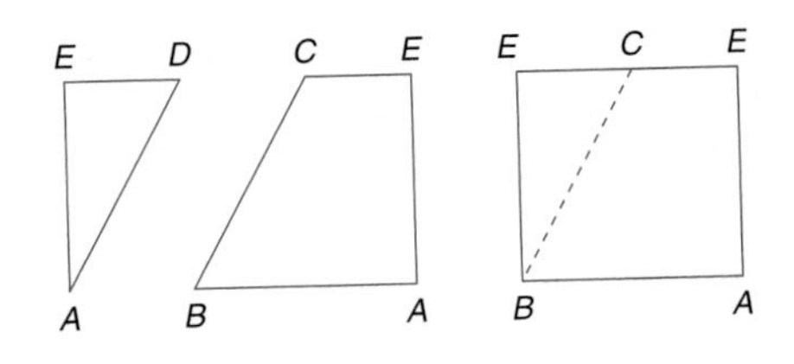

Figure 11.79c

Now let's move the triangle to the left side of $ABCE$ and attach it to $\overline{CB}$ so that $\overline{DA}$ and $\overline{CB}$ coincide. Note that $\overline{DA}$ and $\overline{CB}$ are equal, since the opposite sides of a parallelogram are equal. Therefore, we have Figure 11.79c.

The resulting figure is a rectangle, and the area of a rectangle is equal to the product of its base and its height. Recall that the term *base* is another name for the length of a rectangle, and the term *height* is another name for the width of a rectangle. The area of a parallelogram is equal to the product of its base and height, because the area of a rectangle is equal to the product of its base and height, and Figures 11.79a–c show that the area of a parallelogram is the same as the area of a rectangle. That is, $A = bh$.

Note that the discussion concerning parallelogram $ABCD$ is not a formal proof. It is merely an attempt to show why the area of a parallelogram can be found by multiplying its base by its height.

EXAMPLE 7 ***Finding the Area of a Parallelogram*** Find the area of a parallelogram whose base is 12 centimeters and whose height is 6 centimeters.

Solution The area of a parallelogram is equal to the product of its base and height. Therefore, the area of the given parallelogram is

$$12 \times 6 = 72 \text{ square centimeters}$$

EXAMPLE 8 ***Finding the Area of a Parallelogram*** The area of a parallelogram is 140 square feet and the base is 20 feet. What is its height?

Solution Using the formula $A = bh$, we have $A = 140$ and $b = 20$. Substituting these values in the formula, we have

$$140 = 20h$$

Solving for h,

$$\frac{140}{20} = \frac{20h}{20} \quad \textit{dividing both sides by 20}$$

$$7 = h$$

The height is 7 feet.

NW ***Now Work Problem 21.***

Note of Interest

James A. Garfield (1831–1881) was the twentieth president of the United States and the second president assassinated (during the fourth month of his presidency). Before becoming president, he discovered another proof of the Pythagorean theorem based on the area of a trapezoid (see the accompanying figure).

By using sides a and b as the lengths of the bases and side $(a + b)$ as the height, find the area of the trapezoid. Now find the area of the trapezoid by using the sum of the areas of the embedded right triangles. Set the two areas equal to each other. Using some algebra, you shall see that $a^2 + b^2 = c^2$.

President Garfield was educated at Geauga Seminary, Hiram College, and Williams College, graduating in 1856. Later he became professor of ancient languages at Hiram College.

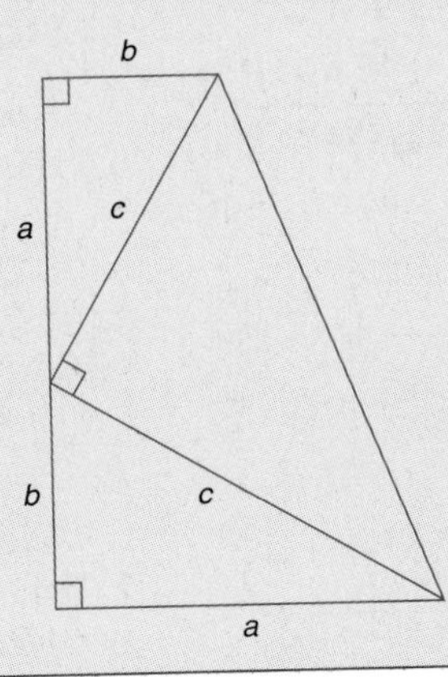

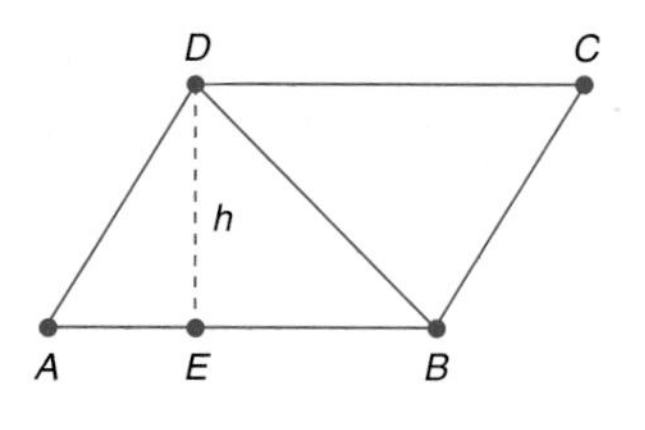

Figure 11.80

Area of Triangles

Now that we are able to find the area of a parallelogram, we can determine a method for finding the area of a triangle. Consider parallelogram $ABCD$ in Figure 11.80, with diagonal $\overline{DB}$ and height $\overline{DE}$.

We know that the area of parallelogram $ABCD$ is equal to the product of its base and height; that is, $A = bh$. What is the area of triangle ABD?

If we were to trace triangle ABD on a piece of paper and then take the paper and turn it so that the tracing of triangle ABD fitted over triangle DBC, we would see that the two triangles are exactly the same; that is, they are *congruent*. Two figures that can be made to coincide are said to be *congruent*. They match exactly in shape and size. The two triangles are congruent; together they form a parallelogram and we already know how to determine the area of a parallelogram. Because the parallelogram is formed by two congruent triangles, the area of one triangle (ABD) is equal to one-half of the area of the parallelogram. Therefore, the area of triangle ABD is $\frac{1}{2}bh$. We can state formally that

The area of a triangle is equal to one-half the product of its base and height; that is, $A = \frac{1}{2}bh$.

EXAMPLE 9 ***Finding the Area of a Triangle*** Find the area of a triangle whose base is 12 inches and whose height is 6 inches.

Solution The area of a triangle is equal to one-half the product of its base and height. Therefore, the area of the given triangle is $\frac{1}{2} \times 12 \times 6 = 36$ square inches.

EXAMPLE 10 ***Finding the Area of a Triangle*** The area of a triangle is 70 square centimeters and the base is 20 centimeters. What is the height?

Solution Using the formula $A = \frac{1}{2}bh$, we have $A = 70$ and $b = 20$. Substituting the values in the formula, we have

$$70 = \tfrac{1}{2}(20)h$$

Solving for h,

$$70 = 10h$$

$$\frac{70}{10} = \frac{10h}{10} \quad \textit{dividing both sides by 10}$$

$$7 = h$$

The height is 7 centimeters.

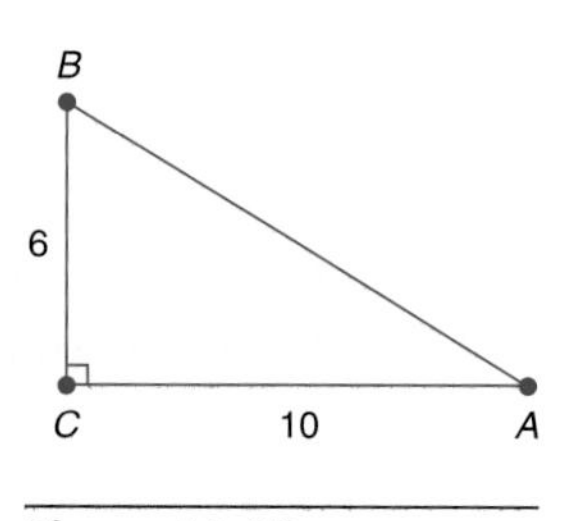

Figure 11.81

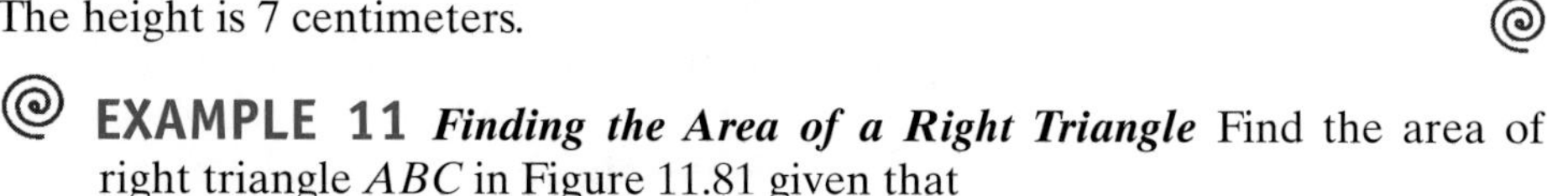

EXAMPLE 11 ***Finding the Area of a Right Triangle*** Find the area of right triangle ABC in Figure 11.81 given that

$$m(\overline{AC}) = 10, \quad m(\overline{BC}) = 6$$

and $\angle BCA$ is a right angle.

Solution The base and height are not given directly in this problem, but the triangle is a right triangle, and therefore the legs of the triangle are its base and height. The base is $\overline{AC}$, whose length is 10, and the height is $\overline{BC}$, whose length is 6. Using the formula $A = \frac{1}{2}bh$, we have

$$A = \tfrac{1}{2}(10)(6)$$

$$A = 30 \text{ square units}$$

Note that the lengths of the sides were not given in terms of a specific measurement. Therefore, the area is described in terms of *square units*.

NW ***Now Work Problems 25 and 27.***

Area of Trapezoids

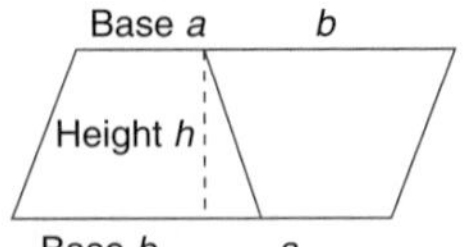

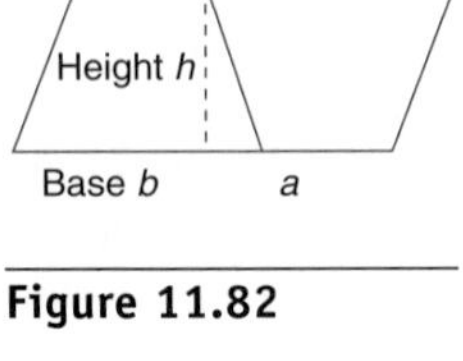

Figure 11.82

The area of a trapezoid can be determined if we think of the trapezoid as half of a parallelogram (see Figure 11.82).

The base of this parallelogram is $a + b$ (which is the sum of the bases of the trapezoid). The area of the parallelogram is $(a + b)h$. Hence, the area of the trapezoid is $\frac{1}{2}(a + b)h$.

The area of a trapezoid is equal to one-half the height times the sum of the bases; that is,

$$A = \frac{1}{2}h(a + b)$$

EXAMPLE 12 ***Finding the Area of a Trapezoid*** Find the area of a trapezoid whose bases are 8 centimeters and 22 centimeters and whose height is 10 centimeters.

Solution The area of a trapezoid is equal to one-half the height times the sum of the bases. Therefore,

$$A = \frac{1}{2}h(a + b)$$
$$A = \frac{1}{2}(10)(8 + 22)$$
$$A = 5(30)$$
$$A = 150 \text{ square centimeters}$$

NW ***Now Work Problem 29.***

Circles

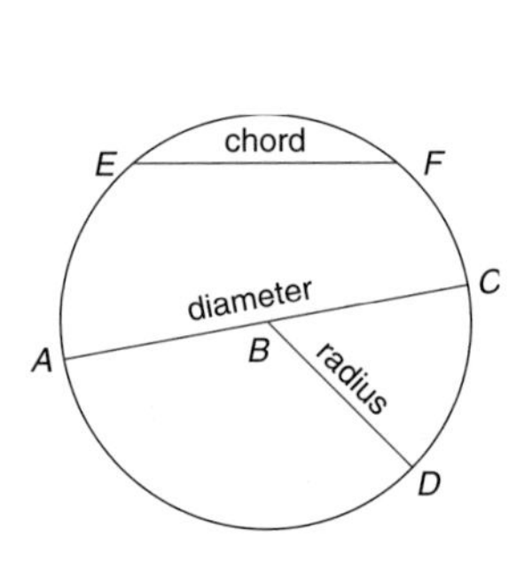

Figure 11.83

Thus far we have discussed the perimeter and area of certain polygons. It is also important to be able to do the same for another common geometric figure, the circle. The distance around a circle is called the **circumference** of a circle. Figure 11.83 is a circle whose center is at point B. A *chord* is a line segment that joins two points on a circle. $\overline{EF}$ is a chord. A **diameter** is a chord that passes through the center of the circle. $\overline{AC}$ is a diameter. A **radius** is a line segment that has the center and a point on the circle as end points. $\overline{BA}$, $\overline{BC}$, and $\overline{BD}$ are radii.

The constant ratio of the circumference (C) of a circle to its diameter (d) is represented by the Greek letter π (pi). That is, $C/d = \pi$

The value of π, correct to the nearest ten-thousandth, is 3.1416. The fraction $\frac{22}{7}$ gives a value of π correct to within 0.04 of one percent. Pi is a decimal that never ends and has no repeating pattern:

$$\pi = 3.141592653589793238\ldots$$

It has been shown that the value of π to 10 decimal places is sufficiently accurate to give the circumference of a circle as large as the earth's equator correct to a small fraction of an inch.

For any circle, the circumference is equal to π ***times the diameter. That is,***

$$C = \pi d$$

Since a diameter is composed of two radii, we can also say: For any circle, the circumference is equal to two times π times the radius. That is,

$$C = 2\pi r$$

EXAMPLE 13 ***Finding Circumference*** Find the circumference of a circular swimming pool if its diameter is 10 meters. Use 3.14 for π.

Solution The circumference of a circle is equal to π times its diameter. Therefore,

$$C = \pi d$$
$$C = (3.14)(10)$$
$$C = 31.4 \text{ m}$$

EXAMPLE 14 ***Finding Circumference*** Find the circumference of a circle that has a radius of 21 cm. Use $\pi = \frac{22}{7}$.

Solution The circumference of a circle is equal to two times π times its radius. Hence,

$$C = 2\pi r$$
$$C = 2 \cdot \frac{22}{7} \cdot 21$$
$$C = \frac{924}{7} = 132 \text{ cm}$$

NW ***Now Work Problem 37.***

The area of a circle is defined as the area enclosed by the circle.

For any circle, the area is equal to π times the radius squared. That is,

$$A = \pi r^2$$

If the radius of a given circle is 10 cm and we wish to find its area, then we have

$$A = \pi r^2$$
$$\text{let } \pi = 3.14\text{:} \quad A = (3.14)(10^2)$$
$$A = 3.14(100)$$
$$A = 314 \text{ cm}^2$$

Note: The notation cm^2 represents the label "square centimeters" and is a common notation for expressing square units. Similarly, 9 m^2 means 9 square meters and 32 mm^2 means 32 square millimeters.

EXAMPLE 15 ***Finding the Area of a Circle*** The radius of a circular rug is 3 m; find its area. Use $\pi = 3.14$.

Solution The area of a circle is equal to π times the radius squared. Therefore,

$$A = \pi r^2$$
$$A = (3.14)3^2$$
$$A = (3.14)9$$
$$A = 28.26 \text{ m}^2$$

EXAMPLE 16 ***Finding the Area of a Circle*** What is the area of a circle whose diameter measures 18 dm? Use $\pi = 3.14$.

Solution We must still use the formula $A = \pi r^2$, so first recall that the diameter of a circle is composed of two radii. Hence, for this problem, since $d = 18$, r must equal 9. Therefore,

$$A = \pi r^2$$
$$A = (3.14)9^2$$
$$A = (3.14)81$$
$$A = 254.34 \text{ dm}^2$$

NW ***Now Work Problem 43.***

Exercises for Section 11.6

NW **1.** Find the perimeter of each polygon.

a.

7 m
5 m
5 m
7 m

b.

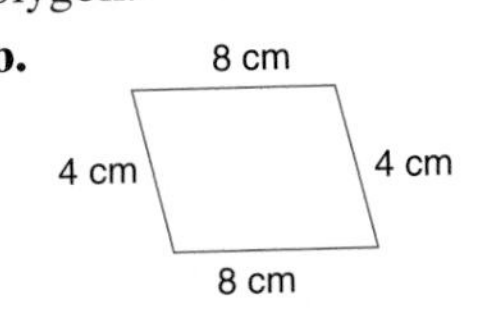

c.

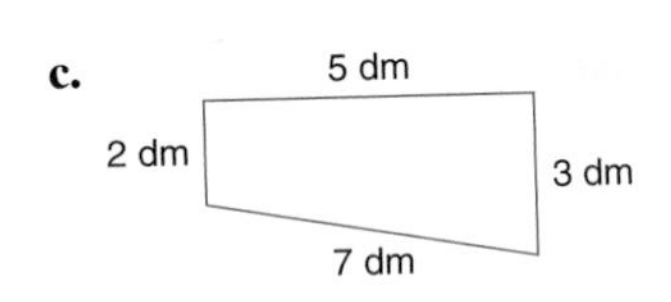

2. Find the perimeter of each polygon.

a.

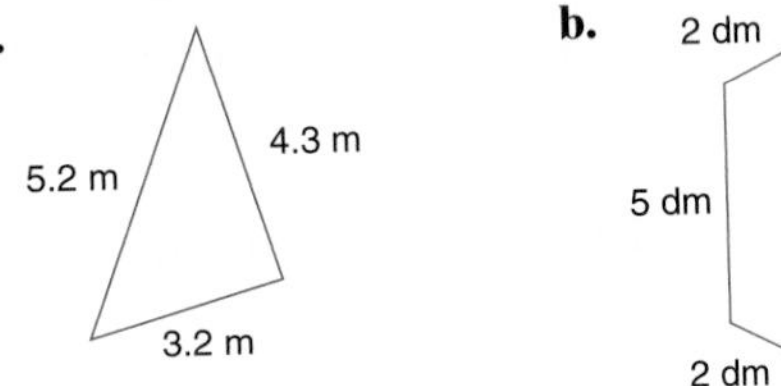

b.

2 dm
2 dm
5 dm
5 dm
2 dm
2 dm

c.

4 cm
4 cm
4 cm
4 cm

NW **3.** Find the perimeter of a rectangle whose length is 12 meters and whose width is 7 meters.

4. Find the perimeter of a square whose side is 5.5 centimeters in length.

5. Find the perimeter of a rhombus whose side is 1.6 meters in length.

6. Find the perimeter of a parallelogram whose sides are 10 inches and 15 inches.

7. Find the perimeter of a parallelogram whose sides are 32 centimeters and 55 centimeters.

8. Find the perimeter of a rectangle, given that its width is 6 and the length of its diagonal is 10.

9. Find the perimeter of a rectangle, given that its length is 4 and the length of its diagonal is 5.

10. Find the perimeter of a rectangle, given that its length is 9 and the length of its diagonal is 15.

11. The area and perimeter of a certain square are the same number. What is the length of a side of the square?

12. Find the area of a rectangle whose length is 11 meters and whose width is 6 meters.

NW **13.** The area of a rectangle is 24 square inches and the length is 8 inches. What is the width?

14. The area of a rectangle is 72 square centimeters and the width is 6 centimeters. What is the length?

NW **15.** Find the area of a rectangle, given that its length is 15 and the length of its diagonal is 17.

16. Find the area of a rectangle, given that its width is 3 and the length of its diagonal is 5.

NW **17.** Find the area of a square whose diagonal measures 5 inches.

18. The length of a diagonal of a square is 8 centimeters. Find the number of square centimeters in the area of the square.

19. If the length of a diagonal of a square is 4 inches, find the number of square inches in the area of the square.

20. Find the area of a parallelogram whose base is 14 inches and whose height is 8 inches.

NW **21.** The area of a parallelogram is 300 square meters and the base is 30 meters. What is its height?

22. The area of a parallelogram is 30 square feet and the height is 4 feet. What is the length of the base?

23. Find the area of a triangle whose base is 13 and whose height is 7.

24. The area of a triangle is 62.5 square centimeters and the base is 10 centimeters. Find the height.

NW **25.** The area of a triangle is 37 square feet and the height is 10 feet. Find the base.

26. Find the area of a right triangle whose legs measure 5 inches and 6 inches.

NW **27.** Find the area of a right triangle whose legs measure 5 and 7.

28. Find the area of a trapezoid whose bases are 8 and 12, and height is 5.

NW **29.** Find the area of a trapezoid whose bases are 10 and 20, and height is 6.

30. The area of a trapezoid is 200, and the bases are 15 and 25. Find the height.

31. The bases of a trapezoid are 14 centimeters and 6 centimeters. Both nonparallel sides are 5 centimeters. Find the area.

32. The bases of an isosceles trapezoid are 9 centimeters and 21 centimeters. The nonparallel sides are each 10 centimeters. Find the area.

*33. Find the area of an isosceles right triangle whose hypotenuse is 10 meters in length.

*34. The area of a square is 16 square units. Find the length of a diagonal.

35. Find the area and perimeter of each of the following figures. Let $\pi = 3.14$.

a.

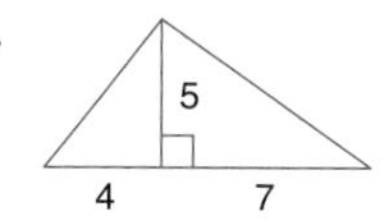

b.

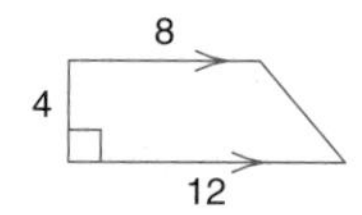

c.

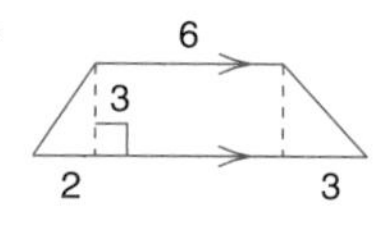

d.

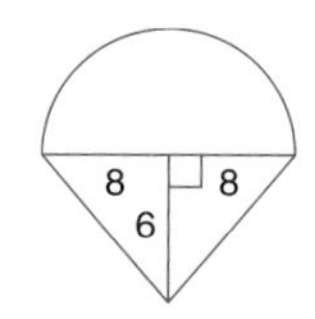

e.

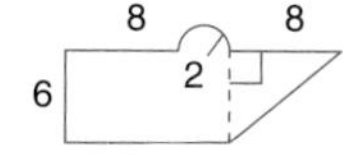

For Exercises 36–48, let $\pi = 3.14$.

36. The diameter of a tree trunk is 20 cm. Find its circumference.

NW 37. The diameter of a wheel is 30 cm. Find its circumference.

38. The radius of a circle is 3.5 mm. Find its circumference.

39. The minute hand of a clock is 15 cm. What is the distance that the tip of the minute hand moves in one hour?

40. The minute hand of a wristwatch is 14 mm. What is the distance that the tip of the minute hand moves in two hours?

41. How much fencing would be needed to enclose a circular garden that has a diameter of 10 m?

42. A circular rug has a diameter of 3 m. Find the area.

NW 43. The floor of a circus tent is circular with a radius of 30 m. What is the area of the floor?

44. Find the area of a circle whose diameter is 20 dm.

45. Find the area of a circle whose radius is 5 mm.

46. What is the area of a circular garden that has a diameter of 8 m?

47. Find the area of a circle if its circumference is 314 m.

*48. If the radius of a circle is increased by 1 cm, by how many cm is its circumference increased?

Writing Mathematics

49. What is the difference between a square and a rectangle?

50. How is a rectangle different from a parallelogram?

51. What is the difference between a trapezoid and a parallelogram?

52. In your own words, explain the meaning of π (pi).

Challenge Exercises

53. A circle and a square each have a perimeter of 140. Which has the greater area and by how much?

54. Find the radius of a circle whose area is numerically equal to its circumference.

55. The area of a rectangle, whose width is 3, equals the area of a square, whose side is 6. Find the perimeter of both figures.

56. The area of parallelogram $ABCD$ is equal to the area of a square whose side is 6. If $m(\overline{AB}) = 12$, find the length of an altitude drawn to side $\overline{AB}$.

57. The circumference of a circle is 18π centimeters. If the length of an arc of this circle is 3π centimeters, find the number of degrees in the measure of the arc.

58. Which of the following statements is not always true?
   a. A square is a rhombus.
   b. A square is a rectangle.
   c. A parallelogram is a polygon.
   d. A trapezoid is a parallelogram.

Research/Group Activities

59. The *Great Pyramid of Cheops* was constructed on a plan such that the perimeter of the four sides had the same ratio to its height as the ratio of the circumference of a circle to its radius, that is, 2π. Research the Great Pyramid of Cheops, noting its many astounding properties, and prepare a report for class.

60. Another formula for finding the area of a triangle is attributed to Heron (a.k.a. Hero) of Alexandria (first century A.D.). It states

$$A = \sqrt{s(s-a)(s-b)(s-c)}$$

where $s = \frac{1}{2}(a + b + c)$, a, b, c are sides of the triangle. Research the life and mathematical contributions of Heron. Prepare a report. Be sure to include an example of the given formula.

Just for fun

Are you good at counting squares? How many squares are there in this figure?

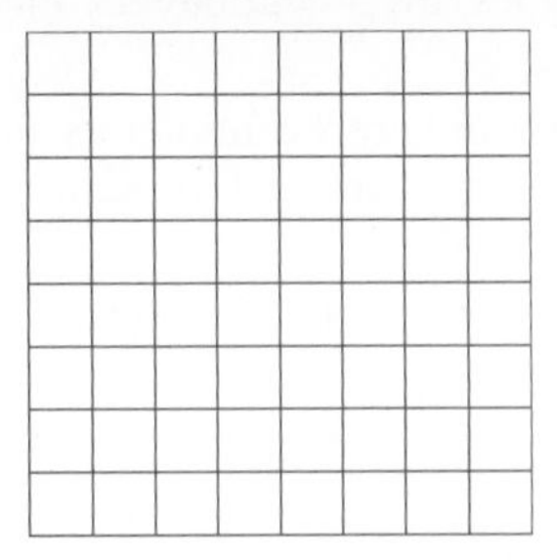

11.7 SOLIDS

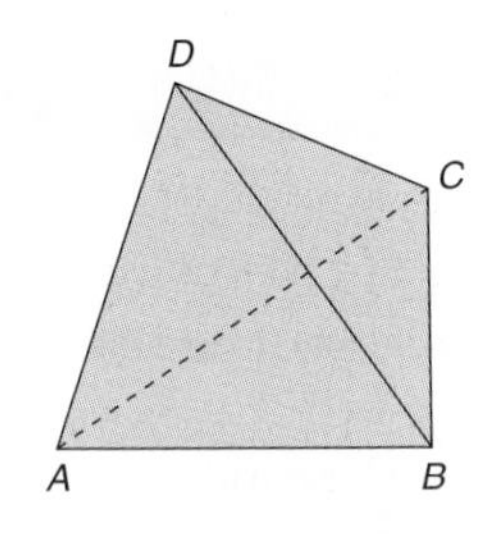

Figure 11.84

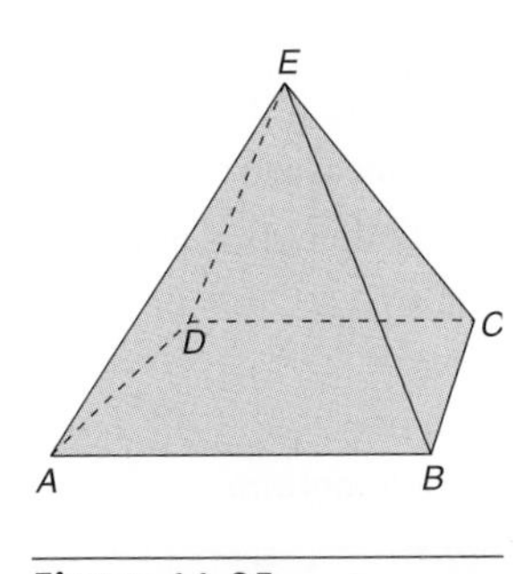

Figure 11.85

Introduction

In previous sections we discussed lines and planes. Two distinct points A and B determine a unique line segment AB, whereas any three distinct points A, B, and C that are not on the same line determine a unique plane. What happens if we have any four distinct points A, B, C, and D that are in a space, but not on the same plane? Figure 11.84 shows the outcome for this situation. The object shown in Figure 11.84 is called a **tetrahedron.** The four points A, B, C, D are the **vertices** of the tetrahedron. The six line segments $\overline{AB}$, $\overline{BC}$, $\overline{CD}$, $\overline{DB}$, $\overline{AD}$, and $\overline{AC}$ are the **edges** of the tetrahedron. The four triangles, ABD, ABC, BCD, and ACD, are called the **faces** of the tetrahedron. The solid in Figure 11.84 is called a tetrahedron because it has *four* faces. A **polyhedron** is a space figure or solid bounded by polygonal regions. For example, Figure 11.85, in Example 1 shows a different polyhedron, called a square pyramid.

EXAMPLE 1 ***Identifying the Parts of a Square Pyramid*** Figure 11.85 contains the square pyramid $ABCDE$. Identify (name) its

a. vertices **b.** edges **c.** faces

Solution

a. The vertices are the end points of the line segments—that is, A, B, C, D, E.

b. The edges are the line segments $\overline{AB}$, $\overline{BC}$, $\overline{CD}$, $\overline{DA}$, $\overline{AE}$, $\overline{BE}$, $\overline{CE}$, $\overline{DE}$.

c. The faces are the polygonal regions—namely, triangles ABE, BCE, CDE, ADE and square $ABCD$.

NW ***Now Work Problem 5.***

The famous Euler formula for polyhedrons states that

For any single polyhedron

$$V - E + F = 2$$

where V is the number of vertices, E is the number of edges, and F is the number of faces.

Let's verify this formula for the square pyramid in Figure 11.85.

$$V = 5, \quad E = 8, \quad F = 5 \quad \text{(see Example 1)}$$

Historical Note

The French mathematician Pierre de Fermat (1601–1665) is famous for his contributions to number theory. He was the son of a leather merchant and received his early education at home. Later in life he was a counselor for the local parliament at Toulouse. He devoted most of his leisure time to the study of mathematics.

There are a number of mathematical problems that are still mysteries to mathematicians. *Fermat's last theorem* is one of them. He wrote "If n is a number greater than 2, there are no whole numbers a, b, c such that $a^n + b^n = c^n$." It can be shown that when $n = 2, x^2 + y^2 = z^2$; for example, (3, 4, 5), (5, 12, 13), and (8, 15, 17). Yet, no one has ever been able to find positive whole numbers such that $x^3 + y^3 = z^3$ or $x^4 + y^4 = z^4$. Fermat said that he could prove that no such numbers could be found.

Although many prominent mathematicians have tried, at the start of 1997 no one had yet found a correct proof to the problem. Fermat's Last Theorem has the distinction of being the mathematical problem for which the greatest number of incorrect proofs have been published.

In 1908 Paul Wolfskehl bequeathed in his will a prize of 100,000 Marks to anyone who could prove Fermat's Last Theorem. Andrew Wiles, a mathematician from Princeton University, collected the Woldskehl Prize worth $50,000 on June 27, 1997. It had taken him seven years working in isolation to solve the problem. Finally, Fermat's Last Theorem had been officially solved.

Hence,

$$\begin{aligned} V - E + F &= 2 \\ 5 - 8 + 5 &= 2 \quad \textit{substituting} \\ 10 - 8 &= 2 \\ 2 &= 2 \end{aligned}$$

Surface Area of Solids

Thus far in our treatment of polyhedrons we have discussed the square pyramid where all the faces are triangles and the base is a square. A rectangular pyramid has a rectangle for a base and its other faces are triangular regions. Similarly, a pentagonal pyramid has a pentagon as its base and its other faces are triangular regions. In addition to these solids, another type of polyhedron is a **prism**, which has two parallel and congruent bases. Its other faces are formed by parallelograms, which are usually rectangles or squares. If the edges of the prism are perpendicular to the base, then the prism is called a **right prism**. Examples of prisms are shown in Figure 11.86.

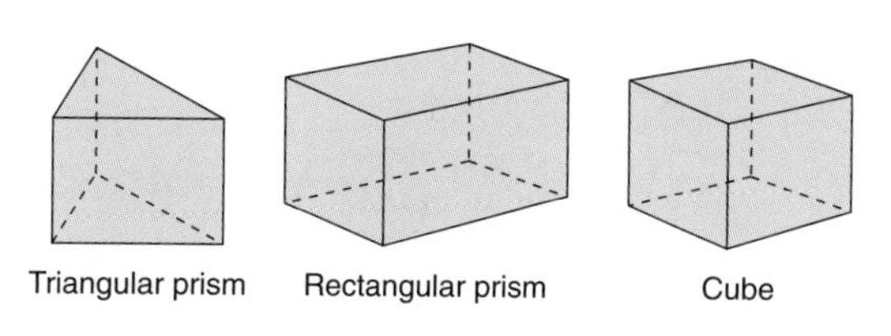

Figure 11.86

Notice that in all of these solids, their faces are composed of polygons. As a result, if we were to find the area of all the faces of a solid and then add them together, we will have the **surface area** of the solid. Thus, the surface area of a solid is simply the area of the solid's faces, which is its outer surface.

The total surface area of a polyhedron is the sum of the areas of all the faces.

To find the surface area of a solid, we apply the area formulas for polygons we developed in the previous section. The surface area formulas for the triangular prism, rectangular prism, and cube of Figure 11.86 are provided in Table 11.1.

TABLE 11.1 Surface Area Formulas for the Triangular Prism, Rectangular Prism, and Cube

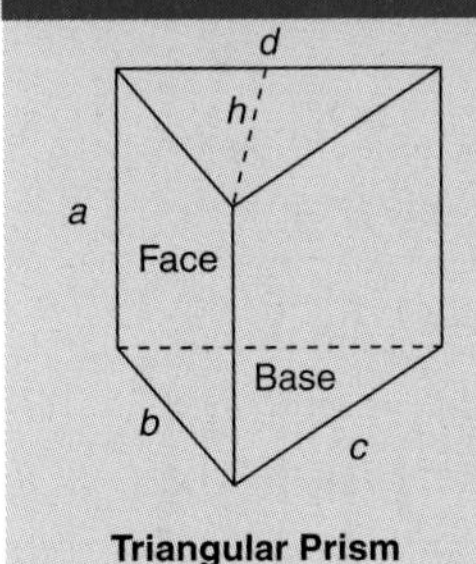

Triangular Prism

A **triangular prism** has 2 *triangular* bases and 3 *rectangular* or *square* faces:

$$\text{Surface Area} = \frac{1}{2}dh + \frac{1}{2}dh + ab + ac + ad$$

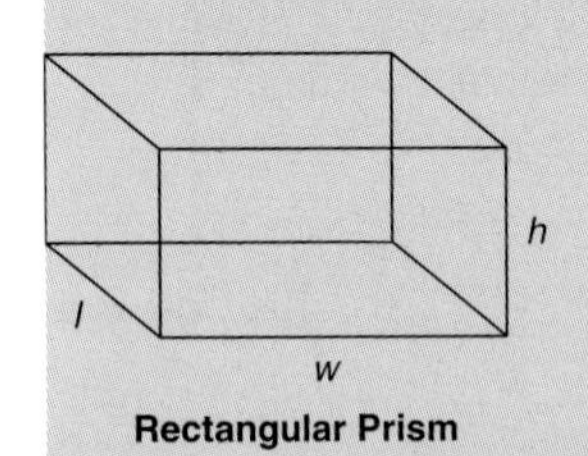

Rectangular Prism

A **rectangular prism** has 6 *rectangular* sides:

$$\begin{aligned}\text{Surface Area} &= lh + lh + wh + wh + lw + lw \\ &= 2lh + 2wh + 2lw\end{aligned}$$

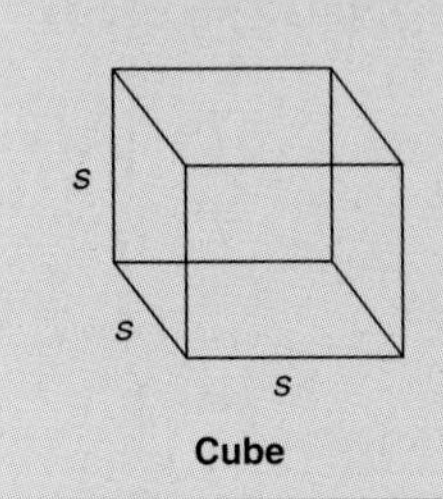

Cube

A **cube** has 6 *square* sides all of which are of equal length:

$$\begin{aligned}\text{Surface Area} &= ss + ss + ss + ss + ss + ss \\ &= s^2 + s^2 + s^2 + s^2 + s^2 + s^2 \\ &= 6s^2\end{aligned}$$

EXAMPLE 2 ***Finding the Surface Area of a Prism*** Find the surface area of the prism in Figure 11.87.

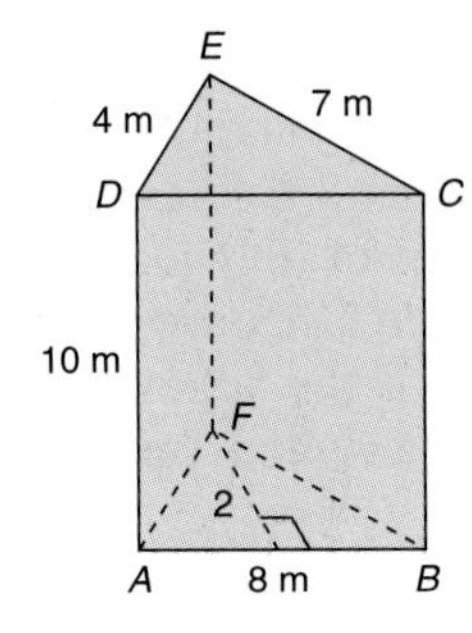

Figure 11.87

Solution We find the area of each face, and then find the total:

$$\begin{aligned}\text{Area } ABF &= \tfrac{1}{2}(8)(2) = 8 \\ \text{Area } DCE &= \tfrac{1}{2}(8)(2) = 8 \\ \text{Area } ABCD &= (8)(10) = 80 \\ \text{Area } ADEF &= (4)(10) = 40 \\ \text{Area } BCEF &= (7)(10) = 70 \\ \text{Total surface area} &= 206 \text{ m}^2\end{aligned}$$

NW ***Now Work Problem 9. (Surface Area Only)***

Three additional geometric solids are cylinders, cones, and spheres. These solids contain circular surfaces and hence their surface area formulas will involve the formula for the area of a circle.

A **cylinder** (can) has two congruent circular bases. The total surface area is the sum of the areas of the bases and the area of the curved surface. The area of a circular base is πr^2. Hence, the total area of the two circular bases is $2\pi r^2$. If the cylinder is opened up, the curved surface flattens out to form a rectangle as shown in Figure 11.88. The length of the rectangle is $2\pi r$ because that is the circumference of the circular base; it still has the same length when it is "opened" up. The area of the rectangle (curved surface) is $2\pi rh$. Therefore,

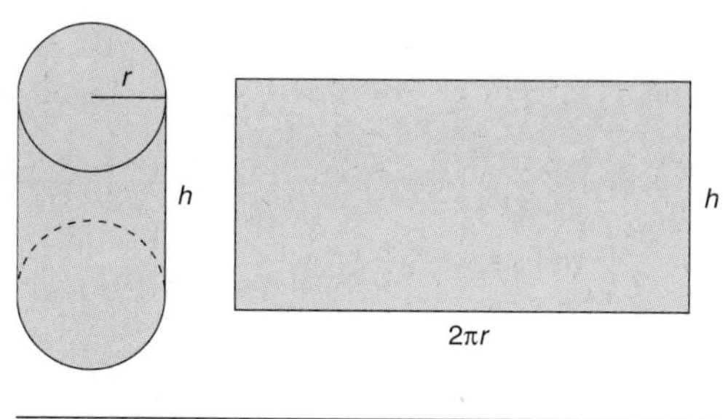

Figure 11.88

The total surface area of any cylinder is equal to 2 times π times the radius squared, plus 2 times π times the radius times the height:

$$\textbf{Surface area of a cylinder} = 2\pi r^2 + 2\pi rh$$

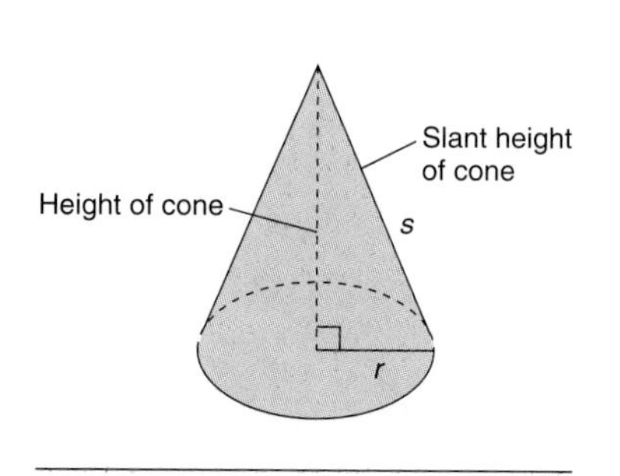

Figure 11.89

A **cone** is similar to a cylinder except it has one circular base and a slant height. Thus, its outer surface contains a circular base and a "slanted" curved surface. An illustration of a cone is shown in Figure 11.89. With slight modification to the surface area formula of a cylinder, we can get the surface area formula of a cone. As a result,

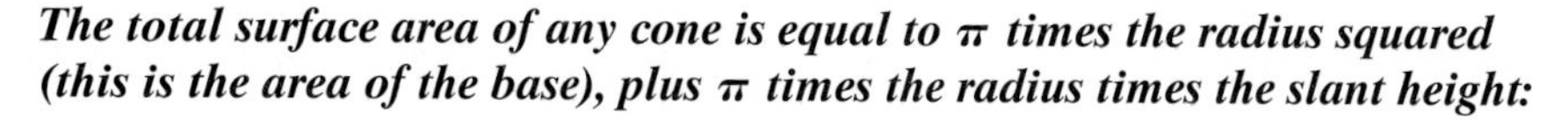

The total surface area of any cone is equal to π times the radius squared (this is the area of the base), plus π times the radius times the slant height:

$$\textbf{Surface area of a cone} = \pi r^2 + \pi rs$$

A **sphere** is a solid surface that resembles a baseball or basketball. Formally, a sphere consists of a set of points in space such that all the points are of equal distance from a fixed point called the *center*. The distance between the center to any point on the sphere is the *radius* of the sphere. The surface area formula for a sphere is given here without further discussion. (Additional information on the development of this formula may be found in a geometry or calculus textbook.)

The surface area of any sphere is equal to 4 times the area of a circle with the same radius:

$$\textbf{Surface area of a sphere} = 4\pi r^2$$

A summary of the surface area formulas for cylinders, cones, and spheres is given in Table 11.2.

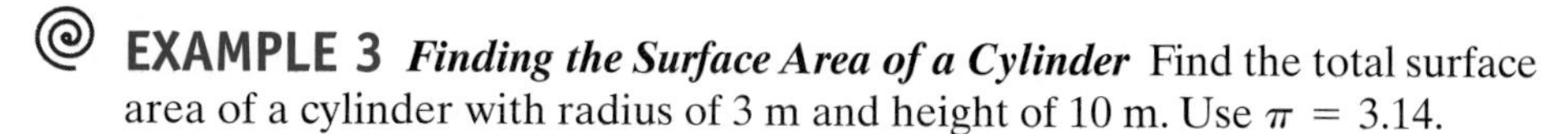

EXAMPLE 3 ***Finding the Surface Area of a Cylinder*** Find the total surface area of a cylinder with radius of 3 m and height of 10 m. Use $\pi = 3.14$.

Solution The formula for the total surface area of a cylinder is

$$A = 2\pi r^2 + 2\pi rh$$

TABLE 11.2 Surface Area Formulas for Cylinder, Cone, and Sphere

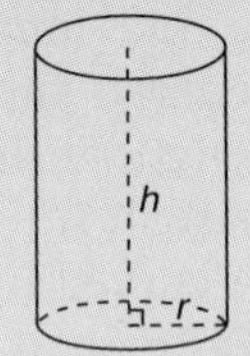

Circular Cylinder

A **circular cylinder** has *two circular* bases and a curved surface that flattens into a rectangle:

$$\text{Surface area} = \pi r^2 + \pi r^2 + 2\pi rh$$
$$= 2\pi r^2 + 2\pi rh$$

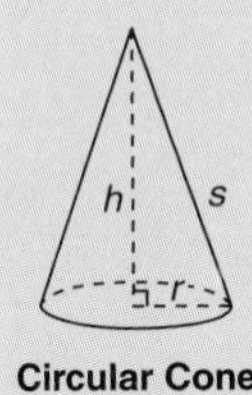

Circular Cone

A **circular cone** has *one circular* base and a slant height:

$$\text{Surface area} = \pi r^2 + \pi rs$$

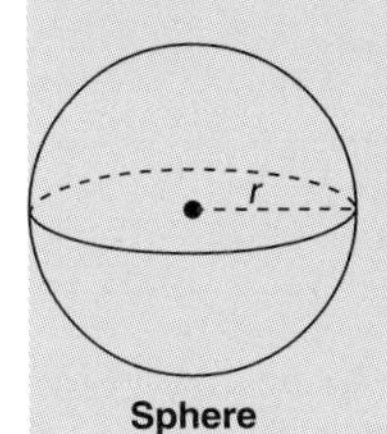

Sphere

A **sphere** is a set of points in space that is of equal distance from some fixed point called its center. The distance between the center to any point on the sphere is the radius:

$$\text{Surface area} = 4\pi r^2$$

Substituting,

$$A = 2(3.14)(3^2) + 2(3.14)(3)(10)$$
$$A = (6.28)9 + (6.28)(30)$$
$$A = 56.52 + 188.40$$
$$A = 244.92 \text{ m}^2$$

NW ***Now Work Problem 13. (Surface Area Only)***

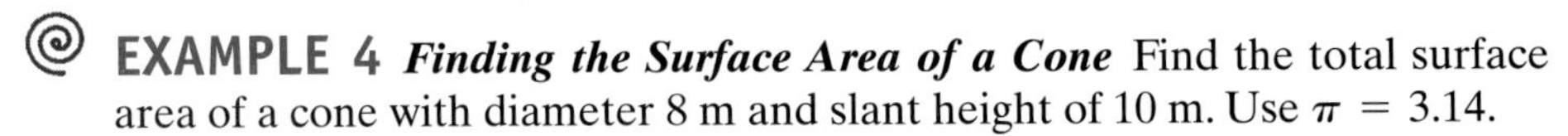

EXAMPLE 4 ***Finding the Surface Area of a Cone*** Find the total surface area of a cone with diameter 8 m and slant height of 10 m. Use $\pi = 3.14$.

Solution The formula for the total surface area of a cone is

$$A = \pi r^2 + \pi rs$$

Note that $d = 8$; therefore, $r = 4$. Substitution yields

$$A = (3.14)(4^2) + (3.14)(4)(10)$$
$$A = (3.14)(16) + (3.14)(40)$$
$$A = 50.24 + 125.60$$
$$A = 175.84 \text{ m}^2$$

NW ***Now Work Problem 17. (Surface Area Only)***

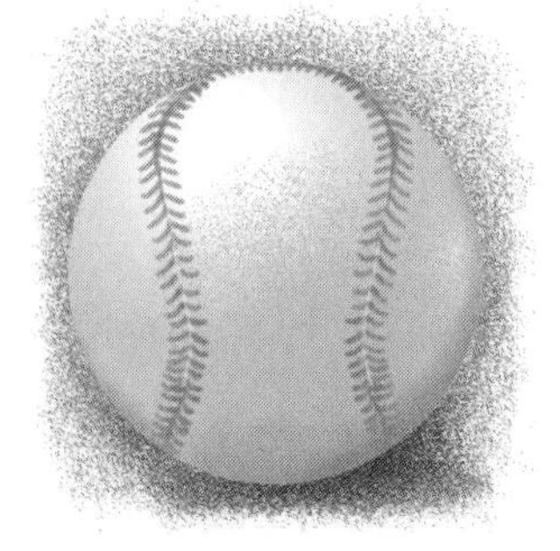

EXAMPLE 5 ***Finding the Surface Area of a Sphere*** Find the surface area of a baseball, given that its radius is 3.3 cm. Use $\pi = 3.14$. Express your answer to the nearest tenth.

Solution The formula for the surface area of a sphere is

$$A = 4\pi r^2$$

Therefore,

$$A = 4(3.14)(3.3)^2$$
$$A = (12.56)(10.89)$$
$$A = 136.7784$$
$$A = 136.8 \text{ cm}^2$$

NW ***Now Work Problem 21. (Surface Area Only)***

Volume of Solids

In addition to surface area, we can also measure the **volume** of a solid, which is the amount of space that is enclosed by the solid. Volume is measured in cubic units. Consequently, when we find the volume of a solid, we are measuring the number of cubic units contained by the solid.

We begin with volume formulas for the triangular prism, rectangular prism, and cube. Note from Table 11.1 that each figure respectively contains a triangular base, rectangular base, and a square base. Note further, that if we were to find the area of the base for one of these solids, and then create a "stack" of these areas that is the height of the solid, we will have a measure of the amount of space contained within the solid.

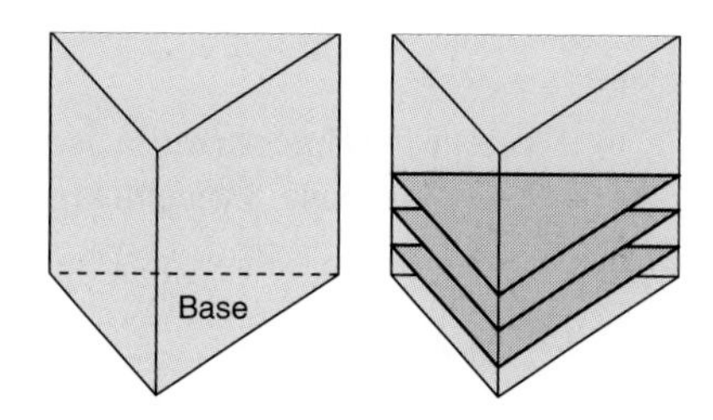

Thus, in general, the volume of these figures is the product of their base and height. Because these bases are not the same, we will let B = the area of the base. As a result,

The volume of any prism is equal to the area of the base times the height:
$V = Bh$, where B = area of the base

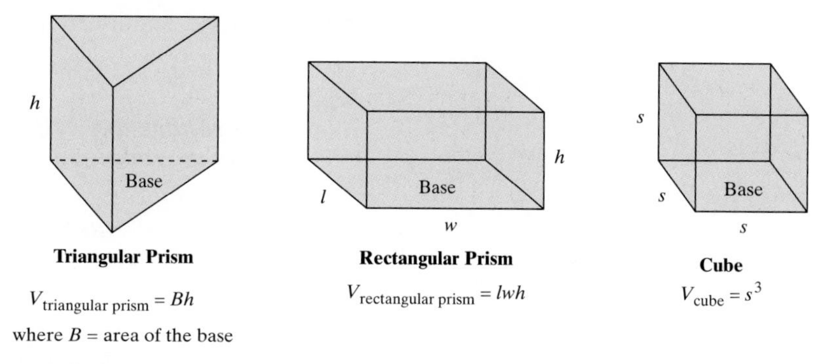

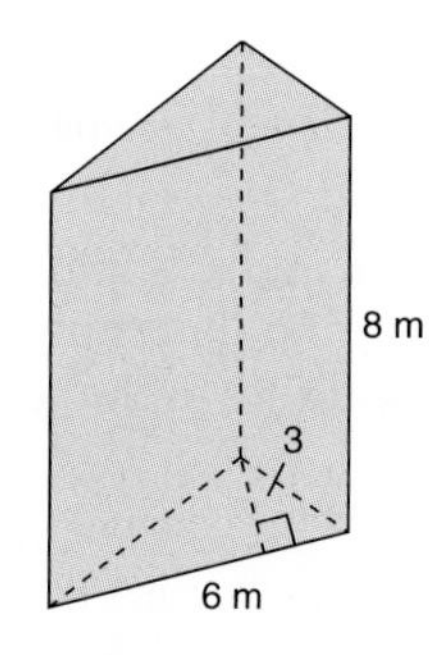

Figure 11.90

EXAMPLE 6 ***Finding the Volume of a Prism*** Find the volume of the prism shown in Figure 11.90.

Solution The base of the prism is a triangle. The formula for the area of a triangle is

$$A = \tfrac{1}{2}bh$$

Therefore,

$$A = \tfrac{1}{2}(6)(3)$$
$$A = (3)(3)$$
$$A = 9$$

Note that this is the area of the base. Now,

$$V = Bh$$
$$V = (9)(8)$$
$$V = 72 \text{ m}^3$$

NW ***Now Work Problem 9. (Volume Only)***

The volume of a pyramid can be found in a similar manner as that of a prism. Its formula is as follows:

The volume any pyramid is equal to one-third the area of the base times the height:

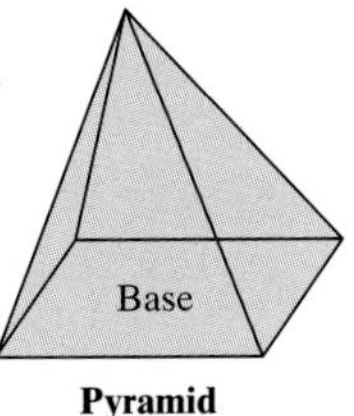

Pyramid

This formula is applicable whether the pyramid has a triangular or rectangular base. For example, the Great Pyramid of Egypt has a square base of 227 m on a side and its height is 144 m. To find its volume we use the formula

$$V = \tfrac{1}{3}Bh$$

The area of its base is $(227)(227) = 51{,}529$; hence, we now have

$$V = \tfrac{1}{3}(51{,}529)(144)$$
$$V = \tfrac{1}{3}(7{,}420{,}176)$$
$$V = 2{,}473{,}392 \text{ m}^3$$

The volume of a cylinder is found in the same fashion as a prism. For any prism, its volume is equal to the area of the base times its height. But the base of a cylinder is a circle, and its area is πr^2. Hence,

The volume any cylinder is equal to π times the radius squared times the height:

$$V_{cylinder} = \pi r^2 h$$

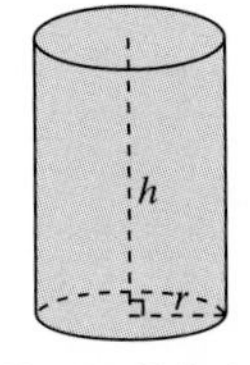

Circular Cylinder

A typical coffee can has a diameter of 10 cm and a height of 13 cm. To find its volume using the formula $V = \pi r^2 h$, we must first determine r. Since $d = 10$, r must equal 5, because $2r = d$. Therefore, substituting in the formula, we have

$$V = (3.14)(5^2)(13)$$
$$V = (3.14)(25)(13)$$
$$V = 1020.5 \text{ cm}^3$$

The volume of a cone is found in a similar manner to that of a pyramid. Recall that the formula for the volume of a pyramid is $V = \frac{1}{3}Bh$, where B is the area of the base. Therefore,

The volume of a cone is equal to one-third times π times the radius squared times the height:

$$V_{cone} = \frac{1}{3}\pi r^2 h$$

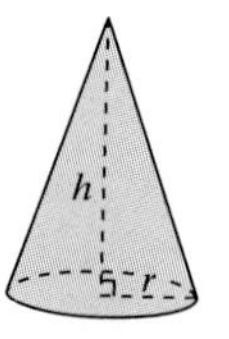

Circular Cone

Note that πr^2 is the area of a circle, which is the base of a cone. The volume of a cone is one-third the volume of a cylinder with base and height the same size as in the cone.

EXAMPLE 7 ***Finding the Volume of a Cone*** Find the volume of a cone whose height is 10 cm and whose base has a radius of 4 cm. Use $\pi = 3.14$. Express your answer to the nearest tenth.

Solution The formula for the volume of a cone is

$$V = \frac{1}{3}\pi r^2 h$$

Substituting,

$$V = \frac{1}{3}(3.14)(4^2)(10)$$

$$V = \frac{1}{3}(3.14)(160)$$

$$V = \frac{1}{3}(502.4)$$

$$V = 167.46\overline{6}$$

$$V = 167.5 \text{ cm}^3$$

NW ***Now Work Problem 17. (Volume Only)***

The last volume formula we will consider is that of a sphere. The volume of a sphere is two-thirds the volume of a cylinder with the same radius and a height equal to twice the radius. Recall that the volume of a cylinder is equal to $\pi r^2 h$. Now if h is equal to $2r$, we have

$$V = \pi r^2(2r) \quad \textit{substituting for h}$$

$$V = 2\pi r^3 \quad \textit{volume of a cylinder}$$

The volume of a sphere is equal to two-thirds of this. That is,

$$\tfrac{2}{3} \text{ of } 2\pi r^3 = \tfrac{2}{3}(2\pi r^3) = \tfrac{4}{3}\pi r^3$$

Therefore,

The volume of a sphere is

$$V_{sphere} = \frac{4}{3}\pi r^3$$

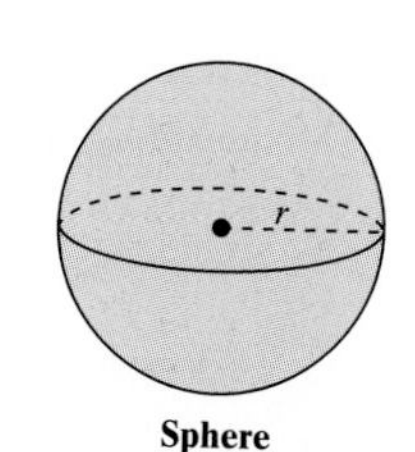

Sphere

EXAMPLE 8 ***Finding the Volume of a Sphere*** Find the volume of a baseball, given that its radius is 3.3 cm. Use $\pi = 3.14$. Express your answer to the nearest tenth.

Solution The formula for the volume of a sphere is

$$V = \tfrac{4}{3}\pi r^3$$

Substituting,

$$V = \frac{4}{3}(3.14)(3.3)^3$$
$$V = \frac{4}{3}(3.14)(35.937)$$
$$V = \frac{4}{3}(112.84218) = \frac{451.36872}{3}$$
$$V = 150.45624$$
$$V = 150.5 \text{ cm}^3$$

NW ***Now Work Problem 21. (Volume Only)***

In this section we have discussed a variety of geometric solids and how to find their surface area and volume. The total surface area of a polyhedron is the sum of the areas of all the faces. Following is a list of the formulas that we used in this section.

Name	Surface Area	Volume
Cylinder	$A = 2\pi r^2 + 2\pi rh$	$V = \pi r^2 h$
Cone	$A = \pi r^2 + \pi rs$	$V = \frac{1}{3}\pi r^2 h$
Sphere	$A = 4\pi r^2$	$V = \frac{4}{3}\pi r^3$
Prism	sum of the areas of the faces	$V = Bh$
Pyramid	sum of the areas of the faces	$V = \frac{1}{3}Bh$

Exercises for Section 11.7

1. What is the least number of vertices that a polyhedron may have?
2. What is the least number of edges that a polyhedron may have?
3. What is the least number of faces that a polyhedron may have?
4. **a.** How many faces does a tetrahedron have?
   b. How many vertices does a tetrahedron have?
   c. How many edges does a tetrahedron have?
   d. Verify that $V - E + F = 2$ for a tetrahedron.
5. (NW) **a.** How many faces does a cube have?
   b. How many vertices does a cube have?
   c. How many edges does a cube have?
   d. Verify that $V - E + F = 2$ for a cube.
6. **a.** How many faces does a triangular prism have?
   b. How many vertices does a triangular prism have?
   c. How many edges does a triangular prism have?
   d. Verify that $V - E + F = 2$ for a triangular prism.
7. Figure 11.91 is called an octahedron.

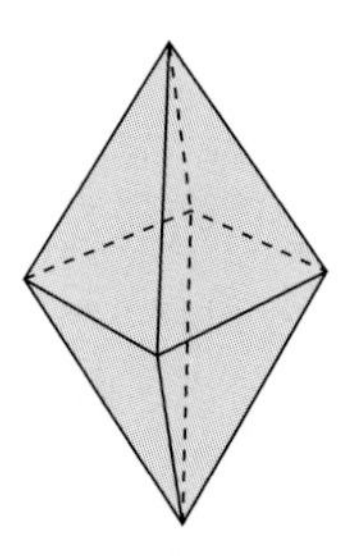

Figure 11.91

   a. How many faces does it have?
   b. How many vertices does it have?
   c. How many edges does it have?
   d. Verify that $V - E + F = 2$ for an octahedron.

For Exercises 8–18, find the total surface area and volume of each polyhedron. Let $\pi = 3.14$. Express decimal answers to the nearest tenth.

8.

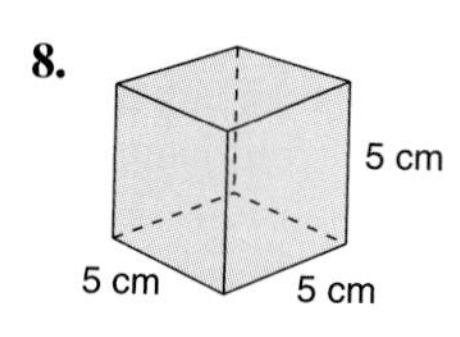

NW **9.**

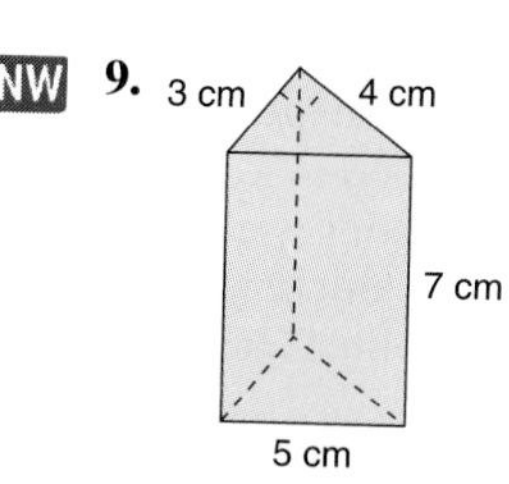

10.

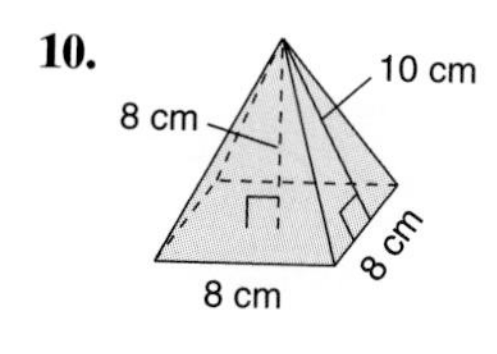

11.

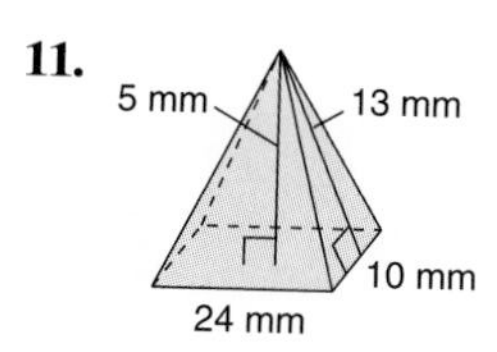

12.

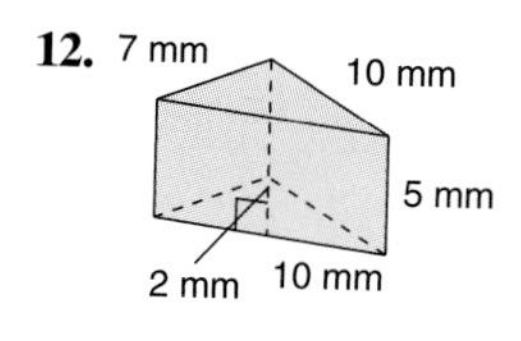

NW **13.**

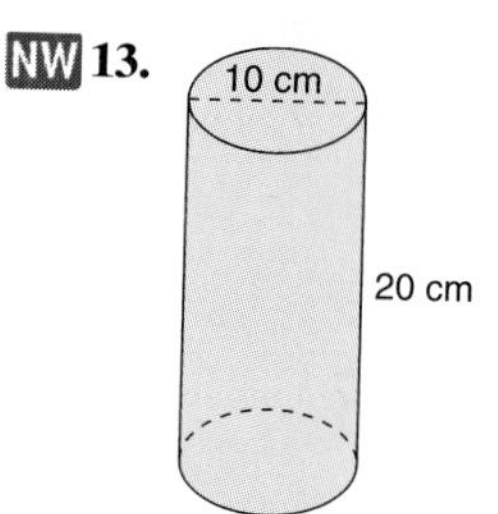

14.

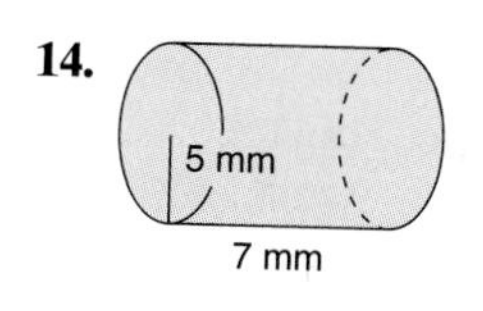

15.

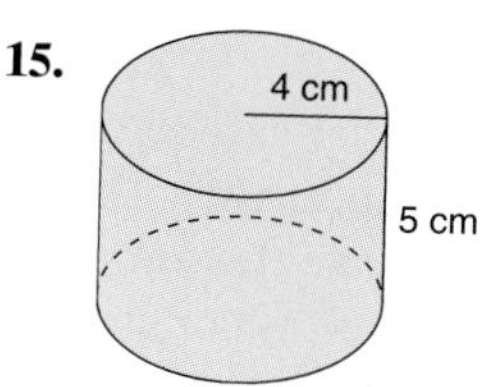

16.

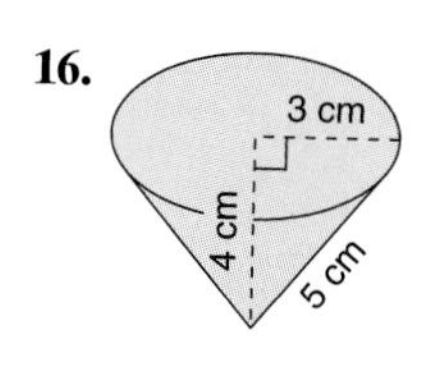

NW **17.**

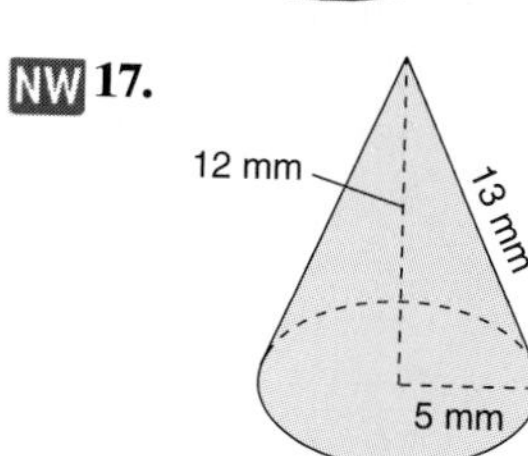

18.

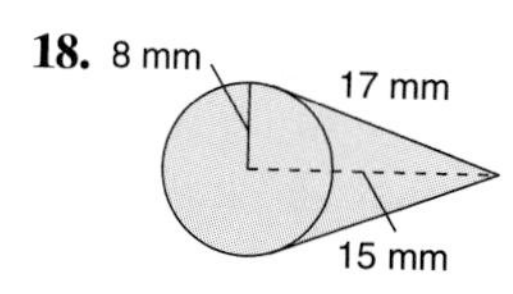

***19.** Will the contents of a can whose base has a radius of 5 cm and a height of 20 cm fit into a canister whose base measures 5 cm by 6 cm and whose height is 20 cm? (Let $\pi = 3.14$.)

***20.** Will the contents of a can whose base has a radius of 10 cm and a height of 10 cm fit into a canister whose base measures 15 cm by 15 cm and whose height is 15 cm? (Let $\pi = 3.14$.)

NW **21.** Find the surface area and volume of a basketball, if its radius is 13 cm. Let $\pi = 3.14$. Express your answer to the nearest tenth.

22. Find the surface area and volume of a marble, if its radius is 0.53 cm. Let $\pi = 3.14$. Express your answer to the nearest tenth.

23. Find the surface area and volume of a ball, if its radius is 9.5 cm. Let $\pi = 3.14$. Express your answer to the nearest tenth.

24. Find the surface area and volume of a pearl, if its diameter is 5 mm. Let $\pi = 3.14$. Express your answer to the nearest tenth.

25. Find the surface area and volume of a spherical container that has a radius of 0.3 m. Let $\pi = 3.14$. Express your answer to the nearest tenth.

***26.** The moon has a radius of approximately 1080 miles. What are its surface area and volume? Let $\pi = 3.14$. Express your answer to the nearest tenth.

Writing Mathematics

27. What is a tetrahedron?

28. What is a prism?

29. In your own words, describe the major characteristics of a pyramid.

Challenge Exercises

30. What is the cost of concrete for a driveway that is 50 feet long, 14 feet wide, and 8 inches deep, if the concrete sells for \$70 per cubic yard? Express your answer to the nearest dollar.

31. If the radius of a sphere is doubled, then what is the corresponding change in its volume?

32. A cylinder whose radius is 6 inches and whose length is 10 inches is $\frac{2}{3}$ full of sand. How many cubic inches of sand are in the cylinder? Let $\pi = 3.14$. Express your answer to the nearest tenth.

33. An oil drum, 5 feet high, is filled with 770 cubic feet of oil. What is the radius of the drum in feet? Let $\pi = 3.14$. Express your answer to the nearest tenth.

34. The larger of two similar pyramids has 8 times the volume of the smaller. If the smaller pyramid has a height of 5 feet, what is the height of the larger pyramid?

35. A conical sand pile has a 12-foot diameter, and it is 4 feet high. How many cubic feet of sand are contained in the pile? How many cubic yards of sand are contained in the pile? Let $\pi = 3.14$. Express your answer to the nearest tenth.

Group/Research Activities

36. A polyhedron is said to be *regular* if its faces are congruent regular polygons and its angles are all congruent. There are regular polygons of all order, but there are only five different regular polyhedra. Identify (name them) these five different polyhedra and provide examples of each.

37. Suppose that a new sidewalk is to be installed on your campus connecting two major buildings. You and some of your classmates are to compute the cost of concrete for such a sidewalk. Select two buildings on your campus and use existing walks as a guide for dimensions. Find the number of cubic yards needed for the new sidewalk and check with a local supplier on the cost of concrete.

Just for fun

A farmer who had three and one-quarter piles of potatoes in one row and four and three-eighths piles in another row decided to put them together. How many piles did he then have?

11.8 CONGRUENT AND SIMILAR TRIANGLES

Congruent Triangles

Congruent figures are figures that can be made to coincide. This means that it is possible to place one figure upon the other and have the two exactly match. Hence, two congruent figures have the same shape and the same size. The symbol for congruence is $\cong$ and is read "is congruent to." Therefore, $A \cong B$ is read "A is congruent to B." Sheets of paper in your notebook are congruent to each other.

Figure 11.92 contains two triangles that are congruent. If made to coincide, which parts would match? If we mentally turn ΔDEF we can see that

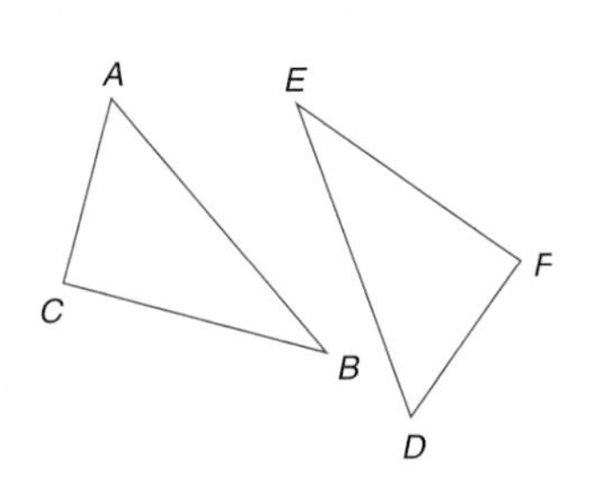

Figure 11.92

$$\angle F = \angle C, \quad \angle A = \angle D, \quad \text{and} \quad \angle B = \angle E$$

Also,

$$\overline{AC} = \overline{DF}, \quad \overline{AB} = \overline{DE}, \quad \text{and} \quad \overline{CB} = \overline{FE}$$

In general, corresponding parts of congruent polygons are **congruent**.

If certain parts of one triangle are congruent to parts of another triangle, the two triangles will be congruent. These conditions are summarized in the following postulates:

1. **Side-Angle-Side (S.A.S.)** Two triangles are congruent if measures of two sides and the included angle are equal, respectively, to measures of two sides and the included angle of the other.

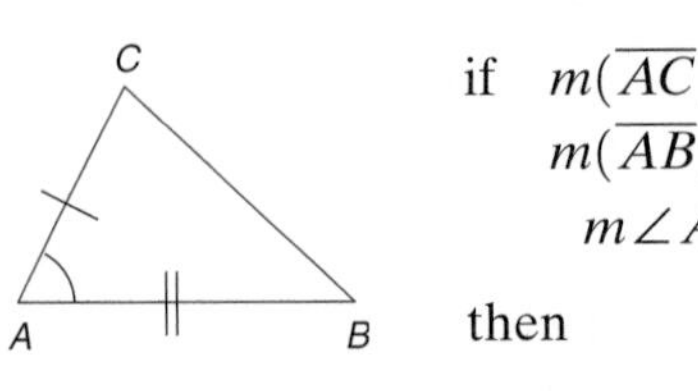

$$\text{if} \quad m(\overline{AC}) = m(\overline{DF})$$
$$m(\overline{AB}) = m(\overline{DE})$$
$$m\angle A = m\angle D$$

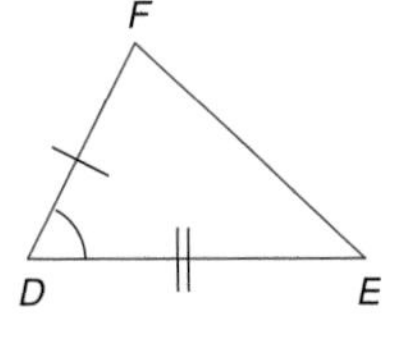

then

$$\Delta ABC \cong \Delta DEF$$

2. **Angle-Side-Angle (A.S.A.)** Two triangles are congruent if two angles and the included side of one are congruent, respectively, to two angles

and the included side of the other.

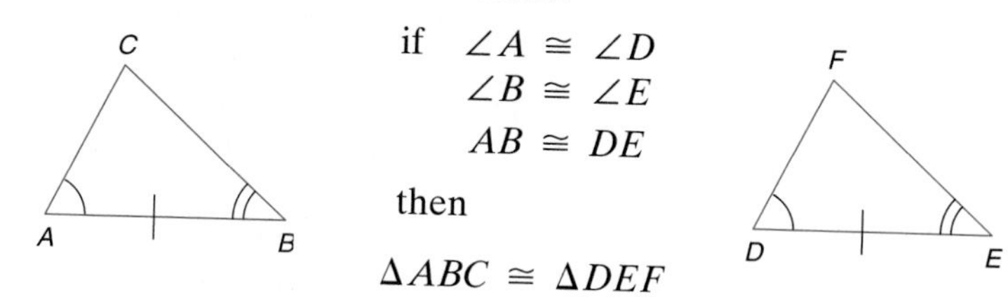

if $\angle A \cong \angle D$
$\angle B \cong \angle E$
$AB \cong DE$

then

$\Delta ABC \cong \Delta DEF$

3. **Side-Side-Side (S.S.S.)** Two triangles are congruent if the three sides of one are congruent, respectively, to the three sides of the other.

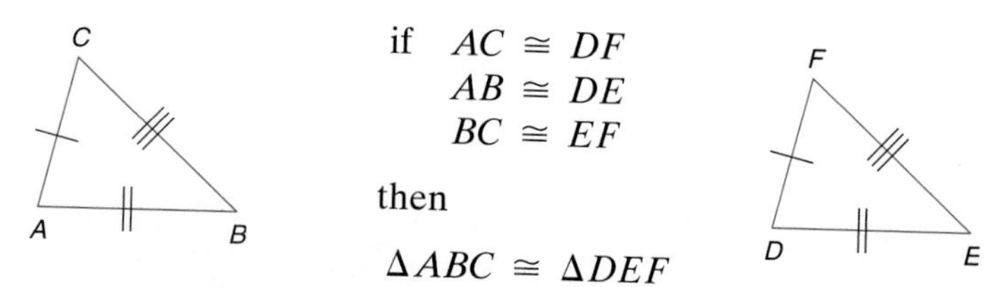

if $AC \cong DF$
$AB \cong DE$
$BC \cong EF$

then

$\Delta ABC \cong \Delta DEF$

EXAMPLE 1 ***Finding the Length of a Side of Congruent Triangles*** In triangles ABC and DEF, angles A and B are congruent, respectively, to angles D and E. If side AB is represented by $3a - 6$ and side DE by $a + 8$, for what value of a are the triangles congruent?

Solution These triangles will be congruent if A.S.A. = A.S.A. Hence, in this case AB must equal DE.

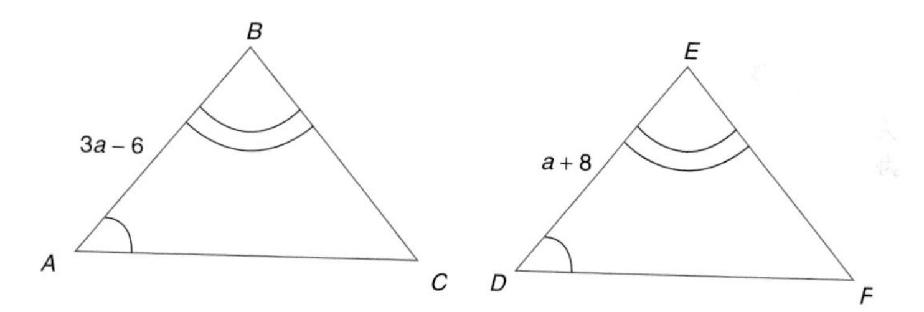

That is,

$$\begin{aligned} 3a - 6 &= a + 8 \\ 3a &= a + 14 \\ 2a &= 14 \\ a &= 7 \end{aligned}$$

If $a = 7$, then the triangles are congruent.

NW ***Now Work Problem 9.***

Similar Triangles

Similar polygons are polygons with the same number of sides that have their corresponding angles equal and their corresponding sides in proportion. The symbol $\sim$ represents the word *similar*. Some examples of similar polygons are shown in Figure 11.93.

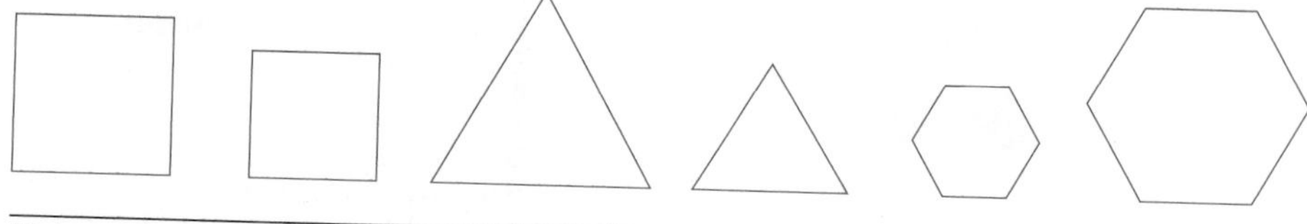

Figure 11.93

Note that polygons are similar if and only if their corresponding angles are equal *and* their corresponding sides are in proportion. Both conditions must be met. If we consider a square and a rectangle, they would have corresponding angles equal, but their corresponding sides would not be in proportion.

But when we consider only triangles we have a special case. It can be shown that

If the three angles of one triangle are congruent with the three corresponding angles of another triangle, then the triangles are similar.

It also follows that if the measures of two angles of one triangle are equal to the measures of two angles of a second triangle, then the two triangles are similar. Why? The measures of the third angles must be equal. Hence, the triangles are similar.

Math Connections

Fractal Geometry

The study of geometry traditionally involves the length, area, and volume of various geometric figures such as circles, triangles, and squares. However, many objects in nature, such as clouds, coastlines, trees, mountains, and weather patterns, represent more complicated, *irregularly* shaped objects that do not lend themselves to traditional study. How, then, can we construct realistic geometric models of these natural shapes? We do this by *fractal geometry*.

The term *fractal* comes from the Latin word *fractus*, which means "broken or fragmented." The mathematician Benoit Mandelbrot first coined this term because nature does not ideally assume the classic geometric shapes. Fractals are geometric figures, just like rectangles, circles, and squares, but fractals have special properties that those figures do not have. A distinguishing characteristic of *regular* fractals is the concept of *self-similarity*. From an intuitive perspective, if the parts of a figure can be shown to be small replicas of the overall figure itself, then the figure is called self-similar.

To facilitate an understanding of the self-similarity concept, let's construct a fractal called the *Sierpinski triangle*. We begin by constructing an equilateral triangle and shading the area of the triangle. This is shown below.

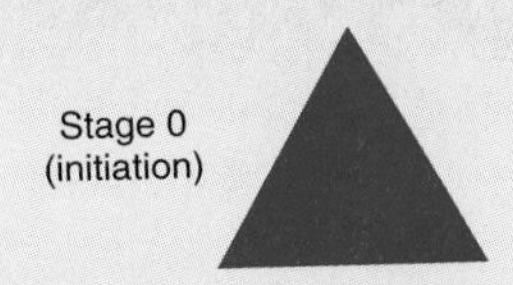

We next connect the midpoints of each side and construct a new triangle, which we shade white. Thus, we transformed the initial triangle into three new subtriangles, T_1, T_2, and T_3, with sides half as long as the first triangle.

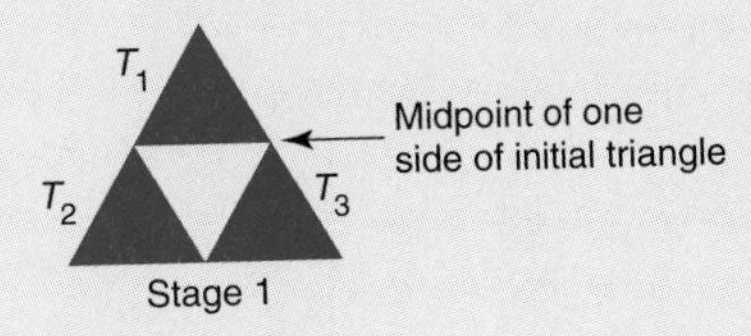

We continue this process. That is, for each dark shaded subtriangle, we connect the midpoints of each side and construct new triangles that are shaded white. This transformation creates three new subtriangles from each of the three previous subtriangles. Thus, we now have a total of nine shaded subtriangles

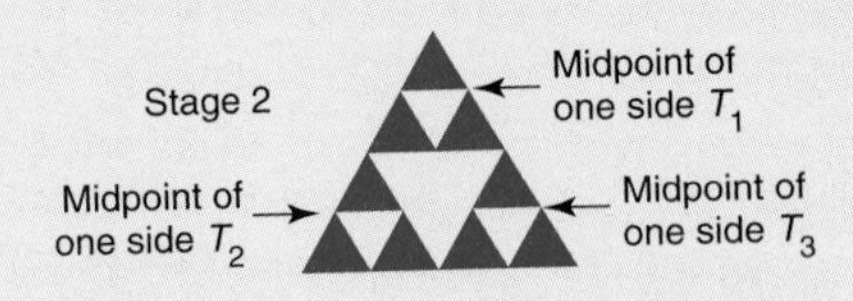

Once again, for each dark shaded subtriangle, we connect the midpoints of each side and construct new triangles that are shaded white. As before, this transformation creates three new subtriangles from each of the previous subtriangles. Since we had nine shaded subtriangles in the previous iteration, we now have 27 shaded subtriangles

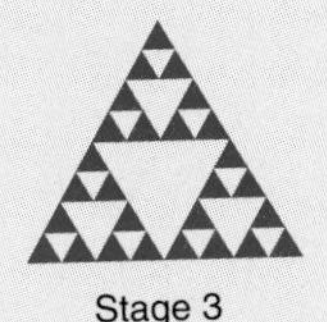

By continuously repeating this process without end, a Sierpinski triangle emerges.

Note how the concept of self-similarity becomes apparent as a result of this construction. If we take any part of the Sierpinski triangle—that is, any piece at all of any size that contains some shaded part—this piece is indeed a replica of the entire triangle first shown in Stage 0. Note also the manner in which the Sierpinski triangle was developed. Once we identified the initial figure, we then defined a rule that was applied over and over again. Each successive figure was based on applying the rule to the previous figure. This process, formally called *recursion*, is another attribute of fractals. (*Note:* Recursion is discussed in more detail in Chapter 1.)

As another illustration, consider the *Koch snowflake curve*, which was named after the Swedish mathematician Helge von

(continued)

(continued)

Koch (1870–1924), who first discovered its properties. The recursion process for developing this fractal is as follows:

The Koch Snowflake Curve

1. ***Stage 0 (Initiation):*** Begin with an equilateral triangle.
2. ***Stages 1–n:*** Apply the following rule:
   a. Scale the size of the current triangle by one-third.
   b. Place copies of the newly scaled triangle at the middle of each side of the previously created triangles.

The application of this process is shown below.

By continually appending new triangles to the middle part of each side, we get a Koch curve. A portion of this curve is shown below. An interesting aspect of this fractal is that its perimeter is of infinite length but its area is finite. That is, the Koch curve is a line of infinite length surrounding a finite area. Specifically, the area is 8/5 of the area of the initial equilateral triangle.

Since their introduction, fractals have been used to study natural phenomena ranging from coastlines to weather patterns to blood supply (e.g., consider the branching of arteries, veins, smaller blood vessels, and capillaries). Finally, images of some of the most famous fractals can be viewed at the Web site **http://library. thinkquest.org/26242/full/fm/fm.html**.

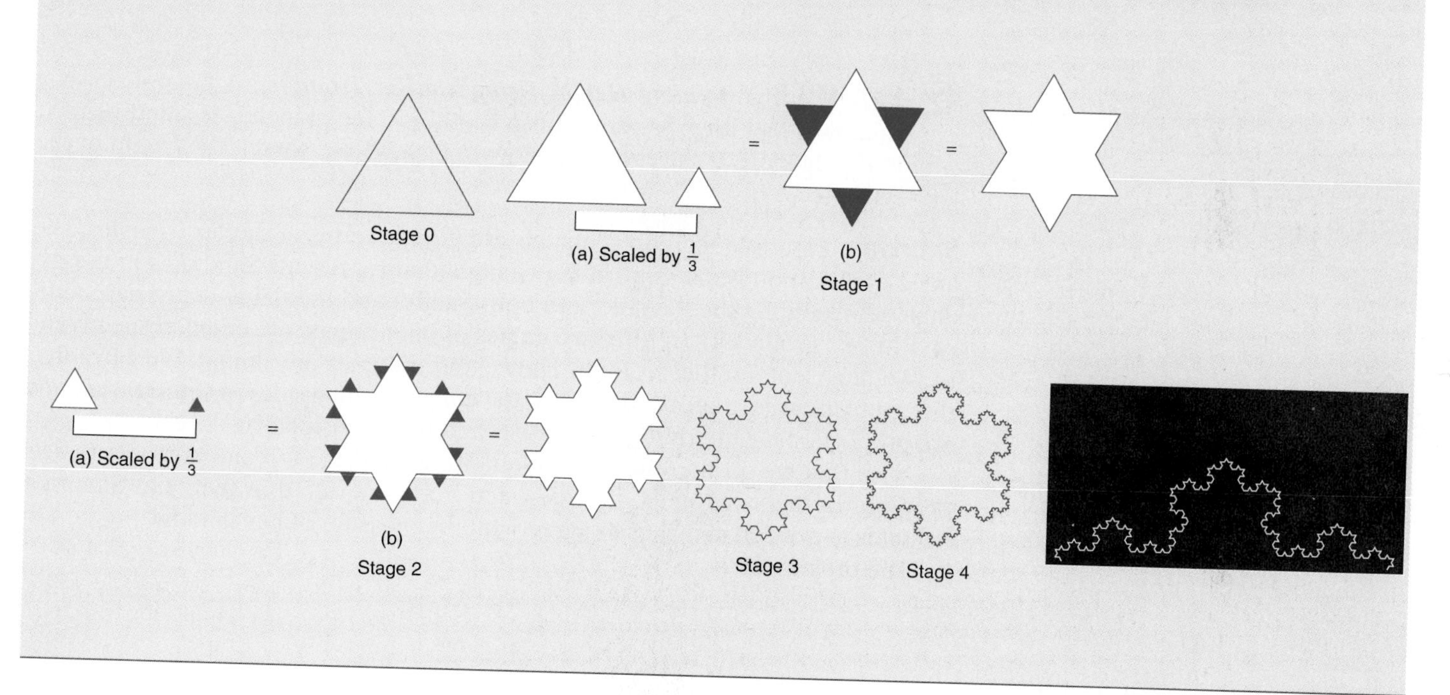

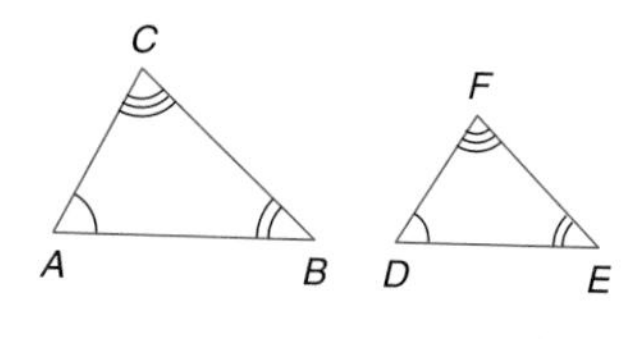

Figure 11.94

Triangle ABC is similar to triangle DEF in Figure 11.94. We write $\Delta ABC \sim \Delta DEF$. In similar triangles the corresponding angles are congruent or equal and the lengths of the corresponding sides are proportional. That is, we can write

$$\angle A \cong \angle D, \quad \angle B \cong \angle E, \quad \angle C \cong \angle F, \quad \text{and} \quad \frac{AB}{DE} = \frac{BC}{EF} = \frac{CA}{FD}$$

In dealing with similar triangles, we will use any two of the three ratios to find a missing side. For a detailed discussion of ratios and proportions, see Section 12.2.

We are given that the two triangles in Figure 11.94 are similar. Suppose the lengths of AC, CB, and AB are 4, 3, and 5, respectively, and the lengths of DF and FE are 16 and 12, respectively. We can find the length of DE by using a proportion and solving it. That is,

$$\begin{aligned} \frac{4}{16} &= \frac{5}{x} & \text{or} \quad \frac{3}{12} &= \frac{5}{x} & \\ 4x &= 5(16) & 3x &= 12(5) & \textit{cross-multiplying} \\ 4x &= 80 & 3x &= 60 & \\ x &= 20 & x &= 20 & \end{aligned}$$

The length of DE is 20.

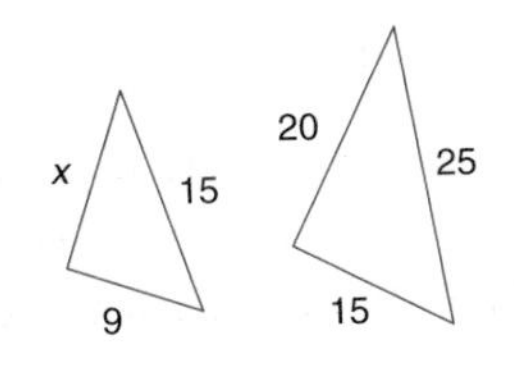

Figure 11.95

EXAMPLE 2 ***Finding the Length of a Side of Similar Triangles*** The triangles in Figure 11.95 are similar. Find the missing length.

Solution

$$\frac{x}{20} = \frac{15}{25}$$

$$25x = 15(20) \quad \textit{cross-multiplying}$$

$$25x = 300$$

$$\frac{25x}{25} = \frac{300}{25} \quad \textit{dividing both sides by 25}$$

$$x = 12$$

EXAMPLE 3 ***Finding Height Using Similar Triangles*** A flagpole casts a shadow 5 feet long. Sandra, whose height is 5 feet 4 inches, is standing next to the flagpole. If she casts a shadow 16 inches long, what is the height of the flagpole?

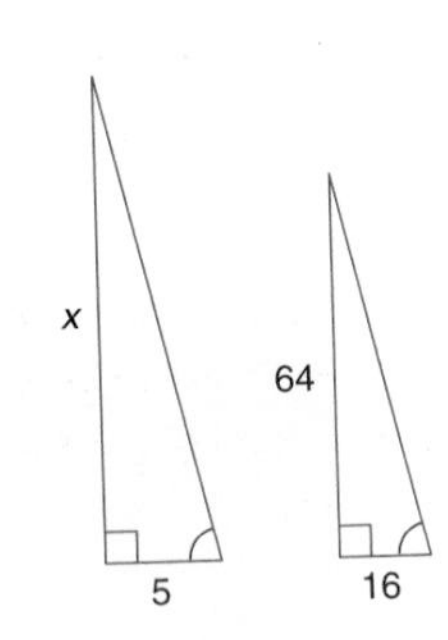

Figure 11.96

Solution Figure 11.96 represents the situation in terms of similar triangles. The angle of elevation is the same in the triangle that represents Sandra as in the triangle representing the flagpole, and both Sandra and the flagpole are perpendicular to the ground. Therefore, two angles of one triangle are equal, respectively, to two angles of another triangle. Hence, the triangles are similar. We can solve this problem using a proportion since the corresponding sides of similar triangles are proportional. One ratio could be x:5, and the other is 64:16.

Note that we set up ratios of like quantities, and the top and bottom of each ratio are in the same units. That is, x and 5 are measured in feet, and Sandra's height is expressed in inches to agree with the measure of her shadow.

The proportion is

$$x:5 = 64:16$$

$$\frac{x}{5} = \frac{64}{16}$$

$$16(x) = 5(64)$$

$$16x = 320$$

$$x = 20 \text{ feet}$$

The height of the flagpole is 20 feet.

NW ***Now Work Problem 31.***

Exercises for Section 11.8

Determine whether each statement (1–8) is true or false.

1. The corresponding parts of congruent polygons are equal.
2. Congruent polygons have the same size and shape.
3. S.A.S. = S.A.S. is a way of proving triangles similar.
4. Two triangles are congruent if three angles of one triangle are congruent with three angles of another triangle.
5. If two right triangles have the legs of one congruent with the legs of the other, the triangles are congruent.
6. If two right triangles have a leg and the adjoining acute angle of one congruent with the corresponding parts of the other, the triangles are congruent.
7. If two isosceles triangles have the legs of one congruent with the legs of the other, the triangles are congruent.
8. If two isosceles triangles have a leg and the base of one congruent with the corresponding parts of the other, the triangles are congruent.
9. **NW** In triangles *MNO* and *RST*, angles *M* and *N* are congruent, respectively, with angles *R* and *S*. If side *MN* is represented

by $6x - 12$ and side RS is represented by $2x + 16$, for what value of x are the triangles congruent?

10. In triangles ABC and DEF, angles A and B are congruent, respectively, with angles D and E. If side AB is represented by $4x - 3$ and side DE by $2x + 9$, for what value of x are the triangles congruent?

11. In triangles ABC and DEF, angle $A \cong$ angle D and side $AB \cong$ side DE. If AB is represented by $3x + 7$, DE by $2x + 12$, AC by $4x - 1$, and DF by $3x + 3$, are the triangles congruent? Why?

12. In Figure 11.97, $\overline{DBE}$ bisects $\overline{ABC}$ and $\angle A \cong \angle C$. Which postulate could be used to prove $\Delta ABE \cong \Delta CBD$?

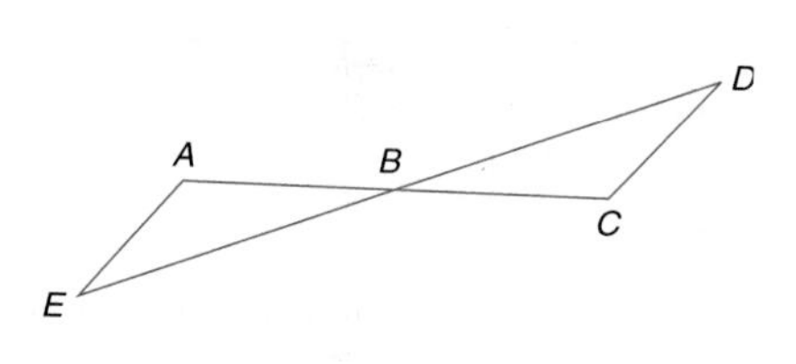

Figure 11.97

Triangles ABC and DEC in Figure 11.98 are congruent. Give the measures of the segments and angles in Exercises 13–18.

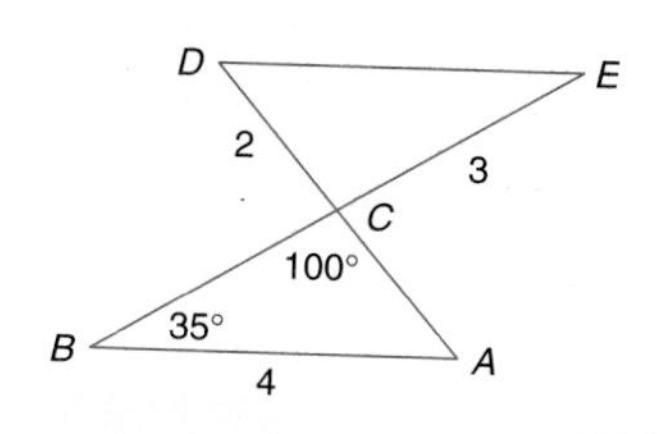

Figure 11.98

13. $\overline{DE}$ **14.** $\overline{CB}$ **15.** $\overline{AC}$

16. $\angle A$ **17.** $\angle D$ **18.** $\angle ECD$

In Figure 11.99, $\Delta ABC \cong \Delta DEF$. Give the measures of the segments and angles in Exercises 19–24.

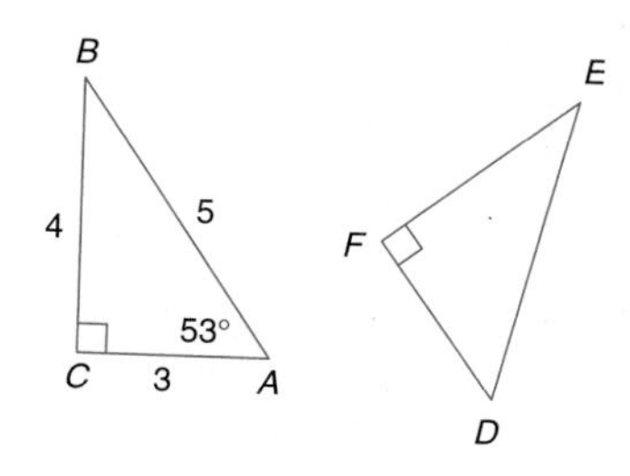

Figure 11.99

19. $\overline{FE}$ **20.** $\overline{DE}$ **21.** $\overline{FD}$

22. $\angle E$ **23.** $\angle B$ **24.** $\angle D$

In each of the Exercises 25–30, the two triangles are similar. Find the missing length.

25.

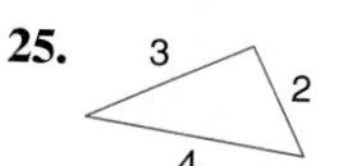

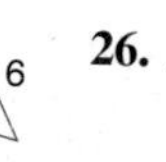

26.

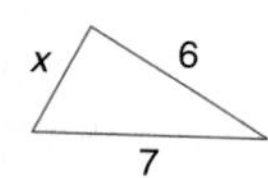

27.

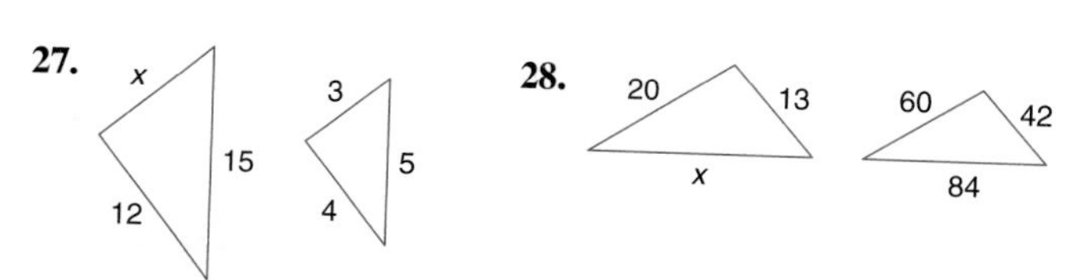

29.

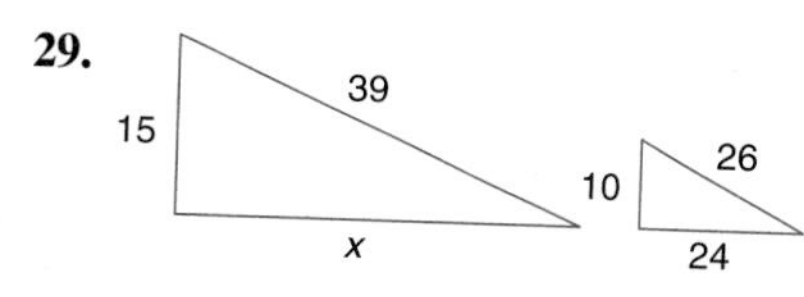

30.

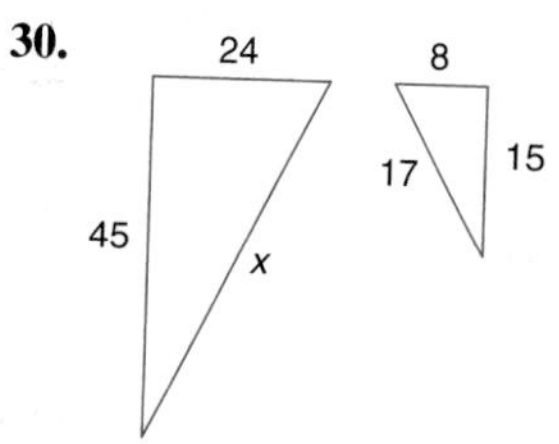

NW **31.** A vertical flagpole casts a shadow 9 m long at the same time that a nearby vertical pole 4 m high casts a shadow 3 m long. Find the height of the flagpole.

32. A vertical rod 5 m high casts a shadow 4 m long. At the same time a nearby tree casts a shadow 20 m long. Find the height of the tree.

33. A tree casts a shadow 40 feet long. At the same time, a nearby student 5 feet 6 inches tall casts a shadow 8 feet long. Find the height of the tree.

34. A vertical pole 3 m high casts a shadow 2.5 m long. At the same time, a nearby signal tower casts a shadow 15 m long. Find the height of the signal tower.

***35.** Two triangles are similar. The sides of one are 6, 8, and 10. If the perimeter of the second triangle is 36, find the lengths of its sides.

***36.** Two triangles are isosceles. The vertex angle (the angle opposite the base) in one measures 50°. An exterior angle at one end of the base of the other measures 110°. Are the triangles similar?

37. The triangles are similar in Figure 11.100. Find the distance across the river.

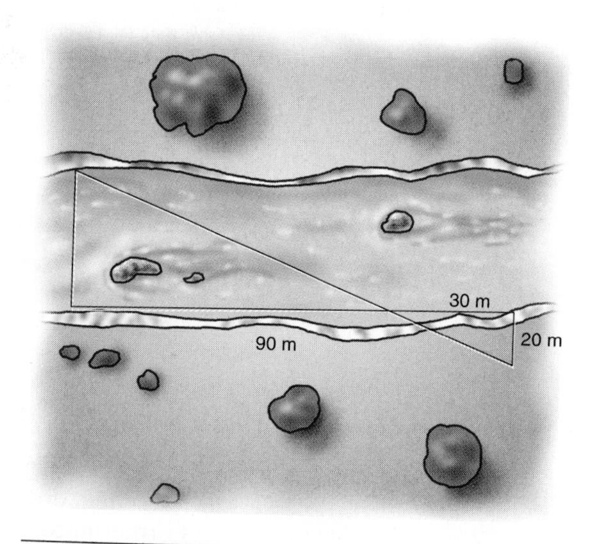

Figure 11.100

38. From the similar triangles in Figure 11.101, find the length of the lake.

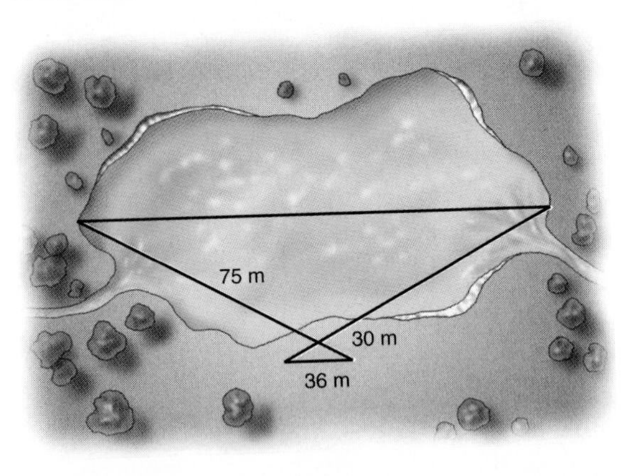

Figure 11.101

Writing Mathematics

39. What are congruent triangles?

40. In your own words, define similar triangles.

Challenge Exercises

41. The bases of a trapezoid are 20 feet and 30 feet, respectively, and the altitude measures 2 feet. Find the altitude of the triangle formed by the longer base and the two nonparallel sides extended until they meet.

42. The bases of a trapezoid are 12 meters and 5 meters, respectively, and the altitude measures 7 meters. Find the area of the triangle formed by the longer base and the two nonparallel sides extended until they meet.

43. The ratio of the areas of two similar triangles is 4 : 9. If the longest side of the smaller triangle is 10 meters, find the length of the longest side of the larger triangle.

Group/Research Activities

44. Archimedes (287–212 B.C.), considered to be one of the greatest mathematicians ever, was one of the first to demonstrate the relationship of the circumference and diameter of circles and how to compute π as accurately as possible. Write a paper on the life of Archimedes and his many contributions to the development of mathematics.

45. Research the history of π and write a paper on the chronology of π.

46. Today the landscape is dotted with numerous cell-phone towers. You and members of your group are to select such a tower and determine its height by means of similar triangles. Keep a log or diary as you and your classmates perform this task. Report back to your class with details and the height of the tower.

Just for fun

What has an inside on the outside?

11.9 NETWORKS

A set of line segments or arcs is called a **graph**. If it is possible to move from any point in the graph to any other point in the graph by moving along the line segments or arcs, then we say the graph is **connected**. A **network** is a connected graph.

A network is **traversable** if it can be drawn by tracing each line segment or arc exactly once without lifting the pencil from the paper. Figure 11.102 shows some examples of networks that are traversable. The end points of the arcs or line segments are called **vertices**.

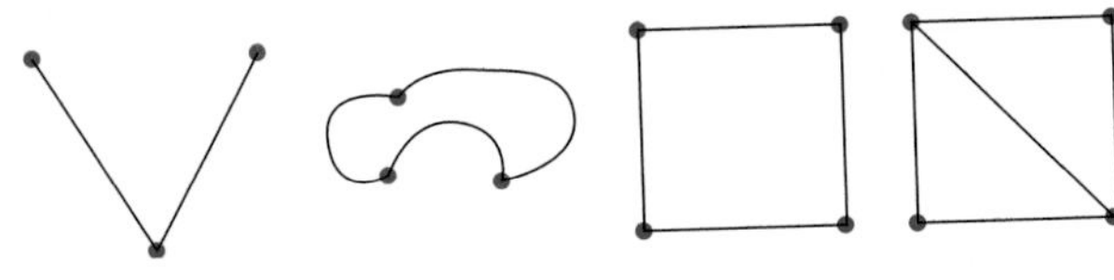

Figure 11.102

A **closed network** divides a plane into two or more regions. A **simple network** is one that does not cross itself. A network that is simple and closed is traversable. Furthermore, you can start at any vertex of a simple closed network to traverse the network. The vertex chosen for the initial point will also be the terminal point.

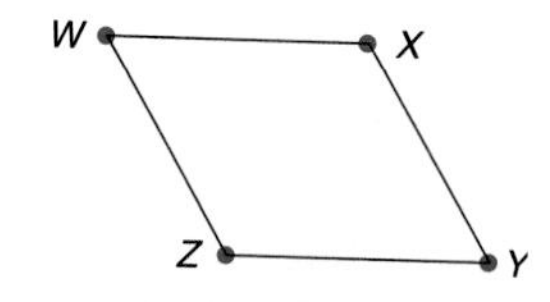

Figure 11.103

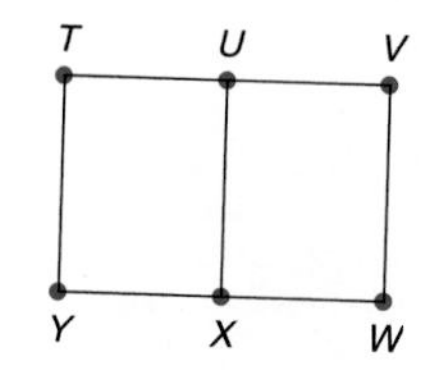

Figure 11.104

Consider the network in Figure 11.103. It is simple and closed. Therefore, we can begin at any vertex and traverse the network.

The network in Figure 11.104 is closed, but not simple. It can also be traversed, but we must begin at either U or X. If we start at U, then we end at X; if we start at X, then we end at U.

A famous puzzle is largely responsible for beginning the study of network theory. This puzzle, the "Seven Bridges of Königsberg," first attracted attention during the 1700s. Königsberg (now Kaliningrad, Russia) was a town in Prussia built on both sides of the Pregel River. Located in the river were two islands, connected to each other and to the city by seven bridges, as shown in Figure 11.105.

Figure 11.105

The problem associated with these bridges and islands was to determine if a person could start at a given point in the town of Königsberg and follow a path that would cross every bridge once and only once on a continuous walk through the town. The citizens of Königsberg tried many routes but found that—no matter where they started, or what path they chose—they could not cross each bridge once and only once. However, it was not until Leonhard Euler (1707–1783), a Swiss mathematician, became interested in the problem that it was proved that each bridge could not be crossed once and only once on a continuous walk through the town. To prove this, Euler analyzed the problem by transforming it into a network similar to that shown in Figure 11.106.

Euler called the points where the paths of the network came together *vertices*. Furthermore, he classified the vertices of a network as **odd** or **even,** depending on whether an odd or even number of paths passed through the vertex. For example, in Figure 11.104, vertex T is even because two paths pass through it, and vertex X is odd because three paths pass through it. In Figure 11.106 all of the vertices are odd.

Euler proved that any network containing only even vertices is traversable by a route beginning at any vertex and ending at the same vertex. He also showed that a network that has exactly two odd vertices is traversable, but the traversing route must start at one of the odd vertices and end at the other.

Finally, Euler showed that if a network has more than two odd vertices, then it is not traversable. This means that the network in Figure 11.106 is not traversable.

Figure 11.106

Therefore, Euler proved that each of the bridges of Königsberg could not be crossed once and only once on a continuous walk because the equivalent network in Figure 11.106 has four odd vertices.

EXAMPLE 1 ***Identifying Vertices of a Network*** For each network, identify the even and odd vertices.

a.

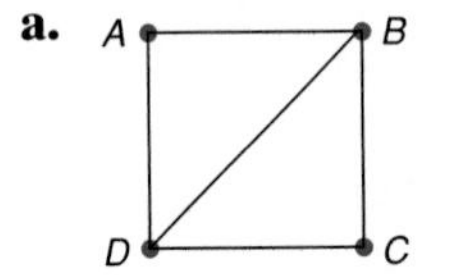

b.

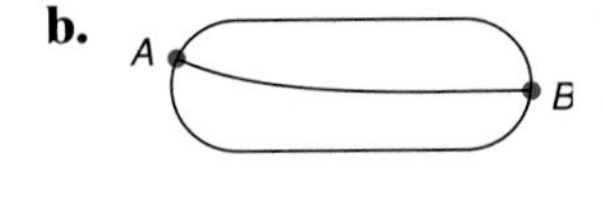

c.

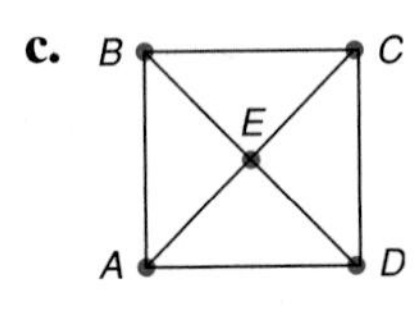

Biography: Sophie Germain

Sophie Germain is considered to be one of the truly outstanding women mathematicians. She decided at an early age to study mathematics, but for a different reason than most people would guess.

Germain was born in Paris on April 1, 1776. In 1789, the Bastille fell and a decade of revolution followed. The turmoil that ensued caused Sophie to be confined to her house, and as a result, she spent many hours reading in her father's library. Here she read of the violent death of Archimedes as he studied a mathematical figure in the sand. Sophie was extremely impressed with Archimedes' absorption in the subject, as he had been oblivious to the events around him. She resolved to study mathematics.

She obtained lecture notes from different professors at the Ecole Polytechnique, even though the school did not accept women. Sophie corresponded with many mathematicians, including the famous Lagrange, but she signed herself M. Leblanc. After a period of some time, Lagrange learned of her true identity and openly praised her work in number theory and mathematical analysis.

In 1815, Germain was awarded a prize by the Institute de France. She was also recommended for an honorary degree from the University of Göttingen, but died in 1831 before the degree could be conferred.

Solution Recall that an even vertex is one that is an end point of an even number of arcs or line segments, and an odd vertex is one that is an end point of an odd number of arcs or line segments.

a. Even vertices: A and C; odd vertices: B and D
b. Even vertices: none; odd vertices: A and B
c. Even vertices, E; odd vertices, A, B, C, and D

EXAMPLE 2 ***Determining if a Network is Traversable*** Determine whether the networks in Example 1 are traversable. If the network is traversable, find the possible starting points.

Solution

a. The network is traversable; B and D are the possible starting points.
b. The network is traversable; A and B are the possible starting points.
c. The network is not traversable because it has more than two odd vertices.

NW ***Now Work Problems 1 and 5.***

Exercises for Section 11.9

*For Exercises 1–10, find (**a**) the number of even vertices, (**b**) the number of odd vertices, (**c**) whether the network is traversable, and (**d**) the possible starting points if the network is traversable.*

NW **1.**

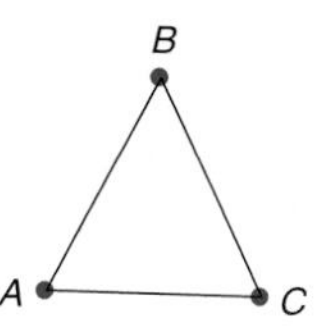

2.

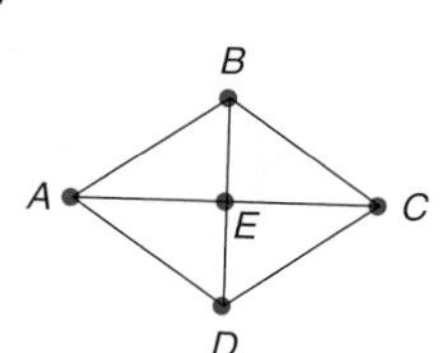

3.

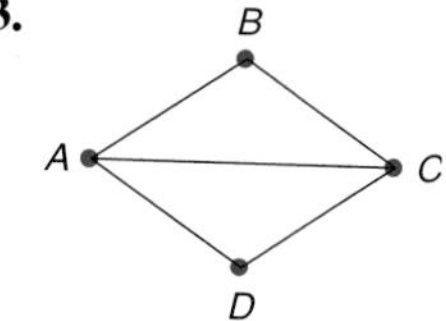

4.

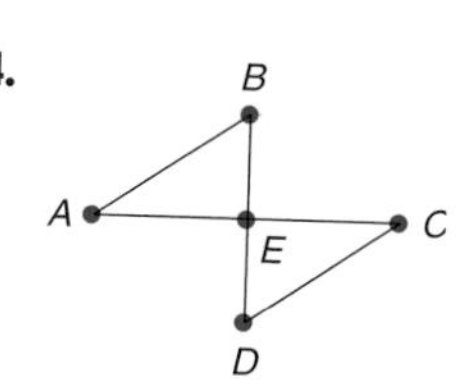

NW **5.**

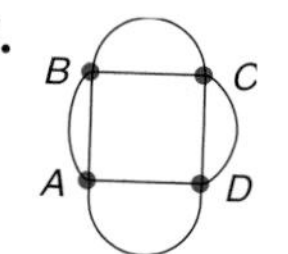

6.

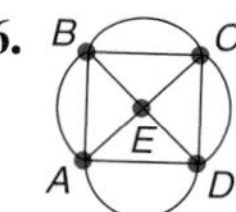

7.

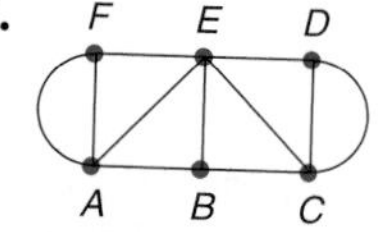

8.

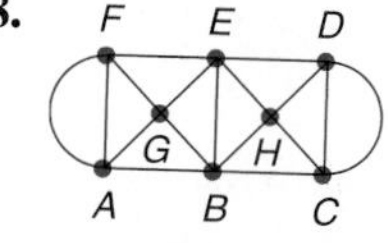

9.

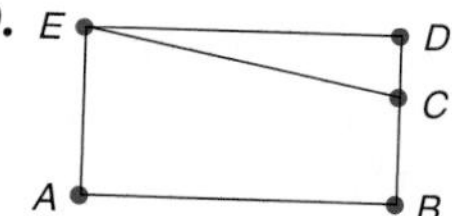

10.

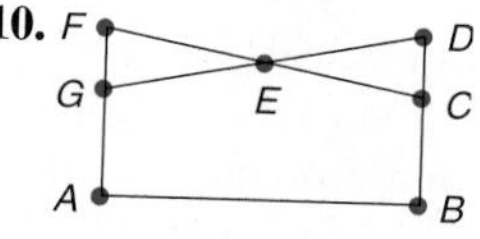

11. Is it possible to walk through a house with the floor plan given in Figure 11.107 and pass through each doorway exactly once?(*Hint:* Use a network.)

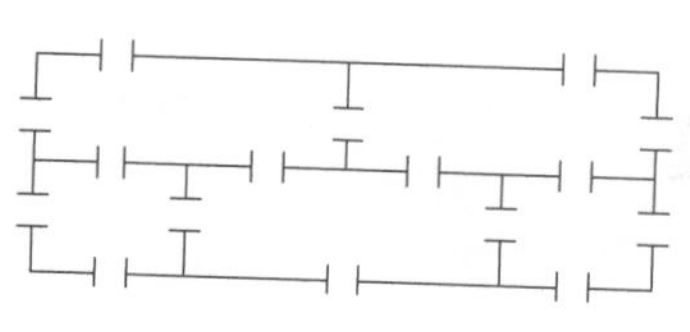

Figure 11.107

12. Is it possible to walk through a house with the floor plan given in Figure 11.108 and pass through each doorway only once?(*Hint:* Use a network.)

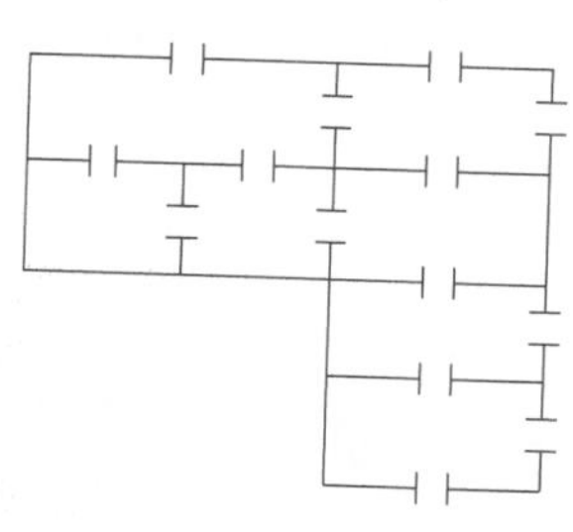

Figure 11.108

Writing Mathematics

13. What is a network?

14. Explain what is meant by a network being traversable.

15. What is a simple network?

16. Explain Euler's rules for determining whether a network is traversable or not.

Challenge Exercises

17. Consider a cube and the network formed by the edges of the cube. Is this network traversable? Why or why not?
18. Construct a network that contains four vertices and is traversable from exactly two points.
19. Construct a network that contains six vertices and is traversable from exactly two points.

Research/Group Activities

20. The Königsberg Bridge problem is an example of a larger problem that many industries face today; that is, the problem of finding efficient ways to route the delivery of goods and services. These industries want to provide speedy service at a minimum cost. Research this topic and write a paper explaining why and how various industries such as airlines and delivery companies need efficient routes or networks.
21. Telephone companies use *minimum networks*. They need to construct or lay phone lines (networks) so that one can go from any point to any other point and the total cost of the network is as small as possible. Write a paper on *minimum networks*.

B.C. by permission of Johnny Hart and Creators Syndicate, Inc.

Just for fun

Can you cut a hole in a standard-size piece of paper in such a way that your entire body can pass through it? Take a standard-size piece of paper and cut it as indicated by the broken lines.

If you have cut correctly, you should be able to pass your body through the hole in the paper. With some practice, you can use a smaller piece of paper and make more cuts.

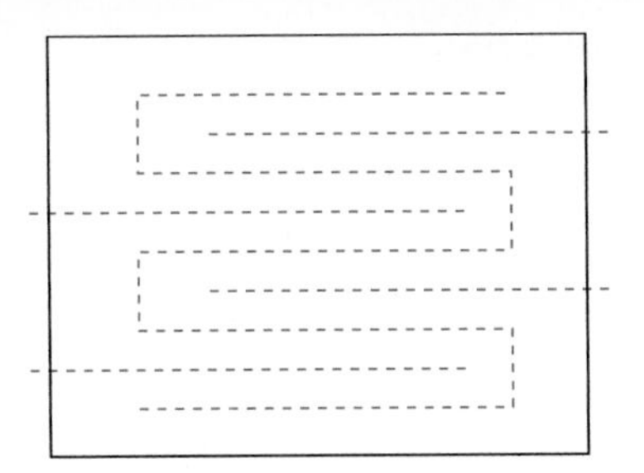

CHAPTER REVIEW MATERIALS

Summary/Chapter 11

11.2 Preliminary Concepts: Points, Lines, and Angles

- In geometry, points and lines are considered undefined terms. Although we do not define these terms, we may still describe properties about them.
- A geometric *point* has no dimension; that is, it has no length, breadth, or thickness. A point does indicate position, however.
- A geometric *line* also has no width, but it does have length; that is, it can be extended infinitely in either direction. We may think of a line as a set of points.
- *Collinear points* are points that lie on the same line.
- Some sets of points commonly used in geometry include a *half-line*, a *ray*, a *line segment*, and an *open line segment*. These are summarized in the following table:

Description	Diagram	Notation
point P	•	P
line PQ	P Q	$\overleftrightarrow{PQ}$
half-line PQ	P Q	$\overset{\circ\!\rightarrow}{PQ}$
ray PQ	P Q	$\overrightarrow{PQ}$
ray QP	P Q	$\overrightarrow{QP}$
line segment PQ	P Q	$\overline{PQ}$
open line segment PQ	P Q	$\overset{\circ\!-\!\circ}{PQ}$

11.3 Planes

- A *plane* is a flat surface such as a white board, a table top, desk top, ceiling, or floor.
- If a line is drawn through any two points of a plane, then the points of the line will also be points of the plane.
- A *unique plane* is determined by any three noncollinear points. In other words, if we are given three distinct points that are not on the same line, then there is one and only one plane that can contain these three points.
- *Parallel planes* are two planes that are parallel to each other; that is, they will never intersect each other. An example of two parallel planes is a floor and ceiling.

11.4 Angles

- An *angle* is the union of two rays that have a common endpoint called a *vertex*.
- An angle is generated by rotating a ray about a point. The two sides of the generated angle are referred to as the *initial side* and the *terminal side*. The amount of rotation from an angle's initial side to its terminal side is referred to as the *measure of the angle*, and the path of a rotation resembles a circle.
- Angles are usually measured in *degrees* (denoted°), and one complete revolution forms a *round angle*, which is equal to 360°.
- A *right angle* is formed by one-fourth of a complete revolution and hence contains 90°, and a *straight angle* is formed by one-half of a complete revolution and hence contains 180°.
- An *acute angle* is an angle whose measure is greater than 0°, but less than 90°; an *obtuse angle* is an angle whose measure is greater than 90°, but less than 180°; and a *reflex angle* is an angle whose measure is greater than 180°, but less than 360°.
- If the sum of the measures of two angles is equal to 360°, then the two angles are *conjugate angles*; if the sum of the measures of two angles is equal to 180°, then the two angles are called *supplementary angles*; and if the sum of the measures of two angles is equal to 90°, then the two angles are called *complementary angles*.
- A *transversal* is a line that intersects two or more lines in different points. When two lines are cut by a transversal eight angles are formed.
- Angles that are on the opposite side of a transversal and do not have the same vertex are called *alternate angles. Alternate interior angles* are on opposite sides of the transversal and "inside" the two lines; *Alternate exterior angles* are on opposite the transversal but "outside" the two lines.
- Two angles that are (1) on the same side of the transversal, (2) do not share a common vertex, and (3) have one angle that is an interior angle and the other that is an exterior angle are called *corresponding angles.*
- Two angles that have a common side between them and a common vertex are called *adjacent angles*.
- When two straight lines intersect they form *vertical angles*, which are two nonadjacent (or opposite) angles.
- If the sides of an angle form a straight line, then the angle is called a *straight angle*.

11.5 Polygons

- A *broken line* is a set of connected line segments. A *simple closed broken* line is one that starts and stops at the same point and does not intersect itself.
- A *polygon* is a simple closed broken line in a plane. The connected line segments that make up a polygon are called the *sides* of the polygon, and the endpoints of these line segments (i.e., the points at which the line segments are connected) are called the *vertices* of the polygon.
- Polygons are classified by their number of sides. For example, a *triangle* is a polygon that has three sides, and a quadrilateral is a polygon that has four sides. Other polygons include *pentagon* (5 sides) and *decagon* (10 sides).
- Triangles may be classified according to the characteristics of their sides. These include a *scalene triangle* (no sides are equal), an *isosceles triangle* (two sides are equal), and an *equilateral triangle* (all three sides are equal).
- Triangles may also be classified according to the characteristics of their angles. These include an *acute triangle* (all three angles are acute), *obtuse triangle* (one angle is obtuse), *right triangle* (one angle is right), and *equiangular triangle* (all angles are of equal measure).
- Quadrilaterals may classified according to the characteristics of their sides. The quadrilaterals discussed in this chapter included trapezoid, parallelogram, rhombus, rectangle, and square.
- A *trapezoid* has two parallel sides, and an *isosceles trapezoid* has two parallel sides and two nonparallel sides that are of equal length.
- In a *parallelogram*, the opposite sides are parallel, the opposite sides are of equal length, the opposite angles are of equal measure, two consecutive angles are supplementary, and the diagonals bisect each other.
- In a *rhombus*, both pairs of opposite sides are parallel, all the properties of a parallelogram hold. In addition, all sides are of equal length (i.e., *equilateral*), and the diagonals are perpendicular to each other.
- In a *rectangle*, all the properties of a parallelogram hold. In addition, all angles are of equal measure (i.e., *equiangular*), and the diagonals are of equal length.
- In a *square*, all the properties of a rectangle and rhombus hold.
- A *regular polygon* is a polygon that is both equilateral and equiangular. Thus, a square is considered to be a regular polygon.

11.6 Perimeter and Area

- The *perimeter of a polygon* is the sum of the lengths of its sides, and the *circumference of a circle* is the distance around the circle. Perimeter and circumference are always measured in linear units (e.g., inches, feet, yards, centimeters, and meters).
- The *area of a polygon* represents the number of square units that are contained in the region enclosed within the sides of the polygon, and the *area of a circle* is the area of the plane surface enclosed within the circle. Area is always measured in square units.
- Common *perimeter and circumference formulas* are as follows: triangle: $P = a + b + c$; parallelogram: $P = 2a + 2b$; rectangle: $P = 2l + 2w$; square: $P = 4s$; and circle: $P = 2\pi r$ or $P = \pi d$.
- Common *area formulas* are as follows: triangle: $A = \frac{1}{2}bh$; trapezoid: $A = \frac{1}{2}h(b_1 + b_2)$; parallelogram: $A = bh$; rectangle: $A = lw$ or bh; square: $A = s^2$; and circle: $A = \pi r^2$.

11.7 Solids

- A *polyhedron* is a simple closed surface in space comprised of polygonal regions. Each polygonal region is called a *face*, and when two faces intersect they form a boundary (i.e., a line segment) called an *edge*. The point at which two or more edges intersect is called a *vertex*.
- The relationship among the number of vertices, edges, and faces of a polyhedron may be expressed via *Euler's formula*: $V - E + F = 2$, where $V =$ the number of vertices of the polyhedron, $E =$ the number of edges of the polyhedron, and $F =$ the number of sides of the polyhedron.
- Two special kinds of polyhedra are *pyramids* and *prisms*.
- A prism is classified according to the polygonal regions forming its bases. If the bases are triangles then the solid is called a *triangular prism*; if the bases are rectangles then the solid is called a *rectangular prism*; and if the bases are squares then the solid is called a *square prism*, or *cube*.
- Solids that involve circular regions are *cylinders, cones*, and *spheres*.
- Common *surface area formulas* for solids are as follows: pyramid: add the area of all faces/bases; prism: add the area of all faces/bases (e.g., $SA_{\text{rectangular prism}} = 2lh + 2wh + 2lw$ and $SA_{\text{cube}} = 6s^2$); cylinder: $SA = 2\pi r^2 + 2\pi rh$; cone: $\pi r^2 + \pi rs$; and sphere: $SA = 4\pi r^2$.
- Common *volume formulas* for solids are as follows: triangular prism: $V = Bh$, where $B =$ area of base; rectangular prism: $V = lwh$; Cube: $V = s^3$; pyramid: $V = \frac{1}{3}Bh$ where $B =$ area of the base; cylinder: $V = \pi r^2 h$; cone: $V = \frac{1}{3}\pi r^2 h$; and sphere: $\frac{4}{3}\pi r^3$.

11.8 Congruent and Similar Triangles

- *Congruent figures* are figures that coincide; that is, they have exactly the same shape and size. Two polygons are congruent when their corresponding sides and angles are also congruent. The symbol for congruency is $\cong$ and is read as *is congruent to*.
- Congruency between triangles may be shown using *side-angle-side, angle-side-angle*, or *side-side-side* strategies, whichever is appropriate for the given problem.
- *Similar polygons* are polygons whose corresponding angles are equal, and whose corresponding sides are in proportion. The symbol for similarity is $\sim$ and is read as *is similar to*.
- If the measures of the three angles of one triangle are equal to the respective measures of the three angles of the second triangle, then the two triangles are similar. Similarly, if the sides of two triangles are respectively proportional, then the triangles are similar.

11.9 Networks

- A *network* consists of a set of points, called *vertices*, and a set of line segments or curves, called *edges*, that connect the points.
- A network is *traversable* if it can be drawn by tracing each line segment or arc exactly once without lifting the pencil point from the paper.
- Any network that has only even vertices is traversable. If a network has exactly two odd vertices, then it is traversable if we begin at one of the odd vertices and end at the other. A network that has more than two odd vertices is not traversable.

Key Terms and Concepts

Review Exercises

11.2 Preliminary Concepts: Points, Lines, and Angles

For Exercises 1–10, use Figure 11.109 with the indicated points to find each of the following:

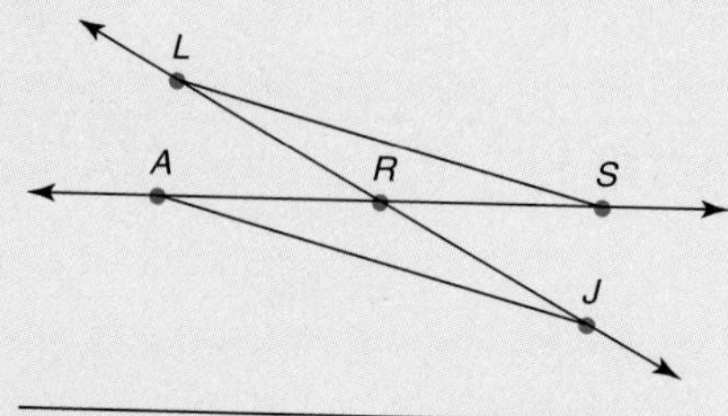

Figure 11.109

1. $\overleftrightarrow{AR} \cap \overleftrightarrow{LJ}$
2. $\overrightarrow{AR} \cap \overrightarrow{SR}$
3. $\overrightarrow{RS} \cap \overrightarrow{RA}$
4. $\overrightarrow{RS} \cup \overrightarrow{RA}$
5. $\overline{LS} \cap \overline{AJ}$
6. $\overrightarrow{RA} \cap \overline{LS}$
7. $\overrightarrow{RA} \cup \overrightarrow{RL}$
8. $\overrightarrow{RS} \cup \overrightarrow{RJ}$
9. $\angle SRJ \cap \angle ARJ$
10. $\angle SRJ \cap \angle LRA$

11.3 Planes

For Exercises 11–20, use Figure 11.110 with the indicated points to find each of the following:

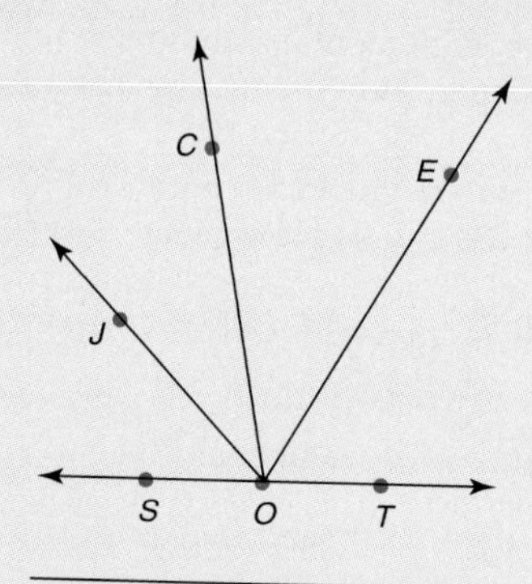

Figure 11.110

11. $\overrightarrow{OS} \cup \overrightarrow{OJ}$
12. $\overrightarrow{OT} \cup \overrightarrow{OE}$
13. $\overrightarrow{OS} \cap \overrightarrow{OJ}$
14. $\overline{OT} \cap \overline{ST}$
15. (interior $\angle COT$) $\cap$ (interior $\angle EOT$)
16. (exterior $\angle COT$) $\cap$ (interior $\angle JOS$)
17. $\angle JOS \cap \angle EOT$
18. $\angle COT \cap \angle COS$
19. (exterior $\angle EOT$) $\cap$ (exterior $\angle JOS$)
20. (interior $\angle JOE$) $\cap$ (interior $\angle COE$)

11.4 Angles

21. Find the measures of the complement and supplement of angles with each of the following measures:

    a. 52° **b.** 46° **c.** 68°
    d. 76°30′ **e.** 13°55′ **f.** 44°44′44″

22. Two angles are complementary, and one angle measures 36° less than the other. How many degrees are there in each angle?
23. Two angles are supplementary, and one angle measures 42° more than twice the other. How many degrees are there in each angle?

11.5 Polygons

24. The lengths of two sides of a triangle are 3 and 6. The third side may be (choose one)

    a. 1 **b.** 2 **c.** 3 **d.** 5

25. Which set of numbers may represent the lengths of the sides of a right triangle?

    a. {3, 4, 6} **b.** {5, 6, 7}
    c. {6, 7, 8} **d.** {6, 8, 10}

26. Find the length of hypotenuse $\overline{AB}$ in right triangle ABC, given that $m(\overline{AC}) = 9$ and $m(\overline{BC}) = 12$.
27. Given isosceles triangle ABC with $m(\overline{AB}) = m(\overline{AC})$ and $m\angle BAC = 48°$, what are the measures of $\angle ABC$ and $\angle ACB$?

11.6 Perimeter and Area

28. The area of a rectangle is 72 cm^2 and the width is 5 cm. What is the length?
29. Find the area of a square whose diagonal measures 10 inches.
30. Find the area of a parallelogram whose base is 13 m and whose height is 7 m.
31. The area of a parallelogram is 200 m^2 and the base is 25 m. What is the height?
32. Find the area of a right triangle whose legs measure 5 cm and 12 cm, respectively.
33. Find the circumference of a circle if its diameter is 10 cm. Use 3.14 for π.
34. Find the area of a circle whose diameter measures 18 m. Use 3.14 for π.
35. The diameter of a wheel is 30 cm. Find its circumference and area. Use 3.14 for π.

11.7 Solids

*For Exercises 36–38, find **(a)** the total surface area and **(b)** the volume of each polygon. Let $\pi = 3.14$.*

36.

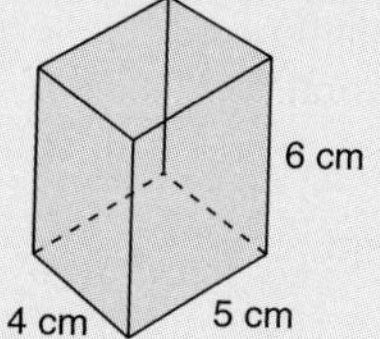

37.

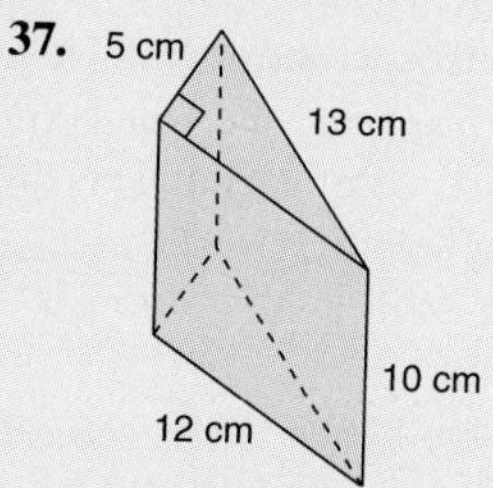

38.

15 cm
10 cm

39. Find the surface area and volume of a cone if its radius is 5 mm, its height is 12 mm, and its slant height is 13 mm. Let $\pi = 3.14$.

40. Find the surface area and volume of a ball if its radius is 12 cm. Let $\pi = 3.14$. Express your answer to the nearest tenth.

11.8 Congruent and Similar Triangles

41. In triangles ABC and DEF, angles A and B are equal, respectively, to angles D and E. If side AB is represented by $4x - 6$ and DE by $2x + 18$, for what values of x are the triangles congruent?

42. A vertical pole 8 m high casts a shadow 6 m long. At the same time a nearby tree casts a shadow 30 m long. Find the height of the tree.

11.9 Networks

*In Exercises 43–46 find (**a**) the number of even vertices, (**b**) the number of odd vertices, (**c**) whether the network is traversable, and (**d**) the possible starting points if the network is traversable.*

43.

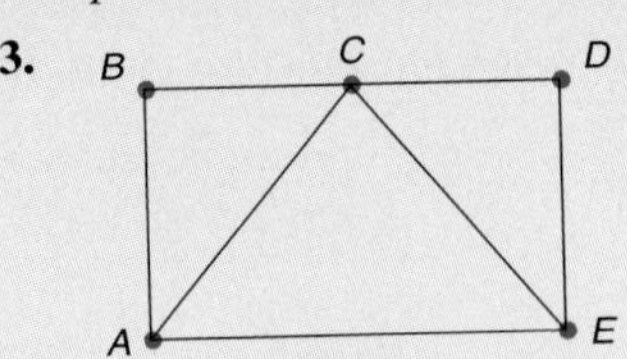

44.

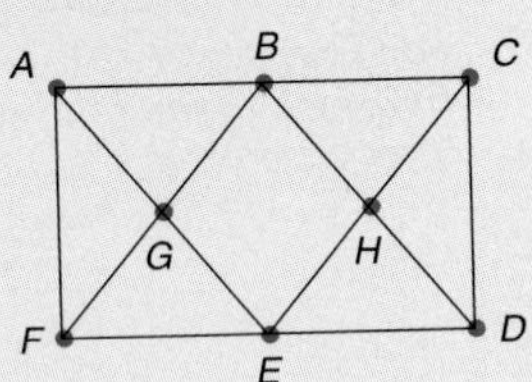

45.

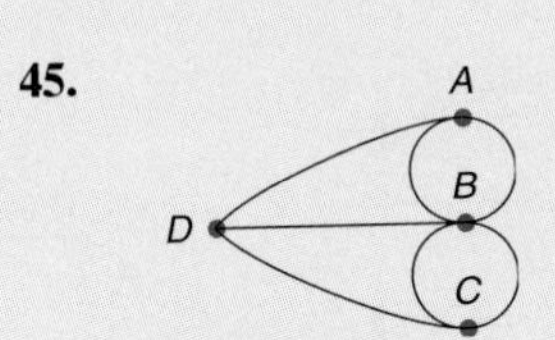

46.

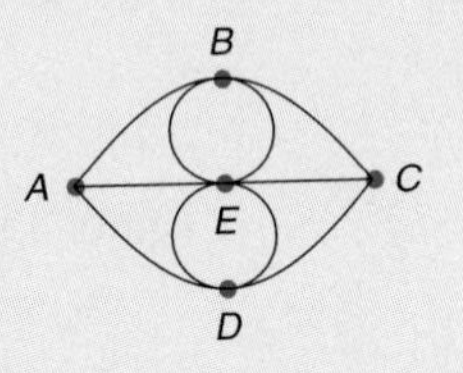

Chapter Quiz

Determine whether each statement (1–20) is true or false.

1. The sum of the squares of two sides of a triangle is equal to the square of the third side.
2. Two rectangles can have equal areas and unequal perimeters.
3. Two rectangles can have equal perimeters and unequal areas.
4. A parallelogram is a rhombus if the diagonals of the parallelogram are equal.
5. Two adjacent angles are angles that have a common vertex.
6. A parallelogram is a polygon whose opposite sides are equal in length.
7. A square is a rectangle with adjacent sides equal in length.
8. A rectangle is a parallelogram with four right angles.
9. Parallel lines are lines that lie in the same plane and do not intersect however far they are extended.
10. An acute triangle is a triangle in which one angle is acute.
11. If the opposite sides of a quadrilateral are not parallel, then the quadrilateral is a trapezoid.
12. If two angles of a triangle are equal in measure, the sides opposite these angles are of equal length.
13. An isosceles trapezoid has both bases equal.
14. Pythagoras is known as the "father of geometry."
15. The area of a triangle is determined by the product of its base and height.
16. An angle is the union of two rays that have a common end point.
17. A solid cone is an example of a polyhedron.
18. A prism is a polygon with two parallel and congruent bases.
19. A scalene triangle is a triangle in which two sides are of equal length.
20. If two lines do not lie in the same plane, they are skew lines.
21. Two angles are complementary, and one angle measures 40° more than the other. How many degrees are there in each angle?
22. Two angles are supplementary, and one angle measures 30° less than twice the other. How many degrees are there in each angle?
23. Find the length of hypotenuse $\overline{AB}$ in right triangle ABC, given that

$$m(\overline{AC}) = 24 \quad \text{and} \quad m(\overline{BC}) = 10$$

24. The area of a rug is 35 ft^2 and the width is 5 ft. What is the length?
25. Find the area of a square book cover whose diagonal measures 12 inches.
26. Find the area of a parallelogram whose base is 15 cm and whose height is 8 cm.
27. Find the area of a window that is a triangle with a base of 12 inches and a height of 6 inches.
28. Find the area of a table top that is a trapezoid, with bases of 36 inches and 26 inches and a height of 18 inches.
29. Find the surface area and volume of a ball if its radius is 10 cm. Let $\pi = 3.14$. Express your answers to the nearest tenth.
30. A pole casts a shadow 40 feet long. At the same time a nearby fence whose height is 6 feet 9 inches casts a shadow 15 feet long. Find the height of the pole.

Just for fun

Cut a strip from a standard-size piece of paper; the strip of paper should measure approximately $\frac{1}{2}$ by 11 inches. Give the paper a half-twist and then tape or glue the two ends together. You have constructed a Möbius strip. It should resemble (a).

The Möbius strip has some interesting properties. For example, it has only one surface. You can demonstrate this by drawing a continuous line, or shading one side, all the way around without lifting your pencil. You will note that this will mark or shade the entire surface.

Next, cut the strip in the middle, as you normally would to obtain two loops, indicated by the broken line in (b). If you have done everything correctly, you should get one bigger loop.

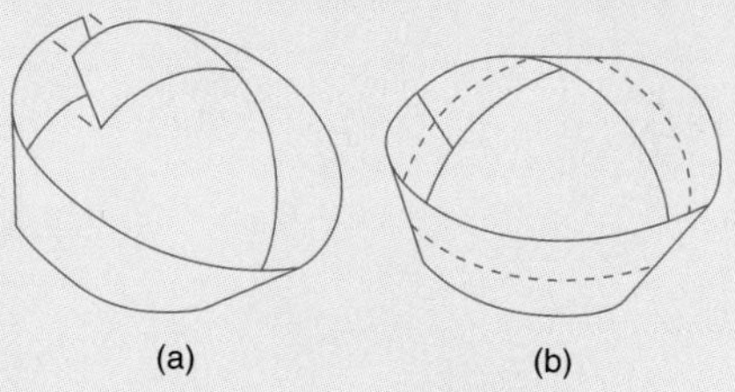

Chapter 12

Chapter Outline

Consumer Mathematics

After Studying This Chapter, You Will Be Able to Do the Following:

1. Express the relationship between two quantities as a **ratio**, and solve a **proportion** for the missing term.
2. Convert a **percent** to a **decimal** or **fraction**, and convert decimals or fractions to percents.
3. Find the **markup** on an item when given the cost or selling price, and determine the **markdown** and **sale price** when given the original retail price and the percent of markdown.
4. Use the formula $I = Prt$ to compute **simple interest,** and compute **compound interest** or **compound amount** using a table.
5. Find the **effective annual interest rate** when money is compounded annually, semiannually, or quarterly.
6. Determine the annual **premium** for the following types of life insurance policies: **term** (5-year), **straight life, limited-payment life** (20-year), and **endowment** (20-year).
7. Find the **true annual interest rate** when an item is purchased on the installment plan, and determine the **finance charge** and **total price** of an item.
8. Determine the monthly payments for **principal** and **interest** for mortgages of various lengths at various interest rates.

Note: *indicates optional material.

Notation Frequently Used in This Chapter

$a \div b$, $\frac{a}{b}$, $a{:}b$	ratio of a to b
A	amount
P	principal
r	interest rate
t	time
n	number of payment periods per year

Visit the PH Companion Website at www.prenhall.com/setek

OVERVIEW

We all have to manage money. We will obtain an item on time (charge it), buy a car, borrow money, purchase a home, and acquire life insurance. Many people put their savings in investments of one type or another. To use your money wisely, you need to make intelligent decisions. A knowledge and understanding of the basics of consumer mathematics will enable you to shop for an item, borrow money, or invest money more wisely.

12.1 INTRODUCTION

Most of us have probably heard the phrase *caveat emptor* at one time or another. It means "let the buyer beware," which can also be interpreted as "the buyer buys at his or her own risk." Unfortunately, many businesses maintain this attitude when dealing with customers. As consumers, we must be wise and discerning shoppers. That is, we must look for the best buy when we purchase an item. When buying items on credit, purchasing insurance, or taking out a loan, consumers must have an understanding of *decimals, percents, simple interest, compound interest, and effective rate of interest*. When faced with several alternatives—for example, buying on the installment plan or paying cash—we must make an intelligent decision based on our own particular situation. The topics in this chapter are designed to give you the information necessary to be a more intelligent shopper and a wiser consumer.

12.2 RATIO AND PROPORTION

Ratio

The **ratio** of two quantities a and b is the quotient or indicated quotient obtained by dividing a by b. The ratio of a to b is written as

$$a \div b \quad \text{or} \quad \frac{a}{b} \quad \text{or} \quad a:b$$

For example, to indicate the ratio of 3 to 5, we could use

$$3 \div 5 \quad \text{or} \quad \frac{3}{5} \quad \text{or} \quad 3:5$$

A ratio provides us with a way of comparing two numbers by means of division. The ratio of one number to another number is the quotient of the first number divided by the second number. Therefore, the ratio of 12 to 3 is $12 \div 3$, or 4 to 1. That is,

$$12 \text{ to } 3 = \frac{12}{3} = \frac{4}{1} = 4 \text{ to } 1$$

EXAMPLE 1 ***Ratio in Simplest Form*** Express each ratio in simplest form.

a. 12 to 36 **b.** 49:14 **c.** $\frac{14}{12}$ **d.** $15 \div 75$

Solution

a. $12 \text{ to } 36 = \frac{12}{36} = \frac{1}{3}$ **b.** $49:14 = \frac{49}{14} = \frac{7}{2}$

c. $\frac{14}{12} = \frac{7}{6}$ **d.** $15 \div 75 = \frac{15}{75} = \frac{1}{5}$

EXAMPLE 2 ***Ratio in Simplest Form*** Find the ratio of 210 minutes to 3 hours.

Solution The quantities compared by a ratio must represent objects measured in the same units. In this case, we have minutes compared to hours. However, to form a ratio, either both quantities should be in terms of minutes, or both should be in terms of hours. We choose to convert both to minutes. Since there are 60 minutes in an hour, we have

$$\frac{210}{180} = \frac{21}{18} = \frac{7}{6}$$

EXAMPLE 3 ***Ratio in Simplest Form*** Express the ratio of $1\frac{1}{2}$ to $2\frac{1}{4}$ in simplest form.

Solution The ratio of one number to another number is the quotient of the first number divided by the second number. Therefore, the ratio of $1\frac{1}{2}$ to $2\frac{1}{4}$ is the same as

$$1\frac{1}{2} \div 2\frac{1}{4} \quad \text{or} \quad \frac{3}{2} \div \frac{9}{4}$$

This problem involves division of rational numbers. (See Section 8.5, Examples 8 and 9.) Hence,

$$1\frac{1}{2}:2\frac{1}{4} = 1\frac{1}{2} \div 2\frac{1}{4} = \frac{3}{2} \div \frac{9}{4} = \frac{3}{2}\cdot\frac{4}{9} = \frac{12}{18} = \frac{2}{3}$$

NW ***Now Work Problem 3.***

EXAMPLE 4 ***Various Ratios*** A baseball team played 20 games and won 15 of them.

a. What is the ratio of the number of games won to the number of games played?

b. What is the ratio of the number of games lost to the number of games played?

c. What is the ratio of the number of games won to the number of games lost?

Solution

a. The ratio of games won to games played is 15:20.

$$15:20 = \frac{15}{20} = \frac{3}{4}$$

b. Games lost = Games played − Games won
Games lost = 20 − 15 = 5

The ratio of games lost to games played is 5:20.

$$5:20 = \frac{5}{20} = \frac{1}{4}$$

c. The ratio of games won to games lost is 15:5.

$$15:5 = \frac{15}{5} = \frac{3}{1}$$

Proportion

A **proportion** is the equality of two ratios. That is, a proportion is a statement that says two ratios are equal. For example,

$$\frac{1}{2} = \frac{4}{8}$$

is a proportion. It states that the two ratios $\frac{1}{2}$ and $\frac{4}{8}$ are equal. If the ratios $a:b$ and $c:d$ are equal, then we can form a proportion by writing

$$a:b = c:d \quad \text{or} \quad \frac{a}{b} = \frac{c}{d}$$

We can read this proportion by saying that "a is to b as c is to d." In this proportion, b and c are called the **means**, whereas a and d are called the **extremes**.

In any proportion, the product of the means equals the product of the extremes:

$$\textbf{If } \frac{a}{b} = \frac{c}{d}, \textbf{ then } ad = bc.$$

For example, in the proportion $\frac{1}{2} = \frac{4}{8}$, 2 and 4 are the means, whereas 1 and 8 are the extremes.

$$\text{If } \frac{1}{2} = \frac{4}{8}, \text{ then } 1 \cdot 8 = 2 \cdot 4, \text{ or } 8 = 8.$$

Extremes *Means*

Many times we know three terms of a proportion and need to find the fourth term. For example, suppose that in a certain mathematics class, the ratio of the number of men to the number of women is 4:3. If there are 12 women in the class, how many men are in the class?

To solve this problem, note that the ratio 4:3 is the same as the ratio of the number of men to the number of women (12). We can therefore set up the proportion

$$\frac{x}{12} = \frac{4}{3}$$

where x represents the number of men in the class. By applying the rule that the product of the means equals the product of the extremes, we have

$$\frac{x}{12} = \frac{4}{3}$$

$$3x = 12 \cdot 4 \quad \textit{cross-multiplying}$$

$$3x = 48$$

$$\frac{3x}{3} = \frac{48}{3}$$

$$x = 16 \text{ men}$$

 **EXAMPLE 5** ***Solving a Proportion*** A car travels 400 miles on 20 gallons of gasoline. At this rate, how many gallons of gasoline will be consumed on a trip of 900 miles?

Solution Let x = the number of gallons used. We can then write the proportion

$$\frac{20}{x} = \frac{400}{900}$$

Solving this proportion,

$$\begin{aligned} 400 \cdot x &= 20 \cdot 900 \qquad \textit{cross-multiplying} \\ 400x &= 18{,}000 \\ \frac{400x}{400} &= \frac{18{,}000}{400} \\ x &= 45 \text{ gallons} \end{aligned}$$

EXAMPLE 6 ***Solving a Proportion*** Find the value of x in each of the following proportions:

a. $4:8 = 3:x$ **b.** $3:7 = x:28$ **c.** $x:6 = 10:12$

Solution To find the value of x in each proportion, we use the property that, for any proportion, the product of the means equals the product of the extremes.

a. $4:8 = 3:x$

$$\frac{4}{8} = \frac{3}{x}$$

$$4 \cdot x = 8 \cdot 3$$

$$4x = 24$$

$$x = 6$$

b. $3:7 = x:28$

$$\frac{3}{7} = \frac{x}{28}$$

$$3 \cdot 28 = 7 \cdot x$$

$$84 = 7x$$

$$12 = x$$

c. $x:6 = 10:12$

$$\frac{x}{6} = \frac{10}{12}$$

$$12 \cdot x = 6 \cdot 10$$

$$12x = 60$$

$$x = 5$$

EXAMPLE 7 ***Height of a Flagpole*** A flagpole casts a shadow 5 feet long. Sandra, whose height is 5 feet 4 inches, is standing next to the flagpole. If she casts a shadow 16 inches long, what is the height of the flagpole?

Solution We can solve this problem using a proportion. Let x represent the height of the pole. Therefore, one ratio could be the height of the pole to its shadow, $x : 5$. Similarly, the ratio of Sandra's height to her shadow is $64 : 16$. Note that we set up ratios of like quantities and that the top and bottom of each ratio are in the same units. The proportion is

$$\begin{aligned} x:5 &= 64:16 \\ \frac{x}{5} &= \frac{64}{16} \\ 16 \cdot x &= 5 \cdot 64 \\ 16x &= 320 \\ x &= 20 \text{ ft} \end{aligned}$$

NW ***Now Work Problems 11 and 13.***

Exercises for Section 12.2

1. Express each ratio as a fraction.
   a. 3 to 2 b. 4 to 7
   c. 8:5 d. 5:6
2. Express each ratio as a fraction.
   a. 5 to 7 b. 6 to 13
   c. 13:6 d. 15:8

NW 3. Express each ratio in simplest form.
   a. 12 to 4 b. 14:28 c. $\frac{18}{12}$
   d. 36:9 e. $3\frac{1}{2}$ to $4\frac{2}{3}$ f. $1\frac{1}{2}:2\frac{1}{4}$
4. Express each ratio in simplest form.
   a. 14 to 56 b. $\frac{14}{12}$ c. 2:18
   d. 14:49 e. $3\frac{1}{3}:2\frac{2}{3}$ f. $2\frac{1}{3}:2\frac{5}{6}$
5. A mathematics class has 30 students in it. There are 18 women and 12 men in the class.
   a. What is the ratio of men to women?
   b. What is the ratio of women to men?
   c. What is the ratio of the number of men to the number of students in the class?
6. The perimeter of a rectangle is 42 meters and the width is 9 meters. Find the ratio of the length of the rectangle to its width.
7. The perimeter of a rectangle is 44 meters and the length is 19 meters. Find the ratio of the width of the rectangle to its length.
8. A final examination has 40 true–false questions and 60 multiple-choice questions.
   a. What is the ratio of true–false questions to multiple-choice questions?
   b. What is the ratio of multiple-choice questions to true–false questions?
   c. What is the ratio of true–false questions to the total number of questions on the examination?
9. Find the value of x in each of the following proportions:
   a. $x:7 = 6:21$ b. $3:x = 14:28$
   c. $5:1 = x:6$ d. $1:7 = 5:x$
10. Find the value of x in each of the following proportions:
   a. $1:2 = 4:x$ b. $5:7 = x:21$
   c. $2:x = 8:12$ d. $x:6 = 4:8$

NW 11. Find the value of x in each of the following proportions:
   a. $3:8 = 12:x$ b. $6:7 = x:21$
   c. $2:x = 18:12$ d. $x:6 = 5:8$
12. Find the value of x in each of the following proportions:
   a. $x:7 = 35:49$ b. $2:x = 8:44$
   c. $7:9 = x:18$ d. $5:13 = 35:x$

NW 13. **Height of a tree** A tree casts a shadow 6 feet long. Gerry, whose height is 5 feet 6 inches, is standing next to the tree. If she casts a shadow 16 inches long, what is the height of the tree?
14. **Typing rate** Marlene can type at a rate of 55 words per minute. At this rate, how long will it take her to type a report that contains approximately 1100 words?
15. **Ratios** A certain athletic squad has 40 members. There are 18 seniors, 6 juniors, 12 sophomores, and 4 freshmen on the team.
   a. What is the ratio of juniors to seniors?
   b. What is the ratio of seniors to freshmen?
   c. What is the ratio of sophomores to seniors?
   d. What is the ratio of freshmen to the total number of team members?
16. **Hydrogen** When decomposed by an electric current, 36 grams of water yield 4 grams of hydrogen. How much hydrogen can be obtained from 50 grams of water?
17. **Alloy** An alloy is composed of two parts tin and one part lead by weight. How much of each is needed to make 60 kilograms of this alloy?
18. **Distance on a map** On a map, a line segment 3 inches long represents a distance of 18 miles. Using the same scale, how many miles long is a road that measures 2 inches on the map?
19. **Reading rate** Addie takes 3 minutes to read an article of 350 words. At the same rate, how many minutes will it take her to read another article of 875 words?
20. If one-half of a number is 30, then how much is two-thirds of the same number?
21. **Length of a shadow** A student who is 6 feet tall stands next to a flagpole that is 50 feet tall. If the shadow cast by the student is 9 feet long, how long is the shadow cast by the flagpole?

22. **Gear ratio** What is the gear ratio of a bicycle that has 20 teeth on the rear sprocket and 46 teeth on the front chainwheel?
23. **Math assignment** Sally takes 35 minutes to do a certain math assignment. Hugh takes 1 hour to do the same job. What is the ratio of their job times?
24. Find the value of x in each of the following proportions:
    a. $3:4 = 12:x$ **b.** $7:10 = x:5$
    c. $6:(5 - x) = 7:3$ **d.** $(x - 6):4 = 4:5$
25. Find the value of x in each of the following proportions:
    a. $x:4 = 10:16$ **b.** $6:x = 16:7$
    c. $5:0.3 = 20:x$ **d.** $6:9 = (x - 2):12$
26. If for every three males in a mathematics class there are two females, how many males are there in the class if there are eight females?

Writing Mathematics

27. What is a ratio? Give at least two examples.
28. What is a proportion? Give at least two examples.

Challenge Exercises

29. **Salt** If 8 liters of sea water are evaporated, then approximately 140 grams of salt will remain. Write three equal ratios for liters of sea water to grams of salt.
30. **Fresh water** A desalting plant produces 56 million liters of fresh water every seven days. How many liters of fresh water are produced in three days?
31. **Better buy** At Bill's Bargain Store, a certain baseball cap is selling 5 caps for $26.25. At Harry's Hat Outlet, the same cap is on sale for 3 for $16. Which store offers the better price? Why?
32. **Profit sharing** Scott invests $5500 and Joe invests $4500 in a partnership. If profits are to be shared in the ratio of their investments, what should Joe receive if Scott's share of the profits is $2100? (Express your answer to the nearest dollar.)
33. **Advertising** The circle graph in Figure 12.1 shows how Jilrac Industries spent $75,000,000 for television advertising. How much money was spent on

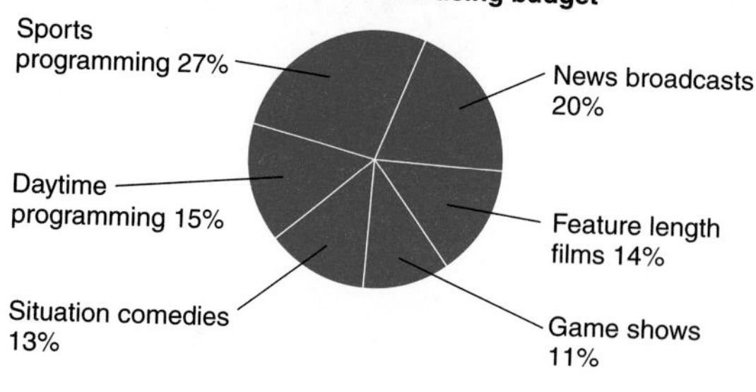

Figure 12.1

    a. news broadcasts?
    b. sports programming?
    c. daytime programming?
    d. feature length films?
    e. situation comedies?
    f. game shows?
34. **Survey** Sheila polled 500 students on campus to determine their favorite type of television program. Complete the following table. Express each answer to the nearest tenth of a percent.

Type of Program	Number of Votes	Percent of Votes
News	99	**a.**
Sports	137	**b.**
Feature length films	87	**c.**
Game shows	55	**d.**
Situation comedies	88	**e.**
Dramas	34	**f.**

Group/Research Activities

35. Use a tape measure to find the circumference of any circle. (Use a can, jar, clock, etc.) Next, measure its diameter. Find the ratio of the circumference to its diameter. Express your answer as a decimal. You should obtain approximately 3.14, which is the value of π (pi) to two decimal places.
36. A profit of $100,000 from the sale of a business is to be allocated among the four partners in a ratio of 4 : 3 : 2 : 1. Determine the amount of profit for each partner.

Just for fun

How long will it take to cut a 12-foot log into 1-foot lengths, allowing 2 minutes for each cut?

12.3 PERCENTS, DECIMALS, AND FRACTIONS

Percents

The concept of percent is one that occurs daily in our lives. For example, when we read the newspaper, listen to the radio, or watch television, we might encounter such statements as

"There is a 30 percent chance of rain tomorrow."
"The sales tax is 7 percent."
"Forman's Department Store's fall sale on women's fashions will feature savings of 20 percent and more on all items in stock."
"The Consumer Price Index rose one-tenth of 1 percent last month."
"First Federal Savings Bank offers loans at 6 percent."

We can describe a **percent** as a ratio with a denominator of 100. That is, a percent is the ratio of a number to 100. The word *percent* means "per one hundred." The symbol for percent is %. Hence, "20 percent" can be written as "20%." Because a percent is the ratio of any number to 100, 20% means the ratio of 20 to 100; that is, $20\% = \frac{20}{100}$. Similarly,

$$18\% \quad \text{means} \quad 18:100, \quad \text{or} \quad \frac{18}{100}$$

$$6\% \quad \text{means} \quad 6:100, \quad \text{or} \quad \frac{6}{100}$$

$$\frac{1}{2}\% \quad \text{means} \quad \frac{1}{2}:100, \quad \text{or} \quad \frac{\frac{1}{2}}{100}$$

EXAMPLE 1 ***Changing a Percent to a Fraction*** Express each percent as a fraction in simplest form.

a. 20% **b.** $33\frac{1}{3}\%$ **c.** $\frac{1}{2}\%$

Solution

a. $20\% = \frac{20}{100} = \frac{1}{5}$

b. $33\frac{1}{3}\% = \frac{33\frac{1}{3}}{100} = \frac{\frac{100}{3}}{100} = \frac{100}{3} \div \frac{100}{1} = \frac{100}{3} \cdot \frac{1}{100} = \frac{1}{3}$

c. $\frac{1}{2}\% = \frac{\frac{1}{2}}{100} = \frac{1}{2} \div \frac{100}{1} = \frac{1}{2} \cdot \frac{1}{100} = \frac{1}{200}$

NW ***Now Work Problems 3 and 5.***

Decimals

It is sometimes necessary to express a percent as a decimal, such as in finding the sales tax on a purchase of $15 if the sales tax is 7%. To convert a percent to a decimal, we use a technique similar to that used for converting a percent to a fraction. Seven percent means $\frac{7}{100}$, and this is the same as 0.07. Therefore, 7% = 0.07. The 7% sales tax on $15 is found by multiplying 15 by 0.07:

$$\begin{array}{r} 15 \\ \times\ 0.07 \\ \hline 1.05 \end{array}$$

The sales tax is $1.05.

EXAMPLE 2 ***Changing a Percent to a Decimal*** Express 15% as a decimal.

Solution

$$15\% = \frac{15}{100} = 0.15$$

To convert a percent to a decimal, we can drop the percent sign and move the decimal point two places to the left.

We can do this because a percent is the ratio of a number to 100, and to divide a number by 100 is the same as moving the decimal point two places to the left. For example, $125 \div 100 = 1.25$. Therefore we can convert 15% to a decimal directly:

$$15\% = 0.15$$

Note that we drop the percent sign and move the decimal point two places to the left. When the number in the percent is a whole number, the decimal point is understood to be to the right of the last digit.

EXAMPLE 3 ***Changing a Percent to a Decimal*** Express each percent as a decimal.

a. 3% **b.** 18% **c.** $\frac{1}{2}\%$

Solution

a. $3\% = 0.03$ **b.** $18\% = 0.18$ **c.** $\frac{1}{2}\% = 0.5\% = 0.005$

Note: To express $\frac{1}{2}\%$ as a decimal, we first had to convert $\frac{1}{2}$ to a decimal: $\frac{1}{2} = 0.5$. For a review of converting fractions to decimals, see Section 8.6.

NW ***Now Work Problem 7.***

Thus far we have converted percents to fractions and to decimals. Next we want to consider changing fractions and decimals to percents.

When expressing a decimal as a percent, it is important to remember that one decimal place to the right of the decimal point represents tenths, two decimal places represent hundredths, three decimal places represent thousandths, and so on. For example,

$$0.5 = \frac{5}{10}, \quad 0.12 = \frac{12}{100}, \quad \text{and} \quad 0.125 = \frac{125}{1000}$$

A percent is the ratio of a number to 100. Therefore, to convert a decimal to a percent, we must obtain an equivalent expression with a denominator of 100. To express 0.15 as a percent, we first rewrite 0.15 as $\frac{15}{100}$. Next we drop the denominator (100) and add a percent sign. Therefore,

$$0.15 = \frac{15}{100} = 15\%$$

To express 0.5 as a percent, we first express it as $\frac{5}{10}$. However, $\frac{5}{10}$ does not have a denominator of 100. Hence, we use a proportion to find an equivalent ratio. That is,

$$\begin{aligned}\frac{5}{10} &= \frac{x}{100}\\ 10 \cdot x &= 5 \cdot 100\\ 10x &= 500\\ x &= 50\end{aligned}$$

Therefore, $0.5 = \frac{5}{10} = \frac{50}{100} = 50\%$.

After examining these examples, you may have discovered another way to express a decimal as a percent:

To write a decimal as a percent, move the decimal point two places to the right and add a percent sign.

For example, $0.15 = 15\%$ and $0.5 = 50\%$. In each case, the decimal point has been moved two places to the right and a percent sign has been added.

EXAMPLE 4 ***Changing a Decimal to a Percent*** Express each decimal as a percent.

a. 0.07 **b.** 0.1 **c.** 3.2 **d.** 0.003

Solution To convert a decimal to a percent, move the decimal point two places to the right and add a percent sign.

a. $0.07 = 7\%$ **b.** $0.1 = 10\%$

c. $3.2 = 320\%$ **d.** $0.003 = 0.3\%$

NW ***Now Work Problem 11.***

Fractions

To change a fraction such as $\frac{1}{4}$ to a percent, we can make use of the process for converting a decimal to a percent because we can express $\frac{1}{4}$ as a decimal. To change a fraction to a decimal, we divide the numerator by the denominator. That is,

$$\frac{1}{4} = 1 \div 4 = 0.25$$

Now that we have a decimal, we can express it as a percent by moving the decimal point two places to the right and inserting a percent sign. Hence,

$$\frac{1}{4} = 1 \div 4 = 0.25 = 25\%$$

EXAMPLE 5 ***Changing a Fraction to a Percent*** Express each fraction as a percent.

a. $\frac{1}{2}$ **b.** $\frac{1}{8}$ **c.** $1\frac{3}{4}$

Solution To change a fraction to a percent, first change the fraction to a decimal, and then change the decimal to a percent.

a. $\frac{1}{2} = 0.5 = 50\%$ **b.** $\frac{1}{8} = 0.125 = 12.5\%$

c. First we rewrite $1\frac{3}{4} = \frac{7}{4}$ and then proceed as before:

$$1\frac{3}{4} = \frac{7}{4} = 1.75 = 175\%$$

NW ***Now Work Problem 17.***

EXAMPLE 6 ***Finding the Amount, Given the Percent*** If there are 30 questions on an exam and a student answered 80% of them correctly, how many did she answer correctly?

Solution To calculate the solution, we express 80% as a decimal, 80% = 0.80, and multiply 30 by 0.80. That is,

$$30 \times 0.80 = 24$$

NW ***Now Work Problem 21.***

Exercises for Section 12.3

For Exercises 1–6, express each percent as a fraction in simplest terms.

1. **a.** 15% **b.** 25% **c.** 75%

2. **a.** 20% **b.** 40% **c.** 36%

NW **3.** **a.** $4\frac{1}{2}\%$ **b.** $2\frac{1}{3}\%$ **c.** $6\frac{1}{4}\%$

4. **a.** $7\frac{3}{4}\%$ **b.** $8\frac{1}{2}\%$ **c.** $20\frac{1}{3}\%$

NW **5.** **a.** 6.5% **b.** 2.3% **c.** 150%

6. **a.** 13.7% **b.** 3.6% **c.** 250%

In Exercises 7–9, express each percent as a decimal.

NW **7.** **a.** 17% **b.** 3% **c.** $4\frac{1}{2}\%$

8. **a.** 12% **b.** 4% **c.** $5\frac{1}{2}\%$

9. **a.** 6.5% **b.** 300% **c.** 0.25%

In Exercises 10–13, express each decimal as a percent.

10. **a.** 0.05 **b.** 0.32 **c.** 0.5

NW **11.** **a.** 0.09 **b.** 2.14 **c.** 0.9

12. **a.** 0.005 **b.** 0.314 **c.** 5.12

13. **a.** 1.125 **b.** 0.010 **c.** 3.01

In Exercises 14–19, express each fraction as a percent.

14. **a.** $\frac{1}{4}$ **b.** $\frac{2}{5}$ **c.** $\frac{3}{8}$

15. **a.** $\frac{3}{4}$ **b.** $\frac{4}{5}$ **c.** $\frac{5}{8}$

16. **a.** $1\frac{1}{2}$ **b.** $\frac{1}{25}$ **c.** $\frac{3}{25}$

NW **17.** **a.** $2\frac{3}{5}$ **b.** $1\frac{1}{8}$ **c.** $\frac{7}{8}$

18. **a.** $\frac{1}{3}$ **b.** $1\frac{4}{5}$ **c.** $\frac{3}{16}$

19. **a.** $\frac{2}{3}$ **b.** $2\frac{3}{4}$ **c.** $\frac{1}{16}$

20. It rained on 40% of the days of November. On how many days did it rain in November?

21. NW Forty percent of the students in a certain mathematics class are females. If there are 50 students in this class, how many are males?

22. Thirty percent of the students in a certain statistics class are males. If there are 30 students in this class, how many are females?

Writing Mathematics

23. Explain how to convert 4.7% to a decimal and then do it.

24. Explain how to convert 0.025 to a percent and then do it.

25. Explain how to convert $1\frac{7}{8}$ to a percent and then do it.

Challenge Exercises

26. Fifteen is what percent of 640? (Express your answer to the nearest tenth of one percent.)

27. What percent of 43 is 94.7? (Express your answer to the nearest tenth of one percent.)

28. What percent of 1.45 is 0.003? (Express your answer to the nearest tenth of one percent.)

29. Find 450% of 18.

30. Find 0.7% of 35.

Research/Group Activities

The following table indicates the ten most frequently used words in written English, and their frequency. (Note: According to most sources there are approximately 50,400 different words in the English language.)

the	*of*	*and*	*to*	*a*
7%	3.6%	2.9%	2.6%	2.3%
in	*that*	*is*	*was*	*he*
2.1%	1.1%	1%	0.98%	0.95%

Select a story from the front page of a newspaper, and find the percent of use of each word listed. How does your percent compare to the percent given in the table?

31. *the* **32.** *and* **33.** *in* **34.** *was* **35.** *to*

36. Write a short paragraph in which more than 2% of the words are the word *is*.

Just for fun

Three sisters met for dinner at a restaurant. Their bill was $30. They divided the amount equally, and each paid $10. The cashier discovered that an error had been made in tabulating their bill. It should have been $25 instead of $30. The cashier informed the waiter of this error and gave him $5 to return to the sisters. On the way back to the table, the waiter decided to keep $2 and return $1 to each sister.

If each sister received $1 back, then each paid $10 − $1, or $9 for dinner, which is a total of $27! The waiter kept $2, which yields a total of $29! Where is the missing dollar?

12.4 MARKUPS AND MARKDOWNS

Markup

The price that we pay for an item when we buy it from a retailer is the retail price, or **selling price**. The amount that a retailer pays for goods is called the **cost** of the item. The difference between the *selling price* of an item and the *cost* of that item is the retailer's **profit margin**. For example, if a color television has a selling price of $400 and it cost the dealer $300, then the profit margin is

$$\$400 - \$300 = \$100$$

The $100 represents a profit margin, but it is not all profit. Out of this $100, the dealer has to meet such expenses as utility costs (heat, light, phone), employees' wages, insurance premiums, taxes, rent or mortgage payments, and so on. Another term commonly used to describe this profit margin is **markup**. Markup is the difference between the selling price and the cost of an item. That is,

$$\text{Markup} = \text{Selling price} - \text{Cost}$$

Using this equation, we can derive two other equations:

$$\text{Selling price} = \text{Markup} + \text{Cost}$$
$$\text{Cost} = \text{Selling price} - \text{Markup}$$

If the selling price of a ring is $90 and the cost is $50, then the markup is $90 − $50 = $40. The amount of markup on an item tells you how much of the price you pay goes to the retailer to cover overhead (the cost of running the business) and profit, and how much goes to the manufacturer or wholesaler. Markup can also be given as a percentage of the cost, or as a percentage of the selling price. A **percent markup on cost** tells you the amount by which the cost of an item was increased to obtain the price you pay, the selling price. A **percent markup on selling price** indicates the amount of the selling price that the retailer retains for overhead and profits.

Most retailers work with percent markups because they deal with large lots of merchandise. A markup of 50% of cost can be applied to a whole group of items whose individual selling prices might vary widely. Then the total markup for the group is easily figured as a percentage of the total cost, without any need to count the items in the group or figure individual markup costs or selling prices.

EXAMPLE 1 ***Finding the Selling Price*** The pro shop at the National Golf Club sells a certain brand of golf clubs at prices based on a markup of 40% of the cost. If the cost of a set of these golf clubs is \$150, what is the selling price?

Solution Because the markup is determined by the cost, we find 40% of \$150:

$$40\% \text{ of } \$150 = 0.40 \times \$150 = \$60.00 = \text{markup}$$
$$\text{Selling price} = \text{Markup} + \text{Cost}$$
$$\text{Selling price} = \$60 + \$150 = \$210$$

EXAMPLE 2 ***Finding the Percent Markup on the Cost*** Another pro shop, at Shinnecock Hills Golf Club, sells a different brand of golf clubs for \$300 per set. If the cost of a set of these golf clubs is \$200, what is the percent markup on the cost?

Solution Markup = Selling price − Cost. Therefore,

$$\text{Markup} = \$300 - \$200 = \$100$$

Now we must find what the percent markup is, based on the cost. That is, we must find what percentage \$100 (the markup) is of \$200 (the cost). To do this, we form a ratio, markup : cost. Therefore,

$$\frac{\text{Markup}}{\text{Cost}} = \frac{\$100}{\$200} = \frac{1}{2} = 0.5 = 50\%$$

The markup (\$100) is 50% of the cost (\$200).

Check:

$$50\% \text{ of } \$200 = 0.50 \times \$200 = \$100 = \text{markup}$$

To find what percentage one number is of another, we form a ratio between the two numbers and convert the ratio (fraction) to a percent. For example, suppose you took a quiz and had 12 out of 15 questions correct. To find the percentage of correct answers, we form a ratio, 12 : 15, and convert it to a percentage:

$$\frac{12}{15} = \frac{4}{5} = 0.8 = 80\%$$

Recall that we can also do this by means of a proportion. That is,

$$\frac{12}{15} = \frac{x}{100}$$
$$15 \cdot x = 12 \cdot 100$$
$$15x = 1200$$
$$x = 80, \quad \text{and} \quad \frac{80}{100} = 80\%$$

EXAMPLE 3 ***Percentage***

a. What percentage of 108 is 27?
b. What percentage of 60 is 48?
c. What percentage of 60 is 90?

Solution

a.
$$\frac{27}{108} = \frac{x}{100}$$
$$108 \cdot x = 27 \cdot 100$$
$$108x = 2700$$
$$x = 25, \quad \text{and} \quad \frac{25}{100} = 25\%$$

b.
$$\frac{48}{60} = \frac{x}{100}$$
$$60 \cdot x = 48 \cdot 100$$
$$60x = 4800$$
$$x = 80, \quad \text{and} \quad \frac{80}{100} = 80\%$$

c.
$$\frac{90}{60} = \frac{x}{100}$$
$$60 \cdot x = 90 \cdot 100$$
$$60x = 9000$$
$$x = 150, \quad \text{and} \quad \frac{150}{100} = 150\%$$

Alternate Solution

a. Express $\frac{27}{108}$ as a decimal and convert the resulting decimal to a percent:

$$\frac{27}{108} = \frac{1}{4} = 0.25 = 25\%$$

b. $\frac{48}{60} = \frac{4}{5} = 0.8 = 80\%$

c. $\frac{90}{60} = \frac{3}{2} = 1.5 = 150\%$

NW ***Now Work Problems 1d and 1f.***

EXAMPLE 4 ***Finding the Percent Markup on the Cost*** Al's Appliance Outlet sells a particular television for $110. If the set costs Al $80, what is the percent markup on the cost?

Solution Markup = Selling price − Cost. Therefore,

$$\text{Markup} = \$110 - \$80 = \$30$$

Since the percent markup on cost is equal to the ratio of markup to cost, we have

$$\frac{\$30}{\$80} = \frac{3}{8} = 0.375 = 37.5\%$$

The markup ($30) is 37.5% of the cost ($80).

Check:

$$37.5\% \text{ of } \$80 = 0.375 \times \$80 = \$30 = \text{markup}$$

NW ***Now Work Problem 5.***

EXAMPLE 5 ***Finding the Cost*** An entertainment system sells for \$320. The markup is 60% of the cost. What is the cost of the system?

Solution Selling price = Markup + Cost. Since the markup is 60% of the cost, we can also state that

$$\text{Selling price} = (60\% \text{ of Cost}) + \text{Cost}$$

This means that the selling price is 60% of the cost plus the full cost. But the full cost is 100% of the cost. Therefore,

$$\text{Selling price} = (60\% \text{ of cost}) + (100\% \text{ of cost})$$

Thus, the selling price is 160% of the cost, so \$320 represents 160% of the cost. To find the cost, we divide \$320 by 160%, or 1.60:

$$\frac{\$320}{160\%} = \frac{\$320}{1.60} = \$200 = \text{Cost}$$

Check:

$$60\% \text{ of } \$200 = 0.60 \times \$200 = \$120 = \text{markup}$$
$$\text{Selling price} = \text{Markup} + \text{Cost}$$
$$\$320 = \$120 + \$200$$
$$\$320 = \$320$$

Thus far we have discussed markup in terms of percentage of cost. However, there are many businesses that figure their markup on the selling price. Suppose a coat costs a retailer \$50 and the markup is 20% of the selling price. What is the selling price for this particular coat? Recall that

$$\text{Cost} = \text{Selling price} - \text{Markup}$$

Then

$$\text{Cost} = \text{Selling price} - (20\% \text{ of Selling price})$$

Note that the selling price is 100% of the selling price. Hence,

$$\text{Cost} = (100\% \text{ of Selling price}) - (20\% \text{ of Selling price})$$

This means that

$$\text{Cost} = 80\% \text{ of Selling price}$$

To find the selling price, we divide \$50 by 80%, or 0.80:

$$\frac{\$50}{80\%} = \frac{\$50}{0.80} = \$62.50 = \text{selling price}$$

We can check our work. The selling price is \$62.50 and the cost is \$50. Therefore, the markup is \$62.50 − \$50.00 = \$12.50. We now find 20% of \$62.50 and check to see if it is \$12.50:

$$20\% \text{ of } \$62.50 = 0.20 \times \$62.50 = \$12.50 = \text{markup}$$

The markup is \$12.50, which is 20% of the selling price, \$62.50.

EXAMPLE 6 ***Finding the Selling Price*** A bike costs a retailer \$90. The markup is 25% of the selling price. Find the selling price.

Solution

$$\begin{aligned}\text{Selling price} &= \text{Markup} + \text{Cost}\\ &= (25\% \text{ of Selling price}) + \text{Cost}\end{aligned}$$

This means that the cost is $100\% - 25\% = 75\%$ of the selling price. That is, \$90 is 75% of the selling price. Therefore, to find the selling price, we divide \$90 by 75%, or 0.75:

$$\frac{\$90}{75\%} = \frac{\$90}{0.75} = \$120 = \text{Selling price}$$

Check:

The markup is $\$120 - \$90 = \$30$. Hence, 25% of \$120 should be \$30:

$$25\% \text{ of } \$120 = 0.25 \times \$120 = \$30 = \text{markup}$$

The solution checks.

EXAMPLE 7 ***Finding the Selling Price*** A color television costs a retailer \$300. The markup is 30% of the selling price. Find the selling price.

Solution

$$\begin{aligned}\text{Selling price} &= \text{Markup} + \text{Cost}\\ &= (30\% \text{ of selling price}) + \text{Cost}\end{aligned}$$

Therefore, the cost is $100\% - 30\% = 70\%$ of the selling price. That is, \$300 is 70% of the selling price. To find the selling price, we divide \$300 by 70%, or 0.70:

$$\frac{\$300}{70\%} = \frac{\$300}{0.70} = \$428.57 = \text{Selling price, to the nearest cent}$$

Check:

The markup is

$$\$428.57 - \$300.00 = \$128.57$$

Therefore, 30% of \$428.57 should be \$128.57:

$$\begin{aligned}30\% \text{ of } \$428.57 &= 0.30 \times \$428.57\\ &= \$128.57 = \text{markup, to the nearest cent}\end{aligned}$$

The solution checks.

NW ***Now Work Problem 11.***

Markdown

Many times, retailers cannot sell everything at the retail selling price. This happens with defective merchandise, overstocked items, discontinued models, unpopular styles, and so on. A retailer still wants to sell the merchandise in stock, but, since it cannot be sold at the original price, it must be reduced in price. Therefore, the merchandise is sold at a new, lower price called the **sale price**. The change, or difference, between the original price and the sale price is called the **markdown**:

$$\text{Markdown} = \text{Original price} - \text{Sale price}$$

Note of Interest

"A penny saved is a penny earned" is a saying that most of us have heard at one time or another. But some people in Congress want to do away with this saying. The Price Rounding Bill was introduced in Congress in 2001 and it would require prices for most goods and services to be rounded to the nearest nickel. This would eliminate the need for pennies.

The total purchase price including coupons and sales tax would be rounded to the nearest nickel. For example, a tab of $2.42 would be rounded to $2.40, whereas a tab of $2.43 would be rounded to $2.45.

Special interest groups such as vending machine makers are urging Congress to pass this bill. It would virtually eliminate pennies from cash register trays and make room for one-dollar coins. Putting coin testers in vending machines is cheaper and easier than putting bill changers in vending machines.

Legislation has been introduced to produce a large gold-colored $1 coin made of 90% copper. To make sure that the coins would be used, unlike the ill-fated Susan B. Anthony $1 coin, the U.S. mint would stop making $1 bills, 18 months after the first coins are made.

This in turn would leave an empty slot in cash register bill trays. According to proponents of the bill, this would be filled by the $2 bill.

If the bill becomes law, pennies would not be entirely eliminated. Prices such as $9.99 would still exist. Payments to the exact cent would still be made by credit card, check, or electronic transfer of funds. Finally, pennies could still be used to pay for items, if you still had some, but you could not use more than 25 of them.

Markdown can be expressed in terms of a dollar amount, and it can also be expressed in terms of percent reduction.

Percent markdowns are usually a better indicator of savings than the dollar amount of the markdown. For example, a $10 markdown on an item that ordinarily sells for $100 is a saving of 10%, but a $10 markdown on a $20 item is a saving of 50%.

If a retailer is selling coats for $30 that were originally priced at $40, we can find the amount of markdown by subtracting $30 from $40. That is,

$$\text{Markdown} = \text{Original price} - \text{Sale price}$$

In this case, the markdown is $\$40 - \$30 = \$10$. But what is the percent reduction? To find the percent reduction, we must find what percent $10 is of $40. Therefore, we divide $10 by $40:

$$\frac{\$10}{\$40} = \frac{1}{4} = 0.25 = 25\%$$

The original price was reduced by 25%.

If we want to find the percent markdown based on the sale price, we divide the markdown ($10) by the sale price ($30):

$$\frac{\$10}{\$30} = \frac{1}{3} = 33\frac{1}{3}\%$$

The markdown of $10 is $33\frac{1}{3}\%$ of the sale price.

EXAMPLE 8 ***Finding the Markdown and Sale Price*** Al's Appliance Outlet had a clearance sale on last year's color television sets. A certain set originally selling for $480 was advertised at a reduction of 20%.

a. What was the dollar markdown?

b. What was the sale price?

Solution

a. We find 20% of the original price, $480:

$$20\% \text{ of } \$480 = 0.20 \times \$480 = \$96 = \text{markdown}$$

b. $\text{Sale price} = \text{Original price} - \text{Markdown}$

$$\text{Sale price} = \$480 - \$96 = \$384$$

NW ***Now Work Problem 15.***

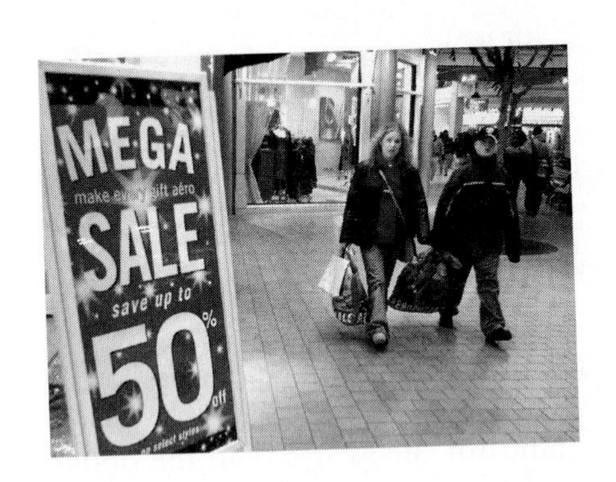

EXAMPLE 9 ***Finding the Percent Markdown*** During a clearance sale, a pair of boots that originally sold at $50 was reduced to $40.

a. What was the percent markdown based on the original price?

b. What was the percent markdown based on the sale price?

Solution

$$\begin{aligned} \text{Markdown} &= \text{Original price} - \text{Sale price} \\ &= \$50 - \$40 = \$10 \end{aligned}$$

a. To find the percent markdown based on the original price, we divide $10 by $50:

$$\frac{\$10}{\$50} = \frac{1}{5} = 0.20 = 20\%$$

b. To find the percent markdown based on the sale price, we divide $10 by $40:

$$\frac{\$10}{\$40} = \frac{1}{4} = 0.25 = 25\%$$

NW ***Now Work Problem 17.***

Exercises for Section 12.4

For Exercises 1 and 2, express your answers to the nearest tenth of a percent.

1. a. What percent of 48 is 24?
b. What percent of 48 is 12?
c. What percent of 30 is 40?
NW d. What percent of 24 is 16?
e. What percent of 216 is 54?
NW f. What percent of 60 is 96?

2. a. What percent of 86 is 43?
b. What percent of 72 is 9?
c. What percent of 81 is 27?
d. What percent of 8 is 3?
e. What percent of 59 is 99?
f. What percent of 49 is 175?

3. Find the dollar amount of markup, to the nearest cent.

	Cost	Percent Markup on Cost
a.	$50.00	25%
b.	$300.00	$33\frac{1}{3}\%$
c.	$25.00	30%
d.	$175.00	$12\frac{1}{2}\%$

4. Find the dollar amount of markup, to the nearest cent.

	Cost	Percent Markup on Cost
a.	$45.00	5%
b.	$39.95	18%
c.	$215.00	$11\frac{1}{2}$%
d.	$59.95	$16\frac{1}{2}$%

NW **5.** Find the markup and the percent markup on the cost. Express each percentage to the nearest tenth.

	Selling Price	Cost
a.	$50.00	$20.00
b.	$100.00	$75.00
c.	$13.50	$10.50
d.	$19.95	$15.95

6. Find the markup and the percent markup on the cost. Express each percentage to the nearest tenth.

	Selling Price	Cost
a.	$220.00	$110.00
b.	$8.00	$6.00
c.	$14.75	$10.00
d.	$99.99	$75.49

NW **7.** A coat sells for $125. The markup is 40% of the cost. What is the cost of the coat, to the nearest cent?

8. A pair of boots retails for $60. The markup is $33\frac{1}{3}$% of the cost. What is the cost of the boots?

NW **9.** A television sells for $358. The markup is 30% of the cost. What is the cost of the television, to the nearest cent?

10. Buying a lawnmower A lawnmower costs a retailer $100. The markup is 25% of the selling price. Find the selling price, to the nearest cent.

NW **11. Buying a snowblower** A snowblower costs a retailer $310. The markup is 40% of the selling price. Find the selling price, to the nearest cent.

12. Buying a basketball A sporting-goods dealer pays $18.50 for each basketball that she buys. If she wants a markup of 35% on the selling price, what should she set as the selling price?

13. Find the markup and cost, to the nearest cent.

	Selling Price	Percent Markup on Selling Price
a.	$400.00	40%
b.	$10.98	20%
c.	$49.95	$12\frac{1}{2}$%

14. Find the markup and the cost, to the nearest cent.

	Selling Price	Percent Markup on Selling Price
a.	$60.00	$33\frac{1}{3}$%
b.	$99.99	25%
c.	$299.95	$12\frac{1}{2}$%

NW **15. Refrigerator sale price** Al's Appliance Outlet had a clearance sale on refrigerators. A certain model, originally priced at $895, was advertised at 20% off. What was the dollar markdown? What was the sale price?

16. Markdown percentage A bike shop advertised one of its bikes for $150. After some time had passed, the bike had not sold, so the dealer lowered the price to $120. What was the percentage of markdown on the original price?

NW **17. Markdown percentage** Dinah's Donut Shop sells fresh doughnuts for $7.00 per dozen. Day-old doughnuts are sold for $4.00 per dozen. What percentage (to the nearest tenth) is the markdown on the original price?

18. Suit sale price A department store sells men's suits for $199.95. During a clearance sale, the price is reduced 25%. What are the dollar markdown and the sale price, to the nearest cent?

19. Used car price During an inventory clearance sale, a used-car dealer reduced the price of all the cars on his lot by $300. If a certain car was originally priced at $1195, what was the sale price? What percent (to the nearest tenth) is the markdown on the sale price?

20. Coat price A coat that was originally priced to sell for $72.50 was reduced to $60.00. What percent (to the nearest tenth) was the markdown on the original price? What percent is the markdown (to the nearest tenth) on the sale price?

21. Lawnmower sale price Monro Hardware paid $150 each for a shipment of lawnmowers. Each mower was marked up by 50% of the cost. At the end of the season, the mowers that had not been sold were reduced 30%. Find the regular selling price of a mower, and the sale price at the end of the season.

22. **Snow tire sale price** Hollywood Service Station paid $60 each for a shipment of snow tires. Each tire was marked up by 50% of the selling price. At the end of the winter, the tires that had not been sold were reduced 40%. Find the regular selling price of a snow tire, and the sale price at the end of the season.

23. **Sales tax** For each of the items in Exercises 15, 17, 19, and 21, find the amount of sales tax, and the total amount due, if a sales tax of 7% is charged on each item at the sale price.

24. **Sales tax** For each of the items in Exercises 16, 18, 20, and 22, find the amount of sales tax, and the total amount due, if a sales tax of $7\frac{1}{2}$% is charged on each item at the sale price.

Writing Mathematics

25. Define *markup.*

26. Define *markdown.*

27. Explain the difference between a percent markup on cost and a percent markup on selling price.

Challenge Exercises

28. Terzo's Pro Shop purchased a certain brand of bowling ball for $102.60 each and sells them for $153.90 each.
    a. What is the percent of profit on each bowling ball if the markup is based on cost?
    b. What is the percent of profit on each bowling ball if the markup is based on the selling price?

29. Ketes agreed to grant an 8% discount on a sale. This discount amounted to $84. What was the original marked price?

30. A watch is on sale for $150 after a markdown of 30% off the regular price. Find the regular selling price.

31. An auto supplies store marks the cost of each article in code on the price tag. If the code words MOST BID AGE are used to represent the digits 0 through 9, how should the store mark a tool that costs $78.94?

Group/Research Activities

32. A store usually sells a certain coat for $200, and advertised a sale on these coats with a 20% markdown. Coats that remained on the rack after the sale were further reduced an additional 30%.
    a. What was the new sale price?
    b. What percent markdown was the new sale price compared to the regular selling price?

33. A retailer typically purchases items from a wholesaler using *trade discounts* and/or *cash discounts*. Research these two discounts and prepare a report for class explaining how each is used in business.

Just for fun

The Bills are in first place and the Dolphins are in fifth, and the Patriots are midway between them. If the Jets are ahead of the Dolphins and the Colts are immediately behind the Patriots, then who is in second place?

12.5 SIMPLE INTEREST

Borrowing money is a necessary transaction for most people and businesses. The borrower may use the money to do something that otherwise would be impossible, whereas the lender charges *interest* and therefore makes a profit on the transaction. Whenever merchandise is bought on credit, the customer pays **interest** on the unpaid balance. Most department stores and credit card companies charge

1.75% to 2% **interest per month** on the unpaid balance, after 30 days. Banks also pay interest to people who deposit their money in savings accounts.

If a person borrows \$1000 from a bank, then the \$1000 is called the *principal*. Similarly, if a person has \$1000 in a savings account, this is also called the principal. The amount of money on which interest is paid is always called the **principal**.

The amount of interest depends on the principal; that is, the interest is a certain percentage of the principal. This percent is called the *rate of interest*, or **rate**. Unless otherwise noted, the rate of interest is an annual one.

When money is borrowed, the borrower agrees to pay back the principal and the interest within a specified period of time. For example, an auto loan may be given for a period of 3 years. At the end of 3 years, both the principal and the interest have been paid, so the interest for such a loan is computed for a period of 3 years. This period is called the **time** of the loan. The interest due depends on three things: the principal, the rate of interest, and the time.

Interest is calculated by multiplying the principal times the rate of interest (expressed as a decimal) times the time (in years). That is,

$$\textit{Interest} = \textit{Principal} \times \textit{rate} \times \textit{time} \quad \text{or} \quad I = Prt$$

This formula determines the *simple interest*. **Simple interest** is the cost of borrowing money computed on the original principal only. This formula can also be used to find the simple interest earned on an investment. If a person borrows \$500 for a period of 1 year at 12% simple interest, we can find the amount of interest by means of the formula $I = Prt$. In this case, we have $P = \$500$, $r = 0.12$ (because $12\% = 0.12$), and $t = 1$ year. Then

$$\text{I} = Prt = \$500 \times 0.12 \times 1 = \$60$$

Therefore, the borrower would have to pay \$60 interest on the loan. The total amount that must be repaid is \$560 (the principal plus the interest). The amount repaid is equal to the principal plus the interest.

$$A = P + I$$

No savings bank or savings and loan can give higher rates on a 6-month deposit of \$10,000 or more.

6-MONTH C.D. RATE
Rates effective July 6-12, 2003

2.50%
PER YEAR

2.52%
EFFECTIVE ANNUAL YIELD

EXAMPLE 1 ***Finding the Interest*** Irene borrows \$3000 for 3 years at a simple interest rate of 13%. How much interest does she pay? What is the total amount that must be repaid?

Solution We use the formula $I = Prt$ with $P = \$3000$, $r = 0.13$, and $t = 3$:

$$\begin{aligned} I &= Prt \\ &= \$3000 \times 0.13 \times 3 \\ &= \$1170 \end{aligned}$$

The total amount that must be repaid is

$$\begin{aligned} A &= P + I \\ &= \$3000 + \$1170 \\ &= \$4170 \end{aligned}$$

NW ***Now Work Problem 7.***

EXAMPLE 2 ***Finding the Interest*** Carl invested \$12,000 for 6 months at a simple interest rate of 10%. How much interest did he earn?

Solution This problem is similar to Example 1, but note that the investment was for a period of time less than 1 year. To use the simple interest formula, we must express 6 months as a fraction of a year:

$$6 \text{ months} = \frac{6}{12}, \quad \text{or} \quad \frac{1}{2}\text{ year}$$

Now we have $P = \$12{,}000$, $r = 0.10$, and $t = \frac{1}{2}$:

$$\begin{aligned} I &= Prt \\ &= \$12{,}000 \times 0.10 \times \frac{1}{2} \\ &= \$600 \end{aligned}$$

NW ***Now Work Problem 3.***

EXAMPLE 3 ***Finding the Interest*** How much will \$5000 earn in 2 years at $9\frac{1}{2}\%$ simple interest?

Solution We use the formula for simple interest with $P = \$5000$, $r = 9\frac{1}{2}\% = 9.5\% = 0.095$, and $t = 2$:

$$\begin{aligned} I &= Prt \\ &= \$5000 \times 0.095 \times 2 \\ &= \$950 \end{aligned}$$

NW ***Now Work Problem 5.***

Amount

Thus far we have used the formula $I = Prt$ to find the simple interest and the formula $A = P + I$ to find the *amount* due on the ending date of the loan or investment. (*Note:* The amount is also known as the *maturity value*.) These two formulas can be combined into one formula that will enable us to find the amount or maturity value in one step. Consider the formulas

$A = P + I$ and $I = Prt$

$A = P + Prt$ *substituting Prt for I*

$A = P(1 + rt)$ *by means of factoring, an algebraic manipulation*

This is the desired formula that will enable us to find the amount in one step.

If Marilyn borrows \$1500 for 2 years at 12% simple interest, what amount must she repay? Using the formula $A = P(1 + rt)$, we have

$$\begin{aligned} A &= P(1 + rt) \\ A &= \$1500(1 + 0.12 \times 2) \\ A &= \$1500(1 + 0.24) \\ A &= \$1500(1.24) \\ A &= \$1860.00 \end{aligned}$$

If we need to find the interest, we would subtract the principal (\$1500) from the amount (\$1860). In this case, the interest is $\$1860 - \$1500 = \$360$.

EXAMPLE 4 ***Finding the Amount*** What is the amount that must be repaid on a \$2400 loan for 18 months at 14% simple interest? How much interest will be paid for the loan?

Solution The amount that must be repaid is

$$A = P(1 + rt)$$

$$A = \$2400\left(1 + 0.14 \times \frac{18}{12}\right) \quad \textit{Note: Time should be expressed in terms of years.}$$

$$A = \$2400(1 + 0.14 \times 1.5)$$

$$A = \$2400(1 + 0.21)$$

$$A = \$2400(1.21)$$

$$A = \$2904$$

The interest is obtained by subtracting the principal from the amount repaid:

$$I = A - P$$

$$I = \$2904 - \$2400$$

$$I = \$504$$

NW ***Now Work Problem 23.***

Exercises for Section 12.5

1. Find the simple interest on a \$2500 loan for 3 years at 9%.
2. Find the simple interest on a \$3000 loan for 2 years at 8%.
3. (NW) Find the simple interest on a \$2500 loan for 6 months at 8.5%.
4. Find the simple interest on a \$4000 loan for 9 months at 9%.
5. (NW) Find the simple interest on a \$6000 loan for 6 months at $12\frac{1}{2}\%$.
6. A merchant borrowed \$2000, agreeing to repay the principal and 12% simple interest at the end of 6 months. Find the amount of interest and the total sum that must be paid.
7. (NW) Luke borrowed \$3000 at $10\frac{1}{2}\%$ simple interest for a period of 3 years. How much interest will he pay? What is the total amount that he must pay?
8. To help pay her tuition bill, Sally borrowed \$1100 at $9\frac{1}{2}\%$ simple interest for 18 months. How much interest will she pay? What is the total amount she will pay?

9. How much simple interest will \$5000 earn in 3 years at a rate of $8\frac{1}{2}\%$? How much interest will it earn in 3 months?
10. How much simple interest will \$6000 earn in 4 years if the rate of interest is 9%? How much will it earn in 4 months?
11. Connie has a balance due of \$240 on her charge account. The rate is $1\frac{1}{2}\%$ simple interest *per month* on the unpaid balance. If Connie decides not to pay anything toward her balance this month, and she does not charge anything to her account next month, what will be her new balance due next month?
12. Sam has a balance due of \$300 on his charge account. The rate is $1\frac{1}{2}\%$ simple interest per month on the unpaid balance. If Sam pays \$180 toward his balance this month, and does not charge anything else next month, what will be the balance due on his charge account next month?
13. Benny deposited \$10,000 in a 26-week certificate of deposit that paid 3.9% simple interest. How much interest will he receive at the end of 26 weeks?
14. If a certificate of deposit pays 5.2% simple interest, how much interest will \$20,000 earn at the end of 26 weeks?
15. Find the interest earned on \$1850 invested for 30 months in an account that pays 4.1% simple interest per year.
16. Bob borrowed \$12,000 from the Friendly Finance Company for 18 months. If the finance company charges 12% simple interest per year, find the total amount that Bob will have to pay back.
17. Find the simple interest on a \$20,000 loan for 3 years at $8\frac{1}{2}\%$. How much interest will it earn in 3 months?
18. How long must \$10,000 be kept at 5% simple interest to become \$12,000?
19. Find the time required for \$35,000 to earn \$1750 at 6% simple interest.
20. What principal will yield \$9280 (principal and interest) in 2 years at 8% simple interest?

*21. Mr. Gilligan has $10,000 and invests $3000 for 1 year at 8% simple interest. At what rate of simple interest should the remainder be invested, if the total income from the $10,000 is to be $660?

*22. An investment firm invested $20,000 for 1 year at simple interest. One portion was invested at 6% and the remainder at 8%. How much money was invested at each rate of interest, if the total interest was the same as it would have been if all the money had been invested at 7%?

NW 23. If a certificate of deposit pays 4% simple interest, find the maturity value of a $5000 certificate at the end of 26 weeks. How much interest will it earn?

24. Maria deposited $8000 in a 2-year certificate of deposit that paid 4.5% simple interest. Find the maturity value of the certificate.

25. Allen borrowed $12,000 for 3 years at 6.4% simple interest. What amount must be repaid? What amount of interest will he pay?

26. Bonnie borrowed $4000 for 2 years at 5.9% simple interest. What amount must she repay? What amount of interest will she pay?

Writing Mathematics

27. Define simple interest.

28. Write the formula for simple interest, and explain what each letter represents.

29. Write the formula for finding the amount (or maturity value), and explain what each letter represents.

Challenge Exercises

30. At a simple interest rate of 5% what principal will earn $720 in four years?

31. What principal would have a maturity value of $936 in 6 months if the simple interest rate is 8%?

32. How long will it take $800 to earn $24 in simple interest if the simple interest rate is 9%?

33. A loan office made a loan of $1000 to be repaid in one month in the amount of $1050. What was the annual interest rate?

34. For each of the following problems **(a–d)**, use the simple interest formula and find the missing item.

	Interest	Principal	Rate	Time
a.	$600	$3000	4%	—
b.	$900	—	9%	2 years
c.	—	$5000	5%	4 months
d.	$150	$3000	—	6 months

35. Jose has a balance due of $300 on his credit card. The rate is 1.5% simple interest per month on the unpaid balance. If Jose pays $180 toward his outstanding balance this month, and he does not charge anything else on his credit card, what will be the balance due on his statement next month?

Group/Research Activity

36. The simple interest formula typically measures time in terms of years. But it is often the case that time in the simple interest formula is given in *days*. This, in turn, leads to two methods of computing the interest, *exact interest* or *ordinary interest* (also called Banker's rule). Research these two methods of interest and prepare a report. Be sure to discuss the differences of each, who uses which method, and the advantages or disadvantages of each.

Just for fun

How much is a billion dollars? To give you some idea, suppose that you could spend $1000 a day every day. How many years would it take you to spend $1,000,000,000? Assume that each year contains 365 days.

12.6 COMPOUND INTEREST

If a person borrows $1000 at 8% simple interest for a period of 1 year, then at the end of that year $1080 must be repaid:

$$I = Prt = \$1000 \times 0.08 \times 1 = \$80$$

$$A = P + I = \$1000 + \$80 = \$1080$$

If the borrower did not pay back any of the loan or the interest by the end of the first year, and if he or she wanted to continue the loan for another year at the same rate, then he or she would owe \$1080 plus the interest on \$1080, which is \$86.40. That is, \$1166.40 would have to be repaid:

$$\begin{aligned} I &= Prt \\ &= \$1080 \times 0.08 \times 1 \\ &= \$86.40 \end{aligned} \qquad \begin{aligned} A &= P + I \\ &= \$1080 + \$86.40 \\ &= \$1166.40 \end{aligned}$$

This is an example of *compound interest*. For this example, we would say that \$1000 was loaned for a period of 2 years, with interest *compounded annually*. Banks pay compound interest on their savings accounts. Most banks pay interest that is compounded quarterly, and some banks pay interest compounded daily.

When the interest due at the end of a certain period is added to the principal and that sum earns interest for the next period, the interest paid is called **compound interest**. The interest for each succeeding period is greater than the previous one, because the principal keeps increasing.

The Cost of a Wait-and-See Investment Attitude

Investing money for retirement can be costly for those who delay their decision to save. To illustrate this point, consider the following scenario involving two 22-year-old college roommates, Milt and Robert, who, in May 2003, graduated from college and are about to begin a teaching career in August.

- On January 1, 2004, Milt invests \$2000 per year into an IRA stock mutual fund. Each January 1, for the next five years, Milt continues making \$2000 contributions. After six years (2004–2009), Milt stops investing any more money into his IRA and leaves the fund untouched until 2048, 45 years after making his first contribution.
- On January 1, 2010, after six years of delay, Robert begins investing \$2000 per year into a similar IRA stock mutual fund. Robert continues investing \$2000 per year each January 1, for the next 37 years.

Thus, Milt's total contribution is \$2000/year × 6 years = \$12,000 and Robert's total contribution is \$2000/year × 38 years = \$76,000. If both Milt's and Robert's accounts do not carry any sales charge, and if both accounts yield a 12% annual rate of return compounded semiannually with dividends and capital gains reinvested, which account do you think will result in a higher aggregate? Surprisingly, it is Milt's. The following table based on our assumptions demonstrates this. In short, Milt only had to invest \$12,000 over six years to receive more than \$1.7 million, whereas Robert invested \$76,000 over 38 years and still ended up with less money than Milt. Although the differences in the final amounts are relatively small (\$1,732,789 versus \$1,693,728), the bottom line is that it can cost you more money to earn less if you delay.

This example illustrates the power of compound interest. Albert Einstein was so impressed with compound interest that he called it one of the greatest discoveries of humankind.

Year	Year-End Value of Milt's Account	Year-End Value of Robert's Account
1	\$2,247	\$0
2	4,772	0
3	7,609	0
4	10,797	0
5	14,379	0
6	18,403	0
7	20,678	2,247
8	23,233	4,772
9	26,105	7,609
10	29,332	10,797
15	52,528	33,714
20	94,070	74,765
25	168,465	148,254
30	301,696	279,880
35	540,292	515,600
40	967,580	937,740
45	\$1,732,789	\$1,693,728

Note: The percent used in this Math Connections example is for illustrative purposes only. The actual percent that is available is a function of the state of the national economy.

EXAMPLE 1 *Interest*

a. Find the amount of simple interest when \$1000 is invested for 3 years at 9%.

b. Find the amount of *compound* interest when \$1000 is invested for 3 years at 9% compounded annually.

Solution

a.
$$I = Prt$$
$$= \$1000 \times 0.09 \times 3$$
$$= \$270$$

b. Because this is compound interest, we must find the interest at the end of each year and add it to the principal before computing the interest for the next year.

$$\text{1st year:} \quad I = Prt$$
$$= \$1000 \times 0.09 \times 1$$
$$= \$90$$

Amount at end of 1 year = \$1000 + \$90 = \$1090.

$$\text{2nd year:} \quad I = Prt$$
$$= \$1090 \times 0.09 \times 1$$
$$= \$98.10$$

Amount at end of 2 years = \$1090 + \$98.10 = \$1188.10.

$$\text{3rd year:} \quad I = Prt$$
$$= \$1188.10 \times 0.09 \times 1$$
$$= \$106.93, \text{ to the nearest cent}$$

Amount at end of 3 years = \$1188.10 + \$106.93 = \$1295.03. The total interest is equal to the total amount accrued at the end of 3 years minus the principal:

$$\text{Interest} = \text{Amount} - \text{Principal}$$
$$= \$1295.03 - \$1000 = \$295.03$$

The highest denomination of currency in circulation is the U.S. Federal Reserve Bank note for \$10,000. It bears the picture of Salmon P. Chase, former Secretary of the Treasury. It is believed that there are some 400 of these bills in circulation. No bills higher than \$100 have been issued since 1969.

Note that in Example 1 the interest earned in 3 years on \$1000 at 9% simple interest was \$270, whereas the interest earned on the same amount over the same time period at 9% *compounded annually* was \$295.03—a difference of \$25.03. This points up the advantage of compound interest over simple interest when money is invested at a given rate over a number of interest periods. Imagine what the difference would be for a large investment such as \$100,000 with an interest rate in the neighborhood of 20%! The compound interest is greater because it is computed more often, and each time on a larger principal.

The computation in part **b** of Example 1 was somewhat tedious. Banks use computers and accountants use calculators to determine compound interest. But they also use books of tables to quote different interest amounts, loan payments,

TABLE 12.1 Compounded Amount of $1

	Interest Rate per Period										
Period	1%	2%	3%	4%	5%	6%	8%	10%	12%	14%	16%
1	1.010	1.020	1.030	1.040	1.050	1.060	1.080	1.100	1.120	1.140	1.160
2	1.020	1.040	1.061	1.082	1.103	1.124	1.166	1.210	1.254	1.300	1.346
4	1.041	1.082	1.126	1.170	1.216	1.262	1.360	1.464	1.574	1.689	1.811
6	1.062	1.126	1.194	1.265	1.340	1.419	1.587	1.772	1.974	2.195	2.436
8	1.083	1.172	1.267	1.369	1.477	1.594	1.851	2.144	2.476	2.853	3.278
10	1.105	1.219	1.344	1.480	1.629	1.791	2.159	2.594	3.106	3.707	4.411
12	1.127	1.268	1.426	1.601	1.796	2.012	2.518	3.138	3.896	4.818	5.936
14	1.149	1.319	1.513	1.732	1.980	2.261	2.937	3.797	4.887	6.261	7.988
16	1.173	1.373	1.605	1.873	2.183	2.540	3.426	4.595	6.130	8.137	10.748
20	1.220	1.486	1.806	2.191	2.653	3.207	4.661	6.728	9.646	13.743	19.461
24	1.270	1.608	2.033	2.563	3.225	4.049	6.341	9.850	15.179	23.212	35.236
28	1.321	1.741	2.288	2.999	3.920	5.112	8.627	14.421	23.884	39.204	63.800

and so on. Table 12.1 can be used to compute compound interest. It shows the amounts that must be paid at the end of different interest periods if $1 is invested at compound interest at one of the given interest rates.

Table 12.1 is an example of a compound-interest table. It will enable us to do the problems in this section. It should be noted that there are tables that are much more detailed. Interest in such tables may be carried out to four or five decimal places, and values are given for many more interest rates and numbers of periods. Table 12.1 shows the amount of interest that $1 will accumulate when interest is paid at the indicated rate and compounded for the indicated number of interest periods. For example, the entry in the third row under the 8% column of Table 12.1 is 1.360. This indicates that if $1 is placed in an account that pays 8% and is compounded for each of four interest periods, then the total accumulation in the account will be $1.36. That is, $1 will grow to $1.36 in four interest periods at 8% per period. The amount accumulated in this way is called the **compound amount**. To find the compound amount on other amounts of money, we multiply the principal by the compound amount for $1. For example, to find the compound amount if $1000 is deposited at 6% compounded annually for 10 years, we first

Revisiting Table 12.1

Table 12.1 indicates the compound amount of $1 for various interest rates for certain periods. For our purpose this table is sufficient. But situations may exist when solving a compound-interest problem that requires more interest periods than the table provides. For example, suppose the table being used covers only 20 periods and we need to find the compound amount for $1000 for six years at 8% compounded quarterly. The interest rate per period is 2% ($8\% \div 4$) and there are 24 periods (6×4). We can find the rate for 24 periods by multiplying any two of the rates, under the 2% column, whose periods add to 24. For example, the rate for 10 periods (under the 2% column) is 1.219 and the rate for 14 periods (under the 2% column) is 1.319. Next, we multiply the two rates, $1.219 \times 1.319 = 1.607861 \approx 1.608$ (rounded off), which is the rate for 24 periods (under the 2% column). Next, we multiply the result by $1000 to find the compound amount: $\$1000 \times 1.608 = \$1608$.

Please note that this is an illustrative example for doing a compound interest problem with a table that has a limited number of interest periods. But the method is sound and will provide an excellent approximation to the correct answer. Let's try it again, using the rate for 12 periods (under the 2% column) twice ($12 + 12 = 24$). The rate for 12 periods is 1.268, and $1.268 \times 1.268 = 1.607824 \approx 1.608$ (rounded off). The compound amount is $\$1000 \times 1.608 = \$1608$. This process can be extended and enables us to approximate the compound amount for problems that have 40 or more periods.

find the compound amount for $1 using these figures for time and interest rate. Using the table, we obtain $1.791. The compound amount for $1000 is therefore

$$\$1000 \times 1.791 = \$1791$$

EXAMPLE 2 ***Compound Amount*** Find the compound amount on deposit when $1200 is deposited for 6 years at 8% compounded annually.

Solution From Table 12.1, the compound amount for $1 at 8% for six interest periods is $1.587. Therefore, to find the compound amount for $1200 at 8% for six periods, we multiply $1200 times 1.587:

$$\text{Compound amount} = \$1200 \times 1.587 = \$1904.40$$

NW ***Now Work Problem 3.***

We can find the total compound interest by subtracting the principal from the compound amount. That is,

$$\text{Compound interest} = \text{Compound amount} - \text{Principal}$$

In Example 2, the compound amount is $1904.40 and the principal is $1200. Hence,

$$\text{Compound interest} = \$1904.40 - \$1200 = \$704.40$$

We noted earlier that banks usually compound interest more often than once a year. There are some banks that compound interest semiannually, others that compound interest quarterly, and others that compound interest daily. If a bank offers 6% interest compounded semiannually, then there are two interest periods every year, and the bank pays 3% interest every 6 months.

To find the compound amount when $1000 is deposited for 4 years at 6% compounded semiannually, we must first determine the number of interest periods and the interest rate per period. Because the principal is deposited for 4 years compounded semiannually, there are eight interest periods. The rate of 6% per year is the same as 3% per half-year. Using Table 12.1, we find 8 under the period column and go across to the 3% column. The entry is 1.267. Now we multiply 1.267 by $1000:

$$\text{Compound amount} = \$1000 \times 1.267 = \$1267$$

EXAMPLE 3 ***Compound Amount*** Find the compound amount on deposit when $500 is deposited for 5 years at 8% compounded quarterly.

Solution Interest compounded quarterly means that there are four interest periods per year, so for 5 years there are 20 interest periods. Eight percent interest per year is the same as 2% every quarter-year. The compound amount for $1 at 2% for 20 interest periods is $1.486. Therefore,

$$\text{Compound amount} = \$500 \times 1.486 = \$743$$

NW ***Now Work Problem 13.***

EXAMPLE 4 ***Compound Interest and Compound Amount*** Find the compound amount and the compound interest on $2000 invested at 12% compounded quarterly for 6 years.

Solution Number of interest periods per year = 4

$$\text{Total number of interest periods} = 6 \times 4 = 24$$

$$\text{Interest rate each interest period} = \frac{1}{4} \text{ of } 12\% = 3\%$$

Note of Interest

The formula used to calculate compound interest is $A = P(1 + \frac{r}{n})^{nt}$, where A is the compound amount, P is the principal, r is the annual rate of interest, t is the time in years, and n is the number of interest periods per year. If your calculator has an exponential key, you may wish to try this formula and use your calculator to solve problems in this section.

The compound amount for \$1 at 3% for 24 interest periods is \$2.033. Therefore,

$$\text{Compound amount} = \$2000 \times 2.033 = \$4066$$
$$\text{Compound interest} = \$4066 - \$2000 = \$2066$$

NW ***Now Work Problem 19.***

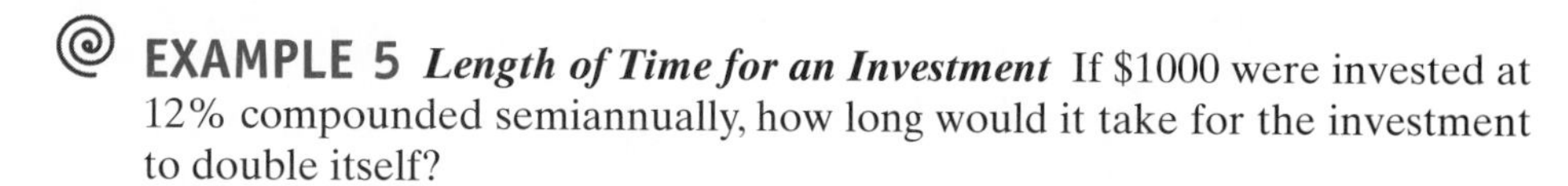

EXAMPLE 5 ***Length of Time for an Investment*** If \$1000 were invested at 12% compounded semiannually, how long would it take for the investment to double itself?

Solution The rate per interest period is $\frac{1}{2}$ of 12%, or 6%. Examining the column headed 6% in Table 12.1, we see that \$1 will double itself in 12 periods; that is, the compound amount for \$1 at 6% for 12 periods is \$2.012. The interest is compounded semiannually, so 12 periods is $12 \div 2 = 6$ years. Therefore, \$1000 will double itself in 6 years when invested at 12% compounded semiannually. The compounded amount would be

$$\$1000 \times 2.012 = \$2012$$

(*Note:* This compound amount is a little more than twice \$1000, but the answer to the original question is correct to the nearest year.)

NW ***Now Work Problem 21.***

Exercises for Section 12.6

For Exercises 1–6, use Table 12.1 in this section to find the compound amount for each investment if interest is compounded annually.

1. \$700 at 6% for 4 years
2. \$900 at 8% for 6 years
3. (NW) \$2500 at 8% for 10 years
4. \$3300 at 10% for 2 years
5. \$10,000 at 14% for 2 years
6. \$12,000 at 16% for 6 years

For Exercises 7–12, use Table 12.1 in this section to find the compound amount for each investment if interest is compounded semiannually.

7. \$500 at 8% for 10 years
8. \$800 at 6% for 6 years
9. \$1000 at 4% for 3 years
10. \$2500 at 12% for 3 years
11. \$5000 at 12% for 5 years
12. \$10,000 at 16% for 2 years

For Exercises 13–18, use Table 12.1 in this section to find the compound amount for each investment if interest is compounded quarterly.

13. (NW) \$500 at 8% for 2 years
14. \$750 at 12% for 3 years
15. \$1000 at 16% for 5 years
16. \$2500 at 8% for 6 years
17. \$10,000 at 12% for 6 years
18. \$20,000 at 16% for 7 years
19. (NW) Find the compound amount and the compound interest on \$3000 invested for 5 years at 12% compounded quarterly.
20. Find the compound amount and the compound interest on \$2500 invested for 10 years at 16% compounded semiannually.
21. (NW) If \$1000 were invested at 6% compounded semiannually, how long would it take for the investment to double itself?
22. If \$1000 were invested at 6% compounded annually, how long would it take for the investment to double itself?
23. **Inflation** In 2003, the price of a "basic" automobile was \$16,000. What can we expect the price of this same type of car to be in the year 2013, if we assume that the annual rate of inflation is 6%?

    (*Hint:* Use the compound-interest table.)
24. **Inflation** A pair of tickets for "good seats" at a NFL game cost approximately \$200 in 2003. If we assume the annual rate of inflation is 8%, what can we expect the price of these same type tickets to be in the year 2015?
25. **Inflation** A typical weekly magazine had a cover price of \$3.95 in 2003. If we assume the annual rate of inflation is

8%, what can we expect the price of this same magazine to be in the year 2015?

26. **Inflation** In 2003, *The New York Times* sold for $1.00 at the newsstand. What can we expect the price of this same newspaper to be in the year 2013, if we assume that the annual inflation rate is 6%?

27. **Compound amount** The sum of $10,000 is deposited in each of four banks, each paying 8% interest per year. The first bank compounds interest annually, the second bank semiannually, and the third bank quarterly. The fourth bank pays simple interest. If no further deposits or withdrawals are made, how much will be in each account at the end of 4 years?

*28. **Finding the principal** A father wishes to give his daughter $20,000 at the age of 25. If the daughter is now 13 years old, how much money must her father invest at 8% interest compounded semiannually in order to have $20,000 when his daughter is 25?

29. **Future value** A grandmother deposits $1000 in an account that pays 6% compounded annually when her granddaughter is born. What will be the value of the account when the granddaughter reaches her twenty-fourth birthday, assuming that no other deposits or withdrawals are made during this time?

*30. **Finding the principal** A grandfather wishes to have $30,000 available for his grandson's college education. If the grandson is now 4 years old, how much money must the grandfather invest at 6% interest compounded semiannually in order to have $30,000 when his grandson is 18?

Writing Mathematics

31. Explain the difference between compound interest and simple interest.

32. Explain what is meant by the *compound amount*.

Challenge Exercises

33. A certain magazine has a cover price of $4.95 today. What can we expect the price of this same magazine to be in 12 years, if we assume the annual rate of inflation is 6%?

34. A complete new video system will be available to the general public in three years. It will cost approximately $6000. The Fratangelo family is looking forward to purchasing this new system. What amount of money should be invested today at 6% compounded semiannually, to yield $6000?

35. Suppose you are offered $10,000 today or $10,500 in one year. Which should you accept if money can be invested at 8% compounded quarterly?

Research/Group Activities

36. Two terms that are frequently used in discussing compound interest are *future value* and *present value*. Research these two topics and prepare a report for class. Be sure to explain the differences between the two, and also explain how each is computed.

37. Records indicate that at one time there existed 12 different denominations of United States paper currency. List the different denominations and also whose portrait appears on each bill. Prepare a report to be presented in class. Be sure to discuss why some presidents were chosen and others chosen were not presidents.

Note of Interest

Benjamin Franklin is credited with the phrase, "A penny saved is a penny earned." Apparently he appreciated the effects of compound interest. Upon his death in 1790, he left 1000 pounds to the city of Boston on the condition that the money could not be touched for 100 years. His original bequest of 1000 pounds was equivalent to approximately $4600 and it compounded to $322,000 for the city of Boston by 1890.

Just for fun

Before the start of the American Revolutionary War there were 13 British Colonies known as the Thirteen Original Colonies. They are represented in the flag of the United States by 13 stripes, six white and seven red. What are the names of the Thirteen Original Colonies?

12.7 EFFECTIVE RATE OF INTEREST

In our previous discussions pertaining to interest, we have seen that interest rates are usually stated on a yearly basis, even though the period of payment is not a year. For example, an interest rate of 8% compounded quarterly means that we use an interest rate of 2% for each interest period. Similarly, an interest rate of 6% compounded semiannually means that we use an interest rate of 3% for each period.

To find the compound amount on deposit when \$1 is deposited for 1 year at 8% compounded quarterly, we can use Table 12.1 from Section 12.6 to find that the amount is \$1.082. That is, \$1 will compound to \$1.082 in 1 year with interest compounded quarterly. The compound amount on deposit when \$1 is deposited for 1 year at 6% compounded semiannually is \$1.061. That is, \$1 will yield \$1.061 in 1 year with interest compounded semiannually.

Effective Rate

We have chosen these two examples to help illustrate the *effective annual interest rate*, also referred to as **effective rate**. For example, an interest rate of 8% compounded quarterly is equivalent to an effective rate of 8.2% compounded annually. This is because, as noted in the preceding paragraph, \$1 compounds to \$1.082 in 1 year at 8% compounded quarterly. Therefore, the total interest earned in 1 year is

$$\$1.082 - \$1.00 = \$0.082$$

This is an effective rate of 8.2% since

$$8.2\% \text{ of } \$1.00 = \$0.082$$

Thus the interest earned at 8% compounded quarterly is the same as the interest earned at an annual interest rate of 8.2%.

Similarly, a 6% interest rate compounded semiannually is equivalent to an effective rate of 6.1% compounded annually. (Note that \$1 compounds to \$1.061 in 1 year at 6% compounded semiannually.)

If interest is paid more than once a year, the effective annual rate is greater than the stated annual rate. If interest is paid annually, then the effective rate is the same as the stated annual rate. In the preceding examples, the effective rate was greater than the stated annual rate.

Effective rate is often used to compare interest rates that are compounded at different intervals. For example, if bank A offers 5% interest on its deposits compounded quarterly and bank B offers $5\frac{1}{2}$% interest on its deposits compounded semiannually, we can find the effective annual rate for each bank to determine which offers the better investment. We can also compare effective rates when borrowing money. In this case, we would select the lowest effective rate. Many banks publish pamphlets listing the various types of accounts they offer, together with the stated interest rate and the effective annual yield. This is also a common practice in newspaper advertisements.

We can determine the effective annual rate by means of the formula

$$E = (1 + r)^n - 1$$

where

E = effective rate

n = number of payment periods per year

r = interest rate per period

We shall now use this formula to find the effective annual rate for 6% compounded semiannually:

$$E = (1 + r)^n - 1$$

The interest rate is compounded semiannually, so there are $n = 2$ payment periods per year. Therefore, $r = 6\% \div 2 = 3\% = 0.03$. (Note that r is expressed without a percent sign.) Substituting these values in the formula, we have

$$E = (1 + 0.03)^2 - 1$$
$$E = (1.03)^2 - 1$$
$$E = 1.0609 - 1$$
$$E = 0.0609, \quad \text{or } 6.1\% \quad \text{(to the nearest tenth)}$$

Note: The symbol [calculator] indicates problems more readily done with a calculator.

EXAMPLE 1 ***Effective Rate*** What is the effective rate, if money is invested at 8% compounded quarterly?

Solution The number of payment periods is four, because interest is compounded quarterly. The interest rate per period is $8\% \div 4 = 2\%$. Therefore, $r = 0.02$. Substituting these values in the formula, we have

$$E = (1 + r)^n - 1$$
$$E = (1 + 0.02)^4 - 1$$
$$E = (1.02)^4 - 1$$
$$E = 1.0824 - 1$$
$$E = 0.0824, \quad \text{or } 8.2\% \quad \text{(to the nearest tenth)}$$

This answer checks with the one we obtained earlier in this section using the compound-interest table.

In Example 1 we rounded our answer to the nearest tenth so that it could be compared with the first illustrative example in this section. However, all other problems in this section will be rounded to the nearest hundredth of a percent. Also, it is recommended that any calculation involving the use of this formula be done on a calculator. (See section 1.3 re: rounding)

EXAMPLE 2 ***Effective Rate*** What is the effective rate, if money is invested at 7% compounded semiannually?

Solution We substitute $n = 2$ and $r = 7\% \div 2 = 3.5\% = 0.035$ into the formula:

$$E = (1 + r)^n - 1$$
$$E = (1 + 0.035)^2 - 1$$
$$E = (1.035)^2 - 1$$
$$E = 1.071225 - 1 = 0.071225$$
$$E = 7.12\%$$

NW ***Now Work Problems 3 and 5.***

EXAMPLE 3 ***Better Offer*** Bank A offers its depositors an interest rate of 5% compounded quarterly and bank B offers its depositors a rate of $5\frac{1}{2}\%$ compounded semiannually. Which bank makes the better offer?

Solution We must compare the effective annual rate for each bank.

Bank A: $n = 4, r = 5\% \div 4 = 1.25\% = 0.0125$

$$E = (1 + r)^n - 1$$
$$E = (1 + 0.0125)^4 - 1$$
$$E = (1.0125)^4 - 1$$
$$E = 1.0509 - 1 = 0.0509$$
$$E = 5.09\%$$

Bank B: $n = 2, r = 5\frac{1}{2}\% \div 2 = 2.75\% = 0.0275$

$$E = (1 + r)^n - 1$$
$$E = (1 + 0.0275)^2 - 1$$
$$E = (1.0275)^2 - 1$$
$$E = 1.0558 - 1 = 0.0558$$
$$E = 5.58\%$$

The effective rate of bank B is greater than that of bank A by 5.58% − 5.09% = 0.49%. Therefore, bank B offers a better interest rate to its depositors.

 ***Now Work Problem 15.***

Exercises for Section 12.7

Answers involving fractional parts of a percentage should be rounded to the nearest hundredth of a percent.

1. What is effective annual interest rate?
2. What is the effective rate if money is invested at 12% compounded semiannually?
3. (NW) What is the effective rate if money is invested at 8% compounded semiannually?
4. What is the effective rate if money is invested at 10% compounded semiannually?
5. (NW) What is the effective rate if money is invested at 12% compounded quarterly?
6. What is the effective rate if money is invested at 10% compounded quarterly?
7. What is the effective rate if money is invested at 6% compounded quarterly?
8. What is the effective rate if money is invested at 7.5% compounded quarterly?
9. What is the effective rate if money is invested at 7.75% compounded quarterly?
10. What is the effective rate if money is invested at 5% compounded **(a)** annually? **(b)** semiannually? **(c)** quarterly?
11. What is the effective rate if money is invested at 6% compounded **(a)** annually? **(b)** semiannually? **(c)** quarterly?
12. What is the effective rate if money is invested at 9% compounded **(a)** annually? **(b)** semiannually? **(c)** quarterly?
13. Which is the higher interest rate: 5% compounded quarterly or 5.5% compounded semiannually?
14. Which is the higher interest rate: 6.5% compounded quarterly or 6.8% compounded semiannually?
15. (NW) Marlene invested her money at 6.4% compounded quarterly, whereas Pamela invested her money at 6.6% compounded semiannually. Who receives the better interest rate?
16. Bob invested his money at 5.5% compounded quarterly, whereas Larry invested his money at 5.75% compounded semiannually. Who receives the better interest rate?
17. Ruth invested her money at 6.75% compounded quarterly, whereas Julia invested her money at 6.9% compounded semiannually. Who receives the better interest rate?

Writing Mathematics

18. Write the formula used for finding the effective rate of interest, and explain what each letter represents.
19. How does a person use effective rate of interest?

Challenge Exercises

20. Some banks compound interest daily. **(a)** Compare the effective rate for 6% compounded quarterly with the effective rate for 6% compounded *daily*. Assume a 360-day year. (*Note:* You will need a calculator with an exponent key.) **(b)** What is the difference in interest paid in 1 year on $1,000,000 between the two methods in part **a**?
21. Rachel invested her money at 5.5% compounded quarterly, whereas Jillian invested her money at 5.55% compounded semiannually. Find the effective rate of interest for each and determine who receives the better interest rate.

Research/Group Activities

22. Select at least three different banks in your area. Determine the stated interest rates and the stated effective annual yield for both a 6-month and a 12-month certificate of deposit. Next, calculate the effective rate of interest for each certificate of deposit. Are the bank's quotes accurate? Report your findings to your classmates.

23. Choose three different banks in your area and determine the fees and interest rates (paid or charged) for **(a)** a checking account, **(b)** a debit card, and **(c)** a credit card. Determine which bank offers the overall best deal for a student. Report your findings to your classmates.

Just for fun

Mario and Roy arranged an unusual race. They agreed that the man whose car crossed the finish line first would be the loser. The man whose car crossed the finish line second would be the winner. How should Mario and Roy drive to have such a race?

Math Connections

Internet Financial Planning Tools

Prior to the advent of the Internet, consumers relied on professionals such as financial planners, bankers, and mutual fund account executives to explain the costs associated with various investments or purchases. Today, however, there are hundreds of Internet-based calculator tools that we can use to determine the costs of mutual fund fees, home mortgages, home improvement loans, and life insurance. Following are a few selected sites that were still "up and running" when this textbook was published.

Home Mortgages

Two sites that are especially useful for buying a home or for refinancing an existing mortgage are the *Countrywide* site at **http://www.countrywide.com** and the *Mortgage Bankers Association* site at **http://www.mbaa.org**. Both sites provide a calculator that will help determine such things as how much of a mortgage you can afford, estimated closing costs, down payment amounts, and monthly principal and interest payments. They will also provide 15- versus 30-year mortgage comparison costs and amortization schedules.

Retirement

There are many sites that will help you estimate if you are saving enough for retirement. Perhaps the most premiere site is the *Quicken Retirement Planner* at **http://www. quicken.com**. This site contains several different calculators, including a 401(k) calculator that helps you determine the future value of your 401(k) as well as your tax savings; an investment calculator that helps you calculate the effect of inflation on your investments; a savings calculator that helps you determine how long you need to save to reach your savings goal; and a family calculator that helps you determine whether your spouse can afford to stop working.

Another site that is also useful is the *American Savings Education Council* at **http://www.asec.org**. This site contains a "ballpark estimate" worksheet designed to help individuals quickly identify approximately how much savings they will need to live comfortably in retirement. A Spanish language version of this worksheet is also available.

Investing

To help identify and keep track of investment costs such as mutual fund fees, the *Securities and Exchange Commission* site at **http://www.sec.gov** provides a mutual fund cost calculator as one of its investor information interactive tools. The site also provides a link to the American Savings Education Council's ballpark estimate retirement calculator.

Another useful site, the *FinanCenter* at **http://www. financecenter.com**, provides a host of financial planning tools, including mortgage, home equity, automobile, credit card, stock, bond, retirement, Roth IRA, budgeting, savings, paycheck planning, and life insurance calculators. Spanish language versions of these calculators are also available. Among the investment tools are the stock calculator, which can help you determine investment costs such as stock commission fees and taxes so that you know what your true profit will be when you sell a stock; and the bond calculator, which can help you compare different types of bonds and determine the effect of interest rate changes on your bonds' value.

12.8 LIFE INSURANCE

Almost all consumers purchase insurance at some point in their lifetimes. The types of insurance that a person may obtain include car insurance, health insurance, homeowner's insurance, and life insurance. Many people obtain all of these types of insurance to protect themselves, their homes, and their families.

Insurance

The basic concept of insurance is the sharing of risks or losses. That is, a person who buys an insurance policy is agreeing to share the risks involved with other people buying a similar policy. A policy is a contract between the insured person, or **policyholder**, and the insurer, or **underwriter**. The fee that an insured person pays to the insurer is called a **premium**. Premiums can be paid monthly, quarterly, semiannually, or annually. In this section we shall consider life insurance. There are three basic types of life insurance policies: (1) *ordinary life* insurance, which is also known as *whole life* or *straight life*; (2) *term*; and (3) *endowment*. An **ordinary life** insurance policy provides protection for the life of the insured person. That is, it offers permanent protection. The insurer will pay the face value of the policy to a designated person, called the **beneficiary**, when the insured person dies, regardless of when this occurs. Premiums for a **whole life** policy can be paid in two different ways: A **straight life** policy is one in which the policyholder pays premiums until death; a **limited-payment life** policy is one in which the policyholder pays premiums for a certain number of years. Whether the policy is a straight life policy or a limited-payment policy, it is in effect for the insured person's lifetime, and the beneficiary receives the face value of the policy whenever the insured person dies. Annual premiums for a limited-payment life policy are higher than those for a straight life policy. A **term insurance policy** is one that provides protection for a specified term, or number of years. The insurer will pay the face value of the policy to a beneficiary if the insured person dies during the term or period of time stated in the policy. At the end of the term, the policy is no longer in effect and is worthless. A common term is 5 years. For example, if a 20-year-old college student purchased a 5-year term life insurance policy, then the insurance company is liable for payment of the face value of the policy if the insured dies within the 5-year period. Once the term of the policy expires, the insurance company is no longer liable. Usually term insurance can be renewed for another term, but at a higher rate.

Note of Interest

The highest life insurance payout is believed to be over $18,000,000. This was for a man in Oklahoma who paid more than $300,000 in premiums the year before his death.

Biography: Nathaniel Bowditch

Nathaniel Bowditch (1773–1838) was America's first actuary (an actuary is a professional who plans insurance policies and premiums). In 1803, he was elected president of Essex Fire and Marine Insurance Company and later became president of Massachusetts Hospital Life Insurance Company. This is quite a feat for a man who was self-educated in surveying, algebra, and Latin. When he was 20 years old, he read Newton's *Principia* and found an error in it. From 1795 through 1802 he took four long ship voyages, ending up as ship's captain in 1802 because of his skill as a navigator. As captain he instructed his crew in making astronomical observations. Also in 1802, he published the New American Practical Navigator, which became the standard navigational guide of the U.S. Navy. Revised editions are still being published by the Navy's Hydrographic Office; the 71st edition was published in 1995. From 1814 to 1820 he translated Laplace's *Mecanique Céleste* (a four-volume set) from French to English, adding his own extensive commentary and publishing it with his own money.

An **endowment insurance policy** is similar to a term policy in that it insures the life of the insured for a specified term or number of years. But if the insured is living at the end of the term of the policy, then the face value of the policy will be paid to the policyholder. Hence, the use of the word *endowment*. A common term for an endowment policy is 20 years.

Premiums vary for the different types of life insurance policies. Term insurance is the cheapest, followed by straight life, then limited-payment; endowment is the most expensive. Different rates are also applied in each individual case. Typical annual premium rates per $1000 of life insurance are given in Table 12.2. A person who is going to buy life insurance should investigate plans and prices thoroughly because different companies offer different prices and options. It should be noted that premiums are usually lower for females than males because women live longer than men. (For the sake of convenience in making computations, we shall assume that the rates are the same for males and females of the same age.)

TABLE 12.2 Annual Premium Rate per $1000 of Life Insurance

Age	5-Year Term	Straight Life	Limited-Payment (20 year)	Endowment (20 year)
20	$6.99	$15.48	$25.96	$49.32
25	7.57	17.50	28.83	49.73
30	8.60	20.15	32.10	50.17
35	9.98	23.84	35.92	51.13
40	12.44	28.37	40.74	52.69
45	15.59	33.90	46.57	55.33
50	19.89	41.23	53.01	59.00
55	27.09	50.88	59.73	64.55
60	—	63.95	70.00	—

Insurance rates are determined by using data about large samples of the population. The most important pieces of information used are the average death rates for different age groups. It stands to reason that a 20-year-old male should not be charged the same premium as a 65-year-old male, as the 65-year-old will pay fewer premiums. The average death rate of 20-year-old males is about 2 per 1000, and they can expect to live for 50 years or more. The average death rate of 65-year-old males is about 32 per 1000, and they can expect to live for another 13 years. This kind of information can be found in tables called **mortality tables**. People who help construct these tables, which determine what premiums an insurance company should charge, are called **actuaries**.

EXAMPLE 1 ***Straight Life Policy*** Tom Jones is 20 years old and wishes to purchase $10,000 worth of life insurance. Determine the annual premium of a straight life policy.

Solution Using Table 12.2, we see that the premium for a 20-year-old person is $15.48 per $1000 of straight life insurance. The policy is worth $10,000; hence, we have

$$10 \times \$15.48 = \$154.80 \text{ annual premium}$$

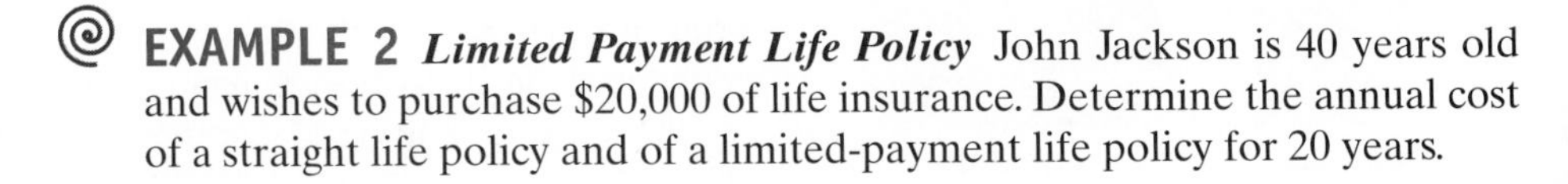

EXAMPLE 2 ***Limited Payment Life Policy*** John Jackson is 40 years old and wishes to purchase $20,000 of life insurance. Determine the annual cost of a straight life policy and of a limited-payment life policy for 20 years.

Solution From the table, the straight life premium per $1000 for a 40-year-old is $28.37. Therefore, the annual cost on a $20,000 policy is

$$20 \times \$28.37 = \$567.40$$

The limited-payment life (also called 20-payment life) premium per $1000 for a 40-year-old is $40.74. Therefore, the annual cost on a $20,000 policy is

$$20 \times \$40.74 = \$814.80$$

NW ***Now Work Problems 1a and 1b.***

EXAMPLE 3 ***Comparing Total Cost of Policies*** If John Jackson (see Example 2) lives for 25 years after he purchases his insurance policy, which policy (straight or limited-payment) will cost more? How much more?

Solution If John Jackson lives for 25 years after purchasing a straight life policy, he will pay 25 annual premiums. Hence, the cost is 25 times $567.40, the annual straight life premium:

$$25 \times \$567.40 = \$14{,}185.00$$

On a 20-year limited-payment life policy, John Jackson will pay 20 annual premiums. Therefore, the cost is 20 times $814.80, the annual limited-payment life premium:

$$20 \times \$814.80 = \$16{,}296.00$$

The limited-payment life policy will cost more. It will cost

$$\$16{,}296 - \$14{,}185 = \$2111 \text{ more}$$

NW ***Now Work Problem 11.***

Thus far we have discussed only annual premiums. But premiums may also be paid semiannually, quarterly, or monthly. These rates are a percentage of the annual rate and may be found in Table 12.3.

TABLE 12.3 Premium Rates

Payment Period	Percentage of Annual Premium
Semiannual	51%
Quarterly	26%
Monthly	9%

EXAMPLE 4 ***Various Payment Methods*** Susan Simons is 30 years old and wants to purchase a $15,000 straight life policy. Determine the annual cost if she pays

a. annually
b. semiannually
c. quarterly
d. monthly

Solution

a. Using Table 12.2, we see that the premium for a 30-year-old person is $20.15 per $1000 of straight life insurance. Therefore, the annual premium is $15 \times \$20.15 = \$302.25$.

b. The semiannual premium is 51% of the annual premium (see Table 12.3), or 0.51 × \$302.25 = \$154.15. The annual cost is 2 × \$154.15 = \$308.30.

c. The quarterly premium is 26% of the annual premium (see Table 12.3), or 0.26 × \$302.25 = \$78.59. The annual cost is 4 × \$78.59 = \$314.36.

d. The monthly premium is 9% of the annual premium (see Table 12.3), or 0.09 × \$302.25 = \$27.20. The annual cost is 12 × \$27.20 = \$326.40.

NW ***Now Work Problem 5.***

From the previous example, we can see that the annual cost of a policy increases as the number of payments increases. But many people would prefer to pay a smaller amount more often as opposed to paying a large amount once a year.

Exercises for Section 12.8

Use Tables 12.2 and 12.3 for these exercises (1–14).

1. Find the annual premium for each of the given insurance policies.

	Face Value of Policy	Age at Issue	Type of Policy
NW **a.**	\$25,000	25	straight life
NW **b.**	20,000	20	20-year limited-payment life
c.	15,000	30	5-year term
d.	50,000	35	20-year endowment
e.	36,000	40	straight life

2. Find the annual premium for each of the given insurance policies.

	Face Value of Policy	Age at Issue	Type of Policy
a.	\$30,000	30	5-year term
b.	20,000	35	20-year limited-payment life
c.	40,000	40	straight life
d.	35,000	35	20-year endowment
e.	22,000	55	5-year term

3. Term life insurance Laurie Adams is 25 years old and wishes to purchase a \$20,000 5-year term life insurance policy. Determine the annual cost if she pays

a. annually **b.** semiannually
c. quarterly **d.** monthly

4. Limited-payment life insurance Carlos Lopez is 30 years old and wishes to purchase a \$25,000 limited-payment (20-year) life insurance policy. Determine the annual cost if he pays

a. annually **b.** semiannually
c. quarterly **d.** monthly

NW **5. Straight life insurance** Irene Gilligan is 35 years old and wishes to purchase a \$30,000 straight life insurance policy. Determine the annual cost if she pays

a. annually **b.** semiannually
c. quarterly **d.** monthly

6. Endowment life insurance Joan Zambriski is 20 years old and wishes to purchase \$20,000 worth of life insurance. She decides to purchase a 20-year endowment policy. Determine the annual cost if she pays

a. annually **b.** semiannually
c. quarterly **d.** monthly

7. Premium savings Paul Thomas is 35 years old and wishes to purchase a \$50,000 endowment (20-year) life insurance policy. How much will Paul save over the 20-year period if he pays his premium annually as opposed to monthly?

8. Term life insurance Stan Simon is 20 years old and wishes to purchase \$10,000 worth of life insurance. Determine the annual cost of a 5-year term policy. If Stan renews his policy when it expires, what is the difference in total cost between the first and second policies?

9. Difference in total cost Jan Peters is 25 years old and wishes to purchase \$12,000 worth of life insurance. Determine the annual cost of a 5-year term policy. If Jan renews her policy when it expires, what is the difference in total cost between the first and second policies?

10. Straight life insurance Jane Lark is 30 years old and wishes to purchase \$20,000 worth of life insurance. Determine the annual cost of a straight life policy and also of a limited-payment life policy for 20 years. If Jane lives for 30 years after she purchases her policy, which policy would cost more? How much more?

NW **11. Difference in total cost** Carl Ronold is 35 years old and wishes to purchase \$25,000 worth of life insurance. Determine the annual cost of a straight life policy and also of a limited-payment life policy for 20 years. If Carl lives for 20 years after he purchases his policy, which policy would cost more? How much more?

12. **Total cost of life insurance** Rich Nelson obtained a $20,000 limited-payment life policy for 20 years when he was 40 years old. Determine the annual cost of this policy. What is the total cost if Rich Nelson lives to be 65 years old? If he had obtained the same policy when he was 25 years old, what total would he have paid for the insurance?

13. **Total cost of life insurance** Lois Carson obtained a $20,000 straight life insurance policy at age 40. Determine the annual cost for this policy. If she died 18 years later, how much more did her beneficiary collect than Lois Carson paid in premiums?

14. **Total cost of life insurance** Tom Tillard is 25 years old and just purchased a $30,000 straight life policy. If he lives to age 65, how much could he have saved by purchasing a 20-year limited-payment life policy with the same face value?

Writing Mathematics

15. Explain what the term *insurance* means to you.

16. Following are some terms used in this section. Provide a definition for each.

    a. underwriter **b.** policyholder
    c. premium **d.** beneficiary

17. Explain the difference between term life insurance and straight life insurance.

18. There exist many different types of insurance, other than life insurance. List at least five other types of insurance that a person might want to purchase.

Challenge Exercises

19. How much is the total quarterly premium paid by a husband, age 35, and his wife, age 30, if each has a straight life insurance policy for $40,000?

20. How much is the total monthly premium paid by a husband, age 30, and his wife, age 25, if each has a $50,000 endowment (20-year) life insurance policy? How much would they save over the 20-year period if they paid the total premium annually instead of monthly?

Research/Group Activities

21. Another form of life insurance available from some companies is *universal life insurance*. Select two different insurance companies and investigate universal life insurance. Prepare a report for class comparing universal life insurance to the types of policies covered in this section. Be sure to include a brief history of the development of universal life insurance.

22. Insurance rates are determined by using data about large samples of the population. The most important information used is the average death rates for different age groups. This kind of information can be found in tables called *mortality tables*. Consult an almanac or the Internet to find

    a. Years of life expected at birth for you and a classmate
    b. Life expectancy for you and a classmate, given your present age

Just for fun

An Amtrak train leaves Miami for Boston traveling at the rate of 80 miles per hour. Two hours later, another train leaves Boston for Miami traveling at the rate of 60 miles per hour. When the two trains meet, which train is closer to Miami?

12.9 INSTALLMENT BUYING

Credit may be thought of as allowing someone to purchase an item without cash and trusting the person to pay for the item at a later specified date. Many consumers purchase goods or services without cash. That is, they obtain possession of something immediately and agree to pay for it at a later date. For example, turning on the lights and using electricity or using the telephone are two examples of using credit. The consumer has used the service or product, but typically will not pay for it until the end of the month.

Installment Purchases

Many items are purchased on an installment plan. That is, the buyer obtains possession of an item immediately and agrees to pay the purchase price plus an additional charge in a series of regular payments, usually monthly. If an entertainment system that sells for $1,000 can be purchased on an installment plan for $200 down and 12 monthly payments of $70 each, we find the total cost of the system when it is purchased on credit by multiplying the monthly payment times the number of payments to obtain the total amount paid in monthly payments, and then we add the down payment.

$$\$70 \times 12 = \$840 \quad \textit{total amount of monthly payments}$$
$$\$840 + 200 = \$1040 \quad \textit{total cost}$$

To find the amount of interest paid (also called service charge, finance charge, or carrying charge), we subtract the cash price from the total installment cost:

$$\$1040 - \$1000 = \$40 \text{ interest}$$

This amount does not reflect the true interest rate. But, because the dollar amount is relatively small, consumers are more than willing to pay it.

Consider the case of Sam Larson, who borrows $1200 from the Friendly Finance Company. The finance company advertised an interest rate of 10% simple interest. Sam wanted to borrow the money for six months and repay the loan in six monthly payments. Hence, we have

$$I = Prt$$
$$I = \$1200 \times 0.10 \times \frac{6}{12}$$
$$I = \$60$$

The total amount that Sam has to repay is $1200 + $60, or $1260. If he is going to repay the loan in six equal monthly payments, he will pay $210 each month.

It should be noted that the Friendly Finance Company computed the interest as though Sam Larson owed the $1200 for the entire six months. But, he did not! Sam owed $1200 for only one month. At the end of the first month, he paid the finance company $210. Of this amount, $200 was applied toward the principal, the other $10 is interest. Because each payment includes $200 of the principal, the amount that Sam owes decreases by $200 each month. The following list shows what Sam owes each month:

$$\begin{aligned} & \$1200 && \textit{(original amount, owed for the first month only)} \\ \$1200 - \$200 = {} & \$1000 && \textit{(amount owed for second month only)} \\ \$1000 - \$200 = {} & \$\ 800 && \textit{(amount owed for third month only)} \\ \$800 - \$200 = {} & \$\ 600 && \textit{(amount owed for fourth month only)} \\ \$600 - \$200 = {} & \$\ 400 && \textit{(amount owed for fifth month only)} \\ \$400 - \$200 = {} & \underline{\$\ 200} && \textit{(amount owed for sixth month only)} \\ \text{Total} = {} & \$4200 && \end{aligned}$$

Biography: Antoine-Augustin Cournot

Antoine-Augustin Cournot (1801–1877), a French mathematician and economist, pioneered the field of mathematical economics. During his life he was a university official at Lyons and Grenoble and held positions in the French government. He also translated and wrote a large number of books on economics, mathematics, statistics, and philosophy. His masterpiece was the publication of *Researches Into the Mathematical Principles of the Theory of Wealth,* in 1838. It is in this work that he establishes the law of diminishing returns, gives interpretations of supply-and-demand as "functions or schedules," and establishes the idea of equilibrium under the conditions of monopoly, duopoly, and pefect competition. When his book was first published it seemed a failure. His ideas were too far advanced for the times. His masterpiece is now considered a classic in its field.

Thus, Sam's debt to the finance company is equivalent to owing \$4200 for one month only. The interest that Sam paid was \$60. Using the formula $I = Prt$, we can find the value of r:

$$I = Prt$$

$$\$60 = \$4200 \times r \times \frac{1}{12}$$

$$60 = \$350 \times r$$

$$\frac{60}{350} = r$$

Solving this equation for r (correct to the nearest thousandth),

$$r = 0.171$$

$$r = 17.1\% \qquad \textit{(to the nearest tenth of a percent)}$$

Therefore, Sam paid about 17.1% interest on his loan, *not* 10% as advertised.

True Annual Interest Rate

In 1969, Congress passed a **Truth-in-Lending Act** that requires all sellers to reveal the **true annual interest rate** that they charge and the total finance charge, which includes additional fees such as service charges. The true annual interest rate is also known as the **annual percentage rate** and is often quoted as APR.

The Truth-in-Lending Act does not establish any maximums on interest rates or finance charges. It does not regulate interest charges. By law, however, before a consumer signs a credit agreement, the creditor must state, in writing, the interest charge and the APR. Hence, the law does make consumers aware of the cost of credit so that they are able to compare terms. This is particularly helpful to buyers who might pay back loans over different periods of time. In 1980, Congress passed the Truth-in-Lending Simplification and Reform Act, which further defined consumer rights.

There is a formula that can be used to compute the true annual interest rate on installment loans. It is

$$i = \frac{2nr}{n + 1}$$

where

i = the true annual interest rate

n = number of payments

r = the stated rate of interest

Applying this formula to the previous example, where $n = 6$ and $r = 10\% = 0.10$, we have

$$i = \frac{2nr}{n + 1}$$

$$i = \frac{2 \times 6 \times 0.10}{6 + 1} = \frac{12 \times 0.10}{7}$$

$$i = \frac{1.2}{7}$$

$$i = 0.171 \qquad \textit{(to the nearest thousandth)}$$

$$i = 17.1\% \qquad \textit{(to the nearest tenth of a percent)}$$

EXAMPLE 1 ***Finding the True Rate of Interest*** Stella Frisco purchased a stereo system for \$1,000. She purchased the system on the installment plan, paying \$200 down and agreeing to pay the balance in 12 monthly payments. The finance charge was 10% simple interest on the balance. What is the true annual rate of interest?

Solution Using the formula $i = \frac{2nr}{n+1}$, we have $n = 12$, $r = 10\% = 0.10$.

Therefore,

$$i = \frac{2nr}{n+1}$$

$$i = \frac{2 \times 12 \times 0.10}{12+1} = \frac{24(0.10)}{13}$$

$$i = \frac{2.4}{13} = 0.185 \quad \textit{(to the nearest thousandth)}$$

$$i = 18.5\% \quad \textit{(to the nearest tenth of a percent)}$$

NW ***Now Work Problem 3.***

EXAMPLE 2 ***Determining the Finance Charge*** A giant-screen television has a cash price of \$1500 and can be purchased on the installment plan with a 12% finance charge. The television will be paid for in 12 monthly payments. If there was no down payment, find

a. the amount of the finance charge
b. the total price
c. the amount of each monthly payment
d. the true annual interest rate.

Solution

a. The finance charge, or interest, is determined by the formula

$$I = Prt$$

$$I = \$1500 \times 0.12 \times 1$$

$$I = \$180 \quad \text{(finance charge)}$$

b. The total price is \$1500 + \$180 = \$1680.

c. The monthly payment is \$1680 ÷ 12 = \$140.

d. The true annual interest rate is determined by the formula

$$i = \frac{2nr}{n+1}$$

$$i = \frac{2 \times 12 \times 0.12}{12+1} = \frac{24(0.12)}{13}$$

$$i = \frac{2.88}{13} = 0.222 \quad \textit{(to the nearest thousandth)}$$

$$i = 22.2\% \quad \textit{(to the nearest tenth of a percent)}$$

NW ***Now Work Problem 1.***

EXAMPLE 3 ***Finding the Total Cost*** Kevin Friday bought a used car for \$3000, which was financed at \$140 per month for 24 months. There was no down payment. Find

a. the total price
b. the finance charge
c. the true annual interest rate

Solution

a. The total price is determined by multiplying the number of payments (24) by the amount of each payment (\$140).

$$24 \times \$140 = \$3360 \quad \text{total price}$$

b. The finance charge is the total price minus the cash price. That is,

$$\$3360 - \$3000 = \$360 \quad \text{finance charge (or interest)}$$

c. To find the true annual interest rate we use the formula $i = \frac{2nr}{n+1}$, but then we need to determine r, the stated interest rate. We do this by using the formula $I = Prt$, where $I = \$360$, $P = \$3000$, and $t = 2$ years (24 monthly payments). Substituting, we have

$$\begin{aligned} I &= Prt \\ \$360 &= \$3000 \times r \times 2 \\ 360 &= 6000 \times r \\ \frac{360}{6000} &= r \\ r &= 0.06 = 6\% \end{aligned}$$

Now we can use the true annual interest rate formula,

$$i = \frac{2nr}{n+1}, \quad \text{where } n = 24 \text{ and } r = 0.06$$

Substituting in the formula, we have

$$i = \frac{2 \times 24 \times 0.06}{24 + 1} = \frac{48(0.06)}{25}$$

$$i = \frac{2.88}{25} = 0.115 \quad \textit{(to the nearest thousandth)}$$

$$i = 11.5\% \quad \textit{(to the nearest tenth of a percent)}$$

NW ***Now Work Problem 9.***

EXAMPLE 4 ***Determining the Monthly Payment*** Nikki Jesske borrowed \$800 from her local credit union for eight months at 6% interest. She agreed to repay the loan by making eight equal monthly payments. Find

a. the finance charge
b. the total to be repaid
c. the monthly payment
d. the true annual rate of interest

Solution

a. The finance charge is the interest on the loan and it is determined by $I = Prt$, where $P = \$800$, $r = 0.06$, and $t = 8$ months. Substituting, we have

$$I = Prt$$
$$I = \$800 \times 0.06 \times \frac{8}{12}$$
$$I = 48 \times \frac{8}{12} = \$32$$

The finance charge is \$32.

b. The total to be repaid is $\$800 + \$32 = \$832$.

c. The monthly payment $= \frac{\$832}{8} = \$104$.

d. The true annual interest rate is determined by the formula $i = \frac{2nr}{n+1}$, where $n = 8$ and $r = 0.06$. Substituting, we have

$$i = \frac{2 \times 8 \times 0.06}{8 + 1} = \frac{(16)(0.06)}{9}$$
$$i = \frac{0.96}{9} = 0.107 \quad \textit{(to the nearest thousandth)}$$
$$i = 10.7\% \quad \textit{(to the nearest tenth of a percent)}$$

NW ***Now Work Problems 5 and 7.***

EXAMPLE 5 ***Determining the Finance Charge*** Joseph Mooney purchased a television for \$500. He paid \$50 down and agreed to pay the balance in 24 equal monthly payments at 12% simple interest.

a. What was the amount of each payment?

b. What was the true annual interest rate (correct to the nearest tenth of a percent)?

c. How much more did Joseph pay for the television by purchasing it on the installment plan?

Solution

a. The down payment of \$50 is subtracted from the price of \$500 to find the amount financed.

$$\text{Amount financed} = \$500 - \$50 = \$450$$

Next we compute the interest on \$450 at 12% for two years. Using the formula $I = Prt$, where $P = \$450$, $r = 0.12$, and $t = 2$, we have

$$I = Prt$$
$$I = \$450 \times 0.12 \times 2$$
$$I = \$108$$

Hence, the amount to be repaid is $\$450 + \$108 = \$558$. The number of payments is 24 and therefore each payment is equal to $\$558 \div 24$ or \$23.25.

b. The true annual interest rate is determined by the formula $i = \frac{2nr}{n+1}$, where $n = 24$ and $r = 0.12$. Substituting, we have

$$i = \frac{2 \times 24 \times 0.12}{24 + 1} = \frac{(48)(0.12)}{25}$$

$$i = \frac{5.76}{25} = 0.230 \quad (\textit{to the nearest thousandth})$$
$$i = 23.0\% \quad (\textit{to the nearest tenth of a percent})$$

c. The total amount Joseph paid for the television consists of the down payment, plus the balance, plus the finance charge. That is,

$$\$50 + \$450 + \$108 = \$608$$

The original cost of the television was \$500. The amount paid for the television was \$608. Hence, Joseph paid \$608 − \$500 = \$108 more for the television by purchasing it on the installment plan.

NW ***Now Work Problem 15.***

Exercises for Section 12.9

For Exercises 1 and 2, find the total cost and the finance charge for each of the given purchases.

	Amount Financed	*Down Payment*	*Cash Price*	*Number of Payments*	*Monthly Payment*
NW **1.**	\$ 500	\$25	\$ 525	24	\$25
2.	\$1000	\$75	\$1075	36	\$35

For Exercises 3 and 4, find the true annual interest rate (correct to the nearest tenth of a percent) for each of the given purchases.

	Amount Financed	*Finance Charge*	*Number of Monthly Payments*
NW **3.**	\$ 300	\$ 27	12
4.	\$3000	\$810	36

NW **5. Buying a television** Carl Thomas purchased a television set advertised for \$500. He bought the television on the installment plan, paying \$100 down and agreeing to pay the balance in 12 monthly payments. The store used a finance charge of 12% simple interest on the balance.

a. What was the amount of each payment?

b. What was the true annual interest rate (correct to the nearest tenth of a percent)?

6. Purchasing a used car A used car is advertised for \$3000. It may be purchased on the installment plan by paying \$400 down and agreeing to pay the balance plus 12% simple interest on the balance in 24 monthly payments.

a. What would be the amount of each payment?

b. What would be the true annual interest rate (correct to the nearest tenth of a percent)?

7. Acquiring furniture An easy chair is advertised for \$450. NW It may be purchased on the installment plan by paying \$100 down and agreeing to pay the balance plus 18% simple interest on the balance in six monthly payments.

a. What is the amount of each payment?

b. What is the true annual interest rate (correct to the nearest tenth of a percent)?

8. Acquiring furniture The Jaxsons purchased a new table priced at \$500. They put no money down and agreed to pay for it in 24 monthly payments. The store stated a finance charge of 12% simple interest.

a. What was the amount of each payment?

b. What was the true annual interest rate (correct to the nearest tenth of a percent)?

c. How much more will the Jaxsons pay for the table by purchasing it on the installment plan?

9. Obtaining a computer A Notebook PC is advertised for NW \$600. It may be purchased on the installment plan by paying \$50 and agreeing to pay the balance plus 18% simple interest on the balance in 24 monthly payments.

a. What is the finance charge?

b. What is the amount of each payment (correct to the nearest penny)?

c. What is the total cost of the Notebook PC?

d. What is the true annual interest rate (correct to the nearest tenth of a percent)?

10. Purchasing supplies Kristine Stewart purchased \$500 worth of items from Benny's Supply House. She put no money down and agreed to pay for the items in 10 equal monthly payments. The Supply House stated a finance charge of 8% simple interest.

a. What was the amount of each payment?

b. What was the true annual interest rate (correct to the nearest tenth of a percent)?

c. How much will Kristine pay for the items in total by purchasing them on the installment plan?

11. **Purchasing appliances** Elizabeth O'Rourke bought an assortment of household cleaning appliances priced at $1200. She put $100 down and agreed to pay the balance in 36 equal monthly payments at 12% simple interest.
    a. What was the amount of each payment?
    b. What was the true annual interest rate (correct to the nearest tenth of a percent)?
    c. How much more will Elizabeth pay for the cleaning appliances by purchasing them on the installment plan?

12. **Buying a coat** Maria Raffel purchased a $350 coat at Kaufmann's Department Store. She paid $25 down and agreed to pay the balance in 12 equal monthly payments. The finance charge was 14% simple interest.
    a. What was the amount of each payment?
    b. What was the true annual interest rate (correct to the nearest tenth of a percent)?
    c. How much more will Maria pay for the coat by purchasing it on the installment plan?

13. **Buying a suit** Fred Gates purchased a suit, with a cash price of $220, on the installment plan. He put $20 down and agreed to pay $34.50 per month for six months. Find
    a. the total price
    b. the finance charge
    c. the true annual interest rate (correct to the nearest tenth of a percent)

14. **Borrowing money** Jane Edgar borrowed $1200 at 8% for 18 months from her credit union. If she agreed to repay the loan in 18 equal monthly payments, find
    a. the total amount repaid
    b. the finance charge
    c. the true annual interest rate (correct to the nearest tenth of a percent)

NW 15. **Borrowing money** Mary Ragusa borrowed $800 from a local bank. She agreed to repay the loan by making 12 monthly payments of $72 each. Find
    a. the total amount repaid
    b. the finance charge
    c. the true annual interest rate (correct to the nearest tenth of a percent)

16. **Remodeling a room** Mike Goho remodeled his family room for $7991 on the installment plan. He made a down payment of $991 and agreed to pay $192.50 per month for 48 months. Find
    a. the total price
    b. the finance charge
    c. the true annual interest rate (correct to the nearest tenth of a percent)

17. **Purchasing a used car** Bill Ketes purchased a $4000 used car on the installment plan. He made a down payment of $250 and agreed to pay $150 per month for 30 months. Find
    a. the total price
    b. the finance charge
    c. the true annual interest rate (correct to the nearest tenth of a percent)

18. **Buying a washing machine** Pam Dretto bought a new washing machine, with a cash price of $600, on the installment plan. She made a down payment of $75 and agreed to pay $69 per month for 8 months. Find
    a. the total price
    b. the finance charge
    c. the true annual interest rate (correct to the nearest tenth of a percent)

19. **Purchasing a boat** Stanley Grabowski purchased a new boat advertised at $8999. He made a down payment of $1000 and agreed to pay $190.35 per month for 60 months. Find
    a. the total price
    b. the finance charge
    c. the true annual interest rate (correct to the nearest tenth of a percent)

20. **Buying a camper** Thomas Walters bought a used camper priced at $8000. He made a down payment of $1000 and agreed to make 48 monthly payments of $205.10 each. Find
    a. the total price
    b. the finance charge
    c. the true annual interest rate (correct to the nearest tenth of a percent)

21. **Obtaining a loan** Marion Starks borrowed $6000 at 10% for 2 years. She agreed to repay the loan by making 24 equal monthly payments. Find
    a. the finance charge
    b. the total to be repaid
    c. the monthly payment
    d. the true annual rate of interest (correct to the nearest tenth of a percent)

22. **Borrowing money** Mike Berdinka borrowed $5000 for 4 years at 12% from a local bank. He agreed to repay the loan by making 48 equal payments. Find
    a. the finance charge
    b. the total to be repaid
    c. the monthly payment (correct to the nearest penny)
    d. the true annual rate of interest (correct to the nearest tenth of a percent)

23. **Obtaining a loan** Jill Johnson borrowed $900 at 8% for 18 months from the credit union. He agreed to repay the loan by making 18 equal monthly payments. Find
    a. the finance charge
    b. the total to be repaid
    c. the monthly payment
    d. the true annual rate of interest (correct to the nearest tenth of a percent)

24. **Acquiring a Television** A television set is advertised for $290 cash, or on the installment plan, for no money down and $30 per month for 12 months. Find
    a. the total to be repaid
    b. the finance charge
    c. the true annual rate of interest (correct to the nearest tenth of a percent)

Writing Mathematics

25. Explain what is meant by the term *credit.*
26. What is a *finance charge?*
27. What is an *annual percentage rate?*

Challenge Exercises

28. Julia Ratigan wants to increase her inventory for the holidays. Julia estimates that on January 1, she will be able to pay back $25,000. If Julia wants the money on October 1, and the bank is willing to loan her the money at a rate of 8.5%, how much should she borrow on October 1?
29. David Rogachefsky purchased a used car for $6000 on the installment plan. He made a down payment of $1000 and agreed to pay $200 per month for 36 months. Find **(a)** the total price of the car, **(b)** the finance charge, and **(c)** the true annual interest rate.

Research/Group Activities

30. Sometimes an installment loan is paid off, or paid in full, before the date of the last payment. This is called a loan payoff, and it is the amount paid by the borrower to the lender to repay a loan before it is due. When the full amount is paid prior to the maturity date the borrower is usually entitled to a refund on the unearned finance charges due to an early payoff. A common method used to calculate this refund is the *Rule of 78*. Research the *Rule of 78* and prepare an example for class illustrating how the rule is used to determine the final payment.
31. Two common ways of purchasing items on credit are installment buying (discussed in this section) and *open-end credit*. Credit cards are examples of open-end credit because there is no fixed number of payments. Credit card companies typically calculate the finance charges using the *average daily balance* method as opposed to the *unpaid balance* method. Prepare a report comparing these two techniques for calculating finance charges.

Just for fun

Ask a person to write his or her age on a piece of paper. Next have the person perform the following calculations: (a) Multiply the number by 2; (b) add 5 to the result; (c) multiply the new number by 50; (d) subtract the number of days in a non-leap year; (e) add any change in his or her pocket less than one dollar; (f) add 115 to the result; and (g) ask the person to read you the resulting number. The first two digits are the person's age, and the last two digits are the amount of change. (*Note:* The person should be 10 or older for this sequence of calculations.)

12.10 MORTGAGES

The largest purchase that a person is likely to make in a lifetime is a house. Typically, most people purchase houses by making a specified down payment and borrowing the balance from a bank or other lending institution. The borrower agrees to make regular payments on the principal and interest until the loan is paid off.

Historical Note

According to historical records, the first bank was founded in 1148. It was the Bank of San Giorgio in Genoa, Italy. Before that time, the lending of money was a personal matter, but the Romans did make some attempt to control money lending by passing certain regulations.

The first bank in the United States was founded in 1781 in Philadelphia. It was the Bank of North America, and it stayed in business only until 1784. Later, in 1791, Congress issued a charter for another bank in Philadelphia called the First Bank of the United States. After that, state governments began to issue charters to banks. The National Banking System was then established by the federal government, and it attempted to guide and regulate banking in the United States.

On December 23, 1913, Congress established the Federal Reserve System, the central banking system of the United States, to improve supervision of banking. Today, the primary function of the system is to promote a flow of credit and money that will determine orderly economic growth, a stable dollar, and a balance of international payments.

This process is called **amortizing** the loan. Mortgages on new homes are usually for a period of 15, 20, 25, or 30 years. Mortgages on older homes sometimes run for shorter periods of time.

The monthly payments necessary to amortize a loan (pay off the principal and interest) are compiled in tables called **amortization tables**. Table 12.4 lists typical monthly payments per \$1000 of a mortgage. The amount that must be paid monthly for principal and interest is given for selected interest rates and different periods of time.

TABLE 12.4 Monthly Mortgage Payment per \$1000

	Length of Mortgage		
Rate (%)	15 Years	20 Years	30 Years
6.0	\$ 8.44	\$ 7.16	\$ 6.00
6.5	8.71	7.46	6.32
7.0	8.99	7.76	6.66
7.5	9.27	8.06	6.99
8.0	9.56	8.36	7.34
8.5	9.85	8.68	7.69
9.0	10.14	9.00	8.05
9.5	10.44	9.32	8.41
10.0	10.75	9.65	8.78
10.5	11.05	9.98	9.15
11.0	11.37	10.32	9.52
11.5	11.68	10.66	9.90
12.0	12.00	11.01	10.29
12.5	12.33	11.36	10.67

Note: This table lists payments per \$1000 for "traditional" fixed-rate mortgages. Many lending institutions offer another type of mortgage: variable-rate mortgages, where the interest rate may be raised or lowered depending on the economic climate.

If a person assumes a \$30,000 mortgage for 20 years at $12\frac{1}{2}\%$, then according to Table 12.4, the monthly payment for each \$1000 is \$11.36. The mortgage is for \$30,000, so we multiply \$11.36 by 30 to find the total monthly payment. Hence, the monthly payment on this mortgage is $30 \times \$11.36$ or \$340.80. It should be noted that actual payments are higher than shown because the mortgagee usually has to make monthly payments on property taxes and insurance. For example, if property taxes were \$1200 a year, then an additional 100 ($\$1200 \div 12$) would have to be paid each month. Also, life and fire insurance might cost \$20 per month. Therefore, the total monthly payment would be $\$340.80 + \$100 + \$20 = \$460.80$.

EXAMPLE 1 ***Monthly Mortgage Payment*** The Woodsons obtained a \$60,000 mortgage for 30 years at 8%. What is their monthly payment for principal and interest?

Solution The monthly payment per \$1000 at 8% for a 30-year mortgage is \$7.34. The mortgage is for \$60,000. Hence, we multiply \$7.34 by 60:

$$60 \times \$7.34 = \$440.40 \quad (\textit{monthly payment})$$

EXAMPLE 2 ***Total Mortgage Interest*** How much total interest will the Woodsons (Example 1) pay on their mortgage?

Solution Their monthly payment is $440.40 and the mortgage is for 30 years. Hence, there will be $30 \times 12 = 360$ payments. Then

$$360 \times \$440.40 = \$158{,}544 \quad \textit{(total payment)}$$
$$\$158{,}544 - \$60{,}000 = \$98{,}544 \quad \textit{(total interest)}$$

NW ***Now Work Problem 11.***

EXAMPLE 3 ***Comparing Amounts of Interest*** How much interest would the Woodsons save if they assume the same $60,000 mortgage at 8% for 20 years instead of 30 years?

Solution The monthly payment per $1000 at 8% for a 20-year mortgage is $8.36. The mortgage is for $60,000. Therefore,

$$\text{monthly payment} = 60 \times \$8.36 = \$501.60.$$

The mortgage is for 20 years. Hence, there will be $12 \times 20 = 240$ payments. Then

$$240 \times \$501.60 = \$120{,}384 \quad \textit{(total payment)}$$
$$\$120{,}384 - \$60{,}000 = \$60{,}384 \quad \textit{(total interest)}$$
$$\text{Interest on 30-year mortgage} = \$98{,}544$$
$$\text{Interest on 20-year mortgage} = \underline{\$60{,}384} \quad \textit{(subtracting)}$$
$$\$38{,}160 \quad \textit{(interest saved)}$$

EXAMPLE 4 ***Comparing Amounts of Interest*** The Robinsons decided to purchase a home that is priced at $95,000. They made a down payment of 20% and borrowed the remainder. Bank A will give them a 30-year mortgage at 8%. Bank B will give them a 20-year mortgage at 9%. Which mortgage should the Robinsons assume so that they will pay the smaller amount of interest? How much will they save?

Solution First, we must determine how much the Robinsons will need to borrow. Their down payment is

$$20\% \text{ of } \$95{,}000 = 0.20 \times \$95{,}000 = \$19{,}000$$
$$\text{Mortgage} = \$95{,}000 - \$19{,}000 = \$76{,}000$$

Bank A: 30-year mortgage at 8%

$$\$7.34 \times 76 = \$557.84 \quad \textit{(monthly payment)}$$
$$12 \times 30 = 360 \text{ payments;} \quad 360 \times \$557.84 = \$200{,}822.40 \quad \textit{(total payment)}$$
$$\$200{,}822.40 - \$76{,}000 = \$124{,}822.40 \quad \textit{(total interest)}$$

Bank B: 20-year mortgage at 9%

$$\$9.00 \times 76 = \$684 \quad \textit{(monthly payment)}$$
$$20 \times 12 = 240 \text{ payments;} \quad 240 \times \$684 = \$164{,}160 \quad \textit{(total payment)}$$
$$\$164{,}160 - \$76{,}000 = \$88{,}160 \quad \textit{(total interest)}$$

The Robinsons will save money by assuming the mortgage with Bank B. They will save $\$124{,}822.40 - \$88{,}160 = \$36{,}662.40$ in interest payments.

NW ***Now Work Problem 13.***

Note that in both cases in Example 4 the interest is more than the face value of the loan. This is the reason banks lend money—to make money! Homebuyers should be aware of the costs involved in assuming a mortgage. A disadvantage of a mortgage is the total amount of interest to be paid on such a loan. But people are willing to do this for a variety of reasons. Advantages to consider include (1) the satisfaction of owning a home; (2) the increasing value of a home usually exceeding the rate of inflation (a house that sells for $60,000 this year would sell for $63,000 next year with a 5% annual inflation rate); (3) repayment of the loan with cheaper dollars because of inflation; and (4) protection against inflation provided by owning a home. Another thing to consider is that the annual interest paid on a mortgage loan may be used as an itemized deduction on an income tax return for that year, which results in an income tax saving. Also, at some point in time the mortgage will be paid off. There will be no more mortgage payments to make!

Exercises for Section 12.10

For Exercises 1–10, find the monthly payment for principal and interest for each mortgage. Use Table 12.4 in this section.

	Amount of Mortgage	Interest Rate (%)	Term of Mortgage (years)
1.	$50,000	7	20
2.	80,000	6	30
3.	75,000	8	30
4.	95,000	7	20
5.	20,000	7.5	15
6.	30,000	10.5	30
7.	25,000	8.5	20
8.	45,000	10	20
9.	55,000	7.5	30
10.	22,000	12	15

NW **11. Monthly payment and total interest** The Smiths assumed a $40,000 mortgage for 20 years at 10.5%. What is their monthly payment? How much total interest will the Smiths pay on their mortgage?

12. Monthly payment and total interest The Garcias assumed a $30,000 mortgage for 30 years at 12%. What is their monthly payment? How much total interest will the Garcias pay on their mortgage?

NW **13. Comparing loans** The Donovans need to borrow $30,000 to buy a house. Bank A will give them a 30-year mortgage at 9%. Bank B will give them a 20-year mortgage at 10.5% Which mortgage should the Donovans assume so that they will pay the smaller amount of interest? How much will they save?

14. Comparing loans The Sullivans need to borrow $40,000 to buy a house. Bank A will give them a 30-year mortgage at 9.5% Bank B will give them a 20-year mortgage at 11% Which mortgage should the Sullivans assume so that they will pay the smaller amount of interest? How much will they save?

15. Purchasing a condo The Seteks purchased a condominium for $100,000. They made a down payment of 20% and borrowed the remainder. If they assumed a 8% mortgage for 30 years, what is their monthly payment? How much total interest will the Seteks pay on their mortgage?

16. Buying a house The Nennos purchased a house for $150,000. They made a down payment of 25% and borrowed the remainder. If they assumed a 7.5% mortgage for 20 years, what is their monthly payment? How much total interest will the Nennos pay on their mortgage?

17. Buying a house Jim and Maria Raffel purchased a house for $125,000. They made a down payment of 20% of the purchase price, and obtained an 8%, 30-year mortgage for the balance. Find their monthly payment and the total interest the Raffels will pay on their mortgage.

18. Purchasing a condo Norma and Bob Caryl purchased a condominium for $90,000. They made a down payment of 15% and borrowed the remainder. If they assumed a 9% mortgage for 20 years, what is their monthly payment? How much total interest will the Caryls pay on their mortgage?

19. Expanding a business Griffin's Hardware will add on to its existing store to accommodate an expanding business.

The owner negotiated a $50,000 mortgage loan at 10% for 20 years. What is the monthly payment? How much total interest will Griffin's Hardware pay on the mortgage?

20. **Buying a warehouse** Josco Industries purchased a new warehouse for $175,000. The company made a down payment of 20% and obtained a 30-year mortgage at 10.5%. What is the monthly payment? How much total interest will Josco Industries pay on the mortgage?

Writing Mathematics

21. What is the difference between a mortgage and a mortgagee?

22. What does it mean to amortize a loan?

Challenge Exercises

23. The Berdinkas purchased a house priced at $120,000. They made a down payment of 10% and agreed to $792.72 per month for 30 years. What is the total amount the house will actually cost them? How much interest will they pay?

24. Bill Ketes purchased a building for $67,200 which he immediately rented to LIPCO Inc. for $640 per month. Bill kept the building for five years. All expenses for this period totaled $14,880. Bill then sold the building for $70,400. What was his net gain from this property?

Research/Group Activities

25. Select at least three different banks in your area and determine what types of mortgages are available. Information should include current interest rates, amount of down payment, length of mortgage, and availability. Present these findings to your classmates.

26. Is it better to rent or to own? Research this topic and determine the advantages and disadvantages of renting or owning a residence. Prepare a report for class.

Math Connections

Inflation

One of the most frequently used terms in any discussion of finances is *inflation*. But what exactly is inflation? In simple terms, inflation is the decaying of money—it describes the decline in the value of money relative to the goods or services that current consumer dollars can purchase. As prices increase, the purchasing power of money declines. Inflation generally results from two entities: (1) specific *current economic conditions* (e.g., What is the present state of factory orders? or What is the present state of productivity?) and (2) the *anticipation of future developments* (e.g., What impact will the current winter weather have on next summer's crops? or What will the current Mideast conflict have on future oil production?). These two actions—the present state of the economy coupled with what "experts" *think* will happen in the future—cause the demand for goods and services to exceed the supply available at their existing prices.

Inflation initially has a positive effect on businesses because overall revenue rises as a result of price increases. Revenue increases in turn lead to increases in profits because the cost of doing business (e.g., employee wages, production costs, etc.) lags behind the increase in prices. Consumer spending also might increase initially because some people begin to adopt the mind-set of "buy now because it will cost more later." For example, the potential for higher real estate prices might attract buyers.

Eventually, though, these temporary gains are short-lived and inflation begins to disrupt normal economic activities, particularly if the pace of inflation fluctuates. For example, as inflation increases, so do interest rates. This leads directly to an increase in business costs, a decrease in consumer spending, and a diminished value of stocks and bonds. As mortgage interest rates increase, home prices escalate as well and this has a negative impact on housing construction. Inflation also erodes the real purchasing power of current incomes and accumulated financial assets, which can result in a pronounced reduction in personal consumption. This is especially true if consumers are unwilling to tap into their savings (assuming they have any) or to borrow money (and thereby increase their personal debt) to purchase goods or services. For example, if home mortgage rates are hovering around 7%, you might consider buying a new house. However, if interest rates skyrocketed to 12%, you probably would not consider making such a purchase. Businesses also suffer on at least two fronts: Overall economic activity declines, which reduces or restricts profits, and employees begin to demand relief from inflation through *cost-of-living pay raises*. In short, inflation contributes significantly to widespread economic uncertainty.

One of the most widely used measures of inflation is the **Consumer Price Index** (CPI), which is a measure of the average change over time in the prices of various goods and services paid by urban consumers. The CPI can also be used to adjust other aspects of the economy, such as retail sales and hourly and weekly earnings, and to translate these aspects into inflation-free dollars. The CPI frequently serves as the basis for adjusting a person's Social Security eligibility and provides the necessary information for providing cost-of-living wage adjustments to millions of American workers.

To experiment with the effect inflation has on the purchasing power of a dollar, go to the Web site **http://woodrow.mpls.frb.fed.us/research/data/us/calc**, which is sponsored by the Woodrow Federal Reserve Bank of Minneapolis. This site features a CPI calculator that enables you to compare the value of a dollar between any two years (currently from 1913 to 2003). For example, if in 1970 we bought goods or services for $3000, in 2003, the same goods or services would cost us $14,211.34.

Just for fun

Whose picture is on the front of a $10 bill? What is pictured on the back of it?

CHAPTER REVIEW MATERIALS

Summary/Chapter 12

12.2 Ratio and Proportion

- The *ratio* of two quantities a and b is the quotient or indicated quotient obtained by dividing a by b. The ratio of a to b is written as $a \div b$, a/b, $a:b$.
- A *proportion* is defined to be an equality of two ratios. In any proportion, the product of the means equals the product of the extremes. That is, if $\frac{a}{b} = \frac{c}{d}$, then

12.3 Percents, Decimals, and Fractions

- A *percent* is a ratio with a denominator of 100. The symbol for percent is %.
- To convert a percent to a decimal, drop the percent sign and move the decimal point two places to the left.
- To change a decimal to a percent, move the decimal point two places to the right and add a percent sign.
- To change a fraction to a percent, express the fraction as a decimal, and then change the decimal to a percent.

12.4 Markups and Markdowns

- *Markup* is the difference between the selling price and the cost of an item.
- *Markdown* is the difference between the original selling price and the sale price.

12.5 Simple Interest

- *Simple interest* is found by the formula, $I = Prt$ where I = interest, P = principal, r = rate of interest, t = time in years.

12.6 Compound Interest

- *Compound interest* occurs when interest due at the end of a certain period of time is added to the principal, and both the principal and the interest from the first period earn interest for the next period.
- Table 12.1 in Section 12.6 shows the amount that $1 will accumulate when interest is paid at the indicated rate for the indicated number of interest periods.

12.7 Effective Rate of Interest

- *Effective annual interest rate* is the annual interest rate that gives the same yield as a nominal interest rate compounded several times a year.
- We can determine the effective annual rate by means of the formula $E = (1 + r)^n - 1$ where E = effective rate, n = number of payment periods per year, and r = interest rate per period.

12.8 Life Insurance

- A *whole life insurance policy* is one in which the insurance company will pay the face value of the policy to the beneficiary when the insured person dies, regardless of when this happens.
- A *term insurance policy* is one where the insurance company will pay the face value of the policy to a beneficiary when the insured person dies, provided this occurs during the period of time (the *term*) stated in the policy.
- An *endowment policy* is similar to a term policy in that it insures the life of the insured for a specific term. If the insured is living at the end of the term, then the face value of the policy will be paid to the policyholder.

12.9 Installment Buying

- Installment buying allows a consumer to obtain possession of an item immediately by agreeing to pay the purchase price plus an additional charge in a series of regular payments.

- *True annual interest rate* also know as *annual percentage rate (APR)* is determined by the formula

$$i = \frac{2nr}{n+1}$$ *where i = the true annual interest rate, n = number of payments, r = the stated interest rate*

12.10 Mortgages

- The monthly payments necessary to amortize a loan (pay off the principal and interest) are compiled in tables called *amortization tables.* Table 12.4 lists typical monthly payments per $1000 of mortgage for various interest rates and periods of time.

Key Terms and Concepts

amount, *680*
APR, *699*
compound amount, *685*
compound interest, *683*
decimal, *666*
effective (annual) rate, *689*
fraction, *668*
markdown, *674*
markup, *670*
mortgage, *706*
percent, *666*
premium, *693*
principal, *679*
proportion, *662*
ratio, *660*
simple interest, *679*
underwriter, *693*

Review Exercises

12.2 Ratio and Proportion

1. Express each ratio in simplest form.

a. 48 : 120 **b.** $\frac{54}{36}$

c. 10 to 25 **d.** $1\frac{1}{2} : 2\frac{1}{4}$

2. Do the ratios form a proportion? Write yes or no.

a. $\frac{9}{5} : \frac{2}{3}$ **b.** $\frac{20}{24} : \frac{5}{6}$

c. $\frac{15}{19} : \frac{90}{119}$ **d.** $\frac{0.4}{0.8} : \frac{0.6}{1.2}$

3. Find the value of x in each proportion.

a. $2 : x = 5 : 10$ **b.** $x : 6 = 5 : 10$

c. $4 : 3 = x : 6$ **d.** $2 : 8 = 6 : x$

12.3 Percents, Decimals, and Fractions

4. Express each percent as a fraction in simplest form.

a. 38% **b.** $12\frac{1}{2}\%$

c. 2.3% **d.** 125%

5. Express each percent as a decimal.

a. 48% **b.** $10\frac{1}{2}\%$

c. 4.7% **d.** 0.25%

6. Express each decimal as a percent.

a. 0.82 **b.** 1.3

c. 0.78 **d.** 2.134

7. Express each fraction as a percent.

a. $\frac{3}{5}$ **b.** $\frac{1}{8}$

c. $2\frac{1}{5}$ **d.** $\frac{3}{16}$

8. During a certain television program, pollsters found that 80 out of 200 people surveyed were watching the program. Predict the number of viewers out of 50,000 people.

9. Twenty-three percent of a certain group of mathematics students had previously studied algebra. If 46 students in this group had studied algebra, how many students are in the group?

12.4 Markups and Markdowns

10. Cost of a suit A suit sells for $125. The markup is 30% of the cost. What is the cost of the suit, correct to the nearest cent?

11. Price of a television A portable television costs a retailer $60. If the markup is 25% of the selling price, find the selling price, correct to the nearest cent.

12. Cost of a couch Find the markup and the cost, correct to the nearest cent, of a new couch that retails for $400, if the percent markup on the selling price is 40%.

13. Sale Price of a top coat A clothing store sells topcoats for $124.99. During a clearance sale the price is reduced by 25%. What is the dollar markdown and what is the sale price, correct to the nearest cent?

12.5 Simple Interest

14. Find the amount of simple interest on a $5000 loan for 3 years at 11%.

15. Find the amount of simple interest on a $10,000 loan for 9 months at 12%.

16. How much interest will $10,000 earn in 4 years at 4.5% simple interest? How much in 4 months?

17. Repaying a loan What is the amount that must be repaid on a $6000 loan for 42 months at 9% simple interest? How much interest will be paid for the loan?

12.6 Compound Interest

18. Compound amount Use Table 12.1 in Section 12.6 to find the compound amount and the compound interest on $4000 invested at 6% compounded semiannually for 10 years.

19. Compound interest Use Table 12.1 in Section 12.6 to find the compound amount and the compound interest on $5000 invested at 12% compounded quarterly for 6 years.

12.7 Effective Rate of Interest

20. What is the effective rate if money is invested at 5% compounded

a. annually **b.** semiannually **c.** quarterly?

21. What is the APR?

12.8 Life Insurance

22. Life insurance Lewis Scott is 25 years old and wants to purchase a $30,000 straight life policy. How much will Lewis save if he pays his premium annually instead of monthly? (Use Tables 12.2 and 12.3 in Section 12.8.)

23. Insurance premiums Use Table 12.2 in Section 12.8 to find the annual premium for each type of insurance policy listed.

	Face Value of Policy	Age At Issue	Type of Policy
a.	$20,000	30	5-year term
b.	$15,000	25	straight life
c.	$30,000	35	20-year limited-payment life
d.	$40,000	40	20-year endowment

12.9 Installment Buying

24. Purchasing a used car Theresa Seager purchased a used car for $5000 on the installment plan. She made a down payment of $1000 and agreed to pay $165 per month for 36 months. Find

a. the total price

b. the finance charge

c. the true annual interest rate (correct to the nearest tenth of a percent)

25. Buying a digital camera A digital camera is advertised for $600. It may be purchased on the installment plan by paying $50 down and agreeing to pay the balance plus 12% simple interest on the balance in 24 monthly payments.

a. What is the finance charge?

b. What is the amount of each payment?

c. What is the total cost of the digital camera?

d. What is the true annual interest rate (correct to the nearest tenth of a percent)?

12.10 Mortgages

26. Find the monthly payment for the principal and interest for each of the following mortgages. (Use Table 12.4 in Section 12.10.)

	Amount of Mortgage	Interest Rate (%)	Term of Mortgage (years)
a.	$30,000	6	15
b.	$45,000	10.5	30
c.	$60,000	12	30
d.	$55,000	6.5	20

27. Mortgage payment The Stones assumed a $50,000 mortgage for 30 years at $12\frac{1}{2}$%. Use Table 12.4 in Section 12.10 to find their monthly payment. How much total interest will the Stones pay on their mortgage?

28. Comparing mortgages The Rohlins need to borrow $75,000 to purchase a house. Bank A will give them a 30-year mortgage at 8%. Bank B will give them a 20-year mortgage at 10%. Which mortgage should the Rohlins assume so that they will pay the smallest amount of interest? How much will they save? (Use Table 12.4 in Section 12.10.)

Just for fun

True or false: Arabic numerals were invented by the Arabs.

Chapter Quiz

For Exercises 1–4, find the value of x in each proportion.

1. $\frac{7}{3} = \frac{21}{x}$ **2.** $\frac{3}{x} = \frac{10}{12}$

3. $8 : x = 12 : 18$ **4.** $6 : 24 = 18 : x$

For Exercises 5–10, complete the table by expressing each of the given entries in equivalent forms.

	Fraction	Decimal	Percent
5.	$\frac{3}{5}$		
6.	$\frac{7}{8}$		
7.		1.25	
8.		0.035	
9.			74%
10.			5%

11. **Markup** A coat sells for $200. The markup is 25% of the cost. What is the cost of the coat?

12. **Markup** An auto alarm system costs a retailer $150. If the markup is 40% of the selling price, find the selling price.

13. **Markdown** An appliance store normally sells a certain model television for $399.99. During a recent sale the price was reduced by 40%. What was the dollar markdown and what was the sale price?

14. **Simple interest** Find the amount of simple interest on a $5400 loan for 9 months at 8.5%.

15. **Insurance** Sharon Thomas is 30 years old and wants to purchase a $25,000 straight life policy. How much will Sharon save if she pays her premium annually instead of quarterly? (Use Tables 12.2 and 12.3 in Section 12.8.)

16. **Simple interest** purchased a camera for $200. She paid $50 down and agreed to pay the balance plus 12% simple interest on the balance in 24 monthly payments. Determine Mary's monthly payment.

17. **APR** If Carlos purchases a car and the dealer states that he can pay for it in 48 monthly payments with an interest rate of 12% simple interest, find the true annual interest rate (correct to the nearest tenth of a percent).

Determine whether each statement (18–30) is true or false.

18. If socks are on sale, three pairs for $12.99, then two pairs of socks should cost $8.68.

19. Given that 1 centimeter = 0.3937 inch, the 9 is in the tenths position.

20. If a jacket that normally sells for $49.95 is marked down 20%, and all transactions are subject to a 7% sales tax, then the final amount due on the jacket is $43.46.

21. If a television sells for $320 and the markup is 60% of the cost, then the cost is $200.

22. If money is invested at 5% compounded quarterly, then the effective rate is 5.5%.

23. The true annual interest rate is the same as the effective annual interest rate.

24. The compound interest on $5000 invested at 12% compounded quarterly for 6 years is $5165.

25. If money is invested at 12% compounded semiannually, then it will double itself in approximately 6 years.

26. James Sparks is 25 years old and wishes to purchase $10,000 worth of life insurance. The annual cost of a 5-year term policy is $175.

27. If you assume a mortgage at 6.5% per year for 30 years, you will pay less interest than if you assume a mortgage at 9% per year for 20 years.

28. If you assume a $50,000 mortgage at 9% for 30 years, then you will pay approximately $100,000 in interest during the life of the loan.

29. Mary purchased a camera for $200. She paid $50 down and agreed to pay the balance plus 12% simple interest on the balance in 24 monthly payments. Her monthly payments were $7.75.

30. The Truth-in-Lending Act requires the disclosure of the finance charge and the annual percentage rate on credit purchases.

Chapter 13

Chapter Outline

An Introduction to the Metric System

After Studying This Chapter, You Will Be Able to Do the Following:

1. Describe the basic characteristics and advantages of the **metric system of measurement.**
2. Identify and use the following prefixes in the metric system: **kilo**, **hecto**, **deka**, **deci**, **centi**, **milli**.
3. Name the basic units of length, volume, and weight (mass) in the metric system and describe their origins and/or relationship.
4. Convert between the various metric units for measuring length, volume and weight (mass).
5. Use conversion tables to convert our customary units to metric units, and vice versa.
6. Describe the **Celsius** temperature scale, and convert Fahrenheit degree readings to Celsius and vice versa.
7. Make approximate conversions from metric measurements to customary measurements and vice versa without computation.
8. Develop an understanding of the metric measurements and be able to determine which metric unit best fits the given item.

Note: *indicates optional material.

Notations Frequently Used in This Chapter

m	meter
L	liter
g	gram
°*C*	degree Celsius
k	kilo, 1000
h	hecto, 100
da	deka, 10
d	deci, 0.1
c	centi, 0.01
m	milli, 0.001

Visit the PH Companion Web site at www.prenhall.com/setek

OVERVIEW

It is therefore declared that the policy of the United States shall be to coordinate and plan the increasing use of the metric system in the United States and to establish a United States Metric Board. . . .

—Metric Conversion Act

13.1 INTRODUCTION

In 1821, John Quincy Adams finished a 4-year investigation on the metric question and published a comprehensive report for the Congress. It was the first United States Metric Study. Adams's report considered the metric system as an alternative for adoption because he believed it approached "the ideal perfection of uniformity applied to weights and measures." But the time was not right; then, most of our trade was with inch-pound England, and the metric system had not yet established itself as the international measurement system. Today the world has committed itself to the metric system, and even in the United States its use is increasing.

The Metric Conversion Act of 1975 (amended in 1988) and a 1991 Presidential Executive Order provide both the rationale and the mandate for a transition to the use of metric units. Federal agencies are developing and implementing metric transition plans, cooperating on mutual concerns, and working with industry and user groups to establish realistic schedules for change.

The **metric system of weights and measures** is an international system that has been formalized as the International System of Units (SI). Many books and pamphlets dealing with the metric system approach the topic from strictly a metric viewpoint. We have not used this approach here. The purpose of this chapter is to introduce you to the metric system and its advantages and to help you understand how the system works.

Our customary system of measures is a "little of this and a little of that"—a hodgepodge. It probably does not seem like a bad system, or one that is awkward to work with, but that is simply because you have used it all your life. However, it can be a confusing system. If you buy a quart of strawberries, you are buying a different amount than if you buy a quart of milk, because one is a dry quart and the other is a liquid quart. A dry quart is 16% greater in volume than a liquid quart. Ounces are also confusing because an ounce can mean volume (for example, the number of ounces in a quart) or weight (as the number of ounces in a pound).

Pound measurement is even more ambiguous. Pound can mean weight, as in a pound of coffee, or it can mean force, as in pounds required to snap a cable or rope. Which is heavier, a pound of hamburger or a pound of gold? This question is not as silly as it seems. A pound of hamburger is weighed by *avoirdupois* weight and contains 7000 grains, but a pound of gold is weighed by *troy* weight and contains 5760 grains. Therefore, a pound of hamburger is heavier. These inconsistencies do not occur in the metric system because each quantity has its own unit of measurement and no unit is used to express more than one quantity.

The metric system is a system of measurement that has basic units of measure for length, width, area, volume, and weight that are in a decimal relationship to each other. For example, 1 meter = 10 decimeters = 100 centimeters. On the other hand, the United States uses a system of measurement that does not relate units in a decimal manner. For example, 12 inches = 1 foot, 3 feet = 1 yard, 16 ounces = 1 pound, and so on. You will find that it is much simpler to multiply or divide by 10 than it is to use 12 for one calculation, 3 for another, 16 for another, 5280 for yet another (5280 feet = 1 mile).

Why is the United States going metric after having lived with the English system for over 225 years? The answer is that the United States cannot afford not to. The European Union adopted a directive that requires all exporters to the European Union nations to indicate the dimensions of their products in metric units. Practically every country in the world presently uses the metric system or is in the process of converting to it. All of our major trading partners belong in one category or the other.

Our neighbors to the north and south, Canada and Mexico, are "metric" countries. Canada announced a commitment to the metric system, and on April 1, 1975 began its conversion to the metric system by giving weather reports with Celsius temperature readings. Today these countries have gone completely metric. Making the conversion to metric on a timetable—that is, doing it abruptly—is known as a "hard conversion." The method that the United States is using, a slow and steady approach, is known as a "soft conversion." In 1981 President Reagan informed the American National Metric Council that he welcomed their efforts to expand metric usage, but emphasized that this was an area for private initiative, not government pressure.

The important element for consideration is that world trade is conducted in metric measures and all the developed and industrialized countries of the world design, produce, and distribute in metric measures for global markets. In addition, the physical sciences (many of which are vital to modern technologies and industrial progress) are researched and conducted in metric measures all over the world.

The United States is the only industrialized nation in the world that does not fully embrace and use the metric system for all facets of design, manufacturing, or delivery of our goods and services.

Originally, the meter was not immediately welcomed by the general public for some of the same reasons that apply today. It was inconvenient for people to learn a new system. But it was gradually accepted, and by 1875 all of Europe (except England) had adopted the metric system.

Throughout the history of the United States many people, including Benjamin Franklin, have recommended that we adopt the metric system. In 1790, Thomas Jefferson recommended that the United States adopt a uniform system of weights and measures that uses a decimal system. Alexander Graham Bell spoke before a Congressional committee in 1906 and stated that the metric system should be provided for the whole of the United States. On December 23, 1975, President Gerald Ford signed the Metric Conversion Act into law. This act established a national policy in support of metric measurement in the United States. Shortly after this, the Bureau of Alcohol, Tobacco, and Firearms of the Treasury Department established a policy that all distilled spirits will be sold in standard metric sizes. This took effect on January 1, 1980, and now the labels on wine and liquor bottles state metric measures only. One immediate economic benefit was to reduce the number of standardized bottles from 16 to 7. Hence, manufacturers do not have to keep on hand as many bottle sizes (inventories) to correspond with various products. The United States fastener industry is also in the process of replacing 59 customary sizes with 25 metric sizes for threaded fasteners. On February 14, 1994, the Federal Trade Commission instituted a rule that many consumer products must include metric measurements. The rule covers such items as soap, tissues, foil wrap, plastic bags, detergents, shampoos, batteries, and mops.

We are already using the metric system in many ways, and it is in greater usage than is commonly realized. Movie cameras use 8- or 16-millimeter film, and 35 millimeters is a popular size for photographic film. Skis are sold in centimeter lengths. Most of the items that we purchase at the supermarket are now labeled in both customary and metric units. For example, a can of pepper whose net weight is 4 ounces is also labeled 113 grams, and a jar of mayonnaise contains 32 fluid ounces, 1 quart, or 0.95 liter. Soft drinks are now sold in liter, 2-liter, and 3-liter bottles.

The automobile industry has been one of the largest industries to use the metric system of measurement. Foreign cars are constructed using the metric system, and American-made autos have most parts made using the metric system. Approximately one-half of the cars being used in this country require some metric tools. Engine displacement is now being described in metric terms; for example, a car might have a 1.6-liter engine. Speedometers on new cars contain gauges that give readings in miles per hour and kilometers per hour. Some oil companies are now selling gasoline by the liter. Many other automotive items are sold using metric measurements, such as tires and spark plugs. Road signs give both metric and English information.

Temperatures are often given in terms of degrees Celsius. Precipitation figures will be given in centimeters (or millimeters), and wind speeds will soon be given in

Note of Interest

Japan converted to the metric system between 1951 and 1962. In fact, by 1962 the usage of the metric system was so widespread in Japan that the government established a fine of $140 for needless use of the old system of weights and measures.

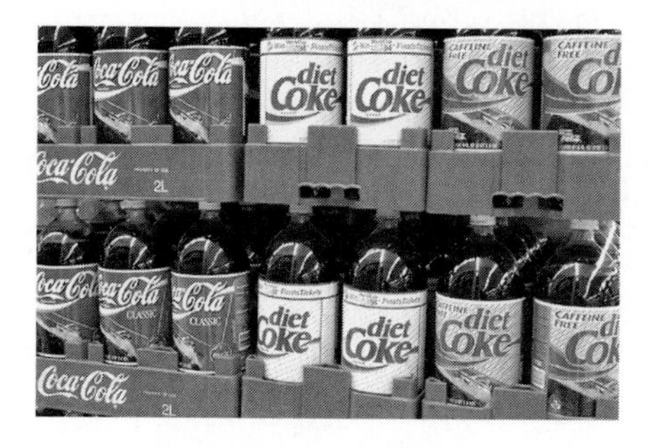

Most items in the supermarket are now labeled in both customary and metric units.

Road signs like this one, showing both miles and kilometers, are a common sight.

kilometers per hour instead of miles per hour, as the National Weather Service has a conversion plan for weather reports and anticipates its use in the near future. International track events, such as the Olympic Games, are now conducted using the metric system. The 100-yard dash changed to a slightly longer race, the 100-meter dash. Similarly, the 220-yard dash became the 200-meter dash, and the mile run became the 1500-meter run. Many American high schools and colleges now conduct their track meets using metric measurements.

Obviously, we will not stop using all of our customary units of measure. Even when the metric system is adopted, the height of horses will probably still be measured in terms of *hands* and depths at sea will still be measured in *fathoms.* Wood may still be sold by the *cord.* In a game like football, distances will probably always be given in yards; it will never be "first down, 9.14 meters to go." It is also doubtful that the Indianapolis 500-mile race will become the Indianapolis 804.7-kilometer race.

Conversion to the metric system will be made in those areas where it is advantageous to do so. Conversion to the metric system will make many computations easier; the metric system is also easier to learn. Conversion will help our economic situation in world markets because countries that already use the metric system will find American products more acceptable; this will benefit the entire population of the United States.

Just for fun

The United States system of measurements uses two different miles, the statute mile and the nautical mile. Which is longer and by how much?

13.2 HISTORY OF SYSTEMS OF MEASUREMENT

Regardless of what period of time we examine, people have always developed some method for weighing and measuring things. History tells us that early humans used a measuring system for making weapons, building places to live in, and even making clothing. The **cubit** is one of the first recorded units of measurement. Noah supposedly built his ark 300 cubits long, 50 cubits wide, and 30 cubits high. According to the Egyptian records (4000 B.C.), a cubit was the distance from the tip of the middle finger of the outstretched hand to the point of the elbow (approximately 19 inches). It is interesting to note that the side of one of the pyramids is 500 cubits long. Volume was measured by filling a container with seeds and then counting the number of seeds in the container. Stones were commonly used for determining the weight of heavy objects.

People first used parts of the body or things that were easily obtained, such as seeds and stones, as measuring instruments. A common brick is supposed to be a span long and one-half span wide. A **span** is the distance from the tip of the little finger to the tip of the thumb of an outstretched hand. A **palm** is the distance across the base of the four fingers that form your palm. A **digit** is the thickness or width of the middle of the middle finger, approximately three-quarters of an inch. The Romans used the idea of the digit to invent the inch. According to the Romans, an **inch** was the thickness or width of a thumb. There are no reliable facts on how the **foot** was invented.

According to most accounts, King Henry I of England decreed that a **yard** was the distance from the tip of his nose to the end of the thumb on his outstretched hand. It was by means of such royal commands that many standards of measurement were determined. For example, Queen Elizabeth I changed the measure of the mile from 5000 feet to 5280 feet. She did this because 1 furlong equaled 220 yards (660 feet), and if 1 mile equaled 5280 feet, then a mile would equal 8 furlongs. Thus, a partial list of English measures about 1500 A.D. was

$$\begin{aligned} 12 \text{ inches} &= 1 \text{ foot} \\ 3 \text{ feet} &= 1 \text{ yard} \\ 5 \text{ feet} &= 1 \text{ pace} \\ 125 \text{ paces} &= 1 \text{ furlong} \\ 8 \text{ furlongs} &= 1 \text{ mile} \\ 12 \text{ furlongs} &= 1 \text{ league} \end{aligned}$$

England became a world power, and by means of trade and colonization, the English system became established in many parts of the world. However, the need for greater uniformity and a single, worldwide, coordinated measurement system was recognized over 300 years ago. As a result, other systems were also being developed in other countries.

Bariel Mouton, vicar of St. Paul's Church in Lyons, France, proposed in 1670 that a standard unit of length be 1 minute of arc of a great circle of the earth. (A great circle is any circumference of the earth; the meridians that pass through the poles of the earth are examples of great circles.) Another proposal was made by Jean Picard, a French astronomer, who proposed a unit of length that was the length of a pendulum that takes 1 second to swing back and forth. But since a pendulum swings faster at the north and south poles than it does at the equator, nothing ever became of Picard's proposal.

After Mouton's proposal, not much was done toward developing a standard unit of measurement for over 100 years. In 1790, at the request of the French government, the French Academy of Sciences devised a new system of measurement. The new basic unit of length was a portion of the meridian of the earth, similar to the unit proposed by Mouton. The new unit was called a *meter*, which was taken from the Greek word *metron,* "to measure."

Since the scientists wanted a unit similar in length to a yard, they chose a portion of the meridian that was approximately the same length. But they also wanted the unit to be part of a base 10, or decimal system. Therefore, they calculated the distance from the North Pole to the equator along the meridian that runs through Dunkirk and Paris, and then took one ten-millionth ($\frac{1}{10{,}000{,}000}$) of that distance as the standard unit of measure, the **meter**. The French Academy of Sciences recommended this unit because all future calculations could be done using the decimal system. There would be no need to divide by 5280, multiply by 16, and so on; all quantities that were larger or smaller than the meter could be converted by multiplying or dividing by 10 or powers of 10. Recall that it is quite easy to multiply or divide a number by 10: We simply move the decimal point to the right or left.

A meter is about 39.37 inches long—a little longer than a yard. By keeping this in mind, you will be able to visualize how long a meter is. It will also help to give you an idea of the size of other metric measurements. The metric unit used

Note of Interest

Measurements of length, weight, volume, and temperature are important to all of us. But many other kinds of measurement are also important in today's society. For example, measurements of force, work, power, and energy all provide necessary information to scientists and manufacturers. Other measurements that may be of interest to someone are related to heat, light, sound, and electricity. Most of these are measured with special units.

A *calorie* is the amount of heat needed to raise the temperature of one gram of water one degree Celsius. A *large calorie* (the nutritionist's calorie) is the amount of heat needed to raise the temperature of 1000 grams (1 kg) of water one degree Celsius.

The brightness or intensity of a light is measured in *foot-candles*. One *foot-candle* is the intensity of light of a standard one-inch-thick candle at a distance of one foot from the flame. One of the smallest units of length is the *angstrom*. It measures the *wavelength* of light. An *angstrom* is a metric unit equivalent to 1/100,000,000 of a centimeter.

A *light-year* is the astronomer's unit of length. It is the distance light travels in a year (about six trillion miles).

The intensity of sound (loudness) is measured in units called *decibels*. A sound that can hardly be heard by a normal ear has an

(continued on next page)

intensity of zero decibels. A very loud sound, one that usually hurts a normal ear, has an intensity of approximately 120 decibels.

Electricity involves a variety of measuring units. The rate of flow of electric charges is measured in *amperes;* the difference in the pressure of electricity between two points in a closed electrical circuit in *volts*. Resistance to the flow of current is measured in *ohms,* and electrical power in *watts.*

for determining mass (weight) is called a *gram,* and the metric unit used for determining volume is called a *liter.* We shall examine grams and liters in greater detail later in this chapter.

Before we proceed with a study of the metric system, we must first develop a familiarity with a set of prefixes used throughout the system. Listed in the table are the most common prefixes and their meanings. For example, a **kilometer** is 1000 meters and a **centimeter** is $\frac{1}{100}$ of a meter.

Prefix	Symbol	Meaning
kilo	k	1000
hecto	h	100
deka	da	10
deci	d	$\frac{1}{10}$ or 0.1
centi	c	$\frac{1}{100}$ or 0.01
milli	m	$\frac{1}{1,000}$ or 0.001

To help you remember these prefixes, observe that prefixes containing an *i* (*deci, centi, milli*) are all fractional parts of one unit. The prefix *deci* should remind you of a word involving 10, such as *decade* or *decimal. Centi* should remind you of a *century,* which is 100 years. The prefix *milli* should remind you of *millennium,* a period of 1000 years.

The prefixes *deka, hecto,* and *kilo* are prefixes that indicate multiplication by 10, 100, and 1000, respectively. A *kilowatt* is a unit of measure used in electricity and is equivalent to 1000 watts. An easy way to remember the correct multiple for *hecto* is to notice that *hecto* and *hundred* both begin with the same letter, *h*. The prefix *deka* is similar to *deci* and should also remind you of words involving 10.

We did not list two prefixes, *micro* and *mega,* as they are not frequently used in the metric system. These prefixes refer to 1 million. A *megaton* is 1 million tons. A *micrometer* is one-millionth of a meter. The prefix *micro* may remind you of *microfilm,* the very small film that libraries use to keep copies of printed matter.

Regardless of the fact that we do not yet know the meaning of gram or liter, we should be able to give an interpretation of the following terms: kilogram, decigram, hectoliter, and milliliter. The key clues are the prefixes. The prefixes have the same meaning for all metric measures. Therefore, a kilogram is 1000 grams, whereas a decigram is $\frac{1}{10}$ of a gram or 0.1 gram. Similarly, a hectoliter is 100 liters, and a milliliter is $\frac{1}{1000}$ of a liter or 0.001 liter.

 **EXAMPLE 1** ***Finding Equivalent Metric Expressions*** Using the prefixes as a guide, find equivalent expressions in grams, meters, or liters for each of the following:

a. centigram **b.** kilometer
c. deciliter **d.** hectometer

Solution

a. The prefix *centi* tells us that 1 centigram = 0.01 gram.

b. The prefix *kilo* indicates that 1 kilometer = 1000 meters.

c. The prefix *deci* indicates *ten* and the *i* tells us that it is a fractional part; that is, 1 deciliter = 0.1 liter.

d. The prefix *hecto* indicates that1 hectometer = 100 meters. 

Math Connections

Evolutionary Changes of the Meter Standard

As discussed in the main body of the text, one meter was initially defined as one ten-millionth of the length of the earth's meridian along a quadrant, which is one-fourth the circumference of the earth. Interestingly, the first prototype, which was a bar made of a platinum-iridium alloy, was actually short by two-tenths (0.2) of a millimeter because researchers miscalculated the flattening of the earth due to its rotation.

Over the years a meter's definition and its physical representation have been modified by the General Conference on Weights and Measures (CGPM), which is the primary intergovernmental treaty organization representing approximately 50 countries. The CGPM is responsible for the International System of Units (SI) and has the responsibility of ensuring that the SI is widely disseminated. The CGPM is also responsible for modifying the SI as necessary so that it reflects the latest advances in science and technology. A brief summary of the changes CGPM has made in defining the meter follows:

- In 1889, the first CGPM sanctioned a new international prototype. It was a bar made of a platinum iridium alloy that had lines inscribed at each end. The distance between these lines defined the meter. This bar was kept in a chamber at the International Bureau of Weights and Measures (BIPM), located outside Paris, France, at a constant temperature and pressure to keep the bar from expanding or contracting.
- In 1960, the eleventh CGPM changed the meter's definition from a mechanical-based one to one that used optical wavelengths. This new definition was a function of the radiation that corresponded to the transition between two different levels of the krypton-86 atom. A special krypton-86 electrical discharge lamp was designed for this purpose and the meter was defined accordingly as a wavelength of krypton-86 radiation. Specifically, 1,650,763.73 wavelengths of the orange-red light of the heated krypton gas equaled one meter.
- In 1983, the seventeenth CGPM once again modified the meter's definition. As technology advanced it was inevitable that new length standards based on more precise atoms or molecules would be proposed. As a result, the CGPM replaced the 1960 radiation-based definition of a meter by one based on time. This new standard defines the meter in terms of the SI second and the defined value for the speed of light in a vacuum. Specifically, the definition is given as follows:

The metre is the length of the path travelled by light in vacuum during a time interval of $\frac{1}{299{,}792{,}458}$ ***of a second.***

This 1983 standard, which is the current standard, defines the meter experimentally. Thus, anyone with a properly equipped science laboratory can confirm the length of the meter without going to France to check it against the international standard. It is also interesting to note that although the meter's definition has changed over the years to reflect advances in science and technology, the original international prototype of the meter is still kept at the BIPM under the conditions specified in 1889.

 **EXAMPLE 2** ***Identifying Prefixes*** What prefix can be used to indicate each number?

a. 0.001 **b.** 10 **c.** 0.1 **d.** 0.01

Solution

a. 0.001 is the same as $\frac{1}{1000}$ and is indicated by *milli.*

b. 10 is indicated by *deka.*

c. 0.1 is the same as $\frac{1}{10}$ and is indicated by *deci.*

d. 0.01 is the same as $\frac{1}{100}$ and is indicated by *centi.*

NW ***Now Work Problem 5.***

Exercises for Section 13.2

1. List two reasons why the United States should convert to the metric system.
2. **a.** List examples of U.S. conversion to the metric system.
   b. What is the basis for the metric system?
   c. Write to your Congressional representative urging a progressive conversion to the metric system. Incorporate ideas that would be conducive to change.
3. List two reasons why the measurement system in the United States is a confusing system.
4. Name the three basic units of measure in the metric system.
5. (NW) What prefix can be used to indicate each number?
   a. 10 **b.** 0.1 **c.** 1000
   d. 0.01 **e.** 100 **f.** 0.001

6. Write the prefix for each symbol.
   a. k b. d c. da
   d. c e. h f. m

7. Given the fact that 1 meter is approximately equal to 39.37 inches, how many inches are contained in each of the following?
   a. 1 kilometer b. 1 hectometer
   c. 1 dekameter d. 1 decimeter
   e. 1 centimeter f. 1 millimeter

8. Given the fact that 1 liter is approximately equal to 1.06 liquid quarts, find the equivalent metric measure.
   a. 1060 quarts b. 106 quarts
   c. 10.6 quarts d. 0.106 quarts
   e. 0.0106 quarts f. 0.00106 quarts

NW 9. Complete each of the following:
   a. 1 kilometer = _____ meters
   b. 1 dekameter = _____ meters
   c. 1 decimeter = _____ meters
   d. 1 millimeter = _____ meters
   e. 1 centimeter = _____ meters
   f. 1 hectometer = _____ meters

10. Complete each of the following:
   a. 1 milligram = _____ grams
   b. 1 centigram = _____ grams
   c. 1 hectogram = _____ grams
   d. 1 kilogram = _____ grams
   e. 1 decigram = _____ grams
   f. 1 dekagram = _____ grams

11. Complete each of the following:
   a. 1 meter = _____ centimeters
   b. 1 liter = _____ milliliters
   c. 1 meter = _____ decimeters
   d. 1 gram = _____ centigrams
   e. 1 gram = _____ milligrams
   f. 1 liter = _____ deciliters

12. Complete each of the following:
   a. 1 kilometer = _____ hectometers
   b. 1 hectometer = _____ dekameters
   c. 1 dekagram = _____ decigram
   d. 1 kiloliter = _____ dekaliters
   e. 1 deciliter = _____ liters
   f. 1 kilogram = _____ hectograms

13. Complete each of the following:
   a. 2000 meters = _____ kilometers
   b. 32 kilograms = _____ grams
   c. 3 grams = _____ milligrams
   d. 40 dekaliters = _____ liters
   e. 3 hectoliters = _____ liters
   f. 5 kilometers = _____ meters

14. Complete each of the following:
   a. 1000 grams = _____ kilograms
   b. 1000 grams = _____ hectograms
   c. 20 liters = _____ dekaliters
   d. 50 liters = _____ hectoliters
   e. 40 meters = _____ dekameters
   f. 20 kilometers = _____ meters

15. Determine whether each statement is true or false.
   a. A liter is larger than a kiloliter.
   b. A liter is smaller than a milliliter.
   c. There are 500 milligrams in a gram.
   d. There are 100 grams in a kilogram
   e. A centimeter is one-hundredth of a meter.

16. Determine whether each statement is true or false.
   a. There are 1000 liters in a milliliter.
   b. There are 100 liters in a centiliter.
   c. A gram is larger than a milligram.
   d. A gram is smaller than a kilogram.
   e. A kilometer is one-thousandth of a meter.

Writing Mathematics

17. Describe the basic characteristics of the metric system.
18. Explain how to convert 25 dekameters to meters and then do it.
19. Explain how to convert 350 centigrams to grams and then do it.

Challenge Exercises

20. List at least 10 different *confusing* units of measure that are used in the United States, and describe why each measure is confusing. For example, did you know that there are two different *mile* measures? Do you know what a *rod* measures?
21. In this section we have encountered the six most common prefixes used in the metric system. There exist a total of 20 prefixes. Prepare a table showing all 20. Be sure to include the prefix, symbol, English name, and the corresponding power of 10.

Group/Research Activities

22. There are seven base units for the International System of Weights and Measures. Determine the seven categories and the corresponding measure that is used.

23. Prepare an exhaustive list of the metric measures that you or your classmates encounter in everyday living experiences. Be sure to include the measure, what type it is (length, area, volume, etc.), and where it was found.

Just for fun

Can you unscramble these 10 words to make the correct metric units of measurement?

1. ETMER	2. MAGR
3. TERIL	4. AGKDMRAE
5. OLTERILKI	6. ICERMTEED
7. OLMAGRKI	8. RITLLIEILM
9. NITREMECTE	10. OERLMTEKI

13.3 LENGTH AND AREA

The basic unit of length in the metric system is the meter. A meter is slightly longer than a yard: 1 meter $\approx$ 39.37 inches. (Recall that the symbol $\approx$ means approximately equal.) The symbol for meter is a lowercase *m*. Most likely there is a *meter stick* in your classroom. If there is, examine it in order to get an idea of how long 1 meter is.

Figure 13.1 shows a ruler. Note that one edge is marked in inches and the other edge is marked in centimeters. A centimeter is 0.01 meter; that is, 1 meter = 100 centimeters. Each small division on the metric edge of the ruler is 1 millimeter; 10 millimeters = 1 centimeter. The symbol for centimeters is *cm*, and the symbol for millimeter is *mm*.

Suppose we want to find the length of line segment AB in Figure 13.2. Using inches, segment AB is approximately $3\frac{3}{16}$ inches long. But if you measure it using the metric edge, you will find that its length is 8 centimeters. This illustrates another reason why the metric system is favored over our customary units of measurement: The metric system eliminates fractions such as $\frac{1}{4}$, $\frac{3}{8}$, $\frac{5}{16}$, and $\frac{1}{32}$.

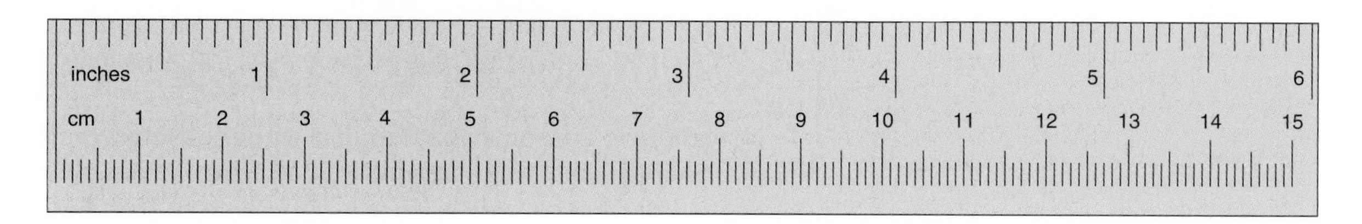

Figure 13.1

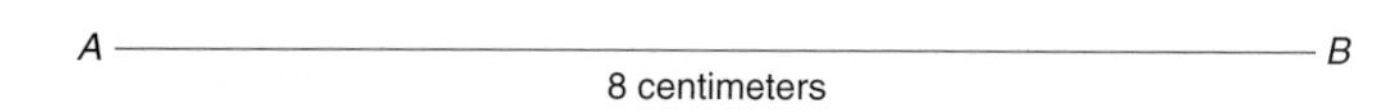

Figure 13.2

Granted, a line segment that is 2 inches long (see Figure 13.3) is easy to measure in terms of inches, but it is also not difficult to measure using the metric system. A line segment 2 inches long is a little longer than 5 centimeters; in fact, it is 51 millimeters in length. Note that we do not have to use a fraction to express this length.

A ———————— B
2 inches

Figure 13.3

EXAMPLE 1 ***Measuring in Millimeters*** Measure each of the given line segments to the nearest millimeter (mm).

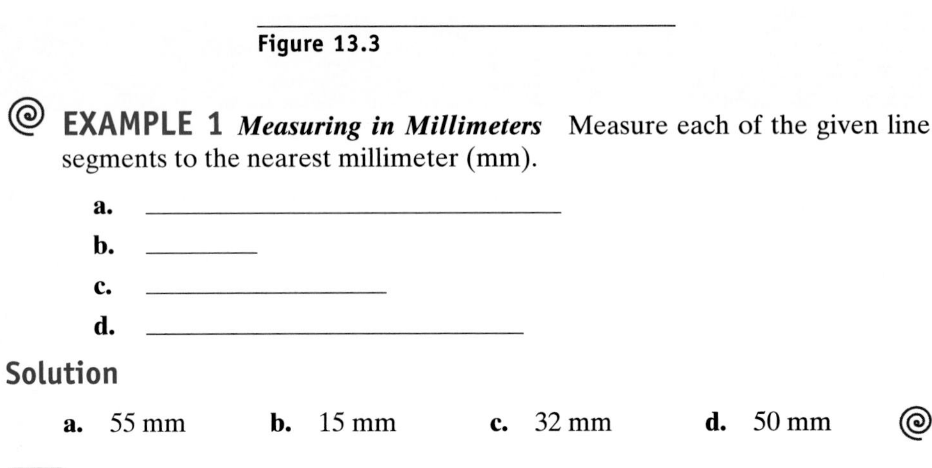

Solution

a. 55 mm **b.** 15 mm **c.** 32 mm **d.** 50 mm

NW ***Now Work Problem 1.***

Although the meter is the basic unit of measurement in the metric system, many lengths (and thicknesses) are measured in terms of parts of a meter, such as decimeter (dm), centimeter (cm), and millimeter (mm). In measuring greater lengths and distances, we can use the dekameter and hectometer, but the kilometer is the most commonly used unit for longer lengths. For example, all of the races in the Olympic Games are described in meters—100-meter dash, 200-meter dash, 400-meter dash, 800-meter run, 1500-meter run, 5000-meter run, and so on—but the distance between two cities is measured in kilometers. The symbol for kilometer is *km*. The distance from Los Angeles to New York is 4690 kilometers; from Dallas to Chicago, 1506 kilometers.

The following table summarizes the units related to the meter. Note that the first part of the symbol indicates the prefix and the second part (m) indicates meter.

Symbol	Word	Meaning
km	kilometer	1000 meters
hm	hectometer	100 meters
dam	dekameter	10 meters
m	meter	1 meter
dm	decimeter	0.1 meter
cm	centimeter	0.01 meter
mm	millimeter	0.001 meter

Now that we are somewhat familiar with the metric prefixes, we can use the following handy chart to find equivalent measures of length.

$\times 10$
km hm dam m dm cm mm
$\div 10$

The chart gives a technique for changing from one metric length measure to another. To change from one unit to another, for each step to the right, multiply by 10 to obtain units that are smaller. You can also think of this as moving the decimal point one place to the right for each step to the right. For each step to the left, divide by 10 to obtain fewer units that are larger. You can also think of this as moving the decimal point one place to the left for each step to the left.

Suppose we want to convert 1.5 kilometers to dekameters. That is, 1.5 km = _____ dam. Using the chart, we start at kilometers and move two places to the right to obtain dekameters. Hence, we multiply by 10×10 (100) or simply move the decimal point two places to the right. That is,

$$1.5 \text{ km} = 150. \text{ dam}$$

To convert 42.7 decimeters to hectometers, that is, 42.7 dm = _____ hm, we start at decimeters and move three places to the left to obtain hectometers. Therefore, we divide by 10 for each step to the left or, in other words, move the decimal point three places to the left. That is,

$$42.7 \text{ km} = 0.042\,7 \text{ hm}$$

EXAMPLE 2 ***Converting from One Metric Measure to Another*** Complete each of the following:

a. 22 m = _____ cm **b.** 31.4 mm = _____ cm
c. 15 mm = _____ dm **d.** 4.1 m = _____ mm
e. 4.2 km = _____ m **f.** 77 dm = _____ hm

Solution

a. Using the chart in this section, we start at meters and move two places to the right to obtain centimeters. Hence, we move the decimal point two places to the right. That is,

$$22 \text{ m} = 2200. \text{ cm}$$

Note: If a decimal point is not indicated, it is understood to be immediately to the right of the last digit in the given whole number.

b. Using the chart, we start at millimeters and move one place to the left to centimeters. Hence, we move the decimal point one place to the left. That is,

$$31.4 \text{ mm} = 3.14 \text{ cm}$$

c. 15 mm = 0.15 dm **d.** 4.1 m = 4100 mm

e. 4.2 km = 4200 m **f.** 77 dm = 0.077 hm

NW ***Now Work Problem 5.***

How long is a kilometer? From the prefix *kilo,* we know that 1 kilometer = 1000 *meters*. But does that give you any idea how long it is? Let's compare it with something that is familiar. How does it compare with a mile? A meter is 39.37 inches long; therefore, a kilometer is 39,370 inches long. A mile contains 5280 feet and 1 foot = 12 inches. Therefore,

$$1 \text{ mile} = 5280 \times 12 = 63{,}360 \text{ inches}$$

A mile is longer than a kilometer. In fact, a mile is approximately 1.6 kilometers.

Math Connections

An Application of Fibonacci Numbers

On page 46 it was mentioned that scientists and mathematicians study the many application of the Fibonacci sequence. Here is one such application that is connected to the golden ratio. Page 48 shows that the ratio of two sequential Fibonacci numbers is approximately 1.618, which means that if you wish to get to the next Fibonacci number all you have to do is multiply by 1.618. For example, 13 is a Fibonacci number, the next one is $13 \times 1.618 \approx 21$. But recall that to convert miles to kilometers you multiply by 1.6, which is the same factor used to arrive at the next Fibonacci number from a given one. This suggests a simple way to convert miles to kilometers with out multiplying. For example, suppose that you want to convert 15 miles to kilometers. First rewrite 15 as the sum of two Fibonacci numbers (thanks to Zeckendorf's theorem):

$$15 \text{ miles} = 2 \text{ miles} + 13 \text{ miles}$$

Remember that writing the next consecutive Fibonacci number has the same effect as multiplying by 1.6; thus,

$$\begin{aligned} 15 \text{ miles} &= 2 \text{ miles} + 13 \text{ miles} \\ &= 3 \text{ km} + 21 \text{ km} = 24 \text{ km} \end{aligned}$$

Check: $15 \text{ miles} \times 1.6 = 24 \text{ km}$

If you take a Fibonacci number and divide it by its successor, you get approximately 0.6. That is, to get a preceding Fibonacci number, we multiply by 0.6, but this is the same factor that is used to convert kilometers to miles. So now we have a simple method to convert kilometers to miles by using the Fibonacci sequence and adding. For example,

$$\begin{aligned} 18 \text{ km} &= 5 \text{ km} + 13 \text{ km} \\ &= 3 \text{ mi} + 8 \text{ mi} \quad \text{(unlike before, you write the} \\ &= 11 \text{ mi} \quad \text{preceding Fibonacci number)} \end{aligned}$$

Check $18 \text{ km} \times 0.6 = 10.8 \text{ mi} \approx 11 \text{ mi}$

Example Practice: Convert each given measurement to the indicated measurement using the Fibonacci sequence.

[*Hint*: Fibonacci sequence: 0, 1, 2, 3, 5, 8, 13, 21, 34, 55, 89, 144, 233, 377 ...]

a. 10 mi = _____ km **b.** 25 mi = _____ km

c. 36 km _____ mi

As another illustration, a football field is 120 yards long (including the end zones); therefore, a football field is

$$120 \times 36 = 4320 \text{ inches long}$$

Nine football fields have a total length of 38,880 inches. Because a kilometer is 39,370 inches in length, a kilometer is 490 inches longer than nine football fields placed end to end. A kilometer is approximately 0.6 mile.

Keeping in mind that 1 mile $\approx$ 1.6 kilometers, we can also discuss speed in terms of kilometers per hour. For example, if a cyclist is pedaling his bike at the rate of 5 miles per hour (mi/hr), his speed is 8 kilometers per hour (km/hr): Because 1 mile = 1.6 kilometers, we have $5 \times 1.6 = 8$ kilometers per hour. If an automobile travels at the rate of 40 miles per hour, then the auto is also traveling at 64 (40×1.6) kilometers per hour.

If a person never drives faster than 90 kilometers per hour, how fast is this in miles per hour? In order to convert kilometers to miles, we simply multiply the number of kilometers by 0.6 (1 kilometer $\approx$ 0.6 mile). Therefore, 90 kilometers per hour is approximately the same as $90 \times 0.6 = 54$ miles per hour. (*Note:* Although most of our conversion factors are approximate, we will use the equals sign for convenience.) Also, we shall use Table 13.1 to obtain the conversion factors.

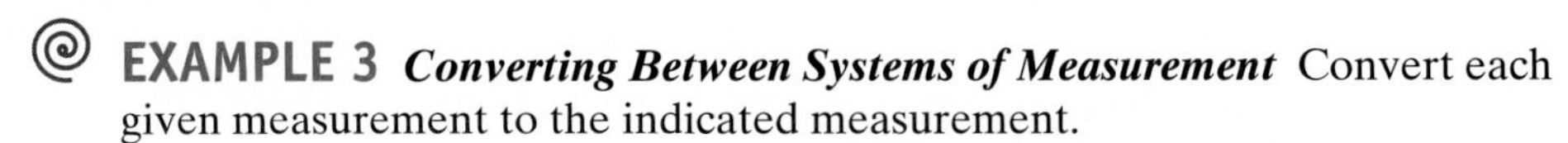

EXAMPLE 3 ***Converting Between Systems of Measurement*** Convert each given measurement to the indicated measurement.

a. 10 mi = _____ km **b.** 25 mi = _____ km

c. 150 km = _____ mi **d.** 25 km = _____ mi

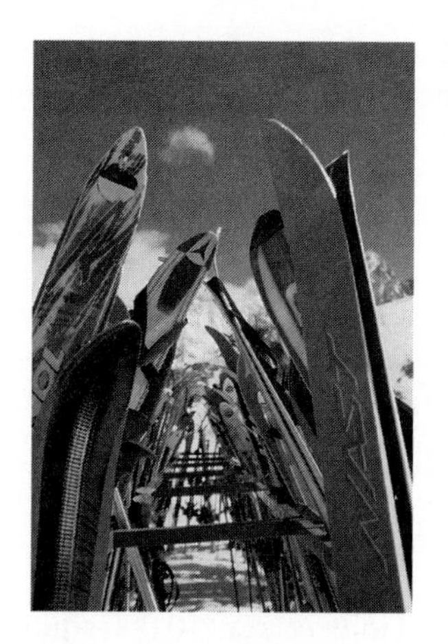
Skis are sold in centimeter lengths.

TABLE 13.1 Metric Conversion Factors—Length and Area

Symbol	When You Know	Multiply by	To Find	Symbol
LENGTH		*To Metric*		
in	inches	2.5	centimeters	cm
ft	feet	30	centimeters	cm
yd	yards	0.9	meters	m
mi	miles	1.6	kilometers	km
		From Metric		
mm	millimeters	0.04	inches	in
cm	centimeters	0.4	inches	in
m	meters	3.3	feet	ft
m	meters	1.1	yards	yd
km	kilometers	0.6	miles	mi
AREA		*To Metric*		
in^2	square inches	6.5	square centimeters	cm^2
ft^2	square feet	0.09	square meters	m^2
yd^2	square yards	0.8	square meters	m^2
mi^2	square miles	2.6	square kilometers	km^2
	acres	0.4	hectares	ha
		From Metric		
cm^2	square centimeters	0.16	square inches	in^2
m^2	square meters	1.2	square yards	yd^2
km^2	square kilometers	0.4	square miles	mi^2
ha	hectares	2.5	acres	

Note of Interest

The term *acre* is an Anglo-Saxon word that literally means "the amount of land that can be plowed by a yoke of oxen in one day." It was coined sometime before the tenth century; however it wasn't until the reign of Henry VIII that the unit *acre* was finally agreed to be 40 poles long by 4 poles wide. Of course, this begs the question, what is a pole? Well, one rod is equal to 16.5 feet and one pole is equal to one square rod. So one pole is 272.25 feet squared. The area of one *acre* is 160 (40×4) times 272.25, or 43,560 square feet. An author lives on a plot of land that is half an acre; how many square feet is that?

Solution (Use Table 13.1)

a. Since 1 mi = 1.6 km, 10 mi = $10 \times 1.6 = 16$ km. (*Note:* A better approximation of 1 mile is 1.61 kilometers, but we shall use 1 mi = 1.6 km.)

b. 25 mi = $25 \times 1.6 = 40$ km

c. Because 1 km = 0.6 mi, 150 km = $150 \times 0.6 = 90$ mi

d. 25 km = $25 \times 0.6 = 15$ mi

NW *Now Work Problems 7 and 9.*

EXAMPLE 4 ***Kilometers and Miles*** Convert each speedometer reading to the indicated measurement.

a. 30 mi/hr = ______ km/hr

b. 45 mi/hr = ______ km/hr

c. 100 km/hr = ______ mi/hr

d. 120 km/hr = ______ mi/hr

Solution (Use Table 13.1) This example is similar to Example 3. To convert miles to kilometers, multiply the number of miles by 1.6. To convert kilometers to miles, multiply the numbers of kilometers by 0.6.

a. 30 mi/hr = $30 \times 1.6 = 48$ km/hr

b. 45 mi/hr = $45 \times 1.6 = 72$ km/hr

c. 100 km/hr = $100 \times 0.6 = 60$ mi/hr

d. 120 km/hr = $120 \times 0.6 = 72$ mi/hr

NW *Now Work Problem 11.*

Area is measured in square units. The floor of a room that measures 8 feet by 10 feet has an area of 80 square feet. This means that 80 squares, each measuring 1 foot by 1 foot, will cover the surface of the floor. Area is sometimes referred to as *surface area*.

A square whose measurements are 1 centimeter by 1 centimeter is said to have an area of 1 square centimeter; this is denoted by 1 cm^2 (see Figure 13.4). The square centimeter is used to find the area of relatively small regions, such as the area of this page. Larger regions are measured in terms of square meters (m^2). Since 1 meter = 100 centimeters, a square meter contains

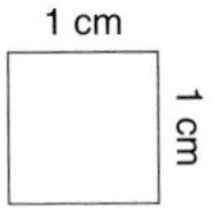

1 cm

1 cm

Figure 13.4

Area = 1 cm^2.

$$10\,000 \text{ square centimeters } (100 \times 100 = 10\,000)$$

(*Note:* We use a space instead of a comma in numbers of 10,000 or more when discussing metric measurements.)

The area of very large regions is measured in *hectares* (10 000 square meters). Land that is measured in terms of acres can also be measured in hectares. Since 1 hectare = 10 000 square meters, a hectare is the area of a square that measures 100 meters on each side. It is highly unlikely that the hectare will be used in measuring the area of land when the United States converts to the metric system. Land will probably continue to be sold in terms of acres for two reasons: (1) Land cannot be shipped overseas—that is, we will not export land to metric countries as we do machinery and other products; and (2) it would be impractical to change all of the property deeds in the United States so that the area would be in terms of hectares. However, the other metric measures of area, such as the square centimeter and the square meter, will be used. Another metric area measure is the *are* (pronounced *air*). It is a square measuring 10 meters on each side; therefore its area is 100 square meters. *Note:* 100 ares = 1 hectare.

You may find the metric conversion factors in Table 13.1 helpful. *Remember, these are all approximate conversions.*

Exercises for Section 13.3

NW **1.** Measure each of the given line segments to the nearest millimeter.

a. ________

b. ____________________________

c. ___________________

d. ____

e. _________________________

f. ___________________

2. Complete each of the following:

a. 7.6 m = ______ mm

b. 25 cm = ______ mm

c. 200 mm = ______ cm

d. 10.2 km = ______ m

e. 30.3 dam = ______ m

f. 56 m = ______ cm

3. Complete each of the following:

a. 18 cm = ______ dm

b. 37 dam = ______ dm

c. 3.2 km = ______ cm

d. 702 km = ______ dam

e. 423 cm = ______ hm

f. 8.14 cm = ______ mm

4. Complete each of the following:

a. 53 dm = ______ dam

b. 203.2 cm = ______ m

c. 123 cm = ______ km

d. 52.4 hm = ______ km

e. 624 mm = ______ cm

f. 2.07 m = ______ cm

NW **5.** Complete each of the following:

a. 413.7 cm = ______ dm

b. 0.45 cm = ______ mm

c. 4.7 dam = ______ cm

d. 4378 cm = ______ km

e. 30.2 dam = ______ km

f. 985 mm = ______ cm

6. Distance from Miami to Atlanta If the distance from Miami to Atlanta is 1070 kilometers, how many miles is it?

NW **7. Distance from Seattle to New Orleans** If the distance from Seattle to New Orleans is 2625 miles, how many kilometers is it?

8. **Distance from Mexico City to Chicago** If the distance from Mexico City to Chicago is 2082.5 miles, how many kilometers is it?

NW 9. **Distance from Los Angeles to New York City** If the distance from Los Angeles to New York City is 4690 kilometers, how many miles is it?

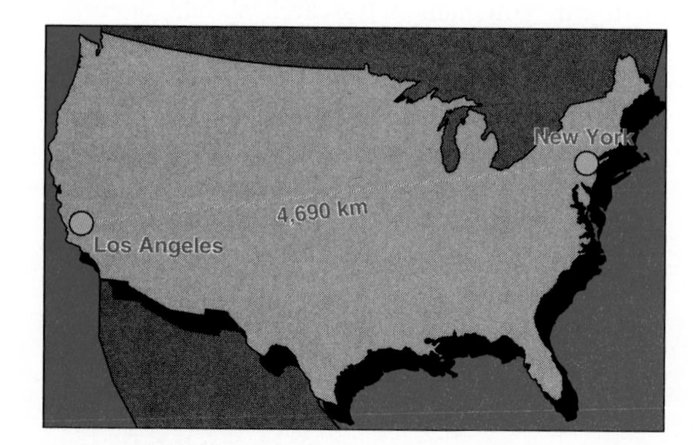

10. **Speed limit** The speed limit in some states is 55 miles per hour. What is the speed limit in kilometers per hour?

NW 11. **Speeding ticket** The speed limit in a certain town is 75 kilometers per hour. If radar records Larry's speed as 50 miles per hour, should he get a ticket?

12. **Distance from earth to the sun** The distance from the earth to the sun is approximately 93 million miles. How many kilometers is it?

13. **Distance from the sun to Mercury** Mercury is the closest planet to the sun; its distance from the sun is 58 million kilometers. How many miles is it?

14. **Perimeter** Find the perimeter of each polygon shown. Your answer should be in terms of the indicated unit.

a. 2 cm, 2 cm

b. 1 cm, 2 cm, 2 cm, 3 cm

c. 4 cm, 4 cm, 3 cm

d. 3 cm, 2 cm, 1 cm, 3 cm

15. **Area** A square that measures 2 centimeters by 2 centimeters has what area?

16. **Area** A hectare is the area of a square that measures _____ meters on each side.

17. **Area** One mile is approximately 1.6 kilometers. Approximately how many square kilometers are in 1 square mile?

18. Which is the longer measurement in each case?
   - **a.** 1 inch or 1 centimeter
   - **b.** 1 yard or 1 meter
   - **c.** 1 mile or 1 kilometer
   - **d.** 6 inches or 13 centimeters
   - **e.** 6 feet or 2 meters
   - **f.** 1 foot or 35 centimeters

19. Which measurement of area is greater in each case?
   - **a.** 1 square inch or 1 square centimeter
   - **b.** 1 square foot or 1 square meter
   - **c.** 1 square yard or 1 square meter
   - **d.** 1 square mile or 1 square kilometer
   - **e.** 1 acre or 1 hectare
   - **f.** 1 square meter or 1 are

For Exercises 20–25, choose the most sensible answer.

20. Width of a newspaper:
   a. 0.38 m **b.** 3.8 m **c.** 0.38 km

21. Length of a shovel:
   a. 0.6 m **b.** 1.6 m **c.** 2.6 m

22. Height of a bowling pin:
   a. 0.39 mm **b.** 3.9 cm **c.** 0.39 m

23. Length of a new pencil:
   a. 19 mm **b.** 19 cm **c.** 30 cm

24. Length of a watch band:
   a. 20 cm **b.** 30 mm **c.** 30 cm

25. Length of a shopping trip:
   a. 5 mm **b.** 5 cm **c.** 5 km

26. What are your metric measurements?
   - **a.** height: feet _____ inches _____; meters _____ centimeters _____
   - **b.** waist: inches _____; centimeters _____
   - **c.** neck: inches _____; centimeters _____
   - **d.** wrist: inches _____; centimeters _____
   - **e.** biceps: inches _____; centimeters _____
   - **f.** foot length: inches _____; centimeters _____

Writing Mathematics

27. Carlos ran 200 yards in the same amount of time that Louie ran 200 meters. Who ran faster? Explain your answer.

28. Rachel drove her car 80 miles in the same amount of time that Jillian drove her car 80 kilometers. Who drove faster? Explain your answer.

Challenge Exercises

29. The United States is not presently using the metric system as its standard system of measurement. Why do you think this is so?

30. Determine which metric unit of length you would use to express the following:

a. The length of a newborn baby

b. The distance between Boston and Providence

c. The length of a baseball bat

d. The diameter of a golf ball

e. The wing span of a butterfly

f. The thickness of a penny

Research/Group Activities

31. There are approximately 27 different units of length in the United States system of measurement. How many of them can you name? Research should enable you to come close to completing the list.

32. We have units of measure for length, area, volume, mass, and temperature. But what about time? How is time measured? Research this concept and prepare a report on time measurement. Include in your report information about atomic clocks.

Note of Interest

NASA's Mars Climate Orbiter

On December 11, 1998, NASA launched the Mars Climate Orbiter. Carrying only two atmospheric instruments, a camera and radiometer, the $125 million spacecraft was designed to study the Martian climate and weather patterns. Throughout the Orbiter's nine-month journey to Mars, however, two separate NASA groups responsible for the mission—the Orbiter's spacecraft team in Colorado and the mission navigation team in California—exchanged information using different units of measurement. Specifically, the Colorado spacecraft team electronically transmitted trajectory data using English units to the mission navigation team, which expected metric units. These data were critical to the maneuvers required to place the spacecraft in the proper Mars orbit. Because the data were sent without units, the mission team interpreted them in metric units.

According to reports stemming from the investigation, the Orbiter was supposed to have approached Mars at an altitude of approximately 150 kilometers, or about 90 miles. Well, if the mission navigation team received "90" from the spacecraft team without any units, this would have been incorrectly interpreted as 90 kilometers, which would have placed the Orbiter at a much lower altitude than required. After reviewing the last six to eight hours of data as the Orbiter approached Mars on September 23, 1999, it was discovered that the actual altitude was 60 kilometers, or about 36 miles. In short, NASA missed the mark by approximately 100 kilometers. Since the lowest survivable altitude was believed to be approximately 85 kilometers, NASA speculates that the spacecraft either burned up in the atmosphere or crashed to the surface. "People sometimes make errors," said Dr. Edward Weiler, NASA's Associate Administrator for Space Science. "The problem here was not the error, it was the failure of NASA's systems engineering, and the checks and balances in our processes to detect the error. That's why we lost the spacecraft."

Just for fun

Given the following equivalences, how many inches are contained in a distance that measures 2 miles 3 furlongs 4 rods 5 yards 2 feet?

$$\begin{aligned} 12 \text{ inches} &= 1 \text{ foot} \\ 3 \text{ feet} &= 1 \text{ yard} \\ 5\tfrac{1}{2} \text{ yards} &= 1 \text{ rod} \\ 40 \text{ rods} &= 1 \text{ furlong} \\ 8 \text{ furlongs} &= 1 \text{ statute mile} \end{aligned}$$

13.4 VOLUME

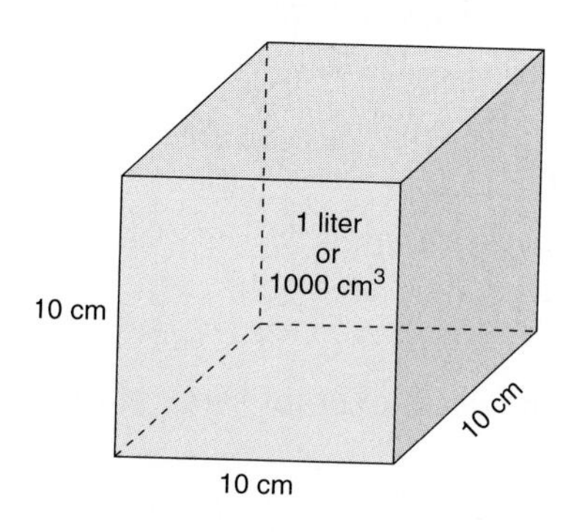

Figure 13.5

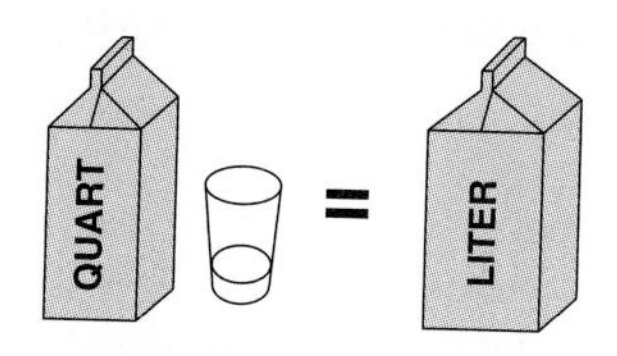

Figure 13.6

One liter of milk contains slightly more than a quart of milk.

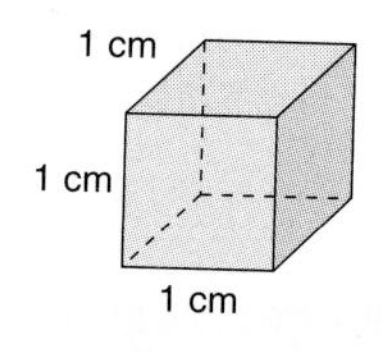

Figure 13.7

One milliliter (1 cubic centimeter).

Volume is the measure of how much a container can hold—that is, its capacity. Unfortunately, our system of measuring volumes and weights is quite confusing. For example, some soft-drink bottles contain 16 ounces, and some cans of coffee contain 16 ounces, but these are two different kinds of ounces. The soft-drink bottle contains 16 **fluid ounces**, which is equivalent to 1 pint. The 16 ounces in the can of coffee are units of weight, equivalent to 1 pound.

In the metric system, the *liter* is the basic unit used to measure capacity. A **liter** is defined as the volume of a cubic decimeter. In other words, a liter is the capacity of a cube (box) that is 1 decimeter long, 1 decimeter wide, and 1 decimeter high (see Figure 13.5). One liter, 1 cubic decimeter, and 1000 cubic centimeters all represent the same volume.

One of the reasons that volume is easier to work with in the metric system than in our present system is that one set of units is used for all volume measures, whether liquid or dry. Using our present system, the volume of a pint of blueberries is different from the volume of a pint of cream. But in the metric system, cream is sold in liter containers and blueberries are sold in liter boxes.

We shall compare a liter to a unit of volume with which you are already familiar, a quart (Figure 13.6). A liter contains a little more than a quart. One liter is approximately 1.06 liquid quarts, and one liquid quart is approximately 0.95 liter.

Two other common units of volume measure in the metric system are the **milliliter** and the **cubic meter**. Recall that the prefix *milli* means $\frac{1}{1000}$ or 0.001; hence, a milliliter is $\frac{1}{1000}$ of a liter. Because a liter contains 1000 cubic centimeters, a milliliter is 1 cubic centimeter. We can think of a milliliter as a cube whose length, width, and height each measure 1 centimeter. The milliliter and the liter are the two most commonly used units of volume.

The milliliter is a small unit of volume measure (Figure 13.7). For example, 5 milliliters = 1 teaspoon, and 15 milliliters = 1 tablespoon. Most liquid prescriptions obtained at pharmacies are sold in milliliters.

A cubic meter (m^3) is used to measure large volumes. One cubic meter is equivalent to 1.3 cubic yards. Items such as large amounts of sand and concrete are sold by the cubic yard at present. In the metric system, they would be sold by the cubic meter. Extremely large quantities of liquids would be measured in terms of the **kiloliter**, which is equivalent to 1000 liters. The capacity of fuel oil trucks and gasoline trucks, for example, would be expressed in kiloliters.

Remember that a liter is defined as the volume of a cube that is 10 centimeters long, 10 centimeters wide, and 10 centimeters high. The volume of 1 liter is 1000 cubic centimeters. The following table lists metric volume measures and their equivalent measure in liters.

Symbol	Word	Meaning
kL	kiloliter	1000 liters
hL	hectoliter	100 liters
daL	dekaliter	10 liters
L	liter	1 liter
dL	deciliter	0.1 liter
cL	centiliter	0.01 liter
mL	milliliter	0.001 liter

We can use a conversion chart, similar to the one in the previous section, to find equivalent measures of volume:

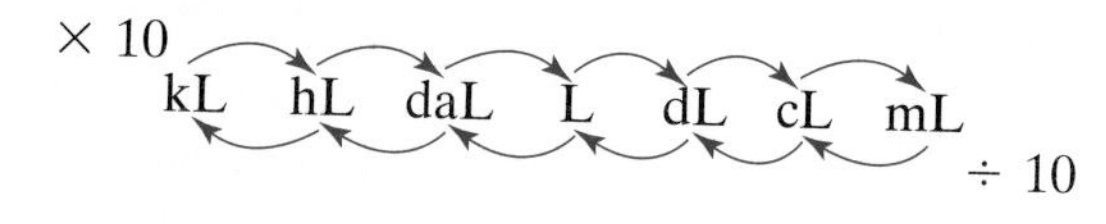

Recall that for each step to the right we multiply by 10, or move the decimal point one place to the right. For each step to the left, divide by 10, or move the decimal point one place to the left.

EXAMPLE 1 ***Finding Equivalent Metric Expressions*** Complete each of the following:

a. 642 mL = ______ L **b.** 74.3 L = ______ cL
c. 5 kL = ______ daL **d.** 15 L = ______ kL
e. 4000 mL = ______ L **f.** 253 L = ______ hL

Solution

a. To change milliliters to liters, move the decimal point three places to the left. That is,

$$642 \text{ mL} = 0.642 \text{ L}$$

b. To change liters to centiliters, move the decimal point two places to the right. That is,

$$74.3 \text{ L} = 7430 \text{ cL}$$

c. 5 kL = 500 daL **d.** 15 L = 0.015 kL

e. 4000 mL = 4.000 L = 4.0 L **f.** 253 L = 2.53 hL

NW ***Now Work Problem 1.***

You may find the metric conversion factors in Table 13.2 helpful.

How much does a barrel contain? Several different sized barrels have been established by law or usage. For example, the taxes on fermented liquors are based on a barrel of 31 gallons. But some states recognize a liquid barrel as 31.5 gallons. Federal law recognizes a 40-gallon barrel for "proof of spirits." A barrel of crude oil or petroleum products contains 42 gallons. Hence, a barrel may contain from 31 to 42 gallons, depending on where it is used and how it is used.

TABLE 13.2 Metric Conversion Factors—Volume

Symbol	When You Know	Multiply by	To Find	Symbol
		To Metric		
tsp	teaspoons	5	milliliters	mL
tbsp	tablespoons	15	milliliters	mL
fl oz	fluid ounces	30	milliliters	mL
c	cups	0.24	liters	L
pt	pints	0.47	liters	L
qt	quarts	0.95	liters	L
gal	gallons	3.8	liters	L
ft^3	cubic feet	0.03	cubic meters	m^3
yd^3	cubic yards	0.76	cubic meters	m^3
		From Metric		
mL	milliliters	0.03	fluid ounces	fl oz
L	liters	2.1	pints	pt
L	liters	1.06	quarts	qt
L	liters	0.26	gallons	gal
m^3	cubic meters	35	cubic feet	ft^3
m^3	cubic meters	1.3	cubic yards	yd^3

Remember, these are all approximate conversions. It should be noted that some texts list a cup as equivalent to 250 milliliters, but according to the National

Institute of Standards and Technology (NIST), 1 cup = 0.24 liter, which is 240 milliliters.

One of the areas where the metric system can provide some help is in the kitchen. For example, a cake recipe that calls for

$2\frac{1}{4}$ cups of flour	$\frac{1}{3}$ cup of shortening
$1\frac{1}{2}$ cups of sugar	1 cup of milk
3 teaspoons of baking powder	1 tablespoon of flavoring
1 teaspoon of salt	2 eggs

would call for the following in the metric system:

540 milliliters of flour	80 milliliters of shortening
360 milliliters of sugar	240 milliliters of milk
15 milliliters of baking powder	15 milliliters of flavoring
5 milliliters of salt	2 eggs

Note that in the first recipe we have to use fractions, as well as different units of measure. However, in the metric recipe, everything (except the eggs) is measured in the same unit, milliliters. If the metric recipe called for fractional amounts, they would indeed be in decimal notation, because the metric system uses only powers of 10.

Using the metric conversions in Table 13.2, we can convert any recipe to a metric recipe. Since 1 cup = 0.24 liter = 240 milliliters, $2\frac{1}{4}$cups = $2\frac{1}{4} \times 240 =$ 540 milliliters. The other conversions are done in a similar manner, using Table 13.2.

EXAMPLE 2 ***Converting to Metric Measure*** Convert the given recipe to a metric recipe.

Blueberry Pie Filling

$\frac{1}{2}$ cup of sugar	$2\frac{1}{2}$ tablespoons of flour
1 tablespoon of lemon juice	$\frac{1}{2}$ teaspoon of cinnamon
$2\frac{1}{3}$ cups of drained blueberries	1 tablespoon of shortening
$\frac{1}{3}$ cup of blueberry juice	

Solution

$$\begin{aligned}
\tfrac{1}{2}\text{cup of sugar} &= \tfrac{1}{2} \times 240 \\
&= 120 \text{ milliliters of sugar} \\
1 \text{ tablespoon of lemon juice} &= 15 \text{ milliliters of lemon juice} \\
2\tfrac{1}{3} \text{ cups of blueberries} &= 2\tfrac{1}{3} \times 240 \\
&= 560 \text{ milliliters of blueberries} \\
\tfrac{1}{3} \text{ cup of blueberry juice} &= \tfrac{1}{3} \times 240 \\
&= 80 \text{ milliliters of blueberry juice} \\
2\tfrac{1}{2} \text{ tablespoons of flour} &= 2\tfrac{1}{2} \times 15 \\
&= 37.5 \text{ milliliters of flour} \\
\tfrac{1}{2} \text{ teaspoon of cinnamon} &= \tfrac{1}{2} \times 5 \\
&= 2.5 \text{ milliliters of cinnamon} \\
1 \text{ tablespoon of shortening} &= 15 \text{ milliliters of shortening}
\end{aligned}$$

NW ***Now Work Problem 13.***

Remember that in the metric system, one type of unit—the liter or some multiple of the liter—is used to measure both liquid and dry volume. The liter and the milliliter are the two most commonly used units of volume. One liter is defined to be the volume of a cube that is 10 centimeters long, 10 centimeters wide, and 10 centimeters high. A milliliter is the volume of a cube that is 1 centimeter long, 1 centimeter wide, and 1 centimeter high.

EXAMPLE 3 ***Finding the Volume of an Aquarium*** How many liters will an aquarium hold if it is 70 centimeters long, 0.5 meter wide, and 500 millimeters high?

Solution We must convert all measurements to the same unit before finding the volume. Recall that volume = length × width × height. Because we want to express the answer in terms of liters, we shall convert all measurements to centimeters because the volume of 1 liter is 1000 cubic centimeters.

$$\begin{aligned} \text{length} &= 70 \text{ cm} \\ \text{width} &= 0.5 \text{ m} = 50 \text{ cm} \\ \text{height} &= 500 \text{ mm} = 50 \text{ cm} \end{aligned}$$

Therefore,

$$V = l \times w \times h$$
$$V = 70 \times 50 \times 50$$
$$V = 175\,000 \text{ cm}^3 \qquad \textit{Note: } 1L = 1000 \text{ cm}^3$$
$$V = 175 \text{ L}$$

NW Now Work Problem 9.

It should be noted that in the United States a common abbreviation for cubic centimeters is cc, as opposed to the metric notation cm^3. This is particularly true in medicine and other areas where the cubic centimeter has been used as a unit of measurement. But because 1 cubic centimeter is the same as 1 milliliter, the milliliter usage is becoming more and more common, particularly with regard to prescriptions.

Exercises for Section 13.4

NW **1.** Complete each of the following:
- **a.** 12 hL = ______ L
- **b.** 25 dL = ______ mL
- **c.** 3500 liters = ______ kL
- **d.** 100 mL = ______ cL
- **e.** 650 cL = ______ L
- **f.** 45 hL = ______ daL

2. Complete each of the following:
- **a.** 42 liters = ______ mL
- **b.** 640 mL = ______ L
- **c.** 25 kL = ______ hL
- **d.** 6 kL = ______ hL
- **e.** 34 daL = ______ hL
- **f.** 75 dL = ______ daL

3. Complete each of the following:
- **a.** 73.2 mL = ______ dL
- **b.** 2314 mL = ______ L
- **c.** 3.14 kL = ______ daL
- **d.** 14.95 mL = ______ L
- **e.** 0.49 daL = ______ cL
- **f.** 7.2 liters = ______ mL

4. Complete each of the following:
- **a.** 324 mL = ______ L
- **b.** 42.4 daL = ______ L
- **c.** 53 liters = ______ mL
- **d.** 5.3 L = ______ mL
- **e.** 62.2 kL = ______ daL
- **f.** 0.5 kL = ______ L

5. Arrange the following measurements of volume in descending order beginning with the largest: dekaliter, liter, hectoliter, milliliter, kiloliter, deciliter, centiliter.

6. Which has the greater volume in each pair?
- **a.** 1 qt or 1 liter
- **b.** 1 gal or 3 L
- **c.** 2 pt or 1 liter
- **d.** 1 tsp or 2 mL
- **e.** 1 c or 1 liter
- **f.** 2 fl oz or 20 mL

7. Which has the greater volume in each pair?
   a. 2 gal or 2 kL b. 3 qt or 3 liters
   c. 3 pt or 50 cL d. 2 c or 2 liters
   e. 3 tsp or 3 mL f. 5 fl oz or 5 mL
8. **Volume of a box** Find the volume of a box that is 1 meter long, 40 centimeters wide, and 50 centimeters high. (*Hint:* Convert all measurements to the same unit before finding the volume, where volume = length × width × height.) Express your answer in liters.

NW 9. **Volume of a container** Find the volume of a box that is 80 centimeters long, 0.5 meter wide, and 50 millimeters high. Express your answer in liters.

10. **Volume of an aquarium** How much water will an aquarium hold if it is 1 meter long, 60 centimeters wide, and 600 millimeters high? Express your answer in cubic meters.

11. **Volume of a Freezer** What is the storage capacity of a food freezer whose inside measurements are 1.5 meters by 1 meter by 80 centimeters? Express your answer in cubic meters.
12. Convert the given recipe to a metric recipe.

Clam Chowder

1 teaspoon salt	$\frac{1}{2}$ cup water
$\frac{1}{4}$ cup butter	1 pint minced clams
2 cups milk	2 cups diced potatoes
$\frac{1}{4}$ cup minced onions	

NW 13. Convert the given recipe to a metric recipe.

Rice Pudding

$\frac{1}{2}$ cup uncooked rice	$\frac{1}{3}$ cup seedless raisins
$2\frac{1}{2}$ cups milk	$\frac{1}{2}$ tablespoon cinnamon
$\frac{1}{4}$ cup sugar	$\frac{1}{2}$ teaspoon salt

For Exercises 14–20, choose the most sensible answer.

14. Volume of a lawn mower fuel tank:
    a. 500 L b. 10 kL c. 2 L
15. Volume of a bottle of soda pop:
    a. 36 mL b. 360 mL c. 36 L
16. Volume of a can of paint:
    a. 4 mL b. 4 cL c. 4 L
17. Volume of a coffee cup:
    a. 20 mL b. 25 L c. 250 mL
18. Volume of a measuring cup:
    a. 5 mL b. 50 mL c. 500 mL
19. Volume of a soup spoon:
    a. 15 cL b. 15 dL c. 15 mL
20. Volume of a teakettle:
    a. 1 L b. 10 L c. 100 L

Writing Mathematics

21. Explain how the volume of a liter was determined.
22. Explain the advantage of the definition of a liter over the similar definition (quart) in the English system.
23. The medical profession tends to use the notation *cc* for cubic centimeters (milliliter). Why do you think this is the case?

Challenge Exercises

24. Many beverages are now sold by milliliters. For example, wine bottles are commonly sold in 750 milliliter sizes. List some other areas where metric volume measures are used extensively in the United States.
25. Write the following volumes in both milliliters and cubic centimeters.
    a. Find the volume of a box that measures 8 cm × 6 cm × 10 cm.
    b. Find the volume of a box that measures 10 cm × 20 cm × 50 cm.

Research/Group Activities

26. Select a country other than the United States and research the development of the metric system in that country. Prepare a report to be presented in class.
27. In the United States there are a variety of *barrels* established by law and usage. That is, not all barrels contain the same amount of liquid. (See Note of Interest, p. 732.) This is also true of other measures in the U.S. system. Determine the number of liters in
    a. one dry pint
    b. one liquid pint
    c. one dry quart
    d. one liquid quart
    e. one British quart
28. Find the capacity of a gas tank of a specific car. Be sure to state the name and model. Next, find out the price of gasoline at a local service station.
    a. Find the capacity of the gas tank in liters.
    b. What is the price of gasoline per liter?
    c. What will be the cost of filling the gas tank?

Just for fun

In the late 1880s, liquid measure was also known as wine measure because it was used to measure liquors and wines. Given the following equivalences, how many gills are contained in 1 hogshead 1 barrel 20 gallons 3 quarts 1 pint?

4 gills = 1 pint
2 pints = 1 quart
4 quarts = 1 gallon
$31\frac{1}{2}$ gallons = 1 barrel
2 barrels = 1 hogshead

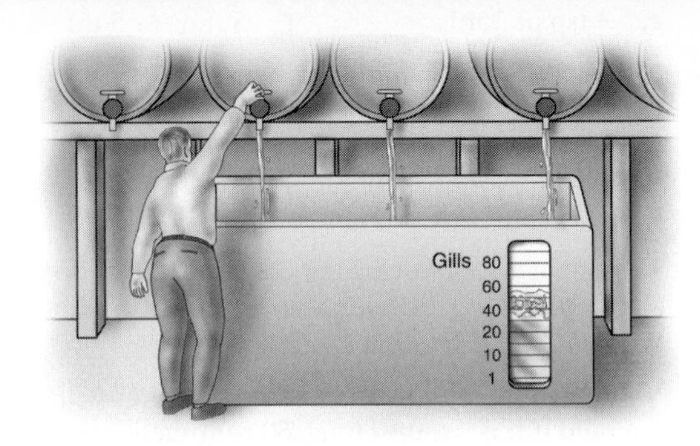

13.5 MASS (WEIGHT)

Figure 13.8

The weight of one nickel is 5 grams.

Weight is a measure of the earth's gravitational pull. (Gravity is the force that holds you on earth.) **Mass** is the measure of the amount of matter—that is, atoms and molecules—that objects are made of. In space, the mass of an object does not change, but its weight does. *Weight* and *mass* are not the same thing, but on earth the mass of an object is always proportional to the weight of the object. Therefore, for this course, we shall assume that weight and mass mean the same thing.

In the metric system, the most common measures of weight (mass) are the kilogram (kg), the gram (g), and milligram (mg). The basic unit of mass in the metric system is the gram. The weight of a common paper clip is approximately 1 gram, whereas a nickel weighs 5 grams (Figure 13.8).

Imagine constructing a leakproof cubic centimeter out of weightless material (see Figure 13.7 in Section 13.4) and filling it with very cold water. The mass (weight) of the water in such a container is 1 gram. Recall that one of the advantages of the metric system is that all of the measures (distance, volume, weight) are related. The mass (weight) of 1 milliliter of water is 1 gram, and a milliliter is the volume of a cube whose length, width, and height each measure 1 centimeter.

Now we can list some metric weight measures and their equivalent measure in grams:

Symbol	Word	Meaning
kg	kilogram	1000 grams
hg	hectogram	100 grams
dag	dekagram	10 grams
g	gram	1 gram
dg	decigram	0.1 gram
cg	centigram	0.01 gram
mg	milligram	0.001 gram

We can use a conversion chart, similar to those in the previous sections, to find equivalent measures of weight (mass):

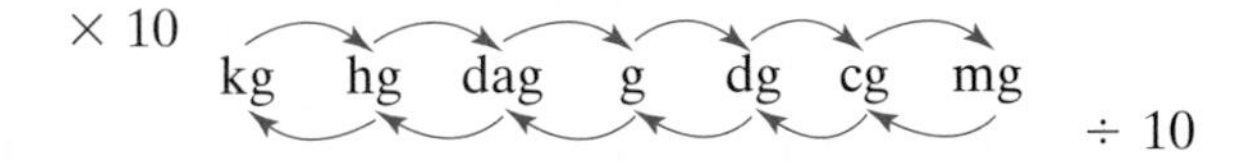

For each step to the right, we multiply by 10, or move the decimal point one place to the right. For each step to the left, divide by 10, or move the decimal point one place to the left.

EXAMPLE 1 ***Finding Equivalent Metric Expressions*** Complete each of the following:

a. 3.5 kg = ______ g **b.** 2000 g = ______ hg

c. 250 mg = ______ g **d.** 3.7 g = ______ dg

e. 4.5 dg = ______ mg **f.** 380 mg = ______ dag

Solution

a. To change kilograms to grams, move the decimal point three places to the right. That is,

$$3.5 \text{ kg} = 3\,500 \text{ g}$$

b. To change grams to hectograms, move the decimal point two places to the left. That is,

$$2000 \text{ g} = 20.00 \text{ hg} = 20 \text{ hg}$$

c. 250 mg = 0.250 g **d.** 3.7 g = 37 dg

e. 4.5 dg = 450 mg **f.** 380 mg = 0.0380 dag

NW ***Now Work Problem 1.***

A kilogram is equivalent to 1000 grams. Therefore, 1000 milliliters—that is, 1 liter—filled with very cold water will approximate the weight of 1 kilogram. One pound is approximately 0.45 kilogram and 1 kilogram is approximately 2.2 pounds. Larger foodstuffs such as meat and fish are weighed in terms of kilograms. A piece of meat that weighs 1 kilogram is ample for four people.

A milligram is equivalent to 0.001 gram. It is an extremely small unit of weight and is used only in medical prescriptions and some areas of science. For example, if you examine the label on a bottle of cold tablets, you will note that the amounts of different substances that each tablet contains are given in terms of milligrams. A common cold tablet contains, among other things, 225 milligrams of aspirin, 30 milligrams of caffeine, and 50 milligrams of ascorbic acid.

The weights of objects that are quite heavy, such as automobiles, are given in *metric tonnes.* One **metric tonne** is equivalent to 1000 kilograms. Because 1 kilogram is approximately 2.2 pounds, a metric tonne (1000 kilograms) is equivalent to 2200 pounds, or 1.1 tons (1 ton = 2000 pounds).

It is interesting to note that 1 gram is the weight of 1 milliliter of water, 1 kilogram is the weight of 1 liter of water, and a metric tonne is the weight of 1 cubic meter of water.

In order to develop a sense of the metric weights discussed, remember that a common paper clip weighs approximately 1 gram and a nickel weighs about 5 grams. A kilogram is equivalent to 2.2 pounds, so a man who weighs 220 pounds weighs 100 kilograms. A milligram ($\frac{1}{1000}$ of a gram) is a minute quantity and is difficult to approximate. A small grain of sand weighs about 1 milligram. The metric tonne is used in measuring the weight of very heavy objects. A metric tonne (1000 kilograms) is equivalent to 2200 pounds.

You may find the metric conversion factors in Table 13.3 helpful. *Remember, these are all approximate conversions.*

TABLE 13.3 Metric Conversion Factors—Mass (Weight)

Symbol	When You Know	Multiply by	To Find	Symbol
		To Metric		
oz	ounces	28	grams	g
lb	pounds	0.45	kilograms	kg
T	tons	0.9	tonnes	t
		From Metric		
g	grams	0.035	ounces	oz
kg	kilograms	2.2	pounds	lb
t	tonnes	1.1	tons	T

EXAMPLE 2 ***Determining Mass (Weight)*** Recall that 1 gram is the mass (weight) of 1 milliliter of water whose temperature is approximately 39° Fahrenheit. Give the mass (weight) in grams of

a. 1 liter of water **b.** 1 cubic meter of water

Solution

a. The volume of 1 liter is 1000 cubic centimeters. A milliliter is the volume of 1 cubic centimeter. Therefore, 1 liter contains 1000 milliliters and 1 liter of water weighs 1000 grams.

As an alternate solution note that a milliliter is $\frac{1}{1000}$ of a liter (recall the meaning of the prefix *milli*). Therefore, 1 liter contains 1000 milliliters and 1 liter of water weighs 1000 grams.

b. A meter contains 100 centimeters, so a cubic meter contains 1 000 000 cubic centimeters. Thus, a cubic meter of water contains 1 000 000 milliliters of water and weighs 1 000 000 grams. Note that

$$1\ 000\ 000 = 1000 \times 1000$$

Therefore, 1 000 000 grams is the same as 1000 kilograms. But 1000 kilograms is equivalent to 1 metric tonne. The weight of 1 cubic meter of very cold water is 1 metric tonne (1 t).

NW ***Now Work Problem 9.***

EXAMPLE 3 ***Converting Between Systems of Measurement*** Convert each of the following to the indicated weight.

a. 16 oz = ______ g **b.** 10 lb = ______ kg

c. 50 kg = ______ lb **d.** 100 g = ______ oz

e. 4000 lb = ______ t **f.** 180 lb = ______ kg

Solution

a. In order to convert ounces to grams, we must multiply the number of ounces by 28. Therefore, 16 oz = 16 × 28 = 448 g.

b. One pound is equivalent to 0.45 kilogram. Therefore,

$$10 \text{ lb} = 10 \times 0.45 = 4.5 \text{ kg}$$

c. One kilogram is equivalent to 2.2 pounds. Therefore,

$$50 \text{ kg} = 50 \times 2.2 = 110 \text{ lb}$$

d. One gram is equivalent to 0.035 ounce. Therefore,

$$100 \text{ g} = 100 \times 0.035 = 3.5 \text{ oz}$$

e. 2000 pounds = 1 ton, so 4000 pounds = 2 tons. Using Table 13.3, 1 T = 0.9 t and

$$4000 \text{ lb} = 2 \text{ T} = 2 \times 0.9 = 1.8 \text{ t}$$

f. One pound is equivalent to 0.45 kilogram. Therefore,

$$180 \text{ lb} = 180 \times 0.45 = 81 \text{ kg}$$

NW ***Now Work Problem 5.***

The metric system is more consistent than our customary system of weights and measures. In the metric system, length, volume, and weight are directly related to each other. For example, the volume of a cube that has length, width, and height of 1 centimeter (0.01 meter) is called a milliliter, and if we fill the milliliter with very cold water (39 °F), it will have a weight of 1 gram.

A liter is the volume of a cube that has a length, width, and height of 10 centimeters. If we fill the liter with very cold water, it will have a weight of 1 kilogram. Similarly, the volume of a cube that has length, width, and height of 1 meter is called a kiloliter, and if we fill the kiloliter with very cold water, it will have a weight of 1000 kilograms or 1 tonne.

Exercises for Section 13.5

NW **1.** Complete each of the following:

a. 52 g = ______ mg **b.** 3.5 kg = ______ g
c. 250 mg = ______ g **d.** 4.3 dg = ______ g
e. 7.2 hg = ______ dg **f.** 4.7 g = ______ mg

2. Complete each of the following:

a. 6300 mg = ______ g
b. 25 dag = ______ g
c. 80 cg = ______ g
d. 20.6 mg = ______ cg
e. 26.5 dag = ______ hg
f. 4.6 kg = ______ mg

3. Complete each of the following:

a. 417 mg = ______ g
b. 342 kg = ______ g
c. 14.3 kg = ______ g
d. 885.9 mg = ______ g
e. 14,359 mg = ______ kg
f. 2.71 kg = ______ g

4. Complete each of the following:

a. 544 g = ______ mg
b. 722 g = ______ kg
c. 423.4 g = ______ kg
d. 267.2 g = ______ mg
e. 315 dg = ______ dag
f. 568 dag = ______ dg

NW **5.** Convert each of the following to the indicated weight:

a. 8 oz = ______ g **b.** 100 kg = ______ lb
c. 500 g = ______ oz **d.** 10 T = ______ t
e. 200 lb = ______ kg **f.** 6000 lb = ______ t

6. Convert each of the following to the indicated weight:

a. 4 lb = ______ kg **b.** 10 kg = ______ lb
c. 100 g = ______ oz **d.** 32 oz = ______ g
e. 10 mg = ______ oz **f.** 5000 lb = ______ t

7. Vitamins A certain vitamin tablet contains 150 mg of vitamin A, 225 mg of vitamin B, and 125 mg of vitamin C. What is the weight of the vitamin tablet in grams?

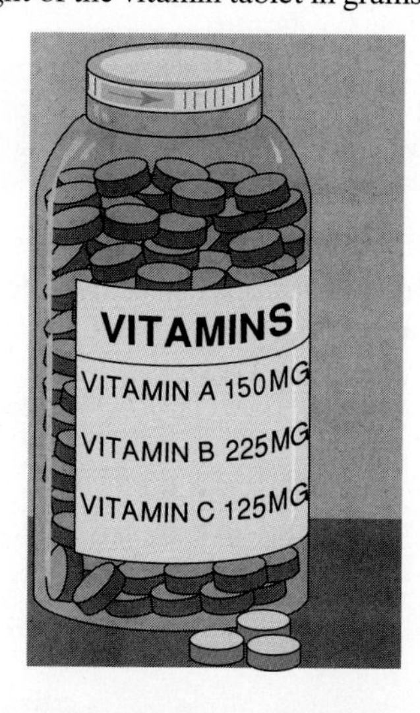

8. **Antacid tablet** A certain antacid tablet contains 248 mg of aluminum hydroxide, 75 mg of magnesium hydroxide, and 41 mg of sodium. What is the weight of the antacid tablet in grams?

NW 9. **Weight** A container is 60 centimeters long, 40 centimeters wide, and 50 centimeters high. It weighs 3 kilograms when it is empty. What will it weigh when it is filled with cold water? Express your answer in kilograms.

10. **Volume** How many liters will an aquarium hold if it is 1 meter long, 60 centimeters wide, and 60 centimeters high? What is the mass of the water in kilograms if it is approximately 39° Fahrenheit?

11. **Swimming pool** A swimming pool is 20 meters long and 9 meters wide. It has a uniform depth of 2 meters.
    - **a.** How many kiloliters of water will it hold?
    - **b.** What is the mass of the water if it is approximately 39° Fahrenheit?

12. Assuming that each of the following is filled with very cold water, give the mass (in grams) of
    - **a.** 2 liters **b.** 1 dekaliter
    - **c.** 5 milliliters **d.** 2 kiloliters
    - **e.** 3 hectoliters **f.** 5 centiliters

13. Which has the greater weight?
    - **a.** 1 lb or 1 kg **b.** 2 oz or 20 g
    - **c.** 10 lb or 5 kg **d.** 4000 lb or 3 t
    - **e.** 40 oz or 2 kg **f.** 280 g or 9 oz

For Exercises 14–19, choose the most sensible answer.

14. Mass of a pound of candy:
    a. 20 g **b.** 120 g **c.** 450 g

15. Mass of a newborn baby:
    a. 3 kg **b.** 30 kg **c.** 300 kg

16. Mass of a grain of sand:
    a. 1 kg **b.** 1 g **c.** 1 mg

17. Mass of an adult person:
    a. 7.5 kg **b.** 75 kg **c.** 750 kg

18. Mass of a nickel:
    a. 5 g **b.** 5 mg **c.** 5 kg

19. Mass of a paper clip:
    a. 1 mg **b.** 1 g **c.** 1 dag

Writing Mathematics

20. How was the mass of one gram determined?

21. Explain the advantage of this definition over the similar definition (ounce) in the English system.

22. Is your weight the same on the moon as it is on earth? Provide reasons for your answer. Is this also true for your mass? Why?

Challenge Exercises

23. Sea water contains about 3.5 grams of salt per 100 milliliters of water. How many grams of salt would be in two liters of sea water?

24. Helium weighs about 0.0002 grams per milliliter. How much does five liters of helium weigh?

25. A biblical measure of weight was the *shekel*. Find its metric equivalent.

26. The heaviest known diamond is the Cullinan, found in South Africa in 1905 and given to King Edward VII. It weighed 3106 carats before it was cut into 105 gems. How many kilograms did the Cullinan weigh? [*Hint*: One gram is the same weight as five carats.]

Research/Group Activities

27. In this chapter, we use the terms *mass* and *weight* interchangeably due to the fact that on earth, the mass of an object is proportional to the weight of an object. But in science they are different. Research these terms and prepare a report to explain the differences between the two terms.

28. A single platinum-iridium bar is sealed in an airtight container in the International Bureau of Weights and Measures in France. It represents the true weight of one kilogram. This bar has some unique characteristics. Find out more about this bar and its properties. Prepare a report to be presented to your class.

Just for fun

In the late 1800s, apothecaries (pharmacists) used the standard apothecaries' weight to make up different medicines. Given the following equivalences, how many grains are contained in 2 pounds 3 ounces 2 drams 2 scruples?

20 grains = 1 scruple
3 scruples = 1 dram
8 drams = 1 ounce
12 ounces = 1 pound

13.6 TEMPERATURE

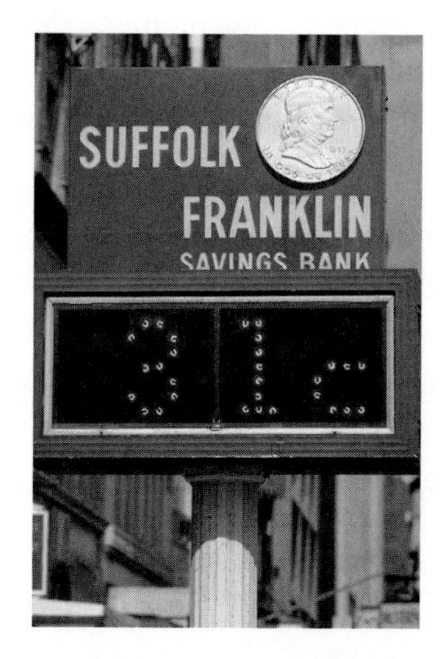

One type of measure with which we are very familiar is temperature. All of us are concerned about weather on any given day, and one of the first questions we ask is "What's the temperature?" Wherever we go, there are time–temperature clocks that indicate how warm or how cold it is. It is even possible to dial a number and hear a recorded message stating the current time and temperature.

Gabriel Fahrenheit (1686–1736), a German physicist, used a brine solution (salt, water, and ice) to devise a scale for the mercury thermometer such that the boiling point of water was 212° and the freezing point of water was 32°. Zero degrees was the lowest point on the thermometer because that was the coldest temperature he could get with the brine solution. However, there are colder temperatures. A temperature of $-81°$ was once recorded in Canada, $-90°$ in Russia, and $-127°$ in Antarctica.

Shortly after Fahrenheit developed his scale, Anders Celsius (1701–1744), a Swedish astronomer, developed another scale. Celsius developed his scale so that the boiling point of water was 100° and the freezing point of water was 0°. You may be familiar with the Celsius thermometer as the centigrade thermometer. Recall that *centi* means $\frac{1}{100}$ or 0.01; there are 100 intervals between 0° centigrade and 100° centigrade. Because the scale that Celsius developed was so convenient, the centigrade thermometer was adopted by scientists, and because 100 is a power of 10 ($100 = 10^2$), all of the countries using the metric system have also adopted it. This thermometer was formerly known as the centigrade thermometer, but, in honor of the man who developed the scale, it is now officially known as the Celsius thermometer, and each unit is called a degree Celsius (°C).

Pictured in Figure 13.9 are a Celsius thermometer and a Fahrenheit (°F) thermometer with indicated temperatures.

As shown in Figure 13.9, the boiling point of water is 212°F or 100°C, and the freezing point of water is 32°F or 0°C. Most of us will be able to remember these comparative temperatures on the Celsius thermometer. But we would like to be able to interpret other Celsius thermometer readings as well. When someone tells us that their body temperature is 37°C, we should be aware that this is normal, or 98.6°F. Note also that Figure 13.9 indicates that a room temperature of 68°F is the same as a room temperature of 20°C. It is obviously important to identify a degree measurement as degrees Celsius or degrees Fahrenheit.

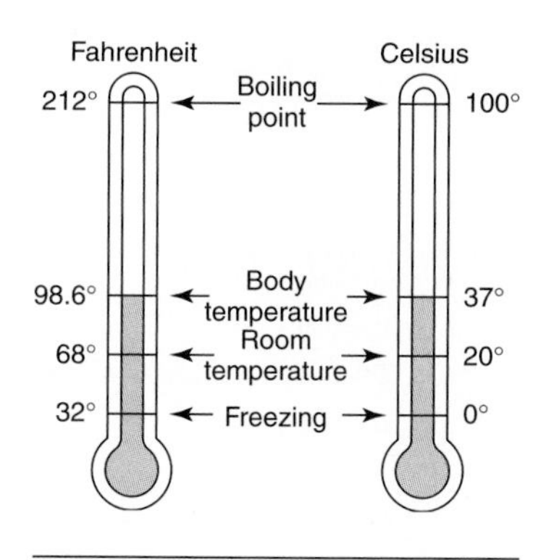

Figure 13.9

The following two formulas will enable you to convert from Fahrenheit to Celsius and vice versa. When you know the Fahrenheit temperature and wish to convert to Celsius, you should use the formula

$$C = \frac{5}{9}(F - 32) \qquad (C = °C; F = °F)$$

When you know the Celsius temperature and wish to convert to Fahrenheit, you should use

$$F = \frac{9}{5}C + 32$$

Let's work out a problem using the first formula. Suppose we wish to find the temperature on the Celsius thermometer equivalent to 32°F. From our previous discussion, you should know that the answer will be 0°C. Using the formula,

$$C = \frac{5}{9}(F - 32)$$

$$C = \frac{5}{9}(32 - 32) \qquad \textit{substituting 32 for F}$$

Historical Note

Gabriel Fahrenheit believed that the lowest temperature that could be reached was the temperature of a mixture of ice and salt. He immersed the end of his thermometer in such a mixture. When the mercury would go no lower, he marked the place on the thermometer and called it zero.

Fahrenheit supposedly put the end of the thermometer in his assistant's mouth and when the mercury stopped rising, he called this point 100°. The space between 0° and 100° was then divided into 100 units. A freezing mark of 32° was obtained because that is where the mercury stopped when the thermometer was put into freezing water. Using the same scaling technique, Fahrenheit discovered that 212° was the temperature of boiling water.

$$C = \frac{5}{9}(0) \qquad \textit{subtracting 32 from 32}$$

$$C = 0 \qquad \textit{multiplying by } 0$$

Therefore, 32°F = 0°C.

Let's try another example using the same formula. Suppose the given temperature is 212°F and we wish to find the equivalent temperature on the Celsius thermometer. From our previous discussion, you should know that the answer will be 100°C. Using the formula,

$$C = \frac{5}{9}(F - 32)$$

$$C = \frac{5}{9}(212 - 32) \qquad \textit{substituting 212 for F}$$

$$C = \frac{5}{9}(180) \qquad \textit{subtracting 32 from 212}$$

$$C = \frac{900}{9} \qquad \textit{multiplying 5 times 180}$$

$$C = 100 \qquad \textit{dividing 900 by 9}$$

Therefore, 212°F = 100°C.

Suppose the given Celsius temperature is 100°C and we wish to find the equivalent temperature on the Fahrenheit thermometer. We use the formula,

$$F = \frac{9}{5}C + 32$$

$$F = \frac{9}{5}(100) + 32 \qquad \textit{substituting 100 for C}$$

$$F = \frac{900}{5} + 32 \qquad \textit{multiplying 9 times 100}$$

$$F = 180 + 32 \qquad \textit{dividing 900 by 5}$$

$$F = 212 \qquad \textit{adding } 180 + 32$$

Therefore, 100°C = 212°F.

Note of Interest

This chapter has provided an introduction to the metric system. The adjacent table lists prefixes that provide the multiples in the International System of units.

Prefix	Symbol	Multiple	Equivalent
exa	E	10^{18}	quintillion
peta	P	10^{15}	quadrillion
tera	T	10^{12}	trillion
giga	G	10^{9}	billion
mega	M	10^{6}	million
kilo	k	10^{3}	thousand
hecto	h	10^{2}	hundred
deka	da	10	ten
deci	d	10^{-1}	tenth
centi	c	10^{-2}	hundredth
milli	m	10^{-3}	thousandth
micro	μ	10^{-6}	millionth
nano	n	10^{-9}	billionth
pico	p	10^{-12}	trillionth
femto	f	10^{-15}	quadrillionth
atto	a	10^{-18}	quintillionth

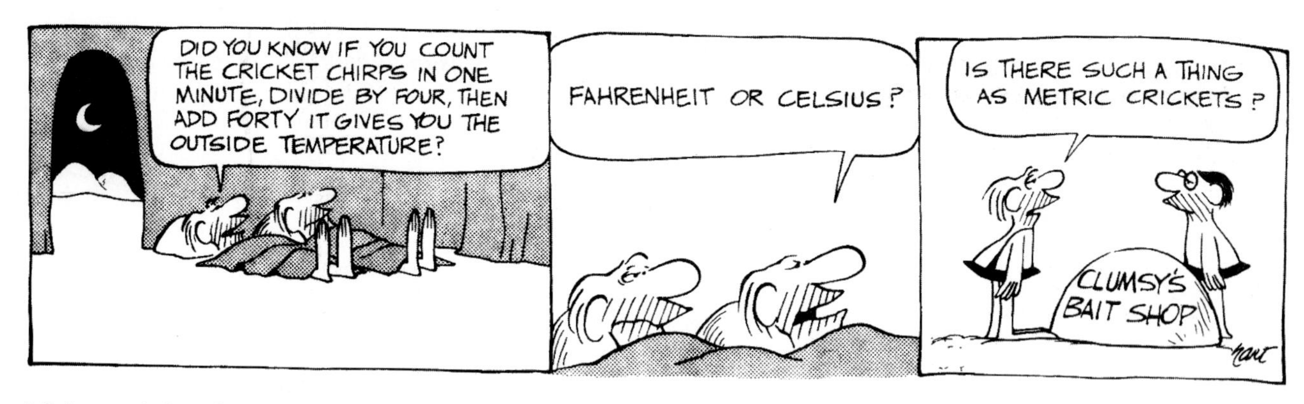

B.C. by permission of Johnny Hart and Creators Syndicate, Inc.

EXAMPLE 1 ***Converting Fahrenheit to Celsius*** Most people set their thermostats at 68°F. Find the equivalent temperature on the Celsius thermometer.

Solution In order to convert Fahrenheit to Celsius, we use the formula $C = \frac{5}{9}(F - 32)$:

$$C = \frac{5}{9}(68 - 32) \quad \textit{substituting 68 for F}$$

$$C = \frac{5}{9}(36) \quad \textit{subtracting 32 from 68}$$

$$C = \frac{180}{9} \quad \textit{multiplying 5 times 36}$$

$$C = 20 \quad \textit{dividing 180 by 9}$$

Therefore, 68°F = 20°C.

EXAMPLE 2 ***Converting Fahrenheit to Celsius*** Usually a person in good health has a body temperature of 98.6°F. Find the equivalent temperature on the Celsius thermometer.

Solution $C = \frac{5}{9}(F - 32)$

$$C = \frac{5}{9}(98.6 - 32) \quad \textit{substituting 98.6 for F}$$

$$C = \frac{5}{9}(66.6) \quad \textit{subtracting 32 from 98.6}$$

$$C = \frac{333.0}{9} \quad \textit{multiplying 5 times 66.6}$$

$$C = 37 \quad \textit{dividing 333 by 9}$$

Therefore, 98.6°F = 37°C.

NW ***Now Work Problem 1.***

EXAMPLE 3 ***Converting Celsius to Fahrenheit*** A recipe in a metric cookbook calls for an oven setting of 175°C. Find the equivalent temperature on the Fahrenheit thermometer.

Solution In order to convert Celsius to Fahrenheit, we use the formula $F = \frac{9}{5}C + 32$:

$$F = \frac{9}{5}(175) + 32 \quad \textit{substituting 175 for C}$$

$$F = \frac{1575}{5} + 32 \quad \textit{multiplying 9 times 175}$$

$$F = 315 + 32 \quad \textit{dividing 1575 by 5}$$

$$F = 347 \quad \textit{adding 315 and 32}$$

Therefore, 175°C = 347°F. (*Note:* We could also divide 175 by 5, which equals 35, and then multiply 35 by 9, which also equals 315 and brings us to the last step.)

EXAMPLE 4 ***Converting Celsius to Fahrenheit*** Convert 37°C to Fahrenheit.

Solution

$$F = \frac{9}{5}(37) + 32 \quad \textit{substituting 37 for C}$$

$$F = \frac{333}{5} + 32 \quad \textit{multiplying 9 times 37}$$

$$F = 66.6 + 32 \quad \textit{dividing 333 by 5}$$

$$F = 98.6 \quad \textit{adding 66.6 and 32}$$

Therefore, 37°C = 98.6°F.

NW ***Now Work Problem 3.***

Besides being able to convert from Fahrenheit to Celsius and vice versa, you should develop a "feel" for temperatures that are given in terms of the Celsius scale. For example, a temperature of 30°C is a hot day in most parts of the country and a temperature of 40°C is a scorcher anywhere. Similarly, if someone has a body temperature of 40°C then that person is very sick, and if his temperature rises to 41°C, then he is near death.

The following may help you to remember what it is like when you are given certain Celsius temperatures:

Thirty makes it hot.
Twenty makes it nice.
Ten gives us a cool spot.
Zero gives us ice.

Math Connections

Metric Prefixes in the Computer Age

In Chapter 7, we learned that the base 2 system of numeration (i.e., the binary number system) is the language of computers. During its development, computer scientists observed that the power $2^{10} = 1024$ was approximately equal to 1000 and began using the International System of Units (SI) prefix *kilo* to denote 1024. Thus, computer professionals talked about computer memory in terms of *kilo*bytes, *mega*bytes, and *giga*bytes, and used similar notation to specify data transmission rates such as *kilo*bits per second, *mega*bits per second, and *giga*bits per second. This notation was accepted in this context and for many years it was not an issue.

As computers achieved critical mass, however, many non-computer scientists and professionals suddenly were using computers and speaking "computerese." It was also common for computer professionals to communicate with laypeople, who considered the concept of kilo as equaling 1000 and not 1024. This resulted in some confusion. For example, when discussing computer memory, does one megabyte mean $2^{20} = 1,048,576$ bytes or does it mean $10^6 = 1,000,000$ bytes? When you say you have a 10-gigabyte hard drive, does this mean that your hard drive's storage capacity is 10×2^{30} bytes or does it mean 10×10^9 bytes? This confusion, which also carries over to expressing data transmission rates, is a real concern and has the potential for incompatible standards that can lead to incompatible devices, products, or systems.

To address this issue, the International Electrotechnical Commission (IEC), the leading international organization for worldwide standardization in electrotechnology, approved in December 1998, as an IEC International Standard, names and symbols for prefixes for binary multiples for use in the fields of data processing and data transmission. These prefixes are provided in the following table.

Factor	Name	Symbol	Origin	Derivation
2^{10}	kibi	Ki	kilobinary: $(2^{10})^1$	kilo: $(10^3)^1$
2^{20}	mebi	Mi	megabinary: $(2^{10})^2$	mega: $(10^3)^2$
2^{30}	gibi	Gi	gigabinary: $(2^{10})^3$	giga: $(10^3)^3$
2^{40}	tebi	Ti	terabinary: $(2^{10})^4$	tera: $(10^3)^4$
2^{50}	pebi	Pi	petabinary: $(2^{10})^5$	peta: $(10^3)^5$
2^{60}	exbi	Ei	exabinary: $(2^{10})^6$	exa: $(10^3)^6$

Source: http://physics.nist.gov/cuu/Units/binary.html

Examples and Comparisons with SI Prefixes

- One *kibi*bit: 1 Kibit = 2^{10} bits = 1024 bits
- One *kilo*bit: 1 kbit = 10^3 bits = 1000 bits
- One *mebi*byte: 1 MiB = 2^{20} bytes (B) = 1,048,576 B
- One *mega*byte: 1 MB = 10^6 bytes (B) = 1,000,000 B
- One *gibi*byte: 1 GiB = 2^{30} bytes (B) = 1,073,741,824 B
- One *giga*byte: 1 GB = 10^9 bytes (B) = 1,000,000,000 B

Although the new prefixes for the binary multiples (kibi, mebi, gibi, tebi, pebi, and exbi) are not part of the SI's metric units, they were nevertheless derived from the metric prefixes for ease of understanding and recall. Given this parallel, the first syllable of the name of the binary-multiple prefix is pronounced in the same way as the first syllable of the name of the corresponding SI prefix, and the second syllable is pronounced "bee."

Exercises for Section 13.6

NW **1.** Convert each Fahrenheit temperature to Celsius.

a. 104°F **b.** 50°F **c.** 41°F
d. 95°F **e.** 86°F **f.** 23°F

2. Convert each Fahrenheit temperature to Celsius.

a. 59°F **b.** 68°F **c.** 5°F
d. 122°F **e.** 32°F **f.** 14°F

NW **3.** Convert each Celsius temperature to Fahrenheit.

a. 20°C **b.** 50°C **c.** 85°C
d. 10°C **e.** 0°C **f.** 65°C

4. Convert each Celsius temperature to Fahrenheit.

a. 80°C **b.** 30°C **c.** 5°C
d. 45°C **e.** 55°C **f.** −5°C

5. Convert each temperature to the indicated scale.

a. 14°F = _____°C

b. 15°C = _____°F

c. 25°C = _____°F

d. −13°F = _____°C

6. Convert each temperature to the indicated scale.
   a. 95°F = _____°C
   b. 35°C = _____°F
   c. −10°C = _____°F
   d. −4°F = _____°C

7. **Fever** A sick person has a high fever if his temperature is 40°C. Find the equivalent temperature on the Fahrenheit thermometer.

8. **Recipe** A recipe in a metric cookbook calls for an oven setting of 200°C. Find the equivalent temperature on the Fahrenheit thermometer.

9. **Melting point of iron** Iron will melt at 5432°F. Find the equivalent temperature on the Celsius thermometer.

10. **Highest temperature in Ohio** The highest temperature ever recorded in the state of Ohio is 113°F. Find the equivalent temperature on the Celsius thermometer.

11. **Lowest temperature in South Dakota** The lowest temperature ever recorded in the state of South Dakota is −58°F. Find the equivalent temperature on the Celsius thermometer.

12. **Lowest temperature in the United States** The lowest temperature ever recorded in the United States is −80°F at Prospect Creek, Alaska. Find the equivalent temperature (to the nearest whole degree) on the Celsius thermometer.

13. **World's highest Temperature** A temperature of 136°F is generally accepted as the world's highest temperature ever recorded under standard conditions. This occurred in Northern Africa in 1922. Find the equivalent temperature (to the nearest whole degree) on the Celsius thermometer.

14. **Change in temperature** In Montana, the temperature once dropped from 44°F to −56°F in a period of 24 hours. This is a change in temperature of 100°F. Approximately how much of a change is it on the Celsius scale? (Think carefully!)

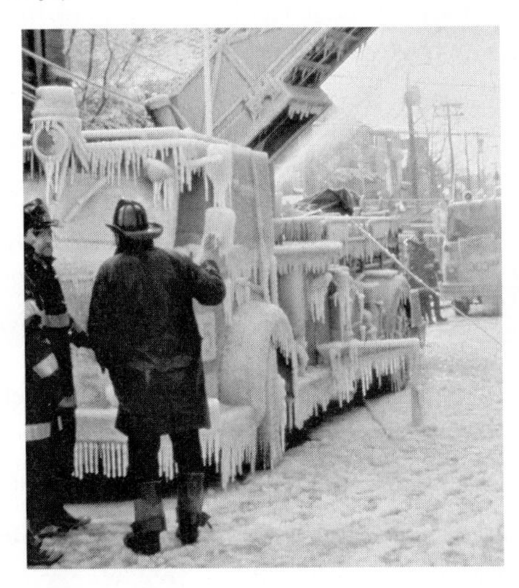

15. **Highest Temperature in the United States** The highest temperature ever recorded in the United States is 134°F, recorded in Death Valley. Find the equivalent temperature (to the nearest whole degree) on the Celsius thermometer.

For Exercises 16–22, choose the most sensible answer.

16. Normal body temperature:
    a. 98°C b. 37°C c. 45°C

17. Room temperature:
    a. 70°C b. 20°C c. 30°C

18. Temperature of freezing rain:
    a. 32°C b. 0°C c. 10°C

19. Temperature of melting wax:
    a. 100°C b. 212°C c. 80°C

20. Temperature of a frozen TV dinner:
    a. 0°C b. −10°C c. 32°C

21. Temperature of an oven when set for baking:
    a. 100°C b. 200°C c. 300°C

22. Ice-skating weather:
    a. −5°C b. 0°C c. 5°C

Writing Mathematics

23. Write the formula for converting Fahrenheit temperatures to Celsius. Explain how to do this conversion.

24. Write the formula for converting Celsius temperatures to Fahrenheit. Explain how to do this conversion.

25. List both the Fahrenheit and Celsius temperatures for each of the following:
    a. water boils
    b. highest temperature ever recorded on earth
    c. body temperature
    d. room temperature
    e. water freezes
    f. mixture of salt and water freezes

26. Until 1948 metric temperatures were given as centigrade and then changed to Celsius to honor its inventor. Even today centigrade is a common usage. Why do you think the word *centigrade* was used?

Challenge Exercises

27. Water boils at 212°F at sea level. For every 550 feet above sea level, the boiling point of water is lowered by 1°F. Rewrite this statement using Celsius readings.

28. Find the Celsius temperature at which methyl alcohol boils.

Research/Group Activity

29. A third temperature scale is the **Kelvin** scale. Research the Kelvin scale and Lord Kelvin, and prepare a report for class. Be sure to compare the characteristics of the Kelvin scale with Fahrenheit and Celsius.

Just for fun

Replace the usual measures (underlined) in the given expressions with the appropriate metric measures.

a. It's third down and there are 10 yards to go.
b. Give them an inch and they will take a mile.
c. The cowboy wore a 10-gallon hat.
d. Did you see the footnote?
e. A miss is as good as a mile.
f. He climbed the rope inch by inch.
g. An ounce of prevention is worth a pound of cure.
h. Have you read "Fahrenheit 451"?
i. Erskine Caldwell wrote "God's Little Acre."
j. A pint's a pound the world around.

CHAPTER REVIEW MATERIALS

Summary/Chapter 13

13.2 History of Systems of Measurement

- In 1790, the French Academy of Sciences devised a new system of measurement. The new basic unit of length was a portion of the *meridian* of the earth. It was called a *meter*, from the Greek word *metron*, "to measure."
- The common prefixes in the metric system are shown in the table.

Prefix	Symbol	Meaning
kilo	k	1000
hecto	h	100
deka	da	10
deci	d	$\frac{1}{10}$ or 0.1
centi	c	$\frac{1}{100}$ or 0.01
milli	m	$\frac{1}{1000}$ or 0.001

13.3 Length and Area

- The basic unit of length in the metric system is the *meter*. 1 meter $\approx$ 39.37 inches.
- A square whose measurements are 1 cm by 1 cm is said to have an area of 1 square cm; this is denoted by 1 cm^2. Larger regions are measured in terms of square meters (m^2). The area of very large regions is measured in *hectares* (10 000 square meters).
- We can make conversions from one metric measure to another by noting the prefixes and using the following chart:

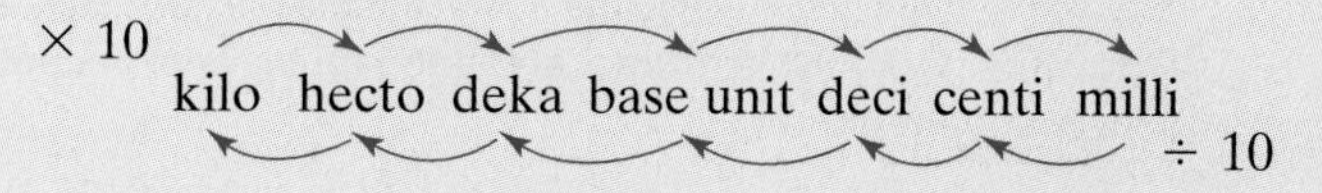

For each step to the right, multiply by 10 or move the decimal point one place to the right. For each step to the left, divide by 10, or move the decimal point one place to the left.

13.4 Volume

- The *liter* is the basic unit of volume in the metric system. It is defined as the volume of a cube that is 10 cm long, 10 cm wide and 10 cm long. A liter contains a little more than a quart.

13.5 Mass (Weight)

- The most common measures of weight (mass) in the metric system are the *kilogram* (kg), *gram* (g), and *milligram* (mg).
- The basic unit of mass in the metric system is the gram. The weight of a common paper clip is approximately one gram.

13.6 Temperature

- The metric system uses the *Celsius* thermometer to measure temperature.
- A room temperature of 20° C is the same as 68 °F, while 0°C = 32°F (freezing) and 100°C = 212°F (boiling).

For more information (free) on the metric system, write to

Office of Metric Programs
U.S. Department of Commerce
Room 4082 Hoover Building
Washington, DC 20230

http://www.doc.gov

or

U.S. Metric Assoc. Inc.
10245 Andasol Avenue
Northridge, CA 91325

or

Director, Metric Program
Building 820, Room 306
National Institute of Standards and Technology
Gaithersburg, MD 20899

(301) 975-3690 phone, (301) 948-1416 fax
e-mail: metric_prg@nist.gov
URL Address: http://www.nist.gov/metric

Note of Interest

Did you ever wonder why alcoholic beverages (a.k.a. spirits) measure the amount of alcohol by proof? It goes back to the eighteenth century, when purveyors of spirits needed a way to measure the amount of alcohol present in their beverages so that tax collectors could determine their share of liquor sales. Some clever soul in Britain figured out that gunpowder would ignite in an alcoholic liquid when water was eliminated. So when the proportion of alcohol to water was high enough, the gunpowder would ignite; this was the *proof* of alcohol in the beverage.

The Cromwell Parliament established the proof system to be eleven parts by volume of alcohol to ten parts water. In the United States, the proof is double that of the alcohol percentage. So liquor being labeled as 80 proof is actually 40% alcohol by volume.

Key Terms and Concepts

Review Exercises

13.2 History of Systems of Measurement

1. Describe the basic characteristics of the metric system and list some of its advantages.

2. Name the prefix that indicates each of the following:

a. 1000 **b.** 10 **c.** 0.1
d. 100 **e.** 0.01 **f.** 0.001

3. How was the length of a meter determined?

4. Complete each of the following:

a. 15 dam = _____ m
b. 30 hm = _____ km
c. 20 mm = _____ cm
d. 25 m = _____ dm
e. 22 km = _____ m
f. 35 m = _____ dam

5. If 1 meter = 39.37 inches, then how many inches are contained in each of the following?

a. 1 kilometer
b. 1 millimeter
c. 1 hectometer
d. 1 centimeter
e. 1 dekameter
f. 1 decimeter

13.3 Length and Area

6. Distance from New York City to Seattle If the distance from New York City to Seattle is 4672 kilometers, approximately how many miles is it?

7. Distance from New Orleans to Montreal If the distance from New Orleans to Montreal is 1583 miles, approximately how many kilometers is it?

8. Convert each of the following to the indicated unit:
   a. 120 cm = ______ in
   b. 450 ft = ______ m
   c. 10 mi = ______ km
   d. 12 m = ______ yd
   e. 7 m^2 = ______ yd^2
   f. 20 in^2 = ______ cm^2

13.4 Volume

9. How is the volume of a liter determined?
10. Complete each of the following:
    a. 3 L = ______ mL
    b. 2 hL = ______ daL
    c. 4 kL = ______ hL
    d. 200 cL = ______ L
    e. 15 cL = ______ mL
    f. 300 dL = ______ daL
11. **Volume of a box** Find the volume of a box (in liters) that is 1 meter long, 30 centimeters wide, and 30 centimeters high.
12. **Price of gasoline** Cynthia paid \$1.59 per gallon for gasoline. Juanita paid \$0.41 a liter. Who obtained the best buy?

13.5 Mass (Weight)

13. How is the weight of a gram determined?
14. Complete each of the following:
    a. 4 kg = ______ g
    b. 8 g = ______ mg
    c. 30 dag = ______ hg
    d. 20 cg = ______ mg
    e. 40 hg = ______ kg
    f. 18 g = ______ cg
15. Assuming that each of the following containers is filled with very cold water, give the mass (in grams) of
    a. 1 liter
    b. 1 cubic meter
    c. 1 centiliter
    d. 1 hectoliter
    e. 45 milliliters
    f. 50 liters

13.6 Temperature

16. **World's lowest temperature** The lowest temperature ever recorded was −127°F, recorded in Antarctica. Find the equivalent temperature (to the nearest degree) on the Celsius thermometer.
17. **Highest temperature in Africa** The highest temperature ever recorded was 58°C, recorded in Africa. Find the equivalent temperature (to the nearest degree) on the Fahrenheit thermometer.

Various Sections

For Exercises 18–29, choose the most sensible answer.

18. The height of a basketball player 6 feet 6 inches tall is approximately
    a. 1 km b. 100 cm
    c. 2 m d. 3000 mm
19. The weight of a football player weighing 200 pounds is approximately
    a. 1 t b. 1000 g
    c. 90 kg d. 100 kg
20. Five gallons of apple cider is approximately equal to
    a. 30 L b. 19 L
    c. 100 cL d. 5 L
21. On a very hot August day in Phoenix, Arizona, the temperature is likely to be
    a. 60°C b. 20°C
    c. 25°C d. 40°C
22. The height of a flagpole is likely to be
    a. 10 m b. 100 m c. 1000 m
23. The distance from home plate to first base on a baseball field is approximately
    a. 27 cm b. 2.7 km c. 27 m
24. The area of a floor tile is likely to be
    a. 9 m^2 b. 90 mm^2 c. 900 cm^2
25. If the weather forecast calls for sunny skies with a high of 20°C, you may plan on
    a. swimming b. hiking c. ice skating

26. If the temperature outside is 35°C, what should you wear?
    a. a heavy sweater b. a ski parka
    c. a T-shirt

27. If you purchased 10 kilograms of groceries at the supermarket, in which of the following should you carry the groceries?
    a. a shopping cart b. a small bag
    c. a trailer

28. What would you be most likely to do if you purchased 400 milliliters of soda pop at the store?
    a. Call your friends to have a party.
    b. Drink it and satisfy your thirst.
    c. Store all the cases in the garage.

29. If your home was located 1 kilometer from your college, how would you get home?
    a. Take a bus. b. Take a plane.
    c. Walk.

30. Convert each of the following to the indicated unit:
    a. 2 tsp = ______ mL
    b. 1 qt = ______ L
    c. 100 liters = ______ qt
    d. 2 oz = ______ g
    e. 200 lb = ______ kg
    f. 95°F = ______ °C

31. Convert the given recipe to a metric recipe.

Sour Cream Cookies	
$\frac{1}{2}$ cup sour cream	$\frac{1}{2}$ pound butter
1 teaspoon vanilla	1 teaspoon baking soda
1 cup brown sugar	$2\frac{1}{2}$ cups flour
2 eggs	
Bake at 350°F	

32. As we complete the metric unit, you are to submit examples of how metrics are used in the everyday world, that is, real examples of the metric measurements, *meters, liters, grams,* and *Celsius.* You may use actual labels from various products, newspaper and magazine articles or ads, photos, and so on. The only stipulation is that the use be real! You need five examples of each of the abovementioned measurements. These are to be affixed to standard notebook paper. Please note: Clearly indicate the metric unit being used by circling it, underlining it, or highlighting it with a marker. Your ability to follow directions and to do neat, orderly work will be considered.

Just for fun

Can you find at least 14 metric terms hidden in this puzzle?

A K E I B C U B I C M E T E R
M I L L I G R A M A B C D R E
I L O U L A E I O U H F M G T
L O C E L S I U S H E I R J E
L L E K L M N O P Q C E T R M
I I N J S T U S C O T T R V A
M T T O N N E W X E A Y I Z K
E E I E A G R A M M R M C C E
T R M B C D E O A F E G H I D
E J E K L M L R D S H T A M E
R O T P Q I G I R N E T E S R
T U E V K A W X T Y A T A R A
Z T R E K L F O U E B L E M U
O H T E N O A D D I R E S K Q
T A D C U B I C M E T E R I S

Chapter Quiz

Indicate whether each statement (1–25) is true or false.

1. A decigram is one-tenth of a gram.
2. A quart is larger than a liter.
3. Water boils at 90°C.
4. A meter is longer than a yard.
5. A mile is longer than a kilometer.
6. A liter contains 500 milliliters.
7. The abbreviation for dekagram is dg.
8. Water freezes at 32°C.
9. A newborn baby weighs about 30 kg.

10. Normal body temperature is 40°C.
11. Metric conversion is voluntary in the United States.
12. A centimeter is equivalent to 100 meters.
13. A hectare is a measure of volume.
14. A liter is defined as the volume of a cube that is 10 cm long, 10 cm wide, and 10 cm high.
15. The liter is used to measure both liquid and dry volume.
16. A milligram is equivalent to 1000 grams.
17. A common paper clip weighs approximately 1 gram.
18. It is most likely to snow at a temperature of 25°C.
19. The height of a basketball player is likely to be 200 cm.
20. The thickness of a dime is about 1 mm.
21. A gallon is larger than 3 liters.
22. A hectare is the area of a square that measures 100 meters on each side.
23. The volume of 1 liter is 1000 cubic centimeters, or 1000 milliliters.
24. A small grain of sand weighs about 1 milligram.
25. A liter is the volume of a cube that has a length, width, and height of 100 centimeters.

Appendix A

TABLE 1 Factorial

n	*n!*
0	1
1	1
2	2
3	6
4	24
5	120
6	720
7	5040
8	40,320
9	362,880
10	3,628,800
11	39,916,800
12	479,001,600
13	6,227,020,800
14	87,178,291,200
15	1,307,674,368,000

TABLE 2 Squares, Square Roots, and Prime Factors for the Numbers 1 through 100

No.	Square	Square Root	Prime Factors	No.	Square	Square Root	Prime Factors
1	1	1.000		51	2601	7.141	$3 \cdot 17$
2	4	1.414	2	52	2704	7.211	$2^2 \cdot 13$
3	9	1.732	3	53	2809	7.280	53
4	16	2.000	2^2	54	2916	7.348	$2 \cdot 3^3$
5	25	2.236	5	55	3025	7.416	$5 \cdot 11$
6	36	2.449	$2 \cdot 3$	56	3136	7.483	$2^3 \cdot 7$
7	49	2.646	7	57	3249	7.550	$3 \cdot 19$
8	64	2.828	2^3	58	3364	7.616	$2 \cdot 29$
9	81	3.000	3^2	59	3481	7.681	59
10	100	3.162	$2 \cdot 5$	60	3600	7.746	$2^2 \cdot 3 \cdot 5$
11	121	3.317	11	61	3721	7.810	61
12	144	3.464	$2^2 \cdot 3$	62	3844	7.874	$2 \cdot 31$
13	169	3.606	13	63	3969	7.937	$3^2 \cdot 7$
14	196	3.742	$2 \cdot 7$	64	4096	8.000	2^6
15	225	3.873	$3 \cdot 5$	65	4225	8.062	$5 \cdot 13$
16	256	4.000	2^4	66	4356	8.124	$2 \cdot 3 \cdot 11$
17	289	4.123	17	67	4489	8.185	67
18	324	4.243	$2 \cdot 3^2$	68	4624	8.246	$2^2 \cdot 17$
19	361	4.359	19	69	4761	8.307	$3 \cdot 23$
20	400	4.472	$2^2 \cdot 5$	70	4900	8.367	$2 \cdot 5 \cdot 7$
21	441	4.583	$3 \cdot 7$	71	5041	8.426	71
22	484	4.690	$2 \cdot 11$	72	5184	8.485	$2^3 \cdot 3^2$
23	529	4.796	23	73	5329	8.544	73
24	576	4.899	$2^3 \cdot 3$	74	5476	8.602	$2 \cdot 37$
25	625	5.000	5^2	75	5625	8.660	$3 \cdot 5^2$
26	676	5.099	$2 \cdot 13$	76	5776	8.718	$2^2 \cdot 19$
27	729	5.196	3^3	77	5929	8.775	$7 \cdot 11$
28	784	5.292	$2^2 \cdot 7$	78	6084	8.832	$2 \cdot 3 \cdot 13$
29	841	5.385	29	79	6241	8.888	79
30	900	5.477	$2 \cdot 3 \cdot 5$	80	6400	8.944	$2^4 \cdot 5$
31	961	5.568	31	81	6561	9.000	3^4
32	1024	5.657	2^5	82	6724	9.055	$2 \cdot 41$
33	1089	5.745	$3 \cdot 11$	83	6889	9.110	83
34	1156	5.831	$2 \cdot 17$	84	7056	9.165	$2^3 \cdot 3 \cdot 7$
35	1225	5.916	$5 \cdot 7$	85	7225	9.220	$5 \cdot 17$
36	1296	6.000	$2^2 \cdot 3^2$	86	7396	9.274	$2 \cdot 43$
37	1369	6.083	37	87	7569	9.327	$3 \cdot 29$
38	1444	6.164	$2 \cdot 19$	88	7744	9.381	$2^3 \cdot 11$
39	1521	6.245	$3 \cdot 13$	89	7921	9.434	89
40	1600	6.325	$2^3 \cdot 5$	90	8100	9.487	$2 \cdot 3^2 \cdot 5$
41	1681	6.403	41	91	8281	9.539	$7 \cdot 13$
42	1764	6.481	$2 \cdot 3 \cdot 7$	92	8464	9.592	$2^2 \cdot 23$
43	1849	6.557	43	93	8649	9.644	$3 \cdot 31$
44	1936	6.633	$2^2 \cdot 11$	94	8836	9.695	$2 \cdot 47$
45	2025	6.708	$3^2 \cdot 5$	95	9025	9.747	$5 \cdot 19$
46	2116	6.782	$2 \cdot 23$	96	9216	9.798	$2^5 \cdot 3$
47	2209	6.856	47	97	9409	9.849	97
48	2304	6.928	$2^4 \cdot 3$	98	9604	9.899	$2 \cdot 7^2$
49	2401	7.000	7^2	99	9801	9.950	$3^2 \cdot 11$
50	2500	7.071	$2 \cdot 5^2$	100	10,000	10.000	$2^2 \cdot 5^2$

Answers to Odd-Numbered Exercises, All Review Exercises & Chapter Quizzes

Chapter 1

Section 1.2

1. Answers will vary.
3. Active observations. Secret Service agents must be aware of everything around them, including what people are wearing or carrying, environmental conditions, and any changes in a particular setting from their established baseline.
5. Recognition observation. The movie theatre patron will most likely leave a lit theatre and return to a darkened one. Thus, she must be able to identify key "landmarks" so she can recognize them when she returns to her seat.
7. Inductive
9. Inductive
11. Deductive
13. Deductive
15. Deductive
17. The pattern is as follows: Each row begins and ends with 1; the "interior" numbers of each row is the sum of the two "adjacent" numbers in the row above it; and there is symmetry among the numbers in each row. That is, the numbers increase in a particular order and then decrease in reverse order. Using this pattern, the next two rows are as follows:

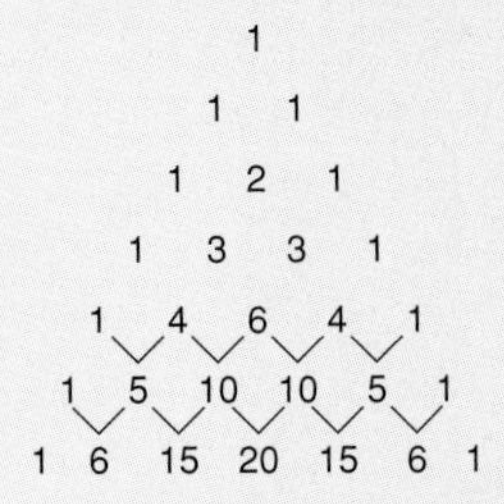

19. The pattern is "double, double, triple, add 3 . . . ". The next three terms are 156, 468, 471.
21. The pattern is "white/black-large pair; white/black-small pair . . . ". The next three terms are

23. Answers will vary. Sample open-ended questions are: When was the last time the car started? What sound does the car make when you try to start it? How long has the car been idle since you last started it? What changes have there been since the last time the car started and now? Sample yes–no questions are: Is the engine cranking but just won't turn over? Were you able to start the car at all today? Is the problem occurring during a "warm" start? Does the car have a manual transmission and, if so, did you depress the clutch pedal all the way to floor? Does the car have an automatic transmission and, if so, did you confirm that the transmission is in "park"?
25. Answers will vary. Sample open-ended questions are: What is the teachers' experience with each medium? What features of the lesson are best served by each medium? What are the trade-offs of using each medium? Sample yes–no questions are: Have you ever used *PowerPoint?* Are a computer and projection equipment conveniently available? Is there an overhead projector available?
27. Answers will vary. Sample open-ended questions are: What programs were you running when this happened? What type of Internet connection do you have? What were you doing just prior to the computer freezing up? What operating system are you running and what Internet service provider and Internet software are you using? Sample yes–no questions are: Is this the first time this has happened? Have you chosen the correct software to run this application? Have you selected the correct modem? Does your computer have sufficient memory to receive information from the Internet?
29. The symbol can be used to represent a subtraction problem; denote a number less than zero; and denote a punctuation mark to connect two words together. In a subtraction problem, it is called "minus sign"; as a number less than zero, it is called a "negative sign"; and to connect two words together, it is called a "hyphen." Examples include $23 - 6 = 17$; -5; and "like-minded."
31. The symbol can be used to denote the omission of letters or figures or to show the possessive case of a word; and to denote a specific unit of measurement. When used in the first instance it is called an "apostrophe"; when used in the second instance it is called "feet." Examples include don't, Tommy's, and $3'$ (i.e., 3 feet).
33. The symbol can be used to represent a punctuation mark to show the sudden, excited, or forceful nature of a phrase; and to denote the product of an integer, n, and all integers less than n up to 1; that is, $n \times (n - 1) \times (n - 2) \times (n - 3) \times \cdots \times 1$. When used as a punctuation mark, "!" is called an "exclamation mark"; when used to denote a special product of integers, it is called "factorial." Examples include: (1) She shouted, "No!"; and (2) $5! = 5 \times 4 \times 3 \times 2 \times 1 = 120$.
35. The symbol can be used in mathematics to represent the base of the natural logarithm. It is an irrational number and its approximation to 5 decimal places is 2.71828. The symbol can also be used to designate the weight printed on packets to denote that it complies with a European Community directive.

37. The symbol can be used as a punctuation mark, called a colon, to direct attention to something that follows it. The symbol can also be used in mathematics to represent a ratio. Examples include: "The answers are as follows: 2, 6, and 9"; and "3 : 4."

39. Answers will vary. Sample answers are: Is it more difficult to ride your bicycle up a hill or into a head wind? Don't forget to wind your watch.

41. Answers will vary. Sample answers are: She took a bow at the end of her performance. The captain walked toward the bow of the ship. She tied a yellow bow around the old oak tree.

43. Answers will vary. Sample answers are: Don't tear your new dress. There was a tear in his eye from watching a very sad movie.

45. Answers will vary. Sample answers are: The album was recorded live. The Ratzells live in Montana.

47. Answers will vary. Sample answers are: From Charles Dickens: "Slipping over stones and refuse on the shore." I refuse to go to the dance unless I get a new dress.

49. To understand the significance of the sign, ask yourself this question: "How much are the tickets if purchased on June 10?" We don't know because the information in the sign only pertains to the price of the tickets before or after June 10.

51. Answers will vary.

53. Answers will vary.

55. One context is from the military in which 2200 hours is equivalent to 10 P.M.

57. One context is Roman numerals in which XXX = 30.

59. 15, 57, 114, 75

61. 2007

63. Answers will vary.

65. If we round \$283 to \$300 and \$64 to \$70, then assuming 5 people in the family (2 parents and 3 children), the estimated airfare is $\$300 \times 5 = \$1500$, and the estimated 5-day pass cost is $\$70 \times 5 = \$350$. Thus, the total estimate is $\$1500 + \$350 = \$1850$. The Albergs probably should plan on \$2000 to account for any price increases as well as for miscellaneous expenses.

67. Inductive reasoning implies that generalizations are being made from observing specific cases; deductive reasoning implies that we are arriving at a logical conclusion based on established facts.

69. $y = 2x + 1$; the next two numbers are 9 and 11.

71. $y = 3x - 3$; the next two numbers are 24 and 30.

73. Answers will vary.

75. Answers will vary.

77. Approximately 10.5 billion

79. Continual decline since 2001–2002.

81. No, rates cannot go much lower; also a function of the economy. Inductive reasoning.

Just for Fun *The average speed of the police car equals the average speed of the speeder.*

Section 1.3

1. **a.** Rounding to nearest whole number: $8 + 4 + 24 = 36$. **b.** Compatible numbers: Answers will vary. One solution is $10 + 5 + 25 = 40$. **c.** Using front-end estimation: $0 + 0 + 20 = 20$.

3. **a.** Rounding to nearest whole number: $2196 \times 23 = 50{,}508$. **b.** Compatible numbers: Answers will vary. One solution is $2000 \times 20 = 40{,}000$. **c.** Using front-end estimation: $2000 \times 20 = 40{,}000$.

5. **a.** Rounding to nearest whole number: $4 \times 43 \times 1 = 172$. **b.** Compatible numbers: Answers will vary. One solution is $4 \times 40 \times 1 = 160$. **c.** Using front-end estimation: $3 \times 40 \times 0 = 0$ (be careful when using numbers less than zero as is the case here).

7. **a.** Rounding to nearest whole number: $4659/16 \approx 291$. **b.** Compatible numbers: Answers will vary. One solution is $4800/16 = 300$. **c.** Using front-end estimation: $4000/10 = 400$.

9. **a.** Rounding to nearest whole number: $(200 \times 4) + 0/23848 = 800$. **b.** Compatible numbers: Answers will vary. One solution is $(200 \times 4) + 0 = 800$. **c.** Using front-end estimation: $(100 \times 3) + 0/20000 = 300$.

11. 320 miles

13. Round 24 hours to 25: $2500/25 = 100$ days ≈ 14 weeks.

15. 6800 pounds

17. 5 P.M.

19. **a.** 0.18 minute/day **b.** 5.4 minutes/month **c.** 72 minutes/year

21. **a.** Approximately 200 **b.** Approximately 48 **c.** $894 \approx 900$ **d.** 15–17-year-olds **e.** 22–25 and 26–30 age groups because they are the largest groups

23. Round 1022 to 1000, 29.7% to 30%, and 63.6% to 64%. **a.** Approximately 940 people had an opinion. **b.** 640 people

25. **a.** Around April; about \$205 **b.** www.addie.com: \$15/share and JB Industries: \$145/share **c.** www.addie.com: \$10,000 investment yields 100 shares. At \$15/share you have \$1500, a loss of \$8500; JB Industries: \$10,000 investment yields 100 shares. At \$145/share you have \$14,500, an increase of \$4500.

27. **a.** 5800 (nearly 6000 people) **b.** about 2000 less

29. First round 833,315 to 800,000. **a.** sedentary **b.** 560,000 **c.** 160,000

31. **a.** $\approx$26–27 miles; **b.** $\approx$24 miles

33. $\approx$800 to 1,000

35. Overestimate to accommodate unexpected price increases or emergencies.

37. Underestimate or else you risk going over the actual price.

39. Answers will vary based on personal opinions. Some will say overestimate so you don't have to pay at the end of the tax year; others will say underestimate so you don't give the government "free" money.

41. Answers will vary.

43. Answers will vary.

45. Answers will vary.

47. They approximate the number of tickets purchased multiplied by the average price per ticket.

49. As long as proper estimation techniques are used and justified, all estimations, regardless of which technique is used, provide good approximations. However, estimates are only approximations and as such can never be a correct answer because their accuracies will vary.

51. Insufficient. The waitress might overestimate your bill (and tip).

53. Sufficient. As long as you are "in the ballpark," then you know that the bill's total is probably a reasonable sum.

55. Sufficient. This is a situation where an exact attendance figure is not necessary.

57. Insufficient. The employer might underestimate your salary.

59. The "5" rule provides a more accurate estimate than the even/odd rule. As an example, consider the number 29. Rounding to the nearest whole number using the "5" rule yields 30, but rounding using the even/odd rule yields 20.

61. Underestimate. The shaded areas of the rectangles do not completely fill the region under the curve. This is because the rectangles are "inscribed."

63. The error in the approximation is related to the width of the rectangles used. The wider the rectangles' width, the less accurate the estimate. The narrower the rectangles' width, the more accurate the estimate.

Just for Fun *17 kg*

Section 1.4

1. ○○○○ △

3. 14; arithmetic progression with a common difference of 3

5. 16; every other term is the square of the term immediately preceding it

7. 3; arithmetic progression with a common difference of $\frac{1}{2}$

9. 8; add the two most recent terms to get the next term

11. 40; level 2 analysis with a common difference of 1

13. 167; level 2 analysis with a common difference of 3

15. 72; level 3 analysis with a common difference of 1

17. 190; level 3 analysis with a common difference of 6

19. -16; level 2 analysis with a common difference of -1

21. 81; geometric progression with a common factor of 3

23. $\frac{1}{81}$; geometric progression with a common factor of $\frac{1}{3}$

25. 3; geometric progression with a common factor of 3

27. -1; geometric progression with a common factor of $\frac{-1}{4}$

29. $\frac{-1}{8}$; geometric progression with a common factor of $\frac{-1}{2}$

31. $x + 2y$; arithmetic progression with a common difference of y

33. $16x$; geometric progression with a common factor of 2

35. x^4; geometric progression with a common factor of x

37. $(x + y)^4$; geometric progression with a common factor of $(x + y)$

39. $16x^5$; geometric progression with a common factor of $2x$

41. The reciprocals are 1, 2, 3, 4. Since a common difference of 1 exists, the original numbers form a harmonic progression.

43. The reciprocals are 2, 4, 6, 8. Since a common difference of 2 exists, the original numbers form a harmonic progression.

45. The reciprocals are $\frac{3}{2}, 2, \frac{5}{2}, 3$. Since a common difference of $\frac{1}{2}$ exists, the original numbers form a harmonic progression.

47. The reciprocals are $\frac{1}{3}, \frac{5}{6}, \frac{4}{3}, \frac{11}{6}$. Since a common difference of $\frac{1}{2}$ exists, the original numbers form a harmonic progression.

49. Zero; The reciprocal of zero is $\frac{1}{0}$, which is undefined.

51. $37 \times 12 = 444$ and $37 \times 15 = 555$

53. 111111111

55. 333333333333333

57. $1234 \times 8 + 4 = 9876$ and $12{,}345 \times 8 + 5 = 98{,}765$

59. $\left(1 + \frac{1}{5}\right) \times 6 = \frac{36}{5}$ and $\left(1 + \frac{1}{6}\right) \times 7 = \frac{49}{6}$

61. Approximately 11%.

63. **a.** The general senior trend is that binge drinking decreased from 1991 to 1993, increased steadily from 1993 to 1998, and decreased for the period 1999–2001. The general eighth grader trend is that binge drinking increased steadily through 1996, decreased for two years, and increased in 1999, and decreased for the next two years.

b. The established pattern for seniors is "down–down–up–up–up–up–up–down–down–down". Based on this pattern, there are two possible scenarios: Either, up–up–up, i.e., 30.2, 30.7, 31.2, or down–down–up, i.e., 29.2, 28.7, 29.

c. The established pattern for eighth graders is "up–up–up–no change–up–down–down–up–down–down." Based on this pattern, we estimate that percentages for 2002–2004 are: 13.2, 12.2, 12.5 (i.e., no change–down–up).

65. Answers will vary.

67. Answers will vary.

69. Answers will vary.

71.

(28)	21	26
23	(25)	27
24	(29)	22

73.

16	2	3	(13)
5	11	(10)	8
9	7	6	12
(4)	14	15	(1)

75. Research

77. Research

79. Zero

81. Overall trend is DOWN!

83. Approximately $20,000,000.

Just for Fun *BOB, BOG, BOO, BOP, BOW, BOX, BOY*

Section 1.5

1. $\left\{1, \frac{1}{\sqrt{2}}, \frac{1}{\sqrt{3}}, \frac{1}{2}, \ldots, \frac{1}{\sqrt{n}}, \ldots\right\}$ where n is a counting number

3. $\left\{0, \frac{1}{2}, \frac{2}{3}, \frac{3}{4}, \ldots, \frac{n-1}{n}, \ldots\right\}$ where n is a counting number

5. $\left\{2, \frac{3}{2}, \frac{4}{3}, \frac{5}{4}, \frac{6}{5}, \ldots, \frac{n+1}{n}, \ldots\right\}$ where n is a counting number

7. $\left\{\frac{1}{2}, \frac{1}{3}, \frac{1}{4}, \frac{1}{5}, \ldots, \frac{1}{n+1}, \ldots\right\}$ where n is a counting number

9. $\left\{2, 1, \frac{2}{3}, \frac{2}{4}, \frac{2}{5}, \ldots, \frac{2}{n}, \ldots\right\}$
where n is a counting number

11. $\left\{\frac{2}{3}, \frac{3}{4}, \frac{4}{5}, \ldots, \frac{n+1}{n+2}, \ldots\right\}$
where n is a counting number

13. $\left\{\frac{1}{2}, \frac{2}{3}, \frac{3}{4}, \ldots, \frac{n}{n+1}, \ldots\right\}$
where n is a counting number

15. $\left\{2, \frac{4}{3}, \frac{6}{5}, \frac{8}{7}, \ldots, \frac{2n}{2n-1}, \ldots\right\}$
where n is a counting number

17. $\left\{\frac{1}{\pi}, \frac{1}{\pi^2}, \frac{1}{\pi^3}, \frac{1}{\pi^4}, \ldots, \frac{1}{\pi^n}, \ldots\right\}$
where n is a counting number

19. $\{1, -1, 1, -1, 1, -1, \ldots, (-1)^n(-1), \ldots\}$

21. **a.** The numerators remain constant while the denominators are increasing;
b. as n increases, the terms of the sequences are getting smaller in value;
c. the terms get smaller and tend to zero.

23. **a.** $1^3 + 2^3 + 3^3 + 4^3 + \cdots + n^3 = \left[\frac{n(n+1)}{2}\right]^2$
where n is a counting number;
b. 3025

25. $37{,}037 \times 3n = (111111)n$
where n is a counting number

27. **a.** $\left(1 + \frac{1}{n}\right) \times (n+1) = \frac{(n+1)^2}{n}$;

b. $\left(1 + \frac{1}{n}\right) + (n+1) = \frac{(n+1)^2}{n}$.

The general rules are the same even though one equation involves multiplying by $(n + 1)$ and the other involves adding $(n + 1)$.

29. $a + (n - 1)d$

31. Let $T_0 = 1$ (This is the initial condition). The recurrence relation is $\{2T_{n-1}\}$, where n is a counting number.

33. An *infinite sequence* is a list of numbers that continues without end; the terms of the sequence are generated by a specific rule. An *infinite series* is the sum of the terms contained within a sequence.

35. Recursion refers to the process where the definition for certain items refers to an earlier version of itself. When applied to sequences, a recursive sequence is one in which the first or the first few values are specified, and all remaining terms of the sequence are then defined in terms of previous values. An example is the Fibonacci sequence, in which the first term is 1 and the second term is 1. Now, all remaining terms are defined by adding the two immediate previous terms. Thus, the third term is equal to the sum of the first two terms $(1 + 1 = 2)$; the fourth term is equal to the sum of the second and third terms $(1 + 2 = 3)$; and the fifth term is equal to the sum of the third and fourth terms $(2 + 3 = 5)$.

37.

1
1 1
1 2 1
1 3 3 1
1 4 6 4 1
1 5 10 10 5 1
1 6 15 20 15 6 1
1 7 21 35 35 21 7 1
1 8 28 56 70 56 28 8 1

39. **a.** The pattern is as follows: The sum of the first 3 terms is 1 less than the 5th term; the sum of the first 4 terms is 1 less than the 6th term; the sum of the first 5 terms is 1 less than the 7th term. **b.** The sum of the first 20 terms must be 1 less than the 22nd term, that is, $17{,}711 - 1 = 17{,}710$.

Just for Fun *Four*

Section 1.6

1. 3 inches
3. 3 pills in 30 minutes
5. \$205 for the modem; \$5 for the cable
7. June 21, 2000
9. 35 students
11. 12 mph
13. 63 games
15. Do not believe him. A liquid that can dissolve anything will be able to dissolve itself.
17. Minimum number of fences is four.

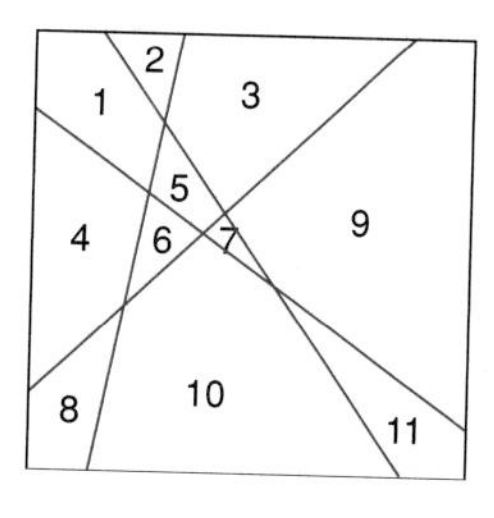

19. 56
21. None remain; the shotgun blast will scare all of the pigeons and they will fly away.
23. If you are on a bridge and you are more than one-half across it. It will take longer to go back than it will take to go forward.
25. Both the Pólya and the IDEAL models are problem solving frameworks that can be used to help guide someone to solve a problem. The first two steps of IDEAL (identify and define the problem) are essentially collapsed into Pólya's first step (understand the problem). The remaining steps of IDEAL are similar to the remaining steps of Pólya.
27. X took one slip of paper, and without opening it, he burned it. He then declared that whatever is written on the remaining slip of paper is what he is *not*.

Just for Fun *Nine: 1. Mercury, 2. Venus, 3. Earth, 4. Mars, 5. Jupiter, 6. Saturn, 7. Uranus, 8. Neptune, 9. Pluto*

Review Exercises for Chapter 1

1. Inductive **2.** Deductive **3.** Inductive

4. Answers may vary. Sample answers are: (1) The symbol $*$ can be used as a symbol of multiplication in computer programs (e.g., $3*4=12$); (2) the symbol can also be used as an *asterisk* in written correspondence as a reference to a footnote.

5. Presumably, the given statement is in reference to age. Assuming this is true, a literal interpretation is that only people who are 18 years old can play the new lottery game. Thus anyone younger or older than 18 is ineligible. Without additional information we cannot make any further interpretations.

6. **a.** One estimate is $13. **b.** Round 8% to 10%, take 10% of $13, which is $1.30, and add this to the $13. Thus, total + tax = $14.30. **c.** Round $14.30 to $15. Now 15% of $15 = $2.25 and 20% of $15 = $3.00. Leave a tip between $2 and $3.

7. 33

8. Round 2095 to 2100. Since "very stressed" is 23%, round it to 20%. Now, 20% of 2100 = 420.

9. Round 1783 to 1800. "Less than satisfied" implies "not very" and "not at all." Round 17.8% of "not very" to 18% and 2.4% of "not at all" to 2%. This sum is 20%. Now, 20% of 1800 = 360.

10. Overall trend is increasing. If the pattern follows the pattern of previous years, then, in 2005, we estimate $55 billion.

11. $10_5 = 5$

12. $10^5 = 100{,}000$

13. $\frac{s}{n} = 10{,}000$ implies 40 dB

14. **a.**

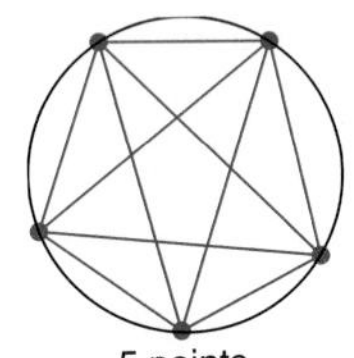

5 points
16 regions

b. Using inductive reasoning, we conjecture that six fully connected points will produce 32 regions. This is an incorrect answer. Only 31 regions are produced, which can be confirmed by a sketch.

15. S

16. 22; arithmetic progression with a common difference of 5

17. $\frac{1}{27}$; geometric progression with a common factor of $\frac{1}{3}$

18. 54; level 2 analysis with a common difference of 2

19. 49

20. 8

21. $\sqrt{4} = 2$

22. $\left\{\frac{1}{x}, \frac{1}{\sqrt[2]{x}}, \frac{1}{\sqrt[3]{x}}, \frac{1}{\sqrt[4]{\sqrt{x}}}, \ldots, \frac{1}{\sqrt[n]{x}}, \ldots\right\}$

23. $\left\{\left(1+\frac{1}{2}\right)\times 3, \left(1+\frac{1}{3}\right)\times 4, \left(1+\frac{1}{4}\right)\times 5, \ldots, \left(1+\frac{1}{n+1}\right)\times(n+2), \ldots\right\}$

24. $\left(1+\frac{1}{n+1}\right)\times(n+2) = \frac{(n+2)^2}{n+1}$

25. 4:50 P.M.

26. There are several solutions. Here is one: $123 - 45 - 67 + 89 = 100$.

Just for Fun *6250*

Chapter 1 Quiz

1. True

2. True

3. True

4. 11, 18

5. $\frac{1}{2}+\frac{1}{2^2}+\frac{1}{2^3}+\frac{1}{2^4}+\cdots+\frac{1}{2^n}=\frac{2^n-1}{2^n}$ where n is a counting number

6. As the number of segments, n, increases by 1, the number of triangles generated is equal to the number of triangles generated previously (for $n-1$ segments) plus n plus 1. For example, for four segments ($n=4$), we add the number of triangles generated when n was 3 (10) to the current value of n plus 1. Thus four segments yield $10+4+1=15$ triangles.

7. (1) Understand the problem; (2) devise a plan; (3) carry out the plan; (4) check your results.

8. One item is $150, the other is $50.

9. A hole will not contain any dirt. Thus the answer is zero.

10. Pedro likes peaches; Claudio likes apples; Susan likes plums; Laura likes bananas

11. 5:40 P.M.; Working backwards strategy.

12. Answers will vary based on rounding strategy. One sample answer is an estimate of $100,000.

13. **a.** ≈$40/share more
b. 1997; ≈$50/share.
c. Answers will vary based on estimation strategy. One sample answer is The Wireless Group stock will increase $20/share while RBN Fiber stock will continue to drop, most likely to below $20/share.

Chapter 2

Section 2.2

1. **a.** True **b.** True **c.** True
d. True **e.** True **f.** False

3. **a.** True **b.** True **c.** False
d. True **e.** False **f.** False

5. **a.** {Ontario, Erie, Huron, Superior, Michigan}
b. {New York, New Jersey, New Mexico, New Hampshire}
c. {Alabama, Alaska, Arizona, Arkansas}
d. ∅
e. {Ohio, New Mexico, Colorado, Idaho}

7. **a.** $\{a, b, c, \ldots, z\}$
b. {Erie, Ontario, Superior, Huron, Michigan}
c. {Ottawa} **d.** {6} **e.** ∅

9. a. $\{x|x \text{ is a day of the week}\}$
 b. $\{x|x \text{ is a vowel in the English alphabet}\}$
 c. $\{x|x \text{ is an odd counting number}\}$
 d. $\{x|x \text{ is a prime number}\}$

11. Yes, it is well defined. The empty set has no elements and there is no difficulty in determining whether any given element belongs to the set.

13. A set is a collection of objects.

15. Equal sets contain the same elements. Equivalent sets contain the same number of elements.

17. For example, $\{x|x \text{ is a counting number less than } 1\}$

19. Research **21.** {Rickey Henderson}

23. {Sammy Sosa, Ken Griffey, Juan Gonzalez}

25. {Eric Young}

27. {Rickey Henderson}

29. {Giants, Dodgers, Cubs, Rangers}

31. {General merchandise stores, Truck rental/moving supplies}

33. {Data processing services, Communication services, Telecommunication services}

Just for Fun *Facetious, abstemious*

Section 2.3

1. a. True b. False c. False
 d. True e. False f. False

3. a. False b. False c. False
 d. True e. True f. False

5. a. $\{15\}, \{20\}, \{15, 20\}, \varnothing$
 b. $\{m, a, t, h\}, \{m, a, t\}, \{m, a, h\}, \{m, t, h\}, \{a, t, h\}, \{m, a\}, \{m, t\}, \{m, h\}, \{a, t\}, \{a, h\}, \{t, h\}, \{m\}, \{a\}, \{t\}, \{h\}, \varnothing$
 c. $\{1\}, \{3\}, \{5\}, \{1, 3\}, \{1, 5\}, \{3, 5\}, \{1, 3, 5\}, \varnothing$
 d. $\varnothing$

7. a. $\{r, i, c\}$ b. $\{m, e, t\}$ c. $\{t, r\}$
 d. $\{m, t\}$ e. $\{m, e, t, r, i, c\}$ f. $\varnothing$

9. a. 16 b. 90¢ c. 0¢

11. 64; 63

13. Current or past Internet search engines or portals

15. Space shuttle names

17. Colors of the rainbow (or color wheel)

19. Given any two sets A and B, if every element in A is also an element in B, then A is a subset of B.

21. 2^n **23.** $2^n - 1$ **25.** $A = B = C$

27. True, transitive property

29. $\mathcal{P}(A) = \{B|B \subseteq A\}$ **31.** No

Just for Fun *There aren't any.*

Section 2.4

1. a. $A \cap B = \{4, 6\}; A \cup B = \{1, 2, 3, 4, 6, 7, 8\}$
 b. $A \cap B = \varnothing; A \cup B = \{f, u, n, t, e, a\}$
 c. $A \cap B = B; A \cup B = A$
 d. $A \cap B = \{s, t\}; A \cup B = \{m, i, s, t, e, r, p, o\}$
 e. $A \cap B = B; A \cup B = A$
 f. $A \cap B = \varnothing; A \cup B = \{1, 2, 3, 4, \ldots\}$

3. a. $\{2, 4, 6\}$ b. $\{3, 8\}$ c. $\{\}$
 d. $\{2, 3, 4, 6, 8\}$ e. $\{2, 3, 4, 6, 8\}$ f. $\{\}$

5. a. $\{0, 8, 9\}$ b. $\{0, 1, 2, 5, 6, 7, 8, 9\}$
 c. $\{0, 1, 2, 5, 6, 7, 8, 9\}$ d. $\{0, 8, 9\}$
 e. $\{3\}$ f. $\{1, 2, 3, 4, 5, 6, 7\}$
 g. $\{0, 1, 2, 3, 5, 6, 7, 8, 9\}$ h. $\{0, 9\}$

7. a. False b. True c. False
 d. False e. False f. False

9. The universal set contains all of the elements being considered in the given problem or discussion.

11. The intersection of two sets is the set of elements common to both sets. A practical example of the concept of set intersection is the intersection of two streets. Their intersection contains common pavement.

13. $\{x|x \in A \text{ and } x \notin B\}$

15. $B' = U - B$

17. No. If $A = \{a, b\}$ and $B = \{c, d\}$, then $A - B = \{a, b\}$ and $B - A = \{c, d\}$.

19. $\{\}$ **21.** Research **23.** {Milk} **25.** {Canola oil}

27. {Sirloin steak} **29.** {Milk, Sirloin steak, Chicken}

Just for Fun *29 hours*

Section 2.5

1. a. Region 2 b. Region 4
 c. Regions 1, 2, and 3 d. Regions 1, 3, and 4
 e. Regions 1, 2, and 3 f. Region 2

3. a. Region V
 b. Regions I, II, III, IV, V, VI, and VII
 c. Regions II, IV, V, VI, and VII
 d. Regions II, IV, and V
 e. Regions I, II, IV, V, and VI
 f. Regions II, III, V, VI and VIII

5. a. $A \cap B$ $(A' \cup B')'$

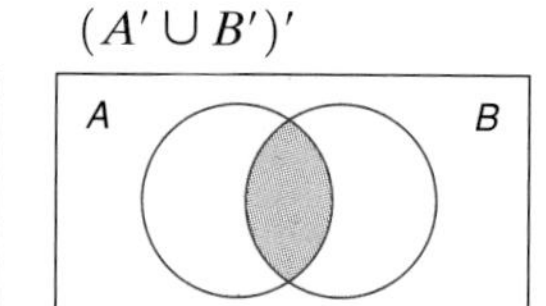

 b. $A \cup B$ $(A' \cap B')'$

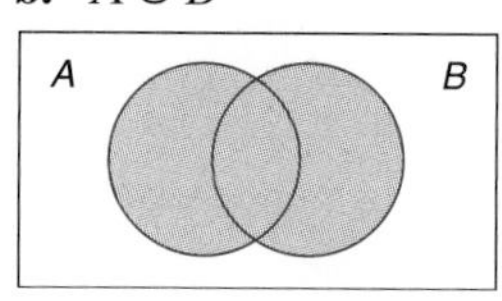

 c. $(A \cup B)'$ $A' \cap B'$

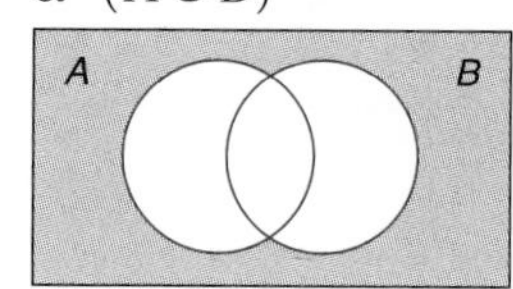

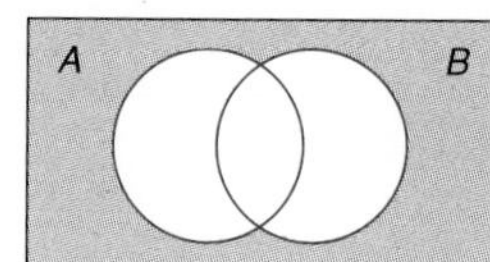

d. $(A \cap B)'$

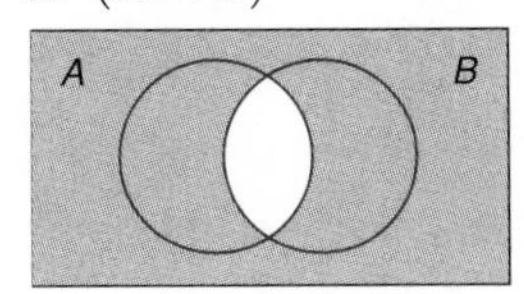

$A' \cup B'$

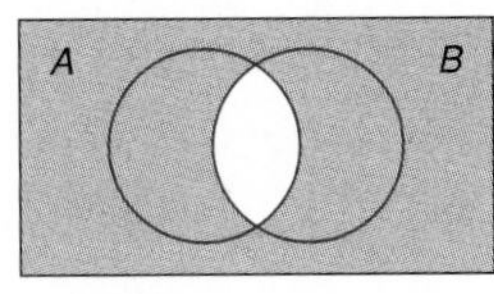

e. $A \cap (B \cup C)$

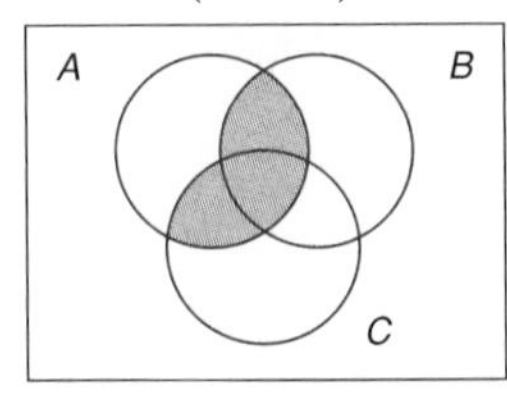

$(A \cap B) \cup (A \cap C)$

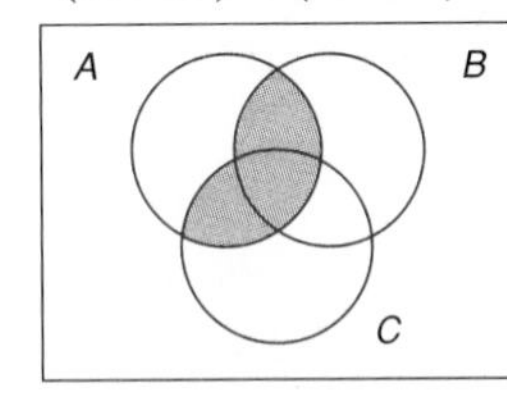

f. $A \cup (B \cap C)$

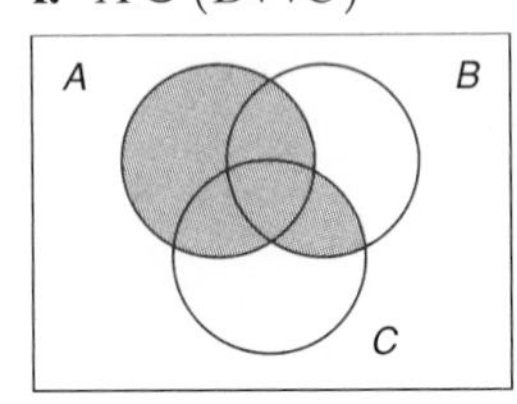

$(A \cup B) \cap (A \cup C)$

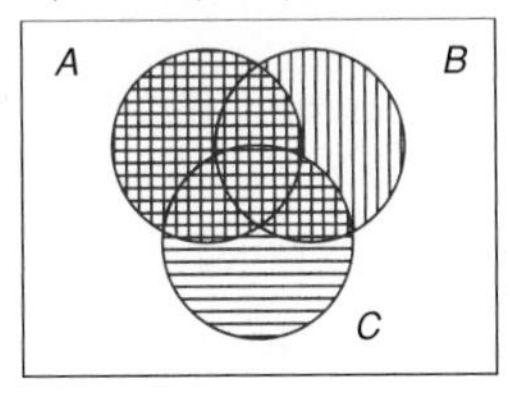

7. **a.** 22 **b.** 26 **c.** 42 **d.** 106
e. 56 **f.** 12 **g.** 3 **h.** 63

9. 6; $i \leftrightarrow a$
$o \leftrightarrow b$
$u \leftrightarrow c$

11. Zero; the two sets do not have the same cardinality.

13. **a.** T **b.** F **c.** T **d.** T
e. F **f.** T **g.** T **h.** T

15. **a.** T **b.** T **c.** F **d.** T **e.** T

17. **a.** 1122 **b.** 1092 **c.** 210 **d.** 1723 **e.** 0
f. 0 **g.** 2505 **h.** 1356 **i.** 1197 **j.** 375

19. A Venn diagram uses a rectangle to represent the universal set and a circle or circles to represent the set or sets under consideration. A Venn diagram provides a picture representation of the set situation.

21. The notation $n(A)$ is used to represent the cardinality of a set.

23.

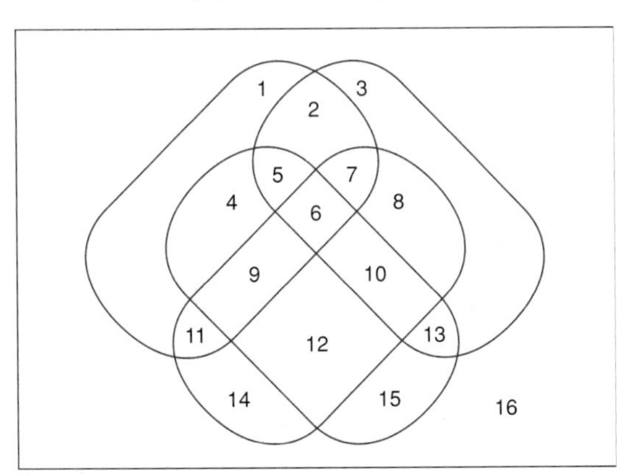

25. One solution is

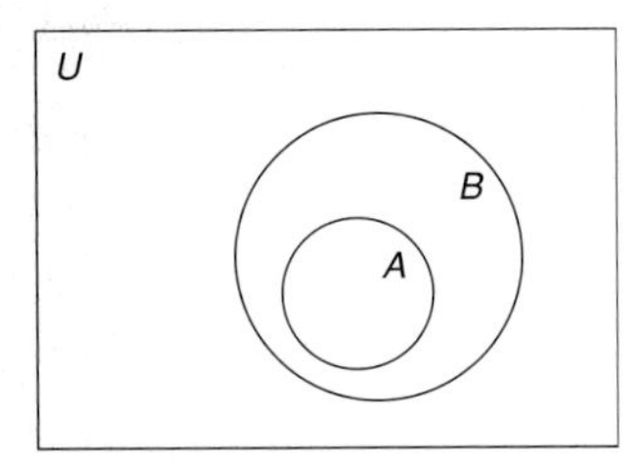

27. The set of real numbers is not countably infinite because it is not denumerable.

Just for Fun

band of gorillas
bed of clams, oysters
bevy of quail, swans
brood of chicks
crash of rhinoceri
down of hares
drift of swine
clowder of cats
colony of ants
congregation of plovers
pod of whales
school of fish
skulk of foxes
flock of sheep
gaggle of geese
herd of elephants
litter of pigs
mob of kangaroos
murder of crows
cast of hawks
sleuth of bears
span of mules
troop of kangaroos
volery of birds
yoke of oxen

Section 2.6

1. **a.** 49 **b.** 31 **c.** 19 **d.** 56

3. **a.** 20 **b.** 55 **c.** 65 **d.** 90

5. **a.** 2 **b.** 18 **c.** 8 **d.** 25

7. **a.** 22 **b.** 49 **c.** 14 **d.** 42

9. **a.** 44 **b.** 11 **c.** 100 **d.** 10

11. **a.** 16 **b.** 54 **c.** 127 **d.** 111

13. **a.** 82 **b.** 19 **c.** 69 **d.** 49

15. **a.** 89 **b.** 6 **c.** 8 **d.** 22

17. **a.** 215 **b.** 270 **c.** 250 **d.** 150

19. Incorrect data: 13 + 6 + 11 = 30 campers, which is more than the 21 stated in the problem.

21. We usually complete region V first because it represents the intersection of all three sets and therefore there is no ambiguity as to which elements belong in that region.

23. Rachel and Ted, Jillian and Andre, Meisha and Al, Sue and Joe

25.

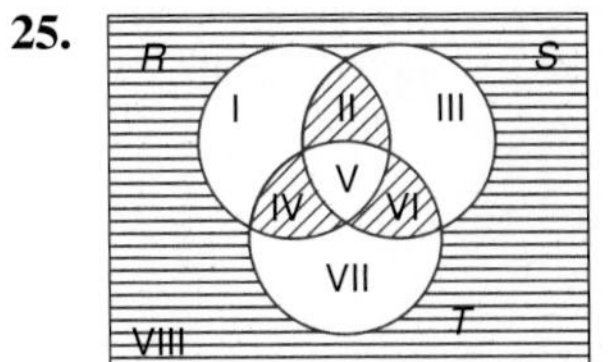

Just for Fun *Joe—accountant; Sharon—attorney; Jack—architect; Sue—author*

Section 2.7

1. $A \times B = \{(5, x), (5, y), (5, z), (10, x), (10, y), (10, z)\}$
$B \times A = \{(x, 5), (x, 10), (y, 5), (y, 10), (z, 5), (z, 10)\}$
$n(A \times B) = 6$

3. $C \times D = \{(4, 3), (4, 5), (4, 7), (6, 3), (6, 5), (6, 7), (8, 3), (8, 5), (8, 7)\}$
$D \times C = \{(3, 4), (3, 6), (3, 8), (5, 4), (5, 6), (5, 8), (7, 4), (7, 6), (7, 8)\}$
$n(C \times D) = 9$

5. $T \times T = \{(t, t), (t, f), (f, t), (f, f)\}$
$n(T \times T) = 4$

7.

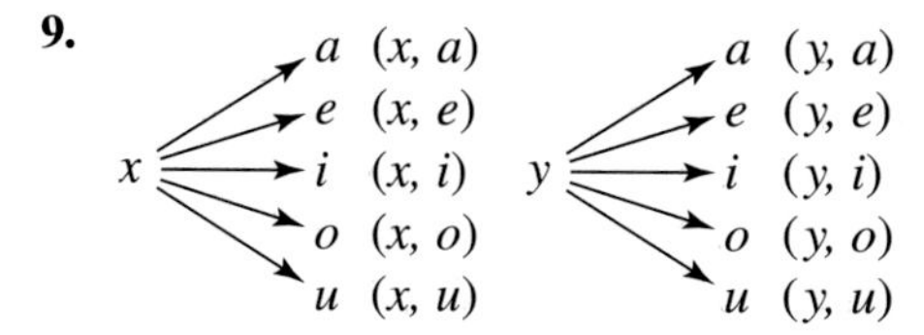

9.

x → a (x, a), e (x, e), i (x, i), o (x, o), u (x, u)

y → a (y, a), e (y, e), i (y, i), o (y, o), u (y, u)

11. **a.** $\{(a, b), (a, c), (a, d), (b, b), (b, c), (b, d)\}$
b. $\{(a, c), (a, d), (a, e), (b, c), (b, d), (b, e)\}$
c. 9
d. $\{(b, c), (b, d), (b, e)\}$
e. $\{(c, a), (c, b), (c, c), (c, d), (d, a), (d, b), (d, c), (d, d), (e, a), (e, b), (e, c), (e, d)\}$
f. $\{(a, c), (a, d), (b, c), (b, d)\}$

13. No, the ordered pairs are different.

15. The Cartesian product, $A \times B$, may be obtained by using the lattice technique or the tree diagram technique.

17. Yes, because of the grouping symbols.

19. **a.** 1 **b.** 2 **c.** 6
d. 24 **e.** 120 **f.** $n!$

Just for Fun *999,999*

Review Exercises for Chapter 2

1. A set is any collection of objects.

2. **a.** {Texas, Tennessee}
$\{x | x$ is a state whose name begins with the letter $T\}$
b. {Huron, Erie, Michigan, Superior, Ontario}
$\{x | x$ is a Great Lake$\}$
c. $\{0, 2, 4, 6, \dots\}$
$\{x | x$ is an even whole number$\}$
d. $\{1, 2, 3, 4, \dots\}$
$\{x | x$ is a positive whole number$\}$
e. $\{1, 2, 3, 4, 5, 6, 7\}$
$\{x | x$ is a counting number less than $8\}$

3. **a.** False **b.** True **c.** False
d. True **e.** False **f.** True

4. **a.** True **b.** True **c.** True
d. True **e.** True **f.** True

5. **a.** True **b.** True **c.** True
d. True **e.** False **f.** False

6. $\{d\}, \{e\}, \{f\}, \{d, e\}, \{d, f\}, \{e, f\}, \{d, e, f\}, \emptyset$

7. **a.** $\{1, 3, 5\}$ **b.** $\{0, 1, 2, 3, 4, 5\}$
c. $\{2\}$ **d.** $\{0, 1, 2, 3\}$
e. U **f.** $\{0, 1, 3, 4, 5\}$
g. U **h.** $\{0, 1, 2, 3, 4\}$

8. **a.** Regions 1, 2, and 3
b. Regions 1, 3, and 4
c. Region 4
d. Regions II, IV, and V
e. Regions I, II, IV, V, and VI
f. Regions I, IV, V, VII, and VIII

9. For example, 1—5, 2—10, 3—15; 6

10. **a.** 11 **b.** 24 **c.** 16 **d.** 5
e. 34 **f.** 38 **g.** 47 **h.** 19

11. **a.** 6 **b.** 14 **c.** 32 **d.** 68

12. **a.** For example, *m—e, a—a, t—s, h—y*
b. $4 \cdot 3 \cdot 2 \cdot 1 = 24$
c. $\{m, a, t, h\}, \{m, a, t\}, \{m, a, h\}, \{m, t, h\}, \{a, t, h\}, \{m, a\}, \{m, t\}, \{m, h\}, \{a, t\}, \{a, h\}, \{t, h\}, \{m\}, \{a\}, \{t\}, \{h\}, \emptyset$
d. $\{(m, e), (m, a), (m, s), (m, y), (a, e), (a, a), (a, s), (a, y), (t, e), (t, a), (t, s), (t, y), (h, e), (h, a), (h, s), (h, y)\}$
e. 16

13. **a.** When $A \subseteq B$ **b.** When $A = B$ **c.** Always
d. When $B = \emptyset$ **e.** Always

14. Given any two sets A and B, if every element in A is also an element in B, then A is a subset of B, denoted by $A \subseteq B$.
If A is a subset of B, and there is at least one element in B not contained in A, then A is a proper subset of B, denoted by $A \subset B$.

15. If a set B contains n elements, then the number of subsets that can be derived from set B is 2^n.
The number of ways a one-to-one correspondence can be established between equivalent sets is $n!$ where n is the number of elements in a given set.
The number of elements in a set A (the cardinality of set A) is denoted by $n(A)$.

Chapter 2 Quiz

1. False **2.** True **3.** True **4.** False
5. False **6.** False **7.** True **8.** False
9. False **10.** False **11.** False **12.** 14
13. 22 **14.** 3 **15.** 18 **16.** 21
17. 10 **18.** 21 **19.** 17 **20.** {i, a}
21. U **22.** {} **23.** {u, l, k, e, m, a, t, h}
24. {u, i, k, e, t, h} **25.** {u, e, a, h, l, k, m, t}

Chapter 3

Section 3.2

1. a. Simple
b. Compound; negation
c. Compound; biconditional
d. Compound; negation
e. Compound; conditional
f. Neither
g. Compound; conjunction
h. Compound; disjunction

3. a. $P \wedge Q$ **b.** $Q \vee P$ **c.** $\sim P \vee \sim Q$
d. $P \rightarrow \sim Q$ **e.** $\sim(\sim P)$ **f.** $Q \leftrightarrow P$

5. a. $P \vee Q$ **b.** $\sim Q \wedge P$ **c.** $\sim(\sim Q)$
d. $Q \leftrightarrow P$ **e.** $P \rightarrow Q$ **f.** $\sim P \wedge \sim Q$

7. a. $P \vee M$ **b.** $G \wedge P$ **c.** $\sim A \rightarrow D$
d. $B \vee \sim W$ **e.** $T \leftrightarrow F$ **f.** $E \rightarrow R$

9. a. I like algebra and I like geometry.
b. If I like algebra, then I do not like geometry.
c. I like algebra or I like geometry.
d. I like algebra or I do not like geometry.
e. I do not like algebra and I do not like geometry.
f. I like algebra iff I like geometry.

11. A statement is a sentence that is true or false, but not both. Questions, commands, or opinions are not statements because we cannot assign a truth value to them.
Question: What time is it?
Command: Stand up!
Opinion: Ms. Scott is the best teacher.

13. A conjunction statement is formed by connecting two simple statements using the word *and* (e.g., Today is Monday and this month is March.)

15. The connective *if* ... *then* ... is used in compound statements known as *conditionals*. (e.g., If the temperature is below 0°C then water will freeze.)

17. The rule *All rules have exceptions* is a paradox because it is self contradictory. The exception to this rule is that there must be a rule that does not have an exception.

19. Answers may vary. **21.** Research

23. $\sim P \wedge F$ **25.** $R \rightarrow \sim C$ **27.** $(H \wedge D), (F \rightarrow M)$

Just for Fun *All of them have 28 days.*

Section 3.3

1. a. None **b.** $P \wedge (Q \leftrightarrow R)$
c. $\sim(P \vee Q \rightarrow R)$ **d.** $\sim P \vee (Q \wedge R)$
e. None **f.** $\sim P \wedge (Q \vee R)$

3. a. $Q \wedge P \rightarrow R$ **b.** $(R \wedge P) \vee Q$
c. $\sim(Q \wedge P)$ **d.** $(Q \wedge R) \vee P$
e. $P \leftrightarrow R \wedge Q$ **f.** $\sim P \wedge \sim R$

5. a. $S \rightarrow E \wedge C$ **b.** $\sim J \wedge \sim C$
c. $\sim(\sim A)$ **d.** $C \rightarrow E \wedge Z$
e. $J \vee (C \wedge P)$

7. a. Algebra is difficult, and logic is easy or Latin is interesting.
b. If algebra is difficult and logic easy, then Latin is interesting.
c. Algebra is difficult, or logic is easy and Latin is interesting.
d. Algebra is difficult, and if logic is easy then Latin is interesting.
e. It is not the case that algebra is difficult and logic is easy.
f. Algebra is not difficult iff logic is easy and Latin is not interesting.

9. We need punctuation marks when writing compound statements so that we can make sense of the statements and understand them.

11. $\leftrightarrow$ biconditional
$\rightarrow$ conditional
$\wedge$, $\vee$ conjunction, disjunction (equal in rank order)
$\sim$ negation

13. No, the first statement is a conditional, while the second statement is a conjunction.

15. $S \vee O$
$O \rightarrow M$
$S \rightarrow \sim E$
$\sim E \vee M$

17. Research **19.** Research

21. Conditional; $B \vee P \vee T \rightarrow M$

23. Conjunction; $D \wedge (\sim S \wedge \sim W)$

25. Disjunction; $(\sim S \wedge \sim W) \vee (\sim S \wedge \sim W)$

Just for Fun *160 ft.*

Section 3.4

1.

~	*P*	∧	~	*Q*
F		**F**	F	
F		**F**	T	
T		**F**	F	
T		**T**	T	

3.

~	(*P*	∧	*Q*)
F	T	T	T
T	T	F	F
T	F	F	T
T	F	F	F

5.

P	∧	~	*P*
T	**F**	F	
F	**F**	T	

7.

P	∨	~	*Q*
T	**T**	F	
T	**T**	T	
F	**F**	F	
F	**T**	T	

9.

~	(*P*	∧	~	*Q*)
F	T	T	F	
F	T	T	T	
T	F	F	F	
F	F	T	T	

11.

~	*P*	∨	(*P*	∧	~	*Q*)
F		**F**	T	F	F	
F		**T**	T	T	T	
T		**T**	F	F	F	
T		**T**	F	F	T	

13.

~	P	⊻	Q
F		**T**	T
F		**F**	F
T		**F**	T
T		**T**	F

15.

~	(~	P	⊻	~Q)
T	F		F	F
F	F		T	T
F	T		T	F
T	T		F	T

17. 4,

T	T
T	F
F	T
F	F

19. A conjunction is false in all cases, except when both statements are true.

21. A disjunction is only false when both statements are false.

23. Answers may vary; one solution is $\sim P \wedge (P \wedge \sim Q)$.

25. $x + x'y$

27. $x'y + xy'$ and $(x' + y')(x + y)$ (The gate implementations are shown.)

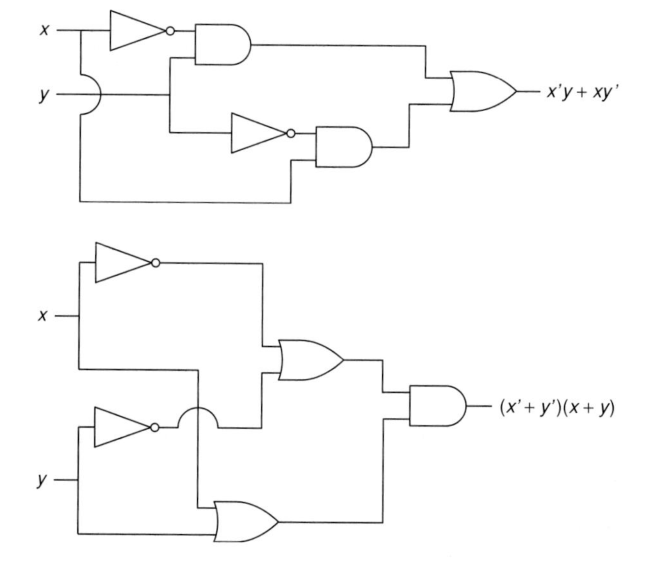

29. Research **31.** True **33.** True

35. False **37.** True

Just for Fun

1	8	6
10	5	0
4	2	9

Section 3.5

1.

P	→	Q
T	**T**	T
T	**F**	F
F	**T**	T
F	**T**	F

3.

~	P	→	~	Q
F		**T**	F	
F		**T**	T	
T		**F**	F	
T		**T**	T	

5.

~	P	→	Q
F		**T**	T
F		**T**	F
T		**T**	T
T		**F**	F

7.

~	P	↔	~	Q
F		**T**	F	
F		**F**	T	
T		**F**	F	
T		**T**	T	

9.

P	∨	Q	→	~	Q
T	T	T	**F**	F	
T	T	F	**T**	T	
F	T	T	**F**	F	
F	F	F	**T**	T	

11.

(P	→	Q)	∨	P	→	Q
T	T	T	T	T	**T**	T
T	F	F	T	T	**F**	F
F	T	T	T	F	**T**	T
F	T	F	T	F	**F**	F

13.

P	∧	Q	↔	P	∨	Q
T	T	T	**T**	T	T	T
T	F	F	**F**	T	T	F
F	F	T	**F**	F	T	T
F	F	F	**T**	F	F	F

15.

(P	∨	Q)	∧	R
T	T	T	**T**	T
T	T	T	**F**	F
T	T	F	**T**	T
T	T	F	**F**	F
F	T	T	**T**	T
F	T	T	**F**	F
F	F	F	**F**	T
F	F	F	**F**	F

17.

(P	∧	Q)	∨	(P	∧	R)
T	T	T	**T**	T	T	T
T	T	T	**T**	T	F	F
T	F	F	**T**	T	T	T
T	F	F	**F**	T	F	F
F	F	T	**F**	F	F	T
F	F	T	**F**	F	F	F
F	F	F	**F**	F	F	T
F	F	F	**F**	F	F	F

19.

P	$\leftrightarrow$	Q	$\vee$	R
T	**T**	T	T	T
T	**T**	T	T	F
T	**T**	F	T	T
T	**F**	F	F	F
F	**F**	T	T	T
F	**F**	T	T	F
F	**F**	F	T	T
F	**T**	F	F	F

21.

$\sim$	$(P$	$\vee$	$Q)$	$\leftrightarrow$	$\sim$	P	$\wedge$	$\sim$	Q
F	T	T	T	**T**	F		F	F	
F	T	T	F	**T**	F		F	T	
F	F	T	T	**T**	T		F	F	
T	F	F	F	**T**	T		T	T	

yes

23.

P	$\wedge$	$\sim$	Q	$\leftrightarrow$	$\sim$	$(\sim$	P	$\vee$	$Q)$
T	F	F		**T**	F	F		T	T
T	T	T		**T**	T	F		F	F
F	F	F		**T**	F	T		T	T
F	F	T		**T**	F	T		T	F

yes

25. **a.** The tide is not out, or we can go clamming.
b. Louise did not drive her new car, or Florence will accompany her.
c. Today is not Friday, or tomorrow is not Sunday.
d. If Chris tosses the ball, then Gary punts.
e. If Carlos was feeling well, then he will not go to New York City.

27. True **29.** True **31.** False
33. True **35.** True **37.** True
39. False **41.** True **43.** True
45. True **47.** True **49.** True

51. A conditional is always true, except when the antecedent is true and the consequent is false.

53. A biconditional is true when both parts are the same; that is, when both parts are true and when both parts are false.

55. A tautology is a statement that is always true. Examples may vary. One example is, Today is Monday or today is not Monday. ($P \vee \sim P$)

57.

$\sim P$	$\wedge$	Q	$\rightarrow$	R	$\wedge$	$\sim S$
F	F	T	**T**	T	F	F
F	F	T	**T**	T	T	T
F	F	T	**T**	F	F	F
F	F	T	**T**	F	F	T
F	F	F	**T**	T	F	F
F	F	F	**T**	T	T	T
F	F	F	**T**	F	F	F
F	F	F	**T**	F	F	T
T	T	T	**F**	T	F	F
T	T	T	**T**	T	T	T
T	T	T	**F**	F	F	F
T	T	T	**F**	F	F	T
T	F	F	**T**	T	F	F
T	F	F	**T**	T	T	T
T	F	F	**T**	F	F	F
T	F	F	**T**	F	F	T

Just for Fun *3*

Section 3.6

1. $\sim(\sim P \wedge \sim Q)$ **3.** $P \vee \sim Q$
5. $\sim(\sim P \vee Q)$ **7.** $P \wedge Q$
9. $P \wedge Q$

11. Larry is not busy, or he can serve on the committee.

13. It is false that I passed the test and that I did not study too much.

15. Pam did not stay late, and Angie did not leave early.

17. It is false that the bus is not late and that my watch is working correctly.

19. Either x is not greater than zero or x is not negative.

21. It is false that the wind did not come up and that we sailed.

23. $A' \cup B'$ **25.** $(A \cup B)'$
27. $(A' \cup B')'$ **29.** $A' \cap B$

31. Equivalent statements are statements that have identical truth values.

33. To construct logically equivalent statements using De Morgan's law, we typically perform three steps: (1) Negate the whole statement. (2) Negate each statement (letter) that makes up the disjunction or conjunction. (3) Change the conjunction connective to a disjunction or the disjunction connective to a conjunction (whichever is given).

35. $\sim P \wedge Q$ **37.** $P \vee (\sim Q \vee R)$

39. Horace did play or we lost.

41. If two does not equal three, then four does not equal six.

43. Either the pond does not freeze over or we will go skating.

45. **a.** $B \wedge M$ **b.** False **c.** $\sim(\sim B \vee \sim M)$; It is false that 34% of women do not expect to get a bachelor's degree or 29% of women do not expect to earn a master's degree.

47. **a.** $M \rightarrow W$ **b.** False **c.** $\sim(M \wedge \sim W)$; It is false that 10% of men expect to get a high school diploma and 25% of women do not expect to get a master's degree.

49. **a.** $W \rightarrow M$ **b.** True **c.** $\sim(W \wedge M)$; It is false that 50% of women expect to get a master's or doctorate degree and 50% of men do not expect to get a master's or doctorate degree.

Just for Fun *They are the same length.*

Section 3.7

1.

Converse:	**a.** $P \rightarrow L$	**b.** $P \rightarrow R$
Inverse:	$\sim L \rightarrow \sim P$	$\sim R \rightarrow \sim P$
Contrapositive:	$\sim P \rightarrow \sim L$	$\sim P \rightarrow \sim R$
Converse:	**c.** $F \rightarrow \sim B$	**d.** $\sim M \rightarrow F$
Inverse:	$B \rightarrow \sim F$	$\sim F \rightarrow M$
Contrapositive:	$\sim F \rightarrow B$	$M \rightarrow \sim F$
Converse:	**e.** $\sim P \rightarrow \sim B$	
Inverse:	$B \rightarrow P$	
Contrapositive:	$P \rightarrow B$	

3. **a.** $P \rightarrow Q$ **b.** $\sim Q \rightarrow \sim P$ **c.** $P \rightarrow Q$
d. $P \rightarrow Q$ **e.** $P \leftrightarrow Q$

5. a. $P \to Q$ b. $\sim P \to \sim Q$ c. $Q \to P$
   d. $P \leftrightarrow Q$ e. $Q \to P$
7. No; the statement is equivalent to "If I pass you, then you will come to class every day."
9. To obtain the contrapositive of a given conditional statement, interchange the antecedent and the consequent, and negate each.
11. Yes, it is a tautology. Truth table verifies it. Answers may vary.
13. a. $C \to A$ b. True c. *Converse*: $A \to C$
    Inverse: $\sim C \to \sim A$
    Contrapositive: $\sim A \to \sim C$
    Negation: $\sim(C \to A)$
15. a. $W \to A$ b. False c. *Converse*: $A \to W$
    Inverse: $\sim W \to \sim A$
    Contrapositive: $\sim A \to \sim W$
    Negation: $\sim(W \to A)$
17. a. $N \to T$ b. True c. *Converse*: $T \to N$
    Inverse: $\sim N \to \sim T$
    Contrapositive: $\sim T \to \sim N$
    Negation: $\sim(N \to T)$
19. a. $F \to A$ b. False c. *Converse*: $A \to F$
    Inverse: $\sim F \to \sim A$
    Contrapositive: $\sim A \to \sim F$
    Negation: $\sim(F \to A)$

Just for Fun

1. Special agent
2. Spy
3. Spy

Section 3.8

1. Invalid 3. Invalid 5. Valid
7. Valid 9. Valid 11. Invalid
13. Invalid 15. Valid 17. Valid
19. A valid argument is one where the conclusion follows logically from the premises.
21. Yes, an invalid argument may have a conclusion that is true. If an argument is valid, it is because of how the conclusion was derived, not whether it is true or false.
23. $S \vee O$
    $O \to M$
    $\underline{S \to \sim E}$ Valid
    $\sim E \vee M$
25. If it rains today, then we will go bowling.
27. If I will not be able to watch my favorite television show, then I will be in trouble.

Just for Fun

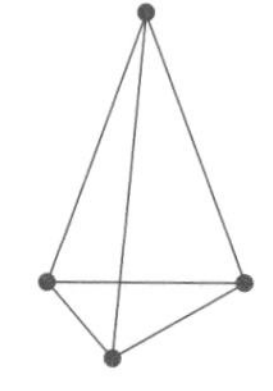

Section 3.9

1. Universal negative

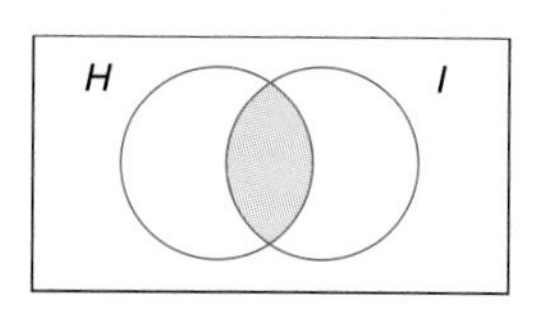

3. Particular negative

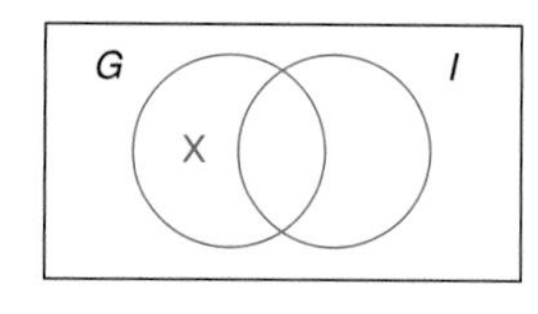

5. Particular affirmative

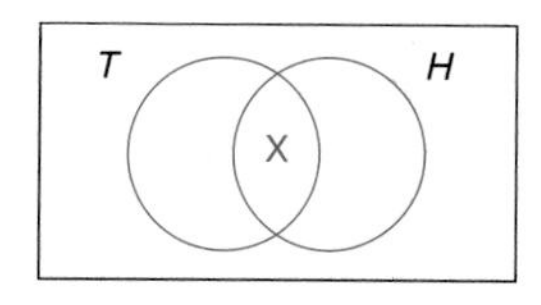

7. Particular negative

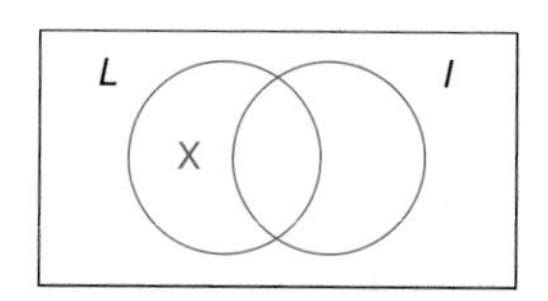

9. Particular affirmative

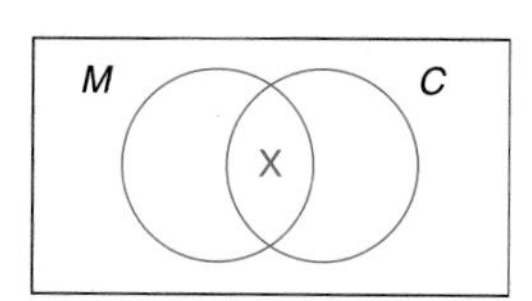

11. Particular negative

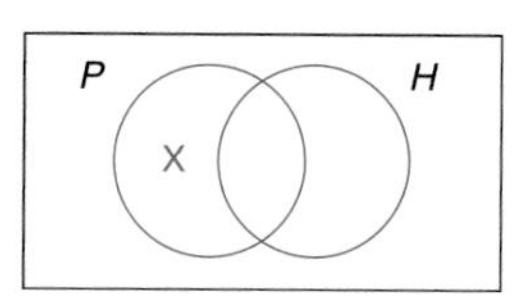

13. Universal negative

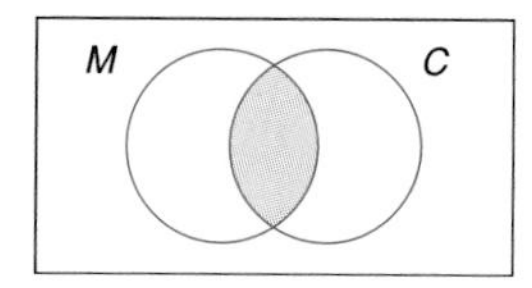

15. Particular negative

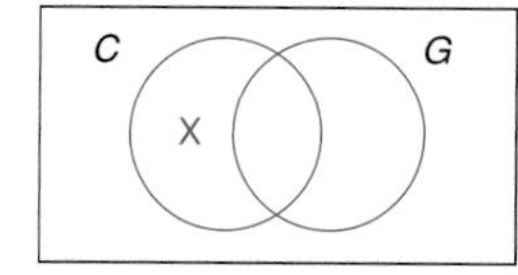

17. Particular negative

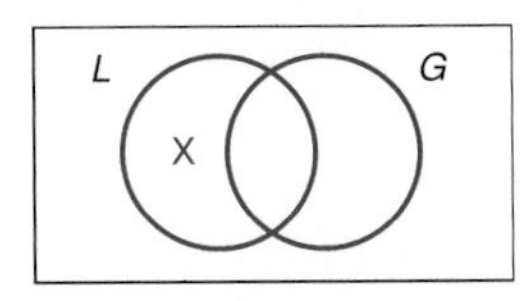

19. Universal negative

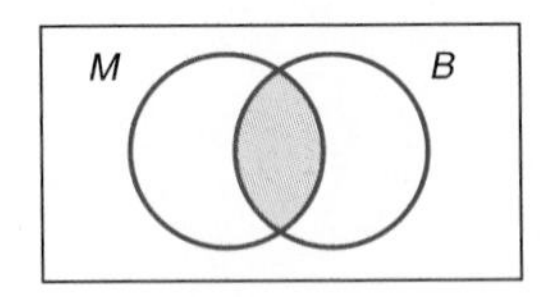

21. Inconsistent **23.** Consistent **25.** Consistent
27. Inconsistent **29.** Consistent **31.** Consistent
33. Consistent **35.** Consistent

37. Consistent statements are statements that can be true together at the same time. Examples may vary.

39. $(\forall x)(Gx)$ **41.** $(\forall x)(\sim Gx)$
43. $(\forall x)(\sim Sx)$ **45.** $(\forall x)(Zx \to Rx)$
47. $\sim(\forall x)(Tx \to Ix)$ **49.** $(\forall x)(Kx \to \sim Cx)$

Just for Fun *11.25 seconds*

Section 3.10

1. Invalid **3.** Valid **5.** Valid
7. Invalid **9.** Valid **11.** Invalid
13. Valid **15.** Invalid **17.** Invalid
19. Valid **21.** Valid **23.** Invalid
25. Invalid

27. Truth has to do with whether a statement is true or false. Validity has to do with the form of an argument. An argument is valid if the conclusion follows logically from the premises. A statement is true or false, not valid or invalid.

29. Some plants are pretty.

Just for Fun *Q and Z*

Section 3.11

1. $P \vee Q$

3. $P \wedge [(P \wedge Q) \vee R]$

5. $[(Q \wedge R) \vee (\sim Q \wedge R)] \wedge P$

7. $[(Q \wedge \sim P) \vee R \vee \sim Q] \wedge P$

9. $(P \wedge \sim Q) \vee [(R \vee P) \wedge (\sim P \vee R)] \vee (Q \wedge P)$

11. P — Q / R

13.

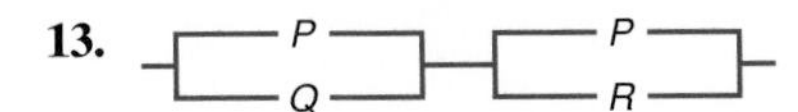

15.

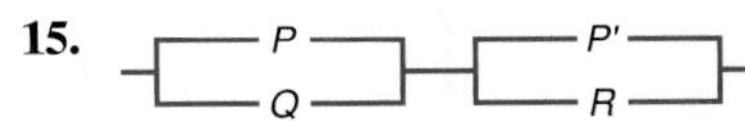

17. P / Q — P / Q'

19. $P \to Q \equiv \sim P \vee Q$ P' / Q

21. **a.** — P —

b. — P —

c. Q / P — R

d. — P — Q / R

23. Complementary switches are switches that always have opposite positions. If a switch is open, then its complement is closed and vice versa. They are typically used in parallel connections

25. $P \wedge \sim Q \wedge R$ — P — Q' — R

Just for Fun *(Meaningless)*

Review Exercises for Chapter 3

1. **a.** Compound; disjunction
b. Neither
c. Simple
d. Compound; conjunction
e. Compound; negation
f. Compound; conditional

2. **a.** $P \wedge Q$ **b.** $\sim Q \to \sim R \vee P$
c. $\sim P \wedge \sim Q$ **d.** $R \to Q \wedge P$
e. $\sim(R \wedge Q) \vee P$ **f.** $P \leftrightarrow Q \vee R$

3. **a.** Bill is golfing, and Addie is sailing or Flo is jogging.
b. Bill is golfing and Addie is sailing, or Flo is jogging.
c. If Bill is golfing or Addie is sailing, then Flo is jogging.
d. Bill is golfing, or if Addie is sailing then Flo is jogging.
e. If Flo is jogging and Addie is sailing, then Bill is not golfing.
f. Bill is not golfing iff Addie is sailing and Flo is not jogging.

4. It is necessary to have a ranking, or dominance, of connectives in symbolic logic in order to translate a symbolic statement correctly. That is, we must be able to identify a statement as a biconditional, conditional, conjunction, disjunction, or negation. Also, dominance of connectives enables us to construct truth tables in a correct and orderly manner.

5. **a.**

~	P	→	Q
F		**T**	T
F		**T**	F
T		**T**	T
T		**F**	F

b.

P	∨	~	Q
T	**T**	F	
T	**T**	T	
F	**F**	F	
F	**T**	T	

c.

P	∨	Q	↔	P
T	T	T	**T**	T
T	T	F	**T**	T
F	T	T	**F**	F
F	F	F	**T**	F

d.

P	∨	Q	→	R
T	T	T	**T**	T
T	T	T	**F**	F
T	T	F	**T**	T
T	T	F	**F**	F
F	T	T	**T**	T
F	T	T	**F**	F
F	F	F	**T**	T
F	F	F	**T**	F

e.

P	$\wedge$	$\sim Q$	$\rightarrow$	Q	$\vee$	R
T	F	F	**T**	T	T	T
T	F	F	**T**	T	T	F
T	T	T	**T**	F	T	T
T	T	T	**F**	F	F	F
F	F	F	**T**	T	T	T
F	F	F	**T**	T	T	F
F	F	T	**T**	F	T	T
F	F	T	**T**	F	F	F

f. $\sim (P \wedge \sim Q) \rightarrow \sim (\sim P \vee Q)$

$\sim$	$(P$	$\wedge$	$\sim Q)$	$\rightarrow$	$\sim$	$(\sim P$	$\vee$	$Q)$
T	T	F	F	**F**	F	F	T	T
F	T	T	T	**T**	T	F	F	F
T	F	F	F	**F**	F	T	T	T
T	F	F	T	**F**	F	T	T	F

6. True **7.** True **8.** True **9.** False

10. a. Today is not Monday or tomorrow is not Sunday.
b. It is false that Neal is first and that David is not second.
c. It is false that Hugh is not painting and not cutting the grass.
d. Norma did not go to the store and Laurie did not go swimming.
e. It is false that mathematics is difficult and that logic is not easy.

11. $\sim P \vee Q \leftrightarrow \sim (P \wedge \sim Q)$ yes

$\sim P$	$\vee$	Q	$\leftrightarrow$	$\sim$	$(P$	$\wedge$	$\sim Q)$
F	T	T	**T**	T	T	F	F
F	F	F	**T**	F	T	T	T
T	T	T	**T**	T	F	F	F
T	T	F	**T**	T	F	F	T

12. *Converse:* **a.** $G \rightarrow P$ **b.** $\sim C \rightarrow P$
Inverse: $\sim P \rightarrow \sim G$ $\sim P \rightarrow C$
Contrapositive: $\sim G \rightarrow \sim P$ $C \rightarrow \sim P$
Converse: **c.** $\sim C \rightarrow \sim A$
Inverse: $A \rightarrow C$
Contrapositive: $C \rightarrow A$

13. No; statement is equivalent to "If I marry you, then I will get a job."

14. a. Possible choices are as follows:
Contrapositive: If I will be gullible then I do not study logic.
Implication: I do not study logic or I will not be gullible.
De Morgan: It is false that I study logic and that I will be gullible.
b. Use the negation of the given sentence: It is false that if I study logic then I will not be gullible.

15. a. $P \rightarrow Q$ **b.** $P \rightarrow Q$ **c.** $\sim Q \rightarrow \sim P$
d. $\sim Q \rightarrow \sim P$ **e.** $\sim P \vee Q$

16. a. $J \rightarrow P$ **b.** False
Converse: $P \rightarrow J$
Inverse: $\sim J \rightarrow \sim P$
Contrapositive: $\sim P \rightarrow \sim J$
Negation: $\sim(J \rightarrow P)$

17. a. $M \rightarrow F$ **b.** True
Converse: $F \rightarrow M$
Inverse: $\sim M \rightarrow \sim F$
Contrapositive: $\sim F \rightarrow \sim M$
Negation: $\sim(M \rightarrow F)$

18. a. $G \rightarrow \sim S$ **b.** False
Converse: $\sim S \rightarrow G$
Inverse: $\sim G \rightarrow S$
Contrapositive: $S \rightarrow \sim G$
Negation: $\sim(G \rightarrow S)$

19. a. $\sim T \rightarrow \sim H$ **b.** True
Converse: $\sim H \rightarrow \sim T$
Inverse: $T \rightarrow H$
Contrapositive: $H \rightarrow T$
Negation: $\sim(T \rightarrow H)$

20. a. Invalid **b.** Valid

21. a. Inconsistent **b.** Consistent
c. Consistent **d.** Inconsistent

22. The validity of an argument has to do with the form of the argument, not the truth or falsity of statements. A valid argument may have a false conclusion if one (or both) premises are false.

23. a. Valid **b.** Invalid
c. Valid **d.** Valid
e. Invalid

24. a. $(P \wedge Q) \vee (\sim P \wedge R)$
b. $(P \vee Q) \vee (R \wedge \sim Q)$
c. $P \wedge ((R \wedge Q) \vee (\sim Q \wedge \sim R))$
d. $P \vee (Q \wedge R) \vee \sim P$

25. a. 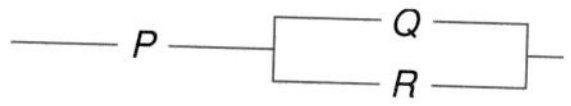
P Q R
b. 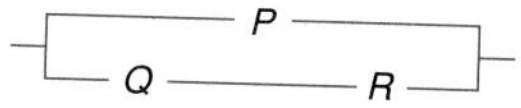
P Q R
c. 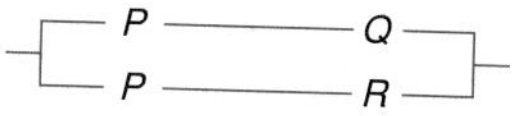
P Q P R
d.
P Q P Q R

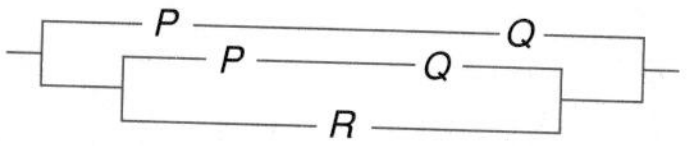

Chapter 3 Quiz

1. False	**2.** True	**3.** True
4. True	**5.** False	**6.** False
7. True	**8.** True	**9.** False
10. True	**11.** True	**12.** False
13. False	**14.** True	**15.** False
16. True	**17.** False	**18.** False
19. False	**20.** False	**21.** True
22. False	**23.** False	**24.** False
25. False		

Just for Fun *Barry—painter; Bob—mason; Bart—carpenter*

Chapter 4

Section 4.2

	a.	b.	c.	d.	e.	f.
1.	$\frac{3}{20}$	$\frac{3}{10}$	$\frac{7}{20}$	$\frac{13}{20}$	$\frac{13}{20}$	0
3.	$\frac{1}{6}$	$\frac{1}{2}$	$\frac{1}{2}$	$\frac{1}{2}$	$\frac{1}{3}$	1
5.	$\frac{1}{52}$	$\frac{1}{52}$	$\frac{1}{13}$	$\frac{1}{2}$	$\frac{1}{4}$	$\frac{1}{26}$
7.	$\frac{3}{13}$	$\frac{11}{26}$	$\frac{4}{13}$	$\frac{1}{52}$	$\frac{1}{26}$	$\frac{1}{52}$
9.	$\frac{5}{7}$	$\frac{4}{7}$	1	$\frac{2}{7}$	$\frac{3}{7}$	
11.	$\frac{1}{6}$	$\frac{1}{6}$	0	$\frac{1}{3}$	$\frac{7}{12}$	$\frac{11}{12}$
13.	$\frac{3}{5}$	$\frac{2}{5}$	0	$\frac{3}{5}$	$\frac{2}{5}$	1
15.	$\frac{1}{3}$	$\frac{5}{18}$	$\frac{5}{9}$	$\frac{4}{9}$	0	$\frac{7}{9}$

17. A sample space is the set of all possible outcomes for a particular experiment.
19. An impossible event is an event that cannot possibly succeed. A certain event is an event that is sure to occur everytime.
23. Research
25. a. $\frac{20}{41}$ b. $\frac{1}{2}$ or 0.5 c. 0.5 theoretical vs. 0.487 actual

Just for Fun *a.* $\frac{0}{1} = 0$
b. $\frac{1}{0}$ *is undefined*
c. $\frac{0}{0}$ *is meaningless*

Section 4.3

1. a. 12
b. H1 T1
H2 T2
H3 T3
H4 T4
H5 T5
H6 T6
c. $\frac{1}{12}$ d. 0 e. $\frac{1}{2}$ f. $\frac{1}{4}$
3. a. 20
b. $1, $5 | $5, $1 | $10, $1 | $20, $1 | $50, $1
$1, $10 | $5, $10 | $10, $5 | $20, $5 | $50, $5
$1, $20 | $5, $20 | $10, $20 | $20, $10 | $50, $10
$1, $50 | $5, $50 | $10, $50 | $20, $50 | $50, $20
c. $\frac{3}{5}$ d. $\frac{3}{5}$ e. $\frac{3}{10}$ f. $\frac{1}{10}$
5. a. $\frac{1}{4}$ b. 0 c. 1 d. $\frac{1}{4}$ e. $\frac{3}{4}$ f. $\frac{1}{2}$
7. a. $\frac{1}{6}$ b. $\frac{1}{18}$ c. 0 d. $\frac{1}{2}$ e. 1 f. $\frac{7}{36}$
9. a. $\frac{2}{5}$ b. $\frac{3}{5}$ c. $\frac{11}{20}$ d. $\frac{9}{20}$ e. $\frac{7}{20}$ f. $\frac{3}{10}$
11. a. $\frac{1}{10}$ b. $\frac{7}{50}$ c. $\frac{7}{50}$ d. $\frac{1}{25}$ e. $\frac{2}{25}$ f. $\frac{23}{50}$
13. The fundamental counting principle states that if one experiment has m different outcomes and another experiment has n different outcomes, then the first and second experiments performed together have $m \times n$ different outcomes.
15. 1,000,000,000

Just for Fun *One possibility is*
$1 + 23 + 4 + 5 + 67 = 100$

Section 4.4

1.

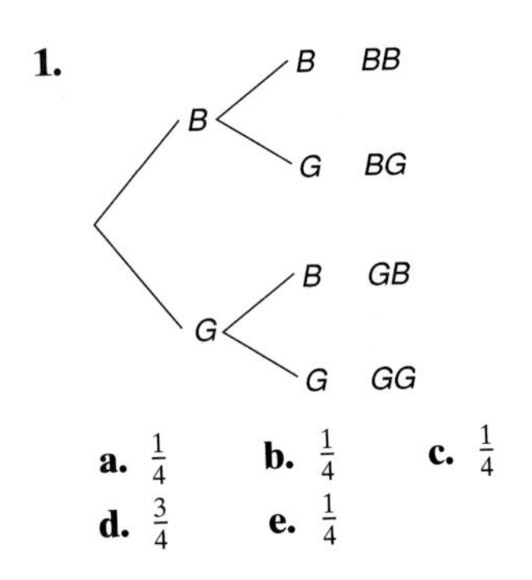

a. $\frac{1}{4}$ b. $\frac{1}{4}$ c. $\frac{1}{4}$
d. $\frac{3}{4}$ e. $\frac{1}{4}$

3.

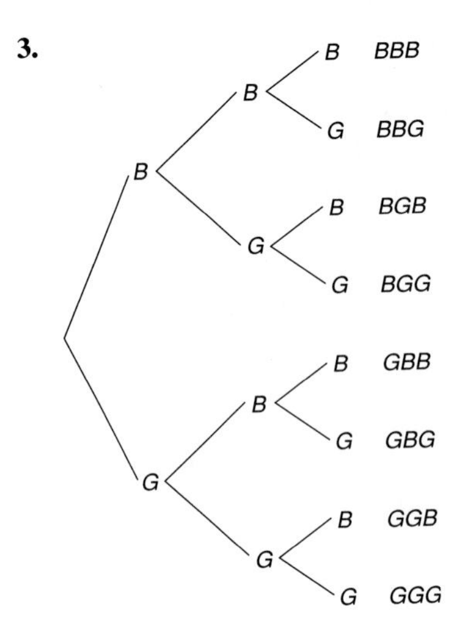

a. $\frac{1}{8}$ b. $\frac{3}{8}$ c. $\frac{7}{8}$ d. $\frac{1}{8}$
e. $\frac{1}{4}$ f. $\frac{1}{2}$

5.

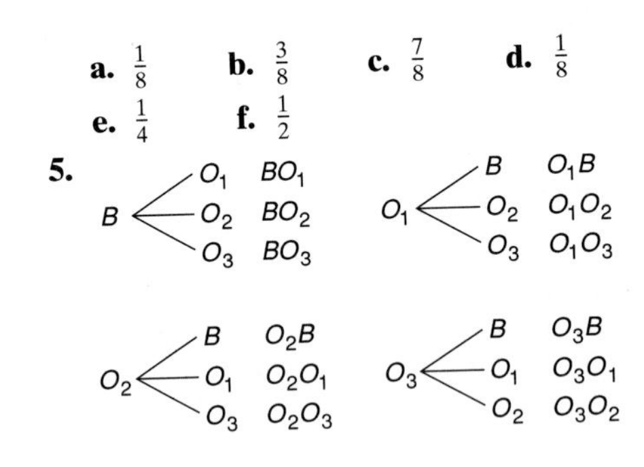

a. $\frac{1}{2}$ b. 1 c. $\frac{1}{4}$ d. $\frac{1}{4}$
e. $\frac{1}{2}$ f. $\frac{1}{2}$

7. a.

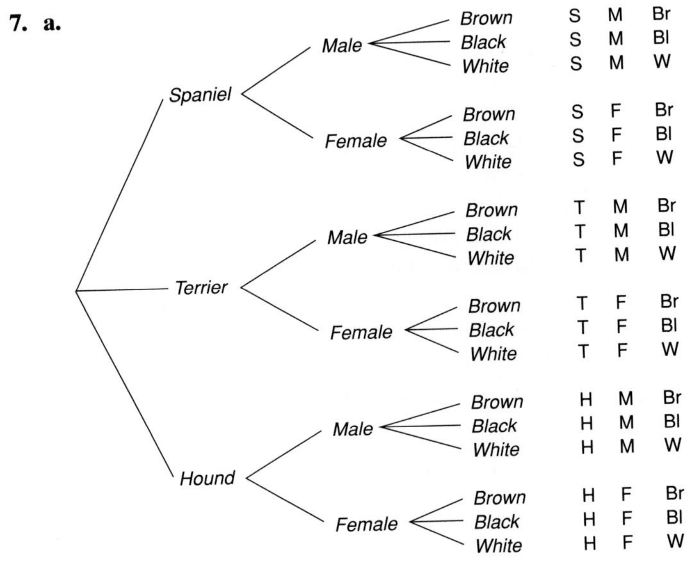

b. $\frac{1}{3}$ c. $\frac{1}{2}$ d. $\frac{1}{18}$ e. $\frac{1}{9}$ f. $\frac{4}{9}$

9. A tree diagram consists of a number of branches that show the possible outcomes for two or more experiments performed together. Read all the branches in the tree, from left to right, and obtain the sample space.

11. a.

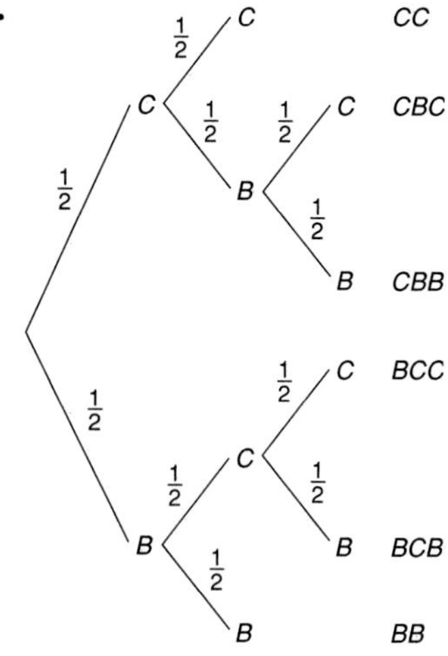

b. $\frac{1}{2}$ **c.** $\frac{1}{4}$ **d.** $\frac{3}{8}$ **e.** $\frac{3}{4}$ **f.** $\frac{3}{8}$

13. a. 8

b.

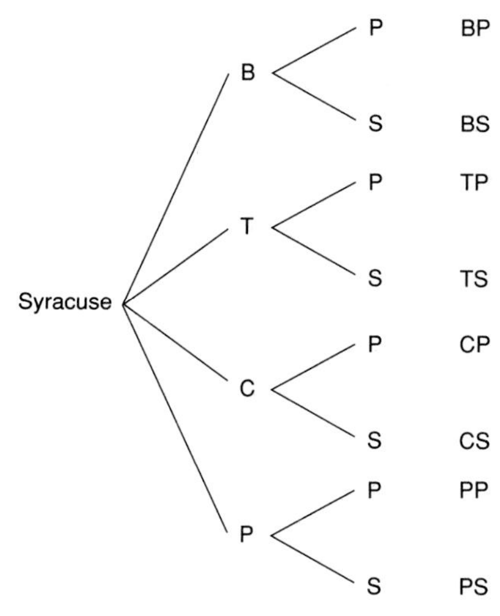

c. $\frac{1}{4}$ **d.** $\frac{1}{2}$ **e.** $\frac{5}{8}$ **f.** 8

Just for Fun *1. horrendous 2. stupendous 3. hazardous*

Section 4.5

1. a. 1 : 5 **b.** 1 : 17 **c.** 1 : 35
d. 35 : 1 **e.** 31 : 5 **f.** 11 : 1

3. a. 1 : 12 **b.** 1 : 3 **c.** 1 : 1
d. 12 : 1 **e.** 3 : 10 **f.** 15 : 11

5. a. 1 : 3 **b.** 3 : 1 **c.** 1 : 1
d. 3 : 1 **e.** 3 : 1 **f.** 1 : 1

7. a. $\frac{8}{11}$ **b.** $\frac{2}{7}$ **c.** $\frac{5}{12}$ **d.** $\frac{5}{8}$

9. $\frac{1}{100}$ **11.** 1 : 5, $3.00

13. $1.00 or 20¢ per ticket

15. 1,000,000; $\frac{1}{1{,}000{,}000}$; 5¢

17. Yes. We can expect to win a total of $36 for every four times we play. Therefore, we would have a profit of $8. A fair price to play is $9.00.

19. $1.20; The cost is too high.

21. Mathematical expectation tells us the expected value or *fair price* to pay to play a game or enter a contest. It also represents the average amount you should expect to win when the game is played a large number of times.

23. 17.5¢; The cost is too high, not worth a first class stamp to enter the contest.

25. Research

27. a. $\frac{153}{20{,}000}$ or 0.00765 **b.** $\frac{19{,}847}{20{,}000}$ **c.** $765

29. Empirical. Death rates change yearly based on number of deaths per year.

Just for Fun *Optical illusion; 3 or 5.*

Section 4.6

1. a. $\frac{5}{32}$ **b.** $\frac{5}{64}$ **c.** $\frac{1}{32}$ **d.** $\frac{1}{16}$ **e.** $\frac{25}{64}$ **f.** $\frac{1}{64}$

3. a. $\frac{7}{228}$ **b.** $\frac{143}{570}$ **c.** $\frac{7}{95}$ **d.** $\frac{11}{76}$ **e.** $\frac{91}{570}$ **f.** $\frac{91}{1140}$

5. a. $\frac{1}{4}$ **b.** $\frac{3}{4}$ **c.** $\frac{3}{26}$ **d.** $\frac{8}{13}$ **e.** $\frac{3}{52}$ **f.** $\frac{29}{52}$

7. a. $\frac{1}{12}$ **b.** $\frac{1}{12}$ **c.** $\frac{7}{12}$ **d.** $\frac{7}{12}$ **e.** $\frac{1}{4}$ **f.** $\frac{3}{4}$

9. a. $\frac{11}{221}$ **b.** $\frac{4}{221}$ **c.** $\frac{1}{221}$ **d.** $\frac{1}{221}$ **e.** $\frac{1}{17}$ **f.** $\frac{1}{17}$

11. a. $\frac{1}{55}$ **b.** $\frac{1}{220}$ **c.** $\frac{1}{22}$ **d.** $\frac{1}{22}$ **e.** $\frac{1}{22}$ **f.** $\frac{14}{55}$

13. a. $\frac{1}{6}$ **b.** $\frac{1}{30}$ **c.** $\frac{2}{6}$ **d.** $\frac{3}{4}$ **e.** $\frac{1}{360}$

15. a. $\frac{1}{5}$ **b.** $\frac{91}{13{,}505}$ **c.** $\frac{2}{25}$ **d.** $\frac{2}{5}$ **e.** $\frac{3}{37}$ **f.** $\frac{4}{5}$

17. a. 0.2 **b.** 0.7 **c.** 0.6 **d.** 0.5 **e.** 0.2 **f.** 0.8

19. No, $P(A \text{ or } B) \neq \frac{7}{12}$

21. Two events are independent if the occurrence of one of them does not affect the probability of the other. Answers may vary.

23. If A and B are not mutually exclusive events, then $P(A \text{ or } B) = P(A) + P(B) - P(A \text{ and } B)$.

25. $\frac{31}{32}$ **27.** $\frac{1}{512}$ **29.** Research

31. a. ≈0.49 **b.** ≈0.63 **c.** ≈0.57 **d.** ≈0.57

Just for Fun **1.** $\frac{1}{4}$ **2.** $\frac{1}{1024}$ **3.** $\frac{3}{4}$ **4.** $\frac{5}{4}$

Section 4.7

1. 380 **3.** 30,240; 100,000 **5.** 120; 216

7. 657,720 **9.** 12

11. 468,000; 421,200; 405,000 **13.** 64; 56

15. **a.** 6 **b.** 120 **c.** 1
d. 210 **e.** 20 **f.** 24

17. 90

19. 362,880

21. 132

23. **a.** 2520 **b.** 50,400 **c.** 5040 **d.** 1260

25. 625

27. 6,400,000,000

29. A permutation is an ordered arrangement.

31. The symbol $n!$ is called *n factorial;*
$n! = n \times (n - 1) \times (n - 2) \times \cdots \times 3 \times 2 \times 1.$

b. 9900

33. 7,503,600

35. **a.** 3,118,752,000 **b.** 59,976,000 **c.** 2,998,800,000

Just for Fun *Hawaii*

Section 4.8

1. **a.** 10 **b.** 10 **c.** 35
d. 35 **e.** 1 **f.** 1

3. 210 **5.** 10; 85¢ **7.** 3003 **9.** 21

11. 1960 **13.** 2970 **15.** 5400 **17.** 108,900

19. A combination is a distinct group of objects without regard to their arrangement.

21. $_nC_r = \dfrac{n!}{(n - r)!r!}$

This is the general formula to obtain the number of combinations of n things taken r at a time.

23. $_{35}C_{11} \cdot {}_{24}C_{11}$

25. 1013; An example is $2^n - (n + 1)$.

Just for Fun *4.5¢*

Section 4.9

1. $\frac{1}{221}$

3. $\frac{11}{850}$

5. **a.** $\frac{7}{306}$ **b.** $\frac{7}{102}$ **c.** $\frac{7}{17}$ **d.** $\frac{28}{153}$ **e.** $\frac{95}{102}$

7. **a.** $\frac{1}{208}$ **b.** $\frac{3}{104}$ **c.** $\frac{21}{52}$ **d.** $\frac{15}{52}$ **e.** $\frac{15}{208}$
f. $\frac{21}{104}$

9. $_{52}C_{13}$; $\dfrac{_{13}C_{13}}{_{52}C_{13}}$ **11.** $\frac{1}{35}$

13. **a.** $\dfrac{_7C_3}{_{20}C_3}$ **b.** $\dfrac{_5C_3}{_{20}C_3}$ **c.** $\dfrac{_8C_3}{_{20}C_3}$
d. $\dfrac{_7C_2 \cdot {}_5C_1}{_{20}C_3}$ **e.** $\dfrac{_5C_2 \cdot {}_8C_1}{_{20}C_3}$ **f.** $\dfrac{_8C_2 \cdot {}_7C_1}{_{20}C_3}$

15. $\dfrac{1287}{2{,}598{,}960}$

17. $\dfrac{1}{108{,}290}$

19. $\dfrac{6}{4165}$

21. No. In just about any game of chance, the probability of winning is very small compared with the probability of losing.

23. $\dfrac{7}{47}$

Just for Fun *The probability this will occur is greater than $\frac{1}{2}$.*

Review Exercises for Chapter 4

1. **a.** $\frac{4}{13}$ **b.** $\frac{6}{13}$ **c.** $\frac{7}{13}$
d. $\frac{9}{13}$ **e.** $\frac{10}{13}$ **f.** $\frac{3}{13}$

2. **a.** $\frac{5}{36}$ **b.** $\frac{31}{36}$ **c.** $\frac{1}{6}$
d. $\frac{2}{9}$ **e.** $\frac{1}{6}$ **f.** 0

3. $\dfrac{4}{10}$ **4.** 24

5.

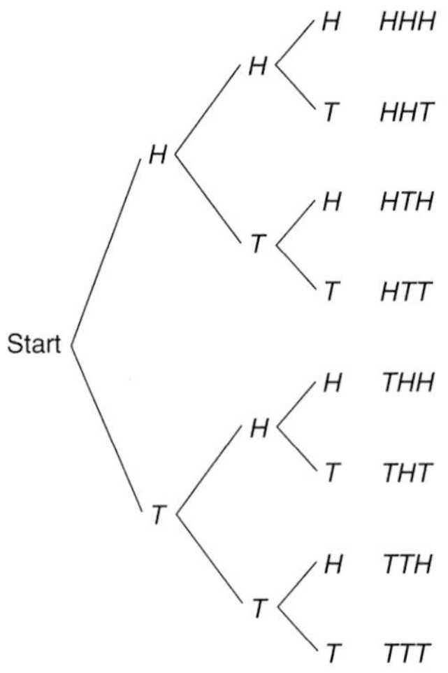

Start — H — H — H: HHH; T: HHT
Start — H — T — H: HTH; T: HTT
Start — T — H — H: THH; T: THT
Start — T — T — H: TTH; T: TTT

a. $\frac{1}{8}$ **b.** $\frac{7}{8}$ **c.** $\frac{1}{8}$

6. **a.** 1 : 12 **b.** 12 : 1 **c.** 3 : 10
d. 4 : 9 **e.** 1 : 51 **f.** 51 : 1

7. $\frac{1}{9}$ **8.** 3 : 8 **9.** $20

10. $4, or 80¢ per ticket

11. **a.** 1000 **b.** $0.50
c. The cost is too high; lose an average of $0.50 for each ticket purchased.

12. The probability that an event will occur is equal to the number of successful outcomes divided by the total number of all possible outcomes. The odds in favor of an event A are found by taking the probability that the event A will occur and dividing it by the probability that the event A will not occur.

13. Independent; it does not matter which ball is chosen first because there is replacement. The occurrence of one event does not affect the occurrence of a second event.

14. Yes; mutually exclusive events cannot happen at the same time. Either a 7 or an 11 may occur, but they cannot occur at the same time.

15. **a.** $\frac{25}{102}$ **b.** $\frac{1}{221}$ **c.** $\frac{13}{51}$
d. $\frac{13}{204}$ **e.** $\frac{40}{221}$ **f.** $\frac{4}{663}$

16. **a.** $\frac{1}{12}$ **b.** $\frac{7}{12}$ **c.** $\frac{1}{78}$
d. $\frac{3}{13}$ **e.** $\frac{1}{156}$ **f.** $\frac{31}{156}$

17. **a.** 0.0728 **b.** 0.18

18. **a.** 7776 **b.** $\frac{1}{1296}$

19. 468,000; 405,000; 302,400

20. 143,640 **21.** 90 **22.** 1260

23. **a.** 720 **b.** 6 **c.** $\frac{5}{2}$ **d.** 12 **e.** 5,527,200 **f.** 1

24. **a.** 10 **b.** 15 **c.** 35 **d.** 35 **e.** 1 **f.** 1

25. 196,000 **26.** 3600

27. **a.** $\frac{1}{30}$ **b.** $\frac{1}{30}$ **c.** $\frac{1}{30}$ **d.** $\frac{1}{5}$ **e.** $\frac{1}{10}$

28. **a.** $\frac{_{26}C_5}{_{52}C_5}$ **b.** $\frac{_{26}C_5}{_{52}C_5}$ **c.** $\frac{_{12}C_5}{_{52}C_5}$ **d.** $\frac{_{40}C_5}{_{52}C_5}$ **e.** $\frac{_{12}C_3 \cdot {_{40}C_2}}{_{52}C_5}$

29. **a.** $\frac{624}{2{,}598{,}960}$ **b.** $\frac{3744}{2{,}598{,}960}$

c. $\frac{5148}{2{,}598{,}960}$ (counting straight flushes)

d. $\frac{5108}{2{,}598{,}960}$ (not counting straight flushes)

30. A permutation is an ordered arrangement of items whereas a combination is an unordered arrangement of items.

Chapter 4 Quiz

1. $\frac{4}{13}$ **2.** $\frac{2}{9}$ **3.** 1 : 5 **4.** $\frac{7}{12}$ **5.** $\frac{3}{4}$

6. $\frac{1}{78}$ **7.** $\frac{3}{13}$ **8.** $\frac{1}{1000}$ **9.** $0.10 **10.** $\frac{3}{4}$

11. True **12.** True **13.** False **14.** False

15. True **16.** True **17.** False **18.** True

19. False **20.** False **21.** True **22.** False

23. False **24.** False **25.** True

Chapter 5

Section 5.2

1. Mean = 5; median = 4; mode = 4; midrange = 6.5

3. Mean = 5.5; median = 5.5; no mode; midrange = 5.5

5. Mean = 7; median = 7; no mode; midrange = 7

7. Mean = 55; median = 55; no mode; midrange = 55

9. Mean = 1755.3; median = 1794; no mode; midrange = 1716.5

11.

	American League	National League
a. mean	50.1	52.6
b. median	50	49
c. mode	47, 52, 56	47, 49
d. midrange	48.5	56.5

e. No, due to the fact that Mark McGwire had two exceptional years for hitting home runs. Also, Barry Bonds had one super season.

13. Mean = 658.9; median = 631; no mode; midrange = 677.5

15. **a.** $30,000 **b.** $25,000; $35,000 **c.** $26,000; $35,000 **d.** $24,500; $35,000 **e.** Median; midrange

17. Mean = 4; median = 3.5; mode = 3; no

19. 1001 **21.** 76.7

23. Mean = $23,125; median = $19,000; mode = $16,000; midrange = $63,000

25. Mean = 13.6; median = 14; mode = 10; midrange = 16

27. **a.** mode **b.** median **c.** mean

29. The mean, median, and mode are all located at the peak of the graph.

31. The *mean* for a set of data is found by determining the sum of the data and dividing this sum by the total number of elements in the set. The *median* for a set of data is found by arranging the data in sequential order and finding the middle number. The *mode* for a given set of data is that item that occurs most frequently. The *midrange* is determined by adding the least value in a given set of data to the greatest value and dividing the sum by 2.

33. Yes, if some of the data have negative values.

35. Harmonic mean ≈ 3.4, arithmetic mean = 5.4; harmonic mean is more representative of this data.

Just for Fun *Put one penny in one cup, four in another, and five in the third, then stack the first cup inside the second.*

Section 5.3

1. $\sigma = 3$ **3.** $\sigma = \sqrt{6}$ or 2.4

5. $\sigma = \sqrt{13.7}$ or 3.7 **7.** $\sigma = \sqrt{26.8}$ or 5.2

9. $\sigma = \sqrt{21.6}$ or 4.6

11. **a.** 4 **b.** 3.5 **c.** 3 **d.** 5 **e.** 4.5 **f.** $\sigma = \sqrt{2.6}$ or 1.6

13. **a.** 34 **b.** 32.5 **c.** Ø **d.** 40 **e.** 40 **f.** $\sigma = \sqrt{134.8}$ or 11.6

15. **a.** 345 **b.** 370 **c.** none **d.** 340 **e.** 320 **f.** $\sigma = \sqrt{11{,}055.6}$ or 105.1

17. **a.** Ruth **b.** Joe **c.** Hugh and Doris

19. The *range* is found by subtracting the smallest value from the largest value for a given set of data.

The *standard deviation* is the square root of the arithmetic mean of the squares of deviations from the mean.

21. The standard deviation takes into account each piece of data, and how large the standard deviation is depends upon the amount of spread that the data exhibits.

Just for Fun *35?*

Section 5.4

1. 94 **3.** 93 **5.** 24 **7.** 8th **9.** 15th

11. **a.** Third **b.** 85 **c.** 60 **d.** 72 **e.** 88 **f.** 87

13. Larry

15. It is false; it is impossible to score at the 100th percentile. Ricci is one of the class and that would exclude Ricci.

17. 0 **19.** 20th

21. **a.** Second **b.** 95 **c.** 14th
d. 65 **e.** 61 **f.** 85

23. A percentile is that value that divides the range of a set of data into two parts such that the given percentage of the data lies below this value.

25. The 50th percentile, the second quartile, and the median are not always equivalent measures.

27. No, it would be too vague and perhaps imply it was close to the 99th percentile.

Just for Fun *qoph, qiviut, there are others.*

Section 5.5

1.

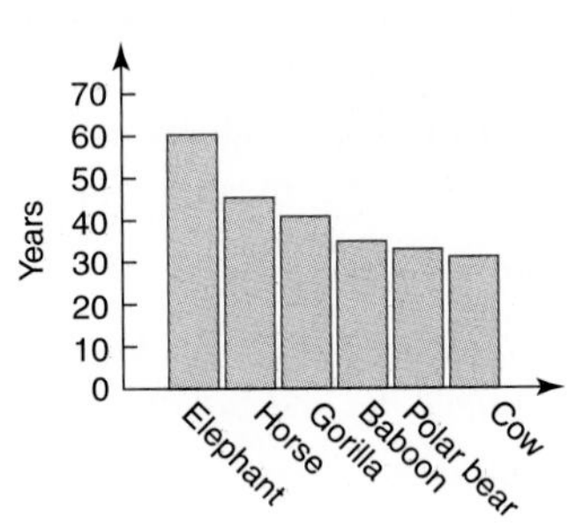

Maximum recorded life spans of six different animals

3.

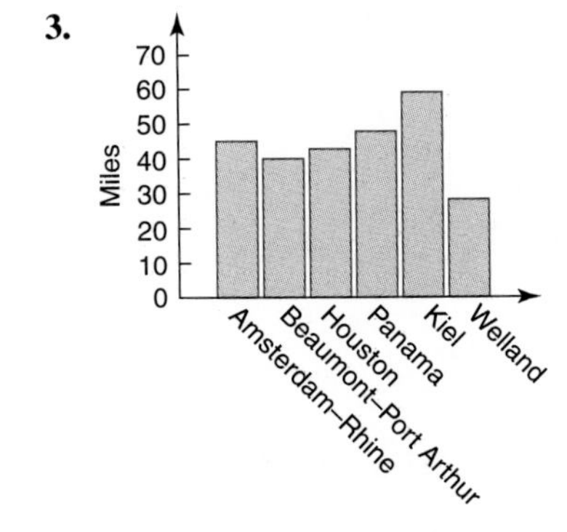

Lengths of six notable canals

5.

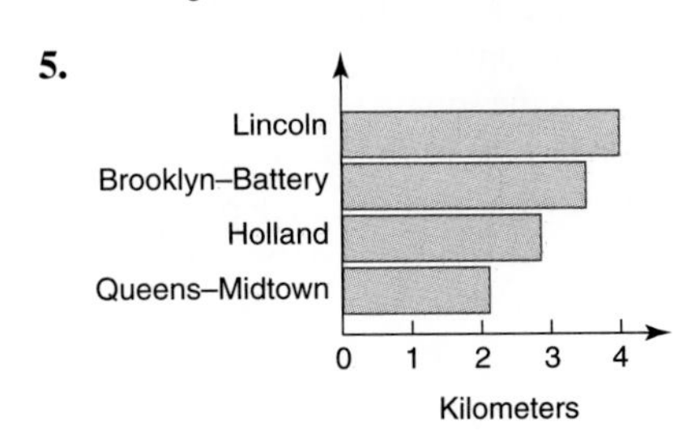

Lengths of notable New York tunnels (for motor vehicles)

7. **a.** 2001 **b.** $262,000
c. 2001, 2004 **d.** $1,362,000

9.

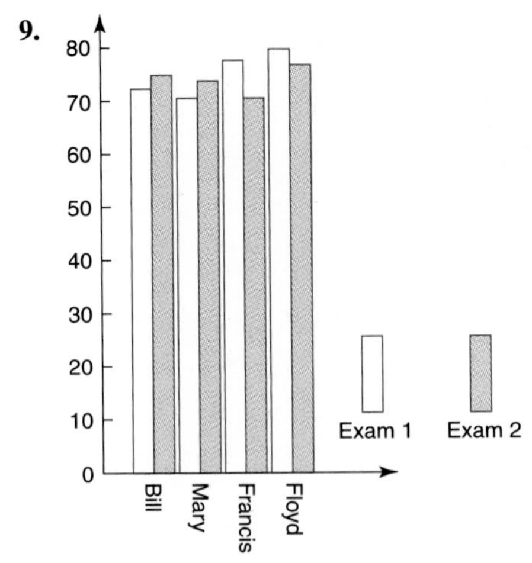

Scores for first two exams

11.

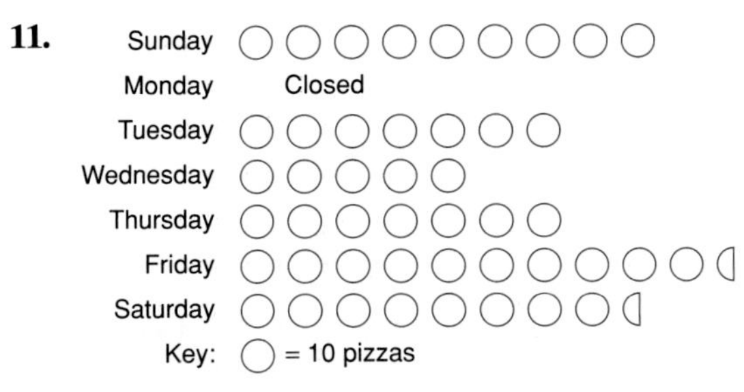

Number of pizzas sold by Phil's Pizza Palace last week

13.

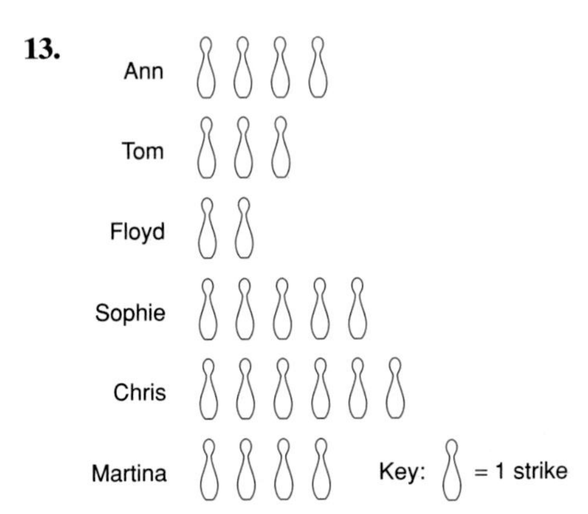

Number of strikes in first game

15. **a.** $175,000 **b.** $225,000
c. $475,000 **d.** $1,050,000

17. Bus = 180°
Car = 120°
Walk = 60°

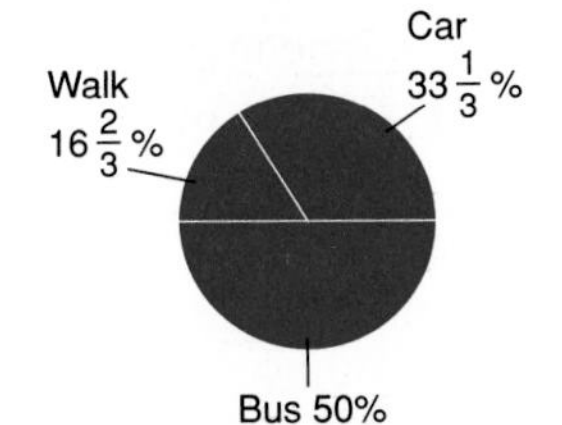

19. Personal income = 90° Sales = 72°
Corporate income = 90° Highway = 36°
Excise = 54° Miscellaneous = 18°

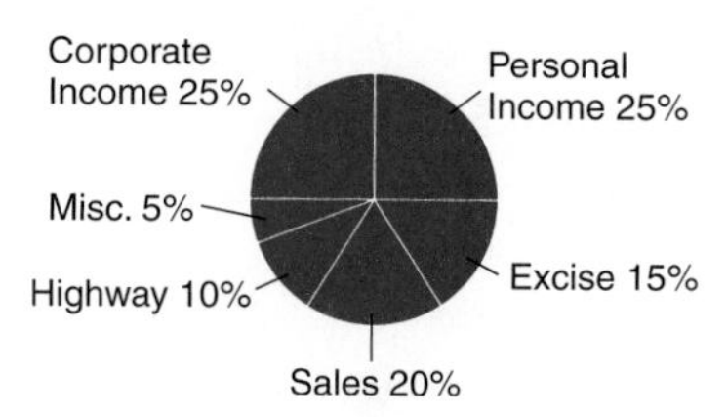

21. Federal aid = 126°   Licenses = 18°
State aid = 54°   Sales tax = 36°
Property tax = 108°   Other = 18°

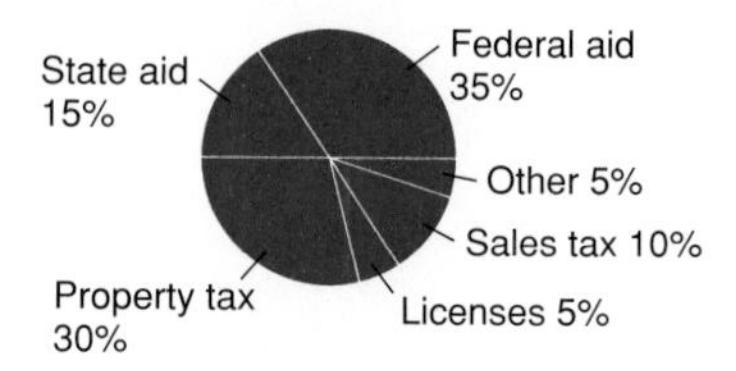

23.

Number	*Tally*	*Frequency*
60	//	2
61	/	1
64	//	2
65	//	2
66	//	2
67	////	4
68	//	2
69	//	2
70	𝍸 /	6
71	//	2
72	//	2
75	/	1

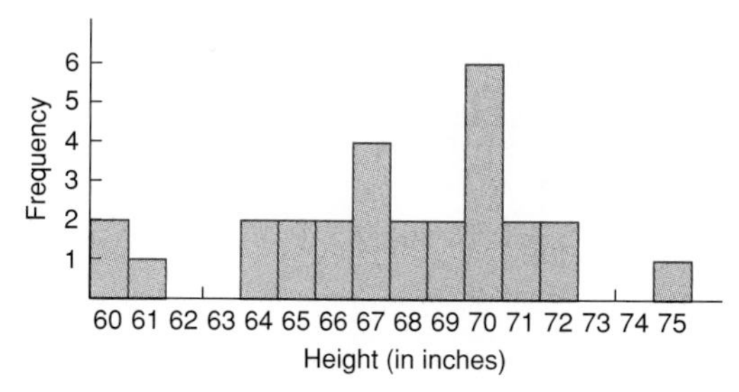

25.

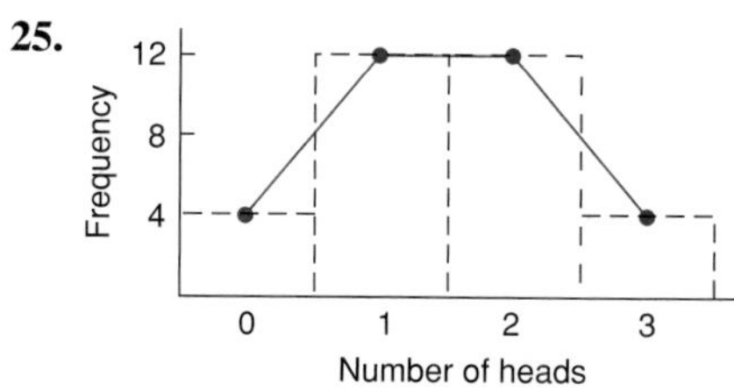

27. **a.**

Interval	*Tally*	*Frequency*
95–99	//	2
90–94	///	3
85–89	𝍸	5
80–84	///	3
75–79	//	2
70–74	𝍸 ////	9
65–69	//	2
60–64	////	4
55–59	///	3

b.

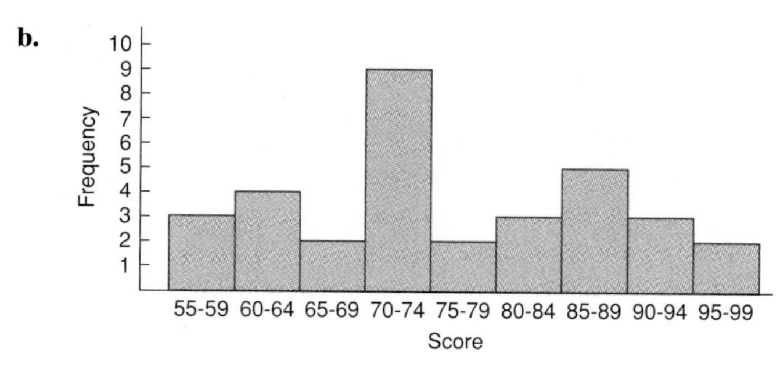

29. **a.**

Interval	*Tally*	*Frequency*
95–99	/	1
90–94	////	4
85–89	////	4
80–84	𝍸 /	6
75–79	𝍸 /	6
70–74	𝍸 /	6
65–69	𝍸	5
60–64	𝍸 /	6
55–59	////	4
50–54	///	3
45–49	//	2
40–44	///	3

b.

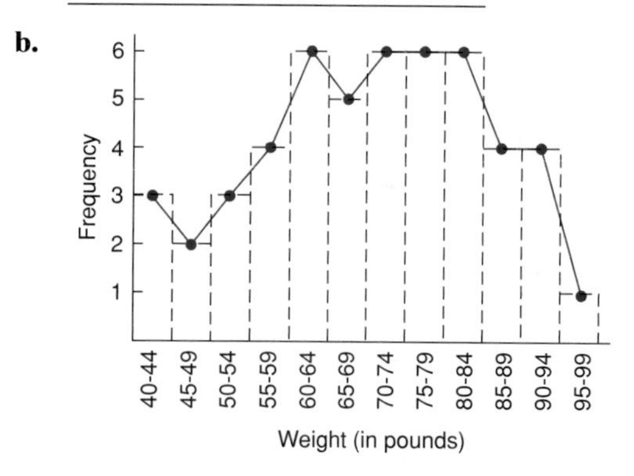

31. No; the graph does not indicate what part or percentage of a dollar is spent.

33. Answers may vary.

35. Graphs used in this section were: vertical bar graph, horizontal bar graph, comparative vertical bar graph, comparative horizontal bar graph, pictogram, circle graph, histogram, and frequency polygon (line graph).

37. A bar graph and a histogram are *not* the same type of graph. The bars of a histogram touch each other and are of uniform width which is not the case for an ordinary bar graph.

39. A range of 5 to 12 classes is desirable to avoid difficulty in interpretation of the data. Hence, too large or too narrow intervals could distort the interpretation.

Just for Fun

Section 5.6

1. **a.** 68.2%   **b.** 95.4%   **c.** 99.7%

3. **a.** 50,000   **b.** 12,000   **c.** 84.1%   **d.** 318

5. **a.** 795 **b.** 115 **c.** ≈6 **d.** 4090
7. **a.** 2.3% **b.** 15.9% **c.** 81.8% **d.** 97.6%
9. **a.** 4.5% **b.** 9.1% **c.** 74.9% **d.** 7.9% **e.** 3.6%
11. **a.** 38.4% **b.** 62.5% **c.** 2.3% **d.** 69.2% **e.** 28.5%
13. **a.** 53.3% **b.** 24.2% **c.** 95.5% **d.** 63.2% **e.** 15.5%
15. **a.** 21.2% **b.** 3.6% **c.** 5.5% **d.** 54% **e.** 17.6%
17. **a.** 0.6% **b.** 6.7%
19. **a.** 38.4% **b.** 61.6%
21. **a.** 0.50 or $\frac{50}{100}$ **b.** 0.308 **c.** 0.48
23. The normal curve is a bell-shaped curve; it is symmetric about the mean and the mean, median, mode all have the same value.
25. A z-score tells us the number of standard deviations a piece of data is located from the mean. A raw score just provides the value of a piece of data.
27. Three items provided in all of the examples in this section are (1) "approximately normally distributed," (2)the mean, and (3) the standard deviation. We need this information in order to proceed with the problem.

Just for Fun

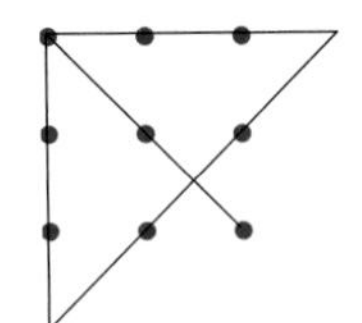

Section 5.7

1. One reason for this discrepancy is the definition of "enrollment." For example, the president might define enrollment as the total number of new students from all classes, freshmen, sophomores, juniors, and seniors. Student affairs office personnel might define enrollment as the total number of new freshmen.
3. Answers may vary.
5. The results of the survey can be misleading because of the distance factor. We are not sure how far people travelled in their commute.
7. The ratio 1 : 3 is misleading because it implies that one computer is available for every three students throughout the school district. But this is probably not the case. For example, in most school districts many more computers are located in junior high schools and high schools as compared with the elementary schools.
9. Answers may vary.
11. An example of unethical practice would be to keep testing or sampling until you get the desired results.
13. Answers may vary.

Just for Fun *Almost 2740 years*

Review Exercises for Chapter 5

1. **a.** Mean, median, mode, midrange **b.** Mode **c.** Median **d.** Mean
2. Mean = 15.4; mode = 12; median = 16; midrange = 15.5
3. Mean = 6; mode = 5; median = 5; midrange = 7.5
4. Mean = 9.6; no mode; median = 9; midrange = 10.5
5. Mean = 55; no mode; median = 55; midrange = 55
6. 72 **7.** **a.** 384 **b.** 76.8
8. Range = 10; midrange = 65; mean = 65
9. Range = 42; midrange = 79; mean = 79.7
10. $\sigma = \sqrt{5}$ or 2.2 **11.** $\sigma = \sqrt{24}$ or 4.9
12. 85 **13.** Julie
14.

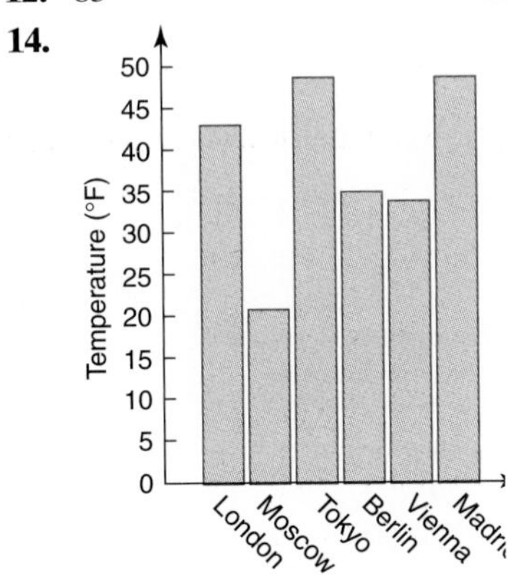

January average high temperatures for selected cities

15.

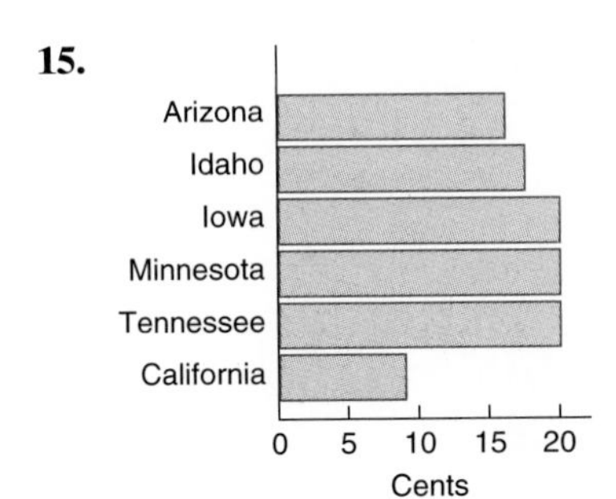

16.

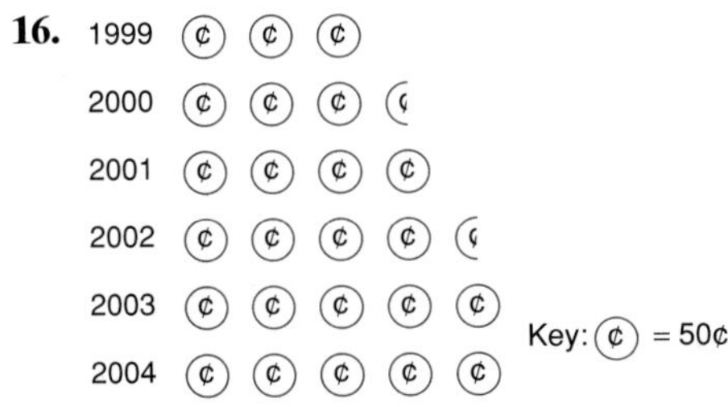

Annual dividend paid to shareholders by Peters Corp.

17.

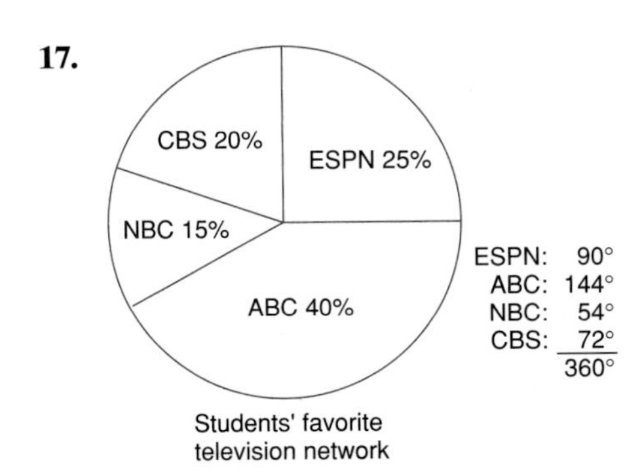

Students' favorite television network

18. a.

Number	*Tally*	*Frequency*
6	𝍸 𝍸 /	11
5	𝍸 ////	9
4	𝍸 𝍸	10
3	////	4
2	𝍸 ///	8
1	𝍸 ///	8

b., c.

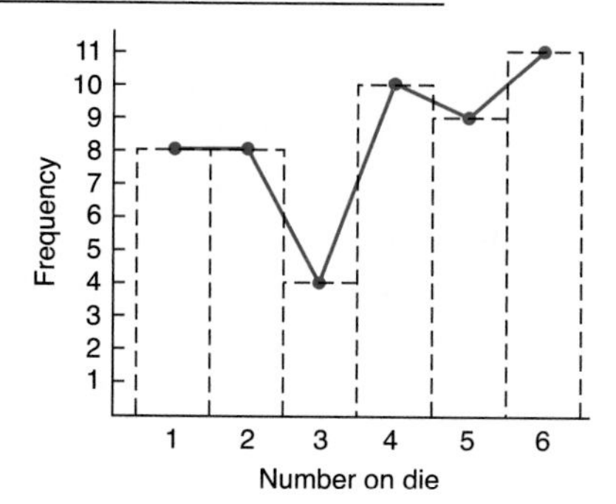

19. a.

Height	*Tally*	*Frequency*
75	//	2
74		0
73		0
72	//	2
71	//	2
70	𝍸 ///	8
69	////	4
68	//	2
67	///	3
66	//	2
65	//	2
64	///	3

b., c.

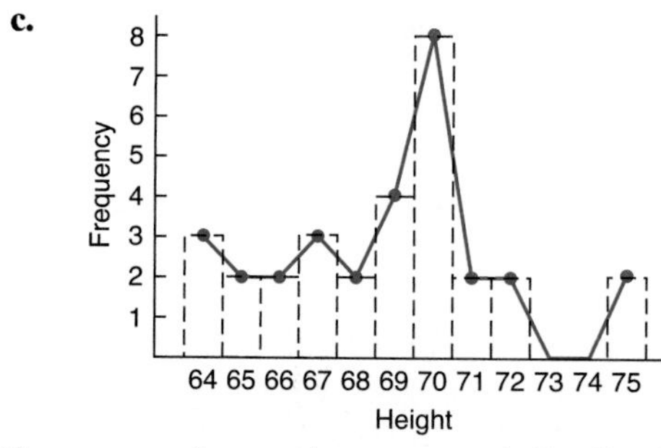

d. No, a normal curve has a normal distribution and the mean, median, and mode all have the same value at the center of the distribution.

20. a. 15.9% **b.** 2.3% **c.** 15.9% **d.** 81.8% **e.** 682

21. a. 5.5% **b.** 34.5% **c.** 60% **d.** 21.2% **e.** 15.7%

22. a. Everybody but Steve **b.** Steve **c.** No one **d.** Steve

23. a. 32.5 **b.** 38 **c.** 46 **d.** 30 **e.** 60

24. a. 25 **b.** 25 **c.** 26 **d.** 18 **e.** 27 **f.** 5.2 **g.** 22 **h.** 28

25. a. Third **b.** 75 **c.** 27 **d.** 22 **e.** Bimodal, 22 and 20

26. a. 6.7 **b.** 6 **c.** 6 **d.** 7.5

27. Answers may vary.

28. Answers may vary.

Chapter 5 Quiz

1. True **2.** True **3.** True **4.** False **5.** False **6.** False **7.** True **8.** False **9.** True **10.** False **11.** False **12.** True **13.** False **14.** True **15.** False **16.** False **17.** True **18.** True **19.** True **20.** False **21.** 4 **22.** 9th **23.** 75% **24.** 36 **25.** Sandra

Chapter 6

Section 6.2

1. a. 6 **b.** 1 **c.** 11 **d.** 10 **e.** 9 **f.** 7

3. a. 9 **b.** 8 **c.** 7 **d.** 6 **e.** 1 **f.** 8

5. a. 10 **b.** 10 **c.** 8 **d.** 11 **e.** 9 **f.** 8

7. a. 12 **b.** 12 **c.** 9 **d.** 4 **e.** 7 **f.** 9

9. a. 2 **b.** 6 **c.** 8 **d.** 9 **e.** 6 **f.** 10

11. a. Closure for addition
b. Commutative property for multiplication
c. Commutative property for addition
d. Associative property for multiplication
e. Inverse element for multiplication
f. Identity element for addition

13. a. True **b.** False **c.** True **d.** False **e.** False **f.** True

15. Yes, it satisfies the required properties.

17. a. 7 **b.** 1 **c.** 5 **d.** 3 **e.** 3 **f.** 2

19. Assuming zero is allowed, when $a = b = c$, or when $a = 9, b = 3, c = 6$.

21. If a binary operation is performed on any two elements of a set and the result is also an element of the set, then that set has *closure* (or is closed) under the given binary operation.

23. An identity element with respect to a binary operation is an element in a set such that when the binary operation is performed on it and any given element in the set, the answer (or result) is the given element. For the set of whole numbers, the additive identity is 0 and the multiplicative identity is 1.

25. a. *B, C* **b.** *A, B, C, D*

27. a.

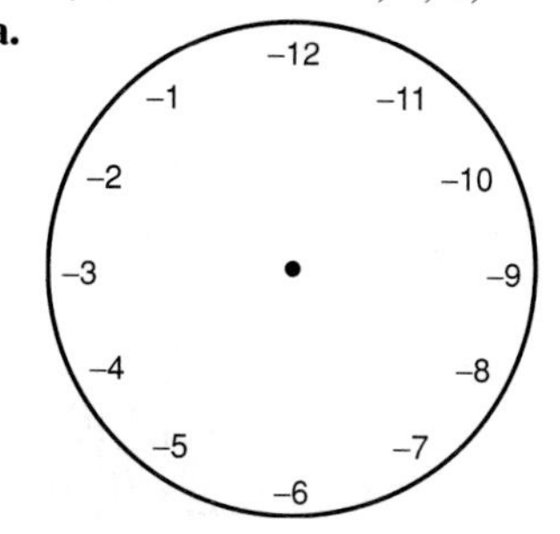

b. $-1, \quad 11 \qquad -7, \quad 5$
$-2, \quad 10 \qquad -8, \quad 4$
$-3, \quad 9 \qquad -9, \quad 3$
$-4, \quad 8 \qquad -10, \quad 2$
$-5, \quad 7 \qquad -11, \quad 1$
$-6, \quad 6 \qquad -12, \quad 12$

Just for Fun *Yes*

Section 6.3

1. Spring **3.** Summer **5.** Fall
7. Winter **9.** Winter **11.** Spring
13. Spring **15.** Fall **17.** No
19. Monday **21.** Wednesday **23.** Monday
25. Friday **27.** Thursday **29.** Thursday
31. Saturday **33.** Tuesday **35.** Saturday
37. a. Thursday **b.** Wednesday **c.** Tuesday
39. Sunday × (Tuesday × Friday)
$\stackrel{?}{=}$ (Sunday × Tuesday) × Friday
Sunday × Wednesday $\stackrel{?}{=}$ Tuesday × Friday
Wednesday = Wednesday
41. Yes **43.** Down **45.** Up **47.** Up
49. Up **51.** Left **53.** Right
55. a. Right **b.** Up **c.** Left
57. Yes
59. A mathematical system is a set of elements together with one or more operations (rules) for combining elements of the set.
61. a. *w* **b.** *z* **c.** *x* **d.** *x*
63.

−	1	2	3	4
1	4	3	2	1
2	1	4	3	2
3	2	1	4	3
4	3	2	1	4

65.

◆	0	1	2	3
0	0	1	2	3
1	1	2	3	0
2	2	3	0	1
3	3	0	1	2

Just for Fun *It is practically impossible.*

Section 6.4

1. a. 4 **b.** 0 **c.** 1 **d.** 2 **e.** 4 **f.** 1
3. a. 3 **b.** 4 **c.** 0 **d.** 2 **e.** 0 **f.** 2
5. a. 3 **b.** 2 **c.** 3 **d.** 4 **e.** 0 **f.** 3
7. a. 2 **b.** 3 **c.** 1 **d.** 4 **e.** 3 **f.** 3

9.

+	0	1	2	3	4	5	6
0	0	1	2	3	4	5	6
1	1	2	3	4	5	6	0
2	2	3	4	5	6	0	1
3	3	4	5	6	0	1	2
4	4	5	6	0	1	2	3
5	5	6	0	1	2	3	4
6	6	0	1	2	3	4	5

a. Yes **b.** 0 **c.** Yes

d. For example,
$(1 + 3) + 5 \stackrel{?}{=} 1 + (3 + 5);$
$4 + 5 \stackrel{?}{=} 1 + 1;$
$2 = 2$
e. Element: 0, 1, 2, 3, 4, 5, 6
Inverse: 0, 6, 5, 4, 3, 2, 1
f. Yes
11. a. 4 **b.** 4 **c.** 2
d. 6 **e.** 3 **f.** 6
13. a. True **b.** False **c.** True
d. True **e.** True **f.** False
15. a. True **b.** False **c.** False
d. False **e.** True **f.** True
17. a. 5 **b.** 4 **c.** 1 **d.** 2 **e.** 6 **f.** 11
19. 51
21. A modulo *m* system consists of *m* elements, 1 through *m*, or 0 through $m - 1$, and a binary operation.
23. a. $10 + 10 = 20, \quad 20 - 12 = 8$
b. $10 + 10 = 20, \quad 20 \div 12 = 1, r = 8$
25. Answers will vary. Two examples: Months of the year, modulo 12; Days of the week, modulo 7.
27. a. 2030 **b.** 2320 **c.** 0810 **d.** 1015

Just for Fun *Five is pretty good, but six is the correct answer.*

Section 6.5

1. *Q* **3.** *R* **5.** *S* **7.** *R* **9.** *S*
11. Yes; there are no new elements.
13. All of them: The inverse of *P* is *P*; the inverse of *Q* is *S*; the inverse of *R* is *R*; and the inverse of *S* is *Q*.
15. ! **17.** ! **19.** ? **21.** ! **23.** ?
25. No; there is a new element in the table, namely, !.
27. None **29.** No **31.** a
33. *d* **35.** *b* **37.** *e*
39. *e* **41.** *e* **43.** *a*
45. Yes; the symmetric table indicates commutativity.
47. *x* **49.** *y* **51.** *z*
53. *z* **55.** *y* **57.** *w*
59. Yes, there are no new elements in the table.
61. All; the inverse of *w* is *w*; the inverse of *x* is *y*; the inverse of *y* is *x*; and the inverse of *z* is *z*.
63. Yes
65. a. Yes **b.** None **c.** None **d.** No
67. a.

*	*a*	*b*	*c*
a	*b*	*c*	*a*
b	*c*	*a*	*b*
c	*a*	*b*	*c*

b. Yes **c.** c
d. All; the inverse of *a* is *b*; the inverse of *b* is *a*; and the inverse of *c* is *c*.
e. Yes
69. To determine whether a system defined by a table is commutative or not, under the given operation, check the diagonal in the table to see if the table is symmetric about it. If it is, then the system is commutative.

71. 4 **73.** 5 **75.** 13

77.
$$7♣(11♦13) \stackrel{?}{=} (7♣11)♦(7♣13)$$
$$7♣(13) \stackrel{?}{=} 11♦13$$
$$13 = 13$$

79. The previous two exercises illustrate the distributive property.

Just for Fun *6 dozen dozen*

Section 6.6

1. a, d

3. An axiomatic system consists of four main parts:
1. Undefined terms
2. Defined terms (definitions)
3. Axioms
4. Theorems

5. Axiom II tells us there is a line, say line AB. Axiom I tells us there are exactly two points on the line, A and B. Axiom IV tells us there is another line (say line CD) that has no points in common with line AB. Axiom I tells us there are exactly two points on the line, C and D. Axiom III tells us that the lines are distinct because for each pair of points there is one and only one line containing them. We have two pairs of points, or at least four points.

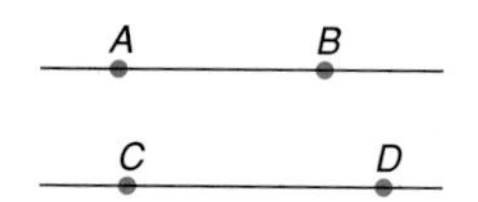

7. **a.** 1. Axiom II ; 2. Axiom IV
b. 1. Axiom I ; 2. Axiom IV
c. 1. Axiom I ; 2. Axiom III
d. 1. Axiom II ; 2. Axiom IV ; 3. Axiom V ; 4. Axiom II
e. 1. Axiom I ; 2. Axiom IV; 3. Axiom V ; 4. Axiom I ; 5. Axiom II

9. An axiom is a statement that is accepted as true without proof.

11. A theorem is a logical deduction made from undefined terms, defined terms, axioms, and sometimes other theorems. Theorems are proved while axioms are accepted as true without proof.

13. **a.** 3
b. Model: $\{a, b\}, \{a, c\}, \{b, c\}$
Axiom 1: There are exactly three sets.
Axiom 2: Every two sets have exactly one element in common.
Axiom 3: Every element belongs to exactly two sets.
Each axiom is satisfied; hence there are three elements.

Just for Fun

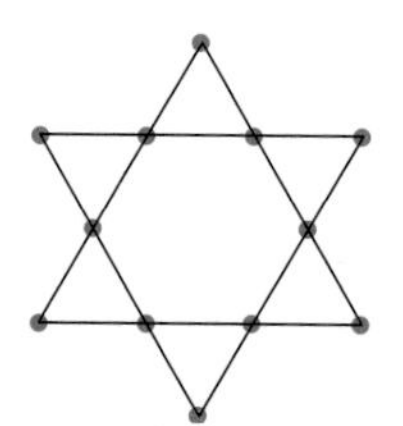

Review Exercises for Chapter 6

1. **a.** 2 **b.** 1 **c.** 3 **d.** 11 **e.** 8 **f.** 5
2. **a.** 10 **b.** 8 **c.** 8 **d.** 2 **e.** 8 **f.** 11
3. **a.** 12 **b.** 8 **c.** 6 **d.** 3 **e.** 6 **f.** 12
4. **a.** 2 **b.** 12 **c.** 6 **d.** 2 **e.** 8 **f.** 4

5. A *mathematical system* is a set of elements together with one or more operations (rules) for combining those elements.

6. **a.** True **b.** False **c.** True **d.** False
7. **a.** True **b.** False **c.** True **d.** False
8. **a.** True **b.** True
9. **a.** False **b.** False
10. **a.** Winter **b.** Summer **c.** Fall **d.** Fall **e.** Fall **f.** Fall
11. **a.** Spring **b.** Summer **c.** Spring **d.** Fall **e.** Winter **f.** Summer
12. **a.** 2 **b.** 1 **c.** 4 **d.** 3 **e.** 2 **f.** 3
13. **a.** 3 **b.** 5 **c.** 3 **d.** 5 **e.** 1 **f.** 1
14. 57
15. The odometer recycles after 100,000 miles; modulo 100,000
16. **a.** $ **b.** ? **c.** π **d.** π **e.** & **f.** $ **g.** ? **h.** π
i. No; there is a new element in the table—namely, π.
j. ¢ **k.** $, ¢ **l.** No

17. An axiomatic system consists of four main parts:
1. Undefined terms
2. Defined terms (definitions)
3. Axioms
4. Theorems

18. We must have three squirrels (axiom I). They must be in a tree (axiom II). If they are all in the same tree, then there must be another squirrel in another tree (axiom IV and II). Therefore, there are at least four squirrels.
If only two of the three squirrels are in a given tree (axioms II and III), then the third squirrel is in another tree (axiom II) together with a fourth squirrel (axiom III).

19.
$$3x + 4 + 2x + 8 + 5x - 4 = 48$$
$$10x + 8 = 48$$
$$10x = 40$$
$$x = 4$$
If $x = 4$, $3x + 4 = 12 + 4 = 16$
$$2x + 8 = 8 + 8 = 16$$
$$5x - 4 = 20 - 4 = 16$$
Therefore, all three sides are equal.

20.
$$5x - 5 + 2x + 9 + 4x - 17 + 3x + 11 = 180$$
$$14x - 2 = 180$$
$$14x = 182$$
$$x = 13$$
If $x = 13$, $m\angle DOC = 35$, and $m\angle COB = 35$, and $\overline{OC}$ bisects $\angle BOD$.

Chapter 6 Quiz

1. True	**2.** True	**3.** True	**4.** False
5. True	**6.** False	**7.** False	**8.** False
9. False	**10.** False	**11.** False	**12.** True
13. True	**14.** True	**15.** True	**16.** 5
17. 8	**18.** 6	**19.** 7	**20.** 8
21. d	**22.** c	**23.** No	**24.** None

25. No, does not satisfy the properties of a group. For example, the system does not contain an identity element.

Just for Fun *The error occurs when we divide by $(a - b)$. If $a = b$, then $(a - b) = 0$, and we cannot divide by zero.*

Chapter 7

Section 7.2

1. VIII **3.** XVI **5.** XXI **7.** CCLXXVI

9. CDLXXIV **11.** MCMLXXXVIII

13. 16	**15.** 151	**17.** 404	**19.** 1512
21. 754	**23.** 1204	**25.** 10,000	**27.** 21
29. 48	**31.** 100	**33.** 1001	**35.** 1100
37. 1500	**39.** 2550	**41.** 13	**43.** 2
45. 20	**47.** 14	**49.** 65	**51.** 1950
53. 1214	**55.** 106,816		

57.

3 + 6 = 9

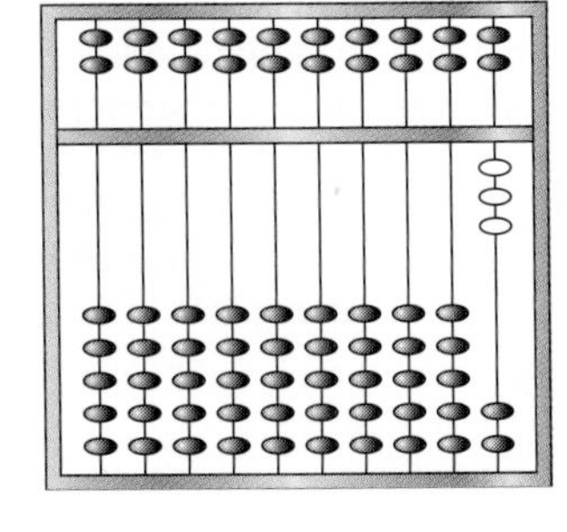

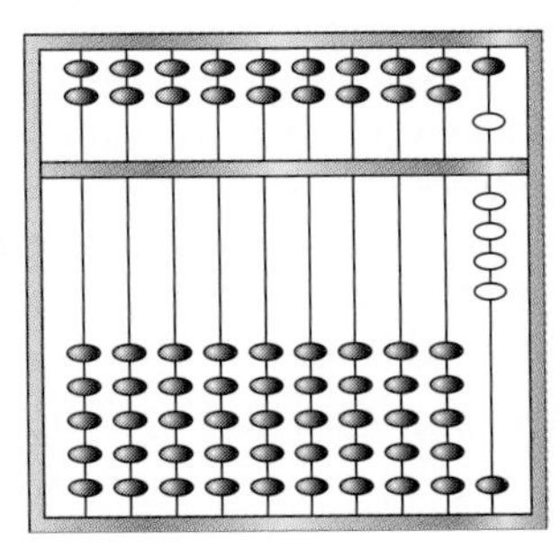

59.

1 + 4 = 5

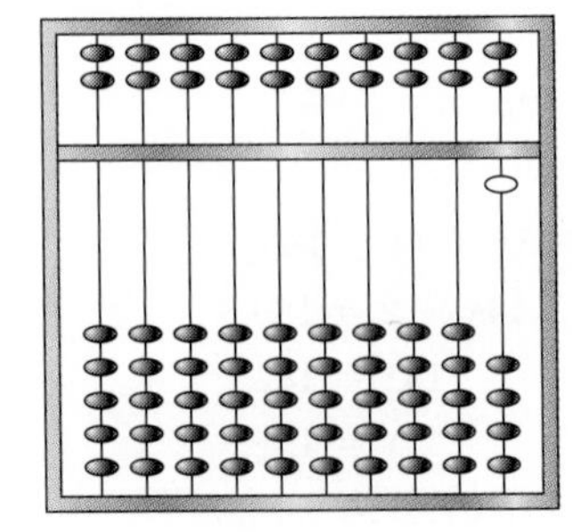

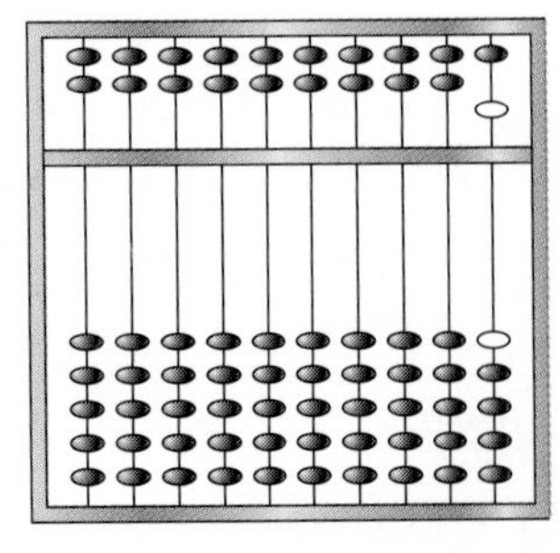

61.

4 + 6 = 10

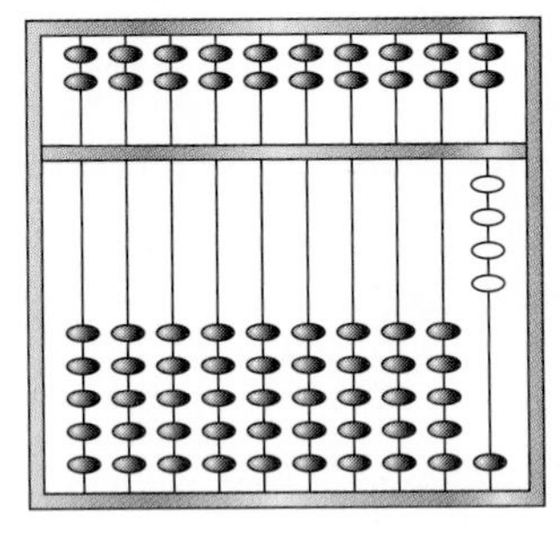

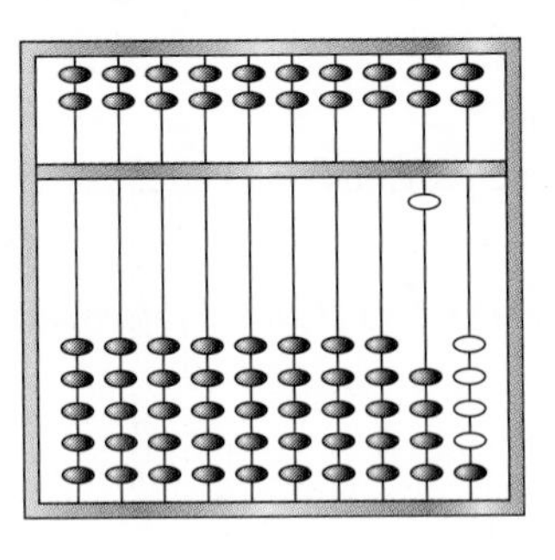

63.

8 + 7 = 15

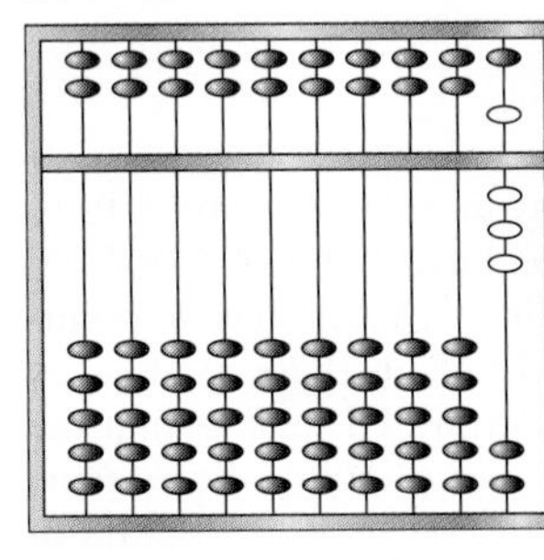

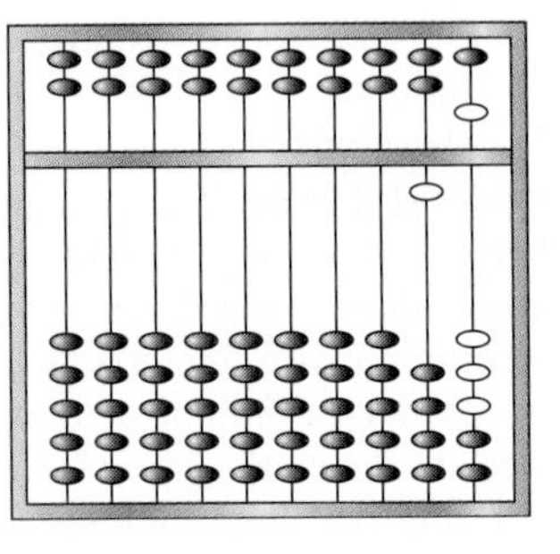

65.

5 + 9 = 14

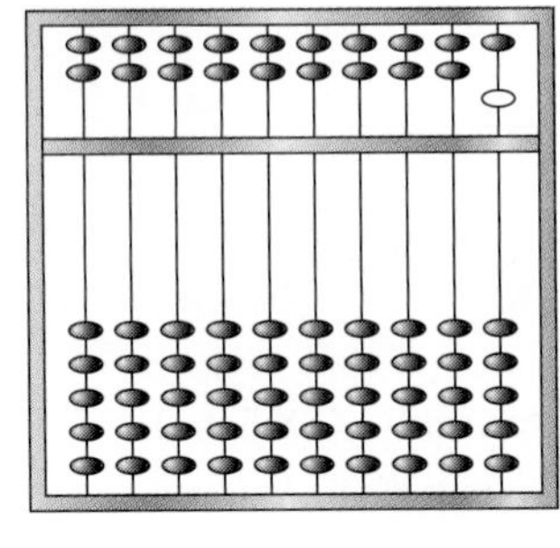

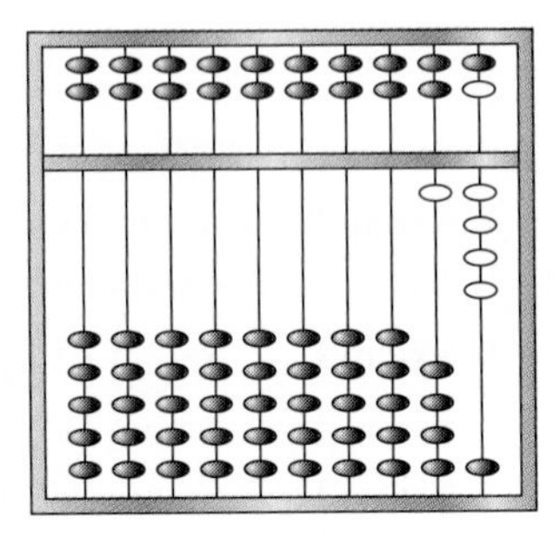

67. Index = 8; Columns = 9, 7; $8 \times 97 = 776$

69. Index = 4; Columns = 9, 7, 5; $4 \times 975 = 3900$

71. The Romans recognized that a symbol did not have to look like the number it represents. Another improvement was compactness. For example, to represent the number twenty-three in the tally system requires 23 "hash" or tally marks. In the Roman system, all we need are five characters, namely, XXIII.

73. It was not a true place-value system because the numerical value of a letter is not dependent on the position it occupies within the numeral. For example, in XI and IX the letter X still represents ten.

75. If we express the numbers 8 and 7 using tally marks, note that these marks can be partitioned into groups of five: 8 = ||||||||; and 7 = |||||||. Now, when we add these tally marks, we get a multiple of five:

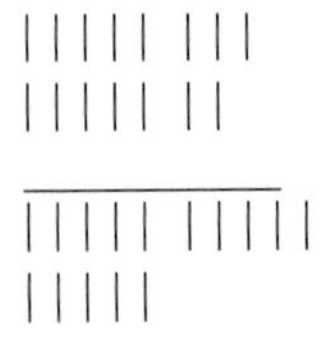

This phenomenon is true for all such cases because the "remainders," that is, the extra group of tally marks, will always add to five when the two numbers are added.

77.

8 − 2 = 6

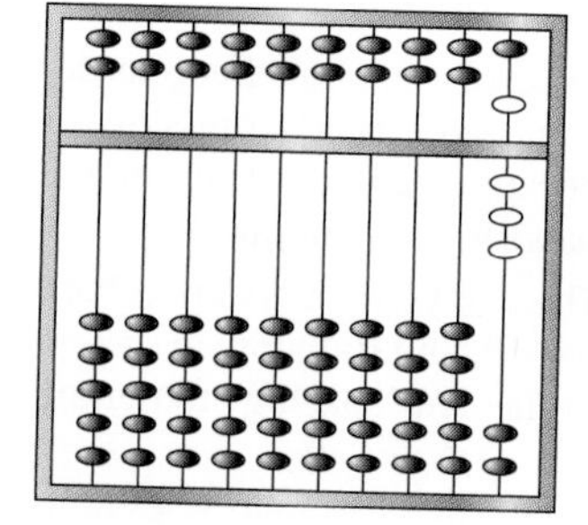
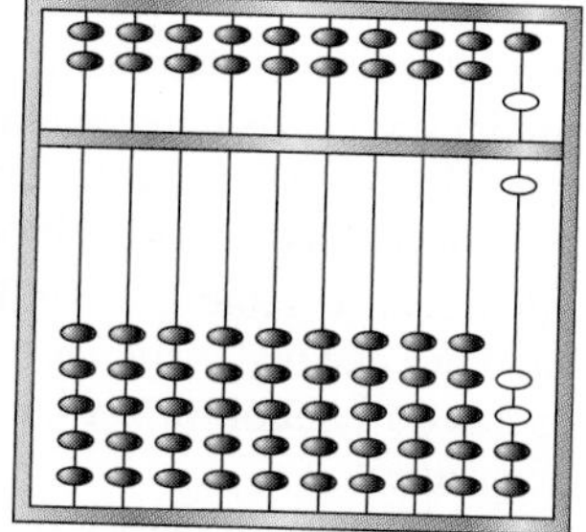

79.

12 − 6 = 6

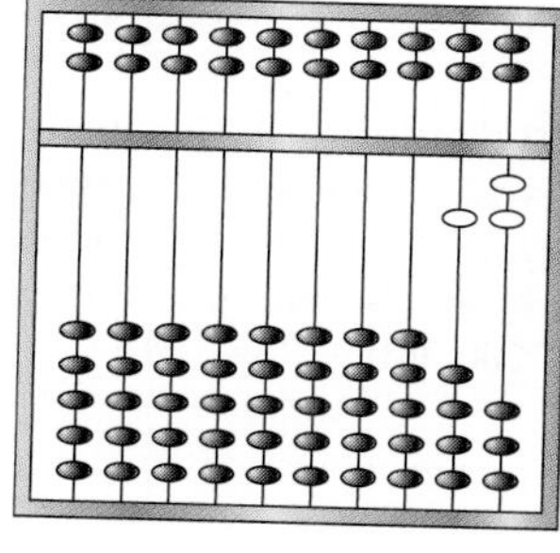
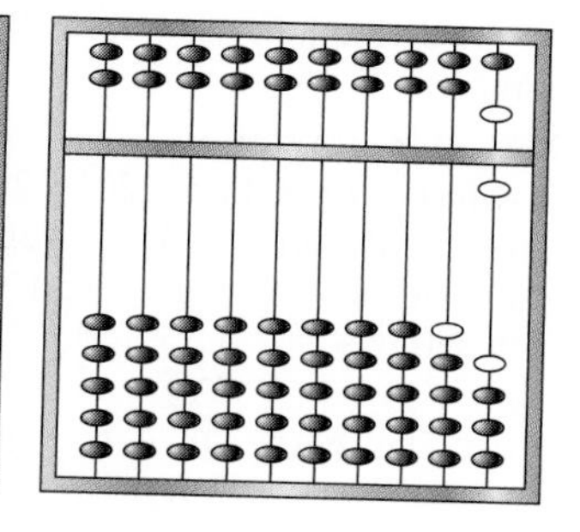

Just for Fun

1. 451 **2.** 7 **3.** 40 **4.** 500 **5.** 40
6. 4 **7.** 10 **8.** 100 **9.** 20,000 **10.** 600

Section 7.3

1. **a.** ∩||||||||| **b.** ∩∩|||
c. ∩∩∩|||| **d.** 𓍢||
e. 𓍢𓍢| **f.** 𓆼𓍢∩∩∩∩||

3. **a.** ∩|||| **b.** ∩∩|
c. 𓍢∩∩∩||
d. 𓆼𓍢𓍢𓍢𓍢∩∩∩∩|||
𓍢𓍢𓍢 ∩∩∩ |||
e. 𓆼𓍢𓍢𓍢𓍢𓍢∩∩∩∩∩||
𓍢𓍢𓍢𓍢 ∩∩∩∩∩||
f. 𓂭|

5. **a.** 22 **b.** 1213 **c.** 1101
d. 12,212 **e.** 1222 **f.** 2212

7. **a.** ∩∩||| **b.** 𓍢∩∩∩∩
∩∩∩∩
c. ∩ **d.** 𓍢

9. **a.** ∩| **b.** 𓍢∩∩∩∩|||||
∩∩ ||||
c. 𓍢∩ **d.** 𓆼𓆼𓆼

11. **a.** |||||| **b.** |||
c. ∩∩∩∩|||||
|||| **d.** 𓍢𓍢𓍢𓍢∩∩∩|||
𓍢𓍢𓍢𓍢∩∩ ||

13. **a.** ΔΓIII **b.** ΔΔIII
c. ΔΔΔIIII **d.** ΔΔΔΔIIII
e. H𐅄ΔΔΔΔΓII **f.** 𐅅𐅄ΔΔΔΔΓIII

15. **a.** ΔΔI **b.** 𐅄ΓII
c. HΔΔΔΓII **d.** X𐅅HH𐅄ΔΔΓI
e. X𐅅HHHH𐅄ΔΔΔIIII **f.** XXI

17. **a.** 17 **b.** 117 **c.** 11,010
d. 656 **e.** 6601 **f.** 5555

19. **a.** 二十一 **b.** 五十七 **c.** 百三十七
d. 千七百七十六 **e.** 千九百八十四 **f.** 二千一

21. **a.** 十六 **b.** 五十四 **c.** 百四十七
d. 八百九十七 **e.** 三千四百七十三 **f.** 四千百七十六

23. **a.** 527 **b.** 315 **c.** 1110 **d.** 8236

25. In a simple grouping system, a symbol's position within a numeral does not affect its numerical value. A simple grouping system also uses different symbols to represent a certain number or group of items. For example, in the Egyptian system, the number 12 is represented as ∩‖, ‖∩, and |∩|. It doesn't matter in what position these symbols are located. They always maintain their numerical values. In a multiplicative grouping system, two symbols are used in tandem. The first represents a basic group and the second is used to denote multiples of this basic group. For example, in the Greek system, five is the basic group and its symbol Γ is used with other Greek symbols to represent multiples of five.

27. Both systems are based on multiples of five, except the Romans used a symbol to represent 50 and 500 whereas the Greeks relied on multiplicative grouping. Also, if we ignore 5 in the Greek system and 5, 50, and 500 in the Roman system, then both systems also involved powers of 10, which is the foundation of our decimal system. Finally, the Greek system used six basic symbols and the Roman system used seven basic symbols. In each case, these symbols did not look like the numbers they represented.

29. The Chinese–Japanese system is more similar to our decimal system than the Greek system because it uses powers of 10 (i.e., 10, 100, 1000) as the basis for its grouping.

31. 1/249

Just for Fun

$$\begin{array}{r} 9567 \\ +\ 1085 \\ \hline 10{,}652 \end{array}$$

Section 7.4

1. **a.** **b.** **c.** **d.**

e.

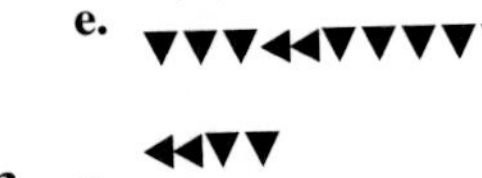

f.

3. **a.** **b.**

c.

d.

e. **f.**

5. **a.** 91 **b.** 142 **c.** 71 **d.** 81

7. **a.** $(2 \times 10^2) + (4 \times 10^1) + (3 \times 10^0)$
b. $(3 \times 10^2) + (7 \times 10^1) + (8 \times 10^0)$
c. $(1 \times 10^3) + (2 \times 10^2) + (3 \times 10^1) + (4 \times 10^0)$
d. $(2 \times 10^3) + (5 \times 10^1) + (1 \times 10^0)$
e. $(1 \times 10^4) + (4 \times 10^2) + (1 \times 10^0)$

9. **a.** $(4 \times 10^2) + (2 \times 10^0)$
b. $(1 \times 10^3) + (4 \times 10^2) + (7 \times 10^1) + (6 \times 10^0)$
c. $(2 \times 10^4) + (1 \times 10^2) + (8 \times 10^1) + (2 \times 10^0)$
d. $(2 \times 10^3) + (3 \times 10^2) + (8 \times 10^1) + (1 \times 10^0)$
e. $(1 \times 10^4) + (5 \times 10^1)$

11. **a.** 240 **b.** 2311 **c.** 1776
d. 4213 **e.** 204 **f.** 40,301

13. Place value implies that the position of a symbol affects its denomination (i.e., its numerical value). Another expression for place value is "position value" because the numerical value of a symbol is based on its position within a numeral.

15. Both are place-value systems. The Babylonian system is base 60 and our decimal system is base 10. The Babylonian system also has no symbol for zero. It relies on "spacing" to denote place value.

17. Answers will vary. Sample answers are as follows: Base 60 is still used in time (e.g., hours are divided into 60 minutes and minutes are divided into 60 seconds), and geometry (e.g., angles are divided into degrees with each degree divided into 60 minutes).

19. One interpretation is straightforward, namely, $(2 \times 60) + (2 \times 1) = 120 + 2 = 122$. Another interpretation is based on the lack of a zero symbol, which can be used as a "placeholder." As a result, we can also interpret this as $(2 \times 3600) +$ "0" $+ (2 \times 1) = 7200 + 0 + 2 = 7202$.

Just for Fun *Turn the page upside down.*

Section 7.5

1. **a.** 7 years 4 months **b.** 2 years 5 months
c. 25 feet 9 inches **d.** 3 feet 11 inches
e. 7 gross 3 dozen 2 units
f. 2 gross 9 dozen 8 units

3. **a.** 10 feet 3 inches **b.** 2 feet 5 inches
c. 6 years 3 months **d.** 10 months
e. 1 great gross 2 gross 4 dozen 3 units
f. 3 great gross 9 gross 8 dozen 10 units

5. **a.** 8 **b.** 24 **c.** 66
d. 79 **e.** 25 **f.** 511

7. **a.** 11_{five} **b.** 34_{five} **c.** 123_{five}
d. 3_{five} **e.** 441_{five} **f.** 3442_{five}

9. **a.** 220_{five} **b.** 2_{five} **c.** 21_{five}
d. 143_{five} **e.** 401_{five} **f.** 13001_{five}

11. **a.** 48 **b.** 64 **c.** 117
d. 187 **e.** 339 **f.** 3457

13. **a.** 36_{twelve} **b.** 45_{twelve} **c.** 50_{twelve}
d. $E5_{twelve}$ **e.** 176_{twelve} **f.** 610_{twelve}

15. **a.** 37_{twelve} **b.** 51_{twelve} **c.** 94_{twelve}
d. 149_{twelve} **e.** 300_{twelve} **f.** 2459_{twelve}

17. **a.** 123 **b.** 142 **c.** 568
d. 1577

19. **a.** 43 **b.** 47 **c.** 22
d. 32 **e.** 19 **f.** 240

21. **a.** 37_{nine} **b.** 66_{nine} **c.** 123_{nine}
d. 160_{nine} **e.** 280_{nine} **f.** 875_{nine}

23. 1915

25. **a.** 210_{five} **b.** 111_{five} **c.** 1012_{five}
d. 200_{five} **e.** 3000_{five} **f.** 10210_{five}

27. $10_{twelve} = (1 \times 12) + (0 \times 1) = 12$

29. Because base 12 requires the numbers zero through eleven. However, we cannot use 10 and 11 to represent ten and eleven, respectively, because the symbols 0 and 1 have already been used to represent zero and one, respectively.

31. Answers will vary. One reason X is an appropriate symbol to represent ten in base 12 is because in the Roman system it also has the meaning of ten. One reason why H is an appropriate symbol to represent eleven is because it looks like 11 if you remove the horizontal bar in H.

33. Base 12 lends itself more easily to division than base 10 because there are more divisors of 12 (i.e., 2, 3, 4, 6) than there are of 10 (i.e., 2, 5).

Just for Fun *1 fathom = 6 feet; 1 hand = 4 inches; 3 barleycorns = 1 inch*

Section 7.6

1. **a.** 41_5 **b.** 41_5 **c.** 110_5
d. 302_5 **e.** 1003_5 **f.** 1010_5

3. **a.** 113_5 **b.** 140_5 **c.** 101_5
d. 311_5 **e.** 1131_5 **f.** 3230_5

5. **a.** 4_5 **b.** 3_5 **c.** 3_5
d. 102_5 **e.** 2424_5 **f.** 424_5

7. **a.** 1243_5 **b.** 1414_5 **c.** 3333_5
d. 10401_5 **e.** 14112_5 **f.** 32343_5

9. **a.** 1234_5 **b.** 424_5 **c.** 4301_5
d. 10111_5 **e.** 22142_5 **f.** 34221_5

11. **a.** 34_5, $R = 1_5$ **b.** 112_5, $R = 2_5$
c. 44_5, $R = 1_5$ **d.** 224_5, $R = 2_5$
e. 22_5, $R = 1_5$ **f.** 22_5

13. In $23_5 + 14_5$ we first add $3_5 + 4_5 = 7_5$. However, because $7 > 5$, we must express 7 in base 5, which is 12_5. Thus, we write 2 and carry one 5. We next add $2_5 + 1_5 + 1_5 = 4_5$. Therefore, the answer is 42_5.

15. In $4_5 \times 3_5$, we multiply 4 by 3 and get 12. However, because $12 > 5$, we must express 12 in base 5. This is 22_5, which means $(2 \times 5) + (2 \times 1) = 12$.

17. **a.** 101_6 **b.** 1060_7 **c.** 15_7 **d.** 155_6

19. **a.** 2036_7 **b.** 2000_6 **c.** 1322_8 **d.** 2146_9

21. **a.** 20211_3 **b.** 22452_6 **c.** 20411_5 **d.** 22312_7

Just for Fun *for, four, fore*

Section 7.7

1. **a.** 2 **b.** 3 **c.** 5 **d.** 7 **e.** 11 **f.** 27

3. **a.** 8 **b.** 9 **c.** 12 **d.** 15 **e.** 16 **f.** 17

5. **a.** 110_2 **b.** 1001_2 **c.** 1101_2
d. 10101_2 **e.** 11001_2 **f.** 100001_2

7. **a.** 101_2 **b.** 110_2 **c.** 100_2
d. 1000_2 **e.** 1001_2 **f.** 1100_2

9. **a.** 1101_2 **b.** 10000_2 **c.** 10101_2
d. 11100_2 **e.** 10100_2 **f.** 11110_2

11. **a.** 10_2 **b.** 101_2 **c.** 10_2
d. 101_2 **e.** 110_2 **f.** 11_2

13. **a.** 1001_2 **b.** 110_2 **c.** 1111_2
d. 10010_2 **e.** 100011_2 **f.** 11001_2

15. **a.** 10101_2 **b.** 1000_2 **c.** 1100_2
d. 10100_2 **e.** 111100_2 **f.** 101101_2

17. **a.** 42 **b.** 60 **c.** 213
d. 171 **e.** 731 **f.** 1679

19. **a.** 34_{16} **b.** $2F_{16}$ **c.** 63_{16}
d. $6B_{16}$ **e.** $A9_{16}$ **f.** CA_{16}

21. **a.** $7D9_{16}$ **b.** $270F_{16}$ **c.** $E71_{16}$
d. $7FD_{16}$ **e.** FED_{16} **f.** $EA6_{16}$

23. **a.** 80_{16} **b.** $D1_{16}$ **c.** 101_{16}
d. $F2_{16}$ **e.** $F7_{16}$ **f.** 163_{16}

25. **a.** $30B_{16}$ **b.** 2973_{16} **c.** $2CDC_{16}$
d. $286E_{16}$ **e.** 4898_{16} **f.** $368D_{16}$

27. **a.** $5B6DA_{16}$ **b.** $5D2B2_{16}$ **c.** $9A0CE_{16}$
d. $C9B406_{16}$ **e.** $8BD496_{16}$ **f.** $D8F1AC_{16}$

29. The base 2 number system has considerable pragmatic applications, especially in the computer world, and hence merits study.

31. Since our base 10 number system only contains the ten symbols 0 through 9, we need to define six new (and unique) symbols for the hexadecimal system.

33. Answers will vary.

35. Answers will vary. One representation is Sunday $= 000_2$, Monday $= 001_2$, Tuesday $= 010_2$, Wednesday $= 011_2$, Thursday $= 100_2$, Friday $= 101_2$, Saturday $= 110_2$.

37. **a.** 1010_2, $R = 1_2$ **b.** 1001_2 **c.** 100010_2, $R = 11_2$

Just for Fun *Let the Yankee fans be Y_1, Y_2, Y_3, and let the Mets fans be M_1, M_2, M_3.*

1. M_1, M_2 go up, leaving Y_1, Y_2, Y_3, M_3.
2. Elevator returns to 1st floor empty.
3. Y_1, Y_2 go up, leaving Y_3, M_3 and joining M_1, M_2.
4. Elevator returns empty.
5. Y_3 and M_3 go up, joining Y_1, Y_2, M_1, M_2.

Alternate solution:

1. M_1, Y_1 go up, leaving M_2, M_3, Y_2, Y_3.
2. Elevator returns empty to first floor.
3. M_2, Y_2 go up, leaving M_3, Y_3 and joining M_1, Y_1.
4. Elevator returns empty.
5. M_3, Y_3 go up, joining M_1, M_2, Y_1, Y_2.

Review Exercises for Chapter 7

1. **a.** LXXVIII **b.** CXXIV **c.** CMXX **d.** MMVI

2. **a.** 519 **b.** 650 **c.** 1400 **d.** 414

3. **a.** 282 **b.** 210 **c.** 143 **d.** 23

4. 753,205

5. $6 + 8 = 14$

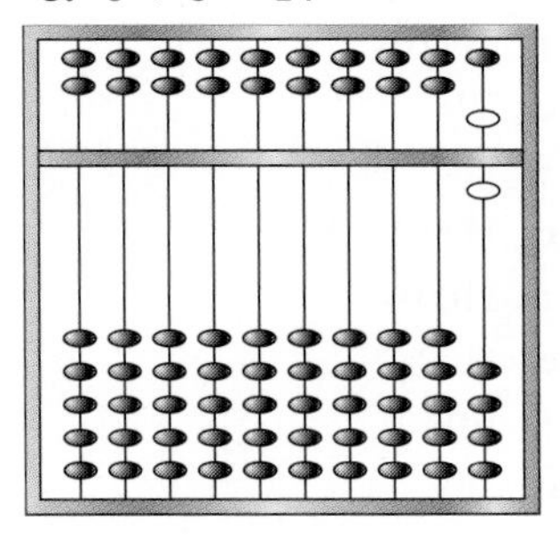

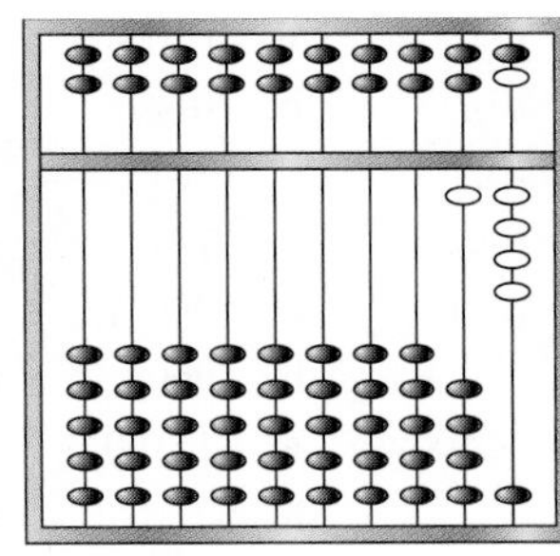

6. $6 \times 29 = 174$

7. **a.** ∩∩∩∩
∩∩∩|||||||||

b. 𐅄ΔΔΓΙΙΙ

c. 七
十
九

8. **a.** ∩∩∩|| **b.** 𓍢𓍢∩| **c.** 𓆼𓍢∩|

9. **a.** 1233 **b.** 12,022

10. **a.** ∩∩∩|||||
∩∩

b. 𓍢𓍢∩∩∩
∩∩

c. ∩∩∩∩|||||
∩∩∩∩||||

d. 𓍢𓍢𓍢𓍢𓍢𓍢𓍢𓍢∩∩∩∩
∩∩∩∩

11. **a.** *Simple grouping:* The position of a symbol does not affect the number represented, and a new symbol is used to indicate a certain number or group of things.

b. *Multiplicative grouping:* Symbols are used for numbers in a basic group, together with a second symbol or notation to represent multiples of the basic group.

12. **a.** Egyptian
b. Greek, Chinese–Japanese

13. ▼◀ ▼▼▼▼
▼▼▼▼

14. *Place value:* The position of a symbol matters.

15. Babylonian, Hindu–Arabic

16. **a.** $(3 \times 10^2) + (4 \times 10^1) + (5 \times 10^0)$
b. $(3 \times 5^2) + (4 \times 5^1) + (2 \times 5^0)$
c. $(1 \times 2^4) + (0 \times 2^3) + (1 \times 2^2) + (1 \times 2^1) + (1 \times 2^0)$
d. $E4_{16} = (E \times 16) + (4 \times 1)$
$= (14 \times 16) + (4 \times 1)$

17. **a.** 56 **b.** 24 **c.** 72
d. 31 **e.** 23 **f.** 707
g. $D5A_{16} = 3418$ **h.** $FFFF_{16} = 65{,}535$

18. **a.** 131_5 **b.** 203_5 **c.** 27_{12}
d. 100011_2 **e.** 111110_2 **f.** 30_{12}
g. $194 = C2_{16}$

19. Base 12 is used when buying many items; for example, we use dozen (12^1), gross (12^2), and great gross (12^3). Modern computers work in base 2, 8, or 16.

20. **a.** 222_6 **b.** 211_3 **c.** 43_5 **d.** 111111_2
e. 1011111_2 **f.** $D9_{16}$

21. **a.** $12_5, 111_2$ **b.** $23_5, 1101_2$ **c.** $113_5, 100001_2$,

22. **a.** 53 **b.** 25 **c.** 53
d. 326 **e.** 121 **f.** 243

23. **a.** 36_9 **b.** 64_7 **c.** 1001_4
d. 10102_3 **e.** 203_6 **f.** 207_8

24. **a.** 1200_6 **b.** 1220_4 **c.** 52_7 **d.** 215_9

25. **a.** 22112_6 **b.** 10011_3 **c.** 16123_7 **d.** 25503_9

26. **a.** 100_5 **b.** 1101_5 **c.** 14_5 **d.** 144_5

27. **a.** 1243_5 **b.** 1102_5 **c.** 20004_5 **d.** $34_5, R = 1_5$

28. **a.** 1011_2 **b.** 1010_2 **c.** 11010_2 **d.** 11_2

29. **a.** 100_2 **b.** 110_2 **c.** 1110_2 **d.** 110111_2

30. **a.** 87_{16} **b.** $1D5_{16}$ **c.** $1A46_{16}$ **d.** $8C70_{16}$

Just for Fun *3 inches*

Chapter 7 Quiz

1. True **2.** False **3.** True
4. True **5.** False **6.** True
7. True **8.** True **9.** False
10. False **11.** 771_{10} **12.** 36_{10}
13. 1705_{10} **14.** 241_5 **15.** 1022_5
16. 2202_3 **17.** 1101111_2 **18.** 1000011_2
19. 10111010_2 **20.** 136_{16} **21.** $7DB_{16}$
22. 1222_5 **23.** 3212_5 **24.** 10010_2
25. 10_2 **26.** 10101_2 **27.** 432_7
28. 32300_6 **29.** 100_{12} **30.** $C83_{16}$
31. 137_{16}

Chapter 8

Section 8.2

1. **a.** Cardinal **b.** Ordinal
c. Nominal **d.** Nominal

3. **a.** Prime **b.** Prime **c.** Neither
d. Composite **e.** Composite **f.** Composite

5. **a.** 2688 is divisible by 2, 3, 4, 8
b. 73,440 is divisible by 2, 3, 4, 5, 8, 9, 10
c. 3,290,154 is divisible by 2, 3

7. 2, 3, 5, 7, 11, 13, 17, 19, 23, 29

9. **a.** $2^2 \times 3^2$ **b.** $2^3 \times 3^2$ **c.** $2^3 \times 3^3$
d. $5^2 \times 19$ **e.** 5^4 **f.** 3×7^2

11. For example, 11 and 13; 17 and 19; 41 and 43; 59 and 61; 71 and 73

13. To say that 3 is a factor of 15 means that 3 divides into 15 without a remainder (i.e., the remainder is zero).

15. One method for determining if a number is prime is to divide the number using prime number divisors. We continue this division process until either we get a zero remainder, which implies that the number is not prime, or until the square of the divisor is greater than the number being tested. For example, to determine if 37 is prime, divide it by 2, 3, 5, and 7. We do not continue dividing 37 by prime numbers greater than 7 (e.g., 11, 13, 17, 19, ...) because $7^2 = 49$, which is greater than 37. Since we did not get a zero remainder in any of the divisions, we conclude that 37 is prime.

17. The fundamental theorem of arithmetic states that every natural number greater than or equal to 2 is either a prime number or can be uniquely expressed as a product of prime numbers.

19. Seven: 1951, 1973, 1979, 1987, 1993, 1997, 1999.

21. The remainders have the pattern 0, 2, 0, 2, 0, 2, ... The general statement is, "Adding 1 to the set of prime numbers greater than or equal to 5 but less than or equal to 19 and then dividing this sum by 6 yields a remainder that consists of the pattern 0, 2, 0, 2, 0, 2, ... "

23. 15

Just for Fun *Let x, y, and z be any natural number. This implies that our number is xyzxyz, which expressed in expanded notation is* $(x \times 10^5) + (y \times 10^4) + (z \times 10^3) + (x \times 10^2) + (y \times 10^1) + (z \times 10^0)$. *Rearranging the terms we get* $(x \times 10^5) + (x \times 10^2) + (y \times 10^4) + (y \times 10^1) + (z \times 10^3) + (z \times 10^0)$. *Simplifying we get* $100100x + 10010y + 1001z$. *Notice that the coefficient of each term is divisible by* 13. *In other words,* 13 *is a divisor of* 100100, 10010, *and* 1001. *In fact, we can factor* 1001 *from each term leaving* $1001(100x + 10y + z)$. *Since 13 is a factor of this product, then 13 must be a factor of any natural number formed as xyzxyz.*

27. A common factor is a number that divides evenly (i.e., zero remainder) into two or more numbers. For example, 2 is a common divisor of 4 and 12. The greatest common divisor is the largest divisor that two or more numbers have in common. For example, of the common divisors of 4 and 12, namely, 1, 2, and 4, the greatest common divisor is 4.

29. One method for finding the GCD between two or more numbers is to list the divisors of each number and then identify the largest they have in common. One method for finding the LCM is to continually list the multiples of each number until you find the first common multiple. Thus, the primary difference between the two methods is we use *factors* for finding the GCD and *multiples* for finding the LCM. Furthermore, the GCD involves finding the *largest* common factor whereas the LCM involves finding the *smallest* common multiple.

31. We first find the least common multiple between the denominators, 9 and 12. This is 36. We then rewrite $\frac{1}{9}$ and $\frac{5}{12}$ as equivalent fractions that have a denominator of 36, $\frac{1}{9} = \frac{4}{36}$ and $\frac{5}{12} = \frac{15}{36}$. We now add the fractions by adding their numerators and placing the sum over the common denominator. Thus, $\frac{1}{9} + \frac{5}{12} = \frac{4}{36} + \frac{15}{36} = \frac{19}{36}$.

33. The GCD of 120 and 460 is 20.

Euclidean Algorithm

q	r
—	460
3	120
1	100
5	20
	0

Repeated Subtraction:
- $460 - 120 = 340 - 120 = 220 - 120 = 100$
- $120 - 100 = 20$
- $100 - 20 = 80 - 20 = 60 - 20 = 40 - 20 = 20 - 20 = 0$

35. The GCD of 472 and 564 = 4.

Euclidean Algorithm

q	r
—	564
1	472
5	92
7	12
1	8
2	4
	0

Repeated Subtraction:
- $564 - 472 = 92$
- $472 - 92 = 380 - 92 = 288 - 92 = 196 - 92 = 104 - 92 = 12$
- $92 - 12 = 80 - 12 = 68 - 12 = 56 - 12 = 44 - 12 = 32 - 12 = 20 - 12 = 8$
- $12 - 8 = 4$
- $8 - 4 = 4 - 4 = 0$

37. The GCD of 3280 and 5425 = 5.

Euclidean Algorithm

q	r
–	5425
1	3280
1	2145
1	1135
1	1010
8	125
12	10
2	5
	0

Repeated Subtraction:
- $5425 - 3280 = 2145$
- $3280 - 2145 = 1135$
- $2145 - 1135 = 1010$
- $1135 - 1010 = 125$
- $1010 - 125 = 885 - 125 = 760 - 125 = 635 - 125 = 510 - 125 = 385 - 125 = 260 - 125 = 135 - 125 = 10$
- $125 - 10 = 115 - 10 = 105 - 10 = 95 - 10 = 85 - 10 = 75 - 10 = 65 - 10 = 55 - 10 = 45 - 10 = 35 - 10 = 25 - 10 = 15 - 10 = 5$
- $10 - 5 = 5 - 5 = 0$

Section 8.3

1.	**a.** 2	**b.** 14	**c.** 3
	d. 26	**e.** 3	**f.** 12
3.	**a.** 4	**b.** 2	**c.** 1
	d. 1	**e.** 9	**f.** 6
5.	**a.** 5/6	**b.** 7/9	**c.** 3/5
	d. 2/3	**e.** 147/152	**f.** 1/2
7.	**a.** 1/3	**b.** 1/4	**c.** 1/6
	d. 1/19	**e.** 1/3	**f.** 1/10
9.	**a.** 56	**b.** 28	**c.** 120
	d. 156	**e.** 990	**f.** 9879
11.	**a.** 24	**b.** 30	**c.** 45
	d. 40	**e.** 42	**f.** 105
13.	**a.** 35/36	**b.** 23/36	**c.** 11/12
	d. 19/24	**e.** 6/55	**f.** 43/60
15.	**a.** 61/77	**b.** 3/4	**c.** 73/120
	d. 7/55	**e.** 19/72	**f.** 13/45

17. 18 inches

19. 3-foot square tiles

21. There are six ways in which this can be done. He can have groups of size 1, 2, 3, 4, 6, and 12.

23. 15 wood blocks

25. 48 minutes; Mark will make 4 round trips and Andrew will make 3 round trips.

Just for Fun

$4 = 2 + 2; 6 = 3 + 3; 8 = 5 + 3;$
$10 = 5 + 5; 12 = 7 + 5; 14 = 7 + 7;$
$16 = 13 + 3; 18 = 13 + 5; 20 = 13 + 7;$
$22 = 11 + 11; 24 = 19 + 5; 26 = 19 + 7;$
$28 = 23 + 5; 30 = 23 + 7$

Section 8.4

1. **a.** -1 **b.** -3 **c.** -3
d. 2 **e.** 1 **f.** 3

3. **a.** -2 **b.** -3 **c.** -14
d. 11 **e.** -1 **f.** -4

5. **a.** -5 **b.** 0 **c.** 25
d. 0 **e.** -35 **f.** 9

7. **a.** 20 **b.** -24 **c.** -24
d. 16 **e.** 40 **f.** 126

9. **a.** 1 **b.** 7 **c.** 5
d. 20 **e.** 30 **f.** -189

11. **a.** 15 **b.** 19 **c.** 10
d. 13 **e.** 18 **f.** 21

13. **a.** -60 **b.** -20 **c.** -27
d. -28 **e.** 7 **f.** -24

15. **a.** $>$ **b.** $<$ **c.** $=$
d. $<$ **e.** $=$ **f.** $>$

17. **a.** Even: $12 = 2 \cdot 6$ **b.** Odd: $17 = 2 \cdot 8 + 1$
c. Even: $-6 = 2(-3)$ **d.** Odd: $101 = 2 \cdot 50 + 1$

19. 10,833 **21.** Aristotle

23. $(2k + 1) + 2n = (2k + 2n) + 1 = 2(k + n) + 1$, which is odd.

25. The set of natural numbers is the set of counting numbers $\{1, 2, 3, \dots\}$. The set of whole numbers is the same as the set of natural numbers except it includes zero $\{0, 1, 2, 3, \dots\}$. The set of integers can be thought of as the set of positive and negative natural numbers with zero that is $\{\dots, -3, -2, -1, 0, 1, 2, 3, \dots\}$.

27. $4 - (-1) = 4 + (+1) = 4 + 1$. Subtracting (or taking away) the opposite of 1 is the same as adding 1.

29. The general rule for dividing integers is to divide without regard to sign. If the sign of the numbers is the same (i.e., both positive or both negative), then the sign of the quotient is positive; otherwise, the sign of the quotient is negative.

31. The distributive property for multiplication over addition says that if we are multiplying a sum of numbers by a factor, then we can distribute this factor to each of the numbers being added. That is $a(b + c) = ab + ac$. For example, $3(5 + 8) = 3(5) + 3(8) = 15 + 24 = 39$.

33.
$2 \times 3 = 6$
$2 \times 2 = 4$
$2 \times 1 = 2$
$2 \times 0 = 0$
$2 \times -1 = -2$
$2 \times -2 = -4$

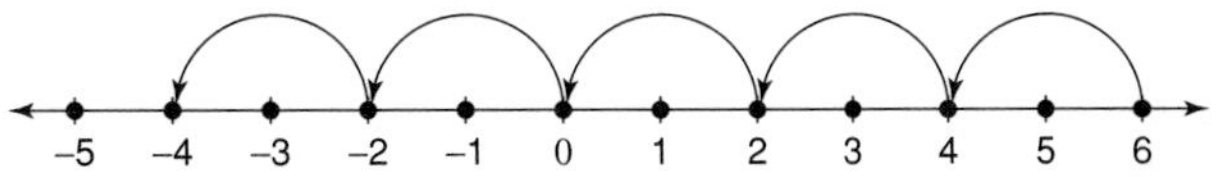

−5 −4 −3 −2 −1 0 1 2 3 4 5 6

35. 9 **37.** 108 **39.** 9 **41.** 1 **43.** -9

45. Research **47.** **a.** 10° **b.** 18° **c.** The "greater" danger occurs at 10° F with a wind speed of 40 mph.

49. "The bigger the negative number . . . " is an incorrect phrase. We may correct it by stating, "The *smaller* the negative number."

Just for Fun *It's a fake! No authentic coins would be dated* B.C.

Section 8.5

1. **a.** $\frac{3}{8}$ **b.** $\frac{1}{9}$ **c.** $\frac{27}{43}$ **d.** $\frac{27}{224}$

3. **a.** $\frac{33}{35}$ **b.** $\frac{11}{12}$ **c.** $\frac{17}{72}$ **d.** $\frac{5}{33}$ **e.** $\frac{1}{80}$ **f.** $\frac{5}{9}$

5. **a.** $\frac{8}{35}$ **b.** $\frac{12}{55}$ **c.** $\frac{32}{91}$ **d.** $\frac{14}{9}$ **e.** $\frac{27}{44}$ **f.** $\frac{14}{13}$

7. **a.** $\frac{27}{44}$ **b.** $\frac{14}{65}$ **c.** $\frac{2}{15}$ **d.** $\frac{9}{10}$ **e.** $\frac{9}{10}$ **f.** $\frac{50}{63}$

9. **a.** $\frac{15}{8}$ **b.** $\frac{40}{3}$ **c.** $\frac{4}{15}$ **d.** $\frac{3}{10}$ **e.** $\frac{44}{75}$ **f.** $\frac{6}{11}$

11. **a.** $\frac{17}{32}$ **b.** $\frac{17}{12}$ **c.** $\frac{7}{24}$ **d.** $\frac{27}{12}$

13. **a.** False **b.** True **c.** False
d. False **e.** False **f.** True

15. **a.** False **b.** True **c.** False
d. True **e.** False

17. **a.** -2 **b.** $10\frac{1}{3}$ **c.** $-2\frac{11}{15}$ **d.** $-3\frac{53}{63}$
e. $\frac{19}{32}$ **f.** -6 **g.** $4\frac{2}{5}$ **h.** $3\frac{1}{2}$
i. -10 **j.** $-\frac{1}{3}$ **k.** $\frac{10}{17}$ **l.** $2\frac{6}{25}$

19. $8\frac{11}{12}$ pounds

21. 35 minutes

23. $18,000

25. $1\frac{1}{6}$ hours (or about 70 minutes)

27. 87 miles

29. One way is to cross multiply. If the cross products are equal, then the fractions are equivalent. Thus, $\frac{3}{8}$ and $\frac{5}{7}$ are not equivalent because their cross products, namely, $3 \times 7 = 21$ and $5 \times 8 = 40$, are not equal.

31. A complex fraction is any fraction that contains a fraction as its numerator, denominator, or both numerator and denominator. An example is $\frac{4}{\frac{2}{3}}$.

33. $A = -1, B = \frac{2}{5}, C = \frac{6}{5}$

35. $A = -\frac{1}{3}, B = \frac{2}{15}, C = \frac{2}{5}$ **37.** Research **39.** $\frac{7}{50}$

Just for Fun $\frac{1}{2}, -\frac{1}{2}$

Section 8.6

1. **a.** 0.375 **b.** 0.3125 **c.** $0.\overline{6}$
d. $0.\overline{21}$ **e.** $0.\overline{09}$ **f.** $0.\overline{405}$

3. **a.** 0.875 **b.** $0.\overline{4}$ **c.** $0.\overline{81}$
d. 0.0625 **e.** $0.\overline{714285}$ **f.** $0.\overline{4705882352941176}$

5. **a.** 1/8 **b.** 1/400 **c.** 7/9
d. 34/99 **e.** 691/110 **f.** 1

7. **a.** $\frac{3}{4}$ **b.** $\frac{7}{8}$ **c.** $\frac{8}{9}$ **d.** $\frac{5}{11}$ **e.** $\frac{235}{999}$ **f.** $\frac{1379}{300}$

9. **a.** $\frac{5}{12}$ **b.** $\frac{7}{24}$ **c.** $\frac{9}{40}$ **d.** $\frac{37}{48}$ **e.** $\frac{69}{88}$ **f.** $\frac{277}{352}$

11. **a.** $\frac{5}{16}$ **b.** $\frac{9}{18}$ **c.** $\frac{1}{16}$ **d.** $\frac{19}{20}$ **e.** $\frac{61}{112}$ **f.** 1.9995

13. 0.005

15. A rational number is a fraction that represents a ratio of two integers (with one restriction: Zero cannot be in the denominator). A decimal is also a fraction but is represented as powers of 10 using a decimal point. For example, $\frac{5}{8} = 0.625$.

17. A repeating nonterminating decimal is a rational number such that when you divide the denominator into the numerator, the quotient does not yield a zero remainder (nonterminating) but has a pattern that continually repeats. For example, $\frac{1}{3} = 0.33\overline{3}$ and $\frac{4}{7} = 0.571428\overline{571428}$ are repeating, nonterminating decimals.

19. Two numbers Dr. Shapiro could have been dividing are 99 and 13501.

21. **a.** $\frac{4}{11}$ **b.** 11 marbles

Just for Fun 1135; *that is,* $11 + 3 + 5 = 19$

Section 8.7

1. **a.** Rational **b.** Rational **c.** Rational
d. Irrational **e.** Irrational **f.** Rational

3. **a.** Rational **b.** Rational **c.** Rational
d. Rational **e.** Rational **f.** Irrational

5. **a.** Terminating decimal
b. Repeating decimal
c. Nonterminating, nonrepeating decimal
d. Terminating decimal
e. Nonterminating, nonrepeating decimal
f. Repeating decimal

7. **a.** True **b.** True **c.** True
d. False **e.** False

9. **a.** False **b.** False **c.** True
d. False **e.** False **f.** True

11. The difference between a rational number and an irrational number is that an irrational number is a nonterminating, nonrepeating decimal.

13. A perfect square is a number that represents a product formed by multiplying a number by itself. For example, 100 is a perfect square because $100 = 10 \times 10$. Similarly, 64 is a perfect square because $64 = 8 \times 8$.

15. The set of whole numbers less than or equal to 2.

17. The set of real numbers greater than 0 but less than or equal to 9.

19. All real numbers greater than -3.

Just for Fun *We use* $\frac{22}{7}$ *only for sake of convenience. It is a close approximation of* π.

Section 8.8

1. 2.40×10^2 **3.** 1.25×10^5 **5.** 4.30×10^6
7. 7.392×10^2 **9.** 2.31×10^{-1} **11.** 2.345×10^{-4}
13. 1.2×10^{-3} **15.** 9×10^{-8} **17.** 4×10^9
19. 1×10^{-7} **21.** 0.0041 **23.** 314,200
25. 48,590,000 **27.** 0.000048 **29.** 8.84×10^7
31. 3.24×10^5 **33.** 2.057×10^2 **35.** 4.35
37. 3×10^3 **39.** 5.0×10^1 **41.** 1.5×10^3
43. 2.0×10^2 **45.** 10^{-2} **47.** 1×10^7
49. 48

51. Scientific notation is a method for expressing very large or very small numbers. It consists of two parts: a "base" that consists of significant digits, and a power of 10. For example, $0.00000000039 = 3.9 \times 10^{-10}$. The 3.9 is the "base" and 10^{-10} is the power of 10.

53. Place a decimal point between the 3 and 9 and count how many digits come before the decimal point. This now becomes the power of 10. Thus, $0.000000000000000000000396 = 3.96 \times 10^{-22}$.

55. The quotient rule for exponents says to divide numbers written in exponential form, if the bases are the same, then keep the common base and subtract the exponents. For example, $\frac{10^9}{10^2} = 10^{9-2} = 10^7$.

57. 0.000000006435

Just for Fun *The Englishman, Mr. Dunn*

Review Exercises for Chapter 8

1. **a.** Composite **b.** Prime **c.** Prime
d. Prime **e.** Composite **f.** Composite

2. **a.** $2 \times 3 \times 7$ **b.** 5×23 **c.** $3 \times 5^2 \times 7$
d. $2 \times 3 \times 29$ **e.** $3 \times 5 \times 47$ **f.** $2 \times 11 \times 101$

3. **a.** 6 **b.** 1 **c.** 24 **d.** 6 **e.** 3 **f.** 18

4. **a.** 252 **b.** 352 **c.** 200
d. 1800 **e.** 300 **f.** 3300

5. **a.** -4 **b.** -1 **c.** -1
d. -4 **e.** 16 **f.** -13

6. **a.** 8 **b.** -15 **c.** -6
d. -8 **e.** -4 **f.** 56

7. **a.** 8 **b.** 4 **c.** 9
d. 4 **e.** 2 **f.** 36

8. **a.** $\frac{7}{12}$ **b.** $\frac{55}{63}$ **c.** $\frac{37}{143}$
d. $\frac{5}{21}$ **e.** $\frac{1}{10}$ **f.** $\frac{19}{40}$

9. **a.** $\frac{2}{15}$ **b.** $\frac{4}{21}$ **c.** $\frac{6}{55}$
d. $\frac{7}{6}$ **e.** $\frac{6}{5}$ **f.** $\frac{10}{11}$

10. **a.** $\frac{27}{32}$ **b.** $\frac{3}{20}$ **c.** $\frac{20}{3}$
d. $\frac{25}{28}$ **e.** $\frac{3}{2}$ **f.** $\frac{1}{4}$

11. **a.** 0.875 **b.** 0.4375 **c.** $0.\overline{18}$
d. $0.\overline{351}$ **e.** $0.\overline{384615}$ **f.** $0.\overline{428571}$

12. **a.** $\frac{3}{4}$ **b.** $\frac{213}{1000}$ **c.** $\frac{157}{50}$
d. $\frac{46}{99}$ **e.** $\frac{247}{99}$ **f.** $\frac{1373}{333}$

13. **a.** Rational **b.** Rational **c.** Irrational
d. Irrational **e.** Irrational **f.** Rational

14. **a.** False **b.** False **c.** True
d. True **e.** True **f.** True

15.

	a.	b.	c.	d.	e.	f.
I:	yes	no	no	no	no	no
II:	yes	no	yes	no	no	no
III:	yes	yes	yes	no	no	no
IV:	yes	yes	yes	yes	yes	no
V:	no	no	no	no	no	yes
VI:	yes	yes	yes	yes	yes	yes

	g.	h.	i.
I:	yes	no	no
II:	yes	no	no
III:	yes	no	no
III:	yes	no	yes
IV:	no	yes	no
VI:	yes	yes	yes

16. **a.** 2.84×10^9 **b.** 5.46×10^{-8}
c. 1×10^7 **d.** 4.8×10^1

Just for Fun *The number of letters in the word equals the digit.*

Chapter 8 Quiz

1. False **2.** False **3.** False
4. False **5.** False **6.** True
7. True **8.** False **9.** True
10. True **11.** True **12.** False
13. False **14.** True **15.** True
16. $\frac{49}{40}$ **17.** $\frac{8}{63}$ **18.** $\frac{9}{10}$
19. $\frac{44}{15}$ **20.** $\frac{1}{[illegible]}$ **21.** 3
22. 144; 24 **23.** $0.\overline{142857}$ **24.** $\frac{1373}{333}$
25. Irrational **26.** Rational **27.** Rational
28. 8.84×10^{10} **29.** 1.1×10^2 **30.** 8.0×10^{17}

Chapter 9

Section 9.2

1. $b + 3$ **3.** $\frac{x - y}{a}$ **5.** $\frac{a - b}{a + b}$

7. $5x - y$ **9.** $xy - 4b$

11. $4xy$ is a single term; 4, x, and y are factors of the term; 4 is the coefficient of the term.

13. The terms are $4s$ and $7t$; the first term consists of the factors 4 and s, the coefficient is 4; the second term's factors are 7 and t, where 7 is the coefficient.

15. The terms are $5x^2$, $-3x$, and 6; the first term consists of the factors 5 and x^2; 5 is the coefficient. The second term's factors are -3 and x; -3 is the coefficient, 6 is the constant.

17. The terms are $8x^3y^2z$ and $2x^2y$. The first term consists of the factors 8, x^3, y^2, and z, 8 is the numerical coefficient. The factors of the second term are 2, x^2, and y; 2 is the numerical coefficient.

19. Commutative property of addition

21. Distributive property of multiplication over addition

23. Associative property of multiplication

25. Identity property of addition

27. Identity property of multiplication

29. $7(4 + 2)$ **31.** $\pi(x + y)$ **33.** 7 **35.** -26

37. 25 **39.** -4 **41.** 25 **43.** 81 **45.** 14

47. 14 **49.** 10 **51.** $9a + 3b + c$ **53.** $-\frac{1}{4}$

55. 6655 **57.** 16 **59.** **a.** 0.4 amps **b.** 240 amps

61. **a.** 143 **b.** IQ decreases **63.** 35 feet

65. **a.** 2450 **b.** No

67. Terms are individual quantities separated by plus or minus signs. Items being multiplied are factors. The expression $xy + c$ has two terms while x and y are factors.

69. To evaluate an expression is to substitute a specific number for a variable, or variables, and perform the indicated operations.

71. $\frac{2x + 3}{12}$ **73.** $x(2x + 9)$

75. 2 is a coefficient, π is a coefficient, r is a variable.

77. Problem is the square root of -4. **b.** $b^2 - 4ac \geq 0$

Just for Fun *Write down the names of the numbers, that is, one, two, three, four, and so on.*

Section 9.3

1. $\{3\}$ **3.** $\{-1\}$ **5.** $\varnothing$ **7.** $\{40\}$

9. $\{8\}$ **11.** $\{1, 2, 3, 4\}$ **13.** $\{x | x > 1\}$ **15.** $\{1, 2\}$

17. $\{0, 1, 2, 3, 4\}$ **19.** $\{-1, 0, 1, 2, 3\}$

21.
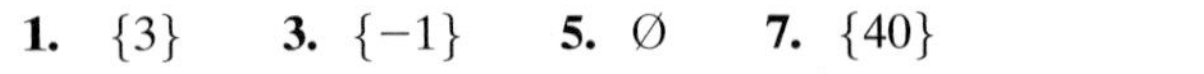

23. 6 7 8 9 10 11 12

25. 0 1 2 3 4 5 6

27. 1 2 3 4 5 6 7

29. –5 –4 –3 –2 –1 0 1

31. 0 1 2 3 4 5 6

33. –4 –3 –2 –1 0 1 2

35. –3 –2 –1 0 1 2 3

37. $45 \leq d \leq 70$ 45 70

39. Statements such as $x + 1 = 3$ cannot be classified as true or false until we replace x by a number; they are called open sentences.

41. $-2 \leq x < 2$ is read "-2 is less than or equal to x and x is less than 2." Hence, -2 is shaded in and 2 is not and all the numbers between them are included.

–2 2

43. $x > 0$ 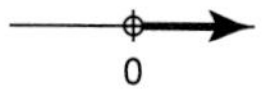

45. $x > 3$ 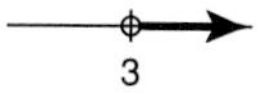

47. Answers may vary. One possibility is $x + 4 < 7$.

49. No solution, or Ø.

Just for Fun 47

Section 9.4

1. $5x$ **3.** $3x$ **5.** $2y$
7. $8x$ **9.** $14y$ **11.** $5x$
13. $2x$ **15.** $5x$ **17.** $5y + 2$
19. $10y + 4$ **21.** $z + 2$ **23.** $3y + 1$
25. $12x + 2$ **27.** $3x - 6$ **29.** $9z + 3$
31. 1 **33.** $7x + 2$ **35.** $15z + 2$
37. x **39.** $7z - 2$ **41.** $23x$
43. $6x + 1$ **45.** $y + 2$ **47.** $-7y + 20$
49. $x = 2$ **51.** $x = 12$ **53.** $y = 6$
55. $y = 4$ **57.** $x = 3$ **59.** $y = 1$
61. $x = 2$ **63.** $x = -6$ **65.** $y = 6$
67. $x = 15$ **69.** $x = 10$ **71.** $x = -10$
73. $z = 3$ **75.** $x = 0$ **77.** $x = 12$
79. $x = 1$ **81.** $x = 5$ **83.** $x = 2$
85. $x = 2$ **87.** $x = 3$ **89.** $x = 16$

91. To simplify an algebraic expression by combining like terms means to reduce the number of terms in an expression. For example, $3x + 4y + 2x + 7y$ becomes $5x + 11y$; the number of terms is reduced from four to two.

93. $3x$ and $4t$ are not like terms because they have no common literal factor.

95. $x = -6$ **97.** $x = -3$ **99.** $x = 2$

101. $5x - x = (5 - 1)x = 4x$

103. a. $0.07x + 4.95, x > 0$; $0.99 + 0.10x, x > 20$
b. 132 minutes

105. $M = \dfrac{F}{A}$

107. $C = \dfrac{5}{9}(F - 32)$

109. $t = \dfrac{A - P}{Pr}$

Section 9.5

1. 2 **3.** 7 **5.** 10 and 11
7. Angela is 37, Richard is 46
9. 14 **11.** Length = 19′; width = 14′
13. Length = 78 m; width = 36 m
15. 13 and 15
17. 8 and 10
19. 12 dimes, 26 quarters
21. 7 nickels, 6 dimes, 3 quarters
23. 6,000 at \$51, 18,000 at \$63, 12,000 at \$75
25. 1st base angle = 70°, top angle = 35°, 2nd base angle = 75°
27. 250 nickels, 286 dimes, 310 quarters
29. 9 m
31. Four less than twice a number is 10.
33. Twenty-seven more than five times a number is 42.
35. If x is an even integer, then $x + 2$ must be the next consecutive even integer because even integers differ by two.
37. $B = \dfrac{P}{48}$, +1 if there is a remainder, and P = total number of people **b.** 18
39. a. $S = 82{,}750 + 0.30b$ **b.** \$89,884; \$7134

Just for Fun 1

Section 9.6

Note: Answers may vary. The following are some possible solutions.

1. $(5, 0), (0, 5), (3, 2)$
3. $(3, 0), (0, -3), (4, 1)$
5. $(0, -6), (-3, 0), (2, -10)$
7. $(0, 10), (2, 16), (-4, -2)$
9. $(0, -4), (4, 2), (2, -1)$
11. $(0, -3), (-5, 0), (5, -6)$
13. $(0, 0), (1, 1), (3, 3)$
15. $(0, 0), (2, -1), (4, -2)$
17. $(2, 5), (5, 7), (8, 9)$
19. $(1, -3), (4, -5), (7, -7)$
21. $(0, 3), (4, -2), (8, -7)$
23. No
25. No
27. Yes, $p = 22 \cdot 3 + 15 \cdot 1 = 66 + 15 = 81$
29. Whole numbers, cannot purchase fractional or negative amounts of shirts and belts.
31. We can use an equation to determine the relationship between two variables. That is, given the value of one variable, we can use the equation to determine the corresponding value of the other variable.
33. $C = 37 + 23(W - 1)$ **b.** $C = \$2.21$; $C = \$2.67$
35. $x + y = 2$ **37.** Research
39. 121 **41. a.** 94 **b.** 2

Just for Fun *Tierce, hogshead, pipe, butt, and tun are names of casks (barrels). They were originally used in measuring amounts of beer and ale.*

Section 9.7

1.

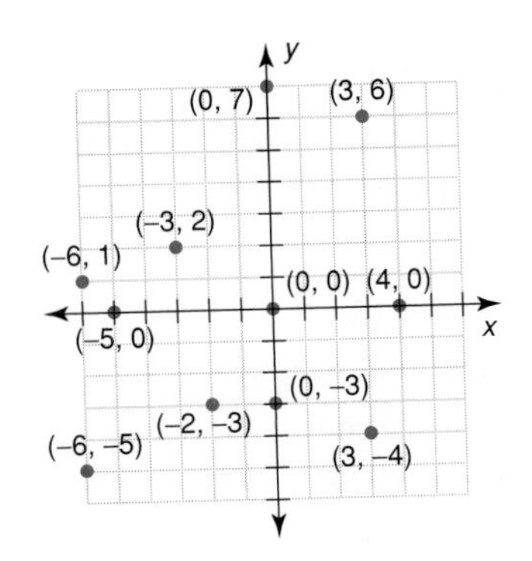

3.

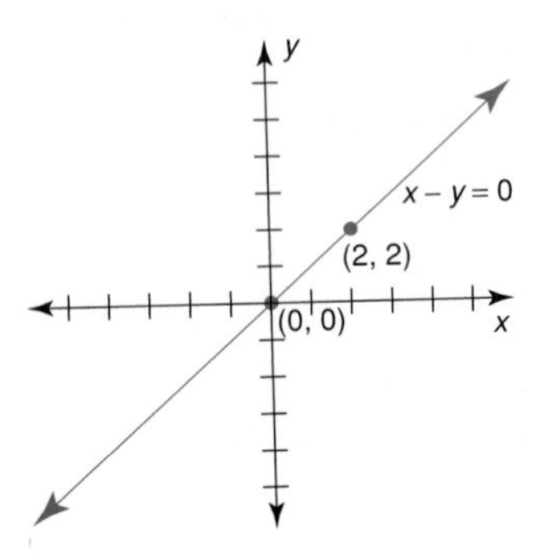

5.

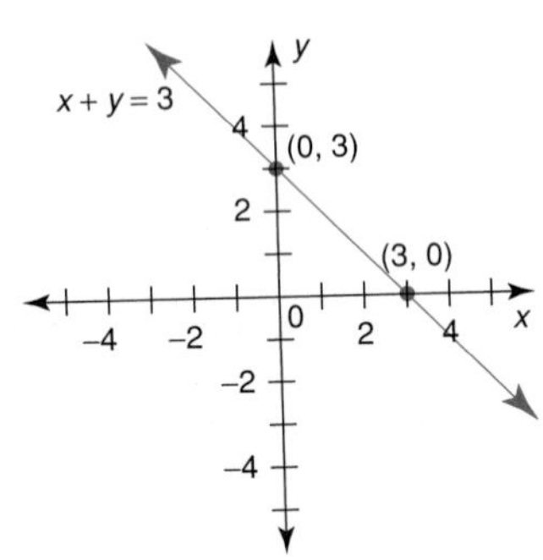

7.

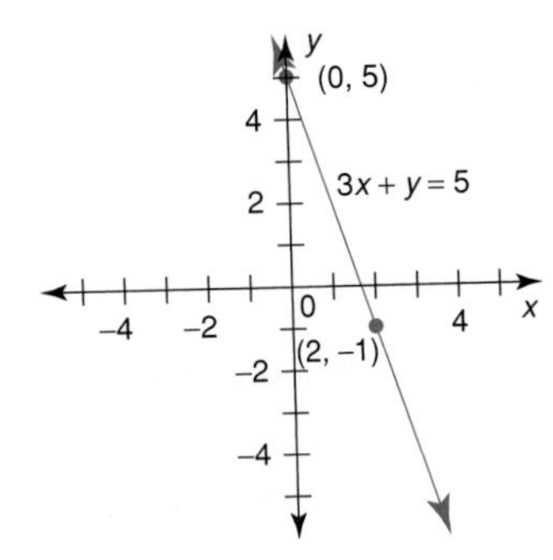

9.

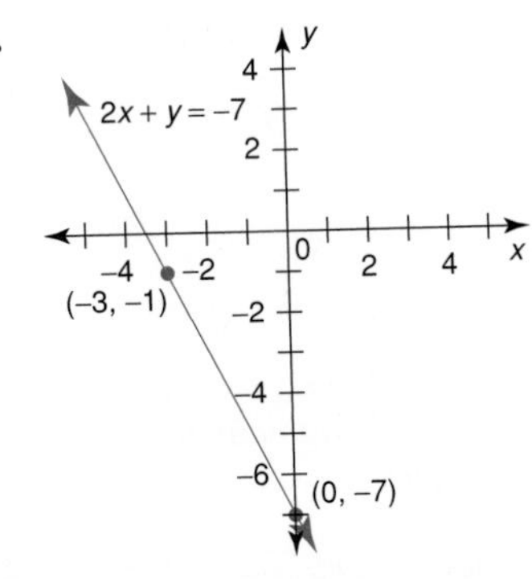

11.

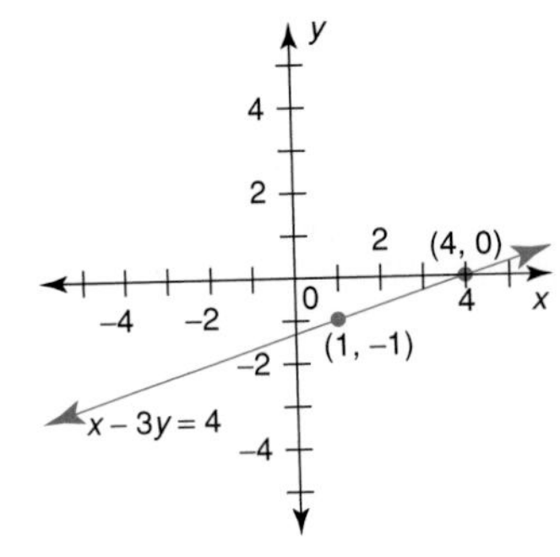

13.

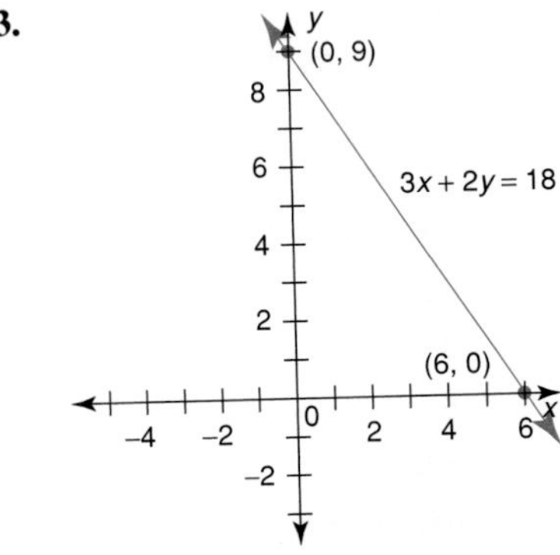

15.

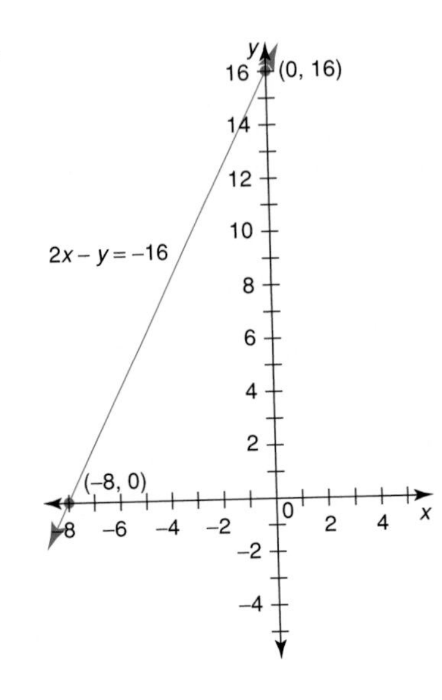

17.

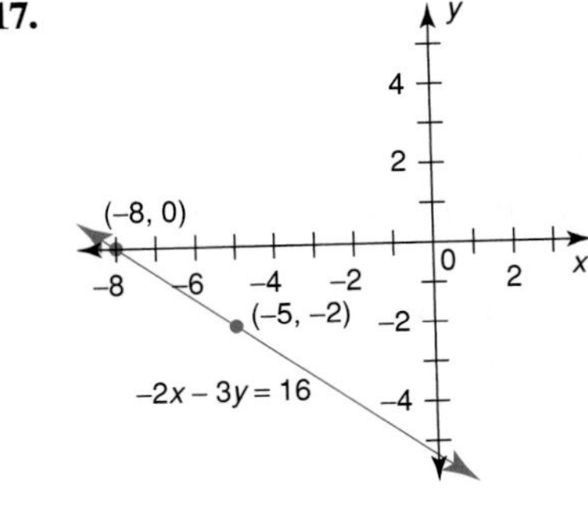

19.

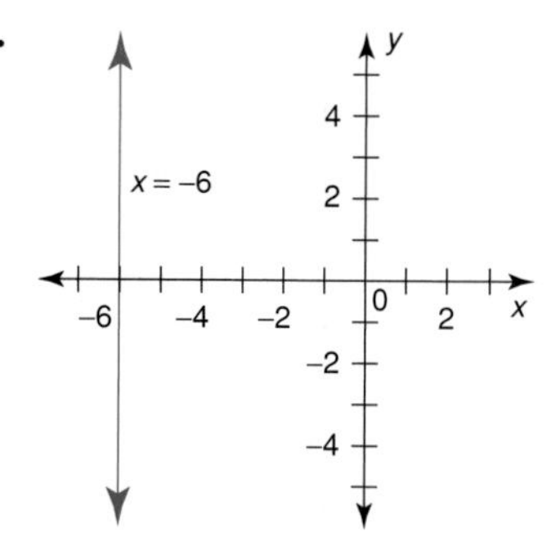

21.

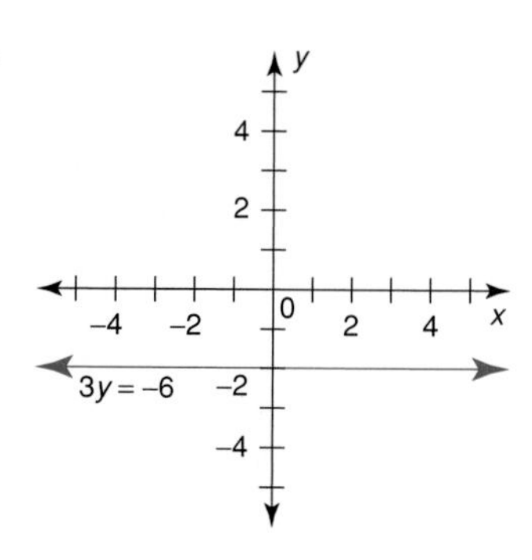

23.

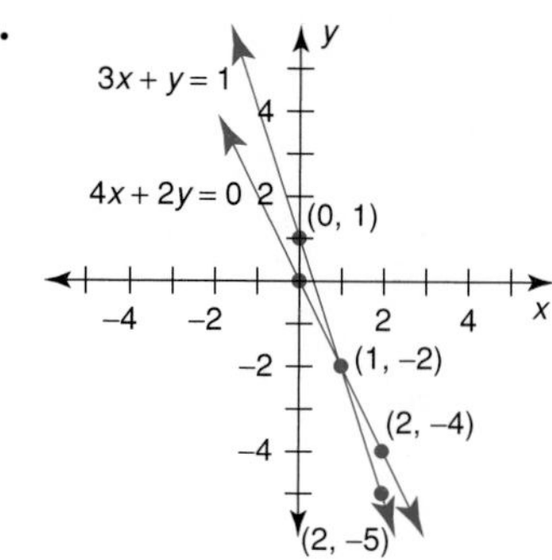

25.

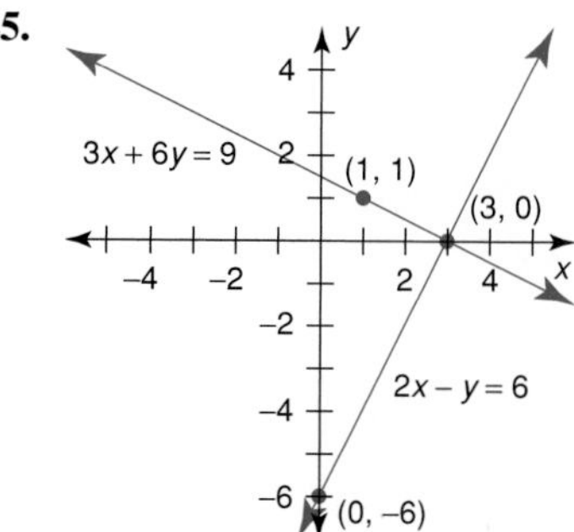

27.

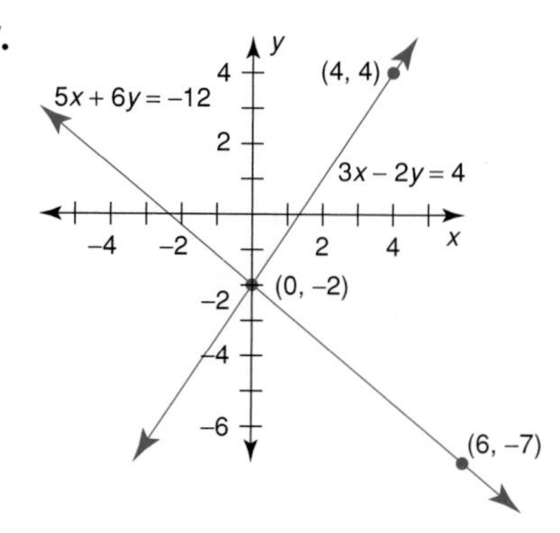

29.

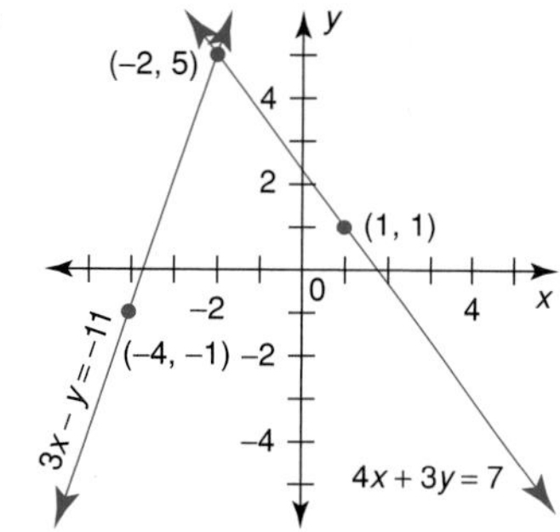

31. The Cartesian plane can be thought of as a grid and it is divided into four quadrants by a horizontal line, the x-axis, and a vertical line, the y-axis. These axes intersect at a common point called the origin, where $x = 0$ and $y = 0$, or $(0, 0)$.

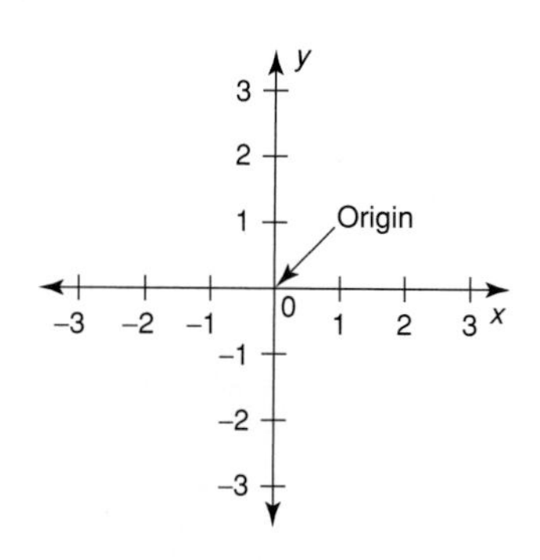

33. To plot $(-3, 1)$ on the Cartesian plane, start at the origin and move 3 units to left (negative direction) along the x-axis, then proceed 1 unit up (in a positive y-direction).

35. A *consistent* system of linear equations has a common point of intersection. An *inconsistent* system of linear equations has no common point of intersection. A system of equations is said to be *dependent* when any solution of one equation is a solution of the other.

b. Answers will vary; here are examples of each.

Consistent	Inconsistent	Dependent
$3x + 2y = 6$	$3x + 2y = 12$	$x - y = 6$
$2x + 3y = 13$	$3x + 2y = 16$	$-x + y = -6$

37. $\approx(1.75, 1.5)$

39. $\approx(17.5, 4.5)$ No, shot will hit corner.

Just for Fun *12*

Section 9.8

1. $x - y \geq -2$

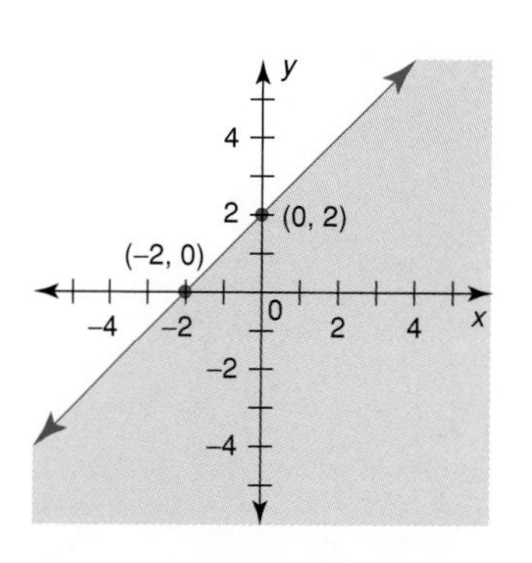

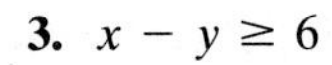

3. $x - y \geq 6$

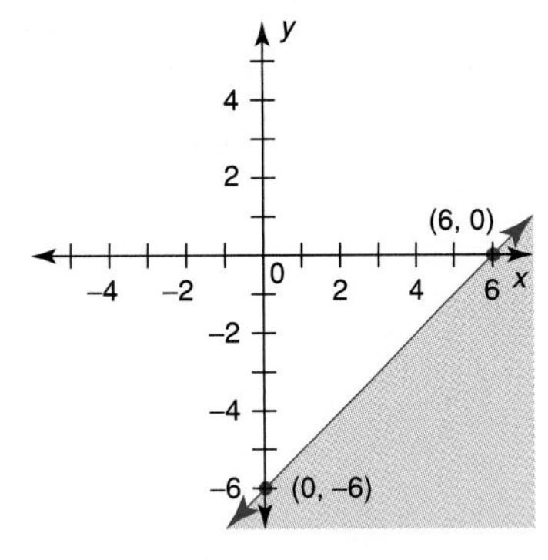

5. $x - y < -6$

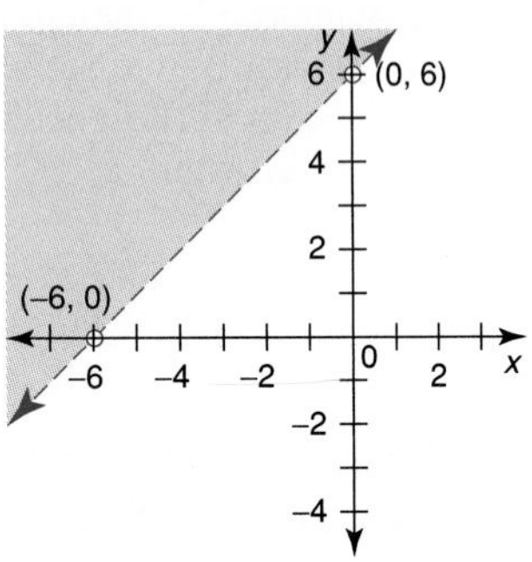

7. $3x - y < 4$

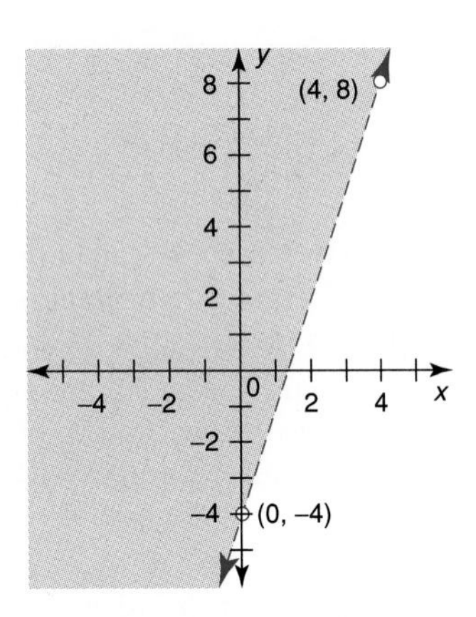

9. $3x - y > 0$

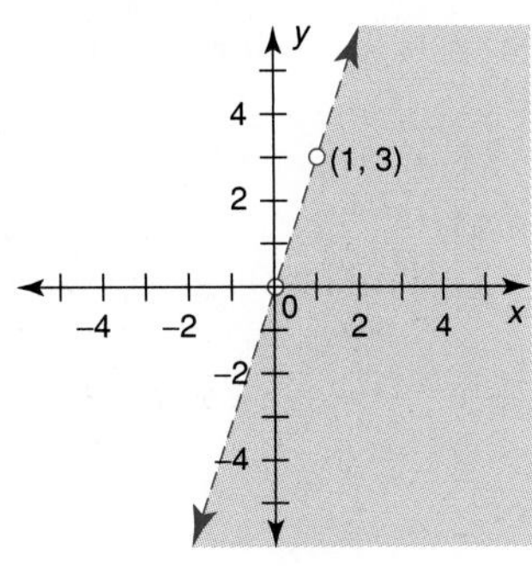

11. $2x + y \leq -5$

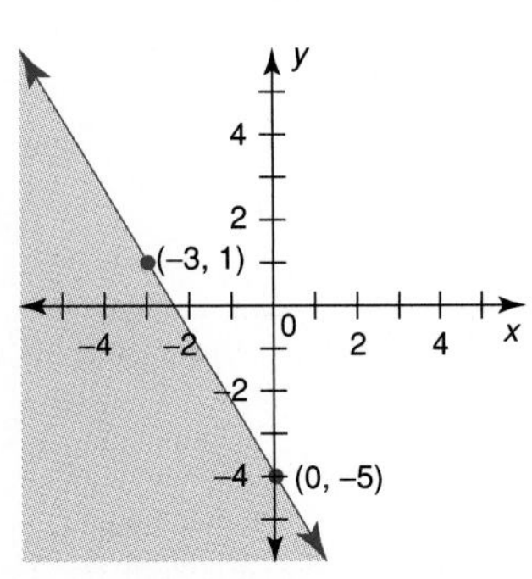

13. $-3x + 2y < 6$

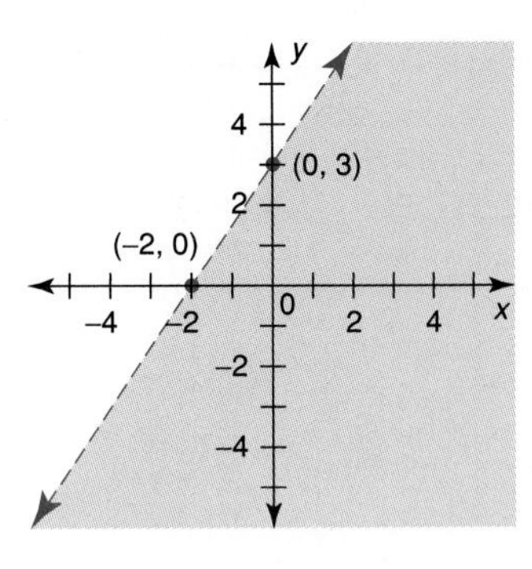

15. $5x - 4y \geq 20$

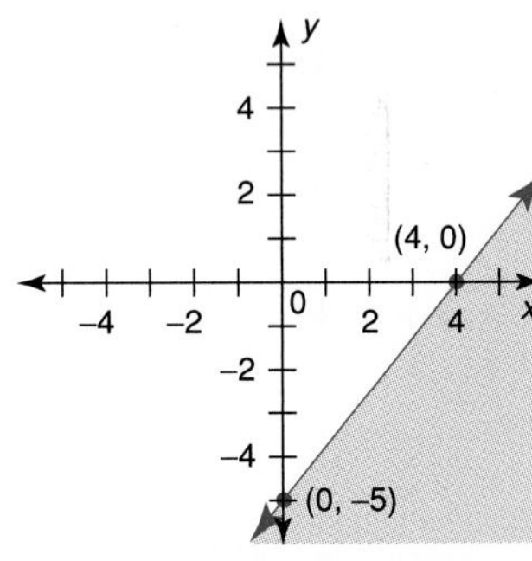

17. $-3x + 4y > -24$

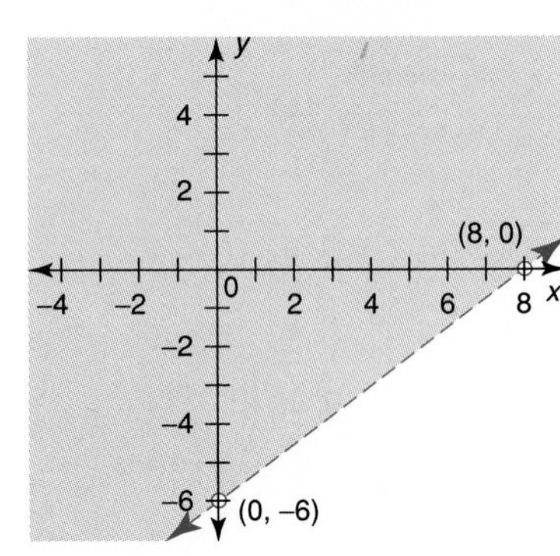

19. $2x + 6y \leq 0$ and $x + 2y > 3$

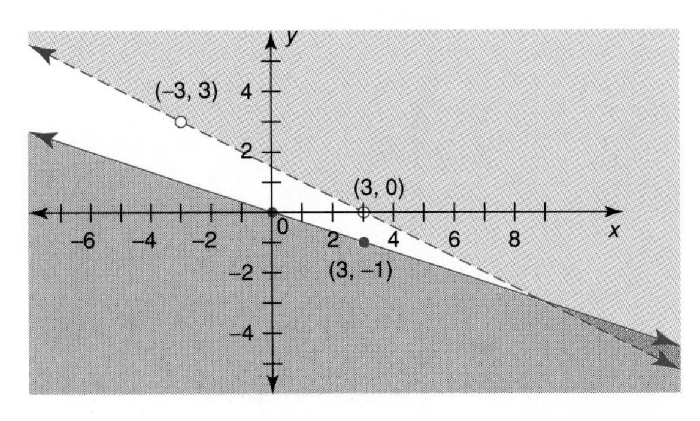

21. $3x + y > -3$ and $-2x + y \leq 4$

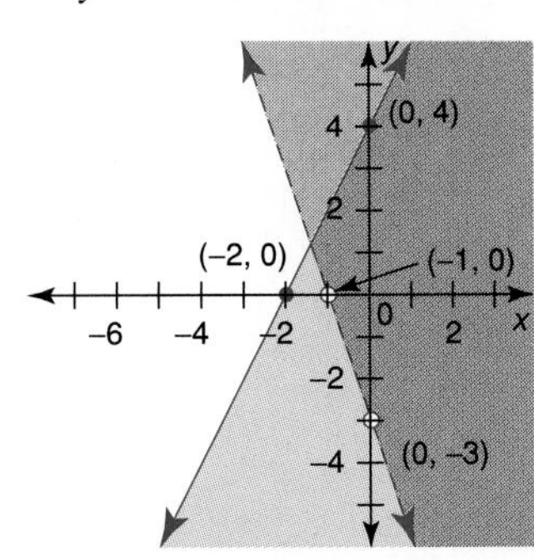

23. The boundary of the half-plane is obtained by graphing the equation derived from the inequality. It is a dashed line for a strict inequality and a solid line for $\geq$ or $\leq$.

25. The solution set for a system of linear inequalities is obtained by graphing both inequalities on the same Cartesian plane. The solution is the region where the half-planes intersect.

27. The solution set will be located in the first quadrant of the Cartesian plane. These restrictions on x and y are necessary when only positive solutions are required.

Just for Fun *None; peacocks do not lay eggs.*

Review Exercises for Chapter 9

1. Variable **2.** Term **3.** Constant **4.** Coefficient

5. a. 3 **b.** $8x^2, -3x, 2$

c. $8, -3, 1$ **d.** $8 \cdot x \cdot x, -3 \cdot x, 1 \cdot 2$

6. Commutative property of multiplication

7. Distributive property

8. Associative property of multiplication

9. 15 **10.** −15 **11.** −270 **12.** 10 **13.** \$20,000

14. a. $\{4\}$ **b.** $\{-6\}$ **c.** $\{6\}$ **d.** $\{1, 2, 3\}$

e. $\{0, 1, 2, 3, 4, 5, 6, 7\}$ **f.** $\{-2, -1, 0, 1, 2\}$

g. $\{1, 2, 3\}$

15. a.

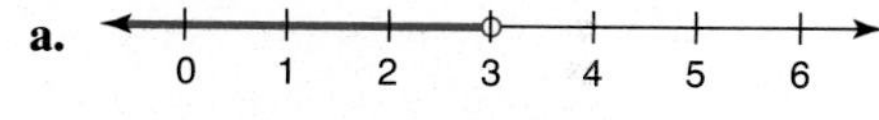

b. −3 −2 −1 0 1 2 3

c. −3 −2 −1 0 1 2 3

d. −3 −2 −1 0 1 2 3

e. −3 −2 −1 0 1 2 3

f. −3 −2 −1 0 1 2 3

16. a. $x = 6$ **b.** $y = 2$ **c.** $z = 2$

d. $y = 3$ **e.** $x = 10$ **f.** $z = \frac{6}{4}$

17. 23 and 25 **18.** Diego is 26 and Andy is 23.

19. $n = 10$ **20.** $w = 5$ m; $l = 10$ m

21. 10 quarters; 5 dimes **22.** $n = 5$

23. $w = 8$ m, $l = 24$ m

24. 10 nickels, 15 dimes, 25 quarters

25. 350

26. a. $x - y = 2$

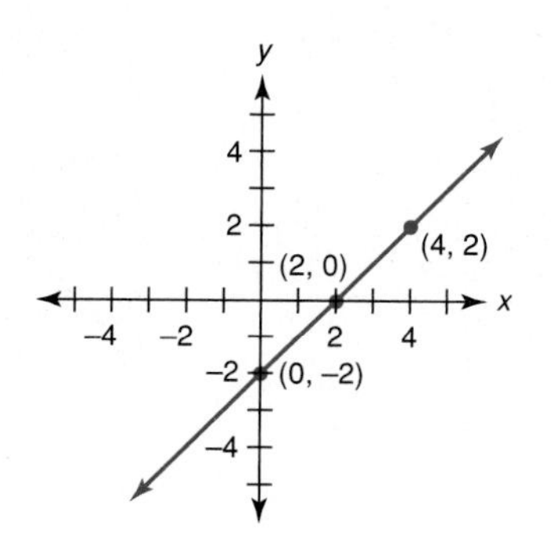

b. $2x + y = 4$

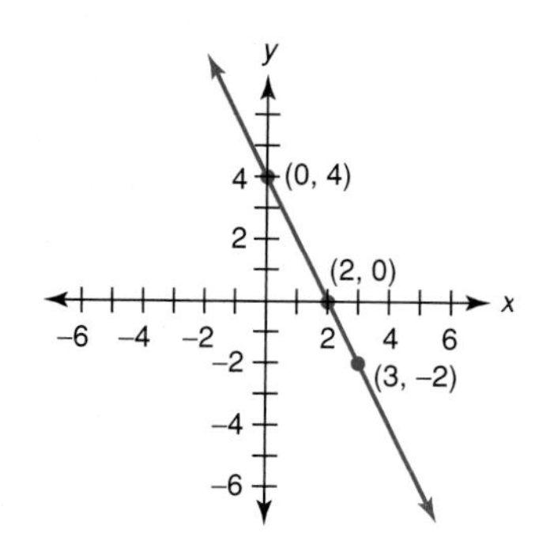

c. $2x - y = 4$

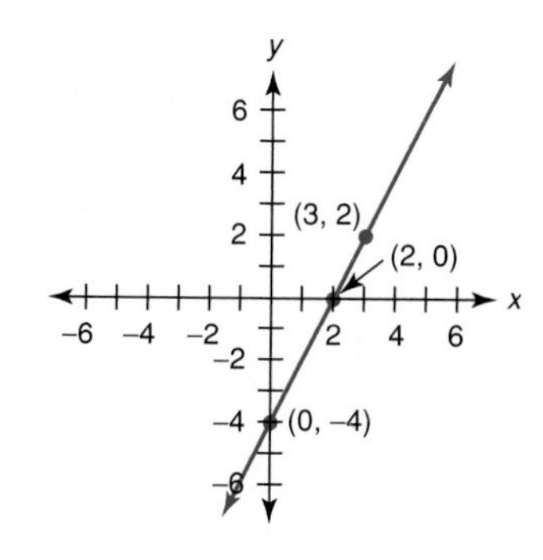

d. $-2x + y = 6$

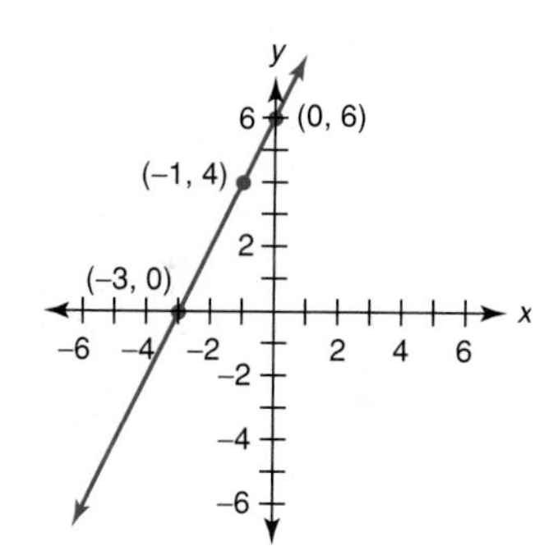

e. $3x - 5y = 15$

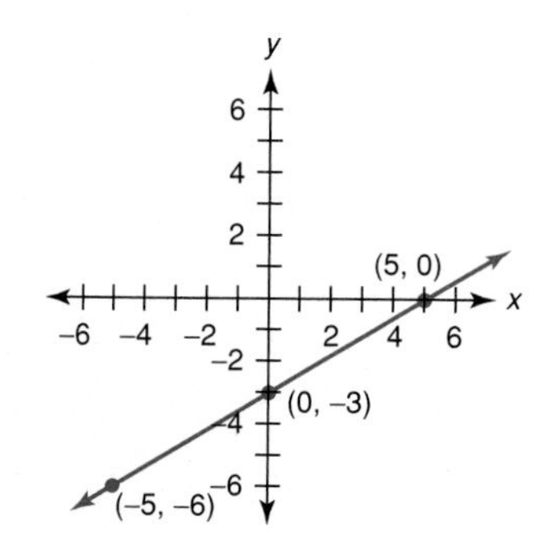

f. $y = x$

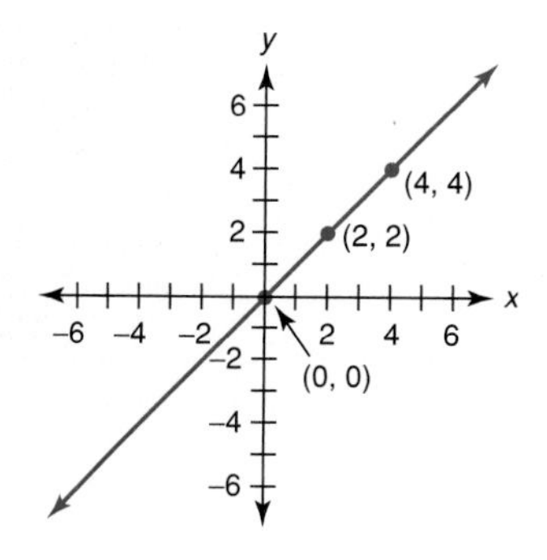

27. a.

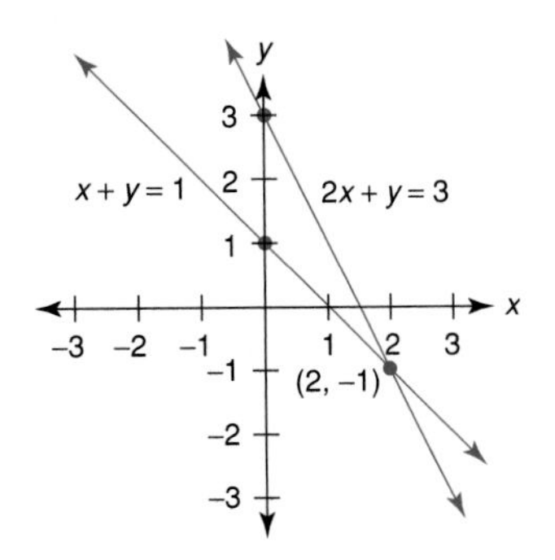

b.

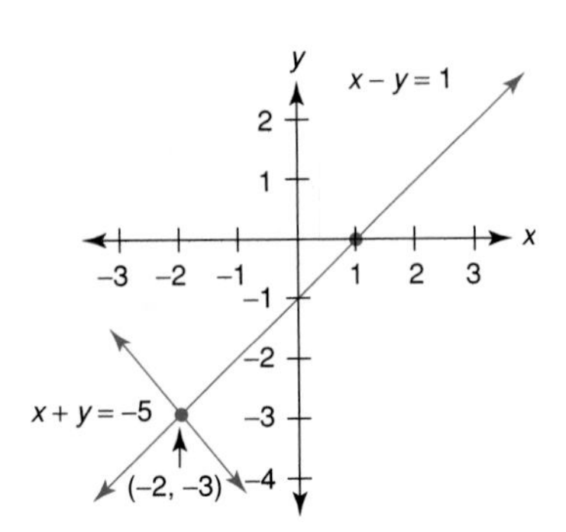

c.

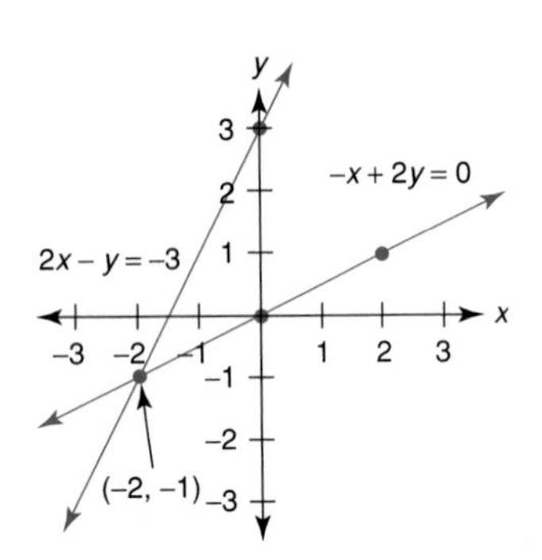

28. a. $x + y > 7$

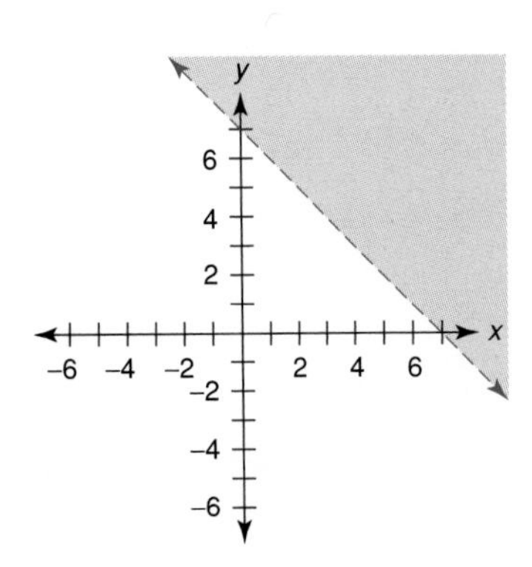

b. $x - y \le 2$

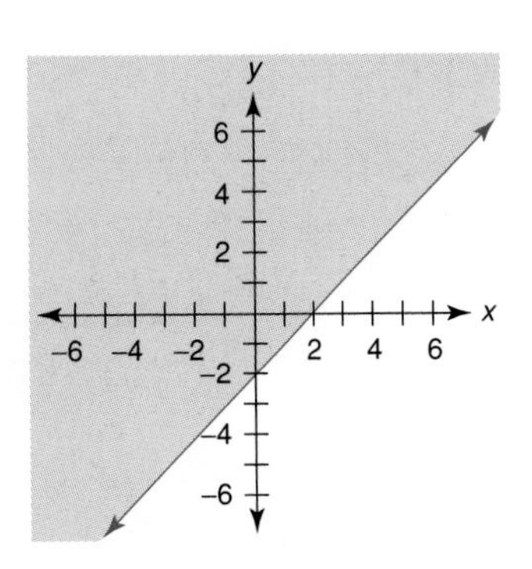

c. $-x + y > 2$

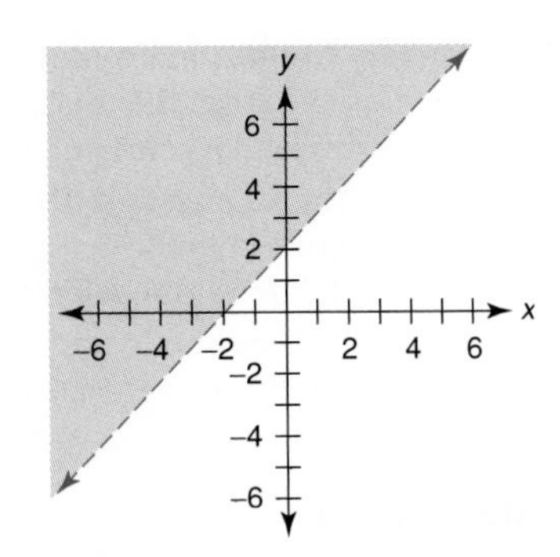

d. $3x + 5y > -15$

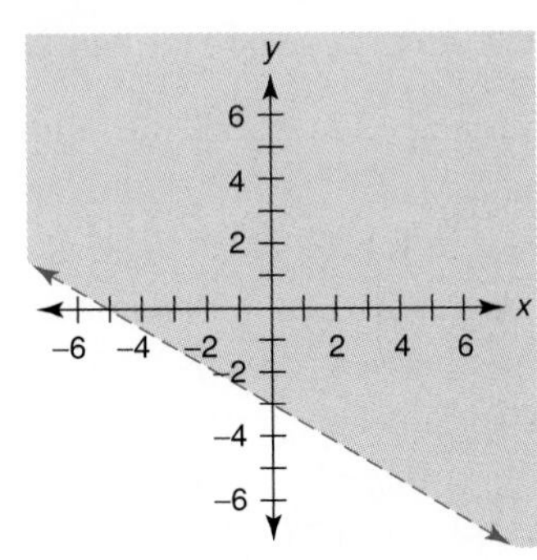

e. $x - 2y \le 0$

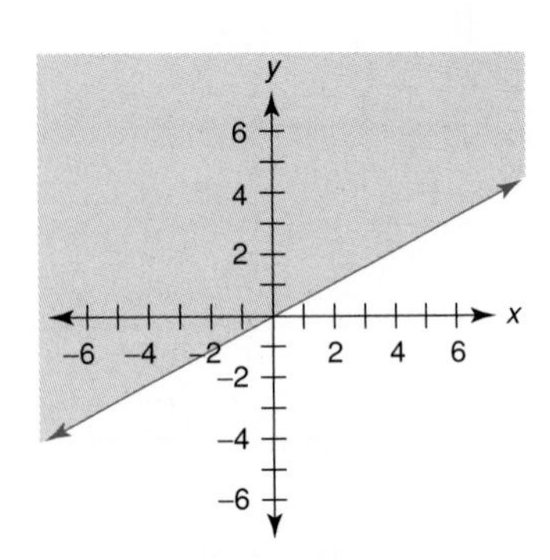

f. $-3x - 2y < -6$

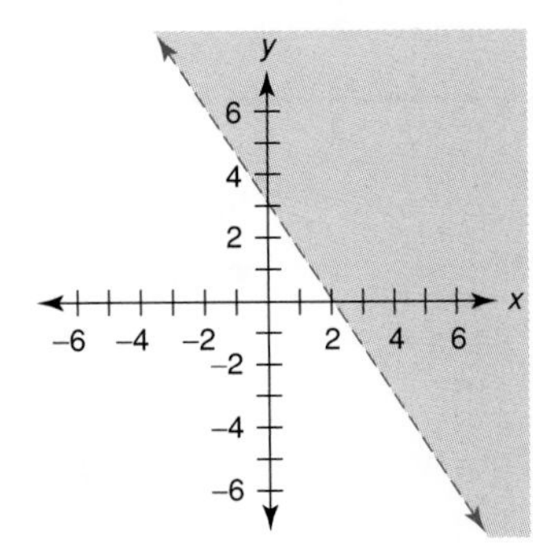

Just for Fun *two hours*

Chapter 9 Quiz

1. False **2.** True **3.** True
4. False **5.** False **6.** True
7. False **8.** False **9.** True
10. False **11.** True **12.** False
13. False **14.** True **15.** False
16. $\{2\}$ **17.** $\{2\}$ **18.** $\{4\}$ **19.** $\{x|x < 1\}$
20. $\{7\}$ **21.** $\{10\}$ **22.** $\{4\}$ **23.** $\{6\}$
24. Mary is 16 and Monique is 11.

25. L = 9 ft., W = 6 ft.
26. 5 quarters, 10 nickels, 4 dimes
27.

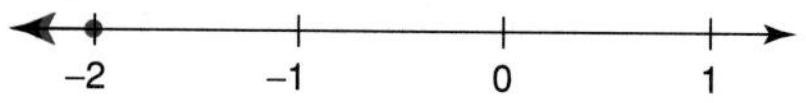

28.

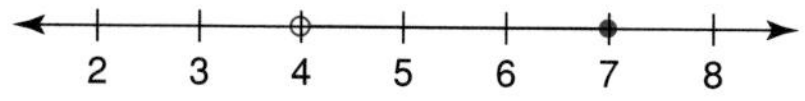

Chapter 10

Section 10.2

1. -3 **3.** 1 **5.** $-\frac{1}{4}$
7. undefined **9.** $-\frac{3}{2}$ **11.** 0
13. **a.** 6 **b.** -3 **c.** $\frac{2}{3}$ **d.** 0
15. **a.** $-\frac{1}{6}$ **b.** $\frac{1}{3}$ **c.** $-\frac{3}{2}$ **d.** undefined
17. $m\overleftrightarrow{AB} = 3$, $m\overleftrightarrow{CD} = 3$; therefore, $\overleftrightarrow{AB} \| \overleftrightarrow{CD}$
19. $m\overleftrightarrow{MN} = \frac{1}{3}$, $m\overleftrightarrow{OP} = -3$; therefore, $\overleftrightarrow{MN} \perp \overleftrightarrow{OP}$
21. $x = -8$ **23.** $x = -20$
25. **a.** No, $m\overline{AD} = \dfrac{y}{x}$ and $m\overline{CD} = \dfrac{y}{2x}$. Since $\dfrac{y}{x} \neq \dfrac{y}{2x}$, the lines are not parallel
b. Cannot tell until we have specific values for x and y; see part d
c. Yes, $m\overline{AB}$ = undefined and $m\overline{AD}$ = undefined. Both lines are vertical, hence parallel
d. Yes, $m\overline{BC} = -1$ and $m\overline{CD} = 1$. Slopes are negative reciprocals (i.e., the product of the slopes is -1), thus the sides are perpendicular
27. The slope of a line is the ratio of the change in vertical distance to the change in horizontal distance.
29. The slope of any horizontal line is zero. There is zero change in the vertical distance.
31. **a.** $m\overline{AB} = \dfrac{5}{2x-7}$, $m\overline{DC} = \dfrac{10}{x-3}$
b. $\dfrac{5}{2x-7} = \dfrac{10}{x-3}$, $x = 3\dfrac{2}{3}$
33. Research
35. Slope of Cook Ins. = 100; slope of TriState Ins. = 80. Cook Ins. rates rise faster as the amount of coverage increases.
37. Slope of digital cameras = 3; slope of traditional cameras = $-\frac{8}{3}$. The number of digital cameras sold is increasing where as the number of traditional cameras sold is decreasing.

Just for Fun $\dfrac{7+7}{.7}$

Section 10.3

1. $x - y + 1 = 0$ **3.** $6x + y + 12 = 0$
5. $3x - 5y - 20 = 0$ **7.** $3x + 4y + 10 = 0$
9. $x - 7y + 1 = 0$ **11.** $y = 3$
13. $x - 3y + 3 = 0$ **15.** $2x + y = 0$
17. $4x + 3y + 12 = 0$ **19.** $x + y + 5 = 0$
21. $x - 2y + 1 = 0$
23.

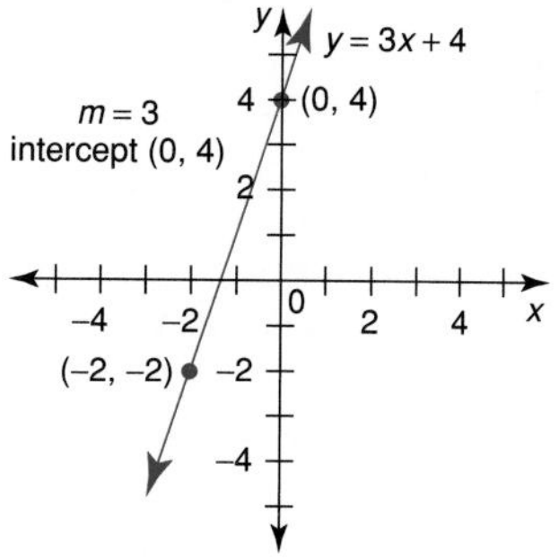

25.

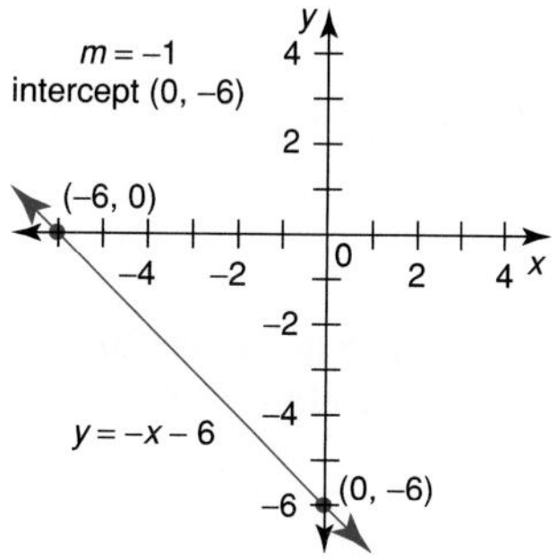

27.

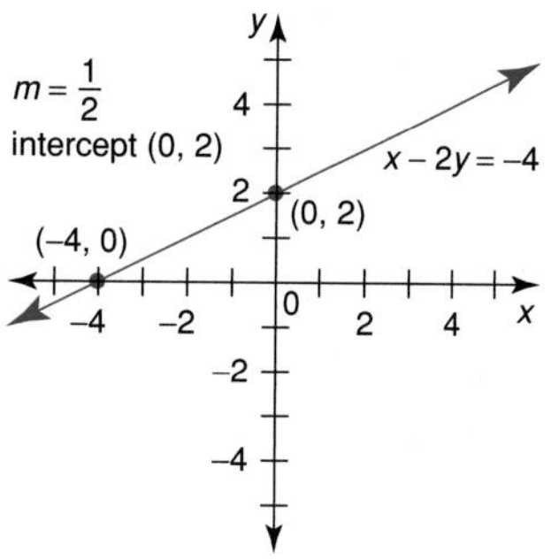

29.

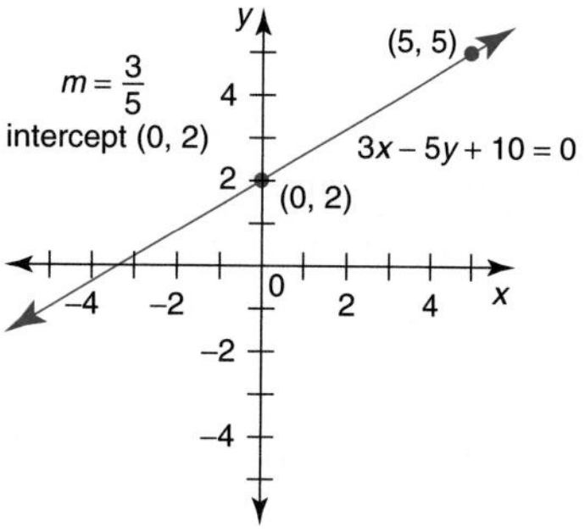

31.

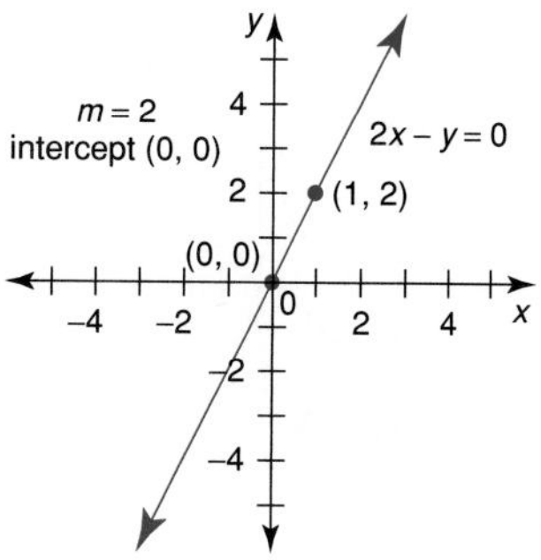

33. $2x - y + 4 = 0$

35. $3x - 2y + 7 = 0$

37. a. 2 **b.** $2x - y + 4 = 0$
c. $3x - y + 3 = 0$ **d.** $5x - 3y + 5 = 0$

39. a. $y = 4$ **b.** $x = 4$
c. $x + 4y - 12 = 0$ **d.** $x - 4y + 12 = 0$

41. $y = mx + b$. We must find values for m and b. First, find the slope of the line and substitute it in the formula for m. Next, substitute the given values for x and y, and solve the equation for b. Finally, write the final answer in the form $y = mx + b$, using the determined values for m and b.

43. a. $y = 2$ **b.** $x = -2$ **c.** $y = x - 3$
d. $y = -x + 3$ **e.** $x = 3$

45. Research

47. *$C = 7x + 0$; $C = 3x + 39$*
$C = \$1.40$; $C = \$0.99$

49. $Y = -0.02W + 75$; $Y = -0.008W + 32$
$Y = 25$ mpg; $Y = 12$ mpg

Just for Fun

Step	Amount Left After Each Step: 8-Gallon	5-Gallon	3-Gallon
1. Fill 5 gal from 8 gal.	3	5	0
2. Fill 3 gal from 5 gal.	3	2	3
3. Empty 3 gal into 8 gal.	6	2	0
4. Empty 5 gal into 3 gal.	6	0	2
5. Fill 5 gal from 8 gal.	1	5	2
6. Fill 3 gal from 5 gal.	1	4	3
7. Empty 3 gal into 8 gal.	4	4	0

Section 10.4

1. Maximum value of P is 16 at $(3, 2)$.

3. Maximum value of P is 26 at $(5, 2)$.

5. Maximum value of P is 20 at $(10, 0)$.

7. Maximum value of P is 50 at $(6, 5)$.

9. Maximum value of P is 3 at $(5, 6)$.

11. 4 lawn mowers and 3 snow blowers

13. 70 acres of potatoes and 30 acres of cauliflower

15. 40 cans of Brand X and 20 cans of Brand Y

17. The objective function in a linear programming problem is the function that is to be maximized or minimized.

19. The minimum value is 19 at $(1, 3)$.

Just for Fun *Two of a set of triplets.*

Section 10.5

1. $\{-2, 4\}$ **3.** $\{-3, 4\}$ **5.** $\{-2\}$

7. $\{-5, 3\}$ **9.** $\{-9, 4\}$ **11.** $\left\{-4, \frac{3}{2}\right\}$

13. $\left\{-\frac{5}{2}, 3\right\}$ **15.** $\left\{-\frac{2}{3}, \frac{5}{2}\right\}$ **17.** $\left\{-\frac{2}{3}, 8\right\}$

19. $\left\{-\frac{3}{2}, -\frac{1}{2}\right\}$ **21.** $\{1 + \sqrt{3}, 1 - \sqrt{3}\}$

23. $\{5 + 3\sqrt{3}, 5 - 3\sqrt{3}\}$

25. $\left\{\frac{-1 + \sqrt{5}}{2}, \frac{-1 - \sqrt{5}}{2}\right\}$

27. $\left\{-\frac{\sqrt{10}}{2}, \frac{\sqrt{10}}{2}\right\}$

29. $x = \dfrac{-b \pm \sqrt{b^2 - 4ac}}{2a} \quad a \neq 0$

31. The standard form for a quadratic equation in x can be written as $ax^2 + bx + c = 0$, $a \neq 0$.

33. $\left\{\frac{1}{3}, \frac{3}{2}\right\}$

Just for Fun **a.** *The tunnel of Corti is a part of the ear.*
b. *The Island of Reil is a part of the brain.*
c. *McBurney's Point is the point of incision for an appendectomy.*

Section 10.6

1. $y = x^2 - 1$

x	y
−2	3
−1	0
0	−1
1	0
2	3

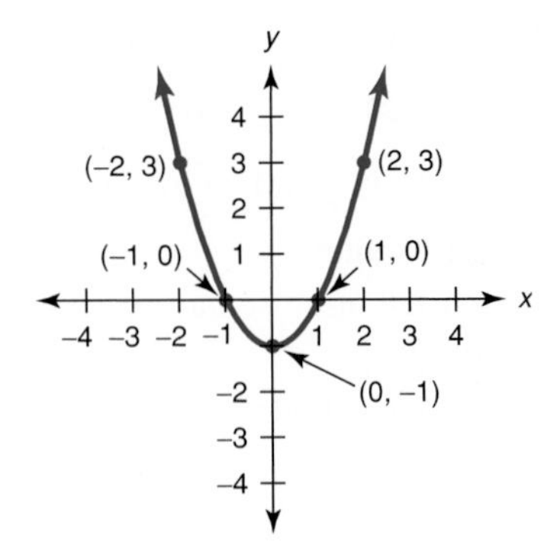

3. $y = -x^2 + 2$

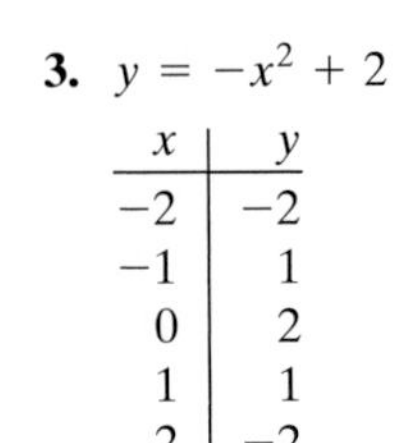

x	y
−2	−2
−1	1
0	2
1	1
2	−2

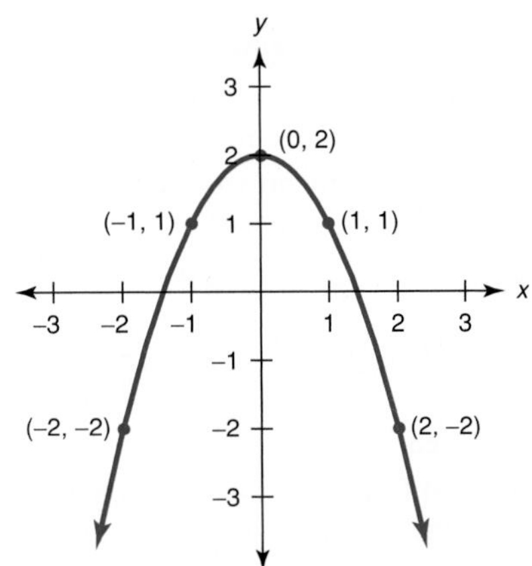

5. $y = -x^2 + 6x$

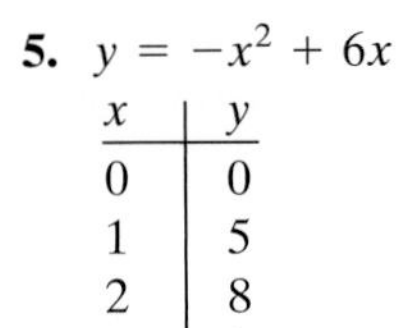
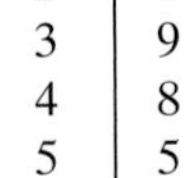

x	y
0	0
1	5
2	8
3	9
4	8
5	5
6	0

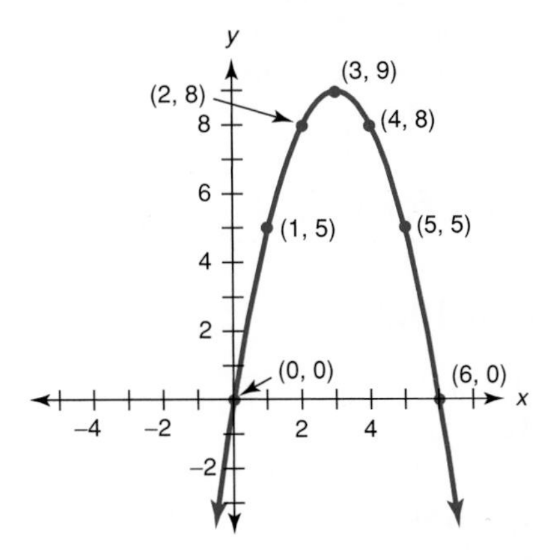

7. $y = x^2 + 2x + 1$

x	y
−4	9
−3	4
−2	1
−1	0
0	1
1	4
2	9

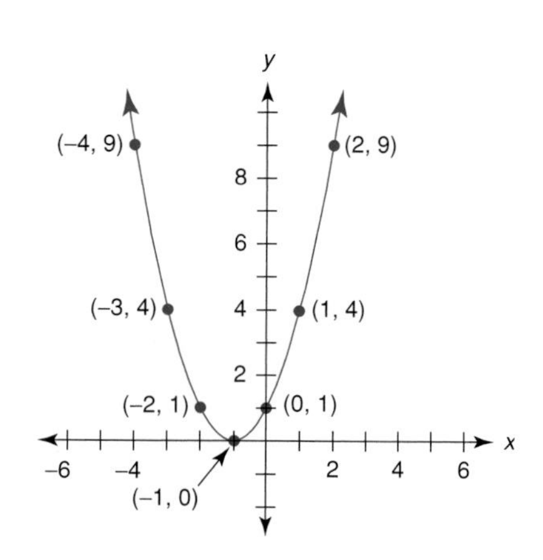

9. $y = x^2 - 4x + 3$

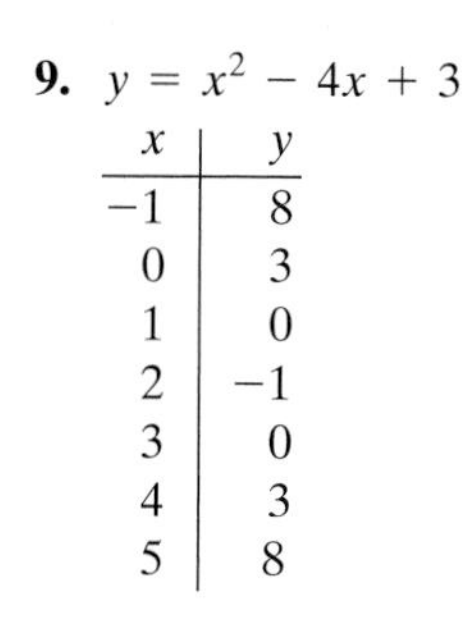

x	y
−1	8
0	3
1	0
2	−1
3	0
4	3
5	8

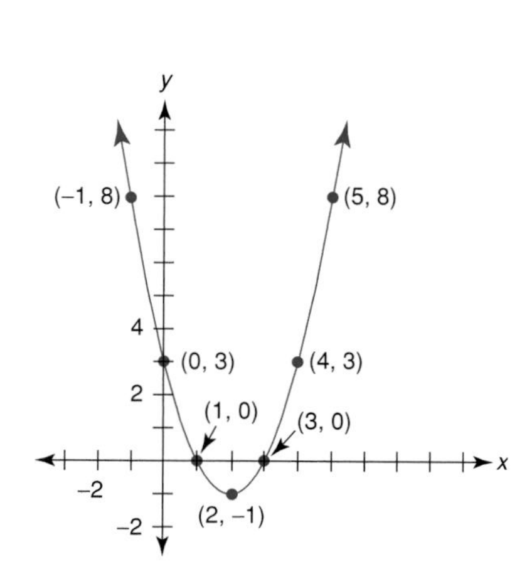

11. $y = -x^2 + 6x - 9$

x	y
1	−4
2	−1
3	0
4	−1
5	−4

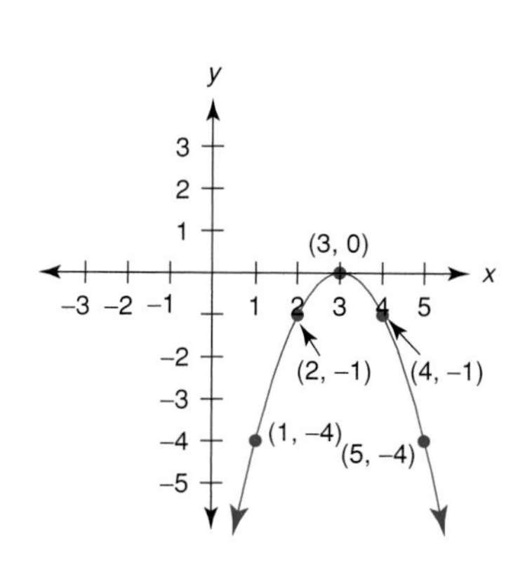

13. $y = x^2 - 6x + 5$

x	y
0	5
1	0
2	−3
3	−4
4	−3
5	0
6	5

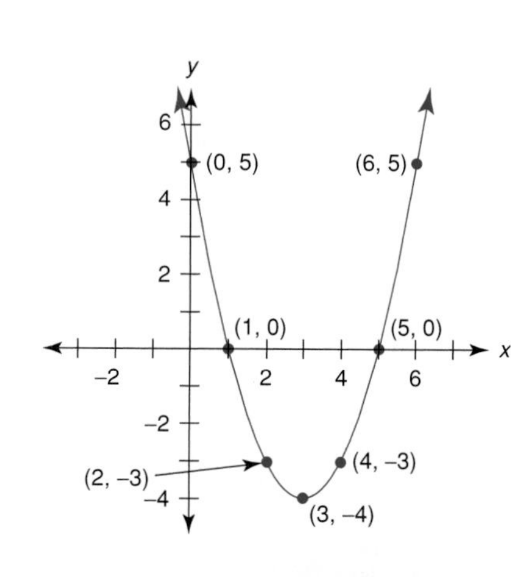

15. $y = -x^2 + 4$

x	y
−2	0
−1	3
0	4
1	3
2	0

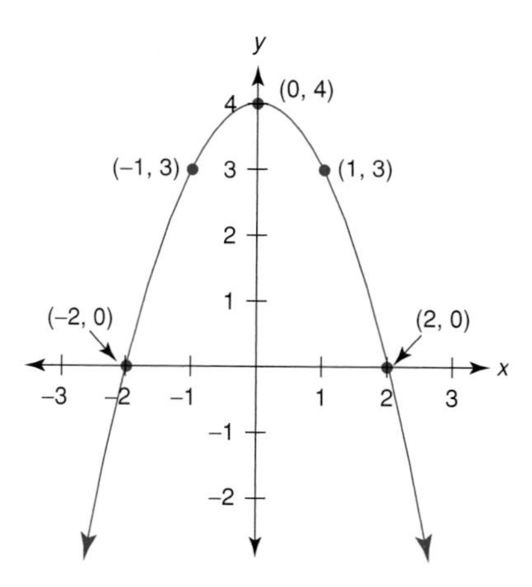

17. $y = x^2 + 2x$

x	y
−3	3
−2	0
−1	−1
0	0
1	3

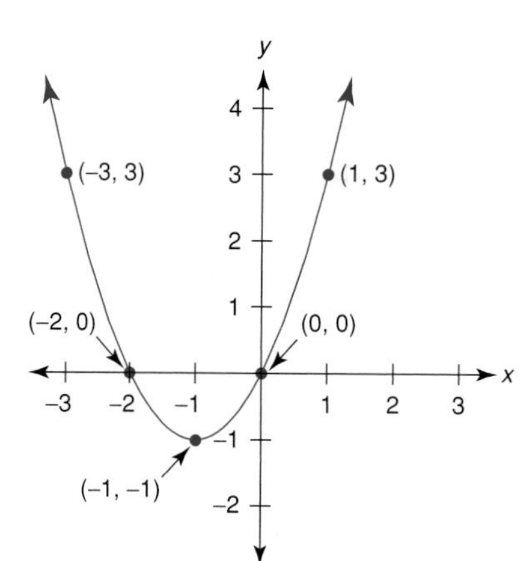

19. $y = x^2 + 3x + 2$

x	y
−3	2
−2	0
−1	0
0	2

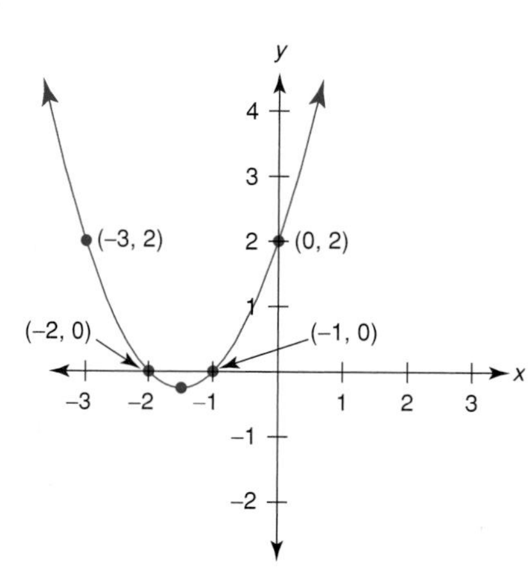

21. A functional relationship is when one variable depends on (is a function of) a second variable, e.g., outdoor weddings are a function of the weather. Thus outdoor weddings and weather form a functional relationship.

23. A function cannot have the same first coordinate and a different second coordinate for any two ordered pairs. Thus $x = y^2$ is not a function because if violates this rule, e.g., (1,1), (1,-1).

25. A linear equation is of the first degree while a quadratic equation is of the second degree.

27. $y = -2x^2 + 11x - 12$

29. 225 ft

Just for Fun *The box full of $10 gold pieces; the value of gold is determined by weight, not denomination.*

Review Exercises for Chapter 10

1. a. $\frac{3}{7}$ **b.** -1

2. a. $x + y - 1 = 0$ **b.** $3x + 2y + 6 = 0$
c. $9x - y - 47 = 0$ **d.** $2x - 3y + 5 = 0$

3. a. $m = -2$, y-intercept $(0, 5)$
b. $m = -1$, y-intercept $(0, 3)$
c. $m = \frac{-2}{3}$, y-intercept $(0, -2)$
d. $m = \frac{1}{3}$, y-intercept $(0, 1)$

4. a.

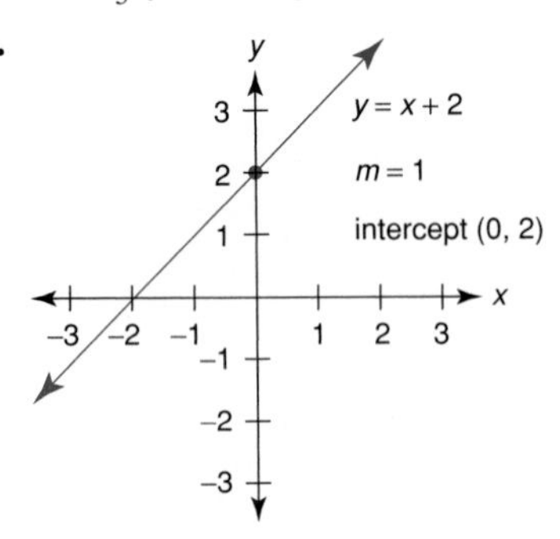

b.

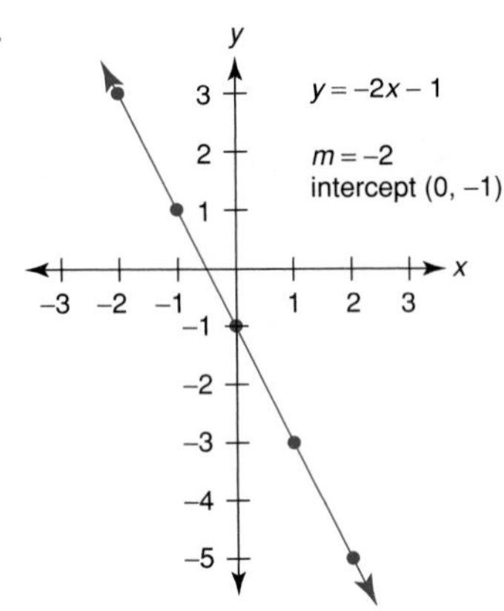

c.

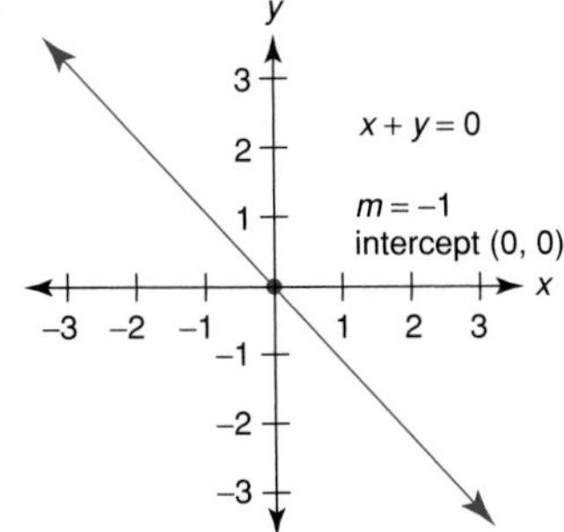

d.

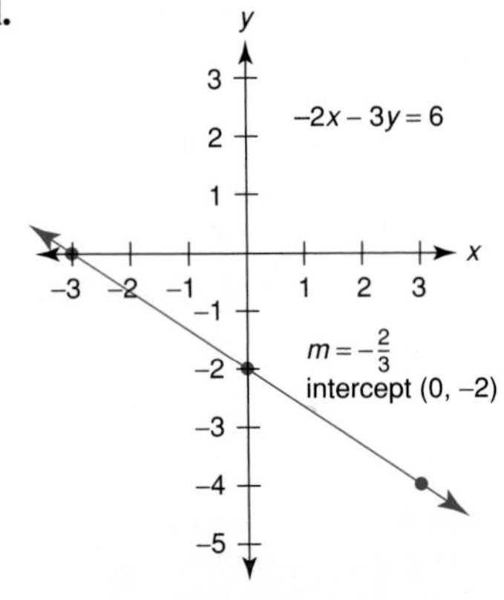

5. a. $m\overline{PQ} = 1, m\overline{PR} = 1$
b. Points P, Q, and R lie on the same line.
c. $x - y - 2 = 0$
d. $(2, 0)$

6. a. $x = 3$
b. $x - y - 1 = 0$
c. $m\overline{AB} = \frac{2}{3}, m\overline{AC} = \frac{2}{3}$. Slopes are equal and both contain point A; therefore, A, B, and C lie on the same straight line.
d. $2x - 3y = 0$

7. Maximum value of P is 18 at $(3, 4)$.

8. Four couches and three recliners; maximum profit is $255.

9. Thirty standard and five deluxe models; maximum profit is $260.

10. $\{2, -\frac{2}{3}\}$ **11.** $\{2, -1\}$ **12.** $\{1, -1\}$

13. $\{4, 0\}$ **14.** $\{\frac{1}{2}, \frac{1}{4}\}$

15. $\{-1 + \sqrt{3}, -1 - \sqrt{3}\}$

16. $\{1, -\frac{1}{2}\}$ **17.** $\{2, -5\}$

18. a. $y = x^2$

x	y
−2	4
−1	1
0	0
1	1
−2	4

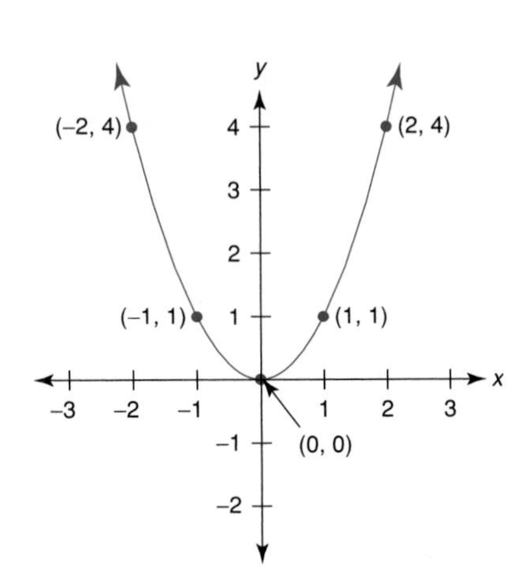

b. $y = x^2 - 4$

x	y
−3	5
−2	0
−1	−3
0	−4
1	−3
2	0
3	5

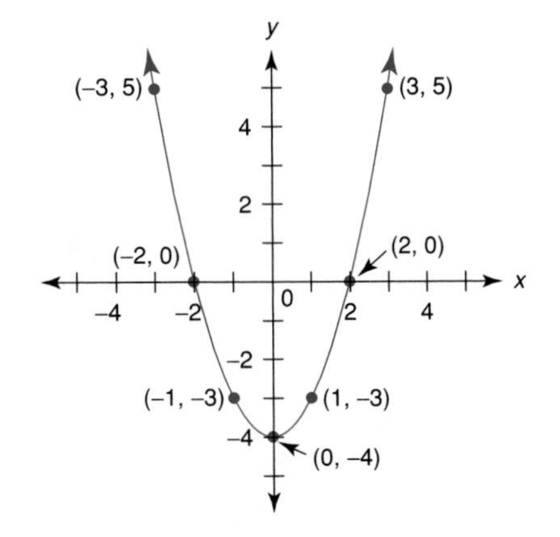

c. $y = -x^2 + 2$

x	y
−2	−2
−1	1
0	2
1	1
2	−2

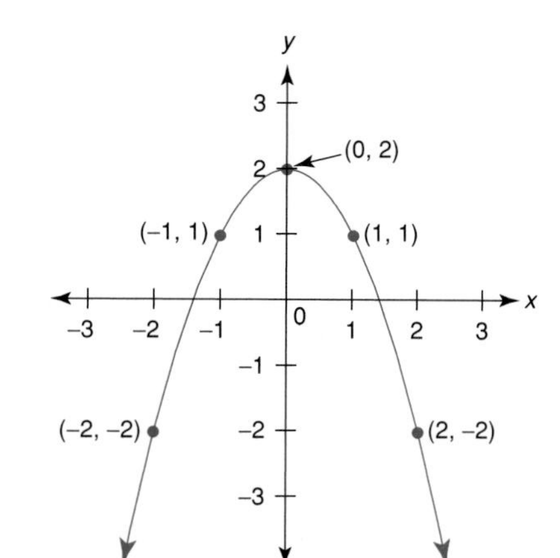

d. $y = x^2 - 6x$

x	y
0	0
1	−5
2	−8
3	−9
4	−8
5	−5
6	0

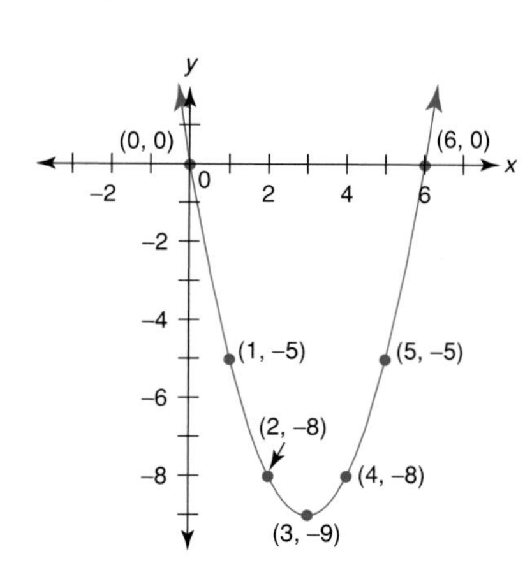

e. $y = x^2 + 2x - 3$

x	y
−4	5
−3	0
−2	−3
−1	−4
0	−3
1	0
2	5

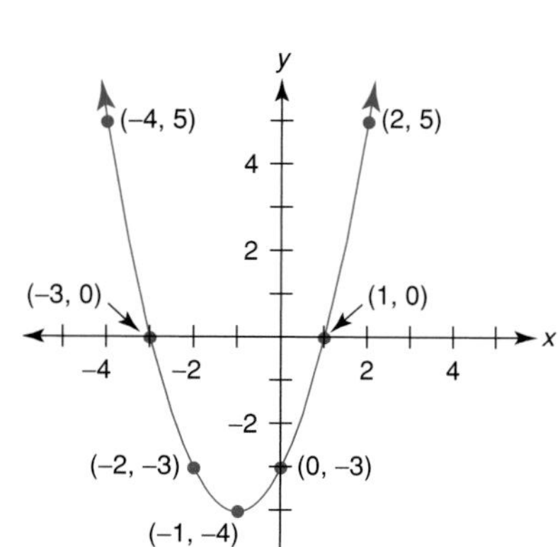

f. $y = -x^2 + 6x - 5$

x	y
0	−5
1	0
2	3
3	4
4	3
5	0
6	−5

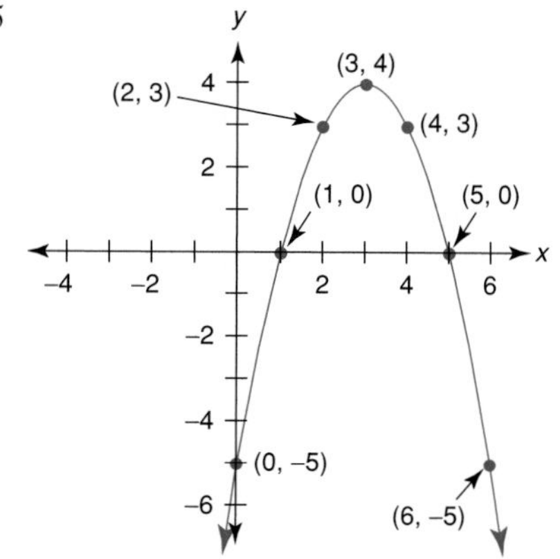

Just for Fun *Zero*

Chapter 10 Quiz

1. True **2.** True **3.** False **4.** True
5. False **6.** True **7.** False **8.** True
9. True **10.** False **11.** $\frac{10}{3}$
12. $3x - y + 11 = 0$ **13.** $x - 2y + 2 = 0$
14. $x + 2y + 8 = 0$ **15.** $m\overline{PQ} = 1, m\overline{PR} = 1$
16. Points P, Q, and R lie on the same line.
17. $x - y - 2 = 0$
18. $(2, 0)$
19. $y = -x^2 + 4x$

x	y
−1	−5
0	0
1	3
2	4
3	3
4	0
5	−5

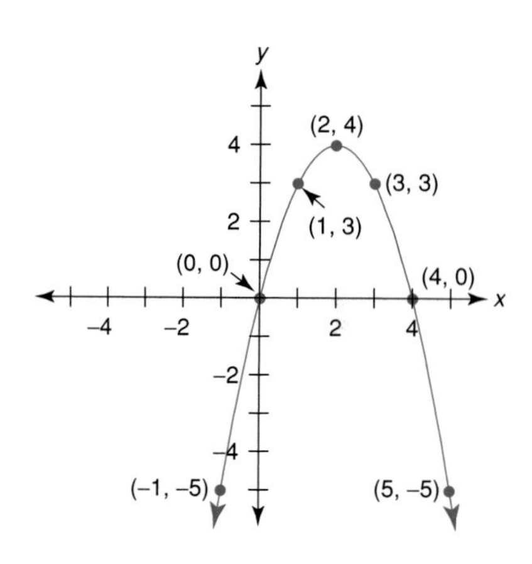

20. $(2, 2)$
21. Maximum value of P is 13 at $(3, 2)$.
22. Maximum value of P is 56 at $(0, 7)$.
23. $\{4, -1\}$
24. $\left\{\frac{1}{2}, \frac{-2}{3}\right\}$
25. $\left\{\frac{1}{3}, \frac{-3}{2}\right\}$
26. $\{5, -3\}$
27. $\{2, -5\}$
28. $\left\{\frac{1}{3}, -3\right\}$
29. $\{4, 0\}$
30. $\left\{\frac{2}{3}, -1\right\}$

Chapter 11

Section 11.2

1. $\{I\}$ **3.** $\overline{QK}$ **5.** $\overline{QI}$
7. $\overleftrightarrow{QK}$ **9.** $\{I\}$ **11.** $\{I\}$
13. $\overline{QU}$ **15.** $\{E\}$ **17.** $\{E\}$
19. $\{E\}$ **21.** $\angle WET$ **23.** $\overleftrightarrow{BT}$ or $\angle BET$
25. $\overline{BE}$ **27.** $\overrightarrow{EW}$ **29.** $\{A\}$
31. $\overleftrightarrow{ZP}$ or $\angle ZAP$ **33.** $\emptyset$ **35.** $\angle ZAW$
37. $\overrightarrow{AM}$ **39.** $\{A\}$ **41.** triangle WAZ

43. A ray has only one end point and may be extended indefinitely in only one direction from that end point.

45. An angle is the union of two rays that have a common end point.

47. An *acute* angle is an angle that is narrower than a right angle. A *right* angle is an angle whose sides are perpendicular. A *straight* angle is an angle whose sides form a straight line. An *obtuse* angle is wider than a right angle and narrower than a straight angle. A *round* (or perigon) angle is formed when a side of the angle is rotated one complete revolution. A *reflex* angle is greater than a straight angle and less than a round angle.

Just for Fun

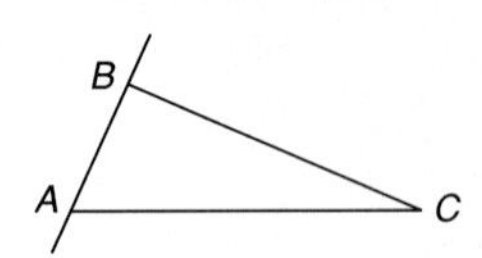

Section 11.3

1. $\{D\}$ **3.** $\overset{\circ\!-\!\circ}{RN}$ **5.** $\angle RDF$

7. $\overset{\circ\!\rightarrow}{DN}$ **9.** $\angle TON$

11. $\overset{\circ\!\rightarrow}{OE}$ **13.** Interior $\angle EON$

15. Ø **17.** Interior $\angle EON$

19. $\angle COT$ **21.** $\angle SOJ$

23. Interior $\angle SOC$ **25.** $\overset{\circ\!\rightarrow}{OT}$

27. $\{O\}$ **29.** Exterior $\angle JOC$

31. True **33.** False

35. False **37.** True

39. A plane has two dimensions, length and width.

41. Parallel lines are on the same plane and do not intersect. Parallel planes are two distinct planes that do not intersect.

43.

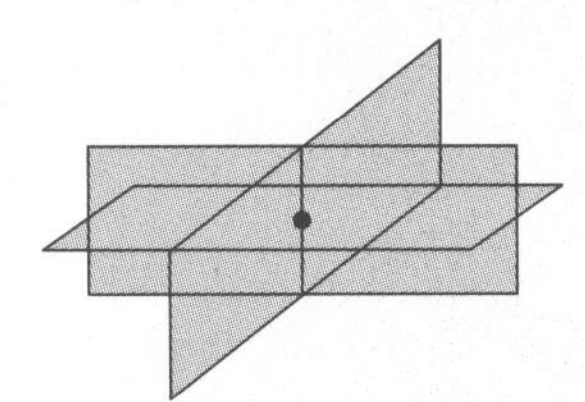

Just for Fun *Irene would receive a larger salary.*

Section 11.4

1. 6: $\angle ABD$, $\angle ABE$, $\angle ABC$, $\angle DBE$, $\angle DBC$, $\angle EBC$

3. *Note*: Answers may vary; some possible solutions are

a. $\angle FOE$, $\angle EOR$ **b.** $\angle LOE$, $\angle NOR$
c. $\angle LOF$, $\angle FOR$ **d.** $\angle WOE$, $\angle LOR$
e. $\angle FOE$, $\angle EOR$ and $\angle LON$, $\angle NOF$
f. $\angle LON$, $\angle SOR$ and $\angle NOF$, $\angle BOS$
g. $\angle FOE$, $\angle EOR$ and $\angle LON$, $\angle NOF$
h. $\angle NOL$, $\angle LOS$ and $\angle FOE$, $\angle BOE$

5. a. 100° **b.** 80° **c.** 110°
d. 59°40′ **e.** 141°30′ **f.** 79°9′35″

7. 60° and 120° **9.** 30° and 60° **11.** $x = 28$

13. 45° **15.** True **17.** False

19. False **21.** True

23. $m\angle 1 = 130°, m\angle 2 = 50°, m\angle 4 = 130°,$
$m\angle 5 = 130°, m\angle 6 = 50°, m\angle 7 = 50°,$
$m\angle 8 = 130°$

25. $m\angle 1 = 155°, m\angle 2 = 25°, m\angle 3 = 25°,$
$m\angle 4 = 155°, m\angle 6 = 25°, m\angle 7 = 25°,$
$m\angle 8 = 155°$

27. $m\angle 3 = 100°, m\angle 5 = 80°$

29. $m\angle 4 = 120°, m\angle 8 = 120°$

31. Vertical angles are two angles such that each side of one is an extension through the vertex of a side of the other.

33. Supplementary angles are two angles whose sum is 180°. Complementary angles are two angles whose sum is 90°. Conjugate angles are two angles whose sum is 360°.

35. 70°

Just for Fun *Tetrahedron (pyramid)*

Section 11.5

1. b, e **3.** a, b, d, e

5. a. True **b.** False **c.** True
d. False **e.** True

7. 13 **9.** 39 **11.** 5 miles **13.** 7 feet **15.** 3

17. a. True **b.** False **c.** False
d. True **e.** False

19. Polygon, triangle, right triangle, hypotenuse

21. a. Rectangle, square **b.** Rhombus, square

23. b **25.** $x = 35$ **27.** $m\angle F = 45°$

29. a **31.** b

33. A broken line is a set of connected line segments.

35. A simple closed broken line is one that does not intersect itself.

37. $x = 5$ **39.** 26 **41.** 50 m

Just for Fun

Nine: *Mercury, Venus, Earth, Mars, Jupiter, Saturn, Uranus, Neptune, Pluto*

Section 11.6

1. a. 24 m **b.** 24 cm **c.** 17 dm

3. 38 m **5.** 6.4 m **7.** 174 cm

9. 14 **11.** 4 **13.** 3 in.

15. 120 square units

17. 12.5 in.^{2} **19.** 8 in.^{2} **21.** 10 m

23. 45.5 square units

25. 7.4 ft

27. 17.5 square units

29. 90 square units

31. 30 cm^2 **33.** 25 m^2

35. a. 27.5; $\sqrt{41} + \sqrt{74} + 11$
b. 40; $24 + \sqrt{32}$

c. 25.5; $17 + \sqrt{18} + \sqrt{13}$

d. 148.5; 45.1

e. 102.3; 50.3

37. 94.2 cm **39.** 94.2 cm **41.** 31.4 m

43. 2826 m^2 **45.** 78.5 mm^2 **47.** 7850 m^2

49. A rectangle is a parallelogram with all of its angles right angles. A square is a rectangle with two adjacent sides equal.

51. A trapezoid is a quadrilateral with only two sides parallel. A parallelogram has both pairs of opposite sides parallel.

53. Circle, by 334.7 square units

55. Perimeter of rectangle = 30, perimeter of square = 24

57. 60°

Just for Fun *204*

Section 11.7

1. 4 **3.** 4

5. **a.** 6 **b.** 8 **c.** 12

d. $8 - 12 + 6 = 2$
$14 - 12 = 2$
$2 = 2$

7. **a.** 8 **b.** 6 **c.** 12

d. $6 - 12 + 8 = 2$
$14 - 12 = 2$
$2 = 2$

9. 96 cm^2, 42 cm^3

11. 539.7 mm^2, 400 mm^3

13. 785 cm^2; 1570 cm^3

15. 226.1 cm^2; 251.2 cm^3

17. 282.6 mm^2; 314 mm^3

19. No, 1570 cm^3 vs. 600 cm^3

21. 2122.6 cm^2; 9198.1 cm^3

23. 1133.5 cm^2; 3589.5 cm^3

25. 1.1 m^2; 0.1 m^3

27. A tetrahedron is a four faced polyhedron (triangular pyramid).

29. A pyramid is a polyhedron with one face a polygon and the other faces are triangles with a common vertex.

31. The volume is increased eightfold.

33. $r = 7.0$ ft

35. $V = 150.8 \text{ ft}^3$; $V = 5.6 \text{ yd}^3$

Just for Fun *One*

Section 11.8

1. True **3.** False **5.** True **7.** False

9. $x = 7$ **11.** No, $\overline{AC} \neq \overline{DF}$, so s.a.s. ≠ s.a.s.

13. 4 **15.** 2 **17.** 45° **19.** 4

21. 3 **23.** 37° **25.** 9 **27.** 9

29. 36 **31.** 12 m **33.** 27.5 ft

35. 9, 12, 15 **37.** 60 m

39. Congruent triangles are triangles that can be made to coincide. They are identical in size and shape.

41. 6 ft

43. 15 m

Just for Fun *An angle*

Section 11.9

1. **a.** 3 **b.** 0 **c.** Yes **d.** *A*, *B*, or *C*

3. **a.** 2 **b.** 2 **c.** Yes **d.** *A* or *C*

5. **a.** 4 **b.** 0 **c.** Yes **d.** *A*, *B*, *C*, or *D*

7. **a.** 2 **b.** 4 **c.** No **d.** None

9. **a.** 3 **b.** 2 **c.** Yes **d.** *E* or *C*

11. No, there are four vertices of odd order.

13. A network is a connected graph.

15. A simple network is one that does not cross itself.

17. A network formed by the edges of a cube is not traversable because it has more than two odd vertices.

19.

Review Exercises for Chapter 11

1. $\{R\}$ **2.** $\overline{AS}$ **3.** $\{R\}$

4. $\overleftrightarrow{AR}$ or $\angle ARS$ **5.** ∅ **6.** ∅

7. $\angle LRA$ **8.** $\angle SRJ$ **9.** $\overrightarrow{RJ}$

10. $\{R\}$ **11.** $\angle SOJ$ **12.** $\angle TOE$

13. $\{O\}$ **14.** $\overline{OT}$ **15.** Interior $\angle EOT$

16. Interior $\angle JOS$ **17.** $\{O\}$ **18.** $\overrightarrow{OC}$

19. Interior $\angle JOE \cup$ exterior $\angle SOT$

20. Interior $\angle COE$

21. **a.** 38°, 128° **b.** 44°, 134° **c.** 22°, 112°
d. 13°30′, 103°30′ **e.** 76°05′, 166°05′
f. 45°15′16″, 135°15′16″

22. 63°, 27° **23.** 46°, 134° **24.** d

25. d **26.** 15 **27.** 66°

28. 14.4 cm **29.** 50 in.^2 **30.** 91 m^2

31. 8 m **32.** 30 cm^2 **33.** 31.4 cm

34. 254.3 m^2 **35.** 94.2 cm; 706.5 cm^2

36. **a.** 148 cm^2 **b.** 120 cm^3

37. **a.** 360 cm^2 **b.** 300 cm^3

38. **a.** 628 cm^2 **b.** 1177.5 cm^3

39. 282.6 mm^2; 314 mm^3

40. 1808.6 cm^2; 7234.6 cm^3

41. $x = 12$ **42.** 40 m

43. **a.** 3 **b.** 2 **c.** Yes **d.** *A* or *E*

44. **a.** 4 **b.** 4 **c.** No **d.** None

45. **a.** 0 **b.** 4 **c.** No **d.** None

46. **a.** 3 **b.** 2 **c.** Yes **d.** *A* or *C*

Chapter 11 Quiz

1. False **2.** True **3.** True **4.** False

5. False **6.** False **7.** True **8.** True

9. True **10.** False **11.** False **12.** True
13. False **14.** False **15.** False **16.** True
17. False **18.** False **19.** False **20.** True
21. $25°, 65°$ **22.** $70°, 110°$ **23.** 26
24. 7 ft **25.** 72 in^2 **26.** 120 cm^2
27. 36 in^2 **28.** 558 in^2
29. 1256 cm^2, 4186.7 cm^3 **30.** 18 ft

21. 30
23. Move the decimal point two places to the left, and drop the percent sign. 4.7% = 0.047.
25. Convert $1\frac{7}{8}$ to a decimal, 1.875, move the decimal point two places to the right, and add a percent sign. $1\frac{7}{8} = 1.875 = 187.5\%$.
27. 220.2%
29. 81

Chapter 12

Section 12.2

1. a. $\frac{3}{2}$ **b.** $\frac{4}{7}$ **c.** $\frac{8}{5}$ **d.** $\frac{5}{6}$
3. a. $\frac{3}{1}$ **b.** $\frac{1}{2}$ **c.** $\frac{3}{2}$ **d.** $\frac{4}{1}$
e. $\frac{3}{4}$ **f.** $\frac{2}{3}$
5. a. $\frac{2}{3}$ **b.** $\frac{3}{2}$ **c.** $\frac{2}{5}$
7. $\frac{3}{19}$
9. a. 2 **b.** 6 **c.** 30 **d.** 35
11. a. 32 **b.** 18 **c.** $\frac{4}{3}$ **d.** $\frac{15}{4}$
13. 24.75 ft
15. a. $\frac{1}{3}$ **b.** $\frac{9}{2}$ **c.** $\frac{2}{3}$ **d.** $\frac{1}{10}$
17. 20 kg lead, 40 kg tin
19. 7.5 min
21. 75 ft
23. $\frac{7}{12}$
25. a. 16 **b.** 3.5 **c.** $2\frac{3}{7}$ or 2.4 **d.** 9.2
27. The ratio of two quantities, a and b, is the quotient or indicated quotient obtained by dividing a by b. The ratio of a to b is written as $a \div b$, $a : b$, or $\frac{a}{b}$.
Answers may vary.
29. 8 : 140, 4 : 70, 2 : 35
31. Bill's Bargain Store offers the better price; 5 for $26.25 yields a unit price of $5.25 each, while 3 for $16 yields a unit price of $5.33.
33. a. $15,000,000 **b.** $20,250,000 **c.** $11,250,000
d. $10,500,000 **e.** $9,750,000 **f.** $ 8,250,000.

Just for Fun *22 minutes*

Section 12.3

1. a. $\frac{3}{20}$ **b.** $\frac{1}{4}$ **c.** $\frac{3}{4}$
3. a. $\frac{9}{200}$ **b.** $\frac{7}{300}$ **c.** $\frac{1}{16}$
5. a. $\frac{13}{200}$ **b.** $\frac{23}{1000}$ **c.** $\frac{3}{2}$
7. a. 0.17 **b.** 0.03 **c.** 0.045
9. a. 0.065 **b.** 3.0 **c.** 0.0025
11. a. 9% **b.** 214% **c.** 90%
13. a. 112.5% **b.** 1% **c.** 301%
15. a. 75% **b.** 80% **c.** 62.5%
17. a. 260% **b.** 112.5% **c.** 87.5%
19. a. $66\frac{2}{3}\%$ **b.** 275% **c.** 6.25%

Just for Fun *We should subtract the $2 the waiter kept in order to get the cost of the meal ($27 − $2 = $25).*

Section 12.4

1. a. 50% **b.** 25% **c.** 133.3%
d. 66.7% **e.** 25% **f.** 160%
3. a. $12.50 **b.** $100
c. $7.50 **d.** $21.88
5.

	Markup	*Percent Markup*
a.	$30	150%
b.	$25	33.3%
c.	$3	28.6%
d.	$4	25.1%

7. $89.29
9. $275.38 **11.** $516.67
13.

	Markup	*Cost*
a.	$160	$240
b.	$2.20	$8.78
c.	$6.24	$43.71

15. $179; $716
17. 42.9%
19. $895; 33.5%
21. $225; $157.50
23. a. $50.12; $766.12 **b.** $0.07; $1.07
c. $62.65; $957.65 **d.** $11.03; $168.53
25. Markup is the difference between the selling price and the cost of an item.
27. A percent markup on cost tells us the amount by which the cost of an item was increased to obtain the selling price. A percent markup on selling price indicates the amount of the selling price that the retailer retains for overhead and profits.
29. $1050 **31.** AGEB

Just for Fun *Jets*

Section 12.5

1. $675 **3.** $106.25 **5.** $375
7. $945; $3945 **9.** $1275; $106.25
11. $243.60 **13.** $195
15. $189.63 **17.** $5100; $425
19. 10 months **21.** 6%
23. $5,100; $100 **25.** $14,304; $2,304
27. Simple interest is based on the entire amount of the loan for the total time of the loan. Interest is the money the borrower pays for the use of the lender's money.

29. $A = P(1 + rt)$, A = amount, P = principal, r = rate of interest (expressed as a decimal), t = time (in terms of years)

31. $900 33. 60% 35. $121.80

Just for Fun *Approximately 2740 years*

Section 12.6

1. $883.40 3. $5397.50 5. $13,000
7. $1095.50 9. $1126 11. $8955
13. $586 15. $2191 17. $20,330
19. $5418; $2418 21. 12 years
23. $28,656 25. $9.95
27. $13,600; $13,690; $13,730; $13,200
29. $4049
31. Compound interest occurs when the interest due at the end of certain period is added to the principal and that sum earns interest for the next period. The interest for each succeeding period is greater than the previous one, because the principal keeps increasing. This is done for the entire time of the loan. For simple interest, the interest is not added to the principal until the end of the time of the loan.
33. $9.96
35. $10,000 now, because at the rate cited, it can grow to $10,820.

Just for Fun *Connecticut, Delaware, Maryland, Massachusetts, New Hampshire, New Jersey, New York, Pennsylvania, Rhode Island, Virginia, South Carolina, North Carolina, Georgia*

Section 12.7

1. The *effective annual interest rate* is the annual interest rate that gives the same yield as the nominal interest rate compounded several times a year.
3. 8.16% 5. 12.55%
7. 6.14% 9. 7.98%
11. a. 6% b. 6.09% c. 6.14%
13. 5.5% compounded semiannually (5.58% vs. 5.09%)
15. Pamela (6.71% vs. 6.56%)
17. Julia (7.02% vs. 6.92%)
19. Effective rate is used to compare interest rates that are compounded at different intervals.
21. Rachel, 5.61%; Jillian 5.63%. Jillian receives the better interest rate.

Just for Fun *Mario should drive Roy's car, and Roy should drive Mario's car.*

Section 12.8

1. a. $437.50 b. $519.20 c. $129
d. $2556.50 e. $1021.32
3. a. $151.40 b. $154.42
c. $157.44 d. $163.56
5. a. $715.20 b. $729.50
c. $743.80 d. $772.44
7. $4091.60 9. $90.84; $61.80
11. $596; $898; limited payment costs $6040 more.
13. $567.40; $9786.80
15. Answers may vary.
The basic concept of insurance is the sharing of risks or losses. That is, a person who buys an insurance policy is agreeing to share the risks involved, with other people buying a similar policy.
17. Term life insurance is for a specific period of time, such as five or ten years. A straight life insurance policy is one in which the policyholder pays premiums till death.
19. $457.50

Just for Fun *They are the same distance from Miami.*

Section 12.9

1. Total cost $625; Finance charge $100
3. 16.6%
5. a. $37.33 b. 22.2%
7. a. $63.58 b. 30.9%
9. a. $198 b. $31.17 c. $798 d. 34.6%
11. a. $41.56 b. 23.4% c. $396
13. a. $227 b. $7 c. 12.0%
15. a. $864 b. $64 c. 14.8%
17. a. $4750 b. $750 c. 15.5%
19. a. $12,421 b. $3422 c. 16.9%
21. a. $1200 b. $7200 c. $300 d. 19.2%
23. a. $108 b. $1008 c. $56 d. 15.2%
25. Credit may be thought of as allowing someone to purchase an item without cash (or borrowing money) and trusting that person to pay for the item (or pay back the money) at a later date.
27. Annual percentage rate (APR) is the true annual interest rate that a person pays when borrowing money or financing a purchase (credit).
29. a. $8200 b. $2200 c. 28.5%

Section 12.10

1. $388 3. $550.50
5. $185.40 7. $217
9. $384.45 11. $399.20; $55,808
13. Bank B; $15,084 15. $587.20; $131,392
17. $734; $164,240 19. $482.50; $65,800
21. A mortgage is a long term loan where property is pledged as security for payment of the loan. A mortgagee is the holder of a mortgage.
23. $297,379.20; $177,379.20

Just for Fun *Alexander Hamilton; U.S. Treasury Building*

Review Exercises for Chapter 12

1. a. $\frac{2}{5}$ b. $\frac{3}{2}$ c. $\frac{2}{5}$ d. $\frac{2}{3}$
2. a. No b. Yes c. No d. Yes

3. a. 4 b. 3 c. 8 d. 24
4. a. $\frac{19}{50}$ b. $\frac{1}{8}$ c. $\frac{23}{1000}$ d. $\frac{5}{4}$
5. a. 0.48 b. 0.105 c. 0.047 d. 0.0025
6. a. 82% b. 130% c. 78% d. 213.4%
7. a. 60% b. 12.5% c. 220% d. 18.75%
8. 20,000 9. 200
10. $96.15 11. $80
12. Markup = $160; cost = $240
13. $31.25; $93.74 14. $1650
15. $900 16. $1800; $150
17. $7890; $1890
18. $7224; $3224 19. $10,165; $5165
20. a. 5% b. 5.06% c. 5.09%
21. Annual percentage rate, which is also known as the true annual interest rate.
22. $42 per year
23. a. $172 b. $262.50 c. $1077.60 d. $2107.60
24. a. $6940 b. $1940 c. 31.5%
25. a. $132 b. $28.42 c. $732 d. 23.0%
26. a. $253.20 b. $411.75 c. $617.40 d. $410.30
27. $533.50; $142,060 28. Bank B; $24,480

Just for Fun *False; they were invented by people in India.*

Chapter 12 Quiz

1. 9 2. 3.6 3. 12 4. 72
5. 0.6; 60% 6. 0.875; 87.5% 7. $\frac{5}{4}$; 125%
8. $\frac{7}{200}$; 3.5% 9. $\frac{37}{50}$; 0.74 10. $\frac{1}{20}$; 0.05
11. $160 12. $250 13. $160; $239.99
14. $344.25 15. $20.17 16. $7.75
17. 23.5%
18. False 19. False 20. False 21. True
22. False 23. False 24. True 25. True
26. False 27. False 28. True 29. True
30. True

Chapter 13

Section 13.1

Just for Fun *Nautical mile = 6076 feet; statute mile = 5280 feet*

Section 13.2

1. (i) To reduce the number of different sizes; (ii) to make conversion from one unit to another easier; (iii) to create a potential increase in export; (iv) because it is the international system of measurement.
3. The measurement system in the United States is a confusing system because measures such as the quart and pound are inconsistent. Measures of a liquid quart and a dry quart differ in weight and volume; a pound of hamburger is heavier than a pound of gold because one is measured in avoirdupois weight and the latter is measured in troy weight. Also there are too many conversions to remember.
5. a. Deka b. Deci c. Kilo d. Centi e. Hecto f. Milli
7. a. 39,370 b. 3937 c. 393.7 d. 3.937 e. 0.3937 f. 0.03937
9. a. 1000 b. 10 c. 0.1 d. 0.001 e. 0.01 f. 100
11. a. 100 b. 1000 c. 10 d. 100 e. 1000 f. 10
13. a. 2 b. 32000 c. 3000 d. 400 e. 300 f. 5000
15. a. False b. False c. False d. False e. True
17. The metric system is the worldwide accepted standard measurement system. There is only one basic unit of measurement for each quantity (i.e., length, volume, and mass), and they are related. The metric system is based on the number 10 and quantities are in a decimal relationship to each other.
19. 350 cg = 3.5 g; move the decimal point two places to the left.
21.

Prefix	Symbol	Power of 10	Name
yotta	Y	10^{24}	septillion
zetta	Z	10^{21}	sextillion
exa	E	10^{18}	quintillion
peta	P	10^{15}	quadrillion
tera	T	10^{12}	trillion
giga	G	10^{9}	billion
mega	M	10^{6}	million
kilo	k	10^{3}	thousand
hecto	h	10^{2}	hundred
deka	da	10^{1}	ten
deci	d	10^{-1}	tenth
centi	c	10^{-2}	hundredth
milli	m	10^{-3}	thousandth
micro	μ	10^{-6}	millionth
nano	n	10^{-9}	billionth
pico	p	10^{-12}	trillionth
femto	f	10^{-15}	quadrillionth
atto	a	10^{-18}	quintillionth
zepto	z	10^{-21}	sextillionth
yocto	y	10^{-24}	septillionth

Just for Fun

1. Meter *2. Gram*
3. Liter *4. Dekagram*
5. Kiloliter *6. Decimeter*
7. Kilogram *8. Milliliter*
9. Centimeter *10. Kilometer*

Section 13.3

1. **a.** 17 mm **b.** 46 mm **c.** 31 mm
   d. 69 mm **e.** 41 mm **f.** 31 mm
3. **a.** 1.8 **b.** 3700 **c.** 320 000
   d. 70 200 **e.** 0.0423 **f.** 81.4
5. **a.** 41.37 **b.** 4.5 **c.** 4700
   d. 0.04378 **e.** 0.302 **f.** 98.5
7. 4200 **9.** 2814
11. Yes, 50 mi/hr = 80 km/hr
13. 34,800,000 **15.** 4 cm^2 **17.** 2.56
19. **a.** 1 $in.^2$ **b.** 1 m^2 **c.** 1 m^2
    d. 1 mi^2 **e.** 1 ha **f.** 1 are
21. b **23.** b **25.** c
27. Louie ran faster because in running 200 meters, he ran approximately 220 yards, 20 yards more than Carlos in the same amount of time.
29. Answers may vary. It should be noted that the United States does use the metric system as a standard of measurement for exports.

Just for Fun *151,476*

Section 13.4

1. **a.** 1200 **b.** 2500 **c.** 3.5
   d. 10 **e.** 6.5 **f.** 450
3. **a.** 0.732 **b.** 2.314 **c.** 314
   d. 0.01495 **e.** 490 **f.** 7200
5. Kiloliter, hectoliter, dekaliter, liter, deciliter, centiliter, milliliter
7. **a.** 2 kL **b.** 3 liters **c.** 3 pt
   d. 2 liters **e.** 3 tsp **f.** 5 fl oz
9. 20 liters **11.** 1.2 m^3
13. Rice pudding: 120 mL uncooked rice; 600 mL milk; 60 mL sugar; 80 mL raisins; 7.5 mL cinnamon; 2.5 mL salt
15. b **17.** c **19.** c
21. A liter is defined as the volume of a cubic decimeter.
23. Basically, the notation cc is used in the medical profession because of tradition. It is similar to the use of other older measures such as "chord" for wood and "bale" for cotton.
25. **a.** 480 cm^3 or 480 mL **b.** 10 000 cm^3 or 10 000 mL

Just for Fun *3,692*

Section 13.5

1. **a.** 52,000 **b.** 3500 **c.** 0.25
   d. 0.43 **e.** 7200 **f.** 4700
3. **a.** 0.417 **b.** 342 000 **c.** 14 300
   d. 0.8859 **e.** 0.014359 **f.** 2710
5. **a.** 224 **b.** 220 **c.** 17.5
   d. 9 **e.** 90 **f.** 2.7
7. 0.5 **9.** 123
11. **a.** 360 **b.** 360 t
13. **a.** 1 kg **b.** 2 oz **c.** 5 kg
    d. 3 t **e.** 2 kg **f.** 280 g
15. a **17.** b **19.** b
21. The metric units of volume, mass, and capacity have a special relationship. A cube with a volume of 1 cubic centimeter can hold 1 milliliter of water. The mass of 1 milliliter of water is 1 gram. An ounce is defined to be equal to 16 *drams* or 437.5 *grains*.
23. 70 grams **25.** 14.1 grams

Just for Fun *13,120*

Section 13.6

1. **a.** 40 °C **b.** 10 °C **c.** 5 °C
   d. 35 °C **e.** 30 °C **f.** −5 °C
3. **a.** 68 °F **b.** 122 °F **c.** 185 °F
   d. 50 °F **e.** 32 °F **f.** 149 °F
5. **a.** −10 °C **b.** 59 °F **c.** 77 °F
   d. −25 °C
7. 104 °F **9.** 3000 °C **11.** −50 °C
13. 58 °C **15.** 57 °C **17.** b
19. a **21.** b
23. $C = \frac{5}{9}(F - 32)$. First substitute for F, next do the subtraction inside the parentheses. Multiply the difference by 5 and then divide by 9.
25. **a.** 212 °F 100 °C
    b. 136 °F ≈58 °C
    c. 98.6 °F 37 °C
    d. 68 °F 20 °C
    e. 32 °F 0 °C
    f. 0 °F ≈−18°C
27. Water boils at 100 °C at sea level. For every 165 m above sea level, the boiling point of water is lowered by 0.56 °C.

Just for Fun

a. *It's third down and 9 meters to go.*
b. *Give them 2.5 cm and they will take 1.6 km.*
c. *The cowboy wore a 38-liter hat.*
d. *Did you see the 30-cm note?*
e. *A miss is as good as 1.6 km.*
f. *He climbed the rope 2.5 cm by 2.5 cm.*
g. *Twenty-eight grams of prevention are worth 0.45 kg of cure.*
h. *Have you read "Celsius 233"?*
i. *Erskine Caldwell wrote "God's Little 0.4 Hectare."*
j. *A 0.47 liter is 0.45 kg the world around.*

Review Exercises for Chapter 13

1. The metric system is a system of measurement with basic units of measure for length, area, volume, and weight in decimal relationship to each other. The basic unit is international and

relates to units of length, volume, and weight. It is a well-planned, logical system with uniformity, allowing for easier and more precise calculations.

2. **a.** Kilo **b.** Deka **c.** Deci **d.** Hecto **e.** Centi **f.** Milli

3. One ten-millionth of the distance from the North Pole to the equator, along the meridian that passes through Dunkirk and Paris.

4. **a.** 150 m **b.** 3.0 km **c.** 2 cm **d.** 250 dm **e.** 22,000 m **f.** 3.5 dam

5. **a.** 39,370 **b.** 0.03937 **c.** 3937 **d.** 0.3937 **e.** 393.7 **f.** 3.937

6. 2803.2 **7.** 2532.8

8. **a.** 48 in **b.** 135 m **c.** 16 km **d.** 13.2 yd **e.** 8.4 yd^2 **f.** 130 cm^2

9. A liter is defined as the volume of a cube that is 1 decimeter long, 1 decimeter wide, and 1 decimeter high.

10. **a.** 3000 mL **b.** 20 daL **c.** 40 hL **d.** 2 L **e.** 150 mL **f.** 3 daL

11. 90 liters

12. Juanita, $1.56 vs. $1.59 for the same amount of gasoline.

13. The weight of one gram equals the weight of one milliliter of very cold water.

14. **a.** 4000 **b.** 8000 **c.** 3 **d.** 200 **e.** 4 **f.** 1800

15. **a.** 1000 **b.** 1 000 000 **c.** 10 **d.** 100 000 **e.** 45 **f.** 50 000

16. −88°C **17.** 136 °F **18.** c **19.** c

20. b **21.** d **22.** a **23.** c

24. c **25.** b **26.** c **27.** a

28. b **29.** c

30. **a.** 10 **b.** 0.95 **c.** 106 **d.** 56 **e.** 90 **f.** 35

31. **Sour Cream Cookies**

120 mL sour cream	224 g butter
5 mL vanilla	5 mL baking soda
240 mL brown sugar	600 mL flour
2 eggs	Bake at 177 °C

32. Answers may vary.

Just for Fun

A	K	E	I	B	C	U	B	I	C	M	E	T	E	R
M	I	L	L	I	G	R	A	M	A	B	C	D	R	E
I	L	O	U	L	A	E	I	O	U	H	F	M	G	T
L	O	C	E	L	S	I	U	S	H	E	I	R	J	E
L	L	E	K	L	M	N	O	P	Q	C	E	T	R	M
I	I	N	J	S	T	U	S	C	O	T	T	R	V	A
M	T	T	O	N	N	E	W	X	E	A	Y	I	Z	K
E	E	I	E	A	G	R	A	M	M	R	M	C	C	E
T	R	M	B	C	D	E	O	A	F	E	G	H	I	D
E	J	E	K	L	M	L	R	D	S	H	T	A	M	E
R	O	T	P	Q	I	G	I	R	N	E	T	E	S	R
T	U	E	V	K	A	W	X	T	Y	A	T	A	R	A
Z	T	R	E	K	L	F	O	U	E	B	L	E	M	U
O	H	T	E	N	O	A	D	D	I	R	E	S	K	Q
T	A	D	C	U	B	I	C	M	E	T	E	R	I	S

Chapter 13 Quiz

1. True **2.** False **3.** False **4.** True

5. True **6.** False **7.** False **8.** False

9. False **10.** False **11.** True **12.** False

13. False **14.** True **15.** True **16.** False

17. True **18.** False **19.** True **20.** True

21. True **22.** True **23.** True **24.** True

25. False

Glossary

Abacus The earliest recorded mechanical counting device, developed by the Chinese during the twelfth century.

Abscissa The x value in the ordered pair (x, y).

Active observation Consciously studying an object so that one can both identify and describe the object in sufficient detail to distinguish it from other objects that might be present.

Acute angle An angle whose measure is greater than $0°$ and less than $90°$.

Adjacent angles Two angles that have the same vertex and a common side between them.

Algorithm Some special process of solving a certain type of problem, usually a method that repeats some basic process.

APR Annual percentage rate, also known as the true annual interest rate.

Area The measure of the total surface within a closed region.

Arithmetic progression A list of numbers in which successive terms are generated by adding a constant. This constant is called the common difference and can be found by subtracting one number of the sequence by the number that immediately precedes it.

Associative If a system consists of a set of elements $\{a, b, c, \dots\}$ and an operation $*$, the operation is associative if, for all elements in the system, $(a*b)*c = a*(b*c)$.

Axiom A statement that is accepted as true without proof.

Axiomatic system An axiomatic system consists of four main parts: (1) undefined terms, (2) definitions, (3) axioms, (4) theorems.

Bar graph A graph consisting of parallel bars whose lengths are proportional to certain quantities given in a set of data.

Base 5 system A system that groups items by fives.

Biconditional The conjunction of two conditional statements where the antecedent and consequent of the first statement have been switched in the second.

Bimodal A set of data is described as bimodal when the given set of data has two modes.

Binary notation The base 2 system of numeration.

Bottom-up processing An approach to problem solving that relies on intuition or creativity. By using this approach solutions to problems are found through exploration, trial and error, or discovery.

Cardinality The cardinality of a set tells us "how many" elements are in a set.

Cartesian plane The grid or graph divided into four quadrants by the x- and y-axes, named after the French mathematician–philosopher René Descartes.

Cartesian product The Cartesian product of sets A and B is the set of all possible ordered pairs such that the first element of the ordered pair is an element of A and the second element of the ordered pair is an element of B.

Celsius Anders Celsius (1701–1744), a Swedish astronomer who developed a thermometer scale on which the boiling point of water is $100°$ and the freezing point of water is $0°$.

Centi A prefix that means $\frac{1}{100}$, or 0.01. A centimeter is 0.01 meter.

Central tendency A measure of central tendency describes a set of data by locating the middle region of the set.

Circle or pie graph A graph in the form of a circle (pie) that shows the relationship of all the parts to the whole.

Circumference The distance around a circle; the length of a circle.

Closure The characteristic of a system that is closed. A system consisting of a set of elements $\{a, b, c, \dots\}$ and an operation $*$ is closed if for any two elements a and b in the set, $a*b$ is also a member of the set.

Collinear Lying on the same line.

Combination A distinct group of objects without regard to their arrangement.

Commutative In a system consisting of a set of elements $\{a, b, c, \dots\}$ and an operation $*$, the operation is commutative if for all elements a and b in the system, $a*b = b*a$.

Comparative bar graph A comparative bar graph represents more than one set of related data within the same bar graph; also known as a *double bar graph.*

Complement The complement of a set A is the set of all the elements in the given universal set, U, that are not in the set A; it is denoted by A'.

Complementary angles Two angles whose measures sum to that of a right angle.

Complex fraction A fraction that has a fraction in the numerator or denominator or both.

Composite number A number that has two or more prime factors.

Compound amount The total amount of principal and compound interest combined.

Compound interest The interest due at the end of a period is added to the principal, and thereafter earns interest.

Conclusion The statement that follows as a consequence of the hypothesis or premises.

Conditional The connective "*If . . . , then . . .*" is used in compound statements referred to as conditionals.

Congruent angles Two angles that can be made to coincide. Two angles that contain the same number of degrees.

Congruent triangles Triangles that can be made to coincide. They have the identical shape and size.

Conjecture The conclusion resulting from inductive reasoning. A conjecture is neither true nor false, only probable.

Conjunction Two simple statements that are connected using the word *and.*

Consistent (1) Two linear equations are consistent if they intersect. (2) Statements that can be true together.

Contradiction A statement that is always false.

Contrapositive The conditional that results from replacing the antecedent by the negation of the consequent, and the consequent by the negation of the antecedent.

Converse The conditional that results from replacing the antecedent by the consequent and the consequent by the antecedent.

Coordinate A point on the Cartesian plane whose position is determined by the values of its abscissa and its ordinate.

Coplanar Lying in the same plane. Coplanar points are points that lie in the same plane.

Correlation A statistical concept that describes the relationship between two entities. A correlation does not imply a cause-effect relationship.

Deci A prefix that means $\frac{1}{10}$, or 0.1. A decimeter is 0.1 meter.

Decimal Any number written in decimal notation.

Decimal notation Using the decimal system to write a number in standard notation.

Decimal system A system of notation for real numbers that uses the base 10.

Deductive reasoning A reasoning process that uses previously accepted and proven facts as its basis for making conclusions. Deductive reasoning moves from the general to the specific. Contrast with inductive reasoning.

Defined terms Definitions that may use undefined terms or terms that have been previously defined. Definitions should be concise, consistent, and not circular.

Degree (1) A unit of measurement of angles; denoted as °. A right angle contains 90°. (2) A unit of measure of temperature. The temperature is 20° Celsius.

Deka A prefix that means 10. A dekameter is 10 meters.

De Morgan's law (Augustus De Morgan, 1806–1871) Rules that enable us to change a disjunction, conjunction, or the negation of one of these to an equivalent statement.

Density property Between every two rational numbers there is another rational number.

Dependent Two equations whose graphs are the same line. Any solution of one equation is also a solution to the other.

Dependent events Two events are dependent if the occurrence of one affects the occurrence of the other.

Descriptive statistics A branch of statistics that is used to describe data. Examples of descriptive statistics include mean, median, standard deviation, frequency distribution, and histogram. Contrast with inferential statistics.

Disjoint sets Two sets whose intersection is the empty set. Disjoint sets have no elements in common.

Disjunction Two simple statements that are connected using the word *or*.

Distributive property The property of arithmetic and algebra that states

$$a(b + c) = ab + ac$$

for any numbers a, b, and c.

Divisible An integer m is divisible by an integer n if there is an integer q such that $m = nq$.

Duodecimal system The base 12 system of numeration.

Effective annual rate The annual interest rate that gives the same yield as a nominal interest rate compounded several times a year.

Element Any member of a set.

Empty set A set with no elements. Also called the *null set.*

Equally likely events Each outcome of the experiment has the same (or equal) chance of occurring.

Equal sets Two sets that contain exactly the same elements.

Equilateral triangle A triangle in which the measures of all three sides are equal.

Equivalent sets Two sets that contain exactly the same number of elements.

Estimation A process that involves forming an approximate solution or response to a given problem or situation. For example, a person might estimate the cost of a new vehicle.

Expanded notation Writing a number in powers of the base system. For example,

$$978 = (9 \times 10^2) + (7 \times 10^1) + (8 \times 10^0)$$

Expectation Expectation tells us the expected value or "fair price" to play a game. It is found by multiplying the probability that an event will occur by the amount that will be won if that event occurs.

Factorial For a positive integer, n, the product of all the positive integers less than or equal to n. For example, $3! = 1 \cdot 2 \cdot 3$ and $n! = 1 \cdot 2 \cdot 3 \cdot \ldots \cdot n$.

Finite (re: finite sets) A finite set contains a limited number of elements. It is a set that has just n members for some integer n.

Foot A unit of linear measure, equal to 12 inches.

Fraction An indicated quotient of two quantities.

Frequency distribution A table that shows the number of times a datum occurs.

Frequency polygon A line graph that results when the upper ends of the middles of the intervals of a histogram are connected by line segments.

Function A relation in which no two ordered pairs have the same first element and different second elements.

Functional observation A subconscious act of being able to note the presence of an object but being unable to consciously identify or describe it.

Fundamental counting principle If one experiment has m different outcomes and a second experiment has n different outcomes, then the first and second events performed together have $m \times n$ different outcomes. This may be extended if there are other experiments to follow: We would have $m \times n \times r \times \cdots \times t$ outcomes.

Fundamental theorem of arithmetic Every natural number greater than 1 is either a prime or can be expressed as a product of prime factors. Except for the order of the factors, this can be done in one and only one way.

General term A special term that corresponds to a particular sequence and used to generate additional terms of the sequence. Also known as the *nth term.*

Geometric progression A list of numbers in which successive terms are generated by multiplying by a constant. This constant is called the *common factor* and can be found by dividing a succeeding number by the number that immediately precedes it.

Going metric The process of converting a system of weights and measures to the metric system.

Gram A unit of mass in the metric system.

Greatest common divisor (greatest common factor) The largest natural number that divides a given pair of natural numbers with a remainder of zero.

Greatest common factor See *greatest common divisor.*

Great gross One great gross = 12 gross or 12 × 144 or 1728 units.

Gross One gross = twelve 12's or 12 × 12 or 144 units.

Group A set of elements and an operation that satisfy the closure property, identity property, inverse property, and associative property.

Hecto A prefix that means 100. A hectometer is 100 meters.

Hexadecimal system The base 16 system of numeration.

Histogram A histogram consists of a series of bars that are drawn all with the same widths on the horizontal axis and uniform units on the vertical axis, to illustrate a frequency distribution.

Horizontal bar graph A horizontal bar graph consists of a set of parallel bars that are parallel to the horizontal axis.

Hypothesis testing A statistical procedure that is used to determine if the results from a sample can be generalized to a population. See also *inferential statistics*.

Identity element A system consisting of a set of elements $\{a, b, c, \dots\}$ and an operation $*$ has an identity element (e), if for every element a in the system, $a * e = a$ and $e * a = a$.

Iff Abbreviation for the phrase "*if and only if*" (biconditional).

Implication An equivalent statement in logic.

$$P \rightarrow Q \equiv {\sim}P \vee Q$$

Improper fraction A fraction in which the numerical value of the numerator is greater than or equal to the numerical value of the denominator. Each of the following is an example of an improper fraction: $\frac{3}{2}, \frac{7}{3}, \frac{8}{4}, \frac{9}{9}$.

Inclusive or One or the other, or both.

Inconsistent (1) Two lines that have no points in common and therefore never intersect. (2) Statements that cannot be true together.

Independent events Two events are independent if the occurrence of one does not affect the occurrence of the other.

Inductive reasoning A reasoning process that uses the results of specific cases as its basis for making conclusions. Inductive reasoning moves from the specific to the general. Unless proven otherwise, conclusions resulting from inductive reasoning should only be considered probable. Contrast with deductive reasoning.

Inequality A statement that one quantity is less than (or greater than) another.

Inferential statistics A branch of statistics that is used to draw conclusions about a population based on information provided by a sample. Thus, results from sample data are *inferred* to a population. Inferential statistics uses hypothesis testing to determine if such inferences are warranted.

Infinite Becoming large beyond any fixed bound.

Integer An element of the set consisting of the set of whole numbers and their additive inverses

$$\{\dots, -2, -1, 0, 1, 2, \dots\}$$

Intersection The intersection of sets A and B is a set of elements that are members of both A and B.

Invalid An argument in which all the premises are true, and the conclusion may be false.

Inverse The conditional that results from negating the antecedent and negating the consequent.

Inverse element Each element in a system consisting of a set of elements $\{a, b, c, \dots\}$ and an operation $*$ has an inverse if for every element a in the system there exists an element b such that $a * b = e$ and $b * a = e$, where e is the identity element of the system.

Irrational numbers Numbers that when expressed as a decimal are nonterminating and nonrepeating.

Isosceles triangle A triangle in which two sides are of equal length.

Kilo A prefix that means 1000. A kilometer is 1000 meters.

Kilogram A kilogram is equal to 1000 grams.

Knowledge transfer The process of using knowledge acquired from past experience(s) to solve a current problem.

Least common multiple The least natural number that is a multiple of each of two given numbers.

Linear In a straight line, having only one dimension.

Linear equation An equation of the form

$$Ax + By = C$$

where A, B, and C are real numbers. The graph of such an equation is a straight line.

Linear programming A system for solving a system of linear inequalities where the goal is to maximize a value subject to certain restrictions.

Liter A unit of volume in the metric system.

Logically equivalent Two statements that have exactly the same truth tables.

Markdown The change, or difference, between the original price and the sale price of an item.

Markup The difference between the selling price and the cost of an item.

Mathematical system A set of elements together with one or more operations (rules) for combining elements of the set.

Mean The mean for a set of data is found by determining the sum of the data and dividing this sum by the total number of elements in the set.

Median The median can be described as the middle value of a set of data when the data are listed in order.

Meter A unit of length in the metric system.

Metric system The system of measurement in which the meter is the basic unit of length and the gram is the basic unit of mass.

Metric ton A metric ton is 1000 kilograms and equivalent to 2200 pounds.

Midrange The midrange is found by adding the least value in the given set of data to the greatest value and dividing the sum by 2.

Milli A prefix that means $\frac{1}{1000}$, or 0.001. One millimeter is 0.001 meter.

Minute The 60th part of a degree of an angle measurement; denoted as $'$. For example, $1^\circ = 60'$.

Mixed number An alternative but equivalent form of an improper fraction. A mixed number has an integer part and a fraction part, such as $3\frac{1}{4}$ and $7\frac{5}{8}$.

Mode The mode for a given set of data is that number, item, or value that occurs most frequently.

Model A diagram (figure) for a proof that satisfies all of the given axioms.

Modulo A system that repeats itself in a cycle. The 12-hour clock is called a modulo 12 system.

Mortgage A conditional pledge of property as a security for money lent.

Multiplicative grouping system A system that uses certain symbols for numbers in a basic group, together with a second symbol or notation to represent numbers that are multiples of the basic group.

Mutually exclusive events Two events that cannot occur together at the same time.

Napier's rods A calculating device developed by John Napier (1550–1617) that was used to perform multiplication.

Natural number Counting number, or an element of the set $\{1, 2, 3, \ldots\}$.

Negation A statement that is formed by prefixing "*It is false that.*"

Network A set of connected line segments or arcs.

Normal distribution A family of distributions for which the mean, median, and mode all have the same value, and all occur exactly at the center of the distribution.

Number A member of the set of positive integers.

Numeral Any symbol for a number. Symbols such as V and 5 are numerals that represent the number five.

Objective function The function in a linear programming problem that describes the relationship between the variables.

Obtuse angle An angle whose measure is greater than 90° and less than 180°.

Odds A ratio of two probabilities. The *odds in favor* of an event A

$$= \frac{\text{Probability that A will occur}}{\text{Probability that A will not occur}}$$

One-to-one correspondence A relation between two equivalent sets such that each element of either set is paired with exactly one element of the other set.

Order of operations A prescribed procedure for evaluating numerical expressions or simplifying algebraic expressions. This order involves working from left to right and performing all arithmetic operations contained within grouping symbols first, followed by multiplications and divisions second, and additions and subtractions last.

Ordered pair Two corresponding elements, where one element is designated as the first and the other as the second.

Ordinate The y-value in the ordered pair (x, y).

Origin The point where the x- and y-axes intersect, whose value is $(0, 0)$.

Parabola The graph of the equation of the general form

$$y = ax^2 + bx + c \quad \text{or} \quad x = ay^2 + by + c$$

Paradox A contradictory statement.

Parallel Equidistant apart at every point.

Parallelogram A quadrilateral that has both pairs of opposite sides parallel.

Percent A ratio with a denominator of 100; per one hundred.

Percentile The value that divides the range of a set of data into two parts such that a given percentage of the data lies below this value.

Perimeter The sum of the lengths of the sides of a polygon; the length of a closed curve.

Permutation An ordered arrangement.

Perpendicular Intersecting at or forming right angles.

Pi The name of the Greek letter π. The symbol π represents the ratio of the circumference to the diameter of a circle, which approximately equals 3.14159.

Pictogram or picture graph A series of identical pictures with a key or code that specifies the quantity that each picture represents.

Place-value system A system in which the position of a symbol matters; the value that any symbol represents depends on the position it occupies within the numeral.

Point-slope formula The formula

$$y - y_1 = m(x - x_1)$$

used to find the equation of a line when given a point on the line and the slope of the line.

Polygon A simple closed broken line.

Polyhedron A space figure or solid bounded by plane polygons.

Population In statistics, a population represents the primary group to be studied. It is the group to which a researcher applies the results of a statistical experiment.

Premise One of the first two propositions (statements) of a syllogism.

Premium A fee that an insured person pays to the insurer.

Prime number Any natural number greater than 1 that is divisible only by itself and 1.

Principal The most important or most significant. In finance, money put at interest, or otherwise invested.

Prism A polyhedron with two parallel and congruent bases.

Probability If an experiment has a total of T equally likely possible outcomes, and if exactly S of them are considered successful, then the probability that event A will occur, denoted by $P(A)$, is $\frac{n(S)}{n(T)}$.

Proper subset If A is a subset of B, and there is at least one element in B not contained in A, then A is a proper subset of B, denoted by $A \subset B$.

Proportion The equality of two ratios.

Pythagorean theorem The sum of the squares of the legs of any right triangle is equal to the square of the hypotenuse.

Quadratic equation An equation of the form

$$ax^2 + bx + c = 0$$

Quartile The 25th, 50th, and 75th percentiles are the 1st, 2nd, and 3rd quartiles.

Random sample A subset of a population in which every member of the population has an equal and independent chance of being selected. A random sample is representative of the population. See also *unbiased sample.*

Range The range for a set of data is found by subtracting the smallest value from the largest value in the given set of data.

Ratio The ratio of two quantities a and b is the quotient obtained by dividing a by b. The ratio of a to b is written as $a \div b$ or $a : b$ or a/b.

Rational number A number that can be expressed in the form a/b where a and b are integers and $b \neq 0$.

Real number Any rational or irrational number.

Recognition observation Similar to functional observation except that the observer is consciously aware of an object to the point where he or she can identify it.

Rectangle A parallelogram that contains a right angle.

Recursion A concept in which the first element of a sequence is defined, and then the nth element is defined in terms of the preceding elements. The operation or computation *recurs* in the definition. An example of a recursive relationship is the Fibonacci sequence.

Relation A set of ordered pairs.

Replacement set The set of permissible substitutes.

Rhombus A parallelogram that has adjacent sides of equal length.

Right angle An angle whose sides are perpendicular. The measure of a right angle equals 90°.

Roman numeral A system of numeration used by the Romans that consists of seven primary symbols: I (1), V (5), X (10), L (50), C (100), D (500), and M (1000).

Sample In statistics, a sample is a subset or subgroup of a population. Members of a sample are selected using a process known as *sampling*. See also *random sample* and *unbiased sample*.

Sample space The set of all possible outcomes of an experiment.

Scalene triangle A triangle with no two sides of equal length.

Second $\frac{1}{60}$ of a minute and $\frac{1}{360}$ part of a degree.

Sequence A set of numbers that are listed in a specific order. The arrangement of these numbers is based on a general rule or principle.

Series The sum of the terms of a sequence.

Set A collection of objects.

Sexagesimal system A system based on powers of 60.

SI An abbreviation for the International System of Units.

Sieve of Eratosthenes A technique, devised by Eratosthenes (276–194 B.C.) used for determining prime numbers by punching holes in a clay tablet.

Similar triangles Triangles whose corresponding angles are equal in measure.

Simple grouping system A system in which the position of a symbol does not affect the number represented, and a different symbol is used to indicate a certain number or group of things.

Simple interest Money paid for the use of money where the interest due at the end of a certain period is computed on the original principal during the entire period.

Slope of a line A ratio of change in vertical distance to the change in horizontal distance.

Slope-intercept formula The formula

$$y = mx + b$$

denotes the equation of a line when given the y-intercept b and the slope of the line m.

Solution set The set of all solutions of a given equation, system of equations, inequality, etc.

Square A quadrilateral with four sides of equal length and four angles of equal measure.

Standard deviation The square root of the arithmetic mean of the squares of the deviations from the mean.

Standard form A form that has been universally accepted by mathematicians as such, in the interest of simplicity and uniformity. For example, the standard form of a quadratic equation is $ax^2 + bx + c = 0$.

Statement A declarative sentence that is either true or false, but not both.

Stem-and-leaf display A statistical method for displaying data. In a steam-and-leaf display, each score is separated into a stem and leaf. The stem consists of all of the digits of a number except its last; the leaf is the last digit. A stem-and-leaf display provides a visual picture of the entire distribution of scores. When rotated 90 degrees, a stem-and-leaf display yelds a histogram of its corresponding grouped frequency distribution.

Straight angle An angle whose measure is 180°.

Subset Given any two sets A and B, if every element in A is also an element in B, then A is a subset of B, denoted by $A \subseteq B$.

Supplementary angles Two angles whose measures sum to that of a straight angle (180°).

Syllogism A logical argument that contains a major premise, a minor premise, and a conclusion.

Tautology A statement that is always true.

Tetrahedron A four-faced polyhedron.

Theorem Logical deductions that are made from undefined terms, defined terms, and axioms.

Top-down processing A systematic method of breaking a large or complicated problem into smaller but related problems that can be solved more readily. The solution of these smaller problems can then lead to the solution of the original problem.

Trapezoid A quadrilateral that has only two parallel sides.

Traversable A network is traversable if it can be drawn by tracing each line segment or arc exactly once without lifting the pencil from the paper.

Tree diagram A tree diagram consists of a number of "branches" that illustrate the possible outcomes for an experiment.

Unbiased sample A sample that has been selected randomly. An unbiased sample is representative of the population from which it was selected.

Undefined term A term used without specific mathematical definition; it satisfies certain axioms but is not otherwise defined.

Underwriter The company or person that agrees to insure a person or item—that is, the insurer.

Union The union of two sets A and B is the set of elements that are members of A, or members of B, or members of both A and B.

Universal set The set of all objects admissible in a particular problem or discussion.

Valid An argument in which all the premises are true and the conclusion is true. A valid argument is one where the conclusion follows logically from the premises.

Variable A symbol used to represent an unspecified member of some set.

Venn diagram A diagram that uses a rectangle to represent the universal set and a circle or circles to represent the set or sets being considered in the discussion.

Vertex (1) The turning point of a parabola. (2) The common endpoint of two intersecting rays of an angle.

Vertical angles Two angles where the sides of one angle extend through the vertex and form the sides of the other.

Vertical bar graph A vertical bar graph consists of a set of parallel bars that are perpendicular to the horizontal axis.

Volume The measure of the amount of space that is enclosed by a three-dimensional figure.

Whole number An integer that is an element of the set $\{0, 1, 2, 3, \dots\}$.

***x*-intercept** The point at which a line crosses the x-axis.

***y*-intercept** The point at which a line crosses the y-axis.

Z-score The position of a piece of data in terms of the number of standard deviations it is located from the mean.

Photo Credits

Table of Contents p. vii Dorling Kindersley Media Library **p. vii**, Frank Whitney/Getty Images Inc.–Image Bank **p. vii**, Getty Images, Inc. **p. viii**, AP/Worldwide Photos **p. viii**, Francesco Ruggeri/Getty Images Inc.–Image Bank **p. viii**, Index Stock Imagery, Inc. **p. ix**, Art Resource, N.Y. **p. ix**, W. Geiersperger/Corbis/Bettmann **p. ix**, NASA Headquarters **p. x**, Andrea Sperling/Getty Images, Inc. **p. x**, Thomas A. Heinz/Corbis/Bettman **p. x**, AP/Wide World Photos **p. xi**, Grant Faint/Getty Images, Inc.–Image Bank.

Chapter 1 CO Dorling Kindersley Media **p. 3**, Richard Hamilton/ Corbis/Bettmann **p. 8**, William M. Setek, Jr. **p. 11**, Mark Richards/PhotoEdit **p. 13**, Daniel Luna/AP/Wide World Photos **p. 23**, Scott Gries/Getty Images **p. 29**, map, South Florida area: Rand McNally & Company; crowd: Richard Lord/The Image Works; map, San Francisco area: Rand McNally & Company **p. 30**, Derek Trask/Corbis/Stock Market **p. 32**, SuperStock, Inc. **p. 45**, Ray Ellis/Photo Researchers, Inc. **p. 46**, The Granger Collection **p. 55**, AP/Wide World Photos **p. 57**, Getty Images, Inc. **p. 59**, Getty Images, Inc. **p. 61**, William M. Setek, Jr. **p. 62**, Getty Images, Inc.

Chapter 2 CO Frank Whitney/Getty Images, Inc.-Image Bank **p. 72**, Photo Researchers, Inc.; Georg Cantor: Corbis/Bettmann **p. 79**, AFP/Patrick Hertzog/Getty Images, Inc.-Liaison **p. 89**, Getty Images Inc.–Hulton Archive Photos **p. 97**, United Media/United Features Syndicate, Inc. **p. 103**, Nancy P. Alexander/PhotoEdit **p. 106**, David Young-Wolff/PhotoEdit

Chapter 3 CO Getty Images-Photodisc **p. 114**, Topham/The Image Works **p. 117**, Everett Collection, Inc. **p. 118**, AFP/Corbis/Bettmann **p. 122**, Grant LeDuc **p. 124**, Corbis/Bettmnn **p. 126**, Peter Correz/Getty Images, Inc.-Stone Allstock **p. 130**, Tribune Media Services, Inc. **p. 141**, Laima Druskis/Pearson Education/PH **p. 154**, Ernest Edwards/Corbis/Bettmann **p. 171**, Maull and Fox/Gonville and Caius College **p. 177**, Smithsonian Institute/Office of Imaging, Printing, and Photographic Services **p. 178**, Culver Pictures, Inc.

Chapter 4 CO AP/Wide World Photos **p. 199**, Tony Freeman/PhotoEdit **p. 200**, Fred Lyon/Photo Researchers, Inc. **p. 208**, Roy Morsch/Corbis/Stock Market **p. 217**, Creators Syndicate, Inc. **p. 219**, Boissard/Corbis/Bettmann **p. 241**, United Media/United Features Syndicate, Inc. **p. 242**, Bettmann Archive/Corbis/Bettmann **p. 246**, C Squared Studios/Getty Images, Inc.-Photodisc **p. 252**, William M. Setek, Jr. **p. 253**, Library of Congress

Chapter 5 CO Ruggeri, Francisco/Getty Images, Inc.-Image Bank **p. 266**, John Griffin/The Image Works **p. 267**, Amos Morgon/Getty Images-Photodisc **p. 273**, David Blackwell **p. 278**, top: Corbis/Bettmann; bottom: Bob Daemmrich/Stock Boston **p. 280**, Creators Syndicate, Inc. **p. 298**, Hugh Rogers

Chapter 6 CO Index Stock Imagery, Inc. **p. 318**, Lee Snider/The Image Works **p. 320**, Historical Pictures/Stock Montage, Inc.-Historical **p. 323**, Mark C. Burnett/Stock Boston **p. 325**, King Features Syndicate, Inc. **p. 326**, Constance Reid **p. 329**, Getty Images, Inc.-Image Bank **p. 331**, Jon Henley/Corbis/Bettmann **p. 333**, Bryce Flynn/Stock Boston **p. 335**, William M. Setek, Jr. **p. 339**, Corbis/Bettmann

Chapter 7 CO Art Resource, N.Y. **p. 361**, Epigraphical Museum **p. 363**, Jeff Greenberg/Visuals Unlimited **p. 371**, Westmorland, Stuart/Getty Images, Inc.-Image Bank **p. 383**, Murray Alcosser/Getty Images, Inc.-Image Bank **p. 386**, Allan Rosenberg/Getty Images, Inc.-Photodisc **p. 391**, Erich Lessing/Cultu/PhotoEdit **p. 402**, Bill Bachmann/Stock Boston **p. 410**, National Archives and Records

Chapter 8 CO W. Geiersperger/Corbis/Bettmann **p. 433**, William M. Setek, Jr. **p. 441**, Tony Craddock/Photo Researchers, Inc. **p. 446**, Sandved, Kjell B./Photo Researchers, Inc. **p. 455**, The Granger Collection **p. 473**, Stock Montage, Inc./Historical Pictures Collection.

Chapter 9 CO NASA Headquarters **p. 507**, Renee Lynn/Photo Researchers, Inc. **p. 512**, Michael Newman/PhotoEdit **p. 523**, Eureka College Archives **p. 528**, Doug Scott/AGE Fotostock America, Inc. **p. 529**, Dorling Kindersley Media **p. 530**, Robert Brenner/PhotoEdit **p. 538**, Tribune Media Services, Inc. **p. 546**, John Madere/Corbis/Bettmann

Chapter 10 CO Andrea Sperling/Getty Images, Inc.-Taxi **p. 551**, William M. Setek, Jr. **p. 571**, Stanford University **p. 575**, Historical Pictures/Stock Montage, Inc./Historical **p. 582**, top: Greg Meadors/Stock Boston; bottom: Jim Pickerell/Stock Boston

Chapter 11 CO Thomas A. Heinz/Corbis/Bettmann **p. 594**, British Museum, London, UK/Bridgeman Art Library **p. 601**, Frank Siteman **p. 616**, Getty Images, Inc.-Hulton Archive **p. 619**, Ted Curtin/Stock Boston **p. 622**, Creators Syndicate, Inc. **p. 633**, Stock Montage, Inc./Historical **p. 641**, NASA/Johnson Space Center **p. 652**, Creators Syndicate, Inc.

Chapter 12 CO AP/Wide World Photos **p. 663**, AP/Wide World Photos **p. 670**, Michael Newman/PhotoEdit **p. 676**, AP/Wide World Photos **p. 679**, Topham/The Image Works **p. 689**, Robert Rathe/Stock Boston **p. 693**, The Granger Collection

Chapter 13 CO Grant Faint/Getty Images, inc.-Image Bank **p. 716**, Culver Pictures, Inc. **p. 717**, Michael Newman/PhotoEdit **p. 718**, Tom McHugh/Photo Researchers, Inc. **p. 722**, Jack Plekan/Fundamental Photographs **p. 727**, Howard Kingsnorth/Getty Images, Inc.-Stone Allstock **p. 735**, Anne et Jacques Six **p. 736**, Ogust/The Image Works **p. 741**, The Image Works **p. 743**, Creators Syndicate, Inc. **p. 744**, top: Gabe Palmer/Corbis; bottom: Robin Smith/photo library.com **p. 746**, Harvey Eisner

Index

T

U

V

W

X

Y

Z